# PLANT PHYSIOLOGY

FOURTH EDITION

**Frank B. Salisbury**
Utah State University

**Cleon W. Ross**
Colorado State University

Wadsworth Publishing Company
Belmont, California
A Division of Wadsworth, Inc.

Biology Editor: Jack C. Carey
Editorial Assistant: Kathy Shea
Production Editor: Angela Mann
Managing Designer: Carolyn Deacy
Print Buyer: Barbara Britton
Art Editor: Donna Kalal
Permissions Editor: Robert M. Kauser
Designer: Ann Kellejian
Copy Editor: Alan Titche
Technical Illustrator: Cecile Duray-Bito and Alexander Teshin Associates
Cover: Section of Autumn Maple Leaf © Rod Planck/Photo Researchers, Inc.
Section Opener Photograms: Ann Kellejian
Compositor: Thompson Type
Printer: Malloy

*This book is printed on acid-free paper that meets Environmental Protection Agency standards for recycled paper.*



5 6 7 8 9 10 — 96 95 94

**Library of Congress Cataloging in Publication Data**
Salisbury, Frank B.
Plant physiology / Frank B. Salisbury, Cleon W. Ross. — 4th ed.
p. cm.
Includes bibliographical references (p. ) and indexes.
ISBN 0-534-15162-0
1. Plant physiology. I. Ross, Cleon W. II. Title.
QK711.2.S23 1991
581.1—dc20 91-7362

# Preface

Frank B. Salisbury

Cleon W. Ross

As we have worked during the past two and a half years on this fourth edition of our text, we have been impressed by the advances in plant physiology that have taken place since 1984, when we finished work on the third edition. It is humbling to realize how many people have contributed to these advances. In some areas especially (photosynthesis, for example) the depth of our understanding now reaches awesome proportions. In other areas far less is known and we frequently emphasize how much remains to be learned. At the same time, basic principles of our science remain just that: fundamental to a comprehension of its frontiers. Because of this, and in spite of our efforts to pick and choose only the most relevant topics, our book has grown significantly since the last edition. We realize that this will pose problems for teachers such as ourselves who must introduce students to the science in a one-quarter or one-semester course. Yet we want students to have some feeling for the extent of the science so we hope interested students will at least scan those topics that are not assigned in class.

## Purpose of the Book

Our purpose, as in previous editions, is to provide a broad explanation of the physiology of plants (their functions) from seed germination to vegetative growth, maturation, and flowering. We present principles and results of previous and ongoing research throughout the world. Partly because of space limitations, we have concentrated on seed plants and little emphasis is usually given to other organisms (except in Chapter 21 on the Biological Clock). *Plant Physiology* is for students who are curious about what plants do and what physical and chemical factors cause them to respond as they do. Many students will use this information in careers in agronomy, horticulture, forestry, range and seed science, and plant pathology. We hope the book will motivate many others to obtain advanced degrees in plant physiology and to perform research to solve present and future problems.

## Organization and Recent Revisions

We retain the same basic organization as in previous editions. The first eight chapters (Section One) are primarily concerned with physical processes that go on in plants, and this topic is introduced with a chapter that summarizes the structures of plant cells. We expect that you will know the highlights of this topic already. The next seven chapters (Section Two) concern the biochemical processes that occur in plants, including three chapters on photosynthesis. These biochemical processes depend partly on some of the physical processes covered in Section One. Section Three describes plant growth and development and with this edition we have added a chapter on molecular biology and genetic engineering as these topics relate to research and discoveries in plant physiology. This chapter was prepared by two guest authors, Drs. Ray Bressan and Avtar Handa, who are experts in the field. The last two chapters (Section Four) on environmental physiology and stress physiology describe important environmental factors, how they limit growth of various species and yet how some species have adapted physiologically to survive in harsh environments.

Finally, as in our third edition, we have added three appendices. Appendix A describes SI units of measurement, which are being increasingly used in all scientific fields. We hope that it will help students to become familiar with these units and that it will serve as a reliable reference source for authors who are preparing technical papers for publication. Relatively minor changes were made in Appendix B, which covers properties of various kinds of radiation including sunlight and various artificial light sources commonly used by plant researchers. Appendix C summarizes transcription and translation. You may have memorized these principles in an introductory biology course but this appendix may serve as a convenient reference.

Although the overall outline is similar to earlier editions, we have not only added much new material but we have reworked virtually every paragraph in the book, attempting to contemplate and evaluate how new information fits into the old, thereby expanding our comprehension of how plants function. And, with the help of many reviewers and an excellent copy editor, Alan Titche, we have tried to improve the presentation so that students will be better able to understand. We have also worked closely with Angela Mann, an outstanding production editor, and Cecile Duray-Bito, who has prepared some delightful new illustrations for this edition.

The reviewers were of immeasurable help. Our science has become so broad that it is impossible for two authors to keep up with it all. Virtually every chapter was examined in preliminary draft by at least three (and as many as a dozen) experts in the respective fields and the final version represents a distillation of their suggestions plus the fruits of our own efforts to assimilate the literature. We are extremely grateful for that help but of course we accept complete responsibility for the present text.

A summary following this preface describes some of the additions and changes in this edition.

## Format and Features

The basic format (page size, double columns, and headings) is the same as it was for the third edition, but with some design improvements. Notably, second-level headings are set above the paragraphs, which makes them easier to find and emphasizes the book's organization. Third-level headings (which are seldom used) are in *italics* on the left margin, initiating paragraphs.

Plant physiology consists of a complex network of information that can be approached from many vantage points and discussed in various chapters. When we have discussed the same subject from a different vantage point in more than one chapter we have usually included a cross-reference to a Section or Chapter in which it is also discussed. New terms or concepts are listed in **bold-faced** type when they are defined; many such terms are defined in more than one place in the book and they are **bold-faced** each time. Page numbers in the index have a "d" by them if the term is *defined* (**bold-faced**) on that page. Page numbers with an "i" by them in the index refer to pages where the item is *illustrated*. Names of biochemical compounds, enzyme names, and certain other names or terms are usually listed in *italics* when they are first introduced to help you find them again as you read further and review.

Nearly everyone who has commented on our book has expressed enjoyment of and appreciation for the guest essays so we have kept most of those from earlier editions and added several new ones. Some of these deal almost exclusively with personal recollections of exciting times in the scientific lives of their authors; others explain some current and important topic in more detail than seemed warranted in the text. There are also smaller boxed essays on special topics, as in earlier editions.

## Literature Citations and Names of People and Plants

We have added many references (authors' names and year of publication), beginning especially with Chapter 6. They are provided for students who wish to learn more about a subject and to document our source of information for subjects that we think might be controversial. We have often listed recent reviews or recent articles that expand on a given topic and introduce students to the past literature. In addition, some classic papers from the previous century and early in this century are also included in certain chapters; many of these are not easily found elsewhere. We do want students to think of plant physiologists as people, so occasionally and rather arbitrarily we have listed their first names and the places where they work or worked. References for each chapter are listed by author and year of publication in the Reference Section at the end of the book.

Plant names pose another problem. We refer to some species mostly by their common names, as did those who worked with them, whereas we identify other species by their scientific (Latin) names (but without the author who first described the species). Usually we give both at least once in the text.

## Some Thoughts

In spite of the agonies of preparing a major text such as this—and it can be agonizing to check some minor point in the library for an hour to update one sentence or to read proofs for what seems like forever—on balance developing this edition has been an enjoyable learning experience. Some questions posed in our last edition have now been answered by former students and our personal understanding of how plants work has grown substantially. There is much yet to learn and answers will be coming fast as the techniques of biotechnology are applied to an increasing number of problems.

As we wrote in the preface to our third edition: "We hope that our enthusiasm and love for the science of plant physiology is apparent to you and that you will come to share these feelings with us. These are the feelings that motivate the rapid advances now being made in virtually all scientific disciplines."

## List of Responsibilities

Frank B. Salisbury wrote chapters 1, 2, 3, 4, 5, 8, 12, 16, 19, 21, 22, 23, 25, 26, and the appendices. Cleon W. Ross wrote chapters 6, 7, 9, 10, 11, 13, 14, 15, 17, 18, and 20. Ray Bresson and Avtar Handa wrote chapter 24.


## Acknowledgments

We greatly appreciate the efforts of the following typists and helpful assistants: Dawn D. Ross, Sharon Goalen, Nancy Phillips, Glenda Nesbit, Laura Wheelright, and Trish Cozart.

## Reviewers

Reviewers for this Fourth Edition include: Tobias Baskin, University of California, Berkeley; J. Clair Batty, Utah State University; Wade L. Berry, University of California, Los Angeles; J. Derek Bewley, University of Guelph, Canada; Robert Allan Black, Washington State University; Peter Brownell, James Cook University; Bruce G. Bugbee, Utah State University; Michael J. Burke, Oregon State University; Martyn Caldwell, Utah State University; William F. Campbell, Utah State University; John G. Carman, Utah State University; James T. Colbert, Iowa State University; Michael Evans, Ohio State University; Donald R. Geiger, University of Dayton; Dr. Govindjee, University of Illinois; Ronald John Hanks, Utah State University; Wolfgang Haupt, Institut für Botanik und Pharmazeutische Biologie der Universität Erlangen-Nürnberg, Germany; John E. Hendrix, Colorado State University; Mordecai J. Jaffe, Wake Forest University; Peter B. Kaufman, University of Michigan; Dov Koller, Hebrew University, Israel; Willard L. Koukkari, University of Minnesota; G. Heinriche Krause, Universität Dusseldorf, Germany; Walter Larcher, Universität Innsbruck, Austria; Wolfram Meier-Augenstein, Universiteit Van Stellenbosch, S. Africa; Anastasios Melis, University of California, Berkeley; Angel Mingo-Castel, Universidat Publica Navarra, Spain; Cary A. Mitchell, Purdue University; Keith Mott, Utah State University; Richard Mueller, Utah State University; Park S. Nobel, University of California, Los Angeles; William H. Outlaw, Florida State University; Robert Pearcy, University of California, Davis; Richard Pharis, University of Calgary; Gregory J. Podgorski, Utah State University; Iffat Rahim, Iowa State University; Fred D. Sack, Ohio State University; John Sager, NASA Kennedy Space Center; Kurt A. Santarius, Universität Düsseldorf, Germany; Ruth Satter, University of Connecticut; Herman Schildknecht, Heidelberg Universität, Germany; Thomas D. Sharkey, University of Wisconsin; Louis F. Sokol, U.S. Metric Association, Inc.; Thomas K. Soulen, University of Minnesota; Daphne Vince-Prue, Goring-on-Thames, England; George W. Welkie, Utah State University; Rosemary White, University of Sydney, Australia; Stephen E. Williams, Lebanon Valley College; Jan A. D. Zeevaart, Michigan State University.


Frank B. Salisbury, Logan, Utah
Cleon W. Ross, Fort Collins, Colorado
February 1991

# Specific Changes in This Edition

## Chapter 1

### Plant Physiology and Plant Cells

This chapter has been greatly expanded and reorganized to give an overview of modern plant-cell biology, emphasizing the endomembrane system and the cytoskeleton. It retains the introduction to plant physiology with the basic postulates that stand as the basis of modern scientific method. The definition of life remains.

## Chapter 2

### Diffusion, Thermodynamics, and Water Potential

Thermodynamic principles should provide the foundation for much of plant physiology but particularly for the study of plant water relations. Dr. Keith Mott, a young plant physiologist at Utah State University, has provided a new, easier-to-grasp approach to thermodynamics.

## Chapter 3

### Osmosis

The basic principles in this chapter have changed little since the last edition but the presentation in this new version has been improved and it reflects some important changes in terminology.

## Chapter 4

### The Photosynthesis-Transpiration Compromise

Stomatal action has been an active field of research since the Third Edition was published, and this is reflected in the new chapter. Stem-flow techniques to measure transpiration are described.

## Chapter 5

### The Ascent of Sap

This traditional topic was thoroughly revised and updated for the Third Edition. Many changes have been made including some new figures for this Fourth Edition, but the basic story remains essentially the same.

## Chapter 6

### Mineral Nutrition

Nickel is added to the list of essential elements. Strategies I and II as adaptations to iron deficiency are explained. Siderophore chemistry is included. New discoveries about the functions of calcium are introduced to be emphasized in later chapters.

## Chapter 7

### Absorption of Mineral Salts

More emphasis is given to modern concepts of ATP-ase and pyrophosphatase pumps, functions of ion channels are introduced, and transport of proteins from cytosol into various organelles is explained.

## Chapter 8

### Transport in the Phloem

This is another field of rapid progress during recent years, particularly in our understanding of phloem unloading and assimilate partitioning and control mechanisms. Recent advances are described.

## Chapter 9

### Enzymes, Proteins, and Amino Acids

Effects of temperature on enzyme activity are expanded and related to plant ecology. Enzyme isoforms are explained and contrasted with isozymes.

## Chapter 10

### Photosynthesis: Chloroplasts and Light

The four major protein complexes of thylakoids are explained, including functions and some aspects of their molecular biology. The Q cycle is introduced into the electron transport process, and the oxidation of water by the "water-oxidizing clock" is explained. How certain herbicides block photosystem II is explained at the molecular level and in relation to resistant mutants. A new model for the electron transport process is presented.

## Chapter 11

### Carbon Dioxide Fixation and Carbohydrate Synthesis

Major updates are on control of photosynthesis and individual enzymes, especially regarding light activation of rubisco and the ferredoxin-thioredoxin system. The outdated LEM system was dropped. The importance of fructose-2,6-bisphosphate in control of sucrose formation is introduced. The ecology of CAM is expanded and biosynthesis and structures of fructans are explained.

## Chapter 12

### Photosynthesis: Environmental and Agricultural Aspects

The status of carbon dioxide in the earth's atmosphere for the past 160,000 years—and especially during recent decades—is described as part of a discussion of the earth's carbon pools and balances. This chapter has also been revised to reflect current thinking in crop ecology and physiological ecology including modern applications of SI units.

## Chapter 13

### Respiration

The importance of fructose-2,6-bisphosphate and of pyrophosphate-dependent phosphofructokinase in glycolysis and its control are explained. The regulation of respiration is expanded and better related to ecology.

## Chapter 14

### Assimilation of Nitrogen and Sulfur

The photorespiratory nitrogen cycle is explained in detail and related to photorespiration described in Chapter 11. Sulfate reduction is updated.

## Chapter 15

### Lipids and Other Natural Products

The role of various secondary products in allelopathy is expanded. Modern information is added to correct late stages in the shikimic acid pathway. How glyphosate acts as an herbicide is explained. Polysaccharide elicitors of fungal invasion are explained briefly.

## Chapter 16

### Growth and Development

This chapter has been revised throughout including the section on the chemistry and physics of cell-wall growth. Several figures were revised. The topics of juvenility and totipotency (tissue-culture studies) have been moved into this chapter.

## Chapter 17

### Hormones and Growth Regulators: Auxins and Gibberellins

This chapter was almost totally rewritten starting with two new figures in the introduction that illustrate possible mechanisms of hormone action. Discussions of the general role of hormones in plants emphasize control by hormone receptors, differential tissue sensitivity, and hormonal control of gene activity. The phosphoinositide cycle, calcium-calmodulin, and protein kinases are introduced with illustrations. A new mechanism for auxin degradation is explained, as is indolebutyric acid as a naturally occurring auxin. Results with genetic mutants are described as these relate to production and action of auxins and gibberellins.

## Chapter 18

### Hormones and Growth Regulators: Cytokinins, Ethylene, Abscisic Acid, and Other Compounds

This chapter expands on the mechanisms of action introduced in Chapter 17, Chapter 24, and Appendix C, especially the role of receptors and new studies with molecular genetics including gene activation and translational control. Some specific additions include evidence for cytokinin control over apical dominance, effects of introducing auxin and cytokinin production genes from TDNA of *Agrobacterium*, ethylene receptors and antagonists, and salicylic acid as a plant hormone that controls heat production important for pollination of *Arum* lilies and that provides resistance to both fungal and viral pathogens. Biochemical reactions of abscisic-acid synthesis are updated and a proposed pathway from violaxanthin (a carotenoid) is illustrated. The probable importance of abscisic acid as a defense compound against salt and cold stress is described and the importance of this hormone in seed maturation is emphasized. Sections on polyamines, brassinosteroids, and jasmonic acid are updated and turgorins are introduced (discussed in Chapter 19).

## Chapter 19

### The Power of Movement in Plants

This chapter has been reorganized, expanded, thoroughly revised, and includes new illustrations in response to a dozen expert reviewers in the field. It has been an active research area during recent years. Phototropism and gravitropism are updated, including evidence that calcium as well as differential sensitivity to auxins may play a role in gravitropism. The turgorins are described as hormones that control leaf movements.

## Chapter 20

### Photomorphogenesis

Major revisions include evidences for two major kinds of phytochromes that differ in chemistry, stability, and function. Light control of genetic activity, including control by phytochrome of its own synthesis, receives much more attention. The importance of phytochrome in plant ecology is expanded, especially in relation to early detection of neighbor plants.

## Chapter 21

### The Biological Clock: Rhythms of Life

Although studies of plant clocks have not been extensive during recent years, this chapter has been revised in various ways, particularly to include a discussion of the spectrum of biological rhythms from the very short to the very long and a description of how ocean tides are controlled.

## Chapter 22

### Growth Responses to Temperature

The discussion of seed physiology has been considerably expanded. Recent advances in our understanding of tuber formation are also included with evidence for a tuber-inducing hormone.

## Chapter 23

### Photoperiodism

Activity in this field has increased during recent years as this chapter points out. The role of phytochrome (in its two forms) in photoperiodism is now more clearly understood and explained.

## Chapter 24

### Molecular Genetics and the Plant Physiologist

This new chapter introduces the rapidly expanding areas of plant molecular biology and genetic engineering to students of plant physiology. The chapter introduces the importance of genetic approaches in understanding plant processes. It then explains how to isolate, clone, purify, and sequence genes. Northern, Southern, and Western blotting methods are explained and illustrated. Transgenic and transformed plants are explained (with two photographs of the latter). Mechanisms for control of gene expression are discussed, with emphasis on *cis* and *trans*-acting factors. Finally, some examples of genes activated by phytochrome, by hormones, by temperature, and by the developmental stage are described.

## Chapter 25

## Topics in Environmental Physiology

This chapter examines principles of plant response to environment such as saturation and limiting factors. As an example, plant adaptations to the radiation environment are outlined (with a new section on effects of sun flecks on understory plants).

## Chapter 26

## Stress Physiology

This is another highly active field of contemporary research. The chapter was extensively revised and reorganized to report the latest developments including a revised discussion of the nature of stress, the critical role of membranes, and heat-shock proteins and other proteins that appear in response to various stresses.

## Appendix A

## The Système Internationale: The Use of SI Units in Plant Physiology

This appendix has been expanded to more thoroughly cover the conventions that plant physiologists should use when reporting their data. It should serve as a valuable reference source as well as a teaching aid.

## Appendix B

## Radiant Energy: Some Definitions

Except for some reorganization and a few important changes in terminology that have become commonplace among plant physiologists during the last few years, this appendix remains mostly as it was in the Third Edition.

## Appendix C

## Gene Replication and Protein Synthesis: Terms and Concepts

This appendix includes only minor changes. Some material about the regulation of protein synthesis was deleted because it is now an important part of Chapter 24.

# Contents in Brief

# Contents in Detail

# PLANT PHYSIOLOGY

# ONE

# Cells: Water, Solutions, and Surfaces

# 1

# Plant Physiology and Plant Cells

**Plant physiology** is the science that studies plant function: what is going on in plants that accounts for their being alive. Plants are not as inanimate as they sometimes appear. (It may be difficult to tell a plastic plant from its real counterpart.) Studying plant physiology will broaden your appreciation for the many things that are happening inside plants. Water and dissolved materials are moving through special transport pathways: water from soil through roots, stems, and leaves to the atmosphere, and inorganic salts and organic molecules in many directions within the plant. Thousands of kinds of chemical reactions are underway in every living cell, transforming water, mineral salts, and gases from the environment into organized plant tissue and organs. And from the moment of conception, when a new plant begins as a zygote, until the plant's death, which could be thousands of years later, organized processes of development are enlarging the plant, increasing its complexity, and initiating such qualitative changes in its growth as formation of flowers in season and the loss of leaves in autumn.

Plant physiology studies all these things.

## 1.1 Some Basic Postulates

Plant physiology, like other branches of biological science, studies life processes that are often similar or identical in many organisms. In this introductory chapter, we state ten postulates or generalizations about science in general and plant physiology in particular. Then, because cell biology is so fundamental to plant physiology, we provide a review of plant cells as the main body of this chapter. Here are the postulates:

**1.** *Plant function can ultimately be understood on the basis of the principles of physics and chemistry*. Indeed, modern plant physiology in particular and biology in general depend upon the physical sciences, which in turn rely on mathematics. Plant physiology is essentially an application of modern physics and chemistry to understanding plants. For that matter, progress in plant physiology has been almost completely dependent upon progress in the physical sciences. Today, the technology of applied physical science provides both the instrumentation upon which research in plant physiology depends and the fundamental knowledge that is applied in interpretation of results.

Furthermore, plant physiologists accept the philosophical statement called the Law of the Uniformity of Nature, which states that the same circumstances or causes will produce the same effects or responses. This concept of cause and effect must be accepted as a working hypothesis (that is, accepted on faith). Although there is no way to prove that the principle always applies everywhere in the universe, there is also no reason to doubt that it does. It is possible that life depends on a spirit or *entelechy*[1] not subject to scientific investigation; but if that is assumed, then by definition we cannot use science to study life. The assumption that plants are mechanistic leads to fruitful research; the contrary assumption, called **vitalism**, has been completely unproductive in science. For example, convictions (yours or ours) about the existence of a Creator may help or hinder your appreciation of plant physiology but cannot play a direct role in the science itself.

**2.** *Botanists and plant physiologists study members of four of the five kingdoms of organisms currently recognized by many biologists* (Table 1-1), *but much discussion in this book is concerned with true plants and, indeed, with a relatively few species of gymnosperms and angiosperms*. Modern biologists consider a five-kingdom approach to a classifi-

[1]A hypothetical agency that is considered inherent in living substance, directing its vital processes but not being discoverable by scientific investigation.

**Table 1-1 A Simplified, Five-Kingdom Outline of the Classification of Organisms.**

VIRUSES: Exhibit properties of life only when present in cells of other organisms; considered by most biologists not to be alive when isolated from living cells.

I. MONERA:[a] prokaryotic organisms (no organized nucleus or cellular organelles), including bacteria, blue-green algae (cyanobacteria), and mycoplasms. (ARCHAEBACTERIA might be a separate kingdom.)

II. PROTISTA: Eukaryotic (true organelles and nucleus), mostly single-celled organisms, including protozoa (single-celled "animals"), some algae,[a] and slime molds.[a] (Some authors include all the eukaryotic algae, even multicellular forms.)

III. FUNGI:[a] The true fungi.

IV. PLANTAE:[a] Most algae and all green plants; true plants include the following, plus some minor groups not mentioned:
- Brown algae[a]
- Red algae[a]
- Green algae[b]
- Mosses and liverworts[a]
- Vascular plants (higher plants)
  - Ferns and relatives[a]
  - Cycads and rare gymnosperms[a]
  - Conifers (common gymnosperms)[b]
  - Flowering plants (angiosperms)[b]
    - Monocotyledons (monocots)
    - Dicotyledons (dicots)

V. ANIMALIA: Multicellular animals.

[a]Studied by plant physiologists
[b]Emphasized by plant physiologists

cation of living organisms to be far superior to the previous attempts to classify all organisms as either plants or animals, but there is still much controversy about the placement of certain groups, such as the slime molds and some of the algae. Suffice it to say that plant physiologists study blue-green algae (or cyanobacteria) and other prokaryotes studied by bacteriologists, various groups of algae, slime molds, true fungi, and representatives of all major groups in the plant kingdom. Nevertheless, our discussions here will strongly emphasize gymnosperms and the flowering plants, only occasionally referring to the other groups.

**3.** *The* ***cell*** *is the fundamental unit of life; all living organisms consist of cells, which contain either membrane-bound nuclei or comparable structures without membranes. Life does not exist in units smaller than cells. Cells arise only from the division of preexisting cells.* Collectively, these three statements are known as the **cell theory**. **Coenocytic organisms** (certain algae, fungi, and slime molds) do not have their organelles (mitochondria, nuclei, and so on) partitioned by membranes into units called cells. Are they exceptions to the theory—or are they multinucleate, single, or few-celled organisms? You decide.

**4.** *Eukaryotic cells contain such membrane-bound organelles as chloroplasts, mitochondria, nuclei, and vacuoles, whereas prokaryotic cells contain no membrane-bound organelles.*

**5.** *Cells are characterized by special macromolecules, such as starch and cellulose, that consist of hundreds to thousands of identical sugar or other molecules; in some macromolecules, such as lignin, groups of molecules may be repeated, or randomness may appear in the distribution of the component molecules.*

**6.** *Cells are also characterized by such macromolecules as proteins and nucleic acids (RNA and DNA) that consist of chains of hundreds to thousands of simpler molecules of various kinds (20 or more amino acids in protein and four or five nucleotides in nucleic acids). These chains include long segments of nonrepeating sequences that are preserved and duplicated (copied) when the molecules are reproduced.* These molecules, so typical of life, contain *information*, much as the sequence of letters in this sentence accounts for the information in the sentence. Information is transferred from cell generation to generation through DNA and from DNA to protein via RNA. The information in a protein bestows upon it certain physical characteristics and the ability to catalyze (speed up) chemical reactions in cells; proteins that catalyze reactions are called **enzymes** and are fundamental to life function.

**7.** *In multicellular organisms, cells are organized into tissues and organs; different cells in a multicellular organism often have different structures and functions.* The tissue-organ concept is more difficult to apply to plants than to animals, but typical plant tissues include, for example, epidermis, cortex, vascular tissues, and pith. The principal organs of a vascular plant are roots, stems, and leaves, which may be modified for various functions (for example, flowers).

**8.** *Living organisms are self-generating structures.* Through the process called **development**, which includes cell divisions, cell enlargement (especially elongation in stems and roots), and cell specialization or **differentiation**, a plant begins as a single cell (the fertilized egg or **zygote**) and eventually becomes a multicellular organism. In contrast to most animals, most plants continue to grow and develop throughout their lives by means of perpetually embryonic (dividing) regions of cells called **meristems**. Though much descriptive information is available, development is probably the least understood phenomenon of contemporary biology (about as mysterious as the functioning of the human brain).

**9.** *Organisms grow and develop within environments and interact with these environments and with each other in many ways.* For example, plant development is influenced by temperature, light, gravity, wind, and humidity.

**10.** *In living organisms, as in other machines, structure and function are intimately wedded.* Clearly, there could be no life functions without the structures of genes, enzymes, other molecules, organelles, cells, and often tissues and organs. Yet the functions of growth and development create the structures. Studies in plant physiology depend strongly upon plant anatomy, **cell biology**,[2] and structural and functional chemistry. At the same time, the structural sciences of plant anatomy and cell biology become more meaningful because of plant physiology.

## 1.2 Prokaryotic Cells: Bacteria and Blue-Green Algae

**Membranes** are the extremely thin layers of material, consisting mostly of lipids and protein, that separate cells, and often cell parts, from their surroundings. We discuss their nature below and especially in Chapter 7. **Prokaryotic cells**, those of bacteria, blue-green algae (cyanobacteria), and mycoplasms, have only the surface membrane that surrounds each cell. Any membranous material found inside such cells is likely to be an inward extension of the cell membrane. **Eukaryotic cells**, on the other hand, contain several kinds of **organelles** ("little organs"), each surrounded by a single- or double-membrane system (or half a unit membrane around lipid globules). The little-studied **Archaebacteria** differ so radically from other prokaryotic cells (as well as from eukaryotes) that it has been suggested that they constitute a separate kingdom of life (see Section 26.6).

The **nucleus** of the eukaryotic cell is surrounded by a double membrane, but the prokaryotic cell has only a central body called a **nucleoid**, which is surrounded by **cytoplasm** but not by a membrane. In bacteria, the nucleoid consists of a single piece of DNA about 1 mm long,[3] closed into a circle and tightly coiled and packed. This is the essential genetic material.

The term *prokaryotic* means "*before* a nucleus" (from the Greek), not *without* a nucleus. Fossil prokaryotes as

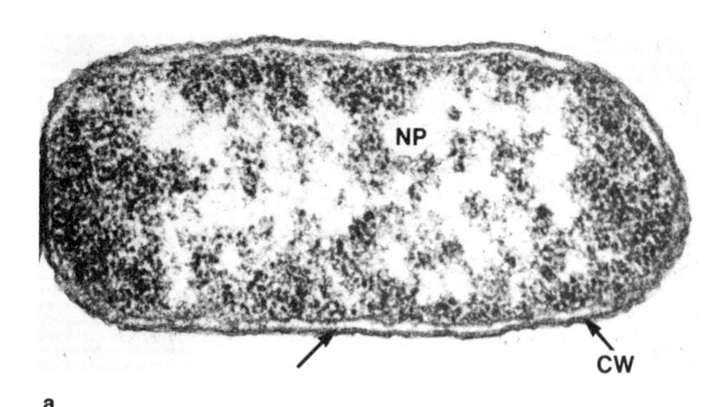


a

prokaryotic (bacterial) cell

plasma membrane
nucleoid
cell wall
capsule
ribosomes
mesosome

b

**Figure 1-1** (**a**) A prokaryotic cell, the bacterium *Escherichia coli*, magnified 21,500 times. The nucleoid (NP), the prokaryote equivalent of a nucleus, occupies the center of the cell, and the cytoplasm surrounding the nucleus is packed with ribosomes. The cell is surrounded by the cell wall (CW), and the plasma membrane (arrow) lies just beneath the cell wall. (Micrograph courtesy of William A. Jensen.) (**b**) An interpretation of a generalized prokaryotic cell. (W. A. Jensen and F. B. Salisbury, 1984.)

old as 3.3 billion years are known, whereas the oldest eukaryotic fossils are less than 1 billion years old. (*Eukaryote* is also from the Greek, meaning "true nucleus.")

Prokaryotic cells are comparatively small, seldom more than a few micrometers long and only about 1 μm thick (Fig. 1-1). Blue-green algal cells are usually much larger than bacterial cells. Also, all blue-green algae carry out photosynthesis with chlorophyll *a*, not found in bacteria, and by metabolic pathways common to plants and algae but not bacteria. Thus the term *cyanobacteria*, which implies that blue-green algae are just another form of bacteria, is perhaps unfortunate, although it has come into wide use.

Most prokaryotic cells are surrounded by **cell walls**. Because they usually lack cellulose, they are chemically different from the typical walls of higher plants. The wall may be anywhere from 10 to 20 nm thick and is sometimes coated with a relatively thick, jellylike **capsule** or slime of proteinaceous material. Inside the wall and tightly pressed against it is the outer membrane of the prokaryotic cell, the **plasma membrane** or **plasmalemma**, which may be smooth or have infoldings that extend into the cell, forming structures called **mesosomes**. Besides controlling what enters and

---

[2] *Cell biology* should properly be called *cytology*, but cytology has become preoccupied with the study of chromosomes; it should be called *cytogenetics*.

[3] Metric and Système Internationale (SI) units are summarized in Appendix A. In this chapter it is important to remember the prefixes that denote decreases of three orders of magnitude in size:

| | | |
|---|---|---|
| 1 *milli*meter (mm) | = 0.001 meter (m) | = $10^{-3}$ m |
| 1 *micro*meter (μm) | = 0.000 001 m | = $10^{-6}$ m |
| 1 *nano*meter (nm) | = 0.000 000 001 m | = $10^{-9}$ m |

Objects smaller than about 200 nm (half the wavelength of blue light, which has the shortest wavelength of visible light) are invisible in the conventional light microscope (they can be *visualized* but not *resolved* in video-enhanced interference light microscopes), but objects as small as 1 to 4 nm can be resolved in electron micrographs.

**Table 1-2 The Components of a Prokaryotic Cell.**[a]

I. CELL WALL (with or without a capsule)

II. PLASMA MEMBRANE or PLASMALEMMA (sometimes with infoldings called mesosomes)

III. NUCLEOID (single circular strand of DNA—the genetic material)

IV. CYTOPLASM (all the substance enclosed by the plasma membrane except the nucleoid)
   A. Ribosomes (sites of protein synthesis; about 15 nm in diameter, which is smaller than those in eukaryotic cells)
   B. Vacuoles (saclike structures, *much* smaller than in plant cells)
   C. Vesicles (small vacuoles)
   D. Reserve deposits (complex sugars and other materials)

V. FLAGELLA (threadlike structures protruding from cell surfaces; capable of beating to cause cell movement; built up from several intertwined, spiral chains of subunits of a protein called *flagellin*; about 15 to 20 nm in diameter, smaller than a single microtubule)

[a]Not all prokaryotic cells have all structures.

*Source*: Modified from W. A. Jensen and F. B. Salisbury, 1984.

leaves the cells, membranes have other important functions. Many enzymatic reactions, including photosynthesis and respiration, take place on the proteins contained in membranes, and the plasma membranes of prokaryotes are thought to play a role in cell replication.

Small spherical bodies, the **ribosomes**, crowd the cytoplasm and are the sites of protein synthesis. They are about 15 nm in diameter, smaller than in eukaryotes. The cytoplasm of the more complex prokaryotes may also contain **vacuoles** (saclike structures), **vesicles** (small vacuoles), and reserve deposits of complex sugars or inorganic materials. In some rare blue-green algae, the vacuoles are filled with nitrogen gas.

Many bacteria are capable of relatively rapid movement, generated by the action of threadlike structures, the **flagella**, that protrude from the cell surface. Prokaryotic flagella are chemically very different from eukaryotic flagella. Table 1-2 summarizes the structures of prokaryotic cells.

## 1.3 Eukaryotic Cells: Protist, Fungal, and Plant

The principal structures of prokaryotic cells are also present in eukaryotic cells, but the latter have several additional structures as well, most of them bound by membranes. A useful fiction in studying eukaryotic

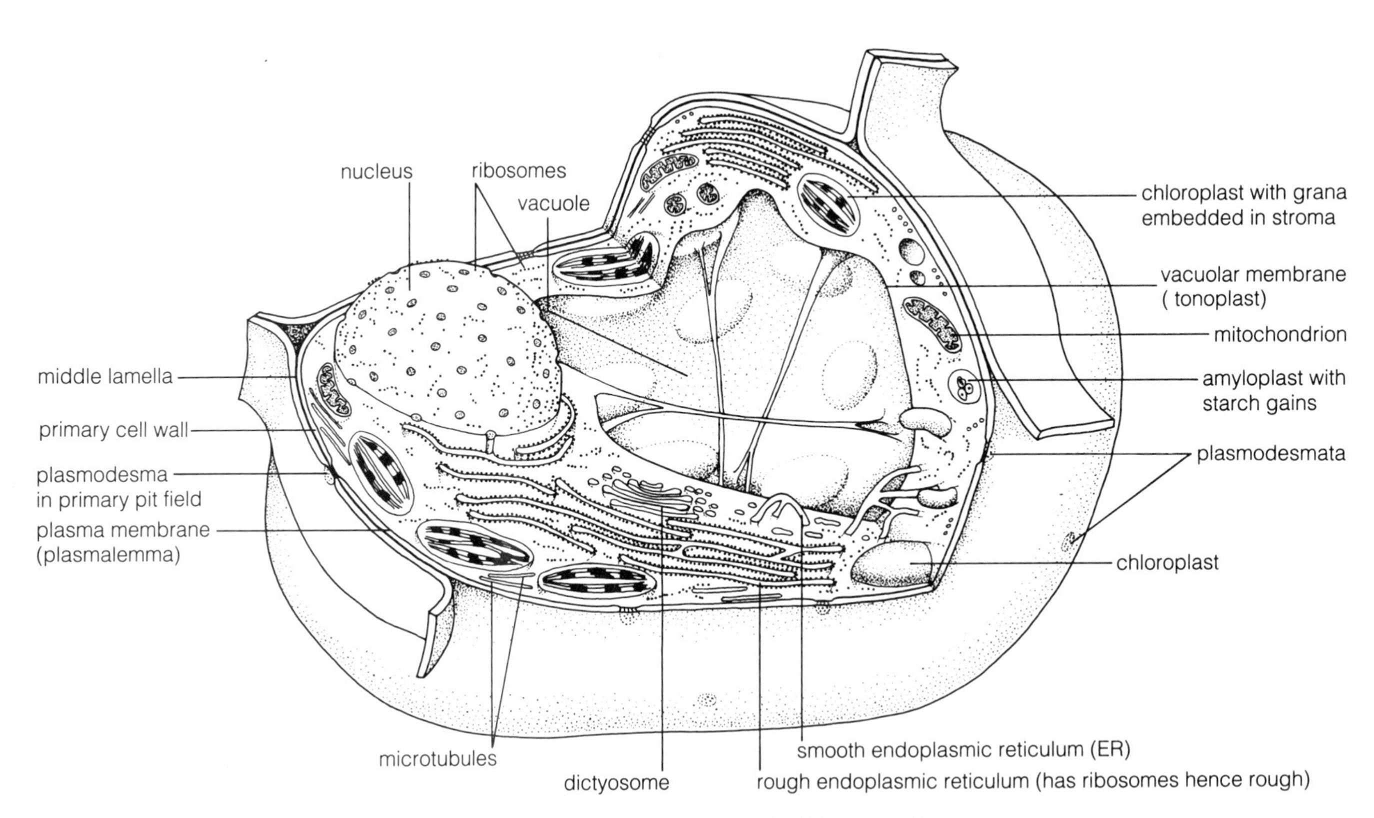


**Figure 1-2** A generalized plant cell. The drawing is based on the appearances of cellular organelles in electron micrographs. (Drawing by Cecile Duray-Bito.)

**Table 1-3 The Components of a Eukaryotic Plant Cell.**

I. CELL WALL[a]
  A. Primary wall (about ¼ cellulose); about 1 to 3 μm thick
  B. Secondary wall (about ½ cellulose + ¼ lignin); may be 4 μm thick or more
  C. Middle lamella (cementing layer between cells, mostly pectin)
  D. Plasmodesmata (strands of plasma membrane penetrating wall); 30 to 100 nm in diameter
  E. Simple and bordered pits

II. PROTOPLAST (contents of the cell exclusive of the cell wall); 10 to 100 μm in diameter
  A. Cytoplasm (cytoplasm + nucleus = protoplasm)
    1. Plasma membrane (plasmalemma); 0.01 μm (10 nm) thick
    2. Endomembrane system
      a. Endoplasmic reticulum (ER); 7.5 nm thick (each membrane; cisternae with two membranes vary in thickness)
      b. Golgi apparatus (consists of dictyosomes; 0.5 to 2.0 μm in diameter; membranes 7.5 nm thick)
      c. Nuclear envelope (two unit membranes); 25 to 57 nm thick
      d. Vacuolar membrane (tonoplast); 7.5 nm thick (see Vacuoles, below)
      e. Microbodies; 0.3 to 1.5 μm in diameter
      f. Spherosomes and protein bodies; 0.5 to 2.0 μm in diameter (surrounded by half of a unit membrane)
    3. Cytoskeleton
      a. Microtubules; 24 to 25 nm thick; core of 12 nm
      b. Microfilaments; 5 to 7 nm thick
      c. Other proteinaceous material
    4. Ribosomes; 15 to 25 nm in diameter (larger than in prokaryotes)
    5. Mitochondria (membrane-bound); 0.5 to 1.0 μm by 1 to 4 μm
    6. Plastids[b] (membrane-bound organelles)
      a. Proplastids (immature plastids)
      b. Leucoplasts (colorless plastids); amyloplasts (contain starch grains, sometimes protein: *proteinoplasts*); elaioplasts (contain fats); etioplasts; and other food-storage plastids
      c. Chloroplasts; 2 to 4 μm thick by 5 to 10 μm in diameter (may contain starch grains)
      d. Chromoplasts (often red, orange, yellow, and other colors)
    7. Cytosol (fluid in which most of above structures are suspended)
  B. Nucleus (cytoplasm + nucleus = protoplasm); 5 to 15 μm or more in diameter (see Nuclear envelope, above)
    1. Nucleoplasm (granular and fibrillar substance of the nucleus)
    2. Chromatin (chromosomes become apparent during cell division)
    3. Nucleolus; 3 to 5 μm in diameter
  C. Vacuoles (from nonexistent to 95 percent of cell volume; sometimes even more)
  D. Ergastic substances (inclusions of relatively pure materials, often in plastids or vacuoles)[a]
    1. Crystals (such as calcium oxalate)
    2. Tannins[b]
    3. Fats and oils (in elaioplasts or lipid globules)
    4. Starch grains (in amyloplasts and chloroplasts; see above)[b]
    5. Protein bodies
  E. Flagella and cilia; 0.2 μm thick, 2 to 150 μm long

[a]Occur in fungal, plant, and some protist cells, but seldom in animals.
[b]Occur only in plant cells and some protistans.

plant cells is the "typical" plant cell, illustrated in Figure 1-2 and summarized in Table 1-3. There is, of course, no such thing as the "typical cell" or the "average teenager." Both are statistical creations, composites of features characteristic of a class but seldom found all together in one individual. Nevertheless, **parenchyma cells** are usually thin-walled, often isodiametric (approximately spherical but with nearly flat faces), living cells that have most features of the typical plant cell. They are found in pith, cortex, mesophyll, and other tissues.

Our understanding of cells has depended to a great extent upon the tools we have had to investigate them. Figure 1-3 shows two kinds of micrographs of cells taken through light microscopes and one taken with the transmission electron microscope. Many features discussed in the remainder of this chapter can be seen in Figure 1-3. Let's begin from the outside of our typical plant cell and progress inward. The exercise will provide insight into how plants, protistans, and fungi differ at the cellular level from each other and from animals, but plant cells are emphasized.

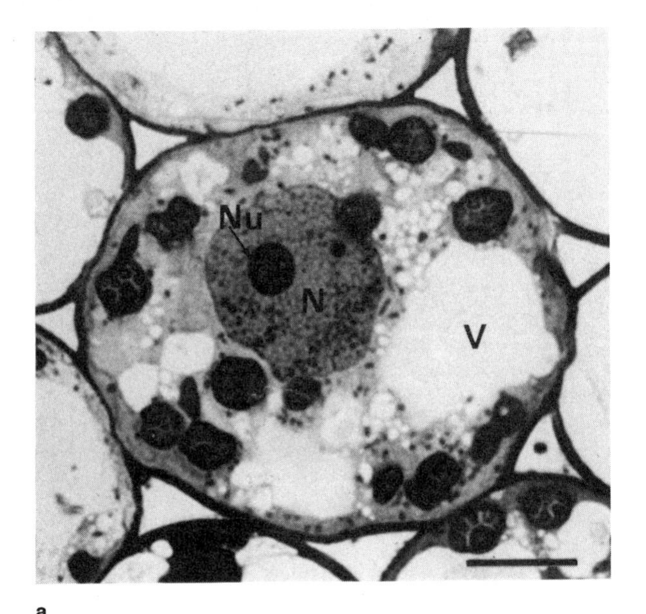


a

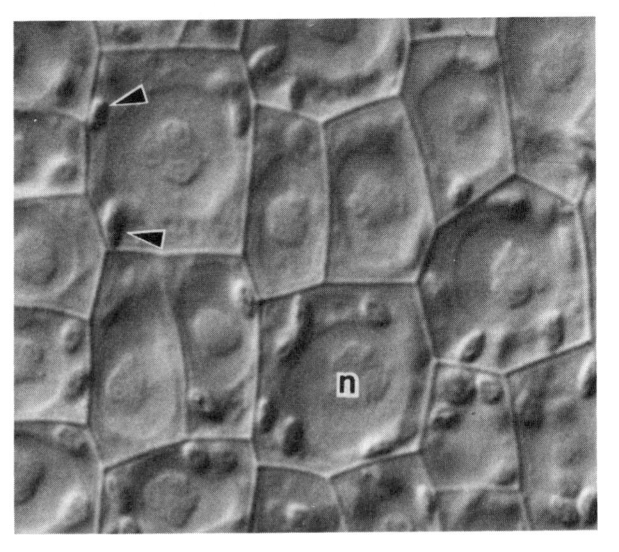


b

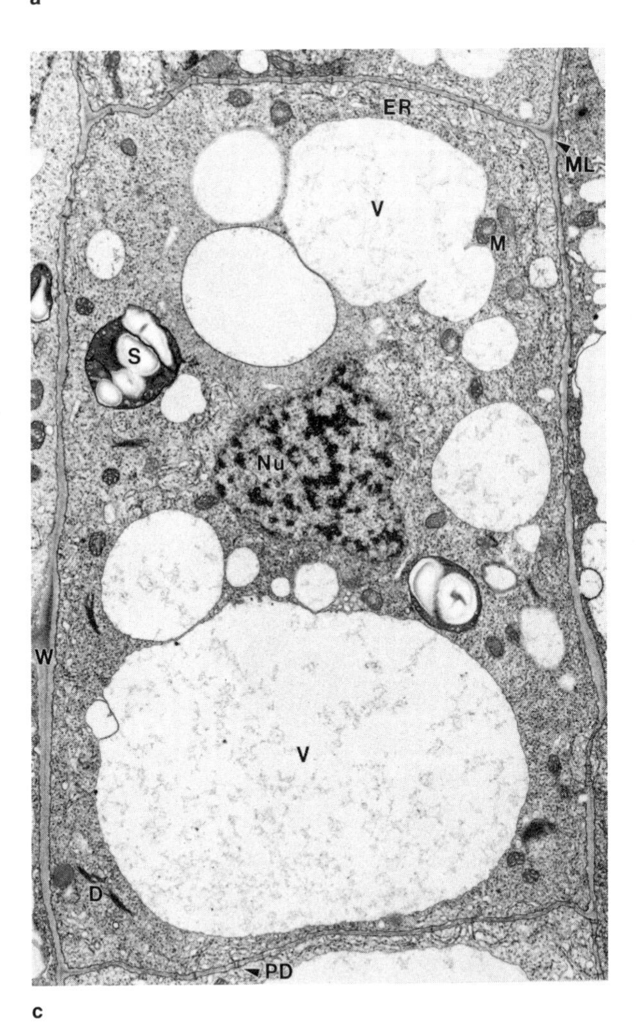


c

**Figure 1-3** Photomicrographs of cells, showing how the techniques of microscopy can influence our visual and thus mental images of cells. (**a**) Micrograph through the light microscope of a parenchyma cell from a maize coleoptile (the sheath that covers the first seedling leaf). The tissue was fixed with glutaraldehyde, sectioned at a thickness of 1 μm, and stained with toluidine blue. The nucleolus (Nu) is prominent in the nucleus (N). Numerous dark-staining amyloplasts (containing starch) are present in the cytoplasm, and the developing vacuole (V) is also prominent. Bar = 5 μm. (**b**) Differential interference contrast (Nomarski) light micrograph of living epidermal cells from the moss *Funaria*. Vacuoles have not yet formed in the cells, but the clear spherical structure in each cell is the nucleus, which contains several nucleoli (n). Young plastids are also visible (arrowheads). 900 ×. (**c**) A transmission electron micrograph of a growing cell from a pea stem. The vacuoles (V) occupy much of the volume of this still-expanding cell. The darker regions in the nucleus (Nu) are condensed chromatin (heterochromatin). Endoplasmic reticulum (ER), mitochondria (M), dictyosomes (D), and plastids containing starch (S) are present throughout the cytoplasm. At this low magnification, membranes are barely visible as dark lines surrounding the cell and its several organelles and vacuoles. Plasmodesmata (PD) in the cell wall (W) connect the protoplasts of adjacent cells. The middle lamella (ML) is especially noticeable where the intercellular spaces will form. 10,000 ×. (Micrographs courtesy of Fred Sack.)

## 1.4 The Cell Wall

Many protistans and most fungal and plant cells are surrounded by a **cell wall**. (Plant sperm and endosperm cells are exceptions.) Indeed, no other feature is more characteristic of fungal and plant cells than the wall. All cells have membranes that enclose their contents, but animal and some protist cells have no walls—*only* membranes, sometimes quite specialized. Young growing cells, some storage cells, the photosynthesizing cells of leaves, all parenchyma cells, and some other cell types have only a **primary wall**, a wall character-

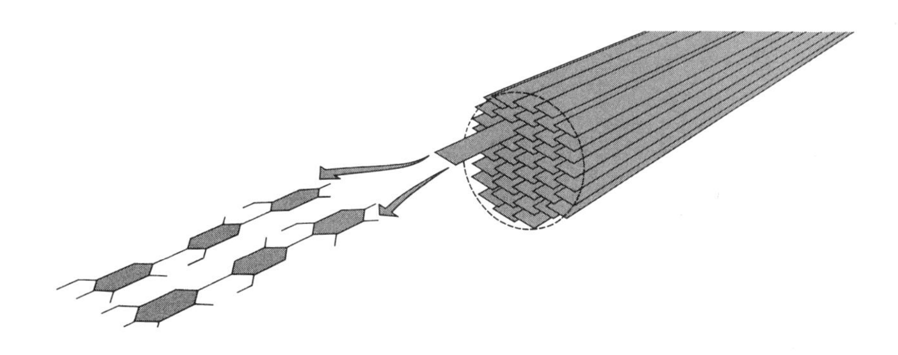

Figure 1-4 A schematic drawing of how cellulose molecules are arranged to form a microfibril of cellulose. Pairs of molecules are held together by hydrogen bonding to form the sheetlike strips, of which there are about 40 in each microfibril. Each strip is held to the ones above, below, and to its sides by hydrogen bonding (discussed in Chapter 2). The hexagons represent glucose molecules in the long-chain cellulose molecules. (From Jensen and Salisbury, 1984.)

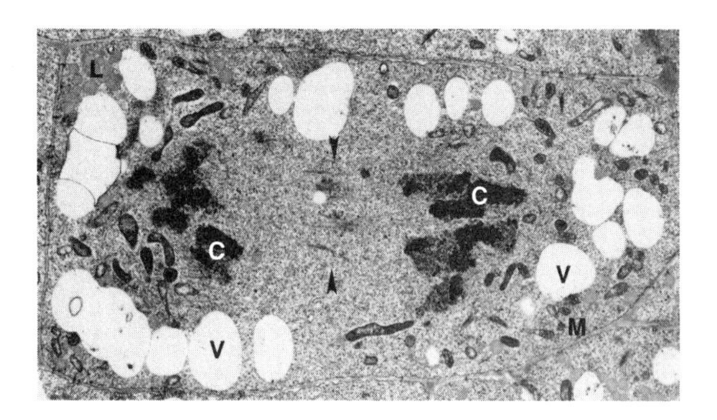


a

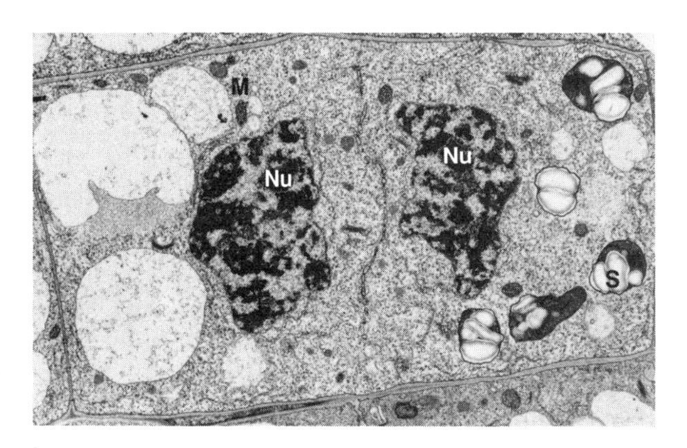


b

c

**Figure 1-5** Formation of the cell plate (phragmoplast) during **cytokinesis** (division of the protoplast after **mitosis**, which is division of the nucleus). (**a**) Early telophase in a maize (corn) root cell. A cell plate is beginning to form (arrowheads) between two groups of chromosomes (C) at the spindle equator. Most other components of the protoplast are distributed around the periphery of the cell. Vacuoles (V) are apparent, as are mitochondria (M). 10,000 ×. (**b**) Late telophase with cytokinesis almost completed in a pea stem cell. The nuclear envelope has reformed around the daughter nuclei (Nu), and the cell plate has almost reached the parent cell walls. Chromosomes have become chromatin in the two nuclei. 9,000 ×. (**c**) Higher magnification view of early cell-plate formation. Microtubules (arrowheads) guide the numerous wall-building vesicles to the midzone, where they coalesce to form the cell plate. 36,000 ×. (Transmission electron micrographs courtesy of Fred Sack.)

ized by being thin and formed while the cell is undergoing growth. The cell wall surrounds the **protoplast**, which includes the plasma membrane and all that it encloses. This membrane is usually pressed tightly against the wall because of the pressure in the fluids inside. Between the primary walls of adjacent cells is the **middle lamella** that cements the two cell walls together. Many mature plant cells, especially those in xylem tissues that have finished growing, have laid down a **secondary wall** between the primary wall and the cell membrane.

## The Primary Cell Wall

Compared with an entire cell, or even with the secondary wall, the primary wall is thin, on the order of 1 to 3 μm thick (about the thickness of an entire bacterial cell). It consists of about 9 to 25 percent **cellulose**. Long, unbranched cellulose molecules are bound together in long cylindrical fibers called **microfibrils**, which are 4.5 to 8.5 nm thick (Fig. 1-4). Because of the parallel arrangement of the cellulose molecules, microfibrils behave like crystals and have as much tensile strength, for their weight, as steel wires in a cable. The primary wall begins to form between cells that are completing cell division; a **cell plate** (or **phragmoplast**) forms where a new cell wall will divide the two daughter cells. This cell plate forms from microtubules (see below) that are separate from those that formed the spindle during separation of the chromatids (Fig. 1-5). These microtubules may guide materials (via Golgi vesicles; see below) to the cell plate, which grows from the center toward the

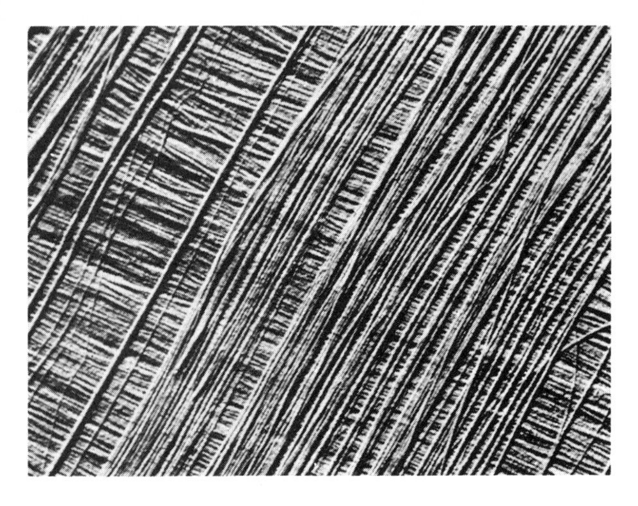

**Figure 1-6** Ordered pattern of cellulose microfibril deposition in a maturing cell wall of a green alga. 21,000 ×. (Courtesy of E. Frei and R. D. Preston, see Frei and Preston, 1961.)

outside by the addition of pectins and other materials, eventually joining with the cell wall that surrounded the mother cell before cell division began. Much of the material that forms the cell plate becomes the pectin-rich middle lamella between the two cell walls of the daughter cells.

In many elongating cells, the cellulose microfibrils are laid down outside of but next to the plasmalemma, roughly at right angles to the long axis of the cell but angling to the right or left to form a helix of low pitch. They are like short threads wrapped around a cylinder (the protoplast). As the cell elongates, the microfibrils slip past each other and are pulled into the long axis of the cell (discussed below and in Chapter 16). Because the most recently deposited microfibrils are still laid down predominantly parallel to the circumference (but often angling in a direction opposite to those previously laid down), the newer and older layers of microfibrils cross each other almost as threads in cloth (but do not go over and under each other as in woven fabric).

The cellulose microfibrils are embedded in a **matrix** of other materials, which are chemically much more complex (Fry, 1986, 1988). Principal among these are the **hemicelluloses**, which form a branching, molecular network filled with water. A typical primary wall may contain 25 to 50 percent hemicelluloses. Closely related are the **pectic substances**, which make up 10 to 35 percent of a primary wall and are highly hydrated. Primary walls also usually have about 10 percent protein, which plays important roles in cell growth (proteins called **extensins**) and in recognition of foreign molecules (proteins called **lectins**). The matrix materials of the primary wall are not crystalline like cellulose, so the matrix cannot be seen in Figure 1-6; it was dissolved away to make the microfibrils more visible in the scanning electron micrograph (also see Fig. 7-10).

The primary wall is admirably adapted to growth. In response to growth-regulating molecules, it "softens" in some way such that the microfibrils both move apart in the longitudinal direction and slide past each other in the water-filled matrix (Chapter 16). This occurs as the protoplast absorbs water, expanding like a balloon and creating pressure against the wall. Thus the wall stretches *plastically* (irreversibly, like bubble gum) rather than *elastically* (like a rubber balloon) as the cell grows. Some primary walls increase their area as much as 20 times during growth. Much new material is added, so they don't become thinner.

When the cell is not growing, even the primary wall resists stretching, thanks to the high tensile strength of its cellulose microfibrils and the crosslinking with the matrix molecules and within the protein network. Yet the wall is porous enough to allow the free passage of water and materials dissolved in the water. The pores (openings between fibrils) are about 3.5 to 5.2 nm in diameter (Carpita et al., 1979; Carpita, 1982), compared with about 0.3 nm diameter for a water molecule and about 1 nm for a sugar molecule. We can think of the cell wall as the medium in which the protoplast functions. Indeed, water and solutes can move throughout the plant in this cell-wall medium called the **apoplast**.

Imagine a cotton (mostly cellulose) cloth bag with a water-filled balloon inside. The cloth is porous, freely allowing the passage of water and solutes dissolved in the water. It also has tensile strength, resisting stretching as one tries to force more water into the balloon. Yet it collapses when the water is released from the balloon. Likewise, if the cell loses water and hence its hydraulic pressure, the primary cell wall collapses (although not as much as a cotton bag; furthermore, the wall with its enzymes is not as inert as cotton cloth). Leaves and young stems are made of cells that have mostly primary walls. They are rigid while the fluid in their cells pushes against their walls, but they wilt when enough water is lost to decrease the internal pressure (Fig. 1-7).

## The Secondary Cell Wall

In many plant cells, especially those in xylem tissues that when mature will provide support for the plant or will conduct fluids under tension (negative pressure), the protoplast secretes a secondary wall after the cell has stopped enlarging. In all but a small portion of the cells in wood and cork, after the secondary wall has been secreted, the protoplast dies, and its contents are removed from the cell so only the wall remains.

In xylem cells that conduct fluids under tension (and sometimes in other cells), secondary wall deposition often produces rings, spirals, or networks (see Fig. 5-6). These beautiful structures prevent collapse from pressures produced by adjacent cells filled with pressurized fluids.

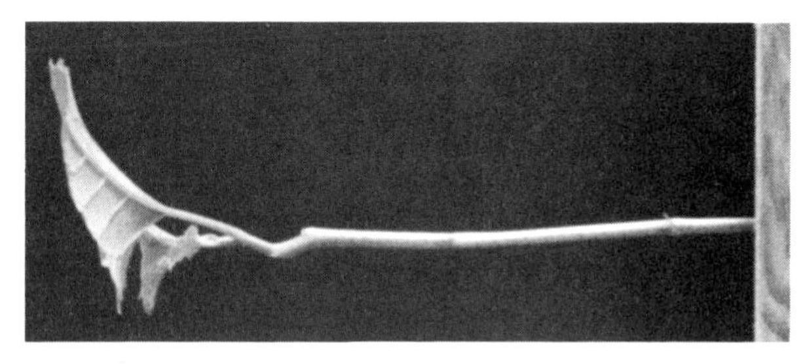

a

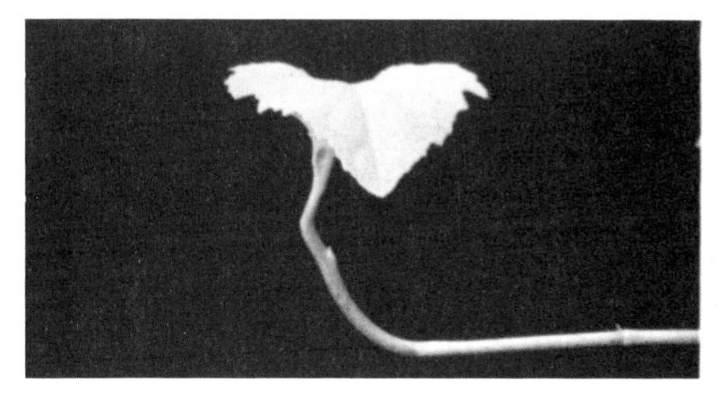

b

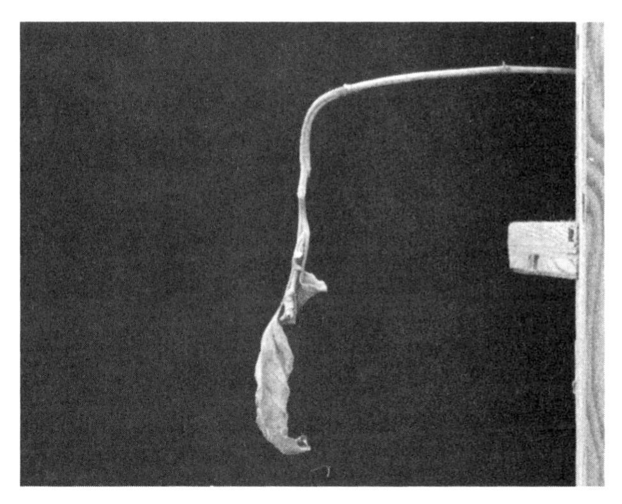

c

**Figure 1-7** How turgor pressure in cells determines the form of nonlignified tissues. (**a**) A cocklebur plant that has just been laid on its side. (**b**) The same plant 18 h later. The upward bend has occurred in the actively growing part of the stem where lignin has not yet been deposited in the cell walls. (**c**) The same plant, 8 days later, after it has dried out and wilted. Note that the lignified part of the stem that did not bend upwards in response to gravity also did not wilt as the plant dried out. (Photographs by F. B. Salisbury.)

Secondary walls are usually much thicker than primary walls; some are several micrometers thick. Secondary walls consist of about 41 to 45 percent cellulose, 30 percent hemicelluloses, and in some cases 22 to 28 percent **lignin** (Chapter 15), which is not easily compressed and resists changes in form; that is, lignin is much more rigid than cellulose. But the combination of

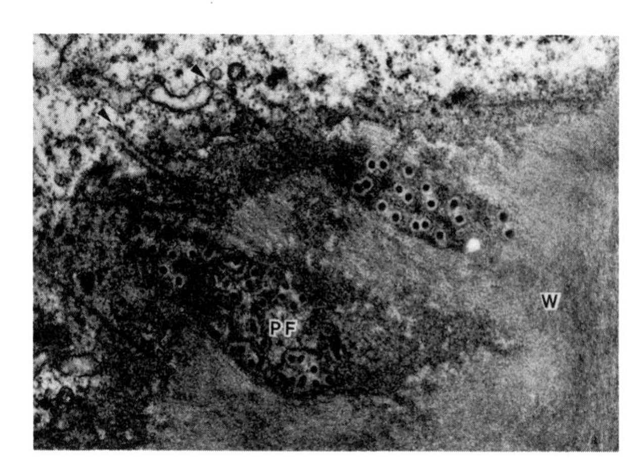


**Figure 1-8** Glancing section through the cell wall (W) and the underlying protoplast of a pea stem cell. There are two primary pit-fields (PF) that contain numerous plasmodesmata. Microtubules (arrowheads) are closely associated with the plasmalemma. 40,000 ×. (Courtesy of Fred Sack.)

stretched cellulose microfibrils embedded in lignin, like the steel rods under tension and embedded in concrete to form prestressed concrete, gives wood its strength. For its weight, wood is one of the strongest materials known. It certainly does not wilt when it loses water!

When a cell that is to have a secondary wall stops enlarging, lignin is first deposited in the already-formed middle lamella, then in the existing primary wall, and finally in the secondary wall, as it forms.

## The Middle Lamella

The pectic substances that cement adjacent cells together form the middle lamella and are ideally suited to their role because they exist as gels. Indeed, we extract them from unripe fruits, in which they are plentiful, and use them in making jams and jellies. Pectins can be broken down by certain enzymes, as happens when many fruits ripen. A green peach, for example, is rock-hard, but as it ripens the tissues become mushy or pulpy.

## Pits, Plasmodesmata, and Other Cell-Wall Features

Primary walls usually have thin areas called **primary pit-fields**. These have a high density of **plasmodesmata** (singular: **plasmodesma**), which are extremely thin strands of cytoplasm that extend through the walls of adjacent cells, connecting the protoplasts of adjacent cells (Fig. 1-8). Plasmodesmata (discussed in Section 7.3) have been seen for decades, but their detailed structures could be understood only after development of the electron microscope. They appear as channels lined

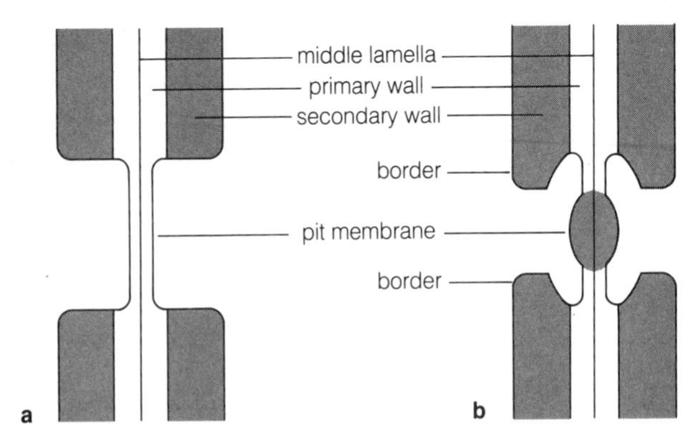


**Figure 1-9** The main characteristics of simple (**a**) and bordered (**b**) pits. (Jensen and Salisbury, 1984.)

by extensions of the plasma membrane of adjacent cells and filled with a tube, about 40 nm in diameter, of the special membrane system we shall discuss below, the endoplasmic reticulum. The tube is called the **desmotubule**. Plasmodesmata may make up about 1 percent of the area of the cell wall, although this varies considerably for different cell types. Plasmodesmata are thought to be of considerable importance because they unite the many protoplasts of a tissue or plant into one functional whole (called the **symplast**). We can calculate that substances like glucose can pass from cell to cell through the plasmodesmata some thousand times faster than across only membranes and cell walls, but particles larger than about 10 nm cannot pass through plasmodesmata, and we are not yet sure whether substances pass through the desmotubule (that is, within the endoplasmic reticulum) or through the space between the desmotubule and the plasmalemma. In any case, the plasmodesmata act as conducting channels for solutes from cell to cell.

Secondary-wall material is not laid down over the primary pit-fields, so the secondary wall includes characteristic openings called **pits** (Fig. 1-9). A pit in a cell wall usually occurs opposite a pit in the wall of the adjacent cell, constituting a **pit-pair**, and the two primary walls and middle lamella between the pits form the **pit "membrane."** There are two kinds of pits: simple and bordered (Fig. 1-9). In bordered pits, the secondary wall arches over the pit cavity. The networklike appearance of some cells with secondary walls (see Fig. 5-6) is caused by the presence of numerous pits.

## 1.5 Eukaryotic Protoplasts

As you can see from Table 1-3, we can divide the contents of protoplasts into three main parts: cytoplasm, nucleus, and vacuole(s). (There are also ergastic substances and organs of locomotion.) All eukaryotic cells have cytoplasm and, when young, at least one nucleus, but the nucleus disappears in sieve elements and some other plant cells as they mature. The entire protoplast is absent from mature xylem elements (tracheids and vessels). The presence of a large vacuole (and of ergastic substances) is unique to plant and fungal cells.

As in the prokaryotes, the **cytoplasm of eukaryotes** is a complex, watery matrix containing many molecular substances, some in colloidal suspension, but membrane-bound organelles are also prevalent. Originally, the term *cytoplasm* was used to designate the matrix surrounding the nucleus. Because of advances in electron microscopy and the discovery of organelles, however, the concept of cytoplasm is evolving and imprecise. Following rather widespread usage, we use the term **cytosol**[4] for the matrix in which cytoplasmic organelles are suspended, realizing that it contains the cytoskeleton discussed below and is much more highly organized than the thin "soup" that might come to mind.

In any case, the cytoplasm, plasmalemma, and nucleus combined can be called **protoplasm**, a term used less often now than formerly. Because most of the chemically functioning parts of the cell occur in the protoplasm, we might think of it as the "living" part of the cell, but chemical changes also occur in the cell wall (for example, the softening that allows growth) and even in the vacuole.

The **vacuole**, which is a volume of water and dissolved materials surrounded by a membrane, usually occupies 80 to 90 percent or more of a mature plant cell. Most mature living plant and fungal cells have a large central vacuole. Small vacuoles (1.0 μm in diameter) are present in certain animal and protist cells, but these seldom resemble the typical vacuoles of plant and fungal cells. A few plant cells also have accumulations of relatively pure nonliving substances such as calcium oxalate, protein bodies, gums, oils, and resins, collectively called **ergastic substances**. These may be in vacuoles, walls, or other parts of the cell.

Each kind of organelle in the cytoplasm is the site of specific chemical processes, as we learn when we separate them by ultracentrifugation and study their biochemical activities. This segregation and organization of processes, typically by membranes, makes the complex chemistry of cells possible partially by increasing efficiency, much as an assembly line does for industrial production. It also allows seemingly incompatible activities, such as the synthesis and breakdown of the

[4]Originally, the term *cytosol* was defined as "that portion of the cell which is found in the supernatant fraction after centrifuging the homogenate at 105,000 × g for 1 hour." (See discussion in Clegg, 1983.) It has now become common practice to use the term with reference to intact cells.

same kinds of molecules, to go on within the same cell at the same time. Cells are highly compartmentalized.

Dynamic action is a living cell's normal condition. Some organelles grow, divide, change shape, contain the enzymes that catalyze thousands of metabolic reactions, and secrete substances through the membrane to the outside wall or the outside world. They take part in growth and specialization of cells and are involved in a myriad of vital activities. Indeed, in many plant cells, the cytoplasm can be seen with the light microscope to stream around the cell periphery and through the cell interior (**cytoplasmic streaming**). The cell is the unit of life, and like all living things it is a changing, vibrant entity.

## 1.6 The Components of Cytoplasm

It may not be possible to construct a completely logical outline of the components of cytoplasm, but the one presented in Table 1-3 comes close. A concept developed over the past decade is that of the **endomembrane system**, which includes the endoplasmic reticulum, the Golgi apparatus, the nuclear envelope, and other cellular organelles and membranes (such as the microbodies, spherosomes, and vacuolar membrane) that have their origins in the endoplasmic reticulum or Golgi apparatus. The plasma membrane is usually considered a separate entity, although it grows by adding vesicles from the Golgi apparatus. Mitochondria and plastids are surrounded by a single smooth membrane and an inner, often convoluted membrane; both membranes appear similar to those of the endomembrane system, but they are apparently self-reproducing and thus are not related to the endomembrane system. Ribosomes are not part of the endomembrane system, nor are microtubules and microfilaments, which form the cytoskeleton. Next we'll consider each of the components of cytoplasm.

### The Plasma Membrane or Plasmalemma

The plasma membranes of eukaryotic and prokaryotic cells are basically similar in nature. In both cases, they regulate the flow of dissolved substances in and out of cells. Osmosis, which functions because water passes through the membranes more rapidly than solutes, regulates the flow of water.

With the electron microscope, the plasmalemma of suitably prepared sections appears as two dark lines separated by a light area. The dark lines are each about 2.5 to 3.5 nm thick and the light area is about 3.5 nm wide, for a total thickness of about 10 nm. This is the **unit membrane**, which is interpreted according to the **fluid mosaic model** that we discuss in Section 7.4. Briefly, consider the membrane to be a lipid bilayer, with the **hydrophilic** ("water-loving") parts of the lipid molecules at the surfaces accounting for the dark lines in electron micrographs, and the **lipophilic** ("fat-loving") parts of the molecules toward the inside of the bilayer accounting for the light space. Protein molecules, which make up as much as 50 percent of the membrane material (but can be much less than this), float in the lipid bilayer, with one or both hydrophilic ends penetrating one or both surfaces of the membrane. The two surfaces of the membrane are typically different, and there is much variety among plasma membranes of different cells and especially among the other membrane systems in the cells: those surrounding various organelles and those that make up the endomembrane system.

### The Endomembrane System

The endomembrane concept was introduced by D. James Morré and H. H. Mollenhauer (1974). Since then it has been discussed and considered by many workers and found to be a valuable concept in our understanding of cells (reviewed by Harris, 1986). The important thing about the endomembrane system is that it plays a major role in the production of cytoplasmic organelles, the deposit of materials within them, and the biosynthesis and transport of material destined to be secreted out of the cell.

*Endoplasmic reticulum (ER)* In very thin sections prepared for transmission electron microscopy, the entire cytoplasm of many eukaryotic cells appears to be filled with a system that consists of two parallel membranes: the **endoplasmic reticulum** or **ER** (Fig. 1-10). High-voltage transmission electron micrographs of thick sections have clarified the three-dimensional aspect of the ER. In many cells, the ER resembles a collapsed sack with layers called **cisternae**; in others, there are thousands of tiny tubes called **tubules**. The tubules are connected to cisternae, with hundreds of tubules extending outward from each cisterna. ER that has many attached ribosomes is called **rough ER**. Typically, it occurs in parallel layers (cisternae). **Smooth ER** lacks ribosomes and often forms tubules, but some of these features could turn out to be fixation artifacts. The tubular forms are especially prevalent in cells that are transporting or secreting lipids or sugars, whereas the cisternal forms with their ribosomes are often engaged in protein synthesis. Actually, it has been shown that one form may change into the other within a few minutes. The transport function could be between cells via plasmodesmata as well as within cells, and it has been suggested that a layer of ER close to the plasmalemma could help regulate what goes in and out of cells. The ER is a highly dynamic component of the cytoplasm, and it may have several functions that are not yet completely appreciated. For example, cellular organelles,

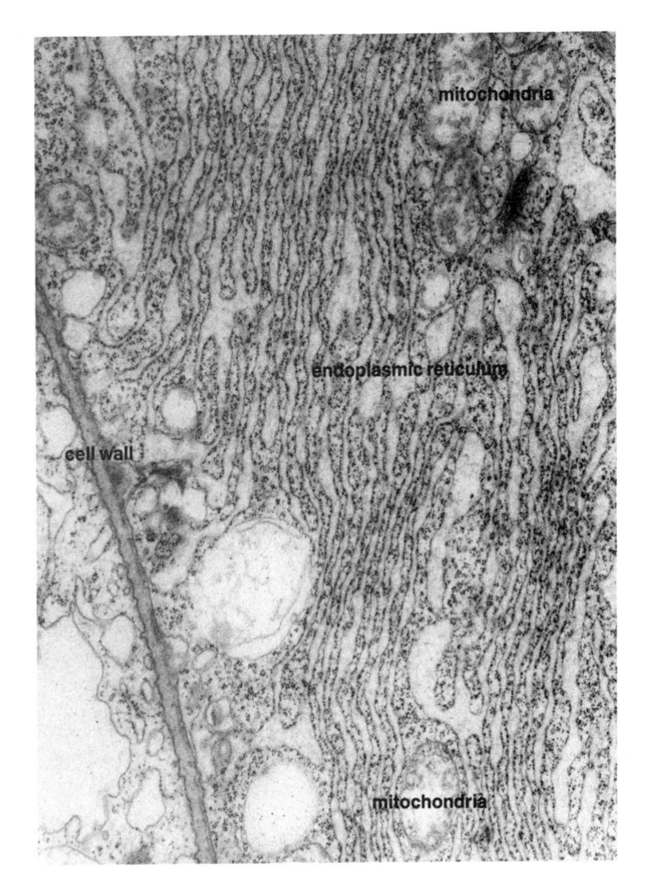


**Figure 1-10** A section through two adjacent cells of the developing fruit of shepherd's purse (*Capsella*) showing extensive rough endoplasmic reticulum (ER) in the upper cell. (Courtesy of Patricia Schulz.)

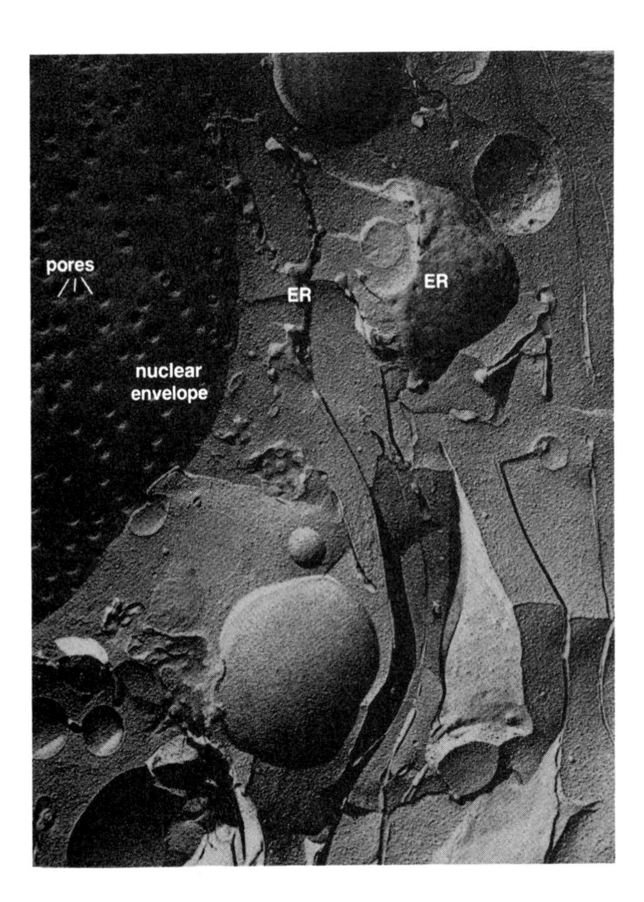


**Figure 1-11** A replica of a fractured onion root-tip cell prepared by freezing the tissue and then causing it to fracture, after which metal is deposited onto the fractured surfaces to make the replica. The fracture lines often split lipid bilayers of membranes. In this splendid photograph, the nuclear envelope is exposed, showing many pores. Fractures in the endoplasmic reticulum are visible in the cytoplasm outside the nucleus. 40,000×. (Micrograph by Daniel Branton.)

sometimes associated with actin filaments (see below), have been seen to move only along paths parallel to the axis of the ER tubules (Allen and Brown, 1988).

Many chemical activities besides protein synthesis are associated with the ER, and metabolism could be the most important ER function. Rough ER becomes more extensive in cells that are especially active in synthesizing protein, but cells that are actively synthesizing oils have extensive smooth ER, suggesting that oils are synthesized by the ER membranes. The ER synthesizes much of itself, including the sterols and phospholipids that are essential parts of all membranes. Because the ER is the source of most of the membrane synthesized within cells, membrane synthesis must be one of its most important roles. ER provides a large surface area within each cell; we can calculate that each cubic centimeter of cells contains about one square meter of ER surface.

Another function of the ER is to transport certain enzymes and other proteins across the plasma membrane and out of the cytoplasm in the process of **secretion**. Some secreted proteins allow the cell wall to stretch plastically during growth, for example. This secretory function is carried out primarily as ER contributes to the formation of dictyosomes, as discussed below.

*Nuclear envelope* Surrounding the nucleus are two parallel unit membranes, together called the **nuclear envelope** (Fig. 1-11). The outer membrane is 7.5 to 10 nm thick, sometimes slightly thicker than the 7.5-nm inner membrane, which is separated from it by the **perinuclear space**, with a thickness of 10 to 40 nm, making a total thickness of 25 to 57 nm for the nuclear envelope. The ER is often seen to be connected to the nuclear envelope, and the perinuclear space is continuous with the space (the **lumen**) between the parallel membranes of the ER and, via the plasmodesmata, perhaps from cell to cell. When the nucleus divides, the nuclear envelope breaks up and disappears; there is some evidence that its parts are maintained within ER-like membranes. In contrast to other membranes, the nuclear envelope (at least the inner membrane) has a high level of strongly bound nucleic acids.

The nuclear envelope is perforated by many **pores**. In surface view, the pores are octagonal and about 70 nm in diameter (Fig. 1-11). The inner and outer membranes are joined to each other to form the margins of the pores, which appear to be lined with material, giving rise to a structure known as the **annulus** (Latin, "ring"), which fills the pore except for a narrow central channel. Sometimes these channels seem to be filled with particles about the right size to be ribosomal subunits caught in transit from nucleus to cytoplasm. Although the pores are thought to allow communication between nucleus and cytoplasm, they are not completely open. There is evidence that particles larger than about 10 nm cannot pass through them. The pores are quite close together and may constitute as much as 20 percent of the surface area of the nucleus. A single nucleus may have as many as 10,000 regularly spaced pores.

*The vacuolar membrane or tonoplast* A single membrane of great importance in plant and fungal cells is the one that surrounds the vacuole, known as the **vacuolar membrane**, or **tonoplast**. It resembles the plasmalemma, but it differs in function and is often slightly thinner (7.5 nm). Whereas the plasmalemma controls the entry and exit of solutes into the cytoplasm, the tonoplast transports solutes into and out of the vacuole and thus controls the water potential (discussed in Chapter 2 and other chapters) of the cell. This is especially important, for example, in the guard cells of the stomatal apparatus. Potassium and other ions are pumped in or out of guard-cell vacuoles; water follows osmotically, causing the cells to swell or shrink, closing or opening the stomates (Chapter 4). The tonoplast is ultimately derived from the ER, but recent evidence suggests that this may be via the Golgi apparatus, as is the case with the plasmalemma. In some cases, ER may swell (dilate) directly to form vacuoles.

*Golgi apparatus* Electron micrographs of suitably prepared tissues show stacks of flattened hollow disks with convoluted margins and surrounded by spherical bodies (Fig. 1-12). Each stack is called a **dictyosome** or **Golgi body**, and the one-to-many dictyosomes in a cell are collectively called the **Golgi apparatus** after the Italian, Camillo Golgi, who discovered them with a light microscope in 1898. The flattened hollow discs are called **cisternae**, a term also applied to flattened sheets of ER, and the spherical bodies around them, which apparently break off from them, are called **Golgi** or **dictyosome vesicles**. Usually, each dictyosome has four to six cisternae located about 10 nm apart, but rarely a dictyosome may consist of only one cisterna — or up to 20 or 30 in many algal cells. Sometimes fine fibrils can be seen bonding between cisternae. Dictyosomes are constantly changing as some of the cisternae grow while others shrink and disappear.

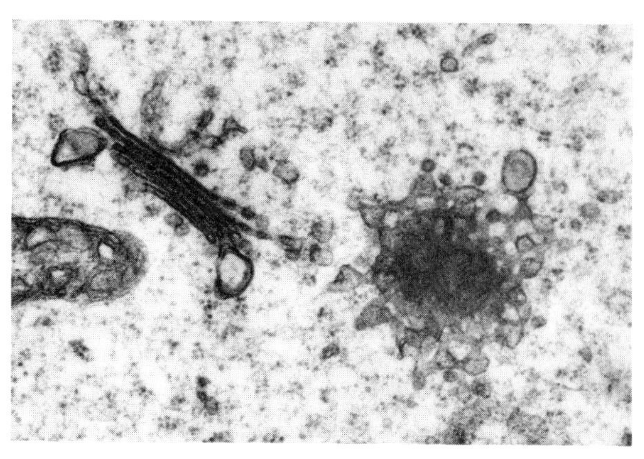

a

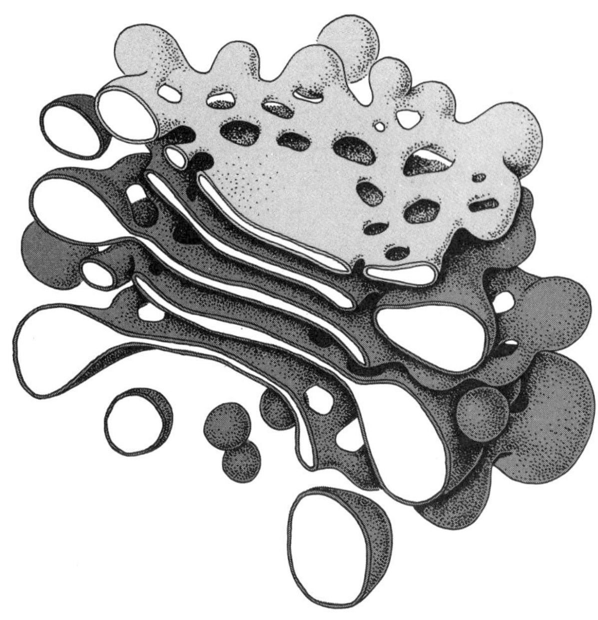

b

**Figure 1-12** Dictyosomes or Golgi bodies. The dictyosomes of a cell collectively form the Golgi apparatus. (**a**) Transmission electron micrograph of two dictyosomes from an *Arabidopsis* (mouse-eared cress) root-cap cell. The dictyosome on the left is seen from its side; the one on the right is seen end on. Large vesicles appear to be budding off from cisterna. 55,000 ×. (Micrograph courtesy of Fred Sack.) (**b**) A three-dimensional drawing of a generalized dictyosome from a plant cell. (From Jensen and Salisbury, 1984.)

The growth and disappearance of the cisternae help explain both the origin and major function of Golgi bodies. Dictyosomes are highly polar, and their membrane thickness and staining characteristics change from one end to the other. One side grows on its forming face (often called the ***cis* face**) when tiny vesicles with thin, lightly staining membranes from the ER fuse

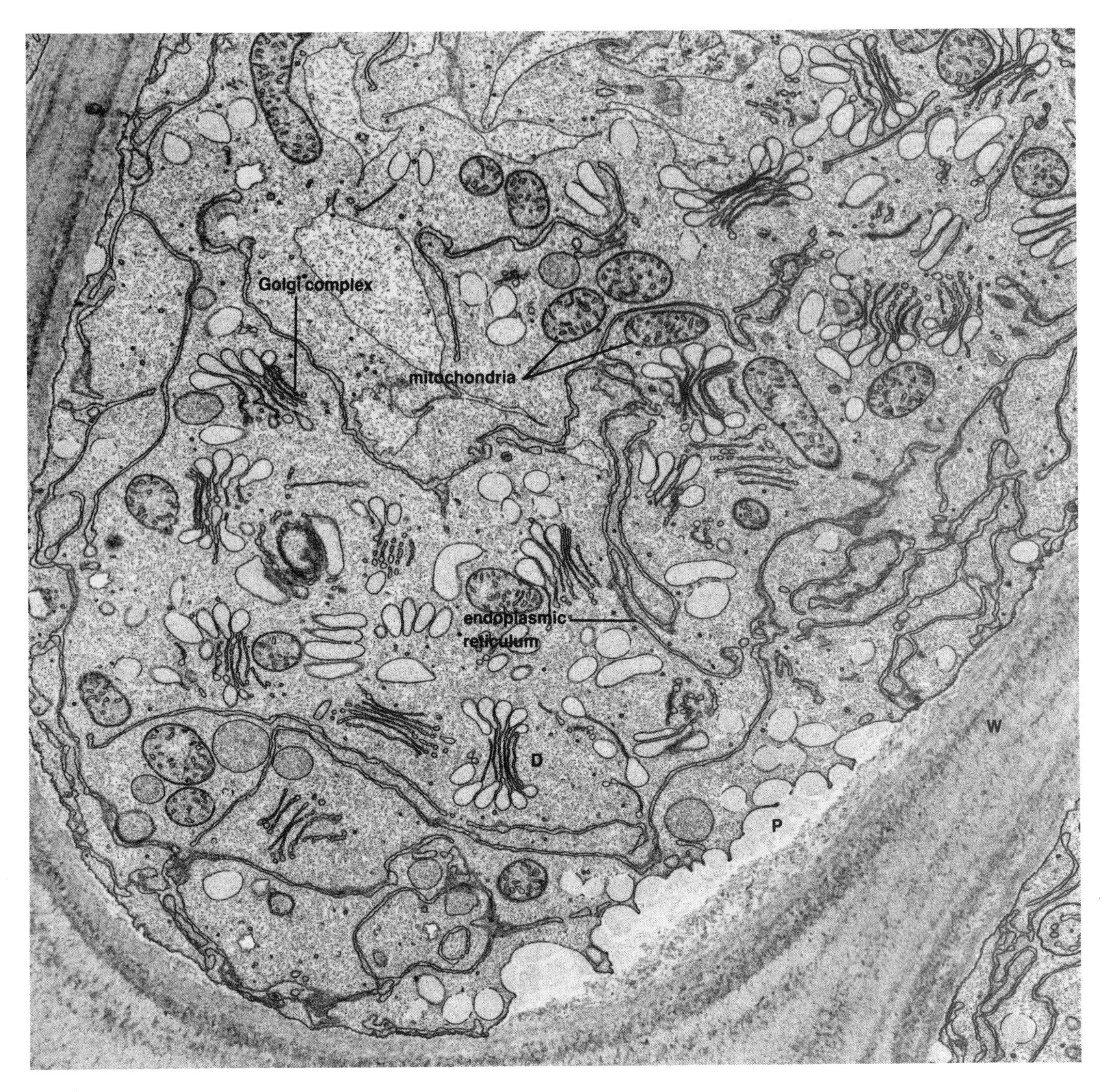


**Figure 1-13** Electron micrograph of a maize root-cap cell showing fusion of secretory products (P) and pinched-off dictyosomes (D). The plasma membrane and cell wall (W) are also visible. (From D. J. Morré et al., 1967.)

with the cisterna on that face (although some workers are not yet convinced that plant dictyosomes form *exclusively* from ER membrane). Such fusion increases the size of each cisterna, and its contents are changed slightly because membranes of the ER are not identical to those of Golgi bodies. Inside the cisterna cavity, newly absorbed or transported compounds may be changed into others. At the opposite side of a dictyosome (the maturing or releasing face, often called the ***trans* face**), other vesicles with new chemical properties are released.

Some of these vesicles are close to the plasma membrane adjacent to the cell wall, and fusion of Golgi vesicles with the plasma membrane increases its surface area during growth. Each vesicle contains matrix polysaccharides and proteins that contribute to the growth of the wall, but cellulose is not present in Golgi bodies and is probably formed in the plasma membrane just before moving into the wall (see discussion below).

Golgi bodies have functions other than contributing to growth of the plasmalemma (or modifying membranes in general) and transporting materials into the cell wall. For example, the slime on the outside of a root cap, which lubricates the root tip as it grows through the soil, is apparently secreted when Golgi vesicles fuse with the plasma membrane as just described (Fig. 1-13). It is likely that individual cisternae act as synthesizing units, although in some cases material may pass from one cisterna to another as synthesis occurs. Golgi bodies are abundant in most secretory cells.

*Microbodies* **Microbodies** are spherical organelles bounded by only a single unit membrane. They range in diameter from 0.5 to 1.5 μm and have a granular interior, sometimes with crystalline inclusions of protein. They also have their origin in the ER and thus are a part of the endomembrane system. Two important kinds are **peroxisomes** and **glyoxysomes**, each of which plays a special role in the chemical activities of plant cells. Peroxisomes break down glycolic acid produced in photosynthesis, recycling other molecules back to the chloroplast, as we discuss in Chapter 11. Glyoxysomes break down fats as they are converted to carbohydrates during and after seed germination. Hydrogen peroxide is a product of this reaction; it is broken down in the glyoxysomes (Chapter 15).

*Oleosomes and protein bodies* **Oleosomes** are spherical (as their previous name **spherosomes** implies) and are bounded by one-half of a unit membrane that is probably derived from the ER. They range in diameter from 0.5 to 2.0 μm. Many spherosomes contain mostly fatty materials and may well be centers of fat synthesis and storage. Some authors distinguish between small spherosomes that are present in nonstorage tissue and **lipid bodies** or **oleosomes** (*ole*, "oil"), but others say this distinction is artificial and use the terms *lipid body* or *oleosome* (depending on the author) exclusively. It has been suggested that lipid bodies form while lipids accumulate in the lipid part of the bilayer of an ER unit membrane, accounting for the "half-unit" membrane around lipid bodies. Lipid bodies or droplets without a surrounding membrane may also occur.

Particularly in the storage cells of developing seeds, protein is accumulated in membrane-bound organelles called **protein bodies**, which are also surrounded by unit membranes probably derived from the ER; they may be thought of as specialized vacuoles. In developing cereal grains, the protein bodies may virtually fill the cells of the subaleurone and outer layer of the starchy endosperm. In legume seeds, which have also been extensively studied, protein bodies form in the cotyledon storage parenchyma. The mechanisms of protein-body formation differ in the two systems, but we will not discuss the details.

*The endomembrane system revisited* We have noted that the ER, the outer membrane of the nuclear envelope, and the tonoplast are often seen in electron micrographs to be part of the same membrane system, although this system is almost certainly different in its different phases. Furthermore, microbodies, lipid bodies, and protein bodies form from the ER, as do dictyosomes, which in turn form the plasma membrane and probably the tonoplast. Thus all are related by connection or origin. The outer membrane of the nuclear envelope apparently forms young parts of the ER, and the ER then grows much more on its own as it synthesizes its own lipids and proteins and contributes to the Golgi-body cisternae and thus the growth of the plasmalemma. We shall further discuss the structures of membranes and some of their functions in Chapter 7.

### The Cytoskeleton

At least as early as the 1960s, thanks to developing techniques with the electron microscope, it was recognized that proteinaceous microtubules and microfilaments occur in virtually all (if not all) eukaryotic cells. Much was learned about the chemistry and structure of these cellular constituents, and functions in motility and cell-wall formation were postulated. It was shown that microtubules, actin microfilaments, and intermediate filaments form three different but closely integrated **cytoskeletal systems**, each with a characteristic distribution in the cell and each helping to determine the form of the cell. During recent years there has been much intense study of filamentous **cytoskeletons** in plants. The cell wall supports plant protoplasts, so the necessity for a cytoskeleton was questioned, and cell walls hindered the search for plant cytoskeletons. But cytoskeletonlike structures were found in isolated protoplasts from several dicot and monocot species (see, for example, Lloyd, 1982; Powell et al., 1982; Seagull, 1989); the process used involved dissolving the cell walls away with enzymes, leaving the protoplasts, and then using strong detergent solutions to remove the membranes and other cytoplasmic materials, leaving the fibrillar cytoskeleton (Fig. 1-14).

Now, after decades of dependence on the electron microscope to study microtubules and microfilaments, a new technique with the light microscope has been responsible for most of the advances. This technique is **fluorescence microscopy**. A substance that reacts specifically with the proteins of microtubules or microfilaments is combined with a fluorescent compound. This makes it possible to see these structures — and the plant cytoskeleton — in intact cells. Perhaps the most important of these substances are antibodies that are produced against specific proteins of microtubules or microfilaments (using techniques described later, page 444), but other fluorescent substances besides antibodies are also important, and the cytoskeleton has been studied in a considerable range of plant cells. This process has been facilitated by a number of new fixation techniques. As Clive W. Lloyd (1987) said, fluorescence

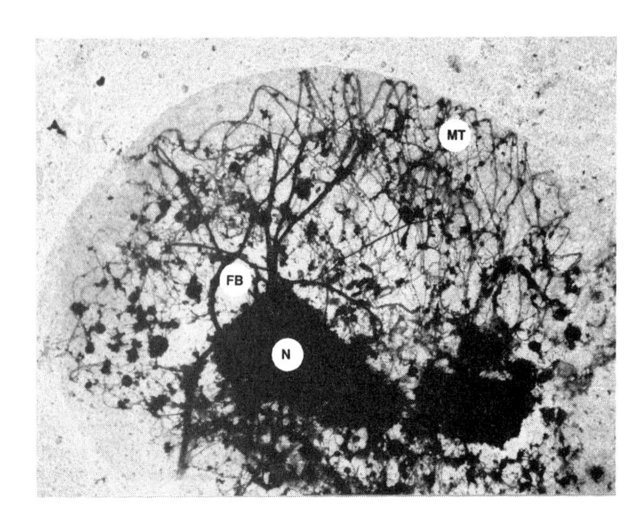


**Figure 1-14** Whole mount of a negatively stained cytoskeleton from a carrot cell protoplast grown in suspension culture. When such cells are extracted in an iso-osmotic, microtubule-stabilizing buffer that contains detergent, cytoskeletons that are insoluble in the detergent remain. When negatively stained, these are seen to consist of cortical microtubules (MT) seen as loops beneath the cell periphery, and of bundles of 7-nm fibrils (FB) that appear to link the nucleus (N) and the cell cortex. The latter cytoskeleton elements were found not to consist of the protein *actin*. (Micrograph courtesy of Clive Lloyd; see Powell et al., 1982.)

microscopy allows "plant cells to be seen—literally as well as metaphorically—in a new light." Before examining the cytoskeleton in detail, we need to consider the nature of microtubules and microfilaments.

*Microtubules* **Microtubules** are long hollow cylinders with an outside diameter of about 25 nm and a central core about 12 nm in diameter. They vary in length from a few nanometers to many micrometers. They consist of spherical molecules of a protein called **tubulin,** which spontaneously assemble under certain conditions to form the long hollow cylinders (Fig. 1-15; see also Figs. 1-5 and 1-8). In cross sections, microtubules consist of 13 subunits in a helical array. These subunits are part of 13 filaments of tubulin, each arranged to form a helix that is part of the wall of the microtubule. Each molecule of tubulin has a molecular weight of 110,000 Daltons (110 kDa) and is a dimer of two different proteins, called $\alpha$ and $\beta$ tubulin, which are arranged alternately along the length of each filament. Thus the microtubule has a polarity, which has important consequences for microtubule formation and action. Microtubules are visualized with the fluorescence microscope primarily by reacting them with antitubulin antibodies, and then with secondary antibodies that have been bonded to fluorescent molecules such as fluorescein. Many of the antibodies used thus far were prepared against molecules of tubulin obtained from animals or yeasts, but there are also antibodies against slime-mold, fungal, and plant tubulin.

Often, microtubules appear in electron micrographs to be surrounded by a narrow clear zone of open space. This could be an artifact of preparation, or it could represent a zone in which the microtubules interact with other components of the cell. Sometimes obvious bridges exist between adjacent microtubules and between units of the plasmalemma; these bridges must be important in maintaining the cytoskeleton structure and might give some rigidity to the cell. A good example is the generative cell of a pollen grain, which has no cell wall to maintain its structure.

Growth and breakdown of the microtubule depend on various factors, such as concentration of the tubulin molecules and presence of $Ca^{2+}$ and $Mg^{2+}$. Breakdown is favored by high $Ca^{2+}$ concentrations, low temperatures, high hydrostatic pressures, and the presence of various drugs, especially colchicine. Heavy water (deuterium oxide) favors microtubule formation. Such treatments can be used to investigate the role of microtubules in various processes such as cell division, for which colchicine has been used for many decades to prevent spindle formation during mitosis. Of course, such treatments could affect processes other than microtubule formation or disappearance. In experiments with isolated tubulin, high concentrations of tubulin produce growth of the microtubule at both ends; lower concentrations cause more growth at one end than the other; and still lower concentrations lead to a net breakdown of the microtubules already present. At a steady-state concentration, growth at one end is compensated for by loss of subunits at the other end. This "treadmilling" means that a given subunit moves along the microtubule, eventually being lost at the dissolving end.

In any case, such observations allow us to imagine how the cell might exercise considerable control over microtubule formation and dissolution. In animal cells, microtubules often arise from recognizable areas of cytoplasm called **microtubule organizing centers (MTOCs).** A good example is the basal body that initiates the growth of flagella, which consist of microtubules. In plants, such centers differ from MTOCs and appear as amorphous material at the spindle poles, in the phragmoplast following cell division, and in corners of growing cells. Their functions are presently not well understood.

The most important functions of microtubules are probably concerned with directed motion, especially of chromosomes in cell division or organelles within cells; with the control of the direction of cellulose microfibrils in cell walls; with cell motion itself; and with flagellar motion.

*Microfilaments* **Microfilaments** are smaller, solid structures, 5 to 7 nm in diameter (Fig. 1-16), that act alone or in coordination with microtubules to produce cellular movements. They also consist of protein, specifically the protein **actin,** which is also a significant

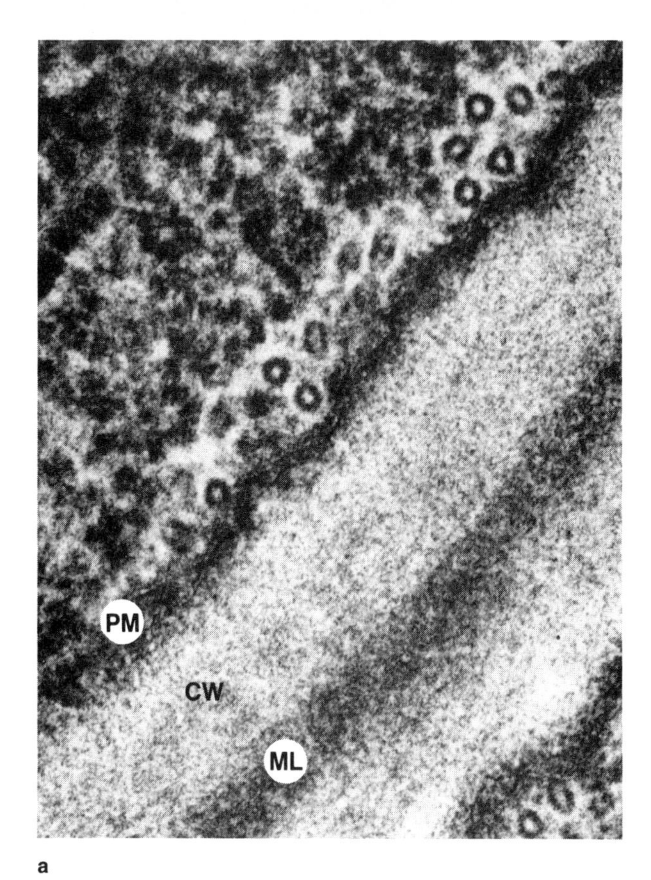


a

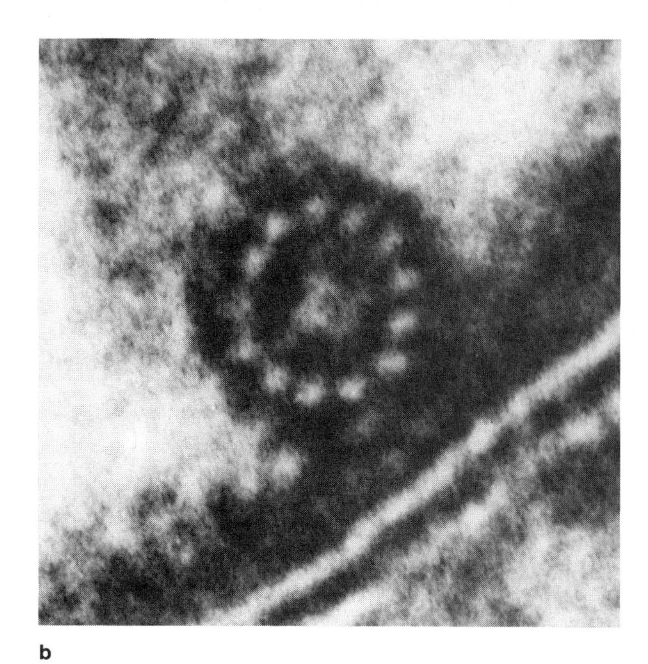

b

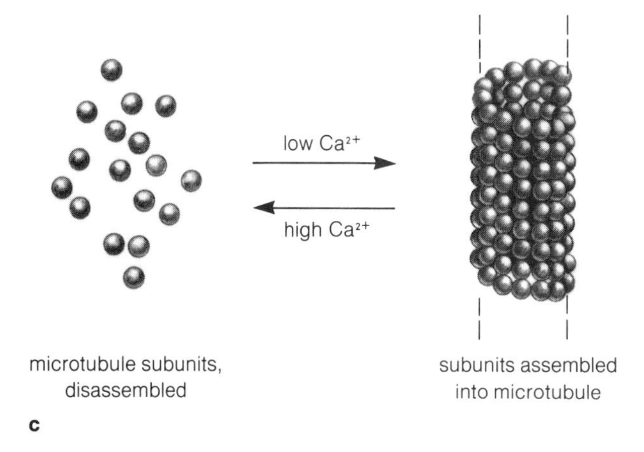


c

**Figure 1-15** Microtubules. (**a**) Part of a cell from a *Juniperus chinensis* root tip showing microtubules in cross section (small circles) next to the plasma membrane (PM). The primary cell wall (CW) and middle lamella (ML) are especially prominent. The microtubules are most abundant in a zone about 0.1 μm thick adjacent to the plasmalemma and are arranged circumferentially around meristematic cells, much as parts of hoops around the inside of a barrel. Along both the side and the end walls of these cells, the arrangement of these microtubules is very similar to that of the cellulose microfibrils (Figs. 1-4 and 1-6) in the adjacent wall. 51,000 ×. (**b**) Higher magnification (650,000 ×) of a microtubule from the juniper root cell. Note the 13 subunits that make up the microtubule, which is about 25 nm in diameter. (**c**) The assembly and disassembly of microtubule subunits in high and low concentrations of calcium ions. (Micrographs courtesy of Myron C. Ledbetter; see Ledbetter, 1965. Used by permission.)

constituent of animal muscle tissue. Activities of microfilaments apparently cause movements such as muscle contraction, cytoplasmic streaming, and **ameboid movements** (movements of single protist, fungal, and animal cells in which protoplasm flows out from the cell, forming a kind of "false foot," or **pseudopod**, followed by the rest of the cell flowing toward the pseudopod and resulting in movement of the cell along a surface). Slime molds, in particular, exhibit ameboid movements.

Microfilaments are visualized in the fluorescence microscope with antiactin antibodies (prepared against animal actin) or with fluorescent analogs of **phallotoxins** (from the fungus *Amanita phalloides*), which bind

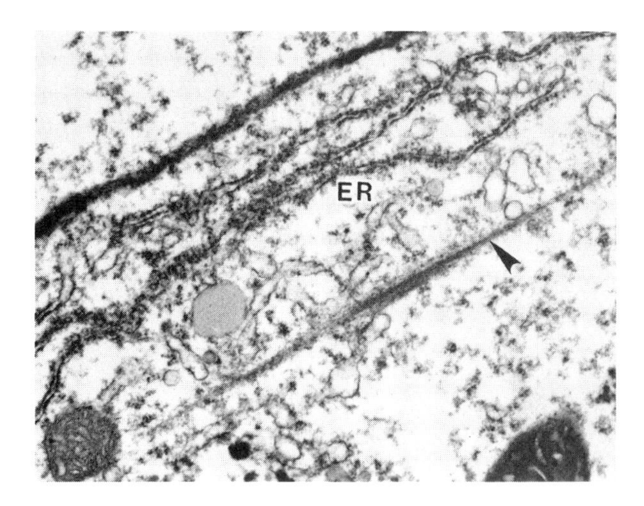


**Figure 1-16** Transmission electron micrograph of actin microfilaments (arrowhead) and endoplasmic reticulum (ER) in a vascular parenchyma cell from a pea stem. 23,000 ×. (Courtesy of Fred Sack.)

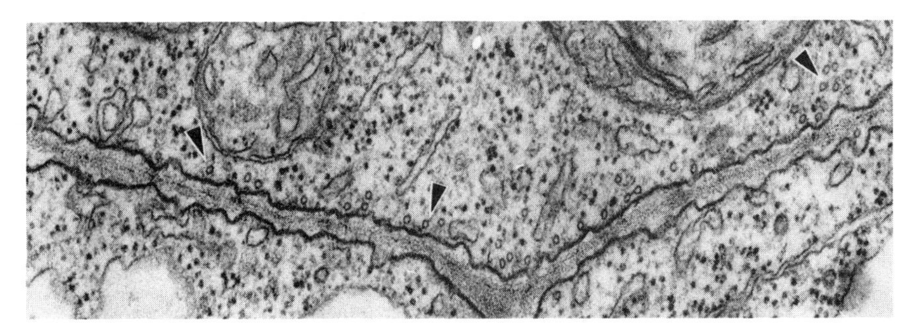

**Figure 1-17** Part of a preprophase band of microtubules in a *Funaria* sporophyte epidermal cell. The microtubules are sectioned across their width and appear as small circles (arrowheads) near the plasmalemma. This band corresponds in position to the regions where the future cell plate will fuse with the parent cell walls as in Figure 1-5b. Fewer microtubules are found near the plasmalemma of the other two cells in the micrograph. 47,000 ×. (Courtesy of Fred Sack.)

specifically with actin (or actinlike) molecules. The phallotoxins have been especially valuable in studies of microfilaments.

*Cytoskeleton dynamics in cell division* Profound changes are seen in the components of the cytoskeletal system during the **cell cycle**, which includes the various steps of cell division as well as the interphase cell.

Before **mitosis** (division of the nucleus) and **cytokinesis** (division of the cell), both the microtubules and the microfilaments undergo radical changes. The microtubules condense in the cell cortex around the cell equator to form a band, which becomes more and more compressed during prophase (Fig. 1-17). This band has associated microfilaments, as indicated by phalloidin staining. The spindle may form from microtubules, and a network of cytoplasmic microfilaments is present throughout division. Near the end of mitosis, the phragmoplast forms from two opposing rings of microtubule bundles at what was the spindle equator (see Fig. 1-5). These rings expand outward (centrifugally) and may guide cell-wall vesicle deposition, forming the cell plate at their midline until they meet the mother-cell walls. Again, microfilaments are associated with these structures. In animal cells, microfilaments form a ring around the equator of the dividing cell, shrinking to divide the cell in half.

Investigators have searched for microtubule organizing centers such as those found in animal cells. It is clear that as microtubules reform following cell division, they often appear to radiate from the nucleus; thus the nuclear envelope must harbor **nucleation sites**, although these are not *organizing centers* in the sense of those found in animal cells because they do not *organize* the arrangement of the microtubules that develop. Such nucleation sites also help to initiate formation of the **spindle**, which forms during mitosis and moves the chromosomes to opposite poles. The sites also help to initiate development of phragmoplast, where the new cell wall between daughter cells will form. All this occurs in spite of the fact that the nuclear envelope breaks down during late prophase before spindle formation and reforms at telophase.

As the microtubules radiate from the nucleus following cell division, they often form an array around the outer layers of cytoplasm close to the plasmalemma; this part of the cell is called the **cortex**. Microtubules also form strands between the cortex and the nucleus and sometimes through strands of cytoplasm that pass through the vacuole.

*Cytoskeleton dynamics in cell-wall formation* The cortical array of microtubules in uniformly elongating interphase cells initially forms a tightly compressed helix (like a compressed bed spring) in which the microtubules are mostly at right angles to the long axis of the cell (but *not* forming closed hoops). The helix can be right- or left-handed, and the handedness can switch from one to the other. As the cell elongates, the helix expands until the microtubules may form angles of 45° with the long axis.

It was noticed some time ago in electron micrographs that the orientation of cortical microtubules is often closely correlated with the orientation of the cellulose microfibrils of both primary and secondary walls. As we noted above, microfibrils are initially at right angles to the major growth axis. Orientation of cellulose microfibrils probably controls the direction of growth because they do not stretch but tend to move away from each other as the cell expands; expansion takes place at right angles to the primary direction of microfibril arrangement (especially that of the microfibrils that have most recently been deposited).

There are other striking correlations between microtubule distribution and cell-wall structures. Microtubules are especially prevalent near isolated regions of secondary wall thickening, such as in xylem elements, and the direction of the spiral of both microfibrils and microtubules can change. Ethylene (a gaseous plant hormone discussed in Chapter 18) causes the microtubules to become more longitudinal, and the microfibrils follow suit, causing the cell to expand instead of elongate. Gibberellins, on the other hand, by promoting cell elongation, cause a more ordered and transverse microtubule and microfibril arrangement. Based on such correlations, it was suggested that the microtubules are involved in wall formation or organization. Although many aspects of cellulose formation are still not understood, much has been learned during recent years about the formation of this most abundant material on earth

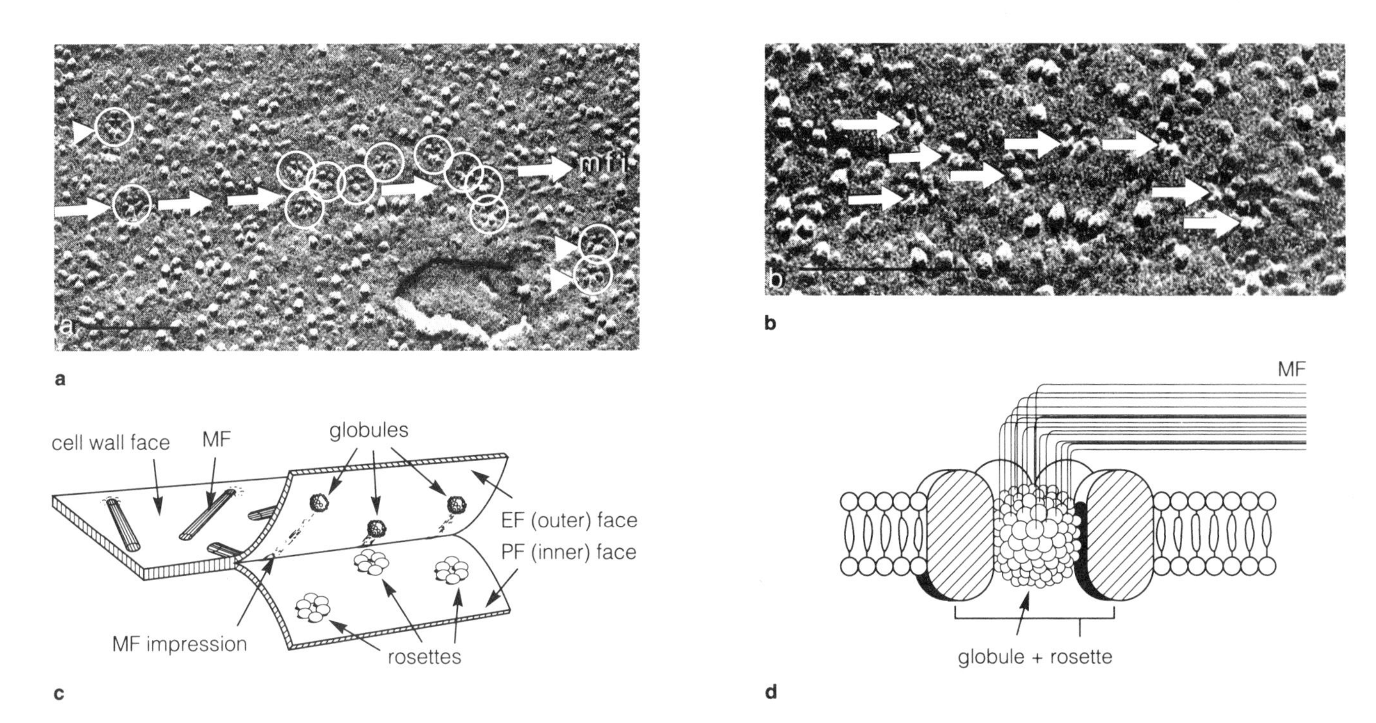


**Figure 1-18** Rosettes and globules in the EF and PF faces of the plasmalemma as they are postulated to synthesize cellulose microfibrils and as they are postulated to be related to the microtubules in the cytoplasm. (**a**) Freeze-fracture of a hypocotyl cell of *Vigna radiata*, plasmatic-fracture (PF) face of the plasmalemma. The rosettes are in circles, and the arrows indicate the direction of the microfibril imprint (mfi). Some rosettes are not in the track. (Bar = 0.1 μm) (**b**) Higher magnification of the aligned rosettes (arrows). (Both photographs courtesy of Werner Herth, see Herth, 1985.) (**c**) Drawing of rosettes and globules found in some cellulosic algae and in lower and higher plants. The linear complexes usually fracture with the EF face but sometimes with the PF face; rosettes fracture with the PF face and globules with the EF face. In double-replica fracturing, globules are sometimes found in complementary location to rosettes, leading to the speculation that rosettes and globules may be part of the same complex, as shown in (**d**). MF = microfibril. (Drawings from Delmer, 1987; used by permission.)

(reviewed by Bacic et al., 1988; Delmer, 1987; Delmer and Stone, 1988).

Information relating to how cellulose microfibrils are synthesized came from studies of the plasmalemma by freeze-fracture techniques, in which the tissue is frozen and then fractured (reviewed by Delmer, 1987). Often the fracture occurs between the inner and outer layers of the plasmalemma. Replicas are made by depositing a thin layer of metal on these membrane faces (evaporating it from a point source), dissolving the tissue away, shadowing the metal with another metal, and observing this preparation with the transmission electron microscope. The two inside faces of the plasmalemma (and other membranes) can be distinguished in such micrographs, and the face of the inner layer (adjacent to the cytoplasm) is called the **PF (inner) face**; the face adjacent to the wall is the **EF (outer) face** (Fig. 1-18). In such studies with algae and then with lower and higher plants, it was possible to see **globules** on the EF face and **rosettes** on the PF face. One hypothesis now holds that these complexes represent the enzymes that synthesize the cellulose microfibrils and that they are some way guided by the orientation of microtubules that often appear in close association with them. The underlying microtubules might provide lanes in which the rosette-globule complexes move as they synthesize cellulose microfibrils in the wall outside the plasmalemma (Fig. 1-18). Accumulating evidence (see, for example, Rudolph and Schnepf, 1988) supports the model of participation of rosettes in cellulose synthesis and also suggests that the rosettes are supplied to the plasmalemma by Golgi vesicles and that their life time is only some 20 minutes.

*Other functions of microfilaments* In addition to the microtubule constituents of the cytoskeleton, there is an extensive network of actinlike cables (groups of microfilaments) in a range of interphase plant cell types. This part of the cytoskeleton is demonstrated mostly with phallotoxins (Seagull et al., 1987). At interphase, microfilaments are organized into three distinct but interconnected arrays: (1) a fine cortical network close to the plasmalemma; (2) large axially oriented cables in the cytoplasm, somewhat farther away from the plasmalemma (the subcortical region); and (3) a "basket" of microfilaments surrounding the nucleus and extending into the strands of cytoplasm that traverse the vacuole. Beginning at the plasmalemma, these arrays disappear as mitosis begins and reappear in reverse order (beginning with the nucleus) after cell division is complete. As with the microtubules, the nucleus seems to be the center for microfilament nucleation and perhaps, in this case, organization as well. In contrast to the microtubules, the microfilaments tend to be parallel with the axis of elongation as cells begin to elongate, but as elongation continues they become increasingly transverse until they are nearly parallel with the microtubules.

Another suggested function of microfilaments, based on their observed orientations, is control of the direction of cytoplasmic streaming. As the microfilament orientation changes, so does the direction of streaming. Microfilaments might also provide the driving force for streaming, and microfilaments and microtubules are also involved in movement and positioning of such organelles as chloroplasts, mitochondria, amyloplasts, and nuclei.

In summary, plant cells have two or more extensive, interacting cytoskeletal networks deeply involved in the dynamics of cell behavior, and knowledge about them is accumulating rapidly.

## Ribosomes

Protein synthesis is a vital cellular function that occurs on the thousands of ribosomes (each about 15 to 25 nm in diameter) that are dispersed in the cytoplasm or associated with rough ER in each cell, always on the cytosol side of the ER double membrane. Identical ribosomes are also attached to the cytosol side of the outer membrane of the nuclear envelope. They are the black dots (often a little larger than a period) so apparent in many electron micrographs of high magnification (for example, Figs. 1-5c, 1-10, and 1-17). Often, ribosomes form a chain like a string of beads, but typically they are in a spiral pattern; these structures are called **polyribosomes** or **polysomes**, and each is held together by a strand of **messenger-RNA** (**mRNA**; discussed in Appendix C). In a polyribosome, the genetic information of the mRNA is translated into protein.

Smaller ribosomes (15 nm, the size of those in prokaryotes) are present in mitochondria and chloroplasts, where they synthesize some of the protein found in these organelles; other chloroplast and mitochondrial proteins are formed on cytoplasmic ribosomes and transported into the organelles (Section 7.11). Nuclei contain no true ribosomes, and modern biochemical studies indicate that nuclei import all of their proteins from the cytoplasm.

In the early days of electron microscopy, researchers attempted to relate cellular organelles to the fractions that were obtained when cells were broken down and centrifuged at various high speeds. One of these fractions contained small particles and membrane-bound vesicles originating from rough ER. The bodies were named **microsomes**, but most of them proved to be vesicles formed from the ER during homogenization (and associated with ribosomes if the ER was rough), so the term *microsome* is now used much less frequently. Nevertheless, it is still encountered in the modern literature (see, for example, Boller and Wiemken, 1986).

## Mitochondria

In the light microscope, **mitochondria** appear as small spheres, rods, or filaments varying in shape and size, usually from about 0.5 to 1.0 μm in diameter and from 1 to 4 μm in length. They were first seen in about 1900. The electron microscope shows them to have a rather elaborate internal structure and frequently an oval shape (see Figs. 1-2, 1-3c, 1-5a, and 1-5b). Most cells have hundreds to thousands of mitochondria, but various algae, including *Chlorella*, have only one large one with a branched and convoluted shape. Mitochondria divide by fission (like a bacterium), and all arise from the mitochondria in the zygote; hence, their membranes are not derived from the endomembrane system. As we have noted, they contain DNA and small (15 nm) ribosomes in the matrix, and thus they synthesize some of their own protein, but they are also dependent on protein synthesized in the cytoplasm under nuclear control. Mitochondrial ribosomes show the same pattern of sensitivity to protein inhibitors as bacterial ribosomes: Cycloheximide inhibits protein synthesis by cytoplasmic ribosomes but not by isolated mitochondria or bacterial cells, whereas chloramphenicol has an opposite effect. Hence, mitochondria are reminiscent of prokaryotes, and it has been suggested that they originated long ago as prokaryotes that invaded eukaryotic cells. But, as noted, mitochondria now depend partially on proteins synthesized in the cytosol.

Mitochondria have a highly folded inner membrane system surrounded by a smooth outer membrane. The inner and outer membranes are quite different, and in most preparations for the electron microscope, neither membrane can be resolved into the typical double structure (the unit membrane) typical of most other membranes. The smooth outer membrane has a large proportion of lipids and is highly permeable to the many compounds that enter and exit mitochondria. The intricately folded inner membrane has several forms, including shelves or tubular protuberances called **cristae**. The tubular forms are common in plant mitochondria. The inner membrane surrounds a **matrix**, and many of the enzymes that control various steps in cellular respiration in particular and metabolism in general occur on the inner membrane or in the matrix. It is likely that over half the cell's metabolism occurs in mitochondria. Mitochondria are further discussed in Chapter 13.

### Plastids

**Plastids** are special structures, bound by double-membrane systems, that occur only in plants and some protistans. Their inner membrane is not folded as in mitochondria, but other membranes arranged in various ways often occur in plastids. They contain DNA and ribosomes, embedded (with the membranes) in a fluid matrix called **stroma**. Plastids are also self-reproducing and thus are independent of the endomembrane system.

Plastid terminology has been confusing, but we can classify them as in Table 1-3: All plastids develop from **proplastids**, which are small bodies found in plants growing in the dark as well as in the light. They divide by fission much as mitochondria (and prokaryotes) do. Colorless plastids in general are called **leucoplasts** (from the Greek: *leucos*, "white" and *plastos*, "formed"). The best known leucoplasts are **amyloplasts**, which contain two or more grains of starch. Other leucoplasts may contain storage proteins (**proteinoplasts**).

There are two kinds of colored plastids: **chloroplasts**, which contain chlorophyll and its associated pigments (see Fig. 1-2), and **chromoplasts**, which contain other pigments (for example, the red pigments of tomato skins). Some authors consider all colored plastids, including chloroplasts, to be chromoplasts.

Chloroplasts contain a system of membranes called **thylakoids**, and these are often connected to form membrane stacks called **grana** (singular: **granum**), which are embedded in **stroma**. The enzymes that control photosynthesis are located on thylakoid membranes and in the stroma, as we discuss in Chapter 10.

## 1.7 The Nucleus

The **nucleus** is, in many respects, the control center of the eukaryotic cell; yet we have seen that it is not an independent organelle but must obtain its protein from the cytoplasm. It is often the cell's most conspicuous organelle, being spherical or elongated and often 5 to 15 μm or more in diameter. It exercises control over cell functions by determining (via mRNA) many of the kinds of enzymes made in the cell, and these determine the chemical reactions that take place and thus the structures and functions of cells. The control lies in the same structure as the genetic or hereditary information and is contained in long fibers of DNA combined with protein, which form a material called **chromatin** (see Figs. 1-3c and 1-5b). This material is duplicated by chemical processes during interphase. Chromatin fibers in plant cells may have a total length of 1 to 10 m, which must fit into a nucleus only 10 μm in diameter—a diameter that is only one millionth the length of the chromatin! Portions of the chromatin, the active genes, must be functional in spite of the packing of the elongated strand into the small nucleus.

During division of the nucleus, the chromatin fibers condense by coiling into elongated, darkly staining bodies called **chromosomes**; these are visible with the light microscope. Apparently, the initial ordering of the DNA to form chromosomes involves formation of a "beads-on-a-string" structure, with the "beads" being molecules of the basic protein, **histone**, which is known to be part of the chromatin. This forms a strand about 10 nm in diameter, which can then be condensed by folding again and again to form the chromosomes. Between divisions the coiling relaxes, and chromosomes cannot be observed in the nucleus. The nucleus also contains a watery, enzyme-filled solution known as **nucleoplasm** in which chromatin, or chromosomes, and nucleoli are suspended. The nucleoplasm is probably as structured as cytosol, containing a cytoskeletal structure that may organize the chromatin and nucleoli.

The nucleus contains one or more (up to about four) roughly spherical bodies, the **nucleoli** (singular: **nucleolus**), each about 3 to 5 μm (up to 10 μm) in diameter. Nucleoli are dense, irregularly shaped, dark-staining masses of fibers and granules suspended in the nucleoplasm. There may also be light-staining areas called **nucleolar vacuoles**, which apparently indicate a highly active nucleolus. Undifferentiated cells such as those in the meristem typically have larger nucleoli than do mature or dormant cells.

A resemblance of the nucleolar granules to cytoplasmic ribosomes is more than coincidence because the subunits of ribosomes, composed mostly of RNA and protein, are made in the nucleolus. Ribosomal RNA is made in the nucleolus, but the protein is synthesized in the cytoplasm and transported into the nucleus, where it is combined in the nucleoli with the RNA to form a subunit that is then transported back to the cytoplasm for final assembly into ribosomes. The subribosomal particles are thought to move in and out through the nuclear pores. Labelling experiments with radioactive tracers suggest that the fibrillar part of the nucleolus is the RNA; a granular part forms the ribosomal subunits. The ribosomal RNA is encoded by special regions of chromosomes called **nucleolar organizing regions**. The nucleolus disappears during mitosis but reappears during telophase as small nucleoli, equal in number to the number of chromosomes with nucleolar organizers. The small nucleoli may condense to form a single nucleolus during interphase.

## 1.8 The Vacuole

As characteristic of plant cells as the cell wall and plastids is the vacuole (reviewed by Boller and Wiemken, 1986). We can consider the vacuole part of the endomembrane system, and we have already discussed the vacuolar membrane or tonoplast as part of that system, but the vacuole is so important that it merits separate consideration. Vacuoles perform several functions, and many of these have come to our attention with the torrent of information that only became available in the 1970s. The information came from studies of isolated vacuoles from higher-plant cells.

### Role in Turgidity and Shape

The form and rigidity of tissues made of cells that have only primary walls (for example, leaves and young stems; see Fig. 1-7) are caused by water and its dissolved materials exerting pressure in the vacuole. The pressure develops by osmosis, which we discuss in Chapter 3.

There is another important aspect of vacuoles that makes plants what they are (Wiebe, 1978a, 1978b). To survive, a plant must absorb relatively large amounts of water, mineral elements, carbon dioxide, and sunlight. Each of these, even sunlight, can be and often is relatively scarce or dilute in the environment. Large surface areas greatly facilitate the absorption of each of these by plants: The surfaces of finely divided roots penetrate large volumes of soil, and the surfaces of leaves capture sunlight and absorb carbon dioxide from the atmosphere. The way for an organism to achieve large surfaces is to begin with relatively large volumes and spread them out into thin layers, like most leaves, or into long narrow structures, like roots or conifer needles. Plants have relatively large volumes because their vacuoles are filled with water, which is much more abundant in the environment than any other constituent of protoplasm. If plant cells contained only protoplasm without vacuoles, as do most animal cells, they could expose only a fraction of their present surface area. It is important for animals to have a compact volume with limited surface and concentrated protoplasm to produce energy and reduce inertia for motion. These two functions of plant vacuoles — maintaining turgor and providing a large volume — are static functions.

### Vacuoles in Storage and Accumulation

The concentration of dissolved materials in vacuoles is high, approximately as high as the concentration of salt in sea water and as high as that of the cytosol (commonly 0.4 to 0.6 M). There are hundreds of dissolved materials, including salts, such small organic molecules as sugars and amino acids, and some proteins and other molecules. Some vacuoles have high concentrations of pigments, which produce the colors of many flowers or the red of red maple leaves (so concentrated in epidermal cells that they mask the green of the chloroplasts). In some plant parts, vacuoles contain materials that would be harmful to the cytoplasm. These include many of the secondary products of metabolism discussed in Chapter 15 (for example, alkaloids, various compounds with attached sugar molecules, and so on). Sometimes, too, vacuoles contain crystals; calcium oxalate crystals are especially common in some species. Still other substances found in vacuoles are noted in the following paragraphs.

In view of all the substances that occur in vacuoles, the vacuole has long been thought of as a kind of dumping ground for cellular waste products and excess mineral ions taken up by the plant. We now know that many of these substances play much more dynamic roles in the vacuole than merely being stored there, although storage, including that of waste products, must be one valid role for the vacuole. Some of these substances are trapped in the vacuole, as their nature changes upon entering into the new environment within the vacuole, which, for one thing, is often more acidic than the cytosol. Neutral red, for example, can pass through the tonoplast as the lipophilic free base, but it becomes ionized by acceptance of a proton in the vacuole, and in this condition it can no longer pass through the tonoplast. $Ca^{2+}$ can be trapped by precipitation with oxalate, phosphate, or sulfate to form crystals, but the vacuole usually contains millimolar concentrations of $Ca^{2+}$.

### Vacuoles as Lysosomes

The enzymes in vacuoles digest various materials absorbed into them, as well as much of the cytoplasm when the cell dies and the tonoplast breaks down. This probably happens when the protoplasts of wood cells break down and die, for example. In this sense, the vacuole is like a **lysosome**, a cellular organelle common in animal and some fungal and protist cells. Lysosomes contain digestive (hydrolytic) enzymes that break down materials they absorb — or these enzymes digest much of the protoplasm following cell death and breakdown of lysosome membranes. The importance of this role for vacuoles is still being investigated because not all enzymes that break down protein in cells, perhaps only about 10 percent in higher plants, are present in vacuoles. In yeast cells, 90 percent of these enzymes are located in the vacuoles.

### Role in Homeostasis

**Homeostasis** is the tendency for various physiological parameters to be maintained at some relatively constant level. Most study of homeostasis has involved animals, and body temperature in birds and mammals is an excellent example of the phenomenon. But a good example in plants is the relatively constant concentration of various substances in the cytosol. The concentration of hydrogen ions (*p*H) provides an excellent example, and the vacuole plays an important role in the maintenance of constant cytosolic *p*H. Excess hydrogen ions in the cytosol may be pumped into the vacuole. The sharp taste of oranges and lemons, which comes from the high concentration of citric acid in their vacuoles, provides a striking example. Such vacuoles may have a *p*H as low as 3.0, whereas the *p*H of the surrounding cytoplasm is between 7.0 and 7.5 (close to neutral). Other organic acids are present in vacuoles of succulent CAM plants (plants with crassulacean acid metabolism; see Chapter 11), which produce the acids at night and then process them photosynthetically during the day. Most vacuoles are slightly acidic (*p*H equals 5 to 6). It has been shown experimentally that when the *p*H around various plant cells is changed drastically, the change is reflected in the *p*H of the vacuole while the cytosol *p*H remains constant.

$Ca^{2+}$ and phosphate ions would be toxic to the cytoplasm if their concentrations became too high. The vacuole absorbs these ions and thus keeps their concentration in the cytosol within suitable limits — sometimes as much as 1,000 times lower in the cytosol than in the vacuole. As we have noted, $Ca^{2+}$ is sometimes trapped in the vacuole by the formation of calcium crystals. (The ER may play a role in regulating cytosolic $Ca^{2+}$.) Phosphate and nitrate provide excellent examples of essential ions stored in vacuoles. If cytosolic phosphate or nitrate levels drop too low, these ions can move from the vacuole into the cytosol. The same is true of sugars, amino acids, and many other stored materials. Thus, the vacuole may be a dumping ground, but it is a warehouse as well.

The total solutes (dissolved substances) in the vacuole determine its osmotic properties (osmotic potential, discussed in Chapter 3) and thus the osmotic properties of the adjoining cytosol as well (cytosol and vacuole are in equilibrium), another example of the role of the vacuole in homeostasis. There are some exceptions, however. Certain compounds such as proline (an amino acid) build up in tissues under water or salt stress (Chapter 26), but the high concentrations develop in the cytosol. They function by protecting the cytosol enzymes from the water or salt stress, so it is appropriate that they should be localized there.

### Metabolic Processes in Vacuoles

A few of the chemical reactions of living cells occur in the vacuole. For example, the last step in the synthesis of ethylene (a gaseous growth regulator in plants) occurs predominantly in the tonoplast of the vacuole, and various sugar transformations also occur there. Some of the secondary metabolites stored in vacuoles also undergo chemical changes there, another finding that illustrates some of the many things learned during recent years by study of isolated vacuoles.

### The Origin of Vacuoles

Young, dividing cells in the growing tips of stems and roots have only minute vacuoles, many in each cell, probably formed from the ER. These grow with the cell, taking up water by osmosis and coalescing with each other until a mature cell often has a vacuole that occupies 80 to 90 percent or more of its volume, with the protoplasm spread in a thin layer between the tonoplast and the plasmalemma.

Some dividing cells have large vacuoles, however. Cambial cells between the bark and wood of a growing stem have large vacuoles and divide along their longitudinal axes to produce both phloem and xylem cells, likewise with large vacuoles. The vacuole is one of the most variable features of plant cells, varying from being almost nonexistent to filling most of the volume of most mature, living plant cells. Even mature vacuoles can drastically change their shape. Vacuoles in guard cells of the stomatal apparatus change from spherical to a tubular network before division, returning to the spherical form thereafter. The variety of vacuolar shapes means that the "typical" cells of Figures 1-2 and 1-3 are not at all typical. Actually, considering the large quantities of wood on earth, the "typical" plant cell is a dead xylem element!

## 1.9 Flagella and Cilia

Flagella and cilia are found mostly in algae, fungi, protozoans, other small (microscopic) animals, specialized cells of other animals, and sex cells of certain lower plants and gymnosperms. These hairlike structures project from cell surfaces and are capable of beating back and forth at high speeds. They are nearly always about 0.2 μm in diameter but vary in length from 2 to 150 μm. The long ones are usually called **flagella** (singular: **flagellum**) and the short ones **cilia** (singular: **cilium**), but there is no sharp dividing line between them. Flagella and cilia are often the sources of locomotion for the cells to which they are attached (one or a few flagella per cell). Cilia may cover the surfaces of adjacent cells, all beating at once in the same direction to propel fluids across their collective surface (as in the human upper respiratory tract).

Electron micrographs reveal flagella and cilia to have a precise internal organization consisting of nine pairs of microtubules that surround two additional microtubules. (Bacterial flagella and cilia do not contain microtubules.) Flagella and cilia grow out of **basal bodies** in the cytoplasm. These have internal structures similar to those of flagella, except that there are nine *triplet* instead of twin groups of microtubules, with none in the center. When a cell with flagella divides, the flagella are often lost, and new ones are regenerated from the basal bodies. The mechanism of movement is not known, but isolated flagella can beat by themselves, so movement is not caused by the basal bodies.

## 1.10 The Plant Cell

It is worth remembering that three features are especially characteristic of plant cells as compared with animal cells: the cell wall with its cellulose; the vacuole (which provides pressure and a large volume and surface area with a minimum of protoplasm); and plastids, especially chloroplasts. Animal cells never have walls, nor do many protist cells, and the walls of prokaryotes and fungi differ in significant ways from those of plants. Vacuoles can be found in all five kingdoms, but the large, central vacuoles we have been discussing are present in practically all cells of plants, fungi, and some protistans. Chloroplasts occur only in plants and in some protistans (depending on how they are classified); in a sense, blue-green algae (cyanobacteria) are structures comparable to chloroplasts, and it is possible that chloroplasts originated when a photosynthesizing prokaryote invaded a eukaryotic cell. Animals and fungi never produce chloroplasts.

Lest it be misleading to end our discussion of plant cells by overemphasizing their uniqueness, we should point out that, after all, plant and animal cells have many more characteristics in common than they have differences. Most of the endomembrane system plus the plasmalemma occur in both plant and animal cells (with the exception of the plant vacuole), and mitochondria, microtubules, microfilaments, ribosomes, and especially the nucleus are similar or identical in all four eukaryotic kingdoms. Both diversity and similarity are features of life.

## 1.11 A Definition of Life

Based upon the postulates in the early part of this chapter and our discussion of cells, we can conclude this chapter by attempting to define life. Of course there are difficulties. Special macromolecular structures exist after an organism has died and might even exist on the surface of a dead world somewhere in the universe, so in that sense they are not characteristic of life. Viruses (which are not cells) display many of the properties of life, but only when they are associated with living organisms. Nevertheless, consider the following statement:

*Life on earth is a distinctive series of functions associated with a unique series of organized structures in which certain macromolecules (proteins, DNA, RNA), having building blocks arranged in nonrepeating but replicating sequences, are capable of reproduction, transfer and utilization of information, and catalysis of metabolic reactions. All this is organized at least to the level of a cell with its surrounding membrane, which level allows the functions of growth, metabolism, response to environment (sometimes called irritability), and reproduction.*

# 2

# Diffusion, Thermodynamics, and Water Potential

How can one organize the topics of plant physiology? We have organized our text around five basic questions, mostly about cells:

1. What are the components of plant cells and what are their functions?
2. How do ions and molecules, especially water, get into and out of cells and move throughout the plant?
3. What happens to the ions and molecules inside cells?
4. How do cells reproduce, grow, and specialize to form the tissues of a multicellular organism?
5. How do plant cells — and plants — interact with their environments?

This book summarizes in an introductory way the current status of answers to these questions. Having discussed the components and functions of cells, the remaining chapters in this first section address the second question, which concerns movement of water and other substances across membranes, throughout the plant, and between the plant and its environment, a topic often called **plant water relations**. Clearly, this subject overlaps that of cellular processes because cellular chemistry influences water relations in many interesting ways. Growth also depends upon water uptake, and much of plant water relations depends upon cells interacting with their environment. Plants are intricately complex and dynamic systems in which any one function interacts with all the others. Put another way, plants are multidimensional systems, but because textbooks are only unidimensional, going from beginning to end, their organization must be to some extent artificial. We will try to integrate the various parts of the book by references to what has been discussed and what is to come, but it will be up to you to build a multidimensional picture in your mind.

During about the past quarter century, it has become fashionable to replace the rather formal terms *hypothesis, theory,* and *law* with the more general term **model**, which is truly an excellent word to describe the products of a scientist's creative thought. This book is, to a great extent, a description of the models that have been developed to explain and predict plant function. For example, how does the water "run uphill" in moving to the top of a tall tree? A physical analog and a conceptual model, continually being modified by new information and new ideas, exist to help us understand this phenomenon.

## 2.1 Plants and Water

Plant physiology is, to a surprising degree, the study of water. Many plant activities are determined by the properties of water and of substances dissolved in water. Thus a brief review of water's properties is a good way to begin our study of plant physiology.

### The Hydrogen Bond, Key to Water's Properties

Most of the unique properties of water can be ascribed to the interesting fact that line segments connecting the centers of the two hydrogen atoms with the center of the oxygen atom do not form a straight line. Instead, they form an angle of about 105°, closer to a right angle than to a straight line (Fig. 2-1). The angle is exact in ice but only an average in liquid water. The two electrons that form the covalent bond between the hydrogen and oxygen atoms are usually closer to the oxygen nucleus, leaving the two hydrogen nuclei (protons) with their positive charges on the surface of the oxygen atom and 105° apart on one side. This provides a slight positive charge to that side of the molecule and an equal negative charge to the other side, with the net charge for the molecule being neutral. Such a molecule is said to be

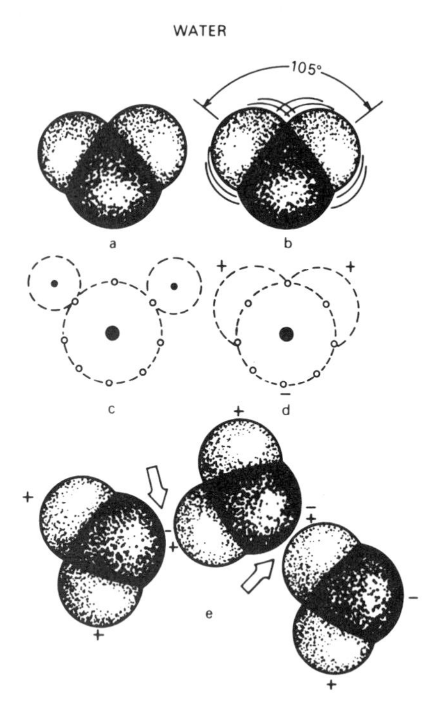


**Figure 2-1** Water molecules. The angle between the two hydrogen atoms (lighter) attached to the oxygen atom (darker) averages 105° (**a** and **b**). The angle is not absolutely stable (**b**) but represents an average sharing of electrons and distribution of charge (**c** and **d**). Attraction of the negative side of one water molecule to the positive side of another produces the hydrogen bond (**e**). Arrows indicate hydrogen bonds.

**polar**. The result is that the positive side of one water molecule is attracted to the negative side of another (Fig. 2-1), forming a relatively weak bond between the polar molecules. Such a bond is called a **hydrogen bond**.

Hydrogen bonds occur in many substances besides water, with those with oxygen and those with nitrogen being especially important in plants. The energy (strength) of the bond between a hydrogen atom in one molecule and some negative part of another molecule varies, depending on the other molecule, from about 8 to 42 kilojoules/mole of bonds (kJ $mol^{-1}$). **Ionic bonds**, in which electrons move from one atom and enter the outer shell of another, have energies that vary from about 582 kJ $mol^{-1}$ for CsI to 1,004 kJ $mol^{-1}$ for LiF, with NaCl being intermediate at 766 kJ $mol^{-1}$. The energies of **covalent bonds**, in which electrons are shared by two atoms, overlap ionic-bond strengths but are usually weaker. Important examples include (all given in kJ $mol^{-1}$) 138 for O—O, 293 for C—N, 347 for C—C, 351 for C—O, 414 for C—H, 460 for O—H (as in water), 607 for C=C, and 828 for C≡C.

**Van der Waals attractive forces** are even weaker than hydrogen-bond forces, having bond energies of about 4.2 kJ $mol^{-1}$. In neutral, nonpolar molecules, these forces result from the fact that electrons are continuously in motion, so the molecule's center of negative charges does not always correspond to its center of positive charges. Thus, as two like molecules approach each other closely, they may induce slight polarizations in each other, with the regions of unlike charge attracting each other. Such forces hold molecules together in the liquid hydrocarbons, for example, but also in membranes and the internal parts of proteins. All these molecular bonds are electronic in nature.

## Liquid at Physiological Temperatures

The higher the molecular weight of an element or a compound, the greater the likelihood that it will be a solid or a liquid at a given temperature, such as room temperature (around 20°C). The lower the molecular weight, the greater the likelihood that a compound will be a liquid or gas. To change from solid to liquid or from liquid to gas—that is, to break the molecular forces that bind molecules to each other—heavier molecules require more energy (heat) than lighter ones. For example, the low–molecular-weight hydrocarbons (methane, ethane, propane, and *n*-butane) are all gases at room temperature. Their respective molecular weights (MW) are 16, 30, 44, and 58 grams per mole. *n*-Pentane (MW = 72) boils at 36°C (96.8°F). *n*-Hexane, *n*-heptane, and *n*-octane, with respective molecular weights of 86, 100, and 114, are liquids at room temperature, and nonadecane (MW = 268) is a solid (melts at 32°C). Ammonia (MW = 17) and carbon dioxide (MW = 44) are gases at room temperature, but water (MW = 18) is a liquid. The explanation is that hydrogen bonds provide a disproportionately high attractive force among water molecules, inhibiting their separation and escape as vapor. The hydrocarbons, on the other hand, have only the relatively weak van der Waals forces among their molecules in the liquid state. Little energy (low temperatures) is needed to drive them into the gaseous phase.

Other liquids with low molecular weights are also polar molecules with hydrogen bonding between them. Good examples are the lower alcohols (methyl = $CH_3OH$, MW = 32) or the lower organic acids (formic = CHOOH, MW = 46; or acetic = $CH_3COOH$, MW = 60). The presence of oxygen and hydrogen atoms makes hydrogen bonding possible in these compounds.

Life as we know it is unthinkable without liquid water. This becomes increasingly evident as we examine other properties of this vital molecule.

## An Incompressible Fluid

For all practical purposes, liquids are incompressible. This means that the laws of hydraulics apply to organisms because they consist largely of water. That a young growing plant is a hydraulic system becomes strikingly evident when the plant wilts (Fig. 2-2). The

**Figure 2-2** Normal (left) and wilted (right) *Coleus* plants. The normal appearance of a plant is dependent upon having sufficient water in the cells to provide turgidity. The wilted plant recovered completely after watering. Note that the youngest leaves on the wilted plant are still quite turgid.

normal form of a plant is maintained by the pressure of water in its protoplasts pressing against its cell walls. Furthermore, plants grow as they absorb the water that causes their cells to expand. Some petals and leaves, such as those of *Mimosa pudica*, the sensitive plant, move or fold up as water moves into or out of special cells at their base. Stomates on leaf surfaces open as water moves into their guard cells and close as water moves out of these cells. Substances are transported in moving fluids in both plants and animals.

## Specific Heat

Almost exactly 1 **calorie**[1] (cal) is required to raise 1 g of pure water 1°C. The SI unit of energy, including heat energy, is the **joule** (abbreviated J; see Appendix A), and the calorie is now defined as 4.184 joules (exactly). The amount of energy required to raise the temperature of a unit mass of a substance 1°C is called its **specific heat**. The specific heat of water varies only slightly over the entire range of temperatures at which water is a liquid, and it is higher than that of any other substance except liquid ammonia. The high specific heat of liquid water is caused by the arrangement of its molecules, which allows the hydrogen and oxygen atoms to vibrate freely, almost as if they were free ions. Thus they can absorb large quantities of energy without much temperature increase. Plants (think of a large succulent cactus) and animals consist largely of water and thus have relatively high temperature stability, even when gaining or losing heat energy.

[1]A kilocalorie, or large Calorie (spelled with a capital C), equals 1,000 small calories and is counted by dieters.

## Latent Heats of Vaporization and Fusion

Some 2,452 joules (586 cal) are required to convert 1 g of water at 20°C to 1 g of water vapor at 20°C. This unusually high **latent heat of vaporization** can again be ascribed to the tenacity of the hydrogen bond and the large amount of energy required for a water molecule to break free from others in the liquid. One consequence is that leaves are cooled as they lose water by transpiration.

To melt 1 g of ice at 0°C, 335 J (80 cal) must be supplied. This is also a high **latent heat of fusion**, caused again by the hydrogen bonds, although ice has fewer per molecule than water. Each $H_2O$ molecule in ice is surrounded by *four* others, forming a tetrahedral structure. (Each oxygen atom attracts two extra hydrogen atoms.) The tetrahedrons are arranged in such a manner that the ice crystal is basically hexagonal, as demonstrated in the pattern of snowflakes. As is usual during conversion from the solid to the liquid state, molecules of water move slightly farther apart during melting. Yet water is extremely unusual because its total volume *decreases* during melting. This is because the molecules are packed more efficiently in the liquid than in the solid. Each molecule in the liquid is surrounded by *five or more* others.

The result of this packing difference is that water expands as it freezes, so ice has a lower density than

**Table 2-1 Viscosities of Fluids.**

| Fluid | Temperature (°C) | Coefficient of Viscosity[a] (centipoises) | Percent of $H_2O$ Viscosity at 20°C |
|---|---|---|---|
| Water | 0 | 1.787 | 177 |
| | 10 | 1.307 | 130 |
| | 20 | 1.002 | 100 |
| | 30 | 0.7975 | 80 |
| | 40 | 0.6529 | 65 |
| | 60 | 0.4665 | 47 |
| | 80 | 0.3547 | 35 |
| | 100 | 0.2818 | 28 |
| Ethyl alcohol | 20 | 1.20 | 120 |
| Benzene | 20 | 0.65 | 65 |
| Glycerine | 20 | 830.0 | 83,000 |
| Mercury | 20 | 1.60 | 160 |
| Machine oil | 19 | 120.0 | 12,000 |

[a]One centipoise = one poise multiplied by 100. One poise is defined as a force of 1 dyne per cm² required to displace a large plane surface in contact with the upper surface of a layer of liquid 1 cm thick over a distance of 1 cm in 1 s. (These are not SI units, but they do show relative viscosities.) A more convenient method of measurement notes the time required for a given volume of liquid to flow by gravity through a tube of given dimensions and then applies a suitable equation.

water. Thus ice floats on the tops of lakes in the winter rather than going to the bottom, where it might otherwise remain without thawing through the following summer. The expansion is also a potential source of damage to plant and animal tissue in freezing weather, as we discuss in Chapter 26. Because water expands upon freezing, increased pressure will make ice melt at a lower-than-normal temperature; that is, increased pressure lowers the melting point. The pressure of the ice-skate blade melts and thus lubricates the ice. With other substances, increased pressure usually raises the melting point.

## Viscosity

Because hydrogen bonds must be broken for water to flow, one might expect water **viscosity** or resistance to flow to be considerably higher than it is. But in liquid water each hydrogen bond is shared on the average by two other molecules, so the individual bonds are somewhat weakened and fairly easily broken. Water flows readily through plants. In ice there are fewer bonds per oxygen atom, so each is stronger. Viscosity of water decreases markedly with increasing temperature (Table 2-1), but this may be physiologically unimportant because water viscosity is low even at cold temperatures.

## Adhesive and Cohesive Forces of Water

Because of its polar nature, water is attracted to many other substances; that is, it wets them. Molecules of proteins and cell-wall polysaccharides, which are also highly polar, provide excellent examples. This attraction between unlike molecules is called **adhesion**. Water wets such substances as its molecules form hydrogen bonds with the other molecules. The attraction of like molecules for each other (again because of hydrogen bonding) is called **cohesion**. Cohesion confers upon water an unusually high **tensile strength**, which is the ability to resist stretching (tension) without breaking. In a thin, confined column of water, such as that in the xylem elements of a stem, this tensile strength can reach such high values that water is *pulled* to the tops of tall trees without breaking (Chapter 5).

Cohesion among water molecules also accounts for **surface tension**: The molecules at the surface of a liquid are continuously being pulled into the liquid by the cohesive (hydrogen-bond) forces, whereas those in the vapor phase are too few and too distant to exert any force on the molecules at the surface. As a result, a drop of water acts as if it were covered by a tight elastic skin; surface tension is what makes a falling drop spherical. The surface tension of water is higher than that of most other liquids.[2]

Surface tension plays many roles in the physiology of plants. For example, under normal pressures, surface tension prevents the passage of air bubbles through the minute pores and pits in cell walls. The water surface of bubbles cannot deform enough to pass through the small openings.

## Water as a Solvent

Water will dissolve more substances than any other common liquid. This is partially because it has one of the highest known **dielectric constants**, which is a measure of the capacity to neutralize the attraction between electrical charges. Because of this property, water is an especially powerful solvent for electrolytes and polar molecules such as sugars. The positive side of the water molecule is attracted to the negative ion or molecular surface of a polar molecule, and likewise the negative side to the positive ion or surface. Water molecules thus form a "cage" around ions or polar molecules, so the ions are often unable to unite with each other and crystallize into a precipitate.

If water contains dissolved electrolytes, then these will carry a charge, and water becomes a good electrical conductor. If water is absolutely pure, however (and pure water is extremely difficult to obtain), then it is a poor conductor. Hydrogen bonding makes it too rigid to carry a charge readily.

The importance of water as a solvent in living organisms becomes quite evident in the first section of this textbook. Osmosis, for example, which will soon

[2]Hydrazine and most metals (including mercury) in the liquid state have higher surface tensions.

be our topic of discussion (see Fig. 2-4 and Chapter 3), depends upon the presence of dissolved materials in the cell's water. We will also be concerned with the movement of dissolved materials by diffusion and bulk flow in plants.

Protoplasm itself is an expression of the properties of water. Its protein and nucleic acid components owe their molecular structures, and hence their biological activities, to their close association with water molecules. Indeed, almost all the molecules of protoplasm owe their specific chemical activities to the water milieu in which they exist. Exceptions are the molecules contained in cellular oil bodies (oleosomes) or the lipid (fatty) portions of membranes, but the oleosomes and the membranes are themselves strongly influenced by the surrounding water.

Water molecules actively enter into the chemistry that is the basis of life. To begin with, water and carbon dioxide are the raw materials for photosynthesis. Indeed, few metabolic processes occur without the utilization or production of water molecules. Nevertheless, water is relatively inert chemically. On balance, it may be more important as the environment for chemical reactions than as a chemical reactant or product.

### Dissociation of Water and the *p*H Scale

Some of the molecules in water separate into hydrogen $[H^+]$ and hydroxyl $[OH^-]$ ions, a process called **dissociation** or **ionization**. The tendency for these ions to recombine is a function of the chances for collisions between them, which in turn depend upon the relative number of ions present in the solution. This **mass action law** may be expressed mathematically by saying that the product of the **molal concentrations** (**m** = moles per kilogram of $H_2O$) equals a constant: $[H^+] \cdot [OH^-] = K$. In dilute solutions, *molal concentrations* are virtually equal to the more convenient **molar concentrations** (**M** = moles per liter of final solution). Near room temperature, $K = 10^{-14}$, so in pure water each $[OH^-]$ and $[H^+] = 10^{-7}$ M. (To multiply $10^{-7}$ by $10^{-7}$, exponents are added to give $10^{-14}$.)

Hydrogen ion concentration is expressed by the ***p*H scale**, on which $pH = -\log[H^+]$. In other words, *p*H equals the absolute value of the hydrogen ion concentration, expressed as a negative exponent of 10. For example, when $[H^+] = 10^{-4}$ M, then $pH = 4$. Neutrality is expressed by $pH = 7$ ($[H^+] = [OH^-]$); decreasing values below 7 indicate increasing acidity, whereas increasing values above 7 indicate increasing alkalinity. The *p*H units are multiples of 10 on a logarithmic scale and should therefore not be added together or averaged. Only a tenth as many $H^+$ need to be added to an unbuffered solution to change the *p*H from 7 to 6 as from 6 to 5.

Water is seldom pure enough to allow equal amounts of hydrogen and hydroxyl ions (that is, neutrality). Presence of dissolved carbon dioxide, as in distilled water, may raise the hydrogen ion concentration so that it reaches $10^{-4}$ M ($pH = 4$). The hydroxyl ion concentration is then $10^{-10}$ M. If tap water contains much dissolved limestone (calcium carbonate), the hydrogen ion concentration can be close to neutral or even slightly basic ($pH = 7$ to 8).

With this discussion of water as background, it is time to return to the question of how water and other materials move into and out of cells. Diffusion is one of the simplest transfer processes in living organisms, but it has enough interesting implications to occupy us for the rest of this chapter.

## 2.2 Diffusion Versus Bulk Flow

The contents of plant cells are under considerable pressure, about as much as the water pressure in a plumbing system: 0.4 to 0.5 MPa (megapascals, equivalent to 60 to 75 pounds per square inch). If we make a small hole through the cell wall and membrane, the cell contents will flow out through the hole until the pressure inside the cell is equal to the pressure outside (atmospheric pressure, probably). **Fluids** are substances, such as liquids or gases, that flow or conform to the shape of their container. When the flow occurs in response to differences in pressure and involves groups of atoms or molecules moving together, it is called **bulk flow**. Sometimes the differences in pressure are established by gravity (the weight of fluid); these are **hydrostatic pressures**. Sometimes the pressure is produced by a mechanical force applied to all or part of the system. In animals, the pump called the heart applies the force. In plants, fluids flow through the vascular tissues by bulk flow in response to pressure differences that are created by diffusion in ways we will discuss in later chapters.

Usually, water and the substances dissolved in it move into and out of cells not by bulk flow but one molecule at a time. The *net* movement from one point to another because of the random kinetic activities or thermal motions of molecules or ions is called **diffusion**. Because diffusion in liquids is slow over macroscopic distances and because the bulk flow of gases and liquids is so common, diffusion is not something that we readily notice. Nevertheless, it is easy to observe diffusion. Carefully place a crystal of dye into a beaker of still water (no stirring or convection currents). As the dye dissolves, you can observe it slowly spreading (diffusing) from its source throughout the liquid. Diffusion in air is much more rapid than in water, as you can observe when a bottle of some strong-smelling substance such as ammonia is opened a meter or two away.

But the diffusive transfer of odor from the bottle to your nose is often aided by air currents (bulk flow).

As we explain below, diffusion often occurs in response to differences in concentration of substances between one point and another. (As the dye begins to dissolve, it is highly concentrated in the water close to the crystal but absent some distance away.) Concentration differences are extremely common in living cells in particular and in organisms in general. For example, as certain organic compounds in the cytosol are taken in and metabolized by a mitochondrion, their concentrations close to the mitochondrion are kept lower than their concentrations close to a photosynthesizing (sugar-producing) chloroplast in the same cell. At the microlevel in cells, diffusion of many substances, including water, occurs constantly and virtually everywhere. Thus, to understand cells it is imperative that we understand diffusion.

The beginning is to understand one of the most fundamental principles of physics: kinetic theory. The basic ideas of kinetic theory are taught early and reviewed often, but the details are easily forgotten. You may never have been taught some of the quantitative aspects of the theory. Hence, we present a brief review.

**Table 2-2 Some Molecular Values for Three Gases.**

| | $H_2$ | $O_2$ | $CO_2$ |
|---|---|---|---|
| Molecular weight of gas (Da) | 2.01 | 32.0 | 44.0 |
| Average velocity at 0°C, meters/second ($m\ s^{-1}$) | 1,696 | 425 | 362 |
| Average velocity at 30°C, ($m\ s^{-1}$) | 1,787 | 448 | 382 |
| Average velocity at 100°C, ($m\ s^{-1}$) | 1,982 | 497 | 424 |
| Mean-free-path (nm) between collisions with other molecules, 0°C, 1 atmosphere (atm) pressure | 112 | 63 | 39 |
| Number of collisions of each molecule per second, in billions ($1 \times 10^9$), 0°C, 1 atm | 15.1 | 6.8 | 9.4 |
| Diameter of each molecule (nm) | 0.272 | 0.364 | 0.462 |
| Number of molecules ($\times 10^{-19}$), 1 atm pressure, in 1 $cm^3$, 0°C | 2.70 | 2.71 | 2.72 |

## 2.3 Kinetic Theory

**Kinetic theory** states that the elementary particles (atoms, ions, and molecules) are in constant motion at temperatures above absolute zero. The *average* energy of a particle in a homogenous substance rises as temperature increases but is constant for different substances at a given temperature. It is instructive when using this model to consider some of the actual velocities and masses of the moving particles. Velocities can be calculated easily for particles in gases, but it is much more difficult to obtain values for liquids and solids. The average velocity ($V_{ave}$) of particles in a gas is calculated by the following formula (see modern texts on statistical thermodynamics):

$$V_{ave} = \left(\frac{8RT}{\pi M}\right)^{1/2} \tag{2.1}$$

where

$V_{ave}$ = average velocity in meters/second ($m\ s^{-1}$)

$R$ = molar gas constant (8.3144 $J\ mol^{-1}\ K^{-1}$; $J = m^2 \cdot kg\ s^{-2}$)

$T$ = absolute temperature in kelvins (K)

$M$ = molecular weight in grams/mole ($g\ mol^{-1}$; also called **daltons, Da**)

$\pi$ = 3.1416

This equation shows that average velocity is proportional to the square root of the absolute temperature; that is, the higher the temperature the faster the motion of the particles. But average velocity is inversely proportional to the square root of the mass, so the smaller the particle the faster it moves at a given temperature.

Applying Equation 2.1 and others produces some impressive numbers, as illustrated in Table 2-2. Average velocities are astonishingly high. The average hydrogen molecule, near room temperature, is moving close to 2 $km\ s^{-1}$, which is 6,433 $km\ h^{-1}$ (3,997 mph)! Even the much heavier $CO_2$ molecule has an average velocity of 1,372 $km\ h^{-1}$. At atmospheric pressure, however, the **mean-free-path** (distance between collisions) is short: only 150 to 400 times the particle diameters. With such high velocities and such short pathways between collisions, the number of collisions of each molecule is enormous: on the order of billions each second. In liquids, for which no one has yet written satisfactory equations (that is, models), the velocities are of the same order of magnitude at room temperatures; but as you would expect, the mean-free-paths are much shorter, so the number of collisions is even greater. In solids, the particles are more or less held in place but vibrate against each other. Realizing the astronomical number of collisions possible in such short time intervals helps us understand how chemical reactions can be so rapid.

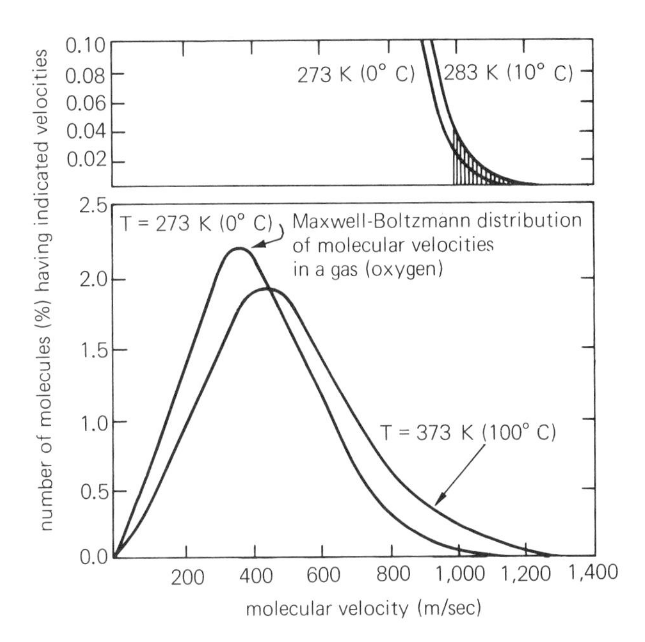


**Figure 2-3** The Maxwell-Boltzmann distribution of molecular velocities in a gas at two temperatures 100°C apart. The curves at the top show the high-velocity portion for a gas at two temperatures 10°C apart. The area under the curves indicated by the vertical lines represents the number of highly energetic particles and approximately doubles in going from the lower to the higher temperature.

Table 2-2 shows that changing the temperature from 0°C to 30°C, which is a large part of the temperature range for life functions, increases the average particle velocities only about 5 percent.

It is important to realize that the actual velocities of individual particles in a homogenous substance vary widely from the average velocity. The velocities are distributed according to the Maxwell-Boltzmann probability function, which produces curves such as those shown in Figure 2-3 for gaseous oxygen. We suspect that this distribution would be similar for liquids and solutes dissolved in them. It is easy to understand why particle speeds vary so widely if we think of the random nature of particle collisions. Unless the collision is perfectly symmetrical, one of the participants gains energy and the other loses it; hence, the speed of a particle will probably change billions of times per second, at virtually every collision. Yet statistically, the particle velocities will be distributed at any instant as in Figure 2-3.

High-speed (high-energy) particles are the ones most likely to cause melting, evaporation, and chemical reactions. As the water molecules with the highest speeds enter the vapor state during evaporation, the average kinetic energy of the remaining molecules is lowered, which is the same as saying that the remaining liquid water is cooled. This is why evaporation is a cooling process. Particles with lower energies (left side of curves in Fig. 2-3) are the first to condense from vapor to the liquid state or to solidify from liquid to solid (freeze). All these processes are important in living plants.

Note in Figure 2-3 and Table 2-2 that *average* particle velocities do not change much even with an increase of 100 K (equivalent to 100°C), yet the number of *high-velocity* particles (under the right-hand tails of the curves) increases considerably even with a 10-K rise in temperature. This is a feature of the shapes of the curves. If the number of these *high-energy* particles doubles with a small temperature increase—and it is these particles that take part in chemical reactions—then we can understand why the reactions double with an increase of just a few degrees. The factor by which a reaction increases with a 10°C increase in temperature is called the $Q_{10}$.[3] If the reaction doubles, $Q_{10} = 2$.

## 2.4 A Model of Diffusion

Let's apply the concept of random molecular motions (kinetic theory) to gain an understanding of diffusion. Imagine two rooms connected by a doorway. One room contains white ping pong balls in free motion, the other contains black ones, also in motion. The imaginary balls lose no energy when they bounce off the walls or each other; the collisions are perfectly elastic. The chances that a black ball will go through the door into the other room in an interval of time depends on the speed and concentration (number per unit volume) of black balls and on the size of the door. At the beginning, the concentration of black balls is higher in one room than the other, but as some black balls go through the door opening, the concentration builds up in the other room. The direction of diffusion before equilibrium of each type of ball will be independent of the other, provided the two do not stick to each other (that is, interact chemically). Gradually, a condition of **dynamic equilibrium** is approached in which the concentration of black balls

[3] $Q_{10}$ values may be calculated when reaction rates are known at any two temperatures:

$$Q_{10} = \left(\frac{k_2}{k_1}\right)^{10/(T_2 - T_1)}$$

or

$$\log Q_{10} = \left(\frac{10}{T_2 - T_1}\right) \log \frac{k_2}{k_1}$$

where

$T_1$ = lower temperature (in K or °C)

$T_2$ = higher temperature (in K or °C)

$k_1$ = rate at $T_1$ = change in measured parameter per unit time at $T_1$.

$k_2$ = rate at $T_2$ = change in measured parameter per unit time at $T_2$.

is the same in both rooms. When dynamic equilibrium is achieved, balls are still passing through the opening, but the chance that a black ball will go in one direction through the opening is the same as the chance that a black ball in the other room will pass through in the opposite direction. The same is true for the white balls. When there is no *net* movement there is no diffusion, only kinetic or molecular motion. Because of this motion, the equilibrium is *dynamic*.

The model just described is nothing more than an expansion in size of what really happens in, say, two compartments connected by an opening, each compartment containing a different gas. The value of the model in this case is that the size and velocities of the "perfect bouncing balls" might be easier to visualize than real molecular-size particles. With the white and black balls in mind, it is not difficult to mentally accelerate their velocities and shrink their sizes to those of molecules given in Table 2-2.

Diffusion (*net* movement of particles or balls) as just described occurs in response to a concentration gradient. **Concentration** is the amount of substance or number of particles per unit volume. A **gradient** occurs when a parameter such as concentration changes gradually from one volume of space to another. (A temperature gradient is easy to visualize; it could be expressed as change in temperature with change in distance: $\Delta T \, \Delta x^{-1}$.)

Gradients in properties other than concentration can also lead to diffusion. This is especially important when we consider the diffusion of water. As noted in the previous discussion, liquid water is virtually incompressible. Because a given amount of liquid water always occupies essentially the same volume, its concentration remains nearly constant at 55.2 to 55.5 M (mol $L^{-1}$). Slight changes occur when substances are dissolved in water and when the temperature of the water changes, but these changes in concentration of water have little effect on the diffusion of the water. Addition of many substances, such as various sugars or salts, causes the water volume to expand (as does increasing temperature), so the water concentration is diminished slightly. Addition of certain other substances actually causes the water volume to shrink slightly, increasing the water concentration. In no cases can these changes in water concentration account in a quantitative way for the observed diffusion of water, which is so important in plants. So our diffusion model based on concentration gradients is too simple.

The science of thermodynamics provides concepts that make it possible to refine the model so that it explains the observed phenomena much more accurately. We cannot teach thermodynamics in a few paragraphs, but a brief overview of some important thermodynamic principles can provide some intuitive feeling for what happens.

## 2.5 Thermodynamics[4]

**Matter** has mass and occupies space, but what is energy? **Energy** occupies no space and has no mass but can transform or act on matter. We observe energy only by observing its effects on matter. **Thermodynamics** (the science that deals with heat and other forms of energy) is a conceptual framework devised during the last century to help us understand heat and machines, especially steam engines. The principles of thermodynamics now apply to all forms of energy and are widely used in almost every field of science.

In thermodynamics the word **system** means that region of space or quantity of matter on which we have focused our interest and attention. A system might be a chlorophyll molecule, a beaker of sugar solution, a photosynthesizing leaf, the Canadian Rockies, or the Milky Way Galaxy. Everything not in the system is called the **surroundings** (environment). The system is separated from the surroundings by a **boundary**, which we usually must imagine. Thermodynamics is often concerned with the energy transfers or **interactions** that take place across the boundary. Nevertheless, it is important to realize that thermodynamics always operates within the limits of a specific known portion of the universe; that is, the system and its surroundings constitute an even larger system that has no interactions with *its* surroundings. In this sense, thermodynamics deals only with closed systems. (Special approaches that we shall not consider here are required to deal with open systems.)

Physiology is the study of processes in organisms. Some (for example, diffusion) are physical; others are chemical. We ask two fundamental questions about processes: *Will the process occur? And how fast will it occur?* Perhaps surprisingly, the two questions have little to do with each other. The tendency for a process to occur does not influence the rate at which it occurs, and vice-versa.

### Enthalpy (*H*) and the First Law of Thermodynamics

The **First Law of Thermodynamics** can be stated in several ways: *In all chemical and physical changes, energy is neither created nor destroyed but is only transformed from one form to another*. Or, *in any process the total energy of the system plus its surroundings remains constant*. Or, *you can't get something for nothing*. Or, *the best you can do is break even*.

The First Law puts some important limitations on what can and what cannot be done. For example, the energy trapped in organic molecules by photosynthesis cannot exceed the light energy absorbed.

[4]We are indebted to Keith Mott for suggestions about the content and organization of Section 2.5.

To study energy transformations and to investigate whether a process will occur, we will make use of **equilibrium thermodynamics**, which provides information about the energy level of a system in some initial state (before the process occurs) and in a final state (after the process has occurred). If the final state has a lower energy level than the initial state, then the process is energetically feasible and can be spontaneous (that is, gives off more energy than it absorbs). Note that equilibrium thermodynamics says nothing about the rate at which a process will occur, only whether it could occur spontaneously or not. A process may be energetically feasible, such as oxidation (burning) of wood, but may not proceed at ambient pressure and temperature because it lacks activation energy (say a burning match). Alternatively, a reaction with a large favorable energy difference may proceed extremely slowly. Life solves the problem of slow but feasible reactions with enzymes (catalysts) that speed the reactions (Chapter 9).

One measure of the energy associated with a particular state is **enthalpy (*H*):**

$$H = E + PV \tag{2.2}$$

where

$E$ = internal energy

$P$ = pressure

$V$ = volume (hence, $PV$ is called the **pressure-volume product**)

Internal energy ($E$) includes the velocity or translational kinetic motions of the particles discussed previously, as well as their rotations and vibrations. Internal energy also includes the electron energies (discussed in Appendix B), which involve energy levels of electrons in molecules and the effects of absorption of radiant energy, and the molecule-electron configurations that collectively we refer to as chemical bonds. Part of the internal energy of gasoline, for example, is the energy locked in the bonds that hold its atoms of carbon and hydrogen together.

We can think of internal energy as existing in two forms. For example, the chemical-bond energy in gasoline is **potential energy**. When the gasoline burns, some bonds break and reform, releasing energy that causes the translation, rotation, and vibration of atoms and molecules (which we observe as an increase in temperature); this atomic and molecular motion is called **kinetic energy**. In an engine, some of this random, disorganized energy is organized and transmitted to the surroundings as **work** by means of moving pistons. We can think of potential energy as being a function of position or condition and kinetic energy as being a function of motion (of objects, molecules, photons, electrons, and so on).

It is impossible to quantify the absolute internal energy of a substance, so it is also impossible to quantify the absolute enthalpy. But *differences* in enthalpy between two states ($\Delta H$) can be quantified by measuring the heat released or gained by the system as it moves between the two states. Processes that release heat are termed **exothermic** when they move from a higher enthalpy to a lower enthalpy; they have a negative change in enthalpy ($-\Delta H$). Processes that gain heat from the surroundings are termed **endothermic**. They occur when the system moves from a lower to a higher enthalpy, so they have a positive enthalpy change ($+\Delta H$).

Because enthalpy is a measure of energy, processes that have a $-\Delta H$ should be spontaneous because they move from a higher energy state to a lower energy state (the extra energy being released as heat), but some spontaneous processes have a $+\Delta H$. In moving from a lower enthalpy to a higher enthalpy, they remove heat from the surroundings. There must be something else besides enthalpy contributing to the overall energy state of a system. That something else is entropy.

## Entropy ($S$) and the Second Law of Thermodynamics

The **Second Law of Thermodynamics** is difficult to put into words because it is very abstract. Nevertheless, it has been stated in several ways: *Any system, plus its surroundings, tends spontaneously toward increasing disorder*. Or, *heat cannot be completely converted into work without changing some part of the system*. Or, *in any energy conversion some energy is transferred to the surroundings as heat*. Or, *no real process can be 100 percent efficient*. Or, *there can never be a perpetual motion machine*. Or, *you can't even break even*.

The consequences of the Second Law are extremely important. For example, photosynthesis will never be 100 percent efficient because some of the light energy driving the process will be converted into heat. Because some of the energy driving *any* process will be converted to or will remain as heat, there will never be a perpetual motion machine. The statement that randomness or disorder must always increase for the system and its surroundings is especially significant. The measure of this randomness is called **entropy** ($S$).

The Second Law says that the entropy or randomness of the universe is always increasing. Therefore, any spontaneous process must result in an increase in entropy somewhere, either in the system or its surroundings. If the process results in an increase in the entropy of the system, we say the process has a positive entropy change ($+\Delta S$). A positive entropy change tends to make a process spontaneous. As with enthalpy, however, there are spontaneous processes that have an unfavorable entropy change ($-\Delta S$) for the system, in which case $\Delta S$ for the surroundings is then positive. Many of the processes of life fall into this category.

Entropy is a valuable concept that leads us to think about the degree of orderliness in the universe and how many changes around us are spontaneously driven by the overall increase in disorder. The concept applies nicely, for example, to our model of diffusion. Think of the energy (and intelligence) required to create the high degree of order of the white and the black balls at the beginning of our thought experiment. Then think about how that order was destroyed—converted to random disorder—by spontaneous mixing as the balls passed through the doorway. Equilibrium was reached at maximum disorder (entropy), and that is one way to define equilibrium.

The two system properties we have defined—enthalpy and entropy—determine the overall energy change as a system moves from one state to another. Processes with a $-\Delta H$ and a $+\Delta S$ will be spontaneous, and processes with a $+\Delta H$ and a $-\Delta S$ will never be spontaneous. But what about a process with a $-\Delta H$ and a $-\Delta S$ or a $+\Delta H$ and a $+\Delta S$? Whether such a process is spontaneous or not will depend on the relative values of $\Delta H$ and $\Delta S$. What is needed is a way to relate $\Delta H$ and $\Delta S$ to each other.

## The Gibbs Free Energy ($G$)

J. Willard Gibbs, working at Yale University in the 1870s, developed a thermodynamic measure that allows us to think of the energy available to do work as it passes across the boundary between a system and its surroundings. This measure is now called the **Gibbs free energy ($G$)**; it is *a measure of the maximum energy available within the system for conversion to work (at constant temperature and pressure).* Because space is limited, we will have to be content with a word description of free energy, an equation that defines it, and applications of the concept of free energy that eventually lead us to a definition of water potential. The water-potential concept is fundamental to the first chapters of this book and appears occasionally in later chapters.

Without derivation, we define the Gibbs free energy ($G$) by combining the enthalpy ($H$) and the entropy ($S$), along with the kelvin temperature ($T$) in the following equation:

$$G = H - TS \tag{2.3}$$

Remembering that the enthalpy ($H$) equals the internal energy ($E$) plus the pressure-volume product ($PV$), we can write Equation 2.3 as:

$$G = E + PV - TS \tag{2.4}$$

As with enthalpy and entropy, we can quantify only differences in free energy ($\Delta G$):

$$\Delta G = \Delta H - T\Delta S \tag{2.5}$$

$\Delta G$ is the quantity that we use to decide whether a process is spontaneous or not. Processes with a negative free-energy change ($-\Delta G$) are energetically feasible and are capable of occurring spontaneously. (Remember that equilibrium thermodynamics does not contain information on rate, and many spontaneous processes do not occur at measurable rates under normal conditions.) Processes with a $+\Delta G$ are not energetically feasible and will not proceed without an energy input.

The definition of free energy is useful conceptually, but in practice it is often difficult to measure changes in enthalpy and entropy. To quantify $\Delta G$ for chemical reactions, we define another term, $\Delta G^0$, the **free energy change under standard conditions**. In this case, standard conditions means one unit of activity of each of the components of the reaction, both reactants and products. For dilute solutions, the **activity** (a kind of "corrected concentration") of a particular component is approximately equal to its molar concentration, so $\Delta G^0$ shows the tendency for a reaction to proceed if each reactant and product has a concentration of approximately 1 M.

Consider a reaction in which certain reactants form certain products:

$$A + B \leftrightarrows C + D \tag{2.6}$$

The **equilibrium constant** for such a reaction equals the product of the concentrations (actually, the activities) of the products divided by the product of the reactant concentrations (activities):

$$K_{eq} = \frac{[C]\,[D]}{[A]\,[B]} \tag{2.7}$$

Note that $K_{eq}$ will be greater than 1 if the reaction proceeds from left to right and less than 1 if the reaction goes from right to left. We can now relate $\Delta G^0$ to $K_{eq}$ with the following equation:

$$\Delta G^0 = -RT \ln K_{eq} \tag{2.8}$$

where

$\Delta G^0$ = standard free energy change in joules (J) or calories (cal)

$R$ = the ideal gas constant (8.314 J $mol^{-1}$ $K^{-1}$; or 1.987 cal $mol^{-1}$ $K^{-1}$)

$T$ = absolute temperature (in kelvins, K)

ln = natural logarithm

$K_{eq}$ = equilibrium constant

The results of the calculation are the same as would be obtained if it were possible to subtract the absolute free energies of the products from those of the reactants. This standard free-energy change is the maximum useful work that can be obtained when one mole of each reactant is converted to one mole of each product. Thousands of experiments have shown that free energy decreases for spontaneous reactions ($-\Delta G^0$) and increases for nonspontaneous reactions ($+\Delta G^0$). Hence, for spontaneous processes, $K_{eq}$ is greater than 1 (activities of products greater than reactants), so $\Delta G^0$ is *negative*, which is another way to define such processes. If the concentration of reactants at equilibrium exceeds that of products, then $K_{eq}$ is less than 1, so its logarithm is thus negative and $\Delta G$ is positive; the reaction is not spontaneous. The farther the reaction proceeds in either direction to reach equilibrium, the more negative or positive $\Delta G^0$ becomes.

Products and reactants seldom reach molar concentrations in cells, so we need a way to determine the actual $\Delta G$ under nonstandard conditions to see whether a reaction is favored or not in the cell. The equation for *actual* reactants [R] and products [P] is (other terms as defined above):

$$\Delta G = \Delta G^0 + RT \ln \frac{[P]}{[R]} \tag{2.9}$$

If there is a high concentration of reactants relative to products, the fraction will be less than 1, making the entire second term negative. If this term is a large enough negative number, $\Delta G$ will be negative in spite of a positive $\Delta G^0$. Therefore, a reaction could be energetically unfavorable under standard conditions (1 M products and reactants) but energetically favorable when reactants are high compared with products. Conversely, a reaction that proceeds to the right (is favorable) under standard conditions might actually proceed to the left when products are high compared with reactants.

An important concept in equilibrium thermodynamics is that $\Delta G$, $\Delta H$, and $\Delta S$ for the transition between two states are independent of the path or route used to go between the two states. As a corollary to this, the values for these parameters are additive for processes occurring in series or at the same time. In other words, if a process can be broken down into several steps that occur in series or simultaneously, then $\Delta G$ for the overall process is equal to the sum of $\Delta G$ values for the individual steps.

This is important in the metabolic processes of living cells. Many of these processes have large positive $\Delta G$ values and can be forced to occur only by coupling them with reactions that have even larger negative $\Delta G$ values, so that the overall process has a negative $\Delta G$ ($-\Delta G$).

## 2.6 Chemical Potential and Water Potential

We can speak of the free-energy change of the total system, or of any one of its components. We note, however, that a large volume of water has more free energy than a smaller one, under otherwise identical conditions. Hence, it is convenient to consider the free energy of a substance in relation to some unit quantity of the substance. The free energy change upon addition of a unit quantity, specifically a gram molecule weight, of substance $i$ is called the **chemical potential** ($\mu_i$). We could then speak of free energy per mole of substance. Chemical potential, like solute concentration and temperature, is independent of the quantity of substance being considered.

For a **solute** (dissolved material) in a **solvent** (the liquid in which the solute is dissolved; in plants, mostly water), the chemical potential is approximately proportional to the concentration of solute. Actually, concentration is usually corrected by some factor that depends upon concentration and other parameters to produce a corrected concentration called the **activity** ($a_i$). Chemical potential of a substance can be calculated with the following relationship:

$$\mu_i = RT \ln a_i \tag{2.10}$$

where

$\mu_i$ = chemical potential of substance $i$

$R$ = gas constant (8.314 J mol$^{-1}$ K$^{-1}$)

$T$ = temperature in kelvins

$a_i$ = activity of substance $i$

A diffusing solute tends to move from regions of high chemical potential (free energy per mole) to regions of low chemical potential.

The chemical potential of water is an extremely valuable concept in plant physiology. In 1960, Ralph O. Slatyer (see his essay on page 48) in Canberra, Australia, and Sterling A. Taylor at Utah State University proposed that the chemical potential of water be used as the basis for an important property of water in plant-soil-air systems. They defined the **water potential** ($\Psi$)[5] of any system or part of a system that contains water, or could contain water, as being equivalent to the chemical potential of the water in that system or system part compared with the chemical potential of pure water at atmospheric pressure and the same temperature; they further suggested that the water potential of the refer-

[5]The symbol for water potential is the *uppercase* Greek letter psi, $\Psi$. The *lowercase* psi, $\psi$, is often used but should be reserved for electric potential.

ence pure water be considered to be zero. In the next chapter we will apply these concepts, but for now suffice it to say that the water potential is negative if the chemical potential of water in the system under consideration is lower than that of the reference pure water, and it is positive if the chemical potential of the water in the system is greater than that of the reference water.

In thermodynamics, the chemical potential of any substance, including water, has units of energy per unit quantity, as in Equation 2.10. Appropriate SI units are joules per kilogram ($J\ kg^{-1}$) or joules per mole ($J\ mol^{-1}$), though calories per mole or per kilogram were often used in the past. (If one is speaking of pure substances, the mole is the correct unit; with mixtures, kilograms must be used.) In 1962, Taylor and Slatyer recommended that the energy terms of chemical potential be divided by the partial molar volume of water, which would then give the water potential in units of *pressure*. (Calculation is given in footnote 4 in Chapter 3, page 49.) Plant physiologists had long been discussing movements of water, including diffusion, in terms of pressure, so this suggestion was accepted and is now almost universally applied. It is still valid to use energy units when speaking of water potential (see, for example, Campbell, 1977), but most plant physiologists and soil scientists now use the following definition of water potential: The **water potential** ($\Psi$) *is the chemical potential of water in a system or part of a system, expressed in units of pressure and compared with the chemical potential (also pressure units) of pure water at atmospheric pressure and at the same temperature and height,*[6] *with the chemical potential of the reference water being set at zero*. This definition can be expressed with the following relationship:

$$\Psi = (\mu_w - \mu_w^*)/V_w \qquad (2.11)$$

where

$\Psi$ = water potential

$\mu_w$ = chemical potential of water in the system under consideration

$\mu_w^*$ = chemical potential of pure water at atmospheric pressure and at the same temperature as the system under consideration

$V_w$ = partial molar volume of water ($18\ cm^3\ mol^{-1}$)

[6]Height is specified to account for the effect of gravity in producing pressure in a standing column of liquid. We will usually ignore this complication by assuming that the reference water is at the same height, but it becomes important in soils and in tall trees, in which water may move in response to gravity.

From Equation 2.11 it is evident that if one is calculating the water potential of pure water, the result will be zero because the chemical potential of the pure water will be compared with itself. If the chemical potential of the water being considered is less than that of pure water (same temperature and atmospheric pressure), its water potential will have a negative value.

As solutes diffuse in response to differences in solute chemical potential, so water diffuses in response to differences in water potential. When water potential is higher in one part (region) of a system than in another, and no impermeable barrier prevents the diffusion of water, it diffuses from the high region of water potential to the low region. The process is spontaneous; free energy is released to the surroundings and the system's free energy decreases. This released energy has the potential to do work, such as osmotically lifting water upward in stems in the phenomenon known as root pressure. The maximum possible work is equivalent to the free energy released, but sometimes no work is done. Then the free energy simply appears in the system and its surroundings as heat or increased entropy. In any case, it is important to remember that equilibrium is reached when the change in free energy ($\Delta G$) or the water potential difference ($\Delta\Psi$) is equal to zero. At this point, entropy for the system and its surroundings will be at a maximum, but entropy change ($\Delta S$) will equal zero.

## 2.7 Chemical- and Water-Potential Gradients

Because gradients in chemical potential or water potential produce the **driving force** for diffusion, it is important to understand the five factors that most commonly produce chemical-potential or water-potential gradients in the soil-plant-air continuum.

### Concentration or Activity

For solute particles in plants (mineral ions, sugars, and so on), the activity (effective concentration) is by far the most common and important factor in establishing the chemical-potential gradients that drive diffusion. In this text, when we discuss the movement of solute particles into and out of cells and throughout the plant, we will be thinking almost exclusively of activity differences: Particles diffuse from regions of high activity to regions of low activity, which means high to low chemical potential.

Water is the common solvent in plants. As we have seen (Section 2.1), water is almost incompressible, so its concentration remains nearly constant, changing only slightly with addition of solutes and changes in temperature. Thus a model based strictly upon water concentration will not explain diffusion of water in plants. Indeed, differences in water concentration can readily be ignored.

### Temperature

Water vapor diffuses from food to the colder freezer coil in a refrigerator. By the same process, *liquid water* or *water vapor* often diffuses from deep in the soil to the surface when the surface is cooled at night, and deeper into the soil during the day. (These processes also often involve evaporation, condensation, or freezing and thawing.) But consider a gas at two different temperatures, separated by a barrier that permits diffusion: The cooler gas diffuses into the warmer gas, opposite to the diffusion in the examples just mentioned. This is because the cooler gas is the more concentrated gas, when pressures are equal, and the concentration difference is more important to diffusion than the slightly higher velocities of the molecules in the warmer gas. (On pages 90–92, John Dacey explains the diffusion of gases into waterlilies.) So temperature effects on diffusion are complex.

Actually, temperature differences are usually ignored in discussions of plant-soil-water relations for a very good reason: *The thermodynamic equations that we have been considering assume constant temperature throughout the system and its surroundings.* But there are ways to estimate the effects of temperature changes, as we shall see in the next chapter. It is important to consider temperature effects because strong temperature gradients may exist in plants. Consider, for example, the plants of the arctic or alpine tundras, sometimes only a few centimeters high, whose roots are in soil close to freezing while their leaves are warmed by the sun to over 20°C.

### Pressure

Increasing the pressure increases the free energy and hence the chemical potential in a system. Imagine a closed container (Fig. 2-4b) separated into two parts by a rigid semipermeable membrane that permits passage only of individual solvent molecules (water, we assume). Bulk flow does not occur, nor can solute particles pass through the membrane. If pressure is applied to the solution on one side of the membrane but not to the other, the water potential of the pressurized side will be *increased* (the $P$ term in Equation 2.4); there will then be a net movement of water molecules by diffusion through the membrane and into the compartment of lower pressure. This effect of pressure is an extremely important consideration in studying plants because the contents of most plant cells are under pressure compared with their surroundings, and because fluids in the xylem can be under tension (negative pressure).

## Effects of Solutes on Solvent Chemical Potential

It has been observed that solute particles decrease the chemical potential of the solvent molecules. This decrease is independent of any effect on solvent (water) concentration, which may be decreased, increased, or not changed, according to the kind of solute and its concentration.[7] Rather, it is a function of the **mole fraction**, which is the number of solvent particles (ions or molecules) compared with the total number of particles:

$$\text{mole fraction of solvent} = \frac{\text{moles of solvent}}{\text{moles of solute} + \text{moles of solvent}} \qquad (2.12)$$

Consider again the closed container in Figure 2-4c. If there is pure water on one side of the membrane and a solution on the other side, a water potential gradient will exist, with water potential being lower on the solution side. Water will diffuse through the membrane from the pure water side into the solution. This special case of diffusion is **osmosis**, which is often the process by which water moves from the soil into the plant and from one living plant cell to another when membranes are crossed.

Of course, a steep chemical-potential gradient exists for the solute particles in our container. If they can penetrate the membrane, they will move from the solution side (where their chemical potential is high) to the other side, where initially their chemical potential is infinitely low. When they become equally concentrated on both sides, there will be no more net movement. Differences in the chemical potential of solutes across cell membranes is a major factor in the movement of ions from the soil into the plant and in transport of ions and nonionized solutes in and out of plant cells. We will see that solutes can also move across membranes *against* chemical potential gradients, but metabolic energy must be used to do so.

---

[7]To see this for yourself, study the many tables by A. V. Wolf, Morden G. Brown, and Phoebe G. Prentiss entitled "Concentrative Properties of Aqueous Solutions: Conversion Tables" in the *Handbook of Chemistry and Physics*, CRC Press, Boca Raton, Fla. In the 1989–1990 edition, the tables are on pages D-221 to D-271. For a range of molar and other concentrations of 99 solutions, including those of many salts and organic compounds—not to mention sea water, human plasma, and urine from humans and various other animals—the tables show such parameters as densities, total water concentration, freezing point depression, osmolality, viscosity, and conductance. Osmolality allows computation of osmotic potentials at any temperature with the van't Hoff equation. (These concepts are discussed in Chapter 3; the van't Hoff equation is Equation 3.2.) It is important to understand that changes in water concentration are quite insignificant, because many textbooks incorrectly explain osmosis in terms of water concentration.

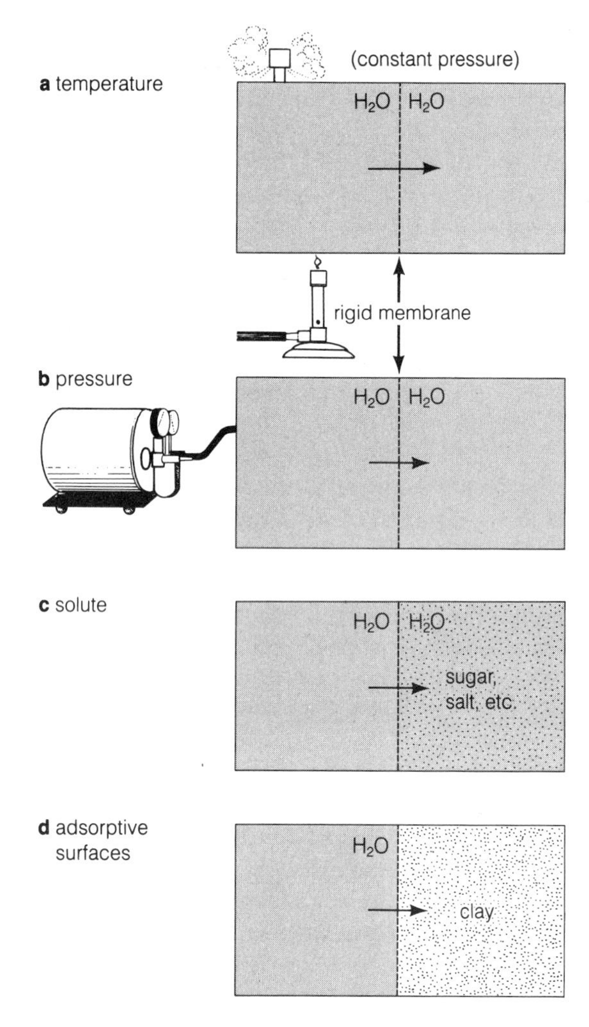


**Figure 2-4** Models of diffusional systems.

It is fundamental that the chemical potential of solutes not be confused with the chemical potential of water or the water potential.

### Matrix

Many charged surfaces, such as those of clay particles in soil, proteins, or cell-wall polysaccharides, have a great affinity for water molecules. These surfaces usually have a net negative charge that attracts the slightly positive sides of the polar water molecules. Because of hydrogen bonding, however, even surfaces that have no net charge, such as starch, also bind water. A material with surfaces that bind water is called a **matrix** (one of several uses of the term). Binding is a spontaneous process that releases free energy ($\Delta G$ is negative). In our imaginary double compartment (Fig. 2-4d), we might have pure water on one side of the membrane and dry protein or clay particles on the other. This condition establishes a steep gradient in water potential, with the water potential being high in the pure water and extremely low in the dry protein or soil. Water molecules will diffuse down the water-potential gradient into the other compartment and become bound to the protein or clay. (In this case, the membrane is required only to keep the water from *flowing* into the protein or clay by bulk flow.) This process of water becoming adsorbed on a matrix is called **hydration**. It is primarily responsible for the first phase of water uptake by a seed prior to germination. This and gravity are the main causes of water flow in soils; hence, we discuss the matric potential in Chapter 3.

## 2.8 Vapor Density, Vapor Pressure, and Water Potential

If a sample of water or a solution is exposed to a closed volume of vacuum or gas (as in Fig. 2-5), water molecules will evaporate into the space until they reach an equilibrium vapor concentration. At equilibrium, water molecules will be condensing back into the liquid phase at the same rate as they are evaporating into the gas phase. A convenient way to express the concentration of water molecules in the **vapor phase** (the gas phase) is as grams per cubic meter ($g\ m^{-3}$). Such an expression is called the **vapor density**. Of course the vapor molecules, colliding with the liquid surface and the walls of the container, exert a pressure. This is called the **vapor pressure**. Note that the vapor density and the vapor pressure are two different ways to express the amount of vapor in equilibrium with the liquid. This amount is independent of the presence of other gases. It will be the same if the volume above the liquid contains air at atmospheric pressure or if the volume was originally a vacuum, in which case the liquid would boil until the **saturation vapor density** or **saturation vapor pressure** had been achieved.

Applying our rule that water potentials are equal in all parts of the system when the system is at equilibrium (which is to say that the water potential gradient equals zero; $\Delta\Psi = 0$), it becomes apparent that factors affecting water potential in the liquid phase will also affect saturation vapor density (or pressure). Figure 2-5 illustrates the three most important effects, and the numbers for vapor density and vapor pressure illustrate the magnitudes of the three factors. Note that vapor density almost doubles when temperatures change from 20°C to 30°C. This effect is much greater than the effect of adding solutes (which lowers vapor density slightly) or of adding pressure (which raises vapor density slightly).

The effect of adding solutes can be calculated by **Raoult's law**, formulated in 1885. This law states that the vapor pressure over perfect solutions is proportional to the mole fraction of solvent:

**Figure 2-5** Illustrating the effects of pressure, temperature, and solutes on vapor density and vapor pressure (and thus indirectly on water potential). Units of vapor density are grams per cubic meter (g m$^{-3}$); vapor pressure units are kilopascals (kPa). These two ways of expressing the amount of vapor are interconvertible with a suitable equation. The equation for calculating water potentials from vapor pressures is given in Chapter 3 (Equations 3.5 and 3.6). Figures for vapor density and pressure of pure water in a closed volume at different temperatures are given in standard tables. Note the relatively large effect of temperature on vapor density or pressure compared with the much smaller effects of pressure and added solutes.

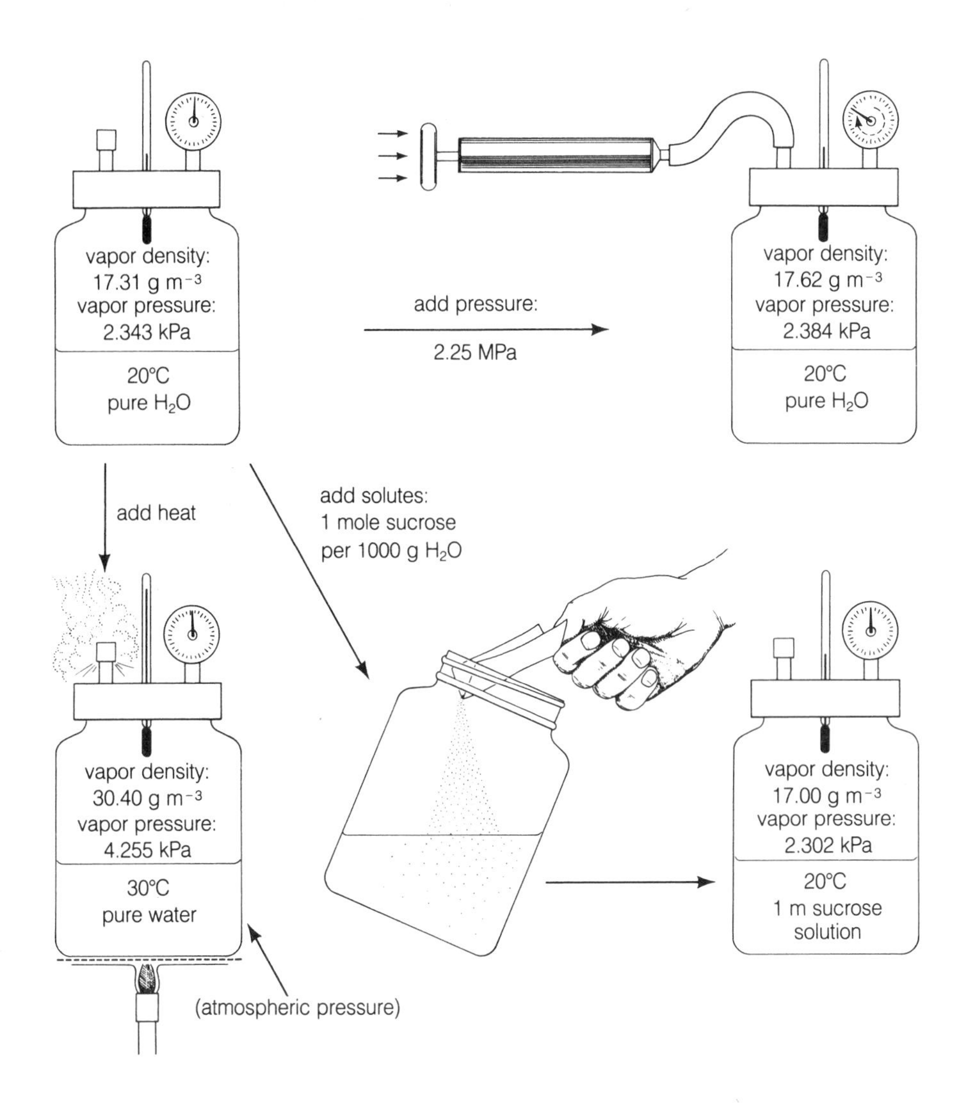


$$p = X_i p^0 \quad (2.13)$$

where

$p$ = the vapor density or pressure of the solution

$X_i$ = the mole fraction of solvent as defined in Equation 2.12

$p^0$ = saturation vapor density or pressure of pure solvent

Perfect solutions (those that exactly follow such laws) are seldom, if ever, encountered in nature, but Raoult's law provides a close approximation for real solutions. Furthermore, it is beautifully simple.

Because the solvent is usually present in quantities much greater than the solute, the actual lowering of the vapor pressure is usually not very great. Consider 1 gram molecular weight (1 mol) of sucrose (342.3 grams) dissolved in 1,000 g (55.508 mol) of water. Using Equation 2.12, the mole fraction of water in this solution is calculated as follows:

$$X_{H_2O} = \frac{55.508}{55.508 + 1.000} = 0.9823\ (98.23\%)$$

At equilibrium, the vapor density or pressure above pure water in a closed container is by definition equal to 100 percent **relative humidity**. At 20°C, this vapor density equals 17.31 g m$^{-3}$, so by Raoult's law the vapor density above a 1.00 molal solution of sucrose equals $17.31 \times 0.9823 = 17.00$ g m$^{-3}$. Reversing the calculation, you see that 17.00 is 98.23 percent of 17.31, so the relative humidity above a 1.00 molal sucrose solution is 98.23 percent. Thus, relative humidity above a solution (at equilibrium) is equal to the mole fraction of solvent expressed as percent.

With modern methods discussed in Chapter 3, it is possible to measure relative humidities to a small fraction of a percent, so vapor pressure above a solution in a closed container can be accurately determined. From this, as we will see, the water potential can be calculated. Water potentials of leaves or other plant materials or soils placed in closed containers can be determined in this way.

## 2.9 The Rate of Diffusion: Fick's First Law

We've examined the tendency for substances to diffuse from regions of high concentration (chemical potential) toward regions of low concentration. But how fast does diffusion take place?

In 1855, the German animal physiologist Adolf Eugen Fick applied concepts of conductive heat transfer to the question of diffusion. This was appropriate because conduction and diffusion are both caused by random thermal motion of the molecules. Consider some species $j$ that has a higher concentration, $C_j$, at one point in a system than at another point. We will use the symbol $J_j$ for the **diffusive flux (flow)** of species $j$. $J_j$ is the amount of $j$ that crosses some area in some unit time (which is a good definition for any **flux** of particles, including the photons in a beam of light). Using conventions of the calculus, we can express the gradient in concentration of $j$ as $\partial C_j/\partial x$, where $x$ is some distance.[8] Fick suggested (and experimentation confirmed) that the diffusive flux is equal to the gradient multiplied by a proportionality constant called the **diffusion coefficient** of species $j$, $D_j$:

$$J_j = -D_j \frac{\partial C_j}{\partial x} \tag{2.14}$$

The negative sign denotes that diffusion in the direction of increasing $x$ takes place from high values of $C_j$ to low values of $C_j$. Equation 2.14 is known as **Fick's first law of diffusion**. The concentration gradient might be expressed (using SI units) as mol $m^{-3}$ $m^{-1}$ (mol $m^{-3}$ denotes concentration and $m^{-1}$ expresses per meter of distance), and the diffusion coefficient would then have units of $m^2$ $s^{-1}$, yielding the diffusive flux in units of mol $m^{-2}$ $s^{-1}$, or amount of substance diffusing across a unit area in a unit interval of time.

We have noted that chemical potential and water potential better describe the tendency for diffusion than does concentration. But the potentials are expressed as energy units or pressure units, so using them in Equation 2.14 does not give diffusion rates directly in units of *amount* of diffusing substance.

[8]The symbol $\partial$ is used in the calculus to indicate a partial derivative, and the expression $\partial C_j/\partial x$ means that the gradient is expressed as some infinitesimal change (a change that is infinitesimally small) in the concentration of $j$ over some infinitesimal (infinitesimally short) distance, $x$. Such an expression of an infinitesimal change is called a **differential**, and it is usually expressed with $d$ instead of $\partial$ (that is, as $dC_j$ or $dx$). The expression $dC_j/dx$ is called a **derivative**, expressed in words as the derivative of $C_j$ with respect to $x$. In the case we are describing, the symbol $\partial$ is used instead of $d$ to show that we are dealing with a *partial derivative* because the gradient in $C_j$ changes with time and other factors besides distance, such as temperature. Thus we are considering the gradient only at one point in time and at one temperature.

Although it involves some assumptions (for example, steady-state flow with no storage in the distance $x$) that we will not discuss here, it is possible to integrate (a procedure used in the calculus) Equation 2.14 from one point in the system to another (say from point 1 to point 2):

$$J_j = \frac{(C_{j1} - C_{j2})\,D_j}{x} \tag{2.15}$$

where

$J_j$ = diffusive flux or flow of species $j$ (mol $m^{-2}$ $s^{-1}$)

$C_{j1}$ = concentration (mol $m^{-3}$) at its highest point, point 1

$C_{j2}$ = concentration (mol $m^{-3}$) at its lowest point, point 2

$x$ = distance (m) from point 1 to point 2

$D_j$ = diffusion coefficient ($m^2$ $s^{-1}$)

The difference in concentrations ($C_{j1} - C_{j2}$) can be expressed simply as $\Delta C_j$. It is the driving force for diffusion. The distance ($x$) can be combined with the diffusion coefficient to give a term called the **resistance** ($r$), and Equation 2.15 can then be expressed as:

$$J_j = \Delta C_j/r \tag{2.16}$$

In this simple form, it is readily apparent that the diffusive flux is proportional to the magnitude of the driving force ($\Delta C_j$) and inversely proportional to the resistance ($r$) encountered between point 1 and point 2 when concentrations are being considered. Resistance has the units of time per unit distance (for example, s $m^{-1}$). Equation 2.16 is analogous to **Ohm's law**, which says that the flow of an electrical current in a wire is proportional to the driving force (difference in voltage between two points) and inversely proportional to the resistance encountered in the wire. We will see that movement (flux) of many things follows a law equivalent or similar to this.

Some workers prefer to think of *permeability* to a diffusing substance rather than *resistance*. The **permeability** ($P_j$) is simply the inverse of the resistance:

$$P_j = \frac{1}{r} \tag{2.17}$$

Units of permeability are distance per unit time (for example, m $s^{-1}$). A medium (for example, a membrane) that has a high resistance to diffusion of solute or solvent has a low permeability. We can write:

$$J_j = P_j \Delta C_j \tag{2.18}$$

We will be concerned with diffusion rates in subsequent chapters. In general, diffusion is much slower than bulk flow. Increasing the temperature increases the average velocity of all the molecular-sized particles and thus increases the rate of diffusion. As we saw in Table 2-2, this effect is not great over the relatively narrow range of kelvin temperatures at which organisms are normally active. The $Q_{10}$ for diffusion of many gases is about 1.03 (meaning that a gas diffuses only 1.03 times faster when temperature is raised 10°C; see footnote 3 in this chapter). Solutes in water, however, have $Q_{10}$ values of 1.2 to 1.4 for diffusion. This is because increasing temperature breaks hydrogen bonds in the water, so solutes can diffuse more rapidly; viscosity of the water (Table 2-1) is decreased while permeability of water to solutes is increased. Because less-massive particles have higher average velocities at a given temperature, they will diffuse more rapidly than larger particles (Equation 2.1), all other factors being equal.

## 2.10 Caveat

In this chapter we have outlined the basic principles of thermodynamics that serve as a foundation for the concepts that relate to water potential. These ideas were developed during the 1960s. By now they have become widely accepted; they are essential to any modern study of plant physiology. The list of references for this chapter (see "*References*" near the end of this book) includes many citations that are not mentioned in the chapter; most of these uncited references are general discussions of plant water relations that can serve as sources of information for further study.

During recent years, a few authors have questioned the value of the water-potential concept. For example, Sinclair and Ludlow (1985) argue that physiological responses of plants are more highly correlated with relative water content[9] than with water potential. They and others (for example, Passioura, 1988) note that water *movement* may not be governed by water potential except in certain situations in which diffusion is paramount; in other cases, especially in soil, gravity is most important, and pressure determines the direction of flow when solutes move with the water, as in the phloem. In short, the situation is more complex than we may have implied in this introductory chapter. Of course, this is typical for most topics in plant physiology. In the meantime, Kramer (1988) defends the use of the concept of water potential, and even its most vocal critics admit that it is the best starting place that we have (see, for example, Schulze et al., 1988).

---

[9]**Relative water content (RWC)** is measured by obtaining a fresh mass of the tissue sample ($W_f$), allowing the tissue to absorb water (usually by floating it on water) until it is saturated, weighing again to get a turgid mass ($W_t$), and weighing it still again after drying in an oven at 85 to 90°C to get a dry mass ($W_d$). Relative water content expresses the water in the original sample as a percentage of the water in the fully hydrated tissue:

$$\text{RWC} = 100\left(\frac{W_f - W_d}{W_t - W_d}\right) \tag{2.19}$$

# 3

# Osmosis

It is an everyday experience to turn on a water faucet or flush a toilet. Thus we are perfectly familiar with water movement as a bulk flow phenomenon — our plumbing systems see to that! But in the world around us, vast quantities of water are moving, usually invisibly, by diffusion. In this and the next few chapters, we'll see how diffusion can establish pressure gradients that result in bulk flow.

It takes some mental effort to visualize this more unfamiliar aspect of the real world. With our mind's eye (there is no other way) we must see those water molecules, flying and bouncing billions of times each second in the vapor state, holding each other in the liquid state with their hydrogen bonding — positive side of one to negative of another — even while their kinetic motions cause some to separate. We must somehow conceptualize the entropy, free energies, and chemical potentials and how these properties can drive the molecules to diffuse down a gradient. We must realize that pressure increases free energies and chemical potentials, whereas solute particles and matric surfaces decrease them.

With these models in mind, we are ready to extend our concepts to the cells of plants. We are ready to discuss osmosis and related matters.

## 3.1 An Osmotic System

A device that measures osmosis is an **osmometer**. This is usually a laboratory device, but a living cell may be thought of as an osmotic system (Fig. 3-1). In both cases, two things are usually present: First, two or more volumes of solutions or pure water are isolated from each other by a membrane that restricts the movement of solute particles more than it restricts the movement of solvent molecules. Second, there is usually some means of allowing pressure to build up in at least one of the volumes. In the laboratory osmometer, pressures typically build up hydrostatically by raising the solution in the tube against gravity, but other means can be used, such as a volume detector (for example, a light beam and photo cell) that can increase pressure in the system (for example, with a piston) as soon as the volume of liquid begins to expand by the first small increment. In the cell, the rigidity of the plant cell wall is responsible for the increase in pressure.

It is important to emphasize the structural differences between the cell membrane and the cell wall. The membrane allows water molecules to pass more rapidly than solute particles; the primary cell wall is usually highly permeable to both. It is the plant cell membrane that makes osmosis possible, but it is the cell wall that provides the rigidity to allow a buildup in pressure. Animal cells do not have walls, so when pressures build up in them, they often burst, as happens when red blood cells are placed in water. Turgid cells provide much of the rigidity of nonwoody plant parts.

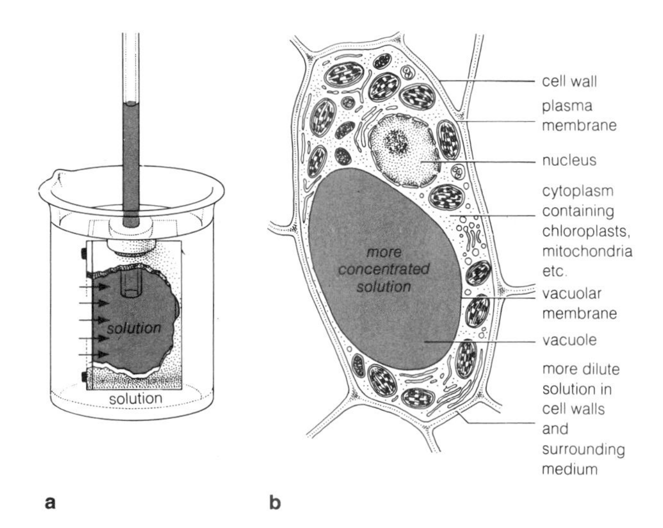


**Figure 3-1** (**a**) A mechanical osmometer in a beaker. (**b**) A cell as an osmotic system.

Consider at first a **perfect osmometer**. In such a device, the membrane is **semipermeable**, allowing ready passage of solvent (water), but *no* passage of solute, and the solution is so strongly confined that movement of water into the osmometer causes no significant increase in solution volume. A nearly perfect osmometer can be constructed in the laboratory, but a cell is never a perfect osmotic system.

As we saw in the previous chapter, restricting the diffusion of solute particles compared with solvent molecules can result in the establishment of a water-potential gradient. If there is pure water on one side of the membrane and a solution on the other side (typically inside the laboratory osmometer or the cell), then the water potential of the solution will be lower than that of the pure water. By convention, water potential of pure water at atmospheric pressure and at the same temperature as the solution being considered is set equal to zero, so *the water potential of an aqueous solution at atmospheric pressure will be some negative number* (less than zero). Hence, water molecules will diffuse from the higher water potential on the outside to the lower water potential in the cell solution; that is, water will diffuse "down" a water-potential gradient into the solution. The result will be a buildup of pressure within the system, either a raising of liquid in the tube of the laboratory osmometer or of pressure upon the cell wall. *Increasing pressure raises the water potential*, so the water potential within the osmotic system will begin to increase toward zero. This is illustrated in Fig. 3-2a and b.

The situation is analogous to the scale of a thermometer, but in this case we are dealing almost exclusively with values below zero. Adding solute decreases the water potential to some level below zero, and adding pressure raises this toward zero.

If pure water is on one side of the membrane (Fig. 3-2b), pressure on the other side will increase until the water potential of the solution is equal to zero; that is, equal to the water potential of the pure water on the other side. *When water potentials ($\Psi$) are equal on both sides, the water-potential difference ($\Delta\Psi$) between the two sides of the membrane is zero, and equilibrium has been achieved* ($\Delta\Psi = \Psi_1 - \Psi_2 = 0$).

If on one side of the membrane there is a solution and on the other side another solution of different concentration, osmosis will still occur (Fig. 3-2b). The more concentrated solution will have the lower (more negative) water potential, so water will diffuse into it from the other solution until its pressure builds up, if it is confined so that that is possible, to the point at which its water potential equals that of the less concentrated solution. If diffusion occurs into a solution that is not confined, it will continue until the more concentrated solution has been diluted to the point at which its water potential equals that of the solution on the other side of the membrane. In either case, both solutions will have a water potential of some negative *but equal* value, and equilibrium will have been reached.

Actually, the process is completely general. There could be pressure on both solutions, or the solution outside the osmometer might be more concentrated (water would move out), but *when equilibrium is achieved, water potential will be equal in all parts of the system*. ($\Psi_1 = \Psi_2 = \Psi_i$, hence, $\Delta\Psi = 0$.) This is not to say that $\Psi = 0$ in all parts of the system; two solutions in equilibrium with each other across a membrane would both have the *same* water potential of some negative number.

## 3.2 The Components of Water Potential

In the preceding paragraphs, we have considered water potential and two of its components: **pressure potential**, as it has been called, which is caused by the addition of pressure and is equal to the real pressure in the part of the system being considered, and **osmotic potential** (also called **solute potential**), which is caused by the presence of solute particles. Because pressure potential is a real pressure, we will simply call it **pressure**.[1] The proper symbol for pressure potential is $\Psi_P$, but $P$ can also be used. The symbol for osmotic or solute potential is $\Psi_s$, but for blackboard discussions the symbol $P$ can be used for pressure and $s$ for solute potential. (The symbols $\pi$ or $\Psi_\pi$ should *not* be used for osmotic *potential*; instead, $\pi$ should be reserved for osmotic *pressure*; see Section 3.6.)

In simple systems at constant temperatures, the water potential results from the combined but opposite actions of the pressure and the osmotic potentials (Fig. 3-2):

$$\Psi = \Psi_P + \Psi_s$$

$$(\Psi = P + s) \qquad (3.1)$$

Pressure can have any value. By convention $P = 0$ at *atmospheric* pressure. An increase of pressure results in a positive pressure, and **tension**[2] (suction or pulling, the opposite of pressure) results in a *negative* pressure.

[1]Some authors think the term *pressure potential* should be abandoned (see, for example, Passioura, 1982, who said that use of the term is "particularly grotesque"), but if we define **potential** (as we do in Chapter 26) as the condition of an environmental parameter in one part of a system that, when compared with the potential in another part of the system, establishes the tendency (the potential) for transfer of the parameter from the one system part to the other, then it seems perfectly logical and consistent to consider pressure potential and osmotic potential as components of water potential (which measures potential for transfer of water).

[2]An archaic use of the term *tension* means pressure, the opposite of its correct meaning. This survives in *hypertension*, which means high blood pressure.

**Figure 3-2** A schematic illustration of various effects on water potential. Black rectangles show the water potential (scale on the left). Wavy lines indicate depression in water potential caused by solutes (the osmotic or solute component of water potential). Unbroken lines suggest the effect of pressure on water potential. (Arrows point up for positive pressure, down for tension.) Zig-zag lines indicate effects of relative humidities below 100 percent on water potential of the atmosphere (discussed later in the chapter and in Chapters 4 and 5). (**a**) The basic effects of solutes and pressures, alone and in combination, on water potential. (**b**) Diffusion of liquid water in response to gradients in water potential, showing how water potential changes as diffusion occurs into an osmometer. (**c**) Water diffusion down a water-potential gradient from the soil through a plant into the atmosphere and from sea water through a plant into the atmosphere. The sea-water example (note *tension* in the xylem sap) is discussed in Chapter 5 (see Figure 5-17).

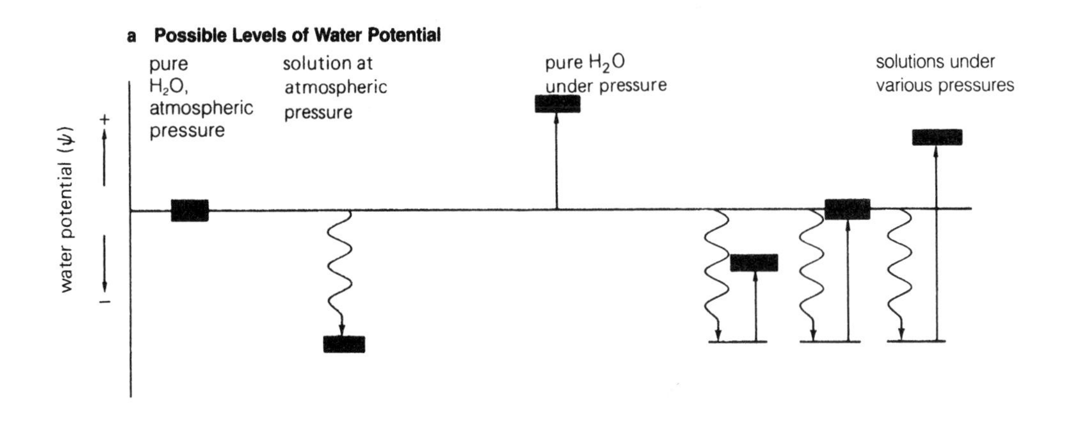


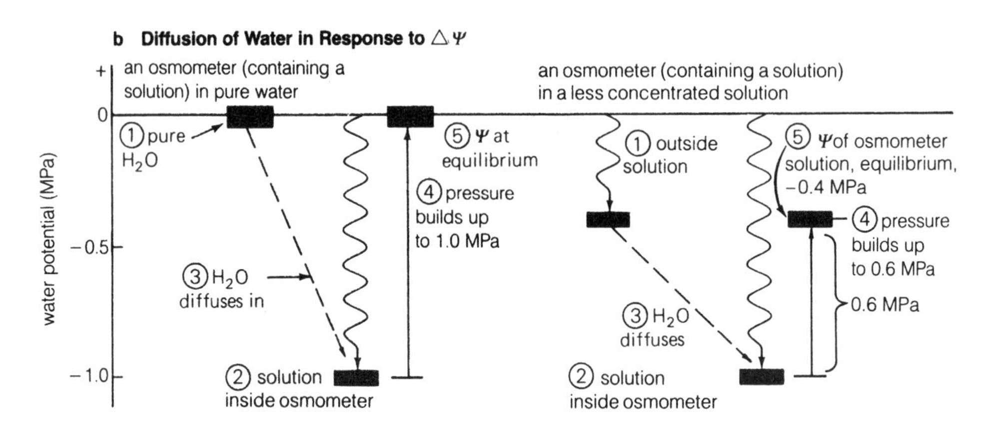


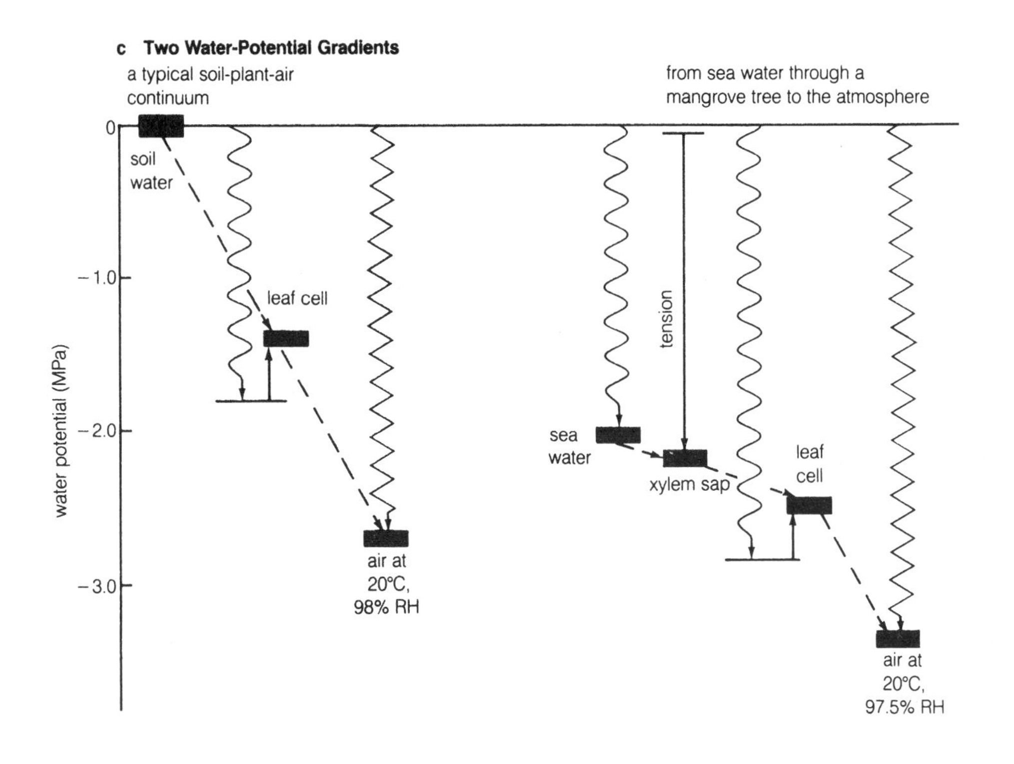

Pressure is usually positive in living cells but is typically negative in dead xylem elements or in soil (positive below the water table). Osmotic potential ($\Psi_s$) is always negative (or zero in pure water) because, in our collective experience, adding solute particles always decreases water potential below that of pure water.

Water potential ($\Psi$) can be negative, zero, or positive because pressure can be positive and very high, and osmotic potential can be zero or negative. We have defined water potential of pure water at atmospheric pressure as zero. In a solution at atmospheric pressure, water potential is negative. In pure water under some external pressure above atmospheric (for example, below the water table), water potential is positive. In a solution under some pressure other than atmospheric, water potential may be negative (osmotic potential is more negative than pressure is positive), zero (pressure equals osmotic potential but is opposite in sign), or positive (pressure is more positive than osmotic potential is negative).

Consider water potentials in the soil-plant-air system. Under most conditions (relative humidity somewhat less than 100 percent), water potential is highest in the soil and lowest in the atmosphere, with intermediate values in various parts of the plant; that is, there is a gradient from the soil, through the plant, to the atmosphere (Fig. 3-2c). But the *components* of water potential vary. In a wet soil above the water table, $P = 0$ and $\Psi_s$ is only slightly negative because the soil solution is dilute, so $\Psi$ is also only slightly negative. Xylem sap is very dilute, so $\Psi_s$ is only slightly negative; but the water is virtually always under *tension* ($P$ is negative), so $\Psi$ is more negative in the xylem than in the soil water, which moves into the plant from the soil. In leaf cells, which contain a more concentrated solution, $\Psi_s$ is quite negative; water moves in and builds up a positive $P$, but water is continuously evaporating from these cells, so $P$ does not increase as much as it otherwise would (that is, equilibrium is not reached), and $\Psi$ in the cells remains more negative than in the xylem. Atmospheric $\Psi$ (not yet discussed) is even more negative, so water tends to evaporate and move out of the leaves and into the atmosphere.

Note that *in land plants* $\Psi$ is virtually never positive (O'Leary, 1970). This is because $\Psi_s$ caused by solutes in cells is always negative, and matric forces also lower $\Psi$ below zero. Pressure in cells can raise $\Psi$ close to zero, but $\Psi$ in those cells never becomes positive. Thus xylem sap, although it is nearly pure water and could have a positive $\Psi$ if it were under pressure, stays under *tension* (negative pressure) as it remains at least close to equilibrium with the living tissues and their negative water potentials. (According to our definition of water potential, which compares the chemical potential of the water under consideration with pure water *at atmospheric pressure* and the same temperature, $\Psi$ increases with depth in a body of nearly pure water. Thus a submerged aquatic plant in near-equilibrium with the surrounding pure water under pressure would also have a positive $\Psi$, but this complication is irrelevant to most of the discussion in these chapters.)

Matric potentials, which also play a role in soil and plants, are discussed in the last section of this chapter. If the soil is dry, matric (and solute) forces may give the soil a negative $\Psi$, so the gradient in $\Psi$ from soil to air is much less steep than in the above example, and water movement through the plant is thus much slower.

## 3.3 Units for Water Potential

Let's look at these components of water potential in relation to the thermodynamic concepts of Chapter 2. The Gibbs free energy indicates the maximum energy available to do work. The free energy per mole is the chemical potential ($\mu$). As originally defined by Slatyer and Taylor (1960), the water potential is the chemical potential of a water solution (free energy per mole) in a system minus the chemical potential of pure water at atmospheric pressure and at the same temperature (see Equation 2.11). The components of water potential consist of solute and matric forces, which decrease the water potential, and pressure, which increases it.

A system's water potential expresses its ability to do work compared with the ability in a comparable quantity of pure water at atmospheric pressure and the same temperature. The osmotic potential of a solution is negative because the solvent water in the solution can do *less* work than pure water. As pressure on the solution increases, the solvent's ability to do work (and thus the water potential of the solution) also increases.

The work is performed by movement of pure water into the solution. In an osmometer, an ideal 1.0 molal[3] sugar (for example, glucose) solution at 28°C has an osmotic potential of $-2.5$ kilojoules per kilogram or $-45\ \mathrm{J\ mol^{-1}}$ ($-10.75\ \mathrm{cal\ mol^{-1}}$), which means that the maximum work that can be done, as pure water comes into equilibrium with the solution in the osmometer, is $2.5\ \mathrm{kJ\ kg^{-1}}$ or $45\ \mathrm{J\ mol^{-1}}$ of solution (see Equation 3.2 below).

[3]**Molality** = moles of solute per kilogram of $H_2O$. As noted in Chapter 2, this expresses osmotic relationships somewhat more accurately than **molarity** = **M** = moles of solute per liter of final solution. The symbol for *molal* is **m**, but that is the same symbol as for *meter*; hence, we will spell out *molal*. Neither *molar* nor *molal* are SI units, but mol kg$^{-1}$ (molal) is a valid SI combination. The liter is not an SI unit, but it can be used with SI units. The recommended symbol for liter is L (not l or ℓ). Note that unit symbols are written in Roman (upright) type; symbols for quantities (for example, length = $l$, concentration = $C$) are written in *italics* (slanted type).

# Pursuing the Questions of Soil-Plant-Atmosphere Water Relations

***Ralph O. Slatyer***

*Ralph O. Slatyer was educated in Australia in the 1950s. Since 1967 he has been a Professor of Biology at the Australian National University in Canberra City. His essay well illustrates the international nature of plant physiology; it also reinforces and expands several topics that we have been considering.*

To me, the most exciting and stimulating part of scientific research is when you make an observation or generate a hypothesis that you think is original and may, in addition, be at variance with accepted phenomena or attitudes.

Of course, in most cases, more careful observation, more thorough reading of the literature, or a critical discussion with colleagues leads to an awareness either that your ideas or observations won't hold up, or that someone else has anticipated you. But occasionally there is the breakthrough, and scientific knowledge and understanding move forward.

In my own case, the first such heady moments came in the mid-1950s when I was investigating the effects of progressive and prolonged water stress on physiological plant responses. At that time, it was widely accepted that the permanent wilting percentage [see Section 5.4] was a soil constant, the soil water content below which no plant growth would occur and further transpiration would cease. This concept had strong empirical support, mainly through painstaking work by Veihmeyer and Hendrickson at the University of California, Davis, on irrigated fruit crops, but dated back to previous works of Briggs and Shantz in the early part of the century. Associated with the concept was the notion that soil water was equally freely available to the plant at water contents down to the permanent wilting percentage, although this had come under challenge, particularly from scientists at the United States Salinity Laboratory at Riverside, California.

My initial research work, associated with my master's and doctoral studies, was concerned with crop and native species responses in arid and semiarid regions, and I repeatedly found evidence that consistently contradicted the established dogma. Accordingly, I set off to spend what was essentially a brief postdoctoral period with Professor Paul Kramer at Duke University. I found him receptive to my ideas and full of encouragement.

Basically, all I had done was to propose that as water stress was progressively imposed by soil water depletion, the turgor pressure of the leaf cells decreased until it reached zero, when the leaf water potential equaled the osmotic potential. I argued that at this point the leaves would be permanently wilted and that growth might be reasonably expected to have ceased. Even with stomatal closure, however, continued transpiration would be expected to continue to deplete soil water until desiccation of the plant itself reached lethal levels. From this argument it followed that the permanent wilting percentage should not be a soil constant but merely the soil water content at which the soil water potential and the plant water potential were balanced, at a level equal to the leaf-cell osmotic potential, so that zero turgor pressure existed.

Somewhat surprisingly, these views, published in both experimental and review papers, were rapidly accepted by the scientific community and have since been reconfirmed in general aspects by numerous investigators. In the process, of course, some of the more specific assertions have required qualification, but overall the dynamic nature of the soil-plant-water interaction and the permanent wilting percentage was clearly established.

This approach also seemed to provide a better basis for interaction among plant and soil scientists interested in plant-environment interactions, and it led to a requirement for a more integrated term to describe the state of water in plants and soils. Through the 1950s, "diffusion pressure deficit," "total soil moisture stress," and related terms were being used by these two groups of scientists who were basically talking about the same thing. This matter finally came to a head, informally, over dinner in a restaurant in Madrid during a UNESCO Plant Water Relations conference, attended by, among others, Sterling Taylor, Wilford Gardner, Robert Hagan, Fred Milthorpe, and myself. We proposed the term "water potential" (already suggested many years earlier by soil scientists), based thermodynamically on the chemical potential of water, as a single term for both soil and plant scientists, to be divided into component potentials, as appropriate. The meeting asked Sterling Taylor and me to draft a letter to *Nature* and a more definitive paper on the subject, and from this rather informal and personal beginning, the new terminology was launched. As far as I know, it is now used almost universally, although it too has been improved by modification and qualification.

I began this brief essay by referring to the excitement of scientific discovery. I conclude by referring to the spirit of cooperation that has existed in that part of the scientific community with which I have been associated. While it has always been a challenge to be first with a new piece of work, my life has been enriched by the warm personal relationships that have been developed between both my immediate colleagues and those further afield, whom one first meets by correspondence, or at conferences, and then by sharing space and facilities in a common laboratory.

The use of energy terms is logical when work is considered, but plant physiologists have traditionally expressed the water-potential concept (under various names during the past century) in pressure units. It is simpler to measure the pressure on the membrane of the laboratory osmometer (or to calculate it, knowing the density of the solution) than it is to measure the amount of energy required to raise the water in the tube. For the 1.0 molal solution, this pressure in a perfect osmometer is equal to 2.5 **megapascals** (**MPa**; 2.5 MPa = 25 bars, 24.67 atmospheres, 18.75 meters of mercury, or 25.49 kg $cm^{-2}$; the pascal is defined as the force of one newton per square meter). In the cell, the work is done by stretching the cell wall. Remember that the actual work is done by the pure water, which has the higher water potential. This is indicated by the fact that the water potential of the solution has a negative sign.

The energy and pressure units for water potential are easy to relate to each other, as was suggested by Taylor and Slatyer in their 1962 paper (see also Kramer, 1983; Slatyer, 1967). When the energy units are divided by the partial molar volume (the volume of 1 mol of $H_2O$, which is 18,000 $mm^3$ $mol^{-1}$) or the specific volume of water ($10^6$ $mm^3$ $kg^{-1}$ = 1 $cm^3$ $g^{-1}$), pressure units are obtained.[4] Thus the osmotic potential of a 1.0 molal glucose solution at 28°C (or the water potential of the same solution at atmospheric pressure) can be expressed in energy terms as $-2.5$ kJ $kg^{-1}$ and in pressure terms as $-2.5$ MPa. With SI units, kilojoules per kilogram have the same numerical value as megapascals.

[4]With SI units, the transformation is straightforward and simple:
Energy units: J $mol^{-1}$
SI definition of the joule: J = newton·meter = N·m
Thus: J $mol^{-1}$ = N·m $mol^{-1}$
Units for the partial molar volume of water: $m^3$ $mol^{-1}$
Units for the specific volume of water: $m^3$ $kg^{-1}$
Hence:

$$\frac{\text{N·m mol}^{-1}}{\text{m}^3\text{ mol}^{-1}} \text{ or } \frac{\text{N·m kg}^{-1}}{\text{m}^3\text{ kg}^{-1}} = \text{N m}^{-2} = \text{pascal}$$

$$= \text{Pa (SI unit of pressure)}$$

Specifically, to convert $-2.5$ kJ $mol^{-1}$ ($\Psi_s$ of a 1.0 molal sugar solution) by using the specific volume of water ($10^6$ $mm^3$ $kg^{-1}$ = $10^{-3}$ $m^3$ $kg^{-1}$):

$$\frac{-2{,}500\text{ J kg}^{-1}}{10^{-3}\text{ m}^3\text{ kg}^{-1}} = -2.5 \times 10^6\text{ N m}^{-2}\text{ or Pa} = -2.5\text{ MPa}$$

Or, to make the conversion by using the partial molar volume of water (18 $cm^3$ $mol^{-1}$ = 18 × $10^{-6}$ $m^3$ $mol^{-1}$)
For water: 1 kg = 1,000 g; molecular weight = 18.015 g $mol^{-1}$
Thus: 1 kg = 55.5 mol

$$\frac{-2.5\text{ kJ kg}^{-1}}{55.5\text{ mol kg}^{-1}} = -0.045\text{ kJ mol}^{-1}$$

$$= -45\text{ J mol}^{-1} = -45\text{ N·m mol}^{-1}$$

$$\frac{-45\text{ N·m mol}^{-1}}{18 \times 10^{-6}\text{ m}^3\text{ mol}^{-1}} = -2.5 \times 10^6\text{ N m}^{-2}\text{ or Pa} = -2.5\text{ MPa}$$

In 1887, J. H. van't Hoff discovered an empirical relationship that allows the calculation of an approximate osmotic potential from the molal concentration of a solution. He plotted osmotic potentials from direct osmometer readings as functions of molal concentration, obtaining the following relationship, the form of which is identical to that of the law for perfect gases:

$$\Psi_s = -CiRT \tag{3.2}$$

where

$\Psi_s = s$ = osmotic potential

$C$ = concentration of the solution expressed as molality (moles of solute per kg $H_2O$)

$i$ = a constant that accounts for ionization of the solute and/or other deviations from perfect solutions

$R$ = the gas constant (0.00831 kg·MPa $mol^{-1}$ $K^{-1}$, or 0.00831 kg·kJ $mol^{-1}$ $K^{-1}$, or 0.0831 kg·bars $mol^{-1}$ $K^{-1}$, or 0.080205 kg·atm $mol^{-1}$ $K^{-1}$, or 0.0357 kg·cal $mol^{-1}$ $K^{-1}$)

$T$ = absolute temperature (K) = degrees C + 273

If $C$, $i$, and $T$ are known for a solution, then osmotic potential ($\Psi_s$) can easily be calculated. For nonionized molecules, such as glucose and mannitol in dilute solutions, $i$ is 1.0, but in other cases $i$ varies with concentration. This is partly because activity varies with concentration and partly because the extent to which a salt or acid ionizes depends upon its concentration. Total $\Psi_s$ for a complex solution such as cell sap is the sum of all osmotic potentials caused by all solutes. It is called **osmolality**.

Consider some examples of simple solutions. For a 1.0 molal glucose solution at 30°C,

$$\Psi_s = -\left(1.0\,\frac{\text{mol}}{\text{kg}}\right)(1.0)\left(0.00831\,\frac{\text{kg·MPa}}{\text{mol K}}\right)(303\text{ K})$$

$$\Psi_s = -2.518\text{ MPa (at 30°C = 303 K)}$$

The same solution at 0°C,

$$\Psi_s = -(1.0\text{ molal})(1.0)\left(0.00831\,\frac{\text{kg·MPa}}{\text{mol K}}\right)(273\text{ K})$$

$$\Psi_s = -2.269\text{ MPa (at 0°C = 273 K)}$$

Note that the osmotic potential is less negative at 0°C than at 30°C, which seems to imply that water would diffuse from the cold toward the warm liquid. That is contrary to our experience (and to the discussion in Chapter 2), and we are incredulous only if we forget that the calculated osmotic potentials are compared with pure water at atmospheric pressure *and at the same temperature*. The fact that the warmer solution has a more negative osmotic potential than the colder one means that pure warm water, at one atmosphere pressure *and at the same temperature as the solution in the osmometer*, would produce a higher equilibrium pressure than the cooler water would produce in a cooler solution.

Because the thermodynamic equations we have presented are valid only at constant temperature, they cannot be used directly to calculate the forces that might be developed by **thermoosmosis**, a situation in which warm water might be separated from cold water by a membrane that permits pressure buildup in the cold water as warm water diffuses in (see Fig. 2-4). The effects of solutes, pressure, and temperature on vapor pressure, however, suggest that high pressures might be built up on the cold side if a perfect thermoosmometer could be built. Figure 2-5 (page 41) shows the relative magnitudes of solute pressure and temperature effects on vapor pressure and thus, by implication, on other properties of solutions related to water potential. The temperature effect is much greater than the solute or pressure effects.

For an application of the van't Hoff equation (Equation 3.2) to an example in which ionization causes $i$ to have a value different from 1.0, consider a 1.0 molal NaCl solution at 20°C. Sodium chloride ionizes 100 percent in dilute solutions, but the value of $i$ is not quite what one would predict from this. Two ions for each formula (NaCl) would suggest that $i = 2$; actual measurements show that $i = 1.8$:

$$\Psi_s = -(1.0 \text{ molal})(1.8)\left(0.00831 \frac{\text{kg MPa}}{\text{mol K}}\right)(293 \text{ K})$$

$$\Psi_s = -4.38 \text{ MPa}$$

Because the van't Hoff equation has the same form as the equation for perfect gases, the **osmotic pressure** (the actual pressure developed in an osmometer rather than the osmotic potential) is the same as would be exerted on the wall of the container if the solute particles existed as a gas in an equivalent volume. At 0°C, the pressure of 1 mole of a perfect gas in a volume of 1 L is 2.27 MPa (2.52 MPa at 30°C), and 2.27 MPa is the osmotic pressure developed by a 1 molal solution of nonionizing solute at 0°C in a perfect osmometer. This interesting relationship, which once caused considerable confusion, is now considered coincidental; it is certainly incorrect to think of solute particles as exerting pressure on the walls of a container as though they were a gas. Pressures on the walls of an open beaker of solution or an ordinary osmometer at equilibrium are hydrostatic pressures only, caused by the weight of the solution. Incidentally, the gas law applies only to "perfect" gases, and the van't Hoff equation is at best only an approximation for ideal solutions.

The van't Hoff equation is not merely coincidence, however, because it can be derived from Raoult's law (Equation 2.13) and other thermodynamic considerations. Its derivation was an early confirmation of developing thermodynamic principles.

Although plant physiologists and many soil scientists have accepted the water-potential conventions described here, physical chemists and most biologists in fields unrelated to plant physiology have not. Rather, they consider the **osmotic pressure** ($\pi$) to be a positive number equal to the real pressure that develops in an osmometer (as noted above). They do not speak of osmotic *potential*. That being the case, the equation for water potential (which goes by several different names) is as follows:

$$\Psi = P - \pi \tag{3.3}$$

Note that, with this convention, water potential of a solution at atmospheric pressure is still a negative number. Some plant physiologists and soil scientists also use this convention, so one must watch out for it when reading the literature.

## 3.4 Dilution

We have neglected one factor that is usually important in a real osmotic system, as contrasted to a perfect osmometer. As water diffuses across the membrane in a real system, it not only causes an increase in pressure but also dilutes the solution. This increases the osmotic potential in the solution (makes it less negative), so the pressure required to reach equilibrium will be less than would have been predicted from the original osmotic potential.

The relationship between water potential and its two primary components during dilution is well illustrated by the **Höfler diagram** (Fig. 3-3). The concept of this diagram was devised by K. Höfler in Austria in 1920. It describes the changing magnitudes of water potential, pressure potential, and osmotic potential as volume changes, assuming that the system expands only by taking up water and that no solutes move out or in. The curve for osmotic potential is derived from the simple dilution relationship, which is a close approximation for dilute molal solutions:

$$\Psi_{s1} V_1 = \Psi_{s2} V_2 \tag{3.4}$$

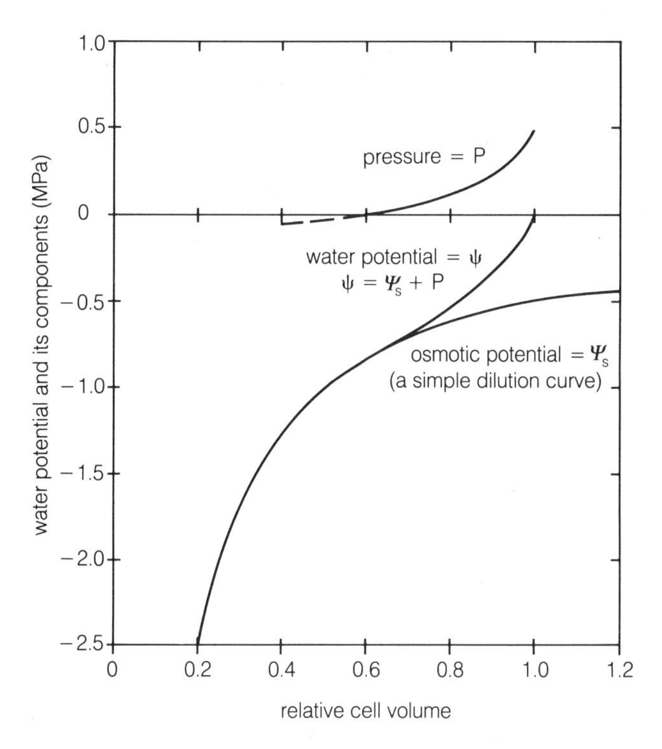


**Figure 3-3** The Höfler diagram. The components of water potential are shown as they change with osmometer (sometimes cell) volume. Osmotic potential is a dilution curve calculated by the relationship $\Psi_{s1}V_1 = \Psi_{s2}V_2$, as described in the text. The pressure curve is arbitrary but expresses the fact that as a cell with zero pressure takes up water, pressure increases slowly at first and then more rapidly (exponentially). The water potential curve is the algebraic sum of the pressure and osmotic-potential curves according to Equation 3.1.

where

$\Psi_{s1}$ = osmotic potential before dilution

$V_1$ = volume before dilution

$\Psi_{s2}$ = osmotic potential after dilution

$V_2$ = volume after dilution

The curve for pressure potential, on the other hand, is more hypothetical. Its shape depends on the tube diameter of the osmometer, or the stretching properties of the cell wall; it is steep if the tube is narrow or the wall is rigid, and less steep if the tube is wide or the wall less rigid. Actually, cell walls stretch easily at first; as pressure increases, their resistance increases somewhat and then becomes rather constant (McClendon, 1982). This is suggested by the pressure line in Figure 3-3. The water-potential curve is the algebraic sum of pressure potential and osmotic potential (Equation 3.1).

The Höfler diagram provides a good way to visualize the principles of Equation 3.1, along with the complications of dilution. It describes what would happen if mature cells were placed in solutions of different osmotic potentials so that water is either gained or lost from the cells, *but the total quantity of solutes within remains constant*. We shall discuss some of these approaches later in this chapter. The Höfler diagram may also describe what happens in some plant cells under normal conditions, at least over short time intervals, but growing cells do not act as suggested by the Höfler diagram. For one thing, the osmotic potential usually remains rather constant in growing cells as they absorb and/or produce solutes within (Chapter 16). Furthermore, when cells grow, their walls become softened (Chapters 16 and 17), so they stretch irreversibly (plastically), and pressures in them often decrease instead of increasing (Rayle et al., 1982).

## 3.5 The Membrane

Membranes exist in wide variety, but osmosis will occur regardless of how the membrane functions, as long as solute movement is restricted compared with water movement (Fig. 3-4). The membrane could consist of a layer of material in which the solvent is more soluble than the solute, which would allow more solvent molecules than solute particles to pass through. A layer of air between two water solutions provides a barrier that completely restricts movement of nonvolatile solutes while allowing water vapor to pass. A third membrane model we can visualize is a sieve with holes of such a size that water molecules could pass but the larger solute particles could not. We shall see in Chapter 7 that water passes rapidly through cell membranes and thus may be slightly soluble in them, and that cell membranes also act like they have pores. It has been suggested that water sometimes passes in the vapor state from soil particles to the root in dry soils.

In 1960, Peter Ray brought an interesting problem to the attention of plant physiologists. Calculations of the thicknesses of certain membranes and the rates of osmotic water movement across them showed that this movement could not occur by diffusion alone; the rates were too high. Ray suggested that the zone of diffusion might be very thin: an interface, say, between the water that is in the pores of the membrane and the solution that is inside the osmotic system. At this interface, the water-potential gradient would be extremely steep, resulting in a rapid diffusion. This rapid movement of water across the interface into the solution would create tension in the water remaining in the pore, pulling it along in a bulk flow (Fig. 3-4d). This fourth membrane mechanism once again illustrates the complexities of nature. Note that the thermodynamic relations (direction and equilibrium) still hold.

The vapor-membrane model is a good example of a truly semipermeable membrane, but all membranes that occur in plants must allow some solute to pass. Such membranes are **differentially permeable** rather

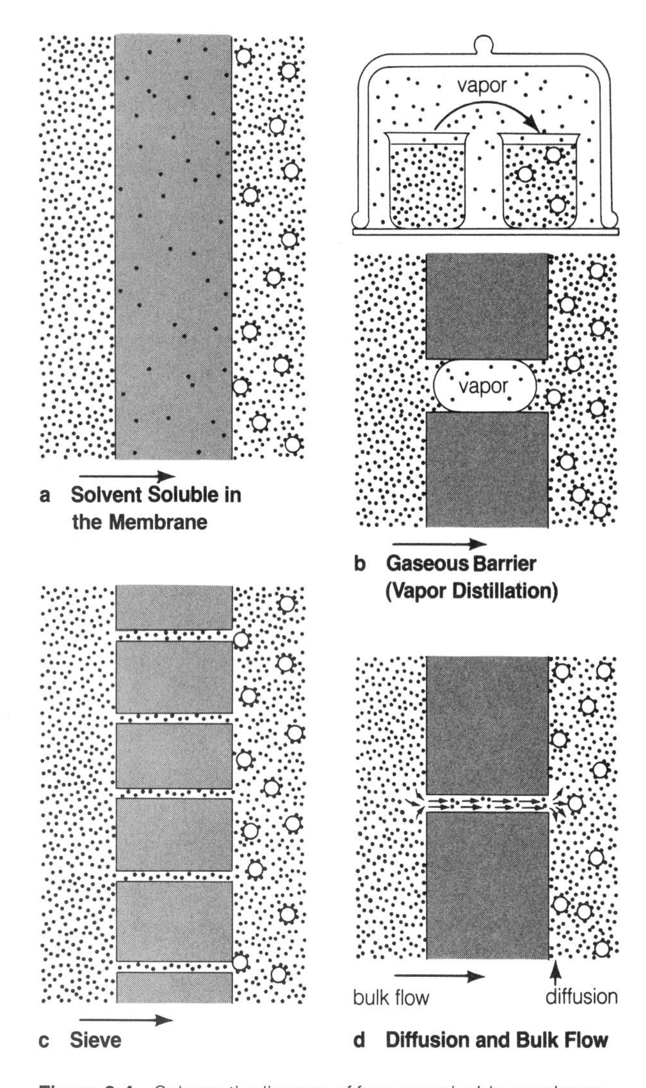


**Figure 3-4** Schematic diagram of four conceivable membrane mechanisms. The black dots represent water molecules about 0.3 nm in diameter, open circles represent sucrose molecules about 1.0 nm in diameter, and the membranes are drawn to the scale of most cellular membranes, at a thickness of 7.5 nm. Note that water concentration is about the same on both sides of the membranes. Water molecules move rapidly through membranes in cells, possibly by mechanisms similar to those represented by models **a** and **c**. This is discussed in Chapter 7. Model **d** is a refinement of model **c**, as discussed in the text. The vapor model **b** may apply in plants and soils in various situations, but no membrane is known to have gas-filled pores.

than truly semipermeable. Though living membranes are permeable to both solvent and solute, they are much more permeable to solvent. The permeability of membranes to solutes introduces a complication into our osmosis model: It determines the rate at which an equilibrium, established by solute concentration and pressure, gradually shifts as osmotic potentials on either side of a membrane change in response to the passage of solute particles.

## 3.6 Measuring the Components of Water Potential

Soon after the water-potential concept was formulated by Otto Renner in 1915, methods were developed for measuring water potential and its components. Newer methods have since been introduced, but the older methods can help us understand the water relations of plants. The newer methods are more useful. We will summarize both methods, not so much as a "cookbook" of useful techniques as an illustration and application of the principles we have been discussing.

### Water Potential

Probably the most meaningful property we can measure in the soil-plant-air system is the water potential. It is not only the final determinant of diffusional water movement, but it is often the indirect determinant of bulk water movement, which occurs in response to the pressure gradients that are set up by such diffusional movement. Furthermore, in principle and in practice, water potential is probably the simplest component of an osmotic system to measure.

Remember that at equilibrium $\Delta\Psi = 0$; that is, $\Psi$ is equal in all parts of the system. Thus, a plant part can be introduced into a closed system, and after equilibrium has been achieved, $\Psi$ may be known or determined for any other part of the system and therefore for the plant part. There are several possibilities for applying this principle, of which three general approaches are shown in Figure 3-5.

In the **tissue volume method** (Fig. 3-5b), a sample of the tissue in question is placed in each of a series of solutions of varying but known concentrations (usually sucrose; sorbitol; mannitol; or, best, polyethylene glycol, PEG; see Fig. 3-6). The best solute for such measurements is one that does not easily pass through membranes or harm the tissue. *The object is to find that solution in which the tissue volume does not change*, indicating no gain or loss in water, which means that the tissue and the solution were in equilibrium to begin with—the water potential of the tissue was and is equal to that of the solution. At atmospheric pressure, when $P = 0$, $\Psi = \Psi_s$. $\Psi_s$ for the solution, with its known concentration, can then be calculated from Equation 3.2.

In practice, there are several ways to determine changes in volume. One is to measure the volume of tissue before placing it in the solution (usually standard volumes are cut), and then to measure the volume (or only the length) after sufficient time for exchange of water (Fig. 3-5b). Volume change can be plotted as a function of solution concentration, which indicates an increase in volume in relatively dilute solutions and a decrease in volume in relatively concentrated ones. On

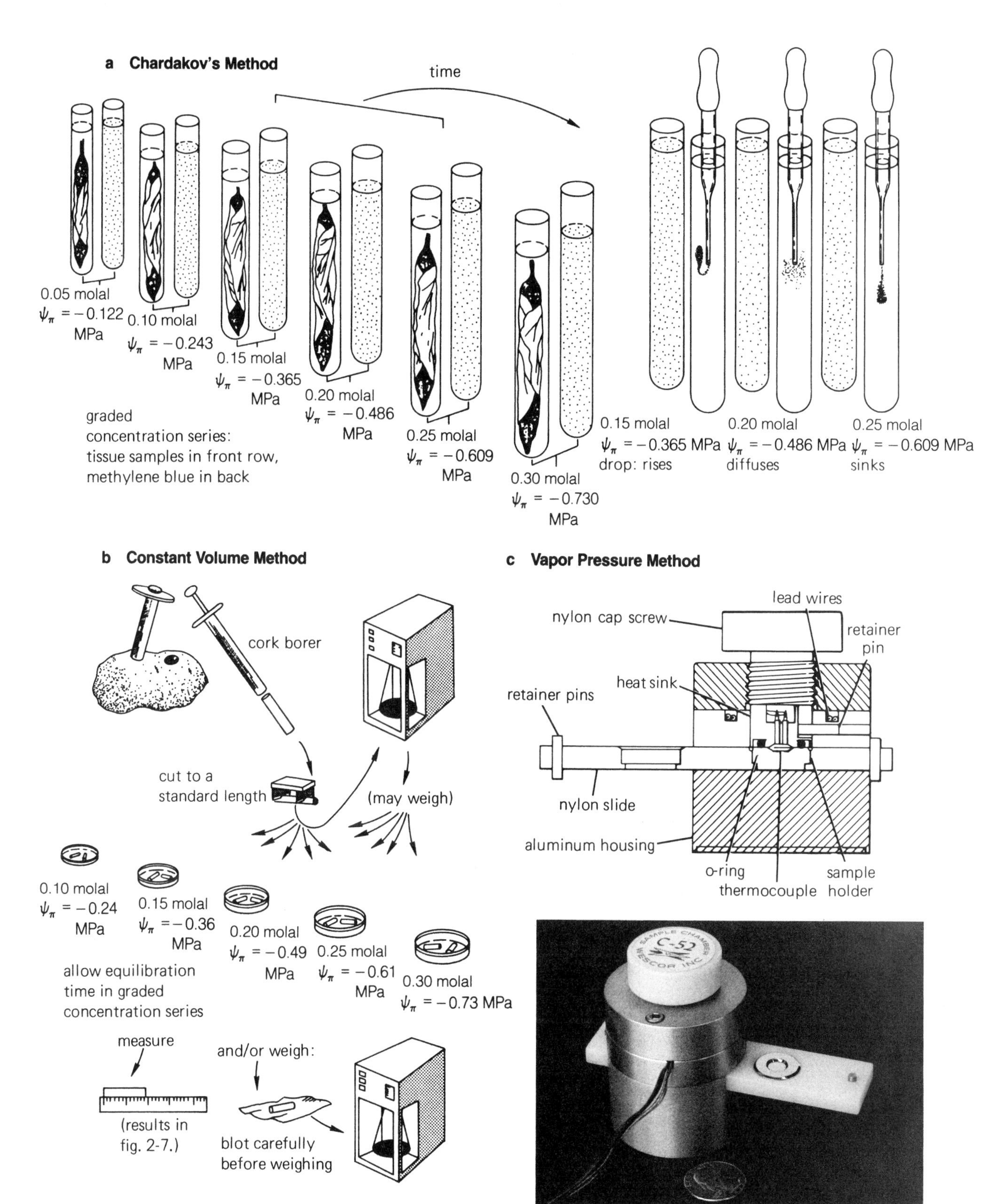


**Figure 3-5** Three different ways to measure water potential. Osmotic potentials are calculated at 20°C by Equation 3.2. The vapor-pressure device is made by Wescor, Inc., Logan, Utah. (Drawing and photograph used by permission.)

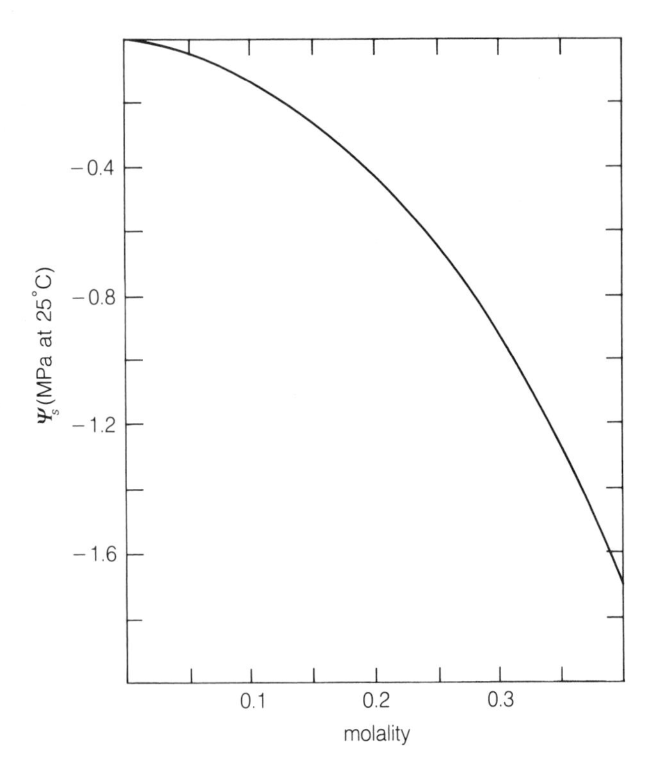


**Figure 3-6** Osmotic potentials for PEG 4000 for various concentrations. PEG 6000 gave almost identical results. (Unpublished data obtained by Cleon Ross with a thermocouple psychrometer; $\Psi_s$ values of PEG solutions cannot be measured correctly by freezing-point depression.)

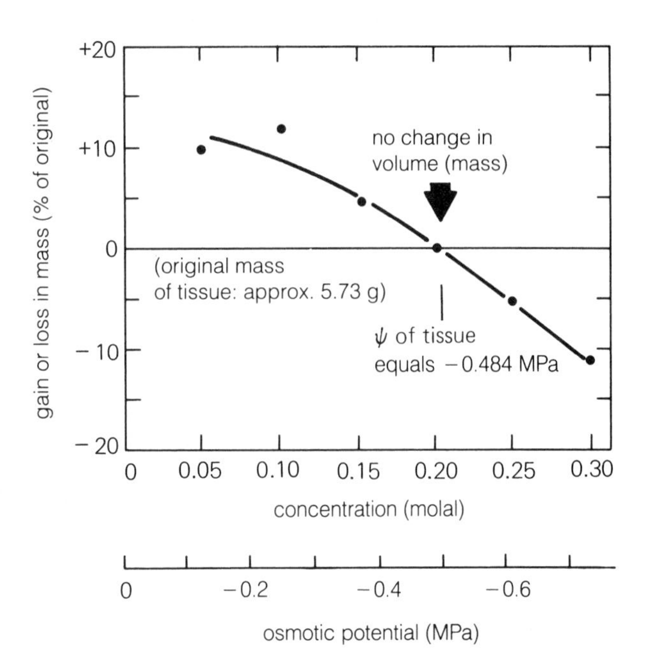


**Figure 3-7** Mass of plant tissue samples as a function of the concentration of solutes with which the samples have been allowed to come into equilibrium. (Data from a student plant physiology laboratory report, Colorado State University.)

such a plot (Fig. 3-7), the point at which the volume curve crosses the zero line indicates the solution that had the same water potential as that of the tissue at the start of the experiment.

In another approach (not illustrated), tissue samples in small closed containers are allowed to equilibrate with the vapor over solutions of known concentration, rather than with the solutions themselves. Thus, the solutions are not contaminated with solutes from the tissue. Mass rather than volumes are usually measured in this method.

Rather than measuring changes in the tissue, one might measure the concentration of the test solution. If it becomes less concentrated, the tissue will have lost water. This faster approach is better than measuring tissue volumes, which often gives $\Psi$ values that are too negative because solutes are absorbed from the test solution by the tissue samples.

In 1948, a Russian scientist, V. S. Chardakov, devised a simple, efficient way to find the test solution in which no change in concentration occurs (Fig. 3-5a). More practical methods are now available for field work, but **Chardakov's method** nicely illustrates the principles we are discussing. Test tubes that contain solutions of graded concentrations are colored slightly by dissolving in them a small crystal of a dye such as methylene blue. (Addition of the dye does not change osmotic potential significantly.) Tissue samples are placed in test tubes that contain solutions of equivalent concentrations but no dye. Time is allowed for exchange of a certain amount of water, but it is not essential that the tissue reach equilibrium with the solution. Sufficient exchange may take place in as little as 5 to 15 min. Then the tissue is removed, and a small drop of the equivalent colored solution is added to the test tube. If the colored drop rises, the solution in which the tissue was incubated has become more dense, indicating that the tissue has absorbed water; in this case, the tissue had a lower (more negative) water potential than the original solution. If the drop sinks, the solution has become less dense, having absorbed water from the tissue; the solution, then, had a lower water potential than the original tissue. If the drop diffuses evenly out into the solution without rising or sinking, then no change in concentration has occurred, and the water potential of the solution equals that of the tissue. When several solutions with different concentrations are used, there is usually one in which the drop neither sinks nor floats. With Equation 3.2 we can calculate its $\Psi_s$ (or we determine $\Psi_s$ empirically). At $P = 0$, $\Psi_s$ is equal to $\Psi$ and thus the average $\Psi$ of the tissue. (See Knipling and Kramer, 1967.)

In the **vapor-pressure method**, tissue is placed in a small, closed volume of air. The water potential of the air comes into equilibrium with the water potential of the tissue, which changes only insignificantly in the

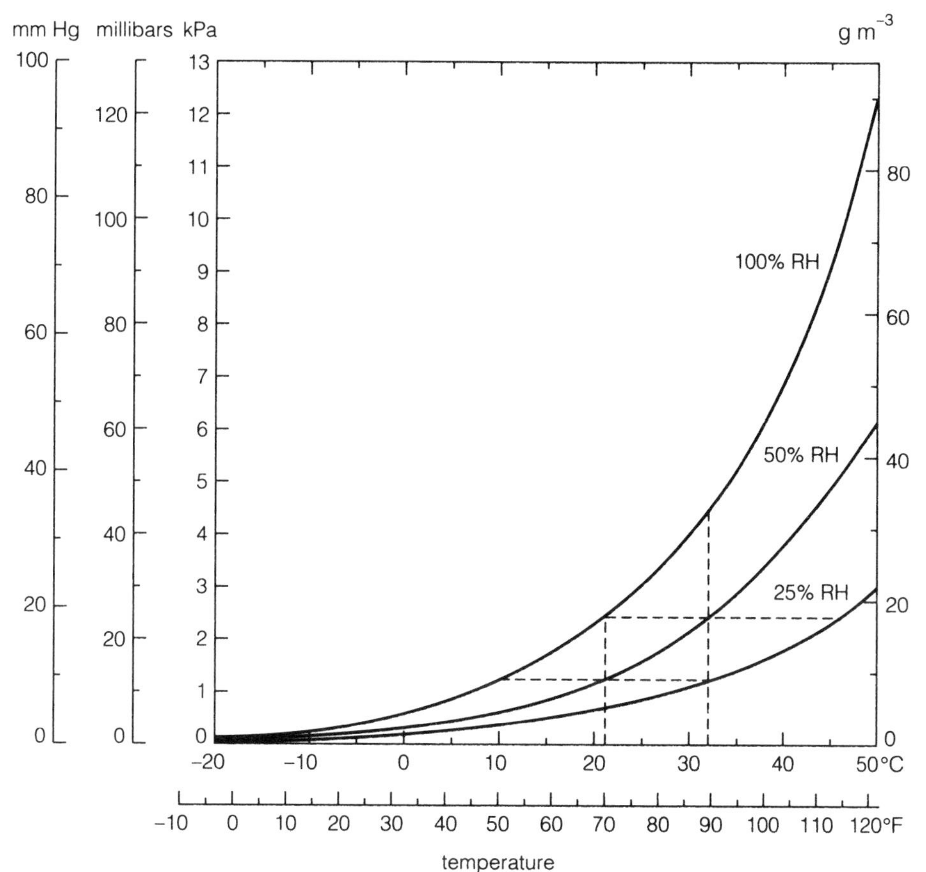


**Figure 3-8** Relationship between moisture in the atmosphere (expressed as both vapor pressure and vapor density) and temperature for saturated air and for air at 50 percent and 25 percent relative humidity. Dashed lines are an example showing that air at 10°C, 100 percent RH has the same amount of moisture as air at 21°C, 50 percent RH, and as air at 33°C, 25 percent RH.

process (Fig. 3-5c). The water potential of the air is then determined by measuring vapor density (humidity) at a known temperature.

Vapor pressure and vapor density were defined in Chapter 2 (page 40). **Relative humidity (RH)** is the amount of water vapor in air at a particular temperature compared with the amount of water vapor that the air *could* hold at that temperature (the **saturation vapor density** or **pressure**). The saturation vapor density approximately doubles for each temperature increase of 10°C (or 20°F). If a volume of air that is saturated with water vapor (100 percent RH) is warmed by 10°C, it can hold about twice as much water as before; so its relative humidity is about 50 percent. This relationship is illustrated in Figure 3-8.

If we know the absolute temperature ($T$ in kelvins), the saturation vapor density or pressure of pure water ($p^0$) at that temperature, the vapor density or pressure in the test chamber ($p$), and the molar volume of water ($V_1$ in L mol$^{-1}$), we can calculate the water potential from the following formula, which is derived from Raoult's law (Equation 2.13):

$$\Psi = -\frac{RT}{V_1}\ln\frac{p^0}{p} \qquad (3.5)$$

The ratio 100 $p/p^0$ is the relative humidity. If we convert to common logarithms and use numerical values for $R$ and $V_1$, Equation 3.5 simplifies to:

$$\Psi \text{ (in MPa)} = -1.06\, T \log_{10}\left(\frac{100}{RH}\right) \qquad (3.6)$$

This equation gives water potential of the air when the temperature and the relative humidity are known.

The measurement of water potential in tissue samples by this method is simple in principle, but in practice a number of difficulties are involved. These difficulties have been solved only since the 1950s, and now this is one of the most commonly used methods. To begin with, if the method is to be sufficiently accurate, temperature must be uniform, at least within one-hundredth of a degree Celsius. This is because slight changes in temperature result in large changes in RH and water potential at constant vapor density.[5]

Another problem involves measurement of the humidity inside the test chamber. An ingenious method was developed by D. C. Spanner (1951) in England and has since been improved. Two thermocouple junctions are built into the test chamber. One of these has a relatively large mass and thus remains close to the temperature of the air in the chamber. The second thermocouple is very small. When a weak current is briefly passed through the two junctions (in the correct direction), the small one cools rapidly by the Peltier effect.[6] As it cools, a minute amount of moisture condenses on it from the air inside the chamber (Fig. 3-9). This point of moisture is then cooled by evaporation and thus acts as a thermocouple "wet bulb," and the difference between the temperature of the wet bulb and that of the dry thermocouple indicates RH and thus the water potential of the air in the chamber. Uniform air temperatures between the two thermocouples are maintained by immersing the chamber in a water bath or by placing it in an aluminum block (or smaller silver block, because silver is the better conductor of heat). In practice, the drop evaporates so rapidly that the actual temperatures cannot be measured. Rather, the system is arbitrarily calibrated using solutions of known concentrations. Typically, measurements (which require less than a minute) are made at regular intervals until they stabilize after an hour or two, indicating that the tissue has reached equilibrium with the air in the chamber.

Still another method for the measurement of water potential in plant stems and leaves (petioles) involves use of a pressure bomb. This is discussed later in this section and again in Chapter 5.

[5]Air at 100 percent RH always has a water potential of zero. If air at 100 percent RH and 20°C is warmed to only 21°C, its RH drops to 94.02 percent and $\Psi = -8.34$ MPa, a value more negative than that found in almost any cell! Hence, the range of relative humidity at which plants "grow well" (at constant $T$) is 99 to 100 percent!

[6]A thermocouple operates on the principle of the **thermoelectric effect**. When a circuit consists of two wires made of different metals (for example, copper and the alloy constantan), and the two junctions between the two wires are at different temperatures, a current flows around the circuit. When the temperature of one junction is known and the current is measured, the temperature of the other junction (the **thermocouple**) can be calculated. The **Peltier effect** is the opposite of the thermoelectric effect. When a current is passed through a circuit of two different metals, the two junctions have different temperatures. One heats and one cools, depending on the direction of the current.

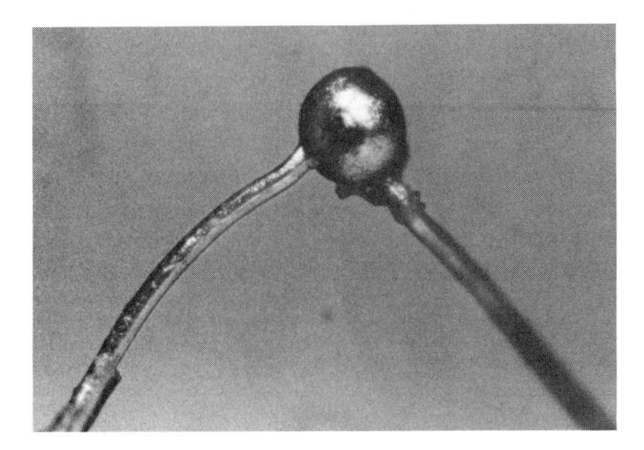

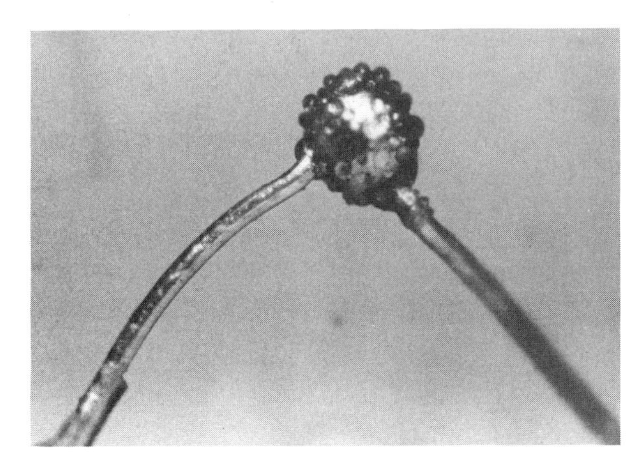

**Figure 3-9** The thermocouple junction (spherical metal object—about 100 μm in diameter) in a psychrometer used to measure atmospheric humidity and thus water potential. Top: the dry junction before the test. Bottom: the wet junction just after Peltier cooling. Drops of water have condensed from the air. (Photomicrographs courtesy of Herman Wiebe.)

## Osmotic Potential

Because the absolute value of osmotic potential (which is negative) is equivalent to real pressure (positive) in a "perfect" osmometer in pure water at equilibrium, the osmotic potential of a solution can be measured directly. Many measurements of this property have been made, particularly during the late 19th century by Wilhelm F. P. Pfeffer (1877), who made nearly perfect, rigid, semipermeable membranes by soaking a porous clay cup in potassium ferrocyanide and then in cupric sulfate, precipitating cupric ferrocyanide in the pores. Columns of mercury were used to determine pressures. Van't Hoff used Pfeffer's data in working out Equation 3.2. With increased understanding of the properties of solutions, it became apparent that other, simpler measurements could be made and the data converted to osmotic potential. Excellent ways of doing this with free liquids have been developed, but no completely satisfactory method is yet available for measuring the osmotic potential of

Figure 3-13 Green and Stanton's experiment. After equilibration in sugar solutions of various osmotic potentials, *Nitella* cells were penetrated by capillary tubes (closed at one end) as illustrated. In the original experiment, atmospheres were the unit of measurement; they are retained here (1 atm = 0.1013 MPa). (Data from Green and Stanton, 1967.)

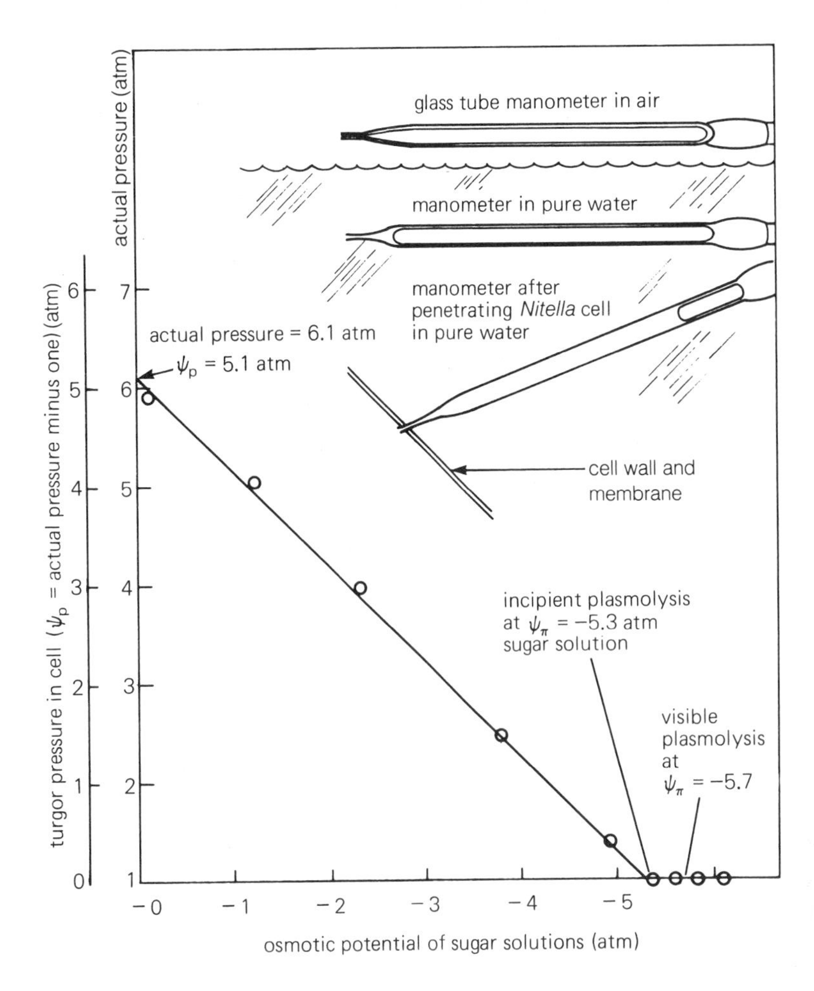


ter potential and osmotic potential have been determined:

$$P = \Psi - \Psi_s \tag{3.9}$$

Paul B. Green and Frederick W. Stanton (1967) described a method for the direct measurement of **turgor pressure** (cell pressure, equal to the pressure in the cells: the protoplast pushing against the wall). They used large cells, those of the alga *Nitella axillaris*. The method has since been greatly elaborated with sophisticated equipment and applied to much smaller, higher-plant cells (see, for example, papers by Nonami et al., 1987 and Shackel, 1987; and a review by Ortega, 1990). Applications of the method in modern studies are mentioned in later chapters.

Green and Stanton made a minute manometer by fusing closed one end of a capillary tube (diameter 40 μm) and fashioning the other end into a tip like that of a syringe needle. If such a tube, with its open end placed in water, is observed with the microscope, water is seen to enter the open end by capillarity, somewhat compressing the air inside the tube. The position of the meniscus inside the open end of the tube is noted and the volume of air calculated. When the open end of the tube punctures a cell, pressure in the cell is transferred to the air in the tube, compressing it further, to an extent indicated by movement of the meniscus (Fig. 3-13). The final pressure in the tube is always equal to the pressure before penetration of the cell, multiplied by the ratio of the original to the final volume (according to Boyle's law). The pressure in the cell before penetration can be approximated by multiplying atmospheric pressure by the ratio of the original volume in the tube to the volume after entrance of water by capillarity. There will be a slight change in pressure within the cell upon penetration by the tube, but even this can be determined by penetrating the cell with a second tube while observing the change in pressure in the first. The method measures actual pressure in the cell, but according to convention, pressure in an open solution at atmospheric pressure is said to be zero. Hence, the actual cell pressure measured by Green and Stanton's method will be about 1 atm (about 0.1 MPa) higher than the conventional *turgor pressure*.

## Matric Potential

Hydrophilic surfaces (for example, those of such colloids as protein, starch, and clay; see box essay) adsorb

## Colloids: Characteristic Components of Protoplasm

Protoplasm is unique not only because it consists of highly complex and special molecules but also because of its physical nature. Because of its high viscosity, protoplasm is a bit like gelatin pudding or sometimes like glue. The physical nature of protoplasm is determined by vast areas of interface between some of those special molecules, especially proteins, and the protoplasmic solutions in which they are suspended. The reactions of life are catalyzed at these enzyme interfaces. Soils are also characterized by huge interfaces between clay (and to a lesser extent, silt and sand) or humus particles and their surroundings. Technology takes advantage of such systems in water softeners, catalytic converters, and numerous other applications.

Crucial to the physical nature of protoplasm are membranes and particles too small to settle out by gravity but larger than the atoms, small molecules, and ions that form true solute particles. When these larger particles are suspended in water, they sometimes form a glue, so they have been termed **colloids**, from the Greek word *kolla*, "glue."

Why do colloids not settle out? Because they are constantly being struck by the surrounding, much smaller water molecules in rapid motion (Chapter 2), and they are small enough that the random velocities of the impacting water molecules do not average out. At any given moment there is a high probability that a colloidal particle is bombarded more strongly on one side than on the opposite side. When colloidal particles are observed in a light microscope by strong illumination from one side, they appear as points of light (the **Tyndall effect**, first noticed by John Tyndall, 1820–1893). They seem to dance around with many random hops per second. The largest (brightest) particles dance less than the smaller (dimmer) particles.

This is **Brownian movement**, discovered by the Scottish botanist Robert Brown in 1827. It is a beautiful, even spectacular, confirmation of kinetic theory. (See an excellent summary by Bernard H. Lavenda, 1985.) This erratic and continuous motion keeps colloids from settling. Indeed, we might define a colloidal particle as one that is not a true solute but is small enough to remain in suspension because of its Brownian movement. With slightly larger particles, there is a much greater chance that the random bombardment on any side will approach an average value for the entire particle. In the contest between kinetic bombardment and gravity, gravity wins, so the particle settles.

The largest particles exhibiting Brownian movement are about 100 to 2,000 nm in diameter, depending on shapes and densities. Because light waves are 385 to 776 nm long (Appendix B), only the largest colloidal particles can cast shadows. The smallest ones *refract* light waves, causing the Tyndall effect, but are not themselves actually visible in a light microscope. The electron microscope, with its electron beams of wavelengths less than 0.1 nm, easily resolves even the smallest colloidal particles, which have diameters of about 10 nm. (Smaller particles are true solutes, but the distinction is not precise.) Many particles in a cell, including the ribosomes and all the single protein molecules that are enzymes, are in the colloidal size range.

Most colloidal particles pass through filter paper, but they cannot pass through cellophane, as can true solute particles. Suspension particles are too large to pass through filter paper.

Although colloidal particles are small, each is large enough to present a surface (a layer of atoms) to surrounding water molecules and solute particles. Because of the small size of colloidal particles, their total surface in a given volume is relatively huge. Imagine a solid cube of material 1 cm on each of its edges. There are six faces, so it has a surface area of 6 $cm^2$. Cut it once. You expose 2 $cm^2$ more of surface. Continue slicing the cube until you have reduced each part to a cube 10 nm long on its edges. Now the total surface area is 6,000,000 $cm^2$ (600 $m^2$). A single cube with the same surface area would be 10 m high with a volume of 1,000 $m^3$! Colloidal particles are seldom cubes, but they are of comparable size.

The reactions of life occur on surfaces, and it is easy to see how relatively large surfaces can exist in a single cell. It is also easy to see how hydration (matric forces) can influence the water milieu of cells and soils.

water, and the tenacity with which the molecules of water are adsorbed is not only a function of the nature of the surface but also of the distance between the surface and the adsorbed water molecules: Those located directly on the adsorbing surface will be held extremely tightly; those at some distance from the surface will be held much less tightly. The adsorption of water by hydrophilic surfaces is called **hydration** or **imbibition**.

The **matric potential ($\tau$)** is a measure at atmospheric pressure of the tendency for the matrix to adsorb additional water molecules. This tendency is equal to the average tenacity with which the least tightly held (most distant) layer of water molecules is adsorbed. Matric potential is expressed in the same energy or pressure units as water potential and may contribute to $\Psi$. A dry colloid or hydrophilic surface, such as filter paper, wood, soil, gelatin, or the stipe of a brown alga, often has an extremely negative matric potential (as low as $-300$ MPa), whereas the same colloid in a large volume of pure water at atmospheric pressure has a matric

**Figure 3-14** A pressure-plate (or pressure-membrane) apparatus used to measure matric potentials of soils and other materials. The wet samples are placed in the circular holders on the pressure plate, a porous plate that allows diffusion of water. After the top has been secured in place with wing nuts and against a rubber O-ring, pressure is introduced and water begins to diffuse through the plate and out through the small tube into the beaker (center of photograph). When water has stopped coming out the tube (often after 24 h), the matric potential (a negative value) of the sample is numerically equal to the pressure (a positive value) in the apparatus, as read on the gauges at the bottom of the picture. A higher pressure might then be introduced to expel more water, producing a moisture release curve such as that of Figure 3-15. (Pressure plate in the laboratory of Ray W. Brown, Forestry Sciences Laboratory, Logan, Utah; photograph by Frank B. Salisbury.)

potential of zero (because it is saturated and therefore in equilibrium with the water). In general, when any colloid at atmospheric pressure is in equilibrium with its surroundings, the least tightly held water molecules have the same free energy as the water molecules in the surroundings, so the matric potential of the colloid is equal to the water potential of the surroundings.

In modern discussions, the radius of curvature of the water surfaces between colloidal particles (the **meniscus**; plural, **menisci**) is often mentioned. The meniscus is the basis of the phenomenon of capillarity, discussed in Chapter 5 (see Fig. 5-3). The smaller the radius of curvature of the meniscus, the tighter the water is held to the colloidal or other hydrophilic surface by hydration and the more negative the matric potential.

A commonly used modern way of measuring matric potential is instructive. A hydrated colloid is enclosed in a pressure chamber with a membrane filter that is supported by a screen to withstand high pressure (Fig. 3-14). The pores of the membrane are 2 to 5 nm in diameter, large enough to allow passage of solutes and water but not the colloid. Surface tension prevents the passage of air through the wet membrane. Assume (in the simplest case) that the colloid is wet with pure water only (no solutes). Compressed gas is introduced into the pressure chamber. Increasing the pressure raises $\Psi$ of the water that is adsorbed on the colloid toward zero. When $\Psi$ of the molecules farthest from the surface of the colloid reaches zero, those molecules begin to diffuse out through the membrane. (You can also think of the air as pressing on the menisci and making their radii of curvatures smaller.) Further increases in pressure on the colloid result in additional but smaller increments of water movement from the colloid out through the membrane.

When all water movement stops, the water on the colloid under pressure will be in equilibrium, through the membrane, with the pure water at atmospheric pressure and the same temperature on the outside of the membrane ($\Psi = 0$; that is, $\Psi$ of the least tightly adsorbed water molecules on the colloid under pressure = 0). If $\Psi = 0$ and if pressure and matric potential are the only components influencing water potential on the colloid side of the membrane:

$$\Psi = P + \tau \qquad (3.10)$$

then the negative matric potential will equal the positive pressure potential produced by the compressed air ($-\tau = P$). Because the pressure is known, the negative matric potential is also known.

Usually, after equilibrium has been reached at a given pressure, the water content of the colloid is determined by weighing before and after drying in an oven. The matric potential of the colloid, plotted as a function of its water content, provides the **moisture release curve**

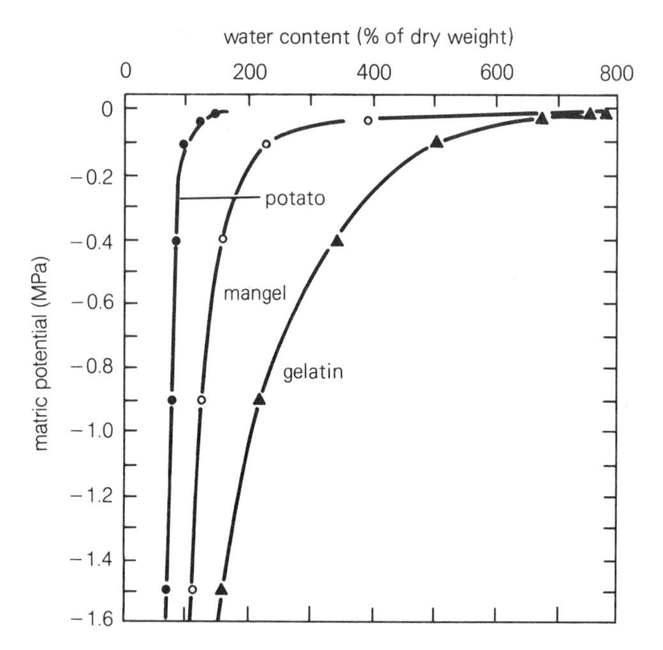


**Figure 3-15** Moisture-release curves for two plant materials and gelatin. (From Wiebe, 1966.)

(Fig. 3-15). Such curves are fundamental in computing soil water flow and soil water-holding capacity.

A test of the assumption that the final pressure in the pressurized membrane apparatus is a measurement of matric potential is to measure the water potential of the colloid at atmospheric pressure by the vapor method described under the heading "Water Potential" above. The two measurements agree closely, indicating the validity of the pressure-chamber approach.

The matric potential is often considered to be a *component* of water potential ($\Psi = P + \Psi_s + \tau$). We don't think this is valid because $\tau$ comes to equilibrium with $\Psi$ as established by $P$ and $\Psi_s$ and thus cannot be added algebraically to $P$ and $\Psi_s$. Nevertheless, matric effects clearly *contribute* to the overall $\Psi$. In a complex system — such as a cell containing molecules of colloidal protein and other hydrophilic surfaces, as well as simple solute ions and molecules — the final water potential will be determined not only by the solute particles and the pressure but also by the proteins and other surfaces. We don't understand the complexities of such systems well enough to account for them with a mathematical model.

Nevertheless, we can imagine some interactions. Some polar solutes, especially ions, are adsorbed on hydrophilic surfaces, influencing their water potential. And adsorption of water molecules on these surfaces decreases the quantity of water in the part of the system not influenced by the matrix, so the solution in this part is more concentrated than it otherwise would be. It is possible that no water molecule in the system will be far enough away from a hydrating surface (for example, that of a protein molecule) to be completely unaffected by such surfaces. This is unlikely in the cytosol, as it turns out (Passioura, 1980, 1982), but it is often true of soil water with its dissolved solutes or of water in the cell wall.

Some modern authors (for example, Passioura, 1980, 1982; Tyree and Jarvis, 1982) suggest that matric effects can be discussed solely in terms of solute and pressure effects. Consider what happens at the molecular level when a hydrophilic surface adsorbs water. In the simplest case, the adsorption may be only a matter of van der Waals forces or hydrogen bonding, which probably extend for only one or two diameters of the water molecule (0.3 to 0.6 nm). But many hydrophilic surfaces in nature, especially those in clays, proteins, and the microstructures of cell walls, have a large unbalanced negative charge. The positive sides of water molecules (see Section 2.1) are strongly attracted to such negatively charged surfaces, and several layers of water molecules can build up on such a surface (Fig. 3-16). These forces are great enough to produce pressures of tens to hundreds of megapascals in the molecules closest to the surfaces. This effect is much like the hydrostatic pressure produced by gravity in deep water, except that the pressure gradient occurs over nanometers instead of hundreds of meters.

This all makes good sense in terms of solute and pressure effects because the water molecules that are attracted to the hydrophilic surface lose much of their energy. That is, extending our thermodynamic discussion to a microscale (which may not be completely valid), we would say that the matric forces, acting like solutes, would produce an extremely negative water potential close to the adsorbing surfaces if this effect were not balanced by the very high positive pressures that exist there. The water potential throughout the layer of adsorbed water molecules is thus constant, as it must be at equilibrium.

If solutes, especially cations, are present, they are strongly attracted to the hydrophilic surface, where their concentration is extremely high (calculated to be at least 2 M). This produces a negative osmotic potential that forms a gradient from extremely negative close to the charged surface to much less negative some distance away. Again, the water molecules strongly diffuse toward the charged surface where solute concentrations are so high, producing a gradient in pressure opposite to the gradient in osmotic potential and keeping the water potential constant throughout (Fig. 3-16).

Clearly, the effects of hydrophilic surfaces on water molecules play an important role in plants. Membranes, proteins, ribosomes, and several components of the cell wall can become hydrated and thus are part of the ma-

Figure 3-16 (a) The electrical double layer of ionic distribution around a charged colloidal particle. Charges next to the surface of the particle are meant to indicate surface charges. These are mostly negative, but the negative charge is interrupted at three points with positive charges. Distribution of ionic charge is statistically opposite to the distribution of surface charge, but a few irregularities are apparent. (b) Pressure increase in water as it approaches a hydrated, charged surface. The lines represent isobars of pressure increasing by 30-MPa intervals as the surface is approached. The pressure increase is caused by attraction between the charged surface and water molecules (by hydrogen bonding) and by the steeply increasing solute concentration, which is caused by mutual attraction of dissolved solutes (ions) and the charged surface.

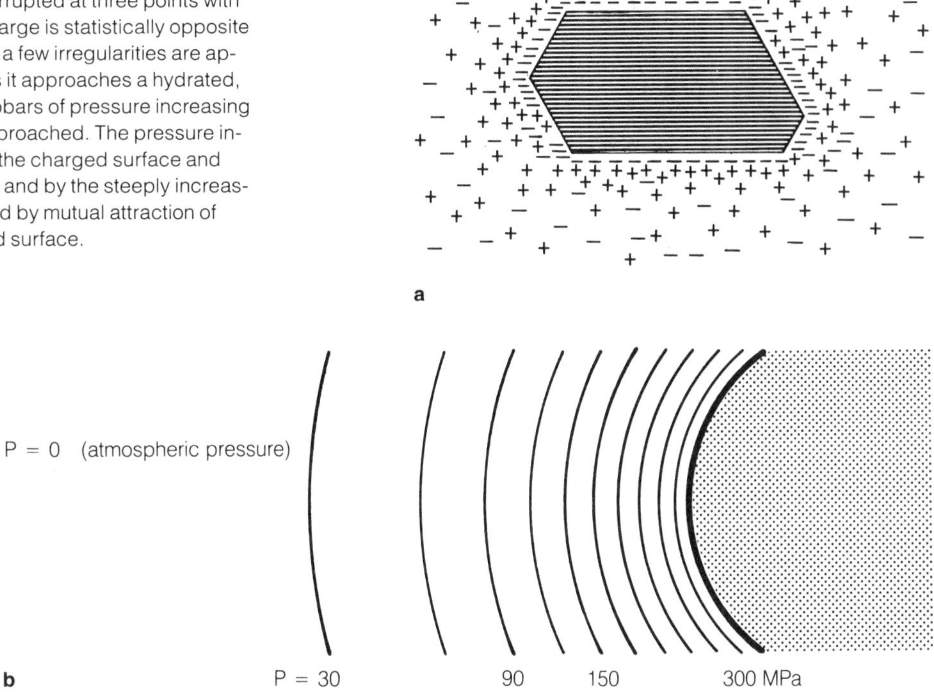


tric component that combines with the osmotic component and pressure to determine water potential. It is important to remember, however, that at *equivalent weights*, small inorganic ions reduce the free energy (and thus the water potential) of an amount of water far more (hundreds to millions of times) than a hydrophilic substance such as protein (because the effect is proportional to the mole fraction and independent of molecular weight). Thus, in the cytosol, the contribution of the matric component to final water potential is probably small and could be negligible compared with that of solute ions and molecules. Yet the matric component is quite significant in the cell wall.

In dry seeds and in the soil, matric effects are especially important. Water uptake into seeds is at first largely a function of matric forces; the surfaces of proteins and some cell-wall polysaccharides must become hydrated before germination can begin. Dry wood is also hydrophilic, and for centuries rocks have been split by a technique that makes use of this property: Holes are drilled along a desired fracture line in the rock, dry wooden pegs are driven into the holes, and water is run over the pegs, which then hydrate, expand, and crack the rock. Growing fungi and perhaps roots and other systems also develop powerful expansion forces that depend on hydration (and osmosis).

Perhaps someday our understanding will advance to the point at which we can calculate the relative contributions of solutes (osmotic potential), pressure, and adsorption of water on hydrophilic surfaces (matric effects) to the true water status of plants. In the meantime, it is important to realize that all three factors play a role but that it is not valid (because of the complications just discussed) to consider matric potential as a component of water potential in the same sense that osmotic potential and pressure are thought of as water-potential components. Nevertheless, such a calculation applied to the soil usually gives reasonable results because matric forces dominate the water potential of soils.

# 4

# The Photosynthesis–Transpiration Compromise

During the summer of 1980, John Hanks, a soil scientist at Utah State University, kept careful records of the amount of water required to grow a crop of sugar beets on the college Greenville farm. To mature the crop, water equivalent to about 620 mm (24.4 in.) of rain was added to the field. About a fourth of this evaporated directly from the soil, but most of the remaining 465 mm passed through the plants into the atmosphere. This evaporation of water from plants (and animals, according to most dictionaries) is called **transpiration**. In plants, it refers to internal water lost through stomates, cuticle, or lenticels. Continuing the calculations, Hanks showed that 465 kg of water were transpired by the sugar-beet plants for each 1 kg of sucrose produced; 230 kg of water were transpired to produce 1 kg of dried **biomass**, including leaves, stems, and roots (Davidoff and Hanks, 1988).

In a 1974 study, Hanks found that 600 kg of water were transpired to produce 1 kg of dry maize (corn), and 225 kg were transpired to produce 1 kg of dry biomass. Thus, of the water that moved through the plant from soil to atmosphere in these examples, only a small fraction of 1 percent became part of the biomass. These figures are typical, although there are substantial differences among species. (See reviews in Hanks, 1982, 1983.)

Why is so much water lost by transpiration to grow a crop? Because the molecular skeletons of virtually all organic matter in plants consist of carbon atoms that must come from the atmosphere. They enter the plant as carbon dioxide ($CO_2$) through stomatal pores, mostly on leaf surfaces, and water exits by diffusion through these same pores as long as they are open. You could say that the plant faces a dilemma: how to get as much $CO_2$ as possible from an atmosphere in which it is extremely dilute (about 0.035 percent by volume) and at the same time retain as much water as possible. The agriculturalist faces a similar challenge: how to achieve a maximum crop yield with a minimum of irrigation or rainfall, a critical natural resource (Sinclair et al., 1984).

Understanding environmental factors and how they influence transpiration from and $CO_2$ absorption into a leaf in the field at various times is a difficult assignment because the factors interact in so many ways. Environmental factors influence not only the physical processes of evaporation and diffusion but also the opening and closing of stomates on the leaf's surfaces, through which over 90 percent of the transpired water and the $CO_2$ pass. Increased leaf temperature, for example, promotes evaporation considerably and diffusion slightly but may cause the stomates to close, or to open wider, depending on species and other factors. At dawn, stomates open in response to increasing light, and the light raises the temperature of the leaf, which makes water evaporate faster. Higher air temperature allows the air to hold more moisture, so transpiration is promoted, and perhaps stomatal aperture is affected. Wind brings more $CO_2$ and blows away the water vapor, causing an increase in evaporation and $CO_2$ uptake, but a little less than expected because increasing $CO_2$ causes stomates to partially close. If the leaf is warmed above air temperature by sunlight, however, wind will lower its temperature, causing a decrease in transpiration. When soil moisture becomes limiting, transpiration and $CO_2$ uptake are inhibited because stomates close. So if one were to visit the Greenville farm, hoping to learn about the extent of transpiration or $CO_2$ uptake at any moment, one would have to be armed with a formidable array of environmental measuring devices, preferably connected to a datalogger and a computer.

## 4.1 Measurement of Transpiration

In this chapter, we emphasize transpiration rather than $CO_2$ uptake. The former process is perhaps more challenging because atmospheric $CO_2$ is fairly constant, whereas humidity varies, and transpiration studies must consider evaporation as well as diffusion. Anyway,

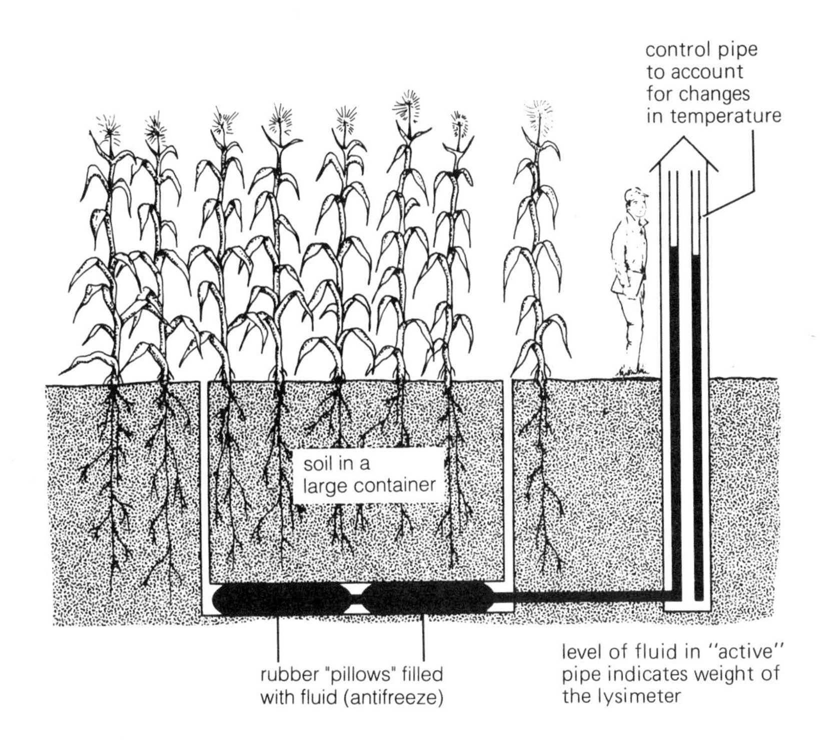


**Figure 4-1** Diagram of a large field lysimeter, operating on a hydraulic principle.

understanding transpiration provides the foundation for understanding $CO_2$ uptake. But note that water-vapor molecules diffuse about 1.6 times as fast as $CO_2$ molecules (because of their lower molecular weight) and that the atmosphere normally contains 10 to 100 times more water vapor than $CO_2$ (at 25°C, $H_2O$ = 3.2 to 32 mmol $mol^{-1}$; $CO_2$ = 0.351 mmol $mol^{-1}$), so that $H_2O$ molecules diffusing out of stomates influence the incoming $CO_2$ molecules (von Caemmerer and Farquhar, 1981).

How could we measure transpiration at the Greenville farm to see if our developing models and calculations were valid? There are many methods; we will discuss three principal approaches: lysimeter, gas exchange, and stem-flow methods.

## Lysimeter or Gravimetric Methods

Two centuries ago, Stephen Hales prepared a potted plant by sealing the pot and soil against water loss while leaving the shoot free to transpire. The potted plant was then weighed at intervals, and because the amount of water used in plant growth (for example, converted to carbohydrate) is only a fraction of 1 percent of the water that is transpired, virtually all the change in weight could be ascribed to transpiration. This is the **lysimeter** method. The only problem is being sure that the potted plant really represents other plants in the field, after being moved from its location in the field to the balance and back, and having its roots confined.

Hanks and others considerably expanded this simple approach. His lysimeters on the Greenville farm were large containers (several cubic meters) full of soil, buried so that their top surfaces were level with the field's surface. The container was placed on a large rubber bag buried beneath it and filled with water and antifreeze that extended into a standpipe above the surface (Fig. 4-1). The level of liquid in the pipe is a measure of the weight of the lysimeter, so it changes with the water content of the soil in the lysimeter and the growing plants, but their weight is small compared with the soil. The amount of soil water is determined by irrigation and rainfall minus the **evapotranspiration**, the combination of evaporation from soil and transpiration from plants. Evaporation from the soil can be estimated in various ways. Lysimeters provide the most reliable field method for studying evapotranspiration, but they are expensive and cannot be moved around on the spur of the moment. Though not universally available, lysimeters are widely used. A more common technique applies the water balance equation to calculate evapotranspiration as the difference between inputs and outputs:

$$E_t = \text{irrigation} + \text{rain} + \text{depletion} - \text{drainage} - \text{runoff} \quad (4.1)$$

where $E_t$ = evapotranspiration, and depletion is the loss from soil water storage. Measurements of soil water storage at the beginning and end of some period gives depletion.

A related method often used in teaching laboratories is to immerse a plant's roots or a detached stem in a closed water reservoir with a measuring device attached. Sometimes the stem is attached directly to a burette, or water loss might be measured as an air bubble moves through a capillary tube connected to the reservoir; such a device is called a **potometer**. It can be

**Figure 4-2** A laboratory cuvette for measuring transpiration.

useful to study relative transpiration rates for brief intervals, but cutting the stem influences transpiration rate, and immersed roots typically become oxygen-deficient, reducing water uptake (Sheriff and McGruddy, 1976).

## Gas Exchange or Cuvette Methods[1]

In this approach, transpiration is calculated by measuring water vapor in a sealed atmosphere that surrounds the leaf. A leaf may be enclosed in a transparent **cuvette** (Fig. 4-2), for example, and humidity, temperature, and volume of gas going into and out of the cuvette can be measured. The information that is obtained depends on which parameters are measured. It is possible to calculate transpiration rate, stomatal conductance, photosynthesis rate, and concentration of $CO_2$ within the leaf. The principles used in the calculations are simple and instructive. They are presently being widely applied in many studies.

The simplest technique seals the leaf in a container and measures relative humidity and temperature of the air in the container at time zero and after some brief time. Absolute humidity (vapor density or pressure) in moles or grams per cubic meter (or in pascals, if pressures are used) is then either measured directly with a dew-point hygrometer, or calculated from relative humidity and temperature by reference to a **psychometric chart** (similar to Fig. 3-8) or standard tables, or calculated by using a calculator and appropriate equations. Knowing the volume of air in the cuvette and the changes in vapor density, the amount of water transpired from the leaf equals the amount of water added to the air. Transpiration is then expressed as grams or moles of water per square meter of leaf per second (g or mol $m^{-2} s^{-1}$). If data are to be compared with photosynthesis measurements, it is logical to use moles instead of grams of water. The problem with this simple approach is that transpiration is affected by humidity, which changes in the cuvette during the measurement.

The first improvement is to pass air through the cuvette, carefully measuring its volume and calculating its vapor density as it enters and as it leaves based on measurements of relative humidity and temperature. Again, the increase in amount of vapor as the air passes through the cuvette represents water transpired from the plant and can be expressed as g or mol $m^{-2} s^{-1}$.

Adding a tiny thermocouple to the system to measure leaf temperature (with an accuracy of 0.1°C) allows the calculation of stomatal conductance, which indicates how open or closed the stomates are—valuable information in many studies. The thermocouple must be small so that it can be pressed against the leaf and thus not respond to the air temperature. The approach is based on Equation 2.16, derived from Fick's First Law of Diffusion, which states that flow is proportional to the driving force and inversely proportional to the resistance (analogous to Ohm's Law). The inverse of resistance to diffusion was called permeability; in the context of gas diffusion, it is called **conductance** ($g_j$, usually in units of $\mu$mol $m^{-2} s^{-1}$). One reason for using conductance instead of resistance is that it is directly proportional to transpiration (or photosynthesis, if $CO_2$ is being measured) rather than inversely proportional.

The driving force is the difference in gas partial pressures ($P_{ji} - P_{jo} = \Delta P_j$; units of Pa) of the gas inside the leaf ($P_{ji}$) and in the atmosphere outside ($P_{jo}$), but amounts can also be used in the calculations (g or mol $m^{-3}$). Absolute humidity outside the leaf (ambient $= P_a$) is calculated, and it is assumed that relative humidity is 100 percent inside the leaf (although it will be slightly lower, perhaps as low as 98 percent). Knowing leaf temperature and relative humidity inside the leaf, we can calculate absolute humidity (vapor density; $P_i$) as noted above. Transpiration ($E$) or diffusive flow ($J_j$; units of $\mu$mol $m^{-2} s^{-1}$) is then expressed as:

$$J_{H_2O} = E = g_{H_2O} \cdot \Delta P_{H_2O} = g_{H_2O}(P_a - P_i) \quad (4.2)$$

Equation 4.2 is then solved for the conductivity:

$$g_{H_2O} = \frac{E}{\Delta P_{H_2O}} \quad (4.3)$$

In the next improvement, a $CO_2$ analyzer is added to the system, measuring $CO_2$ as it enters and as it leaves the cuvette by its absorption of specific wavelengths of radiation in the infrared (IR) portion of the spectrum (Janac et al., 1971; Long, 1982). Recent advances in IR gas analyzers (which can also measure water vapor) allow plant physiologists to measure $CO_2$ changes in a

[1]Bruce G. Bugbee suggested the outline for this section.

few seconds, and future advances may allow measurements in less than a second.

Photosynthesis is the first information that is obtained from such measurements. The decreased $CO_2$ in the air leaving the cuvette represents $CO_2$ removed in photosynthesis, which (as in the transpiration measurement) can be expressed as g or mol $m^{-2}\,s^{-1}$. During recent years there has been much interest in expressing photosynthesis ($A$, which stands for **assimilation**) as a function of $CO_2$ concentration *within* the leaf ($C_i$). But how can internal $CO_2$ concentration be measured or calculated?

Again, we use the equation for flux (Equation 2.10), but this time we apply it to $CO_2$ instead of to $H_2O$. We have measured the $CO_2$ in the air surrounding the leaf (*ambient* $CO_2$ = $C_a$), and we want to know the internal $CO_2$ ($C_i$). As in Equation 4.1, the delta ($\Delta$) signifies *difference* in gas partial pressures or densities ($\Delta P_{CO_2} = C_a - C_i$):

$$J_{CO_2} = A = g_{CO_2} \cdot \Delta P_{CO_2} = g_{CO_2}(C_a - C_i) \quad (4.4)$$

We have calculated $A$, the rate of photosynthesis, already. We have also measured $C_a$, so we are left with two unknowns: $C_i$ and $g_{CO_2}$, but we have calculated $g_{H_2O}$, and we know from principles based on kinetic theory (Equation 2.1) that $CO_2$ molecules diffuse about 0.625 times as fast as $H_2O$ molecules (because of their larger molecular weight). Thus we can calculate $g_{CO_2}$ by multiplying $g_{H_2O}$ by 0.625, and we can solve Equation 4.4 for $C_i$:

$$C_i = C_a - A/g_{CO_2} \quad (4.5)$$

Application of the Equation 4.5 is greatly simplified if amounts are expressed as moles and if SI units are otherwise used. Notice how the units cancel in the following equation if we assume that $C_a = 340\ \mu mol\ mol^{-1}$ (micromoles of $CO_2$ per mole of air), $A = 20\ \mu mol\ m^{-2}\ s^{-1}$ (micromoles of $CO_2$ absorbed by each square meter of leaf surface each second), and $g_{CO_2} = 0.312$ mol $m^{-2}\ s^{-1}$ (moles of $CO_2$ that diffuse across one square meter each second):

$$C_i = C_a - \frac{A}{g_{CO_2}}$$

$$C_i = 340\frac{\mu mol}{mol} - \frac{20\ \mu mol\ m^{-2}\ s^{-1}}{0.312\ mol\ m^{-2}\ s^{-1}}$$

$$C_i = 340\frac{\mu mol}{mol} - 64\frac{\mu mol}{mol}$$

$$C_i = 276\ \mu mol\ mol^{-1}$$

Careful studies have shown that a correction must be made in this calculation. The stream of water vapor going out of the stomates is much larger than the number of $CO_2$ molecules going in (up to 1,000 times as large). This creates a tiny pressure gradient against which the $CO_2$ must diffuse to get into the leaf. If we measure transpiration, we can correct for this, but derivation of the correction factor is quite complicated (von Caemmerer and Farquhar, 1981). Usually, $C_i$ is about 15 percent lower than indicated by Equation 4.5.

Cuvette methods have become important during recent years as instrumentation has improved and microprocessors have become available to rapidly calculate physiological parameters from the sensor outputs.

**Porometers** have come into wide use for field and laboratory measurements of transpiration, thanks again to refined instrumentation and microprocessor calculations. A small chamber (cuvette), often only 1 to 2 cm in diameter, is clamped for a short time onto a leaf surface (usually the lower surface, where most stomates are located), and humidity inside the chamber is monitored. In the initial versions of this device, the rate of change in humidity was used to calculate the rate of transpiration, but this was often subject to error as the increasing humidity slowed transpiration. Most recently, **steady-state porometers** have become commercially available. Air is passed through a drying column and introduced into the chamber at a rate exactly sufficient to maintain humidity in the chamber at its initial value. A microprocessor calculates transpiration from the absolute humidity (relative humidity and air temperature) and the rate at which dry air must be introduced to maintain constant humidity. The resulting data are reliable, but the instrument is expensive.

The cuvette method can be applied on a large scale in the field. A tent of transparent plastic is placed over a number of plants, and this serves as the cuvette. But how can we be sure that the environment surrounding the plants is not influenced by the tent (or a cuvette)? With elaborate instrumentation it is possible to control temperatures, humidities, and gas concentrations inside, although radiation inside the tent is always lower than outside. This is no simple assignment suitable for a basic plant physiology student laboratory; tens of thousands of dollars or the equivalent must be invested in such equipment.

## Stem-Flow Methods

If we could measure the amount of water flowing through the stem, we would have a good approximation of the amount of transpiration, especially in a small herbaceous plant. In a large tree, transpiration may outpace sap flow through the trunk because of the resistance to flow, but when transpiration slows or ceases (for example, at night or during rain), water continues to flow through the stem to make up the deficit (see Chapter 5). And as noted, the amount of water used in photosynthesis is negligible, so most of the flowing sap is transpired. But how can we measure it?

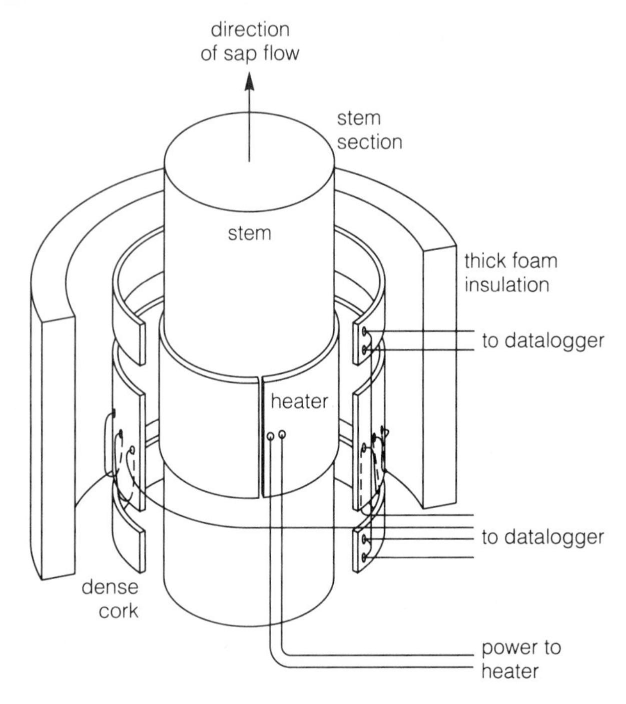


**Figure 4-3** A stem-flow method for measuring transpiration, as discussed in the text. The system shown here is based on schematic drawings of gauges constructed by Baker and van Bavel (1987). In their gauges, the heater, the three strips of dense cork with attached thermojunctions or thermopiles, and the foam insulation are all glued together; they are shown expanded here. The finished gauge is wrapped tightly around the stem so that the upper and lower temperature-sensing junctions are in close contact with the stem. The junctions of the thermopile are glued to the heater. See the paper by Baker and van Bavel (1987) for details of theory, construction, and operation.

In 1932, B. Huber developed a technique in which a pulse of heat was added at a point on a stem, and the temperature was then measured at some point above. The time required for the elevated temperature to arrive at the point above the heater indicated the velocity of sap flow. Knowing the diameter of the stem and other constants, an estimate of transpiration could be made and compared with measurements by lysimeter or cuvette techniques.

A more direct alternative technique measures heat balance rather than the heat-pulse velocity. Cermak et al. (1976), for example, developed a technique in which heat was applied all the way around the trunk of a tree, with thermocouples placed at the heater and below on the trunk. The heater and thermocouples were heavily insulated (for example, with styrofoam). Heat input was automatically and continuously adjusted so that there was a constant temperature difference between the heated segment and the unheated trunk below. As stem flow changed, the required input of heat also changed, and the electrical current required to maintain the constant temperature gradient was recorded. Knowing the specific heat of water and the heat that was added, suitable equations could be used to calculate the rate of sap flow. A digital datalogger can be used to record the data, so a continuous record of transpiration can be obtained. The equations are fundamental enough that the method does not require calibration, and the time for response is essentially instantaneous. The electronics required are not simple, however, and must be provided separately for each gauge.

In a slight modification with herbaceous plants (Fig. 4-3), a flat resistance heater (0.25 mm Inconel foil on 0.05 mm Kapton, varnished) is wrapped around the stem, and thermocouples contact the stem below the heater, at the heater, and above the heater (Sakuratani, 1984; Baker and van Bavel, 1987). The heat input is kept constant, and fluxes of heat out of the system are calculated for measured temperature gradients, so the mass flow rate of water in the stem can be directly calculated. The stem is not injured or penetrated, and no calibration is required. Simultaneous measurements made with this method and lysimeter or cuvette methods were in close agreement (within 10 percent).

Still other methods have been used, but we will not discuss them in detail here. For example, it is possible to combine measurements of net radiation (incoming minus outgoing radiant energy; see Section 4.8), soil heat flow, and temperature and humidity gradients above the plants to calculate evapotranspiration.

## 4.2 The Paradox of Pores

Nature often proves to be more complex than we expect. Suppose we compare the evaporation rate from a beaker of water and from an identical beaker that is half covered, say with metal strips. We would expect evaporation from the second beaker to be about half of that from the first. Now let's cover all but about 1 percent of the second beaker. We will use a thin piece of foil with small holes making up about 1 percent of the total area. Will we measure about 1 percent as much evaporation? Not if the holes have about the same size and spacing as the stomates found in the epidermis of a leaf. We will in fact measure about half as much evaporation (50 percent) as from the open surface.

How can this be? Why isn't evaporation directly proportional to surface area? It certainly seems paradoxical that stomatal openings on the leaf make up only about 1 percent of the surface area, whereas the leaf sometimes transpires half as much water as would evaporate from an equivalent area of wet filter paper. We resolve this apparent paradox by realizing that evaporation is a diffusion process from water surface to atmosphere, and by applying Equation 4.2. Simply stated, diffusion is proportional to the driving force and the conductivity. In our example, the driving force is the same for both beakers: the difference in vapor pressure (or density) between the water surface (where the atmosphere is saturated with vapor) and the atmosphere some distance away (where it must be below saturation if evaporation is to occur).

The different evaporation rates depend on different conductivities to diffusion. Part of the conductivity is a function of the area, and this value is much lower above the beaker covered with porous foil, which is what we expected. But the other part of the conductivity depends on the distance in the atmosphere through which water molecules must diffuse before their concentration reaches that of the atmosphere as a whole. The shorter the distance, the higher the conductivity. This distance can be called the **boundary layer**, and it is much shorter above the pores in the foil than above the free water surface. Molecules evaporating from the free water will be part of a relatively dense column of molecules extending some distance above the surface, whereas molecules diffusing through a pore can go in any direction within an imaginary hemisphere centered above the pore. In the hemisphere, the concentration drops rapidly with distance from the pore, which is to say that the concentration gradient is very steep because the boundary layer is very thin. Of course, if pores are closer together than the thickness of their boundary layers, these hemispheres overlap and merge into a boundary layer.

Many empirical studies were made several decades ago to determine the effects of pore size, shape, and distribution on diffusion rates (see, for example, Brown and Escombe, 1900; Sayre, 1926; and the review in Meyer and Anderson, 1939). Stomates of typical plants proved to be nearly optimal for maximum gas or vapor diffusion. Thus, plants are ideally adapted for $CO_2$ absorption from the atmosphere—but also for loss of water by transpiration. The stomates can close, however, and in most plants they are adapted to close when photosynthesis and $CO_2$ absorption stop (for example, in darkness).

## 4.3 Stomatal Anatomy

Stomates come in considerable variety (Wilkinson, 1979). Figure 4-4 shows drawings of cross sections

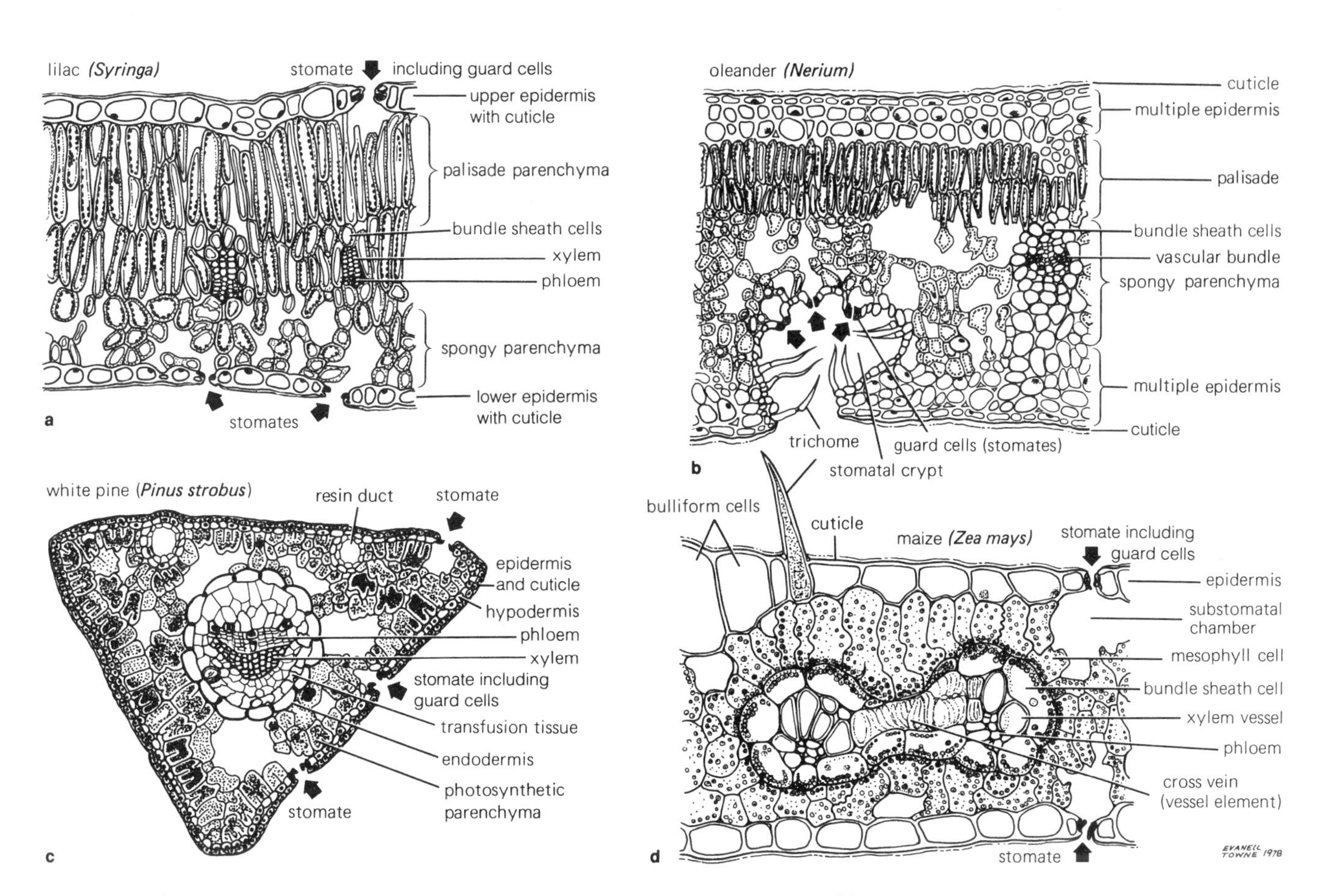


**Figure 4-4** Cross sections through four representative leaves, one with "normal" stomates (**a**), one with stomates deeply sunken in a stomatal cavity (**b**), one a pine needle (leaf) with slightly sunken stomates (**c**), and one a grass leaf with about equal numbers of stomates on both surfaces (**d**). Arrows point to the stomatal pores, but the stomates include the guard cells. Spongy parenchyma and palisade parenchyma (as in **a** and **b**) collectively form the mesophyll. Note details of differing leaf anatomy; pine and grass leaves do not have a palisade layer, for example. Bulliform cells of maize (**d**) shrink in response to water stress, causing the leaf to roll into a cylindrical shape.

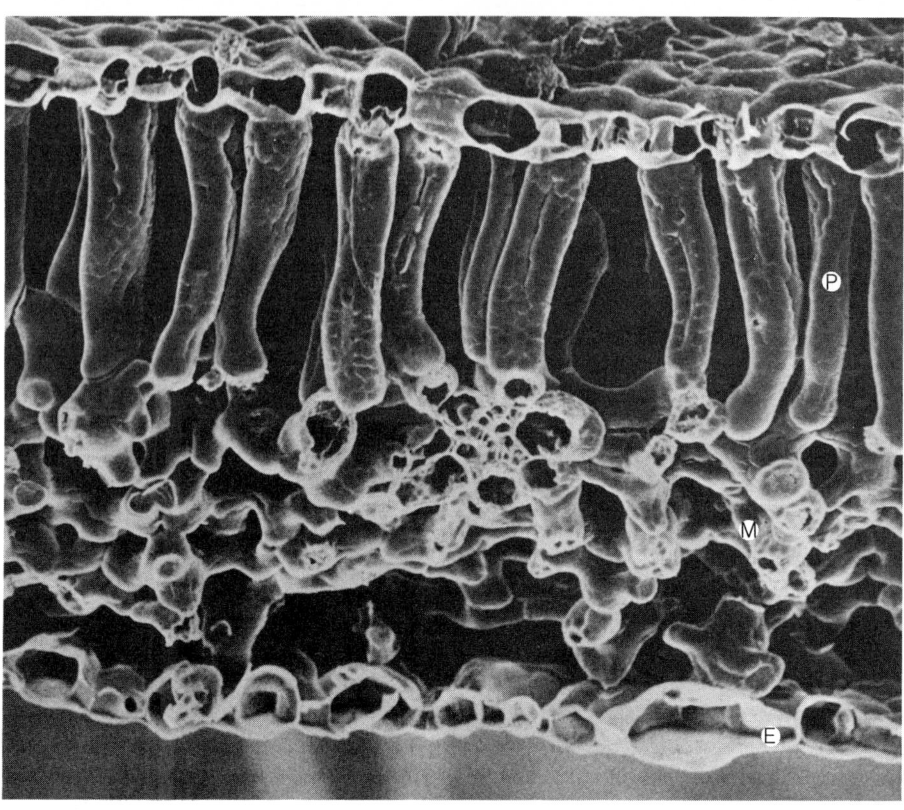


a

through four kinds of leaves. The waxy **cuticle** on leaf surfaces restricts diffusion, so most water vapor and other gases must pass through the openings between the **guard cells**. Some anatomists insist that the term **stomate**[2] refers only to this opening, but others (Esau, 1965; Mauseth, 1988) apply the term to the entire **stomatal apparatus**, which includes the guard cells. The opening is then called the **stomatal pore**. Adjacent to each guard cell are usually one to several other modified epidermal cells called **accessory** or **subsidiary cells** (see Fig. 4-10 below), the number and arrangement depending on the plant family (although different types can occur on a single leaf). Water evaporates inside the leaf from cell walls of the **palisade parenchyma** and **spongy parenchyma**, collectively the **mesophyll**, into the **intercellular spaces**, which are continuous with the outside air when the stomates are open. Carbon dioxide follows the reverse diffusional path into the leaf. Many of the cell walls of mesophyll cells are exposed to the internal leaf atmosphere, although this is seldom evident for palisade cells in drawings and photomicrographs of leaf cross sections. This arrangement becomes much more apparent in sections through the palisade, parallel to the leaf surface, and is also strikingly apparent in scanning electron micrographs such as those in Figure 4-5.

As you can see from Figure 4-4, there is much variability in the anatomy of leaves. In Chapter 11 we discuss a special kind of photosynthesis, called $C_4$ photosynthesis, that is especially prevalent in tropical grasses such as maize (Fig. 4-4d). Leaves with $C_4$ photosynthesis have a special anatomy called **Kranz anatomy** (from the German word for *wreath*) in which the layer of cells around the vascular bundles is especially prominent, with a large number of chloroplasts. Most plants have **bundle-sheath cells**, but they are easy to overlook in species that do not have Kranz anatomy. We'll see in Chapter 11 that they play a special role in $C_4$ photosynthesis.

Stomates sometimes occur only on the lower surfaces of leaves but often are found on both surfaces, often with more on the bottoms. Waterlily pads have stomates only on top, and submerged plants have none at all. Grasses usually have about equal numbers on each side. Sometimes, as in the oleander or pine (Fig. 4-4b and c), the stomates occur in a **stomatal crypt**. Such **sunken stomates** are apparently an adaptation that reduces transpiration.

Figure 4-6 shows scanning electron micrographs of stomates from four species. Typical stomates of dicots consist of two kidney-shaped guard cells; grass and

[2]Most anatomists prefer **stoma** (singular) and **stomata** (plural), but many plant physiologists use the anglicized forms, as we do here.

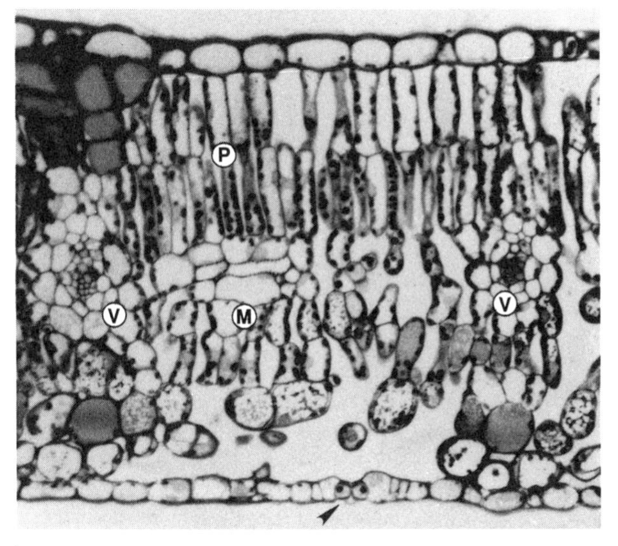


b

**Figure 4-5** (**a**) Transverse view of a mature broadbean (*Vicia faba*) leaf. The internal organization of cells in this leaf is characteristic of many plants, consisting of a layer of palisade cells (P) in the upper half of the leaf, spongy mesophyll cells (M) in the lower half (palisade and spongy mesophyll are collectively called mesophyll), and bounded on both sides by the epidermis (E). Note the large air gaps between both the palisade cells and the spongy mesophyll cells. Most of the surfaces of the cells are exposed to the air; the area of cell wall exposed is 10 to 40 times the surface area of the leaf (Nobel, 1980). The proportion of air volume to cell volume in a leaf can vary from 10 to 80 percent among different types of plants. 420 ×. (Scanning electron micrograph courtesy of John Troughton; see Troughton and Donaldson, 1972.) (**b**) Light micrograph of a cross section of a *Populus* leaf. Note the veins (V) and the layers of palisade (P) and spongy mesophyll cells (M) containing chloroplasts. A stomate is indicated with an arrowhead. 900 × (Courtesy of Fred D. Sack.)

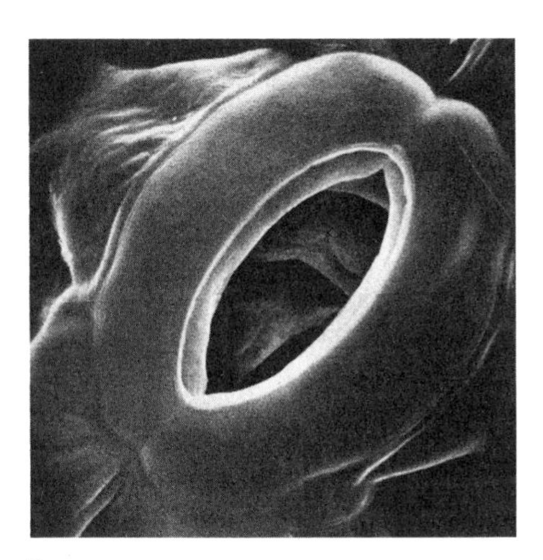

a

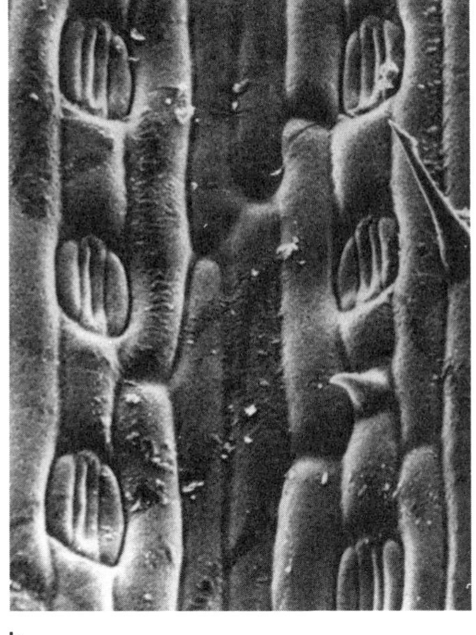

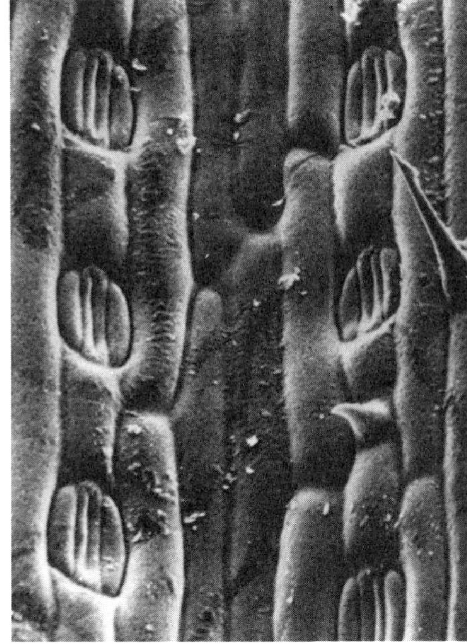

b

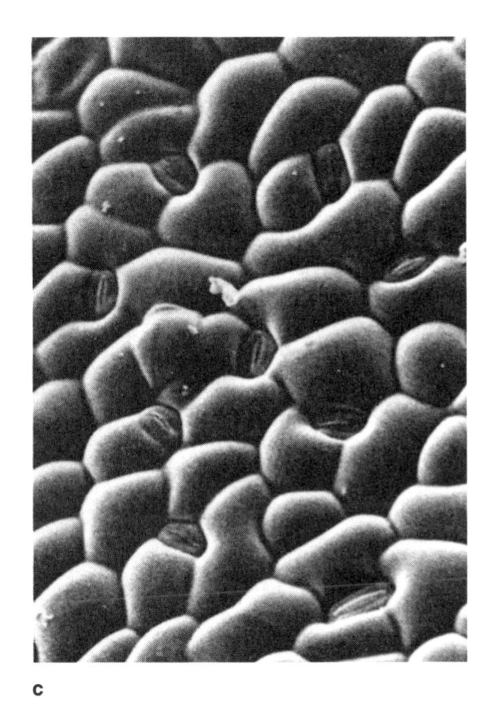

c

d

**Figure 4-6** Stomates. (**a**) Spongy mesophyll cells, as seen through a stomate on the lower surface of a cucumber leaf. 7,900 ×. (Courtesy of John Troughton; see Troughton and Donaldson, 1972.) (**b**) Upper surface of wheat (*Triticum* sp.). Note the characteristic monocot stomates. (**c**) Upper surface of lambsquarter (*Chenopodium rubrum*). (**d**) Upper surface of velvet leaf (*Limnocharis flava*). Note the hairs. (Scanning electron micrographs courtesy of Dan Hess.)

sedge guard cells tend to be more elongate (dumbbell-shaped). Guard cells contain a few chloroplasts, whereas their neighboring epidermal cells usually do not (except in ferns and a few aquatic angiosperms). Typically, there are no (or incomplete) plasmodesmata connecting the protoplasts of guard cells and accessory cells, but there may be plasmodesmata between guard cells and the mesophyll cells below.

Typically, each square millimeter of leaf surface has about 100 stomates, but the number can be ten times that, with a maximum so far recorded of 2,230 (Howard, 1969). Recent studies (for example, Woodward, 1987) suggest that stomatal densities are sensitive to $CO_2$ concentration, with fewer stomates per unit area as $CO_2$ increases. This was shown by laboratory studies and by stomate counts of a given species as a function of increasing elevation. ($CO_2$ partial pressures decrease, along with the other gases in the atmosphere, with increasing elevation.) It was also found by examining herbarium specimens that stomatal densities have decreased by 40 percent over the past two centuries as $CO_2$ in the atmosphere has increased from 280 to over 350 $\mu$mol mol$^{-1}$.

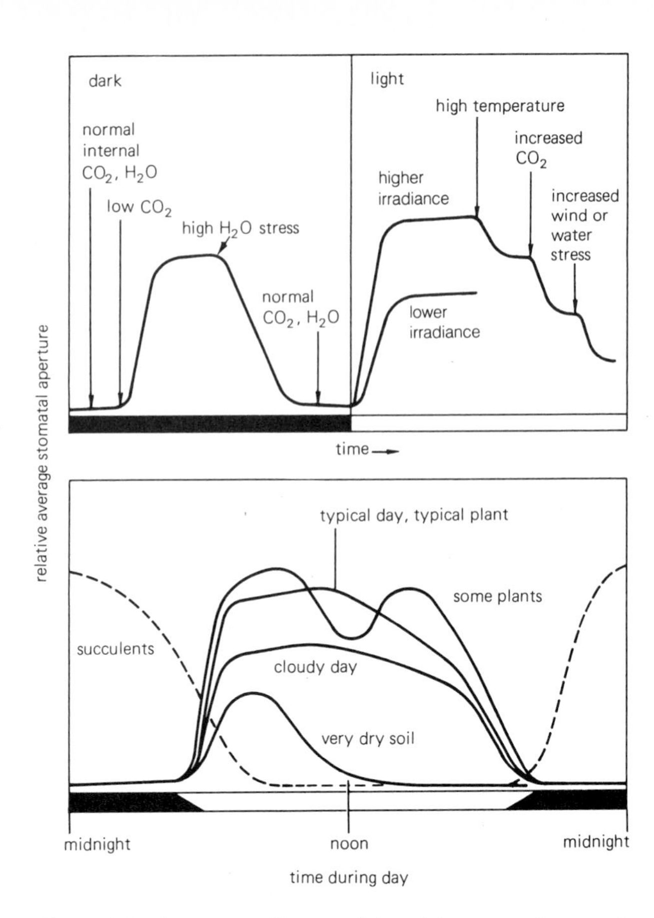


**Figure 4-7** A summary diagram of stomatal response to several environmental conditions. In the top graph, arrows point to times when some environmental parameter was changed as indicated by the label.

## 4.4 Environmental Effects on Stomates

Many factors influence stomatal apertures, and any theory purporting to explain guard-cell action must account for these effects. (For recent reviews of stomatal physiology, see books edited by Biggins, 1987, and Zeiger et al., 1987.)

Stomates of most plants open at sunrise and close in darkness, allowing entry of the $CO_2$ used in photosynthesis during the daytime. Opening generally requires about an hour, and closing is often gradual throughout the afternoon (Fig. 4-7). Stomates close faster if plants are suddenly exposed to darkness. The minimum light level for the opening of stomates in most plants is about 1/1000 to 1/30 of full sunlight, just enough to cause some net photosynthesis. High irradiance levels cause wider stomatal apertures.

Certain succulents that are native to hot, dry conditions (for example, cacti, *Kalanchoe*, and *Bryophyllum*) act in an opposite manner: They open their stomates at night, fix carbon dioxide into organic acids in the dark, and close their stomates during the day (see CAM photosynthesis in Section 11.6). This is an appropriate way both to absorb $CO_2$ through open stomates at night (when transpiration stress is low) and conserve water during the heat of the day.

In most plants, low concentrations of $CO_2$ in the leaves also cause stomates to open. If $CO_2$-free air is blown across leaves even in darkness, then their slightly open stomates open wider. Conversely, high $CO_2$ concentration in the leaves can cause the stomates to close partially, and this occurs in the light as well as the dark. When the stomates are completely closed, which is unusual, external $CO_2$-free air has no effect. In short, stomates respond to intercellular $CO_2$ levels but not to the $CO_2$ concentration at the leaf surface and in the stomatal pore (Mott, 1988). Succulents fix $CO_2$ into organic acids at night, thus lowering internal $CO_2$ concentration, which causes stomatal opening.

### Effects of Light Quality on Stomates

For many years plant physiologists considered several stomatal responses to be responses to the $CO_2$ concentration inside the leaf (Raschke, 1975). Thus the effect of light was explained as an indirect effect via a lowering of the $CO_2$ concentration by photosynthesis. More recently, however, several studies have shown that light has a powerful effect on stomates independent of photosynthesis. Conceivably, the light could be acting on mesophyll cells, which then send some message to the guard cells. Or the photoreceptor for the light could be in the guard cells themselves.

Thomas D. Sharkey and Klaus Raschke (1981a)

studied stomatal action on leaves of five species (cocklebur, cotton, common bean, *Perilla frutescens*, and maize). They were able to vary irradiance while maintaining a constant intercellular $CO_2$ concentration ($C_i$) and to vary $CO_2$ concentration at a constant irradiance. Stomatal conductance indicated stomatal aperture. Results with the first four species clearly showed that the stomatal response to light was mostly direct and only to a small extent a response to changes in intercellular $CO_2$ concentration. At high irradiances, stomates of maize (corn) responded primarily to the lowered intercellular $CO_2$ concentrations. Stomates responded to light even in leaves in which photosynthesis had been reduced to zero by application of an inhibitor (cyanazine). Sharkey and Raschke concluded: "At low light levels, the intercellular $CO_2$ concentration can become the major controlling factor; at high light levels, the direct response to light can overcompensate the $CO_2$ requirement of photosynthesis and cause a rise in the intercellular $CO_2$ concentration." It was possible to observe increasing intercellular $CO_2$ concentrations as light was increasing (because stomates were opening), which is exactly opposite to what one would expect if stomates responded to light only via photosynthetic effects on $CO_2$ concentration.

Sharkey and Raschke further suggested that light absorbed in the guard cells, rather than the mesophyll cells, is primarily responsible for the effect. For one thing, 10 to 20 times as much light energy was required to produce a given stomatal opening response if the leaf was inverted so that the beam of light had to pass through the leaf before reaching the guard cells. If the response were in the mesophyll cells, this would not be expected.

Sharkey and Raschke (1981b) also measured the wavelengths of light that were most effective in causing stomates to open. Blue light (wavelengths between 430 and 460 nm) was nearly 10 times as effective as red light (wavelengths between 630 and 680 nm) in producing a given stomatal opening. There was only a slight response to green light. The wavelengths that were effective in the red part of the spectrum were the same wavelengths that are effective in photosynthesis, and inhibitors of photosynthesis eliminated the response to red light. Thus the red light response is apparently caused by light absorbed by chlorophyll, but the blue light effect is independent of photosynthesis. Actually, as early as 1977, Edwardo Zeiger and Peter Hepler showed that blue light would cause *isolated* guard-cell protoplasts to absorb $K^+$ ions and swell, which in intact stomates (as we discuss below) is what causes stomatal opening.

## Does Photosynthesis Occur in Guard Cells?

If the red-light effect on stomatal opening is via absorption by chlorophyll and photosynthesis, where does this response take place? Do the guard cells photosynthesize? Plant physiologists have investigated these questions for decades. Photosynthesis could be one way for guard cells to sense $CO_2$.

Some workers have been quite convinced that guard cells are capable of photosynthesis and that this photosynthesis plays a significant role in the control of stomatal opening and closing. Photosynthesis was detected in isolated guard-cell protoplasts (Gotow et al., 1988), but the maximum rate was *below* the rate of dark respiration! Three other research groups could detect no photosynthesis in guard cells. Traces of the enzyme called *rubisco,* which fixes carbon dioxide in the first step of photosynthesis (see Chapter 11), was demonstrated immunologically in the chloroplasts of guard cells (Vaughn, 1987; Zemel and Gepstein, 1985), but other careful searches for enzymes that take part in photosynthesis had negative results.

Those who argue against the participation of photosynthesis in guard cells point to the quantitative aspects of the measurements that have been reported (Outlaw, 1989; Tarczynski et al., 1989). Guard cells contain only 3 percent as much chlorophyll as mesophyll cells, so a tiny contamination of mesophyll cells in the guard-cell preparations could account for the observed positive responses. In any case, $CO_2$ levels are apparently not detected only by photosynthesis of guard cells, because stomates are extremely sensitive to $CO_2$ in darkness when there is no photosynthesis. In a careful study, Udo Reckmann, Renate Scheibe, and Klaus Raschke (1990) showed that only 2 percent of the necessary solutes (which we discuss below) could be produced in guard cells of *Pisum sativum* (pea) by photosynthesis and that this was insignificant.

(The controversy about whether photosynthesis occurs significantly in guard cells took place primarily in the published scientific literature, which reminds us that a scientific investigation is not really science unless it is *published* so that other scientists can evaluate it. This fact and some pointers on good scientific writing are discussed in the essay by Page W. Morgan on page 76.)

## Other Effects of Environment on Stomates

Stomates of many (but not all) species are highly sensitive to atmospheric humidity (Tibbitts, 1979). They close when the difference between the vapor content of the air and that of the intercellular spaces exceeds a critical level. A large gradient tends to induce oscillations in opening and closing with a periodicity of about 30 minutes. This is probably because, as the steep vapor gradient induces closing, $CO_2$ in the leaf is depleted, and this in turn leads to opening. The most rapid responses to lowered humidity occur under low irradiances.

The water potential within a leaf also has a powerful effect on stomatal opening and closing. As water

## Must We Write?

### *Page W. Morgan*

*Most of the new findings presented in this book, as well as many early discoveries, are documented with references to the scientific literature (indicated by the name(s) of the author(s) and the year of publication). Publication, as it turns out, is an absolutely essential part of the scientific process, as Page Morgan explains in this essay. Dr. Morgan is a Professor of Plant Physiology in the Department of Soil and Crop Sciences at Texas A&M University. For many years, he has also been an Associate Editor (for plant development and growth regulators) of the journal* Plant Physiology, *in which the findings described in this book were often published. Prof. Morgan tells us how the system works and gives a few pointers on manuscript preparation for the scientific literature.*

Scientists must write! Knowledge uncommunicated is effectively unknown; discovery is but one step on the path of understanding. Although you may agree with these statements, you may not be aware how often scientists and science students are poorly prepared to write. Are genes for writer's block linked to those for investigative talent? Probably not, but while a student is studying the concepts of science, practicing its techniques, and learning its literature, it may be difficult to find the time to develop good communications skills. These skills take practice to develop. Educators soon learn that many students hate to write. Given my experience in teaching, advising graduate students, and serving as a decision editor for a scientific journal, I would like to discuss writing with the readers of this text.

The first point is that writing is necessary. As a young faculty member I remember hearing our Dean say: "Research is not completed until it is published." He was right. Students must ultimately write what they have discovered or learned. Quizzes, reports, term papers, theses, and dissertations allow others to know and judge your ability and accomplishments. Students who follow a scientific career soon are involved in written communications: justifications for equipment purchases, course outlines, research funding proposals, and manuscripts reporting research findings. Review articles, books, and pieces for the semipopular press flow from the pens (or computer keyboards) of the more prolific. Writing is essential in science!

This discussion is not to advocate "Publish or Perish," but rather "Publish or No One Will Know." Until you mentally wrestle with a question and write down your best answer, your instructor cannot evaluate how well you have learned. Likewise, until scientists have published the results and conclusions from their experiments, no one else can evaluate the significance and usefulness of the work. Only after new facts and ideas are published, discussed, and replicated by further testing can the valid ones ultimately contribute to advancing understanding in science as a whole.

Unfortunately, it's not enough to write; the need is to write well. Consider two students who spend hours in the library researching the same term-paper topic. Equally well-versed in the facts, one writes an uninspired, cataloglike listing of who did what and when. The second student identifies the relevant questions, arrays the facts to support logical deductions, and brings the topic to a sharply focused conclusion. One wrote and the other wrote well. At the professional level the objective for writing a scientific paper is not to get it published, but rather to communicate with other scientists. Everyone has a limited amount of time. Writers are literally competing for the time and attention of potential readers. Regardless of how important a discovery is, it will usually be more rapidly accepted if it is presented in a well-written paper. On the other hand, if the writing is poor and the message obscure, readers will often quit before completing the paper.

---

potential decreases (water stress increases), the stomates close. This effect can override low $CO_2$ levels and bright light. Its protective value during drought is obvious.

High temperatures (30 to 35°C) usually cause stomatal closing. This might be an indirect response to water stress, or a rise in respiration rate might cause an increase in $CO_2$ within the leaf. High $CO_2$ concentration in the leaf is probably the correct explanation for high-temperature stomatal closing in some species because it can be prevented by flushing the leaf continuously with $CO_2$-free air. In some plants, however, high temperatures cause stomatal opening instead of closing. This leads to increased transpiration, which removes heat from the leaf.

Sometimes stomates partially close when the leaf is exposed to gentle breezes, possibly because more $CO_2$ is brought close to the stomates, increasing its diffusion into the leaf. Wind can also increase transpiration, leading to water stress and stomatal closing.

## 4.5 Stomatal Mechanics

Stomates open because the guard cells absorb water and swell. At first, this is puzzling. One might imagine that

Perhaps some "do's" and "don'ts" would be helpful. Most instructors have seen a lot of bad writing; thus, the "don'ts" are easy to list. In my experience, the most common error is that writers fail to clearly state their message. The reader wonders: What do we learn about plant physiology from the experiment? Is it new or unique? Is it useful? The student writing a lab report or the scientist writing for a journal has the responsibility to answer those questions. Another common error occurs when a manuscript appears to be a partially condensed thesis. The message is lost in wordiness and secondary topics, and a 10-page story gets told in 25. Also, the technical language of experimental design and statistics is often used as a writing crutch; as a result, clarity suffers. Readers are usually interested in what the response was, and its magnitude, variability, and reproducibility. Jargon about "significant three-way interactions in complete factorials" prompts drowsiness.

Another common writing error is careless preparation. Most writing tasks, whether something as simple as an essay examination or as complicated as a reviewed journal manuscript, include instructions. Yet time after time writers appear to ignore the instructions and submit carelessly prepared material. The axiom "You never get a second chance to make a good first impression" certainly applies to writing. Although technical aspects of composition are beyond the scope of this essay, students would be well served to develop the habit of reviewing their writing for spelling, punctuation, and sentence structure before it is submitted to others. Writing handbooks list the common mistakes, such as using *which* in place of *that*. Equip yourself with a good writing handbook and a dictionary, and use them!

The positive suggestions for scientific writing must start with the scientific method. There is no substitute for expressing the topic as a question that can be answered. Searching out what is already known is equally indispensible. If a writer sits down to write, for example, about some topic as broad as "tissue-culture studies with soybeans," or if a writer dismisses relevant background knowledge with words such as "little is known about this subject," the paper may already be unsalvageable. Writers should have the primary message of their papers in mind before they write the first word. That message, in fact, should be an answer to a clearly stated question, and it should shape the title, the introduction, and the body of the paper. That way readers can recognize a theme that is consistent throughout the paper.

The purpose of scientific writing is to communicate findings, analyses, conclusions, and theories. *Communicate* is the operative word. Unless the reader understands, the writer has failed. The implications are obvious: Identify the audience and write to that level of understanding. Clarity is often inversely related to sentence length. A barrage of words with obscure meanings makes reading a chore rather than a pleasure. The words selected should aid communication rather than make a statement about the writer's vocabulary. Sentences that can be misunderstood usually will be. Brevity is a virtue when wed to clarity. Writers should write in a direct, simple, logical style; the reward is being read and being understood.

Writers should give serious attention to graphic illustrations. Poor ones frequently appear to be an afterthought. Excellent illustrations of findings or conclusions are often the most effective way to communicate. The saying "One picture is worth a thousand words" makes that point. Good illustrations don't just happen, they take forethought, inventiveness, and work. We train graduate students to give seminars with much emphasis on good slides. That emphasis is intended to carry over into their written communications. Computer graphics make the job easier, but the bottom line is still personal initiative.

In conclusion, the ability to write well is a skill needed by science students and scientists alike. It can be acquired and cultivated if the basic goal of communication is kept clearly in view. Those who can think and speak logically can learn to write logically, and even interestingly and entertainingly. Science will be the better for such writing!

---

the swelling guard cells would force their inside walls together. Stomates function the way they do because of special features in the submicroscopic wall anatomy. The cellulose microfibrils, or micelles, that make up the plant cell walls are arranged around the circumference of the elongated guard cells as though they were radiating from a region at the center of the stomate (Fig. 4-8a). The result of this arrangement of microfibrils, called **radial micellation**, is that when a guard cell expands by absorbing water, it cannot increase much in diameter; microfibrils do not stretch much along their length. But the guard cells can increase in length, especially along their outside walls, and as they do so they swell outwardly. The microfibrils then pull the inner wall with them, which opens the stomate (Mauseth, 1988).

It was noticed as long ago as 1856 that the guard cells of some species were slightly more thickened along the concave wall adjacent to the stomatal opening (Fig. 4-8b). Since then, many authors have suggested that the thickening was responsible for the opening when the guard cell takes up water. Donald E. Aylor, Jean-Yves Parlange, and Abraham D. Krikorian (1973) at the Connecticut Agricultural Experiment Station reinvestigated a 1938 discovery by H. Ziegenspeck (see Ziegenspeck, 1955). They showed with balloon models (Fig. 4-9) and

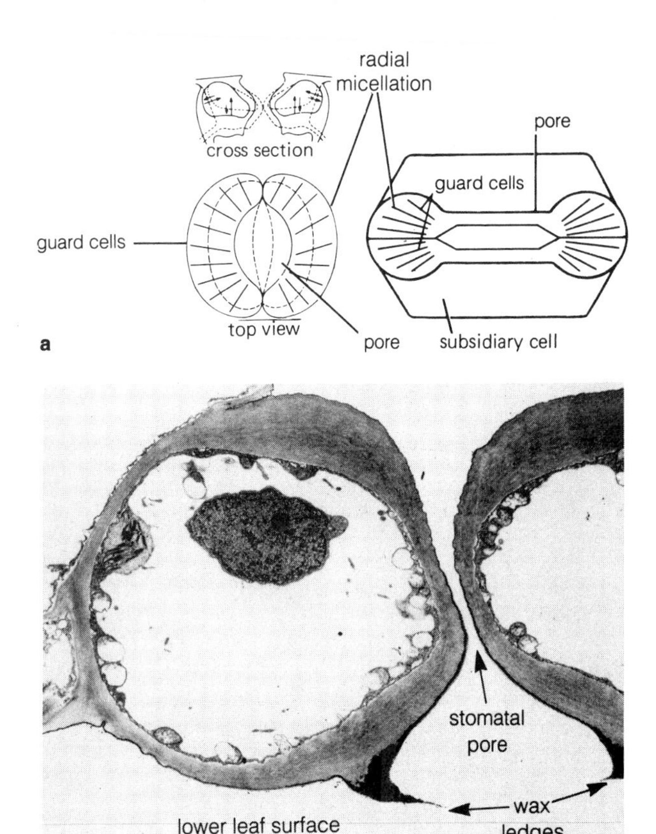


**Figure 4-8** Stomatal structures. (**a**) Schematic drawing of two stomates, showing radial micellation. Left, a dicot leaf (dashed lines show guard cells after loss of water, hence closing); right, a monocot (grass) stomate. (**b**) Transmission electron micrograph of a cross section of tobacco guard cells showing the stomatal pore, wax ledges on the outside of the leaf, and differential thickness of guard-cell walls. 9,000 × (Courtesy of Fred D. Sack.)

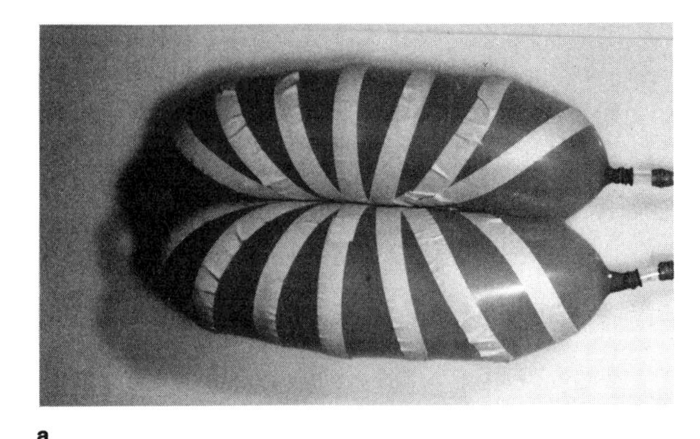

**Figure 4-9** Two balloons representing a guard-cell pair. (**a**) Balloons in their "relaxed" state with masking tape applied to represent both the "radial micellation" and the thickening along part of the ventral walls. (**b**) The balloon pair in an inflated state. Balloons were glued together at the ends with rubber cement before inflating (which weakened the rubber and caused eight pairs to burst when inflated before achieving success with the pair shown!).

mathematical modeling that radial micellation is much more important in stomate opening than is the thickening of the inside walls, and that radial micellation is as important in grass as in dicot guard cells. In grass stomates, however, the ends swell while the middle remains narrow. This pushes apart the midregions of the two guard cells. The resultant dumbbell shape is present only when the guard cells are turgid. Actually, when the guard mother-cell in grasses undergoes cytokinesis to form the two guard cells, the division is incomplete, and the two protoplasts are connected and thus continuous at their ends (Mauseth, 1988).

## 4.6 Stomatal Control Mechanisms

What is it that causes guard cells to absorb water so that the stomates open? For many decades, the prime suspect has been some osmotic relationship. There are at least three possibilities: If the osmotic potentials of the guard-cell protoplasts become more negative than those of the surrounding cells, water should move into the guard cells by osmosis, causing an increase in pressure and swelling. Or a decrease in the resistance of the guard-cell wall to stretching, which would decrease pressure inside and thus allow the uptake of more water, could cause swelling. Or the surrounding subsidiary cells might shrink, again releasing pressure on the guard cells.

By inserting microhypodermic needles filled with silicon oil into individual cells, Mary Edwards and Hans Meidner (1979) directly measured the pressures in epidermal, subsidiary, and guard cells. When they punctured a cell in a way that caused no leakage, the pressure inside forced some cell sap into the needle, pushing back the silicon oil. This activated a highly sensitive pressure transducer that applied pressure to the oil until the meniscus between it and the cell sap was pushed back to the cell surface (as revealed in a microscope). The pressure transducer gave an electrical signal that was calibrated in terms of real pressures. Edwards and Meidner found that such a pressure drop in subsidiary cells did indeed contribute to the swelling of guard cells and the opening of stomates.

This is only a contributing effect, however, because many other measurements have shown that the osmotic potential of guard cells becomes more negative when stomates open. For example, G. D. Humble and Klaus Raschke (1971) measured values of −1.9 MPa for the osmotic potential of broadbean (*Vicia faba*) guard cells with closed stomates and −3.5 MPa when the stomates were open. Because the guard cells nearly doubled in volume during opening, this increase in solute concentration occurs in spite of dilution. The increased solute concentration results in an osmotic transfer of water from the accessory cells to the guard cells. We can now restate the question more precisely: *What causes the change in the osmotic potential in guard cells that results in stomatal opening?*

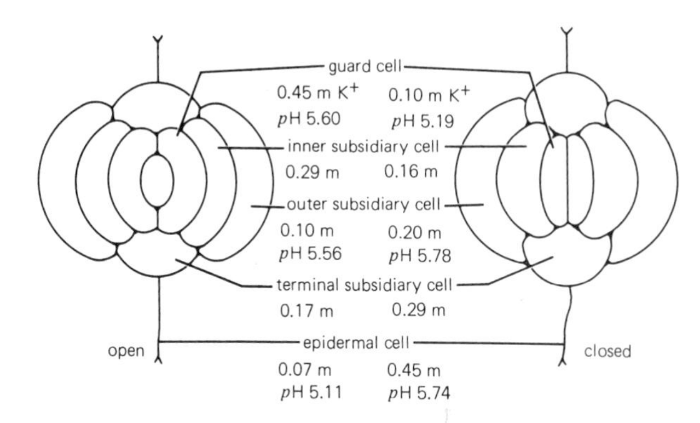


**Figure 4-10** Quantitative changes in $K^+$ concentrations and the *p*H values of the vacuoles in several cells making up the stomatal complex of *Commelina communis*. Values are given for the open and closed conditions of the stomatal pore. Lowercase m stands for molal concentrations. Note buildup of $K^+$ and increase in *p*H (fewer $H^+$ ions) in guard cells as stomate opens; outer and terminal subsidiary cells and epidermal cells show opposite responses. (Data of Penny and Bowling, 1974, 1975.)

## Guard-Cell Absorption of Potassium Ions

Stomates open because guard cells absorb water, and water uptake is caused by more solute and hence a more negative osmotic potential, but what is the solute and where does it come from? Since about 1968 (although the first reports appeared as early as 1943), accumulated experimental evidence has made it clear that potassium ions ($K^+$) move from the surrounding cells into the guard cells as stomates open. Representative data are shown in Figure 4-10.

The quantities of $K^+$ that accumulate in the vacuoles of guard cells during stomatal opening are sufficient to account for the opening, assuming that each $K^+$ ion is associated with a suitable anion to maintain electrical neutrality. Increases of up to 0.5 M in $K^+$ concentration are observed, enough to decrease the osmotic potential by about 2.0 MPa. Stomatal opening and $K^+$ movement into the guard cells are closely correlated in nearly every case investigated. Light causes a buildup of $K^+$ in guard cells and in isolated guard-cell protoplasts, as we noted above (Zeiger and Hepler, 1977). So does $CO_2$-free air. When leaves are transferred to the dark, $K^+$ moves out of guard cells into the surrounding cells, and stomates close. This has been observed in numerous species from all levels of the plant kingdom that have stomates (mosses, ferns, conifers, monocots, and dicots) and for the stomates occurring on various plant organs (leaves, stems, sepals, and moss sporophytes; see, for example, Willmer et al., 1983).

When strips of epidermal tissue are removed from the leaves of *Vicia*, most epidermal cells are broken, but the guard cells remain intact. When these strips are floated on solutions and kept in the light, stomates will not open unless the solutions contain $K^+$. So guard cells must normally obtain $K^+$ ions from adjacent epidermal cells. Stomates also close in response to the application of abscisic acid, a plant growth regulator (see below), which causes loss of $K^+$ from the guard cells. Hence, nearly all the evidence agrees that $K^+$ transport from accessory cells to guard cells is the cause of the more negative osmotic potentials and hence stomatal opening, and that reverse transport is the cause of closing.

Having learned about the $K^+$ flux, we can ask the next question: *What is the mechanism of $K^+$ movement?* Consideration of this question forces us to realize that the science of plant physiology has become increasingly dependent on biochemistry. Now we must add biochemical considerations to our thermodynamic concepts of water relations.

To begin, stomatal opening is not a simple pumping of $K^+$ into guard cells with energy provided by light. Such a mechanism does not account for responses to lowered $CO_2$, even in the dark. Further, it has been observed that increasing the *p*H of guard cells by exposing leaves to ammonium vapor also causes opening, and that stomates sometimes open in response to low oxygen levels. Neither observation is accounted for by a light-driven uptake of $K^+$ into guard cells. Of course, such a mechanism also fails with succulents that open their stomates at night.

In some species, $Cl^-$ ions or other anions accompany $K^+$ into and out of guard cells, but Raschke and Humble (1973) observed that no anion accompanies the movement of $K^+$ into guard cells of *Vicia* leaves. Instead, as potassium ions move into guard cells, an equal number of hydrogen ions move out (see Fig. 4-10 for similar results). Where does the $H^+$ come from? Organic acids are synthesized in the guard cells in response to factors that cause stomatal opening. (As $CO_2$ is used up in photosynthesis, the $pH$ will rise; that could be one factor that triggers the changes that lead to opening.) Malic acid is the most common product under normal conditions. Because hydrogen ions are supplied by organic acids, the $pH$ in the guard cells would drop (they would become more acidic) if $H^+$ were not exchanged for incoming $K^+$. The increased ions ($K^+$ and organic acid anions) make the osmotic potential more negative.

The organic acids are made largely from starch or other carbohydrates stored in guard cells. It has long been known that starch disappears in the guard cells of many plants as stomates open, but careful measurements could never demonstrate that sugar appeared in its place. Thus the increasingly negative osmotic potential is caused by an uptake of $K^+$ and sometimes of $Cl^-$ and/or an accumulation of potassium salts of organic acids, mostly malate. Which of the two processes predominates apparently depends upon the species and perhaps upon the availability of $Cl^-$. But how? Much has been learned and postulated about the biochemical steps involved, steps that we will discuss in later chapters. Starch is apparently broken down to produce a three-carbon compound (phosphoenol pyruvate or PEP); this step is promoted by blue light. The PEP then combines with $CO_2$, producing the four-carbon oxaloacetic acid, which is converted into malic acid. Finally, $H^+$ ions from the malic acid leave the cell, balancing the $K^+$ ions that are entering.

It would at first seem that a balance between loss of $H^+$ and absorption of $K^+$ would not make the osmotic potential more negative. The $H^+$ ions only appear transiently, however, as organic acids are produced from starch. Therefore, there is a net loss of osmotically inactive substances and an increase in active substances as starch disappears and $K^+$ ions are absorbed. (For reviews, see Biggins, 1987; Outlaw, 1983; Permadasa, 1981; Zeiger, 1983; and Zeiger et al., 1987.)

In the succulents with stomates that open at night, $CO_2$ combines with PEP, producing malic acid in the dark in many cells besides guard cells. Again we see that a decrease in $CO_2$ concentration within the leaf is reflected in less dissolved $CO_2$ in the guard cells, in $K^+$ absorption, and in stomatal opening.

## The Effect of Abscisic Acid on Stomates

Another observation of the early 1970s was almost as revolutionary as that of $K^+$ uptake. When the growth regulator abscisic acid (ABA; see Chapter 18) is applied at micromolar ($\mu$M; $10^{-6}$ M) concentrations, it causes stomates to close. Furthermore, when leaves are subjected to water stress, ABA builds up in their tissues. When leaves dry at normally slow rates, ABA builds up before stomates close, suggesting that stomatal closure in response to leaf water stress is mediated by ABA.

To what are the cells in a water-stressed leaf responding? Recall Equation 3.1, which includes the three factors of prime importance to the water relations of plants: solute potential, pressure, and water potential. Changes in any one of these factors could control ABA production. Margaret Pierce and Klaus Raschke (1980) reported that *cell turgor* (pressure) seemed to be in control of ABA production in several species that they studied; changes in solute potential and water potential had little effect by themselves.

During recent years there have been reports that stomates will close even when leaves are *not* experiencing water stress, providing that the *roots* are being stressed (see, for example, Davies et al., 1986; and Schulze, 1986). In one of the most convincing experiments, the root system is divided into two parts, one of which is provided with ample water (so shoots remain turgid) while the other part is allowed to dry out. Stomates often close in response to such a treatment, suggesting that they are getting some kind of signal from the roots. Evidence is accumulating that this signal is ABA. For example, Zhang and Davies (1990) found a close correlation between soil drying, decreased stomatal conductance, and ABA in xylem sap; *leaf* ABA was not as closely correlated with stomatal conductance. Although the experimental results are not questioned, there is controversy over the importance of the phenomenon in nature or in agricultural fields (Kramer, 1988; Passioura, 1988; and Schulze et al., 1988). It would seem at first glance that leaves would virtually always experience water stress before the roots, but on further reflection one realizes that the surface layers of soil might dry out, stressing roots there, while deeper roots are still in moist soil—much like the experiment with divided root systems. In any case, this is currently an active area of research.

One observation seemed counter to the simple picture of water stress leading to ABA production, which in turn leads to stomatal closure. When water stress develops rapidly, as when a leaf is removed and subjected to dry air at warm temperatures, stomates close *before* ABA begins to build up in the leaf tissue. There are at least three possible explanations: First, when water stress develops rapidly, water may evaporate from the guard cells themselves, causing them to lose turgor and thus causing stomates to close. Second, ABA probably occurs in at least three **pools** (a term used when evidence suggests that a substance that occurs in different parts of a tissue or a cell affects some process differently depending on where it is located). The three ABA pools in leaf tissue might be the cytosol of cells, where ABA is apparently synthesized; the chloroplasts, where it accumulates; and the cell walls outside the protoplasts. The ABA could move from one pool to another (eventually into guard cells) before total ABA in the leaf has a chance to build up. (Papers and reviews include Cowan et al., 1982; Outlaw, 1983; Permadasa, 1981; Zeiger, 1983; and Zeiger et al., 1987.)

Third, Harris et al. (1988) have shown that ABA in guard cells represents only 0.15 percent of the total ABA in a leaf. Clearly, changes in guard-cell ABA could easily go undetected if only total leaf ABA is measured. The method used in this study was an enzyme-amplified immunoassay that was claimed to be 100 times more sensitive than other published immunoassays for ABA. It was possible to measure the ABA in individual cells. The investigators found a low but detectable ABA content in guard cells of unstressed (fully hydrated) leaves. When the fresh mass of detached leaves decreased 10 percent by transpiration, guard-cell ABA increased approximately 20-fold. The ABA content of all leaf cells increased in response to such a water stress. ABA concentrations in leaves from stressed cells ranged from 7 to 13 μM. The authors point out that the ABA in the guard cells was probably not restricted to the cell walls because it didn't wash out in 30 min of immersion in water. Yet, ABA did leach out of cells left for 4 h in water, indicating that such losses could account for lower concentrations reported by some other workers.

Before leaving the topic of ABA effects on stomates, it is worth noting that other compounds may also cause stomates to close or open. Cytokinins, for example, are plant growth regulators that, among other things, cause plant cell divisions. They can cause stomatal opening (see reviews by Permadasa, 1981; Zeiger, 1983; and Zeiger et al., 1987). Furthermore, protein synthesis might be essential to keep stomates closed (Thimann and Tan, 1988), again illustrating the complex biochemistry involved in stomatal action.

### Feedback Loops

We have discussed the powerful direct effect of increased light on stomatal opening. In addition, there appear to be at least two **feedback loops** that control stomatal opening and closing. When $CO_2$ decreases in the intercellular spaces and thus in the guard cells, $K^+$ moves into guard cells and stomates open, allowing $CO_2$ to diffuse in, thus completing the first loop. This and the effect of light meet the needs of photosynthesis, but in nonsucculents, it also leads to transpiration. If water stress develops, ABA appears, so the stomates close, completing the second loop. The two loops interact: The degree of stomatal response to ABA depends upon $CO_2$ concentration in the guard cells, and response to $CO_2$ depends upon ABA. One feedback loop provides $CO_2$ for photosynthesis; the other protects against excessive water loss (Fig. 4-11). Stomates, as Raschke (1976) has said, have been "delegated the task of providing food while preventing thirst."

## 4.7 The Role of Transpiration: "What Good Is Transpiration?"

Many philosophers of science would object to such a question, labeling it **teleological**. Such a question assumes that all things in the universe have purpose, an

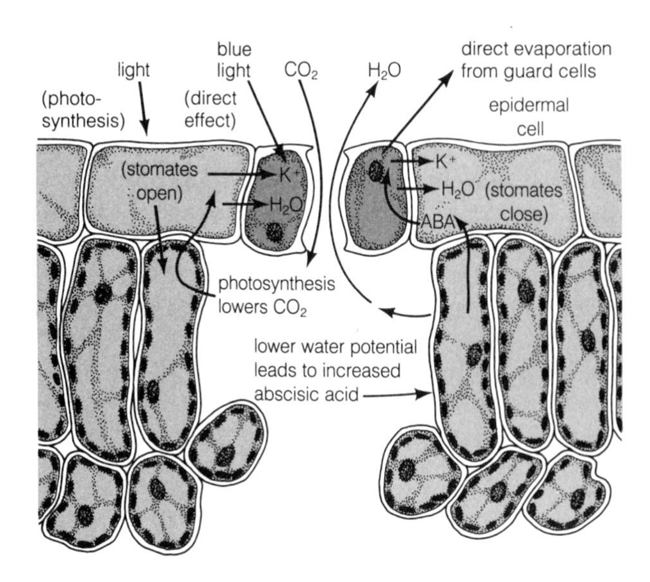


**Figure 4-11** Two important feedback loops, one for $CO_2$ and one for $H_2O$, that control stomatal action. The left part of the drawing illustrates one effect of light: Light promotes photosynthesis, which lowers $CO_2$ levels in the leaf; the leaf's response is to cause more $K^+$ to move into guard cells, and water follows osmotically, causing stomates to open. There is also a direct effect of blue light on guard cells, which causes stomatal opening independently of $CO_2$ levels. The right-hand side shows the effects of water stress: When more water exits in transpiration than can enter from the roots, abscisic acid (ABA) is released or produced from mesophyll cells (or transported from roots), which leads to movement of $K^+$ out of guard cells; water follows osmotically, so stomates close. If the rate of drying is extremely rapid, water is lost from the guard cells directly, bypassing the ABA step but still leading to closure. (Modified from Jensen and Salisbury, 1984.)

assumption that cannot be scientifically demonstrated. The biologist avoids teleology by rewording the question: *What is the selective advantage of transpiration?* Evolutionary theory states that a harmful feature will be eliminated by natural selection if there is some less harmful feature that can be selected in its place. So what is the advantage to a plant of transpiration?

The advantage might be a sort of "win by default." It is essential to the life of a land plant to absorb carbon dioxide from the atmosphere; it seems that the stomatal mechanism has evolved because of this requirement, and the disadvantageous consequence is transpiration. What about oxygen? Because stomates are typically closed at night, they obviously don't need to be open for the plant to absorb the oxygen used in respiration. This is because there is 590 times as much $O_2$ in the atmosphere as $CO_2$, so $O_2$ gets to the cells even if stomates are closed. During the day, when leaves are photosynthesizing and stomates are open, oxygen diffuses out through them. But the stomates' real reason for existence is $CO_2$ absorption. (Water-logged roots, of course, cannot always absorb enough oxygen.)

It has been argued that transpiration is neither essential nor of any advantage to the plant because many plants can be grown through their life cycles in terraria at 100 percent RH, at which transpiration is nearly zero. Indeed, it is a common observation that many plants grow better in atmospheres with a high relative humidity. One of us (F. B. S.) has observed certain alpine *land* plants (for example, *Caltha leptosepala*) growing for days to weeks completely submerged in water. Obviously, no submerged plant can transpire.

Nevertheless, careful investigation and thought have revealed several situations in which transpiration *per se* does seem to be beneficial to the plant. An unavoidable by-product of necessity may have been turned into an advantage. In most of these cases, it is possible for the plant to grow without transpiration, but when it occurs, it seems to confer some benefit, possibly in transporting minerals and maintaining an "optimum turgidity," and certainly by removing large amounts of heat from leaves (Section 4.8, below).

### Mineral Transport

Minerals that are absorbed into the roots typically move up through the plant in the **transpiration stream**, the flow of water through the xylem caused by transpiration. But the transpiration stream is not essential for this movement because minerals move upward in woody stems in spring before leaves appear. The young stems may transpire slightly, but not much compared with transpiration after leaves expand. There is a circulation in plants (see Chapter 8): Solutions move through the phloem tissue from assimilating organs to utilizing organs. Even in the absence of transpiration, the water in these solutions will return to the assimilating organs through the xylem tissue. Such a circulation has been demonstrated with radioactive tracers. So transpiration is not essential for the movement of minerals within the plant. Actually, the rate at which minerals arrive in the leaves is a function only of the rate at which they move into the xylem tissue, providing there is any xylem flow at all. The rate that goods are delivered by an endless belt is a function only of the rate of loading.

Nevertheless, when transpiration occurs, it may aid mineral absorption from the soil and transport in the plant. Calcium and boron in tissues seem to be especially sensitive to the rate of transpiration (reviewed by Tibbitts, 1979). Plants grown in greenhouses with high humidity and air enriched with $CO_2$ (which tends to close stomates) can exhibit calcium deficiency in certain tissues. On the other hand, too rapid transpiration (as in winter homes in which air is especially dry) can lead to a toxic buildup of certain elements.

### Optimum Turgidity Versus Water Stress

Another reason some plants may not grow as well when transpiration is greatly reduced could be that cells function best with some water deficit. There could be an optimum turgidity or water potential for cells, with certain functions being less efficient above and below this level. There is little evidence for this hypothesis, however.

The main concern of plant physiologists and agriculturists is that of not enough water: **water stress**, or a too negative water potential. Plant responses to water stress are studied extensively, largely because of their inhibitory effects on plant yield in natural and agricultural ecosystems. We discuss water-stress effects on plants in Chapter 26. Perhaps the most important point is that cell growth, which depends on absorption of water by cells, is one of the first processes to be affected by water stress. This frequently reduces yield. Other processes such as photosynthesis and synthesis of proteins and cell walls are also adversely affected by water stress.

## 4.8 The Role of Transpiration: Energy Exchange

For years, plant physiologists argued about whether transpiration was necessary to cool a leaf being warmed by the sun. Yes, transpiration is a cooling process, but if transpiration does not cool the leaf, it was argued, other physical processes will—although in the absence of transpiration, it was conceded, leaves might be a few degrees warmer than otherwise. Growth of plants in atmospheres of 100 percent RH were cited to support this view. As we have come to appreciate how a leaf exchanges energy with its environment, these arguments seem much less important. They obscure the important fact that transpiration often plays a highly significant role in leaf cooling.

Evaporation of water is a powerful cooling process. Recall the Maxwell-Boltzmann distribution of molecular velocities (see Fig. 2-3). It is the water molecules with high velocities that evaporate, and as they leave the liquid, the average velocity of the remaining molecules is lower, which is the same as saying that the liquid is cooler. Greenhouses in dry climates are often cooled by evaporative cooling; air is drawn through a wet fibrous pad. When 1 kg of water at 20°C evaporates, it absorbs 2.45 MJ (586 kcal) from its environment; at 30°C the **latent heat of vaporization** is 2.43 MJ $kg^{-1}$ (580 kcal $kg^{-1}$). Vast amounts of water evaporate from plants, and each kilogram of water transpired absorbs 2.4 to 2.5 MJ from the leaf and its environment.

Sometimes transpiration is the only means of net heat transfer to the environment. Consider the large fan-shaped leaf of a native palm tree (*Washingtonia fillifera*) growing at a southern California oasis (Fig. 4-12). Such a leaf, even in full sunlight, is often cooler than the surrounding air, in which case it absorbs heat from the air. And it absorbs more radiant energy from sunlight than it radiates into its environment. It is cooler

**Figure 4-12** Palms (*Washingtonia fillifera*) growing near Palm Springs, California. (Photograph by F. B. Salisbury.)

than air only because it is evaporating large quantities of water.

Investigations of the energy exchange between a plant and its environment provide an interesting example of plant biophysics. In the remainder of this chapter we will consider some of the principles involved, without applying mathematics in the discussion, but the equations are given at the end of the chapter for reference. (See reviews by Gates, 1968, 1971; Nobel, 1983; and Salisbury, 1979.)

## Leaf Temperature

Consider the various factors that influence the temperature of a leaf. While transpiration cools a leaf, condensation of moisture or ice on the leaf (dew or frost) releases the **latent heat of condensation** of water to the leaf and its environment. Incoming radiation warms a leaf, but the leaf is radiating energy to its environment. If leaf temperature is different from air temperature, heat is exchanged first by **conduction** (in which the energies of molecules on the leaf surface are exchanged with those of the contacting molecules of air) and then by **convection** (in which a quantity of warmed air expands, becomes lighter, and thereby rises — or sinks, if it has been cooled). We shall refer to the combination of conduction and convection simply as convection.

If leaf temperature is changing, as is usually the case, the leaf is storing or losing heat. If a thin leaf stores a given amount of heat, its temperature rises rapidly; the same amount stored in a cactus raises the temperature much less, but the cactus stays warm longer. For simplicity, we will consider only a leaf in equilibrium with its environment; that is, at constant temperature. On the order of 1 to 2 percent of the light is converted to chemical energy by photosynthesis, but we can ignore this small amount. Energy produced by respiration and other metabolic processes is also small enough to ignore. Under steady-state conditions, primarily three factors influence leaf temperature: radiation, convection, and transpiration. Each of these is worthy of our consideration.

*Radiation* Appendix B discusses the principles of radiant energy, including those that apply to heat-transfer studies. From the standpoint of leaf temperature, the **net radiation** is important. A leaf absorbs visible (light) and invisible (infrared) radiation from its surroundings and radiates infrared energy. If the leaf is absorbing more radiant energy than it radiates, then the excess must be dissipated by convection or transpiration, or both (or the temperature will rise). At night, leaves often radiate more energy than they absorb. If they cool below air temperature, they will absorb heat from the air and possibly from water that condenses as dew or frost on their surfaces. There are three important things to keep in mind when discussing the net radiation of a leaf: the absorbed wavelengths, the total spectrum of incoming radiation, and the amount of energy radiated by the leaf.

First is the **absorption spectrum** of the leaf. Some energy incident on a leaf is transmitted, some is reflected, and some is absorbed. The energy absorbed depends on its spectrum. Leaves irradiated with white light absorb most blue and red wavelengths and much of the green. Some green is reflected and transmitted, however, which is why leaves appear green. Leaves absorb very little of the near-infrared part of the spectrum; most is either transmitted or reflected. So, if we could see in this part of the spectrum, plants would appear very bright, as when vegetation is photographed with infrared-sensitive film (Fig. 4-13). Virtually all the far-infrared or thermal part of the spectrum is absorbed. If our eyes were sensitive only to that part of the spectrum, vegetation would appear as black as black velvet — except that leaves emit thermal radiation in those wavelengths and would thus appear to glow. The leaf absorption spectrum in Figure 4-14 presents these ideas quantitatively.

Second, radiation sources vary considerably. Figure B-3 in Appendix B shows emission spectra for several sources. The sun and the filament of an incandescent lamp emit **light** (the visible portion of the electromagnetic spectrum) because of their high temperatures. The higher the temperature, the more the peak of the emission spectrum is shifted toward the

a

b

**Figure 4-13** Plants showing high reflectivity in the infrared portion of the spectrum. (**a**) Photograph taken with ordinary panchromatic film, which gives gray tones similar to the color intensities experienced by the human eye. (**b**) Photograph taken with infrared-sensitive film through a dark red filter that excludes most of the visible radiation. Note the brilliant white appearance of the vegetation and compare in particular the clumps of grass. (Logan Canyon, Utah; photographs by F. B. Salisbury.)

blue (see Wien's law in Appendix B). The temperature of the sun's surface is considerably above that of an incandescent filament in a light bulb, so sunlight is richer in blue and green wavelengths than is light from an incandescent lamp.

The sun's radiation is further modified by passing through the atmosphere. Much ultraviolet is removed, and radiant energy is also absorbed by the atmosphere at several discrete wavelengths in the **far-red** (longer than 700 nm, but visible) and infrared parts of the spectrum. Most ultraviolet is absorbed by ozone in the upper atmosphere, and the infrared absorption bands are caused primarily by water and carbon dioxide.

Today, many plants used in physiological research are grown under artificial light sources, including fluorescent, incandescent, and such high-intensity discharge (HID) lamps as mercury vapor, low-pressure and high-pressure sodium, and metal-halide lamps. These each have individual emission spectra (see Fig. B-3) that are absorbed differently by plants.

All objects at temperatures above absolute zero emit radiation (see below). Objects at ordinary temperatures emit most of this as far-infrared wavelengths, so plants receive this radiation from all their surroundings, including molecules of air. The quantity can be a sizable portion (for example, 50 percent) of the total radiation environment.

The radiation absorbed by a plant is a function of the leaf's absorption spectrum and the spectrum of radiation that impinges on it. Hence, the actual percentage of radiation absorbed varies considerably (because both absorption and emission spectra vary), but about 44 to 88 percent may be absorbed in common situations. Absorption is high when plants are irradiated with fluorescent light because the leaf strongly absorbs most of the wavelengths emitted by fluorescent tubes (visible

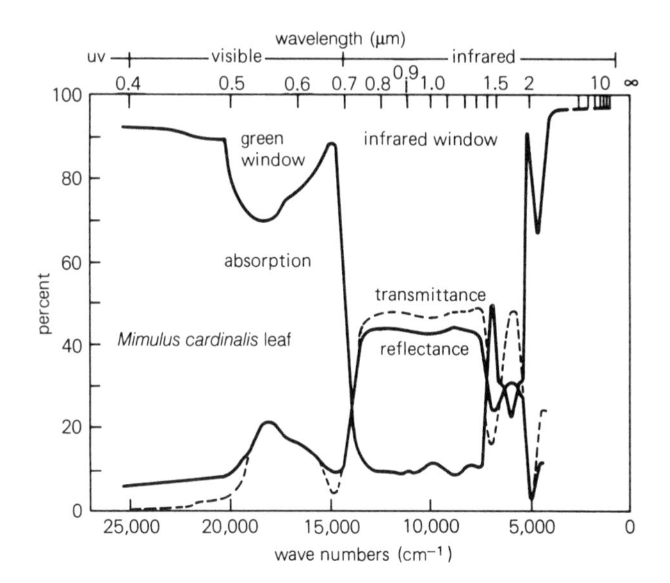


**Figure 4-14** Absorption, transmission, and reflection spectra of a leaf. Note especially the "windows" in the green and in the near-infrared portions of the spectra. The leaves are thin and light green. (From Gates et al., 1965; used by permission.)

light). Absorption is much less when plants are irradiated with incandescent light of equivalent total energy because such light is rich in the near-infrared part of the spectrum that is least absorbed by plants (Mellor et al., 1964).

Third, plants and all objects emit radiant energy in the far-infrared part of the spectrum. The quantity of energy emitted can be calculated by application of the **Stefan-Boltzmann law** (Appendix B), which states that the emitted energy is a function of the *fourth* power of the kelvin (absolute) temperature. Thus, as a leaf's temperature increases in sunlight, the radiant energy it emits also increases. Although on the kelvin scale the normal temperature range of plants (from about 273 K to 310 K) is small, energy emitted over this range varies by about 50 percent, which can be significant. Even when a plant is illuminated by sunlight and is also receiving far-infrared radiation from its environment (for example, from atmosphere, clouds, trees, rocks, and soil), the radiant energy emitted from the leaf is usually more than 50 percent of that absorbed and can reach 80 percent or more.

*Convection* Heat is conducted-convected from the leaf to the atmosphere in response to a temperature difference between leaf and atmosphere. If incoming radiation causes the leaf to be warmer, heat will move from the leaf to the atmosphere. *The temperature difference is the driving force*; the greater the difference, the greater the driving force for convection.

With a given temperature difference, the rate of convective heat transfer is inversely proportional to the resistance to convection. The situation is exactly analogous to the integrated form of Fick's law (Equations 2.16, 3.1, and 3.3) and to Ohm's law, as we have noted. With convective heat transfer, the flow of heat is proportional to the temperature difference between the leaf and the atmosphere and inversely proportional to the resistance to heat flow encountered in the atmosphere.

The resistance to convective heat transfer is expressed by the thickness of the **boundary layer** (also called the **unstirred layer**), which was introduced earlier in this chapter. This layer is the transfer zone of fluid (gas or liquid) in contact with an object (in this case, the leaf) in which the temperature, vapor density, or velocity of the fluid is influenced by the object (Fig. 4-15). For a given temperature difference between the leaf and the air beyond the boundary layer (a given driving force), the convective transfer of heat is more rapid when the boundary layer is thin (steep temperature gradient) and slower when it is thicker (less steep gradient).

Usually there is air movement around a leaf: The more rapid the air movement the thinner the boundary layer. The boundary layer is thinnest next to a leaf's **leading edge** (the edge facing into the wind). If the leaf's surface is parallel to the direction of wind movement, the boundary layer thickens from the leading edge toward the **trailing edge** of the leaf. Small leaves, especially conifer needles, have the thinnest boundary layers and are the most affected by convection. Large leaves, such as those of the desert fan palms, have the thickest boundary layers.

To summarize: The boundary layer is thinnest and offers the least resistance to convective heat transfer for small leaves and high wind velocities. Convective heat transfer is most efficient under such conditions, so smaller leaves have temperatures closer to air temperature than do larger leaves, especially if there is a wind.

*Transpiration* In some ways, transpiration is closely analogous to convective heat transfer; in other ways, it is different. The driving force for transpiration is the gradient in water-vapor density[3] from within the leaf to the atmosphere beyond the boundary layer. The resistance is partially the resistance of the boundary layer. To this extent, convection and transpiration are analogous, but the stomates offer an additional and typically much larger resistance to transpiration. If the stomates are closed or nearly closed, resistance can be very high; if they are open, resistance is relatively low. There are other resistances within the leaves besides that of the stomates, but these usually remain fairly constant. The resistance of the cuticle to passage of water depends

[3]Vapor density (expressed in grams or moles per cubic meter), discussed in Chapter 2 (page 40), can also be called vapor concentration. We have also mentioned vapor pressure, which has pressure units.

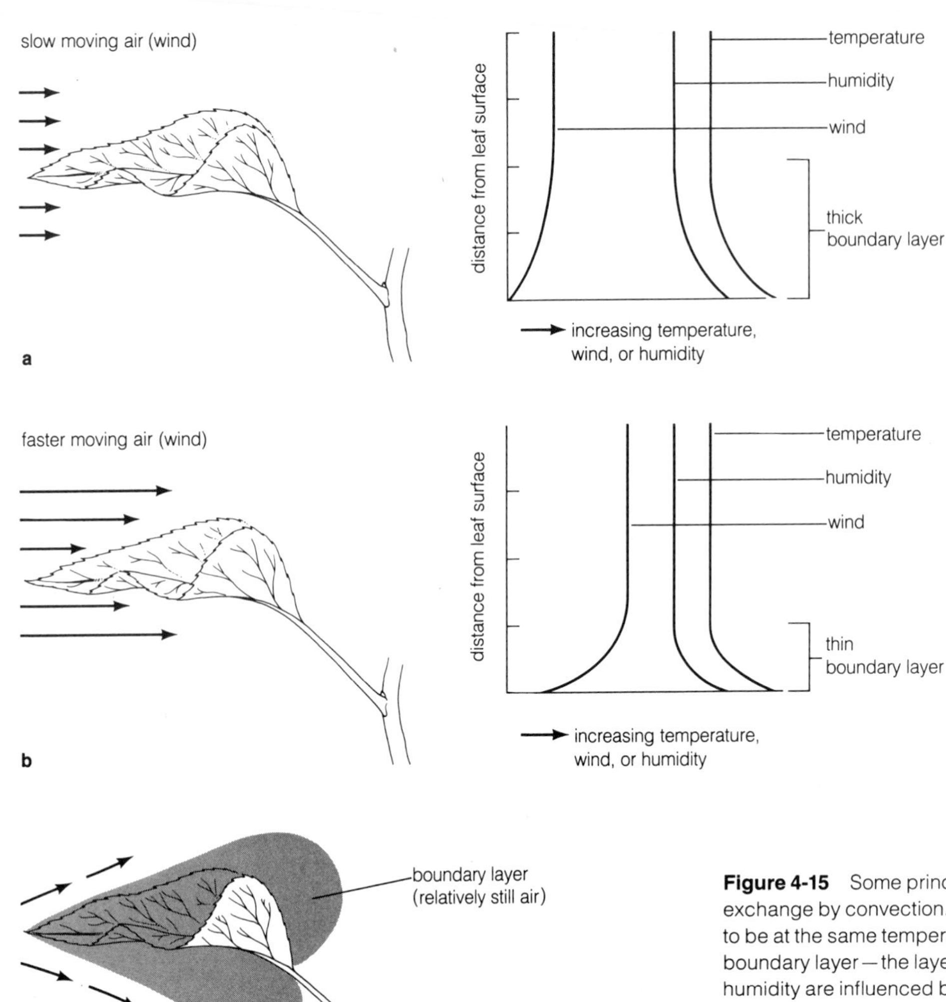


**Figure 4-15** Some principles of the boundary layer and heat exchange by convection. (**a**) and (**b**) The two leaves are assumed to be at the same temperature; only wind velocity is different. The boundary layer—the layer of air in which temperature, wind, and humidity are influenced by the leaf—is represented by the curved parts of the lines in the graphs. The boundary layer becomes thinner with increasing wind speed. (**c**) The boundary layer becomes thicker with distance from the leading edge of the leaf. The shaded area represents a layer of relatively still air. (From Salisbury, 1979.)

upon atmospheric humidity, temperature, and perhaps light or other factors. Because it is always relatively high, it is seldom considered. Note that *there is always some leaf resistance*; that is, the leaf is never simply like a piece of wet paper. And leaf resistance to transpiration can vary over a wide range as environmental factors influence stomatal apertures.

In addition to boundary-layer thickness, the vapor-density gradient is determined by two factors: absolute humidity and leaf temperature. We usually assume that RH in the internal spaces of the leaf approaches 100 percent. Actually, it is somewhat less because at equilibrium the water potential of the internal leaf atmosphere is equal to the water potential of the surfaces from which the water evaporates, and this is usually $-0.05$ to $-3.0$ MPa because it is in equilibrium with the tissue water potential. (If equilibrium is not achieved, the water potential of the leaf atmosphere will be even lower.) Nevertheless, the water potential of the internal leaf is equivalent to an RH of at least 98 percent (see calculations following Equations 3.5 and 3.6). Such high RHs are not common in the atmosphere beyond the boundary layer, so, *even if the leaf is at exactly the same temperature as the atmosphere beyond the boundary layer, under most conditions vapor density is higher inside the leaf.*

A temperature gradient greatly accentuates the vapor-density gradient because maximum vapor density of air is strongly a function of temperature (see Fig. 3-8). Warm air can hold more water than cold air. An examination of Figure 3-8 and Table 4-1 shows, for example, that air at 20°C and an atmospheric humidity of 10 percent establishes a vapor-density difference of about 9.8 g $m^{-3}$ between the leaf and the air, if they are at the same temperature and if the atmosphere within the leaf approaches 100 percent relative humidity. (At 20°C, saturated vapor pressure is about 10.9 g $m^{-3}$, and

**Table 4-1 Vapor-Density Gradients Between Leaves and the Atmosphere When Leaf and Air Temperatures Are the Same or Different and When Atmospheric Humidity Is Different.**

| Conditions | Leaf | Air Beyond Boundary Layer | Difference |
|---|---|---|---|
| Temperature | 20°C | 20°C | none |
| Relative humidity | near 100% | 10% | near 90% |
| Vapor density | 10.9 g m$^{-3}$ | 1.1 g m$^{-3}$ | 9.8 g m$^{-3}$ |
| Temperature | 30°C | 20°C | 10°C |
| Relative humidity | near 100% | 90% | near 10% |
| Vapor density | 20.3 g m$^{-3}$ | 9.8 g m$^{-3}$ | 10.5 g m$^{-3}$ |

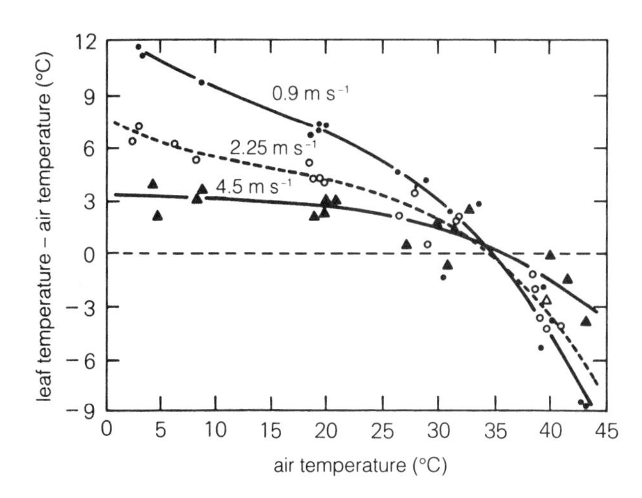


**Figure 4-16** The difference between leaf and air temperature as a function of air temperature for three wind velocities. Total irradiance was 906 watts per square meter (W m$^{-2}$). Curves are third-order polynomials drawn by computer to match the data. (Data from Drake et al., 1970.)

10 percent of this is about 1.1 g m$^{-3}$.) If the leaf is at 30°C, however, and atmospheric humidity is as high as 90 percent (at 20°C), there still is a vapor-density difference of about 10.5 g m$^{-3}$. (At 30°C, vapor density is 20.3 g m$^{-3}$; 90 percent of 10.9 g m$^{-3}$ is 9.8 g m$^{-3}$, which, when subtracted from 20.3 g m$^{-3}$, leaves a gradient of 10.5 g m$^{-3}$.) Thus, if the leaf is warmer than the air (a common phenomenon in sunlight), transpiration can occur into an atmosphere with an RH of 100 percent. As the vapor goes beyond the boundary layer, it may condense to form the minute droplets we see as steam, as in the forest under the sun after a rainstorm, but this is of no consequence to the plant, which has lost water. Remember that, as a rule, the energy source (the driving force) for transpiration is incoming radiation.

## 4.9 Energy Exchanges of Plants in Ecosystems

Application of heat-transfer principles to field situations has provided considerable insight into the function of plant communities. To illustrate, we will consider a few of the principles discussed in previous sections as they apply to plants in the desert, the alpine tundra, and other situations.

### The Desert

In the desert, high temperatures are combined with high radiation, low humidity, and little available water. It is to the plant's advantage to conserve moisture and to maintain a relatively low leaf temperature. Many desert plants—perhaps most—have small leaves, providing thin boundary layers and resulting in efficient convective heat transfer. Their temperatures are thus closely coupled to air temperature, so that at least they are not heated much above air temperature by sunlight. But such a situation might also produce a high rate of transpiration unless leaf resistance is high. Some desert plants have sunken stomates and other features that result in high leaf resistance and retarded transpiration. Some are also light gray and reflect much of the sun's radiation. We have seen that desert succulents conserve water by closing their stomates during the day and fixing $CO_2$ into organic acids at night. But succulent leaves are often large and become much warmer than the air. Their protoplasm can evidently tolerate the high temperatures.

It can be hot in the desert, so an ideal solution might be air conditioning by evaporative cooling, a technique applied widely by people who live in desert cities. A few desert species with roots that reach the water table achieve evaporative cooling. Typically, they have large leaves, as does the palm in its oasis, and their high rates of transpiration result in leaf temperatures several degrees below air temperature. This is possible because the low ambient humidity provides an ample vapor-density difference even when the leaf is 10°C cooler than the air, and because the large leaf has a thick boundary layer, resulting in a low rate of convective heating of the leaf from the surrounding, warmer air. Leaf temperatures of large cocklebur (*Xanthium strumarium*, which is not native to the desert) growing along a ditch bank in southern Arizona were 9 to 11°C cooler than the surrounding air, which was 36°C under a clear, midsummer sky. On the other hand, leaf temperatures of this same species measured in Oregon on a cool summer morning were several degrees above air temperature. Wind-tunnel studies (Fig. 4-16) showed that, to a considerable extent, this plant regulates its leaf temperature by controlling transpiration, staying warmer than cold air and cooler than hot air.

In an oasis or an irrigated field in a dry climate, transpiration is higher than in humid areas, not only because leaves are warmed by high radiation but because they are also warmed by convection. In humid areas, leaves lose more heat by radiation than transpiration; in an oasis, the opposite is the case.

### The Alpine Tundra

Alpine plants are faced with a different situation. Cool, fairly humid, gusty air combines with solar radiation levels that can be extremely high. Water is seldom limiting, but evaporative cooling is of no advantage in a situation in which ambient temperatures are often well below the range considered optimum for metabolic processes. Several hundred measurements of leaf temperatures in the alpine tundra by Salisbury and George Spomer (1964) showed that, when the sun is irradiating the leaves, their temperatures can reach 30°C, which can be as much as 20°C above the temperature of the air just a few centimeters away from the plants. Alpine plants typically have small or finely divided leaves, but they grow in a layer about 10 cm above the ground, where wind velocities are greatly reduced. Many have a cushion or rosette form that results in a thick boundary layer that is determined by the entire plant and nearby soil instead of by individual leaves. High leaf temperatures clearly indicate that transpiration does not provide much cooling under these conditions. Not all the reasons for this are apparent, so integrated field and laboratory research would be in order.

### The Effects of Wind

The ecological problems of transpiration and heat transfer are less challenging in less extreme environments, but careful study of heat transfer can provide some interesting insights. The great variety of leaf shapes in moderate environments is itself suggestive.

Data obtained during the first half of the 20th century often seemed contradictory. Some indicated that wind increased transpiration (as it always increases evaporation from a free surface); others indicated that wind *decreased* transpiration. When radiation loads are relatively low and leaf resistance is also low, transpiration is certainly increased by wind, and when leaf temperature is below air temperature, increasing wind velocity always tends to increase transpiration. But it is now clear that transpiration can be decreased by wind when the radiation-heat load is high, particularly if leaf resistance is also high (that is, stomates are closed). Under such conditions, the leaf temperature can be far above the air temperature, which would cause a high transpiration rate if stomates were open; but the wind cools the leaf by convection, and this cooling is more effective in reducing transpiration than is the wind in decreasing the boundary layer and thus increasing evaporation.

We have now discussed transpiration and energy exchange as they are usually discussed in a basic class, but there are other interesting situations in nature. One of these is the ventilation system in the waterlily, which moves gases in bulk flow through stems in response to pressure gradients (driving forces) set up by special processes of diffusion. In a personal essay (page 90), John Dacey tells how he discovered this fascinating system.

## 4.10 The Heat-Balance Equations

Here, for your reference, are the heat-balance equations that express mathematically the concepts described in Section 4.8 (Nobel, 1983; Salisbury, 1979):

### The Energy-Balance Equation for a Leaf Surface (all values can be expressed as watts per square meter: $W\ m^{-2}$):

$$Q + H + V + B + M = 0 \tag{4.6}$$

where

$Q$ = net radiation (positive if leaf is radiating less energy than the radiant energy absorbed from its surroundings)

$H$ = sensible heat-energy transfer (includes conduction and convection; positive if leaf gains more heat energy than it loses)

$V$ = latent heat-energy flux; the transpiration term (negative when water is vaporizing, positive when condensing or freezing)

$B$ = heat-energy storage (positive when leaf temperature is increasing), and

$M$ = metabolism and other factors (positive when heat is produced)

At constant leaf temperature and ignoring metabolism:

$$Q + H + V = 0 \tag{4.7}$$

### Radiant-Energy Flux *Absorbed* by a Leaf Surface ($Q_{abs}$; $W\ m^{-2}$):

$$Q_{abs} = eQ_v + e'Q_{th} \tag{4.8}$$

where

$eQ_v$ = total absorbed radiation in the photosynthetically active region ($W\ m^{-2}$)

$e'Q_{th}$ = total absorbed (thermal) radiation outside the photosynthetically active region ($W\ m^{-2}$), and

$e$ and $e'$ = leaf emissivities (or absorptivities) in the two spectral regions (dimensionless)

### Net Radiation at a Leaf Surface ($Q$; $W\ m^{-2}$):

Energy *emitted* by a leaf (Stefan-Boltzmann law; see Appendix B) is subtracted from the radiant energy *absorbed* ($Q_{abs}$):

$$Q = Q_{abs} - e'\sigma T^4 \tag{4.9}$$

where

$e'$ = emissivity or absorptivity of the leaf for long-wave (thermal) radiation; typically about 0.95 for living leaves at normal temperatures

$\sigma$ = Stefan-Boltzmann constant ($5.673 \times 10^{-8}$ W $m^{-2}\ K^{-4}$), and

$T$ = absolute temperature of the leaf (K)

Often, the above equation is written (see Monteith, 1973):

$$Q = I_s - rI_s + L_{env} - e'\sigma T^4 \tag{4.10}$$

where

$I_s$ = the solar irradiance incident at the leaf surface ($W\ m^{-2}$)

$r$ = the leaf surface reflection coefficient (decimal fraction), and

$L_{env}$ = the environmental long-wave radiation at the surface ($W\ m^{-2}$)

### Sensible Energy Transfer by Convection at a Leaf Surface ($H$; $W\ m^{-2}$):

$$H = \frac{(T_a - T_\iota)\,c_p\rho}{r_a} = \frac{\Delta T c_p \rho}{r_a} = \Delta T c_p \rho \cdot g_a$$

where

$T_a$ = air temperature (K or °C)

$T_\iota$ = leaf temperature (K or °C)

$\Delta T = T_a - T_\iota$

$c_p$ = specific heat capacity of dry (unsaturated) air ($\approx$ 1,000 J $kg^{-1}\ K^{-1}$; volumetric heat capacity at 20°C, 1 atm = 4.175 MJ $m^{-3}\ K^{-1}$)

$\rho$ = density of dry air (1.205 kg $m^{-3}$ at 20°C and 1 atm)

$r_a$ = boundary-layer resistance (s $m^{-1}$), and

$g_a$ = boundary-layer conductance (m $s^{-1}$)

The **convective transfer coefficient ($h_c$; $W\ m^{-2}\ K^{-1}$),** also called the **heat transfer coefficient** (proportional to the reciprocal of the boundary-layer resistance), may be used to calculate sensible energy transfer:

$$h_c = \frac{c_p \rho}{r_a} \tag{4.12}$$

$$H = \frac{\Delta T c_p \rho}{r_a} = \Delta T c_p \rho \cdot g_a \tag{4.13}$$

$$H = h_c \Delta T \tag{4.14}$$

### Latent Energy Flux of Water Vapor at a Leaf Surface ($V$; $W\ m^{-2}$), the Transpiration Term:

$$V = \frac{(e_\iota - e_a)\,c_p\rho}{\gamma\,(r_\iota + r_a)} = \frac{\Delta p c_p \rho}{\gamma\,(r_\iota + r_a)} = \frac{\Delta p c_p \rho\,(g_\iota + g_a)}{\gamma} \tag{4.15}$$

where

$e_\iota$ = vapor pressure in the leaf; that is, within the substomatal cavity (Pa)

$e_a$ = vapor pressure of the air (Pa)

$r_a$ = boundary-layer resistance (in air) (s $m^{-1}$)

$r_\iota$ = diffusive resistance within the leaf (s $m^{-1}$)

$\gamma$ = psychrometric constant (66.6 Pa $K^{-1}$), and

$g_\iota$ and $g_a$ = leaf and boundary-layer conductivities, respectively (m $s^{-1}$)

# Ventilation in Waterlilies: A Biological Steam Engine

*John Dacey*

*Graduate studies can provide some of the most exciting times in a person's life. This was the case for John Dacey as he unraveled the secrets of ventilation in waterlilies, as he describes in this personal essay. At the time, he was working on a doctoral degree in the Zoology Department at Michigan State University (completed in 1979). Before that, he had grown up in Kingston, Ontario, and he had obtained a combined undergraduate degree in biology and chemistry at the University of King's College. He is continuing his work on production of trace gases in wetlands and marine systems, now at the Woods Hole Oceanographic Institution in Massachusetts. (For further information on this work, see Dacey, 1980, 1981, 1987; Dacey and Klug, 1982a, 1982b.)*

Scientific research as an evolutionary process can be most exciting when it leads to questions that diverge from the expected course. In this essay I recount how my research unfolded; it is an effort to show how successful research depends on a combination of labor, logic, and luck. What began as a study of the role played by sediment gases in generating gas flows in water plants ultimately led to the discovery of a flowthrough ventilation system.

Oxygen is a basic requirement in the metabolism of all plants. Its absence around the roots of aquatic plants inhibits aerobic respiration in roots and allows potentially harmful materials to accumulate. It is generally accepted that the extensive network of internal gas spaces (**lacunae**) in these plants represents an adaptation to this environment, serving primarily to transport $O_2$ to buried roots and rhizomes.

In earlier studies, botanists measured gradients in the concentrations of $O_2$ and $CO_2$ in the lacunae. They found the highest levels of $O_2$ in photosynthesizing leaves and the lowest $O_2$ levels in the roots; vice versa for $CO_2$. They concluded that $O_2$ *diffused* to the roots from photosynthesizing leaves, and $CO_2$ *diffused* from the roots toward the leaves.

That idea is not unreasonable, but it is incomplete. A gas mixture is not only a mixture of randomly moving molecules; it is also a fluid. Just as a gradient in *partial* pressure drives the net diffusion of an individual component in a gas mixture, a gradient in *total* pressure drives a mass flow of the whole gas mixture (i.e., producing a wind). Under a gradient in total pressure, the movements of individual gases are not independent.

In my research I was interested in the relative contributions of diffusion and mass flow in the transport of gases in plants. My approach differed from previous research in two important ways: I measured the total pressure of lacunar gases by manometer, and I measured the concentrations of all principal gases by gas chromatography. This physical perspective made me especially conscious of potential mechanisms for generating pressure gradients.

I was studying a different plant when I saw gas streaming from a submerged waterlily leaf early in spring. The convenience of working with waterlilies quickly became apparent. They are large plants, with petioles up to 2 m long reaching from the sediment to the lake surface. Gas occupies at least half of the total volume of the plant, and I could withdraw samples by syringe. My first waterlily samples showed that methane ($CH_4$) was a significant constituent in lacunar gas, and early measurements confirmed that pressure could be developed in the root lacunae by methane from sediments. This confirmed my original hypothesis, but it soon became apparent that there was a more important process to investigate. Methane might be used as a tracer for gas movement!

Methane is produced in sediments by anaerobic bacteria and diffuses into lacunae in buried roots and rhizomes. As I studied its distribution in the waterlilies, I noticed some surprising patterns. For example, during nighttime in the summer months, $CH_4$ occurred throughout the plant in concentrations one would expect if gases moved principally by diffusion. During daylight, however, $CH_4$ was present in the petioles of older leaves but was absent in the petioles of young floating leaves. I could not explain its disappearance on the basis of diffusion, so I used experimental tracers to investigate the possibility of mass flow. By injecting small quantities of a tracer gas I monitored mass flow in the plant and found that during daylight gas moves from young floating leaves down their petioles to the rhizome. It subsequently moves from the rhizome up the petioles of older leaves to the atmosphere (see figure in this box). The mass flow down young petioles carried away any $CH_4$ tending to diffuse up from the rhizome. When ventilation stopped at night, $CH_4$ could accumulate in the petioles by diffusing from the rhizome.

Mass flow requires a gradient in total gas pressure. Using a sensitive manometer, I found that pressures in young leaves were slightly higher than the surrounding atmosphere (by less than 0.2 kPa). These small pressure differentials were sufficient to move air down the petioles at speeds up to 50 cm/min.

The next problem was to determine what caused the elevated pressures in young leaves. Ventilation stopped in darkness, so it seemed that the phenomenon was light-dependent. I confirmed that by showing that pressures in the

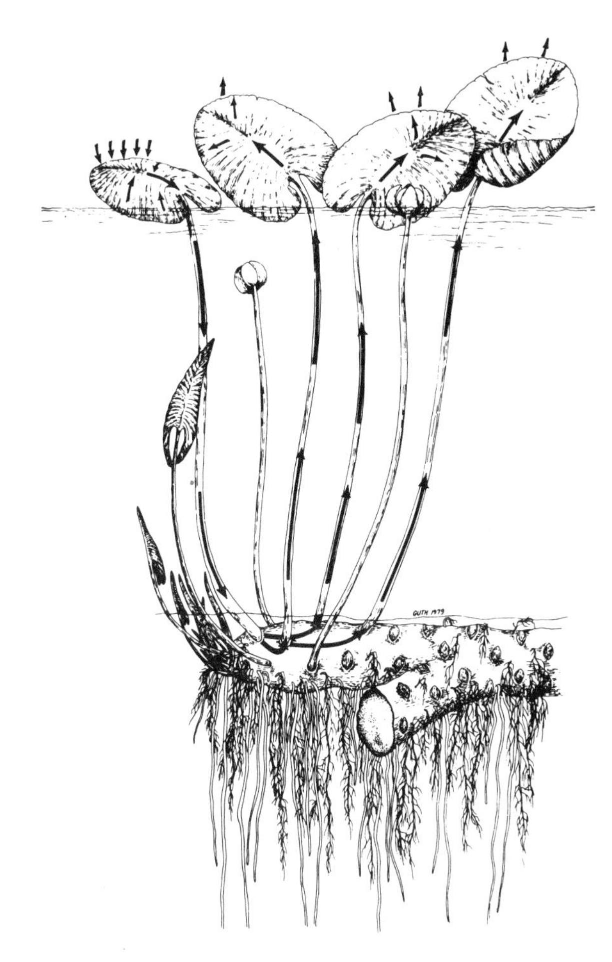

The ventilation system in the waterlily.

lacunae of influx leaves were directly related to the levels of incident light. When I shaded a pressurized leaf, its lacunar gas pressure dropped immediately.

I soon found that photosynthesis did not play a role in pressurization. The pressures were not due to *de novo* gas production by the plant but resulted from movement of air from the surroundings into the lacunae. When I enclosed a pressurized leaf in a transparent bag, the bag tended to collapse, showing that the leaves drew air from the atmosphere against a pressure gradient.

I began to investigate the mechanism of this "pump." Temporarily forgetting the physical perspective I had used up to this point, I hypothesized that the pressurization must be a metabolic process, somehow using light energy to pump $O_2$ into the leaf. Any metabolic pump ought to be influenced by the composition of ambient gas. By varying the composition of gases in the plastic bag, I found that pressurization occurred regardless of the gas composition.

In an effort to understand the mechanism, I looked for differences in gas composition inside and outside the pumping leaves. Not surprisingly, the $H_2O$ vapor pressure was higher inside the leaf. More significantly, the absolute amounts of both $N_2$ and $O_2$ were lower than ambient inside the leaf. There was, therefore, a diffusion gradient for $N_2$ and $O_2$ into the leaf, even though the total gas pressure in the leaf exceeded ambient!

Another key discovery came on the heels of this observation. I found that lacunar gases pressurized in darkness when the leaf was held near a hot object. Pressurization was not dependent on light *per se* but on heat.

It was apparent that the answer lay in physics, not in biochemistry, so I went back to physical principles. Physical theory of gas flow through pores predicts that flow occurs only by diffusion when pores are very small (less than 0.1 $\mu$m at atmospheric pressure). Larger pores allow both diffusion flow and mass flow. The fact that lacunar gas pressures were higher than ambient means that there was no significant mass flow between the lacunae and the atmosphere. Pores separating the lacunae from the atmosphere must have been very small, and diffusion must have been the dominant means of gas exchange.

At diffusive equilibrium, the partial pressures of all gases would be identical inside and outside a leaf with very small pores. There would also be no gradient in the total gas pressure (the sum of partial pressures). A leaf, however, is not in equilibrium with the ambient atmosphere, *since its $H_2O$ vapor pressure almost always exceeds ambient*. There would be net diffusion of $H_2O$ from the leaf. If the pores in the leaf were small enough to prohibit mass flow, the elevated partial pressure of $H_2O$ inside the leaf would increase the total gas pressure in the leaf without influencing the partial pressures of $N_2$ and $O_2$. The total pressure in the leaf would remain elevated as long as water was available inside the leaf to keep its vapor pressure higher than ambient. In daylight, pressure inside the leaf increased as leaves were warmed by the sun. Warming caused more evaporation inside and thus an elevated $H_2O$ vapor pressure. This phenomenon is known in physics as hygrometric pressurization.

As I described earlier, the pressurized gases in young leaves force a mass flow down the petiole to the rhizome. When flow occurs, it tends to lower the total pressure in the lacunae of pressurized leaves. As a result, the partial pressures of all the gases drop proportionately. The partial pressures of $N_2$ and $O_2$ drop below their respective levels in the atmosphere, setting up a gradient that causes those gases to diffuse into the leaf. This diffusion of $N_2$ and $O_2$ moves along a gradient in partial pressure (and incidentally against a gradient in total pressure). Simultaneously, water continues to evaporate into the lacunae, keeping its partial pressure near saturation. The continuous entry of air into the leaf sustains pressurization, thereby allowing it to drive a continuous mass flow.

(*continued*)

There is another relevant pressurization mechanism that has its basis in temperature differences between the lacunar gas and the atmosphere. Very briefly, physical theory predicts that gases diffuse through small pores in a partition more rapidly from the cool side (where gases are more compressed and thus concentrated) than from the warm side (where expansion causes lower concentration). The result is that the warm side tends to be slightly pressurized relative to the cool side (thermal transpiration). This mechanism, along with hygrometric pressurization, tends to elevate the pressure in warm young leaves.

Flowthrough in waterlilies depends on two other factors. First, pores in the older leaves are larger than in younger leaves, so older leaves allow mass flow to the atmosphere. Second, the lacunar system between the young and old leaves is continuous, so pressurized gas in young leaves moves freely by mass flow through the lacunae to the older leaves, where it escapes to the atmosphere.

This system is a flowthrough ventilation system. As much as 22 liters of air a day enter a single floating leaf and flow to the rhizome. This significantly accelerates $O_2$ transport to the roots over what would occur by diffusion alone. The system has the added advantage of also transporting $CO_2$ from the rhizome to the older leaves for use in photosynthesis.

Questions remain about the system: For example, are the flow-limiting pores in the stomatal apertures, or are they in the palisade tissue beneath the stomates? Do similar mechanisms operate in other plants?

It is an essential characteristic of nature to be economical. Waterlily ventilation is a good example of simple design, using heat and the physics of gas behavior to operate a "biological steam engine."

**Figure 5-1** The Dyerville Giant (*Sequoia sempervirens*) located in Humboldt Redwood State Park, California. The tree is so tall and surrounded by so many other trees that two photographs had to be combined to show it. (Photographs by Frank B. Salisbury.)

# 5

# The Ascent of Sap

According to the *Guinness Book of World Records* (McFarlan et al., 1990), "the tallest standing tree in the world is possibly the 'Harry Cole' tree, Humboldt County, California." In July 1988, the tree, a coast redwood (*Sequoia sempervirens*), was 113.1 m (371 ft) tall. For a while, many people in northern California believed that the Dyerville Giant (Fig. 5-1, a "champion tree") in Humboldt Redwood State Park, another coast redwood, was the world's tallest tree, but recent preliminary measurements set its height at 110.4 m (362 ft). Taken as a group, the coast redwoods are the tallest known species of tree. But the *Guinness Book* also says that the tallest tree ever measured was the Ferguson Tree, an Australian eucalyptus (*Eucalyptus regnans*, called mountain ash in Australia) at Watts River, Victoria, Australia. It measured 132.6 m (435 ft) tall and nearly 6 m (18 ft) in diameter at 1.5 m above ground level. It almost certainly measured 143 to 146 m originally. A *Eucalyptus amygdalina* at Mount Baw Baw, Victoria, Australia is believed to have measured 143 m (470 ft) in 1885, and several Douglas fir[1] (*Pseudotsuga menziesii*) trees have reached great heights (for example, the mineral Douglas fir in Washington State, measured in 1905 at 119.8 m or 393 ft).

In any case, water must have moved in some of the tallest trees, from the roots to the uppermost leaves, a vertical distance of well over 120 m. What, asks the plant physiologist, is the mechanism of this movement?

## 5.1 The Problem

Although we tend to take tall trees for granted, the more we cogitate on how water moves uphill at rapid rates to their tops, the more we appreciate the challenge

[1]We have a letter from J. S. Matthews, sent to us by John Worrall of the University of British Columbia, documenting that the big tree of Lynn Valley, British Columbia, said in the *Guinness Book* to have been 126.5 m (415 ft) tall, was a hoax perpetrated by George Cary of the Hoo Hoo Club of lumbermen.

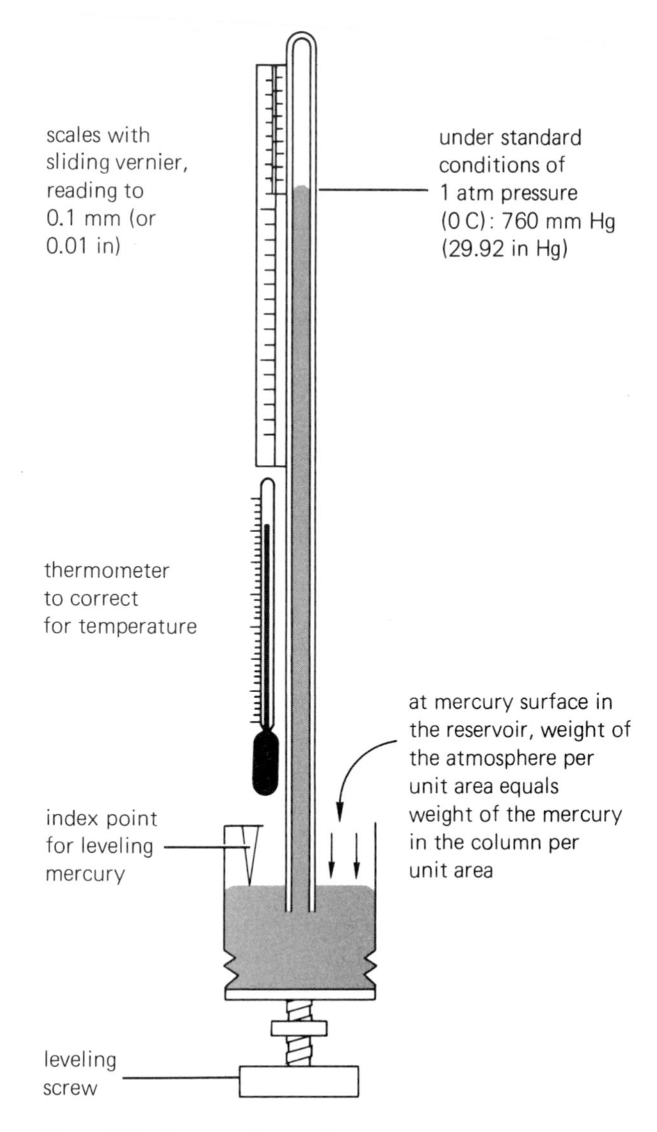


**Figure 5-2** A mercury barometer.

of the problem. A suction pump can lift water only to the **barometric height**, which is the height that is supported by atmospheric pressure from below (about 10.3 m or 34 ft at one **atmosphere**, the normal air pressure at sea level). If a long pipe sealed at one end is filled with water and then placed in an upright position with the open end down and in water, atmospheric pressure will support the water column to about 10.3 m. At this height the pressure equals the vapor pressure of water at its temperature (17.5 mm of Hg or 2.3 kPa = 0.0023 MPa at 20°C); above this height the water turns to vapor. At zero pressure, water will normally **vacuum boil** even at 0°C. (Actually, it boils at 0.61 kPa, its vapor pressure at 0°C.) When the pressure is reduced in a column of water so that vapor forms or air bubbles appear (the air coming out of solution), the column is said to **cavitate** or to undergo **cavitation**. A laboratory barometer (Fig. 5-2) is like our water-filled pipe but contains mercury instead of water. One atmosphere of pressure supports a column of mercury 760 mm high; 0.1 MPa or 1.0 bar supports 10.2 m of water or 750 mm of mercury.

To raise water from ground level to the top of the Harry Cole tree, a pressure at the base of 10.9 atm (1.11 MPa) would be required, plus additional pressure to overcome resistance in the water's pathway and to maintain a flow. If overcoming the resistance requires a pressure about equal to that required to raise the water, a total of about 2.2 MPa (1.1 plus 1.1) is necessary. To raise water to the top of the tallest tree that ever lived (say 150 m), a total of about 3.0 MPa (1.5 plus 1.5) might be required. Clearly, water is not pushed to the tops of tall trees by atmospheric pressure (about 0.1 MPa).

**Root pressures** have been observed in several species. If the stem is cut from a grapevine, for example, and a tube with a mercury manometer is attached, it can be seen that water is indeed sometimes forced from the roots under considerable pressure. Pressures of about 0.5 to 0.6 MPa have been recorded, although in most species values do not exceed 0.1 MPa. Root pressures appear in most plants, but only when ample moisture is present in the soil and when humidities are high; that is, when transpiration is exceptionally low. It is possible to see droplets of water exuded from openings (**hydathodes**), in the tips or edges of grass or strawberry leaves, for example, a phenomenon called **guttation**. When plants are exposed to relatively dry atmospheres or low soil moistures or both, root pressures don't occur because water in their stems is under tension rather than pressure. Root pressures are not found in conifer trees (including the sequoia and the Douglas fir) under any conditions, although slight pressures have been observed in excised conifer roots. Furthermore, rates of movement by root pressure are too slow to account for the total water movement in trees. So we must reject root pressure as the means of moving water to the tops of tall trees, although it does operate in some plants sometimes.

How about capillarity? Most people who are unacquainted with the problem think this is the mechanism that pulls water to the tops of trees. **Capillarity** is the interaction between contacting surfaces of a liquid and a solid that distorts the liquid surface from a planar shape. It causes the rise of liquids in small tubes. It occurs because the liquid wets the side of the tube (by adhesion) and is pulled up, which is evident from the curved **meniscus** at the top of the liquid column. As Figure 5-3 shows, it is simple to calculate that liquids rise higher in tubes of smaller diameter. By the same token, it is easy to calculate that water will rise less than half a meter by capillarity in the xylem elements of plant stems, falling short by a factor of over 300 of accounting for the rise of sap in tall trees! Furthermore, a little re-

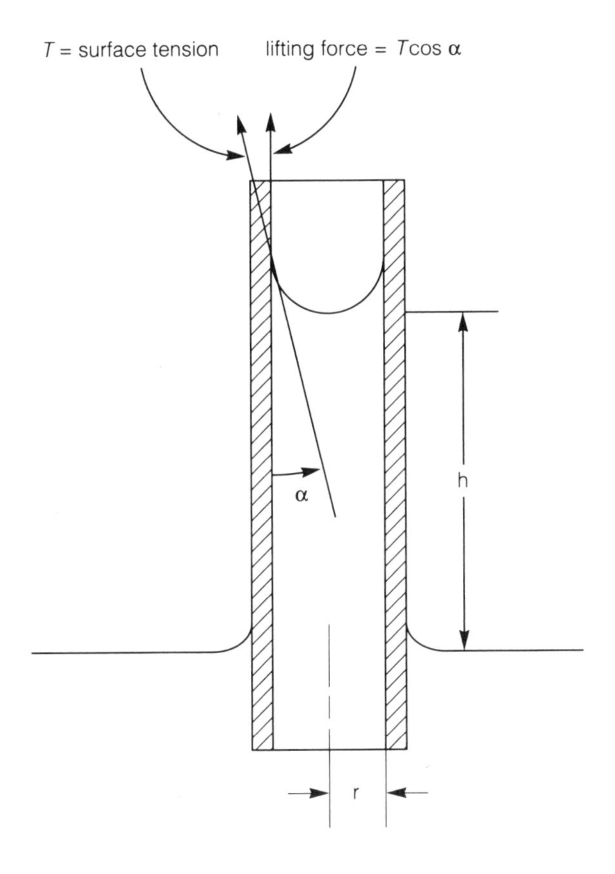


Examples:

| tube radius ($r$) micrometers ($\mu$m) | height ($h$) meters (m) | |
|---|---|---|
| 1.0 | 14.87 | |
| 10 | 1.487 | |
| 100 | 0.1487 | |
| 1,000 | 0.01487 | |
| 40 | 0.3719 | typical tracheid |
| 0.005 | 2975 | approximate pore size in cell walls (height = nearly 3 km) |

total lifting force = $T \cos \alpha \, 2 r \pi$

weight of liquid = $\pi r^2 h d g$

hence,

$T \cos \alpha \, 2r\pi = \pi r^2 h d g$

see drawing for $T$, $\alpha$, $h$, and $r$

for water on glass or a surface with polar groups:

$\alpha = 0$, $\cos \alpha = 1.0$

$\pi = 3.1416$

$d$ = density of liquid
998.2 kg $m^{-3}$ ($H_2O$ at 20°C)

$T = 0.072$ kg $s^{-2}$ ($H_2O$ at 20°C)

$g$ = acceleration due to gravity
(9.806 m $s^{-2}$ at 45° latitude)

solve equation for $h$:

$$h = \frac{T \cos \alpha \, 2r\pi}{\pi r^2 d g} = \frac{2 T \cos \alpha}{r d g}$$

substitute above values for water in glass or xylem elements; give radius in $\mu$m, height is then in m:

$$h = \frac{14.87}{r}$$

**Figure 5-3** The principle of capillarity and the mathematics used to predict the heights a liquid can be expected to reach in a tube of capillary dimensions. The last two examples in the small table show that water would rise only about one-third of a meter in wood with typical-sized tracheids, but that the pores in cell walls are so small that they could, theoretically, support a column of water almost 3 km high. (See Nobel, 1983, for more details.)

flection shows that capillarity cannot raise water in plants at all. Water rises in a small capillary tube because of the open meniscus at the top of the water's surface, but the xylem cells in plants are filled with water; they do not have open menisci. Submicroscopic menisci do exist in the cell walls of plant leaves and other tissues; they are the *holding points* for xylem water, but they are not the source of movement.

During the 19th century it was suggested that water moved up tree trunks in response to some living function or pumping action of stem cells. We can lift water to any height by pumping it over successive intervals, each one less than the barometric height. But careful anatomical study has failed to reveal any pumping cells. Indeed, most water moves in the dead xylem elements, a fact that has been clearly demonstrated by the use of radioactive (tritiated) water as a tracer. Furthermore, in a paper published in 1893, Eduard A. Strasburger (a pioneer investigator of mitosis and meiosis in plants) told how he sawed off trees 20 m tall but left them suspended upright in buckets of copper sulfate, picric acid, or other poisons. The fluid ascended all the way to the leaves, killing the bark as well as the scattered living cells (rays) in the wood. Water continued to move up through the trunk until the leaves died and transpiration stopped. He also scalded long sections of a wisteria vine, but sap continued to rise above 10 m.

The importance of living cells for sap flow in the wood cannot be overemphasized, however. The dead xylem in a plant was built by living cells, and new wood is laid down each year by the living, water-filled cambial cells. Yet the idea that water is pumped up through the trunk by cells along the way must be rejected. So how does water "flow uphill" to the tops of tall trees?

## 5.2 The Cohesion Mechanism of the Ascent of Sap

Near the end of the 19th century a model was formulated to account for the rise of sap in tall trees. One of its elements, the cohesion of water, was not familiar from everyday experience, so the model was quite controversial. As a good hypothesis should, however, it suggested several consequences by which it could be tested. Now, after a century, the numerous data that have accumulated support the model. Most difficulties and criticisms have been laid to rest, but we must still accept a visualization of reality that is not a familiar part of everyday experience—although it is entirely consistent with physics (see reviews by Pickard, 1981; Zimmermann, 1983).

There are three basic elements of the **cohesion theory** for the ascent of sap: the *driving force, hydration* (adhesion), and the *cohesion of water*. The driving force is the gradient in decreasing (more negative) water potentials from the soil through the plant to the atmosphere. Water moves in the pathway from the soil, through the epidermis, cortex, and endodermis, into the vascular tissues of the root, up through the xylem elements in the wood, into the leaves, and finally, by transpiration, through the stomates into the atmosphere. It is the special structure of this pathway (the relatively small diameters and thick walls that prevent collapse of the tubes), the low osmotic potentials of living leaf and stem cells, and the hydration properties of cell walls, especially in leaves, that make the system function. The hydration force between water molecules and the cell walls is caused by hydrogen bonding and is called **adhesion**, which is an attractive force between unlike-molecules.

**Cohesion**, an attraction between like-molecules, is the key. This is the attractive force (also caused by hydrogen bonding; see Chapter 2) between the water molecules in the pathway. In that special environment, these cohesive forces are so great (water has such high tensile strength) that when water is pulled, by osmosis and by evaporation from its holding points in the cell walls at the top of a tall tree, the pull extends all the way down through the trunk and the roots into the soil. Whereas a water column in a vertical pipe of macrodimensions would normally cavitate as just described, cavitation does not stop the flow of sap in the plant because of its highly specialized anatomy.

With this brief preview in mind, we next examine each of the key points, beginning with the pathway and continuing with the nature of the driving force and the role of cohesion. (A book on the topic, highly readable and with many original ideas, was authored by the late Martin H. Zimmermann of the Harvard Forest, Petersham, Massachusetts in 1983.)

## 5.3 The Anatomy of the Pathway

In the preceding chapter on transpiration, we considered the anatomy of the leaf (Figs. 4-4 and 4-5). Important features include the cuticle, the stomatal apparatus, and the considerable intercellular space with exposed moist cell surfaces (the mesophyll cells). Figure 5-4 shows the relationship between leaf cells and nearby

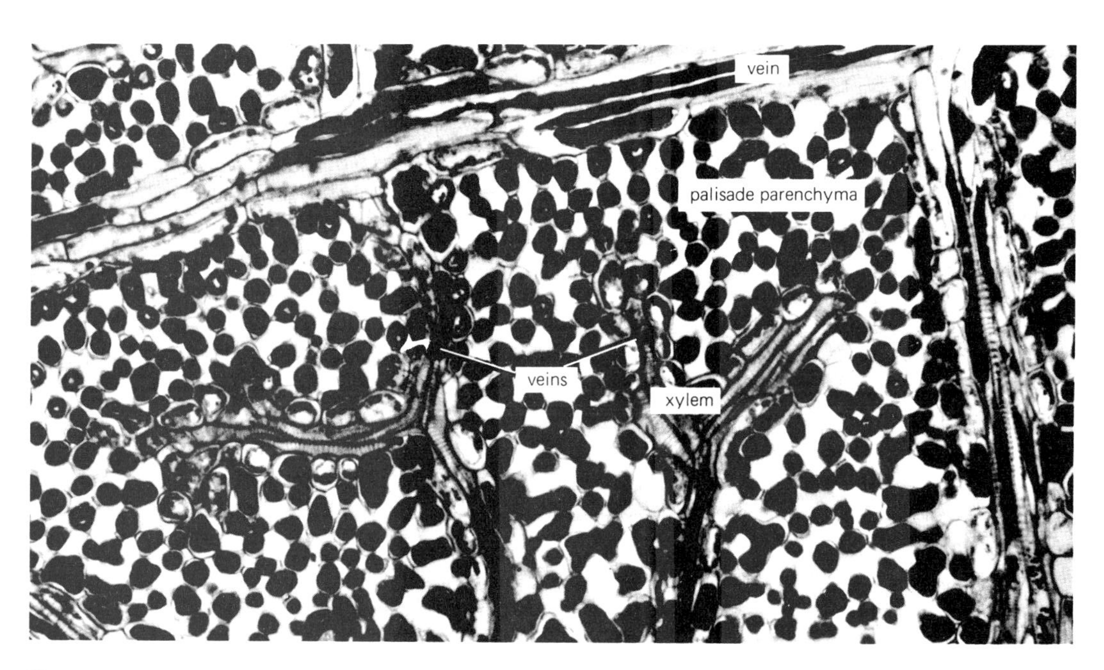


**Figure 5-4** A section cut parallel to the surface of a dicot leaf, showing the close proximity of vascular tissues to the mesophyll cells. The dark bodies are chloroplasts. (Photograph by William A. Jensen.)

vascular elements. It is obvious that no leaf cell is far from a vascular element. The following discussion and figures provide a review of stem and root anatomy. More details are in elementary botany or plant anatomy books (for example, Mauseth, 1988).

## The Vascular Tissues: Stem Anatomy

In this chapter we are concerned primarily with movement of water and dilute solutes through the xylem of stems and roots. Study the tissues called xylem, phloem, and cambium in cross sections of stems and in the three-dimensional drawing of a woody stem in Figure 5-5. In herbaceous stems, **vascular bundles** with their xylem and phloem are "open" in dicots and often "closed" in monocots. Open vascular bundles are "open to growth" because they contain a layer of cambial cells that can produce secondary xylem and phloem; closed vascular bundles lack this cambial layer and are also "closed" in the sense that they are often surrounded by

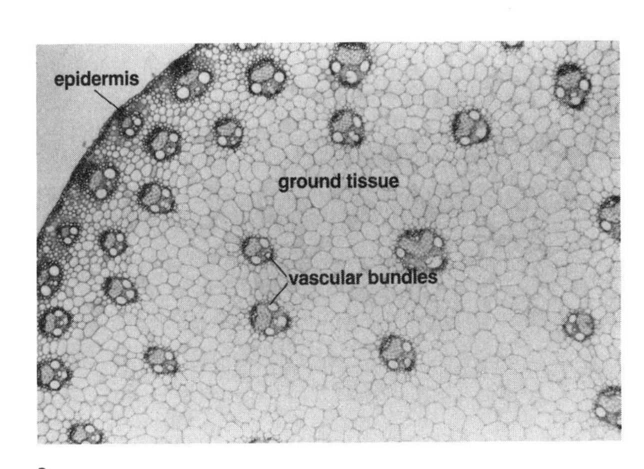


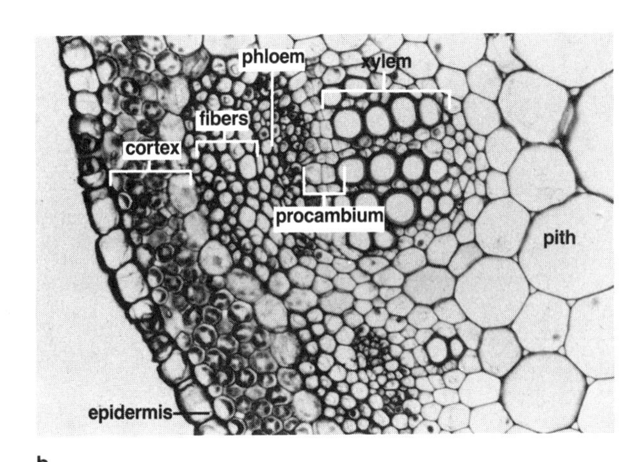


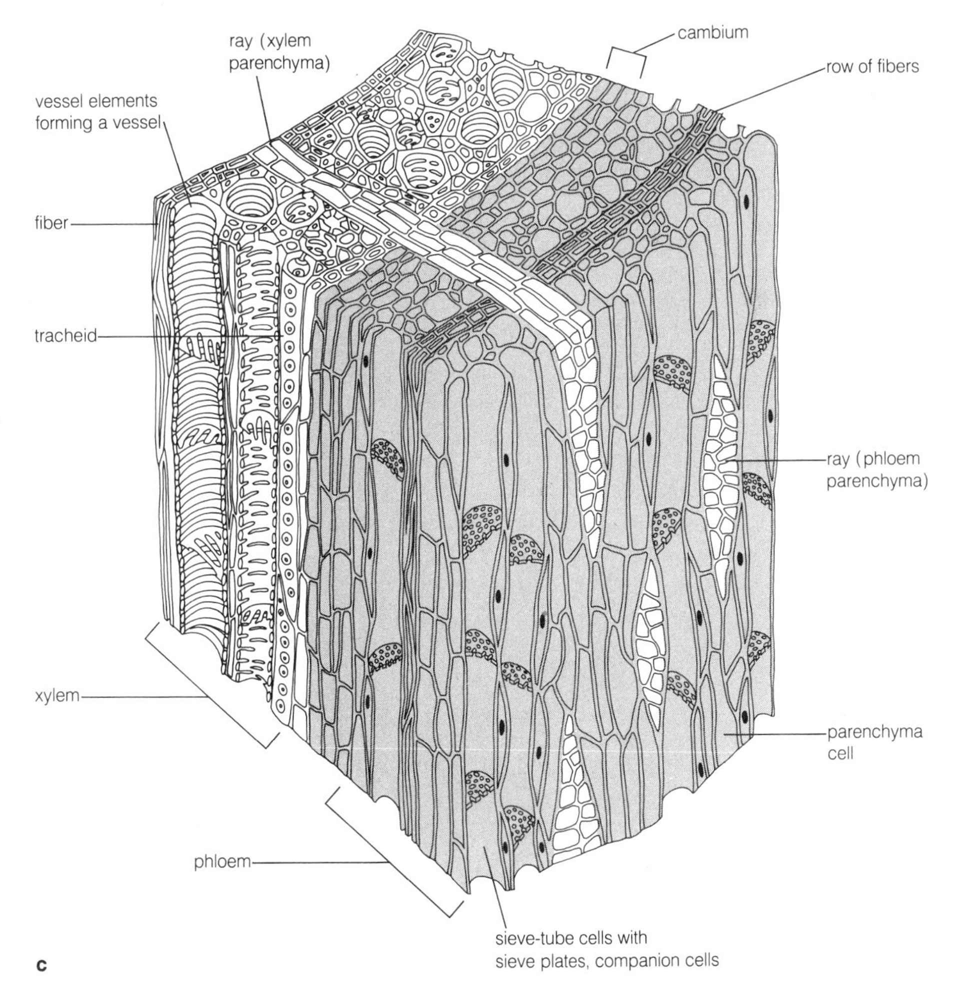


**Figure 5-5** (**a**) A cross section through a typical monocot stem. Note the vascular bundles scattered in a ground tissue of pith. Each is surrounded with a sheath of cells. (**b**) A cross section through a typical herbaceous dicot stem. The vascular bundles form a ring with pith to the inside and cortex (typically with **collenchyma** cells—thickened corners in cross section) to the outside below the epidermis. In both monocot and dicot stems, xylem is usually (but not always) to the inside of the phloem. (**c**) A three-dimensional drawing of a woody dicot stem showing xylem (wood) to the inside of a layer of cambium, and phloem (part of bark) to the outside. (Micrographs courtesy of William Jensen.)

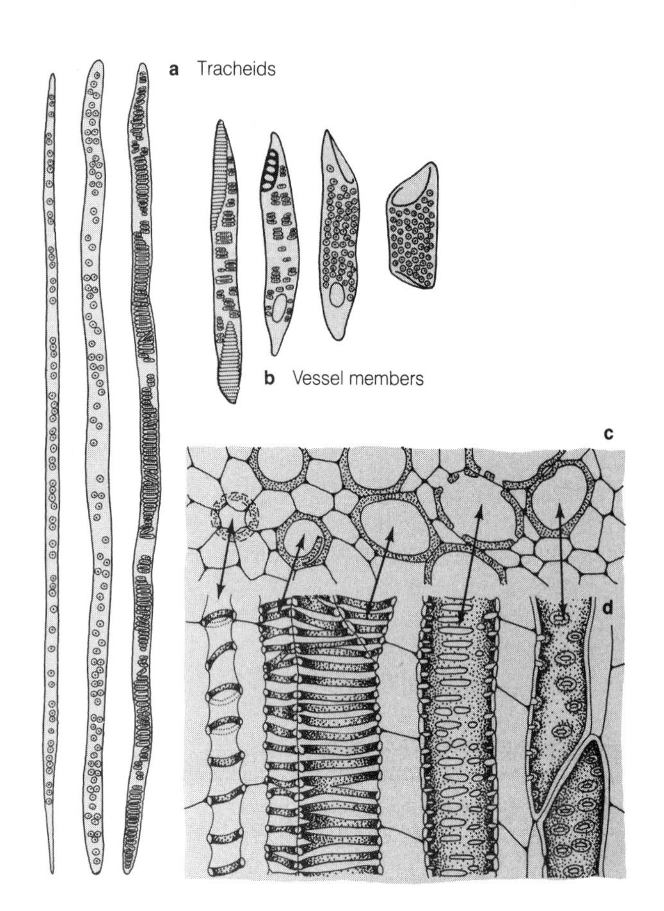


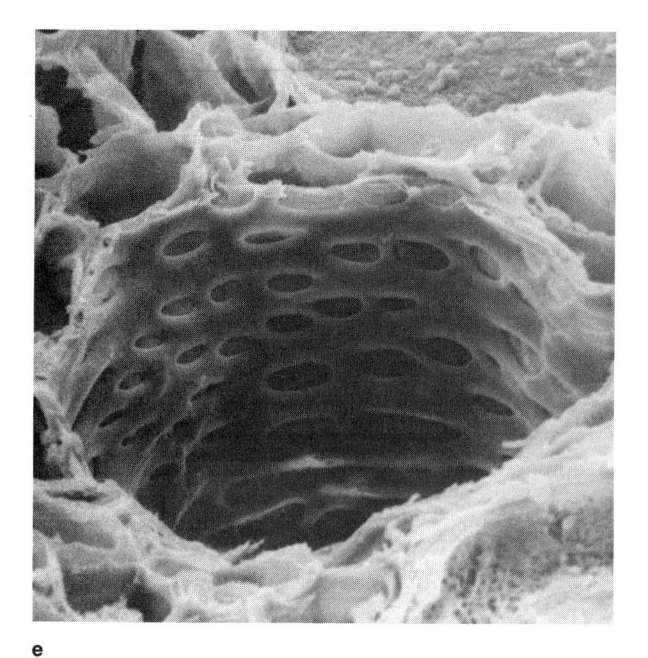


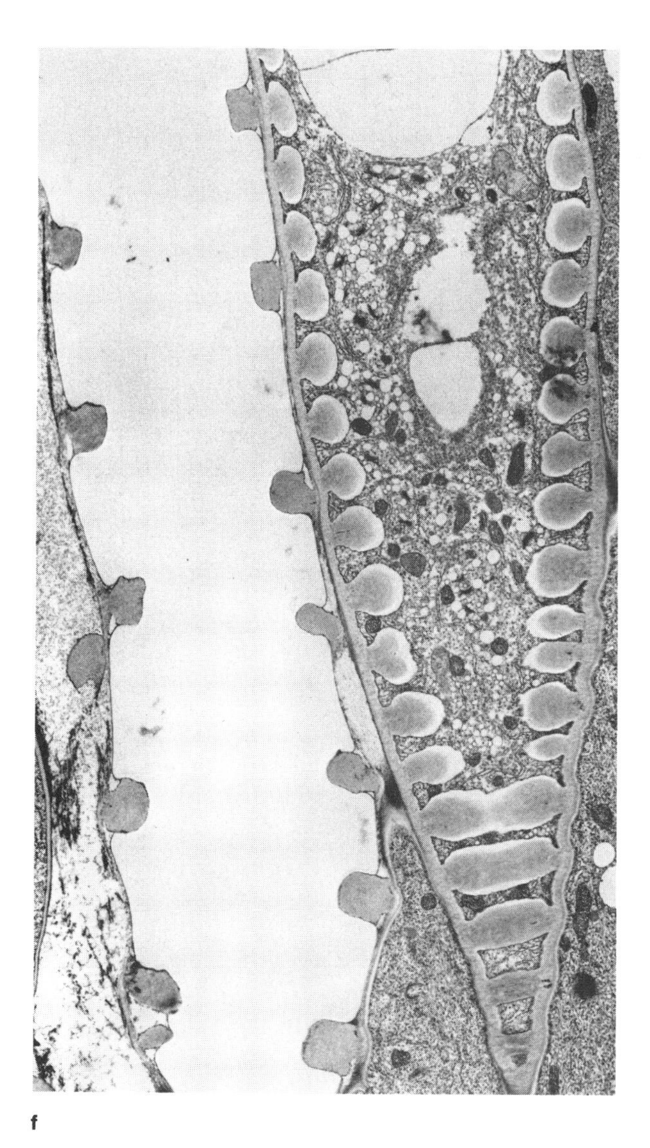


**Figure 5-6** (**a**) and (**b**) Tracheids and vessel members shown isolated from the tissue. Note that the vessels are shorter and thicker than tracheids. Vessel members are shown in cross section (**c**) and in longitudinal section (**d**). (**e**) Tracheid in pine as seen with a scanning electron microscope. Note in all parts of the figure the pits in tracheids and vessels and the wall sculpturing in some of the vessels. In addition to the vessels on the left, (**d**) shows two tracheids with overlapping, tapered ends (right side). (**f**) Longitudinal section through part of two tracheids. The tracheid at the left is mature, empty (dead), and functional; the one at the right still has a protoplast. Both have lignified cell-wall thickenings that are arranged in a helix (but shown here in cross section). The tracheid on the right is sectioned obliquely. 5,000 ×. (**a–d**, from Esau, 1967, pp. 228, 230, and 232. Scanning electron micrograph, **e**, courtesy of John Troughton; see Troughton and Donaldson, 1972. Transmission electron micrograph, **f**, courtesy of Fred Sack.)

a **bundle sheath** of thick-walled fiber cells. Monocot bundles are typically scattered almost randomly in pith (as in maize), but sometimes (in hollow wheat and other grass stems) they form a ring almost as in typical dicot stems. In woody plants the xylem constitutes the wood, which is separated from the bark by a layer of cambium cells. The bark includes phloem and such other tissues as cortex and cork. The phloem is the site of transport for sugars and other products of assimilation, which are highly concentrated in the phloem sap. (Phloem transport is the topic of Chapter 8.) Xylem sap is a dilute solution, almost pure water. (Compositions of xylem and phloem saps are discussed in Chapter 8; see Table 8-1.)

**Xylem** consists of four kinds of cells: **tracheids, vessel elements, fibers,** and **xylem parenchyma**. Fibers and parenchyma cells also occur in phloem. In xylem, especially in woody plants, only the parenchyma cells are living. These occur most abundantly in the **rays** that run radially through the wood of the tree, but some parenchyma cells are scattered throughout the xylem. It is the vertically arranged tracheids and vessel elements that are involved in transport of xylem sap. As a rule, gymnosperms (including conifers and their relatives) have only tracheids, whereas nearly all angiosperms (flowering plants) have both vessel elements and tracheids. (*Gnetum*, traditionally considered a gymnosperm, has fully developed vessels.) Both vessel elements and tracheids are elongate cells, but tracheids are longer and narrower than vessel elements (Fig. 5-6). Both function as dead elements; that is, after they have been produced by growth and differentiation of meristematic cells, they die, and their protoplasts are absorbed by other cells. Before death, however, some changes in the walls occur that are important for water flow through them. One change is the formation of a **secondary wall**, consisting largely of cellulose, lignin, and hemicelluloses, that covers most of the **primary wall** (see discussion in Chapter 1). This wall confers considerable compression strength on the cells and prevents them from collapsing under the extreme tensions that often exist in them. The lignified secondary walls are not as permeable to water as the primary walls, but in forming they leave **pits**, which are round thin places where the cells are separated only by the primary walls. Often the pits are **simple** (a small round hole in the secondary wall); but sometimes, in both vessel elements and tracheids, they are more complex structures called **bordered pits**, in which the secondary walls extend over the center of the pit and the primary walls are swollen in the center of the pit to form a **torus** (Fig. 5-7). In electron micrographs, the primary wall around the torus appears porous. The figure shows that the torus can act as a valve, closing when pressure on one side is greater than pressure on the other.

Tracheid cells have tapered ends that overlap, as shown in Figures 5-5c, 5-6d, and 5-6f. Pits in the tapered

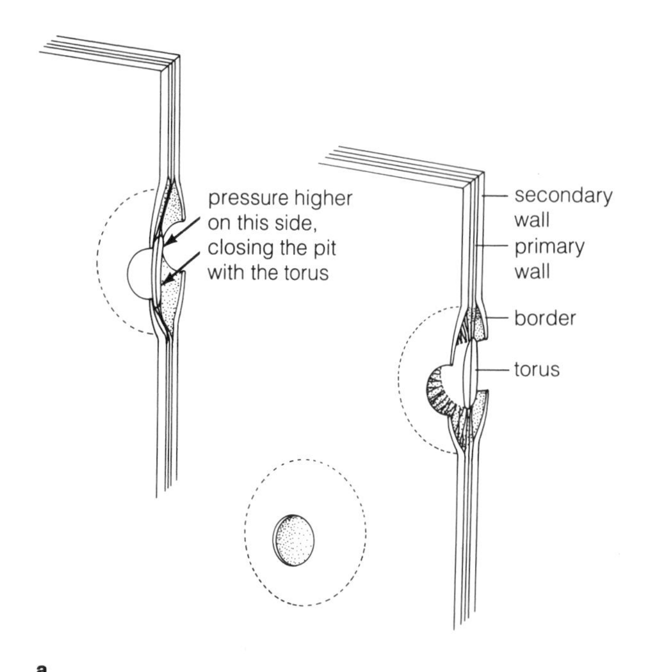


a

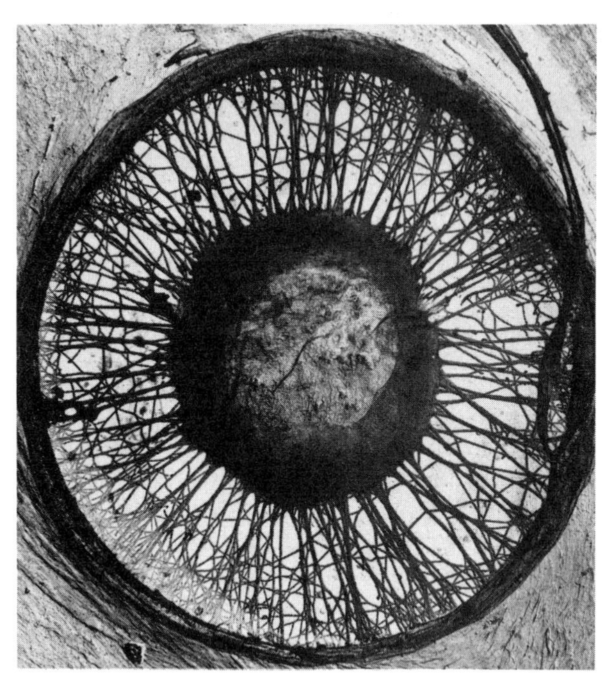

b

**Figure 5-7** (**a**) Diagram of a bordered pit from a pine tracheid. If pressure on one side exceeds pressure on the other side, the high pressure will push the torus so that it plugs the hole, shutting off flow. (**b**) Transmission electron micrograph of a pit from Douglas fir (*Pseudotsuga*). The wood sample was dried by solvent exchange with a low-boiling-point organic solvent that evaporated away in such a manner that pressures on opposite sides of the pit remained equal. The torus is the round structure in the center, and the pit membrane is seen to be highly porous and fibrous. The light circular area on the torus is the area of the pit aperture of the border behind the torus; the dark area is caused by the border, which is itself beyond the plane of focus. 7,800 ×. (Micrograph supplied by Wilfred A. Côté, Jr.; previously unpublished. See Comstock and Côté, 1968, for a description of the methods used.)

part allow water to move upward from one tracheid into the next; the tracheids thus form files of cells. Numerous pits along the sides of tracheids also allow passage of water between adjacent cells; sometimes they exhibit spiral thickenings (Fig. 5-6f) like those that resist compression in the hoses of vacuum cleaners. Vessel elements are typically strengthened by such rings, spirals, or other thickenings, and they also have **perforation plates** at their ends. These plates have openings at which the secondary wall fails to form, and the primary wall and middle lamella dissolve away. These openings allow rapid movement of water. **Vessel elements** (each a single cell) are aligned so that they form long tubes called **vessels** (many cells) that extend from a few centimeters to several meters in some tall trees. Partially because of the perforation plates, but especially because vessels have larger diameters then tracheids, resistance to water flow is usually considerably less in angiosperms than in gymnosperms.

Tracheid diameters are often in the range of 10 to 25 μm, whereas vessel elements are typically 40 to 80 μm in diameter and can be much wider (up to 500 μm = 0.5 mm, probably an upper limit). Rates of nonturbulent flow through small capillaries were studied independently in 1839 and 1840 by Gottfried H. L. Hagen in Germany and by Jean L. M. Poiseuille in France. The rate of flow is reduced by friction (adhesion) between the fluid and the sides of the capillary tube; the molecules of fluid touching the capillary wall do not move at all, and the molecules in the center of the tube move the farthest in a given time. Thus, if at time zero all the molecules in a plane cross section of the capillary could be marked, and if their positions along a longitudinal section could be noted at some later time, they would be found to form a parabolic cone (parabola in longitudinal section) with the peak in the center (Fig. 5-8). Those molecules farthest from the walls of the tube flow the fastest. Indeed, Hagen and Poiseuille worked out an empirical equation for flow rate through capillaries as a function of capillary size and found the rate to be proportional to the *fourth* power of the radius of the capillary (details in Zimmermann, 1983).

Applying the Hagen-Poiseuille equation to tracheids and vessels produces some highly significant results. Consider a tracheid 20 μm in diameter and two vessels, one 40 μm and the other 80 μm in diameter. The relative diameters of the three tubes are 1, 2, and 4, but the relative flow rates will be 1, 16, and 256 (because of the fourth-power function). This means, all other factors being equal, that 256 times as much sap would flow through the 80-μm vessel as through the 20-μm tracheid.

Of course, all other factors are not equal. Applying the Hagen-Poiseuille equation to both situations assumes that friction between the sap and the inner walls of the tracheids and vessels is similar, but this is influenced by the pitting and inner-wall sculpturing of the two kinds of transport elements and by the perforation plates of the vessels. The resistance as sap passes from one tracheid to another or from one vessel or vessel element to another also strongly influences flow rates. Tracheids have higher resistance to water flow because transfer occurs only through pits in the overlapping pointed ends, whereas transfer from one vessel to another often occurs over a considerable distance through *lateral* pitted walls as the two vessels lie in contact next to each other. (Transfer does not occur through vessel ends, which are not aligned.) Vessels are much longer, so water has to pass through pits less frequently as it moves up. In spite of the complications, measurements have shown that flow through large vessels is much more rapid, for a given pressure gradient, than it is through tracheids and small vessels. Indeed, in **ring-porous wood** with large vessels laid down during early spring, often before leaves form, virtually all the sap flow occurs through these vessels and only a small amount through the tracheids and smaller vessels.

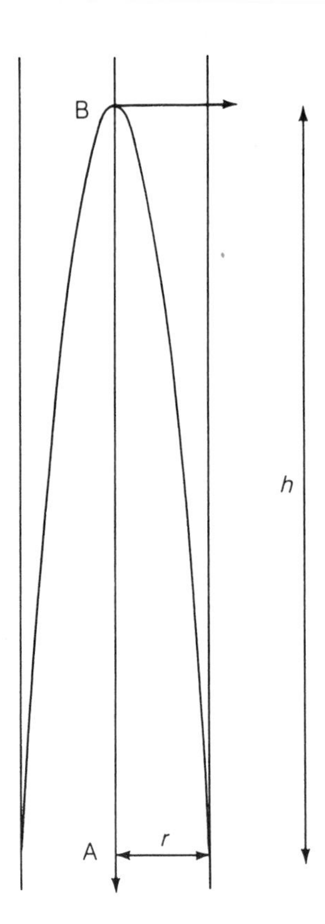


**Figure 5-8** The flow paraboloid in a longitudinal section of a capillary of radius *r*, as calculated by Hagen and Poiseuille in the mid-1800s. If, at time zero, we could label all water molecules on a cross section of the tube at A, we would find them lined up on the surface of a paraboloid at time *t*. The fastest ones, in the center of the capillary, would have covered the distance *h* and reached point B. (From Zimmermann, 1983. Used by permission.)

(*Ring* refers to the circle of "pores" — foresters' term for large vessels in cross section — on the inside of the annual ring.)

Rapid flow through large vessels comes at a price, however, and the price is safety. There is a much greater chance that cavitation will occur in a large vessel than in a small one or in a tracheid, producing an **embolism** (vapor-filled cell). Indeed, in many ring-porous trees (for example, *Castanea, Fraxinus, Quercus,* and *Ulmus*), cavitation occurs in the large vessels by the end of the year in which they are formed, so water transport is mostly dependent on vessels that are less than a year old and on the smaller tracheids and vessels that form a backup system. The wide vessels are so efficient that those of a single growth ring can supply the entire crown with water.

Some vessel elements with spiral thickenings can elongate and grow while conducting water under tension. They grow as the surrounding cells (with contents under pressure) grow and pull them along, their spiral thickenings expanding like coiled springs. Such growth occurs at the base of most grass leaves.

In discussing anatomy of the pathway, we must also mention the cells in the **apical meristems** that produce the **primary xylem**, and the **cambium**, consisting of living cells that divide to produce the **spring wood** and **summer wood** (both **secondary xylem**). These are important features because apical meristems and cambium produce tracheids and vessels filled with sap.

## Vascular Tissues: Root Anatomy

Water enters the plant through its roots (Fig. 5-9). The xylem in the root center is continuous with the xylem in the stem; it is also closely associated with phloem. Cells between the xylem and the phloem form a vascular cambium that produces xylem on the inside and phloem on the outside, leading to growth of the root in diameter.

The xylem and phloem elements are surrounded by a layer of living cells called the **pericycle**. The vascular tissues and the pericycle form a tube of conducting cells called the **stele**. Just outside the stele is a layer

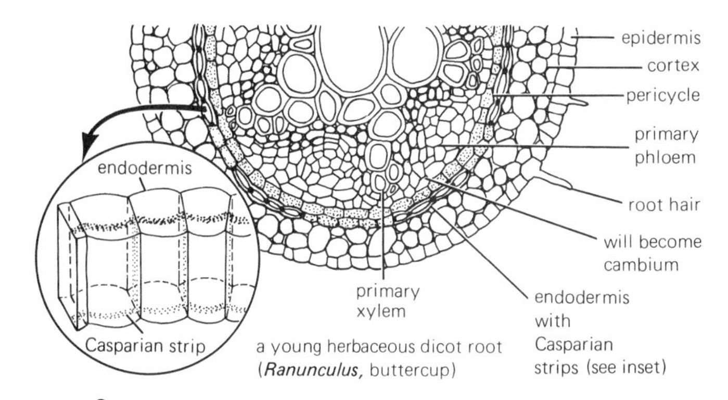


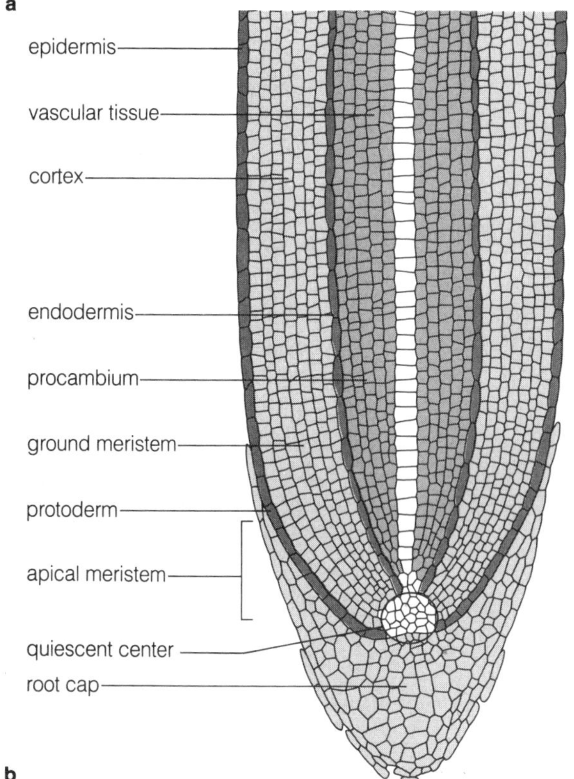


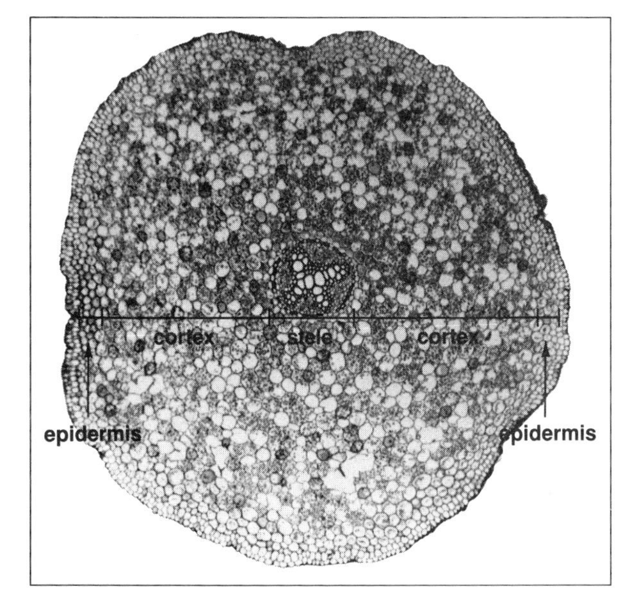


**Figure 5-9** Anatomy of young primary roots. (**a**) A young dicot root in cross section. The inset shows the endodermal layer and the position of the Casparian strips in the radial walls. An exodermal (or hypodermal) layer is not shown but would be the layer just inside the epidermis. (**b**) A longitudinal section of a typical root. The cells surrounding the quiescent center divide to form the tissues of the root. (**c**) A light micrograph of a mature buttercup (*Ranunculus*) root at low magnification to show basic tissue patterns. There can be many variations. For example, there can be more or less cortex compared with stele (compare **a** and **c**), pith in the center instead of xylem (many monocots such as maize), different arrangements of xylem, and so on. (Micrograph courtesy of William A. Jensen; see Jensen and Salisbury, 1984.)

of cells called the **endodermis**. The endodermal cells are especially interesting and important from the standpoint of water (and ion) movement in the plant because their radial and transverse cell walls include thickenings called **Casparian strips** or **bands**, which are impregnated with **suberin**, which, like lignin in secondary walls and the cutin in the cuticle, is quite impermeable to water. (Some Casparian bands also contain lignin.) The endodermal tangential walls (the inside and outside walls parallel to the surface of the root) are usually not impregnated with these substances, although the inner tangential wall sometimes does contain a thin sheet of suberin (a **suberin lamella**).

Outside the endodermis are several layers of relatively large, thin-walled living cells with intercellular air spaces outside their corners. This is the **cortex**. The air spaces form interconnected air channels that seem essential for internal aeration. A layer of somewhat flattened cells on the outside of the cortex forms an **epidermis**. Some epidermal cells develop long projections called **root hairs** (see Figs. 7-3 and 7-5) that extend out among the soil particles surrounding the root, greatly increasing soil-root contact and enhancing water absorption and the volume of soil penetrated.

Plant anatomists and physiologists have long assumed that water with its dissolved substances enters through the epidermis and then moves freely through the cortical cells, both protoplasts (symplast; see below) and cell walls (apoplast; see below) alike, but that it cannot pass through the Casparian strips around the endodermal cells. Instead, water must pass directly through the cells themselves into the symplast (through protoplasts); in the case of a suberin lamella, water can pass inward only through plasmodesmata in the inner wall. Carol A. Peterson, Mary E. Emanuel, and G. B. Humphreys (1981) used special dyes that move only through the apoplast to show that this model is essentially correct for some species. The endodermis prevented dye entry into the stele in most regions of roots, but dye could penetrate to the stele along the margins of secondary roots that had recently emerged from the epidermis of a primary root.

During recent years, the picture has become somewhat more complex by the discovery, just inside the epidermis, of a cell layer, an **exodermis**, that also has cells with Casparian strips and that typically has a suberin lamella. In a brief review of this work, Peterson (1988) outlined the use of sensitive fluorescence techniques that use a number of fabric-brightener dyes as tracers. These dyes bind to cellulose but do not pass through healthy membranes or Casparian bands. Other techniques involve staining the bands with other special dyes that stain the suberin. The suberin lamella obscured the Casparian strips in earlier work, but now it is possible to overcome this problem by pretreating the roots with alkali or by using other techniques. Thanks to these approaches, it was possible to show that only 6 percent of 213 species from 52 families (for example, legumes) had no exodermis, 3 percent had a **hypodermis** (a layer of cells just below the epidermis) with no wall modifications (festucoid grasses such as oats, barley, and wheat), and 88 percent had an exodermis (hypodermis with Casparian bands). These bands mature somewhat farther from the tip than do those in the endodermis, some 5 to 120 mm depending on species and conditions, with more stressful conditions producing bands closer to the tip. There are complications, but it currently appears that most ion uptake and probably water uptake occurs in the epidermis, even in roots without an exodermis. Movement of *ions* through the cortex may be mostly from protoplast to protoplast (symplastic), although *water* can move freely through the membranes and thus may move in the cell walls (apoplast) as well. Future research will be required to determine the significance of the exodermis in uptake of water and its solutes.

The root tips grow through the soil, encountering new regions of moisture. A **root cap** protects the dividing meristematic cells and is continually sloughed off at the forefront and replenished by divisions of these meristematic cells. Because the tissues of the stele and the endodermis are formed as the cells in the meristem divide, enlarge, elongate, and differentiate, the stele will be open at the end where it is being formed. Could water enter through that end, bypassing the endodermal layer? Studies using dyes (for example, Peterson et al., 1981) and radioactive (tritiated) water indicate that such movement is insignificant. Perhaps the cells in the meristematic region are so small and dense and their walls so thin that resistance to water movement is too high. Much water enters through the root hairs and their associated epidermal cells in the region of a young root where xylem vessels are mature and resistance is low.

## The Apoplast-Symplast Concept

In 1930, E. Münch in Germany introduced a concept and terminology that are valuable in our discussion of the pathway of water and solute movement through plants. He suggested that the interconnecting walls and the water-filled xylem elements should be considered as a single system called the **apoplast**. This is, in a sense, the "dead" part of the plant. It includes all cell walls in the root cortex, so, by that definition, endodermal and exodermal walls with Casparian strips are apoplast, but because they are impermeable to water, we usually don't think of them as being part of the apoplast. All the tracheids and vessels in the xylem are part of the apoplast, as are the cell walls in the rest of the plant, including those in leaves, phloem, and other cells in the bark. Except for the Casparian strips, the ascent of sap in a plant could take place entirely in the apoplast, particularly the xylem portion, but perhaps including cell walls of the cortex and even walls of living cells in the leaves.

The rest of the plant, the "living" part, Münch called the **symplast**. This includes the cytoplasm of all the cells in the plant, though some authors would exclude the large central vacuoles. The symplast is a unit because the protoplasts of adjoining cells are connected through the plasmodesmata (see Figs. 1-8 and 7-8).

### The Anatomical Basis of Root Pressure

Based on the apoplast-symplast concept, Alden S. Crafts and Theodore C. Broyer (1938) proposed a mechanism to account for root pressure. A slightly modified version of their model still seems reasonable. Assume that the root is in contact with a soil solution. Ions diffuse into the root via the apoplast (that is, cell walls) of the epidermis. If there is an exodermis, ions must move into the symplast of the epidermis. Otherwise, they might remain in the apoplast of the cortex until they reach the endodermis. In either case, ions pass across cell membranes from the apoplast into the symplast in an active process that requires respiration (presence of oxygen; see Chapter 7). The result is an increase in the concentration of ions inside the cells (within the symplast) to levels higher than those outside (in the apoplast). Because the symplast is continuous across the endodermal and exodermal layers, ions move freely into the pericycle and other living cells within the stele (Fig. 5-9a). This could occur by movement through the plasmodesmata, and the velocity of movement inward could be increased by **cytoplasmic streaming**, the circular flowing of the cytoplasm that is often observed within such cells.

Once inside the stele, ions are actively pumped out of the symplast and into the apoplast (unknown to Crafts and Broyer). The result is a buildup in concentration of solutes within the apoplast in the stele to a level higher than that of the soil solution (but still far below that of the cytoplasm and vacuoles), so the osmotic potential ($\Psi_s$) in the stele is more negative than $\Psi_s$ in the soil. Because water must pass through protoplasts of the endodermal and exodermal layers, these layers act as differentially permeable membranes, and the root becomes an osmotic system. The pressure buildup in this osmotic system must be the cause of root pressure. (Remember that root pressure does not occur in most plants or when plants are actively transpiring.)

## 5.4 The Driving Force: A Water-Potential Gradient

With this picture of the anatomy of the pathway in mind, we return to the mechanism of the ascent of sap. The first question concerns the driving force: Is there a water-potential gradient from the soil through the plant to the atmosphere sufficient to pull water along the pathway?

### Atmospheric Water Potential

The key to understanding is to realize the great capacity that dry air has for water vapor. As the RH of air drops below 100 percent, the air's affinity for water increases dramatically. This is shown by the rapid drop in the water potential[2] ($\Psi$) of increasingly dry air (Fig. 5-10 and Equations 3.5 and 3.6). At 100 percent RH (at all temperatures), $\Psi$ of air equals zero. At 20°C, $\Psi$ of air at 98 percent RH has dropped to $-2.72$ MPa (enough to move a column of water to a height of 277 m); at 90 percent RH, $\Psi = -14.2$ MPa; at 50 percent RH, $\Psi = -93.5$ MPa; and at 10 percent RH, $\Psi = -311$ MPa. Because soil water that is available to plants seldom has a water potential more negative than $-1.5$ MPa, air doesn't have to be very dry to establish a steep water-potential gradient from the soil, through the plant, and into the atmosphere. If the soil is fairly wet, 99 percent RH will establish the water-potential gradient.

Remember that temperature has a marked effect on relative humidity (see Fig. 3-8). Via this effect, temperature strongly influences water potential of the air, but the *direct* effect of temperature on atmospheric water potential is proportional to the kelvin temperature, and this effect is rather small, as indicated by the thin lines in Figure 5-10.

### The Role of Osmosis into Living Cells

To grow, and usually even to remain alive, living cells in plants must be able to obtain water from the apoplast. This means that the water potential in protoplasts must be slightly more negative than that in the surrounding walls and in the tracheids and vessels of the xylem. As we saw in Chapter 3 (Table 3-1), osmotic potentials of plant cells are usually on the order of $-0.5$ to $-3.0$ MPa, although osmotic potentials can be more negative than this: $-4.0$ to $-8.0$ MPa in halophytes. Pressures within living cells are usually on the order of 0.1 to 1.0 MPa, meaning that cell water potentials will range from perhaps $-0.4$ MPa to more negative than $-4.0$ MPa. As noted already, sap in the apoplast, including the

[2]In Chapter 4 we noted that the *rate* of transpiration is a linear function of the vapor-density difference between the inside of the leaf and the atmosphere beyond the boundary layer, when everything else is constant. This has often been experimentally determined (see, for example, Wiebe, 1981). Because the water potential of the air is a *logarithmic* function of relative humidity, the rate of transpiration is not a linear function of the water-potential difference between leaf and air, and it is more difficult to calculate transpiration rates based on $\Delta\Psi$. Nevertheless, it is convenient to use $\Delta\Psi$ when discussing the driving force for water movement from the soil through the plant to the atmosphere.

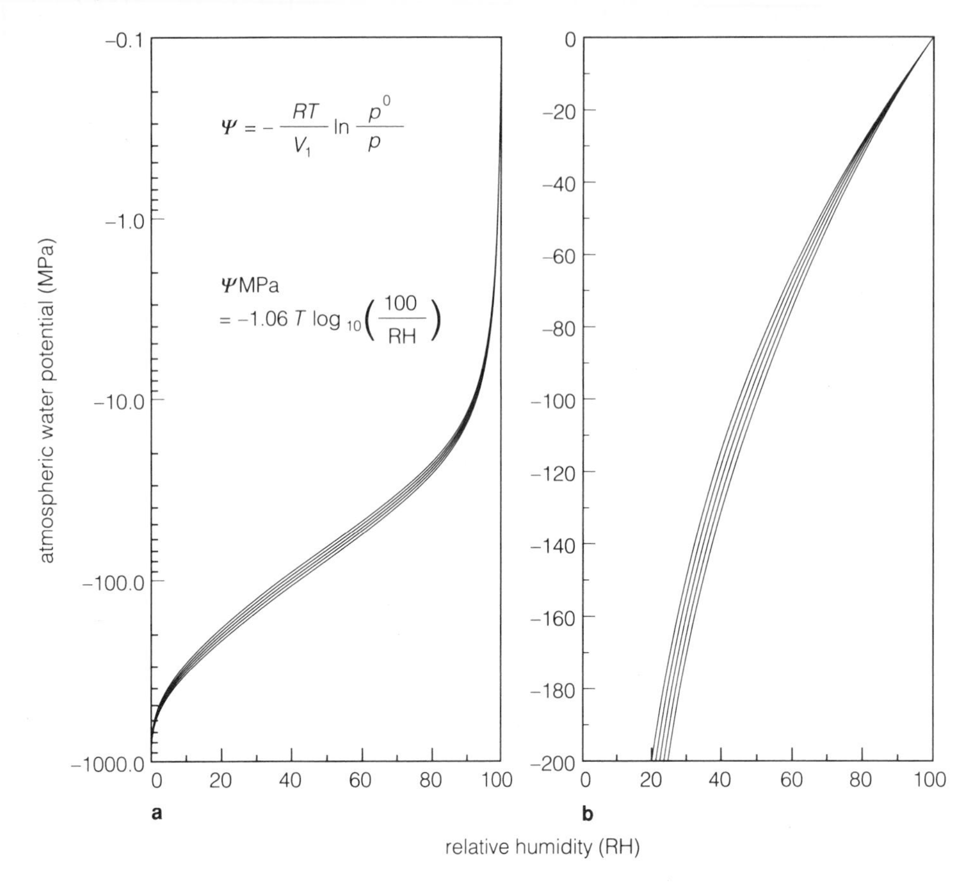


**Figure 5-10** The relationship of atmospheric water potential (20°C) to relative humidity, plotted on a logarithmic scale (**a**) and on a linear scale (**b**). The five lines are for different temperatures: 0°C (bottom line), 10°C, 20°C, 30°C, and 40°C (upper line). The curves were calculated with the equations shown, which are Equations 3.5 and 3.6 on page 55.

tracheids and vessels, is usually quite dilute, having an osmotic potential on the order of −0.1 MPa, or even less negative. But the cohesion theory says that water in the apoplast, particularly in the xylem, is under tension or negative pressure. Because we have reasoned that the osmotic potential of water in the apoplast must be less negative (higher) than that of water in the living cells, we can deduce that tensions in the apoplast will have values no more negative than −0.3 to perhaps −3.5 MPa, which is a fairly wide range.

This line of reasoning leads to another important conclusion: In thinking of how sap gets to the leaves and stems of young plants and to the tops of tall trees, we could begin by ignoring the extremely negative atmospheric water potentials that we just discussed. Even if the atmosphere were always at 100 percent RH throughout the boundary layer around the leaf, so that transpiration did not occur, sap could still be lifted to the tops of small plants and tall trees. The driving force would be the negative water potentials within living cells, these being established by the very negative osmotic potentials and the relatively low positive pressures in cells. If a plant could grow from a seedling to a large tree without ever losing any water by transpiration, water would still be entering the dividing and elongating cells in the meristems and stems and expanding cells of the leaves osmotically, and this movement into living cells would pull columns of water up through the pathway into the plant. As it is, growing cells must compete with transpiration for the water they get. And the transpiration stream brings with it ample amounts of mineral nutrients, if they are available in the soil.

## The Role of Cell-Wall Hydration

Suppose that the atmosphere is quite dry and soil water begins to be depleted. Suppose also that transpiration causes apoplastic $\Psi$ to be much more negative than $\Psi$ in living cells. Then water diffuses out of the cells, they lose pressure, and the soft tissues of the plant wilt. Will the columns of water in the xylem elements then fall back toward the soil, leaving empty tracheids and vessels? No, because in addition to the holding power of negative water potentials in living cells, there is the much greater holding power of hydration within the cell walls of the apoplast.

We can arrive at this conclusion in two ways. First, we can visualize that, as the amount of water diminishes in the cell walls or the xylem elements of leaves, curved menisci begin to form between the cell-wall polysaccharides and in the intercellular spaces. The numbers in the table in Figure 5-3 show that the radii of curvature in these microcapillary pores would be so small that the water would be held with tremendous force, a force strong enough to hold a column of water several kilometers high against gravity. Second, we have discussed matric potential (Sections 2.7 and 3.6) and the forces of hydration or attraction of water molecules to hydrophilic surfaces. Expressed in units of pressure comparable to those we have been discussing, water can be held by hydrophilic surfaces with tensions on the order of $-100$ MPa to $-300$ MPa, whether curved menisci are present or not. Clearly, gravity (the weight of water in the xylem columns) could not remove water against such powerful forces unless the water columns were as tall as mountains. Yet a dry atmosphere can remove water even against the forces of hydration: If the RH is 1 percent at 20°C, the water potential of the air is $-621$ MPa.

Actually, such extreme tensions do not develop in plants. When the water potential in a drying plant drops to some critical level, cavitation and the resulting embolisms occur in the transporting elements. In 1966, John A. Milburn and R. P. C. Johnson published a method of studying such cavitations. They attached an extremely sensitive microphone to the petiole of a detached leaf of castor bean (*Ricinus communis*). As the leaf dried, cavitation of water in a single vessel caused a distinct click, which was recorded. The clicks could be stopped by adding a drop of water to the cut end of the petiole or placing the leaf in a polyethylene bag. Placing the leaf in water for 24 h or (especially) vacuum-infiltrating it led to recovery, from which one could infer that the vessels were again filled with water. The total number of clicks for a leaf (about 3,000) was reduced by about 10 percent if the leaf was allowed to recover and then to wilt again, from which one could infer that 10 percent of the vessels were permanently vapor-locked, probably with air. Commercial applications of this method have appeared on the market; the instrument indicates to a farmer the water status of his crops. Weiser and Wallner (1988) recorded the clicks (**acoustic emissions**) in woody stems that were being frozen; they suggested that the clicks were caused by cavitations in xylem elements as air bubbles were forced out of solution in the freezing water. (See Tyree and Dixon, 1986, for references to refinements in the ultrasonic acoustical technique.)

Zimmermann (1983) has suggested that plants are constructed in such a way that cavitation occurs when water potential reaches a critical low level in a manner that allows refilling of the tracheids or vessels if conditions are right. Zimmermann calls the phenomenon *designed leakage* brought about by *air seeding*. Based on the size of the pores in the porous part of the pit membranes, it is possible to calculate, beginning with the equation of Figure 5-3, the pulling force (equivalent to the *lifting force* of Fig. 5-3) required to pull a meniscus through the pore; that is, to pull an air bubble through the pore. When the pressure in the air spaces on the outside of the pit membrane (probably atmospheric pressure) exceeds the tension inside by that amount, the meniscus will be pulled through the pore (**designed leakage**), allowing a minute quantity of air to enter (**air seeding**). The air would lead to vacuum boiling, and the air and water vapor would immediately and explosively expand to fill the tracheid or vessel (accounting for the cavitation clicks reported by Milburn and Johnson, 1966).

Note that the vessel would not be filled with air, but mostly with water vapor at its vapor pressure for the prevailing temperature. The pressure in the cell would thus be positive but quite low (about 2 to 3 percent of atmospheric pressure at room temperature). Such pressures would immediately seal all the pores, including the one through which the air seeding occurred, because a water meniscus could never be pulled through such a pore by such a small pressure difference across the cell wall. The torus in bordered pits might also contribute to the sealing. Furthermore, when conditions allowed the vapor pressure in the xylem to increase above that of the vapor pressure of water for the particular temperature (as during a rain or by root pressure at night), the cells might refill by condensation of vapor and solution of the minute amount of air. This would be analogous to the restoration of conduction by application of water in the experiments of Milburn and Johnson.

Melvin T. Tyree and John S. Sperry (1988, 1989; Sperry and Tyree, 1988) present data that strongly support this model of designed leakage with reversible embolisms. The capillarity equation of Figure 5-3 (equation for $h$) simplifies to the following form, which gives the maximum pressure difference ($\Delta P$, in MPa) from the pore diameter ($D$, in $\mu$m) and xylem-sap surface tension ($T$, in N $m^{-1}$, close to that of pure water: 0.072 N $m^{-1}$ at 25°C):

$$\Delta P = 4\,(T/D) \qquad (5.1)$$

Tyree and Sperry observed that most vessels in dehydrated stem segments of sugar maple (*Acer saccharum*) embolized when the difference in pressure inside and outside the vessels ($\Delta P$) was about 3 MPa (that is, tensions $< -3$ MPa). This same positive pressure could force air across the pit membranes of hydrated stems. Diameters of pores in pit membranes measured with the scanning electron microscope were within the range predicted by the hypothesis (≥0.4 μm; see also Carpita, 1982).

## Soil Water

Consider a soil, originally fairly dry, that is recently wet from rain or irrigation. Immediately after wetting, the soil is nearly saturated in the wetted zone, and it is said to be at **saturation**. Much of the water (called **gravitational water**) moves down through the pore spaces between the soil particles. If the concentration of ions is dilute (true of most soils), the water potential of this water is quite high, usually about −0.03 MPa. After several hours, or even a day or so, the only water left in the soil is that which can be held against the force of gravity by the adhesion between the water molecules and the soil particles. This remaining water exists in capillary pores that are often wedge-shaped and seldom cylindrical (Fig. 5-11). Soil that contains all the **capillary water** it can retain against gravity is said to be wet to **field capacity**.

As water is removed from the soil by drainage, evaporation, and plant roots, the remaining water is held in ever-thinner films, so its water potential becomes increasingly negative. When soil and root water potentials are the same, the roots no longer remove water from the soil, but water continues to transpire from the shoots, so the plant wilts. Even if transpiration is stopped by enclosing the shoot in an atmosphere with nearly 100 percent RH, it is still not able to obtain enough water to overcome wilting; the amount of soil water is then said to be at the **permanent wilting percentage**. Soil scientists rather arbitrarily consider soil with a water potential of −1.5 MPa to be at the permanent wilting percentage, although most plants, especially those of the desert and those that can grow in salty soils, can remove water when the soil water potential is more negative than −1.5 MPa (see the personal essay of Ralph Slatyer in Chapter 3). As we discussed previously, the minimum plant water potential is apparently determined by the osmotic potential of plant cells and the designed leakage in the xylem. This minimum plant $\Psi$ determines the permanent wilting point.

At such low water potentials, water moves very slowly in soils (thousands of times more slowly than at field capacity), so most plants cannot absorb water rapidly enough to keep up with transpiration. Their roots must grow to the water—but there is not enough water to support root growth! There is very little water held in the soil below the permanent wilting point.

The amount of water in the soil at field capacity and at the permanent wilting percentage depends strongly upon soil texture, soil structure, and the amount and kind of soil organic matter. The **clay** fraction of soil consists of extremely small particles in the colloidal size range and with much hydrophilic surface. These are aggregated into a complex matrix of varying pore size. Water below the field capacity is thus held largely by the attractive forces between water molecules and the surfaces of clay particles and **humus** (soil organic matter decayed to the point that plant structures are not recognizable). Thus, by hydration, water molecules can be held against forces of tens to hundreds of megapascals in both the soil and in plant apoplasts. The apoplasts might be able to compete with the dry soil for water, but if the plant is to live and grow, water must move into living cells. Their water potentials can never be as negative as the water potentials of soil that is much dryer than the permanent wilting percentage, so such soils not only cause plants to wilt but also limit their growth and ultimately lead to death of their cells.

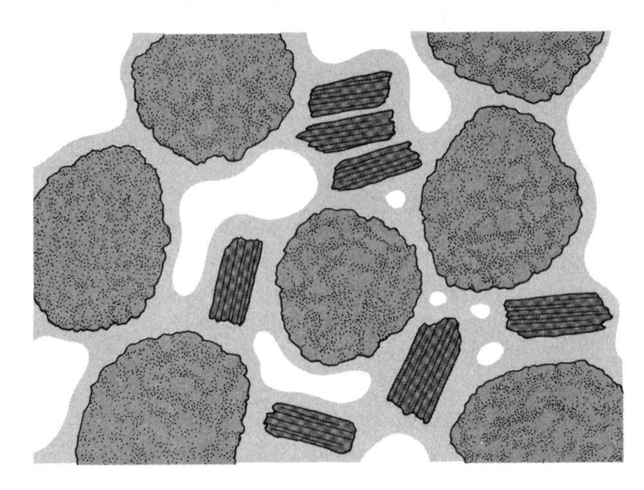

**Figure 5-11** Diagram illustrating the concept of capillary water in soil, in which the large particles represent silt or sand and the smaller particles represent clay. Water is adsorbed to the particle surfaces by hydrogen bonding, hydrating the particles. Forces of hydration extend farther from the more highly charged clay surfaces. Curved surfaces are the menisci that appear in the capillary pores of the soil; they result from surface tension in the water. (Compare with Figure 5-3.) (From Jensen and Salisbury, 1984.)

Although water in a soil that is below the permanent wilting percentage has an extremely low tendency to diffuse, it can be evaporated into a dry atmosphere at the soil surface. It can also be driven off by high temperatures—that is, by supplying a lot of energy. We measure water in a soil by weighing a sample, holding it just above the boiling point for several hours, and finally weighing it again. The difference in weight is that of the water driven off at high temperature.

**Soil texture** refers to the size of the primary particles that make up a soil. A soil with a given texture has a specific distribution of particle sizes. The various sizes are named according to two systems, as shown in Table 5-1. Practically all soils are mixtures of sand, silt, and clay. Soils with about 10 to 25 percent clay and the rest about equal parts of sand and silt are called **loams**.

**Table 5-1 Size Limits of Soil Particles as Used in Classification of Soil Texture.**

| U. S. Department of Agriculture Scheme | | International Scheme | |
|---|---|---|---|
| Name of Separate | Diameter (range, in mm) | Fraction | Diameter (range, in mm) |
| Very coarse sand | 2.0 –1.0 | I | 2.0 –0.2 |
| Coarse sand | 1.0 –0.5 | | |
| Medium sand | 0.5 –0.25 | | |
| Fine sand | 0.25–0.10 | II | 0.2 –0.02 |
| Very fine sand | 0.10–0.05 | | |
| Silt | 0.05–0.002 | III | 0.02–0.002 |
| Clay | <0.002 | IV | <0.002 |

See standard textbooks on soils.

Soil structure is as important as soil texture. **Soil structure** is the arrangement of secondary soil particles or aggregates. If it weren't for this aggregation, most soils would not have enough large pores for good water flow or root penetration. Soil structure can be changed by soil compaction, which can be a serious problem where heavy farm machinery is used.

Figure 5-12 shows soil water potential as a function of the amount of water in clay, loam, and sandy soils. The figure shows that soils with much clay can hold more water that is available to plants. Soils rich in clay and humus (or soils of intermediate texture) may hold the most water, but if the structure is poor, there could be less air space between the particles. Because plants require oxygen to support their root respiration, such soils may not be ideal for plant growth. Root penetration may be restricted in dense clay soils. Loams, sandy loams, silt loams, and clay loams, which make up many of our good agricultural soils, hold adequate water, have ample air space, and are easily penetrated by roots.

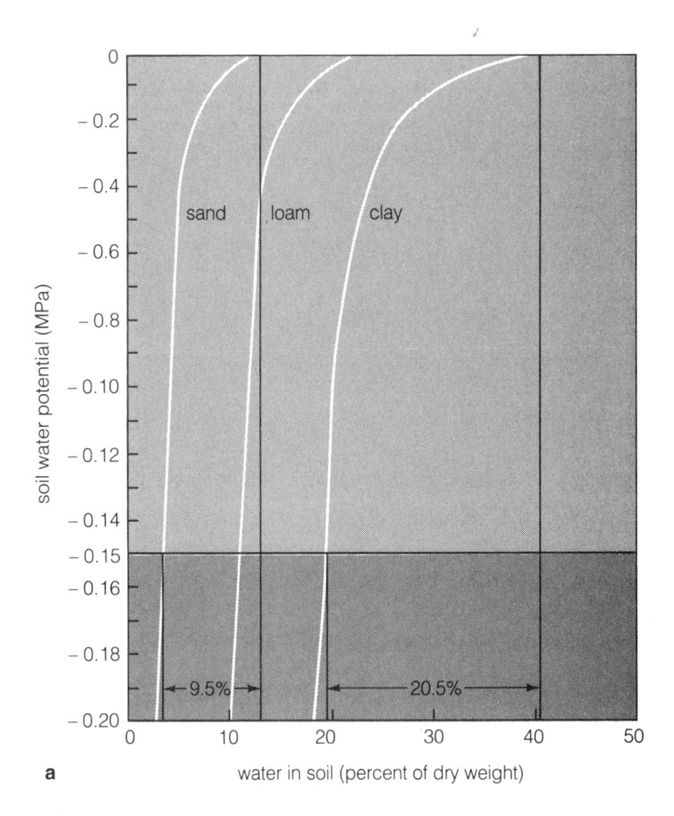


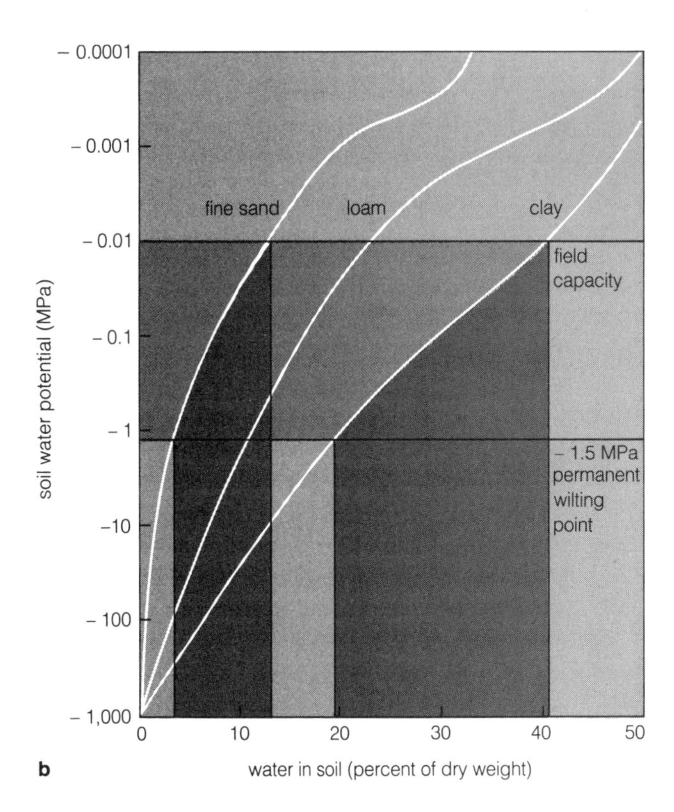


**Figure 5-12** Soil water potential as a function of the amount of water in clay, loam, and sandy soils. Soil water between field capacity ($\Psi = -0.001$ MPa) and the permanent wilting point ($\Psi = -1.5$ MPa, arbitrarily) is considered to be available to plants. (**a**) Curves plotted on a linear scale. As water potential becomes more negative (downward on the ordinate), water is held more tightly by the soil particles; that is, it becomes less available to plants. Note that the water available to plants in the fine sand is only about 9.5 percent of the dry weight of the sand, whereas available water in the clay amounts to about 20.5 percent of the clay's dry weight. (**b**) Curves plotted on a logarithmic scale. The curves represent exactly the same data as in (**a**), but the logarithmic scale allows a much wider range of water potentials to be shown. (From Jensen and Salisbury, 1984.)

## 5.5 Tension in the Xylem: Cohesion

The second question concerns cohesion: Will it hold the water columns together? An Irish botanist, Henry H. Dixon (1914), formulated the hypothesis that the tensions created by transpiration, osmotic uptake of water by living cells, and the hydration of the cell walls (all these drawing water from the pathway within the plant) were relieved by an upward movement of water from below, with the columns of water being held together by cohesion (Fig. 5-13). Dixon and John Joly (1895), a physicist colleague, began to develop the idea as early as 1894, as did E. Askenasy in 1895 (see Askenasy, 1897) and Otto R. Renner (1911). Dixon (1914, p. 103) even cited work by Berlhelot published in 1850. But because the concept was outlined by Dixon in book form with a vast body of supporting data in 1914, it is often called the **Dixon cohesion theory**.

The cohesion question breaks down into subquestions, five of which form the subheadings of the following discussion. Because these subquestions of the cohesion hypothesis suggested several experimental approaches, enough data have now accumulated to place the model on a solid footing.

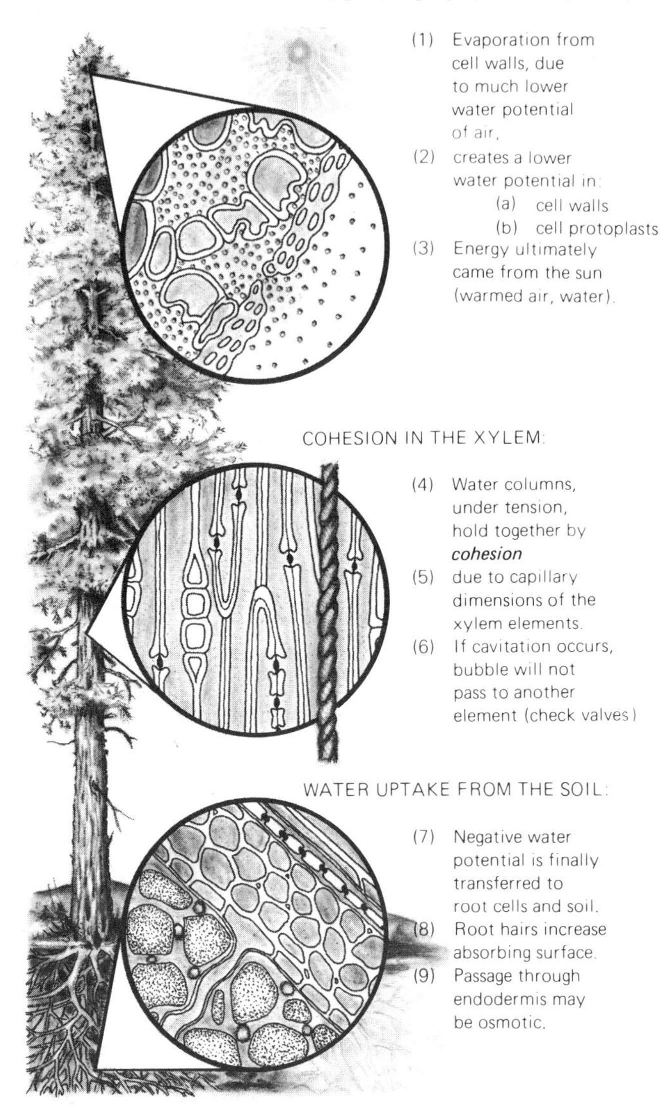


**Figure 5-13** A summary of the cohesion theory of the ascent of sap.

### Does Water Have a High Enough Tensile Strength?

The problem is whether water can sustain tensions of up to $-3.0$ MPa without cavitating. Determination of the tensile strength of water was a difficult task. Three approaches are worth discussing, and the third seems conclusive enough to provide a final answer.

First, knowledge about hydrogen bonding in water suggests that potential cohesive strength under ideal conditions is extremely high, enough to resist a tension of several hundred MPa (Apfel, 1972; Cooke and Kuntz, 1974; Oertli, 1971). But our theories of liquids remain imperfect, and some assumptions must be made.

Second, several experimental measurements suggest high tensile strengths. The force required to separate steel plates held together by a water film has been measured. Fern annuli that are pulled apart by water under tension have been studied. Glass tubes were sealed while full of hot, expanded water, and then cavitation was observed as the water cooled and contracted. These and other methods have produced values for the tensile strength of water, and even of tree sap with dissolved solutes and gases, on the order of $-10$ to $-30$ MPa, though a few workers measured only lower values, on the order of $-0.1$ to $-3.0$ MPa. These methods, designed to measure the tensile strength of water, also of necessity measure the ability of the solid surface in contact with the water to act as a seeding source for cavitation. The slightest bubble of air on the surface expands explosively to produce an embolism, or air might be pulled through pores, as in Zimmermann's designed leakage. Hence, results probably reflect interface conditions instead of tensile strengths of water. The system must work as well as it does in plants because these interface characteristics allow water tensions on the order of $-1.5$ to $-3.0$ MPa.

Third, Lyman Briggs, former director of the National Bureau of Standards, published some especially impressive experimental results in 1950. He used capillary glass tubes bent in the form of a Z (Fig. 5-14). Centrifuging these tubes causes a tension on the water at the center, and the tension present when the water column breaks can easily be calculated. Briggs found that the greatest tensions appeared in the tubes of smallest diameter. With rather fine capillaries, values as negative

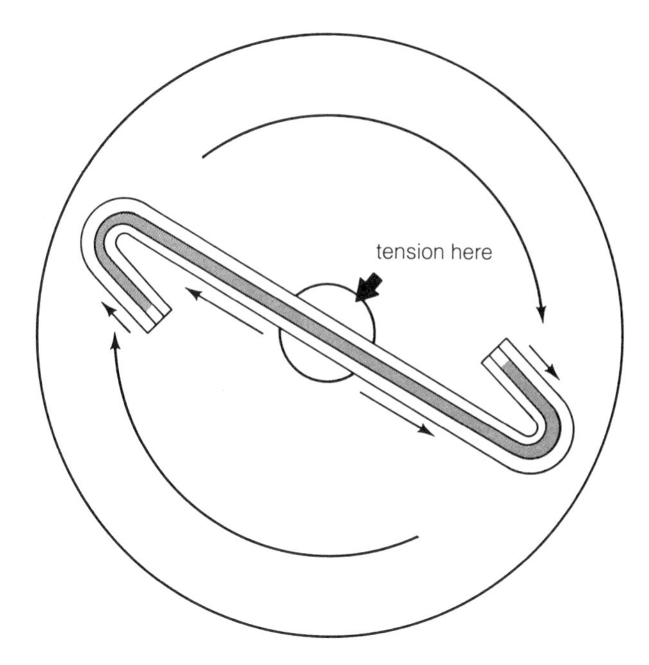


**Figure 5-14** Method of measuring the cohesive properties of water with a centrifuged Z-tube. Small arrows indicate the direction of centrifugal force and the principle of balancing. The shape of the Z-tube prevents water from flying out either end of the tube.

as −26.4 MPa were measured, but even with capillary tubes of 0.5 mm diameter (the size of the largest vessels), air-saturated tap water did not cavitate under tensions of −2.0 MPa. Cavitation did occur when the center of the Z-tube was frozen with dry ice; because air is virtually insoluble in ice, freezing expelled the dissolved air.

It is not obvious why the most negative pressures appeared in the narrowest tubes, but it could simply be that the larger tubes had the most surface and the most volume and therefore were statistically most likely to have points where cavitation could begin. More subtle reasons might involve the degree of curvature of internal tube surfaces; could flatter surfaces be more likely to harbor points of cavitation? Such possibilities need investigation, but Briggs's results support the observation that cavitation is more likely to occur in large vessels than in small tracheids.

## What Are the Flow Velocities? Are the Water Columns Really Continuous?

Results of the best and most recent experiments clearly indicate that the water columns in the xylem are continuous, and observed flow velocities in stems strongly agree. The most elegant method to measure flow velocities is the heat-pulse technique (Section 4.1), introduced by H. Rein (1928) for measurement of blood-flow velocities in animals. B. Huber and E. Schmidt (1936) made many measurements of sap velocities with the method, which consists of heating the liquid briefly at a point along the stem and then measuring (with a sensitive thermocouple) the time of arrival of the heat pulse a few centimeters downstream. There are many problems. We've seen (Fig. 5-8) that velocities depend on distance from the wall of the capillary, for example, but heat-pulse methods can at least give comparative information. As expected, velocities in wide-vessel trees such as ash are much greater than those in narrow-vessel trees such as birch. When velocities are measured at different points along the trunk, it can be seen that they increase first at higher points along the trunk in the morning as transpiration begins, strongly suggesting that transpiration pulls the sap up the trunk. Some reported peak velocities (at noon) range from 1 to 6 m $h^{-1}$ for narrow-vessel trees to 16 to 45 m $h^{-1}$ for trees with wide vessels. Slowest velocities occur in conifers.

## Are the Columns Really Under Tension?

In 1965, Per Scholander and his colleagues at the Scripps Institute of Oceanography in California published the elegantly simple and satisfactory **pressure bomb method** (see Fig. 3-12), which provided a way to measure tension in stems.[3] Scholander reasoned that if the water in the xylem of a stem is under tension, then the pressure outside compresses the cell walls of the xylem. Therefore, when the stem is cut, allowing pressure inside to equal pressure outside, the cell walls should expand, and the water columns in the xylem elements should recede from the cut surface. If the pressure difference is reestablished by increasing the pressure on the outside of the stem until the pressure difference is the same as it was before the stem was cut, the water should move back exactly to the cut. The method of testing this consists of cutting a twig from a tree or a stem with leaves from an herbaceous plant, placing it in a pressure bomb, and increasing the gas pressure on the branch until the water in the xylem can be observed with a hand lens to return to the cut surface. The pressure in the bomb should then be equivalent to the absolute value of the tension in the stem before the cut.

Scholander and his colleagues measured tensions in stems under a variety of conditions. Results for different environments are shown in Figure 5-15. Tensions were always observed, and they varied from a few tenths of a megapascal below zero to more negative than −8.0 MPa, the limit of the instrument. (The highest tensions were in a creosote bush, a desert shrub less than 2 or 3 m tall.) Measurements varied considerably

[3]It is an interesting historical footnote that Henry Dixon, who developed the cohesion theory, understood the principle of the pressure bomb and constructed glass models (see pp. 142–154 in Dixon, 1914). After two rather serious explosions, he abandoned this approach!

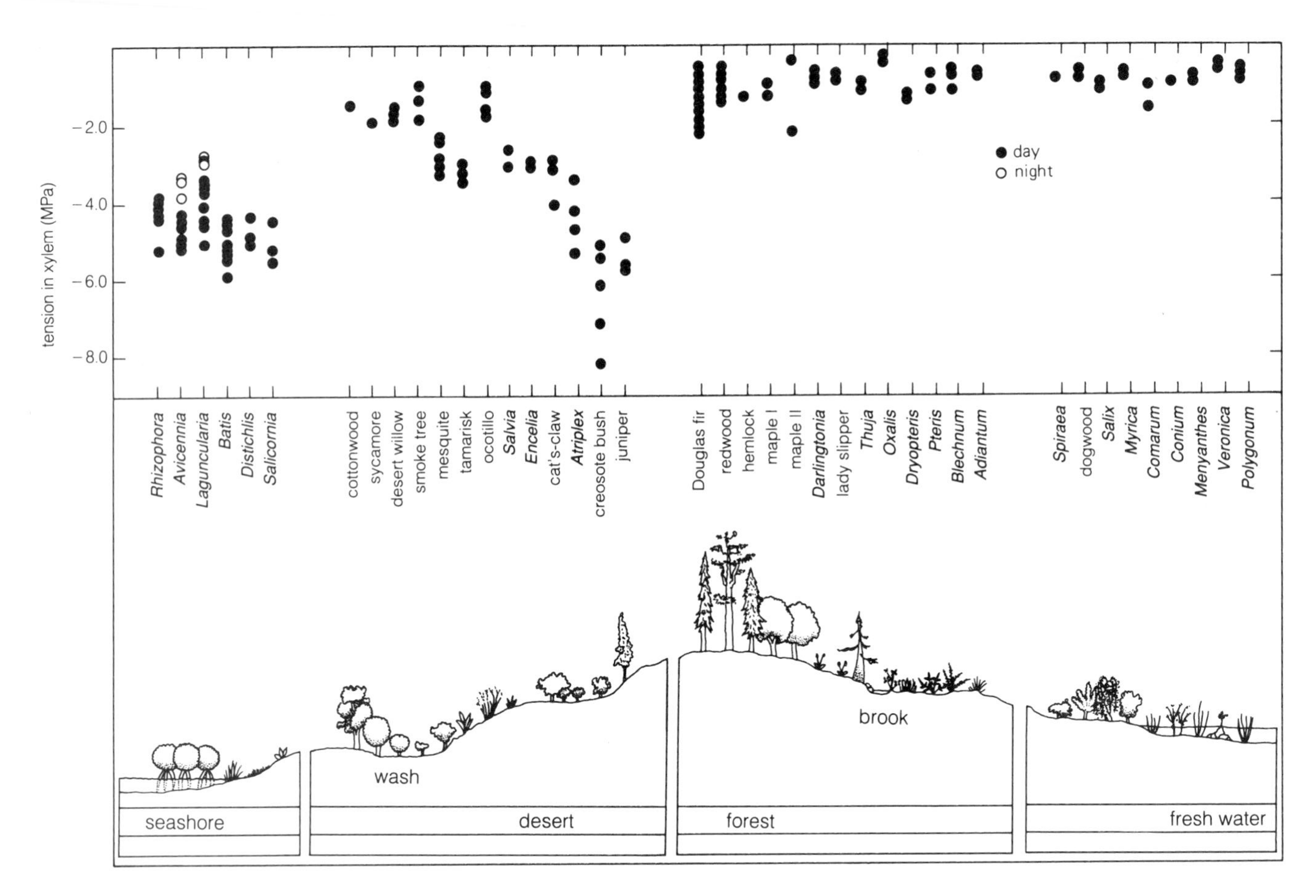


**Figure 5-15** Negative sap pressures in a variety of flowering plants, conifers, and ferns. Most measurements were taken with a pressure bomb during the daytime in strong sunlight. Night values in all cases are likely to be several tenths of a megapascal higher (less negative). (From Scholander et al., 1965; used by permission.)

for each plant, but trends are clear: Forest and freshwater species have tensions with the least negative values, and desert and seashore plants, growing in soils that are likely to be salty, have more negative values. As might be expected, values for tensions are less negative at night (less transpiration). Root pressures were never observed under these conditions, even at night.

In another interesting study (Fig. 5-16), a high-powered rifle was used to knock twigs off a tall Douglas fir tree at 27 and 79 m above the ground. Twigs were quickly placed in a pressure bomb. As might be expected, tensions varied with time of day, being most negative around noon when light levels were highest and humidities were low. The difference in height between the two samples was about 52 m. The hydrostatic pressure difference for 52 m would be 0.51 MPa, which is close to the observed value of about 0.5 MPa. If water is to flow, the gradient must be greater than that established by gravity, so it is surprising that the measured gradient was so close to the hydrostatic gradient. Zimmermann (1983) suggests that this may be because twigs had to be used, and they probably (according to many studies cited by Zimmermann) had much more negative water potentials than xylem in the trunk. Sometimes, especially in small herbaceous plants, pressure gradients are much steeper (for example, 0.08 MPa $m^{-1}$ in tobacco; Begg and Turner, 1970). Reverse gradients in a tall eucalyptus during a rain, indicating downward water flow in the xylem, were reported by Daum (1967) and by Legge (1980).

You can observe tensions in plant stems simply by immersing the stem of a transpiring plant in a dye solution and cutting through the stem. The dye instantly moves a considerable distance both up and down the stem inside the xylem elements and then stops. As with the pressure bomb experiments, the cut suddenly releases the tension on the walls of the xylem tubes, allowing them to expand and pull in dye solution. (It also increases $\Psi$ of the xylem water nearly to zero, allowing water to diffuse osmotically into adjacent cells.)

Renner (1911) performed an elegantly simple experiment (see Renner, 1915). He attached a leafy branch to

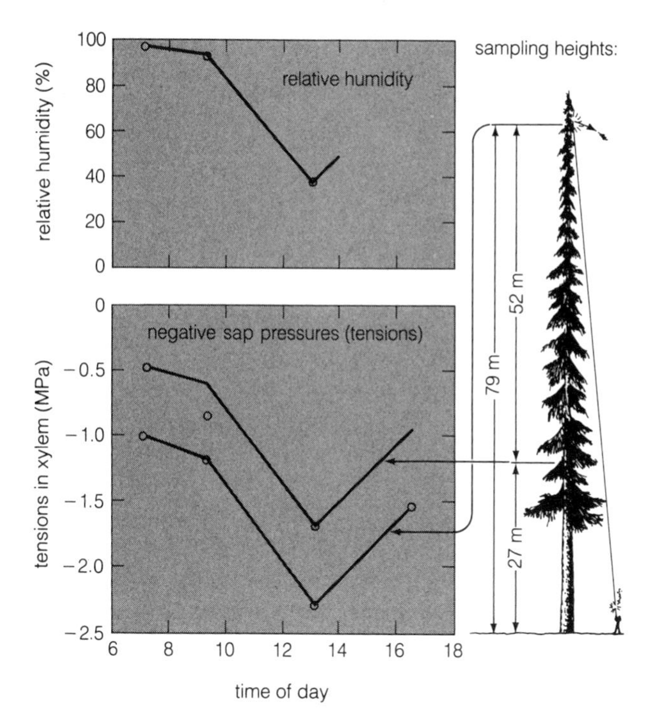



**Figure 5-16** Negative sap pressures (tensions) showing hydrostatic gradients in a tall Douglas fir tree as a function of time of day and height. Pressures were measured by shooting down twigs and placing them in a pressure bomb (see Figure 3-12). The circles represent measured values. The weight of a column of water 52 m high (the distance between the two sampling heights) produces a pressure at the base of 0.503 MPa. This is the expected hydrostatic gradient. Thus the points taken from the highest sampling level are connected (bottom plot in lower graph), and the plot above (for the lower sampling level) is drawn parallel and 0.5 MPa above (0.5 MPa less negative). Two of the three points fall almost on this calculated curve and the third point is not far away, demonstrating that the calculated hydrostatic gradient did exist in the tree. But why is no resistance component apparent? See text for discussion. Note that pressures become more negative as relative humidity drops (top graph). (Data from Scholander et al., 1965; used by permission. The data are typical of several measurements that were made. Figure from Jensen and Salisbury, 1984.)

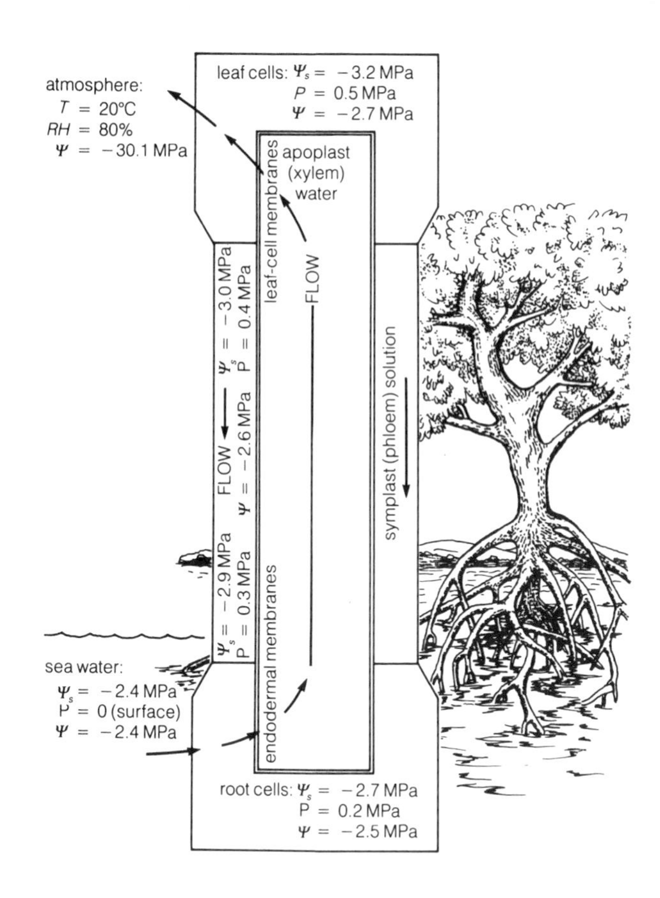



**Figure 5-17** Water relations of a mangrove tree growing with its roots immersed in sea water. The diagram indicates the "essential" parts of the mangrove tree in this context. The endodermal (and exodermal?) membranes keep all but negligible amounts of salt out of the xylem, and the leaf-cell membranes maintain a high solute concentration in the cells. The result is that water in the xylem must be under considerable tension both day and night to remain in equilibrium with sea water, and leaf cells have such a negative osmotic potential that they absorb water from the xylem in spite of its tension and low water potential. Only osmosis keeps the leaf cells from collapsing. (Data based on Scholander et al., 1965, but their hypothetical numbers have been modified to better match the discussions in this chapter and in Chapter 8.)

a burette to measure water uptake and constricted the stem with a clamp to produce a high resistance. After measuring uptake under these conditions, he cut off the leafy end and applied suction with a pump that produced a tension measuring about −0.1 MPa. Only about one-tenth as much water moved in response to the pump as in response to the leafy branch. We are compelled to conclude that the leaves exerted a pull of some −1.0 MPa.

Except for plants showing guttation water, suggesting that root pressures are producing positive pressures in the xylem, water in the xylem of land plants in summer must nearly always be under tension. This is certainly what pressure-bomb measurements have shown (as in Fig. 5-15). Scholander (1968) studied a situation in which tensions in the xylem must always, day and night and in all weather conditions, have values at least as negative as −3.0 MPa. Tropical mangrove trees grow with their roots bathed in sea water, which has an osmotic potential of about −3.0 MPa. As water enters the tree through the roots, salts are excluded, probably by the endodermal (or exodermal?) layer, so the xylem sap is almost pure water with an osmotic potential of nearly zero. (Several other halophytes also have the ability to exclude salt.) The reason that water in the xylem sap remains in the tree and does not move out through the roots into the sea water is that the sap has a water potential of −3.0 MPa or lower (Fig. 5-17). This is achieved by constant xylem tensions of −3.0 MPa or below. Scholander measured the predicted tensions with his pres-

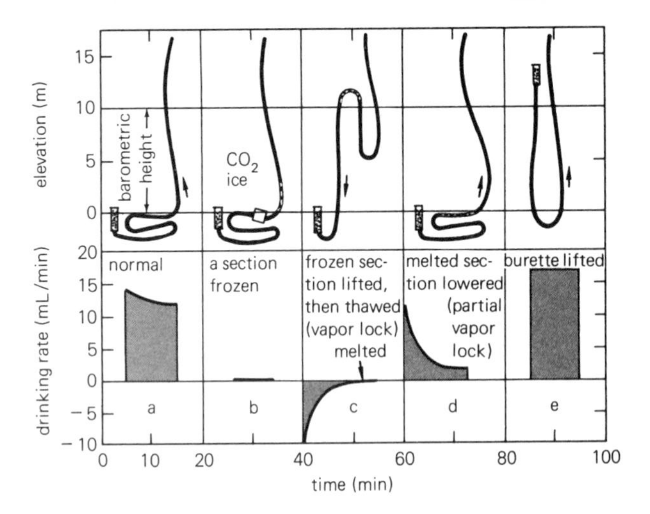


**Figure 5-18** Scholander's experiments with tropical rattan vines (*Calamus* sp.). (**a**) The vine is cut off under water and a burette is attached, allowing measurement of the rate of water uptake. If the burette is stoppered, water continues to be taken up anyway until a vacuum is created in the burette and the water boils. (**b**) To freeze the water in the vine, the burette first had to be taken off so that air entered all the xylem elements, vapor-locking the system. Then after freezing, the vapor-locked portion (about 2 m) was cut off under water and the burette attached again. There was still no water uptake, indicating that freezing had indeed blocked the system. (**c**) If the vapor-locked portion was hoisted above the barometric height and allowed to thaw, some water ran out, but there was no uptake, indicating that the system was now vapor-locked. (**d**) If the vapor-locked, thawed portion was lowered to the ground, there was a rapid initial uptake as vapor condensed to water, breaking the vapor lock, but then uptake was slower than originally because some air had been excluded by freezing. (**e**) When the burette was elevated 11 m, rate of water uptake returned to the original level, indicating that the vapor lock had now been completely eliminated. (From Scholander et al., 1961.)

sure bomb. By this ultrafiltration mechanism, the mangroves avoid a lethal buildup of salt in their leaves.

Obviously, if the xylem sap in mangroves has a water potential of −3.0 MPa, the protoplasts of leaf cells must also have water potentials at least as negative as −3.0 MPa to remain turgid. If they have a turgor pressure of about 0.5 MPa, then their osmotic potentials must be about −3.5 MPa or lower or they would lose water. Osmotic potentials of −3.5 MPa have been measured in mangrove leaf cells. Ordinarily, trees that don't have their roots in salt water have similar, if somewhat less spectacular, water relations. Tensions normally exist in their conductive xylem tissues, but their turgid leaf (and other living) cells have osmotic potentials even more negative than those of the xylem sap with its negative pressures. In any case, the importance of osmosis in the plant is clearly demonstrated. Without the differentially permeable membranes around the living cells and the highly negative osmotic potentials of the sap and cytosol inside, the high tensions in the xylem system would lead to collapse (wilting) of the living tissue. Without osmosis, plants would collapse.

## What If the Columns Cavitate?

Perhaps the most important question raised by plant physiologists about the cohesion hypothesis concerns what would happen if the continuous columns of water should in some way be broken. Say, for example, that a wind bends the stem, further stretching the columns of water and causing them to cavitate, or that the water in the tree trunk freezes, forming bubbles of gas. Or what would happen if some columns were broken by sawing part way through the trunk? Investigators have taken many approaches to solve these and other problems. Figure 5-18 illustrates other elegant experiments of Scholander and his coworkers. They nicely support the cohesion theory.

Some answers to our questions are found in the anatomy of xylem. Pathways of water have been studied by the use of dyes. A hole can be drilled in the trunk of a tree, for example, and dye inserted (summarized by Zimmermann, 1983). Although one must be careful to account for lateral movements caused by the relief of pressure when the hole is drilled, much can be learned about the patterns of vessels by following the dye above and below its insertion.

Another approach to understanding the anatomy is to make micrographs on movie film of consecutive cross sections through a stem. Upon projection, the longitudinal dimension of the stem is represented by time, and movements of individual vessels or vascular bundles (especially in monocots) can be followed. Their positions on various frames can also be measured so that individual vessels can be plotted in three-dimensional drawings. The conclusion of these and the dye studies is that vessels seldom if ever simply lie parallel to each other. When dye is inserted at a point in a trunk, for example, it spreads circumferentially around the trunk as it moves up and down, mostly within a single growth ring (except for the complications of radial transport in the rays). The spread amounts to about 1 to 2°, which equals about 17 to 35 mm per meter of trunk. Studies of individual vessels or vascular bundles show the same thing: a twisting and moving apart with progression up (or down) the stem. Usually the anatomy is highly complex. The conclusion is that water entering the xylem of any root is spread throughout the entire trunk until it reaches virtually all the branches that form the crown. Thus, if a given root is damaged, no part of the crown suffers from lack of water. (This is much less the case for small herbs.)

With this information about the twisting and spreading of vessels in the xylem, the results of saw-cut experiments seem less mysterious. At least as long ago as 1806 (Cotta, cited in Hartig, 1878 and in Zimmermann, 1983), researchers made saw-cuts halfway through a tree trunk at different levels but from opposite sides. Often, one finds that water transport is not interrupted and that the tree survives (see more recent experiments of Greenridge, 1958; MacKay and Weatherby, 1973; Preston, 1952; and Scholander et al., 1961), although the rate of water uptake decreases, and there is a much steeper tension gradient across the section of stem with the cuts.

With xylem anatomy in mind, it is not too difficult to imagine how the path of water might continue uninterrupted around such saw-cuts. In ring-porous wood with long and wide vessels, it is necessary to consider the rest of the transport system, because all the large, long vessels are interrupted by the saw-cuts. Remember that there are always much shorter vessels in such trees, and there are also the tracheids that are shorter still. Water can move circumferentially in the part of the trunk between cuts by zigzagging up and down from vessel to vessel or tracheid to tracheid.

### What about Air Excluded from Solution in the Stem by Freezing?

Microscopic observation shows that air blockage occurs when some trees are frozen, just as when water was frozen in the spinning Z-tube. Inability to restore the water columns in large vessels in the spring may be the factor that excludes certain trees and especially vines from cold climates. How do trees that grow in such regions manage? Observations have shown that the blocked pathways are replaced or restored in such trees. But how?

Several explanations have been proposed. Ring-porous trees apparently use the "throw-away" method. Those species have such large and efficient vessels that a single growth increment of the trunk is sufficient to provide the crown with water. In this case, vessels are formed before leaves emerge in the spring, and the vessels of previous years are not used. A second method involves refilling by root pressure in the spring, which is quite clearly seen in grapevines.

An ingenious explanation was proposed by E. Sucoff (1969), who studied the mathematics of bubbles in liquid. He showed that large bubbles expand more easily than small bubbles, especially if the liquid is under tension. Imagine a northern tree thawing in the spring. As the ice melts, the tracheids become filled with liquid containing the many bubbles of air that had been forced out by freezing. As melting continues and transpiration begins, tension begins to develop in the xylem. Sucoff showed that a large bubble reaches a critical point at which it expands explosively under tension as water turns to vapor in a fraction of a second (similar to the cavitation clicks already mentioned). This is confined to the tracheid in which it occurs because of its anatomy, but it sends a shock wave to surrounding tracheids, driving their small air bubbles back into solution. What about the tracheid in which bubble expansion occurred? It would be vapor-locked and forever lost to sap movement. Study of wood in the spring indicated that about 10 percent of the tracheids were indeed filled with vapor, but the remaining 90 percent are ample to handle sap movement.

## 5.6 Xylem Anatomy: A Fail-Safe System

We have seen that water transport occurs in response to negative water-potential gradients from soil through the plant to the atmosphere and that this depends upon the anatomy and physical characteristics of xylem tissue. Plants, especially tall trees, are apparently designed to allow sufficient flow in response to the pressure gradients to prevent (or at least usually to avoid) cavitation in the transport elements (but allowing it under some circumstances, as in Zimmermann's "designed leakage"), to bypass cells that do become vapor-locked, and sometimes even to restore (and often to replace) such vapor-locked cells. Zimmermann (1983) discussed other safety features in the system; here are some examples:

The wall sculpturing in vessel elements and the various kinds of perforation plates might keep bubbles that do form (as in winter freezing) from coalescing; small bubbles are much easier to dissolve than large ones. Furthermore, it can be shown that in many plants the highest resistance to water flow occurs at the leaf trace or in the base of the petiole. This means that as water stress builds during drought, cavitation occurs first in the leaves, so they may wilt and die; but the water system in the trunk remains relatively intact. It may be possible to produce new leaves but not a new trunk, especially in palm trees, for example, in which there is no secondary growth of xylem in the trunk.

Some species have special cells called **tyloses** that grow into vapor-locked tracheids and vessels. These act as especially effective seals against water loss and pathogen invasion. Sometimes gums or resins are secreted into nonfunctional xylem cells. In many trees, these changes account for the heartwood that often develops in the center of the trunk. Usually, the heartwood contains many cells filled with air or vapor, but sometimes solutes build up in the heartwood, and these draw in water osmotically, so the solution in the heartwood (called wetwood) is under pressure at the same time that xylem sap is under tension in the rest of the trunk.

Some water is stored in the trunk of a tree because of the elasticity of the xylem cells. When transpiration is reduced at night or during rain, tension in the xylem

## Studying Water, Minerals, and Roots

*Paul J. Kramer*

*Paul J. Kramer, one of the deans of the water relations of plants, has pondered the roles of water and minerals in plants since 1931 in his laboratory at Duke University, where he is Professor of Botany, producing many significant data and ideas, not to mention graduate students who are now carrying on the work all over the world.*

My entrance into the field of the water relations of plants was somewhat accidental. In 1928, when I was a young graduate student at Ohio State University searching for a thesis topic, Professor E. N. Transeau showed me a paper by Burton E. Livingston in which it was claimed that osmosis plays a negligible role in water absorption. Transeau suggested that, as Livingston presented little evidence for his view, I might investigate the problem, and I have been working in that field ever since.

Most textbook writers of those days assumed that osmosis was somehow involved in water absorption, but their discussions were so vague that it was quite impossible to understand how absorption really occurred. My early research was done very simply with T-tubes, pipettes, rubber tubing, a vacuum pump, a pressure chamber built from pipe fittings, tomato and sunflower root systems, and papaya petioles. It was intended to test ideas of Atkins, Renner, and other early workers, and the results indicated that root systems can function as osmometers, and transpiring shoots can absorb water through dead root systems. This research led me to support the neglected view of Renner that two mechanisms are involved in water absorption. According to this view, the root systems of plants growing in moist, well-aerated soil function as osmometers when transpiration is slow, resulting in the development of root pressure and the occurrence of guttation. When rapid transpiration lowers the pressure or produces tension in the xylem sap, however, water is pulled in through the roots, and osmotic movement is negligible. Thus, in transpiring plants, most or all of the water enters passively, and the roots act merely as absorbing surfaces. Further research demonstrated that factors such as cold soil and deficient aeration reduce absorption by increasing the resistance to water flow through roots, rather than by inhibiting some mysterious kind of active absorption mechanism. This early research contributed to the development of a relatively clear explanation of how water is absorbed and of how certain environmental factors affect the rate of absorption. Looking back, those early experiments seem prosaic, but at the time they were very exciting.

During the 1950s, I began to realize that many of the contradictory reports in the literature concerning the relationship between soil moisture and plant growth resulted from the fact that one cannot accurately predict plant water stress from measurement of soil water stress. Holger Brix, John

---

is relaxed, but water may continue to be absorbed, relieving the tension as the cells expand elastically. This has often been observed experimentally. For example, Schulze et al. (1985) measured water flow in the trunk of larch (*Larix*) and spruce (*Picea*) trees with a heat-balance technique and concurrently measured canopy transpiration with porometers and cuvettes (methods described in Section 4.1). Transpiration began 2 to 3 h earlier than xylem water flow in the morning, decreased at noon before maximum sap flow was observed, and stopped in the evening 2 to 3 h earlier than did sap flow. They attributed the different rates to stored water. Some 24 percent of the water transpired each day by larch and 14 percent of that lost by spruce trees was stored in the wood, most of it in the crown instead of the trunk. A classical method is to measure the circumference (or radius) of a tree; the circumference contracts slightly during periods of high transpiration (high xylem tension) and expands at night or during rain (Daum, 1967).

Zimmermann (1983) discussed the role of capillarity in water storage. Remember that much of the trunk is filled with gas (perhaps 10 to 15 percent in actively conducting wood and more in heartwood). Although much of this may be water vapor, air (oxygen) must be present to support respiration of the living ray and other parenchyma cells in the wood. We have seen that, because of the small pore sizes in the cell walls, air does not penetrate into the vessels or tracheids, but the air spaces must contain capillary water with a water potential that is in equilibrium through the water-filled pores in the walls with the water in the conducting cells. When the water in the conducting system is under considerable tension (say $P = -1.5$ to $-3.0$ MPa), the menisci of the water in the air spaces have small radii of curvature; that is, there is a minimum of such water present. As tension in the conducting elements relaxes, however, water can diffuse into the air spaces, where menisci will then have larger radii of curvature (Fig. 5-19). This way, even more water can be stored in the wood than its elasticity would suggest.

Boyer, and others began working as students in my laboratory with the recently introduced thermocouple psychrometers, and I started a campaign to educate plant scientists about the necessity of measuring plant water potential. As they were finally beginning to realize the importance of water in relation to plant growth, this idea was generally accepted. I notice, however, that even today workers in water relations sometimes reduce the value of their research by failing to measure the degree of water stress to which their plants are subjected.

My work on water absorption naturally led to an interest in roots as absorbing organs. With the aid of a zoologist colleague, Karl Wilbur, I measured uptake of phosphorus by mycorrhizal roots. Then Herman Wiebe, as a graduate student in my laboratory, did some interesting experiments with radioactive tracers, which showed that the region of most rapid salt absorption and translocation to shoots often is several centimeters behind the root tip, rather than near it, as previously claimed. This conclusion was viewed with considerable skepticism at first, but we have had the pleasure of seeing it verified by more recent research, including work in R. Scott Russell's laboratory at Letcombe, England. I also concluded that considerable salt and water are absorbed by mass flow through suberized roots. This view has not yet been generally accepted, though evidence supporting it has been available for over 30 years. Perhaps this illustrates the difficulty of getting well-entrenched old ideas replaced by new ones.

Recently, Dr. Edwin Fiscus and other investigators in my laboratory have been studying various anomalies in water conduction, especially the reports of decrease in resistance with increase in rate of flow. Most of these reports involve the strict application of Ohm's law (flow being proportional to driving force and inversely proportional to resistance) without taking into account such anatomical and physiological facts as the role of phyllotaxy (leaf arrangement on the stem) in controlling the water supply to individual leaves or the fact that, as the rate of water flow through roots increases, the driving force changes from primarily osmotic to primarily mass flow. Thus, a simple application of Ohm's law is inadequate to explain water absorption without consideration of the driving forces involved.

I am glad that I entered plant physiology at a time when it was still possible to be a generalist. Thus, in addition to my work on water relations, salt absorption, and root systems, I was also able to carry on research on the physiology of woody plants. To me, research in whole plant physiology on the borderline between physiology and ecology, in what perhaps can be called environmental physiology, has been extremely interesting because of the variety of problems that it has presented. I also think that it has enabled me to contribute more to a general understanding of how plants live and grow than I could have contributed by concentration in a narrow area of research. In any event, there has never been any danger of boredom!

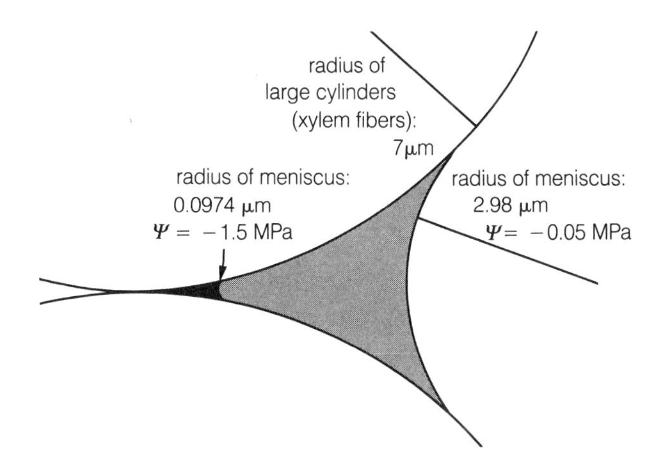


**Figure 5-19** Capillary storage of water in the air spaces between cells. The large portions of circles represent wood fibers or other xylem elements. If these circles have radii of 7 μm, then the black area represents water stored when water potential in the xylem is about −1.5 MPa, and the shaded area represents water stored when xylem water potential is about −0.05 MPa. The menisci have curvatures in equilibrium with the water potentials ($\Psi$) as shown.

All in all, our understanding of xylem anatomy combined with our application of the cohesion theory of the ascent of sap has greatly increased our appreciation for land plants as water-conducting systems. In Chapter 8 we will see that this is only part of the story: Plants are also anatomically constructed for highly efficient transport of the solutes produced in photosynthesis and other metabolic processes (in that case, under positive pressure). But first we must learn more about the minerals required by plants and about the membranes that so strongly influence their movement.

Note added in proof: We are aware that expressing tensions as a negative number might be construed as a double negative: "Negative tension" = positive pressure. It would be confusing, however, to change the sign on numerical values depending on whether we used the term *tension* or *negative pressure*. Hence, we think of these two terms as being completely synonymous and interchangeable, and we use the negative sign before numerical values expressing either term.

# 6

# Mineral Nutrition

What elements must a plant absorb in order to live and grow? Can a plant grow when provided only with elements in inorganic form (mineral salts)? Or do plants, like animals, require vitamins? If only minerals are required, then which ones, in what forms, and in what amounts? How can we know when a plant is lacking some essential element? How can we best provide the limiting element to overcome its deficiency? What do essential elements do within a plant to make them essential? These are some questions of **mineral nutrition**, an important subscience of plant physiology. Because we must properly "feed" plants before we can feed ourselves, these are important questions. Answers obtained so far have greatly improved agriculture during the past century and a half, but still more improvements are needed. Answers to the questions of mineral nutrition also add to our basic understanding of plants, for plant growth requires the incorporation of elements into the materials of which plants are made, and 15 to 20 percent of nonwoody plants is made from such elements; the rest is water.

## 6.1 The Elements in Plant Dry Matter

One approach to determine which elements are essential and how much of each is required is to chemically analyze healthy plants to ascertain what they are made of. When freshly harvested plants or plant parts are heated to 70 to 80°C for a day or two, nearly all the water is driven off; the remaining material is the so-called **dry matter**. Principal components of dry matter are cell-wall polysaccharides and lignin, plus protoplasmic components, including proteins, lipids, amino acids, organic acids, and certain elements such as potassium that exist as ions but form no essential part of any organic compound.

Principal elements in a dry shoot system of maize (corn; reported in 1924) are listed in Table 6-1. Oxygen and carbon were by far the most abundant elements on a weight basis (about 44 percent of each), and hydrogen ranked third. This is approximately the same distribution of elements as in carbohydrates—including cellulose, the most abundant compound in wood. In other species the carbon content is as high as 51 percent and the oxygen as low as 35 percent, values less similar to the composition of a carbohydrate (Williams et al., 1987). Smaller amounts of nitrogen were found in maize, followed by several other elements in even lower concentrations. Also included are two elements, aluminum and silicon, that are believed to be nonessential for most higher plants. More will be said of these elements later, but note that plants absorb and accumulate many nonessential elements from soil solutions. At least 60 elements have been found in plants, including gold, lead, mercury, arsenic, and uranium. Had a complete elemental analysis of the maize plant been made, trace amounts of numerous other elements could have been found, some of which are essential to maize and other species.

A modern analysis of the leaf closest to the young maize cob (the "flag leaf") shows concentrations of three additional essential elements: zinc, copper, and boron (Table 6-1). These results were obtained from leaves of a well-fertilized and highly productive maize field. It emphasizes that leaves generally contain significantly more nitrogen, phosphorus, and potassium than do whole shoot systems. Finally, Table 6-1 also lists data for 11 elements in growing sweet-cherry leaves. Note that although the contents of nitrogen and sulfur are similar to those of the maize leaf, there are substantial differences in phosphorus, potassium, and calcium. These differences show that various species absorb solutes in varying amounts, especially when grown in different soils. Soils are composed largely of aluminum, oxygen, silicon, and iron, yet plants by no means reflect this composition, partly because they absorb carbon and much of their oxygen from the air. Other reasons are that most of the above-mentioned soil elements are present as insoluble minerals and that roots exhibit considerable selection over the rates at which elements are absorbed (see Chapter 7).

## 6.2 Methods of Studying Plant Nutrition: Solution Cultures

Beginning in about 1804, scientists began to appreciate that plants require calcium, potassium, sulfur, phosphorus, and iron. Then, about 1860, three German plant physiologists (W. Pfeffer, Julius von Sachs, W. Knop) recognized how difficult it is to determine the kinds and amounts of elements essential to plants growing in a medium as complex as a soil. They therefore grew plants with their roots in a solution of mineral salts (a **nutrient solution**), the chemical composition of which was controlled and limited only by the purity of chemicals then available. Growing plants in this way is referred to as **hydroponics, soil-less culture**, or **solution culture** (Fig. 6-1). Other investigators later showed that

**Table 6-1 Elemental Analysis of Selected Plant Parts.**

| Element | Maize Shoot[a] (% of dry wt.) | Maize Leaf[b] (% of dry wt.) | Cherry Leaves[c] (% of dry wt.) |
|---|---|---|---|
| Oxygen | 44.4 | — | — |
| Carbon | 43.6 | — | — |
| Hydrogen | 6.2 | — | — |
| Nitrogen | 1.5 | 3.2 | 2.4 |
| Potassium | 0.92 | 2.1 | 0.73 |
| Calcium | 0.23 | 0.52 | 1.7 |
| Phosphorus | 0.20 | 0.31 | 0.15 |
| Magnesium | 0.18 | 0.32 | 0.61 |
| Sulfur | 0.17 | 0.17 | 0.15 |
| Chlorine | 0.14 | — | — |
| Iron | 0.08 | 0.012 | 0.0058 |
| Manganese | 0.04 | 0.009 | 0.0044 |
| Copper | — | 0.0009 | 0.0006 |
| Boron | — | 0.0016 | 0.003 |
| Zinc | — | 0.003 | 0.001 |
| Silicon | 1.2 | — | — |
| Aluminum | 0.89 | — | — |
| Undetermined | 7.8 | — | — |

[a]Data of Latshaw and Miller, 1924. The shoot system included leaves, stem, cob, and grains.
[b]Unpublished 1982 data of P. Soltanpour, F. Moore, R. Cuany, and J. Olson, Colorado State University Soil Testing Laboratory.
[c]Data of Sanchez-Alonso and Lachica, 1987. Values are for middle leaves of new branches sampled on June 14.

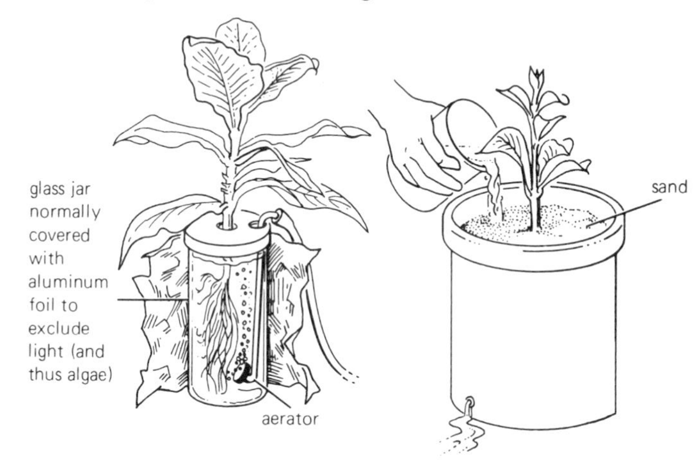


**a** hydroponic culture

**b** slop culture (nutrients applied by hand)

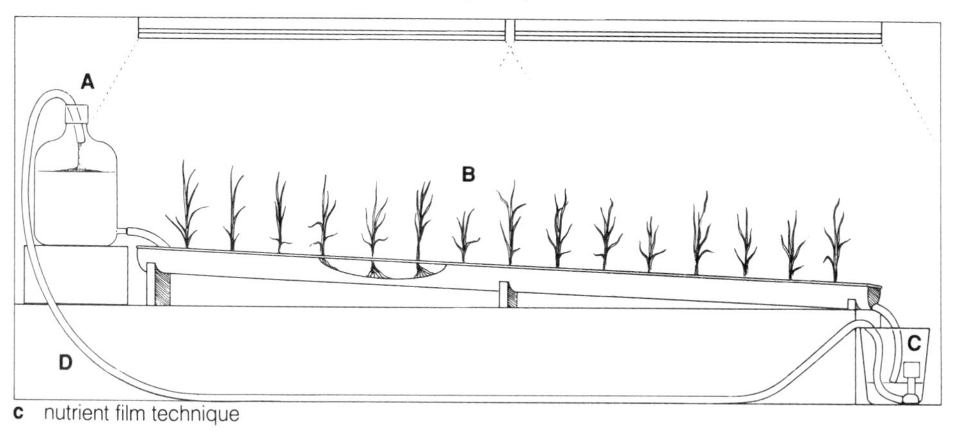


**c** nutrient film technique

**Figure 6-1** Three methods for growing plants with nutrient solutions: (**a**) hydroponic culture (note aerated roots), (**b**) slop culture using sand, and (**c**) nutrient-film technique. Reservoir A contains nutrient solution that drains down trough containing plants B. Plants can be supported in many ways. Unabsorbed solution flows into container C that has a pump to force solution through tube D back to reservoir. (Method **c** from M. W. Nabors, 1983; for practical use of the nutrient-film technique, see Cooper, 1979.)

many plants grew much better if the roots were aerated, as shown in Figure 6-1a. (A history of hydroponics is given by Jones, 1982, and an evaluation of its potential commercial future is given by Wilcox, 1982, and Jones, 1983.)

As techniques were developed to purify salts and the water for hydroponic cultures, more exact control over the elements made available to plants became possible. This proved especially important for several elements required only in very small (trace) amounts. Furthermore, requirements for molybdenum, nickel, copper, zinc, and boron are often difficult to demonstrate for species with large seeds, because large seeds sometimes contain enough of these elements to grow mature plants. In such cases, deficiency symptoms are more easily observed in the second generation grown from seeds taken from parents that have been grown without the added elements. To demonstrate essentiality of nickel, for example, seeds of barley were taken from plants grown for three generations without nickel (Brown et al., 1987). Techniques to measure concentrations of elements in plants, soils, and nutrient solutions have also improved greatly in the past 20 years. **Atomic absorption spectrometers** are now used to measure metals and some nonmetals. Even more valuable (and expensive) are **optical emission spectrometers** in which elements are vaporized at temperatures above 5,000 K. Such high temperatures temporarily excite electrons from their ground-state orbits into higher orbits, and as these electrons move back to their original ground states electromagnetic energy is emitted at wavelengths that are different for each element. These wavelengths are measured and the energy quantified by the spectrometer. Concentrations of more than 20 elements in a single solution can be measured with great sensitivity in less than a minute (Soltanpour et al., 1982; Alexander and McAnulty, 1983).

In spite of advantages for mineral nutrition studies, the solution-culture technique has disadvantages. One is the need for root aeration. Another is the need to replace or supplement the solution every few days if maximum growth is to be achieved; this is because the solution's composition changes continuously as certain ions are absorbed more rapidly than others. This selective uptake not only depletes certain ions but also causes undesirable *p*H changes. Owners of commercial greenhouses sometimes use recirculating solutions that flow in a thin layer through troughs around the roots of valuable crops, including lettuce and tomatoes (Jensen and Collins, 1985). Such solutions are pumped from tanks in which the *p*H and solution composition can be monitored and adjusted automatically (Fig. 6-1c). The pump forces solution across the roots; then, when the pump is temporarily turned off, the solution is drained downhill, leaving a well-aerated film of nutrient solution on the root surfaces (the **nutrient-film technique**—see Cooper, 1979; Graves, 1983; Jones, 1983; Resh, 1989).

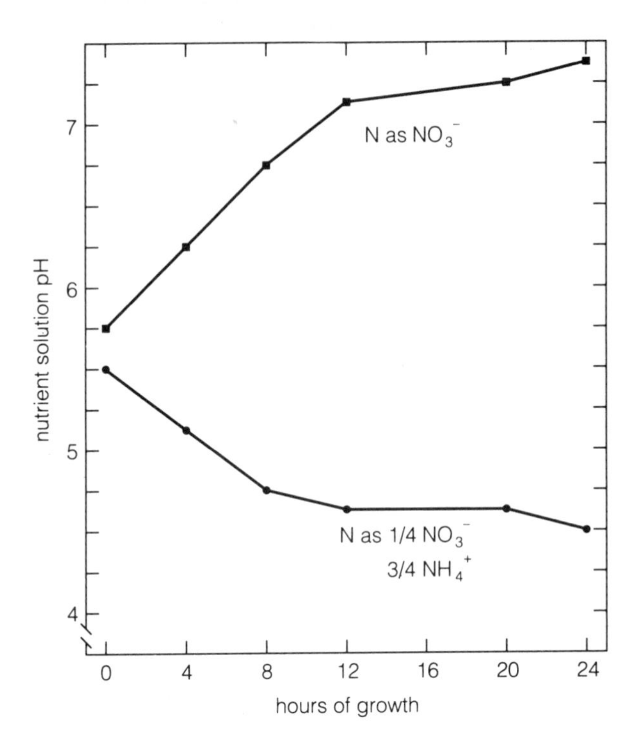


**Figure 6-2** Changes over time in nutrient solution *p*H; nitrogen was provided entirely as nitrate or mostly as ammonium. Rapidly growing sunflower plants were kept in 1-L containers of nutrient solution in a greenhouse during late May. (Unpublished data of C. W. Ross.)

To avoid some problems of liquid cultures, many physiologists use washed white-quartz sand or a mineral called *perlite* (expanded pumice) as a medium for the roots. Nutrient solutions are simply poured or allowed to drip onto these media at suitable intervals in excess amounts to ensure leaching of the old solution through holes in each container (Fig. 6-1b). The technique is convenient but unsuitable for detailed studies with certain elements needed only in trace amounts, because the sand and perlite contribute unknown amounts of certain essential elements. For some studies in which ion uptake measurements by roots are necessary, solution cultures provide the only adequate method. This method is described in Chapter 7 (see Fig. 7-13). In addition to solution cultures applied to roots, fruit trees and some other horticultural crops are often fertilized with foliar sprays (Swietlik and Faust, 1984).

Numerous useful formulations of nutrient solutions were devised from studies of the composition of plants and from other studies in which various concentrations of elements were provided to growing plants. Hewitt (1966) and Jones (1983) described many of these formulations and the ways to prepare and use them effectively. Two such recipes by pioneers of mineral nu-

**Table 6-2 Two Nutrient Solutions for Hydroponic Culture.**

| Hoagland and Arnon's Solution No. 2[a] | | | Evans's Modified Shive's Solution[b] | | |
|---|---|---|---|---|---|
| Salt | mM | mg/L (ppm) | Salt | mM | mg/L (ppm) |
| $KNO_3$ | 6.0 | 235 K<br>196 N as $NO_3$ | $Ca(NO_3)_2 \cdot 4H_2O$ | 5.0 | 140 N as $NO_3$<br>200 Ca |
| $Ca(NO_3)_2 \cdot 4H_2O$ | 4.0 | 14 N as $NH_4^+$<br>160 Ca | $K_2SO_4$ | 2.5 | 216 K<br>160 S |
| $NH_4H_2PO_4$ | 1.0 | 31 P<br>49 Mg | $KH_2PO_4$ | 0.5 | 49 Mg<br>16 P |
| $MgSO_4 \cdot 7H_2O$ | 2.0 | 64 S | $MgSO_4 \cdot 7H_2O$ | 2.0 | |
| Fe-chelate[c] | | | Fe-versenate | | 0.5 Fe |
| $MnCl_2 \cdot 4H_2O$ | 0.009 | 0.5 Mn; 6.5 Cl | KCl<br>$MnSO_4$ | | 9.0 Cl<br>0.25 Mn |
| $H_3BO_3$ | 0.046 | 0.5 B | $H_3BO_3$ | | 0.25 B |
| $ZnSO_4 \cdot 7H_2O$ | 0.0008 | 0.05 Zn | $ZnSO_4 \cdot 7H_2O$ | | 0.25 Zn |
| $CuSO_4 \cdot 5H_2O$ | 0.0003 | 0.02 Cu | $CuSO_4 \cdot 5H_2O$ | | 0.02 Cu |
| $H_2MoO_4 \cdot H_2O$ | 0.0001 | 0.01 Mo | $Na_2MoO_4$ | | 0.02 Mo |

[a]From Hoagland and Arnon, 1950.
[b]From Evans and Nason, 1953.
[c]They made a stock solution of iron chelate at 5 g/L final concentration, then added 2 mL of this to each liter of nutrient solution twice weekly.

trition in the United States are listed in Table 6-2; one is by Dennis R. Hoagland and Daniel I. Arnon, and the second is a modification of one by John Shive (modified by Harold J. Evans). Both have necessary elements in amounts that allow good growth of many higher plants, but a solution ideal for one species is seldom ideal for another. Note that the Evans solution in Table 6-2 contains all its nitrogen as nitrate, but nitrate is often absorbed so fast that there are rapid rises in the nutrient solution *pH*, because absorption of nitrate (and other anions) is accompanied by absorption of $H^+$ or excretion of $OH^-$ to maintain charge balances. At high *pH* values, iron and some other elements precipitate as hydroxides and are then unavailable to the roots (see Section 6.5). The *pH* problem can be minimized by supplying part of the nitrogen as an ammonium salt (for example, $NH_4H_2PO_4$, as in the Hoagland solution) because absorption of $NH_4^+$ and other cations occurs simultaneously with absorption of $OH^-$ or transfer of $H^+$ from the root to the surrounding solution (Fig. 6-2).

Nearly all nutrient solutions are more concentrated than soil solutions. For example, the phosphorus concentration of the Evans solution in Table 6-2 is 500 $\mu$M, whereas about three-fourths of the determinations in one survey of 149 soil solutions (Reisenauer, 1966) gave phosphorus levels less than 1.5 $\mu$M. Over half of these soil solutions had potassium concentrations less than 1.25 mM, whereas the Evans solution is 5.5 mM in $K^+$. Many minerals in the soil are not in solution but are adsorbed on negatively charged surfaces of clay and organic matter or precipitated as insoluble salts. They dissolve only slowly as they are removed by plants from solution or as they are lost by leaching. Plants can grow well in solutions having concentrations of essential elements as low as those dissolved in the soil solution in which they normally grow, provided that the solutions are replenished often enough to maintain those concentrations or that dilute solutions in large volumes flow over their roots from a recirculating tank (Asher and Edwards, 1983; Wild et al., 1987). The higher concentrations typically used in nonflowing solutions avoid the need to change the solution more than once a day or once every several days, depending on growth rates. Of course, the concentration must be low enough to avoid plasmolysis of the root cells. Most solutions have osmotic potentials no more negative than $-0.1$ MPa, so this is not a problem.

## 6.3 The Essential Elements

The nutrient solutions of Table 6-2 include 13 elements believed to be essential for all angiosperms and gymnosperms, although in fact the nutrient requirements of only 100 or so (mostly cultivated) species have been

**Table 6-3 Essential Elements for Most Higher Plants and Internal Concentrations Considered Adequate.**

| Element | Chemical Symbol | Form Available to Plants[a] | Atomic Wt. | Concentration in Dry Tissue mg/kg | (%) | Relative No. of Atoms Compared to Molybdenum |
|---|---|---|---|---|---|---|
| Molybdenum | Mo | $MoO_4^{2-}$ | 95.95 | 0.1 | 0.00001 | 1 |
| Nickel[b] | Ni | $Ni^{2+}$ | 58.71 | ? | ? | ? |
| Copper | Cu | $Cu^+$, **$Cu^{2+}$** | 63.54 | 6 | 0.0006 | 100 |
| Zinc | Zn | $Zn^{2+}$ | 65.38 | 20 | 0.0020 | 300 |
| Manganese | Mn | $Mn^{2+}$ | 54.94 | 50 | 0.0050 | 1,000 |
| Boron | B | $H_3BO_3$ | 10.82 | 20 | 0.002 | 2,000 |
| Iron | Fe | $Fe^{3+}$, **$Fe^{2+}$** | 55.85 | 100 | 0.010 | 2,000 |
| Chlorine | Cl | $Cl^-$ | 35.46 | 100 | 0.010 | 3,000 |
| Sulfur | S | $SO_4^{2-}$ | 32.07 | 1,000 | 0.1 | 30,000 |
| Phosphorus | P | **$H_2PO_4^-$**, $HPO_4^{2-}$ | 30.98 | 2,000 | 0.2 | 60,000 |
| Magnesium | Mg | $Mg^{2+}$ | 24.32 | 2,000 | 0.2 | 80,000 |
| Calcium | Ca | $Ca^{2+}$ | 40.08 | 5,000 | 0.5 | 125,000 |
| Potassium | K | $K^+$ | 39.10 | 10,000 | 1.0 | 250,000 |
| Nitrogen | N | **$NO_3^-$**, $NH_4^+$ | 14.01 | 15,000 | 1.5 | 1,000,000 |
| Oxygen | O | $O_2$, $H_2O$, $CO_2$ | 16.00 | 450,000 | 45 | 30,000,000 |
| Carbon | C | $CO_2$ | 12.01 | 450,000 | 45 | 35,000,000 |
| Hydrogen | H | $H_2O$ | 1.01 | 60,000 | 6 | 60,000,000 |

[a]The more common of two forms is indicated by boldface type.
[b]From Brown et al., 1987.
Source: Modified after Stout, 1961.

investigated much. Adding O, H, and C (from $O_2$, $H_2O$, and $CO_2$) brings the total to 16 elements. A seventeenth, nickel, was not known to be essential when these and other nutrient solution recipes were prepared, but sufficient nickel has always been present as a contaminant of one or more salts used to make those solutions. With these 17 elements and sunlight, most plants can synthesize all the compounds they require. But is it possible that plants require some organic molecules such as vitamins synthesized by microorganisms that normally grow on roots, stems, or leaves? There have been repeated claims that certain plants grow faster or yield more if they are given exogenous vitamins, especially B vitamins, by spraying or dipping vegetation or occasionally from the soil (reviewed by Oertli, 1987; Buchala and Pythoud, 1988; Samiullah et al., 1988), but most of these claims remain unsubstantiated. Furthermore, some plants have been grown well under sterile conditions in plastic or glass enclosures from which all microorganisms were excluded. Plants really are **autotrophic** and make all the organic molecules they need, even though some associated microbes are beneficial (for example, in mycorrhizae, Chapter 7, and root nodules, Chapter 14). Often, these microbes are essential to plants in nature, though not in solution cultures, because they perform roles that allow plant survival in the face of competition and harsh environmental conditions (Rovira et al., 1983; Quispel, 1983).

There are two principal criteria by which an element can be judged essential or nonessential to any plant (Epstein, 1972): *First, an element is essential if the plant cannot complete its life cycle (that is, form viable seeds) in the absence of that element. Second, an element is essential if it forms part of any molecule or constituent of the plant that is itself essential in the plant* (for example, nitrogen in proteins and magnesium in chlorophyll). Either criterion is sufficient to demonstrate essentiality, and most elements in our list of 17 have met both. Historically, the first criterion was the main one used, but improved chemical analyses have generally led to agreement on both criteria.

Although these two criteria are widely accepted by mineral-nutrition experts, other criteria are often considered. Daniel Arnon and Perry Stout (1939) suggested that a third criterion should also be used: If an element is essential, it must be acting directly inside the plant and not causing some other element to be more readily available or antagonizing the effect of another element. This criterion has not been nearly as useful as the other two, but it has been applied in a few cases. One case concerns the initial conclusion that selenium is essential for plants. It was later found that the growth-promoting effects of selenium resulted from the ability of the selenate ion to inhibit the absorption of phosphate, which otherwise was absorbed by the plants in toxic amounts.

We should also emphasize that many investigators consider an element to be essential if deficiency symptoms appear on plants grown without addition of the element to the nutrient solution, even though such plants form viable seeds. The assumption, which seems reasonable, is that plants that contained none of the element (that is, the element was absent from seeds, dust in the air, and nutrient solutions) would develop deficiency symptoms so severe that they would die before they formed viable seeds. Use of this criterion has led to evidence (mentioned later) that sodium and silicon are essential for certain species.

It is usually easier to show that an element is essential than that it is not. Researchers therefore often state that if an element in question is necessary, it must be required in concentrations that are lower than the sensitivity limits of their detecting instruments. For example, it was reported that if vanadium is in fact essential for lettuce or tomato plants, the amount needed is less than 20 μg/kg of dry tissue (20 ppb by weight). Because of such problems, it is likely that a few more nutrient elements, needed in barely detectable amounts, will eventually be added to our list.

Table 6-3 lists the 17 elements currently believed essential to all higher plants, the molecular or ionic form in the soil or air most readily absorbed by plants, the approximate adequate concentration in the plant, and the approximate number of atoms of each element needed relative to molybdenum. About 60 million times as many atoms of hydrogen than of molybdenum are required, a dramatic difference. This difference reflects the importance of hydrogen in thousands of essential compounds, whereas molybdenum acts catalytically and in only a few compounds (enzymes). The first eight elements listed are often called **trace elements** or **micronutrients** (needed in tissue concentrations equal to or less than 100 mg/kg of dry matter), and the last nine are the **macronutrients** (needed in concentrations of 1,000 mg/kg of dry matter). The internal concentrations deemed "adequate" should be considered only as useful guidelines because of variability among plant species and ages. Many similar types of data are given by Shear and Faust (1980) for horticultural fruit and nut crops and by Joiner et al. (1983) for ornamental greenhouse crops.

Besides these essential elements, some species require others. For many years evidence has existed suggesting that sodium is required by numerous desert species, such as *Atriplex vesicaria*, common to dry inland regions of Australia, and *Halogeton glomeratus*, a common introduced weed of salty arid soils in the western United States. Peter F. Brownell and Christopher J. Crossland (see Brownell and Crossland, 1972 and Brownell, 1979) investigated and reviewed the sodium nutrition of 32 species and concluded that those having the C-4 photosynthetic pathway (Section 11.3) probably do require $Na^+$ as a micronutrient. (Dr. Brownell's essay in this chapter expands on these findings.) Numerous C-4 species developed severe **chlorosis** (lack of chlorophyll) in leaves, and sometimes they developed **necrosis** (dead tissues) in the leaf margins and tips. The reasonable assumption is that if tissue levels were reduced even more by elimination of all contaminating sources of sodium, these plants would show more pronounced deficiency symptoms, and death would soon follow. On this basis we could say that sodium is almost certainly essential for C-4 species, especially when grown under the relatively low $CO_2$ concentrations that exist in normal air. Figure 6-3 illustrates the importance of sodium at low (but not at high) $CO_2$ concentrations for *Amaranthus tricolor*. Furthermore, certain species that fix $CO_2$ in photosynthesis via the crassulacean acid metabolism pathway common in succulents (Section 11.6) also grow faster with sodium, and for them $Na^+$ is probably also essential.

**Figure 6-3** Growth of *Amaranthus tricolor*, a species having the C-4 photosynthetic pathway, without (– Na) or with (+ Na) added sodium at atmospheric (Normal) or elevated (High) $CO_2$ levels (– Na, <0.08 μM Na; + Na, 0.01 mM Na; High $CO_2$, 1,500 μL $CO_2$ $L^{-1}$). (Photograph courtesy Peter Brownell, Mark Johnston, and Christopher Grof; also see Johnston et al., 1984.)

Silicon is another element that increases growth of some plants, and the possibility that it is essential has been studied by many investigators (reviewed by Lewis and Reimann, 1969 and Werner and Roth, 1983; and Marschner, 1986). Maize accumulates silicon to the extent of 1 to 4 percent of its dry weight (Table 6-1), as do numerous other grasses, whereas rice and *Equisetum arvense* (a horsetail or scouring-rush) contain up to 16

# The Function of Sodium as a Plant Micronutrient

***Peter F. Brownell***

*Peter Brownell is a native of Australia, recently retired from the James Cook University of North Queensland in Townsville where, beginning in 1971, he taught plant physiology to students in a master of science course in tropical agriculture. As he tells us in this essay, he has long been interested in the role of sodium in plant nutrition and it was some of his early studies that convinced other plant physiologists that sodium is indeed an essential element—at least for some plants. Those plants turned out to be the ones that use C-4 photosynthesis (which we discuss in Chapter 11). Professor Brownell is now studying the biochemical steps in the process where sodium plays its role.*

In 1954, when Professor J. G. Wood of the Botany Department at the University of Adelaide suggested a Ph.D. project to determine whether sodium and/or chlorine were essential micronutrients for plants, little did I think that I would still be working in this field with my students some 35 years later.

Sodium has long been regarded as a possible *macro*nutrient. In 1945, Harmer and Benne arranged species into groups depending upon their responses to sodium with insufficient or sufficient supply of potassium. It was clear that a certain amount of potassium not replaceable by sodium was needed by all species, but there was no evidence for sodium being an essential element. Its main role was in its ability to substitute for some of the potassium needed for maximum growth.

The first suggestion that sodium might be needed as a *micro*nutrient element was probably made by Pfeffer in the late 19th century. Little attention was given to this possibility until the early 1950s, when Professor Wood suggested that sodium and/or chlorine could be essential for plants in very small amounts. At that time no experiments had been reported in which these elements had been carefully excluded from the plants' environment. Soon after the beginning of our study, Clarence Johnson and Perry Stout visited Adelaide with the news that their team at the University of California, Berkeley, had discovered chlorine to be an essential micronutrient for tomatoes. In the same year, also at Berkeley, Allen and Arnon found sodium to be an essential micronutrient element for the cyanobacterium *Anabaena cylindrica*. This was the first time that sodium had been shown to be essential for plant life. At this stage, we decided to invest all our efforts in determining whether higher plants might also require sodium.

Two developments greatly assisted the investigation. The first was the introduction of the flame photometer, followed soon after by Alan Walsh's invention of the atomic absorption spectrophotometer. This enabled us to measure sodium at low concentrations, accurately and rapidly. The second was the availability of plastic materials virtually free of sodium.

One species chosen for this work was *Atriplex vesicaria* (bladder salt-bush), which grows in the southern arid areas of Australia. Salt-bushes accumulate sodium and chlorine at high concentrations (up to 23 percent NaCl on a dry-weight basis) compared with other plants growing in similar habitats. The Australian salt-bushes also possess bundle sheath anatomy, which is now known to be a feature of C-4 plants [see Chapter 11 for description of C-4 plants]. We were very fortunate to have chosen a C-4 species, unknowingly, for our early work.

We took much care to exclude sodium from the environment of the plants. In early experiments, carried out in a conventional glass house, the amounts of sodium recovered in plant material plus that remaining in the culture solutions were several times greater than the amounts of sodium known to have been supplied as impurities of the water, culture media, and seeds. On the other hand, there was no detectable increase in sodium in later experiments conducted in a small greenhouse supplied continuously with air filtered to exclude dust that contained sodium. The water—which was distilled twice, the final stage in a silica still—had a sodium concentration of less than 0.0002 mg $L^{-1}$. Salts of culture solutions were generally purified by up to six recrystallizations in platinum or silica vessels. Some components of the culture solution were prepared from reagents redistilled in silica. For example, ammonium chloride was made up from silica-distilled ammonia and hydrochloric acid. The final concentration of sodium in a complete culture solution prepared from these salts was less than 0.0016 mg $L^{-1}$, which is only a little over a one-hundredth of the sodium which would have been derived from untreated analytical-reagent salts. When plants were grown under these conditions, they showed greatly reduced growth, chlorosis, and necrosis of the leaves. Plants receiving 1 mg $L^{-1}$ sodium in their cultures, irrespective of the salt (the anion), showed normal growth. In some cases, growth of 45-day-old sodium-treated plants was over 20 times that of untreated plants. The response was specific to sodium, as no other group 1 elements were effective. It was very hard to leave such a project, and this is probably one reason why I am still working on it!

Two major questions arose from these results: What is the function of sodium in *Atriplex vesicaria*, and Do other plant species require sodium as a micronutrient element?

The functions of the essential transition elements, including iron, copper, zinc, manganese, and molybdenum, that can take part in oxidation/reduction reactions have generally been discovered incidentally during other metabolic studies. For example, from research on photosynthesis, a function for copper became obvious when it was shown to be a component of plastocyanin. Similarly, a function for molybdenum became apparent from studies of nitrogen metabolism; molybdenum was found to be a component of the nitrate reductase and nitrogenase enzymes. Our knowledge of the functions of other essential micronutrient elements—boron, chlorine, and sodium—has lagged behind because no direct effects on enzyme systems have been observed. Information of their functions had to be acquired by other means, often with little information from which to start.

We were puzzled when we attempted to answer the second question regarding whether or not there was a general requirement by higher plants for sodium. From the pattern observed with other essential elements, it was expected that sodium would be required by all higher plants. Surprisingly, of the 30 species examined, including halophytes, other chenopods, and other nonendemic species of *Atriplex*, only the 10 Australian species of *Atriplex* were found to require sodium. At this time, these differences in response could not be correlated with any obvious differences among the species studied. It still seemed possible that all higher plants might require sodium, but those plants that had grown normally without added sodium might require extremely small amounts compared with the Australian *Atriplex* species. However, after the discovery of the C-4 photosynthetic pathway by Hal Hatch and Roger Slack in 1966, it seemed likely that only plants having the C-4 pathway required sodium. Chris Crossland and I tested this possibility with C-4 plants from different families and found that they all responded to small amounts of sodium, as did the Australian species of *Atriplex*. We were excited at this discovery because it gave us a clue to the possible function of sodium in these plants. It suggested that sodium was required for the operation of the C-4 appendage in transporting $CO_2$ to the bundle sheath cells, where it is reduced to carbohydrates [as described in Chapter 11]. In support of this, we found that the signs of sodium deficiency were alleviated in plants grown in atmospheres with elevated $CO_2$ concentrations. However, plants that had received the sodium treatment showed little or no response to the high $CO_2$ treatments, which suggested that under conditions of sodium deficiency transport of $CO_2$ to the bundle sheath cells is decreased, thus limiting the rate of photosynthesis. When the atmospheric concentration of $CO_2$ in which the plants are grown is elevated to about 1,500 $\mu L\ L^{-1}$, $CO_2$ enters the bundle sheath cells by diffusion, thus bypassing the C-4 system.

Chris Crossland and I also found responses to small amounts of sodium by a crassulacean-acid metabolising (CAM) plant, *Bryophyllum tubiflorum*, when it was grown under certain conditions that encourage activity of the CAM pathway [see Chapter 11 for description of CAM]. There are strong similarities in the photosynthetic metabolism of these plants with those of C-4 plants.

Ross Nable and Mark Johnston found high levels of alanine and pyruvate and low levels of phosphoenolpyruvate in sodium-deficient C-4 plants. The pyruvate formed from the C-3 compounds returning from the bundle sheath is converted to phosphophenol pyruvate. The major steps in this process involve the transport of pyruvate into the mesophyll chloroplast, the enzymatic conversion of pyruvate to phosphoenolpyruvate within the stroma, and the provision of energy required for the conversion reaction. No effect of sodium nutrition on the activity of pyruvate phosphate dikinase, the enzyme catalyzing the conversion of pyruvate to phosphoenolpyruvate, has been found.

Recently, Jun-ichi Ohnishi and Ryuzi Kanai discovered a sodium-induced uptake of pyruvate into the mesophyll chloroplasts of *Panicum miliaceum*. This immediately suggested a role for sodium in C-4 plants. However, Mark Johnston and Chris Grof have obtained evidence for damage to the light-harvesting photosystem, the source of energy for pyruvate transport and/or the regeneration of phosphoenolpyruvate. In sodium deficiency, they found lower chlorophyll a/b ratios and lowered photosystem II activity, with altered ultrastructure in the mesophyll chloroplasts. With the discoveries of the light- and sodium-activated membrane translocator system and of the damage to the energy-producing machinery in the mesophyll chloroplasts, research on the actual function of sodium has reached an exciting phase. We still have a difficult question to answer: What is the primary function of sodium? Is it needed to maintain the integrity of the light-harvesting and energy-transducing systems in the mesophyll chloroplasts, or for transport of pyruvate into the mesophyll chloroplasts? If the latter is the case, then damage to the light-harvesting system observed in mesophyll chloroplasts of sodium-deficient plants could have been caused by excess energy that normally would have been used to convert pyruvate to phosphoenolpyruvate.

An intriguing feature of the hypothesis of sodium's involvement in the transport of pyruvate across a membrane is that a similar sodium-requiring system has been demonstrated in species of cyanobacteria for the uptake of bicarbonate ions. In 1967, Professor Don Nicholas and I found the activity of nitrate reductase to be many times greater in sodium-deficient compared to normal cells of *Anabaena cylindrica*. We were unable to obtain a similar effect of sodium

(*continued*)

## The Function of Sodium *(continued)*

deficiency on nitrate reductase in C-4 plants. The effect of sodium nutrition on nitrate-reductase activity in *Anabaena* might have been a consequence of some earlier effects of the sodium treatment, perhaps those related to the transport of inorganic carbon.

It is unlikely that a lack of sodium will ever limit plant growth in nature. However, as sodium has an important role in C-4 photosynthesis, this continues to be a challenging and exciting project. It has been extremely rewarding to have been able to demonstrate the essentiality of another element, to show that its need is restricted to C-4 plants, and to have evidence that directs us to focus our attention on the metabolism of the mesophyll chloroplast. We have had our share of luck (which is necessary for some success in this kind of work), including the choice of a C-4 plant as our experimental material long before we knew what a C-4 plant was, and the timely discovery of this C-4 photosynthetic pathway during the project. Perhaps the greatest good fortune has been to have had excellent colleagues who have been generous with their help and interest. Another real asset has been having enthusiastic students who have shared the disappointments and periods of excitement that are part of such a project.

percent silicon. The contents of most dicots are much lower than those of grasses or *E. arvense*.

As is so often the case in studies of essentiality, it is difficult to completely remove the element from the plants' environment to determine whether they can grow without it. But with silicon the problems are especially serious because it is present in glass and many nutrient salts, and it also exists as particulate $SiO_2$ in the atmosphere. In several species (rice, barley, sugarcane, tomato, and cucumber) the amount of silicon has been reduced enough to create deficiency symptoms (Miyake and Takahashi, 1985). In rice, for example, overall growth was retarded, transpiration was increased about 30 percent, and the older leaves died. In tomatoes, growth rates were lowered about 50 percent, new leaves of nearly mature plants were deformed, and many plants failed to set fruits. Werner and Roth (1983) and others indicated that silicon is generally an essential element. So far as we are aware, however, no dicot grown with limited silicon has failed to produce viable seeds, nor has any essential role for silicon in the plants been shown. Furthermore, soybean plants grown without silicon accumulated unusually high concentrations of phosphorus (Miyake and Takahashi, 1985), so it seems possible that symptoms of silicon deficiency sometimes represent phosphate toxicity. For certain algae (diatoms and some flagellate Chrysophyceae) that are surrounded by a silica-rich sheath, silicon is certainly essential.

Silicon exists in soil solutions as silicic acid, $H_4SiO_4$ [or $Si(OH)_4$], and is absorbed in this form. It accumulates largely as polymers of hydrated amorphous silica ($SiO_2 \cdot nH_2O$), most abundantly in walls of epidermal cells but also in primary and secondary walls of other cells of roots, stems, and leaves and in grass inflorescences (Kaufman et al., 1985; Sangster and Hodson, 1986). It also accumulates intracellularly in specialized epidermal cells called silica cells.

Various functions of silicon in plants have been suggested. When it accumulates in epidermal cell walls, it seems to cause less transpiration and fewer fungal infections. There is evidence that in xylem cells silicon provides rigidity and limits compression, such as that caused by bending in wind. In fact, it is well known that silicon-deficient cereal grain crops are more easily **lodged** (bent down by wind or rain) than those with adequate silicon. There are also claims that the silicates present in grass leaves and inflorescences reduce grazing (herbivory) by animals and insects. Prevention of lodging or herbivory would thus represent an **ecological requirement** for silicon rather than a physiological or biochemical requirement included in the two criteria for essentiality mentioned earlier.

When sheep and cattle eat grasses abundant in silica, they excrete most of it in the urine, but sometimes it forms kidney stones. Silica is also blamed for causing excessive wear on sheep's teeth, and it is implicated in throat cancers of people from northern China and Iran who eat bracts of inflorescences of foxtail millet (*Setaria italica*) and of *Phalaris minor*, respectively.

Cobalt is essential for many bacteria, including cyanobacteria (blue-green algae). It is required for nitrogen fixation by bacteria in root nodules on legumes (Section 14.2). Figure 6-4 illustrates growth of soybeans with and without cobalt and with only atmospheric nitrogen, which was fixed in root nodules. Cobalt concen-

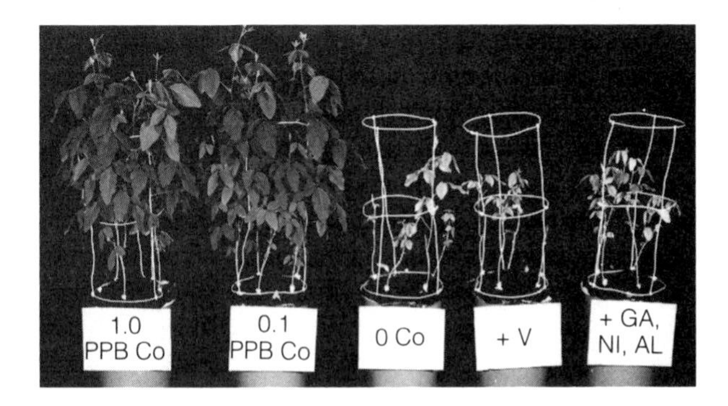


**Figure 6-4** Specific cobalt requirement for nitrogen fixation in soybeans. (From S. Ahmed and H. J. Evans, 1960, *Soil Science* 90:205, with the permission of the publisher. Copyright © 1960 by Williams & Wilkins Company, Baltimore, Maryland.)

trations in the nutrient solution as low as 0.1 $\mu g\ L^{-1}$ were high enough for rapid growth, and neither vanadium, germanium, nickel, nor aluminum could substitute for cobalt. Free-living bacteria that fix nitrogen apart from any symbiotic relationship with plants also require cobalt. Organisms requiring cobalt, including many animals, need it principally because it is a component of the vitamin $B_{12}$ they require. Higher plants and algae are generally believed to contain no vitamin $B_{12}$ and need no cobalt.

In addition to the 17 elements essential for plants, higher animals require sodium, iodine, cobalt, selenium, and apparently also silicon, chromium, tin, vanadium, and fluorine, but apparently not boron (Miller and Neathery, 1977; Mertz, 1981). Another element that was formerly thought to be essential for a few plant species is selenium (see boxed essay entitled "Selenium"). It, too, might eventually prove essential.

## 6.4 Quantitative Requirements and Tissue Analysis

Table 6-3 lists essential elements and their tissue concentrations that appear to be necessary to provide good growth. Such values provide useful guides to physiologists, foresters, orchard managers, and farmers because concentrations of elements in the tissues (especially in selected leaves) indicate more reliably than soil analyses

## Selenium

Selenium is absorbed and accumulated in relatively high concentrations (up to at least 0.5 percent; that is, 5 g per kg dry weight) by certain "accumulator species" of *Astragalus*. Interestingly, although this genus contains about 500 North American species, about 475 of these are nonaccumulators. Certain species of the genera *Stanleya*, *Haplopappus*, and *Xylorhiza* are also notable selenium accumulators.

Because *Astragalus* accumulators live only on seleniferous soils, workers in the 1930s sought to determine whether these plants require selenium. Although it was found that they grew much better on nutrient solutions to which up to nearly 30 mg $L^{-1}$ of selenate ($SeO_4^{2-}$) were added, more recent studies indicate that this growth enhancement occurred because selenate reduced the toxic effects of phosphate; such species are unusually sensitive to phosphate (Bollard, 1983). The dramatic difference in ability to accumulate selenium is illustrated by results from *A. racemosus* and *A. missouriensis* growing side-by-side on a Nebraska soil containing 5 mg of Se/kg soil. *Astragalus racemosus* contained 5.56 mg Se/kg dry weight, whereas *A. missouriensis* contained only 0.025 mg/kg. This genus has apparently evolved in different directions with respect to selenium accumulation.

Such selenium accumulators frequently poison livestock with an often fatal sickness called the alkali disease or the blind staggers. This disease is noted occasionally in certain regions of the western Great Plains of North America, yet seleniferous soils and selenium accumulators are much more widespread geographically (Brown and Shrift, 1982). Toxic forms of Se are certain amino acids in which Se has replaced the sulfur normally present, especially in selenomethionine. Why does the selenium in accumulators not also poison those plants in which it replaces sulfur? We do not know all the answers, but a major one is that accumulators form primarily seleno-amino acids that are not toxic themselves and are not incorporated into certain proteins that are functionally inactive or even toxic (Brown and Shrift, 1982; Bollard, 1983; Anderson and Scarf, 1983).

In bacteria and animals, both of which require selenium, a few essential proteins that contain selenium have been found (Stadtman, 1990). Most such proteins are enzymes that catalyze oxidation-reduction reactions, and the selenium present is essential to their activity. Perhaps similar enzymes occur in selenium-accumulator plants, but so far there is no evidence that this is true. Bollard (1983) concluded that if selenium is essential for *Astragalus* species, it would have to function at tissue concentrations at or less than 0.008 mg/kg of dry matter, *in vivo* levels that are slightly lower than those of molybdenum.

whether plants will grow faster if more of a given element is provided (Wolf, 1982; Bouma, 1983; Moraghan, 1985; Marschner, 1986; Walworth and Sumner, 1988). Figure 6-5 shows an idealized plot of growth rate as a function of the concentration of any given element in the plant (also see Fig. 25-1). In the range of low concentrations called the **deficient zone**, growth increases dramatically as more of the element is provided and its concentration in the plant increases. Above the **critical concentration** (minimal tissue concentration giving almost maximum growth, some say 90 percent of maximum), increases in concentration (fertilizations) do not appreciably affect growth (**adequate zone**). The adequate zone represents **luxury consumption** of the element, during which storage in vacuoles occurs. This zone is fairly wide for macronutrients but is much narrower for micronutrients. Continued increases of any element lead to toxicities and reduced growth (**toxic zone**).

Increasing environmental pollution is bringing much more attention to toxic effects of both essential and nonessential elements (see boxed essay entitled "Metal Toxicity and Resistance"). Increased accumulation of salts is also a problem in many soils, and many approaches, including genetic tolerance, are being sought to relieve the toxicity (Gabelman and Loughman, 1987; Hasegawa et al., 1987; Cheeseman, 1988).

Figure 6-6 shows the average growth responses to various concentrations of calcium for 18 dicots and 11 monocots. The separate curves for these two groups show that critical calcium levels for dicots (about 0.2 percent, dry weight basis) are higher than those for monocots (less than 0.1 percent) and emphasize the approximate nature of the tissue concentrations considered "adequate" in Table 6-3. In general, grasses absorb

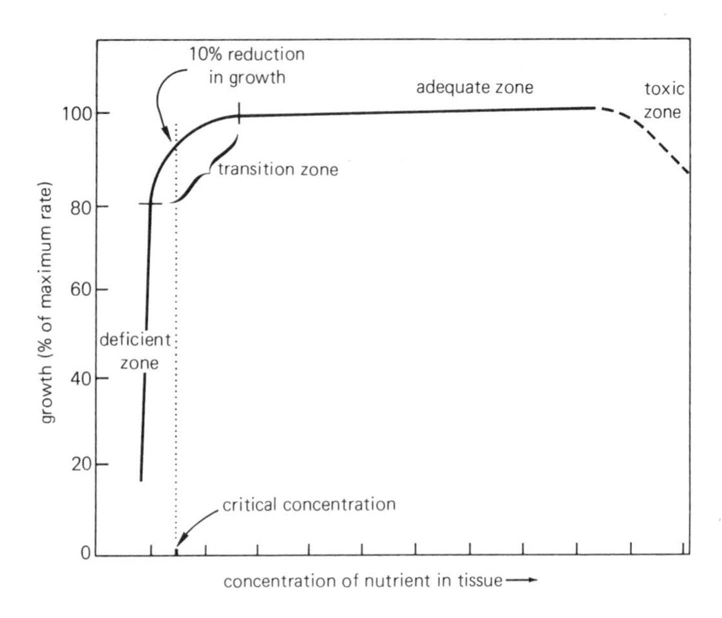


**Figure 6-5** Generalized plot of growth as a function of the concentration of a nutrient in plant tissue. (After Epstein, 1972.)

## Metal Toxicity and Resistance

There is considerable genetic variation in the abilities of various species to tolerate otherwise toxic amounts of nonessential lead, cadmium, silver, aluminum, mercury, tin, and other metals (Woolhouse, 1983). In some species, the elements are absorbed only to a limited extent, so this more accurately represents avoidance rather than tolerance (Taylor, 1987). In other cases the elements accumulate in roots with little transport to shoots. In still others, both roots and shoots contain much higher amounts of such elements than nontolerant species or varieties could live with. This represents true tolerance.

Recently, an important and phylogenetically widespread mechanism of tolerance was discovered (reviewed by Gekeler et al., 1989; Steffens, 1990; and Rauser, 1990). Metals are detoxified by chelation with **phytochelatins**, small peptides rich in the sulfur-containing amino acid cysteine. These peptides generally have two to eight cysteine amino acids in the center of the molecule and a glutamic acid and a glycine at opposite ends. The sulfur atoms of cysteine are almost certainly essential to bind the metals, but other atoms such as nitrogen or oxygen likely also participate.

Phytochelatins are produced in numerous species, but so far they have only been found when toxic amounts of a metal are present. They are also produced when excess amounts of zinc and copper are present, so they can detoxify even essential metals. Their formation therefore represents a true adaptive response to an environmental stress. They act similarly to the far larger metallothionein proteins that detoxify metals in humans and other animals, but in contrast, phytochelatins do not represent direct gene products. Still, genetic control of their production will no doubt prove essential in understanding how various species live on mine wastes and other soils. Molecular-biology studies on metal tolerances by plants have begun and were reviewed by Tomsett and Thurman (1988).

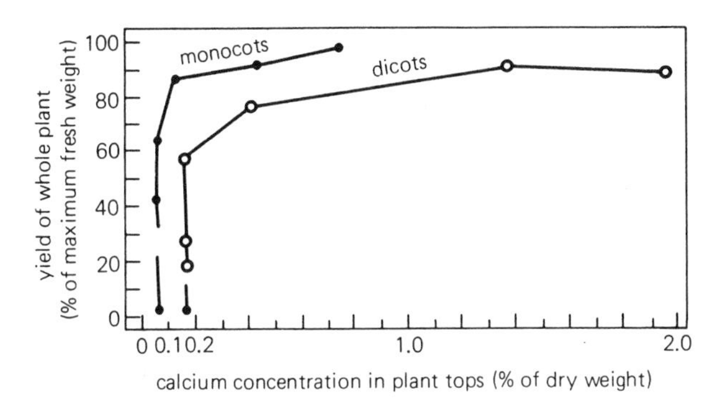


**Figure 6-6** The relation between calcium concentration in tops and relative yields of 18 dicots and 11 monocots after 17 to 19 days' growth in constant calcium concentrations in solution. Each point represents the average values for all species in each plant group given a single $Ca^{2+}$ treatment (0.3, 0.8, 2.5, 10, 100, or 1,000 μM). (From J. F. Loneragan, 1968.)

more potassium and less calcium than do members of the legume family (Leguminosae or Fabaceae) and certain other dicots. However, few comparative data for trees and shrubs exist.

Data in Figure 6-6 emphasize the differences in nutrient requirements among species. Furthermore, graphs similar to those in Figures 6-5 and 6-6 have been effectively used to plan efficient uses of fertilizers for crop plants and forest trees. In the past, only cost prevented soils from being fertilized with nitrogen, phosphorus, or potassium beyond the critical plant-tissue concentrations, but now we know that excess nitrate and some phosphate not absorbed by plants are leached from soils and ultimately appear in lakes and streams. There they cause excessive growth of algae, which leads to the problems of **eutrophication** (nutrient enrichment leading to growth of algae and other plants; upon death, the decomposition of these plants by microorganisms uses so much dissolved oxygen that fish and other animals die). Furthermore, the manufacture of nitrogen fertilizers is one of the most energy-expensive aspects of modern agriculture. Therefore, users of fertilizers should consider not only high yields but also water pollution and world energy requirements.

The abilities of plants to obtain essential nutrients from soils are important in determining where they grow. Although we know much about the mineral nutrition of crop plants, far too little is known about wild species, including forest trees (Chapin, 1987, 1988; *Plant and Soil*, 1983; and Gabelman and Loughman, 1987). Except for carefully fertilized orchards, most trees and native grasses grow on rather unfertile soils, and their soil nutrient requirements are lower than those of crops bred to respond to fertilizers. These lower nutrient requirements result mainly from the ability of native trees, grasses, and herbaceous dicots to absorb nutrients faster than selected crops at low (but not at high) concentrations. Such species are therefore good competitors in their natural environments, where growth is usually slow, but they could not compete in modern agriculture. Nevertheless, thousands of acres of genetically-selected forest trees in the northwestern United States are now fertilized with nitrogen. Of course, leaf fall from deciduous trees in autumn returns some absorbed nutrients to soils. Furthermore, significant amounts of nitrogen, phosphorus, potassium, and magnesium move out of tree leaves into twigs and branches before leaf fall (Ryan and Bormann, 1982; Titus and Kang, 1982). These nutrients are used in new growth the next season. Perennial range grasses similarly conserve minerals by translocation to roots and to lower stem tissues making up the crown in late summer.

## 6.5 Chelating Agents

The micronutrient cations iron, and to a lesser extent zinc, manganese, and copper, are relatively insoluble in nutrient solutions when provided as common inorganic salts, and they are also rather insoluble in most soils. This insolubility is especially marked if the *p*H is above 5, as it is in nearly all soils of the western United States and many other regions with low rainfall. Under these conditions, micronutrient cations react with hydroxyl ions, eventually precipitating insoluble hydrous metal oxides. An example in which the ferric form of iron yields the reddish-brown oxide (rust) is shown in Reaction 6.1:

$$2Fe^{3+} + 6OH^- \rightarrow 2Fe(OH)_3 \rightarrow Fe_2O_3 \cdot 3H_2O \qquad (6.1)$$

Because of this and other reactions that contribute to insolubility, these micronutrients must be held in solution by some other agents. An important kind of agent is called a **ligand** (or **chelating agent** or **chelator**). Reaction of a divalent or trivalent metal ion with a ligand forms a **chelate** (from the Greek, "clawlike"). A chelate is the soluble product formed when certain atoms in an organic ligand donate electrons to the cation. Negatively charged carboxyl groups and nitrogen atoms possess electrons that can be shared in this way. In calcareous soils (those rich in $Ca^{2+}$ and with a *p*H generally of 7 or higher), more than 90 percent of the copper and manganese and half or more of the zinc are probably chelated with microbially produced organic compounds, but what the ligands are isn't known.

Iron deficiency characterized by a lack of chlorophyll (**chlorosis**) is a widespread and worldwide problem in calcareous soils, and it is found in both monocots (mainly grasses) and dicots. It can often be eliminated

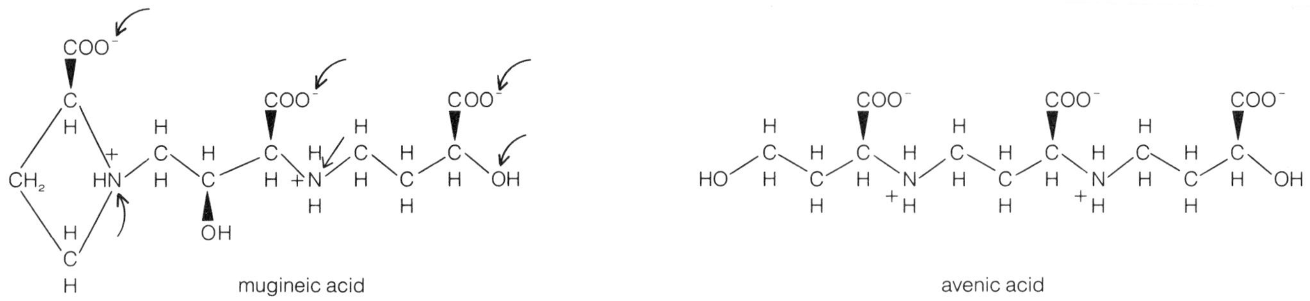

COO⁻ C H H CH₂ HN H C H OH COO⁻ C H H N H C H H COO⁻ C H OH H H H C H

mugineic acid

HO H C H COO⁻ C H N H C H COO⁻ C H N H C H COO⁻ C H OH

avenic acid

**Figure 6-7** Structures of the phytosiderophore ligands mugineic acid and avenic acid. The four oxygen atoms and the two nitrogen atoms of mugineic acid that combine with one $Fe^{3+}$ are indicated by arrows. The atoms of avenic acid that combine with $Fe^{3+}$ have not yet been determined, but note structural similarities in the two acids.

or reduced by adding iron to soils or to leaves in a commercial chelate called *Fe-EDDHA—Fe-ethylenediamine di(o-hydroxyphenyl) acetic acid*, sold under the trade name Sequestrene. Another iron chelate is *Fe-EDTA, Fe-ethylene-diaminetetraacetic acid* (trade name Versenate), but it also chelates $Ca^{2+}$ strongly and therefore is not effective in calcareous soils.

Because iron deficiencies are so widespread, special interest has been paid to what ligands keep iron dissolved in soils and why they sometimes fail to do so. First, realize that $Fe^{3+}$ is far less soluble than is $Fe^{2+}$, so when soils are well aerated the unchelated $Fe^{2+}$ is oxidized to $Fe^{3+}$, which then precipitates as in Reaction 6.1. Of the two forms of unchelated iron, $Fe^{2+}$ is much more easily absorbed by roots, so oxidation strongly removes the available $Fe^{2+}$ form (Lindsay, 1979). There appear to be two major kinds of ligands that form chelates with iron and keep it from totally precipitating: ligands synthesized by microbes and those synthesized by roots. Those produced by roots are excreted into the soil nearby (the **rhizosphere**). To some extent ligand synthesis by roots represents a defense system or strategy against iron deficiency, as described below.

There appear to be two general strategies for iron acquisition by angiosperms (reviewed by Marschner et al., 1986; Romheld, 1987; Chaney, 1988; Brown and Jolley, 1988; Bienfait, 1988; Longnecker, 1988). Gymnosperms have not yet been studied. Strategy I, present in dicots and some monocots, involves release of phenol-like ligands such as caffeic acid (structure given in Fig. 15-11). These ligands chelate mainly $Fe^{3+}$; then this chelated iron moves to the root surface, where it is reduced to $Fe^{2+}$ while still chelated. Simultaneously, roots of iron-stressed strategy-I plants more rapidly form reducing agents (such as NADPH) that carry out the reduction process. Reduction causes loss of $Fe^{2+}$ from the ligand, and $Fe^{2+}$ is immediately absorbed. Also, stressed strategy-I plants more rapidly release $H^+$ ions that favor solubility of both forms of iron, especially $Fe^{3+}$. This defense mechanism often fails in calcareous soils because the soil *p*H is so high and well buffered with bicarbonate ions ($HCO_3^-$). This failure contributes to the physiological disease called **lime-induced chlorosis** (Korcak, 1987; Mengel and Geurtzen, 1988).

Strategy-II plants are represented only by grasses, including cereal grains, so far as is known. They respond to iron-deficiency stress by forming and releasing powerful ligands that chelate $Fe^{3+}$ specifically and strongly. These ligands are called **siderophores** (Greek, "iron-bearers") or, more specifically, **phytosiderophores** (Sugiura and Nomoto, 1984; Neilands and Leong, 1986). Structures of two of the most studied phytosiderophores (*avenic acid* and *mugineic acid*) are in Figure 6-7. Both are iminocarboxylic acids that bind to $Fe^{3+}$ by oxygen and nitrogen atoms as described in Figure 6-7. These and other siderophores are absorbed with the iron still in them, so roots must absorb phytosiderophores and then reduce the iron they contain to $Fe^{2+}$. Presumably, the $Fe^{2+}$ is immediately released and used by the plant, whereas the siderophore might then be either degraded chemically or released from the root to carry in more iron. Future breeding or genetic-engineering techniques with cereals might concentrate on genes that control siderophore formation as a means of improving the plants' ability to grow in calcareous soils.

Once absorbed, divalent metals are kept soluble partly by chelation with certain cellular ligands. Anions of organic acids, especially citric acid, appear to be most important as ligands for transport of iron, zinc, and manganese through the xylem, whereas amino acids seem most important for transport of copper (White et al., 1981; Mullins et al., 1986). Ultimately, much iron, zinc, manganese, nickel, and copper are bound to proteins. In this form they speed electron-transport processes of photosynthesis and respiration and increase the catalytic activity of enzymes. Monovalent cations such as $K^+$ and $Na^+$ do not form stable chelates, but even they are associated loosely by ionic attractions with both inorganic and organic acid anions, including proteins.

## 6.6 Functions of Essential Elements: Some Principles

Essential elements have sometimes been classified functionally into two groups: those having a role in the structure of an important compound and those having an enzyme-activating role. There is no sharp distinction between these functions because several elements form structural parts of essential enzymes and help catalyze the chemical reaction in which the enzyme participates. Carbon, oxygen, and hydrogen are the most obvious elements performing both functions, although nitrogen and sulfur, also found in enzymes, are equally important. Another example of an element with both structural and enzyme-activating roles is magnesium; it is a structural part of chlorophyll molecules and also activates many enzymes. Most of the micronutrients are essential mainly because they activate enzymes (Robb and Peirpont, 1983).

All elements in soluble form, whether free or bound structurally to essential compounds, perform another function by contributing to osmotic potentials, thus aiding buildup of the turgor pressure necessary to maintain form, speed growth, and allow certain pressure-dependent movements (for example, stomatal opening, Chapter 4, and "sleep" movements of leaves, Chapter 19). Abundant, nonbound potassium ions are dominant in this regard, but all ions contribute somewhat to osmotic potentials and, therefore, to turgor pressure. Potassium and perhaps chloride—both monovalent ions—are also necessary elements because they temporarily combine with and activate certain enzymes. No permanent structural roles that would make these elements essential are known, yet they perform transient structural roles.

## 6.7 Nutrient Deficiency: Symptoms and Functions of Elements

Plants respond to an inadequate supply of an essential element by forming characteristic **deficiency symptoms**. Such visually observable symptoms include stunted growth of roots, stems, or leaves and chlorosis or necrosis of various organs. Characteristic symptoms often help determine the necessary functions of the element in the plant, and knowledge of symptoms helps agriculturists and foresters determine how and when to fertilize crops. Several symptoms are described below and are illustrated in books by Gauch (1972), Hewitt and Smith (1975), Grundon (1987), Mengel and Kirkby (1987), Bould et al. (1984), Scaife and Turner (1984), and Robinson (1987), and, for horticultural trees, in an article by Shear and Faust (1980).

Most of the symptoms described appear on the plant's shoot system and are easily observed. Unless plants are grown hydroponically, root symptoms cannot be seen without removing the roots from the soil, so root deficiency symptoms have been less well described. Furthermore, all symptoms differ to some extent according to the species, the severity of the problem, the growth stage, and (as you might suspect) complexities resulting from deficiencies of two or more elements.

The deficiency symptoms for any element depend primarily on two factors:

1. The function or functions of that element
2. Whether or not the element is readily translocated from old leaves to younger leaves

A good example emphasizing both factors is the chlorosis that results from magnesium deficiency. Because magnesium is an essential part of chlorophyll molecules, no chlorophyll is formed in its absence, and only limited amounts are formed when it is present in too low a concentration. Furthermore, chlorosis of lower, older leaves becomes more severe than that of younger leaves. This difference illustrates an important principle: Young parts of a plant have a pronounced ability to withdraw mobile nutrients from older parts, and reproductive organs, flowers, and seeds are especially good at withdrawing nutrients—as mentioned in Chapter 8 and as you might predict if a species is to perpetuate itself. We do not yet understand this withdrawing power, but hormonal relations are involved (Chapters 17 and 18).

Whether withdrawal of an element from a leaf is successful, as with magnesium, depends on the element's mobility in the phloem of vascular tissues. This mobility is determined partly by the solubility of the chemical form of the element in the tissue and partly by how well it can enter the sieve tubes of the phloem. As discussed in Chapter 8, some elements readily move through the phloem from old leaves to younger ones and then to storage organs. These elements include nitrogen, phosphorus, potassium, magnesium, and chlorine. Others such as boron, iron, and calcium are much less mobile, and the mobility of sulfur, zinc, manganese, copper, and molybdenum is usually intermediate. If the element is soluble and can also be loaded into translocating phloem cells, its deficiency symptoms appear earliest and most pronounced in older leaves, whereas symptoms resulting from lack of a relatively immobile element such as calcium or iron appear first in younger leaves. A general guide to deficiency symptoms, partly emphasizing the phloem mobility principle, is given in Table 6-4. However, this table does not include nickel deficiency symptoms, and special attention will be given this element at the end of the next section.

**Table 6-4 A Guide to Plant Nutrient Deficiency Symptoms.**

| Symptoms | Deficient Element |
|---|---|
| Older or lower leaves of plant mostly affected; effects localized or generalized. | |
| Effects mostly generalized over whole plant; more or less drying or firing of lower leaves; plant light or dark green. | |
| Plant light green; lower leaves yellow, drying to light brown color; stalks short and slender if element is deficient in later stages of growth. | Nitrogen |
| Plant dark green, often developing red and purple colors; stalks short and slender if element is deficient in later stages of growth. | Phosphorus |
| Effects mostly localized; mottling or chlorosis with or without spots of dead tissue on lower leaves; little or no drying up of lower leaves. | |
| Mottled or chlorotic leaves; typically may redden, as with cotton; sometimes with dead spots; tips and margins turned or cupped upward; stalks slender. | Magnesium |
| Mottled or chlorotic leaves with large or small spots of dead tissue. | |
| Spots of dead tissue small, usually at tips and between veins, more marked at margins of leaves; stalks slender. | Potassium |
| Spots generalized, rapidly enlarging, generally involving areas between veins and eventually involving secondary and even primary veins; leaves thick; stalks with shortened internodes. | Zinc |
| Newer or bud leaves affected; symptoms localized. | |
| Terminal bud dies, following appearance of distortions at tips or bases of young leaves. | |
| Young leaves of terminal bud at first typically hooked, finally dying back at tips and margins, so that later growth is characterized by a cut-out appearance at these points; stalk finally dies at terminal bud. | Calcium |
| Young leaves of terminal bud become light green at bases, with final breakdown here; in later growth, leaves become twisted; stalk finally dies back at terminal bud. | Boron |
| Terminal bud commonly remains alive; wilting or chlorosis of younger or bud leaves with or without spots of dead tissue; veins light or dark green. | |
| Young leaves permanently wilted (wither-tip effect) without spotting or marked chlorosis; twig or stalk just below tip and seedhead often unable to stand erect in later stages when shortage is acute. | Copper |
| Young leaves not wilted; chlorosis present with or without spots of dead tissue scattered over the leaf. | |
| Spots of dead tissue scattered over the leaf; smallest veins tend to remain green, producing a checkered or reticulating effect. | Manganese |
| Dead spots not commonly present; chlorosis may or may not involve veins, making them light or dark green. | |
| Young leaves with veins and tissue between veins light green. | Sulfur |
| Young leaves chlorotic, principal veins typically green; stalks short and slender. | Iron |

Source: Based on data of McMurtrey (1938) and Grundon (1987).

## Nitrogen

Soils are more commonly deficient in nitrogen than any other element, although phosphorus deficiency is also widespread. Two major ionic forms of nitrogen are absorbed from soils: nitrate ($NO_3^-$) and ammonium ($NH_4^+$), as described in Chapter 14. Because nitrogen is present in so many essential compounds, it is not surprising that growth without added nitrogen is slow. Plants containing enough nitrogen to attain limited growth exhibit deficiency symptoms consisting of a general chlorosis, especially in older leaves. In severe cases these leaves become completely yellow and then tan as they die. They frequently fall off the plant in the yellow or tan stage. Younger leaves remain green longer because they receive soluble forms of nitrogen transported from older leaves. Some plants, including tomato and certain cultivars of maize, exhibit a purplish coloration in stems, petioles, and lower leaf surfaces caused by accumulation of anthocyanin pigments.

Plants grown with excess nitrogen usually have dark green leaves and show an abundance of foliage, usually with a root system of minimal size and therefore a high shoot-to-root ratio. (A reverse ratio often occurs when nitrogen is deficient.) Potato plants grown with superabundant nitrogen show excess shoot growth with only small underground tubers. Reasons for this relatively high shoot growth are unknown, but undoubtedly sugar translocation to roots or tubers is affected in some way, perhaps because of a hormone imbalance. Excess nitrogen also causes tomato fruits to split as they ripen. Flowering and formation of seeds of several agricultural crops are retarded by excess nitrogen.

## Phosphorus

Second to nitrogen, phosphorus is most often the limiting element in soils. It is absorbed primarily as the monovalent phosphate anion ($H_2PO_4^-$) and less rapidly as the divalent anion ($HPO_4^{2-}$). The soil *p*H controls the relative abundance of these two forms, $H_2PO_4^-$ being favored below *p*H 7 and $HPO_4^{2-}$ above *p*H 7. Much phosphate is converted into organic forms upon entry into the root or after transport through the xylem into the shoot. In contrast to nitrogen and sulfur, phosphorus never undergoes reduction in plants and remains as phosphate, either free or bound to organic forms as esters. Phosphorus-deficient plants are stunted and, in contrast to those lacking nitrogen, are often dark green in color. Anthocyanin pigments sometimes accumulate. Oldest leaves become dark brown as they die.

Maturity is often delayed compared to plants containing abundant phosphate. In many species phosphorus and nitrogen interact closely in affecting maturity, excess nitrogen delaying and abundant phosphorus speeding maturity. If excess phosphorus is provided, root growth is often increased relative to shoot growth. This, in contrast to effects with excess nitrogen, causes low shoot-to-root ratios.

Phosphate is easily redistributed in most plants from one organ to another and is lost from older leaves, accumulating in younger leaves and in developing flowers and seeds. As a result, deficiency symptoms occur first in more mature leaves.

Phosphorus is an essential part of many sugar phosphates involved in photosynthesis, respiration, and other metabolic processes, and it is also part of nucleotides (as in RNA and DNA) and of the phospholipids present in membranes. It also plays an essential role in energy metabolism because of its presence in ATP, ADP, AMP, and pyrophosphate (PPi).

## Potassium

After nitrogen and phosphorus, soils are usually most deficient in potassium. Because of the importance of these three elements, commercial fertilizers list the percentages of nitrogen, phosphorus, and potassium they contain (but the last two are actually expressed as equivalent percents of $P_2O_5$ and $K_2O$). As with nitrogen and phosphorus, $K^+$ is easily redistributed from mature to younger organs, so deficiency symptoms first appear on older leaves. In dicots, these leaves initially become slightly chlorotic, especially close to dark **necrotic lesions** (dead or dying spots) that soon develop. In many monocots, such as cereal crops, cells at the tips and margins of leaves die first, and the necrosis spreads basipetally along the margins toward the younger, lower parts of leaf bases. Potassium-deficient maize and other cereal grains develop weak stalks, and their roots become more easily infected with root-rotting organisms. These two factors cause the plants to be rather easily bent to the ground (**lodged**) by wind, rain, or early snowstorms.

Potassium is an activator of many enzymes that are essential for photosynthesis and respiration, and it also activates enzymes needed to form starch and proteins (Bhandal and Malik, 1988). This element is also so abundant that it is a major contributor to the osmotic potential of cells and therefore to their turgor pressure. (See discussion about potassium and stomatal action in Section 4.6.)

## Sulfur

Sulfur is absorbed from soils as divalent sulfate anions ($SO_4^{2-}$). It seems to be metabolized by roots only to the extent that they require it, and much sulfate is translocated unchanged to the shoots in the xylem. Because enough sulfate is present in most soils, sulfur-deficient plants are fairly uncommon. Nevertheless, they have been observed in several parts of Australia, some regions of Scandinavia, southwestern grain-producing regions of Canada, and in scattered parts of the northwestern United States. Deficiency symptoms consist of a general chlorosis throughout the entire leaf, including vascular bundles (veins). Sulfur is not easily redistributed from mature tissues in some species, so deficiencies are usually noted first in younger leaves. In other species, however, most leaves become chlorotic at about the same time, or even first in older leaves. Many crop plants, including the roots, contain about one-fifteenth as much total sulfur as nitrogen (on a weight basis), and this appears to be a useful guide for evaluating nutritional needs (Duke and Reisenauer, 1986).

Most of the sulfur in plants occurs in proteins, specifically in the amino acids cysteine and methionine, which are building blocks for proteins. Other essential compounds that contain sulfur are the vitamins thiamine and biotin, as well as coenzyme A, a compound essential for respiration and for synthesis and breakdown of fatty acids.

Sulfur can also be absorbed by leaves through stomates as gaseous sulfur dioxide ($SO_2$), an environmental pollutant released primarily from burning coal, wood, and oil. $SO_2$ is converted to bisulfite ($HSO_3^-$)

when it reacts with water in the cells, and in this form it both inhibits photosynthesis and causes chlorophyll destruction. Bisulfite is oxidized further to $H_2SO_4$; this acid has been blamed for toxic effects of acidic rainfall (acid rain) in the northeastern United States and adjacent Canadian regions and in many Scandinavian regions.

## Magnesium

Magnesium is absorbed as divalent $Mg^{2+}$. In its absence, chlorosis of the older leaves is the first symptom, as already mentioned. This chlorosis is usually interveinal because for unknown reasons the mesophyll cells next to the vascular bundles retain chlorophyll for longer periods than do the parenchyma cells between them. Magnesium is almost never limiting to plant growth in soils. Besides its presence in chlorophyll, magnesium is essential because it combines with ATP (thereby allowing ATP to function in many reactions) and because it activates many enzymes needed in photosynthesis, respiration, and formation of DNA and RNA.

## Calcium

Calcium is absorbed as divalent $Ca^{2+}$. Most soils contain enough $Ca^{2+}$ for adequate plant growth, but acidic soils where high rainfall occurs are often fertilized with lime (a mixture of CaO and $CaCO_3$) to raise the *p*H. In contrast to $Mg^{2+}$, $Ca^{2+}$ apparently cannot be loaded into translocating phloem cells; as a result, deficiency symptoms are always more pronounced in young tissues (Kirkby and Pilbeam, 1984). Meristematic zones of roots, stems, and leaves, where cell divisions occur, are most susceptible, perhaps because calcium is required to form a new middle lamella in the cell plate that arises between daughter cells. Twisted and deformed tissues result from calcium deficiency, and the meristematic zones die early. In tomatoes, degeneration of young fruits near the blossom ("blossom end rot") is caused by calcium deficiency. Calcium is essential for normal membrane functions in all cells, probably as a binder of phospholipids to each other or to membrane proteins (Section 7.4).

Calcium is receiving renewed attention because it is now recognized that all organisms maintain unexpectedly low concentrations of free $Ca^{2+}$ in the cytosol, usually less than 1 $\mu M$ (reviewed by Hanson, 1984; Hepler and Wayne, 1985; Trewavas, 1986; Leonard and Hepler, 1990). This is true even though calcium is as abundant in many plants, especially legumes, as phosphorus, sulfur, and magnesium. Most calcium in plants is in central vacuoles and bound in cell walls to pectate polysaccharides (Kinzel, 1989). In vacuoles, calcium is frequently precipitated as insoluble crystals of oxalates, and in some species as insoluble carbonate, phosphate, or sulfate. The low, near micromolar concentrations of $Ca^{2+}$ in the cytosol apparently must be maintained in part to prevent formation of insoluble calcium salts of ATP and other organic phosphates. Also, concentrations of $Ca^{2+}$ above the micromolar range inhibit cytoplasmic streaming (Williamson, 1984). Although a few enzymes are activated by $Ca^{2+}$, many are inhibited, and the inhibition provides still further need for cells to maintain unusually low $Ca^{2+}$ concentrations in the cytosol, where many enzymes exist.

Much calcium within the cytosol becomes reversibly bound to a small protein called **calmodulin** (Cheung, 1982; Roberts et al., 1986). This binding changes the structure of calmodulin in such a way that it then activates several enzymes (Chapter 17). The relation of calcium and calmodulin to enzyme activity in plants is now undergoing vigorous research (reviewed by Roberts et al., 1986; Allan and Trewavas, 1987; Poovaiah and Reddy, 1987; Ferguson and Drobak, 1988; Gilroy et al., 1987; Marmé, 1989). More is said about the probable roles of calcium and calmodulin in plant development in later chapters. For now, we emphasize that an enzyme-activating role for $Ca^{2+}$ likely exists mainly when the ion is bound to calmodulin or closely related proteins.

## Iron

Iron-deficient plants are characterized by development of a pronounced interveinal chlorosis similar to that caused by magnesium deficiency but occurring first on the youngest leaves. Interveinal chlorosis is sometimes followed by chlorosis of the veins, so the whole leaf then becomes yellow. In severe cases, the young leaves even become white with necrotic lesions. The reason that iron deficiency results in a rapid inhibition of chlorophyll formation is incompletely known, but two or three enzymes that catalyze certain reactions of chlorophyll synthesis apparently require $Fe^{2+}$.

Iron that has accumulated in older leaves is relatively immobile in the phloem, as it is in the soil, perhaps because it is internally precipitated in leaf cells as an insoluble oxide or in the form of inorganic or organic ferric-phosphate compounds. Direct evidence that such precipitates are formed is weak, and perhaps other unknown but similarly insoluble compounds are formed. One abundant and stable form of iron in leaves is stored in chloroplasts as an iron-protein complex called *phytoferritin* (Seckback, 1982). Entry of iron into the phloem transport stream is probably minimized by formation of such insoluble compounds, although phytoferritin seems to represent a storehouse of iron.

Iron deficiencies are often found in particularly sensitive species in the rose family, including shrubs and fruit trees (Fig. 6-8), and in maize and sorghum. In soils of the western United States, the high *p*H and the

**Figure 6-8** Iron deficiency in apple leaves: O, normal; A–D various levels of deficiency, with D the most severe. (Courtesy of M. Faust.)

presence of bicarbonates contribute to iron deficiency, whereas in acidic soils soluble aluminum is more abundant and restricts iron absorption.

Iron is essential because it forms parts of certain enzymes and numerous proteins that carry electrons during photosynthesis and respiration. It undergoes alternate oxidation and reduction between the $Fe^{2+}$ and $Fe^{3+}$ states as it acts as an electron carrier in proteins. The importance of iron, zinc, copper, and manganese in electron-transport processes in plants was reviewed by Sandman and Boger (1983).

## Chlorine

Chlorine is absorbed from soils as the chloride ion ($Cl^-$) and most of it remains in this form, although more than 130 organic compounds that contain chlorine have been detected in the plant kingdom in trace amounts (Engvild, 1986). One of the more interesting is 4-chloroindoleacetic acid, which seems to be a natural auxin hormone. Most species absorb 10 to 100 times as much chloride as they require, so it represents a common example of luxury consumption. One function of chloride is to stimulate the split (oxidation) of $H_2O$ during photosynthesis (Chapter 10), but it is also essential for roots, for cell division in leaves, and as an important osmotically active solute (Terry, 1977; Flowers, 1988).

Chloride-deficiency symptoms in leaves consist of reduced growth, wilting, and development of chlorotic and necrotic spots. Leaves often eventually attain a bronze color. Roots become stunted in length but thickened, or club-shaped, near the tips. Chloride is rarely if ever deficient in nature because of its high solubility and availability in soils and because it is also transported in dust or in tiny moisture droplets by wind and rain to leaves, where absorption occurs. Because of its presence in human skin, it was necessary for researchers investigating its essentiality to wear rubber gloves.

## Manganese

Manganese exists in three oxidation states ($Mn^{2+}$, $Mn^{3+}$, and $Mn^{4+}$) as insoluble oxides in soils, and it also exists in chelated form. It is absorbed largely as the divalent manganous cation ($Mn^{2+}$) after either release from chelates or reduction in higher-valence oxides at the root surface (Uren, 1981). Deficiencies of manganese are not common, although various disorders such as "gray speck" of oats, "marsh spot" of peas, and "speckled yellows" of sugar beets result when inadequate amounts are present. Initial symptoms are often an interveinal chlorosis on younger or older leaves, depending on the species, followed by or associated with necrotic lesions. Electron microscopy of chloroplasts from spinach leaves showed that the absence of manganese causes disorganization of thylakoid membranes but has little effect on the structure of nuclei and mitochondria. This and much biochemical work indicate that the element plays a structural role in the chloroplast membrane system and that one of its important roles is, like that of chloride, in the photosynthetic split of $H_2O$ (Section 10.6). The $Mn^{2+}$ ion also activates numerous enzymes.

## Boron

Boron is almost entirely absorbed from soils as undissociated boric acid ($H_3BO_3$, more accurately represented as $B(OH)_3$). It is only slowly translocated out of organs in the phloem of many species once it arrives there in the xylem (Raven, 1980). However, in some species it moves out of the phloem much more effectively (Welch, 1986; Shelp, 1988). Deficiencies are not common in most areas, yet several disorders related to disintegration of internal tissues, such as "heart rot" of beets, "stem crack" of celery, "water core" of turnip, and "drought spot" of apples, result from an inadequate boron supply. Plants deficient in boron show a wide variety of symptoms, depending upon the species and plant age, but the earliest symptom is failure of root tips to elongate normally, accompanied by inhibited synthesis of DNA and RNA. Cell division in the shoot apex is also inhibited, as is that in young leaves. Boron plays an undetermined but essential role in elongation of pollen tubes.

Much evidence indicates that it is required only for two major taxonomic groups, vascular plants and diatoms (Lovatt, 1985). In diatoms it forms part of the silicon-rich cell wall.

Biochemical functions of boron in vascular plants remain unclear in spite of much study, partly because we don't know to what extent $B(OH)_3$ is modified in cells and partly because there might be several functions. Probably, much of this weak acid becomes bound as *cis*-diol borate complexes with adjacent hydroxyl groups of mannose and certain other sugars in cell-wall polysaccharides (but not with glucose, fructose, galactose, and sucrose, which have no *cis*-diol arrangements of hydroxyl groups). The biochemical and physiological functions proposed for boron were reviewed by Dugger (1983), Pilbeam and Kirkby (1983), and Lovatt (1985). No specific function is yet certain, but evidence favors special involvement of boron in nucleic-acid synthesis that is so essential to cell division in apical meristems.

## Zinc

Zinc is absorbed as divalent $Zn^{2+}$, probably often from zinc chelates. Disorders caused by zinc deficiency include "little leaf" and "rosette" of apples, peaches, and pecans, resulting from growth reduction of young leaves and stem internodes. Leaf margins are often distorted and puckered. Interveinal chloroses often occur in leaves of maize, sorghum, beans, and fruit trees, suggesting that zinc participates in chlorophyll formation or prevents chlorophyll destruction. The retardation of stem growth in its absence might result partly from its apparent requirement for the production of a growth hormone, indoleacetic acid (auxin). Many enzymes contain tightly bound zinc that is essential for their function; considering all organisms, more than 80 such enzymes are known (Vallee, 1976).

## Copper

Plants are rarely deficient in copper, partly because they need so little of it (Table 6-3). Nevertheless, many Australian soils are extremely deficient in copper (and other micronutrients such as zinc and molybdenum). These soils are widely fertilized with copper and other micronutrients (Donald and Prescott, 1975). Without copper, young leaves often become dark green and are twisted or otherwise misshapen, often exhibiting necrotic spots. Citrus orchards are occasionally deficient, in which case the dying young leaves inspired the name "die back" disease. Copper is absorbed both as the divalent cupric ($Cu^{2+}$) ion in aerated soils or as the monovalent cuprous ion in wet soils with little oxygen. Divalent $Cu^{2+}$ is chelated in various soil compounds (generally unidentified), and these likely provide most copper to root surfaces. Partly because such small amounts are needed by plants, copper readily becomes toxic in solution culture unless its amounts are carefully controlled.

Copper is present in several enzymes or proteins involved in oxidation and reduction. Two notable examples are *cytochrome oxidase*, a respiratory enzyme in mitochondria (Section 13.7), and plastocyanin, a chloroplast protein (Section 10.5).

## Molybdenum

Molybdenum exists to a large extent in soils as molybdate ($MoO_4^{2-}$) salts and as $MoS_2$. In the former, molybdenum exists in the redox (valence) state of $Mo^{6+}$, but in sulfide salts it occurs as $Mo^{4+}$. Probably because only trace amounts are required by plants, virtually nothing is known about the forms in which it is absorbed and the ways it is changed in plant cells. Most plants require less molybdenum than any other element, so molybdenum deficiencies are rare. Nevertheless, they are geographically widespread, especially throughout Australia. Examples of disorders caused by inadequate molybdenum include "whiptail" of cauliflower and broccoli, found, for example, in certain areas of the eastern United States. Symptoms often consist of an interveinal chlorosis occurring first on the older or midstem leaves, then progressing to the youngest leaves. Sometimes, as in the "whiptail" disease, plants do not become chlorotic but develop severely twisted young leaves, which eventually die. In acidic soils, adding lime increases availability of molybdenum and eliminates or reduces the severity of its deficiency. The most well-documented function of molybdenum in plants is as part of the enzyme *nitrate reductase*, which reduces nitrate ions to nitrite ions (Chapter 14), but it may also play a role in breakdown of purines such as adenine and guanine because of its essentiality as part of the enzyme *xanthine dehydrogenase* (Mendel and Muller, 1976; Perez-Vicénte et al., 1988). A third probable function of molybdenum is to form an essential part of an oxidase that converts abscisic acid aldehyde to the hormone ABA (Walker-Simmons et al., 1989).

## Nickel

There is now good evidence that nickel ($Ni^{2+}$) is an essential element for plants (Dalton et al., 1988). It has been known for several years that nickel is an essential part of an enzyme called *urease*, which catalyzes hydrolysis (breakdown using $H_2O$) of urea to $CO_2$ and $NH_4^+$. If urease is essential to plants, then nickel would be deemed essential according to the second criterion for essentiality mentioned earlier in this chapter. But it was not known whether urease is essential because it was not clear whether most or all plants form urea and need urease to hydrolyze urea. Apparently plants generally do form urea and require urease. Although mammals can remove excess urea via the kidneys, they too require nickel and urease.

Legumes of tropical origin, including cowpea (*Vigna unguiculata*) and soybean (*Glycine max*), form ureides in root nodules during nitrogen fixation; ureides are then transported via the xylem to leaves (Section 13.2). They also transfer ureides from old, senescing leaves to developing seeds and younger leaves via the phloem. Utilization of the nitrogen in such ureides by cowpeas and soybeans apparently involves breakdown to urea and then hydrolysis of urea, because without nickel, toxic amounts of urea build up in leaf tips when plants begin to flower (Eskew et al., 1984; Walker et al., 1985). When nickel was carefully removed from nutrient solutions, plants accumulated so much urea in their leaf tips that necrotic (dead) spots appeared. Therefore, ureide breakdown produces urea, and without nickel no urease can be formed to remove the toxic urea.

Breakdown of purine bases (adenine and guanine) occurs via ureides in all plants, so it seemed likely that all plants probably required urease and nickel. Good evidence that nickel is essential for barley has now been provided by Brown et al. (1987). They obtained seeds from plants grown three generations in nutrient solutions in which nickel had been carefully removed with a chelating agent. They found that third-generation seeds were frequently incapable of germinating (were nonviable) and showed various anatomical abnormalities. Thus the first criterion for essentiality has been demonstrated for nickel in barley. Beneficial effects of nickel on growth of oats, wheat, and tomato are also known, and nickel seems essential for certain algae (Welch, 1981; Rees and Bekheet, 1982). Therefore, nickel is likely essential for all plants, and it is the first element to be added to the list of essential elements since chlorine in 1954. We assume that nickel is essential mainly because of its presence in urease, but other functions for it may be discovered.

# 7

# Absorption of Mineral Salts

In previous chapters we explained how water moves into, through, and out of plants and how osmosis is essential to water movement. We found it generally advantageous to ignore the much slower movement of solutes across membranes, a viewpoint that allows simplified explanations of osmosis. In fact, osmosis could not occur unless movement of water was much faster than that of solutes.

Yet solutes do move from cell to cell and from one cellular organelle to another, and this movement is essential for life. Carbon, oxygen, and hydrogen are provided by $H_2O$ and by atmospheric $CO_2$ and $O_2$, but the 14 other elements that are essential to plants are absorbed as ions from the soil by a process aptly called "solution mining." Just as leaves must absorb carbon from a low concentration of $CO_2$ in the atmosphere, roots must absorb essential mineral salts from low concentrations in the soil solution. This chapter concerns the morphological and anatomical properties of roots that allow them to absorb these mineral salts effectively. Properties of membranes, which control absorption rates, are also described. Finally, some theories and hypotheses about how solutes move across membranes are discussed.

## 7.1 Roots as Absorbing Surfaces

Plants solve the problem of absorbing frequently scarce water and mineral elements from soils by producing surprisingly large root systems. Although many plants invest only 20 to 50 percent of their total weight in roots, in some cases (especially when plants are stressed by insufficient water or mineral nitrogen) as much as 90 percent of the total plant biomass is in roots (Fig. 7-1). On the other hand, when wheat plants were grown hydroponically with adequate water and high nitrogen, only 3 to 5 percent of the plants' biomass was in roots (Bugbee and Salisbury, 1988).

The overall shapes of root systems are controlled mainly by genetic rather than environmental mechanisms. Thus grasses have fibrous and highly branched root systems near the soil surface, although the roots of perennial grasses extend more deeply than those of related annual species. Many perennial herbaceous (nonwoody) dicots have a dominant taproot that can extend several meters downward (for example, alfalfa), although the dominant taproot is shorter in most species (for example, carrot, sugar beet, dandelion, and Canada thistle). Other common herbaceous dicots such as soybean and tomato have root systems with a taproot that is difficult to distinguish from branch roots. The same is true for many trees and shrubs, both angiosperms and gymnosperms alike, although tree-root morphology can be complex, especially in pines (Kramer and Kozlowski, 1979). Roots commonly extend outwardly from tree trunks much farther than above-ground branches.

Even though root morphology is genetically controlled, soil environments have influences (Klepper, 1987). For example, when soils are dry, many species invest relatively more biomass in roots, so the root-to-shoot ratio is greater than when soils are moist. Also, branching patterns of roots are more varied than those of shoots. If a topsoil is only a thin layer covering a hardpan of clay or rocks, roots cannot grow deeply but must instead spread laterally near the surface. Essentially, roots grow where they can, and mechanical impedance, temperature, aeration, and availability of water and mineral salts are all important factors. In moist and fertile regions, roots proliferate extensively (Fig. 7-2) until water or nutrients become depleted (Drew, 1975, 1987; Granato and Raper, 1989). After water and nutrients are depleted, roots grow into new soil regions by formation of additional branch or feeder roots. If water is more available deep in the ground, roots generally grow far below the soil surface. Nevertheless, plants adapted to dry regions are not necessarily deep-rooted because shallow root systems take

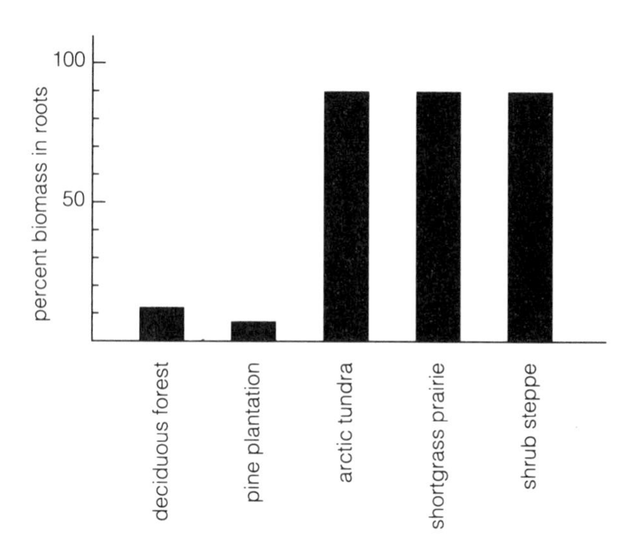


**Figure 7-1** The percent of biomass in root systems of perennial plants in various ecosystems. The deciduous forest was dominated by *Liriodendron tulipifera* (tuliptree), the pine plantation by *Pinus sylvestris* (scotch pine), and the shrub steppe (cool desert) by *Atriplex confertifolia* (shadscale saltbush). The arctic tundra was in Alaska, and the shortgrass prairie was in the nongrazed Pawnee Grassland site in northeastern Colorado. (Redrawn from Caldwell, 1987.)

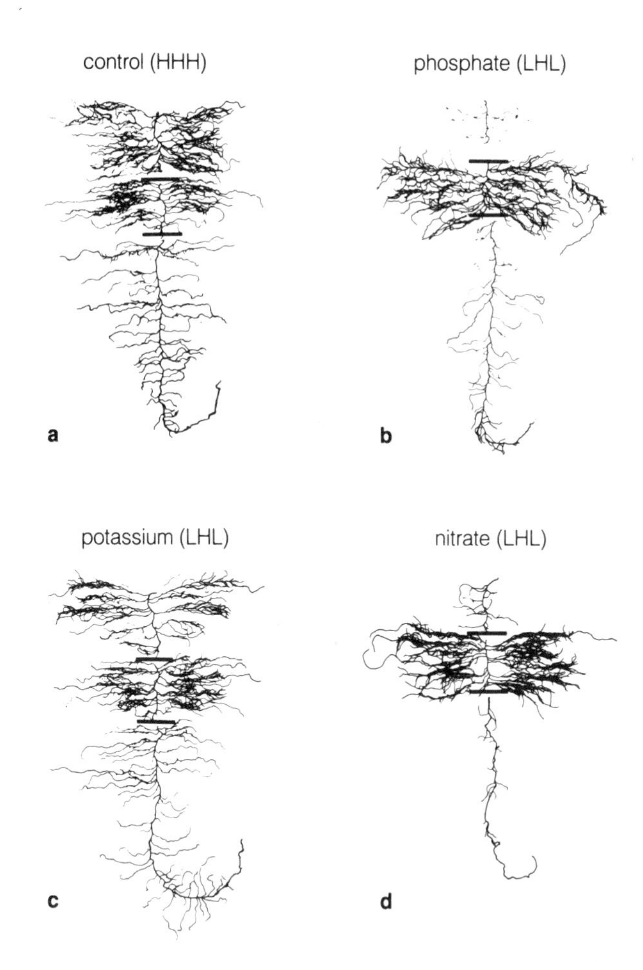


**Figure 7-2** Root proliferation of barley in localized zones of sand fertilized with phosphate, potassium, or nitrate. Portions of root systems (shown separated by line-bars) were grown 21 days in sand compartments separated into three layers by wax barriers through which roots could grow but solution did not flow. Layers were fertilized with nutrient solution containing high (H) or low (L) levels of the particular element. Controls (HHH) received high levels of elements in all three layers. Plants exposed to varying potassium showed little proliferation in the well-fertilized central layer, but the acid-washed sand was found to contribute $K^+$. (From Drew, 1975.)

better advantage of brief intermittent rains. In fact, root systems of some species proliferate both near soil surfaces and at substantial depths with a few long and relatively unbranched connections between; such systems probably represent adaptations to variable climates.

Far too little is known about properties of roots in soils because roots are difficult to observe. Yet careful studies show that branch roots of annual crops elongate for only a few days and that those of perennial species live a year or more before decay occurs. Some desert shrubs replace up to a fourth of their root systems each year, absorbing water and mineral salts from new locations through new roots. Yearly root losses from perennial grasses occur at slower rates than those from perennial shrubs, and retention of old roots for several years contributes to the pronounced ability of grasses to prevent soil erosion. Not much is known about death and replacement of tree roots, but reports summarized by Sutton (1980), which indicated that a 100-year-old Scots pine (*Pinus sylvestris*) had five million root tips and that a mature red oak (*Quercus rubrum*) had 500 million live root tips, show the vastness of such root systems.

The cylindrical and filamentous form of roots is unexpectedly important for absorption of water and solutes from soils. A cylinder has more strength per unit cross-sectional area than do other shapes, and this shape (with a protective root cap) helps growing roots force soil particles aside without breaking the roots. The filamentous form of roots allows exploration of much more soil volume per unit root volume than if roots were spherical or disc-shaped (Wiebe, 1978). Exploration of large soil volumes is important if roots are to grow toward water and ions. When soils are moist (near field capacity), diffusion toward roots is reasonably rapid, but when soils dry to a water potential near 1.5 MPa (a common permanent wilting point), diffusion of water and dissolved ions can decrease 1,000-fold (see Section 5.4). Plants then have difficulty obtaining water and mineral ions for two reasons: limited exploration of the soil by roots and limited diffusion of water and ions into roots.

Besides filamentous roots, **root hairs** contribute to absorption of ions and water. Each root hair is a modified epidermal cell with a filamentous extension up to 1.5 mm long (Dittmer, 1949). Root hairs develop just

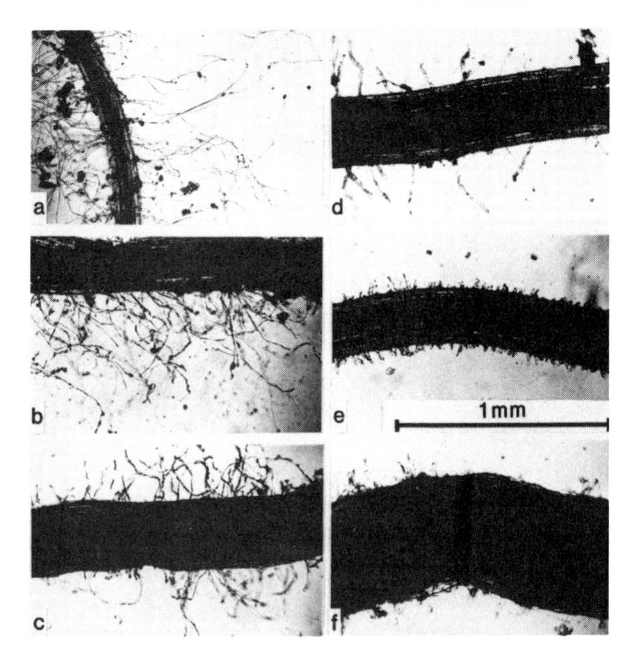


**Figure 7-3** Root hairs of (**a**) Russian thistle, (**b**) tomato, (**c**) lettuce, (**d**) wheat, (**e**) carrot, and the lack thereof in onion (**f**). (From S. Itoh and S. A. Barber, *Agronomy Journal*, 1983, by permission of the American Society of Agronomy.)

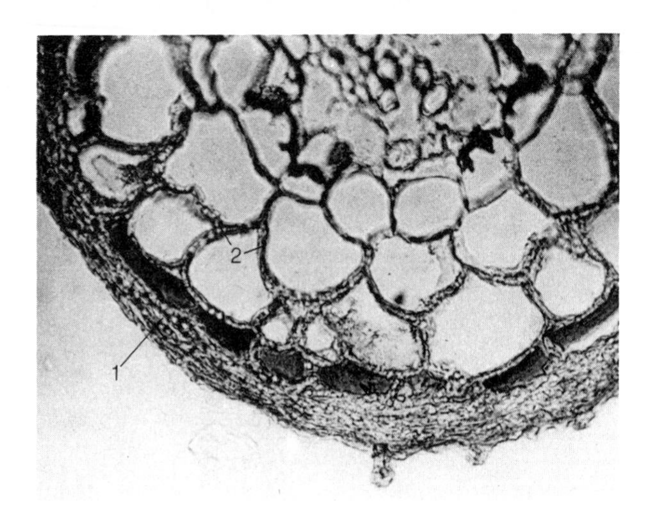


**Figure 7-4** Ectomycorrhiza formed between *Pinus taeda* and the fungus *Thelephora terrestris*. Note the external mantle (1) and the Hartig net (2) between the cells. (Courtesy C. P. P. Reid.)

behind the short region of root elongation close to the tip, and the region containing them is often less than 1 cm long. In the absence of soil but under conditions of adequate moisture and aeration, some plants form an unusually extensive system of root hairs. Nevertheless, the extent of root-hair formation in soils depends on the plant species and is often minimized by microbes and other soil conditions. Figure 7-3 illustrates root hairs of various angiosperms growing in soil. In general, root hairs are more frequent and extend over a greater region of the root when soils are moderately dry rather than wet, but if soils are too dry, root hairs desiccate and die. Sutton's review article (1980) indicates that although some conifers have root hairs, others probably have few or none. The presence of mycorrhizae, especially those of the ecto-type, minimizes root-hair formation in conifers and other species. Mycorrhizae are described next.

## 7.2 Mycorrhizae

We usually learn about root structures from plants grown in greenhouses. But in nature young roots of most species (perhaps 97 percent) look somewhat different because fungi present in native soils infect them and form mycorrhizae. A **mycorrhiza** (fungus-root) is a **symbiotic** (intimate) and **mutualistic** (mutually beneficial) association between a nonpathogenic or weakly pathogenic fungus and living root cells, primarily cortical and epidermal cells. The fungi receive organic nutrients from the plant but improve the mineral and water-absorbing properties of roots. Generally only tender young roots become infected by the fungus. Root-hair production either slows or ceases upon infection, so mycorrhizae often have few such hairs. This would greatly decrease the absorbing surface, were not the soil volume penetrated greatly increased by the slender fungal hyphae extending from the mycorrhizae. The hyphae take over the absorbing functions of root hairs.

Two main groups of mycorrhizae are recognized: the **ectomycorrhizae** and the **endomycorrhizae**, although a rarer group with intermediate properties, the **ectendotrophic mycorrhizae**, is sometimes encountered. In the ectomycorrhizae, the fungal hyphae form a mantle both outside the root and within the root in the intercellular spaces of the epidermis and cortex. No intracellular penetration into epidermal or cortical cells occurs, but an extensive network called the **Hartig net** is formed between these cells (Fig. 7-4). Ectomycorrhizae are common on trees, including members of the families Pinaceae (pine, fir, spruce, larch, hemlock), Fagaceae (oak, beech, chestnut), Betulaceae (birch, alder), Salicaceae (willow, poplar), and a few others. Figure 7-5 shows scanning electron micrographs of two *Pinus contorta* roots, one uninfected and having root hairs and the other infected with an ectomycorrhiza fungus.

Endomycorrhizae consist of three subgroups, but by far the most common are the **vesicular arbuscular mycorrhizae (VAM)**. The fungi present in VAM are members of the Endogonacae, and they produce an in-

a

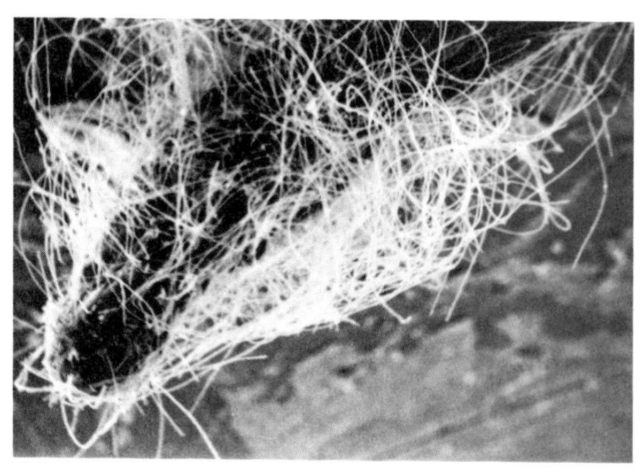

b

**Figure 7-5** (**a**) Scanning electron micrograph of dichotomous roots and root hairs of *Pinus contorta*. Note the absence of a fungal mantle. (**b**) Scanning electron micrograph of ectomycorrhiza of *Pinus contorta* inoculated with *Cenococcum graniforme*. (Courtesy John G. Mexal, Edwin L. Burke, and C. P. P. Reid.)

ternal network of hyphae between cortical cells that extends out into the soil, where the hyphae absorb mineral salts and water (reviewed by Safir, 1987; Hadley, 1988; and Smith and Gianinazzi-Pearson, 1988). Although VAM fungi seem to penetrate directly into the cytosol of cortical cells (where they form structures called vesicles and arbuscules, giving them their name), the hyphae are surrounded by an invaginated plasma membrane of the cortex cell. The VAM are present on most species of herbaceous angiosperms, whether monocot or dicot, annual or perennial crops, or native or introduced species; they also occur in the gymnosperm genera *Cupressus*, *Thuja*, *Taxodium*, *Juniperus*, and *Sequoia* and in ferns, lycopods, and bryophytes.

The fungal partner of both kinds of mycorrhizae receives sugars from the host plant, and plants that are grown in shade and are deficient in sugars predictably have poor mycorrhizal development. Also, plants growing on fertile soils often have mycorrhizae that are less developed than those of wild plants growing on nonfertile soils. The most well-documented advantage of mycorrhizae to plants is increased phosphate absorption, although absorption of other nutrients and water is often increased. The greatest benefit of mycorrhizae is probably increased absorption of ions that normally diffuse slowly toward roots or are in high demand, especially phosphate, $NH_4^+$, $K^+$, and $NO_3^-$. Mycorrhizae offer great advantages to trees growing on nonfertile soils (Fig. 7-6). In fact, without the nutrient-absorbing properties of mycorrhizae, many communities of trees could not exist. For example, some European pines introduced to the United States grew poorly until they were inoculated with mycorrhizal fungi from their native soils. A considerable potential exists for populating certain mine-waste areas, landfills, roadsides, and other infertile soils by introducing plants inoculated with

**Figure 7-6** Growth promotion of six-month-old juniper (*Juniperus osteosperma*) plants by mycorrhizal formation. Plants were grown under identical conditions in a growth chamber. Only the three plants at right had mycorrhizae. (Courtesy F. B. Reeves.)

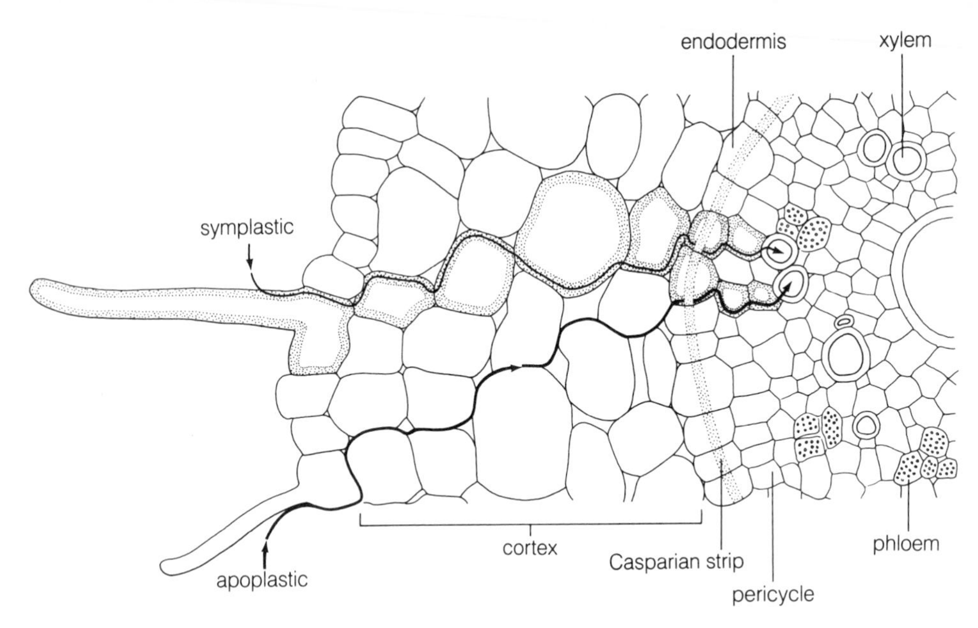


**Figure 7-7** Anatomical aspects of symplastic and apoplastic pathways of ion absorption in the root-hair region. The symplastic pathway involves transport through the cytosol (stippled) of each cell all the way to nonliving xylem. The apoplastic pathway involves movement through the cell wall network as far as the Casparian strip, then movement through the symplasm. Casparian strip of endodermis is shown only as it would appear in end walls (the walls above or below the plane of the section). (Redrawn from K. Esau, 1977.)

fungi capable of forming mycorrhizae (Marx and Schenck, 1983). Greater contributions of mycorrhizae to agriculture and forestry should occur as we understand them better, and such understanding is developing rapidly.

## 7.3 Ion Traffic into the Root

The mineral salts most readily available for roots are those dissolved in the soil solution, even though their concentrations are usually low. As noted in Chapter 6, in one survey of more than 100 agricultural soils at or near field capacity, more than half had dissolved $NO_3^-$ concentrations of less than 2 mM, phosphate concentrations of less than 0.001 mM, $K^+$ concentrations of less than 1.2 mM, and $SO_4^{2-}$ concentrations of less than 0.5 mM (Reisenauer, 1966). Soils on which crops are grown are more fertile than either range or forest soils. Even so, concentrations of such elements in crop plants can reach 10 to 1,000 times those of soil concentrations. Such elements reach roots in three ways: by diffusing through the soil solution, by being passively carried along as water moves by bulk flow into the roots, and by roots growing toward them.

Mineral salts can be absorbed by and transported upward both from root regions containing root hairs and by much older regions many centimeters from the root tip (Clarkson and Hanson, 1980; Drew, 1987). Mycorrhizae have not been investigated nearly as much as nonmycorrhizal roots, but they absorb nutrients rapidly near the tips where fungal hyphae are concentrated and somewhat less rapidly in older regions. Root tips are frequently exposed to higher concentrations of dissolved mineral salts than are older regions because the older regions exist in parts of the soil already explored by growing root tips.

In Section 5.3 we examined the pathway of water movement into young regions of nonmycorrhizal roots in relation to apoplastic and symplastic pathways. The apoplastic pathway essentially involves diffusion and bulk flow of water from cell to cell through spaces between cell-wall polysaccharides. Essential and nonessential mineral salts are carried along in this water. We once believed that the apoplastic pathway always extended from root hairs or other epidermal cells to the endodermis, where the waterproof Casparian strip of the endodermis forced substances to enter endodermal cells across their plasma membranes. This theory meant that plasma membranes of endodermal cells represented the final point at which the root could control entry of any dissolved solute. This theory still seems correct in the sense of *final control* for many (perhaps all) species. Figure 7-7 shows that apoplastic pathway up to the endodermis and a symplastic pathway from a root-hair cell to the endodermis and across it all the way to dead xylem cells with no plasma membrane. However, as mentioned in Section 5.3, roots of many angiosperms have another Casparian strip in the hypodermis (reviewed by Peterson, 1988; Shishkoff, 1987). Peterson (1988) defined a hypodermis with a Casparian strip as an **exodermis**. This Casparian strip develops and matures farther from the root tip (up to 12 cm) than does the comparable strip in the endodermis, so it can exist in moderately old regions of primary roots that have not lost their external cells. This exodermis restricts movement of dyes and sulfate ions into the cortex, so when it is present it must represent an important control point that forces external solutes to be absorbed by the selective plasma membrane of exodermal cells. Once inside the cytosol of the exodermis, ions can move to the xylem from cell to cell via a symplastic pathway.

An ion that is absorbed by an epidermal cell and moves toward the xylem in the symplastic pathway must first cross the epidermis, then probably an exodermis, several cortical cells, the endodermis, and then finally the pericycle. Any such movement from one living cell to another could involve transport directly through each of the two primary walls, the shared middle lamella, and both plasma membranes of adjacent cells. Alternatively, the ion could move through **plasmodesmata**, which are tubular structures that extend through the adjacent cell walls and the middle lamella of nearly all living plant cells (reviewed by Robards, 1975; Gunning and Overall, 1983; Robards and Lucas, 1990; also see Section 1.4). Densities of plasmodesmata (singular, *plasmodesma*) are commonly greater than 1 million per square millimeter! Three electron micrographs of plasmodesmata are shown in Figure 7-8.

At its exterior each plasmodesma consists of a tube of plasma membrane continuous between the two adjacent cells. Within the tube of plasma membrane lies another tube called a **desmotubule** that is a compressed part of the endoplasmic reticulum extending from one cell to the other. Thus there is one tube of membrane inserted within another tube. Research indicates that the central desmotubule is plugged, so that solutes cannot pass directly through but only between it and the plasma membrane; that is, solutes move between one tube and the other. Even though plasmodesmata probably contribute to solute movement across cells, direct movement through other membrane regions is also involved; we shall return to direct movement across membranes later.

Regardless of the pathway through the root from soil to xylem, ions transported upward to the shoot must somehow get into the xylem's dead conducting cells, mainly vessel elements and tracheids. This involves transfer either from living pericycle cells or from still-living xylem cells. Evidence obtained with inhibitors of respiration (especially those that block ATP formation) indicates that transfer into the conducting xylem requires metabolic energy and ATP formation. This apparently means that pericycle or living xylem cells can absorb ions from other living cells on one side and secrete them into dead xylem cells on the other side.

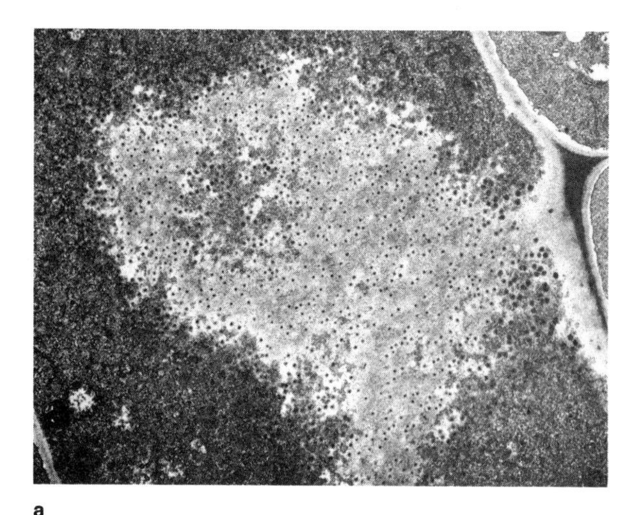


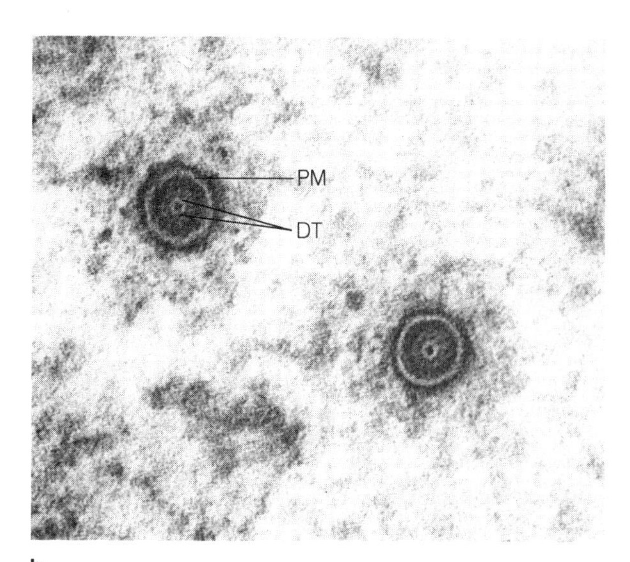


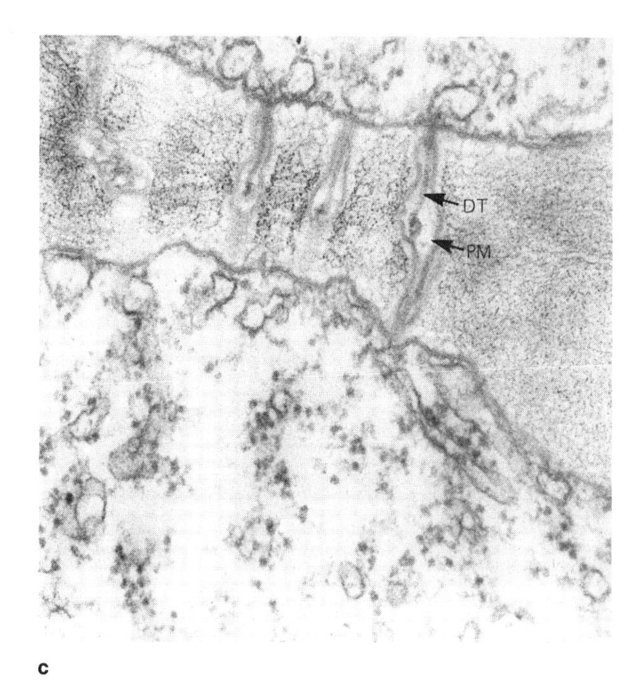


**Figure 7-8** The structure of plasmodesmata. (**a**) Hundreds of plasmodesmata (small black dots) in a primary pit field in the wall of a young barley endodermal cell. (**b**) High magnification of two such plasmodesmata, showing their tubular nature. (**c**) Longitudinal view of plasmodesmata across two adjacent young endodermal cells. PM, plasma membranes; DT, desmotubule. (Courtesy A. W. Robards.)

Most studies on the pathways of ion absorption by roots have involved young and nonmycorrhizal roots, but our views of apoplastic and symplastic pathways for such roots must be modified for older roots and for mycorrhizae. For older roots and nonmycorrhizal roots, information indicating the importance of both pathways is accumulating (Clarkson, 1985; Drew, 1987).

Attention has recently been given both to the ways hyphal strands of mycorrhizal fungi transport ions to mycorrhizal roots and the ways these hyphae obtain organic solutes from the plant. To learn how solutes move from the soil into root cells, Ashford et al. (1989) studied ectomycorrhizae of *Eucalyptus pilularis*, in which a distinct Hartig net is present. They concluded that a solute first enters the fungal cytoplasm and travels through it toward the root by a symplastic pathway. Then, because there are no plasmodesmata or other cytoplasmic connections between the fungal hyphae in the Hartig net and the root cells, the fungus must release the solute into apoplastic space, where root cells can absorb it. The suberin layers on exodermal cells of the root are thought to force early entry of the solute into the cytoplasm of those cells, with little or no penetration of solute through cell walls of the cortex up to the root endodermis. Thus there is first absorption into the symplast of the fungus, then release into an apoplastic space, then absorption into the root symplast, and finally symplastic transport across root cortex cells to the xylem. Ashford et al. (1989) also concluded that solutes (such as sucrose), transferred from the root to the fungal partner, must enter the same restricted apoplastic space that is virtually sealed from other organisms. This arrangement prevents other soil microbes from absorbing either mineral ions being transferred from fungus to root or organic compounds transferred from root to fungus.

Although pathways for ion traffic into roots can vary, ions must always penetrate the plasma membranes of living root cells, even when they are first absorbed by a fungal hypha. Therefore, the plasma membrane represents an important barrier for ion absorption. In fact, the most important function of a membrane around any cell or organelle within a cell is to control the composition inside so that life's processes can occur normally when changes in the surroundings occur. To understand such control we need to understand membranes.

## 7.4 The Nature of Membranes

As noted in Chapter 1, electron micrographs show that most biological membranes are similar, regardless of the kind of cell or organelle they surround (see Section 1.6). They are generally 7.5 to more than 10 nm thick and usually appear in cross-sectional, high-resolution transmission electron micrographs as two dark (electron-dense) lines separated by a lighter (electron-transparent) layer. Figure 7-9 illustrates this three-layered appearance of the plasma membrane and the tonoplast in two potato root-tip cells.

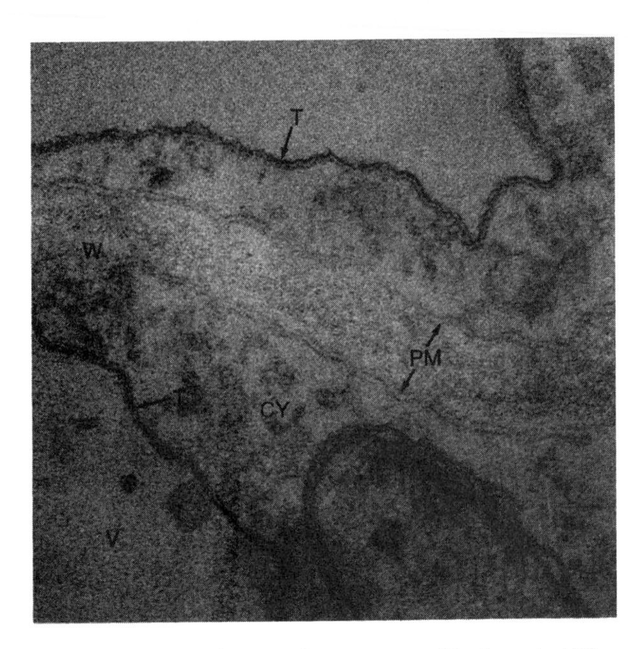

**Figure 7-9** The three-layered appearance of the tonoplast (T) and plasmalemma (PM) in two root-tip cells of potato. The cell wall (W), cytosol (CY), and part of the vacuole (V) are also shown. (Courtesy Paul Grun.)

Every membrane consists largely of proteins and lipids. Proteins usually represent one-half to two-thirds of the membrane's dry weight. Figure 7-10 shows a scanning electron micrograph of a pea root-cell membrane in surface view. When the tissue was sliced for electron microscopy, the knife tore away part of the membrane and neatly exposed the microfibrils of cellulose in the wall beneath (upper two-thirds of Fig. 7-10). The tiny bumps on the membrane (lower one-third of Fig. 7-10) represent protein molecules that extend out of the membrane toward the middle lamella and wall of the adjacent cell.

Some differences in protein and lipid contents exist among plasmalemma, tonoplast, endoplasmic reticulum, and membranes of dictyosomes, chloroplasts, nuclei, mitochondria, and microbodies (peroxisomes and glyoxysomes). The composition of membranes also depends on the species and the environment in which it lives. Nevertheless, the principal lipids of all plant membranes are **phospholipids, glycolipids** (sugar-lipids), and **sterols**.

There are four abundant phospholipids: *phosphatidyl choline, phosphatidyl ethanolamine, phosphatidyl glycerol*, and *phosphatidyl inositol*. There are two abundant glycolipids: *monogalactosyldiglyceride*, with one galac-

**Figure 7-10** Scanning electron micrograph showing part of plasma membrane of pea root cell in surface view. The root section was rapidly frozen and then sectioned by a freeze-fracture technique that splits the membrane between the bilayers. As a result, the half of the bilayer facing the cytosol is absent, and only the portion adjacent to the cell wall is visible. Tiny bumps in the membrane half represent protein molecules (p). In the upper portion of the photograph, cellulose microfibrils (c) in the primary wall are visible. (Courtesy Dan Hess.)

tose sugar, and *digalactosyldiglyceride*, with two galactoses (Fig. 7-11). (Glycolipids exist mainly in chloroplast membranes, in which phospholipids are much less abundant.) Structures of all these lipids have some common features that are important to membrane structure.

First, these lipids have a three-carbon glycerol backbone (shown at the left of each structure in Fig. 7-11) to which two long-chain fatty acids (usually 16 or 18 carbon atoms long) are esterified. Most of these fatty acids have one, two, or three double bonds, commonly in the *cis*-configuration. The melting points of fatty acids increase somewhat with chain length but decrease substantially with the number of double bonds present; thus membranes with fatty acids containing two or three double bonds are much more fluid than those with one or no double bonds. Ecological aspects of this feature are described in Chapter 26.

Every fatty acid is **hydrophobic** (water-fearing), whereas the glycerol backbone (with its oxygen atoms) is more **hydrophilic** (water-loving) because the oxygens can form hydrogen bonds with water. The final part of these lipids (shown at the bottom of each structure in Fig. 7-11) is also a hydrophilic portion. This portion is

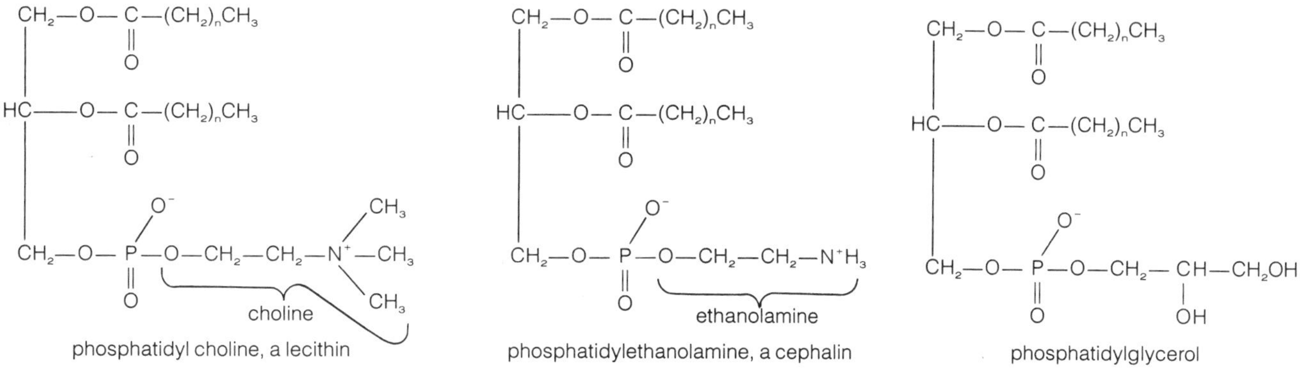

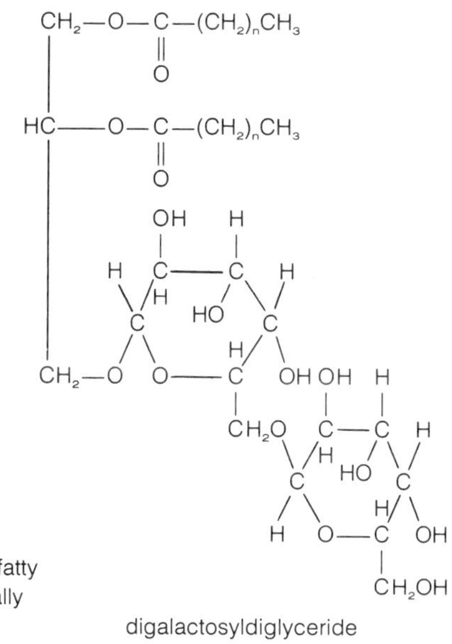

$CH_2$—O—C—$(CH_2)_nCH_3$ … $CH_2$—O—P—O—$CH_2$—$CH_2$—$N^+(CH_3)_3$ (choline)

phosphatidyl choline, a lecithin

$CH_2$—O—C—$(CH_2)_nCH_3$ … $CH_2$—O—P—O—$CH_2$—$CH_2$—$N^+H_3$ (ethanolamine)

phosphatidylethanolamine, a cephalin

$CH_2$—O—C—$(CH_2)_nCH_3$ … $CH_2$—O—P—O—$CH_2$—CH(OH)—$CH_2OH$

phosphatidylglycerol

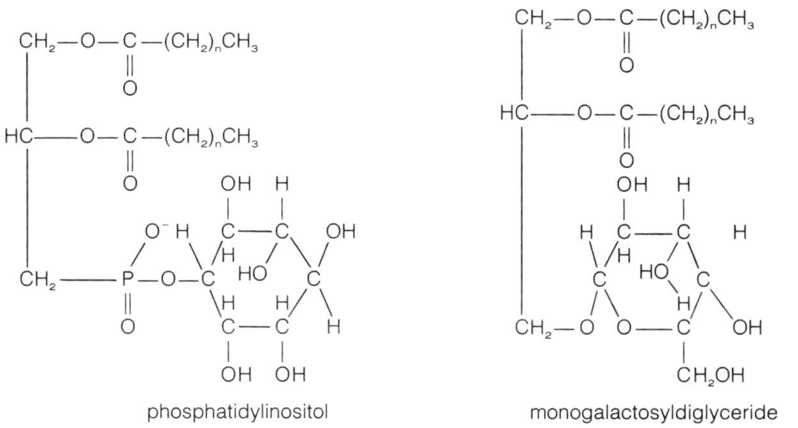

phosphatidylinositol

monogalactosyldiglyceride

digalactosyldiglyceride

**Figure 7-11** Structures of abundant membrane phospholipids and glycolipids. The subscript *n* in fatty acids represents a variable number of $CH_2$ groups or groups containing unsaturated carbons (usually 14 or 16).

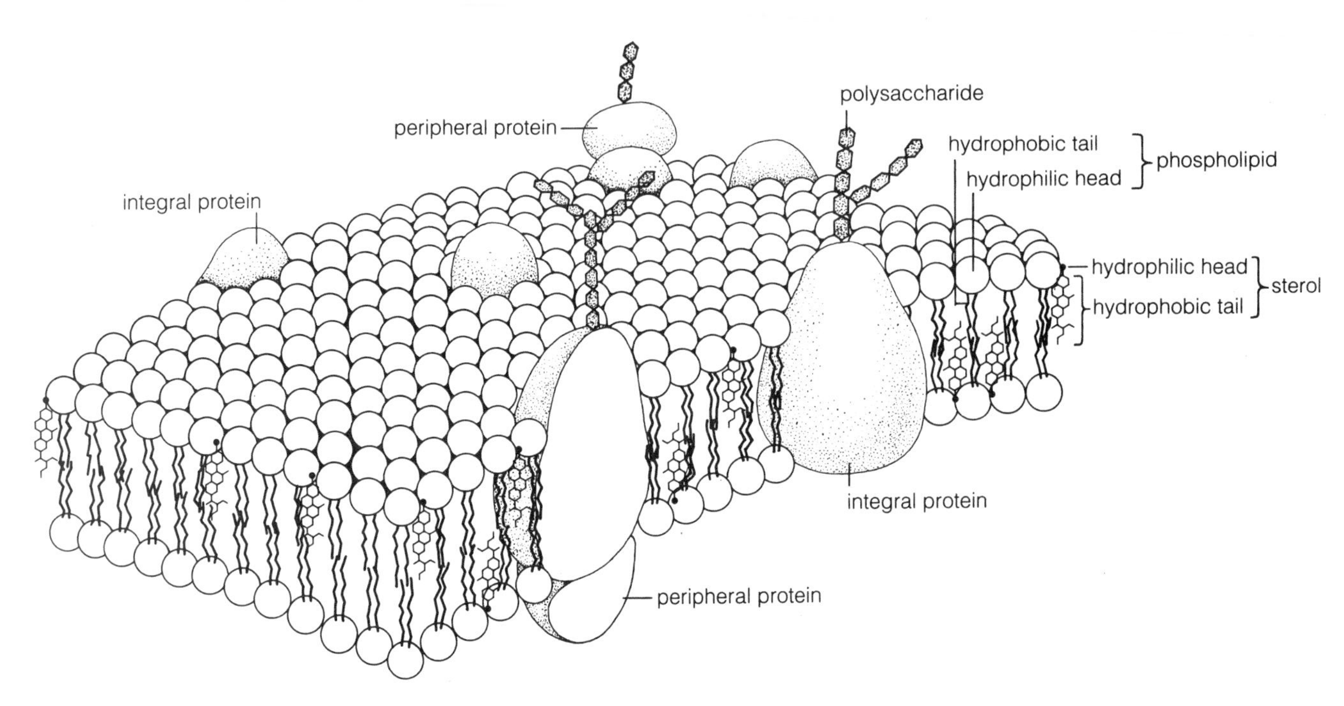


**Figure 7-12** The fluid mosaic model of membrane structure. A bilayer of lipids such as those of Figure 7-11 gives stability to the membrane through association of the lipids' hydrophilic heads with $H_2O$ on either side and association of hydrophobic fatty acids in the interior of the membrane. Sterols also have a relatively small hydrophilic head (made of one hydroxyl group) and a longer hydrophobic portion. Proteins are either continuous across the membrane or lie on either side. Most membranes contain substantially more protein than this drawing suggests. Some proteins are glycoproteins with small polysaccharides attached as shown at the top of the model.

hydrophilic because it is either electrically charged or has one or more oxygens that are attracted to water by hydrogen bonding. Molecules with distinct hydrophobic and hydrophilic regions are called **amphipathic molecules**.

In all membranes, the hydrophilic parts of each lipid dissolve in water at one membrane surface or the other, but the hydrophobic fatty-acid parts repelled by water are forced toward the interior part of the membrane. In this hydrophobic interior they associate with each other by van der Waals forces. The hydrophilic and hydrophobic forces in phospholipids and glycolipids cause the membrane to become a bilayer, as emphasized in the **fluid mosaic model** of membrane structure shown in Figure 7-12. Sterols are also amphipathic because they have both a long hydrophobic part that is rich in carbon and hydrogen and a small hydrophilic part (one hydroxyl group shown as circles in Fig. 7-12). Structures of major sterols and fatty acids are shown in Chapter 15 (Fig. 15-7 and Table 15-1), but Figure 7-12 shows how they probably fit into the membrane structure.

The amount of sterol in membranes varies greatly with species. For example, considering only the plasma membrane, the ratio of sterols to phospholipids in barley roots was reported to be 2.2, but for spinach leaves the ratio was only 0.1 (Rochester et al., 1987). The main function of sterols in membranes is apparently to stabilize the hydrophobic interior and prevent it from becoming too fluid as the temperature rises.

The proteins in membranes are of three known types: catalytic proteins, proteins that make up solute channels, and proteinaceous carriers. Catalytic proteins (enzymes) use energy to pump protons ($H^+$) across membranes; the most abundant of these proteins are enzymes that catalyze hydrolysis of energy-rich ATP to less energy-rich ADP and $H_2PO_4^-$. These ATP-hydrolyzing enzymes are called **ATPases**. Every membrane in every organism apparently has at least one kind of ATPase, whose major function in membranes is to release energy in ATP to transport ions and other solutes across the membrane against a free-energy gradient. Proteins in membranes that pump ions against such a gradient are called **ion pumps**.

Several kinds of proteins make up **solute channels**. Each channel has a hole between the protein molecules by which solutes diffuse across the membrane. Some channels allow only one ion, say $K^+$, to diffuse across, whereas others are less specific and allow both $K^+$ and malate to move across simultaneously; still others are permeable to several solutes. An important property of solute channels is that they are gated, so the holes can be open, closed, or partly open depending on cellular conditions.

There is much indirect evidence for **proteinaceous carriers** in plant membranes; these probably function first by combining with a specific solute on one side of a membrane and then by rotating and releasing the solute on the other side. Thus far only a sucrose carrier in plant membranes has in fact been identified (Lemoine et al., 1989), so little can be said about how proteinaceous carriers function in plants.

Because plants usually absorb solutes quite selectively and because carriers are expected to provide much more selectivity than is apparent in some of the indiscriminate channels found to date, both carriers and channels probably exist in plant membranes. Collectively we call pumps, channels, and carriers **transport proteins** (Sussman and Harper, 1989).

Another essential membrane component is $Ca^{2+}$, without which membranes lose their ability to transport solutes inward and also become leaky for solutes they already contain. The function of $Ca^{2+}$ is not well understood, but it probably bonds hydrophilic portions of phospholipids to each other and to negatively charged parts of proteins within the membrane.

The arrangement of proteins, lipids, and $Ca^{2+}$ in membranes is a continuing problem receiving much attention from biologists, chemists, and physicists. The fluid mosaic model depicted in Figure 7-12 has received great support, even though it is general and does not attempt to accurately describe any particular membrane. This model indicates that some protein molecules are imbedded in various places in the fluid bilayer of lipids such that "the proteins float in a lipid sea." The lipids are truly quite fluid; in fact, a phospholipid molecule can move laterally in one-half of the bilayer of a bacterial membrane from one end of the cell to the other in one second. Flip-flop of phospholipids between opposite halves of the bilayer is rare, although sterols flip-flop often (Stein, 1986).

The two sides (faces) of membranes are different because the proteins and lipids on each side differ. Some proteins extend all the way through the bilayer, as shown in Figure 7-12. These are called **integral** or **intrinsic proteins**. They are bound tightly within the membrane and can be removed only with certain detergent solutions that break hydrogen bonds among all components of the membrane. Others, the **peripheral** or **extrinsic proteins**, are more loosely bound to one or the other side of the membrane surface and can be removed with dilute salt solutions or with detergents. No proteins seem to be only partway embedded in the lipid bilayer; that is, they either go all the way through or are stuck to the surface (Chabre, 1987).

Some of the peripheral proteins in the plasma membrane, tonoplast, ER, and dictyosomes contain short, often branched polysaccharides attached at the outer membrane surface (LaFayette et al., 1987; Grimes and Breidenbach, 1987). These proteins are called **glycoproteins**. The polysaccharides of glycoproteins are shown in Figure 7-12 as outwardly protruding branched projections. These polysaccharides usually consist of joined hexose sugars or modified sugars, especially *mannose*, *fucose* (6-deoxygalactose), and *glucosamine*. The main function of these polysaccharides in the plasma membrane is to give it recognition properties.

Specifically, the polysaccharides recognize external proteins and other polysaccharides of various kinds. For example, certain pathogenic bacteria and fungi must penetrate the cytoplasm of plant cells to cause disease, and before penetration they first approach the host plasma membrane and send protein signals (enzymes) or polysaccharide signals to it (reviewed by Boller, 1989; Stone, 1989; and Dixon and Lamb, 1990). Attachment of these signals probably depends on a lock-and-key process involving polysaccharides of the membrane. If the match is correct for the pathogen, attachment can occur and disease may result. If not, the plant will be resistant. Another example involves infection of legume roots by *Rhizobium* bacteria that fix nitrogen (see Section 14.2). Again there is apparently a recognition between surface proteins or polysaccharides of the bacteria and glycoproteins in the plant's plasma membrane. Still other examples include graft compatibility and incompatibility and pollen-stigma interactions during flower pollination (Hodgkin et al., 1988). Finally, it is possible (although unproven) that the polysaccharides of glycoproteins represent recognition or receptor sites for plant hormones delivered from one cell to another. Most likely the main function of glycoproteins in the tonoplast, ER, and dictyosomes as well is to provide recognition of molecules involved in subcellular traffic.

## 7.5 Early Observations about Solute Absorption

A primary goal in studying membranes with electron microscopes or by chemical methods is to understand how they control the movement of solutes across them. Long before electron microscopes were developed or anyone knew how to isolate membranes for chemical analysis, the absorption properties of membranes could be studied. Some observed properties gave important clues about the nature of membranes. We will mention four observations about solute absorption discovered by pioneers in the field, but keep in mind that these observations say nothing about any *mechanism* of solute absorption. Later, we will distinguish between passive and active mechanisms, but for now recognize that total transport (the sum of both mechanisms) was being measured and that in most cases absorption was mainly or entirely passive.

**1.** *If cells are not alive and metabolizing, their membranes become much more permeable to solutes.* If a cell is killed by high temperatures or poisons or if its metabolism is inhibited by low temperatures, nonlethal high temperatures, or specific inhibitors, many solutes in the cell leak out and those outside diffuse in. This is one measure of death, and it represents only passive transport by free diffusion down a free-energy gradient for all solutes involved.

**2.** *Water molecules and dissolved gases, such as $N_2$, $O_2$, and $CO_2$, diffuse passively through all membranes rapidly.* No one knows how water can penetrate membranes far more rapidly than most of the solutes dissolved in it, but that it does is essential for the occurrence of osmosis. Surprisingly, water even diffuses rapidly across artificial membranes formed from phospholipids only. This probably means that water normally diffuses through the hydrophobic lipids of membranes, not through protein channels. The same is probably true for gases. Perhaps the continuous side-to-side movement of fatty acids in membrane lipids forms small and transient holes through which water and gases move.

For $N_2$, rapid diffusion seems inconsequential for most cells. Dissolved nitrogen from air simply moves into and out of cells and their organelles at equal rates and without any noticeable effect. For $O_2$, rapid inward movement allows respiration to occur and is important for all aerobic cells both day and night. For photosynthetic cells, net movement of $O_2$ out into surrounding air is a normal process during daylight, when photosynthesis exceeds respiration. For $CO_2$, rapid movement into photosynthetic cells is crucial during daylight, yet at night $CO_2$ moves out of all cells, whether photosynthetic or nonphotosynthetic, to the surrounding air. For all of these gases, diffusion across any membrane is only passive movement down a free-energy gradient.

**3.** *Hydrophobic solutes penetrate at rates positively related to their lipid solubility.* More-hydrophobic, less-hydrophilic solutes move across membranes more rapidly than those with opposite properties. Consider a few examples. Methyl alcohol ($CH_3OH$) is not much smaller than urea ($H_2N-CO-NH_2$), but methyl alcohol is about 30 times as lipid-soluble and moves into giant cells of the alga *Chara ceratophylla* about 300 times as rapidly as urea. Valeramide (five carbons) is larger than lactamide (three carbons), and valeramide is about 40 times as lipid-soluble and moves into *Chara* cells about 35 times as fast as lactamide. Urea, methyl alcohol, valeramide, and lactamide are presumed to move across the plasma membrane into cells merely by diffusing passively through the lipid bilayer toward a region of lower concentration. Such observations gave initial evidence that membranes are rich in lipids, even before a bilayer was known to exist.

A practical problem of lipid solubility relates to whether a solute can become charged when dissolved in water, because any such charge (whether positive or negative) greatly decreases lipid solubility, increases water solubility, and decreases the permeability of cells to the solute. An important example concerns dissolved $CO_2$, its hydrated form $H_2CO_3$, the major ionic species $HCO_3^-$, and the further-ionized species $CO_3^{2-}$ formed *reversibly* at *p*H values above 8:

$$CO_2 + H_2O \rightleftharpoons H_2CO_3 \leftrightarrow HCO_3^- \leftrightarrow CO_3^{2-} \quad (H^+ \text{ released at each ionization}) \qquad (7.1)$$

At low *p*H values, far more carbon is absorbed from dissolved $CO_2$ than at high *p*H values at which the negatively charged and lipid-insoluble $HCO_3^-$ and $CO_3^{2-}$ predominate.

A related example concerns absorption of the **herbicide** (weed-killer) 2,4-D (*2,4-dichlorophenoxyacetic acid*). This herbicide has an acetic-acid group that releases an $H^+$ to form negatively charged 2,4-D at neutral or high *p*H, but at low *p*H little ionization occurs, and the uncharged 2,4-D molecule remains much more lipid-soluble than the anionic form. Leaves absorb the herbicide far more effectively at *p*H 5 than at *p*H 8. Positive charges are also important. Contrary to the behavior of acidic compounds such as $H_2CO_3$ and 2,4-D, nitrogenous bases (in which the nitrogen attracts an $H^+$ and becomes positively charged at low *p*H) are usually absorbed more rapidly from neutral or slightly basic solutions in which no charge exists. Again, an important reason is their greater solubility in membrane lipids when noncharged. (As we shall explain in Section 7.7, another reason that anions are absorbed slowly is that the cytosol is negatively charged relative to both the cell wall and the external solution, and this charge repels anions.)

**4.** *Hydrophilic molecules and ions with similar lipid solubilities penetrate at rates inversely related to their size.* For ions, the size relevant to penetration rate is that attained after *water of hydration* is attached (Clarkson, 1974). Each ion attracts to itself a different (average) number of rather firmly bound water molecules, depending upon the net charge density at its surface. For example, $Li^+$ (atomic mass 6.9), which has only one full shell of electrons around its nucleus, is 0.12 nm in diameter when nonhydrated and binds about five $H_2O$ molecules. Alternatively, $K^+$ (atomic mass 39.1) has several shells of electrons and a nonhydrated diameter of 0.27 nm, but it binds only about four $H_2O$ molecules. Hydrated $Li^+$ is thus slightly larger than hydrated $K^+$ and diffuses across membranes less rapidly than $K^+$. Divalent cations such as $Mg^{2+}$ and $Ca^{2+}$ have higher charge densities than either $Li^+$ or $K^+$, bind about a dozen $H_2O$ molecules, and are absorbed far

**Table 7-1 Concentrations of Major Ions in Sea Water Compared with Their Concentrations in Vacuoles of Algae Living There.**

| | *Nitella obtusa*[a] — Baltic Sea | | *Halicystis ovalis*[b] | |
|---|---|---|---|---|
| Ion | Vacuole Concentration | Sea Water Concentration | Vacuole Concentration | Sea Water Concentration |
| $Na^+$ | 54 mM | 30 mM | 257 mM | 488 mM |
| $K^+$ | 113 mM | 0.65 mM | 337 mM | 12 mM |
| $Cl^-$ | 206 mM | 35 mM | 543 mM | 523 mM |

[a]Data for *Nitella* are from Dainty, 1962.
[b]Data for *Halicystis* are from Blount and Levedahl, 1960.

more slowly than monovalent cations. Nevertheless, divalent cations such as $Fe^{2+}$ are absorbed more rapidly than trivalent cations such as $Fe^{3+}$. The same principle occurs with anions; thus monovalent $Cl^-$ and $NO_3^-$ are absorbed far faster than divalent $SO_4^{2-}$. Furthermore, monovalent $H_2PO_4^-$ is absorbed faster than divalent $HPO_4^{2-}$ and much faster than trivalent $PO_4^{3-}$. At *p*H 7 (the approximate *p*H of the cytosol), ionization of $H_2PO_4^-$ into $HPO_4^{2-}$ and $H^+$ is about half complete, so nearly equal amounts of monovalent and divalent forms of phosphate ions exist, with essentially no $PO_4^{3-}$. At a *p*H less than 6 (as in cell walls, vacuoles, and acidic soils), monovalent $H_2PO_4^-$ is dominant. Considering that the *p*H of many cell walls is about 5 (even in neutral or slightly alkaline soils), phosphate transport into the cytosol across the plasma membrane usually involves mainly $H_2PO_4^-$. The reversible ionization of $H_2PO_4^-$ is shown below:

$$H_2PO_4^- \leftrightarrow HPO_4^{2-} + H^+ \leftrightarrow PO_4^{3-} + H^+ \qquad (7.2)$$

## 7.6 Principles of Solute Absorption

Next we discuss the four important principles of solute absorption that apply to nearly all dissolved solutes that plants must acquire from their environment or must transport internally from one cell to another. These principles led to the theory that transport proteins, specifically called *carriers*, control absorption. In fact, all the principles mentioned are consistent with that theory. Later we suggest that proteinaceous channels are more likely to account for some of the results than are true carriers, but otherwise the theory that specific transport proteins in membranes control absorption of most essential solutes seems valid.

### Many Solutes Are Accumulated Inside Cells

A remarkable fact about all cells is that they can absorb certain essential solutes so fast and over such long periods of time that concentrations of these solutes become much higher within the cells than in the external solution. We call such absorption **accumulation**. The extent to which the concentration is greater internally than it is externally is called the **accumulation ratio**. For example, slices of storage tissues such as potato tubers placed in nutrient solutions often deplete the concentration of external ions to nearly zero within a day or two. During this time some ions (especially $K^+$) attain internal concentrations over 1,000 times higher than those finally present in the surrounding solution. Plant tissues usually contain at least 1 percent $K^+$ on a dry-weight basis. When alive, such tissues typically contain 80 to 90 percent water, so the living plant contains about 0.1 percent or 25 mM $K^+$. Yet the dissolved $K^+$ even in fertile soils is often no more than 0.1 mM, indicating overall whole plant accumulation ratios of about 250 to 1. Numerous data for potassium accumulation ratios in roots and shoots of five species are given by Asher and Ozanne (1967). Thermodynamic laws explained in Chapter 2 show that free diffusion not involving expenditure of metabolic energy could not be responsible for such a great accumulation. Therefore, we conclude that plant cells use energy for accumulation, and we know from other studies that the main energy-rich compound is ATP.

Accumulation of certain solutes is a universal phenomenon of living cells. As explained in Chapter 8, certain cells in phloem of vascular bundles accumulate sucrose or other carbohydrates in high concentrations and then translocate them elsewhere. This transport is essential if photosynthetic cells are to supply other parts of the plant. Algae living in salty water must accumulate solutes to prevent plasmolysis, and one of their adaptation mechanisms to a saline environment is to increase their internal salt concentrations. Table 7-1

a b

**Figure 7-13** (**a**) and (**b**): An apparatus used for studies of solute absorption by root segments. The root tissues were placed in Plexiglass vials **a** which were then placed into aerated absorption wells, as shown in **b**. To terminate an experiment the Teflon valve is rotated and the solution drawn by vacuum into the chamber below the vials. (From Kochian and Lucas, 1982.)

lists accumulation data for two species of algae living in different sea waters. Cells of both species have large central vacuoles from which solution was obtained for analysis. Note that both algae accumulated $K^+$; *Nitella* also accumulated $Cl^-$ and to a lesser extent $Na^+$, whereas *Halicystis* appeared to be in equilibrium with respect to $Cl^-$ and actually restricted entry of $Na^+$. We conclude from these and other data that whether or not a solute is accumulated depends upon the solute and the species of plant. Also, we emphasize that restriction of sodium is common to most angiosperms and gymnosperms. We will return to that subject later in terms of mechanisms by which restriction occurs. (It does not occur by a sodium-potassium pump, which some general biology textbooks erroneously say occurs in all eukaryotic cells.)

## Absorption of Solutes Is Specific and Selective

That solutes are absorbed and accumulated by selective processes was further indicated by studies in which roots were excised from seedlings and allowed to accumulate ions. Emanuel Epstein pioneered many such studies with excised barley roots in the 1950s and 1960s. His essay in this chapter describes some of that research, and his excellent book (Epstein, 1972) provides many details about how roots absorb mineral salts. Seedlings were often grown from seeds in a dilute solution of calcium sulfate, so most of the mineral salts were provided from seed reserves. Such "low-salt" roots have a high capacity for subsequent absorption of several ions, and this capacity is maintained for several hours even if the roots are cut off from the shoots. Figure 7-13 shows a method by which excised roots can be used in ion-uptake studies. Aeration of excised roots and of those attached to seedlings is required to allow respiration that is essential for normal ion accumulation in most species (note tube in Fig. 7-13b for forcing air through the solutions).

If such roots are provided with a solution containing dilute (about 0.2 mM) KCl and with about 0.2 mM $Ca^{2+}$ to maintain normal membrane functions, the rate of absorption of $K^+$ is unaffected by similar concentrations of $Na^+$ salts. This is true even though $Na^+$ is chemically similar to $K^+$. The process of $K^+$ accumulation is therefore selective and is not influenced by a related ion under these conditions. As expected, several other monovalent and divalent ions also have no influence on $K^+$ uptake. In similar studies, absorption of chloride is unaffected by the related halides fluoride and iodide, as well as by $NO_3^-$, $SO_4^{2-}$, or $H_2PO_4^-$. Calcium ions are essential for this selectivity because without them, $K^+$ absorption, for example, becomes inhibited by low concentrations of $Na^+$.

In spite of this apparent high selectivity, uptake mechanisms can sometimes be "fooled." Potassium absorption is inhibited competitively by $Rb^+$, and penetration of membranes by these two ions is apparently governed by the same mechanisms. Similar competitive results are often obtained with monovalent $Cl^-$ and $Br^-$, with divalent $Ca^{2+}$, $Sr^{2+}$, and sometimes $Mg^{2+}$, and with divalent $SO_4^{2-}$ and selenate ($SeO_4^{2-}$). This selectivity of ion transport by roots also applies to organic compounds such as amino acids and sugars, and it occurs in all parts of the plant. Selectivity supports the theory that proteinaceous carriers in membranes help move solutes into cells because enzymes (which are proteins) are known to recognize selectively and be activated or inactivated by certain ions or molecules.

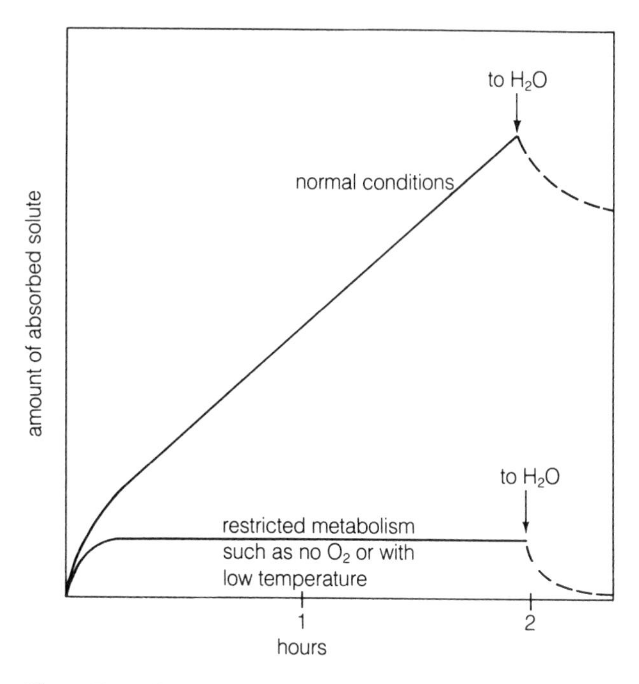


**Figure 7-14** Progress of ion uptake and efflux with time under various conditions. For explanation, see text.

## Absorbed Solutes Often Leak Out Only Slowly

Once ions or organic molecules are absorbed into the cytoplasm or vacuoles of cells, they do not readily leak out (that is, **efflux**, or outward movement, is often slow). Rapid leakage can be induced by damaging the membranes with heat, poisons, or lack of $O_2$, and to some extent by removing $Ca^{2+}$, yet these abnormal situations eventually cause cell death. Slow leakage shows that absorption, *especially in low-salt roots*, is primarily unidirectional **influx**.

If the concentration of an ion in a low-salt tissue is measured while cells are being exposed to that ion, graphs such as those in Figure 7-14 are usually obtained. Under normal conditions of temperature and aeration (upper curve), there is an initial rapid influx of ions, although most of this influx simply represents diffusion into the cell walls rather than actual movement across the plasma membrane. Subsequently, the absorption rate becomes essentially constant, often for as long as several hours.

Now suppose that tissues are removed from the solution at the time indicated by arrows in Figure 7-14 and are then placed in water at room temperature, and that the amount of ion retained inside is measured at various times. Only a small fraction of the ions present leak out rapidly, within a few minutes (top dashed line of Fig. 7-14). Such ion losses occur mainly from cell walls, not from inside cells. Ions present in the cytoplasm and vacuole remain there longer, and they represent by far the majority of ions absorbed. Only if leakage experiments are continued for several minutes or hours can efflux first from the cytoplasm and then from the vacuole be detected, especially in low-salt roots.

The major point here is that most ions can be transported across membranes into cells much faster than they move out. This is further evidence that membrane carriers or unidirectional channels speed only their inward absorption. (We shall explain later, however, that a few ions, especially $Na^+$, $Ca^{2+}$, and $Mg^{2+}$, diffuse inward down a concentration gradient and are transported outward with the aid of ATP-dependent pumps.) Furthermore, the efflux of several ions becomes significant when their internal concentrations become high (non-low-salt roots). For intact plants, efflux is especially active during darkness and in some cases may actually exceed influx (Glass, 1989). A possible explanation is that during dark periods roots are not receiving enough carbohydrates from the leaves to produce the energy they need for rapid influx.

Figure 7-14 (lower curve) also shows the time course of accumulation when respiration is inhibited by the lack of $O_2$, by cold temperatures, or by numerous respiratory poisons. The initial rapid absorption phase is not greatly altered, but the absorption rate then quickly falls to nearly zero (that is, the curve becomes nearly horizontal). When these low-salt roots are transferred to water, nearly all the apparently absorbed ions rapidly diffuse outward into the water. This efflux is of the same magnitude as that from the healthy tissues, and efflux from both healthy and unhealthy types occurs mostly from the cell walls (apoplast) by simple diffusion. Such results illustrate the inhibition of absorption and accumulation by conditions that prevent normal respiration. Respiration and solute absorption are probably strongly related because respiration provides the ATP needed for solute absorption.

## The Rate of Solute Absorption Varies with Solute Concentration

To learn the mechanisms of solute absorption and to understand relations between fertilizer absorption and fertilizer application rates, hundreds of studies relating absorption rates to external concentrations have been made. A major conclusion for most plants in nature is that for nutrients used in abundance (nitrate, ammonium, phosphate, and potassium), diffusion to the root surface is the limiting factor, and thus absorption properties of the roots are of only limited importance for plant nutrition (Nye and Tinker, 1977). However, for crop plants well fertilized with nitrogen, phosphorus, and potassium, and for other essential solutes and for cells other than roots, absorption through membranes is often rate-limiting. Plant physiologists interested in how solutes cross membranes have studied the problem for many years using numerous techniques. Emanuel Epstein and his students investigated the absorption of

# Roots — Mining for Minerals

## *Emanuel Epstein*

*Emanuel Epstein was born in Germany, but all of his college degrees were from the Davis and Berkeley campuses of the University of California. He then spent eight years (1950–1958) with the United States Department of Agriculture at Beltsville, Maryland; since then he has been at U.C. Davis. He is an expert in processes by which ions move across membranes and in salt tolerance in crops. Much of his research is directed toward the plant parts we seldom see: the roots. His work on ion transport in the early 1960s, described in this essay, brought him worldwide acclaim. His application of Michaelis-Menten enzyme kinetics (which we describe in Chapter 9) to ion-transport studies emphasized the carrier concept of transport across membranes. He now uses comparative physiology and molecular-biology techniques to understand salt tolerance.*

My scientific interests have always combined free-ranging curiosity about the workings of nature with a desire that scientific knowledge should make a useful contribution — that is, should have the potential of practical application. Agricultural science fitted this inclination, and a particular, purely emotional liking for trees channeled those interests into pomology, the study of fruit trees, one aspect of the broad field of horticulture. In those days — the late 1930s — agricultural studies at the University of California began on the Berkeley campus, where future "Aggies" took the basic courses in chemistry, physics, mathematics, and biology. They then went to the Davis campus for their specific agricultural studies. I followed that path and upon completion of my undergraduate work stayed on in Davis for a Master's degree in horticulture. This I did under the tutelage of Omund Lilleland. He had observed a characteristic chlorosis of trees in orchards in the foothills of the Sierra Nevada; the chlorosis turned out to be due to manganese deficiency, and thus began my lifelong interest in the mineral nutrition of plants.

At that time (1941), a Master's degree was as far as one could go on the Davis campus. So back I went to Berkeley and started out on the arduous preparation for a Ph.D. in plant physiology. Things did not go smoothly. World War II was raging in Europe; on December 7, 1941, the Japanese attacked Pearl Harbor; and shortly thereafter I fell in love. None of these events was conducive to a single-minded concentration on the job at hand. By the time the dust had settled at the beginning of 1946, I was married and the father of a baby boy, was mustered out of the army, and was at the beginning once again (instead of at the end) of my studies for my Ph.D.!

I joined Dennis R. Hoagland's famous Division of Plant Nutrition at Berkeley as a graduate student. Perry R. Stout of that division was my major professor. I chose him for three reasons: He struck me as an original thinker; he was straightforward, without airs and pretenses; and he was one of the few people in the whole world who, before the war, had done some experiments with plants in which radioactive isotopes of nutrient elements had been used. I wanted to avail myself of that marvelous new tool for the study of mineral plant nutrition.

That was before the days of mail-order radioisotopes. Rather, they came from Ernest Lawrence's 60″ cyclotron up the hill above the Berkeley campus. There were then no units for "Environmental Health and Safety," or I never would have been allowed to go up there, get a piece of copper deflecting plate from the cyclotron, and then do my own target chemistry to separate the zinc-65 and get a pure solution of it for my experiments. I was still interested in micronutrients, and in addition to zinc-65 I used radioactive isotopes of iron and manganese. (I had to determine the half-life of manganese-52 myself because no accurate value had been published, and I did that by means of a quartz-fiber electroscope because the Geiger-Mueller counters we then used were too unstable from day to day.)

Taking courses was important in addition to conducting research. There was then probably no other place in the world as good as the Hoagland Division of Plant Nutrition if one wanted to study that part of plant physiology. Beyond that, two courses were crucial to what I would do: D. M. Greenberg's biochemistry course included a thorough discussion of enzyme kinetics, and G. Ledyard Stebbins's lectures on organic evolution opened my eyes to the origins and genetic diversity of plant life.

One day Dr. Stout introduced me to a visitor, Cecil H. Wadleigh of the U. S. Department of Agriculture. He was talent-scouting to staff a new plant-nutrition laboratory at the main U.S.D.A. experiment station in Beltsville, Maryland. The new laboratory was to specialize in the application of radioisotopes to studies in mineral plant nutrition. I was offered and took the job. It seemed an ideal opportunity to pursue the kind of research I wanted to do and to continue the development of isotopic techniques for doing it. Besides, the income for our family (now four of us) would go from $2,280 a year (partly from the G. I. Bill for veterans, partly from a half-time research assistantship) to $4,600!

In Beltsville I was free to pursue whatever I chose to do — thanks mainly to Sterling B. Hendricks, who left me to my own devices. I decided to study the selectivity between potassium and sodium in their absorption by roots. Selectivity between those elements, I knew, ran like a thread through all biology and was of agricultural significance to the western United States, with its many salt-affected soils.

I taught myself the excised-root technique, which I had never observed in the Berkeley laboratory in which it had been developed by D. R. Hoagland and T. C. Broyer; I used barley roots as they had done. The look-alike elements potassium and rubidium consistently inhibited each other's absorption, but the interference with the absorption of either by sodium was both less, and less consistent. At the same time I confirmed a main tenet of the Hoagland school: The permeability of plant cell membranes to these ions was quite low. Radioions once absorbed by the roots were very slow to exchange with unlabeled ions of the same element in the external solution. How then did they enter in the first place, and how could selectivity be accounted for? There were references in the literature to "binding compounds" or "ion pumps," but there was no clear rationale, let alone any quantitative treatment, for these ideas.

But mine was a "prepared mind" when C. E. Hagen, another recent arrival at Beltsville, asked me whether I thought that the interference by sodium with potassium absorption that I was observing was competitive or not. I did not know, but the question triggered one of those sudden insights that chroniclers of science so often write about but which in actuality are so rare. Entry of ions into cells through an impermeable membrane suggested the need for some agent in the membrane to effect this transport—the formation of an enzyme-substrate complex in enzymic catalysis—and the specificity of that process. If the ion was the "substrate" and if its specific (selective) combination with a carrier in the membrane—the "enzyme"—resulted in a carrier-ion complex, then the kinetics of transport across the membrane might follow the well-known kinetics of enzymic catalysis, or "Michaelis-Menten" kinetics. The difference would be in the process brought about: not the chemical transformation of substrate into product, but the transport of the ion from the outer to the inner side of the membrane. That would explain the paradox of an ion traversing a membrane impermeable to it: It would not be the free ion that would traverse the membrane but rather the carrier-ion complex; once released at the inner side of the membrane, the ion would not be free to diffuse out. The evidence for a rather ion-impermeable membrane was entirely consistent with this scheme. The matter of selectivity could also be accounted for: Just as enzymes possess "active sites" to which the substrate is specifically bound, so the carriers would be expected to have sites to which, say, potassium and its look-alike, rubidium, would bind but sodium would not.

I set to work with a will, using the excised barley-root technique, and eureka!—it worked! Enzyme kinetics could be applied to ion transport. After publishing these findings on the transport of potassium in 1952, I extended the approach to other ions, but the truly definitive work was done in the early 1960s, after I had moved to the (then) Department of Soils and Plant Nutrition at the University of California, Davis—back to my old stomping grounds, where I have stayed ever since.

The key to progress at Davis was my finding that the high selectivity of the transport process for potassium in the presence of sodium depended absolutely on the presence, in the external solution, of calcium ions. Also, I expanded the range of concentrations of the ion whose transport I was studying—potassium to begin with—to cover a very wide range: from 0.002 mM to 50 mM. The rate of potassium absorption by the barley roots followed Michaelis-Menten kinetics nicely from the lowest concentration to about 0.2 or 0.5 mM. At those concentrations, the rate of potassium absorption leveled off; that is, it became nearly independent of the external concentration. Yet at still higher concentrations the rate of uptake rose to levels that were much higher than the theoretical maximum for the low-concentration (high-affinity) mechanism. In addition, this second, high-concentration (low-affinity) mechanism differed drastically from the former one in selectivity and other features. I dubbed the former "mechanism 1" and the latter "mechanism 2," and I have had the satisfaction of seeing an entire body of research and publications—sometimes controversial!—arising from those findings. Many physiologists became convinced by the kinetic evidence for the operation of carriers that a direct search for their biochemical identity was worthwhile.

There is a general lesson to learn from this experience. Specialization is, to be sure, a necessity in scientific research in an age in which ever more is being learned in every aspect of science. But it is important as well to keep in touch with related fields of science because unexpected, fruitful insights may come when information and modes of thinking from different specialties are conceptually connected. Such was the case when I, not an enzymologist, nevertheless knew enough about enzyme kinetics to bring its concepts to bear on the problem of the transport of ions across plant cell membranes. The lesson is to let one's mind's eye roam beyond the immediate, sharply defined focus to see what other knowledge, what other viewpoints, may contribute to the solution of a problem. Narrow-minded specialization is our chief occupational hazard in scientific research; we must deliberately set about combating it.

Ever since taking Ledyard Stebbins's remarkable course

(*continued*)

## Roots — Mining for Minerals *(continued)*

on organic evolution I had been fascinated by the diversity of living beings and particularly by genotypic differentiation within species and among closely related species. For many years my interest had centered on ionic selectivity, especially on selection between potassium and sodium ions and the role of calcium in this process. And as I was living and working in the West, I was keenly aware of salinity of soils and water, "salinity" referring almost without exception to high concentrations of sodium salts — so high as to be deleterious to the growth of most crop species. Yet wild plants inhabit highly saline media: Witness the marine phytoplankton, and on land, the plant denizens of the seashore, salt marshes, and saline deserts. Even among and within crop species there was considerable variation in salt tolerance. Here was yet another opportunity to put together knowledge and approaches of two different disciplines: mineral plant nutrition on the one hand and genetics on the other. Thus beginning in the 1960s, I began educating myself in the genetic control of selective ion transport with emphasis on sodium, potassium, and chloride, and differential responses to salt. I also beat the drums for such a genetic approach by publishing research findings and reviews and emphasizing it in a textbook.

My graduate students and other collaborators conducted fairly large-scale screenings of genetic lines of barley, *Hordeum vulgare*, and wheat, *Triticum aestivum*, finding in both considerable variation in salt tolerance. We failed to find much variability in this regard in the tomato, *Lycopersicon esculentum*, mainly, I now suspect, because we did no adequately large screening. But the failure forced us to consider exotic germplasm. We turned to the Department of Vegetable Crops's Charles M. Rick, the world's foremost authority on tomato genetics and germplasm. He gave us seed of an exotic tomato species he had collected near the seashore on one of the Galapagos Islands, where presumably the plant was exposed to salt. Its fruits were tiny, yellow, and bitter, but in solution culture the plants proved indeed to be salt-tolerant. Fortunately it and the crop species could easily be hybridized, and thus a beginning was made in the development of salt-tolerant tomatoes.

We grew some of our best barley, wheat, and tomato selections in the dunes at Bodega Marine Laboratory on the Pacific coast north of San Francisco. We used seawater and dilutions of it to irrigate the plants and thus provided rather spectacular evidence for our thesis that a genetic dimension should be added to the conventional irrigation-and-drainage techniques for coping with the salinity problem. Indeed, it is our hope and expectation that the immense reservoir of water and mineral nutrients of the world's oceans will one day be put to use along the world's coastal deserts to grow crops specifically developed for seawater culture.

In our current research we focus mainly on mechanisms of salt tolerance, attacking the problem with the tools of comparative physiology — a combination of genetics and plant physiology — with molecular biology not far behind.

It has been, and still is, a fascinating journey. Being a professor and research scientist is, to me, the most important, the most challenging, the most rewarding job a person can have. An extravagant claim? I think not. In all society there are just two growing points: the new, young people who are our students and the new knowledge gained through research. And it is these two sole sources of renewal and growth that a professor in a research university deals with: you readers — our young, new people — and the new knowledge that we generate through research. Nothing is more crucial to the advancement of society, and no occupation more satisfying than that dealing with these two wellsprings of society's future.

specific ions by excised roots as a function of external concentration and time. They developed the idea that proteins in membranes might act like enzymes, to which Michaelis-Menten kinetics usually apply (Chapter 9). To a large extent this idea proved true, which means that membranes contain various proteins that specifically recognize certain solutes, combine with them, and speed their transport inward. Epstein and many others have called these transport proteins **carriers**. Let us investigate kinetic studies in which the rate of absorption is measured as a function of a wide range of external solute concentrations to which excised roots or other plant parts are exposed. In such studies, an apparatus similar to that shown in Figure 7-13 is often used.

There are two possible overall mechanisms for solute absorption. First, simple one-way diffusion across the membrane would cause the rate of absorption to be directly proportional to the external solute concentration. A hypothetical graph illustrating this mechanism is shown in the lower line of Figure 7-15. Results with some organic solutes to which cells have never been exposed in nature show such kinetics. This is an example of passive absorption (described in Section 7.5), and it applies to such solutes as methanol, ethanol, urea, valeramide, and even to atmospheric nitrogen gas. But

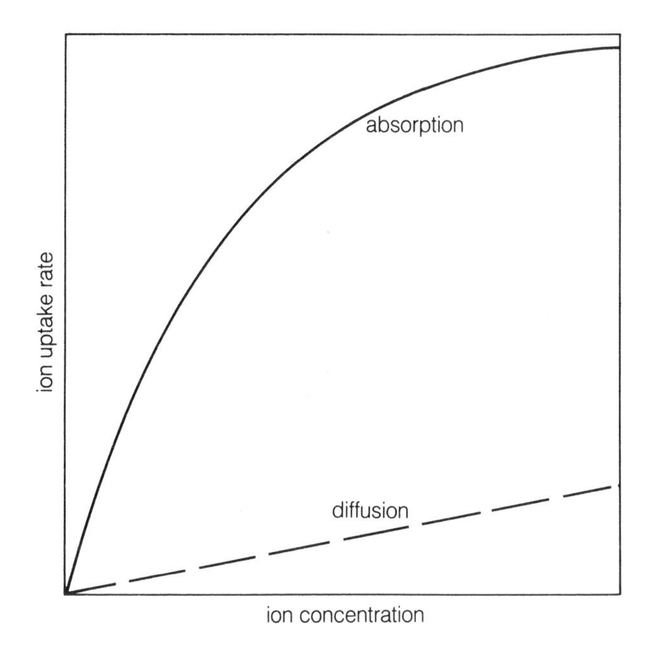


**Figure 7-15** Influence of ion concentration surrounding plant cells on the rate of ion uptake. If free diffusion were responsible for uptake, the rate would be low and essentially proportional to concentration, but actual rates are considerably higher and reflect saturation kinetics.

for solutes (both inorganic and organic) that cells must accumulate, actual absorption rates are much faster, and a linear component of simple diffusion is seldom observed. The upper line of Figure 7-15 illustrates what is usually found. The rate of absorption (jargon: "uptake") increases rapidly as the solute concentration increases in low concentration ranges generally similar to those that exist in soils (up to about 0.1 mM), but at higher concentrations the absorption rate starts to level off. Such findings suggest that a carrier in the membrane transports each solute as fast as it can until it becomes essentially overwhelmed by excess solutes at high solute concentrations.

Other researchers (especially Per Nissen in Norway) have done similar experiments and have conducted careful computer-assisted analyses of kinetic data of Epstein and his colleagues, as well as those of others. Surprisingly, Nissen and coworkers found that the general absorption curve of Figure 7-15 actually consists of numerous increases and leveling-off phases (Nissen, 1986, 1991; Nissen and Nissen, 1983). Epstein and his colleagues had earlier found distinct evidence for two (and for chloride, even three) such kinetic phases (reviewed by Epstein, 1972). Data on phosphate absorption rates over a wide range of phosphate concentrations are plotted in Figure 7-16. Because the concentration range was so great (3 μm to 7 mM), data are plotted logarithmically. Results show a series of curves, each curve increasing and then leveling off before the next curve does likewise. These general results have been obtained with multicellular-layered roots and stems, with unicellular organisms, and even with isolated plastids. Furthermore, they apply to several cations and anions that plant cells normally absorb, as well as to amino acids and glucose. Apparently, the results apply to all plant membranes and to all solutes to which plants have become adapted during evolution. Such results are called **multiphasic kinetics**. We do not yet know how to explain multiphasic kinetics, but it is clear

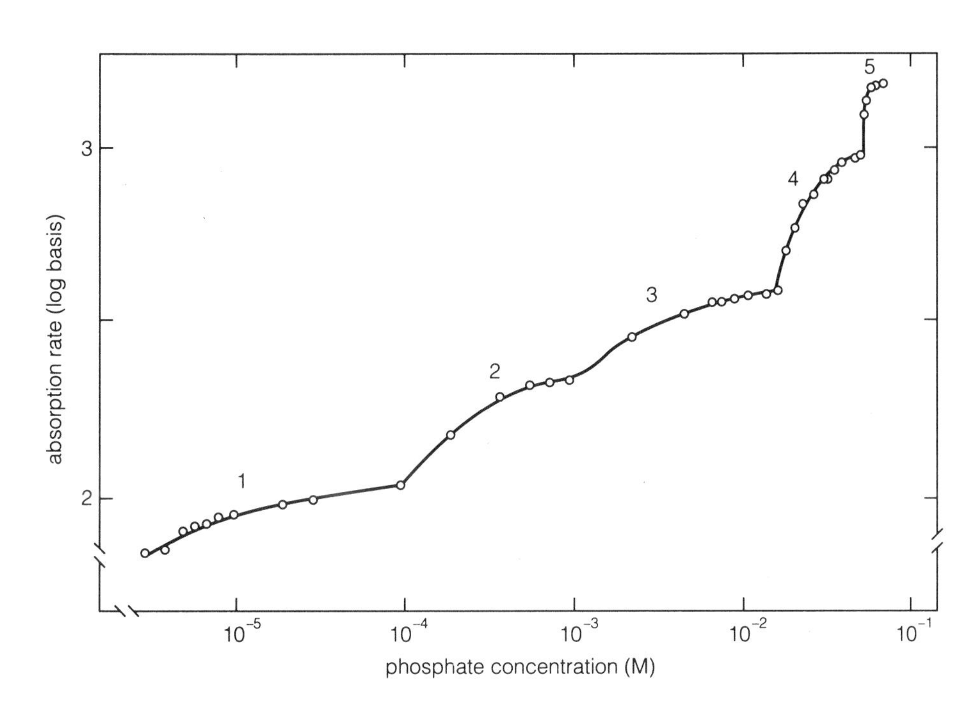


**Figure 7-16** Multiphasic absorption of phosphate by maize root sections. In this experiment, five separate, nearly saturable curves were obtained over a range of phosphate concentrations in the nutrient solution that varied 25,000-fold (from 3 μM to 75 mM). Because of the large range of phosphate concentrations, data are plotted logarithmically. Absorption was for 1-h periods. (Redrawn from Nandi et al., 1987.)

that simple diffusion through lipids or proteins in the membrane is not the answer. Instead, transport proteins (carriers, channels, or pumps) must be involved.

## 7.7 Passive and Active Transport: The Energetics

We emphasized earlier that many solutes that cells absorb are accumulated at higher concentrations within cells than outside them. For example, most cells accumulate potassium, and sieve tubes of the phloem accumulate sucrose before they translocate it (see Section 8.3). But a few solutes that are absorbed rapidly by cells never reach higher concentrations inside than out. A good example is gaseous $N_2$, but it probably moves in and out only by diffusion through the lipid bilayer. Table 7-1 shows that algae (as is also true for most plants) actually restrict movement of sodium so that accumulation ratios for that ion are commonly less than 1.0. Ignoring mechanisms and any transport protein, what do the thermodynamic considerations of Chapter 2 tell us about the feasibility of solute accumulation, nonaccumulation, and restriction?

For any solute that is not charged (mostly gases and sugars), the difference in concentration on the two sides of a membrane is essentially the only factor that determines the gradient in **chemical potential**. But for charged solutes (ions), another important factor is involved: the gradient in **electropotential**. Briefly, this gradient simply concerns the attraction or repulsion of ions resulting from a difference in electrical charge across the membrane. We are accustomed to thinking that the positive and negative charges in any solution of salts must exactly balance, and this is generally true for solutions in which life does not exist. But cells use energy to pump protons, $Na^+$, and $Ca^{2+}$ out into the cell wall (as described in the next section), and this loss of cations causes their cytosol to become slightly negatively charged. This electrical charge can be measured by electrodes placed simultaneously inside and outside the cell. Thousands of such measurements show that the voltage is small, yet the cytosol is negatively charged by some 100 to 150 millivolts (mV).[1] Also, the cytosol is slightly negative relative to the large central vacuole in parenchyma cells by about 30 mV (Sze, 1985). Because the charge in the cytosol relative to the solution outside cells attracts cations and repels anions, the charge's effect must be added to the concentration differences to determine the total chemical-potential gradient for ions.

[1]The amount of charge across the membrane needed to make the cytosol 100 mV negative is tiny, less than one nonbalanced ion in a million anions and cations (Clarkson, 1984).

The total chemical-potential gradient across any membrane is called the **electrochemical gradient**.

A mathematical equation that adds the chemical concentration gradient across a membrane to the electrical gradient (voltage difference) across it is the **Nernst equation**, somewhat simplified as follows:

$$\Delta\mu = \Delta(RT\ln C) + \Delta(zF\psi) \qquad (7.3)$$

Here, $\Delta\mu$ represents the total chemical-potential difference (electrochemical gradient) across the membrane in joules/mole; $RT\ln C$ is the chemical contribution to the electrochemical gradient for the solute, resulting only from the natural log of the concentration difference; and $zF\psi$ is the electrical contribution. R is the gas constant (8.314 J $mol^{-1}$ $K^{-1}$) and $T$ is the absolute temperature in kelvins. $C$ is the ion concentration in moles per liter, and the $\Delta$ sign preceding ($RT\ln C$) means that the difference in concentration across the membrane must be measured; $z$ is the number of charges on the ion (for example, +1 for $K^+$ and −2 for $SO_4^{2-}$); $F$ is the **Faraday constant** (96,400 joules per volt per mole); and $\Delta\psi$ is the electropotential charge difference across the membrane in volts.

If we assume that the temperature is the same on both sides of the membrane (as is usually true) and then convert to log base 10 values, then Equation 7.3 can be simplified for easy use to produce Equation 7.4:

$$\log\frac{C_i}{C_o} = \frac{-zF\Delta\psi}{2.3RT} \qquad (7.4)$$

Here $C_i/C_o$ is the *predicted* ratio of ion concentrations inside and outside the membrane at equilibrium when $\Delta\mu$ = zero. If $\Delta\mu$ is zero or less than zero, the gradient in free energy (electrochemical-potential gradient) allows absorption by **passive transport**, but if it is positive the cell must use energy to transport the ion inward by **active transport**. Therefore, passive and active absorption are defined by the Nernst equation, which considers differences in both concentration and electrical charge.

Table 7-2 lists representative data for excised pea root sections that were allowed to absorb ions from a balanced nutrient solution for 24 h. Tissue concentrations of various ions in the cells that were predicted if no active absorption occurred ($C_i$ values) were calculated from Equation 7.4, and the measured concentrations in the cells represent actual data. After absorption, the $\Delta\psi$ across the root cells was −110 mV. Nearly all ions were accumulated, so the accumulation ratio ($C_i/C_o$) was almost always greater than 1 (column 4), but the interesting comparisons are between predicted and measured concentrations (last column). For $K^+$, predicted and measured concentrations were similar, indicating that the ion was near equilibrium and was

**Table 7-2 Use of the Nernst Equation to Predict Whether Ions Were Absorbed Passively or Actively by Pea Roots.**[a]

| Ion | Solution Conc. | Measured Tissue Conc. | Accumulation Ratio | Predicted $C_i$ | Measured Tissue Conc. / Predicted $C_i$ |
|---|---|---|---|---|---|
| $K^+$ | 1.0 mM | 75 mM | 75 | 72 mM | 1.04 |
| $Na^+$ | 1.0 | 8.0 | 8.0 | 72 | 0.111 |
| $Mg^{2+}$ | 0.25 | 1.5 | 6.0 | 1,310 | 0.00115 |
| $Ca^{2+}$ | 1.0 | 0.80 | 0.80 | 5,250 | 0.000152 |
| $NO_3^-$ | 2.0 | 27 | 14 | 0.0276 | 978 |
| $Cl^-$ | 1.0 | 5.3 | 5.3 | 0.0138 | 384 |
| $H_2PO_4^-$ | 1.0 | 25 | 25 | 0.0138 | 1,810 |
| $SO_4^{2-}$ | 0.25 | 9.5 | 38 | 0.000048 | 198,000 |

[a]Several root segments 1 to 2 cm long (0.20 g fresh wt. total of all segments) were shaken in 50 ml of nutrient solution at 25°C for 48 h. The electropotential difference between solution and root cells averaged 110 mV. Segments were rinsed, frozen, then extracted twice in boiling water to determine the concentration of each ion in the tissues (column 3 above). The accumulation ratio (column 4) is the ratio of measured tissue concentration to nutrient solution concentration. The predicted $C_i$ (column 5) was calculated from the Nernst equation, two examples of which are given below. The last column indicates how far from electrochemical equilibrium each ion was within the tissues. (Data are from N. Higinbotham et al., 1967.)

Sample calculations of $C_i$:

1. For $K^+$:

$$\log C_i/C_o = \frac{(1.0)(96{,}400 \text{ J/V mole})(0.110 \text{ V})}{(2.3)(8.31 \text{ J/mole K})(298 \text{ K})} = 1.86$$

$C_i/C_o = 10^{1.86} = 72$. Because $C_o$ was 1.0 mM, $C_i = 72$ mM

2. For $Ca^{2+}$:

$$\log C_i/C_o = \frac{(2.0)(96{,}400 \text{ J/V mole})(0.110 \text{ V})}{(2.3)(8.31 \text{ J/mole K})(298 \text{ K})} = 3.72$$

$C_i/C_o = 10^{3.72} = 5{,}250$. Because $C_o$ was 1.0 mM, $C_i = 5{,}250$ mM

probably absorbed passively. For $Na^+$, $Ca^{2+}$, and $Mg^{2+}$, the measured tissue concentration was always less than predicted, so these ions could have moved in passively by diffusion, almost surely with aid of a transport protein. (In fact, they must have even been transported out actively, a subject to which we return later.) The point is that, energetically speaking, their inward movement was passive.

Similar results from other experiments for $Mn^{2+}$, $Fe^{2+}$, $Zn^{2+}$, $Cu^{2+}$, and boron as $H_3BO_3$ indicate that their absorption is also passive, even though that passive process depends on energy-dependent production of ATP and its hydrolysis to cause a negative charge inside the cytosol. For all anions, the measured internal concentrations were far higher than those predicted, showing that anions were absorbed actively. Cells probably always use energy to accumulate anions because the negative charge inside the membrane always repels them. This repulsion factor also helps explain why $CO_2$ and $H_2CO_3$ are absorbed faster than both $HCO_3^-$ and $CO_3^{2-}$, and why $H_2PO_4^-$ is absorbed faster than $HPO_4^{2-}$, as mentioned in Section 7.5. The next section concerns how cells use energy to maintain a negative charge in their cytosol.

## 7.8 How ATPase Pumps Transport Protons and Calcium

As indicated above, cells use energy stored in ATP to absorb solutes actively. How do they obtain energy from ATP to do this? Let's first consider ATP and its hydrolysis.

ATP gives up its energy when its terminal phosphate is hydrolyzed off to release ADP and inorganic phosphate ($H_2PO_4^-$ or $HPO_4^{2-}$, collectively abbreviated as Pi), as follows:

$$ATP(Mg) + H_2O \leftrightarrow ADP(Mg) + Pi \quad (7.5)$$

This reaction is catalyzed by an enzyme that is apparently present in every membrane of every living cell. It is called **ATP phosphohydrolase** and is universally abbreviated **ATPase**. It is one of the transport proteins mentioned earlier. Each molecule of ATP and ADP is chelated with one $Mg^{2+}$, which emphasizes one essential function of magnesium for all life. The reaction is strongly exergonic in the forward direction, with the release of approximately 32 kJ (7.6 kcal) for each mole of ATP hydrolyzed. If you mix equal amounts of ATP, ADP, Pi, and $Mg^{2+}$ in water and then add an ATPase to speed the reaction, it will proceed nearly to completion with essentially no ATP left at equilibrium. Nearly all the energy will be released as heat.

*However, in membranes most ATPases instead ensure that much of this energy is used to transport protons from one side of the membrane to the other against an electrochemical gradient*. This transport of $H^+$ provides energy that is then used to transport essential mineral salts. Another generally much-less-active ATPase that pumps $Ca^{2+}$ out of the cytosol at the same time it pumps $H^+$ into the cytosol probably exists in plants (as it does in animals), but we are just starting to learn about it. This **$Ca^{2+}/H^+$ ATPase** is thought always to pump calcium out of the cytosol, either outward to the cell wall via the plasma membrane or inward to the vacuole via the tonoplast (Hepler and Wayne, 1985; Giannini et al., 1987). Furthermore, there is evidence for still another calcium pump in the plasma membrane of some species (Gräf and Weiler, 1989; Kasai and Muto, 1990). This is called a **(Ca + Mg)-ATPase** because it depends on a chelate of $Mg^{2+}$ and ATP to move $Ca^{2+}$ out of the cell. This pump does not move $H^+$ in as $Ca^{2+}$ moves out, and it probably depends on calmodulin for its activity (for discussions of calmodulin, see Sections 6.6 and 17.1).

The mechanism by which ATPases pump $H^+$ and $Ca^{2+}$ from one side of a membrane to the other is not fully understood, yet several facts are known. All ATPases are large enough to span any membrane (often more than once) and so they represent integral proteins mentioned in Section 7.4. Also, part of every ATPase molecule protrudes out of the membrane into the aqueous cytosol, where that part reacts with ATP and water. It is likely that the ATPase contains a potential hole extending across the membrane, but that hole opens only when ATP is being hydrolyzed. Furthermore, only when ATP is being hydrolyzed can the ATPase combine with a proton to move the proton across the membrane through the hole.

There are three major kinds or classes of $H^+$-transporting ATPases known to exist in membranes of various prokaryotic and eukaryotic organisms (reviewed by Blumwald, 1987; Rea and Sanders, 1987; Nelson, 1988; Poole, 1988; Serrano, 1989). These transport only $H^+$. We first summarize some properties of the $H^+$-transporting ATPase in plasma membranes of plants and fungi because this ATPase so strongly controls what first gets into any cell and how fast. This ATPase becomes energized by catalyzing a short-lived transfer of the terminal phosphate from ATP to a part of itself that extends into the cytosol, where ATP is available. Combination of the ATP's terminal phosphate with the ATPase causes the ATPase to become energy-rich, and ADP is released. During formation of the energy-rich state the ATPase changes its shape, combines with a proton, and presumably opens the hole in itself through which the proton can move across the membrane. Immediately thereafter, $H_2O$ is used to hydrolyze phosphate off the ATPase; the proton moves through the hole and is released on the outside of the plasma membrane (cell-wall region). Then the ATPase again assumes its original low-energy shape and is ready to function again. We don't yet understand what changes in shape the ATPase undergoes during these processes and how such changes cause proton transport from one side of the membrane to the other; progress and problems yet to be solved are reviewed succinctly by Pedersen and Carafoli (1987) and by Serrano (1989). An important point about this ATPase is that it requires $K^+$ for its activity, even though it probably does not directly transport $K^+$ across the membrane.

The plasma membrane $H^+$-ATPase is thought to transport out of the cytosol and into the cell wall only one proton for each ATP hydrolyzed. This ATPase is the most energy-wasteful of the three major classes of ATPase known (much of the energy in ATP is lost as heat), yet it has three important effects:

1. *It causes the pH of the cytosol to increase*. This *p*H change is usually small because the cytosol buffers itself against the loss of $H^+$. The cytosol *p*H is usually between 7 and 7.5.

2. *It causes the pH of the cell wall to decrease*. The cell wall has much less buffering capacity than the cytosol, so its *p*H often drops to 5.5 or 5.0. This phenomenon also helps account for the ability of roots to acidify soils, although their release of $CO_2$ in respiration also causes acidification by the formation of carbonic acid, as shown in Reaction 7.1.

3. *It causes the cytosol to become electronegative relative to the cell wall as the cytosol loses $H^+$ but retains $OH^-$*. The cytosol's loss of a cation and retention of an anion accounts for the negative electropotential-voltage difference across the plasma membrane described earlier.

Although the plasma membrane is the first barrier for absorption, most solutes enter the vacuole and so must also penetrate the tonoplast membrane. The tonoplast has an ATPase that pumps protons into the vacuole, which makes the vacuole acidic (often near *p*H 5 but even lower in citrus fruits and some other plant parts). This ATPase differs in several ways from that in the plasma membrane. One important difference is that it transports $H^+$ without combining with a phosphate from ATP. A second difference is that it transports two $H^+$ into the vacuole for each ATP hydrolyzed, so it is more energy-efficient than the plasma membrane ATPase. A third difference is that it is not $K^+$-dependent. How it transports $H^+$ is unknown, but the energy released when ATP is hydrolyzed likely changes its structure so that a hole opens in it through which protons then are forced to move. The membranes of endoplasmic reticulum and dictyosomes also contain an ATPase proton pump much like that of the tonoplast; this pump transports $H^+$ into the interior of organelles and makes them slightly acidic.

In addition to the ATPase pump, the tonoplast has a **proton pumping pyrophosphatase** that uses energy in inorganic pyrophosphate (Rea and Sanders, 1987). Just as with ATP, hydrolysis of a pyrophosphate ester is involved:

$$\mathrm{HO{-}\overset{\overset{O^-}{|}}{\underset{\underset{O}{\|}}{P}}{-}O{-}\overset{\overset{O^-}{|}}{\underset{\underset{O}{\|}}{P}}{-}OH + H_2O \rightarrow 2\,H_2PO_4^-} \tag{7.6}$$

The pyrophosphate actually exists as chelate with $Mg^{2+}$, just as do ATP and ADP. It is likely that the pyrophosphatase pump transports only one $H^+$ into the vacuole for each pyrophosphate hydrolyzed, but this has not yet been proved. Apparently, this pump contributes less to solute absorption by vacuoles than does the ATPase, partly because the PP*i* concentration in the cytosol is less than that of ATP.

Finally, there is still another kind of ATPase that transports protons across chloroplast and mitochondrial membranes. As described in Chapters 10 and 13, it synthesizes rather than hydrolyzes ATP.

## 7.9 How Carriers and Channels Speed Passive Transport

In this section we speculate about how carriers and channels could passively speed movement of solutes across membranes by taking advantage of the electropotential gradient established by an ATPase or pyrophosphatase pump. Carriers are thought to specifically

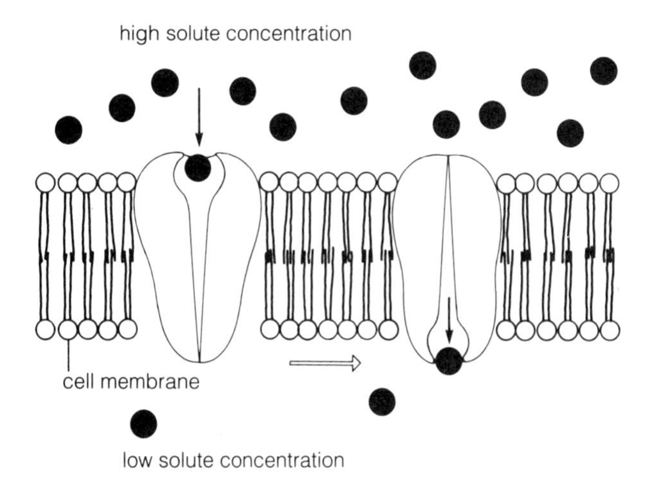


**Figure 7-17** Simplified model of facilitated diffusion or uniport. Here it is assumed that an integral protein in the membrane, a carrier, consists of two equal halves (a dimer) with a hole in its center through which a solute can move down its electrochemical gradient. For simplicity, this example implies that only the concentration difference is important for the establishment of the electrochemical gradient.

bind one or a very few related solutes (much like an enzyme attaches to its substrate). Carriers are also thought to be integral proteins that completely span the membrane. The term *carrier* implies that these proteins pick up solutes and move across the membrane with them, but this almost surely is not true. Rather, they probably undergo a reversible conformational (structural) change that facilitates solute transfer. Figure 7-17 provides a highly schematic explanation of how a carrier might work to move a solute down its free-energy gradient into a cell.

We know much more about channel proteins than about carriers, especially those in animal cells. Work on plant channel proteins didn't begin until about 1984, so we still have much more to learn about them (see reviews by Hedrich et al., 1987; Satter and Moran, 1988; Hedrich and Schroeder, 1989; Schroeder and Hedrich, 1989; and Tester, 1989). Channel proteins are integral proteins. As is true for all proteins, they exist in only one conformational structure of the lowest free energy specified by the particular cell environment at any given time; the conformation varies when the cell environment varies. Some such conformations have a central hole through which solutes can pass at very high rates. Satter and Moran (1988) stated that specific ions can move through open channels as fast as $10^8$ ions per second, which they suggested is three to four orders of magnitude faster than movement via carriers.

Whether channels are open, closed, or partly open depends on the environment. Two main types of channels have been discovered: One has a gating system that

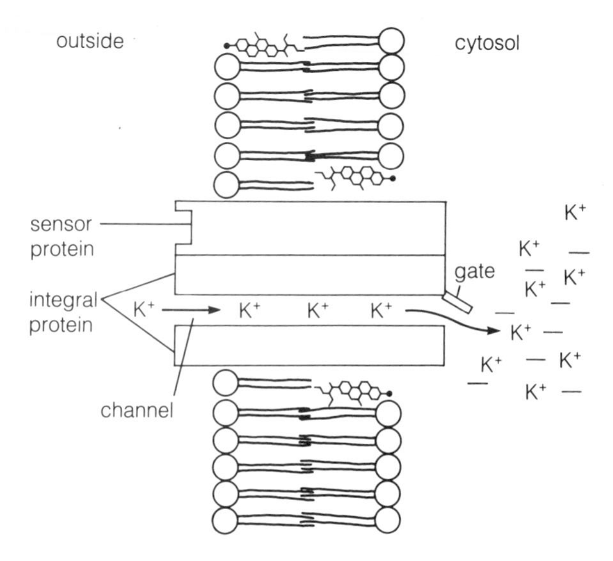


**Figure 7-18** Hypothetical model of an ion channel. Integral proteins that span the membrane contain a central hole through which ions such as potassium can move down their electrochemical gradients. In this model, potassium ions can move from left to right into the cytosol *down* an electrochemical gradient (note excess negative charges at right of channel) but *up* a potassium concentration gradient. The sensor protein embedded near or in an integral protein responds both from outside the cell and from the cytosol to chemical stimuli resulting from light, hormones, or even calcium ions. The channel protein has a gate attached (right center) that closes or opens in unknown ways in response to the voltage gradient across the membrane or to chemicals produced by environmental stimuli.

responds to the voltage gradient across the membrane, and the other responds to external modulator stimuli such as light or hormones. Figure 7-18 gives a model of an ion channel and shows how it might allow movement of $K^+$ ions into a cell.

## 7.10 How Membranes Take Advantage of Proton Pumps for Ion Transport

As first suggested by Peter Mitchell in 1961 (see his review in Mitchell, 1985), the proton pumps described above (largely unknown to him in 1961) provide cells with two usable sources of energy: the *p*H gradient and the electropotential gradient. (This research led to Mitchell's Nobel prize in chemistry in 1978.) One of these two energy sources is usually used to transport anions, cations, or neutral molecules across membranes. (Additional evidence suggests that the plasma membrane can oxidize certain compounds in the cytosol and use the energy released to drive absorption of some solutes. We shall not discuss this topic, but it is reviewed by Bienfait and Lüttge, 1988 and by Crane and Barr, 1989.)

First consider the plasma membrane. Cation absorption is favored by the electropotential gradient (the electropotential is negative inside the cytosol). For $K^+$, $NH_4^+$, $Mg^{2+}$, and $Ca^{2+}$, this absorption usually occurs with the help of either a carrier or a channel. Diffusion through a lipid bilayer is far too slow and cannot account for inhibition by related ions or for multiphasic kinetics of these ions. Only involvement of a transport protein can explain what we know about cation absorption. Researchers have especially sought evidence for a true potassium pump in plants, but none has been found (see reviews by Kochian and Lucas, 1988 and by Lüttge and Clarkson, 1989). Generally, direct inward pumping of $K^+$ by an ATPase occurs only in animals that have a sodium-potassium pump. The simplest explanation for cation absorption into plant cells is a process called **uniport** or **facilitated diffusion**. This process involves specific transport proteins that greatly speed movement of each cation into the cell down its overall free-energy (electrochemical) gradient. For $K^+$ and $NH_4^+$ the ratio of measured concentrations inside and outside the cell indicates the approximate concentration gradients across the plasma membrane because those two monovalent cations are not held tightly to any anion. Chelation of $Mg^{2+}$ and $Ca^{2+}$ by chlorophyll, proteins, ATP, or other ligands in the cytoplasm makes it much more difficult to measure their unbound concentrations inside cells. Nevertheless, facilitated diffusion into cells across the plasma membrane by a "uniporter" might work mechanistically as shown by the simple carrier model in Figure 7-17.

The electropotential gradient will not favor absorption of neutral molecules such as sugars, and it will repel anions such as bicarbonate, nitrate, chloride, phosphate, and sulfate (Dunlop, 1989). For sugars and anions, cells take advantage of the *p*H gradient between the cell wall and cytosol (say, *p*H values of 5 and 7, or a 100-fold difference in $H^+$ concentration). This *p*H gradient and the electropotential gradient favor passive reabsorption of $H^+$, but $H^+$ ions move inward very slowly unless they combine with a carrier or move through a channel. It is important that such carriers will allow $H^+$ to move in only if they can simultaneously combine with an anion and help transport it into the cell. This is an example of **cotransport** or **symport** (Fig. 7-19). Separate carriers are thought to exist for different anions, which explains the specificity of anion absorption mentioned in Section 7.6. Still other carriers move sugars and amino acids into cells by cotransport with $H^+$ (Giaquinta, 1983; Reinhold and Kaplan, 1984; Lemoine et al., 1989). In all these cotransport examples, $H^+$ moves inward down its electrochemical-potential gradient, whereas the anion or neutral molecule is transported actively; that is, an energy-releasing process drives an energy-requiring one.

Passive $H^+$ absorption can also be used to transport cations simultaneously out of cells, as is also shown in

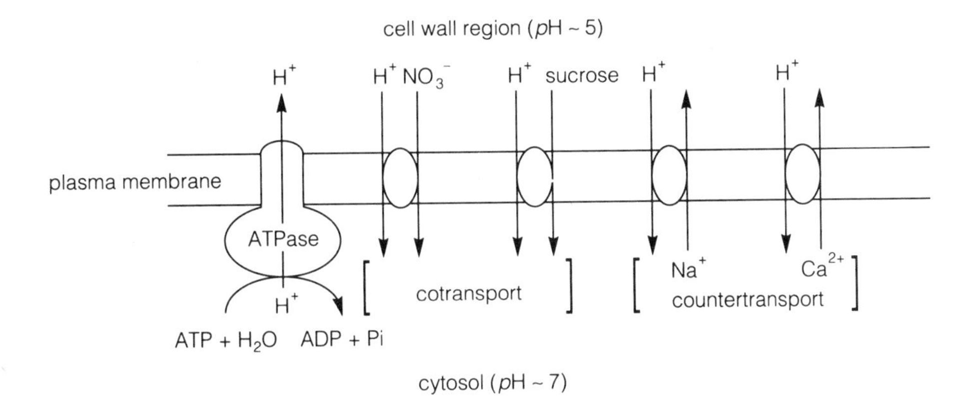


**Figure 7-19** Cotransport and countertransport of solutes across the plasma membrane driven by energy in ATP. The proton-translocating ATPase shown at the left moves $H^+$ out of the cytosol. These protons can return to the cytosol in cotransport via carriers that simultaneously move in anions such as nitrate or sugars such as sucrose. Countertransport (antiport) also involves return of protons to the cytosol, but one or more protons are exchanged for an outgoing cation such as $Na^+$, $Mg^{2+}$, or $Ca^{2+}$.

Figure 7-19. Here a carrier is thought to combine with $H^+$ on the outside and, for example, $Na^+$ on the inside; then it somehow transports them in opposite directions. This is an example of **countertransport** or **antiport**. (It seems impossible for a channel to perform countertransport.) Countertransport of $Na^+$ is an important way that root cells eliminate that ion from their cytosol when they grow in salty soils (Briskin and Thornley, 1985). Furthermore, as we mentioned in Chapter 6, the concentration of $Ca^{2+}$ is kept extremely dilute (often 0.1 to 1.0 μM) in the cytosol of eukaryotes. The energetics described in the Nernst equation strongly favor free diffusion of $Ca^{2+}$ into the cytosol, so energy must be used to remove it (via the $Ca^{2+}$ pumps mentioned in Section 7.8).

Transport of solutes across the tonoplast into the predominant central vacuole uses energy from either the ATPase or the pyrophosphatase pump. This process allows storage of ions and molecules that can be recovered by the cytosol when needed. For $Na^+$, countertransport into the vacuole prevents this ion's accumulation at harmful levels in the cytosol. (Some plants do accumulate and tolerate $Na^+$, a topic described in Chapter 26.) Calcium is also transported into the vacuole by countertransport with $H^+$ (Blumwald and Poole, 1986; Bush and Sze, 1986; Kasai and Muto, 1990). Mechanisms by which anions are absorbed by the vacuole have not been studied much. It seems likely that anions move inward by a uniport process because the vacuole is slightly positively charged relative to the cytosol. However, anions could also move into the vacuole by antiport with protons.

## 7.11 Absorption of Very Large Molecules, Even Proteins, by Organelles

A final kind of transport we wish to mention concerns movement of proteins from one organelle to another within a single cell. We are accustomed to thinking that such large molecules simply cannot penetrate membranes, yet they do. So far as we know, all of the hundreds or thousands of different kinds of proteins in the nucleus must be absorbed by it after they are made on ribosomes in the cytosol. And most of the proteins in chloroplasts and mitochondria are synthesized not within these organelles but also on ribosomes in the cytosol. Proteins are also transported into vacuoles, cell walls, the endoplasmic reticulum, glyoxysomes, and peroxisomes. How can such proteins be absorbed by any organelle, and how are they absorbed by only the proper organelle?

Answers to those questions are largely unknown, but information is accumulating rapidly (Dingwall and Laskey, 1986; della-Cioppa et al., 1987; Jones and Robinson, 1989; Keegstra et al., 1989). What we have learned and suspect can be summarized by a few major conclusions:

1. ATP is usually or always required to provide energy for transport, but it isn't likely that such energy is used mainly to provide a *p*H gradient or an electropotential gradient. A popular hypothesis is that ATP provides the energy needed to unravel the protein's globular structure (to make the polypeptide chain or chains more narrow and linear) so that penetration is made easier (Hurt, 1987).

2. There is great specificity with respect to which organelle absorbs which protein. This specificity results from matching recognition sites in a membrane protein at the surface of the organelle and in the protein to be absorbed. Chloroplasts and mitochondria apparently have no glycoproteins in their outer membranes, so the polysaccharides that give recognition properties to most other membranes (Section 7.4) are probably not involved.

3. Specificity in the protein is confined to only a relatively small part of it, often some 20 to 50 amino acids connected at one end of the molecule. This sequence of amino acids is often referred to for chloroplasts as the **transit sequence**, for mitochondria as the **leader**

**sequence**, and for the endoplasmic reticulum and nucleus as the **signal sequence**. (Presumably, enough new names to match all the cellular organelles will be thought up, or a single inclusive term to represent what seems to be the same general process of recognition will be agreed upon.)

**4.** After the protein is absorbed by the proper organelle, the recognition sequence is split (hydrolyzed) off and the mature functional protein is released. The recognition sequence is probably hydrolyzed even further into its 20 to 50 separate amino acids. The mature protein somehow finds its way to the place in the organelle where it then functions to help maintain life.

## 7.12 Correlations Between Root and Shoot Functions in Mineral Absorption

Except for certain desert species with unusually large root systems, most species invest most of their biomass in shoots (Section 7.1). Therefore, it seems reasonable that absorption of mineral salts should be controlled in part by activities of the shoot. There are two ways of looking at this control. In a *demand sense,* the shoot might increase root absorption of mineral salts by rapidly using them in growth products (for example, proteins, nucleic acids, and chlorophyll). In a *supply sense,* the shoot supplies carbohydrates via the phloem that the root must respire to produce ATP that drives mineral salt absorption (see, for example, Gastal and Saugier, 1989). The shoot probably also supplies the roots certain hormones that affect absorption by roots. In fact, there is much evidence for an interdependence between activities of roots and shoots (reviewed by Wild et al., 1987; Ingestad and Ågren, 1988; Cooper and Clarkson, 1989; and Glass, 1989). For example, excellent

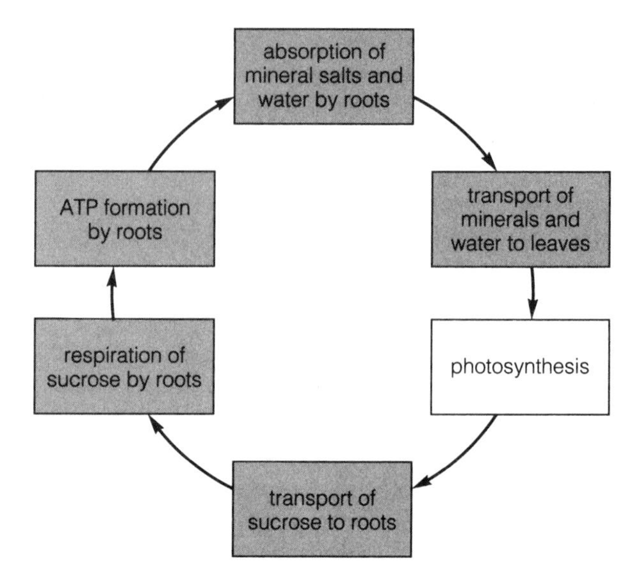


**Figure 7-20** Interrelations of some physiological processes in roots and shoots that affect absorption of mineral salts from the soil. (From Starr and Taggart, 1989.)

correlations between the rate of shoot growth and the rate of absorption of nitrogen, phosphorus, and potassium have been obtained. Respiration rates of roots over time are sometimes highly correlated with rates of photosynthesis (maxima near noon). Root respiration has also correlated with the rate of sugar translocation to the roots. Maximum absorption of nitrate and ammonium ions correlate with maximum photosynthetic rates, except that absorption lags by about 5 h, suggesting the need for carbohydrate translocation and root respiration during the lag period. Correlations don't prove cause and effect, but in a growing plant the interdependence of root and shoot activities seems obvious. Figure 7-20 summarizes this concept in a simple way.

# 8

# Transport in the Phloem

Most higher plants need not only air and sun but also soil and water. The complex structures of roots and shoots meet this need, but there must be an ordered and integrated transfer of materials within the plant. In Chapters 4 and 5 we examined the movement of water with its dissolved minerals from soil through the xylem and ultimately via transpiration out through the leaves. Now we examine the movement of organic molecules—those made in photosynthesis and other processes—from their points of origin to other parts of the plant.

Plant physiologists have devoted considerable effort to the study of this movement or **translocation** of dissolved materials, mostly organic molecules. To begin with, we need descriptive information. In which tissues does the movement take place? How fast? In what form? Then we can ask: What are the mechanisms of translocation, what coordinates supply with demand, and what determines the tissues and organs to which substances are translocated?

The concept of blood circulation has hardly been in doubt since William Harvey lectured on it in the early 1600s, but the mechanisms of solute translocation in plants have only begun to yield to our probing during recent decades, although the problem has been studied by modern scientific methods for well over a century. Why has the elucidation of transport mechanisms in plants been so difficult? Partly because transport occurs within tubes that consist of rows of microscopic cells. In animals, the vascular system is part of the extracellular space: tubes with walls made of cells. The heart, with its macroscopic valves and associated plumbing, is easy to comprehend. (The stumbling block in understanding animal circulatory systems was the microscopic capillary system that returns arterial blood to the veins.)

More subtle, but perhaps even more important than the dimensions of the plant vascular system, is the fact that fluids in phloem sieve tubes are under high pressure, many times higher than that in animal circulatory systems. When cells in the phloem are cut, the pressure is released, and the instantaneous surging of phloem contents alters or destroys the cellular structures that existed before cutting. Only in recent years have rapid freezing, followed by freeze-drying and other techniques, provided help with this problem.

During the late 1970s and the early 1980s most plant physiologists became convinced that a mechanism of phloem transport first suggested in 1926 was indeed basically correct. Of course, the model has been modified since its initial presentation, but in its contemporary form it is supported by a large and convincing body of evidence. Virtually all of the once-disturbing counterevidence has now been satisfactorily explained and laid to rest.

## 8.1 Transport of Organic Solutes

In 1675, the Italian anatomist and microscopist Marcello Malphighi **girdled** a tree by removing a strip of bark from around its trunk (Fig. 8-1). Stephen Hales repeated the experiment in 1727. The bark above the girdle swelled, whereas that below the girdle shrank. Although the tree eventually died, the shoots grew and continued to transpire for some time. Clearly, the transpiration stream was not affected by the girdle and must occur in the wood, but nutrients essential to the life of the bark (and presumably the roots) must move in the bark.

Although this was one of the first experiments in plant physiology, it still makes an interesting demonstration. In its most sophisticated development, it has been combined with radioactive tracers. Bark can be surgically separated from wood, or wood can be removed, leaving the bark virtually intact. Living cells in the bark can also be killed with a treatment such as heat.

**Figure 8-1** The effects of girdling the trunk of a tree by removing bark from around the circumference. Note the swelled bark above the girdle compared with that below. The trunk has been cut to reveal the annual growth rings. Note that an entire year's growth was laid down above the girdle but not below. (Specimen courtesy of Herman Wiebe.)

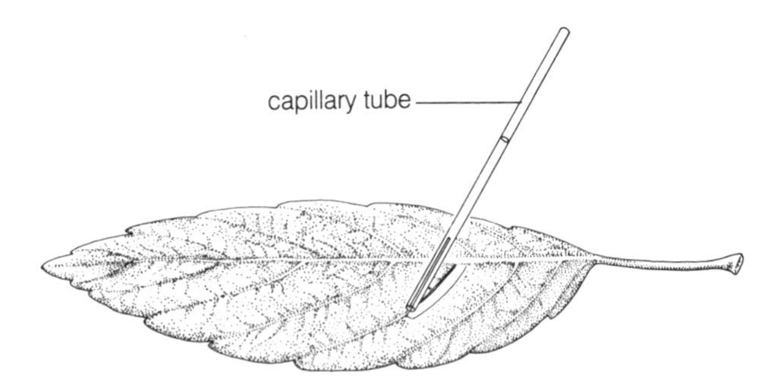


**Figure 8-2** The reverse-flap technique for applying solutions to leaves. Note how the reverse feature makes it unlikely that solution will be pulled from the leaf to relieve tension in the xylem because there is no direct connection between the flap and xylem in the stem. (Courtesy John Hendrix.)

It was long the aim of plant physiologists to measure translocation directly by following the movement of marked materials in the transport system. Early investigators used dyes; indeed, the dye fluorescein moves readily in phloem cells and is still an effective tracer. Viruses and herbicides have also been used, but by far the most important tracers are the radioactive nuclides that became available after World War II. Radioactive phosphorus, sulfur, chlorine, calcium, strontium, rubidium, potassium, hydrogen (tritium), and carbon have been used in these studies. Heavy, stable isotopes such as those of oxygen ($^{18}O$), nitrogen ($^{15}N$), or carbon ($^{13}C$) are also used.

Tracers can be applied by the **reverse-flap technique**, in which a flap including a vein is cut from a leaf, as shown in Figure 8-2. Another widely used approach is to remove some of the cuticle of a leaf by abrasion (which in some species breaks open the epidermal cells). Solutions applied to an abraded leaf penetrate to the mesophyll cells and veins about three times as fast as they would otherwise. Frequently, a leaf in a closed container is exposed to carbon dioxide labeled with carbon-14 ($^{14}C$) or, in recent years, carbon-11 ($^{11}C$), which has a short half-life but emits a more powerful ray than $^{14}C$.[1] The labeled $CO_2$ is incorporated into the **assimilates** (products of assimilation and metabolism) by photosynthesis, and in this form it is exported from the leaf in the translocation stream. It is important to know what happens chemically to the tracer in the plant. As a rule, such inorganic ions as phosphate, sulfate, potassium, or rubidium remain chemically unchanged, and most of the $^{14}C$ in $^{14}CO_2$ is at first incorporated into sucrose or some other sugar; but the radioactive carbon is eventually incorporated into every organic compound in the plant. Sometimes, phosphate and sulfate are incorporated into nucleotides and amino acids.

The radioactive tracer may be detected during transport by touching the plant stem or other part with a radiation detector. Another common method is **autoradiography** (Fig. 8-3). The plant is placed in contact with a sheet of X-ray film for several days to months, and the film is then developed to show the location of the radioactivity. The immediate problem is to immobilize the radioactivity in the plant after harvest. Plants may be placed between blocks of dry ice and then, while still frozen, subjected to a vacuum, allowing the water to sublime away (**freeze-drying** or **lyophilizing**). Another way of arresting movement is to dismember the plant before making the **autoradiogram**.

The girdling experiments and those using tracers lead to three general conclusions and many more specific ones. As noted, removing the bark (containing the phloem) from the stem or trunk (the xylem) had no immediate effect on growth of the shoot or transpira-

[1] $^{14}C$ has a half-life of 5,730 years; $^{11}C$ has a half-life of 20.3 minutes. The positron emitted by $^{11}C$ has an energy about 6.3 times as strong as the beta particle emitted by $^{14}C$; the decay energy of $^{11}C$ is 12.69 times as strong as that of $^{14}C$. The $^{11}C$ must be prepared just before use by, for example, the bombardment of boric oxide with a beam of deuterons accelerated in a Van de Graaff accelerator (Troughton et al., 1974).

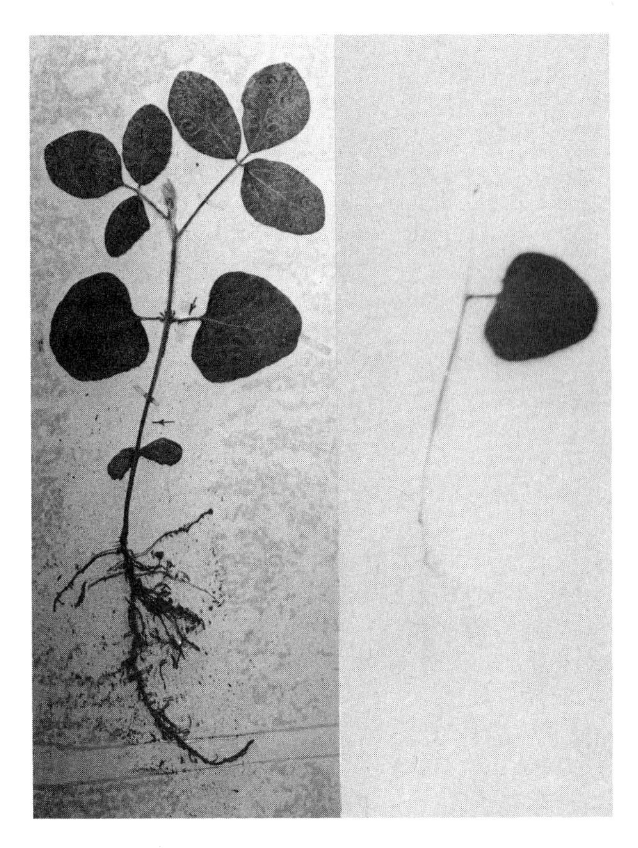

**Figure 8-3** Results of an experiment in which autoradiography is used. The purpose of the experiment was to observe the effect of wilting on translocation. The first true leaf on the right of each soybean plant was held for 1 h in an illuminated chamber containing $^{14}CO_2$, which was converted by photosynthesis into radioactively labeled assimilates. After 6 h, the plants were harvested, dried, pressed (left, in each pair), and placed in tight contact with X-ray film. After 2 weeks of exposure, the film was developed (right, in each pair). The dark areas show where the most $^{14}C$ was located. In both cases, the leaf exposed to the $^{14}CO_2$ had by far the most tracer, but more was moved from the turgid plant (left) than the wilted one (right), which was predicted from the Münch hypothesis discussed in this chapter. (Specimens and films courtesy of Herman Wiebe; see Wiebe, 1962.)

tion from the leaves. In more recent experiments, tritiated water ($^3H_2O$) or water with some radioactive solute is followed through the wood in the transpiration stream, even though the trunk has been girdled. Analysis of the xylem sap reveals that it contains mostly dissolved minerals from the soil plus small amounts of various organic compounds, including sugars and amino acids (Läuchli, 1972; Ziegler, 1975). Hence our first conclusion about translocation: *Water with its dissolved minerals moves primarily upward in the plant through xylem tissues.*

It was clear from Malphighi's and Hales's experiments that assimilates from leaves, including the products of photosynthesis (**photosynthates**), are necessary for the growth of plant parts that cannot photosynthesize, and even for some parts that photosynthesize but only at low levels, such as some stems and fruits. These materials move through the bark in the phloem. Indeed, it is possible to kill a tree by girdling the trunk and making the roots dependent upon their stored food, which runs out after a few weeks to a few years, depending on the species.

We will discuss the anatomy of phloem tissue in a later section. Here it is important to note that detailed studies, especially with radioactive tracers, have shown that assimilates move through the *sieve elements* that form *sieve tubes* in phloem tissue (Fig. 8-4). Hence the second conclusion: *Assimilates, including photosynthates, move over relatively long distances primarily through sieve tubes in the phloem. This is phloem transport.*

Various parts of the plant besides the main stem can be girdled. For example, the bark between a leafy branch and developing fruit can be removed. Again, sugars accumulate in the bark on the side of the leafy branch. Or the girdle can be placed above the leaves on the stem but below a developing shoot tip, in which case sugars still accumulate on the side of the leaves, at the *bottom* of the girdle. Gravity does not govern phloem

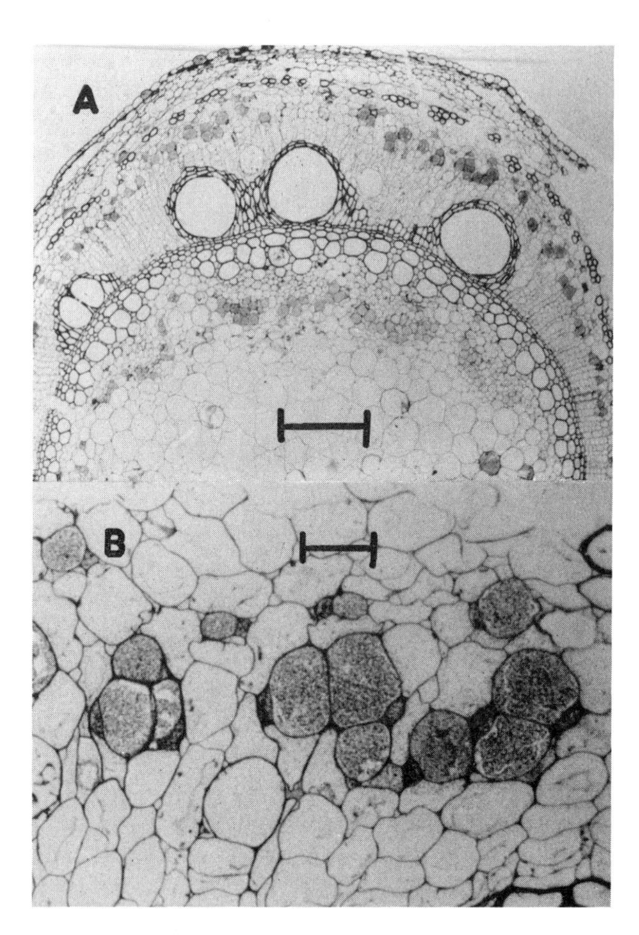


**Figure 8-4** Microautoradiographs of 2-μm Epon sections from a morning glory vine, taken from internodal tissue 100 mm below a leaf that had been allowed to photosynthesize for 6 h in the presence of $^{14}CO_2$. The sections were stained with methyl violet. The black spots are silver grains that appear in the developed photographic emulsion previously poured over the sections and allowed to set. Thus, the spots indicate locations of molecules labeled with $^{14}C$; they are mostly confined to sieve elements. (**A**) Low magnification; bar indicates 200 μm. (**B**) High magnification; bar indicates 20 μm. (Microautoradiographs courtesy of Donald B. Fisher; see Christy and Fisher, 1978.)

transport; the controlling relationship is the relative positions of the *source* and the *sink*. The leaves, with their photosynthetic capacity, typically constitute the **source**, but an exporting storage organ such as a beet or carrot root in the spring of its second year is also a source. Cotyledons and endosperm cells of seeds are sources for germinating seedlings. Any growing, storing, or metabolizing tissue might be a **sink**. Growing fruits, stems, roots, corms, tubers, flowers, or young leaves are examples. Hence the third important point: *Assimilates move from source to sink*.

The first two of the three conclusions, though correct, may obscure the facts that important organic materials are transported in the xylem and that minerals essential to the growth of many sinks are transported in the phloem (Table 8-1). There are sufficient organic

**Table 8-1 Comparison of Xylem and Phloem Sap Compositions from White Lupine (*Lupinus albus*).**

| | Xylem Sap (Tracheal) mg $L^{-1}$ | Phloem Sap (Fruit bleeding) mg $L^{-1}$ |
|---|---|---|
| Sucrose | ND[a] | 154,000 |
| Amino acids | 700 | 13,000 |
| Potassium | 90 | 1,540 |
| Sodium | 60 | 120 |
| Magnesium | 27 | 85 |
| Calcium | 17 | 21 |
| Iron | 1.8 | 9.8 |
| Manganese | 0.6 | 1.4 |
| Zinc | 0.4 | 5.8 |
| Copper | Tr[b] | 0.4 |
| Nitrate | 10 | ND[a] |
| *p*H | 6.3 | 7.9 |

[a] ND = Not present in detectable amount.
[b] Tr = Present in trace amount.
Source: Pate (1975).

and inorganic nutrients in either sap to satisfy the requirements of some sucking insects for their entire life cycles. Neither xylem nor phloem saps provide *balanced* nutrition for the insects. This is reflected in the honeydew produced by the insects. Xylem feeders (for example, cicadas, sharpshooters) produce much watery honeydew (several drops per minute) that is high in inorganics. They remove the available organic compounds and secrete the inorganic materials in their honeydew. Phloem feeders (aphids, mealybugs, scale insects, leaf hoppers) produce smaller amounts of honeydew that is high in organics, especially sticky sugars—and that accumulates on cars and sidewalks below.

## 8.2 The Pressure-Flow Mechanism

The model of phloem transport currently favored by most plant physiologists was proposed in its elemental form by E. Münch in Germany in 1926 (see Münch, 1927, 1930). Much of what we know about translocation has been learned in the process of testing Münch's model. Most tests have been positive, and the negative ones have been reconciled, as we shall see. The story of this model provides a good example of the scientific

comparable plant structures:

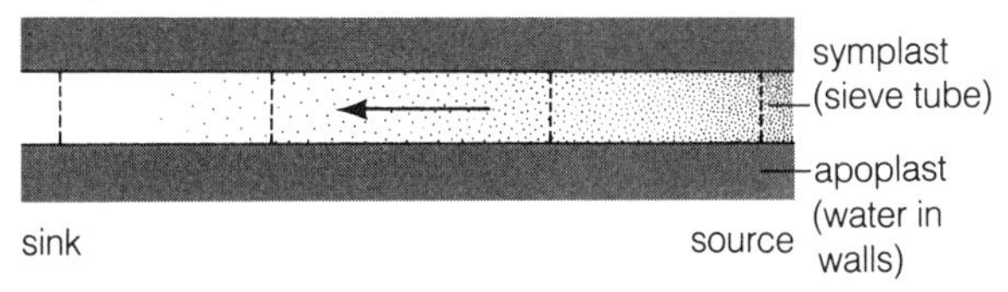


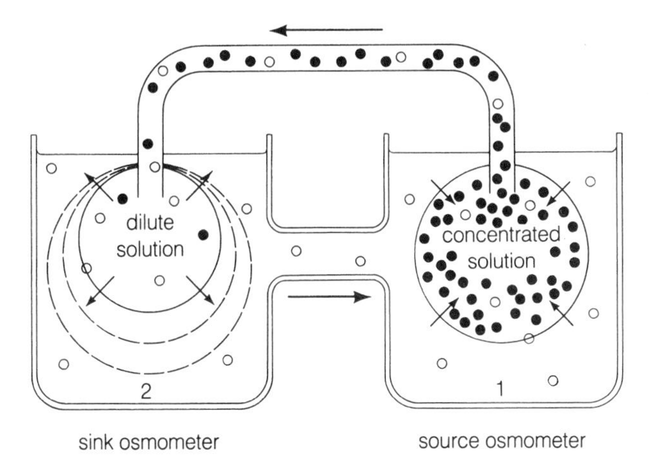


**Figure 8-5** Bottom: A laboratory model consisting of two osmometers and illustrating the pressure-flow theory of solute translocation as proposed by Münch. Note that the concentration of solute present in the largest amount (represented by black circles) will control the rate and direction of flow, whereas more dilute solutes (open circles) will move along in the resulting stream. Dashed lines on the left imply that flow will occur if pressure is relieved by expansion of the osmometer as well as by outward movement of water. Top: A schematic suggestion of how the model might apply to the concentrated solutions in the phloem system (symplast) surrounded by the more dilute solutions of the surrounding apoplast. Solute concentration is maintained high at the source end of the system as sugars and other solutes are moved into the sieve tubes there; concentrations are low at the sink end as solutes move out, which movement also occurs to some extent along the route from source to sink. Lowered concentration of solutes at the sink end allows water to move out in response to the pressure transmitted from the source end (or in response to even higher concentrations of solutes in the apoplast at the sink). Sieve tubes do not expand in analogy to the expanding osmometer of the laboratory model, but growth of storage cells at the sink will cause absorption of water from the apoplast, lowering its water potential and thus facilitating exit of water from sieve tubes there.

approach: model building, testing, modification of the model, and further testing.

Münch's **pressure-flow model** is simple, straightforward, and based on a real model that can be built in the laboratory: two osmometers connected to each other with a tube (Fig. 8-5). The osmometers can be immersed in the same solution or in different solutions, which may or may not be connected but which have approximately the same water potential. The first osmometer contains a solution that is more concentrated than its surrounding solution; the second osmometer contains a solution less concentrated than that in the first osmometer, but either more or less concentrated than the surrounding medium. Water moves osmotically into the first osmometer, and pressure builds up. Because the osmometers are connected, pressure is transferred from the first osmometer to the second (with the velocity of sound, which is basically a pressure-transference phenomenon). Soon the increasing pressure in the second osmometer leads to a more positive water potential than exists in its surrounding medium, so water diffuses out through the membrane. This releases the pressure in the system, and more water diffuses into the first osmometer from its surrounding solution. The result is a bulk flow of solution (water with its solutes) through the tube into the second osmometer. If the walls of the second osmometer stretch, pressure is relieved even if no water moves out, and if the second osmometer is surrounded by a solution more concentrated than that inside, water will diffuse out into the medium even without increasing pressure.

In Münch's laboratory model, bulk flow (also called **mass flow**) ceases when enough solute has been moved from the first osmometer to the second to equalize their pressure potentials. Münch suggested that the living plant contains a comparable system but with advantages. Sieve elements near the source cells (usually photosynthesizing leaf mesophyll cells) are analogous to the first osmometer, but the concentration of assimilates is kept high in these sieve cells by sugars produced photosynthetically in nearby mesophyll cells. Assimilate concentration in the sink end of the phloem system is kept low as assimilates are transferred to other cells, where they are metabolized, incorporated into protoplasm (growth), or stored, often as starch. The connecting channel between source and sink is the phloem system with its sieve tubes (part of the symplast); the surrounding solutions are those of the apoplast (in cell walls and in the xylem). Münch proposed the concepts of the symplast and the apoplast as part of his mass-flow hypothesis.

Note that flow through the sieve tubes is *passive*, occurring in response to the pressure gradient caused by osmotic diffusion of water into sieve tubes at the source end of the system and out of the sieve tubes at the sink end of the system. There is no active pumping of solution by sieve cells along the route, although metabolism is required to maintain these cells in a condition that will allow flow, prevent leakage, and retrieve assimilates that do leak out (see, for example, Giaquinta and Geiger, 1972).

**Cytoplasmic streaming** provides an example of an *active* transport of solutes, as proposed over a century ago by the Dutch botanist Hugo de Vries (1885), one of the discoverers of Mendel's paper on genetics. The cytoplasm streams around the periphery of many cells, and any solute that passes from one cell to another will be accelerated in its transport across the cell by the cytoplasmic streaming in that cell. This must be an important active transport mechanism in many plant tissues; it could be important in moving sugars from leaf mesophyll cells to phloem sieve tubes, for example, or from sieve tubes in storage organs to the storage cells. But because the cytoplasm does not stream in mature sieve elements, cytoplasmic streaming cannot play a role in phloem transport. Streaming is also much slower than sieve-tube transport.

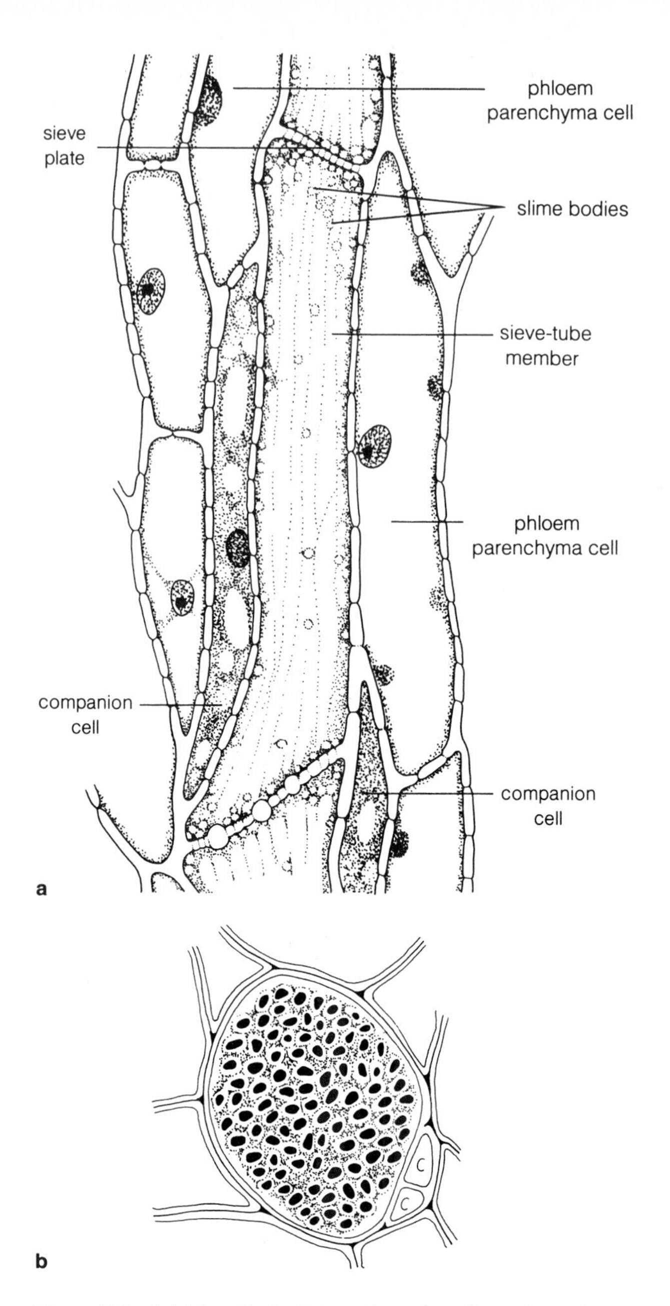


**Figure 8-6** (**a**) A longitudinal view of a mature sieve element, with companion cells and phloem parenchyma cells. (**b**) A face view of a sieve plate; the black areas represent holes in the wall. (From Jensen and Salisbury, 1972.)

## 8.3 Testing the Hypothesis

The Münch hypothesis suggests several ways that it might be tested, as any good model should. We will consider the following approaches: phloem anatomy, rates of phloem transport, the transported solutes, phloem loading and unloading of assimilates, pressure in the phloem, and some complications.

### Phloem Anatomy

A function can often be understood by comprehending the structure in which it occurs. We have noted that an examination of the valves and chambers of the heart makes its function as a pump clear to anyone with some sense of mechanics. Can we discover the mechanism of phloem transport by studying the structure of phloem tissue? Perhaps—if we can understand the structure well enough. Certainly, we cannot hope to understand phloem transport without understanding phloem anatomy.

*Phloem tissue* First are the **sieve elements** or **sieve-tube members**, which are elongated living cells, usually without nuclei, in which transport actually takes place (Esau, 1977; Fahn, 1982; Mauseth, 1988). In angiosperms, they are connected end to end, with pore-filled **sieve plates** between, forming long cellular aggregations called **sieve tubes** (Fig. 8-6; see Fig. 5-5c). In gymnosperms and lower vascular plants, the sieve plates are not as clearly displayed. There are sieve areas with smaller pores on lateral walls and on slanted end walls, so the units are called **sieve cells** instead of sieve elements. Second are the **companion cells** (in angiosperms) or **albuminous cells** (in gymnosperms), which are closely associated with the sieve elements or sieve cells and have relatively dense cytoplasm and distinct nuclei. There are usually many plasmodesmata in the walls between sieve elements and their companion cells, with the plasmodesmatal pores frequently being branched on the side of the companion cell. The exact function of the companion cells remains unknown, but they are always present, viable in functioning phloem and degraded in senescent phloem. They typically have the same osmotic potential (that is, sugar concentration) as the associated sieve elements (Warmbrodt, 1986, and references cited there). Third are the **phloem parenchyma cells**, which are thin-walled cells similar to other parenchyma cells throughout the plant, except that some are more elongated. They may act in storage as well as in lateral transport of solutes and water. Fourth are the **phloem fibers**, which sometimes are grouped in a bundle. As in other tissues, they are thick-walled cells that provide strength.

The vascular anatomy of the minor veins in leaves is especially important to an understanding of phloem transport. The large veins in a leaf branch into smaller veins and eventually into the minor leaf veins. Each minor vein may contain only one vessel representing the xylem and one or two sieve tubes (Fig. 8-7). The vessel is usually above the phloem tissue, and the sieve elements are typically smaller than and surrounded by companion cells. Intermingled with the companion cells are large phloem parenchyma cells, and vascular parenchyma may separate the phloem tissue from the xylem. Companion cells and phloem parenchyma cells sometimes contain chloroplasts, and actively photosynthesizing mesophyll cells (usually consisting of palisade and spongy parenchyma tissues) are often in close contact with the minor vein. Indeed, typically no mesophyll cell in the leaf is separated from a minor vein by more than two or three other mesophyll cells (see Fig. 5-4).

In some species, companion cells have numerous cell-wall ingrowths, which greatly amplify the membrane surface area of the cell (Fig. 8-8). Cells with such wall ingrowths and expanded membrane surfaces are called **transfer cells**. Although most species do not have transfer cells in minor veins of their leaves, such cells might contribute significantly to transfer of assimilates from mesophyll cells to sieve tubes in species that do, which include many legumes and asters (two of the largest families). Brian E. S. Gunning and coworkers (1974) calculated that wall ingrowths in companion cells of *Vicia faba* leaf veins expand the surface area for absorption to over three times the area that would have been available if the companion cells lacked ingrowths.

Transfer cells are not restricted to phloem, but occur throughout the plant. They are found in xylem and phloem parenchyma of leaf nodes (where they are common) and in reproductive structures, such as the interface between the gametophyte and the sporophyte of both lower and higher plants. In all vascular plants studied, they are correlated with active transport processes occurring over short distances, such as those in salt glands, nectaries, and the connections (**haustoria**) between a parasite and its host. At the other, unloading end of the phloem system, transfer cells are clearly involved in movement of unloaded sugars into the endosperm of developing maize seeds (Porter et al., 1985; Shannon et al., 1986). But do they play a role in phloem loading? Only in certain species, it appears, and that has yet to be demonstrated. In a survey of over 1,000 species, it was found that phloem transfer cells associated with sieve elements in minor leaf veins were relatively rare, occurring only in certain herbaceous dicot families. They have been reported (Barnabas et al., 1986) in phloem parenchyma cells of sea grass (*Zostera capensis*), a monocot in the pondweed family (Zosteraceae). The sea grasses are the only angiosperms adapted to the marine environment.

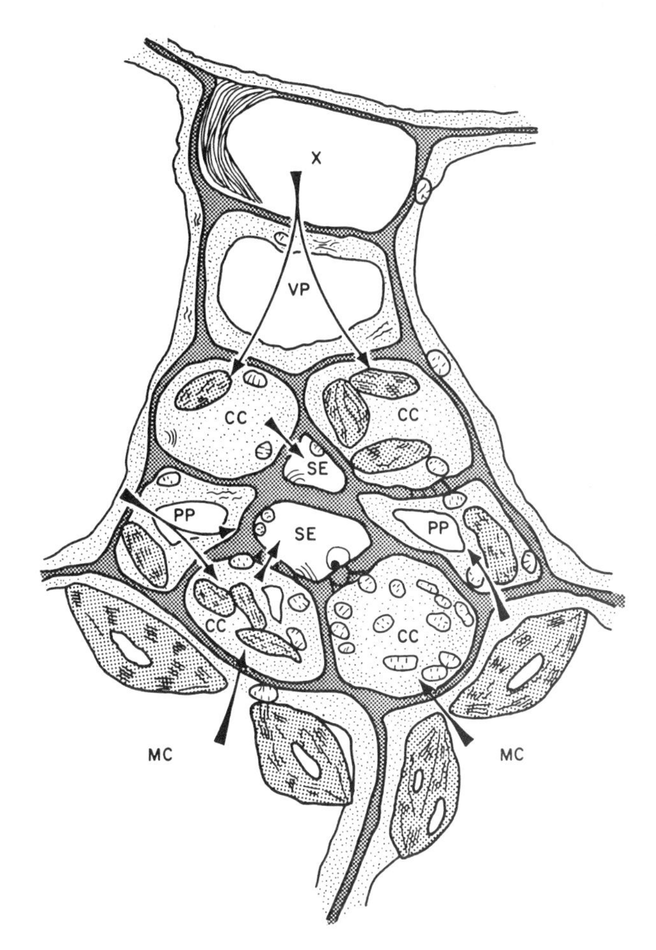


**Figure 8-7** Tracing of an electron micrograph of a cross section of a minor vein from a tobacco leaf. Arrows illustrate possible assimilate entry routes into the sieve-element–companion-cell complex. X = xylem; VP = vascular parenchyma; CC = companion cell; SE = sieve element; PP = phloem parenchyma; and MC = mesophyll cell. Note that sieve elements are much smaller than companion cells, the reverse situation from that found in stems and roots. Solutes moving from the mesophyll cells to the sieve elements could remain in the cytoplasm (symplast), passing through plasmodesmata (not shown), or they could pass through the plasmalemma into the cell walls (apoplastic pathway). (From Giaquinta, 1983.)

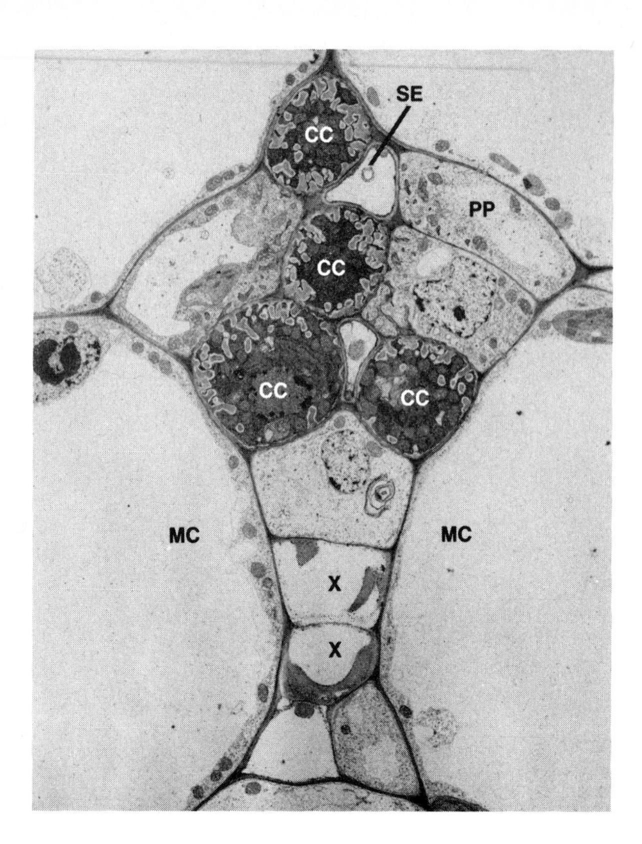


**Figure 8-8** Transfer cells in a cross section of a minor leaf vein of groundsel (*Senecio vulgaris*). In the upper half of the photograph, two sieve elements (**SE**; small, relatively empty-looking cells) are in contact with four companion cells (**CC**; dense cytoplasm, wall ingrowths; these are *transfer cells*) and three phloem parenchyma cells (**PP**; more vacuolated, less dense cytoplasm, wall ingrowths of faces nearest the sieve elements; also transfer cells). Another parenchyma cell separates the phloem region from the two xylem elements (**X**) at lower center. Higher magnifications show that the companion cell transfer cells are connected to the sieve elements by plasmodesmata. The large surrounding cells (**MC**) are leaf mesophyll cells. They are heavily vacuolated. (Transmission electron micrograph courtesy of Brian E. S. Gunning; see Gunning, 1977.)

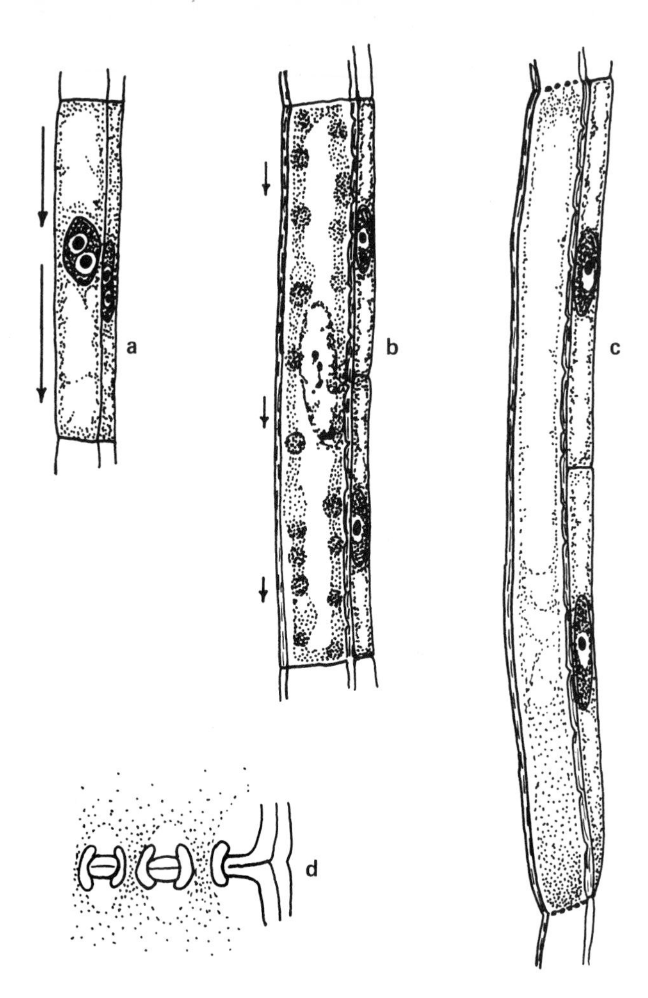


**Figure 8-9** Developing sieve elements. (**a**) Young element with companion cells. Long arrows suggest considerable cytoplasmic streaming. (**b**) Sieve element of intermediate maturity. Slime (P-protein) bodies are evident, the nucleus is beginning to disappear, and cytoplasmic streaming has virtually stopped. (**c**) Mature sieve element. The nucleus and tonoplast are no longer evident. (**d**) Longitudinal section of sieve tube through a sieve plate, showing cytoplasmic connections through the pores.

*Phloem development* Consider the typical development of a secondary sieve element and its companion cell (Fig. 8-9). A single cambial cell divides twice to produce a sieve element and its accompanying companion cell. The companion cell may divide at least once more. The sieve element expands rapidly and becomes highly vacuolated, with a thin layer of cytoplasm pressed against the cell wall. Minute bodies appear in this cytoplasm. They are generally ovoid in shape, some appearing rather amorphous, whereas others have a more fibrillar or stranded appearance. Traditionally, they have been called **slime bodies**, but they apparently consist of phloem protein (*P-protein*, discussed later). As they grow, their boundaries become less defined, and eventually they fuse into a single diffuse mass that is dispersed throughout the cell. At about this time, the nucleus begins to degenerate. Eventually it disappears completely in most sieve elements, although there are a few exceptions in which degenerate or even whole nuclei remain. The tonoplast (vacuolar membrane) also disappears at about this stage, but the plasmalemma remains intact. Cytoplasmic streaming has been observed in some developing sieve elements, but when the elements are mature, this activity ceases.

While these events are taking place, the sieve plate is developing. This process begins with small deposits of a special glucose polymer called **callose**, usually around plasmodesmata. (The box essay "A Review of Carbohydrate Chemistry" further describes callose.) The deposits increase in size until they assume the shape of the final pore. The middle lamella first disappears in the center of the deposit, and then the deposits on both sides of the wall fuse as the wall between disappears, so that the callose-lined pore resembles a grommet. The plasmalemma extends through the pore and is thus continuous from cell to cell. The pores are usually 0.1 to 5.0 μm in diameter, much larger than any solute ion or molecule, or even a virus particle. The sieve elements are usually 20 to 40 μm in diameter and 100 to 500 μm (0.1 to 0.5 mm) long. Of course, there is great variation, depending on species, to most of the information just summarized. For instance, in gymnosperms the albuminous cells and the sieve cells do not arise from a single cambial cell.

Some sieve tubes remain functional for several years. This is probably most striking in perennial monocots, such as palm trees, which have little cambial tissue capable of producing secondary phloem. Some sieve elements laid down at the base of a palm tree apparently must remain functional throughout the life of the tree (which may be well over 100 years), although some new sieve elements are formed by a **primary thickening meristem**.

*Phloem ultrastructure* Companion cells have unusually dense cytoplasm with small vacuoles. Mitochondria are abundant, as are dictyosomes and endoplasmic reticulum. The nucleus is well defined until the associated sieve element finally begins to become nonfunctional. Phloem parenchyma cells, in contrast to companion cells, contain large vacuoles and perhaps fewer organelles (although this may be apparent only because of relatively less cytoplasm). Chloroplasts are often conspicuous in these cells, and plastids occur in both types; indeed, if the vacuole is not well developed in parenchyma cells, it is not easy to distinguish phloem parenchyma from companion cells.

It is the ultrastructure of the sieve tube that is of immediate interest here. Smooth endoplasmic reticulum (ER) occurs as an almost continuous network along the inner surface of and somewhat parallel with the plasmalemma. There are also regions of flattened or convoluted stacks of ER cisternae close to the plasma membrane. In the sieve cells of gymnosperms, the ER exists as a network of smooth tubules close to the plasmalemma but forming massive aggregates at the sieve areas. Mitochondria apparently remain both unmodified throughout the differentiation of the sieve element and capable of carrying out cellular respiration and other metabolic activities (discussed in Chapter 13). Plastids, sometimes containing starch, protein, or both, occur in young as well as mature sieve elements, but their internal membrane systems remain poorly developed. Microtubules are abundant in the cytoplasm of young sieve elements but disappear at intermediate stages of differentiation. Microfilament bundles are frequently seen in differentiating sieve elements and have been reported in mature sieve elements. Walls of mature elements are nonlignified, rich in cellulose, and often thickened. Water and dissolved minerals move freely through them between adjacent cells, depending on permeability of their membranes and velocities of flow within the sieve-tube members.

In addition to the slime bodies in sieve elements, earlier workers had often seen amorphous material in the **lumen** (the part that was originally the vacuole before disintegration of the tonoplast). Most, if not all, of this slime proved to be a fibrillar, proteinaceous material (Lucas and Madore, 1988). The diameter of the fibers is on the order of 7 to 24 nm, and molecular weights vary (in cucurbits, at least) from 14,000 to 158,000. There are several different kinds of this protein, and some of these may be interconvertible. The protein is referred to as **P-protein** (for phloem protein), and slime bodies may be called **P-protein bodies**.

Since the discovery of P-protein, phloem anatomists have considered whether the pores of the sieve plates are filled with the substance. Clearly, the sieve plates themselves offer a considerable resistance to a passive bulk flow of material as postulated in the Münch model. If the pores are partially or completely blocked with P-protein, the resistance could be greatly increased. Perhaps the evidence accumulated from electron micrographs, which suggested that the pores were indeed blocked, led in the 1960s to the serious questioning of the pressure-flow hypothesis and to the formulation of alternative models. Many of these hypotheses suggested that the P-protein played some kind of active role in pumping solution through the pores (see, for example, Fensom, 1972; Peel, 1974). Are the sieve-plate pores open to a bulk flow of solution or occluded by P-protein?

The earliest electron-microscope studies were equivocal because of primitive fixation techniques. Even then, however, some electron micrographs showed occluded pores, whereas others showed open pores. With improving fixation techniques, electron micrographs continued to show predominately occluded pores, but one can always suggest that the release of pressure caused by cutting the phloem for taking samples might well cause surging and a movement of P-protein from the cell periphery to the pores.

Attempts to fix material in ways designed to reduce surge artifacts have led to more observations of open pores. The following techniques have been used: use of rapidly penetrating fixative (acrolein); rapid freezing of whole plants in liquid nitrogen and then transfer to chemical fixative (Fig. 8-10); fixation of isolated sieve tubes from tissue cultures grown in the laboratory (presumably they have low internal pressure); and starving or wilting of plants to reduce pressure in the sieve tubes before sampling and fixation. Furthermore, some plants such as maize (*Zea mays*), duckweed (*Lemna*), and some palms always have unplugged pores, even in the absence of special measures to prevent surging. Carbon-black particles and mycoplasmalike bodies pass

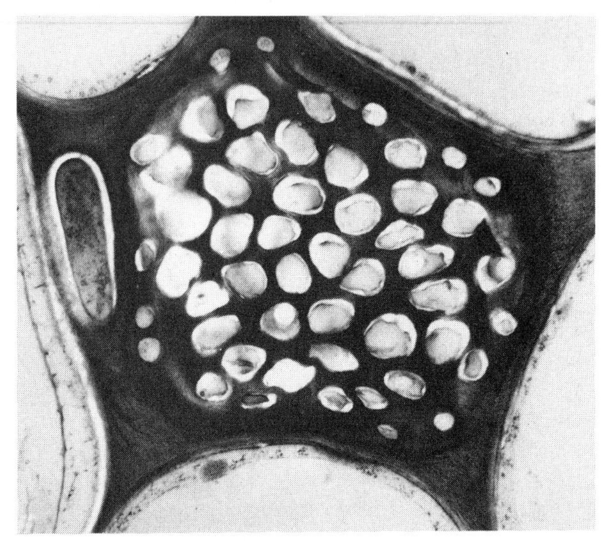

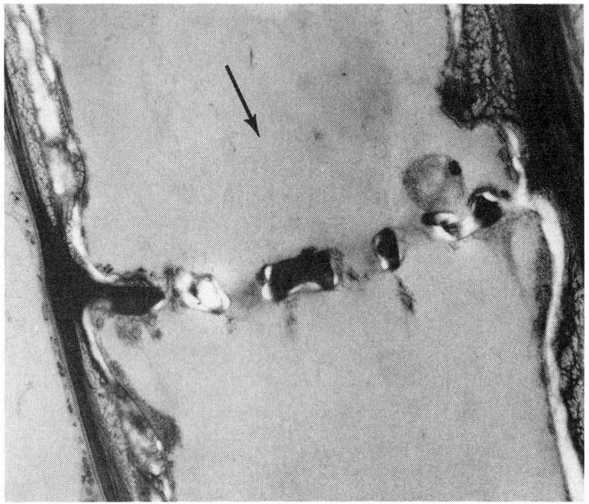

**Figure 8-10** Electron micrographs of the sieve plate in soybean (*Glycine max*), showing rather open pores with relatively small amounts of P-protein fibers. (Top, cross section; bottom, longitudinal section; both 14,000 ×). The petiolar tissue containing $^{14}C$-sucrose was quick-frozen, freeze-substituted in acetone or propylene oxide, and embedded in Epon. The presence of $^{14}C$-sucrose allowed verification of the direction of flow in the sieve tubes. Arrow indicates probable direction of flow. (Micrographs courtesy of Donald B. Fisher; see Fisher, 1975.)

through sieve pores, and virus particles can replace P-protein pore plugs in some specimens (not likely if the P-protein were essential for active transport). (See Hull, 1989, for review of virus movement in plants.)

With these and other evidences, a strong argument can be made that sieve-tube pores are open in normal growing plants. The controversy seems to have died down; a 1986 book (Cronshaw et al.) titled *Phloem Transport* cites 82 papers, none devoted to P-protein in sieve plates. "Most workers agree that there is a mass flow of solution through unplugged sieve-plate pores" (Cronshaw, 1981).

Without some protective mechanism, plants might "bleed to death" when injured by grazing or other means. The surging that occurs when a sieve element is cut causes P-protein to flow to the sieve plate, thereby blocking the pores, and there is evidence that P-protein coagulates when exposed to air. In a few cases the sieve cells collapse upon wounding. (They have relatively soft walls; their original shape is maintained only by turgor pressure.) These mechanisms might also keep potential pathogens from obtaining nutrients. Yet they make it extremely difficult to study and understand uninjured sieve elements.

## The Rates of Phloem Transport

In 1944, Alden S. Crafts and O. Lorenz at the University of California, Davis, weighed 39 pumpkins from August 5 to September 7, estimating that individual fruits gained an average of 482 g of dry material during 792 hours of growth. Thus, compounds that increase the net dry mass moved into each fruit through its peduncle at an average rate of $0.61 \text{ g h}^{-1}$. Based on anatomical studies of stems, Crafts and Lorenz estimated that the average cross section of phloem tissue in the peduncles was $18.6 \text{ mm}^2$; of this, about 20 percent consisted of sieve tubes ($3.72 \text{ mm}^2$). Thus, material was moving through sieve tubes with a mass transfer rate of $0.61 \text{ g h}^{-1} \div 3.72 \text{ mm}^2 = 0.164 \text{ g mm}^{-2} \text{ h}^{-1}$. The **mass transfer rate** is the quantity of material passing through a given cross section of sieve tubes per unit of time.

What about the **velocity** of movement, which is a measure of the linear distance traversed by an assimilate molecule per unit of time? Assuming a specific gravity or density for the dry material of about 1.5 g $\text{cm}^{-3}$ ($1{,}500 \text{ kg m}^{-3}$) and dividing this figure into the rate gives a velocity of about $110 \text{ mm h}^{-1}$. Of course, the material does not move in dry form, but as a solute in water. If its concentration is 10 percent, its velocity will be about 10 times ($100 \div 10$) that calculated for the dry material, or $1{,}100 \text{ mm h}^{-1}$. Although phloem exudates often consist of 10-percent solutions, they are usually more dilute than phloem sap. Osmotic potentials of $-2$ to $-3$ MPa are common in intact sieve tubes, and these values are approximately equivalent to 20- to 30-percent sucrose solutions. If the assimilates were moving in a 20-percent solution, then they were moving five times ($100 \div 20$) as fast as the calculated figure for the dry material, or about $550 \text{ mm h}^{-1}$.

These results are typical of many similar studies, but more meaningful data are now obtained with the short–half-life, energetic $^{11}C$. The isotope, incorporated into $CO_2$, is introduced into the translocation stream via photosynthesis in the leaf, and two or more radiation detectors are placed at intervals along the stem. Radiation from the $^{11}C$ is then measured for brief intervals at different times (Fig. 8-11). ($^{14}C$ has also been used, but it is more difficult to detect.) Results can be expressed as

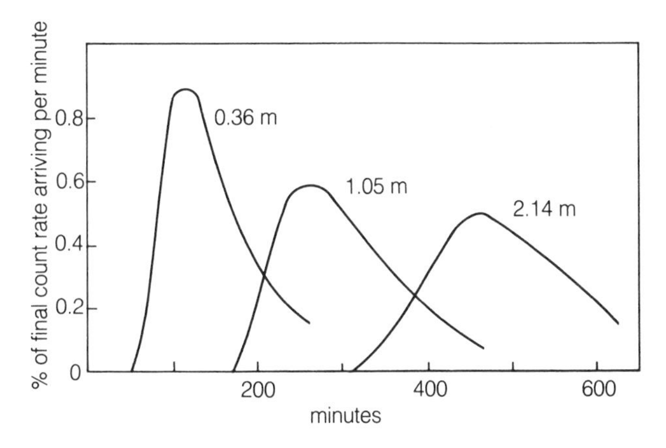


**Figure 8-11** Time profiles of radioactivity arriving at three points along a stem. A single source leaf on a morning glory (*Ipomea nil*) vine was labeled by allowing it to photosynthesize with $^{14}CO_2$ for 5 min, and then radioactivity was continuously measured with a thin-window Geiger tube pressed against expanding sink leaves on the tips of branches at three different distances (shown as meters, m) from the source leaf. All other leaves and the stem above the source leaf were removed. Data are expressed as amount of radioactivity arriving per minute. Note how the profiles broaden and the peaks become lower with increasing distance from the source leaf. Times between the peaks suggest translocation velocities of about 300 to 320 mm $h^{-1}$. (From Christy and Fisher, 1978)

profiles of radioactivity passing each point on the stem as a function of time, and profiles of radioactivity along the stem can also be determined, especially if several detectors are used. Mathematical models were developed from such data, and the data and models can be used to test the Münch or other hypotheses (Minchin and Troughton, 1980). Although there is variation among species, studies such as these generally agree with the older work. In most species, the maximum velocity of transport is 500 to 1,500 mm $h^{-1}$.

It is difficult to appreciate these velocities. If a lone sieve element in an angiosperm is 0.5 mm long (as noted, gymnosperm sieve cells are longer, about 1.4 mm), then at translocation velocities of 900 mm $h^{-1}$ (0.25 mm $s^{-1}$) an entire element would be emptied and refilled in 2 s. At 300 × magnification, an entire phloem element could not be seen at one time in a microscope field, so if one could watch movement of particles through the microscope in sieve elements, velocities would be too rapid to follow easily with the eye; slow-motion moving pictures would be necessary. Without magnification, movement at 1,000 mm $h^{-1}$ is equivalent to the motion of the tip of a 160-mm (6.3-inch) minute hand on a clock. Such movement can easily be detected with the unaided eye.

Geiger and Shieh (1988) developed still another method to study quantities of assimilates fixed and transported throughout the plant. Whole bean plants (*Phaseolus vulgaris*) were supplied with $^{14}CO_2$ of constant specific radioactivity throughout an entire photoperiod. Geiger and Shieh could then measure the gain of tracer carbon in each plant part and deduce the net accumulation of recently fixed carbon from direct fixation, importation, or both. From these data, current and future growth rates could be calculated and projected. By waiting for various time periods after labeling (various **chase periods** during which plants were allowed to photosynthesize with unlabeled $CO_2$, which *chased* the $^{14}CO_2$), they could determine how much of the fixed carbon was lost through respiration.

The following section discusses the compounds, mostly carbohydrates, that are transported in the phloem. If you need a review of carbohydrate chemistry, we suggest that you read the box essay entitled "A Review of Carbohydrate Chemistry" before reading the next section.

## The Transported Solutes

The simplest approach to determine what solutes are contained in phloem sap is simply to cut the phloem and let the sap exude, forming droplets that can then be collected and analyzed. Although bleeding is often rapidly stopped as P-protein and other particulate matter clog the sieve pores, droplets do often form before bleeding stops. Yet these droplets must not be typical of phloem sap; surely they contain some of the P-protein and perhaps other nontransported materials from the sieve tubes. Indeed, Crafts and Lorenz (1944) found a higher percentage of nitrogen (probably in protein) in phloem exudate from peduncles (flower stems) going to pumpkin fruits than in the developed pumpkins. Furthermore, all solutes are usually more dilute in phloem exudate than intact phloem sap. Releasing the pressure in the phloem by cutting lowers the water potential, and water then moves in by osmosis. These effects of cutting on composition and concentration are more noticeable in some species than in others (little effect on composition in many trees, for example), but they are often complications to be reckoned with.

If only we had a miniature hypodermic needle that we could insert into a single sieve tube, carefully extracting some of the contents without a sudden pressure release! In 1953, two insect physiologists at Cambridge, J. S. Kennedy and T. E. Mittler, suggested that, in fact, we have such an instrument. They wondered whether aphids suck to get the phloem sap that nourishes them, using their "hypodermic" mouthpart (**stylet**), or whether the sap is simply forced into their bodies by the pressure in the phloem. When they cut the stylet of a feeding aphid with a sharp razor blade (after administering an anesthetizing stream of $CO_2$), leaving the stylet in place, about 1 $mm^3$ of material exuded from the cut stylet each hour for up to about four days. They said:

(*Text continues on page 176.*)

# A Review of Carbohydrate Chemistry

To understand modern studies on phloem transport, it is essential to know something about the chemistry of the carbohydrates that are transported. Here is a brief review of carbohydrate chemistry, including even a few compounds that have nothing to do with phloem transport but are mentioned to make the review complete. Eight topics are discussed.

*First*, the general formula of **carbohydrates** is $(CH_2O)_n$; that is, for each C there is an $H_2O$ (water, although it does not exist in this form), thus suggesting the name *carbohydrate*. The *n* in the formula means that $CH_2O$ is repeated a certain *number* of times.

*Second*, the basic building blocks of carbohydrates are called **monosaccharides** or **simple sugars** because they are not easily broken down to even simpler sugars. They contain various numbers of carbon atoms (Fig. 8-12) and are named accordingly:

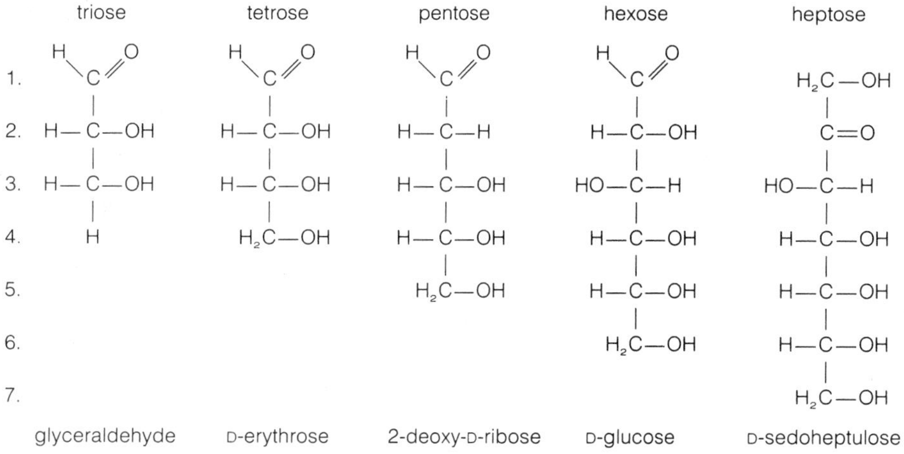

| | triose | tetrose | pentose | hexose | heptose |
|---|---|---|---|---|---|
| 1. | H—C=O | H—C=O | H—C=O | H—C=O | $H_2C$—OH |
| 2. | H—C—OH | H—C—OH | H—C—H | H—C—OH | C=O |
| 3. | H—C—OH | H—C—OH | H—C—OH | HO—C—H | HO—C—H |
| 4. | H | $H_2C$—OH | H—C—OH | H—C—OH | H—C—OH |
| 5. | | | $H_2C$—OH | H—C—OH | H—C—OH |
| 6. | | | | $H_2C$—OH | H—C—OH |
| 7. | | | | | $H_2C$—OH |
| | glyceraldehyde | D-erythrose | 2-deoxy-D-ribose | D-glucose | D-sedoheptulose |

**Figure 8-12** Examples of monosaccharides having three to seven carbon atoms.

*Three-carbon sugars (trioses):* These and similar compounds are important intermediates in the metabolic pathways of photosynthesis and cellular respiration.

*Four-carbon sugars (tetroses):* There are not many of these sugars, although one takes part in photosynthesis and respiration.

*Five-carbon sugars (pentoses):* These compounds are crucial in photosynthesis and respiration. Two pentoses (ribose and deoxyribose) also form key structural components of the nucleic acids, which are essential for all life. Certain gums peculiar to specific plants (both algae and higher plants) consist largely of pentoses, and hemicelluloses found in all plant cell walls are rich in pentoses.

*Six-carbon sugars (hexoses):* These often-discussed sugars take part in many steps of respiration and photosynthesis and constitute the building blocks of many other carbohydrates, including starch and cellulose. Glucose and fructose are key hexoses, but there are several others that occur naturally (Fig. 8-13). For all the mention made of glucose in biology books, in plants the majority of it is bound in polymers and other compounds.

*Seven-carbon sugars (heptoses):* One of the heptoses is an intermediate in photosynthesis and respiration. Otherwise they are seldom encountered.

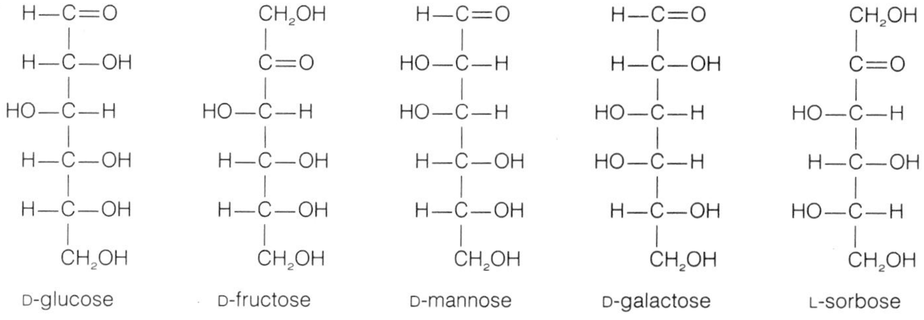

| D-glucose | D-fructose | D-mannose | D-galactose | L-sorbose |
|---|---|---|---|---|
| H—C=O | $CH_2OH$ | H—C=O | H—C=O | $CH_2OH$ |
| H—C—OH | C=O | HO—C—H | H—C—OH | C=O |
| HO—C—H | HO—C—H | HO—C—H | HO—C—H | HO—C—H |
| H—C—OH | H—C—OH | H—C—OH | HO—C—H | H—C—OH |
| H—C—OH | H—C—OH | H—C—OH | H—C—OH | HO—C—H |
| $CH_2OH$ | $CH_2OH$ | $CH_2OH$ | $CH_2OH$ | $CH_2OH$ |

**Figure 8-13** Five important hexose sugars. The aldehyde groups (top of all but D-fructose and L-sorbose) are commonly written as —CHO but are expanded here for clarity.

*Third*, the carbohydrates exhibit **stereoisomerism**. If four different atoms or groups of atoms are attached to a single carbon atom, forming a tetrahedral structure, there are two ways to make the attachment, which result in mirror images. Thus a carbon atom with four *different* things attached can exist as two **stereoisomers**, and molecules with such atoms are said to exhibit stereoisomerism. Two mirror-image isomers of a given compound rotate the plane of plane-polarized light in opposite directions.

Study the hexose sugar, glucose, shown in Figure 8-13. The top (number 1) carbon has only three things attached to it: a hydrogen, an oxygen (by double bond), and the rest of the molecule. The bottom (number 6) carbon has only three *kinds* of things attached to it: two hydrogens, an —OH, and the rest of the molecule. Each of the four carbons in between has *four different kinds* of things attached to it, so each of those four carbons and attached atoms or groups could exist as two stereoisomers (two mirror images). Note in Figure 8-13 that the names of the sugars are prefaced by the letter D (written as a small capital). This designation indicates their stereoisomeric structure and refers to the position of the —OH on the next-to-the-bottom carbon. If it were on the other side, we would use the letter L. (If *only* this —OH changes from one side to the other, a new sugar is produced with a new name; if the name stays the same, as in D-glucose and L-glucose, *all* asymmetric carbons in the two molecules are mirror images of each other.) The chain of carbons in a sugar molecule form a zigzag, but this three-dimensional pattern cannot be represented conveniently on a two-dimensional sheet of paper, so the carbon chain is usually shown as though it were straight.

*Fourth*, the monosaccharides are characterized by the presence of an **aldehyde** group ($-\underset{\underset{O}{\|}}{C}-H$; sugars with the aldehyde are called **aldoses**) or a **ketone** group ($-\underset{\underset{O}{\|}}{C}-$; ketone sugars are called **ketoses**). The aldehyde or the ketone group is highly reactive in alkaline solution; either one is a reducing agent. In a solution containing one or more oxidizing ions, the aldehyde or ketone group becomes oxidized to an acid group ($-\underset{\underset{O}{\|}}{C}-OH$; called a **carboxyl**). The oxidizing ion, of course, becomes reduced. This reaction is the basis for several standard reagents that measure so-called **reducing sugars**. In such reagents, an oxidizing cupric ion ($Cu^{2+}$) is held in solution by some chelating agent (Section 6.5) such as citric or tartaric acid, and the solution is made alkaline with potassium or sodium hydroxide. When a reducing sugar (any of the monosaccharides or their relatives shown in Figs. 8-12 or 8-13) is added to the reagent, the sugars are oxidized to form complex mixtures of sugar acids, and the cupric ions are reduced to cuprous ions ($Cu^{+}$), which in turn form cuprous hydroxide (a yellow precipitate) and then become dehydrated to produce cuprous oxide (a brick-red precipitate).

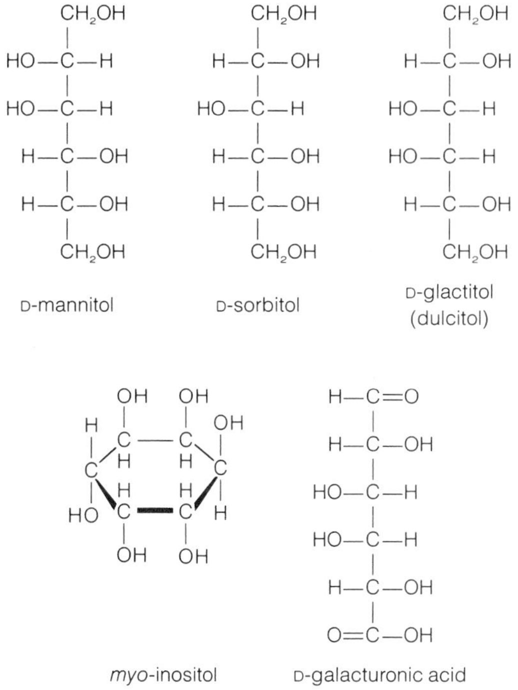

D-mannitol D-sorbitol D-glactitol (dulcitol)

*myo*-inositol D-galacturonic acid

**Figure 8-14** Three important sugar alcohols, a closely related carbocyclic inositol, and a uronic acid. The carboxyl groups at the botton of D-galacturonic acid are usually written —COOH; they are expanded here for clarity. The heavy line in the ring structure of *myo*-inositol signifies that that bond is closer to the viewer, as though the ring were tipped with its top away from the viewer.

*Fifth*, the aldehyde or ketone group of the monosaccharides can be reduced as well as oxidized. When it is reduced, it produces another —OH group where the aldehyde or ketone had been located. In that case, all the carbons then have an —OH group attached. These are called **sugar alcohols** and are important as transported solutes in the phloem of certain species (Fig. 8-14). Note that glucose, fructose, and sorbose can all be reduced to produce the sugar alcohol sorbitol. Note further that when fructose is reduced, the —OH can go on either side of the number 2 carbon; if it attaches to the right side, the product is sorbitol—if to the left side, mannitol.

Some of the sugar acids (produced by oxidation of the aldehyde) occur naturally in plants (for example, galacturonic acid in the pectins found in the middle lamella between the walls of adjacent cells). Various groups can also be attached to the sugars by acetal bonds (for example, the bond

(*continued*)

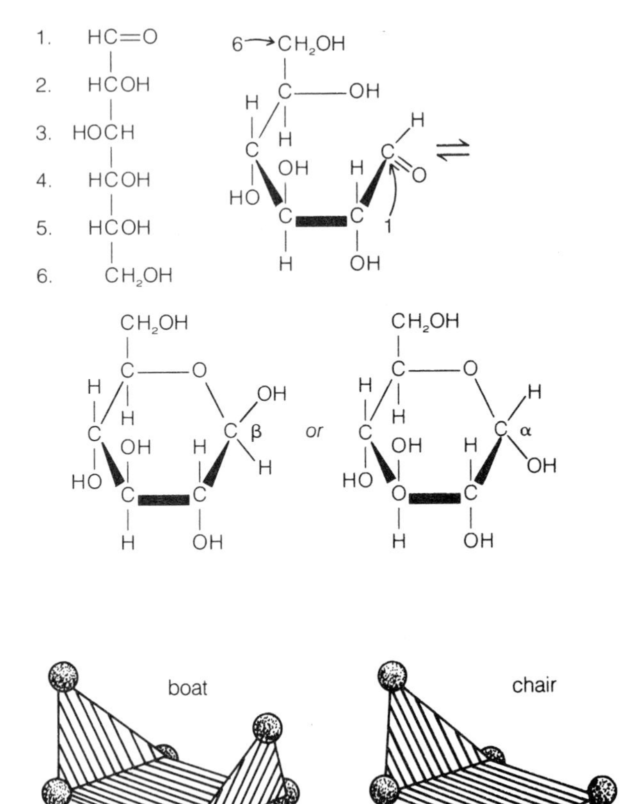


**Figure 8-15** Formation of the ring structures of sugars, specifically D-glucose. Note the α and β forms and the heavy line in the rings, indicating that the ring is tipped with its top away from the viewer. The figure also shows that the six-membered rings naturally assume one of two configurations: a "boat" or a "chair."

connecting glucose and fructose in sucrose; see below), producing **glycosides** (for example, **glucosides** from glucose).

*Sixth*, most sugars in solution form rings rather than straight chains. There are two possible ways for the ring to form. As shown in Figure 8-15, the =O group on the 1-carbon becomes an —OH group after the ring forms. With the ring drawn as shown, the —OH can point either up or down. The two forms are called alpha (α) and beta (β). The ring form of the monosaccharides does not lie in a single plane (that is, it is not flat); rather, it takes the form of either a "boat" or a "chair" (Fig. 8-15).

*Seventh*, these details suggest how the rings can attach to each other in so many ways. This attachment occurs by removal of a water molecule from between two of the sugar molecules (Fig. 8-16). If two monosaccharide molecules (often simply called *units* or *moieties*) hook together, they form a **disaccharide**. Two glucose molecules form the disaccharide maltose—or cellobiose, depending on whether the unit forming the linkage is in the α or the β form. A glucose and a fructose molecule form the disaccharide sucrose (by far the most abundant sugar in the plant kingdom). Three monosaccharide units form a **trisaccharide** and four a **tetrasaccharide**.

The di-, tri-, and tetrasaccharides are referred to collectively as **oligosaccharides**. In some cases, an aldehyde or a ketone group remains potentially exposed in the oligosaccharide, so the compound has the reducing properties of the monosaccharides. (Maltose and cellobiose are examples.) In other cases, the classical example being sucrose, both reducing groups are utilized to form the glycoside linkage between the monosaccharides. In the case of sucrose, the aldehyde of glucose and the ketone of fructose both take part in forming the linkage between the two monosaccharides. Hence, sucrose is a **nonreducing sugar**. Note that the glucose unit forms a six-membered ring (called a **pyranose**), and the fructose forms a five-membered ring (a **furanose**).

A series of nonreducing oligosaccharides that can be thought of as additions of different numbers of galactose molecules to sucrose includes solutes transported in the phloem of several species. This series of oligosaccharides is called

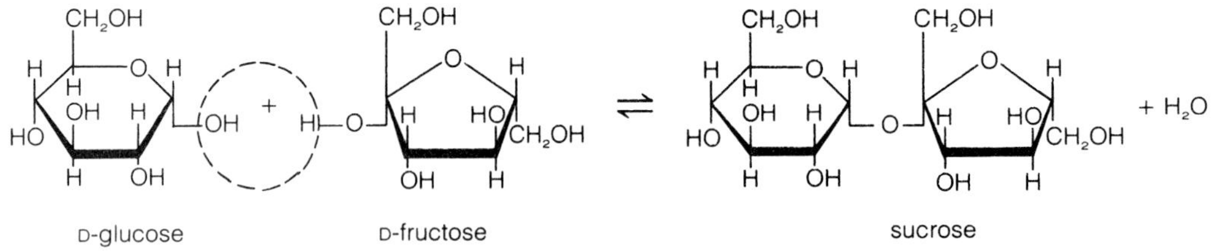

D-glucose D-fructose sucrose + $H_2O$

**Figure 8-16** Formation of the disaccharide sucrose by removal of a water molecule from between two monosaccharides. It is assumed that angles in the rings represent carbon atoms, each with its four bonds. The ring form of D-glucose is called glucopyranose, whereas the ring form of D-fructose is called fructofuranose.

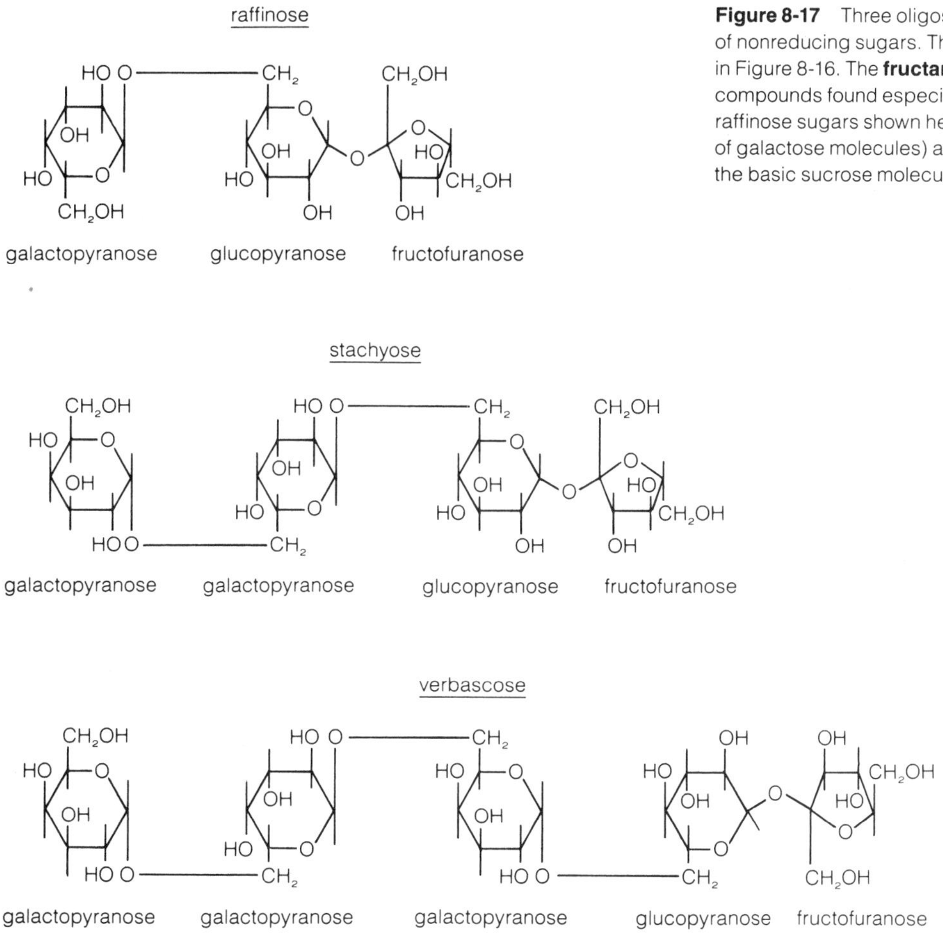

raffinose

galactopyranose glucopyranose fructofuranose

stachyose

galactopyranose galactopyranose glucopyranose fructofuranose

verbascose

galactopyranose galactopyranose galactopyranose glucopyranose fructofuranose

**Figure 8-17** Three oligosaccharides of the **raffinose** group of nonreducing sugars. The fourth member is sucrose, shown in Figure 8-16. The **fructan** group is a series of closely related compounds found especially in monocots. They are like the raffinose sugars shown here, but *fructose* molecules (instead of galactose molecules) are attached to the fructose part of the basic sucrose molecule.

the **raffinose group** (Fig. 8-17). Another group of oligosaccharides and higher polymers, collectively called **dextrins**, are degradation products of starch (Section 13.2). They consist of glucose molecules attached to each other end to end. Because the aldehyde group on one glucose is always free on one end, the dextrins are reducing sugars. Usually dextrins are a mixture of molecules, each with three to several dozen glucose **residues**, as they are sometimes called.

*Eighth*, when many units hook together, they form such **polysaccharides** as starch, fructan, cellulose, callose, hemicellulose, or pectins. When these **complex sugars** and polysaccharides are degraded to **simple sugars**, it is by addition of a water molecule where the original one had been removed. This degradation is called **hydrolysis**. It occurs during seed germination, for example, when starch, which is insoluble, is hydrolyzed to soluble glucose that can be metabolized and transported within the seedling.

**Cellulose** molecules consist of glucose units in the β-ring form (Fig. 8-18). A single molecule of cellulose contains 3,000 to 10,000 glucose units in an unbranched chain. When β-glucose rings are hooked together to form a long chain, the chain proves to be almost perfectly straight (see Fig. 1-4). Glucose units are the basic constituents of **starch**, but in this case they have the α linkage, which produces coiled instead of straight chains (Fig. 8-18). How can we account for the **synthesis** (putting together) of two such different molecules as starch and cellulose from the same units, ultimately the open chains of glucose? The answer lies in the highly specific nature and shape of the enzymes responsible for the synthesis of these polysaccharides. The enzyme that synthesizes cellulose is mechanically configured to form only β bonds from substrate molecules, whereas the starch enzyme combines the same molecules to form α bonds. (Enzymes are discussed in Chapter 9.)

Another series of polysaccharides is **fructan**, which can be thought of as additions of different numbers of fructose molecules to the fructose end of sucrose. They are important in many grasses (Chatterton et al., 1986; Hendrix et al., 1986). Fructan with β (2-1)-**glycosidic linkages** (linkages between

(*continued*)

## A Review of Carbohydrate Chemistry *(continued)*

monosaccharides) is called **inulin** (more common in dicots but also found in onion), whereas those with β (2-6)-glycosidic linkages are called **levans**. A third fructan group contains branched chains with both types of linkage (Section 11.7).

Another important polysaccharide is **callose**, which is composed of β-D-glucopyranose residues linked together by β,1-3-glycosidic linkages as illustrated in Figure 8-19. It is another **glycan**, a compound formed of glucose residues. This interesting linkage produces tightly coiled chains. Callose is important in the formation of the sieve plate, and it also appears almost instantly in various parts of plants that are subjected to mechanical stress (for example, being shaken). It seems to play a role in the healing of damaged tissue. Its chemical structure is very similar to the storage glycans of several algae.

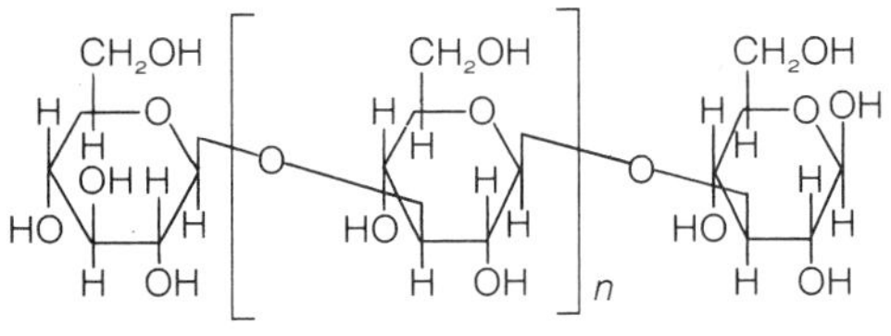

starch

a

cellulose

b

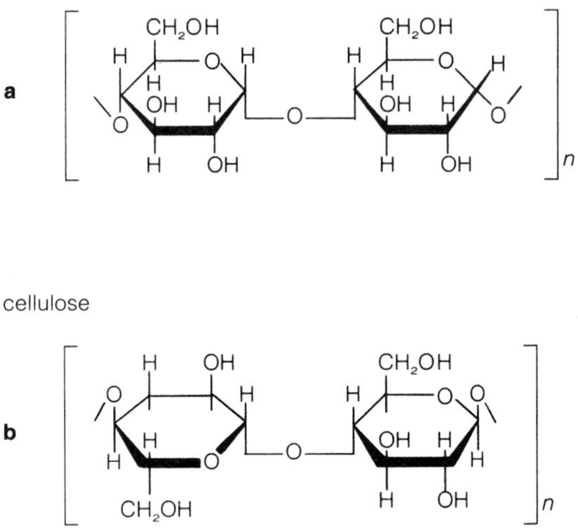

**Figure 8-18** (**a**) The α linkage between glucose residues, as in starch. (**b**) The β linkage between glucose residues, as in cellulose. Note that the residue on the left in the cellulose segment is upside down relative to the one above it in starch.

callose

**Figure 8-19** The molecular structure of callose. Note the interesting 1,3-linkage between β-D-glucopyranose residues, which causes a tight coiling of the molecular callose chain.

"This method of obtaining phloem sap is now in routine use in a study of aphid nutrition and might also be of use to plant physiologists." Since then, the method has indeed been used by plant physiologists.

In many experiments, it is not even necessary to cut off the insect. Normally the phloem sap passes through the insect and forms droplets called **honeydew** on the aphid's body (Fig. 8-20; honeydew secretion stops when the insect is anesthetized with $CO_2$). Often radioactive tracers or dyes can be observed in the honeydew, although its composition is no longer quite the same as it was in the plant, as we have already noted.

As can be seen in Table 8-1, 90 percent or more of the material translocated in phloem consists of carbohydrates. There are species for which this is not necessarily true, with phloem sap containing as much as 45 percent nitrogen compounds, but sugars make up the great bulk of translocated solutes in phloem saps of most species. Furthermore, virtually all the sugars transported in the phloem are nonreducing sugars. Among these, sucrose (common table sugar) is by far the most abundant. Other sugars, if they occur at all, are present only in trace amounts. (Reducing sugars, including glucose and fructose, often found along with sucrose in fruits, are sometimes found in phloem exudates, but they are hydrolysis products of sucrose and are not themselves translocated.)

The major (if not the sole) nonreducing carbohydrates that are transported in higher plants belong either to the raffinose series of sugars—sucrose, raffinose, stachyose, and verbascose (Fig. 8-17)—or to the sugar alcohols: mannitol, sorbitol, galactitol, and *myo*-inositol (Fig. 8-14). Zimmermann and Ziegler (1975) listed phloem-sap compositions for more than 500 species belonging to about 100 dicotyledonous families and subfamilies. Samples of phloem sap were placed on sheets of filter paper in the field and then later chromatographed and stained to identify the various sugars

a

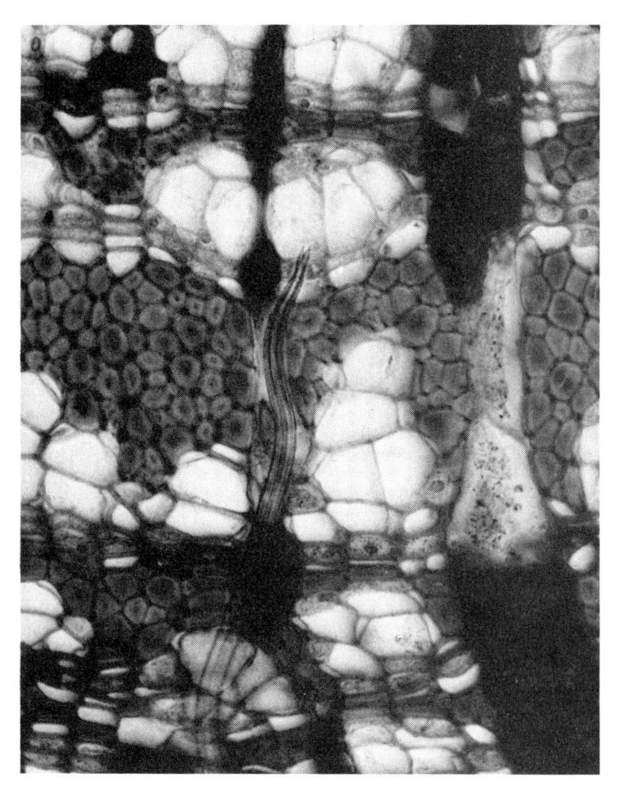

b

**Figure 8-20** Study of translocation in the phloem by the use of aphids. (**a**) An aphid, with a droplet of honeydew, hanging upside down on a branch of a tree. (**b**) A cross section of the tree showing an aphid stylet that has penetrated to a sieve element. (Photographs courtesy of Martin H. Zimmermann; see Zimmermann, 1961. For a series of micrographs of aphid stylets penetrating sieve elements, see Botha et al., 1975.)

and to provide some indication of relative amounts (size and intensity of the stained spots on the chromatogram). This primitive approach could produce some errors, but it did provide a broad picture of translocated solutes.

The list confirms that sucrose is by far the most common transported sugar, although raffinose and stachyose (sometimes verbascose) also appear. *Myo*-inositol occurs in trace to very small amounts in many species, whereas the other sugar alcohols sometimes occur in considerable amounts, but only in certain plant families. Because species in the list are nearly all trees or sometimes woody vines, they might not be representative of flowering plants as a whole, but the list nevertheless provides some interesting facts about certain families. Many families transport only sucrose, with traces of other compounds appearing only rarely; others transport less sucrose than some other sugar or sugars (Table 8-2), but this is not common.

There has been considerable interest in the rose family (Rosaceae; one of the largest families) because it includes many fruit trees and other species of commercial importance and because much sorbitol (Reid and Bielenski, 1974) is transported with a little less sucrose and only traces of raffinose, stachyose, and *myo*-inositol (a few with traces of verbascose). The family consists of several subfamilies, and it is interesting that most of these have that pattern of translocated carbohydrates, except for the Rosoideae, which includes the genus *Rosa* (from which the family takes its name). Members of this subfamily transport no sorbitol, but mostly sucrose with traces of raffinose, stachyose, and *myo*-inositol (verbascose in a few species). Some genera from the subfamilies that do transport mostly sorbitol include *Cotoneaster* (pyracantha), *Crataegus* (hawthorne), *Malus* (apple), *Prunus* (apricot, cherry, and so on), *Pyrus* (pear), *Sorbus* (mountain ash), *Sorbaria* (false spirea), and *Spiraea* (spirea). (See review by Oliveira and Priestley, 1988.)

Fructans, which consist of a glucose molecule plus two to 260 fructose units, occur in several hundred species but are probably not transported in the phloem (Chatterton et al., 1986; Hendrix et al., 1986). They are also nonreducing sugars.

It is highly significant that only nonreducing sugars are translocated, whereas reducing sugars and their phosphate derivatives are not. Although the reasons for this are unclear, nonreducing sugars are less reactive and less labile to enzymatic destruction in sieve elements (Arnold, 1968). Indeed, for perhaps the same reason, reducing sugars are seldom abundant in plant cells. Glucose and fructose, for example, occur most abundantly in cells as their phosphate derivatives, although they do appear as storage sugars in many sweet fruits, probably largely in vacuoles.

In addition to the carbohydrates of phloem sap, much is also known about translocated nitrogenous

**Table 8-2 Some Examples of Less Common Sugars Found in Phloem Saps of Several Woody Families.**

| | S | R | St | V | Aj | M | So | Du | I |
|---|---|---|---|---|---|---|---|---|---|
| Most families | + + + + | + | + | + | | | | | Tr |
| Aceraceae (maple) | + + + + | Tr | Tr | | | | | | Tr |
| Anacardiaceae (cashew) | + + + | Tr | Tr | | | | | | Tr |
| Asteraceae (aster) | + | Tr | Tr | | | | | | Tr |
| Betulaceae (birch) | + + + + | + + | + + | + | | | | | Tr |
| Buddleiaceae (butterfly bush) | + + | + + + | + + + + | + | Tr | | | | + |
| Caprifoliaceae (honeysuckle) | + + + | + + | Tr | | | | | | Tr |
| Celastraceae (staff tree) | + + + | + + | + + + | Tr | | | | + + + | Tr |
| Combretaceae (white mangrove) | + + + | + + | + | Tr | | + + + | | | |
| Fabaceae (legume) | + + + + | Tr | Tr | | | | | | Tr |
| Fagaceae (beech & oak) | + + + + | Tr | Tr | Tr | Tr(?) | | | | + |
| Moraceae (fig) | + + + + | + | + + | Tr | | | | | + |
| Oleaceae (olive) | + + | + + | + + + | + | | + + + | | | Tr |
| Rosaceae (rose) | + + + | Tr | Tr | | | | + + + + | | Tr |
| Verbenaceae (verbena) | + + | + | + + + | Tr | | | | | Tr |

**Key**

S = sucrose
R = raffinose
St = stachyose
V = verbascose
Aj = ajugose
M = D-mannitol
So = sorbitol
Du = dulcitol
I = *myo*-inositol
Tr = trace (could be artifacts)

Source: Based on Zimmermann and Ziegler, 1975.

components of both phloem and xylem (Pate, 1980, 1986). As with carbohydrates, nitrogen components are highly species-specific. In some species much inorganic nitrogen is transported in the xylem as nitrate ($NO_3^-$), which is virtually never present in phloem sap. In other species, nitrogen is transported in the xylem as ureides, amides, or other nitrogen-rich molecules. The same group of organic nitrogen molecules might carry most of the nitrogen in both xylem and phloem channels, but differences in solute composition between xylem and phloem sap have been observed in some species. Alkaloids carry significant amounts of nitrogen in the xylem of certain species, as do certain amino acids that are usually not found in protein. Amino acids, other organic nitrogen compounds, and the reduction and incorporation of nitrate into organic compounds are discussed in Chapters 9 and 14, but structures of the most important organic compounds involved in nitrogen transport are shown in Figure 8-21. Note that such compounds often contain more than one nitrogen atom per molecule.

It is important to note the relative nutritional completeness of sieve-tube sap (alluded to above in our reference to the nutrition of sap-sucking insects). Many plant parts with no or minimal transpiration (for example, meristems, tubers, roots, and some fruits) are almost completely dependent on the phloem for organic and inorganic nutrients during part or all of their growth.

## Phloem Loading

In 1949, Brunhild Roeckl determined the osmotic potentials of photosynthetic cells and sieve sap of *Robinia pseudoacacia* (black locust) using plasmolysis, refractometry, and cryoscopic techniques (Section 3.6). Such measurements have since been repeated by others, and many studies have used radioactive tracers. Typically, the mesophyll cells of trees have an osmotic potential of −1.3 to −1.8 MPa, whereas sieve elements in leaves have an osmotic potential of about −2.0 to −3.0 MPa. Herbaceous plants frequently have somewhat less-negative osmotic potentials in the mesophyll cells. Sugar beets, for example, have an osmotic potential of about −0.8 to −1.3 MPa for mesophyll and phloem parenchyma cells and about −3.0 MPa for the sieve-element–companion-cell complex (Geiger et al., 1973). Because most of the osmotic potential is caused by the presence of sugars in both kinds of cells, it is clear that the sugar concentration is approximately 1.5 to 3 times as high in the sieve elements as in the surrounding mesophyll cells. The process in which sugars are raised

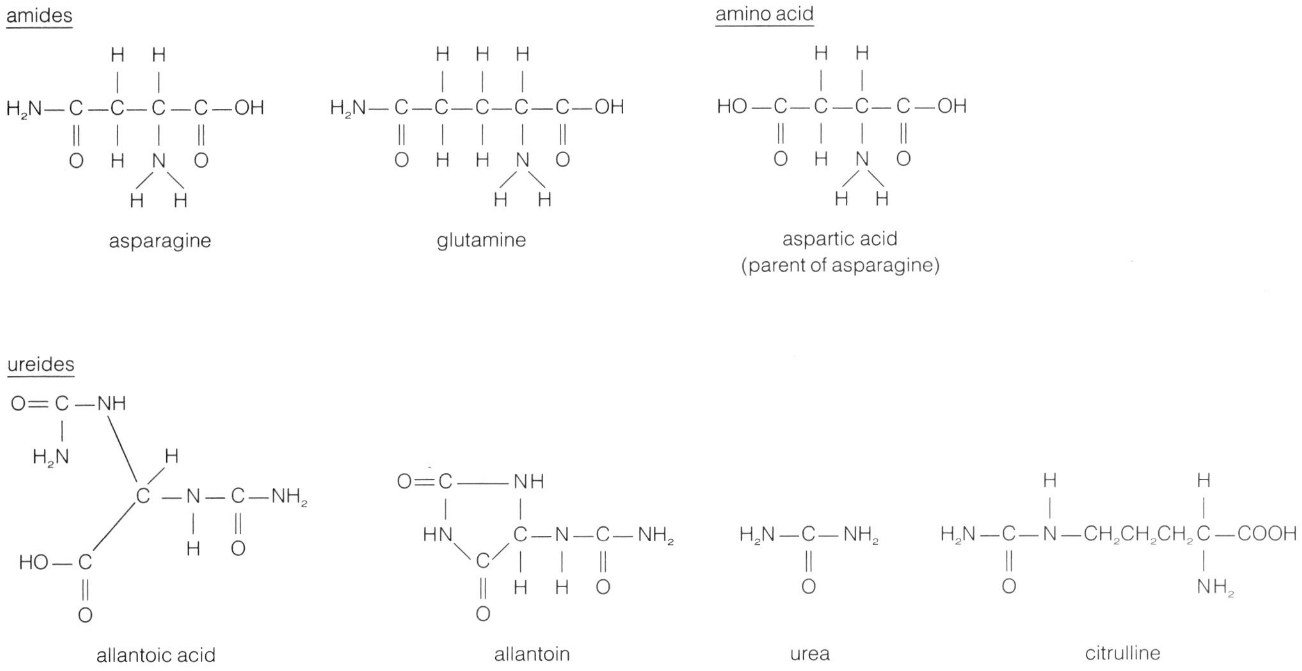

amides

asparagine

glutamine

amino acid

aspartic acid (parent of asparagine)

ureides

allantoic acid

allantoin

urea

citrulline

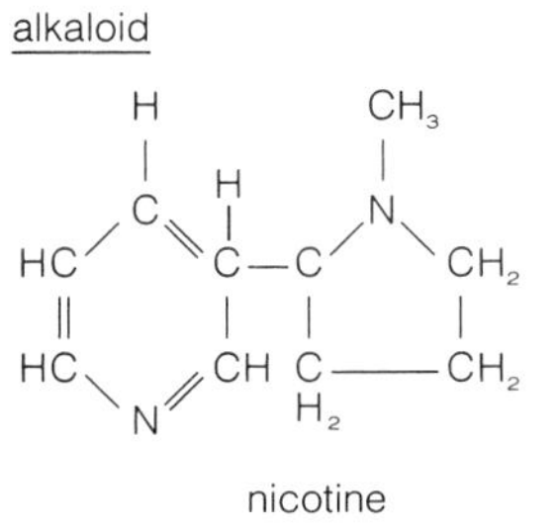

alkaloid

nicotine

**Figure 8-21** Examples of organic nitrogen compounds important in transport of nitrogen in both xylem and phloem of many species. Other nitrogen compounds important in transport are discussed in Chapters 9 and 14.

to high concentrations in phloem cells close to a source such as photosynthesizing leaf cells is called **phloem loading**. During recent years, there has been much interest in phloem loading (Baker and Milburn, 1990; Geiger and Fondy, 1980; Giaquinta, 1983; Lucas and Madore, 1988). Next we will consider some of the high points.

*The pathway of transport* How does sucrose get from the leaf mesophyll cells where it is synthesized to the sieve tubes in the minor leaf veins? Often it moves directly to a minor vein, or it must pass through only two or three cells. Does it move through the apoplast (cell walls outside the protoplast) or through the symplast (cell to cell through plasmodesmata, remaining in the cytoplasm)? Many plants have numerous plasmodesmata between mesophyll cells, and movement from one mesophyll cell to another apparently occurs through the symplast because $^{14}CO_2$ assimilated into carbohydrates in mesophyll cells does not appear in their cell walls.

In many species (for example, broadbean, maize, and sugar beets), plasmodesmata are less common between mesophyll cells and the adjacent companion cells and sieve elements, but there are other species with a direct symplastic continuity between mesophyll or **bundle-sheath cells** (cells forming a sheath around the vascular bundles of many species; see Fig. 5.5a) and adjacent companion cells and sieve tubes (for example, *Cucurbita pepo* and *Fraxinus*). It seemed during the late 1970s and early 1980s that sugar was actively secreted from mesophyll cells into the apoplast of the minor veins (see, for example, Geiger et al., 1974; Giaquinta, 1976, 1983). From the apoplast, sugar could then be absorbed actively, probably into the large companion cells of the minor veins, from which it then passed symplastically into the sieve elements (see Fig. 8-7). Several lines of evidence supported this model (Delrot, 1987). For example, it was difficult to imagine how movement across plasmodesmata could specify the molecules that were loaded into the phloem or how the steep concentration gradients could be generated. But, of course, plasmodesmata are complex structures, still not well understood; they are not just membrane-lined holes in the cell walls. Our inability to understand the mechanics of plasmodesmata does not prove that they are incapable of selective and active transport.

More recent evidence suggests that phloem loading can sometimes occur via the symplast (reviewed by Lucas and Madore, 1988; van Bel, 1987). It is clear that sucrose and other sugars readily leak out of mesophyll as well as phloem cells into the apoplast, but these cells have a powerful ability to *retrieve* these sucrose molecules. Lucas and Madore (1988) suggest that several experiments previously thought to demonstrate phloem loading from the apoplast can be interpreted on the basis of retrieval by mesophyll cells. In a set of experiments that demonstrate a possible symplastic pathway for phloem loading, it was possible to inject a fluorescent dye, lucifer yellow, into the cytosol of a leaf mesophyll cell. Although the dye does not cross membranes, it was easy to follow its movement from cell to cell and even into phloem sieve tubes (Madore et al., 1986).

Apparently, a symplastic pathway occurs in some tissues, whereas others use an apoplastic step. Similar techniques have demonstrated different pathways in different species (Turgeon, 1989; Turgeon and Wimmers, 1988). Some species may well use elements of both pathways. It is equally clear that much remains to be learned about phloem loading.

Active loading of sucrose into the companion cells could produce a strongly negative osmotic potential in those cells, leading to an osmotic entrance of water, which would then pass in bulk flow across the plasmodesmatal connections between the companion cells and sieve elements, carrying the sucrose along with it. Regardless of how the high concentration of sucrose in sieve tubes is produced, it leads to the osmotic uptake of water that produces the high pressures and mass flow. Münch was quite unaware of phloem loading and assumed that the pressures might build up directly in the mesophyll cells, but phloem loading is a highly appropriate modification of his model.

*The selective loading of sugars* Phloem loading has been studied by abrading the leaf surfaces, which destroys the cuticle but ruptures only a few epidermal cells, after which solutions of radioactively labeled sugars are applied. An autoradiogram of leaves made at various times after application of the solutions shows the progress of the loading process. When loading is complete, the minor and major veins are highly radioactive compared with the surrounding interveinal tissue (Fig. 8-22). Using this and other approaches, Donald R. Geiger and his coworkers (1973, 1974) applied various labeled sugars to abraded sugar-beet leaves and studied their absorption into minor veins. In these and other studies, it has become apparent that only those sugars that are transported in the phloem are accumulated into the minor veins, regardless of the pathway. As we have seen, this includes (in various species) the raffinose series of sugars, especially sucrose, as well as sugar alcohols. Such reducing sugars as glucose and fructose are taken up by mesophyll cells, but only small amounts are transferred to the phloem. Presumably, the selectivity of phloem loading is based on mutual recognition by the sugars and a carrier in the plasmalemma that transports the sugars into the cytoplasm.

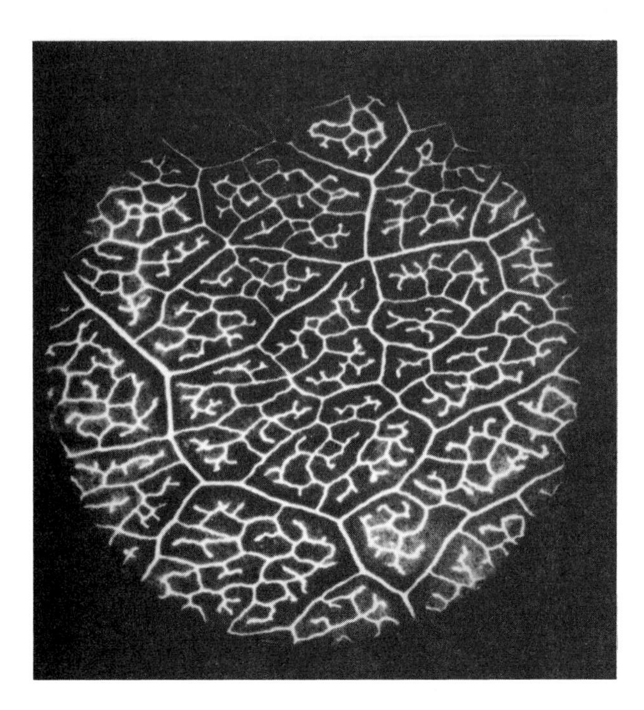

**Figure 8-22** An enlarged positive print of an autoradiograph showing phloem loading in sugar-beet leaves. Light areas are the minor veins that have accumulated the radioactive sucrose. (Courtesy of Donald Geiger; see Geiger et al., 1974.)

Comparable studies with amino acids have shown that certain types are preferentially loaded. Again, these prove to be the compounds that are readily transported in phloem. This is also true of minerals; those that are readily transported in phloem (phosphorus, potassium) are readily loaded; those that are usually not transported (calcium, boron, sometimes iron) are not (Chapter 6). This may be true even for synthetic compounds such as the herbicides 2,4,5-T (relatively immobile in phloem), 2,4-D (intermediate), and maleic hydrazide (most mobile; Field and Peel, 1971; Kleier, 1988; McReady, 1966). But in emphasizing the selectivity of the loading process, we should not lose sight of the fact that many substances can also enter the phloem by passively diffusing in along their own concentration gradients. This is apparently true for several growth regulators, for example.

*Sucrose/proton cotransport mechanism* In many systems, including bacteria, algae, yeast, fungi, and animal cells, transport of organic molecules such as sugars and amino acids is linked with transport of hydrogen ions. Several studies (reviewed by Giaquinta, 1983) suggested that sucrose loading into the phloem may occur by such a *cotransport system*. As noted in Chapter 7, protons are pumped out through the plasmalemma using the energy from ATP and an ATPase carrier enzyme, so the *p*H outside the cell in the apoplast becomes much lower (more acidic) than inside the cell. Protons then diffuse back into the cell, and their movement across the membrane is coupled to a carrier protein that transports sucrose or other sugars into the cell along with the hydrogen ions.

Several evidences for such a cotransport mechanism from the apoplast into phloem cells have been reported, but during recent years it has been shown that these results apply equally well to sugar uptake by mesophyll cells (reviewed by Lucas and Madore, 1988). Consider one example: W. Heyser (1980) used excised leaf strips of *Zea mays* (corn) mounted so that artificial solutions could flow through the xylem vessels. The solution *p*H was measured as it went in and as it came out of the xylem conducting elements. Adding sucrose to the solution raised the *p*H by 0.75 units as the solution moved through the leaf strip. Other sugars, specifically those not loaded into phloem, did not cause this *p*H response. This is what the cotransport hypothesis would predict: $H^+$ ions going into cells with sucrose molecules would leave the solution more alkaline (higher *p*H). But when Fritz et al. (1983) repeated the experiment with $^{14}C$-sucrose and determined its location in the tissue by high-resolution microautoradiography, they found that the majority of the labeled sucrose was in the xylem parenchyma and *not* in the phloem. Thus it appears that the living xylem cells can retrieve sucrose by a cotransport mechanism, but that this has little if anything to do with phloem loading.

*The role of metabolism in transport* Münch's model suggests that sap flow through sieve tubes is a passive phenomenon generated only by the high pressures in the leaf minor veins (or other sources) and the lower pressures at the sink. Thus the model does not immediately suggest that metabolic energy might be required along the pathway *to maintain flow*. Of course, metabolism might be required to maintain the phloem tissues in a condition suitable for transport, to reduce leakage of sugars through the plasma membranes of sieve elements, and to retrieve sugars that do leak out.

Early studies suggested that any inhibition of metabolism (for example, by low temperatures or respiration inhibitors) along the pathway did inhibit transport. This metabolic requirement was often cited by those who presented alternative theories to pressure flow. Indeed, it was suggested that metabolic energy was required along the pathway to move solutes across the sieve plates (for example, by some pumping or peristaltic contraction of P-protein in sieve pores). Hence these alternative ideas were often referred to as *active* theories, as contrasted to the *passive* pressure-flow mechanism.

Studies have shown that the apparent inhibitory effects of low temperature or anoxia (lack of oxygen) in certain species were really only transient effects, and that phloem transport continued after an adjustment period of 60 to 90 min (Geiger and Sovonick, 1975; Watson, 1975; Sij and Swanson, 1973). Thus maintenance of the phloem transport system for bulk flow of sap apparently requires only a minimum of metabolic energy. Of course, metabolic energy is required for phloem loading, as we have seen.

*The development of loading capacity* Young leaves normally act as sinks rather than sources. This is true even after they develop some photosynthetic capability. At a certain time, however, they begin to export carbohydrates through the phloem, although import of carbohydrate may continue for a while through different vascular strands. What accounts for the changeover from an import to an export mode of phloem transport? The development of phloem loading capacity in the minor veins could account for this switch from import to export (Giaquinta, 1983). Once sucrose begins to be actively loaded into the sieve elements, water will enter by osmosis, and flow will begin out of the minor veins; the leaf becomes a source instead of a sink. Robert Turgeon (1989) has recently reviewed the sink-source transition in leaves.

### Phloem Unloading

Removal of sucrose and other solutes from sieve elements at the sink end of the system plays an important role in phloem transport, often determining the sinks into which most translocation occurs. Such solute unloading, which is another highly appropriate modification of Münch's original hypothesis, maintains low phloem turgor pressures at the sink (see Fig. 8-5). Solute unloaded at the sink can then be absorbed into developing fruit or other cells, in which concentrations can reach values as high or higher than in the sieve tubes at the source. We will defer our discussion of unloading to Section 8.4 (Partitioning and Control Mechanisms).

### Pressure in the Phloem

Sieve tubes contain solutions under pressure, as indicated by the feeding habits of aphids. In a few plant species, exudation continues for several hours to many days after phloem is cut. When the top of a sugar palm trunk or its inflorescence is tapped, for example, as much as 10 liters of sugary sap may drip out of the cut sieve tubes per day, and the Palmyra palm in India produces as much as 11 liters of sap per day from cut phloem (reviewed by Crafts, 1961). This sap consists of about 10 percent sucrose and 0.25 percent mineral salts and has probably been diluted by water moving osmotically from the apoplast into the phloem after pressure is released. Pressure exists in sieve tubes, as a pressure-flow mechanism requires.

Münch postulated that a pressure *gradient* must occur in the phloem sufficient to account for flow from source to sink. For many years, pressures in the phloem could not be measured directly, so they were calculated by comparing osmotic potentials in the phloem with water potentials of the surrounding apoplast. At equilibrium (probably seldom achieved), water potential in the sieve tubes would be equal to water potential of the apoplast ($\Psi_i = \Psi_e$, $\Delta\Psi = 0$), so pressure in the sieve tubes ($P_i$) would equal apoplast water potential ($\Psi_e$) minus sieve-tube solute potential ($\Psi_{si}$). ($P_i = \Psi_e - \Psi_{si}$, when $\Delta\Psi = 0$). Gradients in osmotic potential in the sieve tubes from source to sink have often been measured, with the most negative values at the source (see, for example, Housley and Fisher, 1977; Rogers and Peel, 1975; Warmbrodt, 1986). But we saw in Chapter 5 that there is also a gradient in xylem (apoplast) water potential, with the most negative values in the transpiring and photosynthesizing leaves; that is, at the source for much phloem transport. So the existence of a pressure gradient from source to sink in the phloem will depend upon the relative steepness of the osmotic gradient in the sieve tubes compared with the water-potential gradient in the apoplast. Most calculations that take these factors into account suggest that there is a pressure gradient in sieve tubes from source to sink (Fig. 8-23).

Nevertheless, it would be satisfying to measure pressures directly, and several workers have now done this. One approach was to attach a pressure gauge to a cut palm shoot; still another was to apply a pressure cuff similar to those used in measuring blood pressure, increasing pressure in the cuff until wound exudate stopped. Pressures as high as 2.4 MPa were measured. (Normal human blood pressure is about 0.016 MPa.)

H. T. Hammel (1968) inserted a specially prepared hollow needle of microdimensions into the bark of red oak (*Quercus rubrum*) trees. A glass capillary partially filled with dyed water and sealed at one end was attached to the needle. When the needle penetrated a phloem tube, sap moved into the needle and capillary, compressing the gas in the glass tube (as in the method illustrated in Fig. 3-13). Hammel measured pressures at two points on the trunk, one 4.8 m higher than the other, with average values 0 to 0.3 MPa higher for the upper sampling point. Although there was considerable variation, results were what the Münch model predicts.

John P. Wright and Donald B. Fisher (1980) glued glass capillary tubes, sealed at one end, to stylets severed from aphids feeding on phloem sap in weeping willow (*Salix babylonica*). Pressures were calculated by measuring compression of the gas in the capillary tubes as in Hammel's experiments. Stable values up to 1.0 MPa

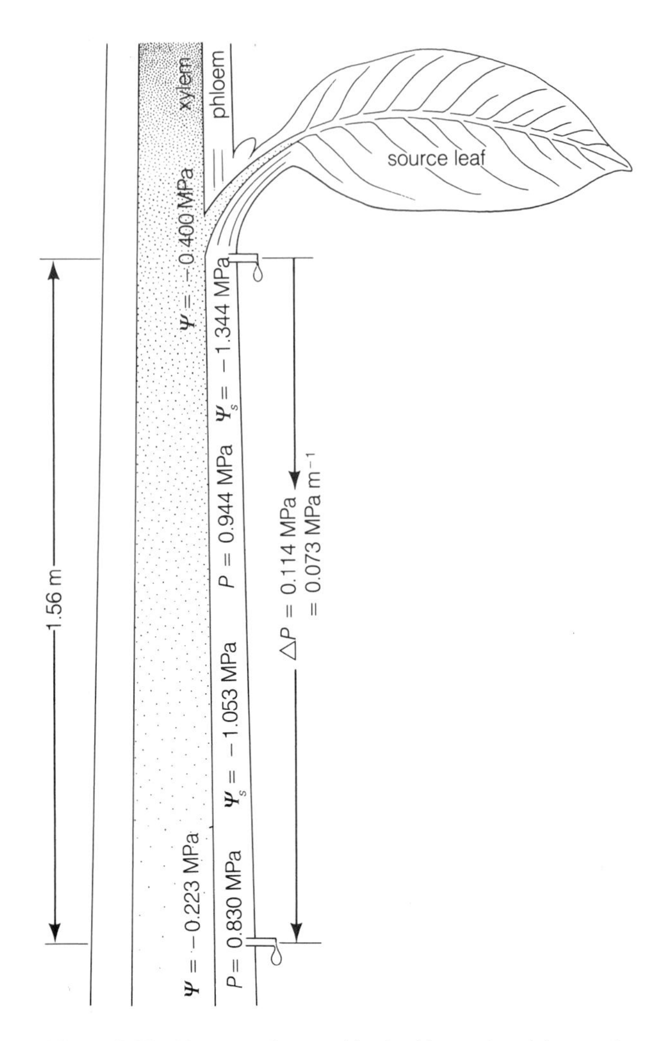


**Figure 8-23** The osmotic quantities in phloem sieve tubes and xylem (apoplast) of a young willow (*Salix viminalis*) sapling. The osmotic potentials ($\Psi_s$) were determined from phloem sap exuded from aphid stylets; water potentials ($\Psi$) of the apoplast (bark samples) were determined with a vapor psychrometer system (see Figs. 3-5 and 3-9); and pressures ($P$) in the sieve tubes were calculated by assuming that water potential of the phloem sap was in equilibrium with that of the surrounding tissues ($P = \Psi - \Psi_s$), including xylem. Note that there is a positive pressure gradient ($\Delta P$) in the sieve tubes from the apex toward the base, even though there is an opposite gradient in the water potential of the apoplast (caused by tension in the xylem plus matric forces). The pressure gradient of about 0.07 MPa $m^{-1}$ is ample to drive a pressure flow of sap through sieve tubes. (Data are averages from several experiments of S. Rogers and A. J. Peel, 1975.)

(accuracy ±0.03 MPa) were measured. They also calculated pressures by measuring sucrose in the phloem exudate (*refractometry*) and leaf water potential by a psychrometric method. The two methods agreed well, when account was taken of the amino acids and $K^+$ ions in the phloem exudate (not measured by refractometry).

## Two Problems with Pressure Flow

The most simplistic view of the mass flow model suggests that substances should move in the phloem not only in the same direction but at the same velocity. Thus, several workers (for example, Biddulph and Cory, 1957; Fensom, 1972) measured the velocity of flow of different tracer substances (for example, $^{14}C$-sucrose, $^{32}P$, and $^{3}H_2O$). Velocities between two points along the transport system were often highest for $^{14}C$-sugars, with $^{32}P$-labeled phosphates moving more slowly and $^{3}H_2O$ moving slowest of all. At first glance, it would seem that if the water moves more slowly than the solutes it is supposed to be carrying, mass flow would have to be rejected.

There are two important complications, however. First, water exchange occurs rapidly along the pathway. Because water moves easily through membranes, much water diffuses out from sieve tubes into surrounding tissues while much water moves from these tissues into the sieve tubes. Sucrose and phosphate do not pass as readily through the membranes along the way, so they might appear to move much faster than the carrier water molecules. Second, it is simplistic to imagine that the phloem transport system consists only of inert tubes. Sieve elements are alive and contain cytoplasm with mitochondria, P-protein, and other substances. Thus solutes can be metabolized or otherwise interact (for example, by adsorption) to different degrees along their transport route.

Another much-discussed potential obstacle to the pressure-flow hypothesis is that it is incompatible with movement of two different substances in opposite directions in the same sieve tube at the same time. Does such true **bidirectional transport** occur? Between the 1930s and the 1970s researchers attempted to answer this question. Numerous experiments suggested that such transport does occur, but so far it has always been possible to devise alternative explanations (reviewed by Peel, 1974).

In the best experiments, two different tracers are applied to different points, and their movement is followed. There is no question that bidirectional movement does occur. Tracers applied to young leaves may move **basipetally** (toward the base) while tracers applied to older leaves below may move **acropetally** (toward the tip), so the two pass each other in the stem, moving in opposite directions. Such bidirectional movement may even occur in a leaf petiole. But do both tracers move in the same vascular bundle or the same sieve tube? Many studies have indicated that they do not. Carol A. Peterson and Herbert B. Currier (1969) abraded the stem surface of several species and applied fluorescein, which was absorbed into intact phloem tissue. After various times they cut sections from above and below the point of dye application. The tracer moved both up and down, but after short time intervals it was never present in the same bundle both above and below a treated area; each bundle and sieve tube translocated the dye in only one direction. With longer times, the dye could move to a node in one bundle, pass laterally to another bundle, and move back down the stem in the opposite direction.

Studies with aphids might settle the question of whether tracers can move in opposite directions in a single phloem cell. For example, fluorescein was applied to one leaf at one node and $^{14}CO_2$ to another at another node; the honeydew from aphids on the stem between was collected on a slowly rotating cellophane disk (Eschrich, 1967; Ho and Peel, 1969). Much of the honeydew contained both tracers.

Are these experiments conclusive? No, for various reasons. For one thing, a stylet inserted into a sieve tube itself becomes a low-pressure sink; substances can flow from both directions toward the stylet. The direction of flow might also change with time in a given sieve tube, perhaps as source and sink roles change, or perhaps in response to more subtle hormonal mechanisms. Thus, if the experiment lasts an hour or more (as is usual), flow might have reversed during the experimental period, so that some honeydew was produced from one source and the rest from another. Furthermore, there is considerable lateral transport between sieve tubes and even between phloem and xylem, especially at the nodes but in the internodes as well; in some species, sieve elements have lateral sieve pores.

Evidence for bidirectional movement contributed strongly to the impetus to find an alternative to the pressure-flow model. To date, however, other evidences for mass flow are so strong that most workers in the field accept the explanations just presented that reconcile *apparent* bidirectional transport with pressure flow. Furthermore, pulse-labeling experiments with $^{11}CO_2$ clearly show a distinct peak of radioactive carbon from one source moving in only one direction along the path of transport (see, for example, Troughton et al., 1974; see also Christy and Fisher, 1978, who worked with $^{14}C$).

## Pressure Flow: A Summary

Let us return to the laboratory model of the Münch system (Fig. 8-5). It is easy to make it work and to list the requisites for suitable function: (1) an osmotic gradient between the two osmometers, (2) membranes that allow the establishment of a pressure gradient in response to the established osmotic gradient, (3) a low-resistance pathway (tube) between the two osmometers that allows flow, and (4) the osmometer with the most negative osmotic potential being immersed in a solution with a higher water potential than that in the osmometer. If these requisites are met in the plant, the system must function as it does in the model. Clearly, the osmotic system (the symplast) with surrounding membranes exists in the plant, and pressures are observed in the transport system, although they must be more widely and accurately measured to see whether the gradient is sufficient to drive flow over long distances, as in trees. The medium with high water potential around the source phloem tissues is the hydrated apoplast.

The sieve plates have offered the biggest problem all along. Will they allow flow rapid enough to account for observed transport rates? Or do they provide too much resistance? Several workers have calculated the resistance based on assumptions that the pores are open (as the best evidence now suggests) or that they are partially occluded. Comparing calculated resistances with measured or calculated pressure gradients suggests that resistance is not too high; known pressure gradients are great enough to produce flow (see, for example, Passioura and Ashford, 1974).

## 8.4 Partitioning and Control Mechanisms

What controls the amounts and directions of phloem transport? It has long been known, for example, that lower leaves transport relatively more to roots than to young leaves, fruits, or seeds, and that the flag leaves of grasses (for example, on wheat) and other upper leaves transport preferentially upward to young stems or to developing fruits and seeds (Fig. 8-24). Why? What is in control?

These questions are of great interest because agricultural yields depend upon the amount of assimilate transported to the harvested organ compared with the amount transported to other organs. Yields of many species have been improved during the past few decades as the **harvest index** (the ratio of the harvest yield to total shoot yield) has been increased, mostly by breeding. This is true for oats, barley, wheat, cotton, soybean, and peanuts, for example (reviewed by Gifford and Evans, 1981; Gifford, 1986). Attempts to increase photosynthetic efficiency of the leaves have so far met with little if any success, but breeding efforts have serendipitously increased the portion of assimilates partitioned to harvested storage organs.

### Photosynthesis and Sink Demand

Several cases are known in which photosynthesis of leaves is strongly influenced by sink demands (Gifford and Evans, 1981). For example, if potato tubers are removed during their development, photosynthesis drops markedly. Short-term responses could be caused by effects on stomatal aperture, but this explanation does not apply to the more lasting effects that are often observed. There are cases in which senescent (aging) leaves can be rejuvenated to full photosynthetic capacity when the sink/source ratio is increased substantially. On the other hand, rapid growth of a sink can sometimes compete with leaves for remobilizable nitrogen, leading to senescence of the leaf and a drop in its photosynthetic capacity.

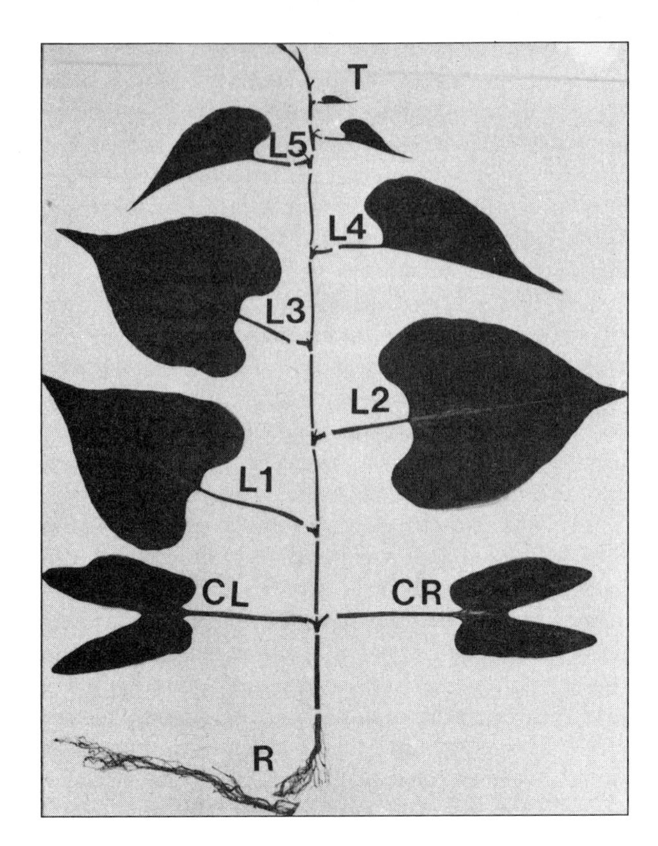


**Figure 8-24** Autoradiogram showing how different leaves on a tall morning-glory plant export assimilates to different sinks. The leaf or cotyledon marked **a** in each radiogram was allowed to photosynthesize in the presence of $^{14}CO_2$ for 24 h prior to harvest. At harvest, plants were dissected into several parts (roots, cotyledonary leaves, true leaves, nodes, internodes, and shoot tip) to prevent movement of assimilates between parts (see above). The parts were mounted, dried with forced air, and held in contact with X-ray film for 24 d. The results show that the cotyledonary leaf (**A**) and the lower true leaves (**B** and **C**) export to the roots (less in **C**) and to the rest of the plant but that upper leaves (**D**, **E**, and **F**) export only to the shoot tip while themselves acting as sinks (**A**, **B**, and **C**). The small leaf in (**F**) does not export at all; it only acts as a sink. (Autoradiograms courtesy of Steven A. Dewey, previously unpublished; see Dewey and Appleby, 1983, for a description of methods.)

How might sink demand regulate photosynthesis in the leaves? The simplest explanation is that when sink demand is low, sucrose piles up in the leaves, causing a product inhibition of photosynthetic reactions. Such a product inhibition has often been reported (see, for example, Blechschmidt-Schneider, 1986; Wardlaw and Eckhardt, 1987), but the situation is more complex. Foyer (1987) and Stitt (1986) have proposed that when sucrose builds up in the mesophyll cells, it leads to synthesis of **fructose-2,6-bisphosphate**, which is known to be an important regulator of sucrose synthesis and of photosynthesis—but this is beyond our discussion here. We will defer its further consideration to Section 13.12.

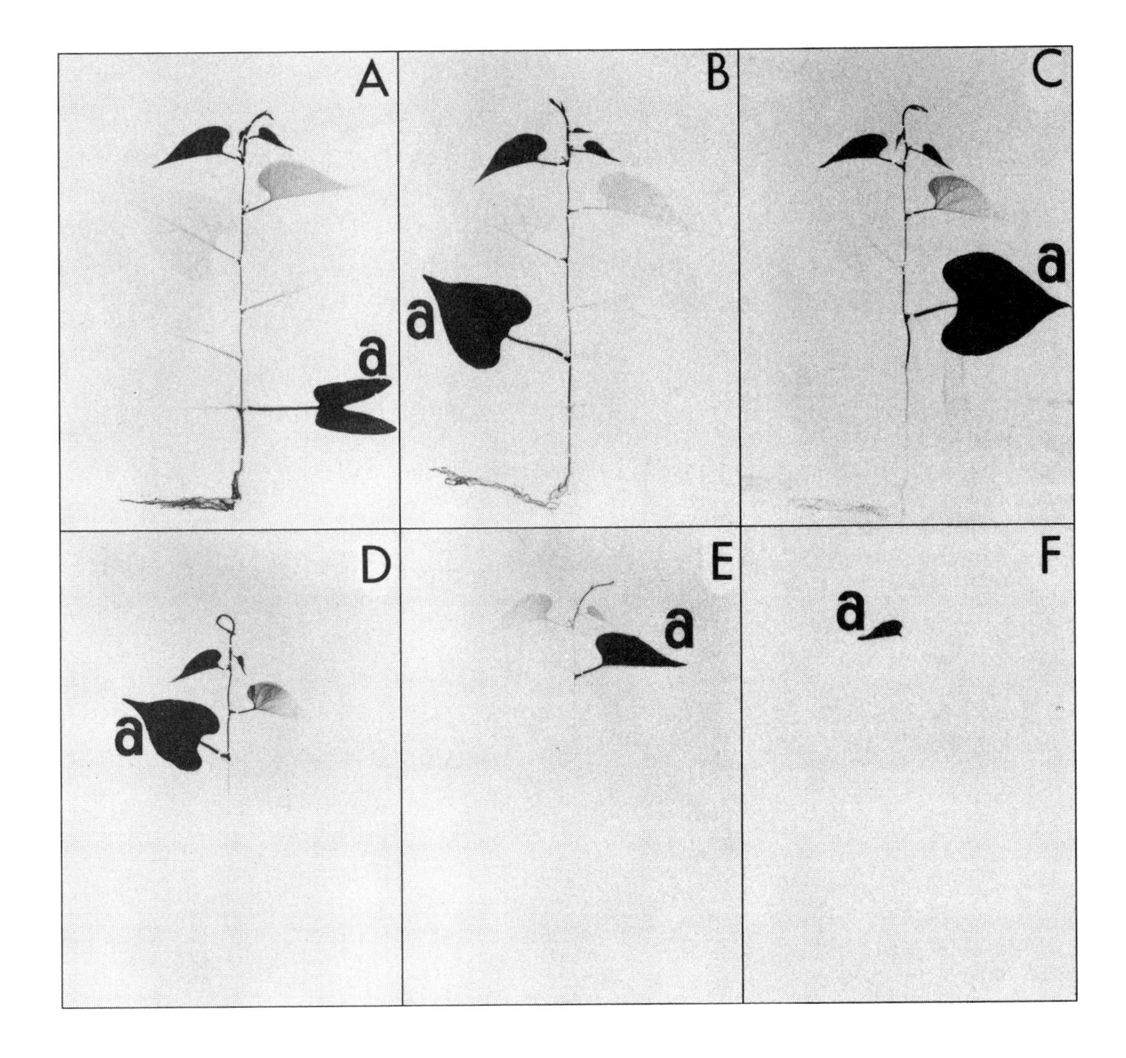


## Metabolically Driven Gradients

In many cases, the sucrose gradient that drives phloem transport is produced by metabolism of sucrose in the sink tissues. This is true of actively growing tissues in which the sucrose is used as an energy source to drive growth. Even in storage tissues, the sucrose may be changed to starch or some other less-soluble product that has a lower effect on osmosis than sucrose.

## Phloem Unloading

We have already noted that unloading occurs from the sieve tubes in the sink regions. According to the pressure-flow model, this would strongly direct the flow of phloem sap toward those regions (Ho, 1988). Removal of sucrose or other solutes from sieve cells makes the osmotic potential in those sieve cells less negative, so pressure transmitted from the source areas raises the water potential even further, and water diffuses out into the apoplast (see Fig. 8-5). Decreasing pressure in the sink end of the sieve tubes increases the pressure gradient between source and sink and leads to further flow toward the sink region. In the sinks, sucrose or other solutes are metabolized (used in respiration, converted to starch, and so on) or actively loaded into storage-cell vacuoles. The unloading process must be a critical control mechanism in carbon partitioning, which is why it has been the object of so much intense research during recent years (Eschrich, 1986; Geiger, 1986; Lucas and Madore, 1988; Thorne, 1985; Turgeon, 1989; Wolswinkel, 1985a; Wyse, 1986; and several other papers cited in Cronshaw et al., 1986).

As with the loading process, sucrose unloading occurs into the apoplast and also via plasmodesmata symplastically to sink cells. In growing and respiring sinks such as meristems, roots, and young leaves, in which the sucrose can be rapidly metabolized as noted above, unloading is typically symplastic. For example, young sugar-beet leaves act as sinks until their photosynthetic apparatus is fully developed, at which point they become sources; sink unloading occurs through the symplast (Schmalstig and Geiger, 1985, 1987). Generally, storage organs such as fruits (for example, grape and orange), roots (sugar beet; Lemoine et al., 1988), and even stems (sugarcane) have sucrose unloaded into the apoplast. Oparka (1986) noted, however, that phloem

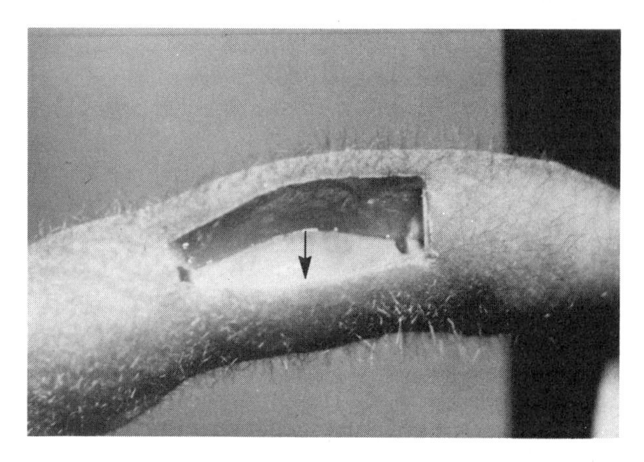

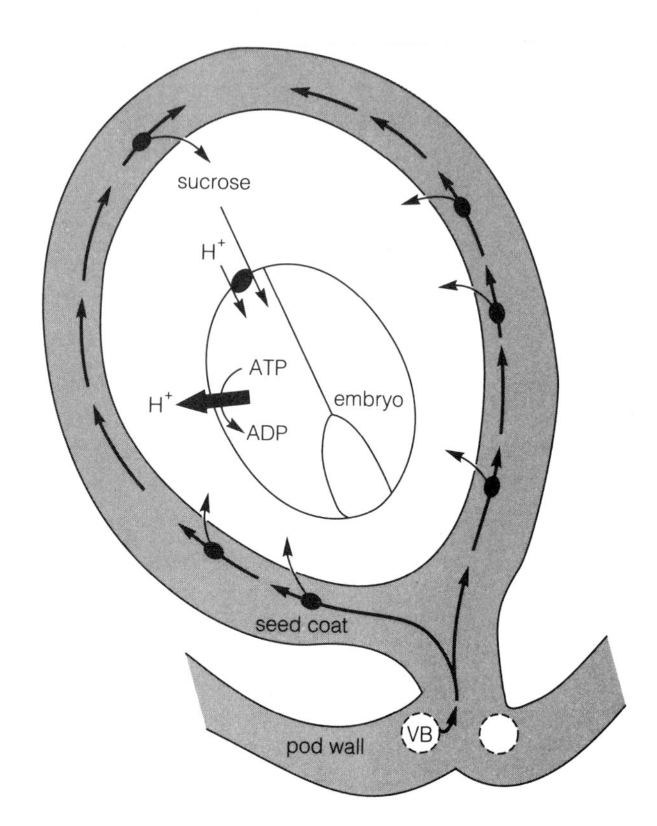


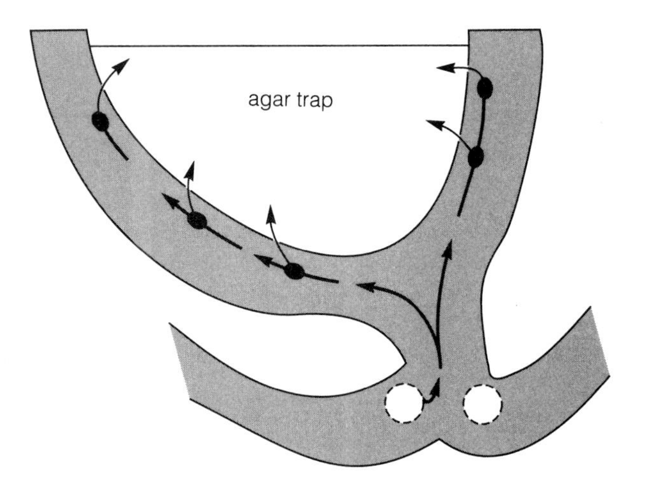


**Figure 8-25** The empty legume ovule method for studying phloem unloading into the apoplast of seed-coat endothelium. The photograph shows a pod with a window cut in its wall and the distal half of the developing seed removed along with the remaining part of the embryo. The cup (arrow) that remains for study is clearly visible. The drawing illustrates the anatomy of both the intact ovule and the ovule cup, which is filled with agar (but not quite to the rim, avoiding contamination of the agar with sap bleeding from the cut seed coat). (Courtesy of John H. Thorne; see Thorne and Rainbird, 1983.)

unloading from sieve tubes to cortical cells in the potato tuber is a symplastic and passive process. For one thing, the phloem is surrounded by an endodermis with Casparian strips, so sucrose must move through plasmodesmata rather than cell walls. The sucrose is converted to starch in the cortical storage cells.

Unloading in most developing seeds is into the apoplast because there is no symplastic connection between phloem of the mother plant and that of the developing embryo. The study of assimilate unloading has been revolutionized by the use of **empty legume ovules** (Gifford and Thorne, 1986; reviews by Thorne, 1986 and Wolswinkel, 1985a). The technique was developed simultaneously and independently in the United States (Thorne and Rainbird, 1983), the Netherlands (Wolswinkel and Ammerlaan, 1983), and Australia (Patrick, 1983). The technique can also be applied to such nonleguminous seeds as maize (Shannon et al., 1986). In his personal essay in this chapter, John H. Thorne tells about the discovery of the empty-ovule technique.

Incisions are made in the pod, exposing a developing seed through a window (Fig. 8-25). The seed is then cut in half, the top (distal) half is removed and discarded, and the embryo inside is carefully removed with a small spatula. The seed-coat cup that remains can then be filled with warm agar (4 percent), which quickly solidifies, or with a buffer solution. Phloem unloading of assimilates occurs unabated into the apoplast of the seed-coat lining (the **endothelium**), and the assimilates, which would normally be absorbed by the developing embryo, diffuse into the agar or buffer trap.

The endothelium can be pretreated with various solutions before the agar or buffer is added. Plants are usually allowed to photosynthesize in $^{14}CO_2$ for some period, and then the radioactivity in the trap can be analyzed after various intervals.

Unloading into the trap is greatly promoted by EGTA and similar compounds, which form **chelates** with divalent cations, especially $Ca^{2+}$. Apparently, removing such cations from the membranes of phloem tissues allows assimilates to leak out. Such metabolic inhibitors as azide, cyanide, and low temperature inhibit unloading into the trap (Wolswinkel, 1985b). The compound PCMBS, which modifies sulfhydryl groups, inhibits sucrose transport across membranes, apparently because it makes the sucrose carrier ineffective (Madore and Lucas, 1987). If the cup is pretreated with this compound before the agar is added, release of sucrose is greatly inhibited. These lines of evidence indicate that assimilate unloading is under metabolic control and may well involve carriers located in membranes.

## Hormone-Directed Transport

Evidence is accumulating that growth regulators (see Chapters 17 and 18) help direct translocation (see, for example, Aloni et al., 1986; Courdeau et al., 1986; Patrick, 1979; Peretó and Beltran, 1987; brief review in Lucas and Madore, 1988). The best studied cases involve remobilization of stored reserves in storage organs such as taproots or the sugarcane stalk parenchyma. The stored assimilates are often directed to new, typically reproductive sink tissues, and the formation of these new sinks is often under the control of growth regulators. There is evidence that the new sinks depend not only on hormones but also on increased concentrations of sucrose. This appears to be the case for flower formation in some species, for example (Bodson and Bernier, 1985).

In most cases, the growth regulators may not only induce the formation of the new growth regions (sinks), but they are released from the new sinks and act as strong mobilizing agents. Application of cytokinins to a leaf, for example, sometimes causes that leaf—specifically, the point of application—to become a sink. Similar observations have been reported for auxin, ethylene, gibberellic acid, and abscisic acid. Combinations of growth regulators can have additive, synergistic, or inhibitory effects (Gifford and Evans, 1981). There is much to learn about growth-regulator effects on translocation and especially on partitioning.

## Turgor Sensing in Sugar Transport

As we noted when we discussed the Münch laboratory model, if the solution surrounding the sink osmometer has a higher solute concentration (more negative $\Psi_s$), this will accentuate flow out of the osmometer and thus transport from one osmometer to the other. Much evidence now indicates that most if not all strong sinks have high concentrations of solutes (usually sucrose) in their apoplasts (Lang et al., 1986; reviews by Wolswinkel, 1985a; Lucas and Madore, 1988). With empty ovules, for example, unloading occurs at rates comparable to those of the intact ovule only if the cup is filled with a solution of fairly high concentration (for example, 400 mM sucrose or mannitol). The strongly negative $\Psi_s$ results in diffusion of water out of the phloem cells and a consequent reduction in phloem turgor (*P*), which makes the pressure gradient from source to sink steeper and thus increases flow. Reduced turgor in phloem cells at the source promotes more rapid phloem loading, which also increases the rate of transport.

## The Control of Fruit and Vegetable Composition

We have learned a great deal about how fruit and vegetable composition is controlled by phloem transport, xylem interchange with phloem, and import of xylem sap (Pate, 1980). Because many developing seeds, fruits, or other food storage organs transpire at a low rate if at all, they essentially subsist on a diet made up of phloem sap. A developing potato tuber, for example, probably does not transpire and may even absorb water directly from the soil, but a head of wheat or a pea pod does transpire, thereby gaining some of its nutrients from the xylem stream. A model for fruit growth based on extensive measurements made by John S. Pate and his coworkers at the University of Western Australia has been developed for the white lupine (*Lupinus albus*; Pate et al., 1977; Pate, 1986). Phloem sap supplies about 98 percent of the carbon, 89 percent of the nitrogen, and 40 percent of the water entering the fruit from the parent plant. The remaining nitrogen and water are supplied by the xylem sap, and the remaining carbon comes both from xylem sap (as part of organic nitrogen compounds) and from photosynthesis in developing pods. Asparagine and glutamine collectively carry 75 to 85 percent of the nitrogen of xylem and phloem saps, and sucrose carries 90 percent of the carbon of phloem. The nitrogen from the amides (asparagine and glutamine) is used to synthesize a wide variety of seed proteins, as shown by studies with $^{15}N$. And much of the nitrogen from the two amides passes through the amino acid arginine, a compound that is almost absent from phloem sap in this plant.

The phloem sap changes significantly in composition while underway from leaves to sinks such as developing fruits or tubers. In white lupine, Pate and his coworkers (1979) found that the phloem sap entering the developing fruits is more dilute in sucrose and much richer in certain amino acids than is the sap exported from leaves. Apparently, as the sap passes through the stem, sucrose is lost by transfer to adjacent tissue (is unloaded), and amino acids are loaded into the phloem. The amino acids must come from pools stored in the stems but originally obtained from xylem sap. Ratios of sucrose to amino acids also change with time. These brief examples suggest the vast possibilities of accomplished and future research for elucidating how the composition of fruits, seeds, and other storage organs is determined by partitioning of solutes in phloem and xylem saps. It is hoped that such research will lead to increased agricultural yields, as well as to a better understanding of whole-plant physiology.

# Discovery of the Empty-Ovule Technique

*John H. Thorne*

*John H. Thorne was the leader of one of the three groups of people who independently and almost simultaneously discovered the empty-ovule technique to study phloem unloading. In his letter accepting our invitation to tell about that experience, he says: "Those years (1980–83) represent one of my most productive professional periods." He is no longer involved in research but enjoys the challenges of the business world at E. I. du Pont de Nemours and Company, in Wilmington, Delaware. Here is his story:*

For many years, efforts to study phloem unloading mechanisms were hampered by the fragile, inaccessible nature of the phloem. Today, however, plant scientists can more readily study phloem unloading with the "empty-ovule" technique. How it was independently developed in three laboratories in different countries is nearly as interesting as the technique itself. After a brief introduction to the technique, I'll give you my perspective on its development.

To gain access to the unloading sites, maternal tissues of attached seeds are cut open and the developing embryo removed from the maternal ovule (seed coat). This exposes the maternal tissues and their phloem unloading sites responsible for the nutrition of the embryo. These can be challenged with buffers, solutes, inhibitors, and so on to characterize photosynthate import processes. The technique was first developed with legume fruits (Thorne and Rainbird, 1983), but studies were later conducted with maize and other species (Thorne, 1985).

My interest in photosynthate translocation into developing seeds began in 1971, while I was a Ph.D. student at Purdue, and continued through research positions at a seed company and the Connecticut Agricultural Experiment Station. There I completed studies of the ultrastructure of soybean fruit vascular tissues and the kinetics, biochemistry, and environmental controls of photosynthate import. Conducting $^{14}C$ washout studies with detached soybean seedcoats in 1979, I first thought of examining unloading in surgically opened, attached seeds.

As I began to publish and speak about my work, I received several letters from John Patrick (Australia) and Pieter Wolswinkel (the Netherlands). Both are great guys who were anxious to compare philosophies on translocation and source/sink interactions. Pieter had worked for several years with *Vicia faba* (broadbean) stem segments parasitized by *Cuscuta* (dodder). His letters told of his frustration that many of his peers neither shared his enthusiasm for the *Cuscuta* technique nor concurred with his conclusions that unloading has an energy-dependent component. I encouraged him to consider working with unparasitized *Vicia* fruit, for discoveries might advance efforts to increase harvest yields and the data would appeal to a wider audience.

John Patrick had worked for several years to demonstrate hormone-directed transport of metabolites and, having left that, was then working with translocation in stems and fruit of *Phaseolus vulgaris* (a bush-bean cultivar). As my papers on the kinetics and metabolism of photosynthate movement within developing soybean seeds (1980) and on vascular ultrastructure of the seed coat (1981) were being reviewed and published, John published a paper on the movement of $^{14}C$ within developing ovules of bush bean. Published in the Australian Journal of Plant Physiology, I first read it many months later when a bundle with the entire 1980 volume arrived in our library. He had the same problem with our journals. It was obvious we were pursuing the same goal.

John wrote to me again. Concerned about duplication of effort, he requested that we choose different areas of study. I indicated that I would leave hormone-directed transport to him but could not reveal the direction of my work beyond that. I was in the process of setting up a new laboratory at du Pont's research facility in Delaware, and their policy on research confidentiality was fresh in my mind.

With Ross Rainbird, a postdoc from John Pate's lab in Australia, I conducted experiments on both photosynthate release from attached, empty seed coats and active uptake into isolated embryos. Papers on both subjects were submitted in 1982, but the empty-seed-coat paper was published in *Plant Physiology* early the next year (Thorne and Rainbird, 1983). It was published as a technique paper, with an extensive discussion of the surgical procedure involved and the impact of various chemical challenges to unloading. A month later, we submitted the bulk of our work characterizing the photosynthate released by the maternal tissues. Roger Gifford's arrival from Australia added considerably to our characterization of the steady-state kinetics of unloading and solute concentrations in the tissues.

John Patrick and Pieter Wolswinkel published their papers on photosynthate release from seed coats a few months later. Only a few weeks separated their dates of submission. John's work, published in *Zeitschrift für Pflanzenphysiologie*, characterized washout kinetics from detached, preloaded seed coats (Patrick, 1983), but Pieter's work was more similar to mine. His paper, published in *Planta*, was the first of several papers utilizing attached fruit (Wolswinkel and Ammerlaan, 1983). Of course, we had all used different legumes in our studies. Thus our letters for the next year or so were filled with discussions of the similarities and differences we saw in our relative systems.

What fond memories I have of those years. The highlight of this five-year period of intense competition came in August, 1985. We were all invited to Asilomar, California, for the International Phloem Transport Conference. Located on the sands of Pebble Beach, this picturesque conference center is the ideal site for long, philosophical discussions. Many an evening was spent talking around a campfire or walking, wine bottle in hand, along the beach. The conference organizers put us together in the same unheated hut for the week to ensure that the three of us resolved our philosophical differences. I think we succeeded.

# TWO

# Plant Biochemistry

# 9

# Enzymes, Proteins, and Amino Acids

Living cells are energy-dependent chemical factories that must follow chemical laws. The chemical reactions that make life possible are collectively referred to as **metabolism**. Thousands of such reactions occur constantly in each cell, so metabolism is an impressive process. Plant cells are especially impressive with regard to the kinds of compounds they can synthesize. Thousands of compounds must be formed to produce the organelles and other structures present in living organisms. Plants also produce a whole host of complex substances called secondary metabolites that probably protect them against insects, bacteria, fungi, and other pathogens. Furthermore, plants produce both vitamins necessary for themselves (and incidentally for humans) and hormones used by cells in different parts of the plant to control and coordinate developmental processes.

Some reactions form large molecules such as starch, cellulose, proteins, fats, and nucleic acids. We call formation of large molecules from small molecules **anabolism** (from Greek, *ana*, "upward," and *ballein*, "to throw"). Anabolism requires an input of energy. **Catabolism** (from Greek, *kata*, "downward") is the degradation or breakdown of large molecules to small molecules, and this process often releases energy. Respiration is the major catabolic process that releases energy in all cells; it involves oxidative breakdown of sugars to $CO_2$ and $H_2O$.

Both anabolism and catabolism consist of **metabolic pathways**, in which an initial compound, A, is converted to another, B; then B is converted to C, C to D, and so on until a final product is formed. In respiration, glucose is the initial compound, and $CO_2$ and $H_2O$ are the final products of a metabolic pathway involving some 50 distinct reactions. Most metabolic pathways have fewer reactions.

How can a cell control whether most of its metabolic pathways will be anabolic or catabolic and which specific ones will operate? For example, all cells contain glucose, which can undergo respiration or can be converted by anabolism into starch in plastids or into cellulose of cell walls. For a cell to function and develop properly, metabolic pathways must be carefully controlled. *The first control is related to energy*. Only certain reactions in metabolic pathways can occur without energy input, whereas other potential reactions require so much added energy that they do not occur and are considered essentially impossible. Thus plants can form the cellulose and starch they require from glucose (with the help of energy from the sun), but they cannot convert that glucose into light energy that would allow them to photosynthesize.

The science of thermodynamics helps us understand the control of metabolism through energy input or output (see Chapter 2). However, thermodynamic laws say nothing about *how fast* possible chemical reactions can occur. For example, these laws state that when $H_2$ and $O_2$ combine to form $H_2O$, much energy will be released, but they cannot predict how rapidly the reaction will occur. In fact, if an inorganic catalyst such as finely ground platinum is present, $H_2$ and $O_2$ react so fast that an explosion occurs! *Therefore, the second type of control in metabolic pathways must be one of rates*. A limited amount of heat release (or heat absorption) in any reaction is tolerable and common, but explosions are unacceptable.

Cells control which metabolic pathways operate, and how fast, by producing the proper catalysts, called **enzymes**, in the proper amounts at the times they are needed. Nearly all chemical reactions of life are too slow without catalysts, and enzymes are much more specific and powerful catalysts than are any metal ions or other inorganic substances that plants could absorb from the soil. Thus enzymes generally speed reaction rates by factors between $10^8$ and $10^{20}$. Compared with man-made catalysts, enzymes are usually more than $10^6$ times as effective catalytically. Enzymes are also much more specific than inorganic or even synthetic organic catalysts in the kinds of reactions they catalyze, so thousands of reactions can be controlled by the formation of the proper compounds needed for life (and without the formation of toxic by-products). Finally, enzymes respond to environmental changes so that control is possible for plants living in a variety of climates. These advantages of enzymes are accompanied by a disadvantage: Enzymes are large protein molecules that an organism forms by anabolism only at the cost of considerable energy. Let us investigate in more detail the nature of enzymes and how they function so effectively.

## 9.1 The Distribution of Enzymes in Cells

Enzymes are not uniformly mixed throughout cells. The enzymes responsible for photosynthesis are located in chloroplasts; many of the enzymes essential for aerobic respiration occur exclusively in mitochondria, whereas still other respiratory enzymes exist in the cytosol. Most enzymes essential to the synthesis of DNA and RNA and to mitosis occur in nuclei. Enzymes that govern the steps in metabolic pathways are sometimes arranged so that a kind of assembly-line production process occurs (Srere, 1987; Gontero et al., 1988). In this case, the product of one reaction is released at a site where it can immediately be converted to a related compound by the next enzyme involved in the pathway, and so on, until the metabolic pathway is completed and a quite different compound is formed.

Such compartmentalization almost surely increases the efficiency of numerous cellular processes for two reasons: First, it helps ensure that the concentrations of reactants are adequate at the sites where the enzymes that act upon them are located. Second, it helps ensure that a compound is directed toward the necessary product and is not diverted into some other pathway by the action of a competing enzyme that can also act upon it elsewhere in the cell. However, this compartmentalization is often not absolute, nor apparently should it be. For example, the membranes surrounding chloroplasts allow outward passage of certain sugar phosphates produced by photosynthesis, which are then acted upon outside the plastids by numerous enzymes that are involved in cell-wall synthesis and respiration, which are essential to growth and maintenance of the plant.

## 9.2 Properties and Structure of Enzymes

### Specificity and Nomenclature

One of the most important properties of enzymes is their specificity. Each enzyme acts on a single **substrate** (reactant) or a small group of closely related substrates that have virtually identical functional groups that are capable of reacting. With some enzymes, specificity appears to be absolute, but with others there is a gradation in their abilities to convert related compounds to products. As explained later, specificity results from enzyme-substrate combinations that may be likened to a lock-and-key arrangement.

More than 5,000 different enzymes have been discovered in living organisms, and the number grows as research continues. Each enzyme is named according to a standardized system, and each is also given a simpler common or trivial name. In both systems the name commonly ends in the suffix *-ase* and characterizes the substrate or substrates acted upon and the type of reaction catalyzed. For example, *cytochrome oxidase*, an important respiratory enzyme, oxidizes (removes an electron from) a cytochrome molecule. *Malic acid dehydrogenase* removes two hydrogen atoms from (dehydrogenates) malic acid. These common names, although conveniently short, do not give sufficient information about the reaction catalyzed. For example, neither identifies the acceptor of the removed electron or hydrogen atoms.

The International Union of Biochemistry lists longer but more descriptive standardized names for all well-characterized enzymes. As an example, cytochrome oxidase is named *cytochrome c:*$O_2$ *oxidoreductase*, indicating that the particular cytochrome from which electrons are removed is the c type and that oxygen molecules are the electron acceptors. Malic acid dehydrogenase is called *L-malate:NAD oxidoreductase*, indicating that the enzyme is specific for the ionized L form of malic acid (malate) and that a molecule abbreviated as NAD is the hydrogen-atom acceptor. Table 9-1 lists six major classes of enzymes based on the types of reactions they catalyze and includes a few examples.

### Reversibility

Enzymes increase the rate at which chemical equilibrium is established among products and reactants. At equilibrium, the terms *reactants* and *products* are arbitrary and depend upon our point of view. Under normal physiological conditions, an enzyme has no influence on the relative quantities of products and reactants that would eventually be reached in its absence. Therefore, an enzyme cannot affect whether the equilibrium state is favorable or unfavorable for formation of a compound.

The equilibrium constant depends upon the chemical potentials (concentrations, approximately) of all compounds involved in the reaction (see Equation 2.7). If the chemical potential of the reactants is very high compared with that of the products, the reaction might proceed only toward product formation because of the chemical law of mass action. Most **decarboxylations**, in which carbon dioxide is split out of a molecule, are ex-

**Table 9-1 The Major Enzyme Classes and Subclasses.**

| Class and Subclass | General Reaction Type |
|---|---|
| Oxidoreductases | Remove and add electrons or electrons and hydrogen. Oxidases transfer electrons or hydrogen to $O_2$ only |
| Oxidases | |
| Reductases | |
| Dehydrogenases | |
| Transferases | Transfer chemical groups |
| Kinases | Transfer phosphate groups, especially from ATP |
| Hydrolases | Break chemical bonds (e.g., amides, esters, glycosides) by adding the elements of water |
| Proteinases | Hydrolyze proteins (peptide bonds) |
| Ribonucleases | Hydrolyze RNA (phosphate esters) |
| Deoxyribonucleases | Hydrolyze DNA (phosphate esters) |
| Lipases | Hydrolyze fats (esters) |
| Lyases | Form double bonds by elimination of a chemical group |
| Isomerases | Rearrange atoms of a molecule to form a structural isomer |
| Ligases or Synthetases | Join two molecules coupled with hydrolysis of ATP or other nucleoside triphosphate |
| Polymerases | Link subunits (monomers) into a polymer such as RNA or DNA |

Source: Modified from S. Wolfe, *Biology of the Cell*, 2nd ed., 1981, p. 45.

amples of such reactions because the $CO_2$ can escape; its concentration (and hence its chemical potential) remains low. **Hydrolytic reactions**, which involve cleavage of bonds between two atoms and addition of the elements of $H_2O$ to those atoms, are also essentially irreversible. For example, hydrolysis of starch to glucose by *amylases*, hydrolysis of phosphate from various molecules by *phosphatases*, and hydrolysis of proteins to amino acids by *proteases* are essentially irreversible processes. Other enzymes using different substrates carry out synthesis of starch and proteins and the addition of phosphate to various molecules. In fact, large molecules such as fats, proteins, starch, nucleic acids, and even certain sugars are synthesized by one series of enzymes and degraded by another. Synthetic and degradative enzymes are often kept separate from each other by membranes or are formed at different times so that competition between degradation and synthesis is minimized.

## Chemical Composition

Nearly every known enzyme has a protein as a major part of its structure, and many contain nothing other than protein. (There are at least two recently discovered exceptions to this in which RNA molecules have catalytic ability, as reviewed by Cech and Bass, 1986, but otherwise all of the several thousand known enzymes are proteins.) However, some proteins appear to have no catalytic function and are not classified as enzymes. For example, proteins of microtubules (**tubulin**) and microfilaments (**actin**) and some of the proteins in ribosomes seem to perform a structural rather than a catalytic function. Other proteins, such as **cytochromes** that transport electrons during photosynthesis and respiration, are not enzymes but rather electron carriers. Furthermore, several storage proteins in seeds also have no known enzymatic function. The role of most seed-storage proteins is to serve as a reservoir of amino acids for the seedling after germination, not to act as enzymes.

Proteins consist of one or more chains (**polypeptide chains**), each of which is usually made of hundreds of amino acids. The composition and size of each protein depend upon the kind and number of its amino-acid subunits. Commonly, 18 to 20 *different kinds* of amino acids are present, with most proteins having a full complement of 20. The *total number* of amino acid subunits varies greatly in different proteins, and so protein molecular weights also vary. Most plant proteins so far characterized have molecular weights of at least 40,000 grams $mole^{-1}$ (also called *Dalton* units, or Daltons, where one **Dalton**, abbreviated Da, is the mass of one hydrogen atom), yet that of *ferredoxin*, a protein involved in photosynthesis, is only about 11.5 kDa and that of *ribulose bisphosphate carboxylase*, another photosynthetic enzyme, is over 500 kDa. The latter is composed of eight small, identical polypeptide chains and eight larger, identical polypeptide chains. The chains of such complex enzymes are frequently held together by noncovalent bonds, often ionic and hydrogen bonds, and can be separated *in vitro*. Nevertheless, if care is taken to prevent chain separation during extraction, even complex enzymes such as ribulose bisphosphate carboxylase can be isolated as homogeneous crystals (Fig. 9-1).

The amino-acid building blocks in proteins can be represented by the general formula:

$$\mathrm{R{-}\overset{\displaystyle H}{\underset{\displaystyle NH_2}{C}}{-}C\!\!\begin{array}{l}{/\!\!/}O\\ \backslash OH\end{array}} \quad \text{or} \quad RCHNH_2COOH$$

The $-NH_2$ is the **amino group** and the $-COOH$ is the **carboxyl group**. These two groups are common to all amino acids, with slight modification of the amino group in proline. R denotes the remainder of the molecule, which is different for each amino acid. Figure 9-2 shows the structures of the 20 amino acids commonly found in proteins. The R groups cause amino acids to

**Figure 9-1** Crystals of pure ribulose bisphosphate carboxylase. Note the dodecahedron shape; the largest crystals are more than 0.2 mm in diameter. (Courtesy S. G. Wildman.)

differ greatly in physical properties, such as water solubility. The aliphatic types in the upper left of the figure and the aromatic types at the lower right are much less soluble in water (are more hydrophobic) than the more hydrophilic basic, acidic, and hydroxylated types (serine and threonine).

Structures of two amides, glutamine and asparagine, that occur in most proteins are included in Figure 9-2. (Technically, amides are amino acids because they have the general amino-acid structure given above. They are amides because the R portion of the amino acid has an amino group connected to a carbonyl carbon.) These amides are formed from glutamic and aspartic acids, the two amino acids that have an additional carboxyl group as part of R. The amides are structural parts of most proteins. They also represent especially important forms in which nitrogen is transported from one part of the plant to another and in which surplus nitrogen can be stored (Sections 8.3 and 14.5).

The union of amino acids and amides into polypeptide chains of proteins occurs through **peptide bonds** involving the carboxyl group of one amino acid and the amino group of the next, as summarized in an oversimplified form in Reaction 9.1. The vertical arrow indicates the peptide bond:

$$HOOC{-}CH_2{-}\underset{H}{\overset{NH_2}{C}}{-}\underset{O}{\overset{}{\underset{\|}{C}}}\overbrace{OH + H}^{H_2O}\overset{H}{N}{-}\underset{H}{\overset{H}{C}}{-}COOH \rightarrow$$

$$HOOC{-}CH_2{-}\underset{H}{\overset{NH_2}{C}}{-}\underset{O}{\underset{\|}{C}}{\downarrow}\overset{H}{N}{-}CH_2{-}COOH \qquad (9.1)$$

When aspartic and glutamic acids, each of which has two carboxyl groups, form peptide bonds with other amino acids, only the carboxyl group adjacent to the amino group participates. The other carboxyl group remains free and gives acidic properties to the protein. When lysine and arginine, each of which has two amino groups, form peptide bonds, the amino group farthest from the carboxyl group is always free. The nitrogen atom of each of these groups possesses two electrons that can be shared by $H^+$ in the cells; as a result, $H^+$ is attracted to these basic nitrogen atoms, giving them a positive charge.

Proteins rich in aspartic and glutamic acids usually have net negative charges in cells because these amino acids lose a $H^+$ ion during dissociation of the carboxyl group that is not involved in a peptide bond. Alternatively, proteins rich in lysine and arginine usually have net positive charges. These charges are important because whether an enzyme is catalytically active or inactive or whether it is bound to another cellular component frequently depends upon whether one of its free amino or carboxyl groups is charged or uncharged. For example, chromosomes contain five major kinds of positively charged proteins called **histones**, which are rich in lysine or arginine. These histones are held to negatively charged DNA by ionic bonds, which help control the structure and genetic activity of the chromosomes. Also, the different net charges on various enzymes allow us to separate them by their chemical and physical properties. Their functions and properties can then be studied without interference from other enzymes.

## Prosthetic Groups, Coenzymes, and Vitamins

In addition to the protein parts of enzymes, some contain a much smaller, organic nonprotein portion called a **prosthetic group**. Prosthetic groups are usually tightly attached to the protein by covalent bonds and are essential to catalytic activity. An example is found among some of the dehydrogenases involved in respiration and fatty-acid degradation. In this case a yellow pigment called a *flavin* is attached to the protein. The flavin is essential to enzyme activity because of its ability to accept and then transfer hydrogen atoms during the course of the catalyzed reaction. Some enzymes contain prosthetic groups to which is attached a metal ion (for example, iron and copper in cytochrome oxidase; see Chapter 13). Other proteins, **glycoproteins** (from the Greek, **glykys**, "sweet"), contain a group of sugars attached to their protein parts. Such attached carbohydrates can contribute to enzymatic action or to protection of the enzyme against temperature extremes, internal destructive agents such as proteases, and pathogens and herbivores (Paulson, 1989). The importance of glycoproteins in the plasma membrane was described in Section 7.4.

**aliphatic amino acids**

$CH_2$—COOH, $NH_2$ — glycine (75.1)

$CH_3$CH—COOH, $NH_2$ — alanine (89.1)

$H_3C$, $H_3C$ CHCH—COOH, $NH_2$ — valine (117.1)

$H_3C$, $H_3C$ CHCH$_2$CH—COOH, $NH_2$ — leucine (131.2)

$CH_3CH_2$, $H_3C$ CHCH—COOH, $NH_2$ — isoleucine (131.2)

**basic amino acids**

$H_2N$, HN CNHCH$_2$CH$_2$CH$_2$CH—COOH, $NH_2$ — arginine (174.2)

$H_2NCH_2CH_2CH_2CH_2$CH—COOH, $NH_2$ — lysine (146.2)

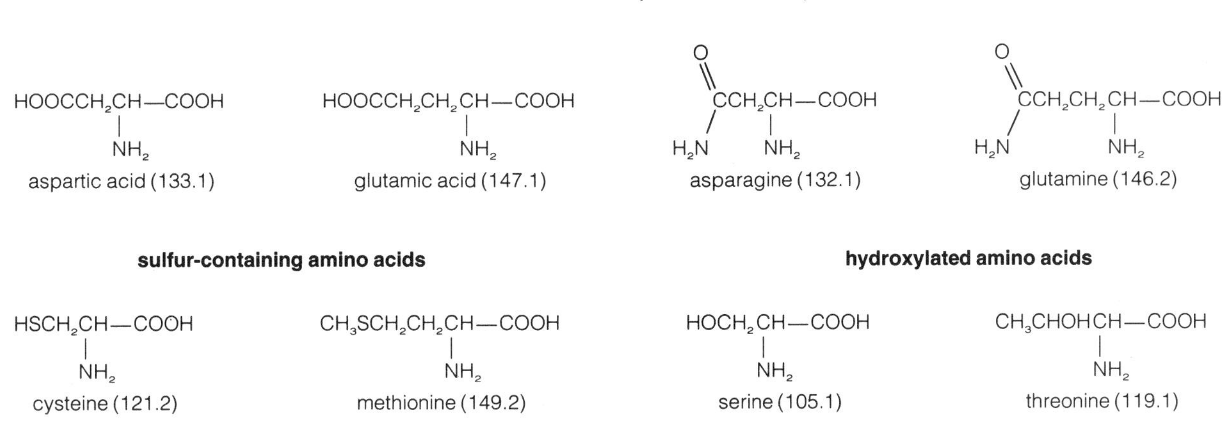


**heterocyclic amino acids**

$H_2C$—$CH_2$, $H_2C$ C—COOH, N H, H — proline (115.1)

CH$_2$CH—COOH, N H, $NH_2$ — tryptophan (204.2)

N—CH, HC C—CH$_2$CH—COOH, N H, $NH_2$ — histidine (155.2)

**aromatic amino acids**

HO—CH$_2$CH—COOH, $NH_2$ — tyrosine (181.2)

CH$_2$CH—COOH, $NH_2$ — phenylalanine (165.2)

**Figure 9-2** Molecular structures of 20 amino acids present in most proteins and their molecular weights (g mol$^{-1}$).

Many enzymes that do not have prosthetic groups require for activity the participation of another organic compound or a metal ion, or both. These substances are usually called **coenzymes**, although the metal ions are often called **metal activators**. In general, coenzymes and metal activators are not tightly held to the enzymes, but often no sharp distinction between coenzymes and prosthetic groups can be made. *Several vitamins synthesized by plants form parts of coenzymes or prosthetic groups required by enzymes in plants and animals, and this largely explains the reason vitamins are essential for life.* Several essential mineral elements also act as enzyme activators, but most mineral elements have a variety of functions (see Section 6.7).

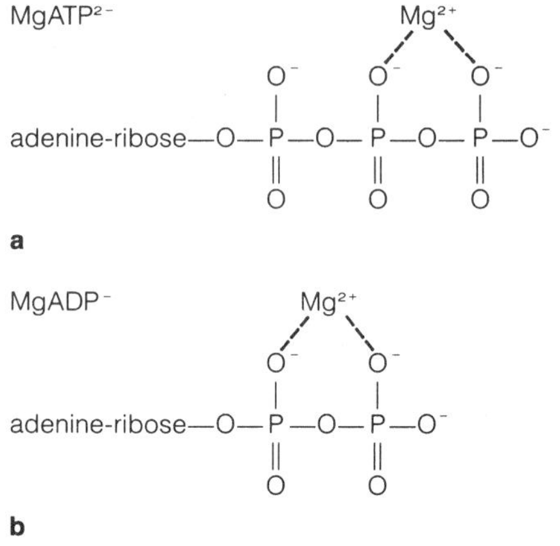

```
MgATP2-            Mg2+
                  /    \
          O-    O-      O-
          |     |       |
adenine-ribose—O—P—O—P—O—P—O-
          ||    ||      ||
          O     O       O
```

a

```
MgADP-        Mg2+
             /    \
           O-      O-
           |       |
adenine-ribose—O—P—O—P—O-
           ||      ||
           O       O
```

b

**Figure 9-3** The $Mg^{2+}$ chelate of (**a**) ATP and (**b**) ADP.

The magnesium ion acts as a metal activator for most enzymes that use ATP or other nucleoside di- or triphosphate as a substrate. A stable chelate between ATP and $Mg^{2+}$ is formed, probably having the structure shown in Figure 9-3a. The enzyme-substrate complex is then a Mg-ATP-enzyme complex. $Mg^{2+}$ also combines with ADP, as shown in Figure 9-3b. Furthermore, $Mn^{2+}$ can combine with ADP or ATP in a similar way, forming a chelate that is often as active as that formed with $Mg^{2+}$. This combination of cations with a substrate rather than with the enzyme is important for other divalent cations, but direct combination of certain enzymes with manganese, iron, zinc, copper, calcium, and potassium also occurs, as with iron and copper in the case of cytochrome oxidase.

## Amino-Acid Sequences

The number of different ways in which amino acids could theoretically be arranged in proteins is staggering. Consider as a simple example a small enzyme with a molecular weight of 13 kDa, consisting of 100 amino acids with an average molecular weight of 130. With the usual 20 different amino acids, the number of possible arrangements would be nearly $20^{100}$ ($10^{130}$). The number of different proteins of all sizes and kinds in nature does not even begin to approach this figure,[1] although estimates of the existence of as many as 40,000 unique proteins in plants have been made. Considering this, how all the various forms of life with different genes can be so different from each other is no longer puzzling. *We know that the arrangements of nucleotides in genes that code for these proteins determine their amino-acid sequences and therefore their functions*, as described in basic biology textbooks and in Appendix C.

Methods used to determine the amino-acid sequence in proteins by first hydrolyzing them into their constituent amino acids are tedious and require the availability of a pure protein. As a result, we currently know the complete sequences for relatively few distinct proteins (probably less than 100).[2] Such studies are important, however, because the results are necessary if we are to learn how enzymes catalyze reactions. Furthermore, comparison of sequences in proteins that have the same function in different organisms provides a powerful tool in evolutionary studies. For example, 38 ferredoxins from angiosperms, a liverwort, some horsetails, ferns, green algae, and certain photosynthetic and nonphotosynthetic bacteria have been sequenced. The similarities and differences have been used, with the help of computers, to construct tentative family trees (Minami et al., 1985; Schmitter et al., 1988). The same kinds of comparisons have been made with cytochrome c molecules from many organisms, including both primitive and advanced plants and animals. Sequences in hemoglobin have already provided several clues about animal evolution. Comparison of one of the histone molecules found in nuclei of both cattle and pea cells showed that the same amino acids occur in 100 of the 102 amino-acid positions present in each. These results suggest that many mutations in genes that control this protein have been eliminated by natural selection during the last 1.5 billion years or so; therefore, this protein and each amino acid in it play an important role in the lives of various organisms.

## Three-Dimensional Structures of Enzymes and Other Proteins

The simplest proteins consist of only one long polypeptide chain, but each such chain is usually coiled and twisted to form somewhat spherical or globular molecules. An example of a rather simple globular protein with 124 amino-acid residues in only one chain is the enzyme *ribonuclease* (which hydrolyzes RNA), shown in Figure 9-4. *The three-dimensional structure of any polypeptide chain or any protein with several polypeptide chains is determined by the kinds of amino acids present and the sequence in which they are arranged*. Each chain and the whole protein then spontaneously attain the lowest

[1] If each of the $10^{130}$ possible kinds of proteins with 100 amino acids occupied $10^3$ cubic nanometers, all $10^{130}$ taken together would fill our known universe $10^{27}$ times!

[2] In the past, determination of the amino-acid sequence in proteins was almost entirely done by hydrolyzing the protein in a stepwise procedure. Now that we understand the genetic code, it is easier to *deduce* the amino-acid sequence from the sequence of bases in a messenger RNA or DNA (gene) that codes for the protein. Base sequences in nucleic acids are easier to determine directly than are amino-acid sequences (see Chapter 24). Using the base-sequence technique, several thousand more proteins have been sequenced indirectly.

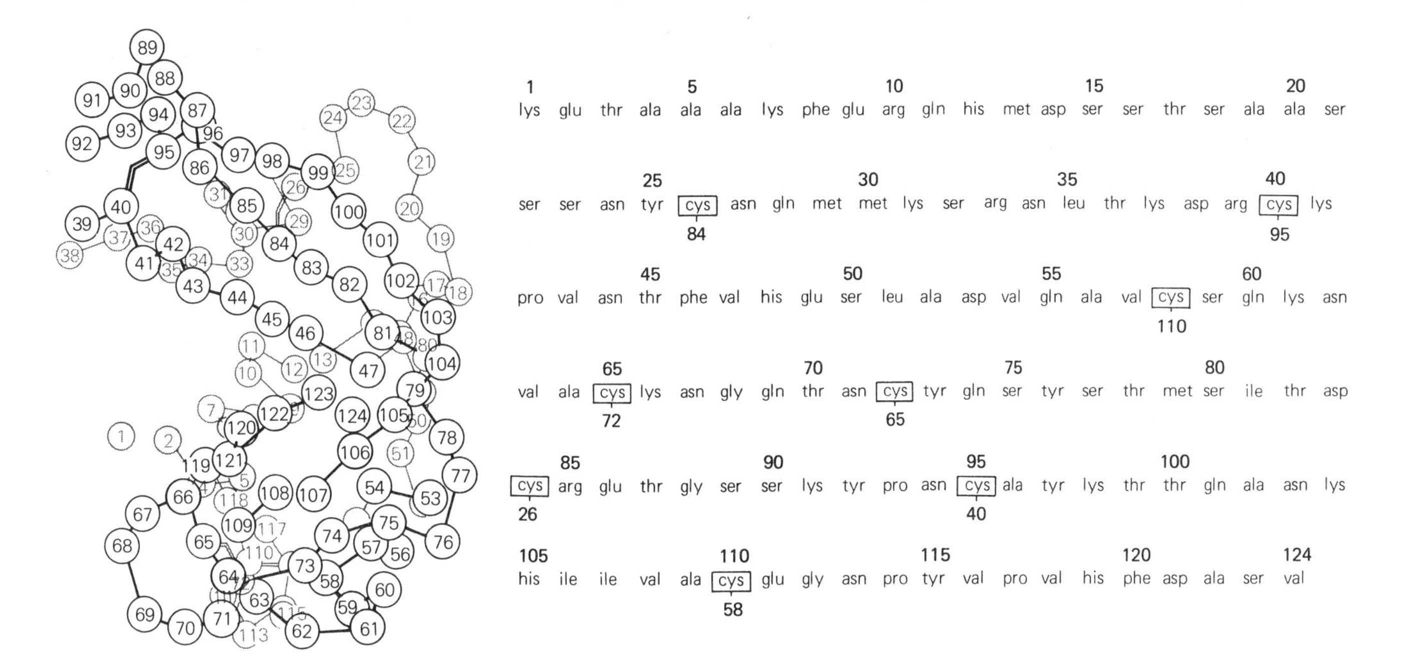


**Figure 9-4** Three-dimensional configuration of ribonuclease A. The active site of the enzyme is in the cleft at the left center of the molecule. The positions of individual amino acid residues are marked by the numbered circles, which correspond to the sequence list above. Gln and asn represent glutamine and asparagine. Disulfide bridges are indicated by a bent heavy line. Cysteine residues joined by disulfide linkages are enclosed in boxes in the sequence list. The number displayed below each box indicates the amino acid residue to which the disulfide linkage is joined. (From R. E. Dickerson and I. Geis, 1969. *The Structure and Action of Proteins*, W. A. Benjamin, Inc., Menlo Park, Calif. Copyright © 1969 by Dickerson and Geis.)

free-energy configuration (shape) consistent with its amino-acid composition and sequence at the existing cellular conditions of *p*H, temperature, ionic strength, and the presence of other nearby molecules. For each chain, this configuration is stabilized only after all the chains in it come together.

In the cytosol, the more-hydrophobic amino acids in a protein (such as valine, leucine, isoleucine, methionine, phenylalanine, and often tyrosine) become concentrated on the inside of the protein's folded structure, where they are shielded from water, whereas the more-hydrophilic amino acids (such as serine, glutamic acid, glutamine, aspartic acid, asparagine, lysine, histidine, and arginine) are more commonly exposed to the surface of the protein, where they are in contact with water of the cytosol. Within membranes, the hydrophobic amino acids of integral proteins usually associate with the hydrophobic long-chain fatty acids, whereas the hydrophilic amino acids of integral proteins congregate on either side of the membrane near water. The folding or coiling of a polypeptide chain in a protein (such that some portions of it stick to other portions) occurs partly because of stabilizing attractive forces between certain R groups that happen to contact each other in a long chain.

Another factor that affects folding or coiling is the repulsion of hydrophobic R groups by water, which brings these hydrophobic groups close together such that they contact each other and not water. Some important kinds of intrachain bonding that result from these forces are shown in Figure 9-5. Note especially the presence of **disulfide bonds (S-S)**. In complex proteins that have more than one polypeptide chain, such as ribulose bisphosphate carboxylase, bonds that hold one chain to another are similar to those that hold separate parts of the same chain together in its three-dimensional structure (Fig. 9-5). Therefore, each polypeptide chain is brought together into a final protein by bonds within each chain that contact other bonds in proximate chains. In some proteins all polypeptide chains are identical, in which case **homopolymers** are made. A good example of a homopolymer is *phosphoenolpyruvic acid carboxylase (PEP carboxylase)*, an enzyme involved in photosynthesis of certain species and in general metabolism of all plants. This enzyme contains four identical polypeptide chains (Huber et al., 1986).

Often the chains in proteins are different, and **heteropolymers** result. Each polypeptide chain in a heteropolymer is coded for by a different gene. Therefore, heteropolymers with two or more different polypeptide chains require the presence of two or more different corresponding genes in the cell. Ribulose bisphosphate carboxylase, mentioned earlier, is a good example of a heteropolymer. For it, the gene that codes for the smaller

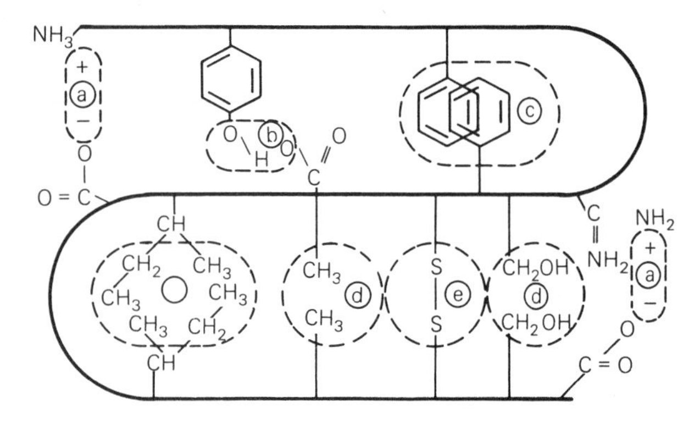

Figure 9-5 Probable types of bonding responsible for holding one polypeptide chain close to another. (a) Electrostatic attraction. (b) Hydrogen bonding. (c) Interaction of nonpolar side-chain groups caused by repulsion of each by water. (d) Van der Waals attractions. (e) Disulfide bonding between former —SH groups of cysteine molecules. (Modified after C. B. Anfinsen, 1959.)

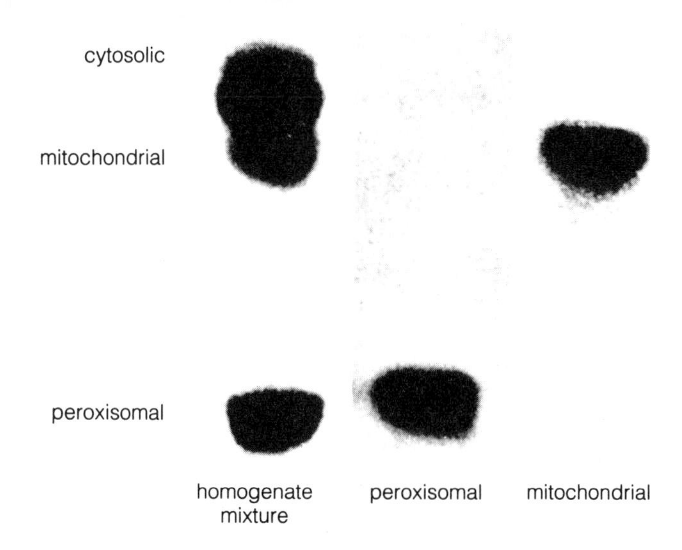


Figure 9-6 Separation of NAD$^+$-malate dehydrogenase isozymes of spinach leaves by starch gel electrophoresis. In the homogenate mixture at left, three isozymes are detectable. Two of these correspond to isozymes extracted from isolated peroxisomes and mitochondria; the third resides in the cytosol. (From Ting et al., 1975.)

polypeptide chain is a nuclear gene, whereas the gene that codes for the larger polypeptide chain is in the chloroplast.

## Isozymes

During the 1940s a technique called **electrophoresis** was developed to separate proteins, and this technique led to an important new discovery about many enzymes. Electrophoresis is essentially the separation of dissolved proteins or other charged molecules in an electrical field. A mixture of enzymes is placed in a buffered solution or in an inert medium such as a layer of starch gel or a column or slab of polacrylamide gel that is wetted with a buffer at a controlled *p*H. The various R groups in each enzyme ionize to an extent controlled by their chemical nature and the *p*H of the buffer chosen. For example, if the *p*H is 7, enzymes that are rich in aspartic and glutamic acid have a net negative charge because of their dissociated carboxyl groups. Alternatively, enzymes that are rich in lysine or arginine are more likely to be positively charged at this *p*H. Each enzyme in the mixture attains a different charge; and if these differences are great enough, the enzymes can be separated by an electrical current that flows from a negative electrode inserted at one end of the solution or gel to a positive electrode placed at the other end. The enzymes migrate in the electrical field, the distance they travel depending upon their net charge and size. After they have migrated, their positions in the gel can be detected, for example, by incubating the gel with the proper substrate and then chemically staining the reaction product.

When a particular enzyme is investigated by electrophoresis, it is often found that more than one stained zone appears in the gel, indicating the presence of more than one enzyme that can act on the same substrate and convert it to the same product. Such enzymes are referred to as **isozymes** or **isoenzymes**. Often, differences among isozymes result from the presence in an organism of more than one gene that codes for each isozyme; nevertheless, the amino-acid sequence differs only slightly from one isozyme to another. This is commonly true for isozymes that have only one kind of polypeptide chain. If only one polypeptide chain is present, two isozymes could result from genetic coding by each of the two allelic genes derived from a different parent. Alternatively, isozymes that are heteropolymers can also differ in the kinds of polypeptide chains they contain. Several other genetic possibilities that result in isozymes exist.

One importance to an organism of having different isozymes capable of catalyzing the same reaction is that their isozymes differ somewhat in their responses to various environmental factors. What this means is that if the environment changes, the isozyme most active in that environment performs its function and helps the organism survive. In addition, one isozyme often exists in one tissue or organ and another in a different tissue or organ with a different function. Different isozymes can sometimes be found even within the same cell. Figure 9-6 illustrates the separation of three isozymes of malate dehydrogenase from various organelles in spinach leaf-mesophyll cells: one isozyme from mitochondria, one from peroxisomes, and one from the cytosol. Each isozyme is exposed to a different chemical

## Plant Proteins and Human Nutrition

Humans and other animals depend on plants for many of their amino acids, so the composition of seed, leaf, and stem proteins is important in diet. We and other animals use these amino acids to build our own proteins and as a food (energy) source. Although adult humans can synthesize most of the amino acids they require from carbohydrates and various organic nitrogen compounds, eight amino acids must be provided in the diet. These are leucine, isoleucine, valine, lysine, methionine, tryptophan, phenylalanine, and threonine. Furthermore, adequate amounts of the sulfur-containing amino acid cysteine can apparently be formed only when sufficient methionine (another S-amino acid) is provided, and we use phenylalanine to synthesize tyrosine, which otherwise would be essential.

Most of the proteins in the human diet come from seed proteins, especially those of the cereal grains rice, wheat, and corn (maize). Approximately two-thirds of the world's population depends on wheat or rice as the principal source of calories and protein. Maize is important in many tropical and subtropical parts of Central and South America. A smaller but still important contribution is made by legume seeds such as beans, peas, and soybeans. Soybeans are an unusually rich, fairly well-balanced protein source; about 40 percent of their dry weight is protein compared to about 12 percent for most cereal grains (Table 9-2).

Compared with most animal proteins, cereal grain proteins are low in lysine, whereas legume seeds are low in methionine. For example, the lysine content of total protein from seeds of 12,561 wheat cultivars averaged 3.14 percent on a weight basis compared with 6.4 percent in whole egg protein (Table 9-2). Bean seed proteins averaged only 1.0 percent methionine compared with 3.1 percent in egg proteins. Plant breeders are making some progress in introducing new cultivars or hybrid species with increased protein contents and increased percentages of essential amino acids. Examples include the opaque-2 and fluory-2 cultivars of maize, both of which are considerably richer in both lysine and tryptophan than are commonly grown cultivars (Harpstead, 1971; Larkins, 1981; Payne and Rhodes, 1982; Doll, 1984).

**Table 9-2 Protein Content and Amino-Acid Composition of Selected Food Legumes and Cereals.**

| | | Amino-Acid Composition (Percent of Total Protein) | | | | | | | | |
|---|---|---|---|---|---|---|---|---|---|---|
| Food | Percent Protein | Lysine | Methionine | Threonine | Tryptophan | Isoleucine | Leucine | Tyrosine | Phenylalanine | Valine |
| Soybean | 40.5 | 6.9 | 1.5 | 4.3 | 1.5 | 5.9 | 8.4 | 3.5 | 5.4 | 5.7 |
| Peas | 23.8 | 7.3 | 1.2 | 3.9 | 1.1 | 5.6 | 8.3 | 4.0 | 5.0 | 5.6 |
| Beans | 21.4 | 7.4 | 1.0 | 4.3 | 0.9 | 5.7 | 8.6 | 3.9 | 5.5 | 6.1 |
| Oats | 14.2 | 3.7 | 1.5 | 3.3 | 1.3 | 5.2 | 7.5 | 3.7 | 5.3 | 6.0 |
| Barley | 12.8 | 3.4 | 1.4 | 3.4 | 1.3 | 4.3 | 6.9 | 3.6 | 5.2 | 5.0 |
| Wheat | 12.3 | 3.1 | 1.5 | 2.9 | 1.2 | 4.3 | 6.7 | 3.7 | 4.9 | 4.6 |
| Rye | 12.1 | 4.1 | 1.6 | 3.7 | 1.1 | 4.3 | 6.7 | 3.2 | 4.7 | 5.2 |
| Sorghum | 11.0 | 2.7 | 1.7 | 3.6 | 1.1 | 5.4 | 16.1 | 2.8 | 5.0 | 5.7 |
| Maize | 10.0 | 2.9 | 1.9 | 4.0 | 0.6 | 4.6 | 13.0 | 6.1 | 4.5 | 5.1 |
| Rice | 7.5 | 4.0 | 1.8 | 3.9 | 1.1 | 4.7 | 8.6 | 4.6 | 5.0 | 7.0 |
| Whole egg | 12.8 | 6.4 | 3.1 | 5.0 | 1.7 | 6.6 | 8.8 | 4.3 | 5.8 | 7.4 |

Source: Data from Orr and Watt, 1957 and Johnson and Lay, 1974.

environment within the cell, and each participates in a different sequence of reactions (metabolic pathway). *Therefore, within individual organelles of any organism and within individual cells or tissues of any organism, the presence of more than one isozyme usually confers advantages to coping with environmental changes.*

### Isoforms of Enzymes: Post-Translational Modifications

After enzymes are synthesized during translation, (Appendix C), they are subject to minor chemical changes that can profoundly affect their catalytic ability. These changes cause production of **isoforms**. Isoforms differ from isozymes by being coded for by the same gene and by having the same amino-acid sequence. They differ only subtly from isozymes. Several kinds of modifications cause isoforms to be produced, including **phosphorylation** (in which a phosphate is esterified to the hydroxyl group of one or more serines or threonines present), **glycosylation** (in which one or more sugars are attached), and **methylation** or **acetylation** (in which one or more methyl or acetyl groups are added). Such modifications can even occur in the same cell at different stages in its development. They allow yet another kind of control over the kind of reactions occurring in a given cell or part of a cell at a particular stage in its life cycle or at the same stage but in a different environment. Lists of enzymes known to be phosphorylated and dephosphorylated (and how such modifications affect their activities) are given by Ranjeva and Boudet (1987) and by Budde and Chollet (1988). In plants, some environmental signal such as light often causes phosphorylation or dephosphorylation of certain proteins, so that the plant responds by modifying a preformed enzyme, which changes the enzyme's activity, and the plant responds better to the environment (Chapters 11, 13, 17, and 20).

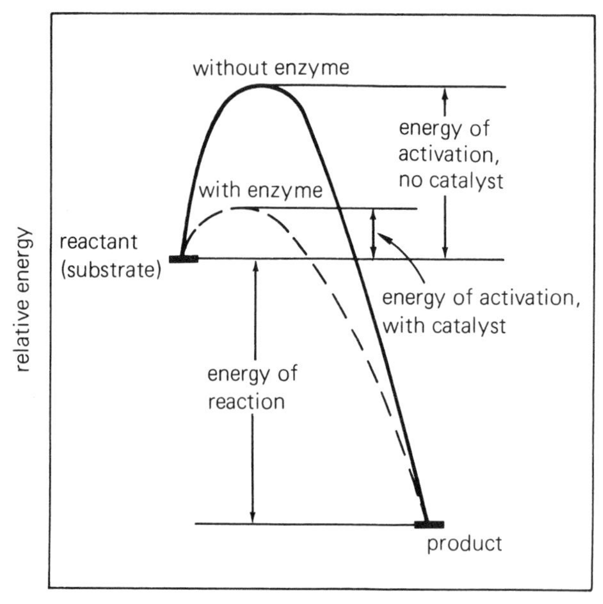


**Figure 9-7** Energy diagram for a metabolic reaction occurring in the presence and absence of an enzyme. Reacting substrate molecules must pass over an "energy hump" (accumulate activation energy) to allow formation of new chemical bonds present in the product, even though the product may be at a lower free energy level than the substrate. A catalyst, such as an enzyme, lowers the activation energy required, thus increasing the fraction of molecules that can react in a given time.

## 9.3 Mechanisms of Enzyme Action

Usually only the most energetic molecules are able to undergo chemical reactions. Such molecules temporarily become more energetic than others of the same kind by being subjected to different numbers and types of collisions. If we could analyze the energies in a population of such molecules, statistical predictions indicate that a distribution similar to the hypothetical values depicted in Figure 2-3 would be found. The curves in that figure show that a temperature rise of 10°C greatly increases the number of molecules that have relatively high energies. Note that the higher-temperature curve is slightly more skewed to the right than the lower curve. Suppose that only those few molecules that occur within the shaded areas of the figure are energetic enough to react in the absence of enzymes. Molecules that have energy equal to or greater than that indicated by the arrow have reached what we call the **energy of activation**. Note that in this example the area under the higher-temperature curve is about twice as large as that under the lower-temperature curve. Therefore, twice as many molecules will react in a given period of time at the higher temperature. As a result, a rise in temperature will double reaction rates, unless the rise is too great and causes cell death. If a doubling occurs, the $Q_{10}$ value is 2.0. ($Q_{10}$ values are explained in Section 2.3.)

But how do enzymes increase reaction rates? Do they cause a shift in the frequency distribution similar to that caused by a temperature increase? The answer is no, but to understand how they act we must consider another aspect of the problem. Figure 9-7 shows energy changes during substrate conversion to products. Here energy is plotted on the ordinate, and the maximum energy that must be obtained again represents the energy of activation. This maximum represents an energy barrier that must be overcome. Note that enzymes lower the activation energy so that a much greater fraction of the substrate molecules has sufficient energy to react without a temperature increase. In other words, enzymes do not shift the curves of Figure 2-3, but in their presence the shaded area of the figure is displaced far to the left.

For an enzyme to decrease the activation energy during the course of the reaction, the enzyme must combine temporarily with the substrate or substrates to form an **enzyme-substrate complex**. The electrical forces within this complex change the shape of the substrate, causing some of the substrate bonds to break; then the bonds rearrange to form products much more rapidly than in the absence of the enzyme.

The enzyme-substrate complex, as first hypothesized by the great organic chemist Emil Fischer in about 1884, assumed a rigid lock-and-key union between the enzyme and substrate (Fig. 9-8a). The portion of the enzyme to which the substrate (or substrates) combines

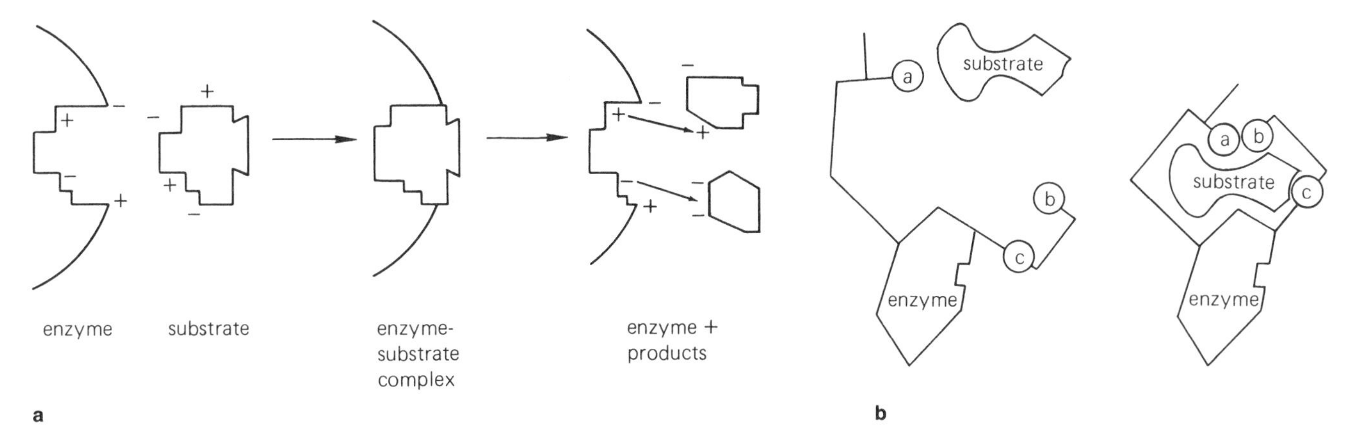


**Figure 9-8** (**a**) The lock-and-key model of the active site, as hypothesized by Emil Fischer. The active site is considered to be a rigid arrangement of charged groups that precisely matches complementary groups of the substrate. (**b**) A modified conception of the active site, as advanced by D. E. Koshland. Here the catalytic groups a and b must be aligned, but the orientation of these groups is altered by the approaching substrate, resulting in a better fit. (From Wolfe, 1981.)

as the substrate undergoes conversion to a product is called the **active site**. If the active site were rigid and specific for a given substrate, reversibility of the reaction would not occur because the structure of the product is different from that of the substrate and would not fit the enzyme well. As contrasted to a rigidly arranged active site, Daniel E. Koshland, Jr. (1973) found evidence that the active site of enzymes can be induced by close approach of the substrate or substrates (or product or products, when the reaction reverses) to undergo a change in conformation (shape) that allows a better combination. This idea is now widely known as the **induced-fit hypothesis** and is illustrated in Figures 9-8b and 9-9 for an enzyme with only one substrate. If two or more substrates are involved, the principle is the same, with each substrate contributing. Apparently the structure of the substrate is also changed during many cases of induced fit, allowing a more functional enzyme-substrate complex.

The kinds of bonds between enzymes and substrates can be covalent, ionic, hydrogen, and van der Waals. The covalent and ionic bonds are most important with respect to the activation energy for a reaction, but the more numerous hydrogen bonds and van der Waals interactions contribute to the structural orientation of the enzyme-substrate complex. Even when strong covalent bonds are formed, they are usually very rapidly broken to release new product molecules. Both covalent and noncovalent bonds between enzyme and substrate are formed between parts of the R portions of amino-acid residues in the enzyme, not between the carbon and nitrogen atoms involved in the peptide bonds. Thus enzymes must have the proper amino acids (composition) and must have them in the right places (sequence).

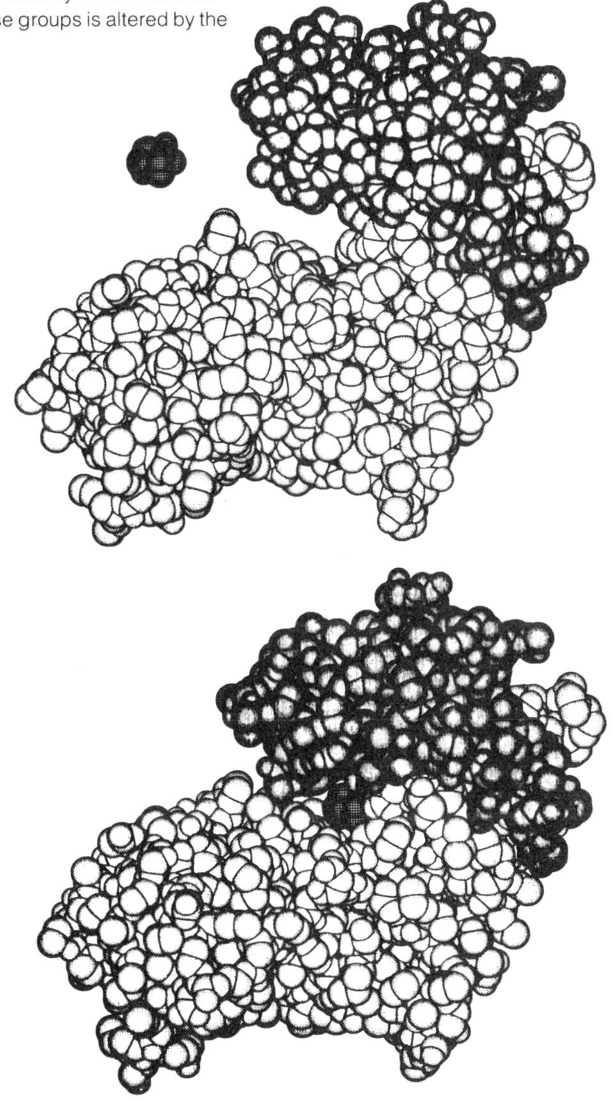

**Figure 9-9** Induced fit of yeast hexokinase and glucose, one of its substrates. Without glucose (top), hexokinase has an open cleft that partially closes when glucose is bound (bottom). (From Bennett and Steitz, 1980, pp. 211–230. With permission from the *Journal of Molecular Biology*, copyright 1980, by Academic Press Inc. [London] Limited.)

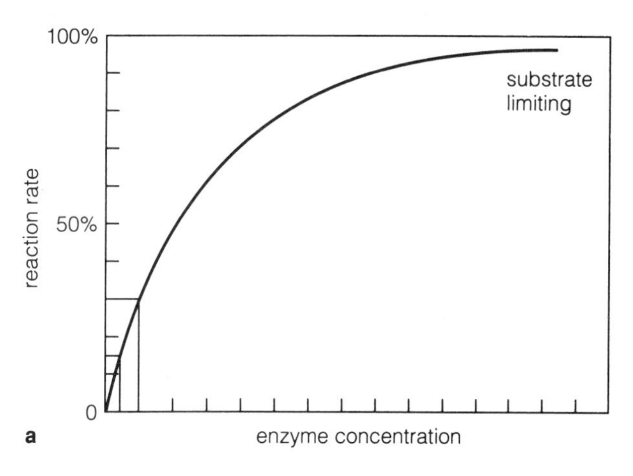


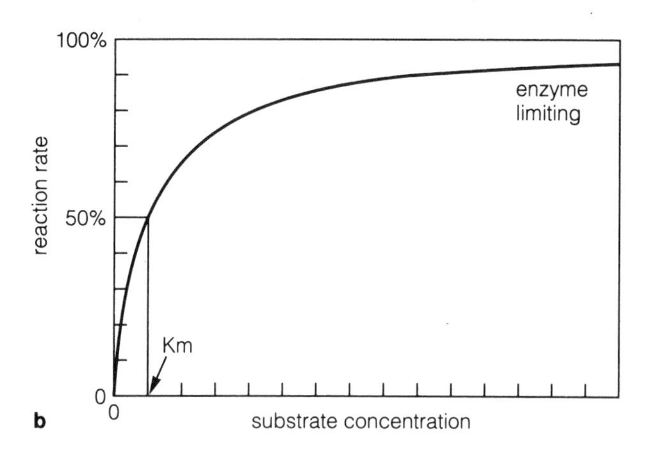


**Figure 9-10** (**a**) Effects of enzyme concentration on rate of reaction when substrate concentration is held constant. (**b**) Effect of substrate concentration on reaction rate when enzyme concentration is held constant.

## 9.4 Denaturation

The previous discussion of enzyme-substrate complexes and the three-dimensional structures of proteins implies that if the structure of an enzyme is altered so that the substrate can no longer bind with it, catalytic activity will be eliminated. Numerous factors cause such alterations and are said to cause enzyme **denaturation**. In many cases, denaturation is irreversible. High temperatures easily break hydrogen bonds and often cause irreversible denaturation; you can't unboil an egg! Extreme heating causes the formation of new covalent bonds between different polypeptide chains or between parts of the same chain, and these bonds are so stable that they break at negligible rates.

Cold temperatures are nearly always maintained during extraction and purification of enzymes to prevent heat denaturation. This is true even though enzymes normally remain undenatured in cells at higher temperatures. We do not fully understand why purification of enzymes at temperatures identical to those at which cells normally exist causes denaturation, but we suspect that extraction and purification procedures remove or dilute substances that normally protect the enzymes. Alternatively, homogenization of cells often releases and allows exposure of enzymes to denaturing substances from subcellular compartments (for example, vacuoles) that *in vivo* are prevented by membranes from contacting such enzymes. A few enzymes are known to be inactivated by low temperatures during purification. Again, a change in structure is the cause.

Oxygen and other oxidizing agents also denature numerous enzymes, often by causing disulfide bridges to be formed in chains in which the —SH groups of cysteine are normally present. Reducing agents can denature in the opposite manner, by breaking disulfide bridges to form two —SH groups. Heavy metal cations such as $Ag^+$, $Hg^{2+}$, or $Pb^{2+}$ can denature enzymes, often by replacing the H on an —SH group; it is for this reason that great concern has developed about the presence of heavy metals in the environment. Many organic solvents also denature enzymes.

When enzymes are dehydrated, they are much less susceptible to heat denaturation than when they are hydrated. This is the main reason why dry seeds and dry fungal or bacterial spores can resist high temperatures and why the presence of steam in autoclaves causes sterilization more quickly than a dry oven at the same temperature. The dry state also prevents enzyme denaturation caused by freezing during winter in seeds, buds, and other parts of perennial shrubs and trees.

## 9.5 Factors Influencing Rates of Enzymatic Reactions

### Enzyme and Substrate Concentrations: Either Can Be Limiting

Catalysis occurs only if enzyme and substrate form a transient complex. The reaction rate depends on the number of successful collisions between them, which in turn depends upon their concentrations. If enough substrate is present, doubling the enzyme concentration usually causes a twofold increase in rate (Fig. 9-10a). With the addition of still more enzyme, the rate begins to become constant because substrate becomes limiting.

Figure 9-10b shows the effect of substrate concentration on the reaction rate when the enzyme concentration is held constant. There is usually an approximately direct proportionality between rate and substrate concentration until the enzyme concentration becomes limiting. At this substrate concentration, the addition of more substrate causes no further rise in the reaction rate because nearly all the enzyme molecules have combined with substrate. When this occurs, no more enzyme active sites are available to cause catalysis. To increase the speed of the reaction then requires the addition of more enzyme.

Figure 9-10b also illustrates another useful fact about enzymes: the substrate concentration required to cause half the maximal reaction rate, a value named the **Michaelis-Menten constant ($K_m$).**[3] $K_m$ values are more or less constants independent of the amount of enzyme present, at least within reasonable limits. The values vary somewhat with *p*H, temperature, and ionic strength, and also with the kinds or amounts of coenzymes present when these are required. Most enzymes studied so far have $K_m$ values between $10^{-3}$ and $10^{-7}$ molar (1 mM and 0.1 $\mu$M), although exceptions exist. If an enzyme catalyzes a reaction between two or more different substrates, it will have a different $K_m$ value for each.

Certain advantages result from knowing the $K_m$ value for an enzyme of interest. First, if we can measure the concentration of the substrate in that part of the cell in which the enzyme occurs, we can predict whether the cell needs more enzyme or more substrate to speed up the reaction. Some studies indicate that most enzymes involved in respiration are usually not saturated by their substrates. The same is probably true with many photosynthetic enzymes in leaves exposed to bright light because $CO_2$ fixation can be increased by increasing the concentration of this gas around such leaves. Second, $K_m$ values represent approximately inverse measures of the affinity of the enzyme for a given substrate; thus the lower the $K_m$ the more stable the enzyme-substrate complex. In cases in which an enzyme can catalyze reactions with two similar substrates (for example, glucose and fructose), the substrate for which the enzyme has the lower $K_m$ is the one most frequently acted upon in the cell. And third, the $K_m$ gives an approximate measure of the concentration of the enzyme's substrate in the part of the cell in which reaction occurs. For example, enzymes that catalyze reactions with relatively concentrated substrates such as sucrose usually have relatively high $K_m$ values for these substrates, and enzymes that react with hormones or other substrates that are present in very low concentrations have much lower $K_m$ values for their substrates. The concepts of the Michaelis-Menten constant and the equations concerning it are applied in the essay by Emanuel Epstein in Chapter 7 and also in Section 19.5, Figure 19-17.

[3]In practice, $K_m$ values can seldom if ever be accurately determined by extrapolating the reaction rate to its maximum because the rate approaches the maximum asymptotically. There are methods to plot the data differently to overcome this problem. The most popular method is the Lineweaver-Burk technique described in many biochemistry textbooks. In this technique a straight line results when one plots the reciprocal of the reaction rate as a function of the substrate concentration; the line intersects the ordinate and is extrapolated downward to the left of the ordinate until it encounters the abscissa at a negative reciprocal substrate concentration. That intersection provides the reciprocal of $K_m$. Naqui (1986) pointed out that the Michaelis-Menten equation (see Fig. 19-17) is a rectangular hyperbola, so analytic geometry can be used to determine both $K_m$ and the maximum reaction rate ($V_{max}$). Gannon (1986) showed that a semi-log plot of the data (in which substrate concentrations are plotted logarithmically) allows easy determination of $K_m$ and $V_{max}$. This method is used in Figure 19-17.

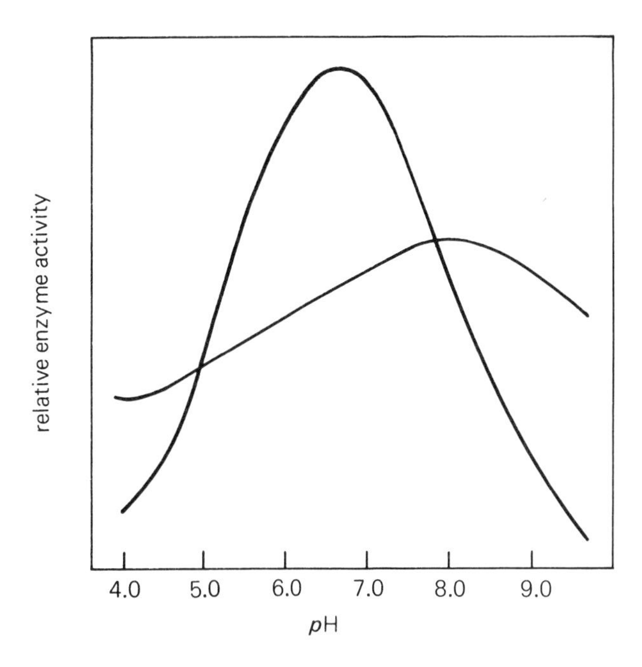


**Figure 9-11** Influence of *p*H upon activity of two different enzymes. The *p*H optimum and shape of the curve vary greatly among enzymes and depend upon reaction conditions.

## *p*H

The *p*H of the medium influences enzyme activity in various ways. Usually there is an optimum *p*H at which each enzyme functions, with decreased activity at higher or lower *p*H values. Sometimes plots of enzyme activity versus *p*H, such as those in Figure 9-11, are almost bell-shaped, whereas for other enzymes the curves may be almost flat. The optimum *p*H is often between 6 and 8, but it can be higher or lower for some enzymes. Extremes in *p*H usually cause denaturation. Many enzymes probably never function *in vivo* at their *p*H optima.

Apart from denaturation effects, *p*H can influence reaction rates in at least two ways. First, enzyme activity often depends upon the presence of free amino or carboxyl groups. These can be either charged or uncharged, depending on the enzyme, but only one form

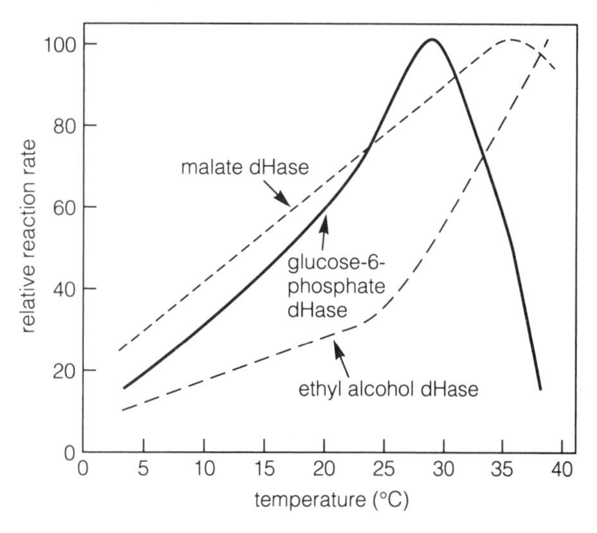


**Figure 9-12** Effects of temperature on rates of reactions catalyzed by three dehydrogenases (dHase) extracted from hypocotyls of soybean seedlings. Note the variable effects of temperature on the activity of each enzyme and that the dehydrogenase for ethyl alcohol functioned continuously better as the temperature rose to 38°C. At higher temperatures even this enzyme is rapidly denatured. (Drawn from data of Duke et al., 1977.)

is presumed effective in a given case. If an uncharged amino group is essential, the *p*H optimum will be relatively high, whereas a neutral carboxyl group requires a low *p*H. Second, the *p*H controls the ionization of many substrates, some of which must be ionized for the reaction to proceed.

## Temperature

As opposed to mammals and birds, plants cannot regulate their temperatures. As a result, all reactions in them are strongly influenced by external temperatures. In general, reactions catalyzed by enzymes increase with temperature from 0°C to 35 or 40°C. $Q_{10}$ values are commonly 2 to 3 over the range from 0°C to 30°C, partly because heat increases the number of molecules that have energy equal to or greater than the energy of activation already discussed. Yet because the reaction rate so strongly depends on catalysis by the enzyme, temperature also affects reactions by changing the shape of the enzyme. The enzyme's shape determines its ability both to combine with the substrate (effects on $K_m$) and to cause catalysis (effects on maximal reaction rate). Various enzymes, even from a single species, often differ greatly in their response to temperature (Fig. 9-12). This means that at any given temperature some enzymes function optimally or nearly so, whereas many others function far less than optimally. An organism's growth and reproduction vary greatly with temperature, and for any given species this may depend on the optimum temperature for action of certain enzymes that control rate-limiting reactions (Burke et al., 1988).

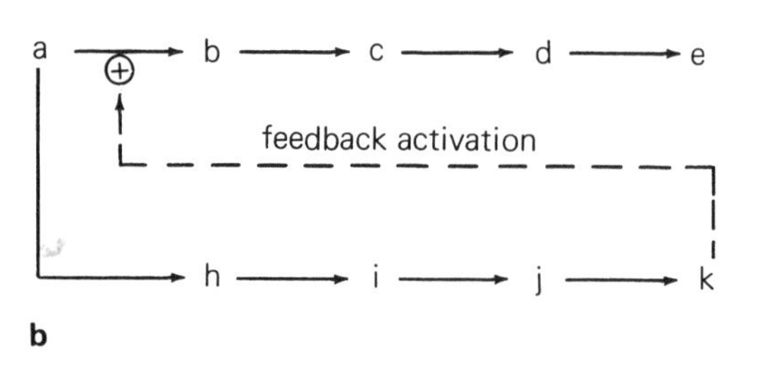


**Figure 9-13** (**a**) Feedback inhibition; (**b**) feedback activation.

Data in Figure 9-13 for malate dehydrogenase and glucose-6-phosphate dehydrogenase illustrate another common observation. If the temperature is raised above a certain value (which varies with the enzyme), the reaction rate begins to decrease. Why is this so? After all, at higher temperatures more substrate molecules will be excited than at 25 or 30°C. The answer is enzyme denaturation. Above 35 or 40°C denaturation of most plant enzymes occurs rapidly, so at high temperatures there is no effective catalyst to lower the activation energy, and not enough substrate molecules have sufficient energy to react without a catalyst.

Low temperatures can also denature certain enzymes, as mentioned in Section 9.4. This phenomenon appears to be especially important in species that are sensitive to chilling, a subject discussed in Section 26.5. It seems reasonable as well to wonder whether differences in temperature optima for enzymes determine the environment in which different species live. For example, the overall temperature optimum for the process of photosynthesis (catalyzed by many enzymes) in alpine and arctic plants is often 10 to 15°C, whereas the optimum for maize plants is near 30°C. So far there are few data that show distinct differences in temperature optima for enzymes extracted from plants that grow at different temperatures (Patterson and Graham, 1987), which probably means that extraction and purification of enzymes separates them from other cellular compounds that in nature (but not *in vitro*) strongly modify their response to temperature.

## Reaction Products

The rate of an enzymatic reaction can be determined by measuring the rate of disappearance of substrate or the rate of product appearance, or both. By either method the reaction is usually observed to proceed more slowly as time passes. This rate decrease is sometimes caused by denaturation of the enzyme while the reaction is being measured, but other factors are also involved. One of the most important factors is the continuous decrease in concentration of substrate or substrates and the accumulation of products. As products accumulate, their concentrations sometimes become high enough to cause appreciable reversibility of the reaction, provided that the relative chemical potentials of products and reactants allow reversibility. In some cases products can inhibit the forward reaction by combining with the enzyme in such a way that further formation of the enzyme-substrate complex is inhibited.

### Inhibitors

Many "foreign" substances can block the catalytic effects of enzymes. Some are inorganic, such as several metal cations, and some are organic. Both types are usually classified according to whether their effect is competitive or noncompetitive with the substrate. **Competitive inhibitors** usually have structures sufficiently similar to the substrate that they are able to compete for the active site of the enzyme. When such a combination of enzyme and inhibitor is formed, the concentration of effective enzyme molecules is lowered, decreasing the reaction rate. The inhibitor itself sometimes undergoes a change caused by the enzyme, but that change is not essential for inhibition. Addition of more of the natural substrate overcomes the effect of a competitive inhibitor. A classic example of competitive inhibition is caused by **malonate** ($^{-}OOC-CH_2-COO^{-}$), the doubly charged anion of malonic acid, on the action of **succinate dehydrogenase**. This enzyme functions in mitochondria to carry out an essential reaction of the Krebs cycle (see Chapter 13). It removes two H atoms from succinate and adds these to its covalently bound prosthetic group flavin adenine dinucleotide (FAD), forming fumarate and enzyme-bound $FADH_2$:

$$\begin{array}{lcl} COO^- & & COO^- \\ | & & | \\ CH_2 & + \text{enzyme—FAD} \rightleftharpoons & CH \quad + \text{enzyme—}FADH_2 \\ | & & \| \\ CH_2 & & HC \\ | & & | \\ COO^- & & COO^- \\ \text{succinate} & & \text{fumarate} \end{array} \qquad (9.2)$$

Malonate combines reversibly with the enzyme in place of succinate, but because hydrogen removal cannot occur, no reaction takes place. In this way, succinate dehydrogenase molecules bound to malonate are unable to catalyze the normal dehydrogenation of succinate, and respiration is poisoned. Interestingly, beans and certain other legumes contain unusually high concentrations of malonate, probably in the central vacuole, where it cannot affect respiration. Another much more powerful inhibitor of succinate dehydrogenase is oxaloacetate ($^{-}OOC-CH_2-CO-COO^{-}$), a normal intermediate in the Krebs cycle (Burke et al., 1982). Oxaloacetate poisoning of succinate dehydrogenase is probably prevented because an unusually low concentration of oxaloacetate exists in mitochondria.

**Noncompetitive inhibitors** also combine with enzymes but at locations other than the active site. This effect is not overcome by simply raising the substrate concentration, rather only by removing the inhibitor. Noncompetitive inhibitors generally show less structural resemblance to the substrate than do competitive inhibitors. Toxic metal ions and compounds that combine with or destroy essential sulfhydryl groups often are noncompetitive inhibitors. For example, excess $O_2$ can oxidize $-SH$ groups that are close to each other, removing the H atom from each and forming new disulfide bridges. This changes the structure of the enzyme so that its active site can no longer combine well with the substrate or substrates. Heavy $Hg^{2+}$ ions can replace the H atom on a sulfhydryl group, forming heavy mercaptides that are often insoluble; $Ag^{+}$ ions act similarly. Mercury and silver cause noncompetitive inhibition of enzymes that is essentially irreversible because once such metals attach to enzymes they are almost impossible to remove.

Most poisons affect plants and animals by inhibiting enzymes. Some of these poisons will be discussed later in relation to the specific processes that are affected. Enzymes are also inhibited noncompetitively by any protein denaturant, such as strong acids or bases, or by high concentrations of urea or detergents, which break hydrogen bonds. A relatively simple treatment of these and other aspects of enzyme kinetics and enzyme inhibition is given by Engel (1977).

## 9.6 Allosteric Enzymes and Feedback Control

We have mentioned that numerous foreign ions or molecules can inhibit enzymatic action, in most cases by altering the configuration of the enzyme so that it cannot effectively form a complex with the substrate. However, several enzymes can also be altered by normal cellular constituents, with resulting decreases or increases in their functions. Such effects are important mechanisms for homeostatic control at the metabolic level; that is, they help organisms produce only the proper amount of the compounds they need. The most common case is inhibition of a particular reaction by a metabolite that is chemically unrelated to the substrate with which the enzyme reacts.

To understand this, consider an example in which a compound A is converted by a series of enzymatic reactions via intermediates B, C, D, and E to an essential

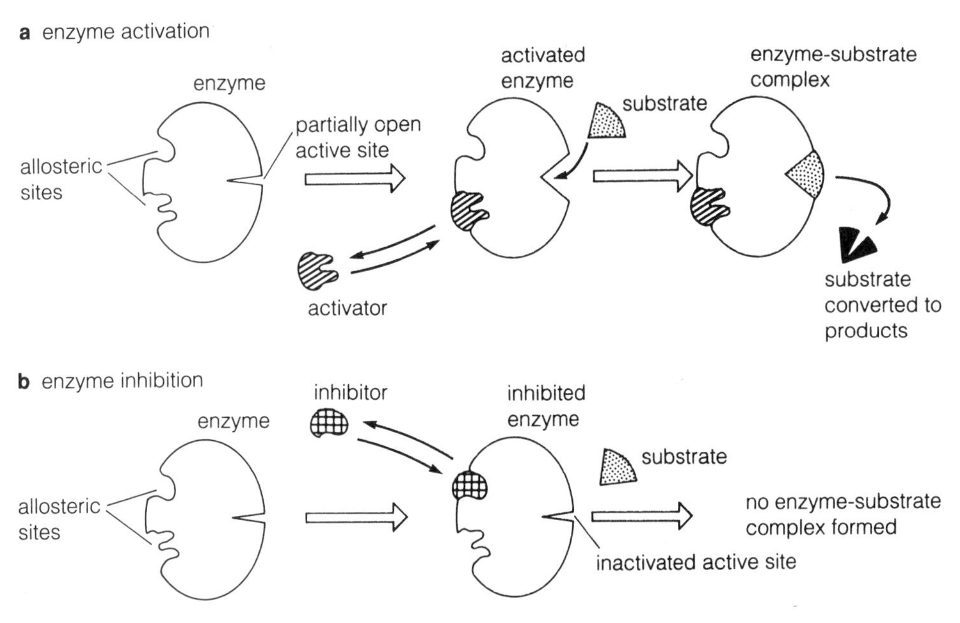


**Figure 9-14** A hypothetical model illustrating how the presence of activators and repressors might influence allosteric enzymes, thus affecting reaction rates. In (**a**), attachment of the activator to the enzyme opens the active site, allowing it to combine with the substrate. In (**b**), attachment of the inhibitor closes the active site, preventing attachment of the substrate.

product F (Fig. 9-13a). After this number of reactions, compound F no longer bears much structural resemblance to A. Nevertheless, F can sometimes reversibly combine with the first enzyme to inhibit its combination with A. This is an example of **feedback inhibition** or **end-product inhibition**. Its advantage is that it provides a rapid and sensitive mechanism to prevent oversynthesis of compound F because the feedback inhibition occurs only after F has built up to a level that is sufficient for cellular needs. Later, when the amount of F in the cell has been reduced (say, by incorporation into a structural component of the cell), F molecules dissociate from enzyme number 1 and allow it to become active again. Cases of feedback inhibition nearly always involve action of a product of a metabolic pathway upon the first enzyme of that pathway. A well-studied example of feedback inhibition in plants occurs in formation of the nucleotide *uridine monophosphate (UMP)*, which begins with aspartic acid and carbamyl phosphate. The pathway requires five enzyme-catalyzed steps, but only the first enzyme, *aspartic transcarbamylase*, is susceptible to feedback control by UMP; no other reaction is blocked in a similar way by reactants and products in the pathway (Ross, 1981).

To understand cases in which a metabolic process increases activity of an enzyme, consider the situation in which another compound, a, is converted by an enzyme to compound b in a series of reactions that eventually leads to e; yet a is also acted upon by a competing enzyme to initiate a second reaction leading to product k (Fig. 9-13b). Here the cellular levels of e and k depend on the relative activities of the first enzymes in their separate pathways of formation. Oversynthesis of k is prevented by its activation of the competing enzyme that converts a to b. Other more complicated kinds of feedback loops are known, primarily in bacteria (Ricard, 1980, 1987).

Enzymes that combine with and respond (either negatively or positively) to small molecules such as F or k are called **allosteric enzymes**. The sites at which combination with the smaller molecules occurs are called **allosteric sites** (*allo* means "other"; that is, different from the active site). Sometimes an allosteric site is on a polypeptide chain that is different from the chain containing the active site. The small molecules undergoing reversible binding to allosteric sites are called **allosteric effectors**. One result of allosteric binding is to lock the enzyme into a different configuration such that its $K_m$ for the normal substrate is either increased or decreased. A second result is a change in the maximum reaction rate without a change in the $K_m$. Figure 9-14 shows a diagram of how a single enzyme with two different allosteric sites could be activated by one allosteric effector and inhibited by another. The activation case is another example of induced fit, but here it is an allosteric effector instead of the substrate that alters the shape of the enzyme so that it now combines more readily with the substrate.

# 10

# Photosynthesis: Chloroplasts and Light

Photosynthesis is essentially the only mechanism of energy input into the living world. The only exceptions occur in chemosynthetic bacteria that obtain energy by oxidizing inorganic substrates such as ferrous ions and sulfur dissolved from the earth's crust, or by oxidizing $H_2S$ released from volcanic action. In addition, thermal vents on the ocean floor put energy into biological systems as heat. Because of its importance to life, we shall devote three chapters to photosynthesis.

Like energy-yielding oxidation reactions upon which all life depends, photosynthesis involves oxidation and reduction. The overall process is an oxidation of water (removal of electrons with release of $O_2$ as a by-product) and a reduction of $CO_2$ to form organic compounds such as carbohydrates. (The reverse of this process—the combustion or oxidation of gasoline or carbohydrates in wood to form $CO_2$ and $H_2O$—is a spontaneous process that releases energy.) The similar yet effectively controlled oxidative process of respiration keeps all organisms alive (see Chapter 13). During both combustion and respiration, electrons are removed from carbon compounds and passed downhill (energetically speaking), and then they and $H^+$ combine with a strong electron acceptor, $O_2$, to make stable $H_2O$. Considered in this way, photosynthesis uses light energy to drive electrons "uphill" away from $H_2O$ to a weaker electron acceptor, $CO_2$. These relations are summarized in Figure 10-1. In this chapter we emphasize the light-harvesting apparatus of plants—the *thylakoid membrane* in chloroplasts—and the way in which it accomplishes this "uphill" transport of electrons.

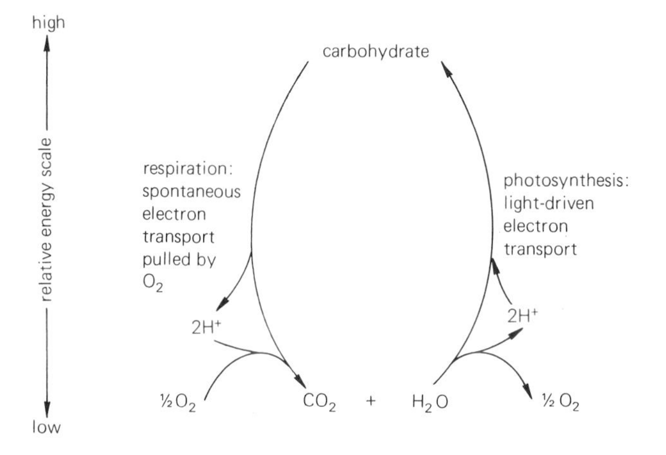


**Figure 10-1** Contrasting energy relations of photosynthesis and respiration.

## 10.1 Historical Summary of Early Photosynthesis Research

Before the early 18th century, scientists believed that plants obtained all of their elements from the soil. In 1727, Stephen Hales suggested that part of their nourishment came from the atmosphere and that light participated somehow in this process. It was not known then that air contains different gaseous elements.

In 1771, Joseph Priestly, an English clergyman and chemist, implicated $O_2$ (although this "dephlogisticated air," as he called it, was not known to be a molecule) when he found that green plants could renew air made "bad" by the breathing of animals. Then a Dutch physician, Jan Ingenhousz, demonstrated that light was necessary for this purification of air. He found that plants, too, made "bad air" in darkness. This surprisingly (to us) caused him to recommend that plants be removed from houses during the night to avoid the possibility of poisoning the occupants! This and earlier pioneering experiments in the early 1700s by Stephen Hales were reviewed by Gest (1988).

In 1782, Jean Senebier showed that the presence of the noxious gas produced by animals and by plants in darkness ($CO_2$) stimulated production of "purified air" ($O_2$) in light. So by this time the participation of two gases in photosynthesis had been demonstrated. Work of Lavoisier and others made it apparent that these gases were indeed $CO_2$ and $O_2$. Water was implicated by N. T. de Saussure when, in 1804, he made the first quantitative measurements of photosynthesis. He found that plants gained more dry weight during photosynthesis than could be accounted for by the amount by which the weight of $CO_2$ absorbed exceeded the weight of $O_2$ released. He correctly attributed the difference to the uptake of $H_2O$. He also noted that approximately equal volumes of $CO_2$ and $O_2$ were exchanged during photosynthesis.

The nature of the other chemical product of photosynthesis—organic matter—was demonstrated by Julius von Sachs in 1864, when he observed growth of starch grains in illuminated chloroplasts. Starch is detected only in areas of the leaf exposed to light. Thus, the overall reaction of photosynthesis was demonstrated to be as follows:

$$nCO_2 + nH_2O + \text{light} \rightarrow (CH_2O)_n + nO_2 \qquad (10.1)$$

In this reaction, $(CH_2O)_n$ is simply an abbreviation for starch or other carbohydrates with an empirical formula close to it. Starch is the world's most abundant photosynthetic product formed by chloroplasts.

A further important discovery was that of C. B. van Niel, who in the early 1930s pointed out the similarity between the overall photosynthetic process in green plants and that in certain bacteria. Various bacteria were known to reduce $CO_2$ using light energy and an electron source other than water. Some of these bacteria use organic acids such as acetic or succinic acid as electron sources, whereas those to which van Niel gave primary attention use $H_2S$ and deposit sulfur as a by-product. The overall photosynthetic equation for these bacteria was believed to be as follows:

$$nCO_2 + 2nH_2S + \text{light} \rightarrow (CH_2O)_n + nH_2O + 2nS \qquad (10.2)$$

When Reaction 10.2 is compared with Reaction 10.1, an analogy can be seen between the roles of $H_2S$ and $H_2O$, and between those of $O_2$ and sulfur. This suggested to van Niel that the $O_2$ released by plants is derived from water, not from $CO_2$. This idea was supported in England in the late 1930s by Robin Hill and R. Scarisbrick, whose work showed that isolated chloroplasts and chloroplast fragments could release $O_2$ in the light if they were given a suitable acceptor for the electrons being removed from $H_2O$. Certain ferric ($Fe^{3+}$) salts were the earliest electron acceptors provided to the chloroplasts, and they became reduced to the ferrous ($Fe^{2+}$) form. This light-driven split (a **photolysis**) of water in the absence of $CO_2$ fixation became known as the **Hill reaction**. Hill and Scarisbrick's work showed that whole cells were not necessary for at least some of the reactions of photosynthesis and that the light-driven $O_2$ release is not mandatorily tied to reduction of $CO_2$.

More convincing evidence that the $O_2$ released is derived from $H_2O$ came in 1941 from work by Samuel Ruben and Martin Kamen (see the historical review by Kamen, 1989). They supplied the green alga *Chlorella* with $H_2O$ containing $^{18}O$, a heavy, nonradioactive oxygen isotope that can be detected with a mass spectrometer. The $O_2$ released in photosynthesis became labeled with $^{18}O$, supporting van Niel's hypothesis. For technical reasons Ruben's experiments could not prove that $O_2$ came entirely from $H_2O$, but subsequent work by Alan Stemler and Richard Radmer (1975) seems to provide such proof. We must therefore modify the summary equation for photosynthesis given in Reaction 10.1 to include two $H_2O$ molecules as reactants:

$$nCO_2 + 2nH_2O + \text{light} \xrightarrow{\text{chloroplasts}} (CH_2O)_n + nO_2 + nH_2O \qquad (10.3)$$

In 1951 it was found that a natural plant constituent—a coenzyme that contains vitamin B (niacin or nicotinamide) and is called **nicotinamide adenine dinucleotide phosphate** (commonly abbreviated **$NADP^+$**)—could also act as a Hill reagent by accepting electrons from water in reactions occurring in isolated thylakoid membranes or in broken chloroplasts. This discovery again stimulated photosynthesis research; because it was already known that the reduced form of $NADP^+$, **NADPH**, could transfer electrons to a number of plant compounds, it was correctly suspected that its normal role in chloroplasts was reduction of $CO_2$. Therefore, *one of two essential functions of light in photosynthesis is to drive electrons from $H_2O$ to reduce $NADP^+$ to NADPH; the other function is to provide energy to form ATP from ADP and* P*i* as described next.

Conversion of ADP and P*i* to ATP in chloroplasts was discovered in the laboratory of Daniel Arnon at the University of California, Berkeley, in 1954 (reviewed by Arnon, 1984). Before that, the only important mechanism known to form ATP was respiration, especially those reactions, called oxidative phosphorylation (see Chapter 13), that occur in the mitochondria. Arnon found that ATP was synthesized in isolated chloroplasts only in the light, and the process became known as **photosynthetic phosphorylation**, or simply **photophosphorylation**. This process of ATP formation by photophosphorylation can be summarized as follows:

$$\text{ADP} + \text{P}i + \text{light} \xrightarrow{\text{chloroplasts}} \text{ATP} + H_2O \qquad (10.4)$$

Photophosphorylation in chloroplasts accounts for much more ATP formation in leaves during daylight than does oxidative phosphorylation in mitochondria of those leaves, and so it is clearly of great quantitative significance. Notice, however, that our summary equation for photosynthesis (Reaction 10.3) says nothing about ATP, NADPH, or $NADP^+$. The reason for this is that once ATP and NADPH are formed, their energy is

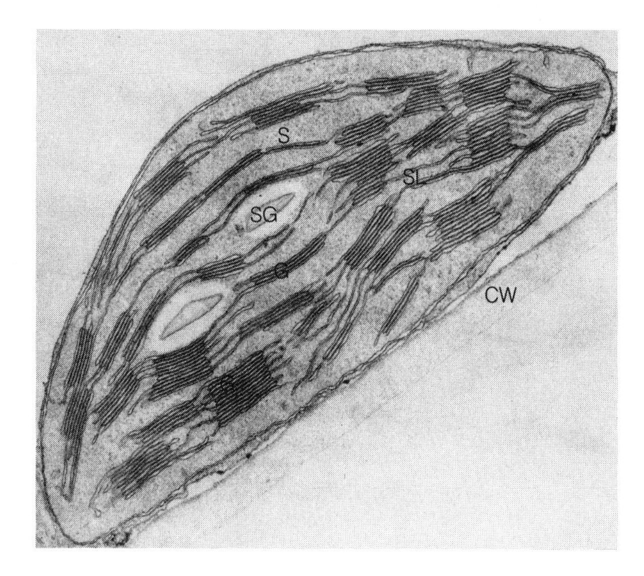


**Figure 10-2** Oat leaf chloroplast. S, stroma; ST, stroma thylakoid; G, granum; SG, starch grain; CW, cell wall. (Courtesy P. Hanchey-Bauer.)

used in the process of $CO_2$ reduction and carbohydrate synthesis, and ADP, P*i*, and $NADP^+$ are again released. So ADP and P*i* are rapidly converted to ATP by light energy, and the ATP is just as rapidly broken down when photosynthesis is occurring at a constant rate. How ATP and NADPH are used to help fix $CO_2$ is a subject of Chapter 11, but in the remainder of this chapter we are concerned with thylakoid membranes in chloroplasts and how they trap light energy and use it to form ATP and NADPH.

## 10.2 Chloroplasts: Structures and Photosynthetic Pigments

Chloroplasts of many shapes and sizes are found in various kinds of plants (Kirk and Tilney-Bassett, 1978; Possingham, 1980; Wellburn, 1987). They arise from tiny **proplastids** (immature, small, and nearly colorless plastids with few or no internal membranes). Most commonly, proplastids are derived only from the unfertilized egg cell; the sperm contributes none. Proplastids divide as the embryo develops, and they develop into chloroplasts when leaves and stems are formed. Young chloroplasts also actively divide, especially when the organ containing them is exposed to light, so each mature leaf cell often contains a few hundred chloroplasts. Most chloroplasts are easily seen with the light microscope, but their fine structure was discovered only by electron microscopy.

Assuming that organisms can be logically segregated into five kingdoms, chloroplasts occur in almost all members of the Kingdom Plantae and in certain algae or algal-like members of the Kingdom Protista. Only colorless, parasitic angiosperms are known exceptions. Cyanobacteria and other photosynthetic bacterial

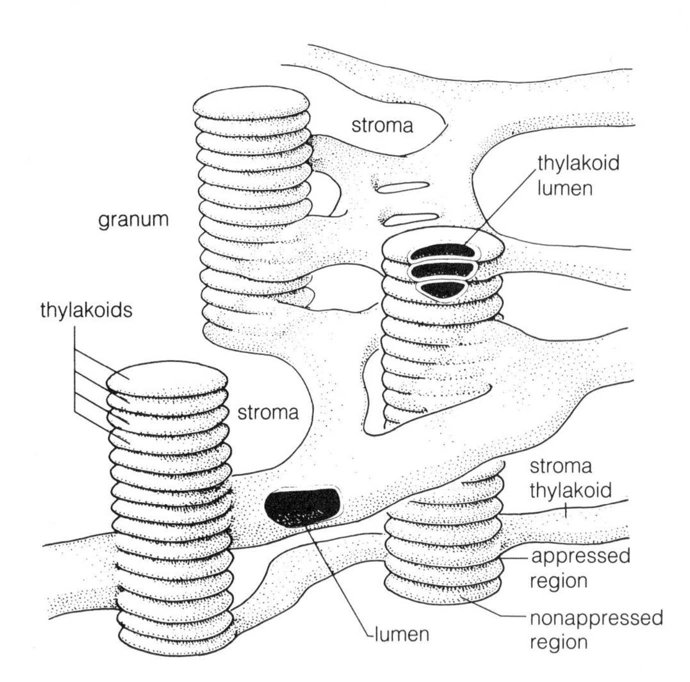


**Figure 10-3** A three-dimensional interpretation of the arrangement of the internal membranes of a chloroplast, emphasizing the relation between stroma thylakoids and grana. Note the lumen in both kinds of thylakoids. (Redrawn from an original by T. E. Weir.)

members of the Kingdom Monera have no chloroplasts, but they still have photosynthetic pigments embedded in specialized membranes, as do chloroplasts. No chloroplasts occur in either the Kingdom Fungi or the Kingdom Animalia. We discuss here only the chloroplasts typical of vascular plants, bryophytes, and many green algae in the Kingdom Plantae.

The structure of a chloroplast from an oat leaf is shown in Figure 10-2. Each chloroplast is surrounded by a double-membrane system or envelope that controls molecular traffic into and out of it. Within the chloroplast one finds the amorphous, gel-like, and enzyme-rich material called the **stroma**, which contains enzymes that convert $CO_2$ into carbohydrates, especially starch. Embedded throughout the stroma are the pigment-containing **thylakoids** (from the Greek, *thylakos*, "sac" or "pouch"), in which energy from light is used to oxidize $H_2O$ and form energy-rich ATP and NADPH, needed by the stroma to convert $CO_2$ to carbohydrates. In certain portions of the chloroplast are thylakoid stacks called **grana** (a single stack is called a **granum**). The region where one granum thylakoid contacts another is called an **appressed region**, and as we explain later these regions of grana carry out somewhat different photoreactions than do nonappressed grana regions and stroma thylakoids, both of which directly contact the stroma. The **stroma thylakoids** are longer thylakoids that connect one granum to another and extend through the stroma. They often extend into and make up part of one or more grana, and in such locations there is no apparent distinction between them and the grana thylakoids. Figure 10-3 depicts a three-dimensional interpretation of the relation between thylakoids of grana and stroma. Note that there is a cavity, usually called a **lumen**, between the two membranes of each thylakoid. This lumen is filled with water and dissolved salts, and it plays a special role in photosynthesis.

The pigments present in thylakoid membranes consist largely of two kinds of green *chlorophylls*, **chlorophyll *a*** and **chlorophyll *b***. Also present are yellow-to-orange pigments classified as **carotenoids**. There are two kinds of carotenoids: the pure hydrocarbon **carotenes** and the oxygen-containing **xanthophylls**. Certain

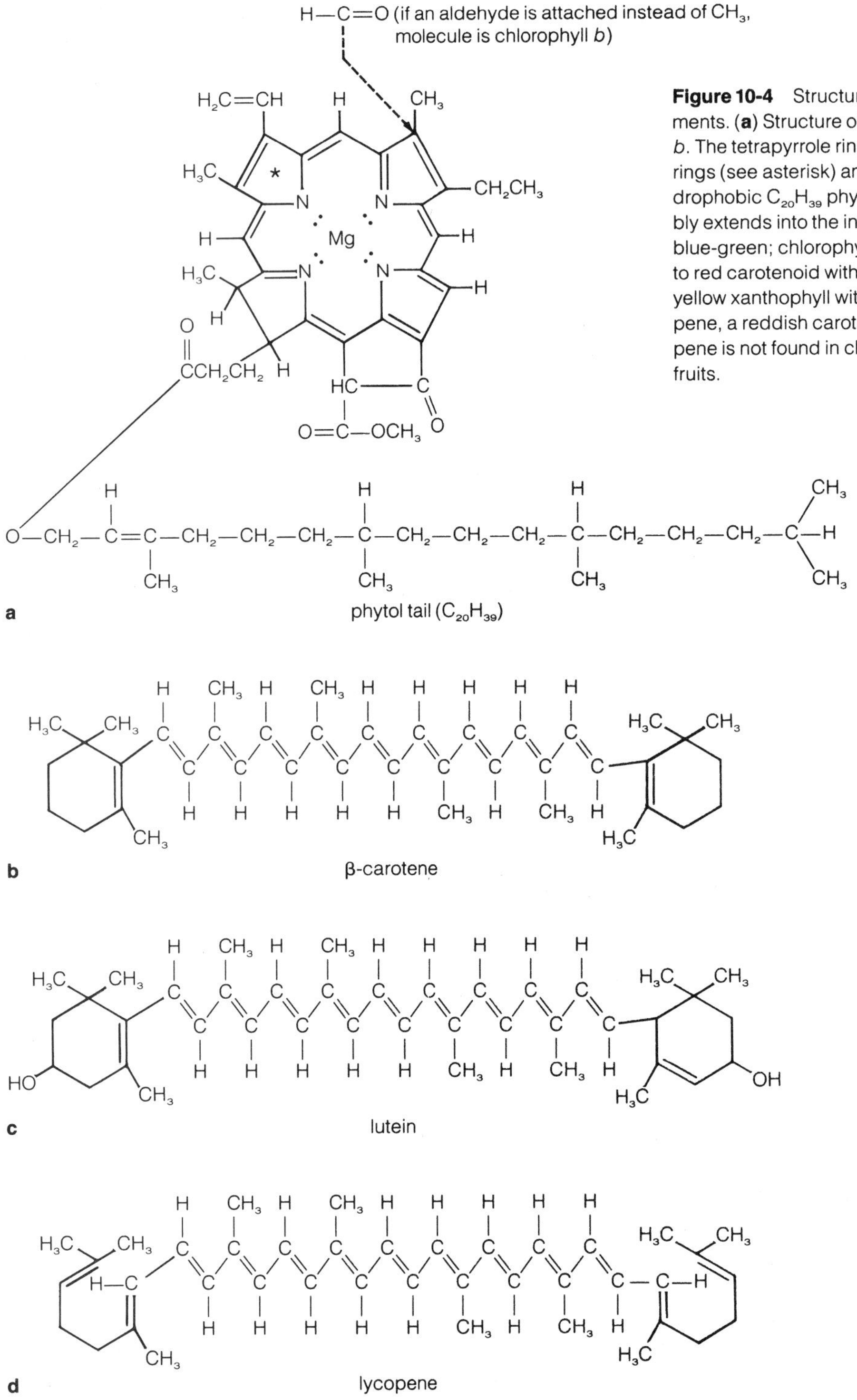

H—C=O (if an aldehyde is attached instead of $CH_3$, molecule is chlorophyll *b*)

a — phytol tail ($C_{20}H_{39}$)

b — β-carotene

c — lutein

d — lycopene

**Figure 10-4** Structures of some chlorophyll and carotenoid pigments. (**a**) Structure of chlorophyll *a* and its relation to chlorophyll *b*. The tetrapyrrole ring structure on top is made of four pyrrole rings (see asterisk) and gives the green color, whereas the hydrophobic $C_{20}H_{39}$ phytol tail common to both chlorophylls probably extends into the interior of the membrane. Chlorophyll *a* is blue-green; chlorophyll *b*, yellow-green. (**b**) β-carotene, a yellow to red carotenoid with the empirical formula $C_{40}H_{56}$. (**c**) Lutein, a yellow xanthophyll with the empirical formula $C_{40}H_{56}O_2$. (**d**) Lycopene, a reddish carotene with the empirical formula $C_{40}H_{56}$. Lycopene is not found in chloroplasts but gives the red color to tomato fruits.

carotenoids (especially *violaxanthin*, a xanthophyll) also exist in the chloroplast envelope, giving it a yellowish color, whereas chlorophylls do not occur in the envelopes. The structures of chlorophylls *a* and *b* and some carotenoids are shown in Figure 10-4. In most plants, including green algae, β-*carotene* and the xanthophyll *lutein* are the most abundant carotenoids in thylakoids.

Electron microscopy gives us no information about how chlorophylls and carotenoids are arranged in thylakoids, but other techniques show that all chlorophylls and most or all carotenoids are embedded within thylakoids and are attached by noncovalent bonds to protein molecules. Altogether, chloroplast pigments represent about one-half of the lipid content of thylakoid membranes, whereas the other half is composed largely of galactolipids with small amounts of phospholipids (for structures, see Fig. 7-13). Few, if any, sterols are present, as contrasted to other plant membranes. The galactolipids and phospholipids make a bilayer typical of membranes described in Chapters 1 and 7. The fatty-acid portions of thylakoid lipids are rich in both linolenic acid (with three double bonds) and linoleic acid (with two double bonds). These unsaturated fatty acids cause thylakoid membranes to be unusually fluid, and certain compounds within them are quite mobile.

Also present in chloroplasts are DNA, RNA, ribosomes, and, of course, many enzymes. All these molecules are present largely in the stroma, where both transcription and translation occur (see Appendix C). Chloroplast DNA (the genome) exists in 50 or more supercoiled double-stranded circles per plastid. Numerous plastid genes code for apparently all of the transfer-RNA molecules (about 30) and ribosomal-RNA molecules (four) used by plastids for translation. About 85 other such genes code for proteins involved in transcription, translation, and photosynthesis, but most proteins in plastids are coded for by nuclear genes (Steinback et al., 1985; Murphy and Thompson, 1988).

## 10.3 Some Principles of Light Absorption by Plants

To learn how light causes photosynthesis, we must learn something about its properties. First, light has both a *wave nature* and a *particle nature*. Light represents only that part of radiant energy with wavelengths that are visible to the human eye (approximately 390 to 760 nanometers, nm). This is a very narrow region of the electromagnetic spectrum.

The particulate nature of light is usually expressed in statements that light comes in **quanta** or **photons**: discrete packets of energy, each having a specific associated wavelength. The energy in each photon is inversely proportional to the wavelength, so the violet and blue wavelengths have more-energetic photons

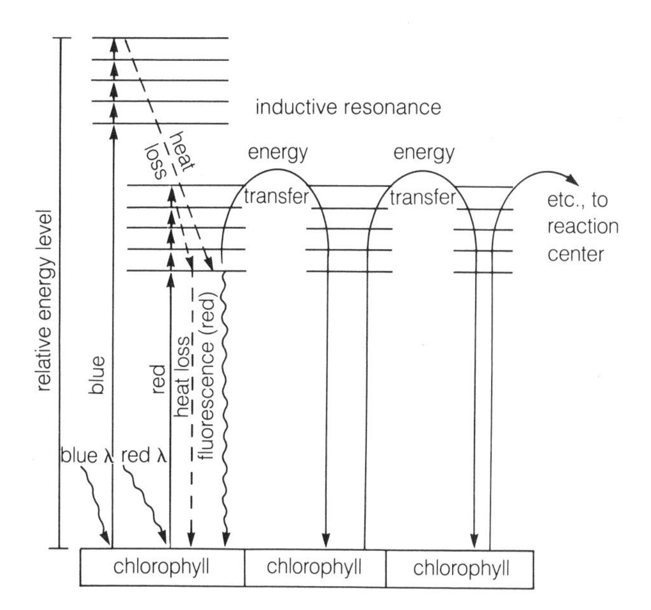


**Figure 10-5** Simplified model to explain how light energy striking a chlorophyll molecule is given up. Note that excitation by blue or red light leads to the same final energy level (often called the first excited singlet). From here, the energy can be lost by decay back to the ground state (heat loss or fluorescence of red light) or can be transferred to an adjacent pigment by inductive resonance. Each time a pigment transfers its excitation energy to an adjacent pigment, the excited electron in the first pigment returns to the ground state.

than the longer orange and red ones. One mole ($6.02 \times 10^{23}$) of photons was formerly called an **einstein**, but the term *einstein* is now discouraged because the mole is an SI unit and the einstein is not. Quantitative and other aspects of light relations are discussed in Appendix B.

A fundamental principle of light absorption, often called the **Stark Einstein Law**, is that any molecule can absorb only one photon at a time, and this photon causes excitation of only one electron. Specific valence (bonding) electrons in stable ground-state orbitals are the ones that are usually excited, and each electron can be driven away from its ground state in the positively charged nucleus a distance corresponding to an energy exactly equal to the energy of the photon absorbed (Fig. 10-5). The pigment molecule is then in an *excited state*, and it is this excitation energy that is used in photosynthesis.

Chlorophylls and other pigments can remain in an excited state for only very short periods, usually a billionth of a second (a nanosecond) or even less. As shown in Figure 10-5, the excitation energy can be totally lost by heat release as the electron moves back to its ground state. This is what is now happening to electrons in the ink of the words you are reading. A second way that some pigments, including chlorophyll, can lose excitation energy is by a combination of heat loss and fluorescence. (**Fluorescence** is light production accompanying rapid decay of electrons in the excited state.) Chlorophyll fluorescence produces only deep-red light, and these long wavelengths are easily seen when a sufficiently concentrated solution of either chlorophyll *a* or *b* or a mixture of chloroplast pigments is illuminated, especially with ultraviolet or blue radiation. In the leaf, fluorescence is weak because the excitation energy is used in photosynthesis.

Figure 10-5 helps explain why blue light is always less efficient on an energy basis in photosynthesis than red light. After excitation with a blue photon, the electron in a chlorophyll always decays extremely rapidly by heat release to a lower energy level, a level that lower energy red light produces without that heat loss when a red photon is absorbed. From this lower level, either additional heat loss, fluorescence, or photosynthesis can occur.

Photosynthesis requires that energy in excited electrons of various pigments be transferred to an energy-collecting pigment, a **reaction center**. We shall explain later that there are two kinds of reaction centers in thylakoids, both of which consist of chlorophyll *a* molecules that are made special by their association with particular proteins and other membrane components. Figure 10-5 illustrates that the energy in an excited pigment can be transferred to an adjacent pigment, and from it to still another pigment, and so on, until the energy finally arrives at the reaction center. There are various theories to explain energy migration within a group of neighboring pigment molecules, the most popular of which states that energy migrates by **exciton transfer** through **inductive resonance**. We won't discuss this theory, but we emphasize that excitation of any one of many pigment molecules in a thylakoid allows momentary collection of the light energy in a chlorophyll *a* reaction center.

Leaves of most plant species absorb more than 90 percent of the violet and blue wavelengths that strike them and almost as high a percentage of the orange and red wavelengths (see Fig. 4-14). Almost all of this absorption is by the chloroplast pigments. In thylakoids, each photon can excite an electron in a carotenoid or chlorophyll. Chlorophylls are green because they absorb green wavelengths ineffectively and instead reflect or transmit them. We can use a spectrophotometer to measure the relative absorbance of various wavelengths of light by a purified pigment. A graph of this absorption as a function of wavelength is called an **absorption spectrum**. The absorption spectra of chlorophylls *a* and *b* are given in Figure 10-6a. These spectra show that very little of the green and yellow-green light between 500 and 600 nm is absorbed *in vitro* and that both chlorophylls strongly absorb violet, blue, orange, and red wavelengths.

Most of the carotenoids (both β-carotene and the xanthophylls) in thylakoids efficiently transfer their excitation energy to the same reaction centers as do chlorophylls, and so they contribute to photosynthesis

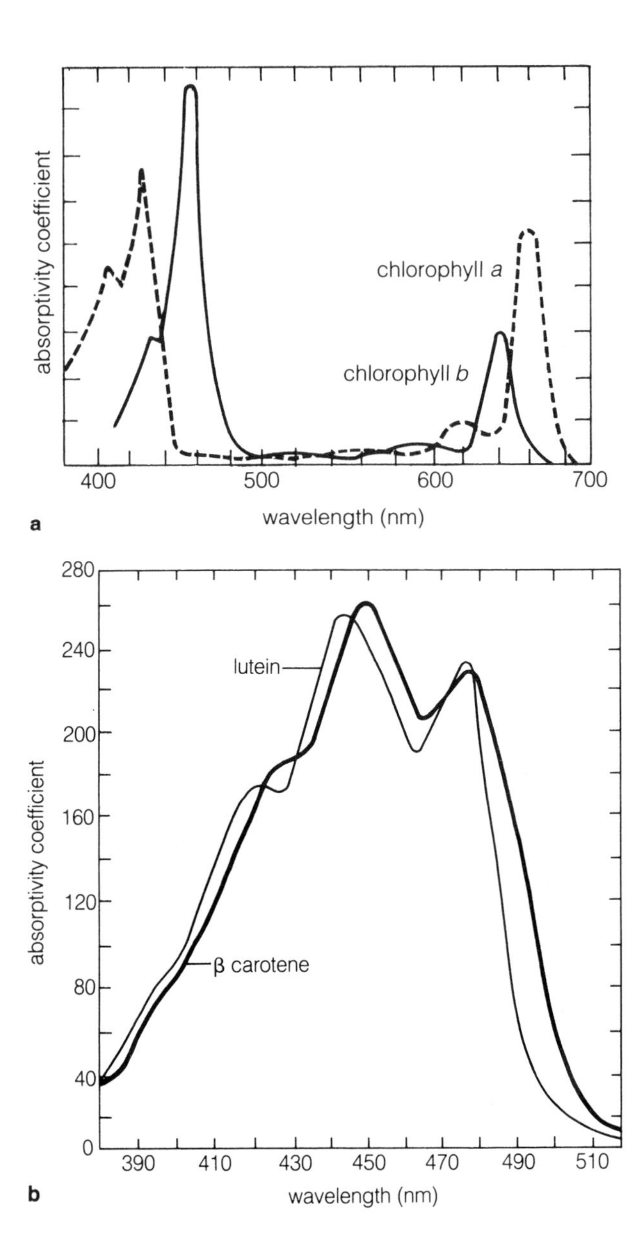


**Figure 10-6** (**a**) Absorption spectra of chlorophylls *a* and *b* dissolved in diethyl ether. The absorptivity coefficient used here is equal to the absorbance (optical density) given by a solution at a concentration of 1 g/L with a thickness (light path length) of 1 cm. (From Zscheile and Comar, 1941.) (**b**) Absorption spectra of β-carotene in hexane and of lutein (a xanthophyll) in ethanol. The absorptivity coefficient used is the same as that described in Figure 10-6a. (Data from Zscheile et al., 1942.)

(Siefermann-Harms, 1985, 1987). The absorption spectra of β-carotene and lutein (a xanthophyll) are in Figure 10-6b. These pigments absorb only blue and violet wavelengths *in vitro*. They reflect and transmit green, yellow, orange, and red wavelengths, and this combination appears yellow or orange to us. *Besides the function of carotenoids as light-harvesting pigments that contribute to photosynthesis, they protect chlorophylls against oxidative destruction by $O_2$ when irradiance levels are high* (see solarization, Section 12.3).

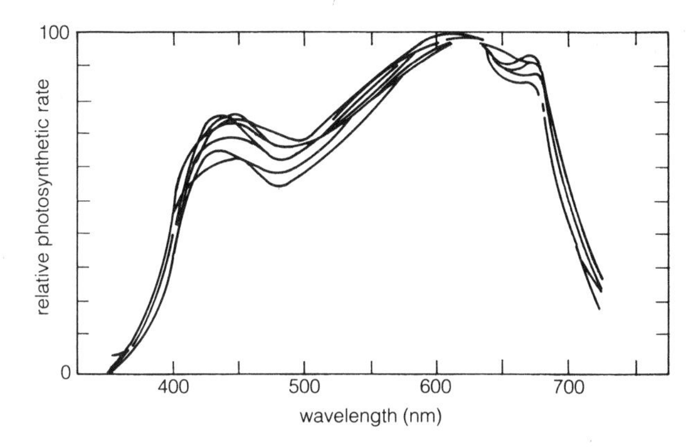


**Figure 10-7** Action spectra of 22 species of crop plants. (From McCree, 1972.)

When we compare the effects of different wavelengths on the rate of photosynthesis, always making sure that we do not add so much energy of any wavelength that the process becomes saturated, we obtain an **action spectrum**. Action spectra for photosynthesis and other photobiological processes help identify the pigment involved because these spectra often closely match the absorption spectrum of any pigment that participates. Light, to be effective, must be absorbed. The relative rates of photosynthesis for several herbaceous dicot and grass species can be plotted as a function of the wavelength striking a unit area of leaf (Fig. 10-7). Similar results were obtained by Inada (1976). All the species show a major peak in the red-light region and a distinct lower peak in the blue-light region, both of which result from light absorption by chlorophylls and carotenoids.[1] Similar results occur with deciduous trees. However, conifers show less response to blue light because their waxy needles reflect more of the blue light and because they contain high amounts of carotenoids, some of which absorb blue light but don't transfer the energy to chlorophylls for photosynthesis (Clark and Lister, 1975). In particular, blue spruce and the blue-green Colorado spruce species photosynthesize little in blue or violet light.

Compared with the absorption spectra of purified chlorophylls and carotenoids, the action of green and yellow light in causing photosynthesis of seed plants and the absorption of these wavelengths by leaves is surprisingly high. Nevertheless, carotenoids and chlorophylls are apparently the only pigments that absorb this light. The main reason that the action spectra are higher than the absorption spectra for yellow and green wavelengths is that although the chance of any such wavelength being absorbed is small, those wavelengths not absorbed are repeatedly reflected from chloroplast to chloroplast in the complex network of photosynthetic cells. With each reflection a small additional percentage of those wavelengths is absorbed, until finally half or more are absorbed by most leaves and cause photosynthesis. This internal reflection does not occur in a spectrophotometer cuvette containing dissolved chlorophyll, and so the absorbance of green wavelengths is very low (Fig. 10-6a). Furthermore, the *in vitro* absorption by these pigments in an organic solvent occurs at shorter wavelengths than when they are present in chloroplast thylakoids.

In living leaves absorption of the carotenoids shifts from the blue portion of the spectrum into the green, and some photosynthesis in the green portion at about 500 nm results from absorption by active carotenoids. Both chlorophylls show only small *in vivo* shifts in the blue region, but chlorophyll *a* shows several shifts in the red region. The association of chlorophyll *a* with thylakoid proteins causes additional peaks to occur in the red region. We are interested in two of these minor peaks, those at about 680 and 700 nm, because they result from chlorophyll *a* molecules in special chemical environments acting as reaction-center pigments. These pigments are abbreviated **P680** and **P700**. Their functions will be discussed in more detail in Section 10.5.

## 10.4 The Emerson Enhancement Effect: Cooperating Photosystems

In the 1950s, Robert Emerson at the University of Illinois was interested in why red light of wavelengths longer than 690 nm is so ineffective in causing photosynthesis (Fig. 10-7), even though much of it is absorbed by chlorophyll *a in vivo*. His research group found that if light of shorter wavelengths was provided at the same time as the longer red wavelengths, photosynthesis was even faster than the sum of the two rates with either color alone. This synergism or enhancement became known as the **Emerson enhancement effect**.

We can think of enhancement as the long red wavelengths helping out the shorter wavelengths, or as the short group helping out the long red wavelengths. We now realize that two separate groups of pigments or photosystems cooperate in photosynthesis and that

[1]Slightly different action spectra with lower blue peaks are obtained when we plot the data as a function of the *energy* in each applied wavelength. This is because although a blue photon absorbed by a photosynthetically active pigment is as effective as any other photon, it contains more energy than others and more of its energy is wasted as heat. Therefore, blue light can be as *effective* in photosynthesis as red but never as *efficient*. In many algae, carotenoids and phycobilin pigments (the red phycoerythrins or the blue phycocyanins) absorb light, causing photosynthesis. The action spectra of these algae are quite different from those of seed plants.

such long red wavelengths are absorbed only by one photosystem, called **photosystem I (PS I)**. The second photosystem, **photosystem II (PS II)**, absorbs wavelengths shorter than 690 nm, and for maximum photosynthesis wavelengths absorbed by both systems must function together. In fact, the two photosystems normally cooperate to cause photosynthesis at all wavelengths shorter than 690 nm, including the red, orange, yellow, green, blue, and violet, because both photosystems absorb those wavelengths. The importance of Emerson's work is that it suggested the presence of two distinct photosystems. Nevertheless, it was R. Hill and F. Bendall (1960) who first clearly proposed how two photosystems might cooperate, and this idea was then substantiated by extensive work of L. N. M. Duysens and his colleagues in the Netherlands. (See Govindjee's essay, "The Role of Chlorophyll *a* in Photosynthesis," in this chapter.) Reaction 10.5 summarizes how PS I and PS II use light energy to oxidize $H_2O$ and cooperatively transfer the two available electrons in it to $NADP^+$, thus forming NADPH:

$$\text{PS II} \xrightarrow{+\text{ light}} \text{PS I} \xrightarrow{+\text{ light}} 2\text{NADPH} + 2\text{H}^+ \quad (2\text{H}_2\text{O} \rightarrow \text{O}_2 + 4\text{H}^+;\ 2\text{NADP}^+) \tag{10.5}$$

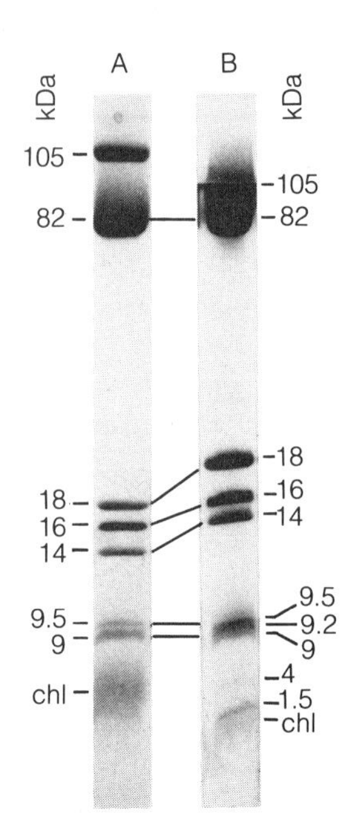


**Figure 10-8** Separation of PS I polypeptides from barley (*Hordeum vulgare* L.) leaves by SDS polyacrylamide gel electrophoresis. Two different gels were used in A and B, which caused polypeptides to migrate somewhat differently. Note that the small polypeptides run farther from top to bottom according to their molecular weights in kDa units. Polypeptides were stained with Coomassie brilliant blue dye. (Courtesy Henrik V. Scheller; see also Scheller et al., 1989.)

## 10.5 The Four Major Complexes of Thylakoids

When thylakoids are separated from isolated chloroplasts and then treated with the proper mild detergents and certain other solutions, four major complexes of proteins can be dissolved. These complexes can be purified by ultracentrifugation, which takes advantage of their different densities. Their proteins can be further resolved into various polypeptide subunits by polyacrylamide gel electrophoresis (see Section 9.2) if the proteins are first heated in a detergent called *sodium dodecyl sulfate* (*SDS*). This SDS treatment causes protein denaturation involving breakage of hydrogen bonds and ionic bonds that normally hold one polypeptide to another. Polyacrylamide gels used to separate polypeptides resulting from SDS treatment are called **SDS gels**. In these gels the large polypeptides move relatively slowly and remain near the top of the gel, whereas smaller polypeptides move progressively farther down toward the anode. Separation of several polypeptides from photosystem I by this powerful technique is shown in Figure 10-8. Using ultracentrifugation and SDS gels, most of the polypeptides from the four major protein complexes have been separated and their molecular weights determined (in kilodaltons, kDa). In most cases partial amino-acid sequences of the polypeptides have been determined and compared with base sequences in genes from the nucleus or chloroplast; this process has allowed identification of the genes that code for these polypeptides.

We first describe the photosystem II complex here because it is more closely involved with oxidation of water, which initiates electron transport in photosynthesis. (Helpful reviews of structure, functions, and locations of the four complexes in thylakoids are those by Glazer and Melis, 1987; Barber, 1987a, 1987b; Irrgang et al., 1988; Renger, 1988; Chitnis and Thornber, 1988; Reilly and Nelson, 1988; Mattoo et al., 1989; Marder and Barber, 1989; Lagoutte and Mathis, 1989; Govindjee and Coleman, 1990; Ghanotakis and Yocum, 1990; and Scheller and Møller, 1990.)

### Photosystem II (PS II)

This photosystem consists of a core complex of six integral (intrinsic) polypeptides connected noncovalently to each other, and it contains the reaction center P680. All of these polypeptides are coded for by the chloroplast genome. Two such polypeptides with molecular weights near 33 kDa and 31 kDa are commonly called D1 and D2, respectively, and these directly bind P680 and certain quinones necessary for oxidation of water. (The D terminology is used because D1 and D2 stain *diffusely* when separated on SDS gels.) Also associated with the PS II core complex and the membrane-lumen interface are three peripheral (extrinsic) polypeptides coded by nuclear genes; these polypeptides are thought to aid binding of $Ca^{2+}$ and $Cl^-$, both of which are essential for photolysis of water. Besides these polypeptides, the core complex contains about 40 molecules of chlorophyll *a*, several molecules of β-carotene, some membrane lipids (mostly galactolipids), four manganese ions, one noncovalently bound iron, one or more $Ca^{2+}$, several $Cl^-$, two molecules of plastoquinone, and two molecules of pheophytin. **Plastoquinones** are special quinones in plastids; as we explain later, they carry two electrons from PS II toward photosystem I, and they also transport $H^+$ from the stroma into the thylakoid lumen. **Pheophytin** is a modified chlorophyll *a* molecule in which two H atoms have replaced the central $Mg^{2+}$. It is now accepted that most PS II is present only in appressed regions of grana thylakoids (see Fig. 10-3). Nonappressed regions of grana and stroma thylakoids have far less PS II.

The P680 in the PS II core complex receives light energy by inductive resonance from a total of about 250 chlorophyll *a* and chlorophyll *b* molecules (present in about equal numbers) and numerous xanthophylls. These pigments are present in the **PS II light harvesting complex**, often called **LHCII**. Each pigment is associated with an integral protein, about 10 chlorophylls, and two or three xanthophylls per protein molecule. Their function is to act as an **antenna system**, absorbing light and passing exciton energy to P680. Probably all of the protein in LHCII is coded for by nuclear DNA and synthesized on cytoplasmic ribosomes, so each protein must be transported into the chloroplast and into the thylakoids (as described in Section 7.11 and reviewed by Smeekens et al., 1990). The overall function of PS II is to use light energy to reduce oxidized plastoquinone (abbreviated **PQ**) to its fully reduced form (**$PQH_2$**) using electrons from water.

Because two $H_2O$ molecules (four electrons) are needed to reduce each $CO_2$ (see Reaction 10.3) and because two light photons are required to oxidize each $H_2O$, we can summarize the function of PS II as follows:

$$2H_2O + 4\text{ photons} + 2PQ + 4H^+ \rightarrow O_2 + 4H^+ + 2PQH_2 \qquad (10.6)$$

It is helpful to show $H^+$ on both sides of the equation because (as we explain later) oxidation of water results in release of $H^+$ in the thylakoid lumen, and reduction of PQ requires $H^+$ taken from the opposite (stroma) side of the thylakoid.

## The Cytochrome $b_6$-Cytochrome f Complex

This complex, abbreviated **cyt $b_6$-f**, consists of four different integral polypeptides, three of which contain iron that undergoes reduction to $Fe^{2+}$ and then oxidation back to $Fe^{3+}$ during electron flow. The first two are *cyt $b_6$* (also called *cyt b563*) and *cytochrome f* (for *frons* or leaf). Each cytochrome contains iron in a heme prosthetic group. Third is a protein with two nonheme iron atoms, in which each of the two irons is connected to two nonprotein sulfur atoms and to two sulfur atoms of cysteine residues in the protein. This protein is an **iron-sulfur protein** (a **2Fe-2S protein**), one of several involved in photosynthetic electron transport. The fourth polypeptide, component IV, has no iron and no known function. The gene for the 2Fe-2S polypeptide is in the nucleus, whereas genes for the other three are in the chloroplast itself.

The cyt $b_6$-f complex exists in roughly equal concentrations in grana and stroma thylakoids. Its major function is to pass electrons from PS II to PS I. It does this by oxidizing $PQH_2$ and by reducing a small, unusually mobile, copper-containing protein called *plastocyanin*. It also causes transport of $H^+$ from the stroma into the thylakoid lumen, and this further aids the separation of electrons and protons in $H_2O$ started by PS II. The stoichiometry of partial reactions catalyzed by the complex is unclear because of the likely presence of a quinone or *Q cycle* that increases the number of $PQH_2$ molecules involved and doubles the amount of $H^+$ transferred into the thylakoid lumen. We won't discuss the Q cycle, but evidence for its function in chloroplasts is given, for example, by Hope and Matthews (1988), and ideas of its general function in photosynthetic bacteria, mitochondria, and chloroplasts are reviewed by Malkin (1988), O'Keefe (1988), and Marder and Barber (1989). Assuming conservatively and for the sake of simplicity that the Q cycle is not operating and instead using the two $PQH_2$ molecules from PS II, we can write the following summary for the function of the cyt $b_6$-f complex, where PC represents plastocyanin with oxidized or reduced copper:

$$2PQH_2 + 4PC(Cu^{2+}) \rightarrow 2PQ + 4PC(Cu^+) + 4H^+ \text{ (lumen)} \qquad (10.7)$$

## Photosystem I (PS I)

This photosystem absorbs light energy independently of PS II, but it contains a separable core complex that receives electrons originally taken from $H_2O$ by the PS II core complex. It was reviewed by Lagoutte and Mathis (1989), Evans and Bredenkamp (1990), and Scheller and Møller (1990). The PS I core complex from barley contains 11 different polypeptides varying in size from about 1.5 to 82 kDa (Fig. 10-8; also see Scheller and Møller, 1990); six are coded for by nuclear genes and five by chloroplast genes. There is probably one of each polypeptide present per P700 reaction center. The two largest polypeptides (about 82 kDa each) are called Ia and Ib because they are similar, because their genes are recognized as a single operon in the chloroplast genome, and because they are bound closely in thylakoids (probably to make a heterodimer protein). These two polypeptides bind the reaction center P700, and with another polypeptide they also bind some 50 to 100 chlorophyll *a* molecules, some β-carotene, and three electron carriers that help transport electrons toward $NADP^+$.

The electron carriers associated with polypeptides Ia and Ib in the core complex, frequently called $A_0$, $A_1$, and X, have now been tentatively identified as follows: $A_0$ is a chlorophyll *a* molecule, $A_1$ is probably another kind of quinone called **phylloquinone** (vitamin $K_1$), and X is an iron-sulfur group similar to that in the cyt $b_6$-f complex, except that X has a 4Fe-4S center instead of the 2Fe-2S center in the cyt $b_6$-f complex. Two other iron-sulfur groups, each with a 4Fe-4S center, are attached to a 9-kDa polypeptide identified as a band in Figure 10-8. All of these Fe-S centers can pick up (via one $Fe^{3+}$) and transfer (via $Fe^{2+}$) only one electron at a time, even though as many as four irons are present. The PS I core complex receives light energy by inductive resonance from about 100 molecules of chlorophyll *a* and *b* (in a ratio of about 4:1) that exist bound to nuclear-encoded proteins in a light-harvesting system of antennae pigments surrounding the core complex. This antenna system is called **LHCI**.

Photosystem I is located exclusively in stroma thylakoids and nonappressed regions of grana that face the stroma. It functions as a light-dependent system to oxidize reduced plastocyanin and transfer the electrons to a soluble form of Fe-S protein called ferredoxin. **Ferredoxin** is a low-molecular-weight protein present as a peripheral protein attached loosely to the stroma side of thylakoids. It contains a 2Fe-2S cluster but picks up and transfers only one electron as one of its irons first gets reduced to $Fe^{2+}$ and then oxidized to $Fe^{3+}$. (See Arnon, 1988, for the history and involvement of ferredoxin in this and other biochemical processes.) An overall reaction for the function of PS I can be written as follows using the four reduced plastocyanins (per two $H_2O$ undergoing oxidation) provided by the cyt $b_6$-f complex and abbreviating ferredoxin as Fd:

$$\text{Light} + 4\text{PC}(\text{Cu}^+) + 4\text{Fd}(\text{Fe}^{3+}) \rightarrow 4\text{PC}(\text{Cu}^{2+}) + 4\text{Fe}(\text{Fd}^{2+}) \qquad (10.8)$$

Electrons from mobile ferredoxin are then commonly used in the final electron-transport step to reduce $NADP^+$ and form (with $H^+$) NADPH. This reaction is catalyzed in the stroma by an enzyme called *ferredoxin-$NADP^+$ reductase* (reviewed by Pschorn et al., 1988). The following reaction shows transfer of the four electrons originally taken from two water molecules:

$$4\text{Fd}(\text{Fe}^{2+}) + 2\text{NADP}^+ + 2\text{H}^+ \rightarrow 4\text{Fd}(\text{Fe}^{3+}) + 2\text{NADPH} \qquad (10.9)$$

### The ATP Synthase or Coupling Factor

The final known complex in thylakoids is a group of polypeptides that converts ADP and inorganic phosphate (P*i*) to ATP and $H_2O$. It is called **ATP synthase** or, because it couples ATP formation to transport of electrons and $H^+$ across the thylakoid membrane, a **coupling factor**. We shall call it by the first name because it better describes its function. It exists, along with photosystem I, only in stroma thylakoids and the nonappressed regions of grana thylakoids (Anderson and Andersson, 1988). It contains two major parts: a stalk called $CF_0$ that extends from the lumen across the thylakoid membrane to the stroma and a spherical (headpiece) part called $CF_1$ that lies in the stroma. A total of nine polypeptides exist in ATP synthase, some of which are coded for by chloroplast DNA and some by nuclear DNA. The structure of the chloroplast ATP synthase is very similar to those in both mitochondria (see Section 13.7) and the bacterium *Escherichia coli*. Its function in photophosphorylation will be described in Section 10.8.

## 10.6 Oxidation of $H_2O$ by Photosystem II: The Supply of Electrons from the Oxygen-Evolving Complex

We emphasized that during photosynthesis electrons are transported from $H_2O$ to $NADP^+$ and are temporarily stored in NADPH molecules before reduction of $CO_2$. The reason light energy is required is that $H_2O$ is thermodynamically quite difficult to oxidize, and $NADP^+$ is moderately difficult to reduce. The stepwise transfer of electrons from $H_2O$ components in the PS II core first to those in the cyt $b_6$-f core, then to those in PS I, and finally to ferredoxin and $NADP^+$ completely across a thylakoid membrane helps stop back reactions that would prevent oxidation of $H_2O$. The question of how electrons are actually removed from $H_2O$ by an **oxygen-evolving complex (OEC)** is partly answered by recent information about the structure of the PS II complex and by information obtained in the late 1960s.

In 1969, Pierre Joliot demonstrated in Paris that chloroplasts first kept in darkness and then given extremely rapid flashes of light release $O_2$ from $H_2O$ in distinct four-flash peaks. Because release of one $O_2$ molecule requires oxidation of two $H_2O$ molecules and removal from them of four electrons, the four-flash periodicity suggested to Bessel Kok that some molecule or molecules must be accumulating a positive charge after each flash, until it has accumulated four positive charges and can get four electrons back in a one-step oxidation of two $H_2O$ molecules. He called these various states $S_0$, $S_1$, $S_2$, $S_3$, and $S_4$, and his important idea was

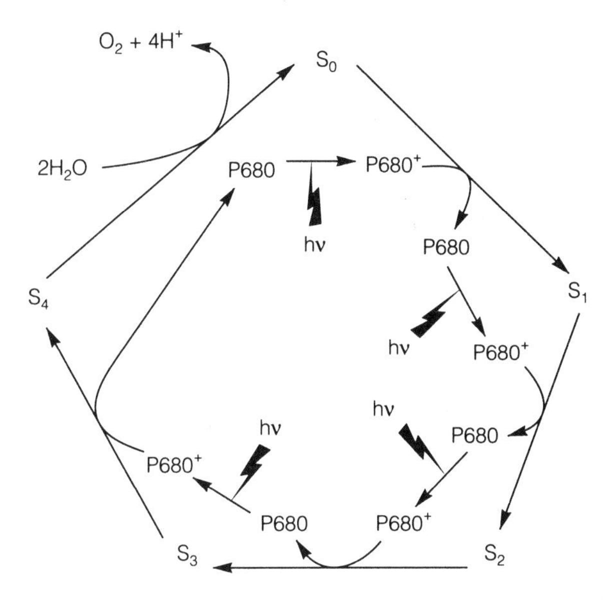


**Figure 10-9** The water oxidizing clock. $S_0$ through $S_4$ probably represent increasing oxidation states of two or more manganese ions, as explained in the text. When light energy is transferred to P680, P680 is oxidized to P680$^+$, and the latter (probably indirectly via a tyrosine in the D1 polypeptide) accepts an electron from $S_0$, $S_1$, $S_2$, or $S_3$ to return to noncharged P680. As shown, four photons (represented by $h\nu$) are used, each photon causing one advancement of the clock. Finally, $S_4$ absorbs four electrons from two $H_2O$ molecules and returns to noncharged $S_0$. (Adapted from various sources.)

that conversion of $S_0$ to $S_1$ involves the loss of one electron, conversion of $S_1$ to $S_2$ the loss of another, and so on, until $S_4$ has four extra positive charges; then $S_4$ gets all four electrons back from two $H_2O$ and returns to $S_0$ with four fewer charges than $S_4$. These changes are depicted in the model presented in Figure 10-9. The model is sometimes called the water oxidizing clock.

But what is the chemical nature of the S states in the clock? We soon learned that P680 becomes oxidized by loss of one electron after a light flash but that P680 cannot be S because P680 can lose only one electron and can accumulate only one positive charge. Further study led to the current understanding that the various S states represent various oxidation states of manganese, including $Mn^{2+}$, $Mn^{3+}$, and $Mn^{4+}$. We know that four Mn atoms are associated with each PS II core system and that all four are essential for $O_2$ release. We suspect that these are somehow bound to each other and to the D1 and D 2 polypeptides on the thylakoid lumen part of PS II where $O_2$ is released. However, it is still not clear how they function or the valence of each in the changing S states. Helpful reviews of this topic are those by Renger (1988), Rutherford (1989), Ghanotakis and Yocum (1990), and Govindjee and Coleman (1990). These reviews also present various hypotheses about the essential roles of $Cl^-$ and $Ca^{2+}$ during oxidation of $H_2O$.

## 10.7 Electron Transport from $H_2O$ to $NADP^+$ Across Thylakoids

Figure 10-10 presents a model in which the three major electron transport complexes (PS II, the cyt $b_6$-f complex, and PS I) cooperate to transport electrons from $H_2O$ to $NADP^+$. We shall trace most of the individual electron-transport reactions within the model. Note that PS II and the cyt $b_6$-f complex are in an appressed thylakoid region, whereas PS I and the CF complex are in a stroma thylakoid or nonappressed region.

As $H_2O$ is oxidized in the OEC, two electrons are released for transport. The first compound to receive them, one at a time, is a tyrosine amino acid in polypeptide D1, and this tyrosine then passes them to P680. However, P680 can accept an electron only if it has just lost one of its own, and this occurs when it has been excited by light energy transferred to it from a light-absorbing pigment in LHCII. So, light causes oxidation of P680, and P680$^+$ then acts as an electron attractant (oxidant) that is sufficiently powerful to pull an electron from the D1 tyrosine, which in turn pulls an electron from a Mn ion of the OEC. P680 gives its electron to pheophytin (Pheo), and Pheo then passes it on to a specialized plastoquinone called $Q_A$ that is attached strongly to D2. $Q_A$ passes the electron to another plastoquinone, called $Q_B$, that is nearby and attached loosely on polypeptide D1. To fully reduce each $Q_A$ and $Q_B$ requires two electrons, so both electrons from $H_2O$ arrive, one at a time, at these molecules. In fact, their reduction also requires addition of two $H^+$, as follows:

$$\text{PQA (quinone: } (H_3C)_2\text{-ring(=O)}_2\text{-}(CH_2-CH{=}\underset{CH_3}{C}-CH_2)_9H) \underset{-2e,\ -2H^+}{\overset{+2e,\ +2H^+}{\rightleftharpoons}} \text{PQAH}_2 \text{ (hydroquinone: } (H_3C)_2\text{-ring(OH)}_2\text{-}(CH_2-CH{=}\underset{CH_3}{C}-CH_2)_9H) \qquad (10.10)$$

The requirement of two electrons and two $H^+$ to reduce plastoquinone (PQ) is important for photosynthetic phosphorylation because the $H^+$ used come from the stroma, but when PQ is later oxidized the $H^+$ are transferred from it into the thylakoid lumen. The overall result of PQ reduction at PS II and then oxidation at the cyt $b_6$-f complex, therefore, is the transport of two $H^+$ from stroma to thylakoid lumen. These $H^+$ are subsequently transported back by ATP synthase in a process that drives ATP formation. Although the PQ called $Q_A$

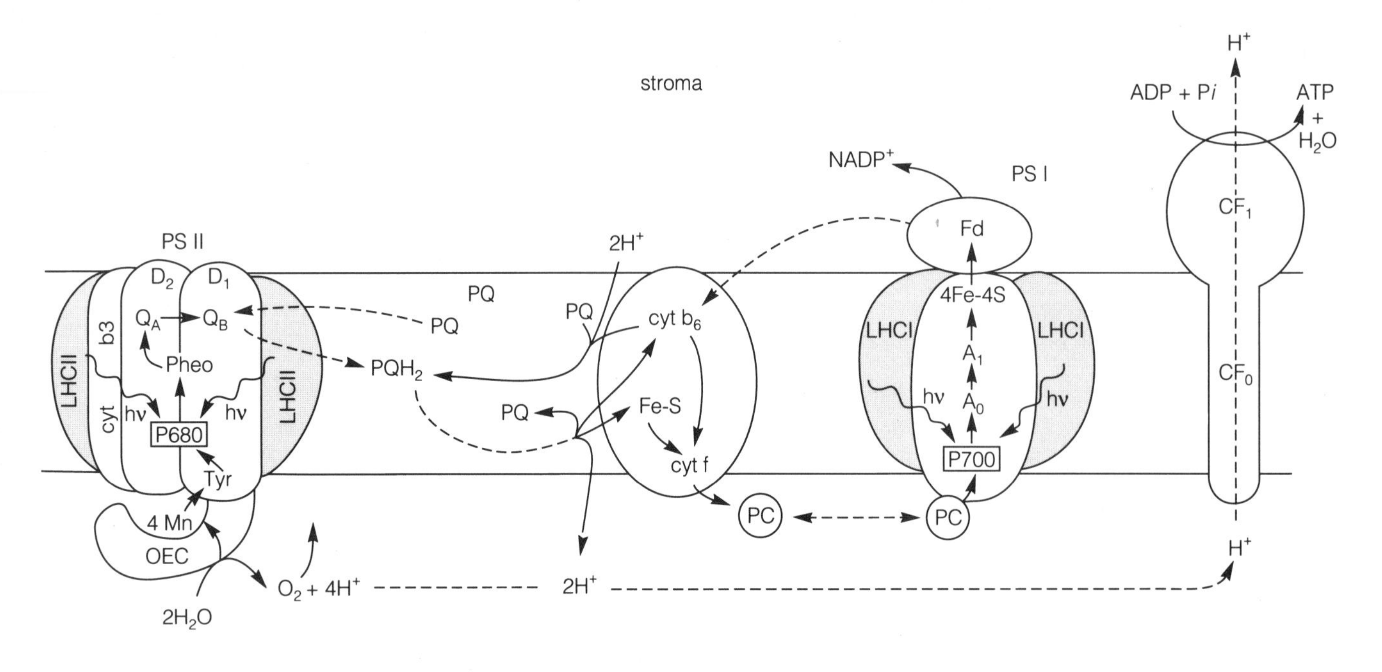


**Figure 10-10** Cooperation of PS II (left), the cyt $b_6$-f complex, and PS I in carrying electrons from $H_2O$ (lower left) in a lumen across a thylakoid membrane to $NADP^+$ in the stroma. These complexes also cooperate to deposit $H^+$ in the lumen and remove $H^+$ from the stroma. The ATP synthase (far right, shown as $CF_0$ and $CF_1$) transports $H^+$ back from lumen to stroma and converts ADP and *Pi* to ATP and $H_2O$. For visual clarity, membrane lipids are omitted, as are most of the polypeptides in each of the four major complexes. Plastoquinones (PQ), plastocyanin (PC), and ferredoxin (Fd) are mobile and carry electrons as indicated by dashed lines. The role of cyt $b_3$ in PS II is still uncertain.

is tightly bound to D2, the PQ called $Q_B$ comes off D1 when $Q_B$ receives two electrons from $Q_A$ and two $H^+$ from the stroma. Then a different oxidized plastoquinone occupies the $Q_B$ site on D1 where another PQ was formerly bound. There are several such quinones dissolved in the thylakoid membrane, and they are highly mobile so that continued replacement is sufficiently rapid. (Certain herbicides kill weeds by occupying the $Q_B$-binding site on D1, which prevents binding of PQ and stops electron transport, thereby blocking photosynthesis; see the boxed essay entitled "Herbicides and Photosynthetic Electron Transport" in this chapter.) For each pair of $H_2O$ molecules oxidized by the OEC, four electrons are transported through the quinones, so two $Q_B$ molecules must be reduced, leave D1, and be replaced. Each of these reduced $Q_B$s (which are now mobile $PQH_2$) carry two electrons to the cyt $b_6$-f complex, where subsequent electron-transport processes occur. It must be noted that as a mobile carrier PQ can interact with a number of different PS II and cyt $b_6$-f complexes in a thylakoid membrane. Thus far we have described the individual reactions summarized earlier in Reaction 10.6.

The cyt $b_6$-f complex can accept only one electron at a time from each mobile $PQH_2$ formed in the PS II complex. These electrons are passed, one at a time, to either the Fe-S protein in that complex or to cyt $b_6$. In either case it is a $Fe^{3+}$ in the protein that accepts the electron as it is reduced to $Fe^{2+}$. The two $H^+$ from each $PQH_2$ are deposited into the thylakoid lumen. Thus from each molecule of $H_2O$ oxidized by the OEC, a total of four $H^+$ are deposited in the lumen: two from oxidation of water and two from $PQH_2$. Oxidation of two $H_2O$ (for release of one $O_2$ or fixation of one $CO_2$) leads to accumulation of eight protons in the lumen.

Each electron in the iron of either cyt $b_6$ or the Fe-S protein is accepted by cyt f, which reduces the iron in it to $Fe^{2+}$. Then cyt f donates an electron to $Cu^{2+}$ in plastocyanin, reducing it to $Cu^+$. When cyt f reduces four plastocyanins, Reaction 10.7 has occurred.

Each mobile PC carries an electron along the thylakoid lumen to PS I. The first molecule to accept an electron from the $Cu^+$ in PC is P700. However, just as is true for P680 in PS II, P700 in PS I cannot accept an electron unless it just lost one. Loss of an electron by P700 to form $P700^+$ occurs when light energy is transferred from light-harvesting pigments in LHCI to P700, thereby sufficiently exciting an electron in P700 so that it can be removed. Therefore, a photon of light absorbed by a PS I antenna pigment causes formation of $P700^+$,

## Herbicides and Photosynthetic Electron Transport

You might suspect that one way herbicides kill weeds is by inhibiting photosynthesis, and for some herbicides this is indeed true. Those that do so generally interfere with some reaction of electron transport. Several urea derivatives, most notably monuron or CMU (3-p-chlorophenyl-1,1-dimethylurea) and diuron or DCMU (3-(3,4-dichlorophenyl)-1,1-dimethylurea), are applied to soil, move in the xylem to leaves, and block electron transport by replacing $Q_B$ on the D1 polypeptide of PS II. Certain triazine herbicides such as simazine and atrazine and certain substituted uracil herbicides, including bromacil and isocil, seem to effect blockage at the same step (reviewed by Schulz et al., 1990). Maize and sorghum are tolerant to triazines (but not to the urea or uracil derivatives) because they contain enzymes that detoxify such compounds.

During the past 20 years more than 40 weedy species have become resistant to the triazines, and this is largely because of a mutation in the chloroplast gene that codes for the amino acid serine at position 264 in D1 (Mazur and Falco, 1989). In the nonmutants this polypeptide binds both $Q_B$ and triazines; the mutant D1 still binds $Q_B$ so that electron transport can proceed, but it no longer binds triazines, monuron, diuron, or the substituted uracils. Several attempts are being made to incorporate the gene for resistance from weeds into crop plants so that crops will be resistant and so that nonresistant weeds can be killed by the herbicides. In one moderately successful example, sexual crosses of the resistant weed *Brassica campestris* with nonresistant *Brassica napus* plants have been made, so commercially important rutabagas, Chinese cabbage, and oil seed rapes may soon be resistant (reviewed by Mazur and Falco, 1989). The review by Schulz et al. (1990) also describes progress in genetic-engineering experiments intended to produce plants resistant to herbicides other than those that act in photosystem II.

which takes an electron from reduced PC, forming oxidized PC and reduced P700. Each electron from P700 moves to $A_0$ (likely still another chlorophyll *a* made special by its chemical environment); then $A_0$ passes its electron first to $A_1$ (probably the phylloquinone vitamin $K_1$ mentioned earlier) and then to one iron in various 4Fe-4S proteins in the PS I core complex mentioned before. These reactions of PS I were summarized by Reaction 10.8. Finally, mobile ferredoxins (Fd in the model) accept one electron each and transfer them to $NADP^+$ to form NADPH in the stroma (Reaction 10.9). The structure of $NADP^+$ shown in Figure 10-11 indicates how its reduction requires two electrons and one $H^+$.

The light-driven reactions by which electrons are transferred across thylakoid membranes to form NADPH are called **noncyclic electron transport** because those electrons do not cycle back to $H_2O$. However, many researchers believe that via a slightly modified pathway light can cause electrons to cycle from P700 through ferredoxin back to certain components of the electron-transport system we just described and then back again to P700. This process is called **cyclic electron transport**. Electrons not donated to $NADP^+$ by ferredoxin can be transported instead to the cyt $b_6$-f complex, probably directly to cyt $b_6$. From here they can move to a plastoquinone, whose full reduction (you recall) takes two electrons and also two $H^+$; both $H^+$ come from the stroma side of the thylakoid. When the electrons then move to the Fe-S protein, each $PQH_2$ releases its two $H^+$ into the thylakoid lumen. These $H^+$ also contribute to the *p*H gradient that drives photophosphorylation. From the Fe-S protein electrons move via the main noncyclic pathway through cyt f and plastocyanin, and then back to P700. Of course, cyclic electron transport in PS I requires energy: one photon per each electron transferred.

In summary, our model (Fig. 10-10) shows that transfer of one electron from $H_2O$ to $NADP^+$ requires two photons because excitation of both photosystems is essential. This explains the Emerson enhancement effect, which first indicated the cooperation of two photosystems. Functions of several other electron-transport components and their physical relationships in thylakoids are also suggested in the model.

Let us now relate this model to the summary equation for photosynthesis (Reaction 10.3). That equation shows that for each molecule of $CO_2$ fixed, one $O_2$ is released and two $H_2O$ molecules are used. The number of photons of light is not specified, but that number is important if we are to be able to calculate photosyn-

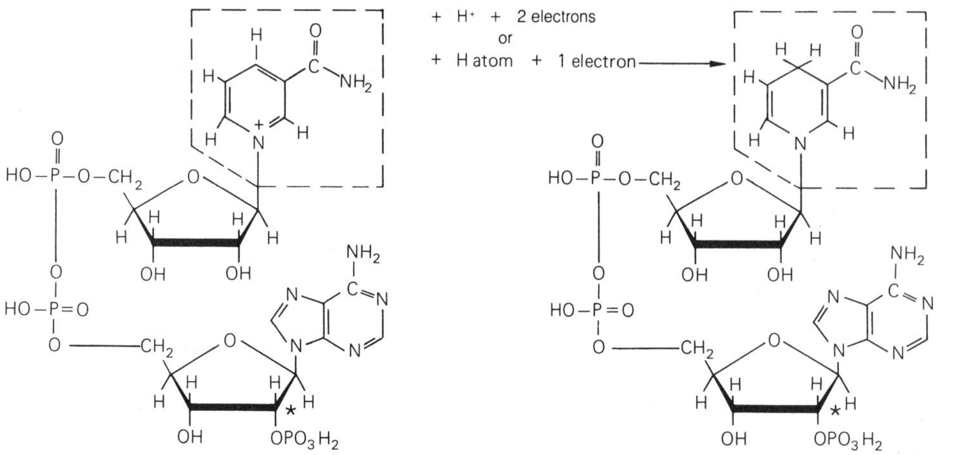

+ $H^+$ + 2 electrons or + H atom + 1 electron

**Figure 10-11** Structures of $NADP^+$ (left) and NADPH (right). The part of the $NADP^+$ molecule undergoing reduction, the nicotinamide ring, is enclosed by a dashed line. One electron is added to the nitrogen atom of the nicotinamide, neutralizing its positive charge, and the second electron is added as part of an H atom to its uppermost carbon atom. The relatively complex $NADP^+$ is a combination of two nucleotides, adenosine monophosphate (AMP; lower half of structure) and nicotinamide mononucleotide (upper half of structure). All **nucleotides** are made of three major parts: (1) a heterocyclic ring, in this case nicotinamide but in other nucleotides a purine or pyrimidine base, (2) the pentose sugar ribose, and (3) phosphate. Phosphate is esterified to the C-5 position of the ribose unit. The two nucleotides in $NADP^+$ are connected in an anhydride linkage between the C-5 phosphate group of each ribose moiety. Notice also that $NADP^+$ contains another phosphate group esterified to the OH group at the C-2 position (see asterisk) of that ribose moiety belonging to AMP. The presence of this additional phosphate is the only way in which $NADP^+$ and NADPH differ from another important electron-carrying coenzyme called **$NAD^+$ (nicotinamide adenine dinucleotide)**. $NAD^+$ and its reduced form, **NADH**, are much less abundant than $NADP^+$ and NADPH in chloroplasts, but they are involved in electron transport during several reactions of respiration (Chapter 13), nitrogen metabolism (Chapter 14), fat breakdown (Chapter 15), and even of a few photosynthesis reactions (Chapter 11).

thetic efficiencies. Our model requires two photons for each electron transported. Each $H_2O$ provides two electrons and two $H_2O$ molecules are required, so our model predicts that a minimum of eight photons would be required to oxidize two $H_2O$ molecules, release one $O_2$, and provide four electrons. These four electrons could reduce two $NADPH^+$, and two NADPH are indeed essential to reduce one $CO_2$ (see Section 11.1). Therefore, based only on NADPH formation, our model shows an eight-photon requirement to accomplish reduction of one $CO_2$ molecule. For leaves of many kinds of plants, 15 to 20 photons per $CO_2$ molecule are usually required (Ehleringer and Pearcy, 1983; Osborne and Garrett, 1983), but probably under the most ideal conditions only 12 photons are required (Ehleringer and Björkman, 1977). Nevertheless, a minimum of nine or 10 photons was required for spinach and pea leaves (Evans, 1987), and there are repeated claims that the green alga *Chlorella* requires only about six photons, as reviewed by Pirt (1986) and by Osborne and Geider (1987). No known model can adequately explain a six-photon requirement, and the stated reviews concern results largely obtained many years ago.

To understand the apparent discrepancy between our eight-photon model and requirements of nine to 12 photons obtained by careful experimenters, we must ask how much ATP is required to cause photosynthesis and how much our model indicates is provided. As is shown in Section 11.2, three ATPs are needed to reduce one $CO_2$ to a simple carbohydrate; more ATP (an uncertain amount) is needed to drive solute accumulation (Section 7.8) and cytoplasmic streaming, and still more is needed to form complex polysaccharides, proteins, and nucleic acids from each reduced $CO_2$. Therefore, our model ought to account for an excess of three ATP produced per two $H_2O$ molecules oxidized and per two NADPH formed. How much ATP is actually produced during photophosphorylation, and how does that process occur? The following section addresses these questions.

## 10.8 Photophosphorylation

Formation of ATP from ADP and P*i* is highly unfavorable from a thermodynamic standpoint. Photophosphorylation can occur only because light energy somehow drives it. To understand how that occurs, note that the ATP synthase or coupling factor ($CF_0$ + $CF_1$) in Figure 10-10 causes ATP formation in the stroma and transport of $H^+$ from the thylakoid lumen to the stroma.

Each process is favored by the other, but ATP formation absolutely requires $H^+$ transport. The $H^+$ ions in the thylakoid lumen arise from oxidation of $H_2O$ and $PQH_2$. These oxidations cause the $H^+$ concentration in the lumen (*p*H 5) to become about 1,000 times as great as that in the stroma (*p*H 8) when photosynthesis is occurring. There is thus a strong $H^+$ concentration gradient toward the stroma, but thylakoids are quite impermeable to $H^+$ and other ions except when transported by ATP synthase. This *p*H gradient across the membrane provides a powerful form of chemical-potential energy largely responsible for driving photophosphorylation.

The idea that *p*H gradients could provide energy for ATP formation in chloroplasts, mitochondria, and bacteria was first proposed in 1961 by Peter Mitchell in England, but his ideas were not accepted by most biochemists for many years. (Mitchell finally received a Nobel prize for chemistry in 1978; see his 1966 and 1985 reviews.) His theory is called the **chemiosmotic theory**, although it bears no clear relation to osmosis, which we described in early chapters. Direct evidence for the chemiosmotic theory was first obtained by photosynthesis researchers G. Hind and Andre Jagendorf at Cornell University in about 1963 (see Jagendorf, 1967). Their work and Mitchell's perseverance sparked thousands of other studies, and now the theory is widely accepted. The theory even explains how **uncouplers** of photophosphorylation work. Uncouplers were so-named because they remove the interdependence (**coupling**) of electron transport and phosphorylation. Many are known, including the relatively simple $NH_3$ and dinitrophenol. Most of them act as "ferryboats," moving into thylakoid channels, picking up a proton, and carrying it back to the stroma, where the *p*H is higher and where the proton is released and reacts with $OH^-$ to form $H_2O$. Still other compounds, such as gramicidin and carbonylcyanide p-trifluoromethoxyphenylhydrazone, may block photophosphorylation by preventing the exit of $H^+$ through $CF_0$. Repeated action of uncouplers destroys the *p*H gradient across the thylakoid membrane and prevents ATP formation, but electron transport is often sped because it is then thermodynamically easier for electron transport to cause a separation of $H^+$ across the membrane.

A continuing problem is that we don't understand the mechanism by which ATP synthase uses the energy released from the "downhill" movement of $H^+$ from channel to stroma to convert ADP and P*i* to ATP. Nevertheless, it is clear that $H^+$ movement through ATP synthase causes structural changes in some of its polypeptides such that they bind ADP and P*i* sufficiently tightly to enable them to react to form ATP and $H_2O$. Even without knowledge of the mechanism, the number of $H^+$ that must be transported to form one ATP can be measured, and that number appears to be three.

Now, consider that the oxidation of two $H_2O$ molecules will directly release four $H^+$ and that four more $H^+$ arise from two $H_2O$ molecules during noncyclic electron transport at the $PQH_2$ oxidation step. Therefore, because eight photons are required by our model to oxidize two $H_2O$ molecules, these eight photons also provide eight $H^+$, nearly enough to form three ATP but not enough to form the more than three that are required to both convert one $CO_2$ into complex compounds and maintain other cellular processes. The process of ATP formation by these reactions in which electrons from $H_2O$ are transported to $NADP^+$, accompanied by $H^+$ transport, is called **noncyclic photophosphorylation**. Formation of additional ATP is thought by most experts to arise from the cyclic pathway described above. Absorption of two such photons causes two electrons to cycle and deposits two $H^+$ in the thylakoid channel when $PQH_2$ is oxidized. No $H_2O$ is split because PS II is not involved, so no NADPH is formed. But ATP is produced by ATP synthase in response to the decreased *p*H in the thylakoid lumen; the formation of ATP by this cyclic electron transport pathway is therefore called **cyclic photophosphorylation**.

If eight photons involving both photosystems produce eight $H^+$ by noncyclic electron transport, and if four additional photons absorbed only by PS I produce four more $H^+$, then the 12-photon total would lead to four ATP molecules. We stated earlier that more than three ATP were needed to convert one $CO_2$ to complex compounds and that measurements with leaves showed that a minimum of nine (or 12) photons were required for each $CO_2$ used. With both cyclic and noncyclic electron transport, our model is consistent with experimental results; so we can rewrite Reaction 10.3 to show approximately how many photons of light are required to convert one $CO_2$ into a product we shall refer to as $(CH_2O)$, as an abbreviation for carbohydrates such as sucrose and starch. In reality, this $(CH_2O)$ represents all organic forms of carbon in the plant. Proteins and nucleic acids that require (on a weight basis) more ATP to be produced than do polysaccharides are more abundant in actively growing cells than in mature cells, in which polysaccharides predominate. Our new equation (Reaction 10.11) therefore shows a minimum requirement of 12 photons, depending on the physiological status of the cells and allowing for some uncertainty in photophosphorylation:

$$CO_2 + 2H_2O + 12\text{ photons} \rightarrow (CH_2O) + O_2 + H_2O \qquad (10.11)$$

Neither ATP nor NADPH appear in this summary because their production is balanced by their use in $CO_2$ reduction. We further note that if the Q cycle operates (see Section 10.5), the number of $H^+$ moved from stroma to lumen during electron transport is doubled, so this could enhance the amount of photophosphorylation (without changing the NADPH output) and could therefore lower the photon requirement to close to eight.

# Role of Chlorophyll *a* in Photosynthesis

*Govindjee*

*Govindjee, who uses one name only (his last name, Asthana, was dropped by his father as a protest against the caste system in India), obtained his Ph.D. (in biophysics) in 1960 from the University of Illinois at Urbana-Champaign (UIUC). Since 1961 he has been on the faculty of the Department of Physiology and Biophysics and Plant Biology at UIUC. His honors include: Past President of the American Society for Photobiology; Distinguished Lecturer of the School of Life Sciences, UIUC; Fellow of the American Association of Advancement of Science and of the National Academy of Science (India). His research interests are in primary photochemistry, use of chlorophyll* a *fluorescence as a probe of photosynthesis, and mechanism of photosystem II reactions.*

My interest in photosynthesis dates back to 1953 when I was a M.Sc. student of Shri Ranjan at the University of Allahabad in India. I had signed up for the special paper in Advanced Plant Physiology and chose to write a term paper on the "Role of Chlorophyll in Photosynthesis." A library search led me to the pioneering papers of Robert Emerson on the action spectra of photosynthesis, complete with all their spicy extensions, such as the controversy raging between Emerson and Otto Warburg regarding the *minimum* quantum requirement for $O_2$ evolution in photosynthesis (eight quanta for Emerson and four for Warburg) and the unusual existence of the "*Red Drop*"—the inefficiency of the light absorbed by chlorophyll *a* in the red region of the spectrum (> 680 nm).

It was not until I had completed my M.Sc. in Botany in 1954 and had been working for awhile on a highly exciting project on the effects of virus infection on amino-acid metabolism with Ranjan's team (M. M. Laloraya, T. Rajarao, and Rajni Varma) that I started correspondence with Emerson. Enthusiastic and personal letters from Emerson, a Fulbright travel grant, and a fellowship in physicochemical biology from the University of Illinois led me to Emerson's laboratory in the basement of the old Natural History Building in Urbana. There my footsteps would cross those of Warburg, who had already visited from Berlin, and those of Eugene Rabinowitch, the physical chemist and prophet of photosynthesis and world peace, who would greet me and tutor me in the biophysics of photosynthesis. Under the stern yet friendly eyes of Emerson, I struggled to learn sophisticated but extraordinarily tedious techniques of manometry (readings were taken with a cathetometer in constantly shaking manometers to a precision of hundredths of a mm of pressure change) and bolometry to measure the absolute quantum yield of $O_2$ evolution and the action spectra of photosynthesis. The high dispersion monochromator that I used for the studies was mammoth and constructed by Emerson himself. Emerson gave us responsibility and granted us independence, and we were infected by his enthusiasm and excitement about his discovery of the *Enhancement Effect*. The tragic air crash in the East River near La Guardia airport in New York on February 4, 1959, that took Emerson's life forced my wife Rajni and me to complete our Ph.Ds. without the benefit of his guidance. We finished the degrees under Rabinowitch, who was very generous and welcomed us as his students.

In the diatom *Navicula minima* I discovered a new band in the action spectrum of the *Emerson Enhancement Effect* that had a peak at 670 nm in the region of chlorophyll *a* absorption. I was quite excited about this discovery because it provided a solution to the discrepancy between (a) Emerson's view that the accessory pigments (such as chlorophyll *b* in green plants, fucoxanthol and chlorophyll *c* in diatoms, and phycobilins in red and blue-green algae) performed one light reaction and that chlorophyll *a* performed another, and that the two cooperated to do photosynthesis; and (b) L. N. M. Duysens's early work in the Netherlands, which showed that light energy absorbed by the accessory pigments was efficiently transferred to chlorophyll *a*. From my results it became clear that chlorophyll *a* played an important role in both photosystems. The idea of two light reactions (we used to call them short-wave and long-wave photoreactions) in series to explain the minimum photon or quantum requirement of eight was already in Rabinowitch's 1945 book, and the idea that one light reaction may reduce a cytochrome molecule and the other oxidize it while reducing $NADP^+$ was also stated in Rabinowitch's 1956 book; that is to say, the basics of the so-called R. Hill and F. Bendall (1960) scheme already existed at Urbana (see discussions by L. N. M. Duysens, Photosynthesis Research 21:61–69, 1989).

In 1957 Bessel Kok had already discovered the existence of the special chlorophyll *a* that he called P700, and he proposed that it was the reaction center chlorophyll *a* of photosynthesis. In our 1965 Scientific American article on the "Role of Chlorophyll in Photosynthesis" (213(1):74–83), Rabinowitch and I speculated on the existence of yet another special chlorophyll *a*, the P680 reaction center of the short-wave system. This reaction center was indeed discovered in 1968 in H. T. Witt's laboratory in Berlin. Most of the chlorophyll *a* molecules in both the pigment systems is for harvesting (or capturing) light energy, so these molecules belong to what is called the "antenna" system. They function to absorb and transfer excitation energy to the reaction center special chlorophyll *a* molecules known as P680 and P700. This role is no different in principle from that of the various accessory pigments such as carotenoids. P680 and P700, which are present in one out of several hundred chlorophyll *a* molecules, are the initiation sites of conversion of light energy into chemical energy. (For a general discussion, see my 1974 article with Rajni, "The Primary Events of Photosynthesis," Scientific American 231(6):68–82.)

The primary reactions of photosynthesis are:

$$\text{P680} + \text{pheophytin } a + h\nu \xrightarrow{3\text{ps}} \text{P680}^{+} + \text{pheophytin } a^{-}$$

$$\text{P700} + \text{chlorophyll } a + h\nu \xrightarrow{\leq 10\text{ps}} \text{P700}^{+} + \text{chlorophyll } a^{-}$$

Here, P680 and P700 are the primary electron donors, and pheophytin *a* (essentially chlorophyll *a* without Mg, but with $H^+$s in the center) and chlorophyll *a* (also called Ao) are the primary electron acceptors. Upon receiving light energy, the above primary charge-separation reactions occur. These are the *true* light reactions of photosynthesis in which light energy is actually converted into chemical energy. Electron transfers from P680 to pheophytin and from P700 to chlorophyll *a* (Ao) are energetically "uphill" processes requiring the input of light energy.

In 1977 V. Klimov and A. A. Krasnovsky and coworkers in the Soviet Union had discovered that pheophytin was the primary electron acceptor of photosystem II, the P680-containing system. It was not until 1989 that we, in collaboration with Mike Wasielewski and Doug Johnson at the Argonne National Laboratory in Illinois and Mike Seibert at the Solar Energy Research Institute in Colorado, were able to measure the incredibly rapid electron transfer between P680 and pheophytin. At 4°C it occurs in 3 picoseconds (or 3,000 femtoseconds; remember that there are more femtoseconds in a second than there are seconds in 30 million years). The nature of the primary electron acceptor of photosystem I, the P700-containing system, has been proposed by several researchers to be a special chlorophyll *a* molecule. In collaboration with James Fenton and working in two different laboratories, we have shown that the primary acceptor of photosystem I is indeed a form of chlorophyll *a* that is reduced within 10 picoseconds.

Next we turn to the matter of chlorophyll *a* fluorescence. Upon absorption of light, the excited antenna chlorophyll *a* molecules have several potential pathways for deexcitation: energy transfer to other chlorophyll *a* molecules, radiationless processes (heat), and light emission (prompt fluorescence). I had been fascinated and intrigued by light emission from plants and animals since I was a student at Allahabad University (1950–1954). When I came to Urbana, Rabinowitch's research group was involved with studies on chlorophyll *a* fluorescence, and the potential of this technique to learn new things about photosynthesis seemed limitless.

In 1958 Steve Brody had not only discovered the new long-wavelength fluorescence emission band (now labeled F730) at 77 K (liquid-nitrogen temperature), now known to emerge from photosystem I, but he had also made the first measurements on the lifetime of chlorophyll *a* fluorescence *in vivo*. I grabbed the very first opportunity to test whether chlorophyll *a* fluorescence could be used as a probe to check the existence of two light reactions in photosynthesis. We excited algal cells with blue and far-red light separately and together, and in 1960 we discovered the quenching effect of far-red light on the red chlorophyll *a* fluorescence produced by blue light; these data were consistent with the two-light reaction scheme of photosynthesis. My fascination for using chlorophyll *a* fluorescence as a probe of photosynthetic reactions has continued until today. There is a long list of our observations that have provided new information on the structure and function of photosystem II and its interaction with photosystem I. I shall give a few examples.

I was particularly excited about the finding in 1963 of a fluorescence emission band at 696 nm (F696), at 77 K, that originated in photosystem II; this is now used to monitor "active" photosystem II. Between 1966 and 1970 the temperature-dependence of these emission bands (F730 and F695) and the well-known F685 band down to 4 K provided information on the mechanism of excitation-energy transfer in photosynthesis; that is, these facts showed the validity of Förster's inductive-resonance transfer mechanism.

Changes in chlorophyll *a* fluorescence yield after a single-turnover, bright actinic flash constitute an extremely powerful probe of the photosystem II reaction. Between a nanosecond and less than a microsecond, the rise in chlorophyll *a* fluorescence monitors electron donation from the electron donor Z (now known to be tyrosine) to the oxidized P680, $P680^+$, whereas during microseconds to milliseconds, the decay in chlorophyll *a* fluorescence monitors electron flow from the primary quinone acceptor of photosystem II to the secondary quinone acceptor of photosystem II, $Q_B$. The fluorescence rise occurs because $P680^+$ is a quencher of chlorophyll *a* fluorescence, and electron flow from Z to $P680^+$ decreases the concentration of $P680^+$. However, the fluorescence decay occurs because $Q_A^-$, formed by the bright flash, becomes reoxidized, and a quencher of chlorophyll *a* fluorescence ($Q_A$) is formed as electrons are transferred to $Q_B$. We have exploited this fluorescence-decay method to understand the site of action of bicarbonate/$CO_2$ in chloroplast reactions in thylakoid membranes.

Warburg had discovered in 1958 that the Hill reaction is drastically diminished if $CO_2$ is removed from photosynthetic systems. He interpreted this observation as support of his hypothesis that $O_2$ in photosynthesis originates in $CO_2$, not in $H_2O$! In 1970 Alan Stemler took up this problem for his Ph.D. thesis after listening to my lecture on the source of $O_2$ in photosynthesis. Starting in 1975, we have successfully used the fluorescence-decay method in establishing with others that one major effect of the removal of bicarbonate ions from the

(*continued*)

## Role of Chlorophyll *a* in Photosynthesis *(continued)*

thylakoid membrane is to drastically slow down (or block) the electron flow from $Q_A$ to the plastoquinone pool, a process that is quite similar to what happens when herbicides such as diuron and atrazine are added to plants [see the boxed essay entitled "Herbicides and Photosynthetic Electron Transport" in this chapter]. This site of bicarbonate effect between $Q_A$ and the plastoquinone pool has been fully confirmed by other biochemical and biophysical measurements. No evidence has yet been found to support Warburg's picture of the involvement of $CO_2$ in $O_2$ evolution. Our current research involves using chlorophyll *a* fluorescence as a probe of reactions involving $Q_A$ in specific herbicide-resistant and site-directed mutants of transformable cyanobacteria to decipher the site of bicarbonate/$CO_2$ binding.

In conclusion, chlorophyll *a* not only plays a role in harvesting light energy, converting light energy into chemical energy, and in acting as both primary electron donors (P680, P700) and as primary electron acceptors (Ao; pheophytin is derived from chlorophyll by the replacement of Mg with $H^+$s in the center), but its fluorescence can be used as a sensitive and a nondestructive probe of the structure and function of photosynthesis.

## 10.9 Distribution of Light Energy Between PS I and PS II

To maintain maximum efficiency of photosynthesis requires that each photosystem receives input of the same number of photons per unit time, so that activation of P680 and P700 can cooperatively drive electrons toward $NADP^+$ (Fig. 10-10). Until about 1980, scientists assumed that the two photosystems existed in about equal ratios in each chloroplast, but then work with various angiosperm species growing in shade or sun (and even cyanobacteria) showed that PS II/PS I ratios varied from 0.43 up to 4.1; the higher ratios were obtained in plants growing in deep shade. The expected ratio of 1.0 seldom occurred (Melis and Brown, 1980). How can efficient photosynthesis occur in plants containing more of one system than the other?

The answer involves adaptation to light, both on a short-term and a long-term scale. On the short term (30 seconds or less), the photosystems adapt so that energy redistribution occurs between PS II and PS I. For energy redistribution from PS II to PS I, there is actual movement of some of the LHCII pigments and proteins so that they now associate with PS I in stroma thylakoids (Barber, 1987b; Anderson, 1986; Anderson and Andersson, 1988). There the LHCII pigments transfer more light energy to PS I and none to PS II. Movement occurs because the proteins of LHCII become more negatively charged by becoming phosphorylated through the actions of ATP and a specific protein kinase that transfers phosphate from ATP to those proteins. Phosphate groups ionize (lose $H^+$) to create the additional negative charges on such proteins. These excess negative charges force some of the LHCII proteins with associated pigments apart, and they are then attracted toward a specific positively charged protein or proteins of PS I in stroma thylakoids.

The role of light in LHCII movement seems to be as follows: Preferential absorption of light by PS II causes reduction of numerous PQ molecules to $PQH_2$ (see Fig. 10-10); then these $PQH_2$ activate the protein kinase. Light absorbed preferentially by PS I causes oxidation of $PQH_2$ molecules, stops phosphorylation, and allows phosphate removal by a phosphatase that hydrolyzes phosphate groups away from mobile LHCII proteins. Loss of phosphate reduces their negative charge so that they then move back to appressed thylakoid regions and donate light energy to PS II. It is still unclear how plastoquinones control activity of the protein kinase and how the resulting changes in charge of LHCII proteins cause them to fit better in one photosystem than in another. Nevertheless, this mechanism certainly increases cooperation of the two photosystems and represents an important adaptation for maximizing photosynthetic efficiency. Long-term adaptations of chloroplasts to various environmental conditions are described in Chapter 12.

# 11

# Carbon Dioxide Fixation and Carbohydrate Synthesis

In Chapter 10 we explained how chloroplasts capture light energy to produce NADPH and ATP. Those two molecules are said by some authors to have reducing power because both help reduce $CO_2$ after it is fixed into the carboxyl group of a three-carbon acid, as explained in this chapter. Strictly speaking, however, it is NADPH that is the reductant; ATP only facilitates reduction. Although plants vary in how they fix $CO_2$ into organic acids, NADPH and ATP are always involved in reduction in such a way that carbohydrates are formed from the $CO_2$ molecules fixed.

The sequence of reactions involving $CO_2$ fixation and formation of carbohydrates by photosynthesis was identified only after radioactive carbon-14 became available in about 1945. Carbon dioxide containing $^{14}C$ was then prepared, and all plant molecules produced from it during photosynthetic experiments were therefore tagged with that isotope. Paper chromatography was developed at about the same time, making separation of photosynthetic products possible.

Labeled molecules on paper chromatograms prepared from alcohol extracts of plants that had fixed $^{14}CO_2$ were detected by autoradiography (see Section 8.1). In this technique, X-ray film is placed in tight contact with chromatograms in darkness for a few days to a few weeks (depending on the amount of radioactivity present); during this time radioactivity exposes the film. When the film is developed, dark spots indicate the locations of radioactive compounds. Identities of the compounds and amounts of radioactivity in each can be determined after cutting out the areas of paper that correspond to each dark spot on the film. At first Geiger-Müller tubes were used to measure radioactivity, but now the far more sensitive liquid-scintillation counters are used. Direct chemical analyses and rechromatography of unknown substances on paper with known compounds (**cochromatography**) were first used to identify radioactive photosynthetic products, but nuclear magnetic resonance (NMR) and mass spectrometry currently provide much more powerful tools for analysis. High-performance liquid chromatography (HPLC) is now a much better and faster separation technique than paper chromatography. These techniques have given us a fairly complete understanding of the pathways of $CO_2$ fixation and carbohydrate synthesis.

## 11.1 Products of Carbon Dioxide Fixation

### The First Product

Paper chromatographic procedures in conjunction with use of $^{14}CO_2$ were applied to the problem of photosynthesis by Melvin Calvin, Andrew A. Benson, James A. Bassham, and others at the University of California, Berkeley, from 1946 to 1953 (see the essays in this chapter by Drs. Calvin and Bassham and the former's personal perspective in Calvin, 1989). These investigators allowed unicellular green algae such as *Chlorella* to attain a constant rate of photosynthesis and then introduced $^{14}CO_2$ into solutions in which the algae were growing. At various times after the introduction of $^{14}C$, algae were dropped into boiling 80-percent ethanol to kill them rapidly and extract any radioactive compounds. Each extract was chromatographed on paper, and autoradiograms were made.

After photosynthesizing for 60 seconds in $^{14}CO_2$ the algae had formed many compounds, as shown by the many dark spots on the film in Figure 11-1 (top). Amino acids and other organic acids had become radioactive, but the compounds containing the most $^{14}C$ were phosphorylated sugars, shown at the lower right of that autoradiogram. To identify the first product formed from $CO_2$, time periods were shortened to 7 s and finally even to as little as 2 s (Fig. 11-1, middle and bottom). When this was done, most $^{14}C$ was found in a phosphorylated three-carbon acid called **3-phosphoglyceric acid (3-PGA)**. This acid was the first detectable product of photosynthetic $CO_2$ fixation in these algae

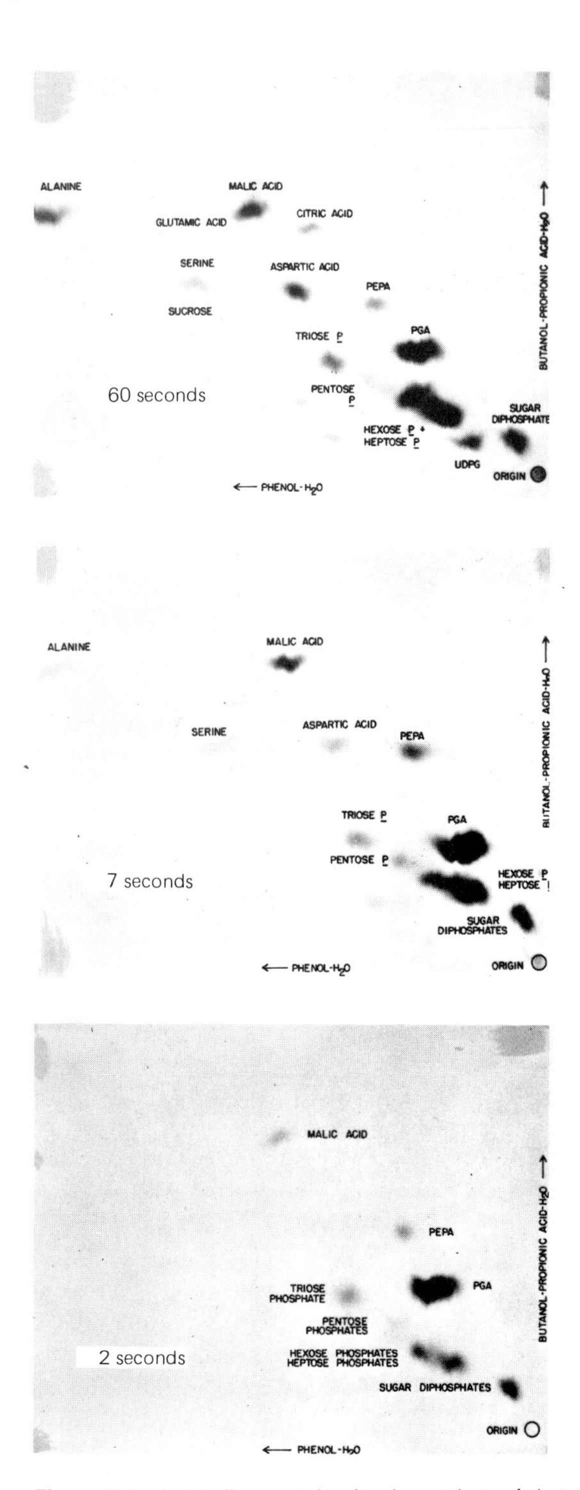


**Figure 11-1** Autoradiograms showing the products of photosynthesis in the alga *Chlorella pyrenoidosa* after various times of exposure to $^{14}CO_2$. (Top) 60 seconds, (middle) 7 seconds, (bottom) 2 seconds. Note the increasing importance of 3-PGA and other sugar phosphates as the exposure time is shortened. (From Bassham, 1965.)

and in leaves of all plants then investigated. (Note that 3-PGA and most plant acids exist largely in the ionized form, without the $H^+$ that exist in the carboxyl groups of true acids. Ionized acids exist as salts with a counter cation, usually $K^+$; hence, 3-PGA exists as the negatively charged 3-phosphoglycerate.) It was expected that the compound with which $^{14}CO_2$ normally combines would accumulate if $^{14}CO_2$ were fed to algae for a short time and then the $CO_2$ supply were suddenly removed. No two-carbon compound was found. The substance that did accumulate was a five-carbon sugar, phosphorylated at each end, called **ribulose-1,5-bisphosphate (RuBP)**.[1] At the same time, there was a rapid drop in the level of labeled 3-PGA, which suggested that ribulose bisphosphate is the normal substrate to which $CO_2$ is added to form PGA.

## The Reaction of $CO_2$ Fixation

A year or so later (1954) an enzyme was found that irreversibly catalyzed the combination of $CO_2$ with RuBP to form two molecules of 3-PGA. This is an unusually important reaction. An unstable intermediate product is formed (shown in brackets in Reaction 11.1) that with the addition of water splits into two 3-PGAs. Thus, if $^{14}CO_2$ is involved (see asterisks* in Reaction 11.1), one of the two 3-PGA products becomes labeled with $^{14}C$ and one remains unlabeled:

$$\begin{array}{c} \mathrm{CH_2OPO_3H^-} \\ | \\ \mathrm{C{=}O} \\ | \\ \mathrm{C^*O_2 + H{-}C{-}OH} \\ | \\ \mathrm{H{-}C{-}OH} \\ | \\ \mathrm{CH_2OPO_3H^-} \end{array} \xrightarrow{Mg^{2+}} \left[ \begin{array}{c} \mathrm{O} \quad \mathrm{CH_2OPO_3H^-} \\ \backslash\backslash \quad | \\ \mathrm{C^*{-}C{-}OH} \\ / \quad | \\ \mathrm{OH} \quad \mathrm{C{=}O} \\ | \\ \mathrm{H{-}C{-}OH} \\ | \\ \mathrm{CH_2OPO_3H^-} \end{array} \right] \xrightarrow{+\ H_2O}$$

$$\begin{array}{c} \mathrm{CH_2OPO_3H^-} \\ | \\ \mathrm{H{-}C{-}OH} \\ | \\ \mathrm{C^*OOH} \\ \text{3-PGA} \end{array} + \begin{array}{c} \mathrm{COOH} \\ | \\ \mathrm{H{-}C{-}OH} \\ | \\ \mathrm{CH_2OPO_3H^-} \\ \text{3-PGA} \end{array} \qquad (11.1)$$

[1]This compound and others with phosphate groups on separate carbon atoms of a molecule were first called diphosphates (thus, ribulose-1,5-diphosphate), but the term *bis*phosphate now distinguishes them from true diphosphates such as ADP in which the two phosphates in a pyrophosphate group are linked only to one carbon.

The enzyme that catalyzes this reaction is commonly called **ribulose bisphosphate carboxylase**, which we abbreviate as **rubisco**. Rubisco is functional in all photosynthetic organisms except a few photosynthetic bacteria. It is important not only because of the essential reaction it catalyzes, but also because it seems to be by far the most abundant protein on earth (Ellis, 1979). Chloroplasts contain approximately half of the total protein in leaves, and about one-fourth to one-half of their protein is rubisco; thus this enzyme is one-eighth to one-fourth of all the protein in leaves, and as such it is important in the diets of animals, including humans. Andrews and Lorimer (1987) estimated that the average human requires the constant activity of about 44 kg of rubisco just to obtain food through photosynthesis.

## 11.2 The Calvin Cycle

Investigations of additional radioactive compounds that are formed rapidly from $^{14}CO_2$ identified other sugar phosphates containing four, five, six, and seven carbon atoms. These included the tetrose (four-carbon) phosphate *erythrose-4-phosphate*; the pentose (five-carbon) phosphates *ribose-5-phosphate, xylulose-5-phosphate*, and *ribulose-5-phosphate*; the hexose (six-carbon) phosphates *fructose-6-phosphate, fructose-1,6-bisphosphate* and *glucose-6-phosphate*; and the heptose (seven-carbon) phosphates *sedoheptulose-7-phosphate* and *sedoheptulose-1,7-bisphosphate*. By noting the time sequences in which each sugar phosphate became labeled from $^{14}CO_2$ and then degrading each to determine which atoms contained $^{14}C$, it was possible to predict a metabolic pathway that related the compounds to each other (Bassham, 1965, 1979).

When $^{14}C$-labeled 3-PGA was degraded, most of the $^{14}C$ was in the carboxyl carbon, as shown by asterisks in Reaction 11.1, but the other two carbons were also labeled. This labeling pattern suggested that the latter two carbons of 3-PGA were not derived from $^{14}CO_2$ directly but were instead formed by transfer of carbon from the carboxyl-carbon atom of 3-PGA by some cyclic process. A cyclic pathway that uses 3-PGA to form the other sugar phosphates previously mentioned and that also converts some of its carbons back to RuBP was soon worked out. These reactions have collectively been named the **Calvin cycle**, the **photosynthetic carbon reduction cycle**, or the **C-3 photosynthetic pathway** (because the first product, 3-PGA, contains three carbons). Calvin was awarded a Nobel prize in chemistry in 1961 for this work.

The Calvin cycle occurs in the stroma of chloroplasts and consists of three main parts: **carboxylation**, **reduction**, and **regeneration**, as explained below and summarized in Figure 11-2. Carboxylation involves addition of $CO_2$ and $H_2O$ to RuBP to form two molecules

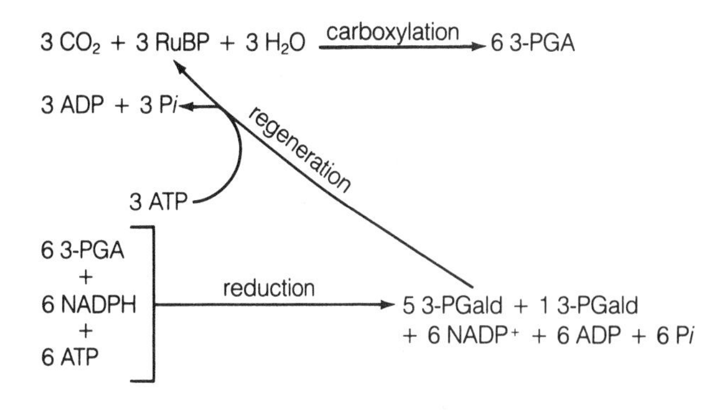


summary:

$$3\,CO_2 + 5\,H_2O + 9\,ATP + 6\,NADPH + 6\,H^+ \longrightarrow 1\ 3\text{-PGald} + 6\,NADP^+ + 9\,ADP + 8\,Pi$$

**Figure 11-2** A summary of the Calvin cycle, emphasizing the carboxylation, reduction, and regeneration phases.

of 3-PGA (Reaction 11.1). In the reduction phase, the carboxyl group in 3-PGA is reduced to an aldehyde group in *3-phosphoglyceraldehyde* (*3-PGaldehyde*), as follows:

$$\underset{\text{3-PGA}}{\mathrm{C(=O)OH{-}HC(OH){-}CH_2OPO_3H^-}} \xrightarrow{ATP \;\to\; ADP} \underset{\text{1,3-bisphosphoglyceric acid}}{\mathrm{C(=O)OPO_3H^-{-}HC(OH){-}CH_2OPO_3H^-}}$$

$$\xrightarrow[H_2PO_4^-\ (Pi)]{NADPH + H^+ \;\to\; NADP^+} \underset{\text{3-phosphoglyceraldehyde}}{\mathrm{C(=O)H{-}HC(OH){-}CH_2OPO_3H^-}} \qquad (11.2)$$

Note that reduction does not occur directly; instead, the carboxyl group of 3-PGA is first converted to an acid anhydride type of ester in *1,3-bisphosphoglyceric acid* (*1,3-bisPGA*) by addition of the terminal phosphate group from ATP. This ATP arises from photophosphorylation (described in Chapter 10), and the ADP released when 1,3-bisPGA is formed is quickly converted back to ATP by additional photophosphorylation reactions. The actual reducing agent in Reaction 11.2 is NADPH, which donates two electrons to the top carbon atom involved in the anhydride ester group. Simultaneously, *Pi* is released from that group to be used again to convert ADP to ATP. The $NADP^+$ is reduced back to NADPH in light-driven reactions described in Chapter 10 (see Fig. 10-10).

# Exploring the Path of Carbon in Photosynthesis (I)

*James A. Bassham*

*James A. Bassham, whom most of his friends call "Al," participated as a graduate student in elucidating the pathway of carbon in photosynthesis and is now a professor of chemistry at the University of California, Berkeley. In 1977 he told us his story:*

When I think back to my days as a graduate student 30 years ago, the first two impressions that enter my mind are those of good fortune and excitement. The good fortune had to do with my becoming involved in that particular research at that time and place.

The chain of events that led me there began with my having to fill out a form of entry to the University of California at Berkeley and deciding to put down chemistry in the appropriate blank space because I had received a chemistry prize in high school. Then, during my freshman chemistry class at Cal one spring morning, our section instructor, Professor Sam Ruben, chose to put aside his usual laboratory lecture notes and tell us a little about his research. He and other scientists had discovered a new radioisotope of carbon, called carbon-14, and were putting it to many uses, of which the most interesting to me was as a tracer to map the pathway of carbon fixation by green plants making sugar and using sunlight. This was certainly more interesting than precipitating various sulfides with hydrogen sulfide on the porch of the freshman chemistry laboratory as part of my training in qualitative analysis. The advent of World War II, however, and three years of service in the U.S. Navy were to put such thoughts from my mind until a later date.

The third link in this chain of circumstances followed my return to Berkeley as a graduate student in chemistry, after the close of the war. Since I had been an undergraduate student at Berkeley, the Chemistry Department was reluctant to admit me as a graduate student, it being their wise policy to encourage former undergraduates to go elsewhere for graduate training. After much discussion, they allowed me to enter for course work only. By some miracle, after the first semester had passed the Dean invited me to stay on and work for my doctorate.

As is usual in such cases, I was directed to interview several professors of organic chemistry so that, it was hoped, I would find one willing to accept me to do research on some project for my thesis. The first person I spoke to was a young professor named Melvin Calvin. By coincidence, the first topic he mentioned to me was the mapping of the path of carbon reduction during photosynthesis using carbon-14 as a label. I soon learned that Dr. Ruben had lost his life in an unfortunate laboratory accident during the war, and that the work on carbon-14 was now being carried forward in the newly formed BioOrganic Chemistry Group of the University of California. Professor Ernest Lawrence, director of the Radiation Laboratory (now the Lawrence Berkeley Laboratory), had invited Melvin Calvin to form a division to explore the uses of carbon-14 in investigations of biochemistry and organic-reaction mechanisms. The work was already under way.

Professor Calvin then proceeded to tell me of several projects involving organic synthesis and reaction mechanisms that were making use of this radioisotope, but he might have saved himself the trouble because my mind was already made up. If at all possible, and if acceptable to him, I was very eager to get going on the work of using carbon-14 to study photosynthesis. He agreed to this and before long escorted me from the old red-brick chemistry building across the court to an even older and shabbier wooden frame building that had been constructed as a "temporary" building about half a century before. As most university colleagues will know, temporary buildings built on university campuses are usually good for at least 50 years.

In this building, called the Old Radiation Laboratory, I soon became acquainted with a small group of people who were to prove instrumental in the work leading to the mapping of the path of carbon in photosynthesis. Of key importance among these was Andrew A. Benson, a young postdoctoral scientist working with Professor Calvin on the photosynthesis project. Andy was an excellent experimentalist and taught me a great deal about how to devise apparatus and techniques

---

Reaction 11.2 represents the only reduction step in the entire Calvin cycle; because both of the 3-PGA molecules produced in Reaction 11.1 are reduced in the same way, this step involves use of two of the three ATPs required to convert one $CO_2$ molecule into part of a carbohydrate. Thus for each $CO_2$ fixed, two NADPH and two ATP are required here. A third ATP is used in the regeneration phase, making the total requirement three ATP and two NADPH for each molecule of $CO_2$ fixed and reduced.

What is regenerated in the regeneration phase is RuBP, which is needed to react with the additional $CO_2$ constantly diffusing into leaves through stomates. This phase is complex and involves phosphorylated sugars

for solving new kinds of problems in the laboratory. Of course, as those familiar with the mapping of the path of carbon in photosynthesis know, he played a very important role in the identification of various sugar-phosphate intermediates, which were to turn out to be the essential compounds in the photosynthetic pathway. There were others who played important roles — visiting scientists, staff scientists, and students — but space will not permit me to describe them all.

Thinking about these former colleagues and the old laboratory brings to mind my second main impression of that time: a sense of great excitement. The Old Radiation Laboratory was an exciting place to work, the research project was fascinating, and the people were all, without exception, enthusiastic about what they were doing. I suppose, to some extent, this is because we had a new and, at that time, almost exclusive technique and an important problem to which to apply it. It went far beyond that, however, and, to a large extent, stemmed directly from the personality of Melvin Calvin. He was in the laboratory every morning, or as soon as he could get free from his teaching duties, asking questions about the latest experiments and laying out a program of new experiments to pursue. Never mind the fact that just the day before he had outlined an experiment that, in our opinion, would take a month or so; he would be in the next morning to ask about the progress we had made with it. As we all know, this is an excellent procedure for stimulating graduate students. I don't mean to imply that he was a taskmaster; he was simply motivated by a tremendous enthusiasm for the project and the results we were getting. When we were fortunate to get a new experimental result, it always took on a new and greater importance after he had examined it.

This sense of achievement was further heightened when we developed techniques for analyzing the radioactive products of photosynthesis by using two-dimensional paper chromatography and radioautography. It was a beautiful analytical tool but suffered from one drawback: It required about two weeks from the original experiment until the films were developed to locate and count the $^{14}C$ in the radioactive spots. Of course, we had to learn to go on doing other experiments without waiting for the results of the previous one. Thus, there was a considerable sense of suspense when the X-ray films were developing in the darkroom and the spots first began to appear before our eyes.

Some other phases of the research were even more painstaking and drawn out. For example, the degradation of molecules following short periods of photosynthesis to locate the radiocarbon within the molecule could take several weeks, or even months, before the final product was obtained.

Driven by our excitement to overcome these difficulties, we tended to work long hours and long weeks. When this reached an intolerable point and we felt the need for some mental rejuvenation, Andy Benson would organize an expedition to the high Sierra, and several of us would dash off to the mountains for a strenuous weekend of climbing 14,000-foot peaks. After such excursions we returned to the laboratory physically exhausted but mentally refreshed. I suspect that Melvin Calvin was always relieved to see us return to the laboratory after these mountain and rock-climbing expeditions with no more serious infirmities than a few sore muscles or blisters on our feet.

About midway through the time when the path of carbon in photosynthesis was being mapped, I had done enough work on my part of the project to justify writing it up as a doctoral thesis. My main reluctance in doing this was the thought that I might then have to depart from this interesting project and find a job somewhere else doing something that was sure to be far less exciting. Fortunately for me, I was invited to stay on in a postdoctoral status and was able to remain a part of the team while the path of carbon in photosynthesis was fully mapped.

The techniques that we developed during that period proved to be extremely valuable for studies of metabolic regulation in plant cells (and even in animal cells) and have shaped my whole scientific career since that time. To a large extent, we have been able to maintain in our laboratory a sense of excitement and cooperation over the years, in spite of two moves and growth from a dozen people to over 100. The person who has been most responsible for maintaining that sense of excitement and purpose is Melvin Calvin.

with four, five, six, and seven carbons, as shown in detail in Figure 11-3. In the final reaction of the Calvin cycle (reaction 13), the third ATP that is required for each molecule of $CO_2$ fixed is used to convert ribulose-5-phosphate to RuBP; the cycle then begins again.

We emphasize that three turns of the cycle fix three $CO_2$ molecules, and there is a net production of one 3-PGaldehyde. Some 3-PGaldehyde molecules are used in chloroplasts to form starch, the major photosynthetic product in most species when photosynthesis is occurring rapidly. Other 3-PGaldehyde molecules are transported out of chloroplasts via an antiport carrier system (see Section 7.10) in exchange for either P*i* or 3-PGA from the cytosol (Heldt and Flügge, 1987; Heldt et al.,

# Exploring the Path of Carbon in Photosynthesis (II)

*Melvin Calvin*

*Melvin Calvin was a busy man (now retired). While I [F.B.S.] was working for the AEC (Atomic Energy Commission, now DOE, the Department of Energy) in 1974, I visited on two or three occasions the laboratory he directed because it was one of the national laboratories supported largely by the AEC. We conducted a formal inspection of the lab that year—and found it to be administered in a unique way based on close trust and cooperation among the directing scientists but designed to frustrate thoroughly a Washington bureaucrat (which I was at that time)!*

*I told Professors Calvin and Bassham that we were planning to revise our text and asked if they would write essays for us, telling about the work on what is now called the Calvin cycle. They agreed, but Calvin was unable to find the time. However, on January 4, 1977, I recorded the following telephone conversation with him:*

*Frank B. Salisbury:* How did you get into science?

*Melvin Calvin:* Well, it was a very practical consideration. When I was still in grade school I was concerned with what my future livelihood would be. I looked around and decided that almost everything that I had contact with (for example, in a food store, the food itself and its processing, the cans and the making of them, the paper and the making of it, the dye stuff on the paper), everything had to do with chemistry. This prompted me to say: "I'll try to understand how the food gets made, how it's processed, and how it eventually reaches the grocery store." I realized that how the food got there was an essential activity for human survival and that I had a good chance of finding a job if I knew anything about it.

*F.B.S.:* So it really was a practical consideration. But what aimed you at biology, especially plant physiology?

*M.C.:* Well, partly it was the food thing, and partly (much later) as I learned more of chemistry, it was the mysteriousness of how a plant could use sunshine to make food.

*F.B.S.:* So the practical approach became a bit of an intellectual challenge as well?

*M.C.:* Yes, exactly. You've got to make a living, but you want to do something interesting; if you can do both, you really have it made.

*F.B.S.:* What got you from the grocery store to Ernest Lawrence?

*M.C.:* That is a long, long trip, that one! I was an undergraduate at Michigan Tech, a graduate student at Minnesota, and then did postdoctorate work in England with Michael Polanyi. It was there that the real interest in photosynthesis, which came long ago, began to be executed. That's when I started working on chlorophyll and chlorophyll analogues: How do they work electronically? What kinds of things were they? Thus began the marriage of the practicality of food production and the intellectual challenge of energy conversion. So when I came to Berkeley, I started working on synthetic analogues of chlorophyll and heme. I was living in an environment in which radioactive isotopes were being turned up every day, and the idea of having the radioactive isotope of carbon was fairly rampant around the place. Work on the chemistry of the production of sugars wasn't something that happened because we had the carbon; the carbon was something we wanted, and we knew what we wanted it for long before we had it. Then Ruben and Kamen came along with both the carbons, carbon-11 and carbon-14. I was busy with porphyrin chemistry during the early war years, and my background in this and metal complexes led to my association with the Manhattan District for the development of methods of uranium purification and plutonium isolation. That's how I got to Ernest. And it came from chlorophyll, whether you like it or not! By that time, Sam Ruben had been killed, and Ernest said, "Well, you ought to do something with radiocarbon." I said, "Yes, I ought to," and I knew exactly what I had to do.

*F.B.S.:* What were your essential insights or acts of creativity?

*M.C.:* You mean in terms of the carbon cycle? Those came after a good deal of work. We knew the experiments that had to be done, and the mechanics of doing them were obvious. It was just a matter of having the material and the time, and with the end of the war we had both. Carbon-14 had appeared just before the war, so the moment we had the time and the opportunity to generate the radiocarbon (which we did in 1944–1945), the work began in earnest, I mean in a serious way! We made carbon-14 at the Hanford and Oak Ridge reactors. Everybody else did, too; we had lots of it. The experiments were easy to design. We did several, and by 1951 we had already mapped a good many of the early compounds on the way to sugar, although actual delineation of the whole sequence didn't occur until somewhat later.

*F.B.S.:* Putting it all together must have been the creative part.

*M.C.:* Yes. The first thing we saw were 3s [three-carbon intermediates], then we saw 5s and 6s and 7s. We didn't see the 4s until toward the end of the line. They were an essential part of the puzzle when we did find them. Putting it all together occurred in the early and middle 1950s. It came in bits and pieces and fits and starts. The first important step was the recognition of phosphoglyceric acid as the first product. We kept looking for a 2-C piece because the first thing we saw was 3, and it was logical that $CO_2$ should add to a 2. But we never did find a 2. As you know, there wasn't one! The reaction was a 5 + 1 that gives two 3s.

I saw that. I can remember being at home and reading some papers in JACS about the mechanism of decarboxylation of β-keto acids and of dicarboxylic acids (malonic acid and acetoacetic acid), completely mechanistic studies in organic chemistry. I then recognized how a $CO_2$ would add to a β-keto sugar. It was the recognition that ribulose diphosphate was the acceptor of the $CO_2$ that allowed me to finally draw the whole thing out. It was done almost all at once because the pieces and parts had been accumulating for several years.

*F.B.S.:* That was the white heat of creation?

*M.C.:* Yes. The actual drawing of the reaction was on a scrap of paper right beside me where I was reading. The article prompted my thinking of the reverse decarboxylation reaction and what I had to do it with. The ribulose diphosphate was the only thing I had to do it with. I tried it with fructose diphosphate, but you come out with one carbon too many. So it had to be a five-carbon and not a six-carbon keto sugar. We had already found the five-carbon sugar, but I didn't know what it was doing there. I know which chair I was sitting in when I put it together! Once you hit it, you know it's right. Everything fits when you hit the right key. And I know when that happened! It was quite an exciting few minutes. The next step was to come back in the lab and pick up the missing pieces, which we did.

*F.B.S.:* You were the director of the lab with several graduate students and postdocs working with you. How did that interaction work?

*M.C.:* Oh, that was wonderful! I enjoyed that immensely. That was the best part of the thing, you know. Every day I would come in and say, "What's new?" We would review and then see what the next experiment had to be. Usually there were anywhere from one to six people involved at any one time.

*F.B.S.:* Were you involved in the laboratory part of the work?

*M.C.:* Oh, yes. It was a joint effort, the design of the experiments. Some of them I would carry out myself. Some were easy! For example, the identification of phosphoglyceric acid I did personally. And I did it by the way the stuff behaved on an ion-exchange column a year or two before we had paper chromatography. One of the graduate students designed a transient experiment with $CO_2$. Alec Wilson was a boy from New Zealand and did a beautiful job! The experiment pointed the finger at ribulose because when you shut off the $CO_2$, the ribulose diphosphate rises. Shutting the light on or off was an easy experiment. But to shut off or turn on the radioactive $CO_2$ wasn't so easy. We had to build a very special apparatus, and he did it. It was a beautiful technical accomplishment.

*F.B.S.:* Two final questions: Just what is the nature of creativity in science? What price do you have to pay, and what are the rewards?

*M.C.:* I have a phrase in response to that to my students. I tell them that it's no trick to get the right answer when you have all the data. A computer can do that. The real creative trick is to get the right answer when you only have half enough data, half of what you have is wrong, and you don't know which half is wrong! When you get the right answer under those circumstances, you are doing something creative! The students learn after a while that this is not a joke. You must be able to sift out the critical points and put it all together in the right way, ignoring what doesn't seem to fit right. If you ignore the right things, you come out with the right answer! But if you pay attention to the wrong things, you don't.

*F.B.S.:* It's not a mechanical thing: If you have only half the data, you've got to sense or feel your way to the right answer.

*M.C.:* One that fits your concept of the physical world. The process involves intuition. You can't do it on a part-time basis, obviously. It is done day and night, winter and summer, under all circumstances, at any given moment. You can't tell where you will be when you are doing it. It does intrude on what some people call their private lives. In fact, I don't see how it can work otherwise. So, you ask for price. The price usually is paid by the person's family in the form of neglect and competition. I don't see any way to beat that.

*F.B.S.:* So what are the rewards? Are they worth it?

*M.C.:* The satisfaction of guessing correctly and then showing that it is correct is really great! You can usually tell when you are right, even before you've proved it; that's when the satisfaction is the greatest. When something is born and you *know* it is right; then you set about for the next 10 years trying to show it's right. It sounds as though you're making the discoveries over a period of 10 years; you're really not. You make them all at once, and then you spend a long time showing that this is the way it is. Even today I have some ideas about how photoelectric charge separation occurs—a conversion of solar energy—and I'm beginning to see physically how to do it. We will simulate the thing in a synthetic system within a decade, I would say, probably less. How soon it will be economic in terms of building useful gadgets is another matter. But it's on its way. That's the great sport.

*F.B.S.:* Now, lastly, your Nobel prize [1961] is the only one that qualifies as plant physiology. How has it affected your life?

*M.C.:* Oh, my! It's made it easier in some ways and more difficult in others. Easier in that I didn't have to spend so much time proving that what I wanted to do was worth doing. On the other hand, the responsibilities that came with it in terms of the students that came here and their expectations, in terms of students everywhere and their expectations of what you can and can't do, is a burden sometimes. I'm beginning to feel it now, a little bit. It's both a privilege and a responsibility.

*F.B.S.:* Most things of value end up that way, don't they?

*M.C.:* Yes. They almost always do. You don't get anything for nothing. Some of our modern-day students don't really understand that. They think someone owes them a living. It's hard to get them to realize that you pay for everything. The price is intellectual, physical, and emotional. It's all three, but it has its compensations.

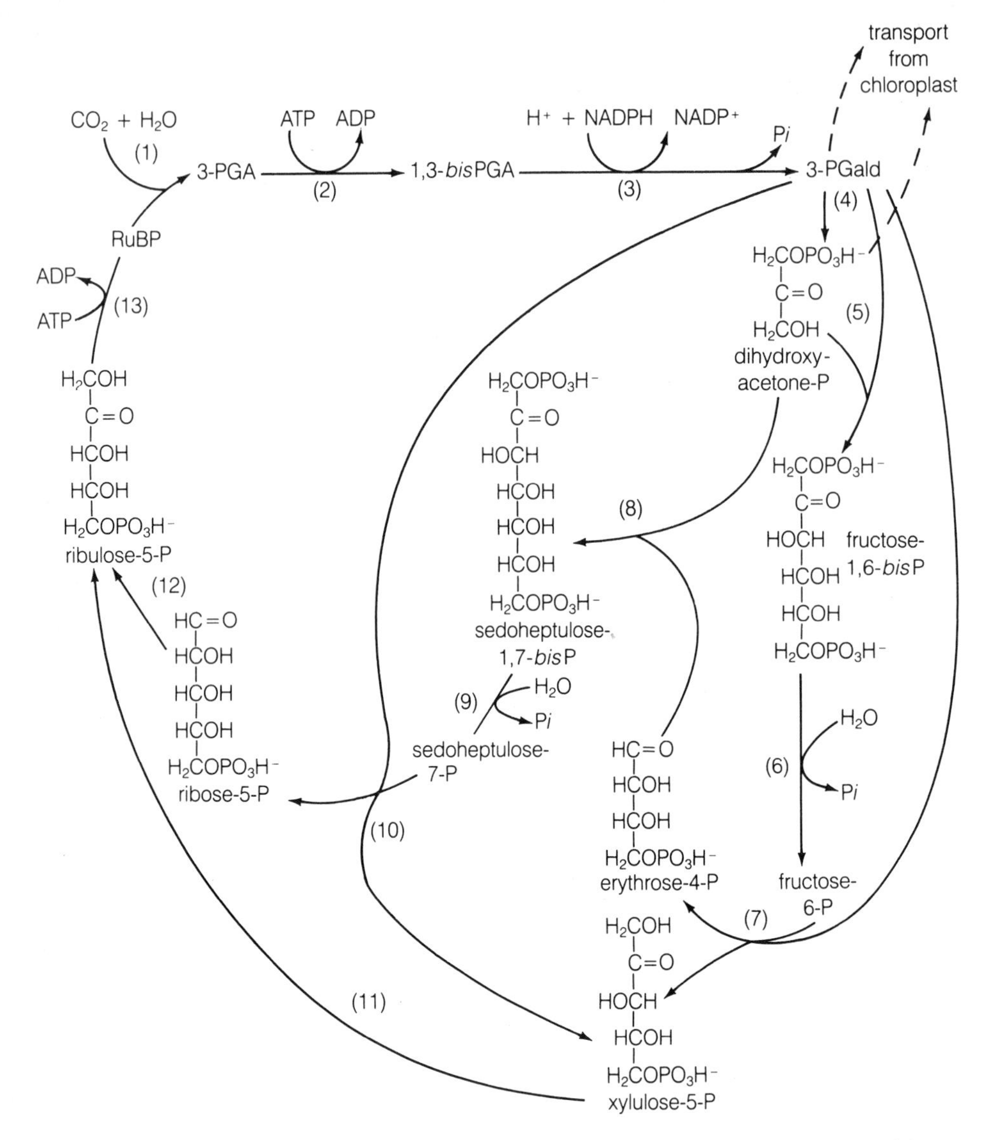


**Figure 11-3** Reactions of the Calvin cycle. Detailed reactions include (**1**), upper left, fixation of $CO_2$ into 3-PGA catalyzed by rubisco, (**2**) phosphorylation of 3-PGA by ATP to form 1,3-*bis*PGA, catalyzed by phosphoglycerokinase, (**3**) reduction of 1,3-*bis*PGA to 3-PGaldehyde, catalyzed by 3-phosphoglyceraldehyde dehydrogenase, (**4**) isomerization of 3-PGaldehyde to form dihydroxyacetone phosphate, catalyzed by triose phosphate isomerase, (**5**) aldol combination of 3-PGaldehyde and dihydroxyacetone-P to form fructose-1,6-*bis*P, catalyzed by aldolase, (**6**) hydrolysis of the phosphate from C-1 of fructose-*bis*P to form fructose-6-P, catalyzed by fructose-1,6-bisphosphate phosphatase, (**7**) transfer of upper two carbons of fructose-6-P to 3-PGaldehyde to form the 5-carbon xylulose-5-P, releasing the 4-carbon erythrose-4-P, catalyzed by transketolase, (**8**) aldol combination of erythrose-4-P and dihydroxyacetone-P to form sedoheptulose-1,7-*bis*P, catalyzed again by aldolase, (**9**) hydrolysis of phosphate from C-1 of sedoheptulose-1,7-*bis*P to form sedoheptulose-7-P, (**10**) transfer of the upper two carbons of sedoheptulose-7-P to 3-PGaldehyde to form ribose-5-P and xylulose-5-P, catalyzed again by transketolase. (Note that this is the fourth reaction the 3-PGaldehyde undergoes within the chloroplast, and that it can also be transported out of the chloroplast, upper right.) In reaction (**11**), xylulose-5-P is isomerized by an epimerase to form another pentose phosphate, ribulose-5-P. This ribulose-5-P can also be formed in reaction (**12**) by a different isomerase that uses ribose-5-P as the substrate. In reaction (**13**), ribulose-5-P is converted to RuBP by ribulose-5-P kinase, allowing $CO_2$ fixation to occur again.

1990). Still others are converted to *dihydroxyacetone phosphate*, a similar three-carbon triose phosphate that can be transferred out of chloroplasts by the same antiport system. This system helps keep the total amount of phosphate in the chloroplast constant but leads to the net appearance of triose phosphates in the cytosol. These triose phosphates are used in the cytosol to form sucrose, cell-wall polysaccharides, and hundreds of other compounds of which the plant is made. Their transport is especially important because the numerous other sugar phosphates of the Calvin cycle are largely held within the chloroplast.

## 11.3 The C-4 Dicarboxylic Acid Pathway: Some Species Fix $CO_2$ Differently

Discovery of the reactions of the Calvin cycle were thought to have elucidated $CO_2$ fixation and reduction in plants, but a new era in photosynthesis research came in 1965 from a discovery made in Hawaii by H. P. Kortschak, C. E. Hartt, and G. O. Burr. They found that sugarcane leaves, in which photosynthesis is unusually rapid and efficient, initially fix most $CO_2$ into carbon 4 of malic and aspartic acids. After approximately one second of photosynthesis in $^{14}CO_2$, 80 percent of $^{14}C$ fixed was in these two acids and only 10 percent was in PGA, indicating that in this plant 3-PGA is not the first product of photosynthesis. These results were quickly confirmed in Australia by M. D. Hatch and C. R. Slack, who found that some grass species of tropical origin, including maize (corn), displayed similar labeling patterns after fixing $^{14}CO_2$ (Hatch, 1977). Other grasses such as wheat, oat, rice, and bamboo, which are not closely related taxonomically to tropical grasses, gave 3-PGA as the predominant fixation product.

The new results of Kortschak et al. (1965) and of Hatch and Slack showed that the primary carboxylation reaction of some species is different from that involving ribulose bisphosphate. Species that produce four-carbon acids as the primary initial $CO_2$ fixation products are now commonly referred to as **C-4 species**; those that initially fix $CO_2$ into 3-PGA are called **C-3 species**. There are, however, a few species with intermediate properties, and these are thought to represent evolutionary transitions from C-3 to C-4 species (Edwards and Ku, 1987; Monson and Moore, 1989; Brown and Hattersley, 1989).

Most C-4 species are monocots (especially grasses and sedges), although more than 300 are dicots. Among the grasses, sugarcane, maize, and sorghum are important agricultural crops, and numerous range grasses (especially in southern latitudes) are also C-4 plants. The C-4 pathway is known to occur in more than 1,000 species of angiosperms distributed among at least 19 families. At least 11 genera include both C-4 and C-3 species. Extensive lists of C-4 species are given in references by Krenzer et al. (1975), Downton (1975), Winter and Troughton (1978), and (for North American grasses) by Waller and Lewis (1979). Numerous C-4 sedges are listed by Hesla et al. (1982). All gymnosperms, bryophytes, and algae and most pteridophytes that have been studied are C-3 plants, as are nearly all trees and shrubs. A few *Euphorbia* tree or shrub species from Hawaii have the C-4 pathway (Pearcy, 1983).

Considering that there are about 285,000 species of flowering plants (monocots and dicots), the presence of the C-4 pathway in roughly 0.4 percent of those investigated might seem hardly worth mentioning. The great attention they have received arose largely because of the economic importance of some of them and because under high irradiance and warm temperatures they can photosynthesize more rapidly and produce substantially more biomass than can C-3 plants. Investigations of the C-4 pathway have also taught us much about limitations to photosynthesis in C-3 plants and have furthered general ecological knowledge about factors that control productivity in various climates (topics covered in Chapters 12 and 25).

The reaction by which $CO_2$ (actually $HCO_3^-$) is converted into carbon 4 of malate and aspartate occurs through its initial combination with **phosphoenolpyruvate (PEP)** to form **oxaloacetate** and P*i*:

$$H^*CO_3^- + \begin{matrix} COOH \\ | \\ C{-}OPO_3H^- \\ \| \\ CH_2 \end{matrix} \rightarrow \begin{matrix} COOH \\ | \\ C{=}O \\ | \\ CH_2 \\ | \\ {}^*COOH \end{matrix} + Pi \tag{11.3}$$

PEP — oxaloacetic acid

This is an irreversible reaction. Oxaloacetate is not usually a detectable product of photosynthesis, but it can be found when precautions are taken to prevent both its rapid conversion to malic and aspartic acids and its unusually high susceptibility to destruction in isolation and chromatography procedures.

**Phosphoenolpyruvate carboxylase (PEP carboxylase)**, a $Mg^{2+}$-requiring enzyme that is apparently present in all living plant cells, is the catalyst involved (see reviews by Andreo et al., 1987; Stiborová, 1988). The reasons it is of special importance in leaves of C-4 species are that an active isozyme of it is unusually abundant in such species (often 10 to 15 percent of total leaf protein) and that a cyclic pathway that maintains a constant and relatively plentiful supply of PEP is also present. In leaves of C-3 species and in root, fruit, and other cells (regardless of the species) that lack chlorophyll, other isozymes of PEP carboxylase are present. Here the major function of the enzyme seems to be to help replace Krebs-cycle acids used in synthetic reactions (see Chapter 13) and to help form malate needed in charge-balancing functions. In all cases, PEP carboxylase exists in the cytosol outside any organelle (including the chloroplast).

The reactions that convert oxaloacetate to malate and aspartate in C-4 plants are as follows:

$$\underset{\text{oxaloacetic acid}}{\text{COOH–C(=O)–CH}_2\text{–*COOH}} + \text{NADPH} + \text{H}^- \longrightarrow \underset{\text{L-malic acid}}{\text{COOH–HC(OH)–CH}_2\text{–*COOH}} + \text{NADP}^+ \qquad (11.4)$$

$$\underset{\text{oxaloacetic acid}}{\text{COOH–C(=O)–CH}_2\text{–*COOH}} + \underset{\text{L-alanine}}{\text{CH}_3\text{–HC(NH}_2\text{)–COOH}} \longrightarrow \underset{\text{L-aspartic acid}}{\text{COOH–HC(NH}_2\text{)–CH}_2\text{–*COOH}} + \underset{\text{pyruvic acid}}{\text{CH}_3\text{–C(=O)–COOH}} \qquad (11.5)$$

Formation of malate in Reaction 11.4 is catalyzed by *malate dehydrogenase*, with the necessary electrons provided by NADPH. Interestingly, malate dehydrogenase is a chloroplast enzyme, which means that oxaloacetate (formed in the cytosol) must move into the chloroplast for reduction to malate. This movement occurs by another chloroplast antiport system in which oxaloacetate is transported in on a carrier that also moves malate out. Formation of aspartate from oxaloacetate occurs in the cytosol and requires another amino acid such as alanine as the source of an amino group. This type of reaction is referred to as **transamination** because transfer of an amino group is involved (see Section 14.3).

It became evident that in C-4 species there is a division of labor between two different kinds of photosynthetic cells: mesophyll cells and bundle sheath cells (Campbell and Black, 1982). Both kinds of cells are required to produce sucrose, starch, and other plant products. One (or occasionally two) distinct layer of tightly packed, often thick-walled and rather gas-impermeable **bundle sheath cells** almost always surrounds the leaf vascular bundles and separates them from the predominant mesophyll cells. This concentric arrangement of bundle sheath cells is described as **Kranz** (German, "halo" or "wreath") **anatomy** (Laetsch, 1974). In contrast to C-3 plants, in which a much less distinct bundle sheath is often also present, bundle sheath cells of many C-4 plants contain thicker walls, far more chloroplasts, mitochondria, and other organelles, and smaller central vacuoles. A comparison of cross-sectional leaf anatomy in representative C-3 and C-4 grasses is shown in Figure 11-4.

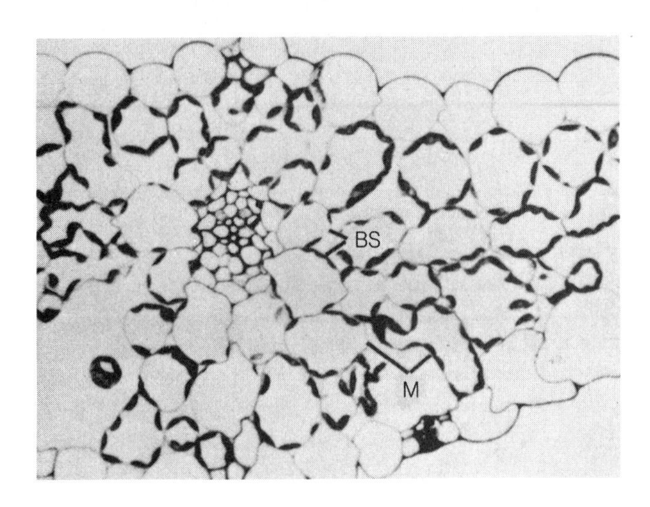


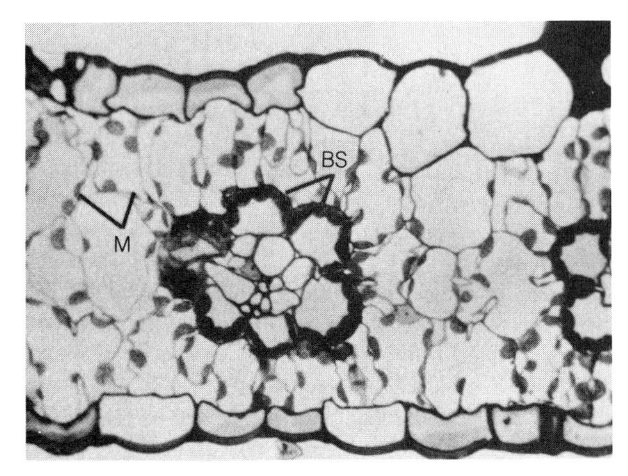


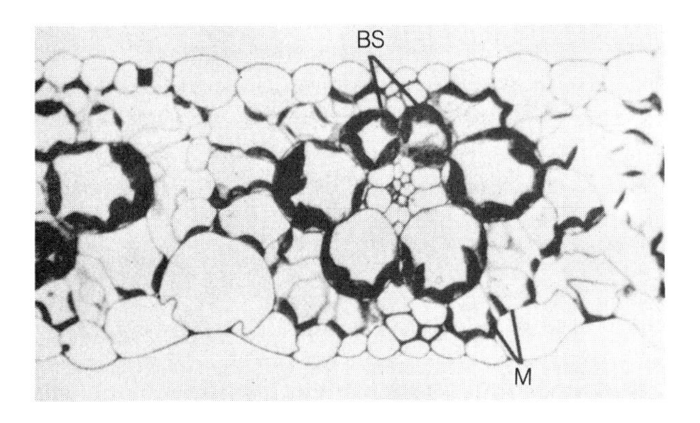


**Figure 11-4** Leaf cross sections of a C-3 monocot (oat, top) and C-4 monocots (maize, middle, and Rhodesgrass, bottom). BS = bundle sheath; M = mesophyll cells. (From Frederick and Newcomb, 1971.)

Chloroplasts of bundle sheath cells frequently contain nearly all the leaf starch, with little present in chloroplasts of more loosely arranged surrounding mesophyll cells, although this distribution of starch is species-dependent (Huber et al., 1990). Studies with segregated bundle sheath and mesophyll cells confirmed earlier suppositions that malate and aspartate are formed in mesophyll cells and that 3-PGA, sucrose, and starch are produced mainly in bundle sheath cells. Rubisco exists only in bundle sheath cells, as do most Calvin-cycle enzymes, so the complete Calvin cycle occurs only in bundle sheath cells. On the other hand, PEP carboxylase occurs mainly in mesophyll cells. *Thus C-4 species really use both kinds of $CO_2$ fixing mechanisms.*

The main reasons that $CO_2$ first appears in malate and aspartate are that after stomatal entry $CO_2$ first penetrates into mesophyll cells, that the activities of PEP carboxylase are high there, and that rubisco is not present there. Most $CO_2$ that has been recently fixed into carboxyl groups of malate and aspartate is rapidly transferred, perhaps via abundant plasmodesmata (see arrows in Figure 11-5), into bundle sheath cells. There these compounds undergo decarboxylation with release of $CO_2$ that is then fixed by rubisco into 3-PGA. The principal source of $CO_2$ for bundle sheath cells is therefore C-4 acids formed in the mesophyll.

Sucrose and starch are ultimately formed from 3-PGA in bundle sheath cells using Calvin-cycle reactions and other reactions not yet mentioned. The division of labor referred to above involves the trapping of $CO_2$ into C-4 acids by mesophyll cells and, after transfer of these acids to bundle sheath cells, decarboxylation and refixation of $CO_2$. The three-carbon acids (pyruvate and alanine) resulting from decarboxylation of C-4 acids are then returned to mesophyll cells, where they are converted to PEP and carboxylated with PEP carboxylase to keep the cycle going. The C-4 acids formed in mesophyll cells seem only to be carriers of $CO_2$ to bundle sheath cells. This idea is illustrated in the model of the C-4 pathway and its relation to the Calvin cycle presented in Figure 11-6.

Two additional aspects of this model require explanation. The first involves the mechanisms by which aspartate and malate are decarboxylated in bundle sheath cells. Surprisingly, three such mechanisms occur (depending on the species), and some species use more than one (Kelly et al., 1989). Two of these mechanisms are shown in Figure 11-6. In the so-called **aspartate formers** (species that form more aspartate than malate), aspartate that moves into bundle sheaths is changed back to oxaloacetate by transamination. Next, this oxaloacetate is reduced to malate by an *NADH-dependent malate dehydrogenase*. (**NADH** is a coenzyme capable of transferring two electrons and is nearly identical in structure to NADPH, as explained in Figure 10-11. The oxidized form of NADH is $NAD^+$.) The malate is oxidatively decarboxylated by a *malic enzyme* that uses $NAD^+$ as the electron acceptor. Pyruvate, $CO_2$, and NADH are the products. The pyruvate is then converted to alanine by another transamination. As alanine moves

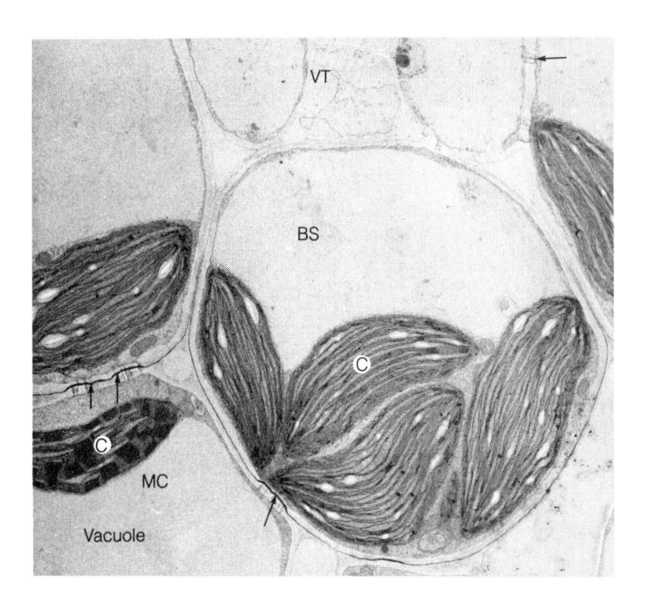


**Figure 11-5** Electron micrograph of adjacent mesophyll cell (MC) and bundle sheath cell (BS) in the C-4 plant crabgrass (*Digitaria sanguinalis*). Note abundant grana and lack of starch in the mesophyll cell chloroplast, but absence of grana and presence of several small starch granules in bundle sheath chloroplasts (C). Arrows mark plasmodesmata where passage of organic acids is suspected to occur. Vascular tissue (VT) is shown at top. (From Black et al., 1973.)

back into mesophyll cells, the nitrogen in it replaces that lost when aspartate was transported to bundle sheath cells.

**Malate formers** (species that form mostly malate) transfer malate into bundle sheaths, where it is also oxidatively decarboxylated to $CO_2$ and pyruvate, but with a malic enzyme that uses $NADP^+$ in strong preference to $NAD^+$ (Fig. 11-6, left center). The NADPH formed by this enzyme helps reduce 3-PGA to 3-PGaldehyde (see Reaction 11.2), as indicated in the model.

The third decarboxylation system, not shown in Figure 11-6, operates mainly in aspartate formers and involves oxaloacetate formed from aspartate in bundle sheaths. In this system, oxaloacetate reacts with ATP (catalyzed by *PEP carboxykinase*) to release $CO_2$, PEP, and ADP. The $CO_2$ is fixed by rubisco and is converted to carbohydrates by the Calvin cycle, and the ADP is converted back to ATP by photophosphorylation.

The second aspect of our model that requires explanation involves the way the three-carbon acids that are transported back to the mesophyll cells regenerate the PEP needed for continued fixation of $CO_2$ there. If pyruvate is transported back, it is absorbed by mesophyll chloroplasts. If alanine is transported back, it is converted to pyruvate by another transamination, and the pyruvate is absorbed by mesophyll chloroplasts. Then, in either case an unusual chloroplast enzyme named *pyruvate, phosphate dikinase* uses ATP to convert the pyruvate to PEP and PP*i*:

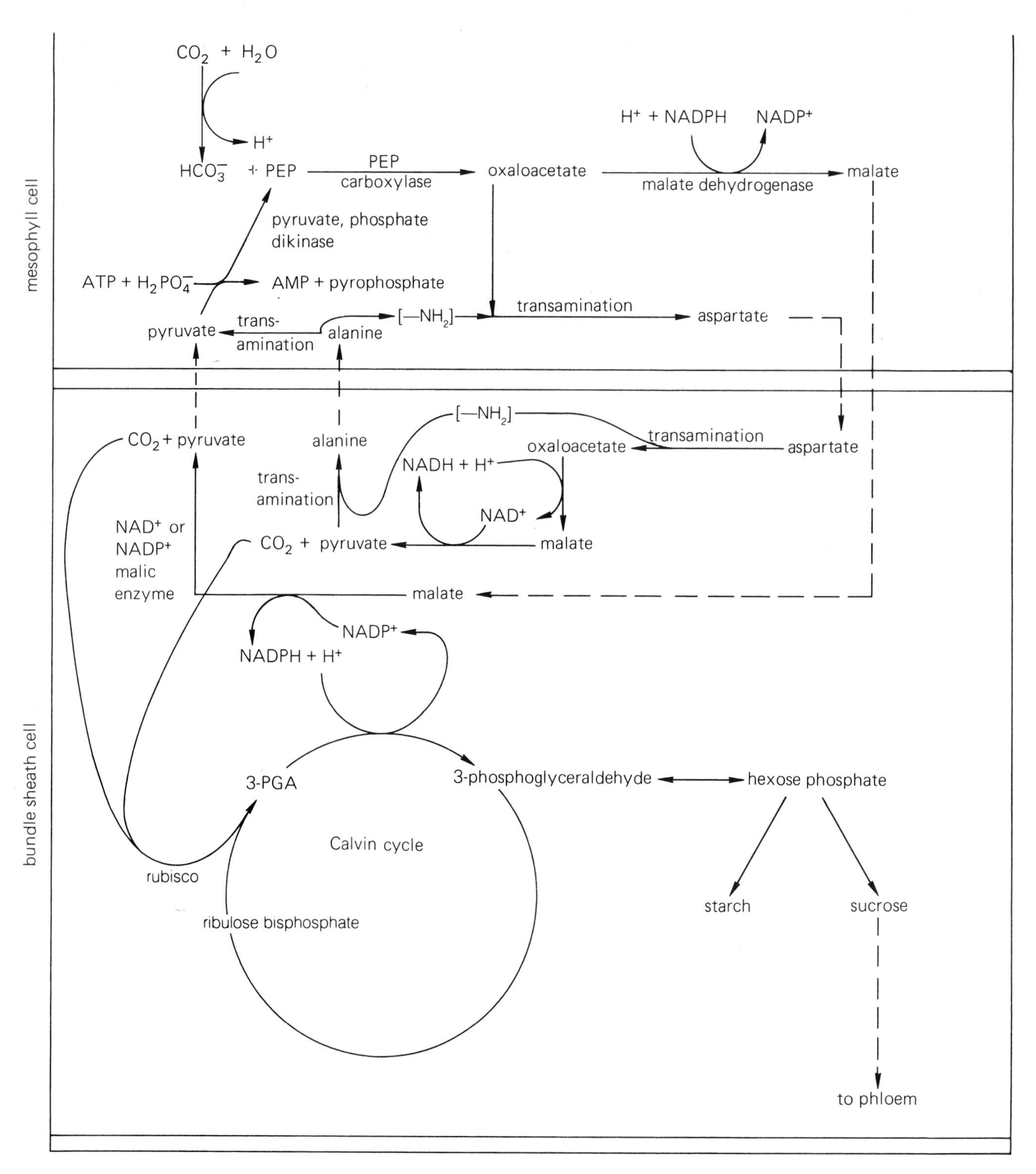


**Figure 11-6** A summary of metabolic division of labor in mesophyll and bundle sheath cells of C-4 plants. $CO_2$ is initially fixed into C-4 acids in the mesophyll; then these acids move into the bundle sheath (probably as $K^+$ salts), where they are decarboxylated. The $CO_2$ thus released is fixed via the Calvin cycle in the bundle sheath chloroplasts. Sucrose and starch are common products, as shown. After decarboxylation of the C-4 acids, three-carbon molecules such as pyruvate and alanine move back to the mesophyll cells, where they are converted to PEP so that $CO_2$ fixation can continue there.

$$
\begin{array}{l}
\text{COOH} \\
| \\
\text{C}{=}\text{O} \\
| \\
\text{CH}_3 \\
\text{pyruvic acid}
\end{array}
+ \text{ATP} + \text{P}i \rightleftharpoons
\begin{array}{l}
\text{COOH} \\
| \\
\text{C}{-}\text{OPO}_3\text{H}^- \\
\| \\
\text{CH}_2 \\
\text{PEP}
\end{array}
+ \text{AMP}
$$

$$
+ \begin{array}{c}
\text{O}^- \quad\quad \text{O}^- \\
| \quad\quad\quad\; | \\
\text{HO}{-}\text{P}{-}\text{O}{-}\text{P}{-}\text{OH} \\
\| \quad\quad\quad\; \| \\
\text{O} \quad\quad\;\; \text{O} \\
\text{pyrophosphate}
\end{array} \qquad (11.6)
$$

Calculations of the energy required to operate the C-4 pathway, with its additional Calvin cycle, indicate that for each $CO_2$ fixed, two ATP in addition to the three needed in the Calvin cycle are required. These two additional ATP are necessary so that PEP synthesis can be maintained for continued $CO_2$ fixation. There is no additional NADPH requirement because for each NADPH used to reduce oxaloacetate in the mesophyll cells, one is also regained during malic-enzyme action in bundle sheath cells.

In spite of the apparent inefficiency with respect to ATP utilization in C-4 species, these plants almost always show more rapid rates of photosynthesis per unit of leaf surface area than do C-3 species *when both are exposed to high light levels and warm temperatures at ambient $CO_2$ levels*. The C-4 species are adapted to and evolved from C-3 species in regions of periodic drought, such as tropical savannas. When temperatures reach 25 to 35°C and irradiance levels are high, C-4 plants are about twice as efficient as C-3 plants in converting the sun's energy into dry matter.

Photosynthesis in C-3 plants is often limited by atmospheric $CO_2$ levels, but C-4 plants are much less limited by $CO_2$ because they effectively pump $CO_2$ into bundle sheath cells when they transport malic or aspartic acid into those cells. This pumping action concentrates $CO_2$ in bundle sheath cells where it is used in the Calvin cycle, so $CO_2$ less often limits photosynthesis in C-4 plants than C-3 plants. At high temperatures $CO_2$ is less soluble in the water of chloroplasts, an effect that lowers photosynthesis of C-3 plants more than it lowers that of C-4 plants. Furthermore, drought stress and the accompanying stomatal closure lower $CO_2$ entry into leaves, again giving C-4 plants an advantage over C-3 plants. We will defer other ecological and environmental aspects of photosynthesis until the next chapter, but it is useful to mention here that the comparatively low photosynthetic efficiency of most C-3 species results largely from light-enhanced loss of part of the $CO_2$ they fix; this loss occurs by a phenomenon called photorespiration. Little or no such loss occurs in C-4 plants that maintain higher levels of $CO_2$ in bundle sheath chloroplasts.

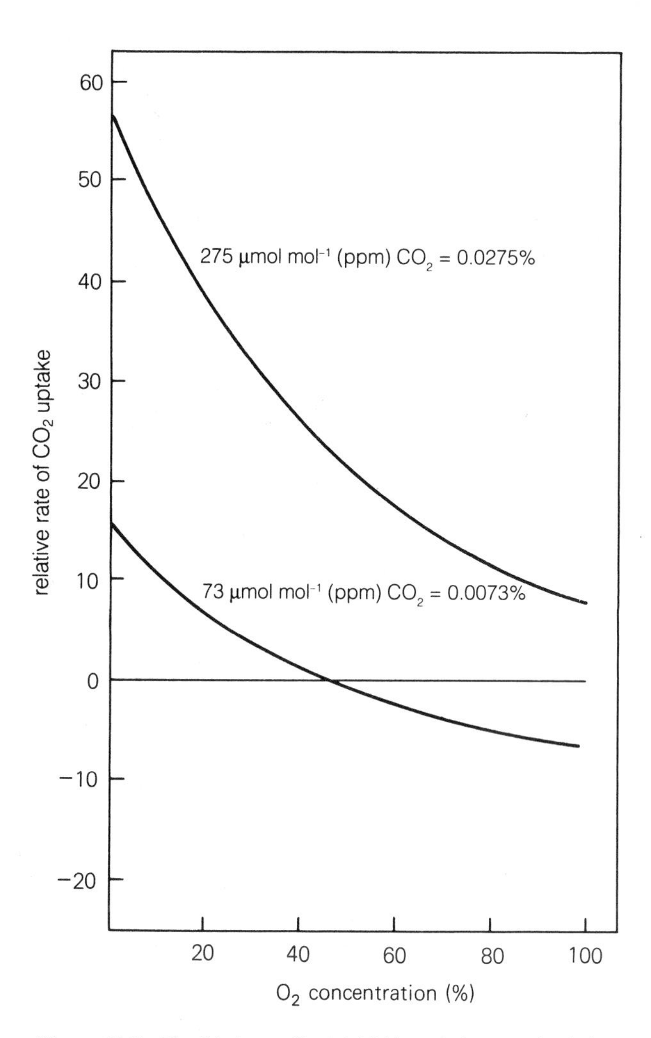


**Figure 11-7** The Warburg effect: inhibition of photosynthesis in soybean (C-3) plants by $O_2$. Normal air contains 20.9 percent $O_2$ and 0.035 percent $CO_2$ (350 µmol mol$^{-1}$). The irradiance level was equal to about one-sixth of maximum sunlight; the temperature was 22.5°C. Negative values represent a net loss of $CO_2$ by respiration. (From Forrester et al., 1966.)

## 11.4 Photorespiration

Otto Warburg, a famous German biochemist who devoted much of his attention to photosynthesis in algae, noted in 1920 that photosynthesis is inhibited by $O_2$. This inhibition occurs in all C-3 species studied since and is termed the **Warburg effect**. Figure 11-7 illustrates this effect for C-3 soybean leaves exposed to two different $CO_2$ concentrations, one a near-normal $CO_2$ concentration and one a much-reduced $CO_2$ level. Note that even the normal $O_2$ concentration of 21 percent is inhibitory compared with zero $O_2$ at both $CO_2$ levels. Furthermore, inhibition by $O_2$ is greater at the lower $CO_2$ concentration. *These and many similar studies show that* *existing atmospheric levels of $O_2$ inhibit photosynthesis in C-3 plants.* In contrast, photosynthesis in C-4 species is not significantly affected by varying $O_2$ concentrations.

To understand these different effects of $O_2$ on C-3 and C-4 species, remember that net $CO_2$ fixation is the amount by which photosynthesis exceeds respiration, because respiration continuously releases $CO_2$. During darkness C-3 leaves respire at a rate that is roughly one-sixth of the photosynthetic rate, yet in light C-3 plants respire much faster than in darkness. Data indicating higher respiration rates for C-3 plants in light than in darkness were obtained with an infrared $CO_2$ analyzer and first published by John P. Decker in the 1950s, but physiologists were slow to accept his conclusions. We now know that total respiration in leaves of C-3 species is often two or three times as rapid in light as in darkness and that under field conditions respiration causes the release of one-fourth to one-third of the $CO_2$ that is simultaneously fixed by photosynthesis (Gerbaud and Andre, 1987; Sharkey, 1988). Respiration in illuminated photosynthetic organs occurs by two processes: the process that occurs in all plant parts even during darkness (see Chapter 13) and a much more rapid process, strictly dependent on light, known as **photorespiration**. The two processes are spatially separated within the cells: Normal respiration occurs in the cytosol and in mitochondria, and photorespiration occurs in a process involving cooperation of chloroplasts, peroxisomes, and mitochondria (Ogren, 1984; Husic et al., 1987).

Loss of $CO_2$ by photorespiration in C-4 species is almost undetectable, which is the principal reason that these species show much higher net photosynthetic rates at high irradiance levels and warm temperatures than do C-3 species. To understand why photorespiration is so much higher in C-3 plants than C-4 plants, we must first understand the chemical reactions of photorespiration.

In 1971, W. L. Ogren and George Bowes at the University of Illinois theorized from various data that carbons 1 and 2 of RuBP were the precursors to glycolic acid, a two-carbon acid. They showed experimentally that $O_2$ could inhibit $CO_2$ fixation by rubisco, thus apparently explaining the Warburg effect. They also showed that rubisco catalyzes the oxidation of RuBP by $O_2$. *Thus rubisco is also an oxygenase.* Details of how it fixes both $O_2$ and $CO_2$ competitively are given by Andrews and Lorimer (1987), Gutteridge (1990), and Keys (1990).

The two products of rubisco action on RuBP and $O_2$ are 3-PGA and *phosphoglycolic acid*, a two-carbon phosphorylated acid. Using heavy oxygen ($^{18}O_2$), it was shown that only one of the $O_2$ atoms is incorporated into phosphoglycolate; the other is converted to water (as an $OH^-$ ion), as follows:

$$
{}^{18}O_2 + \begin{array}{c} CH_2OPO_3H^- \\ | \\ C{=}O \\ | \\ H{-}C{-}OH \\ | \\ H{-}C{-}OH \\ | \\ CH_2OPO_3H^- \end{array} \longrightarrow \left[ \begin{array}{c} CH_2OPO_3H^- \\ | \\ {}^{18}O{-}C{-}OH \\ / \quad | \\ {}^{18}OH \quad C{=}O \\ | \\ H{-}C{-}OH \\ | \\ CH_2OPO_3H^- \end{array} \right] \longrightarrow
$$

ribulose bisphosphate

$$
\xrightarrow{OH^-} \begin{array}{c} CH_2OPO_3H^- \\ | \\ C \\ {}^{18}O {//} \quad \backslash OH \end{array} + \begin{array}{c} COOH \\ | \\ H{-}C{-}OH \\ | \\ CH_2OPO_3H^- \end{array} + {}^{18}OH^- \qquad (11.7)
$$

phosphoglycolic acid 3-PGA

Molecular $O_2$ and $CO_2$ therefore compete for the same rubisco enzyme and for the same RuBP substrate. Oxygen fixation represents about two-thirds of the total $O_2$ absorbed during photorespiration, the remainder coming (as we will show) from oxidation of phosphoglycolate. This competition between $O_2$ and $CO_2$ for rubisco explains the greater inhibition of photosynthesis in C-3 plants at lower $CO_2$ levels (Fig. 11-7). The affinity of rubisco for $CO_2$ is much greater than its affinity for $O_2$, but $O_2$ fixation in all C-3 plants can occur because the $O_2$ concentration in leaves or in cells of algae is much higher than that of $CO_2$. (Atmospheric concentrations of $O_2$ average 20.9 percent by volume and those of $CO_2$ average about 0.0352 percent; see Section 12.1.) At any given time, rubisco enzymes are fixing about one-fourth to one-third as much $O_2$ as $CO_2$. When temperatures are warm, the ratio of dissolved chloroplastic $O_2$ to $CO_2$ is higher than when temperatures are cool, so $O_2$ fixation by rubisco occurs faster, and photorespiration then indirectly slows growth in C-3 species but not in C-4 species.

Photorespiration is light-dependent for several reasons. First, RuBP formation occurs much faster in light than in darkness because operation of the Calvin cycle needed to form RuBP requires ATP and NADPH, both light-dependent products, as explained earlier. Furthermore, light directly causes release of $O_2$ from $H_2O$ in chloroplasts, so chloroplastic $O_2$ is more abundant in light than in darkness, when it must diffuse inwardly through leaf surfaces with closed stomates. Finally, as we explain later, rubisco is activated by light and is inactive in darkness, so it cannot fix $O_2$ (or $CO_2$) in darkness. Thus *photorespiration is essentially absent in C-4 plants for two main reasons: Rubisco and other Calvin-cycle enzymes are present only in bundle sheath cells, and the $CO_2$ concentration in those cells is maintained at such a high level that $O_2$ cannot compete successfully with $CO_2$.* High $CO_2$ concentrations in bundle sheath cells are maintained by rapid decarboxylation of malate and aspartate transferred there from mesophyll cells. If bundle sheath cells are separated from mesophyll cells, their source of $CO_2$ from C-4 acids is removed; they will then photorespire, but not when in an intact leaf with the $CO_2$ pump operating.

Phosphoglycolate formed in Reaction 11.7 represents the source of $CO_2$ released in photorespiration. The pathway by which phosphoglycolate is formed by rubisco and then metabolized to release $CO_2$ during photorespiration is often called the **oxidative photosynthetic carbon cycle** or **C-2 cycle** (Husic et al., 1987; Oliver and Kim, 1990). The pathway is called a "cycle" because some of the carbons in phosphoglycolate molecules are converted back to RuBP via 3-PGA and the Calvin cycle. The phosphate group of phosphoglycolate is first hydrolyzed away by a specific phosphatase found in chloroplasts of C-3 plants, releasing P*i* and *glycolic* acid. The glycolate then moves out of chloroplasts into adjacent peroxisomes.

**Peroxisomes** are small organelles that contain several oxidative enzymes; they and glyoxysomes of fat-rich seeds (see Chapter 15) are the two kinds of plant microbodies (Huang et al., 1983; see Chapter 1). Peroxisomes exist almost exclusively in photosynthetic tissues, and in electron micrographs they often appear to be in direct contact with chloroplasts (Fig. 11-8). In peroxisomes, glycolate is oxidized to *glyoxylic acid* by *glycolic acid oxidase,* an enzyme containing riboflavin as part of an essential prosthetic group:

$$\underset{\text{glycolic acid}}{\underset{\text{OH}}{\underset{|}{CH_2}}\text{—COOH}} + O_2 \rightarrow \underset{\text{glyoxylic acid}}{\underset{\text{O}}{\underset{\|}{HC}}\text{—COOH}} + H_2O_2 \qquad (11.8)$$

Here glycolic acid oxidase transfers electrons (present in H atoms) from glycolate to $O_2$, reducing the $O_2$ to $H_2O_2$ (*hydrogen peroxide*). Nearly all this $H_2O_2$ is then broken down by *catalase* (another peroxisomal enzyme) to $H_2O$ and $O_2$:

$$2H_2O_2 \rightarrow 2H_2O + O_2 \qquad (11.9)$$

Next, glyoxylate is converted to glycine (a two-carbon amino acid) by a transamination reaction with a different amino acid, still in peroxisomes. Then, after transport into mitochondria, two molecules of glycine are converted to one molecule of serine (a three-carbon amino acid), one molecule of $CO_2$, and one $NH_4^+$ ion (Walker and Oliver, 1986). This mitochondrial reaction is the source of $CO_2$ released in photorespiration. It is also important because the $NH_4^+$ released must be reincorporated into amino acids so that glycine formation can continue, and this process requires ATP and re-

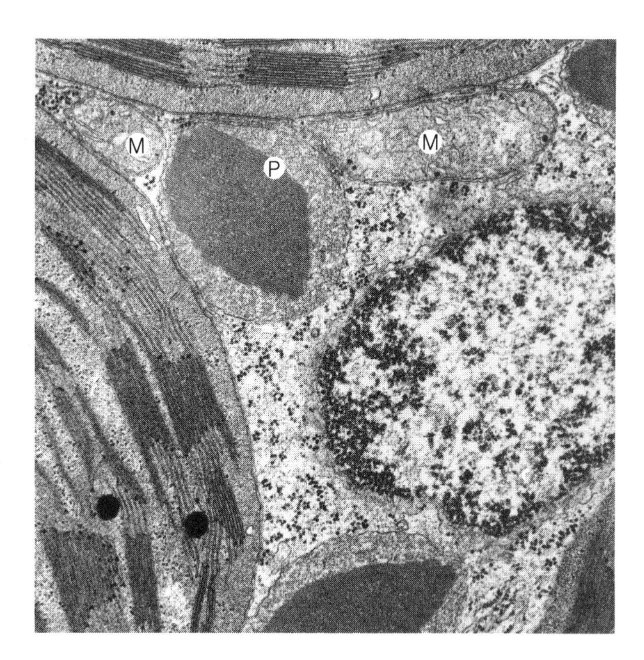


**Figure 11-8** Close association of chloroplasts, peroxisomes (P), and mitochondria (M) in a leaf cell. The crystalline-like matrix in these peroxisomes is due to the enzyme catalase, but many peroxisomes containing catalase show no such matrix. (Courtesy Eugene Vigil.)

duced ferredoxin (see Section 14.4 and Fig. 14-9). Serine is then converted to 3-PGA by a series of reactions that involves loss of its amino group and gain of a phosphate group from ATP. Part of the 3-PGA is converted to RuBP and part to sucrose and starch in chloroplasts. The overall equation for photorespiration (stopping carbon flow at 3-PGA and abbreviating ribulose bisphosphate as RuBP and ferredoxin as Fd) is:

$$2\,\text{RuBP} + 3\,O_2 + 2\,\text{ATP} + H_2O + 2\,\text{Fd}(Fe^{2+}) \longrightarrow CO_2 + 3\,\text{3-PGA} + 2\,\text{ADP} + 3\,\text{P}i + 2\,\text{Fd}(Fe^{3+}) \qquad (11.10)$$

Photorespiration therefore conserves an average of three-fourths of the carbons split off RuBP (as phosphoglycolate) when $O_2$ reacts with it (one $CO_2$ lost for every two two-carbon acids formed and for every three $O_2$ absorbed). Note also from Reaction 11.10 that photorespiration uses rather than produces ATP and $H_2O$ and that it requires a reductant (reduced ferredoxin). Light is necessary to produce the ATP and reduced ferredoxin (see Chapter 10).

But how does photorespiration persist in C-3 plants instead of being eliminated through evolutionary selection pressures? Certainly it reduces net $CO_2$ fixation and growth rates in these plants. The answer is unclear, but some experts have suggested that photorespiration is a means of removing excess ATP and NADPH (or reduced ferredoxin) produced at excessively high irradiance levels (see Ogren, 1984). Because both ATP and

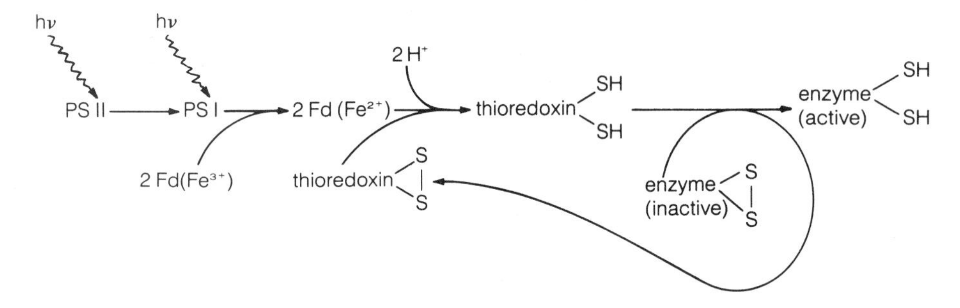


**Figure 11-9** Light activation of enzymes via the ferredoxin-thioredoxin system.

NADPH are needed to regenerate RuBP from 3-PGA formed during $O_2$ fixation, both molecules would certainly be used up in photorespiration without $CO_2$ fixation. This use of excess "reducing power" might prevent high irradiance levels from causing damage to chloroplast pigments (see solarization, Chapter 12).

Other experts have suggested that photorespiration is a necessary consequence of the structure of the enzyme rubisco, a structure that evolved to fix $CO_2$ in ancient photosynthetic bacteria when atmospheric $CO_2$ concentrations were high and those of $O_2$ were low. According to this hypothesis, as $O_2$ accumulated in the atmosphere by the photolysis of $H_2O$ by algae and early land plants, rubisco necessarily began to fix $O_2$ simply because its active site for $CO_2$ could not discriminate effectively between the two similar gases. Whether natural selection or genetic-engineering methods can eventually force modification of rubisco to favor $CO_2$ even more is an interesting and important question. Evidence concerning the selectivity of rubisco for $CO_2$ over $O_2$ indicates that evolution from anaerobic bacteria and cyanobacteria to higher plants is indeed favoring $CO_2$ fixation (Andrews and Lorimer, 1987; Pierce, 1988).

## 11.5 Light Control of Photosynthetic Enzymes in C-3 and C-4 Plants

We emphasized the role of light in providing the ATP and NADPH needed for $CO_2$ fixation and reduction. In addition, light regulates the activities of several chloroplast photosynthetic enzymes. Such enzymes exist in an active form in light and in an inactive or much less active form in darkness. Carbohydrate production from $CO_2$ is, therefore, shut off tightly at night because of enzyme inactivity, closed stomates, and a deficiency of ATP and NADPH.

In C-3 species, five Calvin-cycle enzymes are activated in light: rubisco, 3-phosphoglyceraldehyde dehydrogenase, fructose-1,6-bisphosphate phosphatase, sedoheptulose-1,7-bisphosphate phosphatase, and ribulose-5-phosphate kinase. The function of each enzyme is given in the legend of Figure 11-3. In C-4 species, three enzymes—PEP carboxylase, $NADP^+$-malate dehydrogenase, and pyruvate, phosphate dikinase of mesophyll cells—are also light-activated. In both groups of plants, $CF_1$ of the thylakoid ATP synthase is similarly activated.

Mechanisms of light activation are indirect, and light's energy is not absorbed by the colorless enzymes directly. Instead, light absorbed by photosystems II and I (see Chapter 10) is involved. Most of these enzymes have disulfide (S—S) groups that are reduced to two sulfhydryl groups (—SH plus —SH) when they are light-activated (Buchanan, 1984; Ford et al., 1987; Scheibe, 1987; Crawford et al., 1989). Each reduction causes an important modification in enzyme structure so that the enzyme functions much faster. Reduction occurs using the electrons that are derived from the photolysis of $H_2O$ in photosystem II but are not used to reduce $NADP^+$ to NADPH (Fig. 11-9). Electrons go through PS II to PS I and then to ferredoxin to reduce it from Fd($Fe^{3+}$) to Fd($Fe^{2+}$). Next, electrons move, one at a time, to one or both of two small proteins called **thioredoxins** (Holmgren, 1985). These proteins also contain a disulfide bond that becomes reduced with two electrons. Again, two sulfhydryl groups are produced. Transfer of electrons from ferredoxin to thioredoxin is catalyzed by an enzyme called *ferredoxin-thioredoxin reductase*. Finally, reduced thioredoxin reduces and activates the photosynthetic enzymes. Inactivation in darkness occurs by $O_2$-dependent oxidation of the enzymes, most likely by reversal of the scheme in Figure 11-9 as far as ferredoxin.

Activation of rubisco and PEP carboxylase is generally not as great and occurs differently. Reduction and disulfide-sulfhydryl changes are not involved. For rubisco, three (and in some species, four) effects caused by light are important. First is a rise in *p*H of the chloroplast stroma from about 7 to 8 caused by light-driven $H^+$ transport from the stroma to thylakoid channels (see Fig. 10-10). Rubisco is much more active at *p*H 8 than at a lower *p*H. Second is transport of $Mg^{2+}$ from the thylakoid channels to the stroma accompanying the *p*H change, important because rubisco requires $Mg^{2+}$ for maximal activity. These two effects seem to be much less important factors than the other two, and for the latter to be understood we must explain more about the chemical nature of rubisco.

In all organisms except some photosynthetic bacteria, rubisco exists as a heteropolymer, with eight identical large subunits of about 56 kDa each (coded by a chloroplast gene) and eight small identical subunits of about 14 kDa each (coded by a nuclear gene), making a total of about 560 kDa for the whole enzyme. An active site that binds the substrates RuBP and $CO_2$ and the activator $Mg^{2+}$ exists on each large subunit, yet in some way the small subunits are also essential for catalytic activity (Andrews and Lorimer, 1987). (An exception occurs in purple, nonsulfur bacteria that photosynthesize, because their rubisco contains only eight large subunits and no small ones.) The other two light-activating effects concern how well the active site on each large subunit can bind RuBP.

Rubisco can exist in three different states, two with the active site inactivated and one with the site totally activated. In the first state the enzyme is free, with no bound activators. In the second, still inactive state, a $CO_2$ molecule has been bound, but this is a different $CO_2$ than the one used up in photosynthesis. This activator $CO_2$ binds to the amino group of a certain lysine amino acid in the large subunit to form a carbamate (lys-NH-$CO_2$). Then $Mg^{2+}$ is rapidly bound to the negatively charged carbamate, the enzyme changes its conformation, and it first binds RuBP and then $CO_2$ and $H_2O$ (or $O_2$ when photorespiration occurs). Catalysis follows to form, if $CO_2$ is fixed, two 3-PGA.

A major discovery about understanding rubisco activation in light was made by Somerville et al. (1982). They were studying various mutants of *Arabidopsis thaliana* to learn more about genetic and metabolic control of photorespiration and photosynthesis. One mutant had a defective gene that caused inability of the plant to activate rubisco in light. Then the protein that the nonmutated gene codes for was found to activate rubisco under suitable *in vitro* conditions (Salvucci et al., 1985). This protein, called *rubisco activase*, requires ATP and RuBP for maximal activity, and it functions by catalyzing the carbamylation of rubisco by activator $CO_2$ molecules. Rubisco activase functions in light, probably because ATP from photophosphorylation is then available, but in darkness or dim light rubisco becomes spontaneously decarbamylated and loses activity (reviewed by Salvucci, 1989; Sharkey, 1989; and Portis, 1990). The role of RuBP in activation is not yet clear.

Next, another important discovery about light activation of rubisco was made (Seemann et al., 1985; Gutteridge et al., 1986). A powerful inhibitor of rubisco was

| 2-carboxy-arabinitol-1-phosphate | 2-carboxy-3-keto-arabinitol-1,5-bisphosphate |
|---|---|
| $CH_2OPO_3^{2-}$ | $CH_2OPO_3^{2-}$ |
| HO—C—$CO_2^-$ | HO—C—$CO_2^-$ |
| H—C—OH | C=O |
| H—C—OH | H—C—OH |
| $CH_2OH$ | $CH_2OPO_3^{2-}$ |

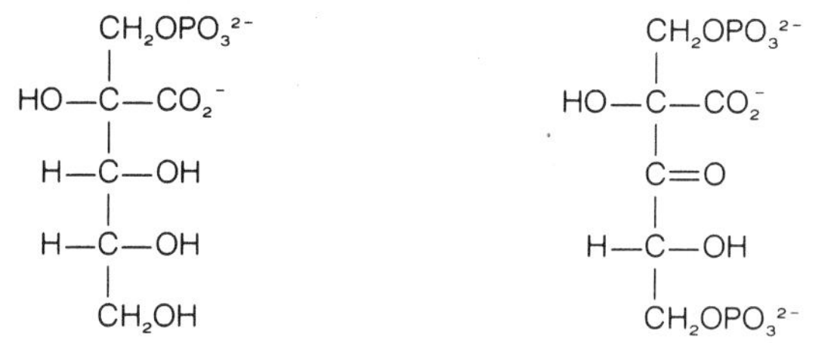

**Figure 11-10** Structures of a nocturnal inhibitor of rubisco (left) and of the normal enzyme intermediate of $CO_2$ fixation before $H_2O$ is added (right).

found to exist in leaves of certain species. The inhibitor, identified as the six-carbon compound *2-carboxyarabinitol-1-phosphate*, or *CA1P* (Gutteridge et al., 1986; Berry et al., 1987), reaches relatively high concentrations at night but is degraded in daylight by a phosphatase enzyme that removes phosphate from carbon 1. The structure of CA1P is similar to that of the enzyme intermediate *2-carboxy-3-keto-arabinitol-1,5-bisphosphate* formed when rubisco attaches $CO_2$ to RuBP to form 3-PGA (Fig. 11-10). At night, CA1P attaches tightly to rubisco, but in daylight rubisco activase helps drive off CA1P, the phosphatase hydrolyzes away the phosphate to form free 2-*carboxyarabinitol*, and carbamylation by rubisco activase begins (reviewed by Salvucci, 1989 and by Servaites, 1990). It is uncertain how light causes the phosphatase to become functional, but there is some evidence that it is activated by the ferredoxin-thioredoxin reductase system. Nevertheless, CA1P seems absent in many (perhaps most) species and shows little change in concentration during light and darkness in other species. It is likely that other nocturnal inhibitors are present in plants that do not use CA1P.

PEP carboxylase of C-4 plants becomes activated in mesophyll cells during daylight and is deactivated at night. This mechanism helps prevent $CO_2$ fixation in C-4 plants at night. How activation and deactivation occur is still controversial. The enzyme is activated by glycine and glucose-6-phosphate, but it is strongly inhibited by malate and to a lesser extent by oxaloacetate and aspartate. The inhibitors act as feedback effectors (see Section 9.6) that control how fast C-4 acids are produced (Stiborová, 1988), yet they are not likely to control day-night changes in enzyme activity. The most likely explanation is that in C-4 plants PEP carboxylase becomes phosphorylated during daylight by **protein kinases**, enzymes that can transfer phosphate from ATP to one or more hydroxyl groups of hydroxylated amino acids (serine or threonine; see Fig. 9-2). This phosphorylation increases the enzyme's activity, but at night the phosphate groups are removed and activity decreases. How light causes phosphorylation in the cytosol where the enzyme exists is unknown.

**Figure 11-11** Miscellaneous succulents. From left to right, *Opuntia, Aloe obscura, Echeveria corderoyi, Crassula argentea, Agave horrida.* (New York Botanical Garden photograph courtesy Arthur Cronquist.)

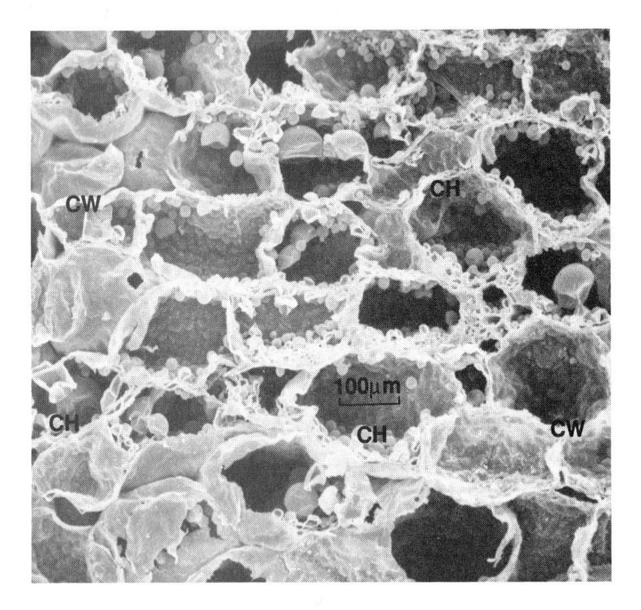


**Figure 11-12** Scanning electron micrograph of mesophyll cells in a mature leaf of the CAM plant *Kalanchoe daigremontiana.* Note unusually large central vacuoles and the organelles (mostly chloroplasts) in a thin layer of cytoplasm. (CW, cell wall; CH, chloroplasts) (From Balsamo and Uribe, 1988.)

Finally, in C-4 plants, both $NADP^+$-malate dehydrogenase and pyruvate, phosphate dikinase are light-activated in mesophyll cells. The dehydrogenase is activated by the ferredoxin-thioredoxin system mentioned above. The mechanism of activation of pyruvate, phosphate dikinase is too complex to describe here, but it depends mainly on light-induced changes in P*i* (an activator) and ADP (a deactivator). Surprisingly, the enzyme becomes phosphorylated by ADP (to form AMP) and dephosphorylated by P*i* (to form PP*i*) (Burnell and Hatch, 1985; Edwards et al., 1985).

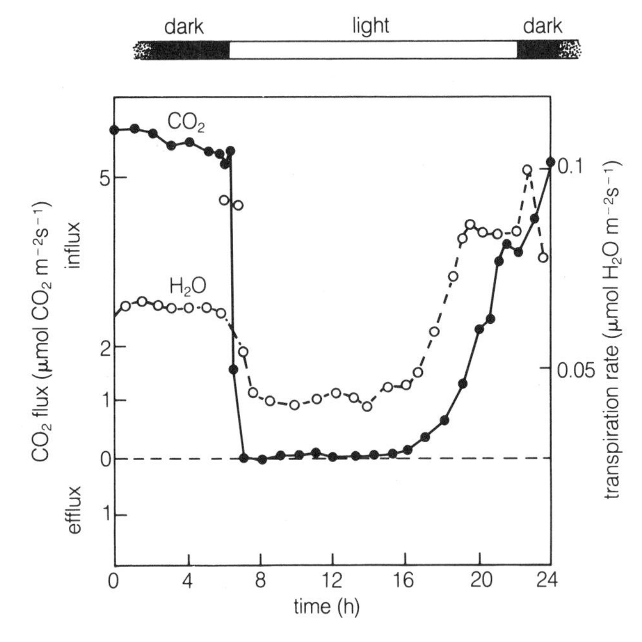


**Figure 11-13** $CO_2$ fixation and transpiration rates of the CAM plant *Agave americana* during alternate light and dark periods. (From Neales et al., 1968.)

## 11.6 $CO_2$ Fixation in Succulent Species (Crassulacean Acid Metabolism)

Numerous species living in arid climates have thick leaves with relatively low surface-to-volume ratios, a thick cuticle, and accompanying low transpiration rates. Such species are frequently referred to as **succulents** (Fig. 11-11). They usually lack a well-developed palisade layer of cells, and most of their leaf or stem photosynthetic cells are spongy mesophyll. These cells have unusually large vacuoles relative to the thin layer of cytoplasm (Fig. 11-12). Bundle sheath cells are present but indistinct (Gibson, 1982).

The metabolism of $CO_2$ in succulents is unusual, and because it was first investigated in members of the family Crassulaceae, it is called **crassulacean acid metabolism (CAM)**. CAM has been found in hundreds of species in 26 angiosperm families (including the familiar Cactaceae, Orchidaceae, Bromeliaceae, Liliaceae, and Euphorbiaceae), in some pteridophytes, and probably in the gymnosperm *Welwitschia mirabilis* (reviewed by Kluge and Ting, 1978; Lüttge, 1987; Ting, 1985; and Ting and Gibbs, 1982). CAM plants usually grow where water is scarce or is difficult to get, including deserts and semi-arid regions, salt marshes, and epiphytic sites (as when certain orchids grow attached to other plants). In these habitats CAM plants (like all plants) must obtain water and $CO_2$, but if they fully open their stomates during daylight and thereby obtain $CO_2$, they transpire

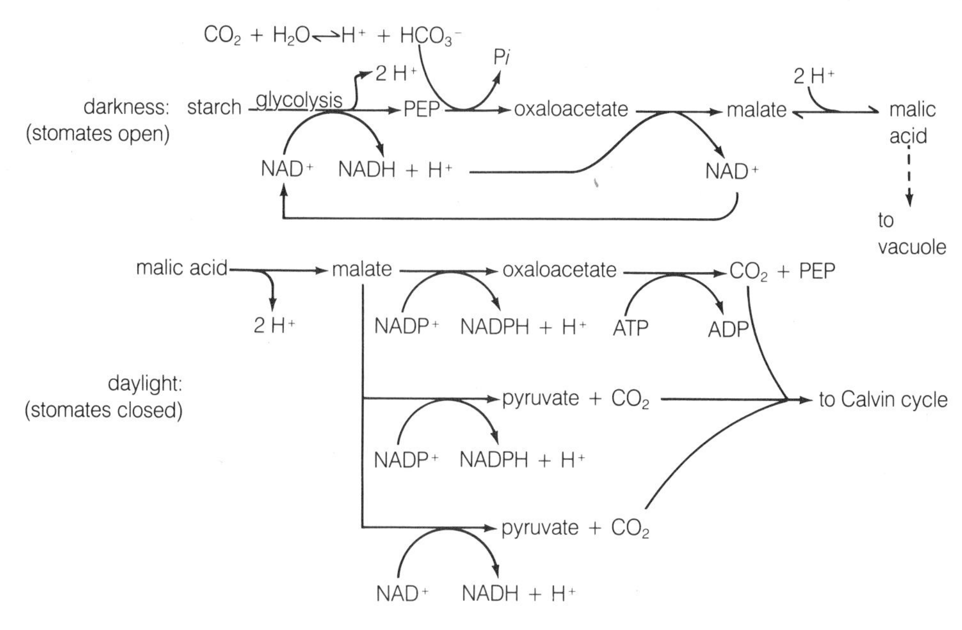


**Figure 11-14** A summary of $CO_2$ fixation in CAM plants.

too much water. As described in Chapter 4, they therefore open their stomates and fix $CO_2$ into malic acid primarily at night, when temperatures are cooler and relative humidities higher. Figure 11-13 illustrates the daily course of transpiration and $CO_2$ fixation in one such species, *Agave americana*, a century plant.

The most striking metabolic feature of CAM plants is formation of malic acid at night and its disappearance during daylight. This formation of acid at night is detectable as a sour taste and is accompanied by a net loss of sugars and starch. The most abundant acid in CAM plants is malic acid, but citric and isocitric acids that are derived from malic acid accumulate to a lesser extent in some species. However, citric and isocitric acids usually show little change in concentrations during daylight and darkness. PEP carboxylase in the cytosol of CAM plants is the enzyme responsible for $CO_2$ fixation into malate at night (in contrast to its low activity in C-4 plants in darkness), but rubisco becomes active during daylight, as in C-3 and C-4 plants. The role of rubisco is identical to its function in bundle sheath cells of C-4 plants—that is, refixation of the $CO_2$ lost from organic acids such as malic acid.

A model that is consistent with what we know about $CO_2$ fixation in CAM plants is shown in Figure 11-14. During darkness, starch is degraded by glycolysis (see Chapter 13) as far as PEP. $CO_2$ (actually $HCO_3^-$) reacts with PEP to form oxaloacetate, which is then reduced to malic acid by an NADH-dependent malate dehydrogenase. The $H^+$ ions from malic acid are transported into the large central vacuole by an ATPase and a pyrophosphatase pump (see Section 7.8), and malate ions follow $H^+$ into the vacuole. Here malic acid accumulates, sometimes even at concentrations of 0.3 M or more, until sunrise. This accumulation makes the osmotic potential of the cells quite negative, so they can absorb water and store it when the plant exists in dry or salty soils.

During daylight, malic acid diffuses passively out of the vacuole and is decarboxylated by one or more of the three mechanisms also present in C-4 plant bundle sheath cells (see Fig. 11-6). The mechanism used depends largely on the species. The $CO_2$ released becomes quite concentrated in the cells and is refixed (without photorespiration) by rubisco into 3-PGA of the Calvin cycle, which then leads to formation of sucrose, starch, and other photosynthetic products. Pyruvate formed by decarboxylation is converted to PEP by pyruvate, phosphate dikinase, as in C-4 plants; PEP is then partly respired, partly converted to sugars and starch by reverse glycolysis, and partly converted to amino acids, proteins, nucleic acids, lipids, and aromatic compounds by reactions to be described in Chapters 12 to 15.

Thus CAM plants, like C-4 plants, first use PEP carboxylase and NADPH-malate dehydrogenase to form malic acid, then decarboxylate that acid to release $CO_2$ by one of three mechanisms, and then refix the $CO_2$ into Calvin-cycle products using rubisco. In C-4 plants a spatial separation between mesophyll and bundle sheath cells aids malate formation and decarboxylation, and both processes occur in daylight. In CAM plants both processes occur in the same cells, one process at night and the other in daylight, and the large central vacuole stores malic acid that would otherwise cause too low a cytosolic *p*H at night. Low permeability of the tonoplast to $H^+$ resulting from malic acid ionization in the vacuole must be especially important in CAM plants because the vacuole *p*H often becomes as low as 4 at night.

An interesting question is what causes rubisco, not PEP carboxylase, to fix $CO_2$ in CAM plants during daylight. Both enzymes are present, both have about equal affinities for dissolved forms of $CO_2$, and the cytosol location of PEP carboxylase should allow it to encounter incoming $CO_2$ before that $CO_2$ reaches rubisco in the chloroplast. Part of the answer is that during daylight CAM plants convert their PEP carboxylase from an active form to an inactive form. The inactive form present during daylight has less affinity for PEP and is strongly inhibited by the malic acid released from the vacuole, but the active form has a higher affinity for PEP and is much less inhibited by the malic acid formed at night. Changes in the activities of still other enzymes that favor $CO_2$ fixation by PEP carboxylase only at night occur.

Although the ability of a plant to perform CAM is genetically determined, it is also environmentally controlled. In general, CAM is favored by hot days with high irradiance levels, cool nights, and dry soils, all of which predominate in deserts. High salt concentrations in soils that lead to osmotic drought also favor CAM. Some species (especially cacti; see Martin and Kirchner, 1987) can remain under drought for several weeks with closed stomates, neither gaining nor losing much $CO_2$ yet still using some light energy for photophosphorylation in daylight (Lüttge, 1987). More commonly, CAM plants are facultative C-3 plants and switch to a greater rate of $CO_2$ fixation by a C-3 photosynthetic mode after a rainstorm in daylight or when night temperatures are high. Stomates then remain open longer during daylight hours.

This switching process suggests that evolution for CAM is driven only by water stress. Nevertheless, CAM also occurs in some primitive underwater plant groups related to club mosses and horsetails, especially in the genus *Isoetes* (quillworts). These frequently live in shallow ponds in which $CO_2$ levels are lower during daylight than at night because of photosynthesis (Keeley, 1988; Keeley and Morton, 1982; and Raven et al., 1988). The occurrence of CAM in such species suggests that genetic selection for CAM has also been driven by diurnal changes in $CO_2$ availability. Finally, CAM can be induced by marked day-night changes in temperatures in alpine species of *Sempervivum* in the central Austrian Alps. Even with adequate moisture, night temperatures near freezing followed by high irradiance levels near noon cause leaf temperature changes so large (up to 45°C) that CAM then occurs (Wagner and Larcher, 1981).

## 11.7 Formation of Sucrose, Starch, and Fructans

In each of the three major pathways by which $CO_2$ is fixed, the principal leaf storage products accumulating in daylight are usually sucrose and starch. Free hexose sugars such as glucose and fructose are usually much less abundant than sucrose in photosynthetic cells, although the opposite is true in many nonphotosynthetic cells. In many grass species (especially those that originated in temperate zones, including the Hordeae, Aveneae, and Festuceae tribes) and also in some dicots, starch is not a major product of photosynthesis; in these plants sucrose and fructose polymers collectively called **fructans** (formerly **fructosans**) predominate in leaves and stems, but starch predominates in roots and seeds. In this section we summarize the formation of sucrose, starch, and fructans from photosynthetic products; their utilization as energy sources during respiration is explained in Chapter 13.

### Synthesis of Sucrose

Sucrose (Fig. 11-15) is especially important because it is so common and abundant in plants and because we consume so much of it as table sugar. It acts as an energy source in photosynthetic cells and is readily translocated through the phloem to growing tissues. It is commercially important in sugar-beet roots and sugarcane stems because of its unusual abundance there.

Synthesis of sucrose occurs in the cytosol, not in chloroplasts where the Calvin cycle occurs. Free glucose and fructose are not important precursors of sucrose; instead, phosphorylated forms of those sugars are. As mentioned in Section 11.1, triose phosphates (3-phosphoglyceraldehyde and dihydroxyacetone phosphate) exported from chloroplasts in photosynthetic cells serve as precursors of hexose phosphates and of sucrose. These triose phosphates (with three carbons) become converted into one P*i* and one fructose-6-phosphate (with six carbons), part of which is changed to glucose-6-phosphate and then to glucose-1-phosphate. (For details of those reactions, which also occur in respiration, see Fig. 13-5.) Glucose-1-phosphate and fructose-6-phosphate contain the two hexose units needed to yield the disaccharide sucrose, but combination of those units is indirect because energy must be provided to activate the glucose unit. This energy is provided by *uridine triphosphate (UTP)*, which is a nucleoside triphosphate similar to ATP except that it contains the pyrimidine base uracil instead of the purine base adenosine. The UTP acts by reacting with glucose-1-phosphate; the two terminal phosphates of UTP are removed together as PP*i*, and the phosphate of glucose-1-phosphate becomes esterified to the remaining phos-

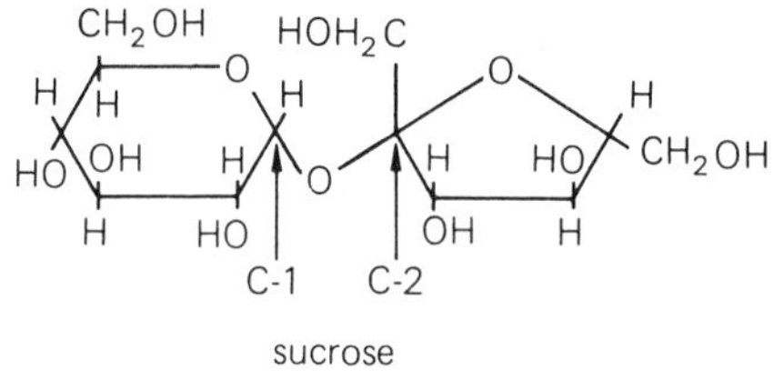

**Figure 11-15** The structure of sucrose, a disaccharide made of a glucose unit (left) and a fructose unit (right) connected between carbons 1 and 2, as shown.

phate in UTP to form a molecule called *uridine diphosphate glucose (UDPG)*. The glucose in UDPG can be considered to be activated because it can readily be transferred to an acceptor molecule such as fructose-6-phosphate.

This reaction and others involved in the major pathway of sucrose formation are summarized by Reactions 11.11 through 11.15 below (Avigad, 1982; Hawker, 1985; ap Rees, 1987). The enzymes that catalyze each reaction require $Mg^{2+}$ as a cofactor, another of the many reasons why magnesium is essential for plants.

$$\text{UTP} + \text{glucose-1-phosphate} \rightleftharpoons \text{UDPG} + \text{PP}i \qquad (11.11)$$

$$\text{PP}i + H_2O \rightarrow 2\ \text{P}i \qquad (11.12)$$

$$\text{UDPG} + \text{fructose-6-phosphate} \rightleftharpoons \text{sucrose-6-phosphate} + \text{UDP} \qquad (11.13)$$

$$\text{sucrose-6-phosphate} + H_2O \rightarrow \text{sucrose} + \text{P}i \qquad (11.14)$$

$$\text{UDP} + \text{ATP} \rightleftharpoons \text{UTP} + \text{ADP} \qquad (11.15)$$

A calculation of the total energy cost to the plant in forming one sucrose molecule can be made by adding the above reactions to produce the following reaction, which shows that only one ATP is needed to form the glycosidic bond that connects glucose to fructose in sucrose:

$$\text{glucose-1-phosphate} + \text{fructose-6-phosphate} + 2H_2O + \text{ATP} \rightarrow \text{sucrose} + 3\ \text{P}i + \text{ADP} \qquad (11.16)$$

Because three ATP molecules are required in the Calvin cycle for each carbon in each hexose of sucrose (36 total ATP), the one additional ATP needed to form the glycosidic bond in sucrose is a small additional requirement.

## Formation of Starch

The major storage carbohydrate of most plants is starch (Jenner, 1982). In leaves, starch accumulates in chloroplasts, where it is formed directly from photosynthesis. In storage organs, it accumulates in **amyloplasts**, where it is formed following translocation of sucrose or other carbohydrate from leaves (see Chapter 1). In plants, starch always exists in one or more starch grains in a plastid. The amounts of starch in various tissues depend on many genetic and environmental factors, but in leaves the level and duration of light are especially important. Starch builds up in daylight when photosynthesis exceeds the combined rates of respiration and translocation; then some of it disappears at night by the latter two processes.

You have eaten many nonphotosynthetic starch-rich organs, such as potato tubers, banana fruits, and the seeds of cereal grains and legumes that are so common in our diets. Besides food crops, most native perennial plants store starch before and during the dormant period, and this starch is used as energy for regrowth the next growing season. In deciduous trees and shrubs, starch is stored largely in amyloplasts of young twigs, in the bark (phloem parenchyma cells), in living xylem parenchyma cells, and also in some root parenchyma storage cells. Many herbaceous perennial grasses and dicots store starch in roots, bases of stems (crowns), or in underground bulbs or tubers. In stems, cortex and pith cells are frequent sites of starch storage, both in annuals and perennials. Sugar maple trees store starch in twig and trunk xylem parenchyma during late summer and early fall, then convert this starch to sucrose in early spring, when it can be collected as maple sugar. Collection occurs by tapping the flow of sap from tree trunks, an osmotic flow resulting from turgor pressure in xylem caused by conversion of relatively few starch grains into many more sucrose molecules.

Two types of starch are present in most starch grains: *amylose* and *amylopectin*, both of which are composed of D-glucoses connected by α-1,4 bonds (Fig. 11-16a). The α-1,4 bonds cause starch chains to coil into helices (Fig. 11-17). Amylopectin consists of highly branched molecules, the branches occurring between C-6 of a glucose in the main chain and C-1 of the first glucose in the branch chain (α-1,6 bonds). The number of glucose units present in various amylopectins ranges from 2,000 to 500,000. Amyloses are smaller and contain a few hundred to a few thousand sugar units, the number depending on the species and environmental conditions (Manners, 1985). Although amylose was formerly thought to be nonbranched, more recent research showed that it, too, contains branches, although far fewer than amylopectin (Kainuma, 1988). Amylose becomes purple or blue when stained with an iodine-potassium iodide solution, a mixture that produces the reactive $I_5^-$ ion (Banks and Muir, 1980). Amylopectin reacts much less intensely with this reagent, and it exhibits a purple to red color. The iodine test is often used

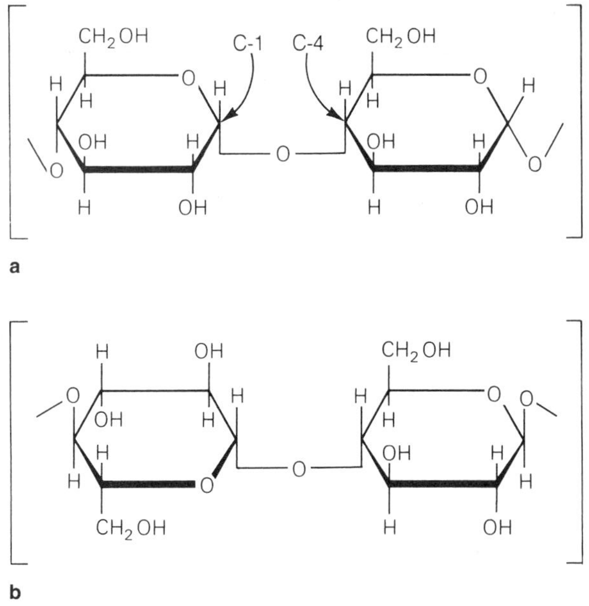

**Figure 11-16** (**a**) The alpha ($\alpha$) linkage between glucose residues, as in starch. (**b**) The beta ($\beta$) linkage between glucose residues, as in cellulose.

by students and researchers to determine whether starch is present in cells. The percentage of amylopectin in starch grains from most species varies from about 70 to 80 percent (Manners, 1985). Potato-tuber starches contain about 78 percent amylopectin and 22 percent amylose. These ratios are similar to the ratios of starches in banana fruits and to those in the seeds of pea, wheat, rice, and field maize.

Starch formation occurs mainly by one process involving repeated donation of glucose units from a nucleotide sugar similar to UDPG called *adenosine diphosphoglucose, ADPG* (reviewed by Beck and Ziegler, 1989; and Preiss, 1982a, 1982b). Formation of ADPG occurs through the use of ATP and glucose-1-phosphate in chloroplasts and other plastids. The following reaction summarizes starch formation from ADPG, in which a growing amylose molecule with a glucose unit having a reactive C-4 group at its end combines with C-1 of the glucose being added from ADPG:

$$\text{ADPG} + \text{small amylose } (n\text{-glucose units})$$
$$\rightarrow \text{larger amylose (with } n + 1 \text{ glucose units)} + \text{ADP} \qquad (11.17)$$

*Starch synthetase*, which catalyzes this reaction, is activated by $K^+$, which is one reason why $K^+$ is essential for plants and is probably why sugars, not starch, accumulate in plants deficient in $K^+$. Various isozymes of starch synthetase occur in different plants and parts of plants.

Branches in amylopectins between C-6 of the main chain and C-1 of the branch chain are formed by various isozymes of enzymes summarily called the *branching* or *Q enzyme*. Surprisingly little is known about how they catalyze branching, but branch bonds do not result from glucose transferred from ADPG. Rather, branching enzymes transfer short units from a growing starch molecule to the same or another starch molecule to form the $\alpha$-1,6 linkage.

Much remains to be learned about starch formation and its control, especially regarding branching. Nevertheless, a few facts about control are important. High light levels and long days favor photosynthesis and carbohydrate translocation, so the long days of summer cause both accumulation of one or more starch grains in chloroplasts and starch storage in amyloplasts of nonphotosynthetic cells. Furthermore, starch formation in chloroplasts is favored by bright light because the enzyme that forms ADPG is activated allosterically by 3-PGA and inhibited allosterically by P*i* (Preiss, 1982a, 1982b, 1984). Levels of 3-PGA increase somewhat in daylight as $CO_2$ is fixed, but levels of P*i* decrease somewhat as it is added to ADP to form ATP during photosynthetic phosphorylation. It was thought for many years that starch storage represented excess photosynthetic product, whereas sucrose represented a more readily available product that could easily be translocated. Although this is somewhat true, sucrose storage is also common (in the vacuole).

Biochemical controls over formation of starch versus sucrose are now being investigated (reviewed by Stitt et al., 1987; Woodrow and Berry, 1988; Stitt and Quick, 1989; Stitt, 1990; Huber et al., 1990; and Hanson, 1990). One important factor is that species that produce primarily sucrose have a light-activated sucrose phosphate synthetase enzyme, whereas starch-formers so far investigated do not. Also, starch-formers less readily hydrolyze sucrose in the vacuole by invertase enzymes (see Section 13.2) than do sucrose-formers. Finally, we note that formation of sucrose versus its use in respiration is controlled by a potent allosteric effector of two enzymes involved both in the glycolysis phase of respiration and in sucrose synthesis (Stitt, 1990). Functions of this effector, *fructose-2,6-bisphosphate*, are described in Section 13.12 and depicted in Figure 13-13.

## Formation of Fructans

Our knowledge of the chemistry and synthesis of various fructans is surprisingly limited considering their importance in plants, especially in stems and leaves of so-called cool-season grasses that dominate pastures in temperate climates. Such C-3 grasses provide most of the feed for cattle and sheep, so the fructans they store are important to animals, including humans, as well as to plants. Besides grasses, fructans exist in certain organs in at least nine other families, including under-

**Figure 11-17** A schematic representation of a small part of starch molecules. Amylose and amylopectin are similar, except that amylopectin is much more branched.

ground storage organs of the Asteraceae (composits such as asters and dandelions) and the Campanulaceae, and in leaves and bulbs of the Liliaceae (lilies), Iridaceae (irises), Agavaraceae, and Amyrillidaceae (reviewed by Meier and Reid, 1982; Pontis and del Campillo, 1985; Hendry, 1987; Pollock and Chatterton, 1988).

Fructans are fructose polymers that are much smaller than the glucose polymers of starch. Fructans usually have only three to a few hundred fructose units. They are quite water-soluble and are synthesized and stored largely or entirely in vacuoles. Most contain one terminal glucose unit, indicating that they are built by adding fructose units onto the fructose of a sucrose molecule. There appear to be four major types of fructans:

1. **Inulins**, which are unusually abundant in tubers of dahlia and Jerusalem artichoke and which also occur in several other species (but not in grasses). Most inulins contain up to about 35 fructose units connected to each other in straight chains by β-2,1 glycosidic bonds (carbon 2 of one fructose connected to carbon 1 of the previous one). Each inulin has glucose at the starting end of the chain. Recently, the long-held theory that inulins are always nonbranched and contain exclusively β-2,1 glycosidic bonds was demonstrated to be erroneous by Carpita et al. (1991). They used modern, highly-sensitive analytical techniques and demonstrated that inulins from roots of chicory (*Chicorium intybus*) contain small amounts of β-2,6 bonds in the main chain and occasional short branches. Whether inulins from other species are similar to those of chicory needs to be investigated by similar methods.

2. **Levans** (formerly also called **phleins**), which are abundant in leaves and stems of many cool-season grasses. The number of fructose units in levans ranges from a few to many — up to 260 in timothy grass (*Phleum pratense*) and up to 314 in orchard grass (*Dactylis glomerata*). In contrast to the inulins, levans contain fructose units connected mainly by β-2,6 glycosidic bonds (carbon 2 of one fructose connected to carbon 6 of the previous one). Each levan has glucose at the starting end of the chain. Some levans have branches, often with only one fructose unit constituting the branch.

3. An unnamed, highly-branched, mixed-linkage fructan common in leaves, stems, and inflorescences of wheat, barley, and certain other cool-season grasses.

The most abundant fructans in wheat stems contain only three or four fructose units, whereas in wheat inflorescences a much larger group is also present (Hendrix et al., 1986). Only the smaller, mixed-linkage group has been analyzed carefully. It contains in its main chains both β-2,1 and β-2,6 bonds, and the bonds where branches occur are of the same kind (Bancal et al., 1991).

**4.** An unnamed, nonbranched group of fructans identified so far only in two species of the family Liliaceae: onion (roots) and asparagus (leaves). This group of nine major fructans consists of relatively small molecules with no more than five fructose units and one glucose unit. However, as opposed to most known fructans in other species, roughly half of the fructans in this group contain the glucose of the sucrose starting unit not on a terminal end but instead bonded on the other end to one, two, or three fructose units. All glycosidic bonds that connect one fructose to another in these fructans are β-2,1 (Shiomi, 1989).

So far, only two major classes of enzymes that form fructans have been identified, both of which produce straight chains without branches. One starter enzyme called **sucrose:sucrose fructosyltransferase (SST)** combines two sucroses to form a glucose-fructose-fructose unit (a three-hexose **kestose**), releasing a glucose from one of the sucroses. The second enzyme (**fructan:fructan fructosyltransferase**, or **FFT**) adds (one at a time) additional fructose units from a kestose or from a fructose unit of an even larger fructan to the terminal fructose formed by SST. FFTs are therefore *chain-elongating enzymes* responsible for formation of fructans that are larger than kestoses. Three different kinds of FFTs have been described, yet none of these forms β-2,6 bonds in the main chain of levans. Furthermore, no enzyme responsible for producing branches has yet been found.

A major role of fructans in plants is carbohydrate storage. In general, leaves of cool-season grasses accumulate fructans in cool temperatures at which photosynthesis exceeds translocation. From an extensive study of nearly 200 species of grasses, Chatterton et al. (1989) concluded that formation of fructans in leaf vacuoles provides an effective sink that allows photosynthesis to continue in cool conditions.

# 12

# Photosynthesis: Environmental and Agricultural Aspects

The amount of photosynthesis occurring on earth is staggering. Estimates of the amount of carbon fixed each year range from about 70 to 120 billion metric tons (equivalent to about 170 to 290 gigatons of dry matter with an empirical formula close to $CH_2O$). Table 12-1 (on page 252) lists estimates of photosynthetic productivity (called **primary productivity**) for specific types of ecosystems. For several decades we thought that roughly two-thirds of this productivity occurred on land and only one-third in seas and oceans; now it appears that measurement methods inhibited growth of phytoplankton, and thus the photosynthesis that occurs in the open oceans was underestimated, perhaps by a factor of two or more (Post et al., 1990). In any case, this vast productivity occurs in spite of the low atmospheric $CO_2$ concentration:[1] only about 0.0352 percent by volume, or 352 $\mu mol\ mol^{-1}$. Within plant canopies, as in a maize field for example, the $CO_2$ content can fall to 260 $\mu mol\ mol^{-1}$ or less during daylight hours, whereas the content there can reach 400 $\mu mol\ mol^{-1}$ during darkness because of respiration by plants and soil microbes. Most of the $CO_2$ used by plants is ultimately converted into cellulose, the major component of wood.

Table 12-2 shows that about 80 percent of the earth's carbon is in sedimentary limestones and dolomites, deposited over geologic time on the bottoms of oceans and seas by marine organisms (Berner and Lasaga, 1989). Most of the rest is in the form of sedimentary organic matter, called **kerogen**, derived from the soft body parts of organisms. Oil shale, coal, and petroleum are only a tiny fraction of the total, but they are also important carbon reservoirs. Over geological time, all this carbon has existed in the atmosphere—but not all at once! Right now the atmosphere has only about one-thousandth of 1 percent of the total. Nevertheless, it contains about 746 billion metric tons, compared with about 560 billion metric tons in living organisms. Over 13 percent of the carbon in the atmosphere is used in photosynthesis each year, and approximately an equal amount exchanges with dissolved $CO_2$ in the oceans.

## 12.1 The Carbon Cycle

The amount of $CO_2$ in the air has remained quite stable at approximately 280 $\mu mol\ mol^{-1}$ for most of the previous thousand years and was relatively stable between 200 and 300 $\mu mol\ mol^{-1}$ for 150,000 years before that (as indicated by analysis of air bubbles trapped in polar ice; see Fig. 12-1a and 12-1b). Beginning about 1850, atmospheric $CO_2$ has increased exponentially (see Fig. 12-1b), to about 352 $\mu mol\ mol^{-1}$ in 1990 (see Fig. 12-1c; Clark, 1982; Post et al., 1990; Rycroft, 1982; Tans et al., 1990; Waterman et al., 1989). $CO_2$ increased at about 1.4 $\mu mol\ mol^{-1}\ y^{-1}$ during the past 15 years, but in 1988 the level rose by more than 2 $\mu mol\ mol^{-1}$, a record jump and over half of 1 percent of the current total (Pieter Tans, personal communication). The main reason for the increase since 1850 is the burning of fossil fuels, but the clearing of land, especially the burning of tropical forests, has also contributed (Stuiver, 1978). Stable ecosystems such as the tropical rain forests add about as much $CO_2$ to the atmosphere (by respiration and decay) as they remove, but when they are cleared and burned the stored carbon in their biomass and much or all of that stored in the soil is transferred from the biosphere to the atmosphere.

Figure 12-2 shows the carbon cycle with the various ways in which $CO_2$ is added to or removed from the atmosphere and transferred between other **compartments** or **pools** in which it exists. Note that in Figure 12-2 and in Table 12-2 (on page 253) the pool sizes vary over several orders of magnitude (and that their sizes

[1]It has been common to refer to $CO_2$ concentrations in units of parts per million (ppm), but ppm is not an SI unit. **Avogadro's law** states that equal volumes of different gases at the same temperature and pressure have the same number of molecules, or moles (Avogadro's number of molecules). The **mole** (symbol **mol**) is a base SI unit (see Appendix A), so an appropriate SI unit for $CO_2$ concentration is $\mu mol\ mol^{-1}$, which, fortunately, is equal to ppm.

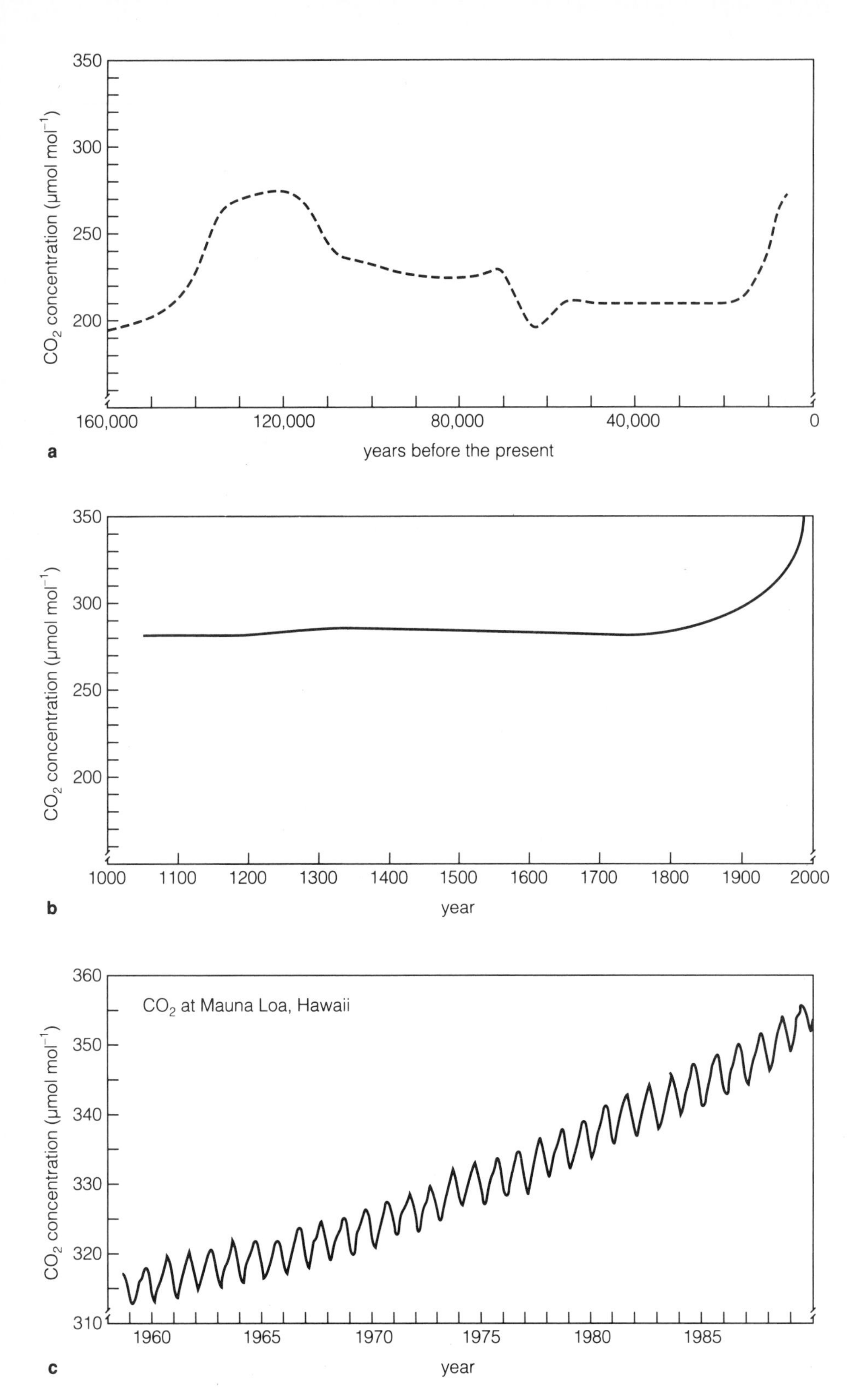


**Figure 12-1** Atmospheric $CO_2$ concentrations during prehistoric and historic times. (**a**) Approximate levels of $CO_2$ during the past 160,000 years as determined by measuring air bubbles trapped in antarctic ice at the Vostok and Bird stations. The curve approximates many individual data points, which show considerable scatter. (**b**) $CO_2$ concentrations during the past 1,000 years (up until 1958), also from air bubbles trapped in ice, and then yearly averages of measurements made at Mauna Loa Observatory in Hawaii since 1958. (**c**) Monthly averages of $CO_2$ measurements made at the Mauna Loa Observatory. Note that rises within years occur in winter, when input to the atmosphere (by respiration, burning, and so forth) exceeds removal; summer decrease is caused largely by photosynthesis. (Data for **a** and **b** are published in various sources; figures here were modified from graphs in Post et al., 1990. Data for **c** were supplied by Pieter Tans, U.S. Department of Commerce, National Oceanic and Atmospheric Administration (NOAA), Environmental Research Laboratories, Boulder, Colorado.)

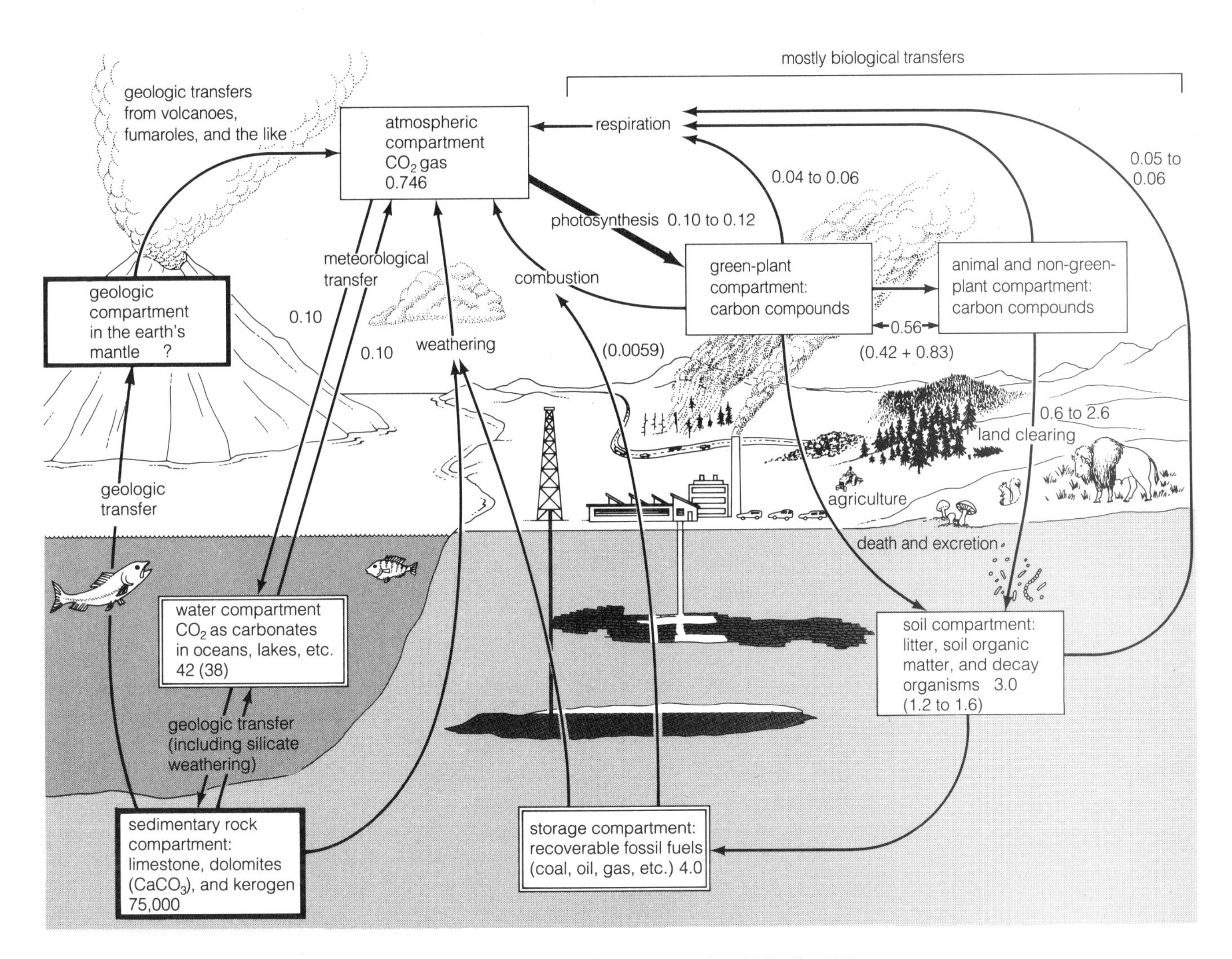


**Figure 12-2** The carbon cycle in nature. The numbers in the boxes (compartments) multiplied by $10^{15}$ represent kilograms of carbon; numbers near arrows (also multiplied by $10^{15}$ to give kilograms) represent annual transfers. Agriculture and land clearing (especially burning of tropical rain forests) are currently adding carbon to the atmosphere, carbon that was stored in plant biomass and as soil organic matter, but amounts of these transfers are only imperfectly known; they occur via combustion and decay (a biological oxidation) and are not shown separately. Because the concentration of atmospheric $CO_2$ can be measured accurately, the amount in the atmosphere is the most accurate figure. (Concentration in $\mu$mol mol$^{-1}$ is multiplied by 2.12 to give gigatons—billions of metric tons—of carbon in the atmosphere.) Numbers not in parentheses are from Berner and Lasaga (1989); numbers in parentheses are from Post et al. (1990); and the differences illustrate the uncertainties in our estimates. (Figure modified from Jensen and Salisbury, 1984.)

are known only imperfectly, with considerable variation among estimates). The atmospheric and biospheric compartments are the smallest; sedimentary rocks, with their carbonates and **kerogen** (sedimentary organic matter), are by far the largest.

In the short run (that is, during our lifetimes), $CO_2$ is added to the atmosphere by the respiration of plants, microorganisms, and animals, by the burning of fossil fuels, and by land clearing. Over geological time (and continuing at present), $CO_2$ is added to the atmosphere in volcanic emissions and soda springs. In the short run, photosynthesis is one of two important mechanisms for removal of $CO_2$ from the atmosphere; the other mechanism is solution of $CO_2$ in the oceans and seas, where solid and dissolved carbonates equilibrate with $CO_2$, a change in one eventually affecting the other, so that the concentration of $CO_2$ on a worldwide scale is buffered by the carbonates in water. There has been much effort to determine the rates of these processes, including photosynthesis (Post et al., 1990), but so far it has been impossible to prepare a balance sheet that will account for all the $CO_2$ added to the atmosphere by the burning of fossil fuels; the atmosphere contains less $CO_2$ than calculations suggest it should.

**Table 12-1 Estimates of Net Primary Productivity and Plant Biomass for Various Ecosystems.[a]**

| Ecosystem Type | Area ($10^6$ km²) | Net Primary Production, per Unit Area (g/m²/yr) | | World Net Primary Production ($10^9$ t/yr) | Biomass per Unit Area (kg/m²) | | World Biomass ($10^9$ t) |
|---|---|---|---|---|---|---|---|
| | | Normal Range | Mean | | Normal Range | Mean | |
| Tropical rain forest | 17.0 | 1,000–3,500 | 2,200 | 37.4 | 6–80 | 45 | 765 |
| Tropical seasonal forest | 7.5 | 1,000–2,500 | 1,600 | 12.0 | 6–60 | 35 | 260 |
| Temperate evergreen forest | 5.0 | 600–2,500 | 1,300 | 6.5 | 6–200 | 35 | 175 |
| Temperate deciduous forest | 7.0 | 600–2,500 | 1,200 | 8.4 | 6–60 | 30 | 210 |
| Boreal forest | 12.0 | 400–2,000 | 800 | 9.6 | 6–40 | 20 | 240 |
| Woodland and shrubland | 8.5 | 250–1,200 | 700 | 6.0 | 2–20 | 6 | 50 |
| Savanna | 15.0 | 200–2,000 | 900 | 13.5 | 0.2–15 | 4 | 60 |
| Temperate grassland | 9.0 | 200–1,500 | 600 | 5.4 | 0.2–5 | 1.6 | 14 |
| Tundra and alpine | 8.0 | 10–400 | 140 | 1.1 | 0.1–3 | 0.6 | 5 |
| Desert and semidesert scrub | 18.0 | 10–250 | 90 | 1.6 | 0.1–4 | 0.7 | 13 |
| Extreme desert, rock, sand, and ice | 24.0 | 0–10 | 3 | 0.07 | 0–0.2 | 0.02 | 0.5 |
| Cultivated land | 14.0 | 100–3,500 | 650 | 9.1 | 0.4–12 | 1 | 14 |
| Swamp and marsh | 2.0 | 800–3,500 | 2,000 | 4.0 | 3–50 | 15 | 30 |
| Lake and stream | 2.0 | 100–1,500 | 250 | 0.5 | 0–0.1 | 0.02 | 0.05 |
| *Total continental* | 149 | | 773 | 115[b] | | 12.3 | 1840[b] |
| Open ocean | 332.0 | 2–400 | 125 | 41.5 | 0–0.005 | 0.003 | 1.0 |
| Upwelling zones | 0.4 | 400–1,000 | 500 | 0.2 | 0.005–0.1 | 0.02 | 0.008 |
| Continental shelf | 26.6 | 200–600 | 360 | 9.6 | 0.001–0.04 | 0.01 | 0.27 |
| Algal beds and reefs | 0.6 | 500–4,000 | 2,500 | 1.6 | 0.04–4 | 2 | 1.2 |
| Estuaries | 1.4 | 200–3,500 | 1,500 | 2.1 | 0.01–6 | 1 | 1.4 |
| *Total marine* | 361 | | 152 | 55.0 | | 0.01 | 3.9 |
| *Full total* | 510 | | 333 | 170[b] | | 3.6 | 1,840[b] |

[a]Units are square kilometers, dry grams or kilograms per square meter, and dry metric tons (t) of organic matter. One metric ton equals 1.1023 English tons.

[b]Totals are rounded to about the same number of significant figures as the data, which are imperfect estimates.

Source: Whittaker, 1975.

Tans et al. (1990) found that their models of the oceans would not accommodate the amounts of $CO_2$ that other workers have assumed were removed by the oceans. They suggest that "a large amount of the $CO_2$ is apparently absorbed on the continents by terrestrial ecosystems." A minute portion of the carbon thus removed by photosynthesis must be fossilized into coal, oil, and natural gas.

Atmospheric $CO_2$ dissolved in water becomes carbonic acid, $H_2CO_3$, which dissolves some of the calcium and magnesium carbonates in rocks and soil to form bicarbonate ions ($HCO_3^-$). Eventually, these ions end up in oceans and seas, where they are converted back to limestone and dolomite by marine organisms, but because a molecule of $CO_2$ is released in the process, this process does not change the atmospheric $CO_2$ level. Carbonic acid also weathers silicate minerals, such as the feldspars found in granites and basalts (represented by the generalized formula $CaSiO_3$), to produce bicarbonate ions. Because the silicate contains no carbon, when these $HCO_3^-$ ions are converted by marine organisms to limestone and dolomite, carbon is removed from the atmosphere and stored in sedimentary rocks. Eventually, even this carbon is returned to the atmosphere, but first the sedimentary rocks are buried to depths of several kilometers, mostly at **subduction zones** where two of the great plates into which the earth's surface is divided collide. One plate slides under the other, carrying the sedimentary rock with it. At great depths, it is heated, and the carbonates are converted to $CO_2$, which finally returns to the atmosphere via volcanoes and soda springs. Over geologic time, these mechanisms have kept the atmospheric $CO_2$ content relatively stable (Berner and Lasaga, 1989).

**Table 12-2 Amount of Carbon Found on the Earth in Various Forms.**

| Form | Carbon Mass ($10^{15}$ kg) | Percent of Total[a] |
|---|---|---|
| Calcium carbonate (limestone) and Ca-Mg carbonate (dolomites); mostly in sedimentary rocks. | 60,000 | 80 |
| Sedimentary organic matter (kerogen) | 15,000 | 20 |
| Oceanic dissolved bicarbonate and carbonate | 42 | 0.05 |
| Recoverable fossil fuels (coal and oil) | 4.0 | 0.005 |
| Dead surficial carbon (humus, caliche, and so on) | 3.0 | 0.004 |
| Atmospheric carbon dioxide | 0.75 | 0.001 |
| All life (plants and animals) | 0.56 | 0.00075 |

[a]Because amounts are known only imperfectly, percentages are rounded to the same number of significant figures.

Source: Modified from Berner and Lasaga, 1989.

**Table 12-3 Maximum Photosynthetic Rates of Major Plant Types Under Natural Conditions.**

| Type of Plant | Example | Maximum Photosynthesis ($CO_2$ fixed, $\mu$mol m$^{-2}$ s$^{-1}$)[a] |
|---|---|---|
| CAM | *Agave americana* (century plant) | 0.6–2.4 |
| Tropical, subtropical, and Mediterranean evergreen trees and shrubs; temperate zone evergreen conifers | *Pinus sylvestris* (Scotch pine) | 3–9 |
| Temperate zone deciduous trees and shrubs | *Fagus sylvatica* (European beech) | 3–12 |
| Temperate zone herbs and C-3 pathway crop plants | *Glycine max* (soybean) | 10–20 |
| Twelve herbacious alpine plants (Austrian alps, 2600 m elev.) | *Ligusticum mutellina* *Taraxacum alpinum* others | 10–24 |
| Tropical grasses, dicots, and sedges with C-4 pathway | *Zea mays* (corn or maize) | 20–40 |

[a]Values are calculated on the basis of one surface of the leaf; for conifers, data are for the optical projection of needles. An extensive list of tree photosynthetic rates was compiled by Larcher (1969). Data for several C-3 and C-4 crops are listed by Radmer and Kok (1977). Values for many C-3 native plants are given by Björkman (1981). Körner and Diemer (1987) report on alpine plants.

Worldwide increases in atmospheric $CO_2$ are of concern because $CO_2$ and some other so-called **greenhouse gases** such as methane absorb long wavelengths of radiant energy more than short wavelengths. The shorter wavelengths predominate in sunlight and penetrate the atmosphere, warming the earth and everything on it. The earth then radiates longer wavelengths (because it is much cooler than the sun; see Wien's Law in Appendix B), which are absorbed by the greenhouse gases—which in turn radiate part of the energy (at long wavelengths) back to earth, further warming it. (Plants in a greenhouse are warmed the same way: Greenhouse glass transmits short and absorbs and then radiates long wavelengths.) This warming of the earth's surface could, within hundreds of years, cause melting of so much polar ice that the oceans would rise and flood many coastal cities (Revelle, 1982). Other accompanying climatic changes, especially in rainfall patterns, would greatly alter both agriculture and natural vegetation. The situation remains uncertain, however, because clouds and atmospheric particulate pollutants (the ever-present haze) reflect the sun's incoming rays, and increases in these pollutants could thus lead to global cooling.

## 12.2 Photosynthetic Rates of Various Species

Photosynthetic rates of species living in such diverse conditions as arid deserts, high mountains, and tropical rain forests differ greatly (Table 12-1). **Leaf photosynthetic capacity**—defined as the photosynthetic rate per unit leaf area when irradiance is saturating, $CO_2$ and $O_2$ concentrations are normal, temperature is optimum, and relative humidity is high—varies by nearly two orders of magnitude (reviewed by Pearcy et al., 1987). Differences result partly from variations in light, temperature, and the availability of water, but individual species show remarkable differences under specific conditions optimum for each. Species growing in resource-rich environments have much higher photosynthetic capacities than those growing where water, nutrients, or light are in short supply. The highest capacities are found among desert annuals and grasses when water is available. Species possessing the C-4 pathway of $CO_2$ fixation generally have the highest photosynthetic rates, whereas slow-growing desert succulents exhibiting crassulacean acid metabolism (CAM) have among the slowest rates. Table 12-3 summarizes the approximate range of maximum photosynthetic rates for a few major groups of plants representing many different species. Relatively few data are available for CAM plants.

Perennial alpine and arctic species make up an interesting group. These plants usually have short growing seasons; alpine species have moderate day lengths and high irradiance levels, and arctic species have long day lengths and low irradiance levels. Their photosynthesis exceeds respiration so much that they can double in dry weight within a month or less, so carbohydrate accumulation is apparently not a survival problem. It has been shown that alpine plants, at least, can utilize $CO_2$ more efficiently than their lowland counterparts (Körner and Diemer, 1987).

## 12.3 Factors Affecting Photosynthesis

Many factors influence photosynthesis: $H_2O$, $CO_2$, light, nutrients, and temperature, as well as plant age and genetics. Which factor most limits photosynthesis in natural or agricultural ecosystems? From Table 12-1 we can conclude that higher plants apparently are most limited by availability of water. Deserts are extremely unproductive, whereas marshes, estuaries, and tropical rain forests are the most productive ecosystems, along with certain irrigated croplands. When water potentials become too negative (that is, when water becomes limiting), cellular expansion is first retarded, so growth is reduced. With only a little more water stress, stomates begin to close and $CO_2$ uptake is restricted. Photosynthesis is then limited by water because of retarded leaf expansion and restricted $CO_2$ absorption. The relationship between water availability and photosynthesis is examined in Chapters 16 and 26, and the importance of leaf nitrogen content to rubisco concentrations and photosynthesis is mentioned in Section 13.4.

Table 12-1 also suggests that two other factors are important in plant ecosystems: First, alpine and arctic tundras have low productivity, mainly because of low temperatures and short growing seasons. Second, oceans are also low in productivity on a unit-area basis (even taking into account the new measurements mentioned above), although there are regions that are highly productive. What limits productivity in open oceans? Obviously water is available, and temperatures are seldom if ever limiting; active algal photosynthesis has been observed under the ice in Antarctica. Sunlight and $CO_2$ are available, especially in surface waters. *Mineral nutrients* are the limiting factor. Many organisms settle to the bottom when they die, taking their minerals with them. Thus surface waters where light and $CO_2$ are most abundant become impoverished in phosphates, nitrates, and other essential nutrients. When special conditions bring upwellings with nutrients to the surface, a profuse bloom of phytoplankton often results. Some plankton species may synthesize toxic chemicals, leading to fish kills, but in regions where upwelling from depths is the rule, such as off the coast of Peru, the plankton provide food for fish. These areas contain some of the world's most important fisheries. But most oceans are "nutrient deserts."

With this short summary of planetary productivity, let us examine in more detail certain factors that influence photosynthesis.

### Light Effects

Figure 12-3 illustrates a rather typical (except for CAM plants) course of $CO_2$ fixation by photosynthesis during daylight and of $CO_2$ release by respiration at night in a plot of alfalfa. Certain interesting facts can be derived from this graph. First, maximum $CO_2$ fixation occurs about noon, when the irradiance is highest. That light often limits photosynthesis is also shown by reduced $CO_2$-fixation rates when plants were exposed briefly to cloud shadows. The figure also shows the relative magnitudes of photosynthesis and dark respiration; in this example the photosynthetic rate reached a maximum of about eight times the nearly constant night respiratory rate. Average ratios for net carbon fixed during daylight compared with night losses by respiration are nearer to 6:1 but vary with plant and environment.

To understand quantitatively how light affects the rate of photosynthesis, we must first examine how much light energy sunlight provides. At the upper boundary of the atmosphere and at the earth's mean distance from the sun, the total irradiance is 1,360 W $m^{-2}$ (the **solar constant**), which includes ultraviolet and infrared wavelengths.[2] This varies about $\pm 2$ percent because of the earth's slightly elliptical orbit. As this radiation passes through the atmosphere to the earth's surface, much energy is lost by absorption and scattering caused by water vapor, dust, $CO_2$, and ozone, so only about 900 W $m^{-2}$ reach plants, depending on time of day and year, elevation, latitude, atmospheric conditions, and other factors. Of this, about half is in the infrared, roughly 5 percent in the ultraviolet, and the rest has wavelengths between 400 and 700 nm and is capable of causing photosynthesis. *When expressed in energy units* (watts or joules per second), this is called **photosynthetically active radiation, PAR** (see McCree, 1981 and Appendix B). Under a cloud-free sky, PAR is commonly about 400 to 500 W $m^{-2}$, depending on the factors just mentioned.[3]

As we discussed in Chapter 10, photosynthesis and other photochemical reactions depend not on the total

[2]Remember that 1 watt (W) = 1 joule per second, so the solar constant can also be expressed as 1,360 J $m^{-2}$ $s^{-1}$; the joule is the SI unit of energy.

[3]Direct sunlight at noon in the northern hemisphere equals about 10,000 foot-candles or 108,000 lux; but the foot-candle and the lux are not energy units but instead are measures of **illuminance**, a subjective description of the ability of the human eye to perceive light (see Appendix B).

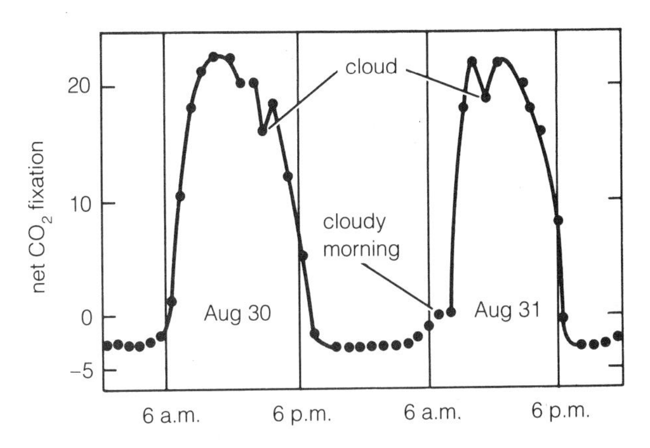


**Figure 12-3** Photosynthesis in an alfalfa plot over a two-day period in late summer. The effect of periods of cloud cover can be noted. Negative $CO_2$-fixation values during hours of darkness indicate the respiration rates. (From data of Thomas and Hill, 1949.)

*energy* in light but on the number of photons or quanta that are absorbed. A highly energetic photon in the blue range of the spectrum has nearly twice as much energy as one in the red range, but the two photons have exactly the same effect in photosynthesis. Hence, when plant physiologists talk about such photochemical reactions as photosynthesis (and others; see Chapter 20), they often express the quantity of light as the number of *photons* in the wavelength regions under consideration: for photosynthesis, 400 to 700 nm, as mentioned above. Expressed this way, the light quantity effective in photosynthesis is called the **photosynthetic photon flux (PPF)**. Its units are moles of quanta (photons) per square meter per second. Sunlight is in the micromolar range; on a clear summer day it equals some 2,000 to 2,300 $\mu$mol $m^{-2}$ $s^{-1}$. (This can be quickly calculated from PAR values by assuming that sunlight in the PAR range has an average wavelength of 550 nm and by knowing that a mole of photons at this wavelength has 217 kJ of energy.[4]) For many studies of photosynthetic productivity, it is appropriate to sum the photons on a daily basis, which brings the units into the molar range: 30 to 60 mol $m^{-2}$ $d^{-1}$ for a monthly average in summer at medium latitudes. On an energy basis (PAR), this is about 6.5 to 13 MJ $m^{-2}$ $d^{-1}$ (megajoules per square meter per day).

Special light sensors have been developed to measure PPF. They include filters that reduce the blue wavelengths before they get to the sensor so that the sensors can be calibrated directly in $\mu$mol $m^{-2}$ $s^{-1}$. They are accurate to within about 10 percent. Because not all wavelengths from 400 to 700 nm are absorbed equally in photosynthesis (see Fig. 10-4), another refinement would be to filter out some of the green wavelengths, which are less effective in photosynthesis, before they reach the sensor. Such meters are available, but so far they have not been widely used; some workers feel that their advantages are minimal, at best.

In any environment the irradiance varies within a plant canopy. About 80 to 90 percent of PPF is absorbed by a representative leaf, although this value varies considerably with leaf structure and age. The remainder (mostly green and far-red wavelengths; see Fig. 4-14) is transmitted to lower leaves or the ground below or is reflected to the surroundings. Of radiation absorbed and potentially capable of causing photosynthesis, more than 95 percent is usually converted to heat; thus less than 5 percent is captured during photosynthesis.

Let us now investigate how varying the irradiance affects photosynthetic rates, first when single leaves are exposed to normal air with about 350 $\mu$mol $mol^{-1}$ $CO_2$. There is, of course, no net $CO_2$ fixed in darkness (except in CAM plants), and in dim light the respiratory loss of $CO_2$ exceeds that used in photosynthesis (Fig. 12-4). Above a certain irradiance, called **light saturation**, increasing light no longer increases photosynthesis. Between darkness and saturation there is an irradiance at which photosynthesis just balances respiration (net $CO_2$ exchange is zero); this is called the **light compensation point**. This point varies with the species, with the irradiance during growth, with the temperature at which measurements are made, and with the $CO_2$ concentration, but in leaves grown in the sun it is usually about 2 percent of full sunlight (approximately the irradiance in a well-lighted classroom: 40 $\mu$mol $m^{-2}$ $s^{-1}$). Only when the irradiance is above the light compensation point can dry-weight increases occur. Differences in light compensation points are caused primarily by differences in rates of dark respiration. When respiration is low, the leaf requires less light to photosynthesize rapidly enough to balance the $CO_2$ being lost, so the light compensation point is then also low.

Figure 12-4 shows responses to irradiance changes exhibited by single leaves of three dicot species after growth in conditions similar to their native habitats. The uppermost curve is for a C-4 perennial shrub species, *Tidestromia oblongifolia*, which grows under unusually hot and arid summer conditions at high light levels in Death Valley, California; the middle curve is for *Atriplex patula* subspecies *hastata*, a C-3 plant that grows along the Pacific coast of the United States; and the lowest curve is for *Alocasia macrorrhiza*, which grows on the floor of a rainforest in Queensland, Australia. The

[4]Two conventions used earlier in discussions of PPF have now been abandoned: We spoke earlier of photosynthetic photon flux *density* (PPFD), but *flux* carries with it the concept of *density*, so *density* is no longer used. A mole of quanta was called the *einstein*, but a mole of anything is still a mole, so the *einstein* is not an acceptable SI unit. Not all workers have adopted the use of PPF; some still use PAR with reference to photon flux.

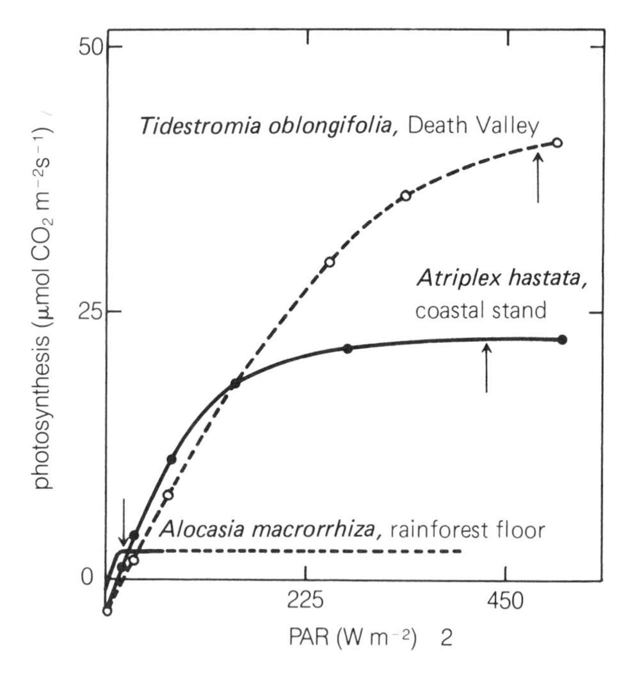


**Figure 12-4** Influence of light on photosynthetic rates in single, attached leaves of three species native to different habitats. Maximum irradiances to which the plants are normally exposed (except for sunflecks that irradiate *Alocasia*) are indicated by arrows. The light compensation points are indicated on the graph where the lines cross the abscissa. (Redrawn from Berry, 1975.)

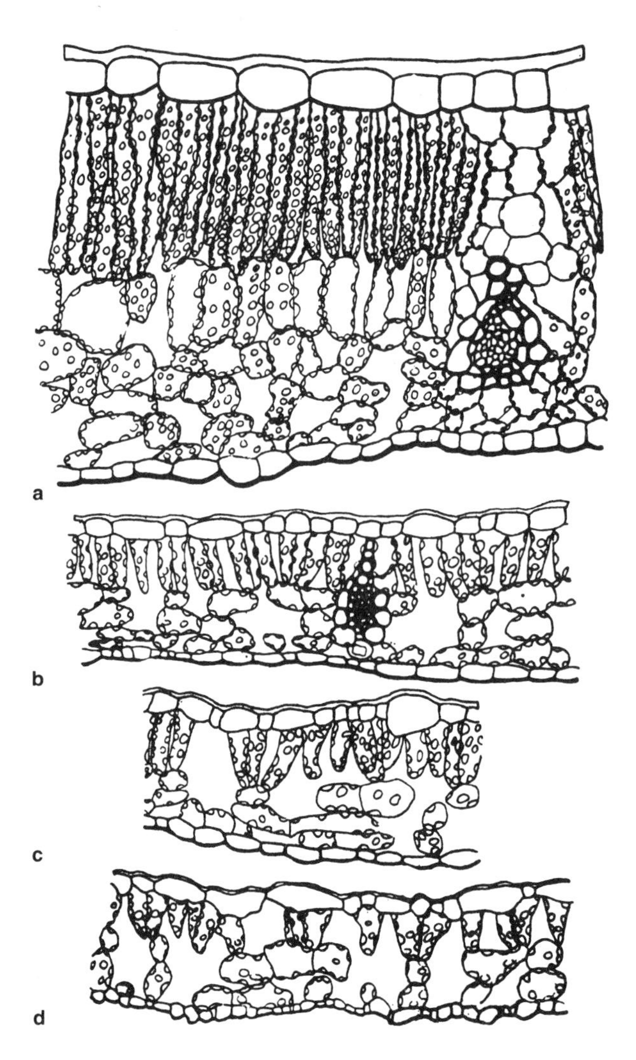


**Figure 12-5** Cross sections of leaves of sugar maple (*Acer saccharum*), an unusually shade-tolerant tree, exposed to different light intensities during growth. (**a**) Sun leaf from south side of isolated tree. Note thick cuticle over the upper epidermis and long palisade parenchyma cells. (**b**) Shade leaf from center of crown of an isolated tree. (**c, d**) Shade leaves from base of two forest trees. All trees were growing near Minneapolis, Minnesota. (From Hanson, 1917.)

energy of PAR received each day by the first and last species during their growth differs by a factor of 300.

The responses of *Alocasia* are typical of many species native to shady habitats (**shade plants**), including most house plants. First, these species exhibit much lower photosynthetic rates under bright sunlight than do crop plants or other species grown in open areas. Second, their photosynthetic responses are saturated at much lower irradiances than are those of other species. Third, under very low irradiance levels they usually photosynthesize at higher rates than do other species. Fourth, their light compensation points are unusually low. These characteristics cause them to grow slowly in their natural shady habitats, yet they survive where species with higher light compensation points could not get enough light and would die. An important complication on the forest floor is the sunflecks that penetrate the tree canopy and may irradiate the shade leaf for anywhere from a fraction of a second (as wind flutters the leaves) to several minutes. As we discuss in Section 25.5, light from sunflecks may account for half or more of the photosynthesis of forest-floor species.

The photosynthetic light responses of single leaves of *Tidestromia oblongifolia* shown in Figure 12-4 are typical of C-4 species native to sunny habitats and of such C-4 crops as maize, sorghum, sugarcane, and millet (the four primary C-4 crops; a few important forage grasses are also C-4 crops). Such leaves show no rate saturation up to and even beyond full sunlight and can have maximum rates more than twice those of most C-3 species (at optimum temperatures for each). For such C-4 crops, rates as high as 40 to 50 $\mu$mol m$^{-2}$ s$^{-1}$ $CO_2$ fixed are not uncommon. The responses of *Atriplex hastata* in Figure 12-4 are representative of many C-3 crop species, such as potatoes, sugar-beets, soybeans, alfalfa, tomatoes, and orchard grass. Individual leaves of these species show photosynthetic light saturation at irradiances one-fourth to one-half that of full sunlight. However, peanut and sunflower are two C-3 species that do not become light-saturated until nearly full sunlight; they also show maximum rates almost as high as those of C-4 crops. Most trees native to temperate climates

**Table 12-4 Some Photosynthetic Characteristics of Three Major Plant Groups.**

| Characteristics | C-3 | C-4 | CAM |
|---|---|---|---|
| Leaf anatomy | No distinct bundle sheath of photosynthetic cells | Well-organized bundle sheath, rich in organelles | Usually no palisade cells, large vacuoles in mesophyll cells |
| Carboxylating enzyme | Rubisco | PEP carboxylase, then rubisco | Darkness: PEP carboxylase light: mainly rubisco |
| Theoretical energy requirement ($CO_2$:ATP; NADPH) | 1:3:2 | 1:5:2 | 1:6.5:2 |
| Transpiration ratio ($H_2O$/dry weight increase) | 450–950 | 250–350 | 18–125 |
| Leaf chlorophyll *a* to *b* ratio | $2.8 \pm 0.4$ | $3.9 \pm 0.6$ | 2.5–3.0 |
| Requirement for $Na^+$ as a micronutrient | No | Yes | Yes |
| $CO_2$ compensation point ($\mu$mol $mol^{-1}$ $CO_2$) | 30–70 | 0–10 | 0–5 in dark |
| Photosynthesis inhibited by 21% $O_2$? | Yes | No | Yes |
| Photorespiration detectable? | Yes | Only in bundle sheath | Detectable in late afternoon |
| Optimum temperature for photosynthesis | 15–25°C | 30–47°C | ≃35°C |
| Dry matter production (tons/hectare/year) | $22 \pm 0.3$ | $39 \pm 17$ | Low and highly variable |
| Maximum on record[a] | 34–39 | 50–54 | |

[a]Monteith, 1978.

Source: Modified from table in Black, 1973.

show maximum rates intermediate between those of typical C-3 crops and shade plants and are often saturated by irradiances as low as one-fourth that of full sunlight.

The more rapid photosynthesis of C-4 species under high irradiances results in a lower water requirement per gram of dry matter produced (a higher **water use efficiency**), but CAM plants have much lower requirements than either C-4 or C-3 species. Often C-4 species can also get by with less nitrogen than C-3 species (Brown, 1978). Table 12-4 compares some of these and other photosynthetic characteristics of C-3, C-4, and CAM plants, most of which are described in Chapter 11 or in the following sections. A good review of the ecophysiology of C-3 and C-4 plants is an article by Pearcy and Ehleringer (1984).

## Adaptations to Sun and Shade

In trees, shrubs, and to some extent in herbaceous plants, many leaves develop in the shade of others and attain during development characteristics that are much like those of true shade plants (Corré, 1983; McClendon and McMillen, 1982). These are called **shade leaves**, as opposed to **sun leaves** that develop in bright light. In dicots, shade leaves are typically larger in area but thinner than sun leaves. Sun leaves become thicker than shade leaves because they form longer palisade cells or an additional layer of such cells (Fig. 12-5). On a weight basis, shade leaves also generally have more chlorophyll, especially chlorophyll *b*, mainly because each chloroplast has more grana than do those of sun leaves. In addition, grana of *Alocasia* and certain other shade plants develop far more thylakoids in the grana, up to 100 per granum (Björkman, 1981). On the other hand, chloroplasts of shade leaves have less total stroma protein, including rubisco, and probably less thylakoid electron transport protein than do sun leaves (Boardman, 1977; Björkman, 1981). Thus shade leaves invest more energy in producing light-harvesting pigments that allow use of essentially all the limited amount of light striking them. Furthermore, chloroplasts in leaves exposed to deep shade become arranged by phototaxis within the cells in patterns that maximize light absorption (Section 20.10). The petioles of dicots also respond to the direction and intensity of light by bending (Section 19.2), causing the leaf blades to move into less shaded regions. All these factors allow net $CO_2$ fixation under low irradiance levels with minimum energy cost to produce and maintain the photosynthetic apparatus.

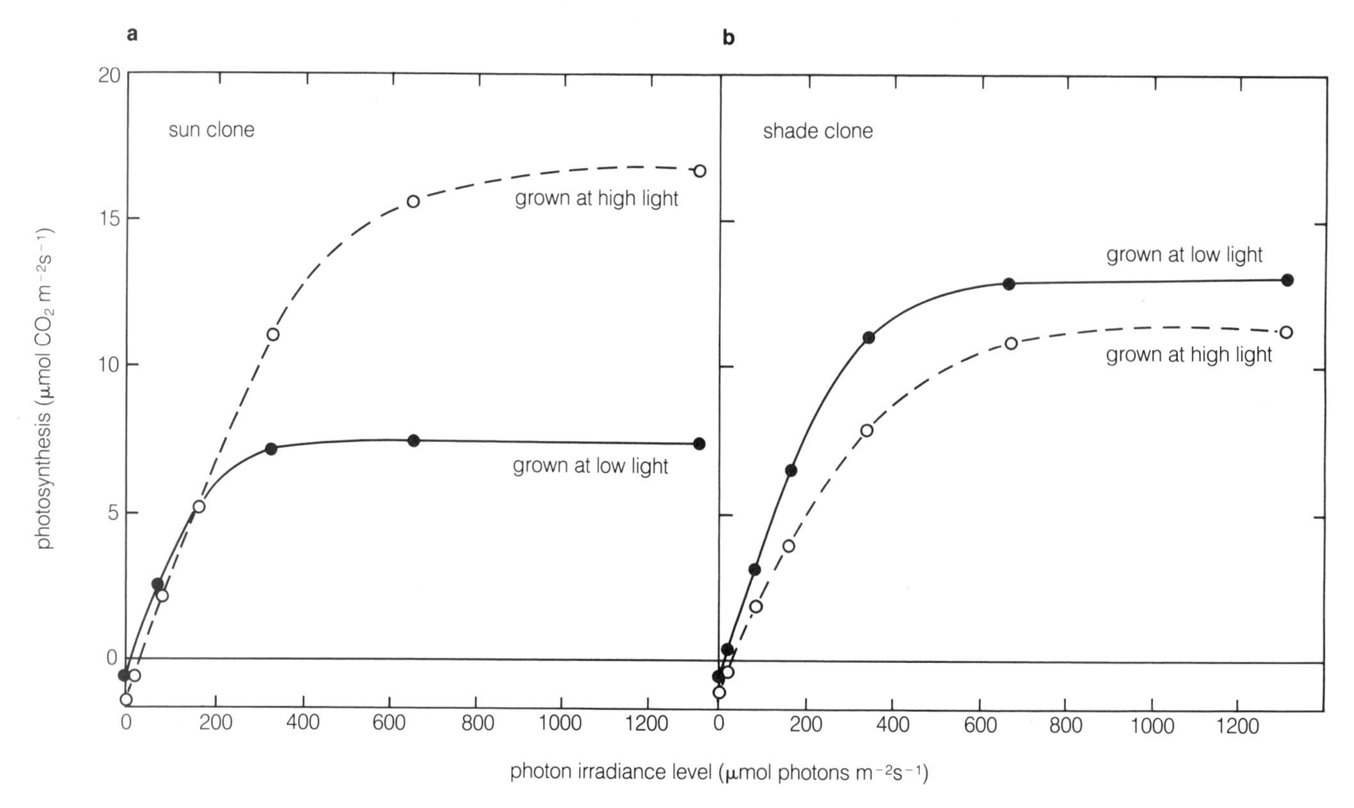


**Figure 12-6** Differences in ability of sun clones and shade clones of *Solidago virgaurea* to adapt photosynthetically to high irradiance levels. Dashed lines represent photosynthetic rates of each type after being grown at high light levels; solid lines represent rates after growth at low levels. (**a**) The sun clone adapted to high light levels during growth; it then required more light to saturate photosynthesis and photosynthesized faster than did plants of the same clone previously grown under low light levels. (**b**) The shade clone behaved differently; it photosynthesized less rapidly after growth in high light levels than in low light levels. (From Björkman and Holmgren, 1963. Used by permission.)

Nevertheless, sun and shade leaves of the same species often exhibit up to a fivefold difference in photosynthetic capacity (Björkman, 1981).

To what extent can sun plants (or sun leaves) adapt to shade and shade plants (or shade leaves) adapt to bright sunlight? Mature leaves show very little adaptation to shade or sun, but whole plants of some species adapt very well to either condition during development, especially to shade. Of course, there are genetic limits to the extent of adaptation. Some plants seem to be obligate shade plants (for example, *Alocasia*); others are obligate sun plants (for example, sunflower, *Helianthus annuus*). But most are facultative shade or sun plants. Facultative C-3 and certain C-4 sun plants adapt somewhat to shade by producing morphological and photosynthetic characteristics similar to those of shade plants (Björkman, 1981). Thus their light compensation points decrease (mainly because they respire much more slowly), they photosynthesize much more slowly, and photosynthesis is saturated at lower irradiance levels. They gradually develop the ability to grow in shade, but this growth is slow.

The reverse adaptation from shade to sun conditions is less common. Shade plants usually cannot be moved to direct sunlight without inhibited photosynthesis and death of the older leaves within several days. Some interesting data were obtained with two different types of *Solidago virgaurea*, one native to open habitats and the other native to shaded forest floors. Clones of each type were made, and these were then grown at high and low irradiance levels. Their photosynthetic responses to various light levels were then measured (Fig. 12-6). Note that the shade clone previously grown under high irradiance levels photosynthesized more slowly than the shade clone grown under low light. Sun clones of the same species, however, photosynthesized much faster after growth under high rather than low irradiances, as expected.

Some conifers are also sensitive to excess light. A dramatic example of this is Englemann spruce (*Picea engelmannii*) in the central and southern Rocky Mountains. When seedlings of this species are transplanted in the open during reforestation work, they usually become chlorotic and die. These symptoms result from a

**Figure 12-7** One method of shading spruce seedlings to prevent solarization during reforestation planting. (Courtesy Frank Ronco.)

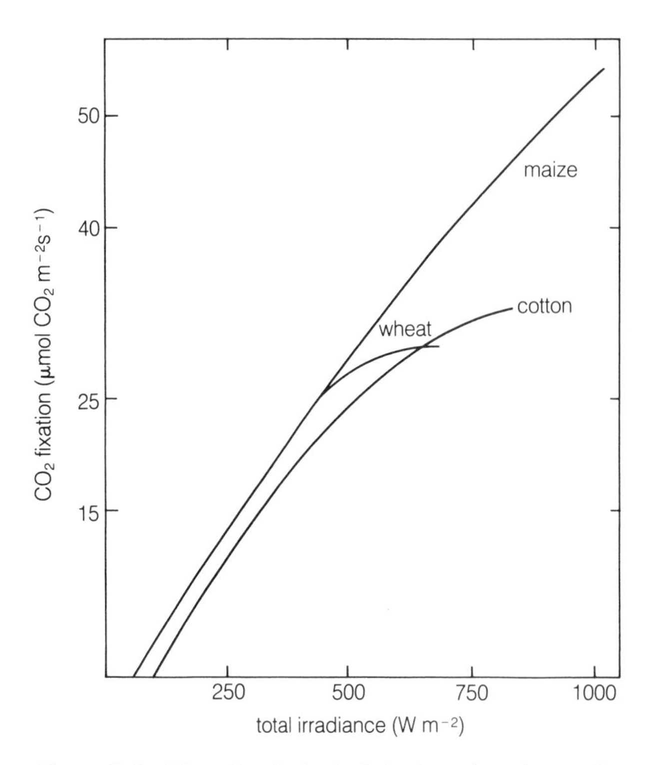


**Figure 12-8** Effect of total solar-radiation intensity at the top of the canopy on net photosynthetic rates in maize, wheat, and cotton plants. [Drawn from data of Baker and Musgrave, 1964 (maize); Puckridge, 1968 (wheat); and Baker, 1965 (cotton).]

phenomenon known as **solarization**, a light-dependent inhibition of photosynthesis followed by oxygen-dependent bleaching of chloroplast pigments. A major function of certain carotenoid pigments is protection against solarization by absorbing excess light energy that is released as heat instead of being transferred to chlorophylls. In Englemann spruce and some other species, this protection is insufficient. If such seedlings are shaded with logs, stumps, or brush, their survival rate is much higher than when unshaded (Fig. 12-7). After maturing for a year or two, they are no longer damaged by high light levels. Light sensitivity is an important factor in plant succession because unusually sensitive species never become established except in the shade of others; these are often climax species that can reproduce in their own shade. Other factors also contribute to shade tolerance, one of which in dicots is the ability to form broad, thin leaves at the expense of reduced root systems.

## Light Effects in a Plant Canopy

The curves of Figure 12-4 show how photosynthesis in single leaves changes with irradiance level, but we may reasonably ask whether a whole plant, an entire crop, or a forest would exhibit similar response curves. Figure 12-8 shows that a stand of maize, a C-4 crop plant, responded in almost a linear fashion to increases in light level up to full sunlight measured on top of the canopy. This curve is much like that for a single leaf of maize or sugarcane (Fig. 12-4), except that it shows even less indication of light saturation. Figure 12-8 also depicts typical results for two C-3 crops, cotton and wheat. For them we see a distinct tendency toward light saturation, but only at much higher levels than in the single-leaf C-3 example of Figure 12-4 (*Atriplex hastata*). (The essay on limiting factors and maximum yields in Chapter 25 describes experiments in which wheat yields were increased to five times the world record in the field by setting virtually all environmental factors at their optimum levels; there was no saturation even at the highest PPF. See box figure on page 561.)

The principal reason for differences in light responses of single leaves and whole plants or groups of plants is that the upper leaves absorb much of the incident light, leaving less for the lower leaves. In this situation, exposure to a higher irradiance may saturate the upper leaves, but more light is then transmitted and reflected toward the shaded leaves below that are not saturated. As a result, single plants, crops, or forests as a whole probably seldom receive enough light to maximize the photosynthetic rate. This is consistent with the alfalfa data of Figure 12-3, in which temporary cloudiness decreased photosynthesis.

Much research has concerned plant and thus canopy **architectures** in relation to productivity. To begin with, it is important to determine **leaf area index (LAI)**, which is the ratio of leaf area (one surface only) of a crop to the ground area upon which that crop stands. LAI values up to 8 are common for many mature crops, depending on species and planting density. Forest trees have LAI values of about 12, and many shaded leaves receive less than 1 percent of full sunlight.

In addition to LAI, it is useful to consider leaf arrangements in the canopy. If all leaves are nearly horizontal and light comes mostly from above, the upper leaves will be exposed to full sunlight; that is, photosynthesis in these leaves will be supersaturated, and much absorbed light will be wasted. Some leaves just below the top of the canopy may be exposed to ideal light levels for photosynthesis, but many leaves below them will have insufficient light. If leaves are nearly vertical, however, as are those of most grasses, and light still comes predominantly from above, light rays will be more or less parallel to the leaf surfaces so that, on a unit-area basis, virtually no leaves will be above the saturation level. Furthermore, light will penetrate deeply into the canopy, so few leaves will be shaded below the light compensation point.

## Availability of $CO_2$

Photosynthetic rates are enhanced not only by increased irradiance levels but also by higher $CO_2$ concentrations, especially when stomates are partly closed by drought. Figure 12-9 illustrates how increasing $CO_2$ levels in the air increase photosynthesis in a C-3 plant at three different irradiance levels. Here the additional $CO_2$ decreased photorespiration by increasing the ratio of $CO_2$ to $O_2$ reacting with rubisco. Photorespiration decreases with increasing $CO_2$-to-$O_2$ ratios, which leads to faster net photosynthesis. Note that at high $CO_2$ concentrations, high irradiance levels increase photosynthesis more than at low $CO_2$ concentrations, and that to saturate photosynthesis a higher $CO_2$ concentration is required at high irradiance levels than at low ones. In contrast, photosynthesis of C-4 species is generally saturated by $CO_2$ levels near 400 $\mu$mol mol$^{-1}$, just above current atmospheric concentrations, even at high irradiance levels in which demands for $CO_2$ are greatest. Some C-4 species are even saturated by present atmospheric $CO_2$ levels (Edwards et al., 1983).

The difference in $CO_2$ requirement between C-4 and C-3 species can be observed easily if $CO_2$ levels are decreased below atmospheric levels. If irradiance levels are above the light compensation points for each, net photosynthesis of C-3 species usually reaches zero at $CO_2$ concentrations between 35 and 45 $\mu$mol mol$^{-1}$ (Bauer and Martha, 1981), whereas C-4 plants continue net $CO_2$ fixation down to levels between 0 and 5 $\mu$mol mol$^{-1}$. The $CO_2$ concentration at which photosynthetic fixation just balances respiratory loss is called the **$CO_2$ compensation point**, a few examples of which are illustrated in Figure 12-10. Note that the value for maize

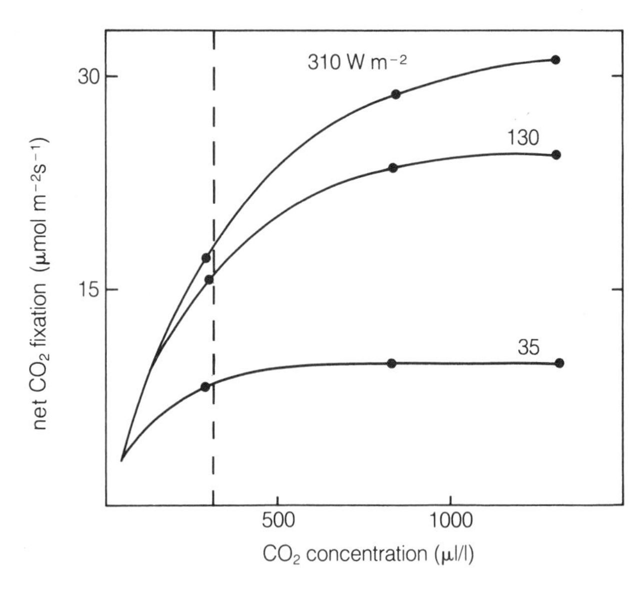


**Figure 12-9** Effects of atmospheric $CO_2$ enrichment on $CO_2$ fixation in sugar-beet leaves. Intact, fully developed leaves from young plants were used. Fixation rates for three different irradiance levels of PAR are shown. The dashed line represents the atmospheric $CO_2$ concentration when the experiment was performed. Higher $CO_2$ levels increased $CO_2$ fixation more at increasing irradiance levels. At the highest level, which is only slightly less than that obtained from full sunlight, the highest $CO_2$ concentration nearly saturated the fixation rate, but at lower irradiances, this rate was saturated by lower $CO_2$ concentrations. Leaf temperatures were between 21 and 24°C. (Redrawn from data of Gaastra, 1959.)

appears to be zero, whereas that for the C-3 species sunflower and red clover is about 40 $\mu$mol mol$^{-1}$.

The difference in $CO_2$ compensation points for C-4 and C-3 species is exhibited dramatically by contrasting responses when a plant of each type is placed in a common, sealed chamber in which photosynthesis can occur (Moss and Smith, 1972); the plants must be grown hydroponically in a soil-less medium such as sand, perlite, or vermiculite to minimize $CO_2$ release by soil microorganisms. Both plants fix $CO_2$ until the $CO_2$ compensation point of the C-3 plant is reached, but the C-4 plant will photosynthesize at still lower $CO_2$ concentrations using $CO_2$ lost by respiration, including photorespiration, from the C-3 plant. As a result, the C-3 plant will usually die within a week or so, but the C-4 plant continues to grow. There is a net transfer of $CO_2$ from one plant to the other (Fig. 12-11).

The lower $CO_2$ compensation points in C-4 than in C-3 species arise from the much lower photorespiratory release of $CO_2$ by C-4 plants. The difference in compensation points essentially disappears if the $O_2$ concentration to which the plants are exposed is decreased from the normal 21 percent down to about 2 percent. In this case the $CO_2$ compensation points of the C-4 species remain the same, but those of the C-3 species also ap-

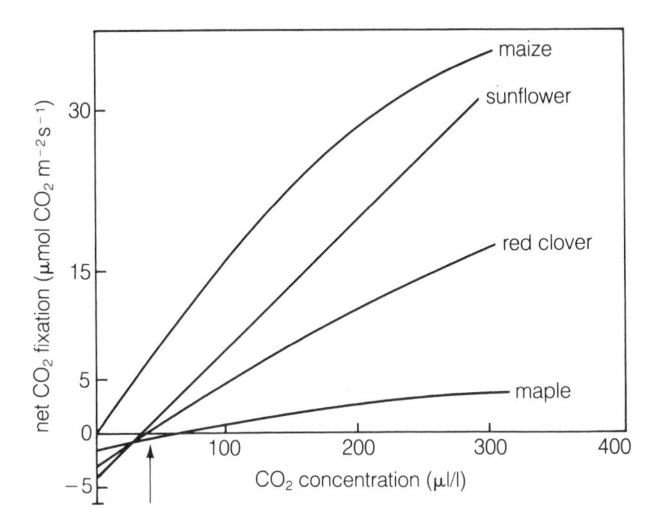


**Figure 12-10** Influence of reduced $CO_2$ concentrations on photosynthetic rate in C-4 (maize) and C-3 plants. Artificial lights providing approximately the same energy as sunlight in the 400–700 nm region were used. Arrow indicates compensation points. (From Hesketh, 1963.)

**Figure 12-11** Left, a C-3 plant (wheat) and right, a C-4 plant (maize) grown in a soil-less medium of perlite inside an airtight chamber in which the $CO_2$ supply became depleted. The maize remained green except at the leaf tips, whereas the wheat leaves were brown and apparently dead. (Courtesy Dale N. Moss.)

proach zero because insufficient $O_2$ is present to compete with $CO_2$ for rubisco; photorespiration becomes negligible.

During the summer growing season, insufficient $CO_2$ is a common cause of suboptimal photosynthesis of C-3 plants, especially for leaves exposed to bright light. Even slight breezes can enhance photosynthesis by replacing $CO_2$-depleted air in the boundary layer around a leaf. Students sometimes ask whether the usual limiting factor in photosynthesis is $CO_2$ or light. The answer is that both can be limiting for C-3 plants and both usually are, but not necessarily for the same leaves. The upper, more illuminated leaves will usually respond to increases in $CO_2$, whereas the lower leaves may be $CO_2$-saturated but will respond to additional light. Thus an increase in either factor increases $CO_2$ fixation of a whole plant or crop. For C-4 plants, light usually limits growth of shaded leaves, unless water or temperature are limiting factors.

Greenhouse crops sometimes lack enough $CO_2$ for maximal growth, and this is especially serious in winter when greenhouses are closed (Enoch and Kimball, 1986). Some growers fertilize the air with $CO_2$ released from high-pressure tanks or other sources such as a clean-burning flame, thereby obtaining increased yields of many ornamental and food crops during the winter months. $CO_2$ levels are usually not allowed to exceed 1,000 to 1,200 $\mu$mol mol$^{-1}$ because such concentrations are frequently toxic or cause stomatal closure, sometimes even reducing photosynthesis (Hicklenton and Jolliffe, 1980). In summer, greenhouses are often cooled with evaporative cooling systems in which outside air is drawn across wet pads in the greenhouse walls. Increased growth in such greenhouses must result in part from increased $CO_2$ levels caused by the incoming fresh air.

## Temperature

The temperature range over which plants can photosynthesize is surprisingly large (Björkman, 1980; Long, 1983; see Section 26.6). Certain bacteria and blue-green algae photosynthesize at temperatures as high as 70°C, whereas conifers can photosynthesize extremely slowly at −6°C or below. In some antarctic lichens, photosynthesis occurs at −18°C, with an optimum near 0°C. In many plants exposed to bright sunlight on a hot summer day, leaf temperatures often reach 35°C or higher, and photosynthesis continues.

The effect of temperature on photosynthesis depends on the species, the environmental conditions under which the plant was grown, and the environmental conditions during measurement. Desert species have higher temperature optima than do arctic or alpine species, and desert annuals that grow during the hot summer months (mostly C-4 species) have higher optima than those that grow there only during winter and spring (mostly C-3 species). Crops such as maize, sorghum, cotton, and soybeans that grow well in warm climates usually have higher optima than do crops such as potatoes, peas, wheat, oats, and barley that are cultivated in cooler regions. In general, optimum temperatures for photosynthesis are similar to the daytime temperatures at which the plants normally grow, except that in cold environments the optima are usually higher than air temperatures—and leaf temperatures in the sun are typically higher than air temperatures. Figure 12-12 illustrates the general relationship for two grasses native to the Great Plains in Wyoming, *Spartina pectinata* (prairie cordgrass), a C-4 plant, and *Leucopoa kingii* (king's fescue), a C-3 plant. *Spartina* grows at a lower elevation (Wheatland site, Fig. 12-12a) than does *Leuco-*

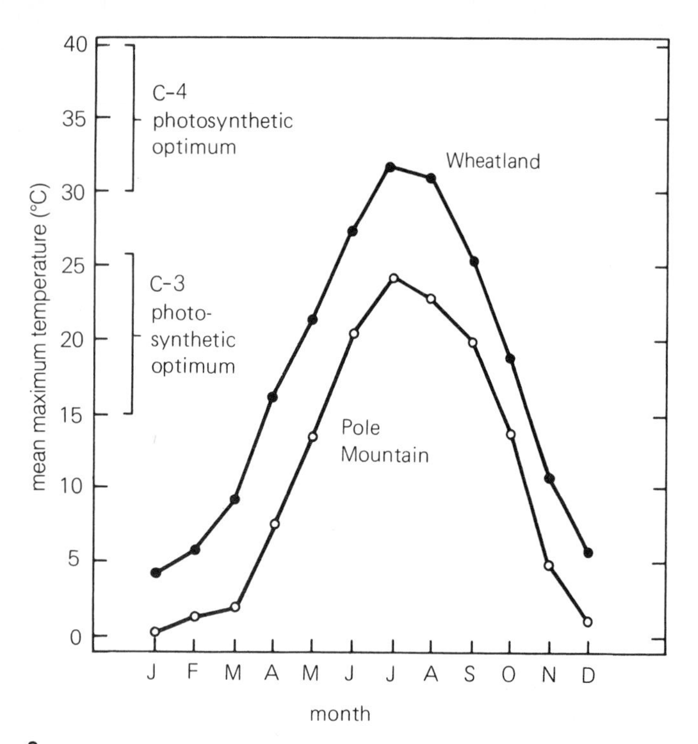


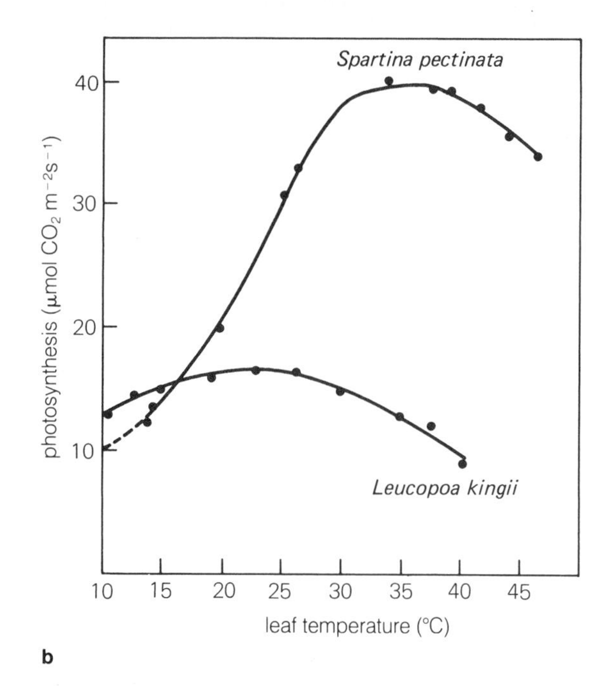


**Figure 12-12** Effect of temperature on photosynthesis in grasses native to the northern Great Plains. (**a**) Mean maximum temperatures during various months at two Wyoming sites. The Wheatland site on the Laramie River has an elevation of 1,470 m; the Pole Mountain site near Laramie is at 2,600 m. Grasses native to the Wheatland site are primarily C-4 species; those at Pole Mountain are mainly C-3 species. The temperature optima for the C-4 species are about 15°C higher than for the C-3 species. (**b**) Temperature-photosynthesis response curves for two species native to the sites in Figure 12-12a. *Spartina*, a C-4 plant, has a higher temperature optimum and greater photosynthetic rates at most temperatures than does *Leucopoa*, a C-3 plant. All measurements were made on whole plants in the field on cloudless days near noon. Stomates of both species remained open at temperatures up to 40°C but partially closed at higher temperatures. (Data of A. T. Harrison.)

*poa* (Pole Mountain site), and the mean daytime temperatures are higher for the Wheatland site. Figure 12-12b shows that the optimum photosynthetic temperature for *Spartina* is near 35°C, compared with about 25°C for *Leucopoa*.

Although there are exceptions, C-4 plants generally have higher temperature optima than do C-3 plants, and this difference is controlled largely by lower rates of photorespiration in C-4 plants. Normal temperature increases have little influence on the light-driven split of $H_2O$ or on diffusion of $CO_2$ into the leaf, but they more markedly influence biochemical reactions of $CO_2$ fixation and reduction. Thus increases in temperature usually increase photosynthetic rates until enzyme denaturation and photosystem destruction begin. However, respiratory $CO_2$ loss also increases with temperature, and this is especially pronounced for photorespiration, largely because a temperature rise increases the ratio of dissolved $O_2$ to $CO_2$ (Hall and Keys, 1983). As a result of $O_2$ competition, net $CO_2$ fixation in C-3 plants is not promoted by increased temperature nearly as much as we might expect. The promoting effect of a temperature rise is nearly balanced by increased respiration and photorespiration over much of the temperature range at which C-3 plants normally grow, so a rather flat and broad temperature response curve between 15 and 30°C often occurs (for example, that of *Leucopoa* in Fig. 12-12b). Because photorespiration is of little significance in C-4 plants, they often exhibit optima in the 30 to 40°C range (Table 12-4). There is also evidence that at high temperatures ATP and NADPH are not produced fast enough in C-3 plants to allow increases in $CO_2$ fixation, so formation of ribulose bisphosphate becomes limiting.

The most dramatic example of photosynthesis at high temperatures in an angiosperm was found by Olle Björkman and coworkers at the Carnegie Institution in Stanford, California, in the shrublike C-4 plant *Tidestromia oblongifolia*. As mentioned before, this plant grows in Death Valley, California, and it does so under the hottest natural environment in the Western Hemisphere. In contrast to species that grow in this area only in winter or spring, *Tidestromia* grows in the hot summer months. It has a remarkable photosynthetic optimum at an air temperature of 47°C (117°F) and, as expected, it proved to be a C-4 plant. Such high temperatures are tolerated well by *Tidestromia* and other comparable species, but the photosystems of most species are destroyed by such excess heat.

Leaves of many species (even when mature) can adapt somewhat to changes in temperature if they are exposed for a few days to different temperatures; this helps plants adjust to seasonal changes (Berry and Björkman, 1980; Osmond et al., 1980; Oquist, 1983). One example of this that involves substantial temperature changes is found in conifers that photosynthesize in both summer and winter. Their temperature optima are higher in summer. Of course, there are genetic limits and considerable genetic variability in the extent of adaptation to temperature, just as to irradiance levels.

### Leaf Age

As leaves grow, their ability to photosynthesize increases until they are fully expanded; then it begins to decrease slowly. Old, senescent leaves eventually become yellow and are unable to photosynthesize because of chlorophyll breakdown and loss of functional chloroplasts. However, even apparently healthy leaves of conifers that persist several years usually show gradually decreasing photosynthetic rates during successive summers. Many factors control net photosynthesis during leaf development (Šesták, 1981).

### Carbohydrate Translocation

One internal control of photosynthesis is the rate at which photosynthetic products such as sucrose can be translocated from leaves to various sink organs. It is often found that removal of developing tubers, seeds, or fruits (strong sinks) inhibits photosynthesis after a few days, especially in adjacent leaves that normally translocate substances to these organs. Furthermore, species that have high photosynthetic rates also have relatively high translocation rates, consistent with the idea that effective transport of photosynthetic products maintains rapid $CO_2$ fixation. Severe infection of leaves by pathogens often inhibits photosynthesis so much that these leaves become sugar importers rather than exporters; the adjacent healthy leaves then gradually photosynthesize much faster, suggesting that enhanced translocation from them has removed some limitation to their $CO_2$ fixation. We do not fully understand the mechanism of these relations (see discussion in Section 8.4), but one factor in some species is buildup of starch grains in chloroplasts when translocation is slow and photosynthesis is fast. Such starch grains press thylakoids unusually close together in chloroplasts and physically prevent light from reaching the thylakoids and causing photosynthesis. Another probable factor is feedback inhibition of photosynthesis by sugars or perhaps other photosynthetic products when translocation is slow (Herold, 1980; Wardlaw, 1980; Gifford and Evans, 1981; Azcón-Bieto, 1983).

## 12.4 Photosynthetic Rates, Efficiencies, and Crop Production

Many crop physiologists, ecologists, and plant breeders are concerned with how environmental factors and plant genotypes can be altered to increase yields of agronomic and forest crops (Evans, 1980; Gifford and Evans, 1981; Johnson, 1981). But crop yields are one thing, and photosynthetic efficiencies are another. **Photosynthetic efficiency** of crop yield is best calculated by dividing the total PAR energy absorbed by a crop from planting to harvest into the total chemical-bond energy of the sucrose produced in photosynthesis. Total radient energy could be used instead of PAR, but it is well known that some 55 percent of solar energy is not used in photosynthesis, so this is ignored. Maximum efficiencies for all species are obtained only at low irradiance levels, not in bright sunlight (during long days) when yields are highest. Considering supplies of PAR to land area on growing crops, overall **biomass-production efficiencies** (including photosynthesis and respiration) are always much below the 18 percent[5] that is the potentially achievable maximum. Many crops, including forest trees and herbaceous species, convert into stored carbohydrates only 1 to 2 percent of the PAR striking the field during the growing season (Wittwer, 1980; Good and Bell, 1980). Much PAR is wasted by striking bare ground between young plants before the canopy has formed; this is true for both C-3 and C-4 plants. As we noted, only 40 to 45 percent of the sun's energy is in the PAR region, so the theoretical maximum efficiency from all the sun's energy is only about 8 percent (45 percent of 18 percent).

At temperatures from 10 to 25°C and normal atmospheric $CO_2$ and $O_2$ levels, efficiencies are about the same for both C-3 and C-4 plants (Ehleringer and Pearcy, 1983; Osborne and Garrett, 1983). Both require about 15 photons of PPF to fix one molecule of $CO_2$. At lower

[5]The potentially achievable maximum of 18 percent is calculated as follows: Assume that 12 moles of photons represent the minimum number of photons needed to fix one mole of $CO_2$ (Reaction 10.8) and that an average photon in the PAR region (400 to 700 nm) has a wavelength of 550 nm. From Planck's equation relating photon energy and wavelength in Appendix B, we can calculate that 1 mol of such photons has an energy of 217 kJ (51.9 kcal). Twelve moles of photons would therefore have an energy of 2.6 MJ. This is the input energy. The output energy—1 mol of fixed carbon in sucrose—has an energy of about 0.47 MJ. Efficiency equals output energy divided by input energy, or 18 percent.

temperatures or with 2 percent $O_2$ or less, photorespiration of C-3 plants is essentially eliminated, and they become more efficient than C-4 plants, requiring only 12 photons per $CO_2$ fixed. Under the same conditions, C-4 plants still require at least 14 photons, in part because they require three ATP molecules for each $CO_2$ to operate the Calvin cycle and two more to operate the C-4 pathway (Section 11.3). The number of photons required by C-4 plants depends upon the mechanism by which they decarboxylate four-carbon acids in the bundle sheath and how effectively $CO_2$ can leak out of the bundle sheath cells (Pearcy and Ehleringer, 1984). The photon requirement varies from 14 to 20.

Photosynthetic efficiencies of crop plants never exceed the theoretical 18 percent of absorbed PAR, and there is no known way by which this can be increased. It has been claimed that dense algal cultures, irradiated with dim red (660 nm) light and corrected for respiratory losses, have achieved a photon requirement as low as 6 to 8 (Emerson and Lewis, 1941; Section 10.7), which, for a photon requirement of 8, is an efficiency of about 33 percent. Demmig and Björkman (1987) studied the quantum requirement of plant leaves by studying oxygen evolution instead of $CO_2$ absorption. This gives a measure of the quantum requirement of the light reactions rather than the entire photosynthetic process. Their values averaged around 9. The difference between these low values and those calculated on the basis of $CO_2$ absorption, which are never lower than 12, is apparently accounted for by the fact that much of the reducing power produced in the light reactions is used for energy-requiring processes other than $CO_2$ fixation. Reduction of nitrate is one good example. Furthermore, because crop plants are not irradiated with dim red light and because respiratory energy is required to maintain life and to drive growth and development, even the calculated 18 percent based on a photon requirement of 12 is not realistic for crops. Correcting for essential respiratory losses reduces the potentially achievable efficiency to about 13 percent at best.

A study with wheat in controlled environments (Bugbee and Salisbury, 1988; see essay in Chapter 25) achieved a life-cycle efficiency of slightly above 10 percent at the lowest irradiance (400 $\mu$mol $m^{-2}$ $s^{-1}$). Considering that some light was wasted during canopy formation and also after leaves began to age (senesce), efficiencies must have reached the potentially achievable 13 percent during the maximum growth phase after the canopy had closed. (Measurements showed that only a few percent of the incoming light was not absorbed — that is, was reflected or transmitted.)

In normal air, increasing temperatures gradually decrease efficiencies of C-3 plants, whereas efficiencies of C-4 plants remain constant. As temperatures rise above 30°C, efficiencies of most C-3 plants become

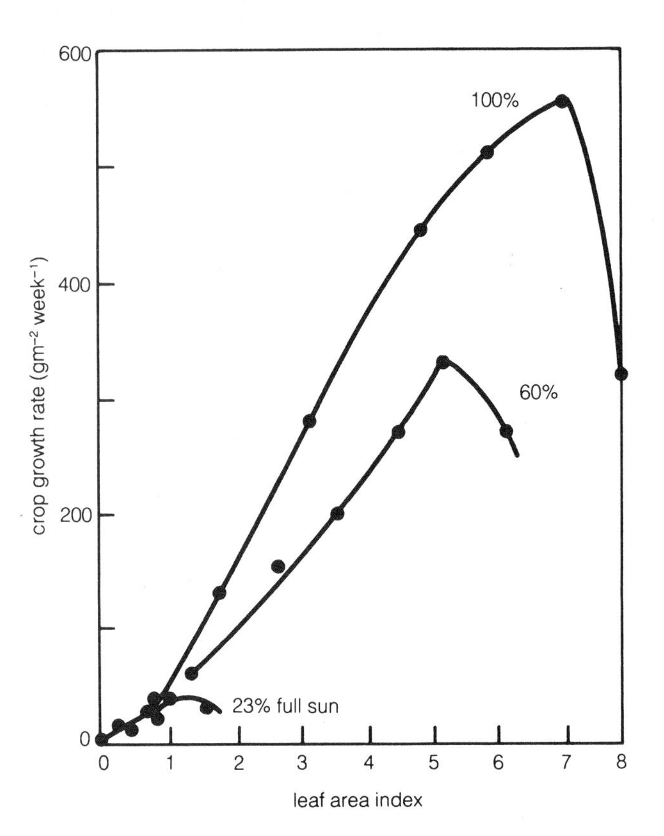


**Figure 12-13** Growth of sunflower plant communities (100 plants $m^{-2}$) at various leaf area indices (LAI) and light levels given as percent of full sunlight. At full sunlight, the optimum LAI is 7; the optimum at 60 percent full sunlight is only 5; and at 23 percent full sunlight, it is only 1.5. (From Leopold and Kriedemann, 1975.)

lower than those of C-4 plants. This efficiency crossover with increased temperature results from lower net photosynthesis in C-3 plants because of faster $CO_2$ loss by photorespiration. The absence of detectable photorespiration in C-4 plants, even above 30°C, gives them a substantial efficiency advantage at high (but not at low) temperatures and especially in nonshaded conditions (Björkman, 1981).

Agriculturists, including foresters, are concerned with productivity of economic plant parts, not with efficiencies or total plant weight, so their goal is to increase both the percentage and the amount of PAR energy that goes to harvestable products. The percentage of weight in the harvestable product compared with aboveground plant biomass is called the **harvest index**. For important crops, including wheat, rice, barley, oats, and peanuts, harvest indices averaging near 50 percent have been reached (Austin et al., 1980; Hargrove and Cabanilla, 1979; Johnson, 1981). For cereal grains, which feed most humans, such high harvest indices have come from breeding programs resulting in cultivars that convert less PAR to leaves and stems and more to seeds. A comprehensive analysis of threefold increases in Minnesota maize yields since the 1930s indicated that introduction of hybrid cultivars accounts for most of that increase (Cardwell, 1982). Photosynthetic efficiencies of very few crops have been improved by breeding (Gifford and Evans, 1981; Zelitch, 1982; Gifford et al., 1984).

Even though a final reduced vegetative growth relative to seeds is advantageous, cultivars that produce extensive leaf cover early in the season are desirable because they intercept more PPF than cultivars that produce relatively more early stem or root growth (Allen and Scott, 1980). Ideally, a crop should form a canopy quickly and then partition its assimilates mostly to the plant part that will be harvested. We discussed canopy architecture above and noted the use of LAI values. Productivity rates increase somewhat with LAI because of increased total light interception, but larger LAI values often cause no more increases and then even decreases on a ground-area basis (Fig. 12-13), probably because of respiratory $CO_2$ loss from heavily shaded leaves and stems. But this depends on leaf arrangements. In the controlled-environment wheat study, LAI values got as high as 30, but this was superoptimal. Such high values were possible because light was penetrating the canopy more or less parallel to the leaves.

Increased stem elongation is often an advantage for plants competing for light, but in a uniformly growing cereal crop no such advantage occurs, and increased grain yields are obtained with dwarf or semidwarf cultivars that allocate relatively more photosynthate to grain than to stems. Another advantage is that such cultivars do not **lodge** (fall over) as easily as do taller cultivars, especially when they are heavily fertilized with nitrogen. Plant breeders have also provided cultivars with other alterations in canopy structure that increase yields. We have noted above the advantages of erect rather than horizontal leaves, and, indeed, yields of the erect-leaf types have often been significantly greater, especially in rice. Photosynthetic capacities have also been increased by increasing the number of leaves in maize and the individual leaf area in wheat (Ho, 1988). It is expected that further cooperation between physiologists and geneticists will bring about increased yields of several other species. Elimination of photorespiration in future agricultural and forest-crop plants appears to be a worthwhile goal (Somerville and Ogren, 1982; Zelitch, 1982), even though no success with crops has thus far been obtained.

# 13

# Respiration

All active cells respire continuously, often absorbing $O_2$ and releasing $CO_2$ in equal volumes. Yet, as we know, respiration is much more than a simple exchange of gases. The overall process is an oxidation-reduction in which compounds are oxidized to $CO_2$ and the $O_2$ absorbed is reduced to form $H_2O$. Starch, fructans, sucrose or other sugars, fats, organic acids, and, under some conditions, even proteins can serve as respiratory substrates. The common respiration of glucose, for example, can be written as follows:

$$C_6H_{12}O_6 + 6\,O_2 \rightarrow 6\,CO_2 + 6\,H_2O + \text{energy} \qquad (13.1)$$

Much of the energy released during respiration—approximately 2,870 kJ or 686 kcal per mole of glucose—is heat. When temperatures are low, this heat can stimulate metabolism and benefit certain species, but usually it is instead transferred to the atmosphere or soil, with little consequence to the plant. Far more important than heat is the energy trapped in ATP because this compound is used for many essential processes of life, such as growth and ion accumulation.

Reaction 13.1, the summary equation for respiration, is somewhat misleading because respiration, like photosynthesis, is not a single reaction. It is a series of 50 or more component reactions, each catalyzed by a different enzyme. It is an oxidation (with the same products as burning) occurring in a water medium near neutral *p*H, at moderate temperatures, and without smoke! This gradual, stepwise breakdown of large molecules provides a means of converting energy into ATP. Furthermore, as the breakdown proceeds, carbon-skeleton intermediates are provided for a large number of other essential plant products. These products include amino acids for proteins; nucleotides for nucleic acids; and carbon precursors for porphyrin pigments (such as chlorophyll and cytochromes) and for fats, sterols, carotenoids, flavonoid pigments such as anthocyanins, and certain other aromatic compounds such as lignin.

Of course, when these products are formed, conversion of the original respiratory substrates to $CO_2$ and $H_2O$ is not complete. Usually only some of the respiratory substrates are fully oxidized to $CO_2$ and $H_2O$ (a catabolic process), whereas the rest are used in synthetic (anabolic) processes, especially in growing cells. The energy trapped during complete oxidation of some molecules can be used to synthesize the other molecules required for growth. When plants are growing, respiration rates increase as a result of growth demands, but some of the disappearing compounds are diverted into synthetic reactions and never appear as $CO_2$. Whether carbon atoms in the compounds being respired are converted to $CO_2$ or to any of the large molecules mentioned above depends on the kind of cell involved, its location in the plant, and whether or not the plant is rapidly growing.

## 13.1 The Respiratory Quotient

If carbohydrates such as sucrose, fructans, or starch are respiratory substrates and if they are completely oxidized, the volume of $O_2$ taken up exactly balances the volume of $CO_2$ released. The ratio of $CO_2$ to $O_2$, called the **respiratory quotient** or **RQ**, is often very near unity. For example, the RQ obtained from leaves of many different species averaged about 1.05. Germinating seeds of the cereal grains and of many legumes such as peas and beans, which contain starch as the main reserve food, also exhibit RQ values of approximately 1.0. Seeds from many other species, however, contain much fat or oil that is rich in hydrogen and low in oxygen. When fats and oils are oxidized during germination, the RQ is often as low as 0.7 because relatively large amounts of oxygen are needed to convert the hydrogen to $H_2O$ and the carbon to $CO_2$. Consider the oxidation of a common fatty acid, oleic acid:

$$C_{18}H_{34}O_2 + 25.5\,O_2 \rightarrow 18\,CO_2 + 17\,H_2O \qquad (13.2)$$

The RQ for this reaction is 18/25.5 = 0.71.

By measuring the RQ for any plant part, information can be obtained about the type of compounds being oxidized. The problem is complicated because at any time several different types of compounds can be respired, so the measured RQ is an average that depends on the contribution of each substrate and its relative content of carbon, hydrogen, and oxygen. In this chapter we emphasize respiration of carbohydrates; the utilization of fats is described in Chapter 15. We first describe some of the major biochemical aspects of respiration; then these will be used to help explain the more physiological and environmental aspects of respiration of various plants and plant parts.

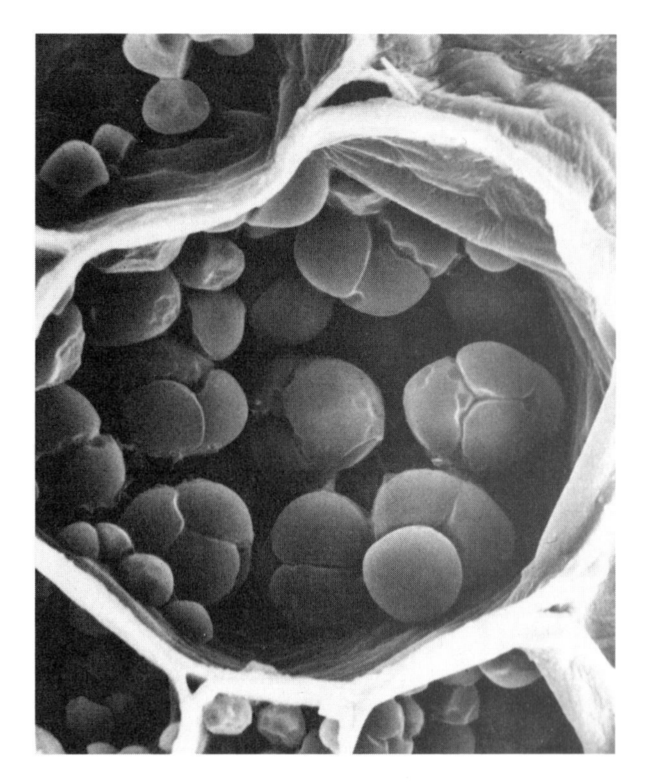

**Figure 13-1** Scanning electron micrographs of starch grains in amyloplasts of a rice stem parenchyma cell. Many amyloplasts in these cells contain four starch grains. (Courtesy P. Dayanandan.)

## 13.2 Formation of Hexose Sugars from Reserve Carbohydrates

### Storage and Degradation of Starch

As described in Section 11.7, starch is stored as water-insoluble granules (grains) that consist of highly branched amylopectin molecules and largely nonbranched amyloses. Starch accumulated in chloroplasts during photosynthesis is the most abundant carbohydrate reserve in leaves of most species. Starch formed in amyloplasts of storage organs from translocated sucrose or other nonreducing sugars is also a principal respiratory substrate for storage organs (Fig. 13-1). Parenchyma cells in roots and stems commonly store starch; in perennial species, starch is stored during winter months and is used in new growth the following spring. Potato tubers are rich in starch-containing amyloplasts, and much of this starch disappears as a result of respiration and translocation of sugars from tuber sections that are planted to obtain a new crop. The endosperm or cotyledon storage tissues of many seeds contain abundant starch, and most of this also disappears during seedling development. Starch storage in various plant parts was reviewed by Jenner (1982).

Figure 13-2a shows the relation of the starch-storing endosperm to the rest of the seed in maize, and a germinating maize seedling is shown in Figure 13-2b. In these instances only some of the glucose molecules derived from starch are totally oxidized to $CO_2$ and $H_2O$. Other glucose molecules are converted into sucrose molecules in the scutellum and are then moved into the growing root and shoot, where some are totally respired and others are diverted into cell-wall materials, proteins, and other substances needed for growth of the seedling.

Most steps in the degradation of starch to glucose can be catalyzed by three different enzymes, although still other enzymes are needed to complete the process. The first three enzymes include an *alpha amylase* ($\alpha$-amylase), a *beta amylase* ($\beta$-amylase), and *starch phosphorylase*. Of these, apparently only alpha amylase can attack intact starch granules, so when $\beta$-amylase and starch phosphorylase are involved they presumably must act on the first products released by $\alpha$-amylase (Stitt and Steup, 1985; Manners, 1985). Some points of attack of these enzymes on amylopectin are shown in Figure 13-3. Alpha amylase randomly attacks 1,4 bonds throughout both amylose and amylopectin, at first causing random pits in starch grains and releasing products that are still large. With nonbranched amylose chains, repeated attack by $\alpha$-amylase leads to *maltose*, a disaccharide containing two glucose units (Fig. 13-4a). Alpha amylase cannot, however, attack the 1,6 bonds at the branch points in amylopectin, so amylopectin digestion stops when branched *dextrins* of short chain lengths still remain. Many $\alpha$-amylases are activated by $Ca^{2+}$, which is one reason that calcium is an essential element.

Beta amylase hydrolyzes starch into $\beta$-*maltose* (Fig. 13-4b); the enzyme first acts only upon the nonreducing ends. The $\beta$-maltose is rapidly changed by mutarotation into the natural mixtures of $\alpha$- and $\beta$-isomers. Hydrolysis of amylose by $\beta$-amylase is nearly complete,

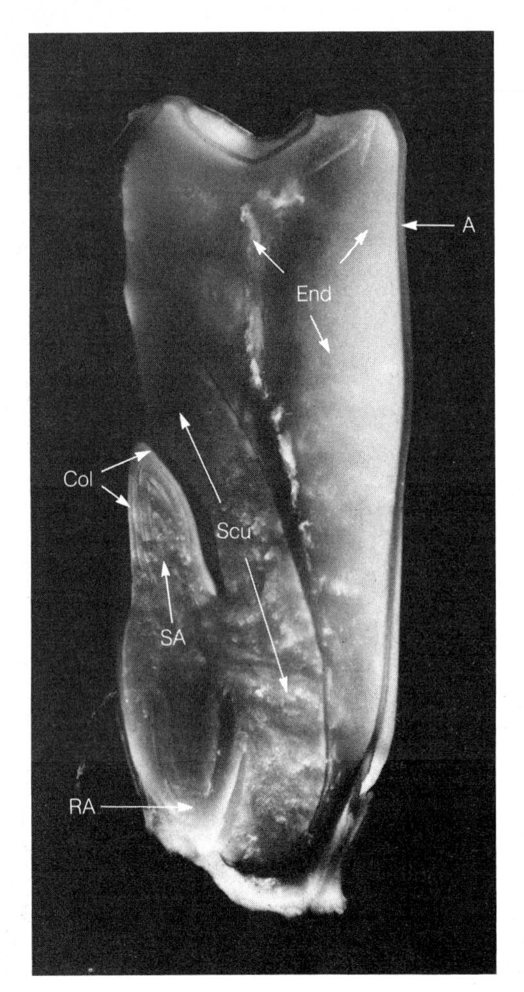


a

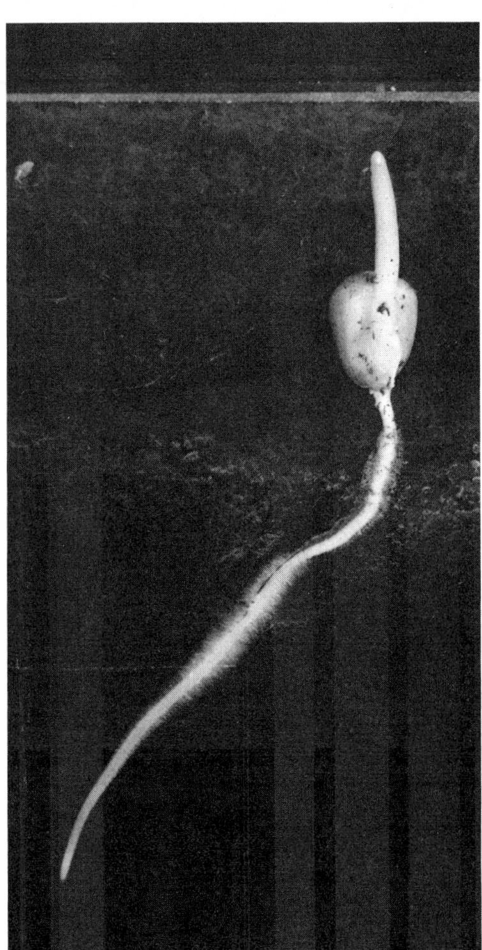

b

**Figure 13-2** (**a**) Longitudinal section of a maize seed, showing the relation of the starch-storing endosperm (End) to the other seed parts: Col, coleoptile; Scu, scutellum or cotyledon; SA, shoot apex; RA, radicle. (From O'Brien and McCully, 1969.) (**b**) Maize seedling being nourished by the endosperm. (From Jensen and Salisbury, 1972.)

β-amylase

H, HO — G¹—⁴G₁—₄G¹—⁴G₁—₄G¹—⁴G₁—₄G¹—⁴₆G₁—₄G¹—⁴G₁—₄G¹—⁴G₁—₄G¹—⁴G — H, OH

nonreducing end

starch phosphorylase

reducing end

α-amylase

₄G⁶—¹G₄—₁G⁴—¹G₄—₁G⁴—¹G₄—₁G⁴—¹G — H, OH

nonreducing end

debranching enzyme

nonreducing end — G¹ — H, OH

**Figure 13-3** Points of degradative attack on amylopectin by alpha amylase, beta amylase, starch phosphorylase, and a debranching enzyme. Shown is part of an amylopectin molecule with one reducing and two nonreducing ends (G represents glucose). Also shown are the α-1,4 bonds between glucoses and (at the single branch point shown here) an α-1,6 bond.

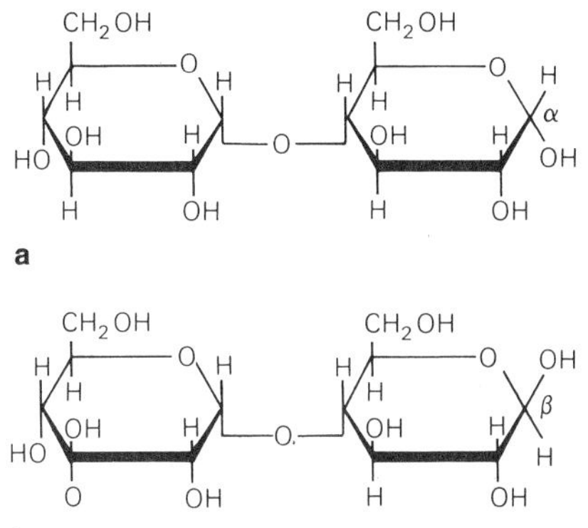

**Figure 13-4** Alpha(α)- and beta(β)-maltose released from starch during action of α- and β-amylases.

but amylopectin breakdown is incomplete because the branch linkages are not attacked. Branched dextrins again remain.

The activity of both amylases involves uptake of one $H_2O$ for each bond cleaved, so they are *hydrolase enzymes* (see Section 9.2). Hydrolytic reactions are not reversible, so no starch synthesis by amylases can be detected. *A general principle is that large molecules are usually synthesized by one series of reactions (pathway) and are broken down by another.* For example, we explained in Section 11.7 how polysaccharide synthesis requires an activated form of a sugar such as ADP-glucose, UDP-glucose, or sometimes even glucose-1-phosphate.

The amylases are widespread in various tissues but are most active in germinating seeds that are high in starch. In leaves, α-amylase is probably of much greater importance than β-amylase for starch hydrolysis. Much of the α-amylase is located inside chloroplasts, often bound to the starch grains that it will attack. It functions during both day and night, although of course during daylight there is a net production of starch from photosynthesis.

Starch phosphorylase degrades starch beginning at a nonreducing end (Fig. 13-3). This degradation does not occur by incorporating water into the products as amylases do, but instead by incorporating phosphate. It is therefore a *phosphorolytic enzyme* rather than a hydrolytic one, and the reaction that it catalyzes is reversible *in vitro*:

$$\text{starch} + H_2PO_4^- \leftrightharpoons \text{glucose-1-phosphate} \qquad (13.3)$$

In spite of the *in vitro* reversibility of this reaction, its only important role seems to be starch degradation. One reason for this is that the P*i* concentration within plastids is often 100 times that of glucose-1-phosphate; under these conditions starch synthesis is negligible. As will become more apparent later, formation of glucose-1-phosphate avoids the need for an ATP to convert glucose into a glucose phosphate during respiration.

Amylopectin is only partially degraded by starch phosphorylase. The reaction proceeds consecutively from the nonreducing end of each main chain or branch chain (Fig. 13-3) to within a few glucose residues of the α-1,6 branch linkages, so dextrins again remain. Amylose, with few such branches, is almost completely degraded by repeated removal of glucose units beginning at the nonreducing end of the chain. Starch phosphorylase is widespread in plants (as are the amylases), and it is often difficult to determine which enzyme digests most of the starch in the cells concerned. The current theory is that α-amylase (or a similar endoamylase) is essential for the initial attack, as mentioned above, and for cereal grain seeds both amylases appear functional but starch phosphorylase does not. For seeds of other species, for leaves, and for other tissues, starch phosphorylase apparently also contributes, especially after the starch grains are partially hydrolyzed by one of the amylases (Steup et al., 1983; Steup, 1988; Manners, 1985; ap Rees, 1988).

The 1,6 branch linkages in amylopectin or branched dextrins that are not attacked by any of the above enzymes are hydrolyzed by various *debranching enzymes*. Plants contain three main types that differ somewhat as to the types of polysaccharides they will attack: a *pullulanase*, an *isoamylase*, and a *limit dextrinase* (Manners, 1985). Action of these enzymes on branched starch chains (Fig. 13-3) provides additional end groups for attack by amylases or by starch phosphorylase, and subsequent action of limit dextrinases on dextrins allows complete digestion of amylopectin into maltose, glucose, or glucose-1-phosphate.

Maltose seldom accumulates during starch digestion because it is hydrolyzed to glucose by a maltase enzyme, as follows:

$$\text{maltose} + H_2O \rightarrow 2\ \alpha\text{-D-glucose} \qquad (13.4)$$

The resulting glucose units are then available for conversion into other polysaccharides as described in Section 11.7 or, as emphasized here, for degradation by respiration.

In summary, the amylases hydrolyze nonbranched amylose chains largely to maltose, whereas starch phosphorylase converts such chains to glucose-1-phosphate. The action of all three enzymes on amylopectin leaves a dextrin, the branch linkages of which must be hydrolyzed by debranching enzymes. Maltose is hydrolyzed to glucose largely by maltase.

All degradation of starch to hexoses presumably occurs within chloroplasts or amyloplasts, yet true respiration of these hexoses begins in the cytosol. As we explained in Section 11.1, hexoses scarcely move out of chloroplasts, and the same is likely true of amyloplasts. If this is so, then hexoses derived from starch must always be converted to triose phosphate (3-PGaldehyde and dihydroxyacetone-P) in plastids, and these molecules must be moved by the phosphate carrier into the cytosol. Here they either could be reassembled to hexose phosphate as described in Section 11.7 or could enter respiration (glycolysis) directly.

### Hydrolysis of Fructans

As mentioned in Section 11.7, the principal carbohydrate food-reserve material in some species—most notably the stems, leaves, and flowers of temperate-region grasses and parts of members of the Asteraceae and other families—is not starch. Instead, fructans predominate (Meier and Reid, 1982; Pontis and del Campillo, 1985; Hendry, 1987; Pollock and Chatterton, 1988; *J. Plant Physiology*, Special Issue, Vol. 134, 1989). But even in these species, fructans are seldom (if ever) abundant in seeds. As usual, starch is the main carbohydrate reserve of seeds. Considering the importance of fructans, surprisingly little is known about their metabolism, although it is known that they are hydrolyzed by β-*fructofuranosidase* enzymes having specificity for the particular β-2,1 or β-2,6 links involved. For example, one such enzyme from the Jerusalem-artichoke tuber successively cleaves fructose units from inulin until a mixture of fructose and the terminal sucrose unit remains:

$$\underset{\text{(fructan)}}{\text{glucose-fructose-(fructose)}_n} + n\text{H}_2\text{O} \rightarrow n \text{ fructose} + \underset{\text{(sucrose)}}{\text{glucose-fructose}} \quad (13.5)$$

Fructose can undergo respiration rather directly, but sucrose must first be split into glucose and fructose as described below.

### Hydrolysis of Sucrose

One important reaction of sucrose degradation is irreversible hydrolysis by *invertases* to free glucose and fructose:

$$\text{sucrose} + \text{H}_2\text{O} \rightarrow \text{glucose} + \text{fructose} \quad (13.6)$$

Invertases exist in the cytosol, the vacuole, and sometimes even in cell walls (Avigad, 1982; ap Rees, 1988; Stommel and Simon, 1990). The cytosol invertase is an alkaline type with a *p*H optimum near 7.5, whereas the other two are acidic invertases with *p*H optima of 5 or less. The cell-wall invertase, when present, hydrolyzes incoming, translocated sucrose into glucose and fructose molecules that are then absorbed by sink cells.

Another common enzyme that can degrade sucrose is *sucrose synthase*, so named because the reaction it catalyzes is reversible and was first thought to be important mainly in the synthesis of sucrose. Sucrose synthase catalyzes the following reaction:

$$\text{sucrose} + \text{UDP} \rightleftharpoons \text{fructose} + \text{UDP-glucose} \quad (13.7)$$

Fructose is then available for respiration, and the glucose in UDP-glucose can be released in one or two ways not shown here.

Evidence indicates that sucrose synthase is the main enzyme that degrades sucrose in starch-storage organs (for example, developing seeds and potato tubers) or in rapidly growing tissues that are converting translocated sucrose to cell-wall polysaccharides. For slow-growing and mature cells, invertase may be the more important enzyme that degrades sucrose and provides glucose and fructose for respiration.

## 13.3 Glycolysis

A group of reactions, collectively called **glycolysis**, degrades glucose, glucose-1-phosphate, or fructose (set free by the preparatory reactions described above) to pyruvic acid in the cytosol. (Several reactions of glycolysis also occur in chloroplasts and other plastids, but the complete pathway might not.) Glycolysis is the first of three closely related phases of respiration; it is followed by the Krebs cycle and electron-transport processes occurring in mitochondria.

The individual reactions of glycolysis, now believed to occur in all living organisms, were discovered between 1912 and 1935 by German scientists interested in alcohol production by yeast and by others concerned with breakdown of animal starch (glycogen) to pyruvic acid in muscle cells (Lipmann, 1975; Cori, 1983). The term *glycolysis*, meaning lysis of sugar, was introduced in 1909 to mean breakdown of sugar to ethyl alcohol (ethanol). However, most cells produce pyruvic acid instead of ethanol when they are normally aerated. Furthermore, the common sugars that are broken down are hexoses, so **glycolysis** *has come to mean degradation of hexoses to pyruvic acid*, although many animal biochemists use the term to mean degradation of glycogen (animal starch) to pyruvate. Plants form no animal starch, so the term *glycolysis* could be misleading to biologists who associate it with glycogen rather than with starch or sucrose. In fact, it has been suggested that botanists should use the term **sucrolysis** because sucrose is the most abundant sugar formed and translocated in plants and is therefore a common supplier of glucose and fructose used in respiration. However, as we emphasized in Section 13.2, starch also supplies glucose for glycolysis, and in amyloplasts this glucose actually reaches the cytoplasm (where glycolysis occurs) largely as triose phosphates, not as sucrose.

The individual reactions of glycolysis, the enzymes that catalyze them, and the particular requirements of the enzymes for metal activators are diagramed in Figure 13-5. The overall process, however, (beginning with glucose) can be summarized as follows:

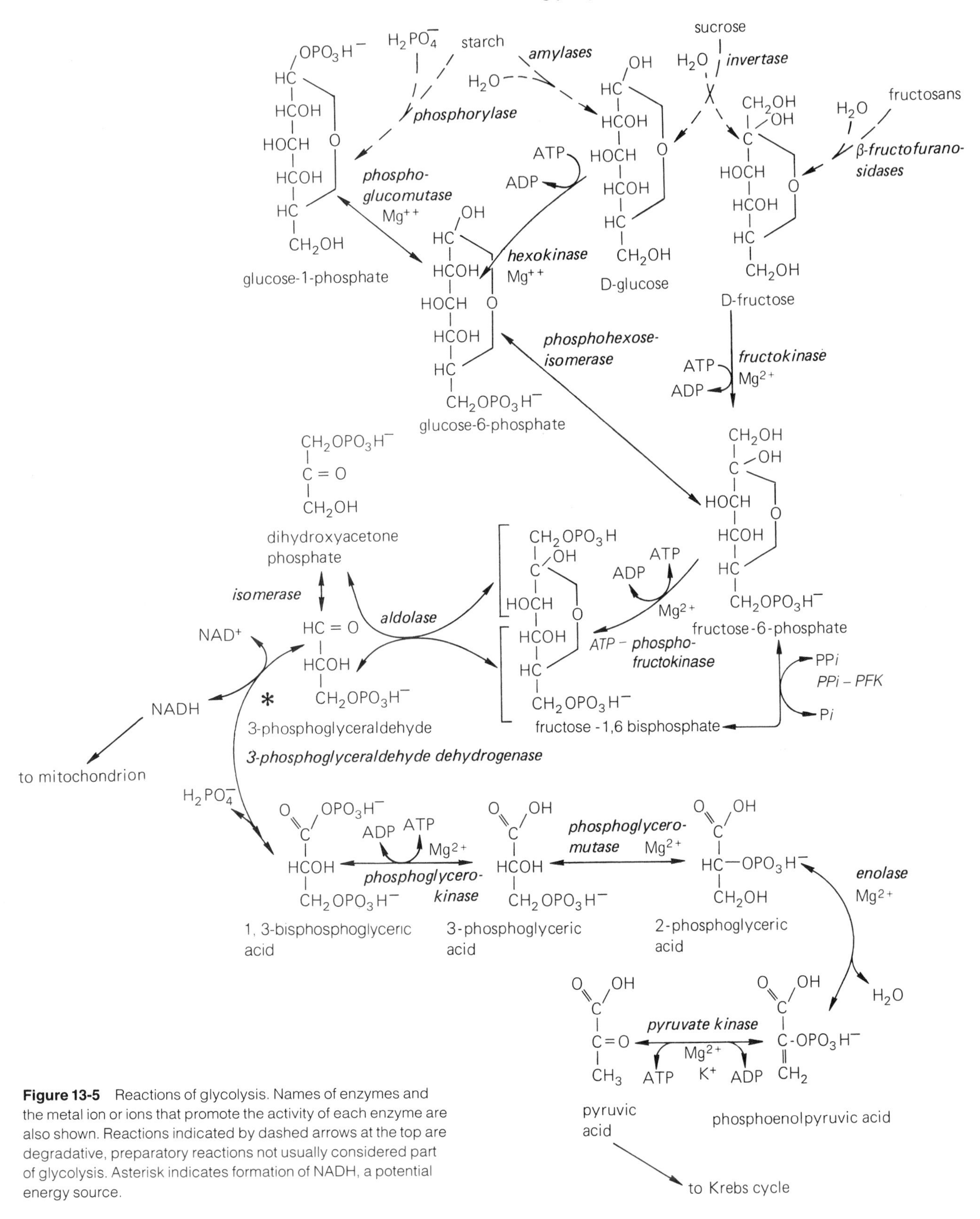


**Figure 13-5** Reactions of glycolysis. Names of enzymes and the metal ion or ions that promote the activity of each enzyme are also shown. Reactions indicated by dashed arrows at the top are degradative, preparatory reactions not usually considered part of glycolysis. Asterisk indicates formation of NADH, a potential energy source.

$$\text{glucose} + 2\,\text{NAD}^+ + 2\,\text{ADP}^{2-} + 2\,\text{H}_2\text{PO}_4^- \rightarrow 2\,\text{pyruvate} + 2\,\text{NADH} + 2\,\text{H}^+ + 2\,\text{ATP}^{3-} + 2\,\text{H}_2\text{O} \quad (13.8)$$

Glycolysis has several functions. *First, glycolysis converts one hexose molecule into two molecules of pyruvic acid, and partial oxidation of hexose occurs.* No $O_2$ is used and no $CO_2$ is released. For each hexose converted, two molecules of $NAD^+$ are reduced to NADH ($+2H^+$). These NADH are important because each can subsequently be oxidized by $O_2$ in a mitochondrion such that $NAD^+$ is regenerated and two molecules of ATP are formed. Furthermore, some of these NADH that do not enter mitochondria are used in the cytosol to drive various anabolic, reductive processes. Inspection of Figure 13-5 shows that NADH is formed at only one step in glycolysis: during the oxidation of 3-phosphoglyceraldehyde to 1,3-bisphosphoglyceric acid (see large asterisk at lower left of Figure 13-5).

*A second function of glycolysis is production of ATP.* Overall glycolysis yields some ATP, but in early steps ATP must be used. When glucose or fructose enters glycolysis, each is phosphorylated by ATP in reactions catalyzed by hexokinase or fructokinase (Fig. 13-5, upper right). Glucose-6-phosphate and fructose-6-phosphate are the products. Subsequently, fructose-6-phosphate is phosphorylated at carbon 1 by another ATP (or UTP) to form fructose-1,6-bisphosphate. The enzyme responsible for this phosphorylation is called *ATP-phosphofructokinase* (*ATP-PFK*).

Note that this reaction represents one of two routes by which fructose-6-phosphate can be converted to fructose-1,6-bisphosphate. The other route, discovered only in the late 1970s and early 1980s (see Carnal and Black, 1983), involves phosphorylation of carbon 1 of fructose-6-phosphate with pyrophosphate as a phosphate donor. The enzyme involved is called *pyrophosphate phosphofructokinase* (*PPi-PFK*). Current evidence suggests that the ATP-PFK route is involved in so-called "maintenance respiration" by cells that are not rapidly growing, differentiating, or adapting to changing environments (Black et al., 1987). This reaction, then, occurs mainly in cells that are mature, or nearly so, and that exist for some time in a moderately constant environment. The PP*i*-PFK route is much more adaptive and can increase or decrease in importance depending on developmental processes and environmental conditions. We shall return to control over biochemical reactions of respiration in Section 13.12, but for now note that conversion of glucose or fructose to fructose-1,6-bisphosphate requires two ATPs if the ATP-PFK route is used and only one ATP if the PP*i*-PFK route is used.

In other reactions of glycolysis, fructose-1,6-bisphosphate is split to form two three-carbon phosphorylated sugars, and these triose phosphates are then oxidized to pyruvic acid. These steps yield two ATPs from each triose phosphate, making a total of four ATPs for each glucose or fructose respired. A yield of four ATPs minus the two (or one) required to form fructose-1,6-bisphosphate leaves a net yield of either two or three ATPs for each hexose used in glycolysis. Reaction 13.8 above lists two ATPs as net products, but the PP*i*-PFK route would increase this to three.

*A third function of glycolysis is formation of molecules that can be removed from the pathway to synthesize several other constituents of which the plant is made.* This function is not apparent in either Figure 13-5 or Reaction 13.8, but it is given special attention in Section 13.11 and in Figure 13-12.

*Finally, glycolysis is important because the pyruvate that it produces can be oxidized in mitochondria to yield relatively large amounts of ATP, much more than is produced in glycolysis.*

## 13.4 Fermentation

Although glycolysis can function well without $O_2$, further oxidation of pyruvate and NADH by mitochondria requires oxygen. Therefore, when $O_2$ is limiting, NADH and pyruvate begin to accumulate. Under this condition plants carry out **fermentation** (anaerobic respiration), forming either ethanol or lactic acid (usually ethanol), as shown in Figure 13-6. The two top reactions in Figure 13-6 consist of a decarboxylation of pyruvic acid to form acetaldehyde, then rapid reduction of acetaldehyde by NADH to form ethanol. These reactions are catalyzed by *pyruvic acid decarboxylase* and *alcohol dehydrogenase*. Some cells contain *lactic acid dehydrogenase*, which uses NADH to reduce pyruvic acid to lactic acid. Ethanol or lactic acid, or both, are fermentation products, depending on the activities of each of the enzymes present. In each case NADH is the reductant, but only under anaerobic conditions is NADH abundant enough to cause reduction. Furthermore, in some plants NADH is used to enable accumulation of other compounds when $O_2$ is limiting, especially malic acid and glycerol (Crawford, 1982; Davies, 1980a). The occurrence of fermentation in various plants under oxygen stress is described in Section 13.13. (Note that lactic acid can accumulate in exercised muscles if their oxygen supply is insufficient; it is this lactic acid that causes "stiffness" after exercise of muscles that are insufficiently conditioned.)

## 13.5 Mitochondrial Structures and Respiration

To understand how pyruvate and NADH produced in glycolysis are oxidized by mitochondria, it is first helpful to understand some of the properties of these organelles. In some respects, mitochondria are similar to chloroplasts, although functionally they are quite different. Each mitochondrion contains circular DNA that has genetic information used to produce a small per-

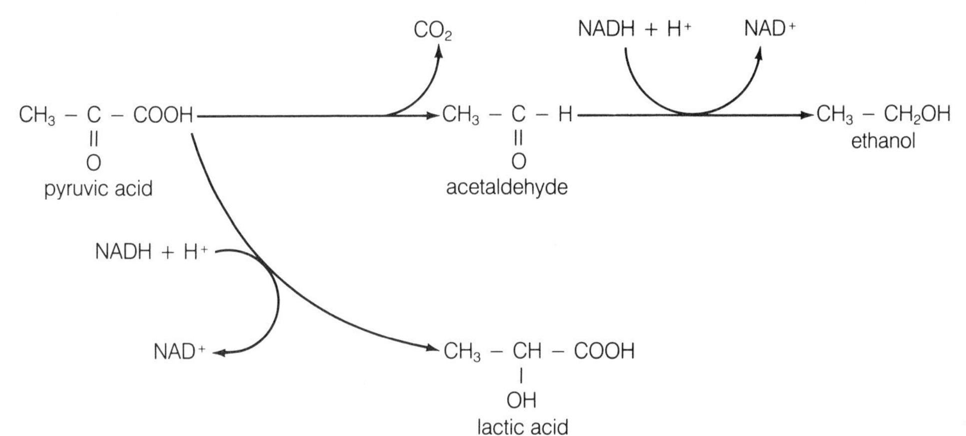


**Figure 13-6** Fermentation of pyruvate to form ethanol or lactic acid.

centage of its enzymes, each is formed mainly or entirely by division of a preexisting organelle, and each is surrounded by a double membrane or envelope with an extensive inner membrane system. Most living plant cells contain a few hundred mitochondria (Douce, 1985).

Mitochondria are only a few micrometers long, not much longer than many bacteria, and although they can be seen with a light microscope, their fine structure is made clear only with the electron microscope. Their general morphology can be seen in thin transmission electron micrographs in earlier chapters (see, for example, Fig. 11-8). Figure 13-7a shows a high-voltage electron micrograph of mitochondria within cells; in this

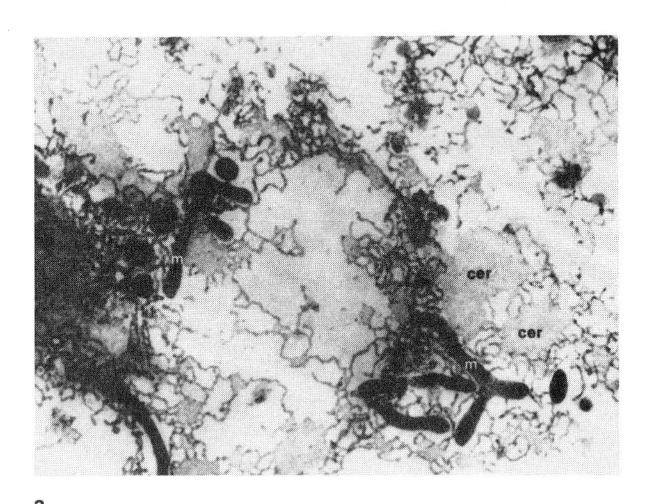


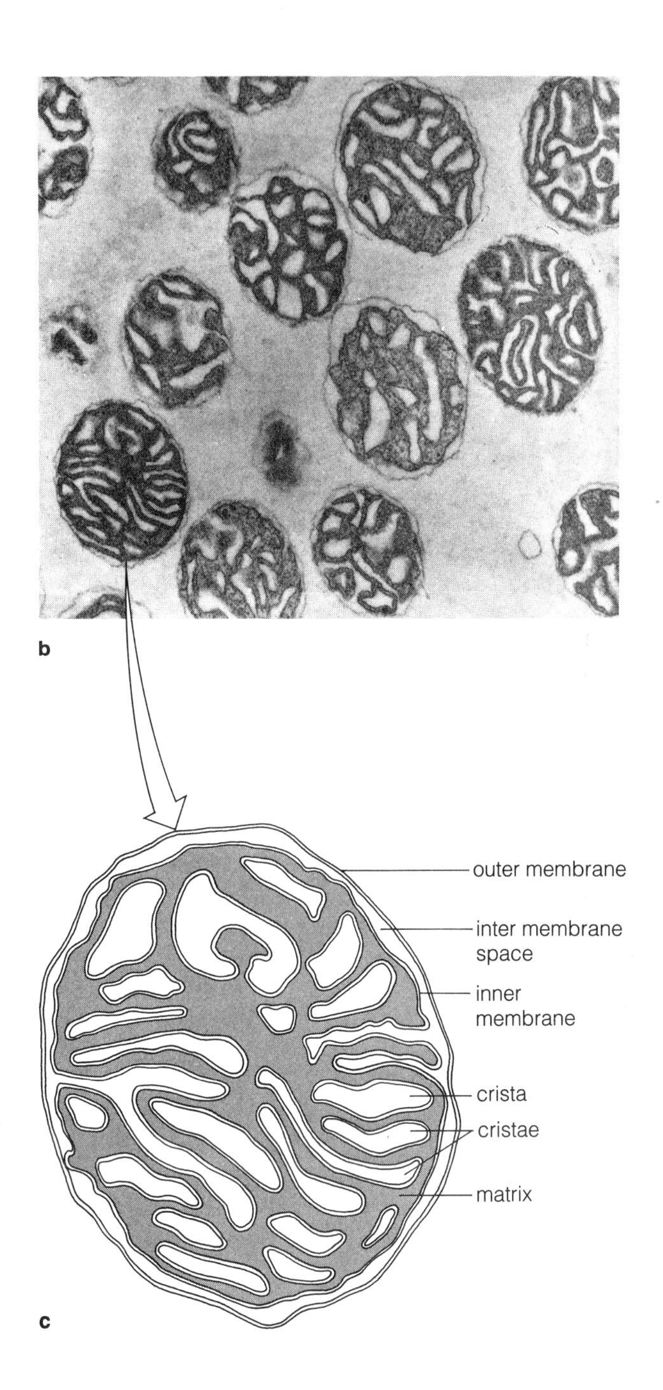


**Figure 13-7** Various views of mitochondria. (**a**) High-voltage transmission electron micrograph of multishaped mitochondria in cotyledon cells of mung bean (*Vigna radiata*). Bar = 1 μm. (From Harris and Chrispeels, 1980.) (**b**) Transmission electron micrograph of isolated plant mitochondria. Cristae appear as light areas, and the matrix appears dark. (From Smith, 1977; photo courtesy W. D. Bonner, Jr.) (**c**) Interpretive drawing of an isolated mitochondrion. (Modified from Malone et al., 1974.)

kind of micrograph one can visualize their three-dimensional, often rod-like shapes. When isolated, mitochondria usually become rather spherical (Figs. 13-7b and 13-7c). However, *in vivo* mitochondria rapidly and reversibly change their shapes from long and narrow and even branched (Fig. 13-7a) to more nearly spherical.

The inner membrane of the mitochondrial envelope is highly convoluted, protruding into the interior matrix in tubelike (narrow or highly dilated tubes) or sometimes even sheetlike patterns in many places. Each such convolution is called a **crista** (plural, **cristae**). In most plant mitochondria, dilated tubular cristae are well developed, but this varies with the type of cell, its age, and its extent of development. In many cells, one crista is fused to another in the interior of the mitochondrion, forming a continuous saclike intermembrane compartment between them (Figs. 13-7b and 13-7c); in other cells still other modifications occur. Regardless of their form, cristae contain most of the enzymes that catalyze steps of the electron-transport system that follow the Krebs cycle, so the increased surface area they provide is of great importance. The Krebs-cycle reactions occur in the protein-rich matrix between the cristae.

## 13.6 The Krebs Cycle

The **Krebs cycle** was named in honor of the English biochemist Hans A. Krebs, who in 1937 proposed a cycle of reactions to explain how pyruvate breakdown takes place in the breast muscle of pigeons. He called his proposed pathway the **citric acid cycle** because citric acid is an important intermediate. Another common name for the same group of reactions is the **tricarboxylic acid (TCA) cycle**, a term used because citric and isocitric acids have three carboxyl groups.

The initial step leading to the Krebs cycle involves the oxidation and loss of $CO_2$ from pyruvate and the combination of the remaining two-carbon acetate unit with a sulfur-containing compound, *coenzyme A (CoA)*, to form *acetyl CoA*:

$$(CH_3{-}\underset{\underset{\displaystyle O}{\|}}{C}{-}SCoA)$$

This and another comparable role of CoA in the Krebs cycle are important reasons why sulfur is an essential element.

This pyruvate decarboxylation reaction also involves a phosphorylated form of *thiamine* (vitamin $B_1$) as a prosthetic group. Participation of thiamine in this reaction partially explains the essential function of vitamin $B_1$ in plants and animals. Besides the loss of $CO_2$, two hydrogen atoms are removed from pyruvic acid during the formation of acetyl CoA. The enzyme catalyzing the complete reaction is called *pyruvic acid dehydrogenase*, but it is actually an organized complex containing numerous copies of five different enzymes, three of which catalyze oxidative decarboxylation of pyruvate and two of which regulate activity of the other three (Miernyk et al., 1987). (We shall return to control of pyruvate dehydrogenase in Section 13.12.) The hydrogen atoms removed from pyruvate are finally accepted by $NAD^+$, yielding NADH. This and other Krebs-cycle reactions are diagramed in Figure 13-8.

The Krebs cycle accomplishes removal of some of the electrons from organic-acid intermediates and transfer of these electrons to $NAD^+$ (to form NADH) or to *ubiquinone* (to form *ubiquinol*[1]). Notice that none of the dehydrogenase enzymes of the cycle uses $NADP^+$ as an electron acceptor. In fact, $NADP^+$ is usually nearly undetectable in plant mitochondria, a situation different from that of chloroplasts, in which $NADP^+$ is abundant and in which there is often less $NAD^+$. Not only are NADH and ubiquinol important products of the Krebs cycle, but one molecule of ATP is formed from ADP and P*i* during the conversion of *succinyl coenzyme A* to *succinic acid*. (In mammals, but not in the few plants so far investigated, formation of ATP at this step requires GDP and GTP, both guanosine nucleotides.) Two additional $CO_2$ molecules (shown boxed in Fig. 13-8) are released in these Krebs-cycle reactions, so there is a net loss of both carbon atoms from the incoming acetate of acetyl CoA. The release of $CO_2$ in the Krebs cycle accounts for the product $CO_2$ in the summary equation for respiration (Reaction 13.1), but no $O_2$ is absorbed during any Krebs-cycle reaction.

*The primary functions of the Krebs cycle are as follows:*

1. reduction of $NAD^+$ and ubiquinone to the electron donors NADH and ubiquinol, which are subsequently oxidized to yield ATP
2. direct synthesis of a limited amount of ATP (one ATP for each pyruvate oxidized)
3. formation of carbon skeletons that can be used to synthesize certain amino acids that, in turn, are converted into larger molecules (see Section 13.11 and Fig. 13-12 for explanation of the kinds of compounds formed from Krebs-cycle intermediates and what prevents the cycle from stopping when intermediates are removed from it)

[1]Most textbooks identify the flavin FAD as the acceptor of electrons and $H^+$ from succinic acid, with $FADH_2$ as the product. FAD and $FADH_2$ are bound to succinic acid dehydrogenase, but they represent transitory intermediate compounds during the overall reduction of membrane-soluble quinone (ubiquinone) to ubiquinol (Cammack, 1987).

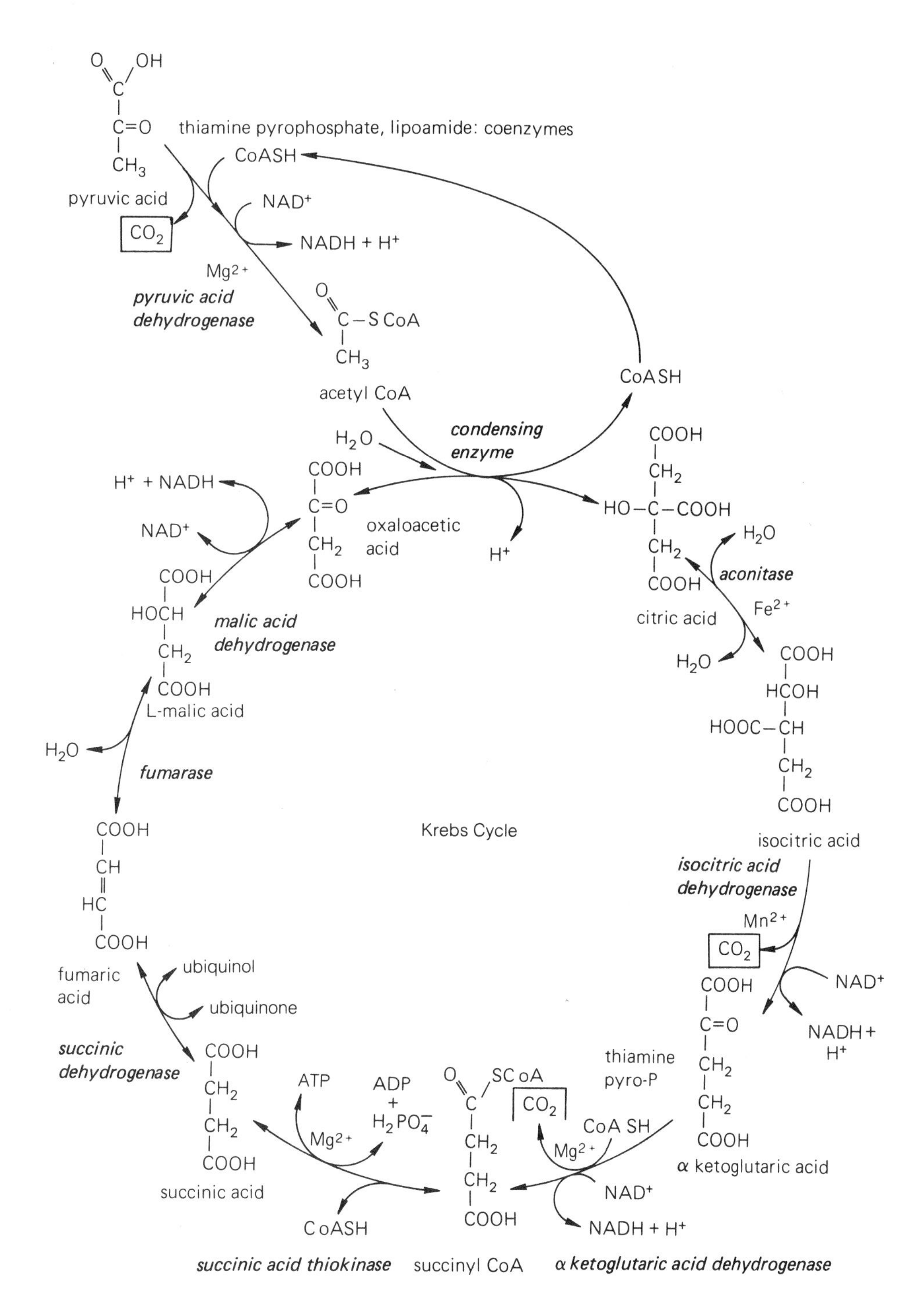


**Figure 13-8** Reactions of the Krebs cycle, including enzymes and coenzymes. Also shown is the oxidative decarboxylation of pyruvic acid.

Considering that two pyruvates are produced in glycolysis from each glucose, the overall reaction for the Krebs cycle can be written as follows:

$$2 \text{ pyruvate} + 8\,NAD^+ + 2 \text{ ubiquinone} + 2\,ADP^{2-} + 2\,H_2PO_4^- + 4\,H_2O \rightarrow 6\,CO_2 + 2\,ATP^{3-} + 8\,NADH + 8\,H^+ + 2 \text{ ubiquinol}$$

## 13.7 The Electron-Transport System and Oxidative Phosphorylation

NADH present in mitochondria comes from three main processes: the Krebs cycle, glycolysis, and (in leaves) oxidation of glycine produced during photorespiration. When NADH is oxidized, ATP is produced. Similarly, the ubiquinol produced by succinic acid dehydrogenase in the Krebs cycle is also oxidized to yield ATP. Although these oxidations involve $O_2$ uptake and $H_2O$ production, neither NADH nor ubiquinol can combine directly with $O_2$ to form $H_2O$. Rather, their electrons are transferred via several intermediate compounds before $H_2O$ is made. These electron carriers constitute the **electron-transport system** of mitochondria. Electron transport proceeds from carriers that are thermodynamically difficult to reduce (those with negative reduction potentials) to those that have a greater tendency to accept electrons (and that have higher, even positive reduction potentials). Oxygen has the greatest tendency to accept electrons, and it ultimately does so, forming $H_2O$. Each carrier of the system usually accepts electrons only from the previous carrier close to it. Carriers are arranged in four main protein complexes in the inner mitochondrial membrane, and there are several thousand electron transport systems in each mitochondrion.

As in the chloroplast electron-transport system, which is involved in the transfer of electrons *from* water molecules, the mitochondrial system involves cytochromes (up to four of the *b* type and two of the *c* type) and a few quinones, especially ubiquinone. Also present are several *flavoproteins* (riboflavin-containing proteins), some iron-sulfur (Fe-S) proteins similar to ferredoxin, an enzyme named *cytochrome oxidase*, and a few other electron carriers not yet identified (Douce, 1985; Moore and Rich, 1985; Douce and Neuberger, 1989). The cytochromes and cytochrome oxidase both contain iron as part of a heme group. The flavoproteins contain either *flavin adenine dinucleotide (FAD)* or the similar *flavin mononucleotide (FMN)* as bound prosthetic groups. Many of these electron carriers have counterparts in chloroplasts, yet each is unique in structure.

The cytochromes and Fe-S proteins can receive or transfer only one electron at a time. Ubiquinone, like the plastoquinone of chloroplasts, receives and transfers two electrons and two $H^+$; the same is true of flavoproteins. This property of ubiquinone and flavoproteins is important in establishing a *p*H gradient from the matrix (*p*H about 8.5) across the inner membrane to the intermembrane space (*p*H near 7), because this *p*H gradient drives formation of ATP from ADP and P*i* according to the Mitchell chemiosmotic theory of Section 10.8 (Mitchell, 1985; Senior, 1988).

In mitochondria, ATP formation from ADP and P*i* is indirectly driven by the strong thermodynamic tendency of $O_2$ to become reduced, and this process is called **oxidative phosphorylation**. As in chloroplasts, phosphorylation is catalyzed by a coupling factor or ATP synthase. This mitochondrial ATP synthase has a connected stalk and headpiece much like that of the thylakoid ATP synthase, and the headpiece extends completely across the inner membrane. The headpiece faces and extends into the matrix, whereas the stalk extends outward toward the intermembrane space between the inner and outer membranes. ATP is formed at or in the headpiece within the matrix and is then transported toward the cytosol by countertransport (Section 7.8) with incoming ADP. The ATP then moves readily across the much more permeable outer membrane into the cytosol, where it performs numerous functions. The outer membrane has **porins**, channels that allow passage of molecules with molecular weights less than about 5 kDa, so nucleotides and many other metabolites easily get through that membrane (Mannella, 1985; Heldt and Flügge, 1987).

Phosphate is also necessary for ATP formation, and it is carried across the much less permeable inner membrane into the matrix by two countertransport systems that simultaneously move either $OH^-$ or a dicarboxylic acid such as malate out of the matrix to the intermembrane space. A similar countertransport system catalyzes exchange of $OH^-$ and pyruvate, and this probably explains how pyruvate from glycolysis gets into the matrix, where it is oxidized by pyruvate dehydrogenase.

Figure 13-9 indicates the major electron-transport pathway, beginning with NADH + $H^+$ formed in the matrix by Krebs-cycle enzymes (upper right). The two electrons and two $H^+$ are passed to a flavoprotein containing FMN, which in turn passes the electrons to a Fe-S protein. The iron in the latter can accept only one electron at a time and accepts no $H^+$; the two $H^+$ are somehow transferred into the intermembrane space. This is the first of four steps in which a pair of $H^+$ is moved from the matrix fully across the inner mitochondrial membrane in concert with the transfer of two electrons. The reduced Fe-S transfers electrons to ubiquinone (UQ), which, with two $H^+$ taken from the matrix, becomes reduced to ubiquinol ($UQH_2$). From $UQH_2$ the electrons move one at a time to various cytochromes *b*, and the two $H^+$ from $UQH_2$ are transferred outwardly to the intermembrane space. Another Fe-S protein then receives and transfers electrons to the $Fe^{3+}$ in cytochrome $c_1$ with a third outward transport of a pair of $H^+$. From cytochrome $c_1$, electrons are received by cytochrome *c*, and then their transfer to $O_2$ to form $H_2O$ is catalyzed by *cytochrome oxidase*. This oxidase contains inseparable *a* and $a_3$ components (Fig. 13-9, bottom) and some other polypeptides that contain, in total, two copper ions that undergo oxidation-reduction between $Cu^+$ and $Cu^{2+}$ forms. The two coppers are involved in electron transport between the iron components of cytochromes *a* and $a_3$. Accompanying oxidation of cytochrome *c* by cytochrome oxidase, another pair of $H^+$ is moved from the matrix to the intermembrane space (Fig. 13-9, lower left), but how this occurs is not clear (Wikström, 1984; Prince, 1988).

Although reduction potentials are not shown in Figure 13-9, the overall $\Delta\psi'_0$ is from NADH at an $\psi'_0$ of $-0.32$ V to $O_2$ at an $\psi'_0$ of $+0.82$ V, a total change of $+1.14$ V. This is the same change that occurs during photosynthetic electron transport from $H_2O$ to $NADP^+$,

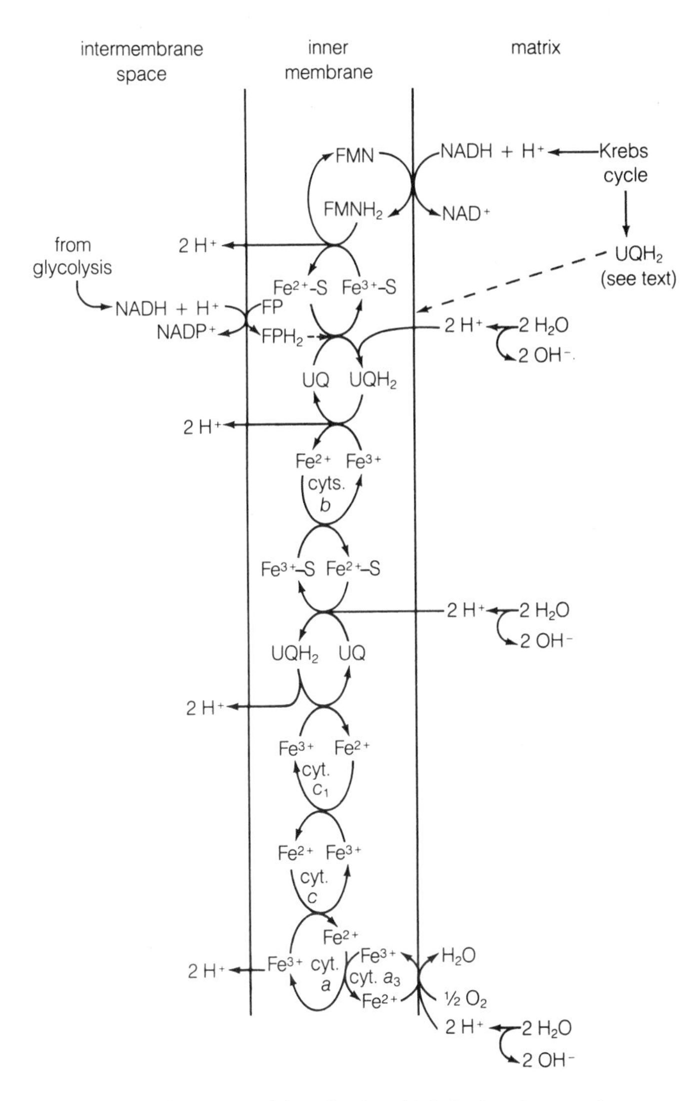


**Figure 13-9** Reactions of the mitochondrial electron-transport system. Abbreviations are in text, except that FP is an oxidized flavoprotein and $FPH_2$ a reduced flavoprotein. The mechanism by which two $H^+$ are transported into the intermembrane space in the last reaction has not been clarified.

but in mitochondria the sign preceding $\Delta\psi'_0$ is positive and 220 kJ of energy are released for each mole of Krebs-cycle NADH oxidized. The transport of four pairs of $H^+$ across the membrane causes a sufficient *p*H gradient to allow formation of three ATP by the ATP synthase (not shown in Fig. 13-9, but compare with Fig. 10-10). Numerous studies with isolated plant mitochondria show that for each Krebs-cycle NADH oxidized, three ATP are formed.

For each NADH formed in glycolysis (Fig. 13-9, upper left) and for each ubiquinol formed in the Krebs cycle by oxidation of succinate, only two ATP are formed. The reason for this is that these NADH and $UQH_2$ molecules donate electrons to the transport chain only after the first pair of $H^+$ in the main pathway has been passed into the intermembrane space, so a smaller *p*H difference across the membrane is created when they are oxidized. For NADH arising from glycolysis in the cytosol, two flavoproteins (FP) containing *NADH dehydrogenase* exist on the outer surface of the inner membrane, one of which is shown in Figure 13-9 (Møller, 1986; Douce and Neuberger, 1989). Furthermore, NADPH (resulting from the pentose phosphate pathway; see Section 13.10) can be oxidized by a similar dehydrogenase (not shown). This ability of plant mitochondria to oxidize cytosolic NADH and NADPH directly is not shared by animal mitochondria. (Animals have a transhydrogenase enzyme that transfers electrons from NADPH to $NAD^+$, forming $NADP^+$ and NADH, and they use special carriers to move electron pairs from NADH into the matrix.) The ubiquinol produced in the Krebs cycle is oxidized similarly in plant and animal mitochondria; its electrons are taken from a cytochrome *b*, and its $H^+$ are therefore moved across the membrane into the intermembrane space. Two ATPs are formed from each ubiquinol arising from succinate in the Krebs cycle.

Oxidative phosphorylation from all mitochondrial substrates is uncoupled from electron transport by numerous uncoupling compounds, just as in chloroplasts (Section 10.8). Most uncouplers neutralize the *p*H gradient by carrying $H^+$ into the matrix, preventing oxidative phosphorylation but still allowing electron transport to occur. Sometimes electron transport occurs even faster, presumably because there is less "back pressure" from the *p*H gradient (that is, $H^+$ transport accompanying electron flow is easier when the $H^+$ concentration into which $H^+$ are transported is lower). Dinitrophenol uncouples in mitochondria even more effectively than in chloroplasts, and so do numerous other compounds. At proper concentrations, dinitrophenol greatly speeds electron transport and respiration because of its uncoupling effect; that is, it minimizes the *p*H gradient across the inner membrane and allows $H^+$ to be transported outward more easily by the ATPase coupling factor. Ammonium ions, which strongly uncouple photosynthetic phosphorylation in chloroplasts by a proton "ferryboat" action like that of dinitrophenol, are somehow much less potent inhibitors of oxidative phosphorylation in mitochondria. In fact, mitochondria can tolerate $NH_4^+$ up to at least 20 mM (Yamaya and Matsumoto, 1985). Part of this tolerance must somehow relate to the abundance of $NH_4^+$ within mitochondria that arises (in leaves) from oxidative decarboxylation of glycine during photorespiration.

Still other compounds inhibit either oxidative phosphorylation or electron transport without uncoupling the two processes. For example, two potent phosphorylation inhibitors are *oligomycin*, an antibiotic produced by a *Streptomyces* species, and *bongkrekic* acid, an antibiotic produced by a *Pseudomonas* species that grows on fungal-infected coconuts called "bongkreks" by native Indonesians (Goodwin and Mercer, 1983). Oligomycin inhibits ATP formation by the ATPase, whereas bongkrekic acid prevents ATP formation by blocking a countertransport system that carries ADP from the intermembrane space into the matrix in exchange for ATP. Without this ADP, no oxidative phosphorylation can occur. *Antimycin A*, also from *Streptomyces*, blocks electron transport at or near the cytochrome *b* to Fe-S protein step (Fig. 13-9). This prevents phosphorylation, but it does not uncouple the two processes.

## 13.8 Energetics of Glycolysis, the Krebs Cycle, and the Electron-Transport System

When a hexose is completely oxidized to $CO_2$ and $H_2O$ using these three processes, Reaction 13.1 describes the overall reaction. However, Reaction 13.1 lists energy as a product, and we now know that much of this energy is trapped in ATP. But how much is in ATP, and how much is lost as heat? To answer this, note that glycolysis yields two ATP and two NADH per hexose used (Reaction 13.8). Each such NADH oxidized by the electron transport system yields two ATP, as described above, so glycolysis contributes a total of six ATP per hexose. The Krebs cycle contributes two ATP per hexose or per two pyruvates (Reaction 13.9) when succinyl CoA is cleaved to succinate and CoASH (Fig. 13-8). This cycle also produces eight NADH per hexose within the mitochondrial matrix; by oxidative phosphorylation each of these NADH yields three ATP, or 24 per hexose. Each ubiquinol from the Krebs cycle yields two ATP by oxidative phosphorylation, or four per hexose (two pyruvates; see Reaction 13.9). The total contribution of the Krebs cycle is then 30 ATP. Adding these 30 ATP to the 6 from glycolysis leads to a total of 36 ATP per hexose completely respired by these processes.

We can also estimate the efficiency of respiration in terms of how much energy in glucose can be trapped in the terminal phosphate bond of ATP. The standard Gibbs free energy change ($\Delta G'_0$) for complete oxidation of one mole of glucose or fructose at *p*H 7 is $-2,870$ kJ ($-686$ kcal), so we shall use this as the energy in the reactants of respiration. Among the products, only the energy in the terminal phosphate of ATP is additional useful energy. The $\Delta G'_0$ for hydrolysis of the terminal phosphate in each mole of ATP is about $-31.8$ kJ ($-7.6$ kcal), or $-1,140$ kJ in 36 moles of ATP. Therefore, the efficiency is about $-1,140/-2,870$, or 40 percent.[2] The remaining 60 percent is lost as heat.

## 13.9 Cyanide-Resistant Respiration

Aerobic respiration of most organisms, including some plants, is strongly inhibited by certain negative ions that combine with the iron in cytochrome oxidase. Two such ions, cyanide ($CN^-$) and azide ($N_3^-$), are particularly effective. Carbon monoxide (CO) also forms a strong complex with iron, preventing electron transport and poisoning respiration. In many plant tissues, however, the poisoning of cytochrome oxidase by such inhibitors has only a minor effect on respiration. The respiration that continues in this situation is said to be **cyanide-resistant respiration** (Lance et al., 1985; Lambers, 1985; Siedow and Berthold, 1986). Several fungi, bryophytes, and algae and a few bacteria and animals are also resistant to cyanide, azide, and CO, but most animals are not (Henry and Nyns, 1975).

The reason respiration can continue when cytochrome oxidase is blocked is that cyanide-resistant mitochondria have an alternative, short branch in the electron-transport pathway at the first step involving ubiquinol ($UQH_2$, Fig. 13-9). This branch or alternative route also allows transport of electrons to oxygen, probably from ubiquinol to a flavoprotein to the oxidase. The terminal oxidase has a much lower affinity for $O_2$ than does cytochrome oxidase, and little or no oxidative phosphorylation is coupled to the alternative pathway; that is, it leads mainly to production of heat, not to production of ATP. This heat production is beneficial to certain plants, as is the case in the pollination ecology of arum lilies such as *Sauromatum guttatum* and *Symplocarpus foetidus* (skunk cabbage). (See Figure 13-10 and the review by Meeuse and Raskin, 1988.)

Recently, the alternative oxidase was isolated from skunk cabbage, and some of its properties were studied. Its activity increased about seven times and the normal electron-transport pathway decreased about ten times when the appendix of the spadix became thermogenic (Elthon et al., 1989). These activity changes force electron transport through the alternative pathway and greatly increase heat production. Yet for most plants, the *rate* of operation of the alternative pathway is unclear because evidence for its existence usually comes during the unnatural conditions when cytochrome oxidase of the normal pathway is poisoned by cyanide, azide, or CO. However, the alternative pathway does commonly operate in plants (Siedow and Musgrave, 1987). Its activity is highest in cells rich in sugars (as after rapid photosynthesis) when glycolysis and the

[2]This efficiency of 40 percent based on $\Delta G'_0$ values is probably somewhat unrealistic for conditions occurring naturally in cells in which $NAD^+/NADH$ ratios might be as high as 30 in the mitochondrial matrix and in which cytosol ATP/ADP ratios might usually be between 4 and 10. A theoretical review by Ericinska and Wilson (1982) indicates that true efficiencies for electron transport and oxidative phosphorylation are as high as 75 percent in liver and heart mitochondria. Use of $\Delta G'_0$ and $\Delta\psi'_0$ values to calculate the efficiency of these processes gives a value of only 43 percent. Thus true efficiencies for the entire process of respiration might be considerably higher than 40 percent.

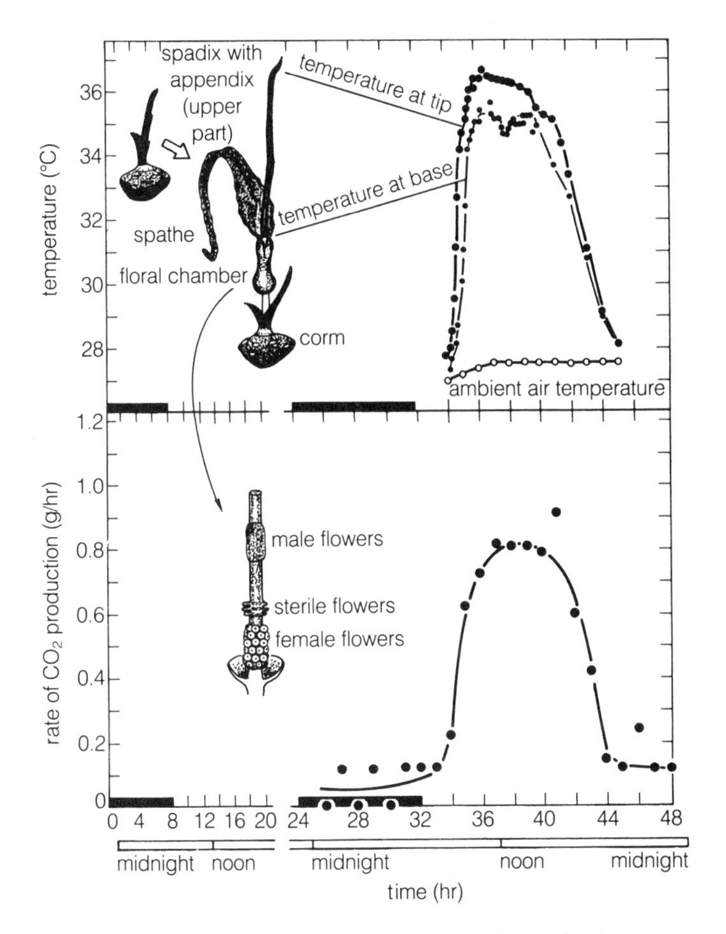


**Figure 13-10** Respiration and temperature of a spadix of *Sauromatum guttatum* as a function of time. *Sauromatum* is a Pakistani and Indian genus in the Araceae family. Growth from the corm to a structure about 50 cm tall may occur in about 9 days (drawings at upper left), with a maximum growth rate of 7 to 10 cm/day. If this occurs in constant light, the spathe remains wrapped around the spadix; but after the "normal" time for flowering has passed, a single period of darkness, if it is long enough (bar on the abscissa—two 8-hour dark periods were given in this experiment), will initiate opening of the spathe and a burst in $CO_2$ production (note extremely large quantities) with a concurrent rise in temperature. The heat apparently serves to volatilize various compounds (especially amines and ammonia), which give an odor of rotting meat. Carrion flies and beetles are attracted and serve in pollination. They enter the floral chamber (lower drawing, somewhat schematic). (Original data. Experiment performed for use in this text by B. J. D. Meeuse, R. C. Buggein, and J. R. Klima of the University of Washington, Seattle.)

Krebs cycle occur unusually rapidly, because then the normal electron-transport pathway cannot handle all the electrons provided to it (Lambers, 1985). Several experts have concluded that the alternative pathway operates largely as an overflow mechanism to remove electrons when the cytochrome pathway becomes saturated by rapid glycolysis and Krebs-cycle activities. Operation of the alternative pathway means that the efficiency of respiration in plants (Section 13.8) is decreased in proportion to the pathway's activity.

## 13.10 The Pentose Phosphate Pathway

After 1950, plant physiologists gradually became aware that glycolysis and the Krebs cycle were not the only reactions by which plants obtain energy from oxidation of sugars into carbon dioxide and water. Because five-carbon sugar phosphates are intermediates, this alternative series of reactions is now usually called the **pentose phosphate pathway (PPP)**. It has also been called the oxidative pentose pathway, the hexose monophosphate shunt, and the phosphogluconate pathway.

Several compounds of the PPP are also members of the Calvin cycle, in which sugar phosphates are synthesized in chloroplasts. The major difference between the Calvin cycle and the PPP is that in the latter, sugar phosphates are degraded rather than synthesized. In this respect, the reactions of the PPP are similar to those of glycolysis. In addition, glycolysis and the PPP have certain reactants in common, and both occur mainly in the cytosol, so the two pathways are greatly interwoven. One important difference is that in the PPP, $NADP^+$ is always the electron acceptor, whereas in glycolysis $NAD^+$ is the common acceptor.

Reactions of the PPP are outlined in Figure 13-11. The first reaction involves glucose-6-phosphate, which can arise from starch breakdown by starch phosphorylase, followed by phosphoglucomutase action in glycolysis; from addition of the terminal phosphate of ATP to glucose; or directly from photosynthetic reactions. It is immediately oxidized (dehydrogenated) irreversibly by *glucose-6-phosphate dehydrogenase* to *6-phosphoglucono-lactone* (reaction 1). This lactone is rapidly hydrolyzed by a *lactonase* into *6-phosphogluconate* (reaction 2); then the latter is irreversibly and oxidatively decarboxylated to ribulose-5-phosphate by *6-phosphogluconate dehydrogenase* (reaction 3). Note that reactions 1 and 3 are catalyzed by dehydrogenases that are highly specific for $NADP^+$ (not $NAD^+$). Furthermore, glucose-6-phosphate dehydrogenase is strongly inhibited noncompetitively (allosterically; see Section 9.6) by NADPH. In chloroplasts, where an isozyme of this enzyme exists and where the PPP also operates during darkness, light inactivates the enzyme, thereby preventing degradation of glucose-6-phosphate and allowing the Calvin cycle to operate faster. One mechanism of deactivation by light is formation of inhibitory NADPH from $NADP^+$ by the thylakoid electron-transport system, and another is the ferredoxin-thioredoxin system described in Section 11.5.

The next reactions of the PPP lead to pentose phosphates and are catalyzed by an *isomerase* (reaction 4) and an *epimerase* (reaction 5), which is a type of isomerase. These and subsequent reactions are similar or identical to some in the Calvin cycle (Fig. 11-3). Important enzymes are *transketolase* (reactions 6 and 8) and *transaldolase* (reaction 7). Note that these last three reactions

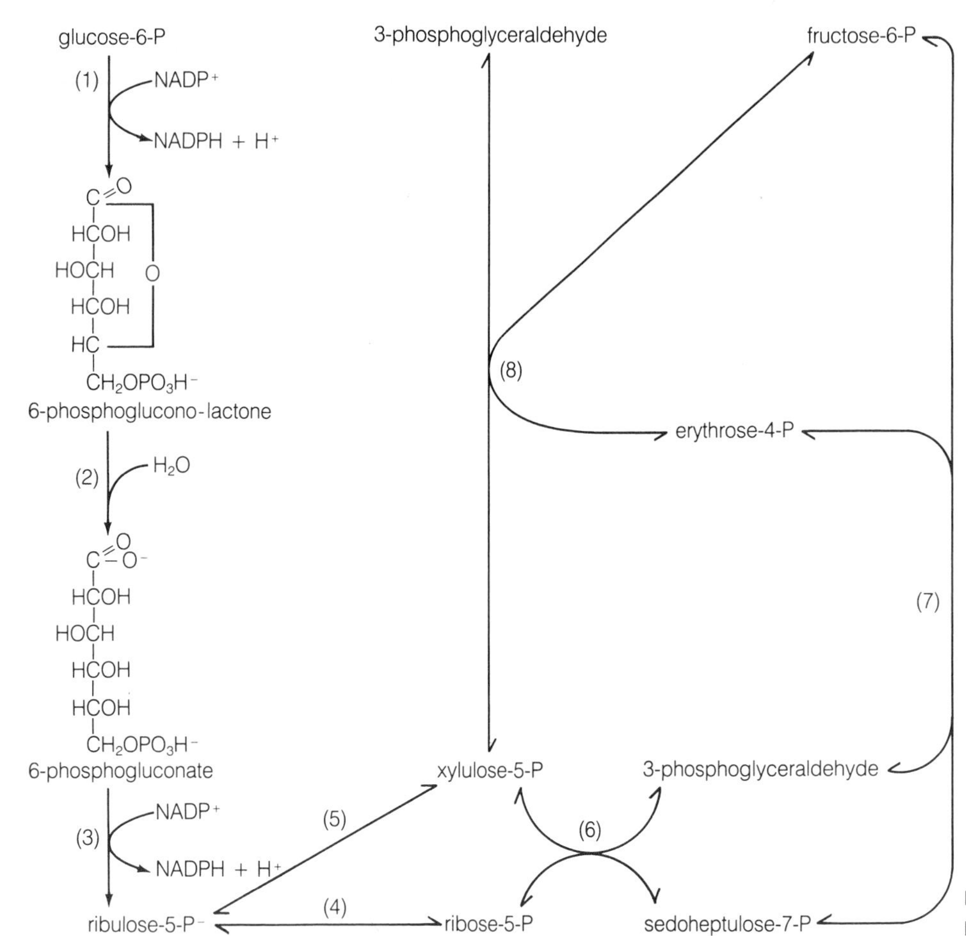


**Figure 13-11** Reactions of the pentose phosphate respiratory pathway.

lead to 3-phosphoglyceraldehyde and fructose-6-phosphate, which are intermediates of glycolysis. As a result, the PPP can be considered an alternative route to compounds subsequently degraded by glycolysis (ap Rees, 1985, 1988).

Three other functions of the PPP are important. *First, NADPH is produced*; this is important because this nucleotide can be oxidized by plant mitochondria to form ATP. Furthermore, NADPH is used specifically in numerous biosynthetic reactions requiring an electron donor. For these reactions (for example, formation of fatty acids and of several isoprenoids described in Chapter 15), NADH is nonfunctional, although for certain other reduction processes it works well. *Second, erythrose-4-phosphate is produced* in reaction 7 or 8, and this four-carbon compound is an essential starting reactant for production of numerous phenolic compounds such as anthocyanins and lignin (Sections 15.6 and 15.7). *Third, ribose-5-phosphate is produced*; this is a required precursor of the ribose and deoxyribose units in nucleotides, including those in RNA and DNA. Clearly, the PPP is just as essential to plants as glycolysis and the Krebs cycle.

## 13.11 Respiratory Production of Molecules Used for Synthetic Processes

Near the beginning of this chapter we stated that respiration is important to cells because many compounds are formed that can be diverted into other substances needed for growth. Many of these compounds are large molecules, including lipids, proteins, chlorophyll, and nucleic acids. ATP is needed to form them, and frequently the electrons present in NADH or NADPH are also required. Another process requiring significant quantities of NADH is the reduction of nitrate to nitrite (Section 14.3). In the preceding section we emphasized the importance of the PPP to the production of NADPH, ribose-5-phosphate, and erythrose-4-phosphate for anabolic reactions. The role of glycolysis and the Krebs cycle in producing carbon skeletons for synthesis of larger molecules is summarized in Figure 13-12. When studying this figure, remember that if carbon skeletons are diverted from the respiratory pathway as shown, not all carbons of the original respiratory substrate (for example, starch) will be released as $CO_2$, and not all

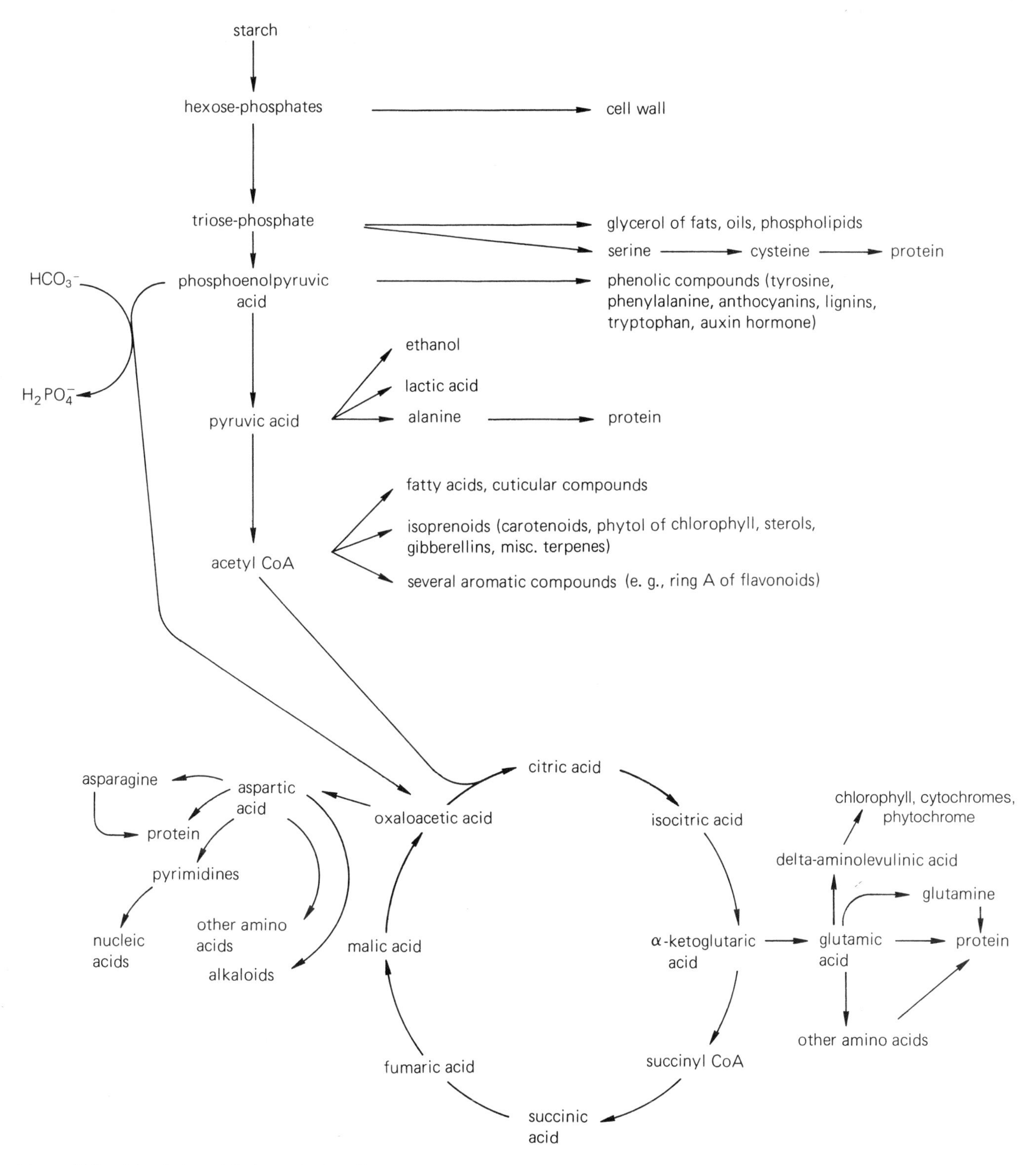


**Figure 13-12** Glycolysis and Krebs cycle simplified to show their roles in formation of some other essential compounds.

electrons normally transferred by NADH or NADPH will combine with $O_2$ to form $H_2O$. Yet some of the substrate molecules must be totally oxidized because use of diverted carbon skeletons to form larger molecules is effective only when oxidative phosphorylation is producing an adequate supply of ATP.

Another important point is that when organic acids of the Krebs cycle are removed by conversion into aspartic acid, glutamic acid, chlorophyll, and cytochromes, for example, regeneration of oxaloacetic acid will be prevented. Therefore, diversion of organic acids from the cycle would soon cause the cycle to stop were it not for another mechanism to replenish oxaloacetate. (Such replenishing or filling-up mechanisms are called **anaplerotic** by biochemists.) In all plants, day and night, there is some fixation of $CO_2$ ($HCO_3^-$) into oxaloacetate and malate by PEP carboxylase and malate dehydrogenase (see Reaction 11.3 and Fig. 13-12, left). These reactions are essential for growth processes because they replenish organic acids converted into larger molecules and allow the Krebs cycle to continue.

## 13.12 Biochemical Control of Respiration

To understand the environmental effects on respiration of various plants and plant parts described in this section, it is useful to learn some of the main biochemical control points and how these controls take place. To help understand control, consider that a photosynthesizing plant must regulate how much carbohydrate is stored in sucrose, starch, and fructans, for example, compared to how much is totally respired and how much is used in growth processes that require formation of membranes and cell walls. One logical control point should be near the start of glycolysis because the hexose phosphates used in glycolysis might instead be used in storage carbohydrates or in cell walls. Indeed, an important control point exists there.

Another important control of metabolism might depend on the ATP, ADP, and P*i* concentrations. Because ATP is the only important product of complete respiration, ADP and P*i* levels should help control how fast ATP is formed. ATP formation is surprisingly fast, even though cell ATP concentrations are only in the millimolar range or less. Pradet and Raymond (1983) calculated that 1 g of actively metabolizing maize root tips would convert about 5 g of ADP to ATP per day! Such high production rates mean that utilization rates are correspondingly high; otherwise, cells would soon be filled with ATP. De Visser (1987) calculated that in each plant cell all the ATP is converted to ADP and P*i* (and back again) one to several times per minute! Utilization of ATP occurs in many ways, and such growth-dependent processes as solute absorption and formation of proteins, starch, sucrose, fructans, cell-wall polymers, nucleic acids, and lipids are among them. This means that there is an important connection between ATP supply and growth. Respiration is needed for growth because it provides the ATP, but at the same time growth uses up the ATP and reforms the ADP and P*i* that respiration requires to make ATP again. Growth and respiration are dependent on each other, yet even nongrowing cells require ATP to maintain themselves and minimize entropy within them.[3] Just how important do ATP, ADP, and P*i* act as control agents? The following discussions deal with this matter and with other problems concerning the ways respiration is controlled.

### Energy Charge

In 1968, David E. Atkinson published an article in which he explained many reasons why ATP, ADP, and AMP should be master controllers of respiration (also see Atkinson, 1977). He realized that ATP has two "high-energy" phosphate bonds, ADP one, and AMP none, and that a highly active mitochondrial and chloroplast enzyme called *adenylate kinase* catalyzes a freely reversible reaction (ATP + AMP $\rightleftharpoons$ 2 ADP) that keeps these nucleotides in equilibrium. He hypothesized that a value he defined as the **energy charge (EC)** that depends on nucleotide concentrations should be important in metabolic control:

$$EC = \frac{(ATP) + 1/2(ADP)}{(ATP) + (ADP) + (AMP)}$$

For nearly all active cells investigated, EC values are between 0.8 and 0.95, but substantially lower values are found in anaerobic or poisoned cells. Atkinson argued that important enzymes of metabolic pathways that *use* ATP (for example, polysaccharide synthesis) should be *activated* allosterically by high EC values. Important enzymes in pathways that *regenerate* ATP should be *inhibited* allosterically by high EC values. He assumed that such enzymes bind two or more such nucleotides on allosteric sites with high affinity, so that response of the enzymes should depend on concentration ratios of nucleotides rather than on the absolute concentration of only one nucleotide.

Numerous results with enzymes from animal and some microbial cells are consistent with measured EC values, but some results are inconsistent. Many investigations concerning the importance of EC in plants have now been made (Pradet and Raymond, 1983; Ray-

[3]Since the 1960s plant physiologists have increasingly recognized that respiration rates of plants or plant parts consist of two major components, one proportional to the rate of growth and the other to its existing size (its dry mass). The first component (sometimes called *growth respiration*) is considered to result from respiration needed to form ATP, NADPH, NADH, and carbon skeletons required to form new plant biomass. The second component (sometimes called *maintenance respiration*) is considered to represent respiration needed to maintain and repair the existing structural system (for reviews, see Farrar, 1985; Amthor, 1989; Williams and Farrar, 1990; and Johnson, 1990). Several models that relate these components and environmental factors have been constructed and are summarized by Amthor (1989).

mond et al., 1987), but only a few plant enzymes respond to EC as Atkinson predicted. Furthermore, EC values in leaves remain constant when plants are switched from light to darkness or vice versa, yet we know that leaf biosynthesis of starch (ATP utilization) occurs only in light and that respiration (ATP formation) is relatively more important than biosynthesis in darkness. We also know that light activates several photosynthetic enzymes rather quickly, independent of EC (Section 11.5). Light also inactivates chloroplast glucose-6-phosphate dehydrogenase, the rate-limiting enzyme of the PPP, as mentioned above. Our conclusion is that plants seem to have important control mechanisms that are different from energy charge, but that EC values are likely important in some cases.

We next mention a few additional control mechanisms, first emphasizing glycolysis, in which the most clear-cut controls of respiration have been discovered.

## Regulation of Glycolysis

Sucrose, starch, and fructans are the major sources of substrates for glycolysis, and none of the enzymes that catalyze hydrolysis of those polysaccharides appears to be regulated allosterically by substrates or products of respiration. However, certain hormones (especially gibberellins) induce hydrolysis of these food reserves into hexoses used in glycolysis. (This subject is covered in Chapter 17.) In general, if hexoses are abundant, glycolysis and other phases of respiration are more rapid than when hexoses are not abundant.

ATP-phosphofructokinase (ATP-PFK) can act as the first enzyme of glycolysis, and it has long appeared to be the enzyme that is most susceptible to important metabolic control (Turner and Turner, 1980). ATP-PFK catalyzes formation of fructose-1,6-bisphosphate (Fig. 13-5). This reaction is the first glycolytic step that involves a hexose phosphate that cannot also be used to form sucrose or starch, so it could represent a control over the entire glycolytic pathway. Activity of ATP-PFK is inhibited by many metabolites, including ATP, PEP, and citric acid, but activity is enhanced by P*i* (reviewed by Copeland and Turner, 1987; Dennis and Greyson, 1987; Raymond et al., 1987). ADP and AMP are usually slightly inhibitory. Inhibition by ATP, PEP, and citric acid, which are formed from or during glycolysis, seems to be a reasonable way to prevent overproduction of these compounds. Activation by P*i* might also be expected because P*i* is used in glycolysis along with fructose-1,6-bisphosphate, but the inhibition by ADP and AMP is unexpected and inconsistent with control by energy charge.

Another important regulator of glycolysis is the $NAD^+$/NADH ratio because $NAD^+$ is an essential substrate for glycolysis, whereas NADH is a product. These substrate/product relations of $NAD^+$ and NADH are also true for the Krebs cycle and the electron-transport system. Because $O_2$ is so important in oxidizing NADH and regenerating $NAD^+$, good aeration favors glycolysis, the Krebs cycle, and the electron-transport system. We discuss aeration effects in Section 13.13.

Much more has been learned recently about control of glycolysis; two discoveries were important (reviewed by Huber, 1986; Copeland and Turner, 1987; ap Rees and Dancer, 1987; ap Rees, 1987; Black et al., 1987; Sung et al., 1988; Stitt, 1990). First was the finding that plants contain PP*i*-PFK and that it (along with the ATP-dependent PFK) catalyzes formation of fructose-1,6-bisphosphate from fructose-6-phosphate (Fig. 13-5). (Nearly all animals lack PP*i*-PFK, but some microbes have it.) Second was the discovery that plants, like most other organisms, contain *fructose-2,6-bisphosphate* (Sabularse and Anderson, 1981). Fructose-2,6-bisphosphate proved to be a potent activator of PP*i*-PFK, thereby favoring fructose-1,6-bisphosphate formation. It also inhibits a cytosolic fructose-1,6-bisphosphatase enzyme that hydrolyzes fructose-1,6-bisphosphate back to fructose-6-phosphate, so this inhibition again favors high amounts of fructose-1,6-bisphosphate and therefore the start of glycolysis. These relations are summarized in Figure 13-13. Note also that if glycolysis is favored, then sucrose formation will be depressed because both processes compete for the same fructose-6-phosphate in the cytosol. This competition is probably a major control mechanism that determines whether sucrose is respired or translocated to other parts of the plant. And it depends on rapidly changing levels of fructose-2,6-bisphosphate in the cytosol that are commonly only 1 to 10 $\mu$M. Future research must explain how environmental changes affect levels of fructose-2,6-bisphosphate.

## Control of Respiration in Mitochondria

As noted above, mitochondrial respiration consists essentially of the Krebs cycle, the electron-transport system, and oxidative phosphorylation. Clearly, there are numerous possible control points in these three interdependent processes. In a review of this subject, Dry et al. (1987) concluded that the major controlling factor was the concentration of ADP in the mitochondria. If this concentration is relatively high, then oxidative phosphorylation (formation of ATP from the ADP and P*i*) is fast, electron transport to oxygen is fast, and the entire Krebs cycle runs faster; within rather wide limits, the EC value is not controlling. This means that the rate of mitochondrial respiration depends on the ability of the mitochondria to transport ATP out into the cytosol, convert ATP back to ADP in biosynthetic and growth processes, and then transport ADP back into the mitochondria, where it is used again to make more ATP. It is often found that cells, organs, or tissues that grow rap-

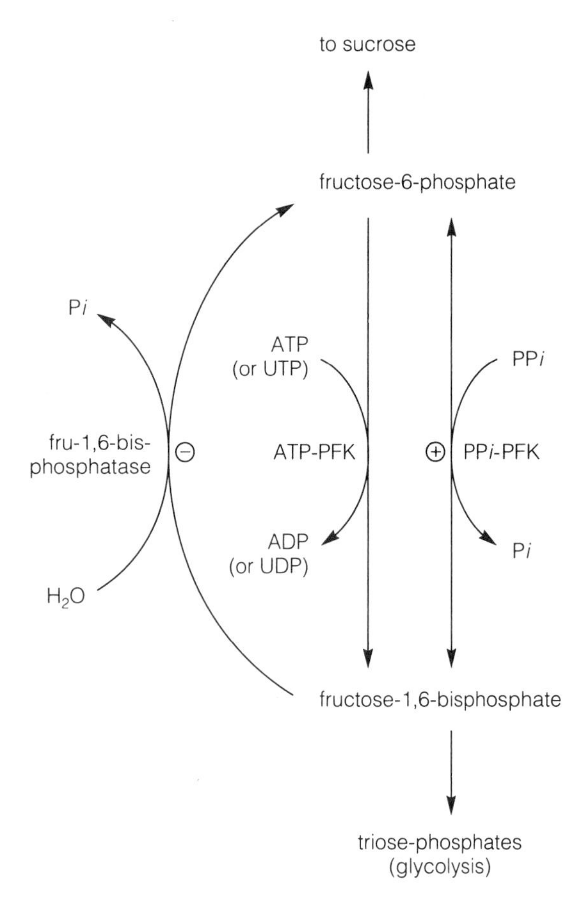


**Figure 13-13** Regulation of glycolysis versus sucrose synthesis by fructose-2,6-bisphosphate. Conversion of fructose-6-phosphate to fructose-1,6-bisphosphate by PP*i*-PFK is activated (note circled plus sign near reversible arrow) by fructose-2,6-bisphosphate, but the reverse reaction catalyzed by fructose-1,6-bisphosphate phosphatase (extreme left) is inhibited by fructose-2,6-bisphosphate (note circled minus sign near arrow). ATP-PFK (center downward arrow) can use either ATP or uridine triphosphate (UTP) as a substrate to phosphorylate fructose-6-phosphate.

idly also respire rapidly, and one important reason is that growth requires so much hydrolysis of ATP back to ADP in energy-dependent reactions.

Evidence favoring another kind of control has appeared recently. This control is regulation of the first step in the Krebs cycle: oxidation of pyruvate by the pyruvate dehydrogenase complex. As mentioned in Section 13.6, this complex contains five different enzymes, two of which regulate activity of the other three. One regulatory enzyme is a kinase, which uses ATP to phosphorylate the hydroxyl group of various threonine amino acid residues in a certain part of the pyruvate dehydrogenase enzyme. This phosphorylation quickly inactivates the enzyme, so the Krebs cycle stops. The second regulatory enzyme, a phosphatase, hydrolyzes phosphate away from the threonines and reactivates the enzyme, so the Krebs cycle can again oxidize pyruvate. Therefore, if the ATP level in mitochondria is high and if the kinase is active, the Krebs cycle shuts down to produce less ATP until some of the phosphates are removed. The EC ratio again is relatively unimportant.

One of the important factors that controls whether pyruvate dehydrogenase is phosphorylated (inactive) or dephosphorylated (active) is the concentration of pyruvate in the mitochondria. If pyruvate is abundant it slows phosphorylation, keeps the dehydrogenase more active, and allows the Krebs cycle to proceed (Budde et al., 1988). In summary, it appears that if mitochondria have accumulated much ATP, this ATP slows the Krebs cycle and therefore all subsequent respiratory processes in mitochondria. However, an abundance of pyruvate formed in glycolysis can partly overcome the effect of high ATP levels and can help keep mitochondrial respiration going.

### Control of the Pentose Phosphate Pathway

We mentioned above that the rate-limiting enzyme of the PPP is the first, glucose-6-phosphate dehydrogenase, and that in chloroplasts the isozyme present is inhibited by NADPH formed in light and is inactivated in light by the ferredoxin-thioredoxin system. Presumably, the isozyme present in the cytosol is also inhibited by NADPH; regardless, this cytosol isoenzyme requires $NADP^+$ as a substrate. Any process that favors conversion of NADPH to $NADP^+$ should therefore speed the PPP. Two such processes are oxidation of NADPH by the electron-transport system and oxidation during biosynthesis of fatty acids and isoprenoid compounds such as carotenoids and sterols.

## 13.13 Factors Affecting Respiration

Many environmental factors influence respiration. The previous descriptions of the individual reactions involved should help us understand how such factors affect the overall rate of respiration and its importance to plant maintenance and growth. Let's investigate how some environmental factors affect the biochemical processes of respiration.

### Substrate Availability

Respiration depends on the presence of an available substrate; starved plants with low starch, fructan, or sugar reserves respire at low rates. Plants deficient in sugars often respire noticeably faster when sugars are provided. In fact, respiration rates of leaves are often much faster just after sundown, when sugar levels are high, than just before sunup, when levels are lower. We mentioned in the previous chapter that lower, shaded leaves usually respire slower than do upper leaves exposed to higher light levels. If this were not true, the lower leaves would probably die sooner than they do. The difference in starch and sugar contents resulting from unequal photosynthetic rates is probably responsible for the lower respiratory rates of shaded leaves.

If starvation becomes extensive, even proteins can be respired. These proteins are first hydrolyzed into their amino-acid subunits, which are then degraded by glycolytic and Krebs-cycle reactions. For glutamic and aspartic acids produced by hydrolysis of proteins, the relation to the Krebs cycle is especially clear because these amino acids are converted to $\alpha$-ketoglutaric and oxaloacetic acids, respectively (see Fig. 13-12). Similarly, the amino acid alanine is oxidized via pyruvic acid. As leaves become senescent and yellow, most of the protein and other nitrogenous compounds in the chloroplasts are broken down. During this process ammonium ions released from various amino acids are combined into glutamine and asparagine (as amide groups), and this prevents ammonium toxicity. These processes are discussed in Chapter 14.

## Oxygen Availability

The $O_2$ supply also influences respiration, but the magnitude of its influence differs greatly among plant species and even within organs of the same plant. Normal variations in $O_2$ content of the air are much too small to influence respiration of most leaves and stems. Furthermore, the rate of $O_2$ penetration into leaves, stems, and roots is usually sufficient to maintain normal $O_2$ levels by the mitochondria, mainly because cytochrome oxidase has such a high affinity for oxygen that it can function even when the $O_2$ concentration around it is only 0.05 percent of that in the air (Drew, 1979).

In more bulky tissues with lower surface/volume ratios, diffusion of $O_2$ from air to cytochrome oxidase in cells near the interior is probably retarded enough to slow respiration rates. One might suspect that in carrot roots, potato tubers, and other storage organs, the rate of $O_2$ penetration would be so low that respiration inside would be primarily anaerobic. Quantitative data on gas penetration into such organs are meager, but measurements do show that the rate of $O_2$ movement through them is certainly much less than in air (Drew, 1979). However, in 1890 the French physiologist H. Devaux showed that central regions of bulky plant tissues do respire aerobically, although slowly (for other reasons). He demonstrated the importance of intercellular spaces for gaseous diffusion.

Microscopic observations show that these intercellular spaces represent significant amounts of the total tissue volume. For example, in potato tubers approximately 1 percent of the volume is occupied by air spaces, and values for roots of from 2 to 45 percent have been observed in various species, the higher values more common among wetland plants (Crawford, 1982). Such intercellular air spaces extend from the stomates of leaves to most cells in the plant, aiding their aerobic respiration. Only tightly packed xylem parenchyma cells and cells in meristematic regions appear to have no access to such air spaces.

We mentioned in Chapter 7 that diffusion of $O_2$ through the intercellular space system from leaves to roots was probably important in moving $O_2$ and other gases through plant tissues more rapidly than might have been expected for organisms with no lungs or hemoglobin to help transport gases. In general, the intercellular air system from leaves to roots is especially important for grasses and sedges because they have hollow stems; indeed, these species are generally more tolerant to flooding than most others (Crawford, 1982). (John Dacey's essay in Chapter 4 explains how bulk flow forces air through petioles of water lilies, and it might also force air down into waterlogged roots.) Nevertheless, flooding for long periods is toxic to nearly all plants, especially when no detectable $O_2$ is present around the roots (**anoxia**, or entirely anaerobic conditions).

Among crop plants, only rice is known to tolerate anoxia for long, although *Echinochloa crus-galii* (barnyard grass), a common weed of rice fields, is also unusually tolerant, as are other species native to wetlands (Barclay and Crawford, 1982). In native plants, tolerance of anoxia is generally greatest when temperatures are low and respiration is slow (as in winter) and when adequate carbohydrates are stored in fleshy rhizomes or roots.

Substantial differences in tolerance also exist in trees (Gill, 1970; Joly and Crawford, 1982). For certain tropical mangroves (*Rhizophora mangle* and *Avicennia nitida*), roots that grow upward out of the water (*pneumatophores*) transfer oxygen to flooded roots. Thus the flooded roots are really not anoxic, but rather **hypoxic** (under reduced oxygen levels). Among conifers, lodgepole pine (*Pinus contorta*) is more tolerant to hypoxia under flooding than is Sitka spruce (*Picea sitchensis*), and part of the difference lies in the greater ability of the pine to transport oxygen to its roots (Philipson and Coutts, 1980). Some species form extensive adventitious root systems when their stems are flooded, and these roots aid absorption of mineral salts and water. Still other species form new roots on the original root system.

Another interesting morphological adaptation of roots to hypoxia is formation of **aerenchyma tissue**. Aerenchyma is produced following collapse and lysis of some mature cortex cells, so it is a tissue with large air spaces (Fig. 13-14). Aerenchyma allows faster diffusion of oxygen from shoots to roots, thereby aiding respiration of hypoxic roots (Kawase, 1981; Jackson, 1985). The cause of aerenchyma formation seems to be ethylene. This gas is produced in trace amounts by many plant parts, especially when they are stressed (Chapter 18), but in waterlogged soils ethylene accumulates because it cannot diffuse away as fast as in aerated soils.

Physiologists are interested not only in how some

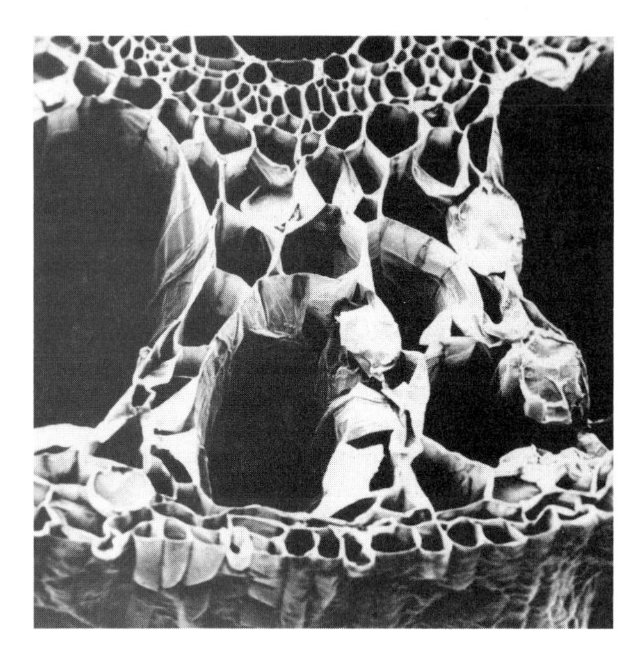

**Figure 13-14** Scanning electron micrograph showing aerenchyma formation in a maize root subjected to hypoxia. (From Campbell and Drew, 1983. Used by permission.)

species tolerate hypoxia better than others (in addition to morphological adaptations), but also how hypoxia injures plants. An important factor in survival of plants in hypoxic, anoxic, or other stressful situations is their ability to express different genes, especially genes that produce enzymes that help overcome the stress metabolically.

Roots commonly respond to hypoxia by faster glycolysis and fermentation. Injurious effects of hypoxia are caused by several metabolic imbalances ultimately resulting from insufficient oxygen (Kozlowski, 1984). One effect is retarded transport of cytokinin hormones from young roots to shoots, and, as explained in Chapter 18, one important source of these hormones for leaves and stems is root tips. Other imbalances include insufficient absorption of mineral salts (especially nitrate); leaf wilting, accompanied by slower photosynthesis and carbohydrate translocation; decreased permeability of roots to water; and accumulation of toxins caused by microbes around the roots (Drew, 1979; Bradford and Yang, 1981).

As we might predict from the preceding biochemical sections, the supply of ATP is limited because the electron-transport system and Krebs cycle cannot function without oxygen. Furthermore, the products of fermentation, especially ethanol (in most species), lactic acid, malic acid, and rarely glycerol, accumulate to some extent. Both ethanol and lactic acid can become toxic, although Kozlowski (1984) suggested that fermentation products usually do not become toxic because they leak out and diffuse away from roots.

Some results with rice and barnyard grass are interesting. The seeds of these plants will germinate under hypoxia or anoxia, but they then do so unusually: by upward protrusion of the coleoptile rather than by downward protrusion of the radicle (Fig. 13-15). Roots hardly develop, but the coleoptile continues to grow under hypoxia. Nevertheless, why do rice and barnyard grass plants tolerate hypoxic conditions unusually well? Some answers are becoming available. Seeds of these species have the unusual ability to produce ATP from a rapid fermentation during anoxia (Cobb and Kennedy, 1987). They use this ATP to synthesize proteins far more effectively than stress-sensitive seeds, which cannot ferment and produce ATP so rapidly. Two of the proteins they synthesize under anoxia are pyruvic acid decarboxylase and alcohol dehydrogenase, both important fermentation enzymes (Morrell et al., 1990). Subsequent elongation of the seedling's shoot system under water soon allows the shoot to emerge into air, become green and photosynthetic, and transfer oxygen to the roots, which then become less hypoxic and grow (Raskin and Kende, 1985).

Many seeds, especially large ones, exhibit fermentation during the normal imbibition of water that leads to germination. Alain Pradet and his colleagues in France studied sensitivity of several cultivated seeds to hypoxia. They classified 12 species into two groups with quite different sensitivities (Al-Ani et al., 1985). Group 1 (lettuce, turnip, sunflower, radish, cabbage, flax, and soybean) were sensitive, and germination would not occur at $O_2$ partial pressures below 2 kilopascals (2 percent $O_2$). Group 2 was much more resistant, and germination was not stopped until the $O_2$ partial pressure decreased below about 0.1 kPa. Group 2 included rice, wheat, maize, sorghum, and pea. There was a poor correlation between ability to germinate and the extent to which hypoxia decreased energy charge, so it is not clear why the seeds differ in sensitivity to hypoxia unless energy charge is a minor controlling factor. No aerenchyma is formed in seeds.

Louis Pasteur discovered a surprising effect of hypoxic and anoxic conditions in his studies of wine production by yeasts over a century ago. When the $O_2$ concentration around yeast and most plant cells is decreased gradually below the atmospheric 20.9 percent, $CO_2$ production from respiration decreases until a minimum is reached, *but lower $O_2$ concentrations cause a rapid rise in $CO_2$ production* (see Fig. 13-16). Pasteur later found that yeast cells grew rapidly in air but used little sugar and produced little ethanol and $CO_2$; under anaerobic conditions they grew slower but used more sugar and produced more $CO_2$ and ethanol. This phenomenon, which became known as the **Pasteur effect**, was subsequently determined by plant physiologists to be caused by inhibited carbohydrate breakdown by oxygen.

A common biochemical explanation of the Pasteur effect involves an allosteric inhibition of phosphofructokinase in the presence of certain compounds formed

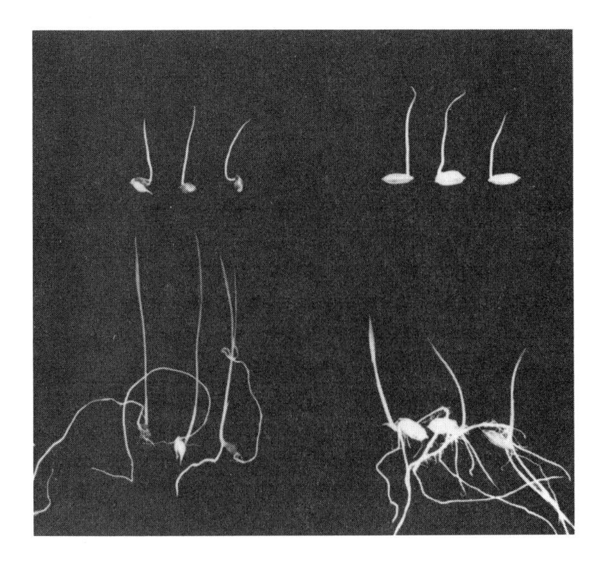

**Figure 13-15** Development of week-old barnyard grass (left) and oat (right) seedlings in an atmosphere of nitrogen (top row) or air (bottom row). Note the failure of root development but coleoptile elongation in the nitrogen atmosphere. (From Kennedy et al., 1980.)

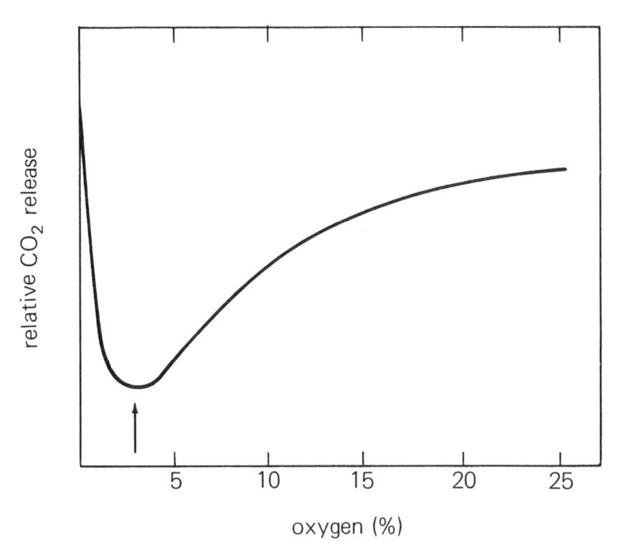


**Figure 13-16** The influence of atmospheric oxygen concentration on $CO_2$ production in apple fruits. On the right side of the arrow, increasing $O_2$ supply increases respiration because of stimulated Krebs-cycle activity, yet anaerobic $CO_2$ release from pyruvate becomes minimal in this region of the curve due to indirect inhibition effects of $O_2$ on glycolysis. At the left of the arrow, the $O_2$ concentration is low enough to allow a very rapid breakdown of sugars to ethanol and $CO_2$. (Redrawn from James, 1963.)

by aerobic respiration. In plants, increased production of ATP and citrate with increased $O_2$ is known to allosterically inhibit ATP-PFK, which should therefore decrease glycolysis and fermentation (in the region to the left of the arrow in Fig. 13-16). But because glycolysis can bypass ATP-PFK when PP*i*-PFK is used (at least in many plants), our explanations of the cause of the Pasteur effect must be reevaluated.

The Pasteur effect no doubt causes decreased carbohydrate reserves in flooded soils because of faster glycolysis under hypoxic conditions. This probably helps explain why plants with swollen rhizomes or thick roots that store carbohydrates can survive anoxia longer, especially at cold temperatures, when respiration is slow.

The Pasteur effect also has some practical importance in storage of fruits and vegetables, especially apples (Weichmann, 1986). Here a goal of storage is to prevent extensive sugar loss and overripening. This is done by carefully decreasing $O_2$ to the concentration at which aerobic respiration is minimized but sugar breakdown by anaerobic processes is not stimulated. Additional $CO_2$ is also added to the air, and in some cases the temperature is lowered closer to the freezing point, which further prevents overripening. As discussed in Chapter 18, $CO_2$ inhibits action of a fruit-ripening hormone, ethylene, and this is a probable explanation for its effectiveness in inhibiting overripening. Low concentrations of $O_2$ also slow ethylene production.

## Temperature

For most plant species and parts, the $Q_{10}$ for respiration between 5 and 25°C is usually between 2.0 and 2.5. With further increases in temperature up to 30 or 35°C, the respiration rate still increases, but less rapidly, so the $Q_{10}$ begins to decrease. A possible explanation for the $Q_{10}$ decrease is that the rate of $O_2$ penetration into cells begins to limit respiration at higher temperatures at which chemical reactions could otherwise proceed rapidly. Diffusion rates of $O_2$ and $CO_2$ are also increased by increased temperature, but the $Q_{10}$ for these physical processes is only about 1.1; that is, temperature doesn't speed diffusion of solutes through water very much.

With even further rises in temperature to 40°C or so, the rate of respiration actually decreases, especially if plants are maintained under such conditions for long periods. Apparently, the required enzymes begin to be denatured rapidly at high temperatures, preventing metabolic increases that would otherwise occur. With pea seedlings, a temperature increase from 25 to 45°C initially increased respiration greatly, but within about two hours the rate became less than before. A probable explanation is that the two-hour period was long enough to partly denature respiratory enzymes.

## Type and Age of Plant

Because there are large morphological differences among members of the plant kingdom, it is to be expected that differences in metabolism also exist. In

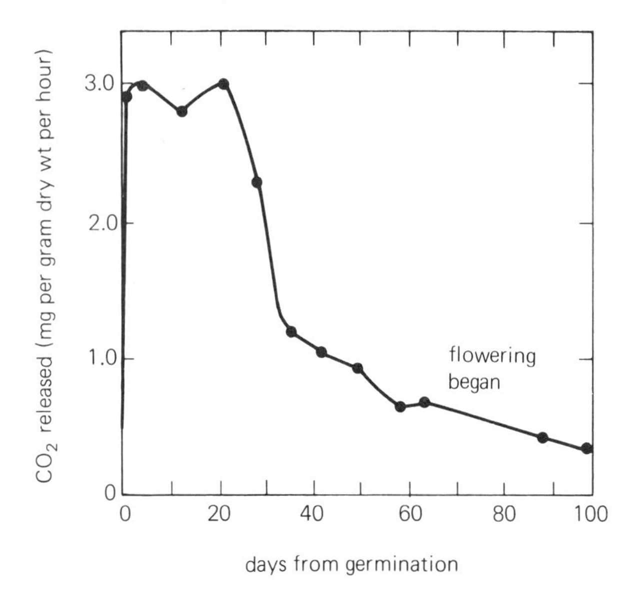


**Figure 13-17** Respiration of whole sunflower plants from germination until maturity. The rate gradually declined after the 22nd day, even though the rate for individual parts, such as inflorescences, increased for a time after that. (Drawn from data of Kidd et al., 1921.)

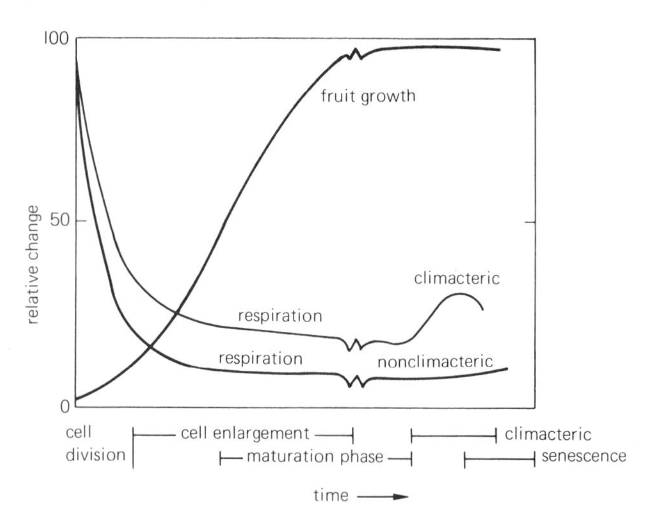


**Figure 13-18** Stages in development and maturation of fruits that undergo the climacteric respiration increase and of those that do not. Discontinuities in the lines indicate that the time scale was changed to show differences in development rates of different fruits. The growth pattern may be single or double sigmoid (see Chapter 16). (From J. Biale, 1964, *Science* 146:880. Copyright 1964 by the American Association for the Advancement of Science.)

general, bacteria, fungi, and many algae respire considerably more rapidly than do seed plants. Various organs or tissues of higher plants also exhibit large variation in rates. One reason that bacteria and fungi have so much higher values than plants, based on dry weight, is that they contain only small amounts of stored food reserves and have no nonmetabolic woody cells. Such dead woody cells contribute to the dry weight and strength of vascular plants but not to respiration. Similarly, root tips and other regions containing meristematic cells with a high percentage of protoplasm and protein have high respiratory rates expressed on a dry-weight basis. If comparisons are made on a protein basis, these differences are smaller. In general, there is a fairly good correlation between the rate of growth of a particular cell type and its respiration rate. This results from several factors, such as the use of ATP, NADH, and NADPH both for synthesis of proteins, cell-wall materials, membrane components, and nucleic acids and for ion accumulation and carbohydrate transport. Consequently ADP, $NADP^+$, and $NAD^+$ become available for use in respiration. Inactive seeds and spores have the lowest (usually undetectable) respiration rates, but here the effect is not entirely because of a lack of growth. Rather, certain changes in the protoplasm, especially desiccation, shut off metabolism. Such seeds and spores generally contain abundant food reserves.

The age of intact plants influences their respiration to a large degree. Figure 13-17 shows how the rate changed in whole sunflower plants from germination until after flowering. The rate is expressed as the amount of $CO_2$ released per amount of preexisting dry weight. The curve is extrapolated to zero time to show the common large burst in respiratory activity as the dry seeds absorb water and germinate. Respiration remained high during the period of most rapid vegetative growth but then fell before flowering occurred. In this and other examples, much of the respiration in mature plants is in young leaves and roots and in the growing flowers.

Changes in respiration also occur during development of ripening fruits. In all fruits, the respiration rate is high when they are young, while the cells are still rapidly dividing and growing (Fig. 13-18). The rate then gradually declines, even if the fruits are picked. However, in many species, of which the apple is a good example, the gradual decrease in respiration is reversed by a sharp increase, known as the **climacteric**. The climacteric usually coincides with full ripeness and flavor of fruits, and its appearance is hastened by production in cells of traces of ethylene that stimulate ripening (Biale and Young, 1981; Brady, 1987; Tucker and Grierson, 1987). Further storage leads to senescence and to decreases in respiration.

Some fruits, including citrus fruits, cherries, grapes, pineapples, and strawberries, do not show the climacteric (lowest curve, Fig. 13-18). Grapefruits, oranges, and lemons ripen on the trees; if removed sooner, their respiration simply continues at a gradually decreasing rate. The advantages and disadvantages of a climacteric are unknown. The biochemical basis for the climacteric respiratory rise is also unclear, but biochemical and molecular-biology techniques should help solve this problem (Tucker and Grierson, 1987).

# 14

# Assimilation of Nitrogen and Sulfur

The importance of nitrogen to plants is emphasized by the fact that only carbon, oxygen, and hydrogen are more abundant in them. Although nitrogen occurs in a vast number of plant constituents, most of it is in proteins. Sulfur is only about one-fifteenth as abundant in plants as nitrogen, but it occurs in many molecules, especially proteins. Both elements are usually absorbed from the soil in highly oxidized forms and must be reduced by energy-dependent processes before they are incorporated into proteins and other cellular constituents. Human metabolic systems cannot duplicate this reduction, just as we cannot reduce $CO_2$. Describing the ways in which nitrate and sulfate are reduced and subsequently combined with carbohydrate skeletons to form amino acids is an important task of this chapter. We shall also discuss fixation of atmospheric $N_2$ and interconversions of nitrogen compounds during various stages of plant development. An excellent summary of many aspects of nitrogen metabolism was given by Blevins (1989).

## 14.1 The Nitrogen Cycle

Nitrogen exists in several forms in our environment. The continuous interconversion of these forms by physical and biological processes constitutes the nitrogen cycle, summarized in Figure 14-1.

Vast amounts of nitrogen occur in the atmosphere (78 percent by volume), yet it is energetically difficult for living organisms to obtain the nitrogen atoms of $N_2$ in a useful form. Although $N_2$ moves into leaf cells along with $CO_2$ through stomates, enzymes are available to reduce only the $CO_2$, so $N_2$ moves out as fast as it moves in. Most of the nitrogen in living organisms arrives there only after either fixation (reduction) by prokaryotic microorganisms, some of which exist in the roots of certain plants, or by industrial fixation to form fertilizers. Small amounts of nitrogen also move from the atmosphere to the soil as ammonium ($NH_4^+$) and nitrate ($NO_3^-$) ions in rain and are then absorbed by roots. This $NH_4^+$ arises from industrial burning, volcanic activity, and forest fires, whereas $NO_3^-$ arises from oxidation of $N_2$ by $O_2$ or ozone in the presence of lightning or ultraviolet radiation. Another source of $NO_3^-$ is the oceans. Wind-whipped white caps produce minute droplets of water called aerosols, from which the water evaporates, leaving ocean salts suspended in the atmosphere. Near coastlines these salts can be brought to the land in rainwater. They are called **cyclic salts** because they eventually cycle through streams back to the oceans.

Absorption of $NO_3^-$ and $NH_4^+$ by plants allows them to form numerous nitrogenous compounds, mainly proteins. Manure and dead plants, microorganisms, and animals are important sources of nitrogen returned to the soil, but most of this nitrogen is insoluble and is not immediately available for plant use. Nearly all soils contain small amounts of various amino acids, produced largely by microbial decay of organic matter but also by excretion from living roots. Even though such amino acids can be absorbed and metabolized by plants, these and other more complex nitrogen compounds contribute little to the plant's nitrogen nutrition in a direct way. They are, however, of great importance as a nitrogen reservoir from which $NH_4^+$ and $NO_3^-$ arise. In fact, as much as 90 percent of the total nitrogen in soils may be in organic matter, although in some cases significant amounts exist as $NH_4^+$ bound to clay colloids.

Conversion of organic nitrogen to $NH_4^+$ by soil bacteria and fungi is called **ammonification**. This process can occur with many kinds of microorganisms, at cool temperatures, and at various *p*H values. Subsequently, in warm, moist soils with near neutral *p*H, $NH_4^+$ is further oxidized by bacteria to nitrite ($NO_2^-$) and $NO_3^-$ within a few days of its formation or its addition as a fertilizer. This oxidation, called **nitrification**, provides

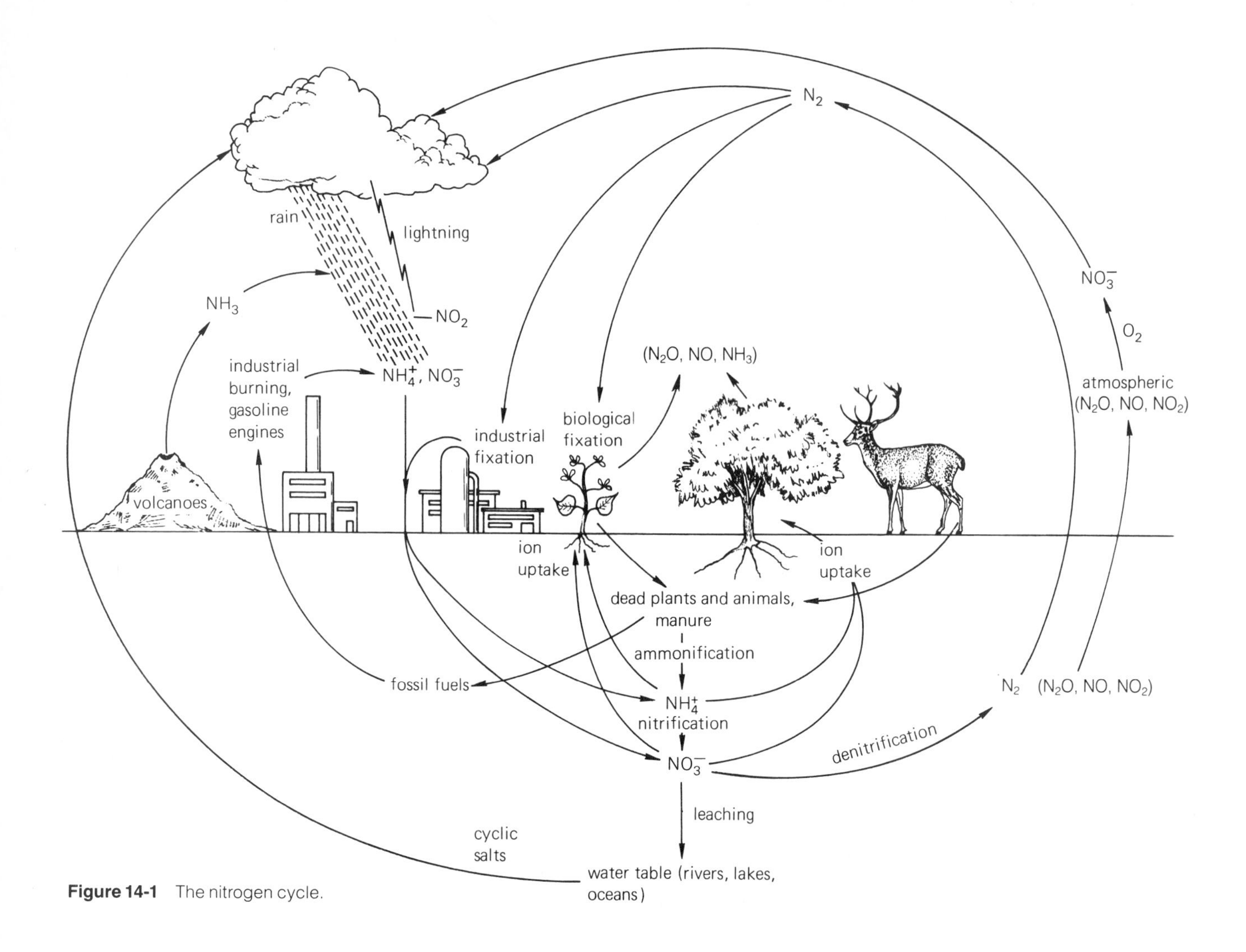


**Figure 14-1** The nitrogen cycle.

energy for survival and growth of such microbes, just as does oxidation of more complex foods for other organisms. Bacteria of the genus *Nitrosomonas* are most important in oxidation of ammonia to nitrite, whereas bacteria of the genus *Nitrobacter* commonly reduce most nitrite to nitrate. In many cold, acidic, or hypoxic soils, however, nitrifying bacteria are less effective and abundant, so $NH_4^+$ becomes a more important nitrogen source than $NO_3^-$. This is true, for example, in arctic soils and anaerobic bogs. Many forest trees absorb most of their nitrogen as $NH_4^+$ because of the low *p*H common to forest soils and probably because other factors contribute to slow rates of nitrification. Because of its positive charge, $NH_4^+$ is adsorbed to soil colloids, whereas $NO_3^-$ is not adsorbed and is much more readily leached.

Nitrate is also lost from soils by **denitrification**, the process by which $N_2$, NO, $N_2O$, and $NO_2$ are formed from $NO_3^-$ by anaerobic bacteria. These bacteria use $NO_3^-$ rather than $O_2$ as an electron acceptor during respiration, thus obtaining energy for survival. Denitrification occurs relatively deep in the soil where $O_2$ penetration is limited, in waterlogged or compacted soils, and in certain regions near the soil surface where the $O_2$ concentration is low because of its especially rapid use in oxidation of organic matter. Furthermore, plants lose small amounts of nitrogen to the atmosphere as volatile $NH_3$, $N_2O$, $NO_2$, and NO, especially when well fertilized with nitrogen (Wetselaar and Farquhar, 1980; Duxbury et al., 1982). Oxidized forms of nitrogen in the atmosphere are important ecologically, because when converted to $NO_3^-$ they contribute $HNO_3$ to acidic rainfall ("acid rain").

## 14.2 Nitrogen Fixation

The process by which $N_2$ is reduced to $NH_4^+$ is called **nitrogen fixation** (reviewed by Dixon and Wheeler, 1986). It is, so far as we know, carried out only by prokaryotic microorganisms. Principal $N_2$-fixers include certain free-living soil bacteria, free-living cyanobacteria (blue-green algae) on soil surfaces or in water, cyanobacteria in symbiotic associations with fungi in

**Figure 14-2** Bog myrtle (*Myrica gale*) plants cultured with (left) and without (right) nodules in a hydroponic solution lacking nitrogen. Plants were grown from seed for five months in the solutions. (From Bond, 1963.)

lichens or with ferns, mosses, and liverworts (Peters, 1978; Peters and Meeks, 1989), and bacteria or other microbes associated symbiotically with roots, especially those of legumes. Their role in nitrogen fixation is of great importance to the food chain in forests, freshwater and marine environments, and even arctic regions. Furthermore, the activities of the roots of nitrogen-fixing plants benefit the roots of surrounding plants, either through excretion of nitrogen from nodules or through microbial decomposition of nodules or even whole plants (Ta and Faris, 1987). This contribution is important in agriculture, in which mixed legumes and grasses often are used as pastures.

About 15 percent of the nearly 20,000 species in the family Fabaceae (Leguminosae) have been examined for $N_2$ fixation, and approximately 90 percent of these have root nodules in which fixation occurs (Allen and Allen, 1981). Important nonlegumes that fix $N_2$ are taxonomically diverse trees and shrubs occurring in eight families and 23 genera, including members of the genera *Alnus* (alder), *Myrica* (such as *M. gale*, the bog myrtle), *Shepherdia, Coriaria, Hippophae, Eleagnus* (autumn olive), *Ceaonothus* (snow brush), and *Casaurina* (Tjepkema et al., 1986; Dawson, 1986). These nonlegumes are typically pioneer plants on nitrogen-deficient soils—for example, *M. gale* in the bog soils of western Scotland and *Casaurina equisetifolia* on sand dunes of tropical islands. Figure 14-2 demonstrates the important role of root nodules in providing nitrogen to *M. gale*. All plants in this figure were five months old, but only the nodulated group on the left could grow well in the nutrient solution lacking nitrogen salts.

There is considerable interest among foresters in selecting and breeding nitrogen-fixing nonlegume trees that can be planted with or before trees of greater economic importance. One goal is to reduce the need for nitrogen fertilizers in the timber industry. Some success has already been obtained using mixed red alder and Douglas fir populations in the Pacific Northwest of the United States. Different alder species have promise for other temperate regions, and species of other genera should be effective in tropical regions (Dawson, 1986).

The microorganisms responsible for nitrogen fixation in roots of many species have been identified. In some tropical trees it is various cyanobacteria, but in most trees or shrubs actinomycetes (filamentous prokaryotes) in the genus *Frankia* carry out this process. In legumes, bacterial species of the closely related genera *Rhizobium, Bradyrhizobium,* and *Azorhizobium* are responsible (Downie and Johnston, 1988; Djordjevie et al., 1987; Quispel, 1988). A particular *Rhizobium* or *Rhizobium*-like species is generally effective with only one legume species. All rhizobia are aerobic bacteria that persist saprophytically in the soil until they infect a root hair (Fig. 14-3) or sometimes a damaged epidermal cell. Root hairs usually respond to invasion first by curling and surrounding the bacterium; curling is caused by unidentified molecules released from the bacteria. Yet another interesting finding is that genes of the rhizobia that control production of the molecules that cause curling are activated first by compounds released by the roots, probably root hairs. In alfalfa, white clover, and broadbean or faba bean, the most abundant compounds are certain flavonoids (described in Section 15.7; Bothe, 1987; Maxwell et al., 1989). These results emphasize that various specific chemical signals are probably sent from the root hair and are somehow recognized by the invading bacteria (Kondorosi and Kondorosi, 1986).

Next, enzymes from the bacteria degrade part of the cell wall and allow bacterial entry into the root-hair cell itself. Then the root hair produces a threadlike structure called the **infection thread**, which consists of an infolded and extended plasma membrane of the cell being invaded, along with new cellulose formed on the *inside* of this membrane. The bacteria multiply extensively inside the thread, which extends inwardly and penetrates through and between the cortex cells.

In the inner cortex cells the bacteria are released into the cytoplasm and stimulate some cells (especially tetraploid cells) to divide. These divisions lead to a proliferation of tissues, eventually forming a mature **root nodule** (Fig. 14-3) made largely of tetraploid cells containing bacteria, as well as some diploid cells without bacteria. Each enlarged, nonmotile bacterium is referred to as a **bacteroid**. A typical root nodule cell contains several thousand bacteroids (Fig. 14-4). The bacteroids usually occur in the cytoplasm in groups, each

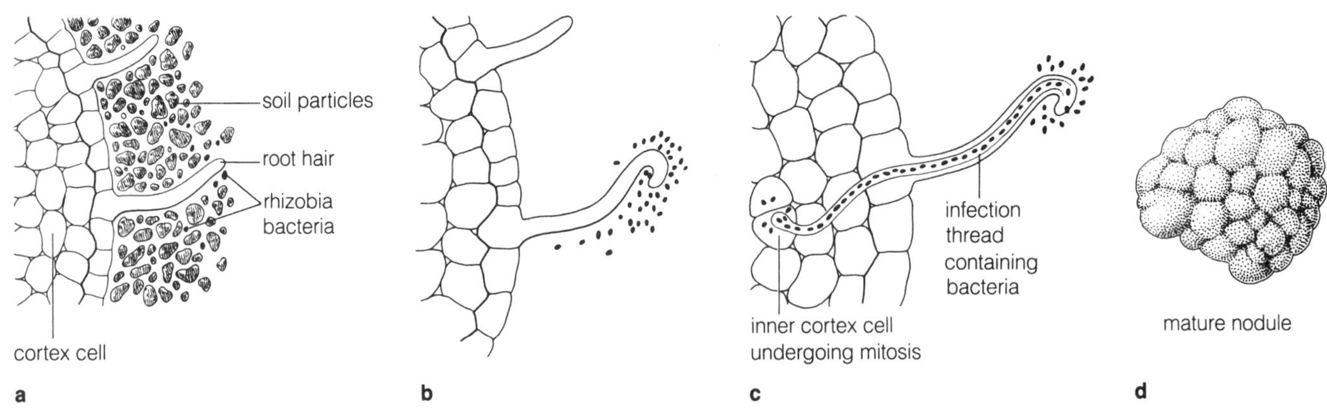

**Figure 14-3** Development of root nodules in soybean. (**a**) and (**b**) *Rhizobium* bacteria contact a susceptible root hair, divide near it, and upon successful infection of the root hair cause it to curl. (**c**) Infection thread carrying dividing bacteria, now modified and apparent as bacteroids. Bacteroids cause inner cortical and pericycle cells to divide. Division and growth of cortical and pericycle cells lead to **d** a mature nodule complete with vascular tissues continuous with those of the root. The three largest nodules shown at the right were 3 to 4 mm in diameter. (Micrograph in **d** from Bergersen and Goodchild, 1973.)

group surrounded by a membrane called the **peribacteroid membrane**. Between the peribacteroid membrane and the bacteroid group is a region called the **peribacteroid space**. Outside the peribacteroid space in the plant cytosol is a protein called **leghemoglobin** (Appleby, 1984; Appleby et al., 1988; Haaker, 1988; Powell and Gannon, 1988). This molecule is red because of a heme group (as in blood hemoglobin) attached as a prosthetic group to the colorless globin protein. Leghemoglobin gives legume nodules a pink color, although it is much more dilute in nodules of nonlegumes. Leghemoglobin is thought to help transport $O_2$ into the bacteroids at carefully controlled rates. Too much $O_2$ inactivates the enzyme that catalyzes nitrogen fixation, yet some $O_2$ is essential for bacteroid respiration.

Nitrogen fixation in root nodules occurs directly within the bacteroids. The host plant provides bacteroids with carbohydrates, which they oxidize and from which they obtain energy. These carbohydrates are first formed in leaves during photosynthesis and then are translocated through phloem to the root nodules. Sucrose is the most abundant carbohydrate translocated, at least in legumes. Some of the electrons and ATP obtained during oxidation by the bacteroids are used to reduce $N_2$ to $NH_4^+$.

## The Biochemistry and Physiology of Nitrogen Fixation

The overall chemical reaction for nitrogen fixation (reduction) is summarized in Reaction 14.1:

$$N_2 + 8\,\text{electrons} + 16\,\text{MgATP} + 16\,H_2O \longrightarrow 2\,NH_3 + H_2 + 16\,\text{MgADP} + 16\,Pi + 8\,H^+ \quad (14.1)$$

As noted, the process requires both a source of electrons and protons and numerous ATP molecules. Furthermore, the production of one $H_2$ formed per $N_2$ reduced seems obligatory (Haaker, 1988; Haaker and Klugkist, 1987). Also required is an enzyme complex called **nitrogenase**, which catalyzes reduction of several other substrates, including acetylene, cyanide, azide, nitrous oxide, and hydrazine. Acetylene reduction to ethylene is often measured as an estimate of nitrogen fixation rates in soils, lakes, and streams, partly because it is so easy to measure ethylene with a gas chromatograph.

The original source of electrons and protons is carbohydrate translocated from the leaves (and then respired by the bacteria). Respiration of carbohydrate in bacteroids leads to reduction of $NAD^+$ to $NADH^+$ or of $NADP^+$ to NADPH, as described in Chapter 13. Alter-

a

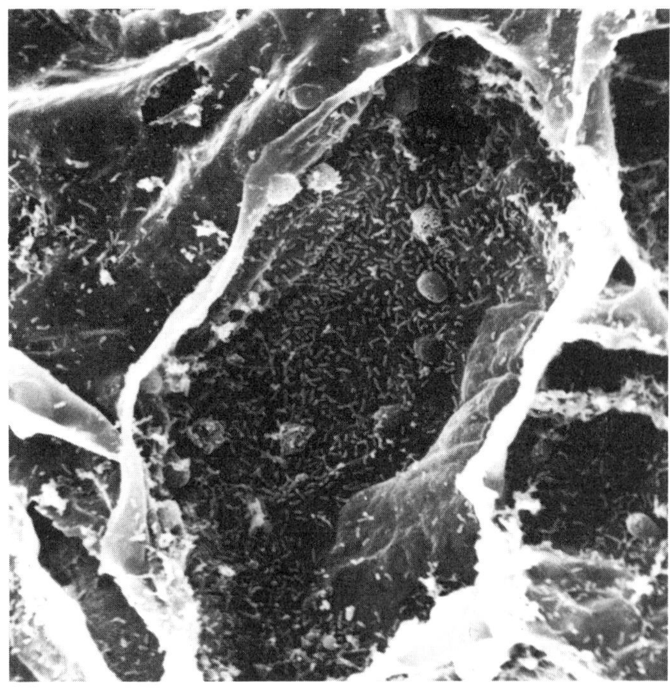

b

**Figure 14-4** (**a**) Transmission electron micrograph of bacteroids within parts of three soybean (*Glycine max*) root nodule cells. Numerous bacteroids (B) are often in groups of four to six, each group surrounded by peribacteroid space (light region), and the peribacteroid space is surrounded by a peribacteroid membrane (PBM). The plant cell walls can be seen, as can one intercellular space filled with air at upper left. (From D. A. Day, G. D. Price, and M. K. Udvardi, 1989.) (**b**) Scanning electron micrograph showing hundreds of bacteroids in a bean (*Phaseolus vulgaris*) root nodule cell. Larger egg-shaped granules probably are starch grains. (Courtesy P. Dayanandan.)

natively, in some nitrogen-fixing organisms oxidation of pyruvate during respiration causes reduction of a protein called *flavodoxin*; then flavodoxin, NADH, or NADPH reduces ferredoxin or similar proteins that are highly effective in reducing $N_2$ to $NH_4^+$.

Nitrogenase accepts electrons from reduced flavodoxin, ferredoxin, or other effective reducing agents as it catalyzes nitrogen fixation. Nitrogenase consists of two distinct proteins, often called Fe protein and Fe-Mo protein. The Fe-Mo protein has two molybdenum and 28 iron atoms; the Fe protein contains four atoms of iron in an $Fe_4S_4$ cluster. Both molybdenum and iron become reduced and then oxidized as nitrogenase accepts electrons from ferredoxin and transfers them to $N_2$ to form $NH_4^+$. ATP is essential to fixation because it binds to Fe protein and causes that protein to act as a stronger reducing agent. The Fe protein transfers electrons to Fe-Mo protein, accompanied by hydrolysis of ATP to ADP. The Fe-Mo protein then completes the transfer of electrons to $N_2$ and to protons to make two $NH_3$ and one $H_2$. The electron transport process (with flavodoxin as electron donor) and use of ATP are summarized in Figure 14-5.

Nitrogen fixation is, as mentioned above, sensitive to too much $O_2$ because the Fe protein and the Fe-Mo protein of nitrogenase are both oxidatively denatured by $O_2$. Leghemoglobin partly controls the availability of $O_2$ in the bacteroid, but complex anatomical features of the bacteroid itself (such as the cortex and endodermis that encircle all vascular bundles and bacteroid-containing cells) seem much more important to maintaining a low $O_2$ level around nitrogenase by acting as a diffusion barrier to air in the soil (Dakora and Atkins, 1989). Apparently, evolution of the symbiotic relation of nitrogen-fixing bacteria and plant roots led to the development of an excellent oxygen protection system — the entire root nodule. In so-called free-living organisms that fix nitrogen (bacteria and cyanobacteria), biochemical modifications appear more important (Becana and Rodríguez-Barrueco, 1989).

The $NH_3$ (probably as $NH_4^+$) is translocated out of the bacteroids before it can be further metabolized and used by the host plant. In the cytosol of bacteroid-containing cells (external to the peribacteroid membrane), $NH_4^+$ is converted into glutamine, glutamic acid, asparagine, and, in many species, nitrogen-rich compounds called **ureides**. The two principal ureides in legumes are *allantoin* ($C_4N_4H_6O_3$) and *allantoic acid* ($C_4N_4H_8O_4$; for structures, see Fig. 8-21); like asparagine ($C_4N_2H_7O_4$), they have relatively high C:N ratios. Each of these three compounds represents a major form of nitrogen translocated from nodules to other parts of

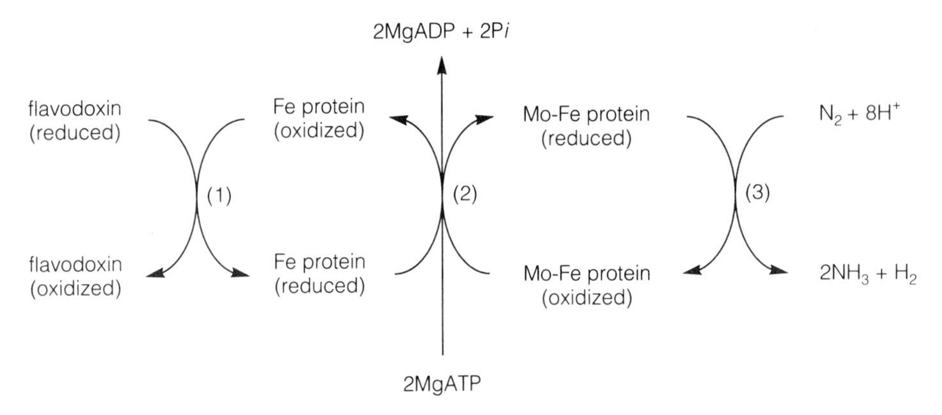


**Figure 14-5** Summary of electron transport from reduced flavodoxin to $N_2$ and $H^+$ in three main stages.

the plant. Asparagine predominates in legumes of temperate origin, including peas, alfalfa, clovers, and lupines. Ureides predominate in legumes of tropical origin, including soybeans, cowpeas, and various beans (Schubert, 1986). In alder, a nonlegume, another ureide called citrulline (see Fig. 8-21) is the major nitrogen compound transported from root nodules.

Asparagine and ureides move from bacteroid-containing cells into pericycle cells adjacent to the vascular bundles that surround the nodule proper. In many species these pericycle cells are modified as transfer cells (Section 8.3), and they seem to actively secrete nitrogen compounds into the conducting xylem cells (Walsh et al., 1989). From there the compounds move into the xylem of the root and shoot to which the vascular bundles of the nodule are connected. They are degraded (largely in the leaves) back to $NH_4^+$, and the nitrogen is converted rapidly into amino acids, amides, and protein (Winkler et al., 1988). Roots of nodulated plants that have not themselves become nodulated seem to receive appreciable nitrogen only after it has moved into the leaves and then back through the phloem along with sucrose. There is, therefore, a cycling process: movement of nitrogen from nodules upward to shoots in the xylem and then the return of excess nitrogen downward to roots via the phloem.

Because of the importance of nitrogen fixation in both nature and agriculture, many ecologists, agronomists, and plant physiologists have studied the environmental and genetic factors that control it. The high costs of nitrogen fertilizers have stimulated even more research in the past few years. In general, the factors that enhance photosynthesis, such as adequate moisture, warm temperatures, bright sunlight, and high $CO_2$ levels, are found to enhance nitrogen fixation (Neves and Hungria, 1987; Sheehy, 1987). Consistent with this, the rate of fixation is usually maximal in early afternoon, when translocation of sugars from leaves to nodules is occurring rapidly. Early afternoon is also a time when transpiration is especially rapid, and the transpiration stream aids removal of nitrogen compounds from roots and root nodules (Pate, 1980).

Several genetic factors control the rates of nitrogen fixation and yields of legumes. One factor concerns how effectively nodulation occurs, and this depends on the genetically controlled recognition process between the bacterial species and the legume species or variety. Researchers are attempting to increase the efficiency of nodulation by altering genes in various rhizobia and by selecting for more compatible host varieties (Rolfe and Gresshoff, 1988; Martinez et al., 1990). Another genetic factor relates to the ability of nitrogenase from all organisms to reduce $H^+$ in competition with $N_2$. As Reaction 14.1 shows, one-fourth of the electrons made available for reduction of $N_2$ are used instead to reduce $H^+$ to $H_2$, and the $H_2$ simply escapes into the soil atmosphere, carrying with it wasted energy. Nevertheless, most *Rhizobium* and closely-related species and free-living bacteria contain a **hydrogenase** enzyme that oxidizes much of the $H_2$ to $H_2O$ before it escapes. During this oxidation, ATP is produced from ADP and P*i*. There is evidence that soybeans and a few other legumes that contain *Rhizobium* strains with an active hydrogenase yield slightly more than legumes that contain a mutant without hydrogenase, probably because of less wasted energy (Eisbrenner and Evans, 1983; Neves and Hungria, 1987; Stam et al., 1987). Perhaps even more effective *Rhizobium* species or strains with active hydrogenases can be found or developed to increase legume yields through genetic-engineering techniques. The goal of incorporating nitrogen-fixing genes into roots of nonlegume crops by genetic-engineering techniques now seems many years in the future, partly because the genes involved in nitrogen fixation and its control are so numerous. Nevertheless, considerable progress in this field is being made.

The stage of growth also influences nitrogen fixation. Three important grain legumes—soybeans, pigeon peas, and peanuts—all show maximum fixation rates after flowering when the demand for nitrogen in the developing seeds and fruits increases. These species, as is common to legumes, contain seeds that are especially protein-rich. In fact, soybean seeds contain 40 percent protein, the highest known percentage of any plant. About 90 percent of the nitrogen fixation in these species occurs during the period of reproductive development, and about 10 percent occurs during the first two months of vegetative growth.

Surprisingly, nitrogen fixation provides only about one-fourth to one-half of the total nitrogen in several mature grain legumes grown on soils of normal fertility; the remaining one-half to three-fourths is absorbed as $NO_3^-$ or $NH_4^+$ from the soil, mainly during the period of vegetative growth. Nevertheless, yields of grain legumes usually cannot be increased with nitrogen fertilizers because nitrogen fixation is decreased as the amount of fertilizer nitrogen absorbed increases. For nitrate fertilizer, this decrease results from inhibition of attachment of rhizobia to root hairs, abortion of infection threads, slowing of nodule growth, inhibition of fixation within established nodules, and more rapid senescence of the nodules when either $NO_3^-$ or $NH_4^+$ is added (Robertson and Farnden, 1980; Streeter, 1988).

The amount of $N_2$ fixed by perennial native and legume crop species during various times in the growing season is probably greatest during reproductive development. The percentage of nitrogen that is derived from fixed $N_2$ in such species is likely to be greater than that in annual grain legumes such as peas, beans, and soybeans because nodules are perennial and fixation should begin earlier than in annuals, in which nodule development must start anew each year. Furthermore, native nitrogen-fixing plants often grow in relatively unfertile soils in which the main input of nitrogen is from fixation. For perennial alfalfa fields, from which each crop is removed as soon as it blooms, the main supply of nitrogen is from nitrogen fixation (Vance and Heichel, 1981).

## 14.3 Assimilation of Nitrate and Ammonium Ions

For plants that cannot fix $N_2$, the only important nitrogen sources are $NO_3^-$ and $NH_4^+$. In general, this is true for all crop plants except legumes (but see box essay on page 296). Crop plants and many native species absorb most nitrogen as $NO_3^-$ because $NH_4^+$ is so readily oxidized to $NO_3^-$ by nitrifying bacteria. However, climax communities of conifers and of grasses absorb most nitrogen as $NH_4^+$ because nitrification is inhibited either by low soil *p*H or by tannins and phenolic compounds (Rice, 1974; Haynes and Goh, 1978). For reviews of the importance of $NO_3^-$ and $NH_4^+$ to various species under various environmental conditions, see Runge (1983) and Bloom (1988). We shall first consider the assimilation of nitrate, because of its abundance in most soils and because it must be converted to $NH_4^+$ in the plant before the nitrogen enters amino acids and other nitrogen compounds.

### Sites of Nitrate Assimilation

Both roots and shoots require organic nitrogen compounds, but in which of these organs is $NO_3^-$ reduced and incorporated into organic compounds? Roots of some species can synthesize all of the organic nitrogen they need from $NO_3^-$, whereas roots of others rely on shoots for organic nitrogen. The evidence for this comes from two kinds of studies. In one, plants are allowed to absorb $NO_3^-$, and then their stems are severed and xylem sap is collected and analyzed to see whether it contains $NO_3^-$ or reduced nitrogen compounds. In the second, the activity of nitrate reductase in roots and shoots (mainly leaves) is compared. The reasonable assumption here is that most $NO_3^-$ reduction occurs at the site (roots or shoots) at which the most nitrate-reductase activity occurs. Each kind of study has problems, but when both agree we are inclined to believe that the results reflect the situation that occurs in nature. Andrews (1986) summarized results for more than 30 species and described some of the factors that affect the location of nitrate reduction (also see Van Beusichem et al., 1987).

Figure 14-7 shows results of one study in which nitrogenous substances were analyzed in the xylem transport stream of several herbaceous species during vegetative growth; plants had their roots in sand watered with a nitrate-containing nutrient solution. (The legumes did not contain root nodules.) None of these plants translocated detectable amounts of $NH_4^+$ to the shoots, but some transported large quantities of organic nitrogen compounds derived from $NH_4^+$, especially amino acids and amides. The cocklebur (*Xanthium strumarium*) and white lupine (*Lupinus albus*) represent extremes among those in this study. Cocklebur roots reduced almost no $NO_3^-$, so they apparently depend upon amino acids translocated in the phloem from the leaves. In the lupine, nearly all $NO_3^-$ was absorbed and converted into amino acids and amides in the roots. Similar results with these two species have been obtained by others (Andrews, 1986). Most conifers and deciduous trees investigated behave like the lupine and translocate little $NO_3^-$ to the shoots; the shoots of these species are usually provided a diet of organic nitrogen. Much more research is needed with plants of different ages and with conifers, other forest trees, and herbaceous plants grown in nature, in which cases the fungal hyphae in mycorrhizae might contribute organic nitrogen compounds to the root cells. Furthermore, the

## Many Grasses Also Support Nitrogen Fixation

Although it was long believed that no nitrogen fixation occurs in cereal grains or other grasses, an assay method developed in the 1960s has helped prove that this is not entirely true. The method depends on the ability of all nitrogen-fixing organisms to reduce acetylene to ethylene, which then can be rapidly measured by gas chromatography. Sealed soil samples or aqueous samples from lakes, ponds, and streams can be collected in the field and assayed later. Positive results with the acetylene reduction method usually correlated well with $^{15}N_2$-fixing ability measured by more laborious and less sensitive methods (use of mass or emission spectrometers).

A few reports of limited fixation by sugarcane and other tropical grasses in the 1960s were verified and widely extended in the 1970s with the new assay. Bacteria living on or near root cells of numerous species show activity (Fig. 14-6). These bacteria reside in a transition zone between soil and root often called the **rhizosphere** (Curl and Truelove, 1986). Occasionally such bacteria even enter the roots, as evidenced in part by the ability of surface-sterilized roots to fix nitrogen. The most prevalent bacteria usually identified with active grass roots are members of the genus *Azospirillum*, although reasonably well-defined associations of sugarcane with *Beijerinckia*, of *Paspalum notatum* (another tropical grass) with *Azotobacter paspali*, of certain wheat cultivars with *Bacillus*, and of rice with *Achromobacter*-like organisms also exist (Stewart, 1982). Even when such bacteria occur only in the rhizosphere, there is loose mutualism involved because some of the nitrogen fixed is absorbed by the roots, and carbohydrates released by the roots nourish the bacteria.

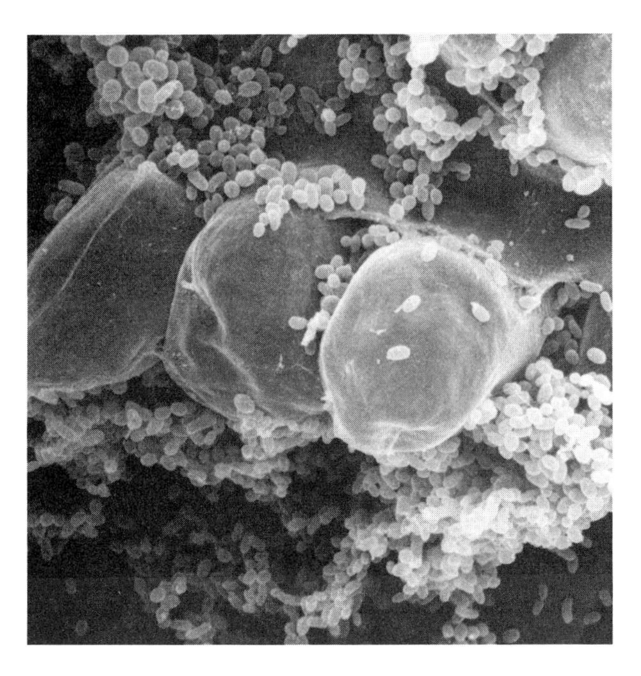

**Figure 14-6** Scanning electron micrograph of part of a sorghum (*Sorghum bicolor*) root covered with nitrogen-fixing (*Azospirillum brasilense*) bacterial cells. Three large epidermal cells were being sloughed off into the rhizosphere. (From Schank et al., 1983; photo by Howard Berg, used by permission.)

There is still disagreement as to how much nitrogen fixation is supported by grasses, because it is impossible to measure rates for whole crops in the field over an entire growing season. Furthermore, it has recently become clear that the acetylene reduction technique so often used should be replaced with $^{15}N_2$-fixation measurements, at least for field studies, because the acetylene reduction method overestimates true nitrogen fixation (Boddey, 1987). Fixation rates with even the most efficient tropical grasses are certainly much less than with legumes and other species that have root nodules harboring nitrogen fixers in much more ideal environments (Van Berkum and Bohlool, 1980). Furthermore, fixation rates with cereal grain crops in the United States are much less than those with tropical grasses. Unless soils are anaerobic (as when moist or wet) for many hours, rates with these cereal grains are barely detectable, perhaps because $O_2$ inactivates nitrogenase. Nevertheless, field inoculations with *Azospirillum* species have reportedly increased dry-matter yields of various crops in Israel, India, the Bahamas, Australia, and Florida (Schank et al., 1983), although the review by Boddey and Dobereiner (1988) indicates that in many cases the *Azospirillum* bacteria contribute unknown plant-growth substances that cause much of the dry-matter increase. Currently, research to increase fixation rates with various grasses is being done. It might be possible to alter aeration properties and genetically alter both bacteria and grasses so that certain native grasses and cereal grains can fix greater fractions of the nitrogen they need. Also, the current genetic variability in grasses presents a possible method of breeding more useful cultivars or varieties. For example, one of four commercial sugarcane varieties in Brazil supported significantly more nitrogen fixation than the other three (Lima et al., 1987). Such cultivars are important to Brazil, the world's leading producer of sugarcane, not only for sugar production, but also to produce ethanol to fuel the more than two million cars that burn 95 percent ethanol. (Note that quite recently, researchers in England, Australia, and China have even induced rice, wheat, and oilseed rape to form small and sparse root nodules with rhizobium bacteria. Results are summarized in *Science*, Research News, Nov. 16, 1990.)

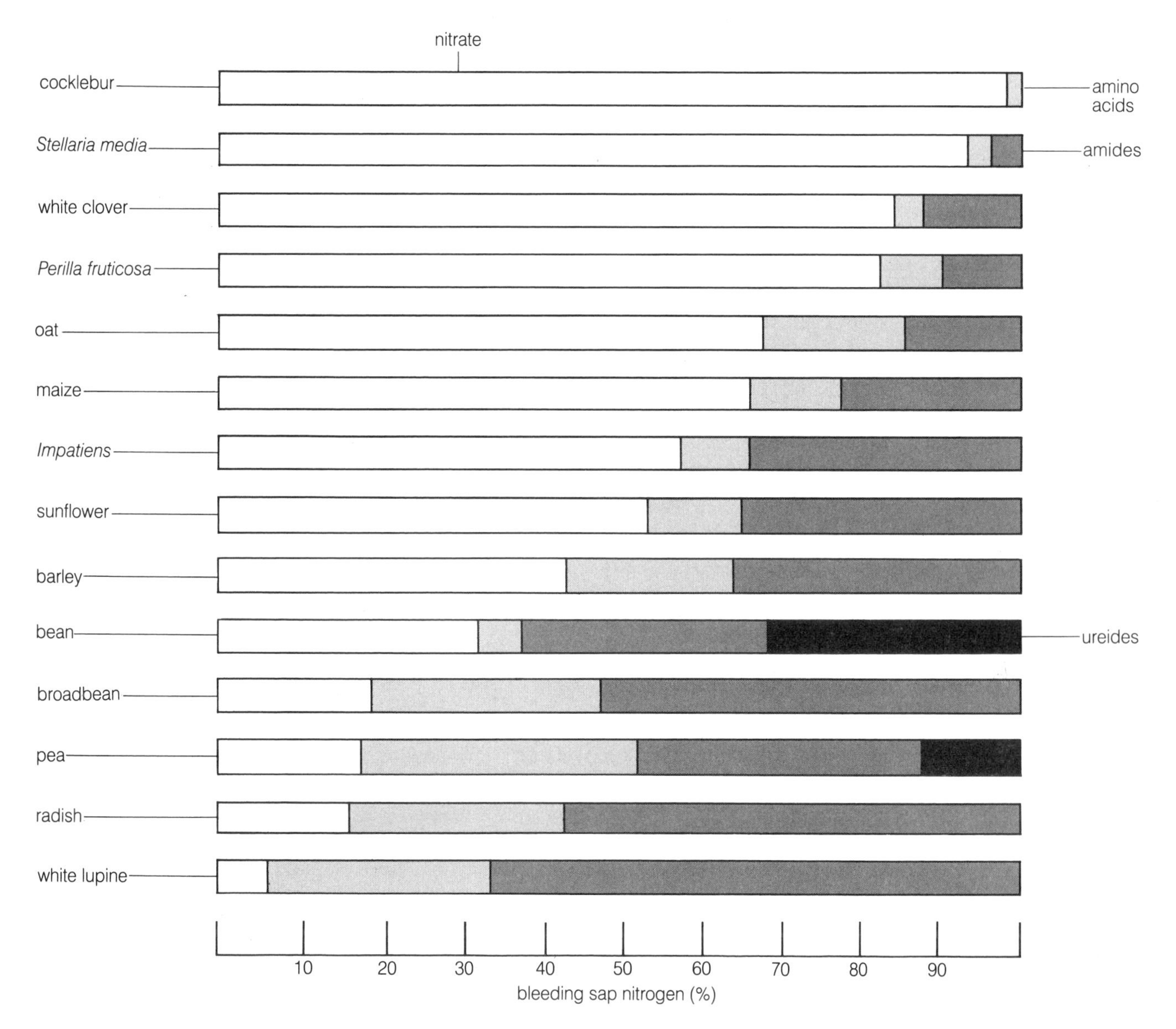


**Figure 14-7** Relative amounts of nitrogen compounds in the xylem sap of various species. Plants were grown with their roots in sterile sand watered with a sterile nutrient solution containing 140 mg/L nitrogen as nitrate (10 mM $NO_3^-$); then the stems were severed to collect xylem sap from the cut stumps. Species at the top of the figure transport primarily nitrate; those at the bottom, mainly amides and amino acids. Two legumes also transport ureides, especially citrulline. (From Pate, 1973.)

kinds of compounds transported to legume shoots are altered when root nodules are present and when nitrogen is fixed into asparagine or ureides (Winkler et al., 1988; Pate, 1989). We have much more to learn about what really occurs in nature.

The relative amounts of $NO_3^-$ and organic nitrogen in the xylem depend upon environmental conditions. Even plants that normally do not translocate much $NO_3^-$ do so if provided with excessive amounts of it in the soil, or if the roots are cold (Andrews, 1986). Under these conditions, reduction of $NO_3^-$ in the roots cannot keep pace with transport to the shoots. Reduction then occurs in leaves and stems, especially during sunny days.

## The Processes of Nitrate Reduction

The overall process of reduction of $NO_3^-$ to $NH_4^+$ is an energy-dependent one summarized in Reaction 14.2:

$$NO_3^- + 8\text{ electrons} + 10\,H^+ \rightarrow NH_4^+ + 3\,H_2O \qquad (14.2)$$

The oxidation number of nitrogen changes from +5 to −3.

Sources of the eight electrons (and 10 $H^+$) required for this reduction process will be described shortly, but note that two more $H^+$ than electrons are used in the reaction. This use of $H^+$ causes the cell's *p*H to rise. Continued *p*H rises would be lethal to plants if they had no way to replace such $H^+$ ions (Raven and Smith, 1976; Raven, 1988). About half are neutralized when $NH_4^+$ is subsequently converted into protein because that process releases one $H^+$ for each nitrogen atom involved. Neutralization of the other half occurs by various mechanisms (Raven, 1988). In algae, aquatic angiosperms, and roots, $H^+$ ions are absorbed from the surrounding medium or $OH^-$ ions are excreted into the medium, either process helping to maintain constant cellular *p*H. In shoots, replacement of $H^+$ ions occurs by production of malic acid and other acids from sugars or starch. This is part of a biochemical *pH stat* (Davies, 1986), and it occurs because PEP carboxylase much more effectively fixes PEP and $HCO_3^-$ into oxaloacetic acid, the precursor of malic acid, as the *p*H rises. It is also believed that some of the malate anions produced during the neutralization process are transported in the phloem back to the roots as $K^+$ salts. This transport prevents the osmotic potential from becoming too negative as organic acid salts accumulate in the shoot cells. In roots, the transported malate is decarboxylated to pyruvate and $CO_2$, whereas the $K^+$ is recirculated with $NO_3^-$ and other anions back to the shoots through the xylem.

Nitrate reduction occurs in two distinct reactions catalyzed by different enzymes. The first reaction is catalyzed by **nitrate reductase (NR)**, an enzyme that transfers two electrons from NADH or, in a few species, from NADPH. Nitrite ($NO_2^-$), $NAD^+$ (or $NADP^+$), and $H_2O$ are the products:

$$NO_3^- + NADH + H^+ \xrightarrow{NR} NO_2^- + NAD^+ + H_2O \quad (14.3)$$

This reaction occurs in the cytosol outside any organelle. NR is composed of two identical subunit polypeptide chains (Campbell, 1988; Solomonson and Barber, 1990), each coded by a nuclear gene. It contains FAD, iron in a heme prosthetic group, and molybdenum, all of which become consecutively reduced and oxidized as electrons are transported from NADH to the nitrogen atom in $NO_3^-$.

NR has been studied intensively because its activity often controls the rate of protein synthesis in plants that absorb $NO_3^-$ as the major nitrogen source. The activity of NR is affected by several factors. One is its rate of synthesis, and another is its rate of degradation by protein-digesting enzymes (proteinases, described in Section 14.5). Apparently, NR is continuously synthesized and degraded, so these processes control activity by regulating how much NR a cell has. Activity is also affected by both inhibitors and activators within the cell. Although it is difficult to separate effects of these factors, abundant levels of $NO_3^-$ in the cytosol clearly increase the activity of NR, largely because of faster synthesis of the enzyme (Rajasekhar and Oelmüller, 1987; Campbell, 1988). This is a case of **enzyme induction**—enhanced formation of an enzyme by a particular chemical. Enzyme induction is widespread in microorganisms, but fewer examples in plants or mammals are known. Induction of NR by $NO_3^-$ is an excellent example of substrate induction because the inducer is also the substrate for the enzyme. Induction of NR is widespread in various parts of various plants. The cells involved apparently conserve energy by not synthesizing NR or the messenger RNA that codes for it until $NO_3^-$ is available; then the enzyme begins to appear within a few hours. How $NO_3^-$ activates the gene needed for NR formation is being investigated.

In leaves and stems, light also increases NR activity when $NO_3^-$ is available. Therefore, there is usually a diurnal (day-night) rhythm in NR activity. How light causes this rhythm is still unclear and seems to vary with the plant and its stage of development. First, in green tissues, light activates one or both photosystems of photosynthesis. This increases (perhaps by providing ATP) transport of stored $NO_3^-$ from the vacuole to the cytosol, where induction of NR then occurs (Granstedt and Huffaker, 1982). Second, light activates the phytochrome system (Chapter 20), which somehow activates the gene that codes for the mRNA that codes for the NR enzyme (Rajasekhar et al., 1988). Finally, light acting through photosynthesis promotes activity of NR because it increases the carbohydrate supply, and NADH necessary for nitrate reduction is produced from these carbohydrates when they are respired (Aslam and Huffaker, 1984). The overall plant response to these light effects is an increase in NR activity and a larger increase in the rate of $NO_3^-$ reduction to ammonium after sunrise, especially in the shoot.

## Reduction of Nitrite to Ammonium Ions

The second reaction of the overall process of nitrate reduction involves conversion of nitrite to $NH_4^+$. Nitrite arising in the cytosol from nitrate-reductase action is transported into chloroplasts in leaves or into proplastids in roots, where subsequent reduction to $NH_4^+$ occurs, catalyzed by **nitrite reductase**. In leaves, reduction of $NO_2^-$ to $NH_4^+$ requires six electrons derived from $H_2O$ by the chloroplast noncyclic electron transport system (Reaction 14.4). During this electron transfer, light drives electron transport from $H_2O$ to ferredoxin (denoted by Fd); then reduced ferredoxin provides the six electrons used to reduce $NO_2^-$ to $NH_4^+$. It is at this step (Reaction 14.5) that net use of two $H^+$ occurs during the

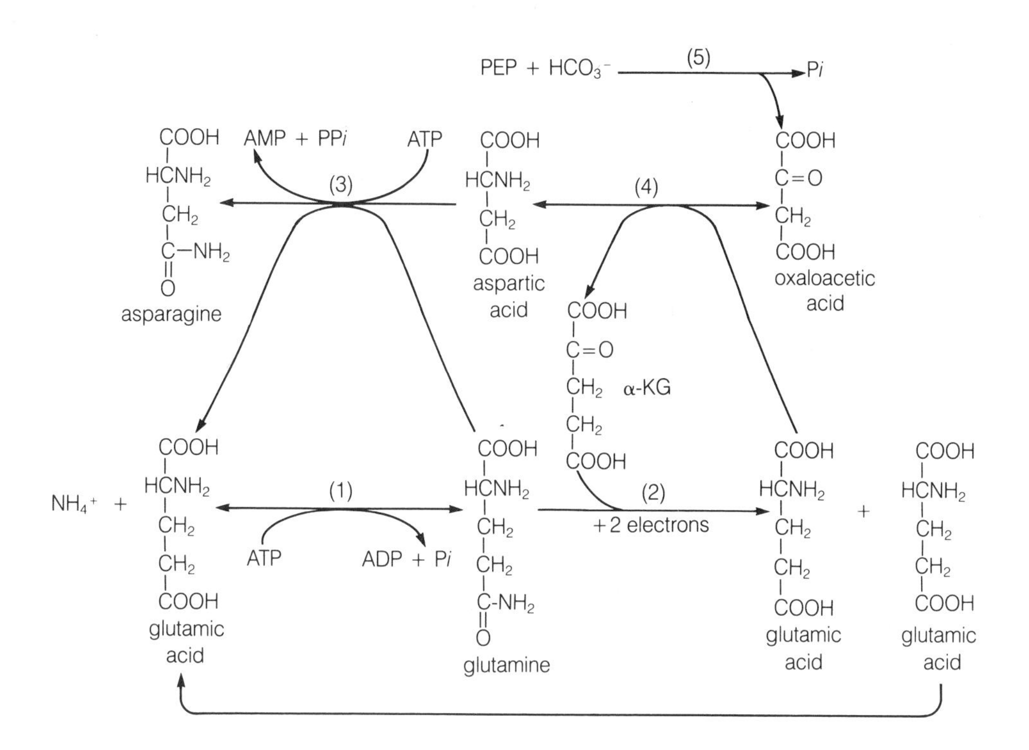


**Figure 14-8** Conversion of ammonium (lower left) into important organic compounds.

overall process of nitrate reduction to $NH_4^+$ (see Reaction 14.6).

$$3\,H_2O + 6\,Fd\,(Fe^{3+}) + light \rightarrow 1.5\,O_2 + 6\,H^+ + 6\,Fd\,(Fe^{2+}) \quad (14.4)$$

$$NO_2^- + 6\,Fd\,(Fe^{2+}) + 8\,H^+ \rightarrow NH_4^+ + 6\,Fd\,(Fe^{3+}) + 2\,H_2O \quad (14.5)$$

$$NO_2^- + 3\,H_2O + 2\,H^+ + light \rightarrow NH_4^+ + 1.5\,O_2 + 2\,H_2O \quad (14.6)$$

Reactions 14.4 and 14.6 show that three $H_2O$ molecules are required to provide the necessary six electrons in reduced ferredoxin (two electrons per $H_2O$ split by light energy), even though two $H_2O$ also appear as products of this overall reaction.

Although reduced ferredoxin is the normal donor of electrons to nitrite reductase in leaves, the reducing substance in roots is still uncertain. When nitrite reductase is studied *in vitro*, it only weakly accepts electrons from NADH, NADPH, or naturally occurring flavin compounds such as $FADH_2$. Reduced ferredoxin will provide electrons for isolated nitrite reductases in roots, but neither proplastids nor other parts of root cells have detectable amounts of ferredoxin. Although we are still unsure how roots reduce $NO_2^-$ to $NH_4^+$, it is clear that a carbohydrate supply from the leaves is necessary and that most reductions occur in proplastids (Bowsher et al., 1989). Furthermore, there is indirect evidence that NADPH derived from the pentose-phosphate respiratory pathway is the active reducing substance. Surprisingly, nitrite reductase is also inducible, as is nitrate reductase, by nitrate. In barley, addition of either nitrate or nitrite to a nutrient solution made available to the roots leads to induction of both nitrate reductase and nitrite reductase in the leaves (Aslam and Huffaker, 1989). However, the experiments indicated that nitrate is the active ion, and that to be effective nitrite had first to be oxidized to nitrate.

## Conversion of Ammonium into Organic Compounds

Whether $NH_4^+$ is absorbed directly from the soil or produced by energy-dependent nitrogen fixation or $NO_3^-$ reduction, it does not accumulate anywhere in the plant. Ammonium is, in fact, quite toxic (Givan, 1979), perhaps because it inhibits ATP formation in both chloroplasts and mitochondria by acting as an uncoupling agent. Except for traces of $NH_4^+$ lost to the atmosphere as volatile $NH_3$ (see Fig. 14-1 and Farquhar et al., 1980), all $NH_4^+$ seems to be converted first into the amide group of glutamine. This conversion and other reactions leading to glutamic acid, aspartic acid, and asparagine are summarized in Fig. 14-8 and described briefly below.

Glutamine is formed by addition of an $NH_2$ group from $NH_4^+$ to the carboxyl group farthest from the alpha carbon of glutamic acid. An amide bond is thus formed (Fig. 14-8, reaction 1), and glutamine is one of two especially important plant amides (for structures, see Fig. 9-2). The necessary enzyme is *glutamine synthetase*. Hydrolysis of ATP to ADP and *Pi* is essential to drive the reaction forward. Because this reaction requires glutamic acid as a reactant, there must be some mechanism to provide it; this is accomplished by reaction 2, catalyzed by *glutamate synthase*. Glutamate synthase transfers the amide group of glutamine to the carbonyl carbon of α-ketoglutaric acid, thereby forming two molecules of glutamic acid. This process requires a reducing agent capable of donating two electrons, which is ferredoxin (two molecules) in chloroplasts and either NADH or NADPH in proplastids of nonphotosynthetic cells. One of the two glutamates formed in reaction 2 is essential to maintain reaction 1, but the other can be converted directly into proteins, chlorophyll, nucleic acids, and so on. Furthermore, some of the glutamate is transported to other tissues, where it is used similarly in synthetic processes.[1]

Besides forming glutamate, glutamine can donate its amide group to aspartic acid to form asparagine, the second important plant amide (Fig. 14-8, reaction 3). This reaction requires *asparagine synthetase*, and the irreversible hydrolysis of ATP to AMP and PP*i* provides the energy to drive it forward. Interestingly, asparagine synthetase is strongly activated by $Cl^-$, which probably helps explain the uncertain role of chlorine in plants (Huber and Streeter, 1985). A continuous supply of aspartic acid must be present to maintain asparagine synthesis. The nitrogen in aspartate can come from glutamate, but its four carbons probably arise from oxaloacetate (Fig. 14-8, reaction 4). Oxaloacetate, in turn, is formed from PEP and $HCO_3^-$ by action of PEP carboxylase (reaction 5).

Perhaps because of its relatively high ratio of nitrogen to carbon compared with most other compounds, glutamine has evolved as an important storage form of nitrogen in most species. Storage organs such as potato tubers and the roots of beet, carrot, radish, and turnip are especially rich in this amide. In mature leaves, glutamine is formed from glutamic acid and $NH_4^+$ produced when protein degradation begins to increase. It is then transported via the phloem to younger leaves or to roots, flowers, seeds, or fruits, where its nitrogen is reused. Finally, glutamine is incorporated directly into proteins in all cells as one of the 20 amino acids, even though technically it is an amide as well as an amino acid (Chapter 9). The other amide, asparagine, performs essentially the same functions as glutamine but less commonly, especially in legumes of temperate origin, in which it is unusually abundant (Ta and Joy, 1985; Sieciechowicz et al., 1988).

[1]Whether small amounts of $NH_4^+$ can be added to α-ketoglutaric acid to form glutamic acid has been under investigation for nearly 40 years (see reviews by Oaks and Hirel, 1985; Joy, 1988; and Rhodes et al., 1989). An enzyme called *glutamate dehydrogenase* can use NADH (or, for some isozymes of the enzyme, NADPH) to reduce α-ketoglutarate to glutamate and release $H_2O$ and $NAD^+$ (or $NADP^+$). Glutamate dehydrogenases exist in chloroplasts and mitochondria, in which their main role is (as the name implies) to dehydrogenate glutamate. This produces α-ketoglutarate, which is used in the Krebs cycle of mitochondria or is involved in the photorespiratory nitrogen cycle described in Section 14.4. Glutamate dehydrogenase probably functions especially rapidly when proteins are being hydrolyzed and glutamate is being released, as, for example, during seed germination and early seedling development and during senescence of leaves. Nevertheless, Yamaya and Oaks (1987) proposed that in leaves this enzyme also contributes to the recapture of some of the $NH_4^+$ lost in photorespiration by adding it to α-ketoglutarate.

## Transamination

When $NH_4^+$ containing heavy isotopic $^{15}N$ is fed to plants or to excised plant parts, first glutamine, then glutamic, and then aspartic acid becomes most rapidly labeled with $^{15}N$, as detected with a mass spectrograph that measures the amounts of $^{15}N$ in each compound. Subsequently, the $^{15}N$ appears in other amino acids. The reason for this labeling sequence is that after glutamate formation (Fig. 14-8, reaction 2), glutamate transfers its amino group directly to a variety of α-keto acids in several reversible **transamination** reactions. An important example of transamination occurs between glutamate and oxaloacetate, producing α-ketoglutarate and aspartate (see Fig. 14-8, reaction 4). The important physiological point about all transamination reactions is that they always accomplish transfer of nitrogen from one compound to another in most organs and cells of most species (Giovanelli, 1980). *Biochemically, all transaminations involve freely reversible donation of an alpha-amino group from one amino acid to the alpha-keto group of an alpha-keto acid, accompanied by formation of a new amino acid and a new alpha-keto acid*. All transaminase enzymes (or aminotransferases) require *pyridoxal phosphate* (vitamin $B_1$) as a prosthetic group, an important reason that that vitamin is essential for life.

The aspartate formed by transamination can transfer its amino group to other α-keto acids to form different amino acids by transamination reactions. Transfer to pyruvate, for example, yields alanine. Alanine and other amino acids can then transfer their amino groups, too, so numerous amino acids are formed by transamination. Reactions by which amino acids other than aspartate are formed are described in biochemical reviews (Miflin, 1980; Lea and Joy, 1983; Bray, 1983). We emphasize, however, that the amino groups of nearly all amino acids in both plants and animals probably passed through glutamine and glutamate on the route from $NO_3^-$, $NH_4^+$, or atmospheric $N_2$. Furthermore, plants synthesize amino acids that animals do not. Besides the 20 amino acids (including the two amides) common in proteins, hundreds of nonprotein amino acids have been identified in the plant kingdom. The functions of such amino acids in the plants containing them are

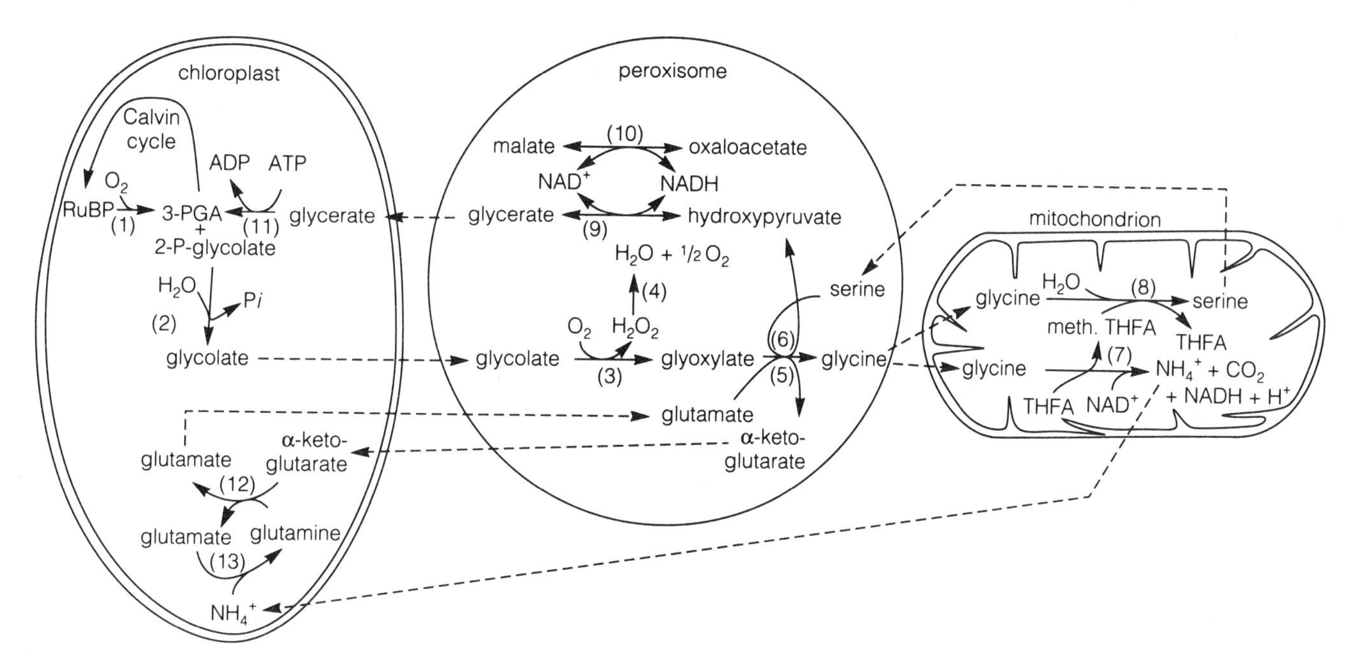


**Figure 14-9** The photorespiratory nitrogen cycle. Chloroplasts, peroxisomes, and mitochondria cooperate to trap ammonium ions lost in the mitochondria during photorespiration. Solid lines represent enzyme-catalyzed reactions; dashed lines represent movement from one organelle to another through the cytosol. Enzymes that catalyze the numbered reactions are (1) rubisco; (2) 2-phosphoglycolate phosphatase; (3) glycolate oxidase; (4) catalase; (5) glutamate (alanine)-glyoxylate aminotransferase; (6) serine (asparagine)-glyoxylate aminotransferase; (7) glycine decarboxylase complex; (8) serine hydroxymethyltransferase; (9) hydroxypyruvate reductase; (10) NAD malate dehydrogenase; (11) glycerate kinase; (12) glutamate synthase; and (13) glutamine synthetase.

mostly unknown, although some are toxic to insects, mammals, and other plants, suggesting ecological defense roles (Bell, 1980, 1981).

## 14.4 The Photorespiratory Nitrogen Cycle

In our description of photorespiration (Section 11.4), we mentioned that two molecules of glycine (each derived from phosphoglycolate when $O_2$ is fixed by rubisco) are converted to one molecule of the three-carbon amino acid serine, one molecule of $CO_2$, and one molecule of $NH_4^+$. So for every $CO_2$ released during photorespiration, an equivalent molar amount of $NH_4^+$ is released from glycine, and this $NH_4^+$ must be recaptured into organic combination. Let us investigate some details of these processes, now commonly called the **photorespiratory nitrogen cycle** (Keys et al., 1978; Givan et al., 1988). The cycle is diagramed in Figure 14-9.

Three organelles are involved (chloroplasts, peroxisomes, and mitochondria), and there is molecular traffic to and from them via the cytosol. As described in Section 11.4 and depicted in the upper left of Figure 14-9, fixation of $O_2$ by rubisco in chloroplasts causes formation of 2-phosphoglycolate, which is then dephosphorylated, and the glycolate product is transported into a peroxisome, where it is oxidized by $O_2$ to glyoxylate. Two peroxisomal transaminase enzymes are then capable of converting glyoxylate to glycine, one enzyme of which uses glutamate (or alanine) and the other commonly serine. In Figure 14-9 (center), reactions with both enzymes are shown.

Glycine now leaves the peroxisome and enters a mitochondrion. Some glycines are oxidized there, each releasing four compounds: the $CO_2$ of photorespiration, $NH_4^+$, NADH + $H^+$ made by sending electrons from glycine to $NAD^+$, and *$N^5$,$N^{10}$-methylene tetrahydrofolic acid* (abbreviated meth. THFA in Fig. 14-9). Tetrahydrofolic acid accepts the methylene group from the α-amino carbon of glycine in this reaction, all steps of which are catalyzed by a multienzyme complex called *glycine decarboxylase* (Walker and Oliver, 1986). Glycines that are not attacked in mitochondria by glycine decarboxylase can accept $H_2O$ and the methylene group of methylene-THFA to form serine, releasing free THFA. This reaction is catalyzed by *serine hydroxymethyltransferase*. Overall, therefore, mitochondria convert two glycines to one $CO_2$, one $NH_4^+$, one NADH, and one serine.

The serine produced from glycine moves back to a peroxisome, where it can donate its amino group in a

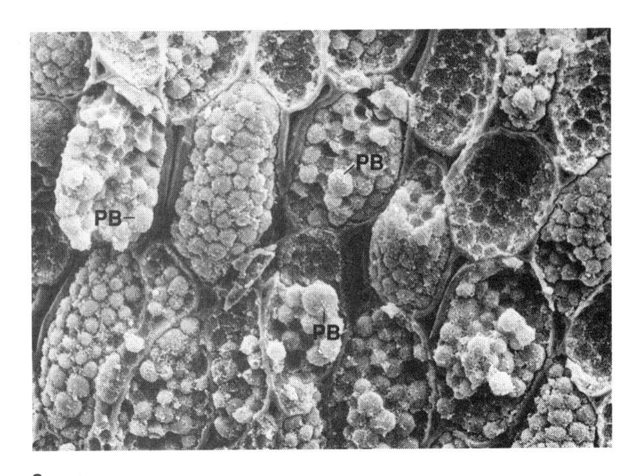


a

b

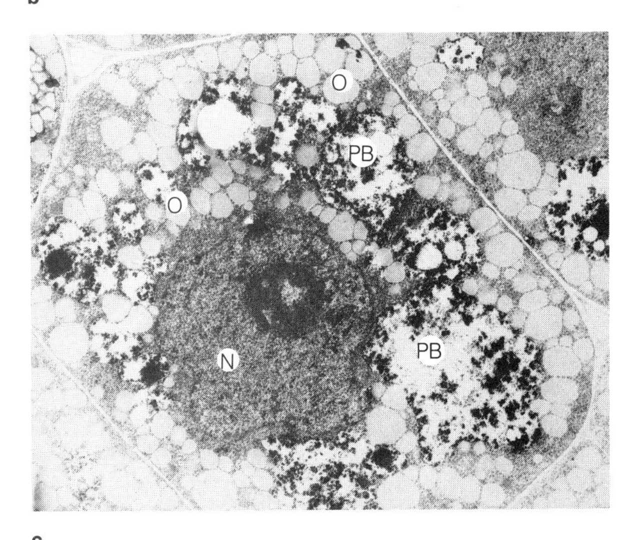


c

**Figure 14-10** (**a**) Scanning electron micrograph of protein bodies (PB) in cells of a seed of *Lupinus albus*. Note that here the PBs nearly fill each cell. (Courtesy Jean Noël Hallet; see Le Gal and Rey, 1986.) (**b**) Electron micrograph of a cortex cell in the radicle of an ungerminated Grand Rapids lettuce (*Lactuca sativa*) seed. The numerous unstained (light grey) structures are oleosomes (O) in which oils and fats are stored, whereas the larger darkly stained structures are protein bodies (PB). White areas in some of the protein bodies identify sites where most of the reserve phosphorus is stored as phytins. Phytins are calcium, magnesium, and potassium salts of phytic acid, myoinositol hexaphosphoric acid. (Courtesy Nicholas Carpita.) (**c**) Digestion of reserve protein in radicle cortex cells of recently germinated lettuce seeds. Protein bodies (PB) surrounding the nucleus were beginning to fuse to form the large vacuole, and most of the protein in them has disappeared. Numerous oieosomes (O) are still visible. (Courtesy Nicholas Carpita.)

transamination with glyoxylate, forming glycine and *hydroxypyruvate* ($CH_2OH-CO-COO^-$). The hydroxypyruvate is then reduced to glycerate in the peroxisome. This glycerate moves back to the chloroplast, where it is converted with ATP and a kinase back to 3-PGA to help keep the Calvin cycle going.

The $NH_4^+$ released by glycine decarboxylase in the mitochondrion (Fig. 14-9, extreme right) moves back to a chloroplast, where it is used to form glutamine by glutamine synthetase. Overall, the reactions of the photorespiratory nitrogen cycle provide a means for recapture of the $NH_4^+$ that is temporarily lost by the oxidative decarboxylation of glycine. Were it not for this cycle, all plants with photorespiration might be poisoned by free $NH_4^+$.

## 14.5 Nitrogen Transformations During Plant Development

### Nitrogen Metabolism of Germinating Seeds

In storage cells in all kinds of seeds, the reserve proteins are deposited in membrane-bound structures called **protein bodies** (Lott, 1980; Pernollet, 1978; Weber and Neumann, 1980). Figure 14-10a is a scanning electron micrograph of protein bodies in a lupine seed, illustrating their three-dimensional appearance. Figure 14-10b is a transmission electron micrograph of these (darkly-stained) bodies in cells of a lettuce seed. In lettuce, protein bodies and oleosomes (oil bodies) are almost the only visible structures. Protein bodies are not pure protein but also contain much of the seed's reserves of phosphate, magnesium, and calcium. The phosphate is esterified to each of the six hydroxyl groups of a six-carbon sugar alcohol called myoinositol (Fig. 14-11). The product of this esterification is called phytic acid, and ionization of $H^+$ from the phosphate groups allows $Mg^{2+}$, $Ca^{2+}$, $Zn^{2+}$, and probably $K^+$ to form salts collec-

myoinositol

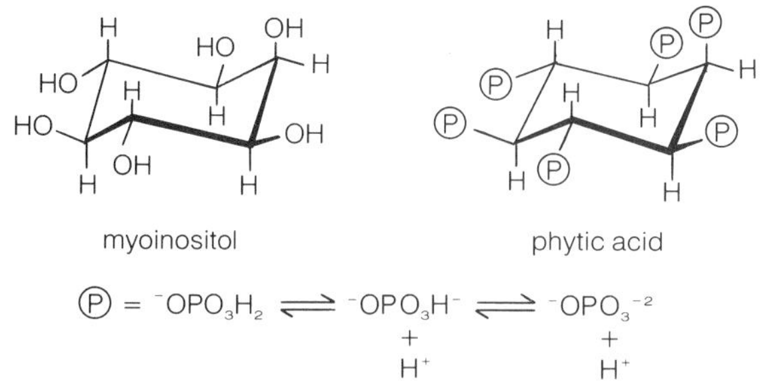

phytic acid

$$\text{Ⓟ} = {}^{-}OPO_3H_2 \rightleftharpoons {}^{-}OPO_3H^{-} + H^{+} \rightleftharpoons {}^{-}OPO_3^{-2} + H^{+}$$

**Figure 14-11** Structures of myoinositol (left) and its hexaphosphate, phytic acid.

tively called **phytin**, or sometimes **phytates** (Maga, 1982; Oberleas, 1983; Raboy, 1990). Phytin is usually bound to proteins in protein bodies.

Imbibition of water by a dry seed sets off a variety of chemical reactions that lead to **germination** (radicle protrusion through the seed coat) and subsequent seedling development. The proteins in protein bodies are hydrolyzed by *proteinases* (*proteases*) and *peptidases* to amino acids and amides (Dalling, 1986; Nielsen, 1988). Note disappearance of most of the protein in the protein bodies of Figure 14-10c, caused by action of proteinases and peptidases. Membranes surrounding disintegrating protein bodies are not destroyed; rather, they fuse to form the tonoplast around the growing central vacuole. Some of the amino acids and amides released during protein hydrolysis in seeds are used to form special new proteins, nucleic acids, and so on in the cells in which hydrolysis occurs, but the majority are translocated through the phloem to growing cells of root and shoot. Release of phosphate and cations from phytin in protein bodies also occurs during or shortly after germination, and some of these ions are also transported to growing regions via the phloem. Soon, the young root system begins to absorb $NO_3^-$ and $NH_4^+$, and nitrogen assimilation for another growing plant starts anew.

## Traffic of Nitrogen Compounds During Vegetative and Reproductive Stages

On a daily basis, measurements on grasses and legumes indicate that in herbaceous plants there is extensive recirculation of nitrogen from roots to leaves and back. This is likely important in directing nitrogen to the strongest sinks to prevent any part of the plant from becoming nitrogen deficient (Millard, 1988). On a seasonal basis the general aspects of protein transformation in various organs of herbaceous plants during maturation were demonstrated over a hundred years ago. Some of these aspects are illustrated in Figure

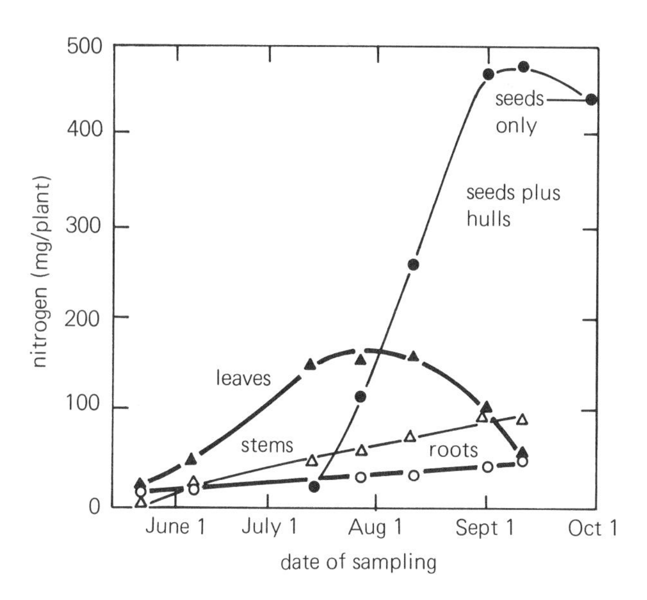


**Figure 14-12** Changes in nitrogen content of various organs of the broadbean (*Vicia faba*) during growth. The extensive accumulation of nitrogen compounds in the fruits (seeds plus hulls) was accompanied by a loss from the leaves and a large uptake from the soil. (Data of Emmerling, 1880.)

14-12, in which changes in amounts of total nitrogen in roots, stems, leaves, and seeds of a broadbean (*Vicia faba*) plant from the seedling stage until maturity are shown.

These changes largely reflect degradation and synthesis of proteins because most of the nitrogen in any plant part is in protein. For leaves, about half of this protein is in chloroplasts. Note that the broadbean leaves actually lost nitrogen during August and September, while the seeds were accumulating it. This transfer of nitrogen compounds from leaves, especially those that are mature, to developing protein bodies in seeds or to fruits via the phloem is typical of both herbaceous and woody plants. The principal organic compounds translocated are glutamine, asparagine, glutamate, and aspartate, and neither $NO_3^-$ nor $NH_4^+$ is translocated in significant amounts in the phloem (Section 8.3). The amount of nitrogen transferred from vegetative organs to fruits of the broadbean was much less than that gained by the fruits during the same period (Fig. 14-12). The additional nitrogen demand of seeds in such legumes is usually satisfied by nitrogen fixed in root nodules during seed development. Nevertheless, the nitrogen demands of protein-rich legume seeds are so great that loss of nitrogen from leaves, especially those near the seeds, is substantial.

Unfortunately for crop production, one of the major leaf proteins that contains this nitrogen is the abundant photosynthetic enzyme rubisco. Photosynthetic activity decreases considerably during fruit and seed production in essentially all crops because rubisco is hydrolyzed by proteinases (Huffaker, 1982). This has

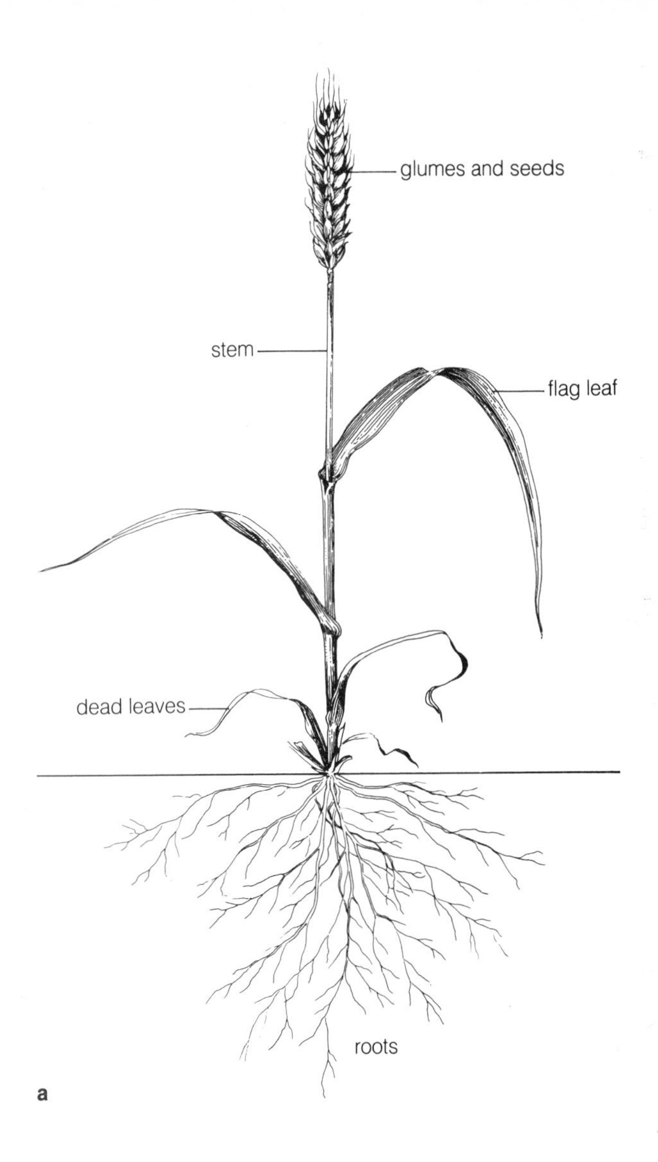


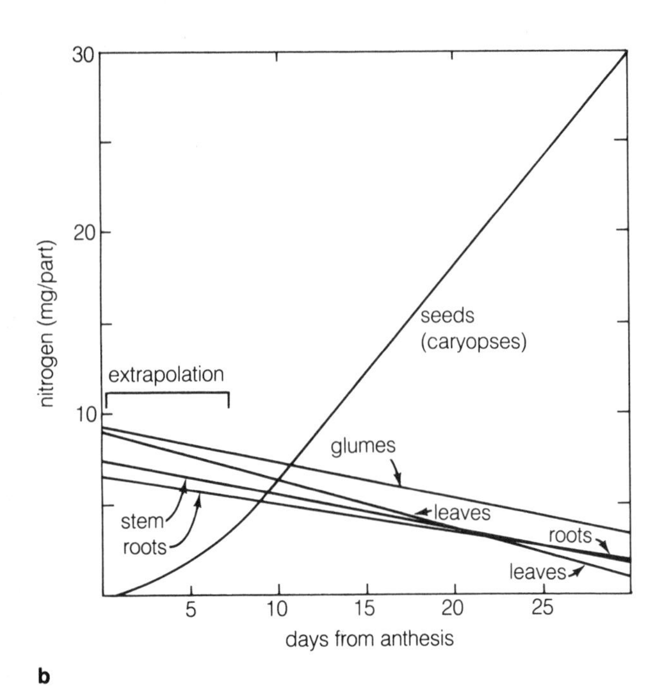


**Figure 14-13** Changes in nitrogen content of various organs during seed development in a wheat plant. Measurements were made beginning about six days after anthesis and continuing until day 30, when seeds were mature. Drawing of plant was made 15 days after anthesis and shows parts analyzed for total nitrogen. On day 15, glumes and flag leaf were green; the leaf just below the flag leaf was yellow at the tip; the second leaf below the flag leaf was half yellow-brown; and other leaves were dead. (From Simpson et al., 1982. Used by permission.)

been called a "self-destruct" phenomenon (at least for leaves), but for plants in soils in which nitrogen is not abundant the hydrolysis of proteins and transport of nitrogen to the seed is essential for seed production. Chlorophyll molecules also disappear from leaves as proteins are degraded, and the nitrogen in these molecules is apparently also transported to reproductive organs.

In cereal grains and many other annuals that do not fix $N_2$, transfer of nitrogen from vegetative parts to seeds is sometimes more extensive than in legumes, even though their seeds contain lower percentages of protein than do legume seeds (Millard, 1988). Wheat leaves, for example, can lose up to 85 percent of their nitrogen (and an equal percentage of phosphate) before they die. Figure 14-13 illustrates nitrogen changes in major parts of wheat plants after flowering begins. This extensive transfer of nitrogen from vegetative organs to flowers and seeds is accompanied by a decrease in the rate of uptake of soil nitrogen as reproductive growth begins. Thus wheat and oats can absorb 90 percent of the nitrogen (and phosphate) needed for maturity before they are half-grown. Again, transport of nitrogen from vegetative organs occurs partly at the expense of rubisco degradation. This degradation is more of a growth limitation in C-3 plants than C-4 plants because C-4 plants contain only about 10 percent as much of this enzyme as C-3 plants (Millard, 1988). Remember that there is essentially no rubisco in mesophyll cells of C-4 plants (Section 11.4).

In perennial herbaceous plants, much of the nitrogen and other elements that are mobile in the phloem moves into the crown and roots after seed demands are satisfied. As a result, these elements are available for the next season's growth, and the decay of dead plant parts returns less to the soil than otherwise would have occurred. We know less about nitrogen relations in woody perennials, but fruits and seeds again are strong sinks for nitrogen. How much of this nitrogen comes directly from the soil via the xylem and how much comes from mature leaves via the phloem is unknown. However, fruits and seeds always have low rates of tran-

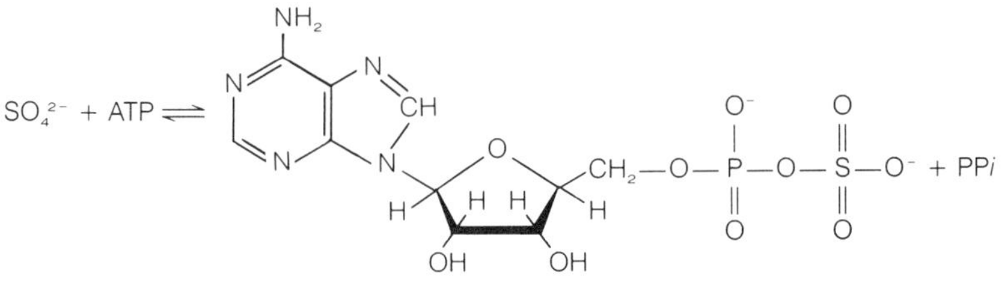

$$SO_4^{2-} + ATP \rightleftharpoons \text{adenosine-5'-phosphosulfate (APS)} + PPi$$

adenosine-5′-phosphosulfate (APS)

$$PPi + H_2O \longrightarrow 2\,Pi$$

$$APS + XSH \longrightarrow AMP + X\text{—}S\text{—}SO_3^-$$

$$XSSO_3^- + 8\,Fd\,(Fe^{+2}) + 7H^+ \longrightarrow S^{2-} + XSH + 8\,Fd(Fe^{+3}) + 3H_2O$$

**Figure 14-14** Four major reactions in reduction of sulfate to sulfide.

spiration compared with mature leaves, so the xylem must be only a limited supplier of mineral salts to these organs. The total composition of fruits and seeds is much more like that of sieve tubes, suggesting that these organs grow largely on a diet of phloem sap. Much of the nitrogen in this sap comes from the leaves, especially leaves near the fruits. In autumn but before leaf fall, deciduous woody plants translocate some of the nitrogen in their leaves to ray parenchyma cells of xylem and phloem in both stems and roots; this transport also occurs in the phloem. In two species each of *Acer, Salix,* and *Populus,* protein bodies in parenchyma cells of the inner bark accumulated in fall and winter, then disappeared in spring (Wetzel et al., 1989). It was proposed that these protein bodies represent bark storage proteins. For apple trees, estimates suggest that up to half of the nitrogen is lost from senescing leaves by transport to other tissues (Titus and Kang, 1982; also see Millard and Thomson, 1989 and Côté et al., 1989). The nitrogen is stored mainly in reserve proteins until new growth begins in spring; then amino acids, amides, and ureides appear in the xylem on their way to young leaves or flower buds. Were it not for this conservation process, losses of nitrogen during leaf fall would cause productivity of trees typically growing on nitrogen-deficient soils to be even lower.

RNA molecules are also degraded in mature and senescing leaves and in seed storage tissues. Hydrolytic enzymes called **ribonucleases** are responsible for this degradation. These enzymes generally release purine and pyrimidine nucleotides in which the phosphate group is attached either to carbon 3 or to carbon 5 of the ribose unit. The nitrogen in these nucleotides is probably translocated to other organs only after further degradation and rearrangement of the nitrogen into glutamate, aspartate, and their amides. Less is known about DNA breakdown in plants, except that DNA is much more stable and much less abundant than RNA. Even senescent leaves from which all chlorophyll and most of the protein and RNA have disappeared still retain much of their DNA. This DNA remains in the leaf as it is shed, and its nitrogen is cycled back to the soil. No one seems to have studied what happens to the DNA that is broken down when sieve-tube elements and conducting xylem cells lose their nuclei.

## 14.6 Assimilation of Sulfate

Except for small amounts of $SO_2$ absorbed by the shoots of plants growing near smokestacks, $SO_4^{2-}$ absorbed by roots provides the necessary sulfur for plant growth. Just as the reduction of $NO_3^-$ and $CO_2$ are energy-dependent reduction processes, so is the reduction of sulfate to sulfide:

$$SO_4^{2-} + ATP + 8\,\text{electrons} + 8\,H^+ \rightarrow S^{2-} + 4\,H_2O + AMP + PPi \qquad (14.7)$$

Sulfate reduction occurs in both roots and shoots of some species, but most of the sulfur transported in the xylem to the leaves is in nonreduced $SO_4^{2-}$. Some transport back to roots and to other parts of the plant occurs through phloem, and both free $SO_4^{2-}$ and organic sulfur compounds are transported (Bonas et al., 1982). We know little about $SO_4^{2-}$ reduction in tissues without chlorophyll, but most of the reactions apparently are the same as those occurring in leaves. ATP is essential in each case. In leaves, the entire process occurs in chloroplasts (Schiff, 1983). In roots, most or perhaps all of the process occurs in proplastids (Brunold and Suter, 1989).

The first step of $SO_4^{2-}$ assimilation in all cells is reaction of $SO_4^{2-}$ with ATP, producing *adenosine-5′-phosphosulfate (APS)* and pyrophosphate (PP*i*). This step is catalyzed by *ATP sulfurylase*. The PP*i* is rapidly and irreversibly hydrolyzed into two P*i* by a pyrophosphatase enzyme, and then the P*i* can be used in mitochondria or chloroplasts to regenerate ATP. These processes are shown in the first two reactions of Figure 14-14.

The sulfur of APS is reduced in chloroplasts by

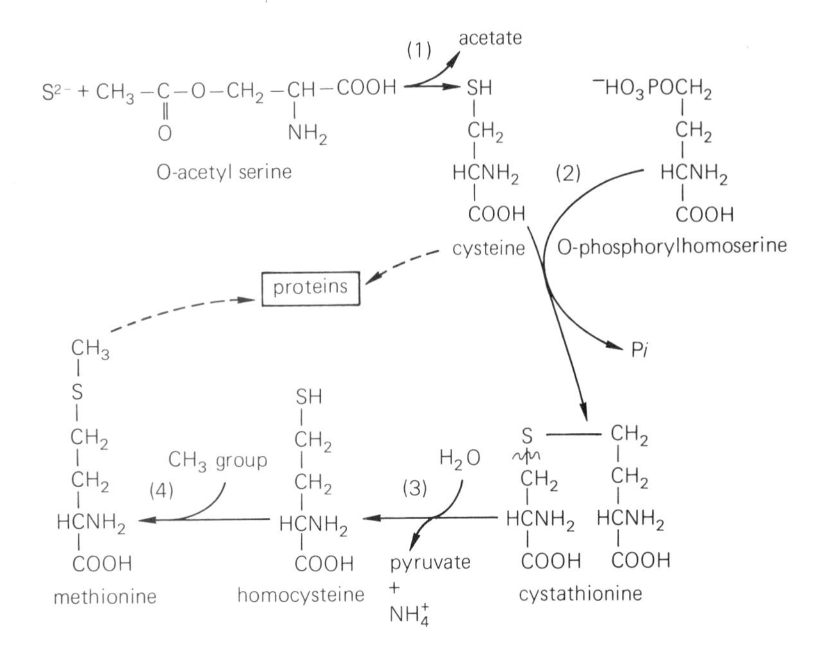


**Figure 14-15** Reactions by which sulfide is converted to cysteine and methionine. Reaction 1, catalyzed by cysteine synthetase, involves replacement of the acetate group in O-acetyl serine by sulfide. Reaction 2, catalyzed by cystathionine synthetase, splits phosphate from O-phosphorylhomoserine and joins the sulfur atom in cysteine to the terminal $CH_2$ group of the homoserine residue. Reaction 3 is catalyzed by cystathionase, an enzyme that hydrolyzes cystathionine between the S and the carbon shown by the wavy line. Pyruvate and $NH_4^+$ are released; these products were formerly part of the cysteine molecule. Homocysteine is converted by methionine synthetase to methionine by receipt of a methyl group in reaction 4. $N^5$-methyltetrahydrofolic acid is the methyl donor for this reaction.

electrons donated from reduced ferredoxin. For proplastids, NADPH represents a reasonable (but unproven) comparable electron donor. The oxidation number of sulfur changes from +6 to −2 during APS reduction, which explains why eight electrons are required (Reaction 14.7). Reduction in chloroplasts is thought to occur as follows (Fig. 14-14): First, the sulfate group of APS is transferred to the sulfur atom of an acceptor molecule by an enzyme called *APS sulfotransferase*. The acceptor molecule has not been identified; likely candidates are **glutathione** (a tripeptide containing glutamate-cysteine-glycine) and a thioredoxin (described in Section 11.5). This acceptor is denoted XSH in Figure 14-14, and after it accepts sulfate from APS it is denoted X-S-$SO_3^-$. Reduction of the sulfur in the $SO_3^-$ group of X-S-$SO_3^-$ by reduced ferredoxin then occurs, producing XSH and free sulfide.[2]

The sulfide (free or bound) resulting from reduction of APS does not accumulate because it is rapidly converted into organic sulfur compounds, especially cysteine and methionine. Reactions by which these two amino acids are formed are shown in Figure 14-15 (Giovanelli et al., 1980). Most of the plant's sulfur (90 percent) is in cysteine or methionine of proteins, but small amounts of cysteine are incorporated into coenzyme A (see Fig. 13-8), and traces of methionine are used to form *S-adenosylmethionine*. One importance of S-adenosylmethionine is that its methyl group can be transferred to help form lignins and pectins of cell walls, flavonoids such as the brightly colored anthocyanins, and chlorophylls; another importance is its role as a precursor of the plant hormone ethylene (Section 18.2). In some species, especially onion, garlic, and cabbage and its close relatives, odiferous **mercaptans** (R-SH)

[2]How the sulfur bound to APS is reduced is controversial. It has been thought since the mid-1970s that this reduction in chloroplasts is rather direct, using electrons donated to APS from reduced ferredoxin. Although reduced ferredoxin is almost surely an electron donor, it is unclear whether the sulfur in APS is the true electron acceptor. In certain well-investigated bacteria and yeasts, APS is not reduced directly but is first converted to **3′,5′-phosphoadenosine-phosphosulfate (PAPS)** with ATP. Then these organisms use two electrons from a reduced thioredoxin to reduce the sulfur in PAPS, releasing free *sulfite* ($SO_3^{2-}$) and *adenosine-3′,5′-bisphosphate*. The sulfur in free sulfite never accumulates but instead is reduced rapidly to the sulfide level of oxidation (valence of −2) as $S^{2-}$, $HS^-$, or $H_2S$) by acceptance of six electrons donated directly from thioredoxin via the enzyme ferredoxin-thioredoxin reductase (see Fig. 11-9).

A report by Schwenn (1989) indicates that in spinach chloroplasts PAPS (not APS) is also the form of sulfur first reduced (by *PAPS reductase*) and that free sulfite is then released. Because chloroplasts from various plant species contain a *ferredoxin-dependent sulfite reductase* that can reduce sulfite using electrons from reduced ferredoxin, Schwenn (1989) proposed that chloroplasts reduce sulfate by the same general processes that occur in bacteria and yeasts and that no XSH (such as indicated in Fig. 14-14) is involved. According to Schwenn's hypothesis, sulfate is converted to APS, APS is converted to PAPS, and then PAPS is reduced via a PAPS reductase that requires reduced ferredoxin and thioredoxin. Finally, free sulfite is released and reduced to the sulfide level of oxidation by gain of six electrons from reduced ferredoxin. This appealing hypothesis for reduction of free sulfite in chloroplasts can be compared to reduction of nitrite in the same organelles, a process that also involves donation of six electrons from reduced ferredoxin. Nevertheless, we are not aware of other work that confirms Schwenn's hypothesis.

such as methyl mercaptan and n-propyl mercaptan, **sulfides** (R-S-R), or **sulfoxides** (general structure below) accumulate (Block, 1985).

$$\begin{array}{c} R—S—R \\ \| \\ O \end{array}$$

Another odiferous compound released in small amounts by leaves of both angiosperms and conifers is $H_2S$ (Grundon and Asher, 1988). Production of $H_2S$ seems energy-wasteful because its formation requires both ATP (in APS formation) and reduced ferredoxin. However, its release begins only when the reduced sulfur (cysteine) supply of the leaf is already adequate—that is, in daylight and when the $SO_4^{2-}$ supply is plentiful—so release might represent a mechanism for maintaining a constant cellular level of cysteine (Rennenberg, 1984). Other control mechanisms of sulfate assimilation by various plants involve inhibition of APS sulfotransferase formation by $H_2S$ or cysteine and inhibition of $SO_4^{2-}$ absorption by cysteine.

Although plants, bacteria, and fungi generally reduce and convert sulfur into cysteine, methionine, and other essential sulfur compounds, mammals cannot. Because of this, we and the animals we eat depend on plants for reduced sulfur, and particularly for the essential amino acids cysteine and methionine. Because we cannot reduce $NO_3^-$, plants are equally essential as providers of organic nitrogen.

# 15

# Lipids and Other Natural Products

In the preceding six chapters we emphasized that plants contain an imposing variety of carbohydrates and of nitrogen and sulfur compounds. Many of the reactions by which these substances are formed have been explained, especially in relation to the importance of light in providing energy to drive the reactions in the shoot system. We have seen that light energy is used to drive reduction of $CO_2$, $NO_3^-$, and $SO_4^{2-}$, processes that humans and other animals cannot accomplish metabolically.

In this chapter we discuss the properties and functions of many other compounds that plants require for growth or survival. Some of these, such as fats and oils, are important food reserves that are deposited in specialized tissues and cells only at certain times in the life cycle. Others, such as waxes and components of cutin and suberin, are protective coats over the plant's exterior or act as water barriers in endodermal and exodermal cells (Section 8.3). Still others help perpetuate the species by facilitating pollination or by defending against other competitive organisms. During the last several years it has been discovered that hundreds of compounds that plants make have ecological roles, opening a new field of scientific endeavor often called **ecological biochemistry** (Harborne, 1988, 1989; Scriber and Ayres, 1988). Such discoveries have explained the functions of many compounds that formerly seemed to be merely waste products of plants.

Besides these compounds, certain plants produce many others, such as rubber, for which no function within them is presently known. Tetrahydrocannabinol, the active compound in marijuana, is another such example. Compounds not required for normal growth and development through metabolic pathways common to all plants are sometimes referred to as **secondary compounds** or **secondary products**. This designation separates them from primary compounds such as sugar phosphates, amino acids and amides, proteins, nucleotides, nucleic acids, chlorophyll, and organic acids, all of which are necessary for life in all plants. The separation is not complete because, for example, a compound such as lignin is considered primary and essential for vascular plants (because of its presence in xylem), but not for algae. In a review, Metcalf (1987) stated that 50,000 to 100,000 secondary plant compounds may exist in the plant kingdom, thousands of which have already been identified.

In a way, plants can be compared to sophisticated organic-chemical laboratories so far as their synthetic abilities are concerned. Modern analytical instruments such as gas chromatographs, high-performance liquid chromatographs, and mass spectrometers now provide valuable tools to separate and identify the compounds plants make. Structures and biosynthesis of hundreds of secondary compounds are summarized in books by Robinson (1980), Vickery and Vickery (1981), and Conn (1981), and in frequent reviews in the journals *Phytochemistry* and *Natural Product Reports*.

We begin by describing the **lipids**, a group of fatty and fatlike substances, rich in carbon and hydrogen, that dissolve in organic solvents such as chloroform, acetone, ethers, certain alcohols, and benzene, but that do not dissolve in water. Among these are the fats and oils, phospholipids and glycolipids, waxes, and many of the components of cutin and suberin.

## 15.1 Fats and Oils

Chemically, fats and oils are very similar compounds, but fats are solids at room temperatures, whereas oils are liquids. Both are composed of long-chain *fatty acids* esterified by their single carboxyl group to a hydroxyl of the three-carbon alcohol *glycerol*. All three hydroxyl groups of glycerol are esterified, so fats and oils are often called **triglycerides**. Except for instances in which an important distinction must be made between fat and

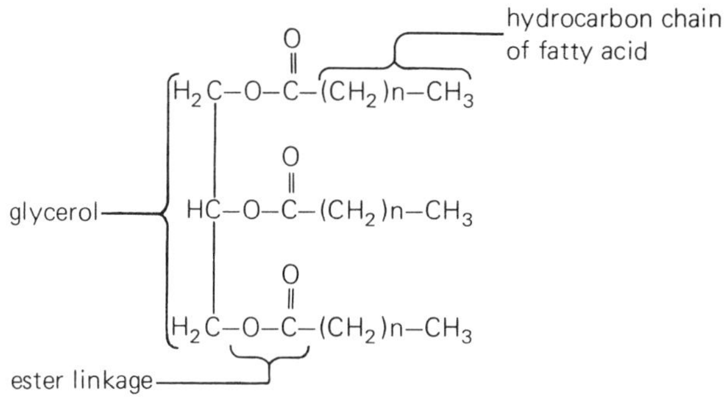

hydrocarbon chain of fatty acid

$$H_2C-O-\overset{O}{\overset{\|}{C}}-(CH_2)_n-CH_3$$

glycerol

$$HC-O-\overset{O}{\overset{\|}{C}}-(CH_2)_n-CH_3$$

$$H_2C-O-\overset{O}{\overset{\|}{C}}-(CH_2)_n-CH_3$$

ester linkage

**Figure 15-1** The general structure of a fat or oil, both of which are triglycerides.

oil triglycerides, we refer to them as fats. The general formula for a fat is given in Figure 15-1.

The melting points and other properties of fats are determined by the kinds of fatty acids they contain. A fat usually contains three different fatty acids, although occasionally two are identical. These acids almost always have an even number of carbon atoms, usually 16 or 18, and some are unsaturated (contain double bonds). The melting point rises with the length of the fatty acid and with the extent of its saturation with hydrogen, so solid fats usually have saturated fatty acids. In the oils, one to three double bonds are present in each fatty acid; these cause lower melting points and make oils liquid at room temperature. Examples of commercially important plant oils are those from the seeds of cotton, corn, peanuts, and soybeans. All of these oils principally contain fatty acids with 18 carbon atoms, including *oleic acid*, with one double bond, and *linoleic acid*, with two double bonds. In fact, these two acids, in the order named, are the most abundant fatty acids in nature.

Table 15-1 lists several important fatty acids, including the number of carbon atoms, structure, degree of unsaturation, position of double bonds, and melting point for each. The most abundant saturated fatty acids are *palmitic acid*, with 16 carbons, and *stearic acid*, with 18 carbons. Coconut fat is a rich source of *lauric acid*, a saturated acid with only 12 carbons. The seven fatty acids listed in Table 15-1 represent about 90 percent of those occurring in lipids of plant membranes (Section 7.4) and about the same percentage of those in commercial oils from seeds (Harwood, 1980, 1989). Seeds of many plants contain a high percentage of fatty acids that are not important in membrane lipids of the same species. Castor beans (*Ricinus communis*), for example, contain *ricinoleic acid* (12-hydroxyoleic acid), which makes up between 80 and 90 percent of the fatty acids in castor oil but which is absent from castor-bean membranes and is rare in other species.

## Distribution and Importance of Fats

Fat storage is rare in leaves, stems, and roots but occurs in many seeds and some fruits (for example, avocados and olives). In angiosperms, fats are concentrated in the endosperm or cotyledon storage tissues of seeds, but they also occur in the embryonic axis. In gymnosperm seeds they are stored in the female gametophyte.

Compared with carbohydrates, fats contain larger amounts of carbon and hydrogen and less oxygen, so

**Table 15-1 Fatty Acids Abundant or Common in Various Plants.**

| Name | Number of Carbons: Number of Double Bonds | Structure | Melting Point (°C)[a] |
|---|---|---|---|
| Lauric | 12:0 | $CH_3(CH_2)_{10}COOH$ | 44 |
| Myristic | 14:0 | $CH_3(CH_2)_{12}COOH$ | 58 |
| Palmitic | 16:0 | $CH_3(CH_2)_{14}COOH$ | 63 |
| Stearic | 18:0 | $CH_3(CH_2)_{16}COOH$ | 71.2 |
| Oleic | 18:1 at C-9, 10 | $CH_3(CH_2)_7\overset{H}{\underset{10}{C}}=\overset{H}{\underset{9}{C}}-(CH_2)_7COOH$ | 16.3 |
| Linoleic | 18:2 at C-9, 10; 12, 13 | $CH_3(CH_2)_4\overset{H}{\underset{13}{C}}=\overset{H}{\underset{12}{C}}-CH_2\overset{H}{\underset{10}{C}}=\overset{H}{\underset{9}{C}}-(CH_2)_7-COOH$ | −5 |
| Linolenic | 18:3 at C-9,10; 12, 13; 15, 16 | $CH_3CH_2\overset{H}{C}=\overset{H}{C}-CH_2\overset{H}{C}=\overset{H}{C}-CH_2\overset{H}{C}=\overset{H}{C}-(CH_2)_7-COOH$ | −11.3 |

[a]Melting points were obtained from Weast, 1988.

**Table 15-2 The Chemical Composition of Some Seeds of Economic Importance.**

| Species | Family | Principal Reserve Tissue | Percent Content[a] Carbohydrate | Protein | Lipid |
|---|---|---|---|---|---|
| Maize *(Zea mays)* | Poaceae (Gramineae) | Endosperm | 51–74 | 10 | 5 |
| Wheat *(Triticum aestivum)* | Poaceae | Endosperm | 60-75 | 13 | 2 |
| Pea *(Pisum sativum)* | Fabaceae (Leguminosae) | Cotyledons | 34–46 | 20 | 2 |
| Peanut *(Arachis hypogaea)* | Fabaceae | Cotyledons | 12–33 | 20–30 | 40–50 |
| Soybean *(Glycine* sp.*)* | Fabaceae | Cotyledons | 14 | 37 | 17 |
| Brazil nut *(Bertholletia excelsa)* | Lecythidaceae | Hypocotyl | 4 | 14 | 62 |
| Castor bean *(Ricinus communis)* | Euphorbiaceae | Endosperm | 0 | 18 | 64 |
| Sunflower *(Helianthus annuus)* | Asteraceae (Compositae) | Cotyledons | 2 | 25 | 45–50 |
| Oak *(Quercus robur)* | Fagaceae | Cotyledons | 47 | 3 | 3 |
| Douglas fir *(Pseudotsuga menziesii)* | Pinaceae | Gametophyte | 2 | 30 | 36 |

[a]The percentages are based on the fresh (air-dry) weights of the seeds.

Source: From Street and Öpik, 1970. A longer list of analyses of whole seeds is given by Sinclair and de Wit, 1975.

when fats are respired, more $O_2$ is used per unit weight. As a result, more ATP is formed, demonstrating that greater amounts of energy can be stored per unit volume of fats than of carbohydrates. Perhaps because of this, most small seeds contain fats as primary storage materials. When these fats are respired, enough energy is released to allow establishment of the seedling, yet the small weight of such seeds often allows them to be scattered effectively by wind. Larger seeds, especially those such as pea, bean, and maize that have been selected by humans for agriculture, often contain much starch and only small amounts of fats, but seeds of conifers and those in nuts are usually fat-rich (Table 15-2). An encyclopedic list of seed compositions from 113 families is given by Earle and Jones (1962).

Fats are always stored in specialized bodies in the cytosol (see Fig. 14-10), and there are often hundreds to thousands of such bodies in each storage cell. These bodies have been called lipid bodies, spherosomes, and **oleosomes** (Latin, *oleo*, "oil"). For a review of terminology, see Gurr (1980). We prefer the term *oleosome*, suggested by Yatsu et al. (1971), because it both correctly indicates that these bodies contain oil and distinguishes them from peroxisomes (Fig. 11-8) and glyoxysomes (see below), which are also spherical bodies. Furthermore, the term *spherosome* has been used for years to describe organelles that contain little if any fat (Sorokin, 1967).

Oleosomes can be isolated from seeds in rather pure form, allowing analysis of their composition and structure. Failure of oleosomes to fuse into one large lipid droplet in cells or when isolated suggests that a membrane surrounds each, yet that membrane frequently cannot be seen in electron micrographs. When it is visible, it usually appears to be only about half as thick (approximately 3 nm) as a typical unit membrane (approximately 8 nm). Apparently, the oleosome membrane is indeed a half-membrane whose polar, hydrophilic surface is exposed to the aqueous cytosol, and whose nonpolar, hydrophobic surface faces the fats stored inside.

An extensive cytological study of oleosome formation during seed development seems to explain how a half-membrane arises (Wanner et al., 1981). Oleosomes apparently originate from two sources: the endoplasmic reticulum and plastids. Fats apparently accumulate between the two layers of phospholipids and glycolipids present in the outer membrane of the plastid envelope or the ER membrane. This accumulation causes separation of the lipid bilayer into two halves, with fats forcing them apart until a distinct oleosome swells and pinches off (Fig. 15-2).

## Formation of Fats

Fats stored in seeds and fruits are not transported there from leaves, but instead they are synthesized *in situ* from sucrose or other translocated sugars. Although leaves produce various fatty acids present in lipids of their membranes, they seldom synthesize fats. Further-

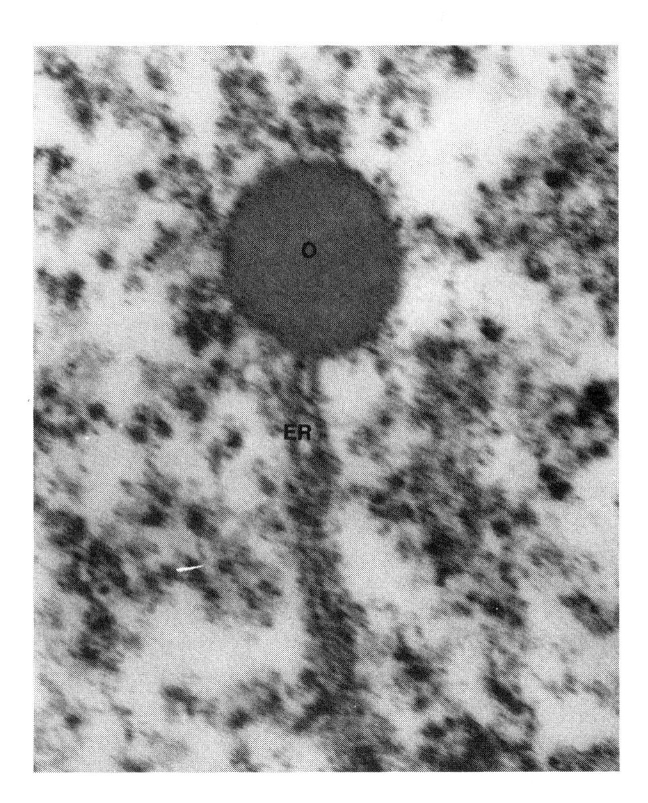


**Figure 15-2** Formation of an oleosome (O) from the endoplasmic reticulum (ER) in a fat-storing cotyledon of a developing watermelon seed. (Courtesy G. Wanner.)

more, both fatty acids and fats are too insoluble in $H_2O$ to be translocated in phloem or xylem.

Conversion of carbohydrates to fats requires production of the fatty acids and the glycerol backbone to which the fatty acids become esterified. The glycerol unit ($\alpha$-glycerophosphate) arises by reduction of dihydroxyacetone phosphate produced in glycolysis (Section 13.3). The fatty acids are formed by multiple condensations of acetate units in acetyl CoA. Most reactions of fatty-acid synthesis occur only in chloroplasts of leaves and proplastids of seeds and roots (Stumpf, 1987; Harwood, 1988; Heemskerk and Wintermans, 1987). The fatty acids synthesized in those organelles are mainly palmitic acid and oleic acid. The acetyl CoA used to form fats in chloroplasts is often produced by a pyruvate dehydrogenase (similar to that in mitochondria), which uses pyruvate made from glycolysis in the cytosol. Another source of acetyl CoA for chloroplasts of spinach and some other plants is free acetate from mitochondria. This acetate is readily absorbed by plastids, converted to acetyl CoA, then used to form fatty acids and other lipids described in this chapter (Givan, 1983; Harwood, 1988, 1989). A summary of the many reactions involved in fatty-acid synthesis is exemplified for palmitic acid (as the CoA ester) in Reaction 15.1:

$$\text{8 acetyl CoA} + 7\,ATP^{3-} + 14\,NADPH + 14\,H^+ \rightarrow \text{palmityl CoA} + 7\,CoA + 7\,ADP^{2-} + 7\,H_2PO_4^- + 14\,NADP^+ + 7\,H_2O \quad (15.1)$$

Subsequently, CoA is hydrolyzed away when palmitic or some other fatty acid is combined with glycerol during formation of fats or membrane lipids.

This summary emphasizes that conversion of acetate units into fatty acids is energy-expensive, because almost two pairs of electrons (2 NADPH) and one ATP are needed for each acetyl group present. In illuminated leaves, photosynthesis provides most of the NADPH and ATP, and fatty-acid formation occurs much faster in light than in darkness. In darkness and in proplastids of seeds and roots, the pentose-phosphate respiratory pathway (Section 13.10) likely provides the NADPH, and glycolysis provides the ATP and the pyruvate from which acetyl CoA is formed.

Although palmitic acid and oleic acid are formed in plastids, most other fatty acids are formed by modifying these acids in the ER. In seeds, all fatty acids they produce can be esterified with glycerol to produce fats that develop into oleosomes directly in the ER (Fig. 15-2). Alternatively, fatty acids can be transported back to proplastids for oleosome formation. Furthermore, the ER of all cells can convert fatty acids into phospholipids needed for growth of the ER itself or of other cellular membranes (Moore, 1984; Mudd, 1980). In leaves, linoleic and linolenic acids are synthesized from oleic acid and (by elongation) from palmitic acid in the ER. Then linoleic and linolenic acids are transported from the ER back to the chloroplasts, where they accumulate as lipids in thylakoid membranes.

The kinds of fatty acids found in membranes and storage fats of plants vary somewhat with the environment in which the plant grows. Temperature is a major controlling factor (Harwood, 1989). At lower temperatures fatty acids are more unsaturated (that is, there are more linolenic and linoleic acids) than at higher temperatures. This unsaturation decreases the average fatty-acid melting points (Table 15-1), makes membranes more fluid, and makes oils rather than solid fats in oleosomes. A popular hypothesis to explain the temperature effect is that the increased solubility of oxygen in water as the temperature drops provides $O_2$ that acts as the essential acceptor of hydrogen atoms for desaturation processes in the ER, therefore causing more unsaturation in the fatty acids present.

## Conversion of Fats to Sugars: β-oxidation and the Glyoxylate Cycle

Breakdown of fats stored in oleosomes of seeds and fruits releases relatively large amounts of energy. For seeds, this energy is necessary to drive early seedling

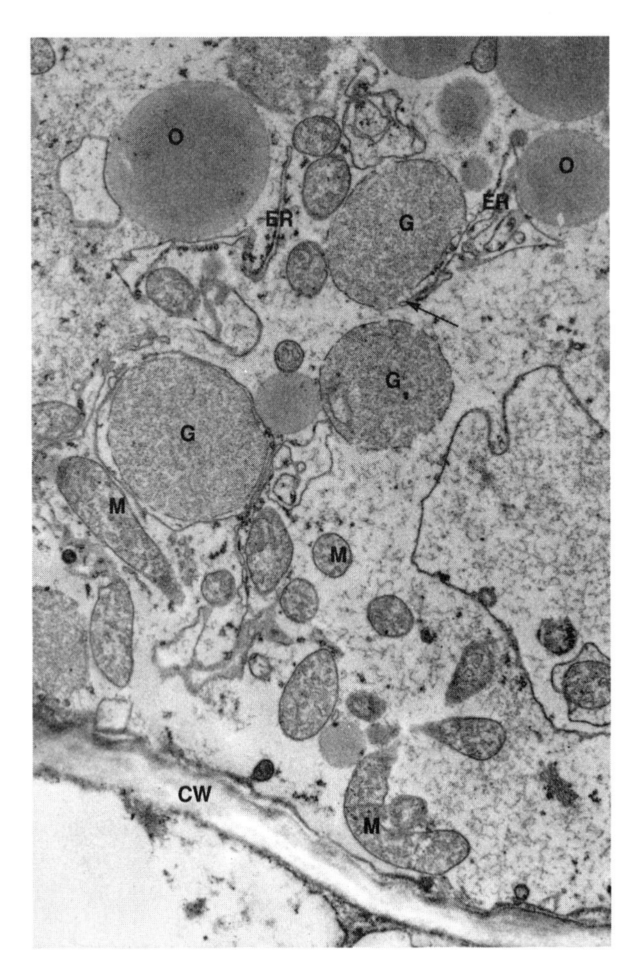


**Figure 15-3** Part of a fat-storing cell in the megagametophyte of a ponderosa pine seed germinated 7 days, showing glyoxysomes (G), fat-storing oleosomes (O), endoplasmic reticulum (ER), and mitochondria (M). Most of the fat has been converted to sugars. Some of the glyoxysomes appear to be connected to the ER, from which they arise; note arrow. (From Ching, 1970.)

development before photosynthesis begins. Because fats cannot be translocated to the growing roots and shoot, they must be converted to more-mobile molecules, usually sucrose. Conversion of fats to sugars is an especially interesting process because it occurs largely in fat-rich seeds and fungal spores and in some bacteria, but not in humans or other animals.

Most reactions necessary to convert fats to sugars occur in microbodies called **glyoxysomes** (Fig. 15-3). Structurally, glyoxysomes are almost identical to peroxisomes of photosynthetic cells (Chapter 11), but many of the enzymes they contain are different (Tolbert, 1981; Huang et al., 1983). In some species, perhaps all, they are formed as **proglyoxysomes** (small glyoxysome precursors) in cotyledons of developing seeds; then during germination and early seedling development they mature into fully functional glyoxysomes (Trelease, 1984). They persist only until the fats are digested; then they disappear. In cotyledons that emerge above ground and become photosynthetic, they are replaced by peroxisomes (Beevers, 1979).

Breakdown of fats begins with action of **lipases**, which hydrolyze the ester bonds and release the three fatty acids and glycerol:

$$\begin{array}{lcl} & (H_2O) \searrow & \\ & H_2C-O-\overset{O}{\overset{\|}{C}}-(CH_2)_nCH_3 & \rightarrow HO\overset{O}{\overset{\|}{C}}(CH_2)_nCH_3 \\ H_2COH & (H_2O) \searrow & \\ HCOH \leftarrow & HC-O-\overset{O}{\overset{\|}{C}}-(CH_2)_nCH_3 & \rightarrow HO\overset{O}{\overset{\|}{C}}(CH_2)_nCH_3 \\ H_2COH & (H_2O) \searrow & \\ & H_2C-O-\overset{O}{\overset{\|}{C}}-(CH_2)_nCH_3 & \rightarrow HO\overset{O}{\overset{\|}{C}}(CH_2)_nCH_3 \\ \text{glycerol} & \text{fat molecule} & \text{free fatty acids} \end{array} \quad (15.2)$$

Most lipase activity is present in the half-membranes of oleosomes (Huang, 1987). How the fatty acids then get into glyoxysomes for further breakdown is not known, but frequently there is direct contact between oleosomes and glyoxysomes (Fig. 15-3).

The glycerol resulting from lipase action is converted with ATP to $\alpha$-glycerolphosphate in the cytosol; the glycerolphosphate is then oxidized by $NAD^+$ to dihydroxyacetone phosphate, most of which is converted to sucrose by reversal of glycolysis. The fatty acids taken into glyoxysomes are first oxidized to acetyl CoA units and NADH by a metabolic pathway called **$\beta$-oxidation** because the beta-carbon is oxidized. Details of $\beta$-oxidation will not be presented here but are summarized below for palmitic acid:

$$\begin{aligned} &\text{palmitate} + ATP^{3-} + 7\,NAD^+ + 7\,FAD + 7\,H_2O \\ &+ 8\,CoASH \rightarrow 8\text{ acetyl CoA} + AMP^{2-} + \text{pyrophosphate}^{2-} \\ &+ 7\,NADH + 7\,H^+ + 7\,FADH_2 \end{aligned} \quad (15.3)$$

Reaction 15.3 is a generalized summary of $\beta$-oxidation that applies whenever the process occurs. In cells other than storage tissues of fat-rich seeds, fatty acids released from membrane lipids also undergo $\beta$-oxidation in peroxisomes. In animal livers, $\beta$-oxidation can occur in mitochondria and in peroxisomes, but plant mitochondria seem unable to carry out this process (Gerhardt, 1986; Harwood, 1988). When $\beta$-oxidation occurs in glyoxysomes or peroxisomes, these organelles have no complete Krebs-cycle enzymes and no electron-transport system to oxidize the $FADH_2$ and NADH products of $\beta$-oxidation.

Oxidation of $FADH_2$ in glyoxysomes and peroxisomes is an energy-inefficient process. This waste occurs because these organelles contain an oxidase enzyme that transfers hydrogen atoms of $FADH_2$ directly to $O_2$, forming $H_2O_2$ (hydrogen peroxide) with release of heat. Each $H_2O_2$ is then degraded to ½ $O_2$ and

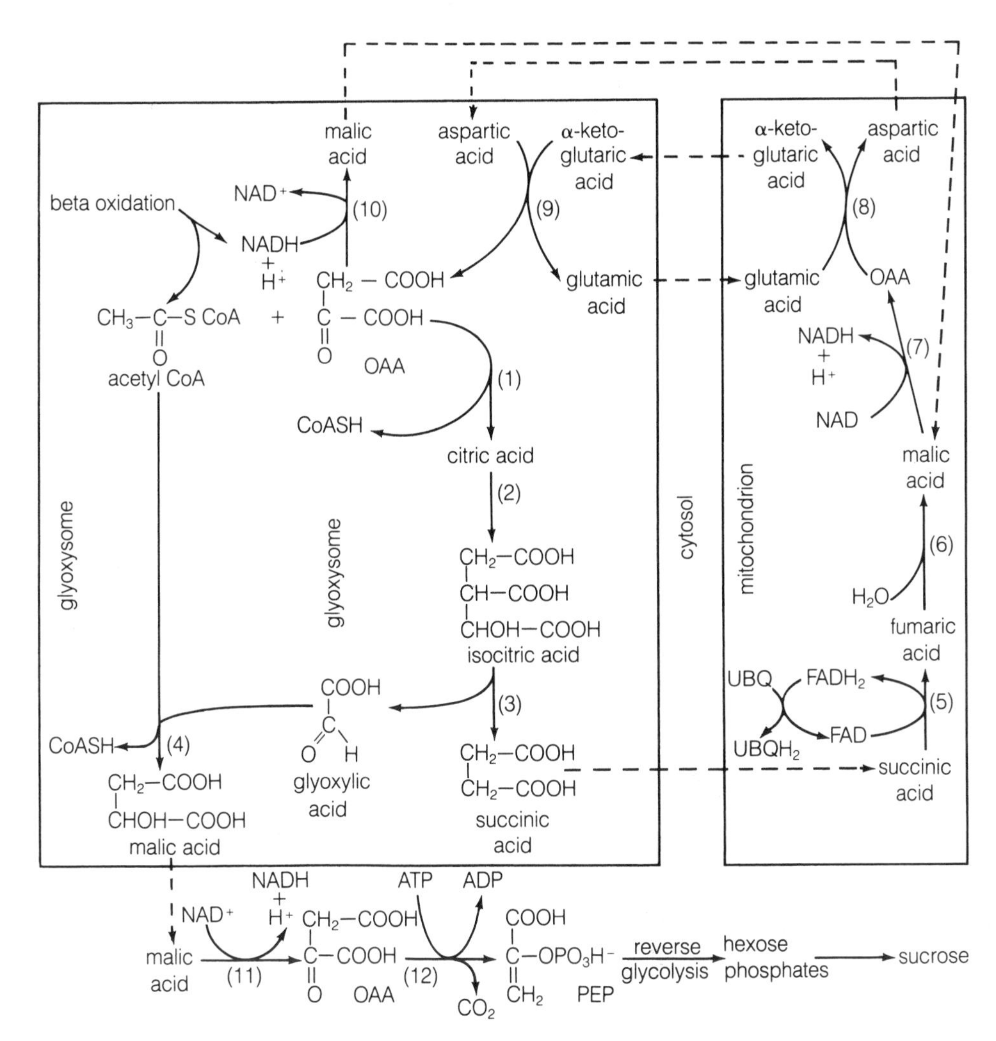


**Figure 15-4** Cooperation of glyoxysomes, cytosol, and mitochondria in converting fatty acids of reserve fats to sucrose by the glyoxylate pathway.

$H_2O$ by catalase, just as occurs in peroxisomes during the glycolate pathway of photorespiration (Section 11.4) and with additional energy loss as heat. Glyoxysomes can process both NADH and acetyl CoA released in β-oxidation, yet they require help of mitochondria and the cytosol to form sugars. Pertinent reactions that occur in the glyoxysomes, namely, conversion of acetate units of acetyl CoA to malic acid, are called the **glyoxylate cycle**. Details of the glyoxylate cycle and additional mitochondrial and cytoplasmic reactions necessary to convert acetate units to sugars are shown in Figure 15-4 and are described briefly below.

Acetyl CoA reacts with oxaloacetic acid to form citric acid, just as in the Krebs cycle (Fig. 15-4, reaction 1). After isocitric acid (six carbons) is formed (reaction 2), it undergoes cleavage by an enzyme unique to the glyoxylate cycle, called *isocitrate lyase*. Succinate (four carbons) and glyoxylate (two carbons) are produced (reaction 3). The glyoxylate reacts with another acetyl CoA to form malate and free coenzyme A (CoASH, reaction 4). This reaction is catalyzed by a second enzyme restricted to the glyoxylate cycle, called *malate synthetase*. This malate is a product of the cycle and is transported to the cytosol, where it is converted to sugars, as we shall explain.

Succinate produced in reaction 3 moves to the mitochondria for further processing. Here it is oxidized by Krebs-cycle reactions 5, 6, and 7 to oxaloacetate (OAA), releasing NADH and reduced ubiquinone ($UBQH_2$). Both NADH and $UBQH_2$ are oxidized with $O_2$ by the mitochondrial electron-transport system to form $H_2O$ and ATP. Reactions 8 and 9 are transaminations between alpha-keto acids and amino acids that require exchange transport of such molecules between mitochondria and glyoxysomes. Their main function seems to be regeneration of the OAA needed to maintain reaction 1 of the glyoxylate pathway (Mettler and Beevers, 1980).

The malate produced by malate synthetase (reaction 4) is more than enough to account for all of the fatty-acid carbons converted to sucrose carbons. This malate is first oxidized to OAA by a cytoplasmic $NAD^+$-malate dehydrogenase (reaction 11); then the OAA is decarboxylated and phosphorylated with ATP to yield

$CO_2$ and phosphoenolpyruvate, PEP (reaction 12). This reaction is catalyzed by an enzyme we have not yet mentioned, called *PEP carboxykinase*. It is likely that ATP produced in the mitochondria during oxidation of NADH and $UBQH_2$ is transported out and drives reaction 12 in the cytosol. Once PEP is formed it can readily undergo reverse glycolysis to form hexose phosphates. Sucrose derived from these hexose phosphates is then transported via the phloem to growing roots and shoots, where it provides much of the carbon needed for growth of those organs in developing seedlings.

An overall summary of the conversion of a fatty acid (palmitic) to a sugar (sucrose) is given by Reaction 15.4:

$$C_{16}H_{32}O_2 + 11\,O_2 \rightarrow C_{12}H_{22}O_{11} + 4\,CO_2 + 5\,H_2O \qquad (15.4)$$

This is a respiration process because $O_2$ is absorbed (during oxidation of $FADH_2$ produced by β-oxidation) and $CO_2$ is released (during conversion of oxaloacetate to PEP). The respiratory quotient (RQ; moles of $CO_2$/ moles of $O_2$) is 0.36, entirely consistent with measurements of RQ values in numerous fat-rich seeds or seedlings. Although one-fourth of the carbon atoms are lost from fatty acids as $CO_2$, the saving of three-fourths is enough for the ecological requirements of species with fat-rich seeds.

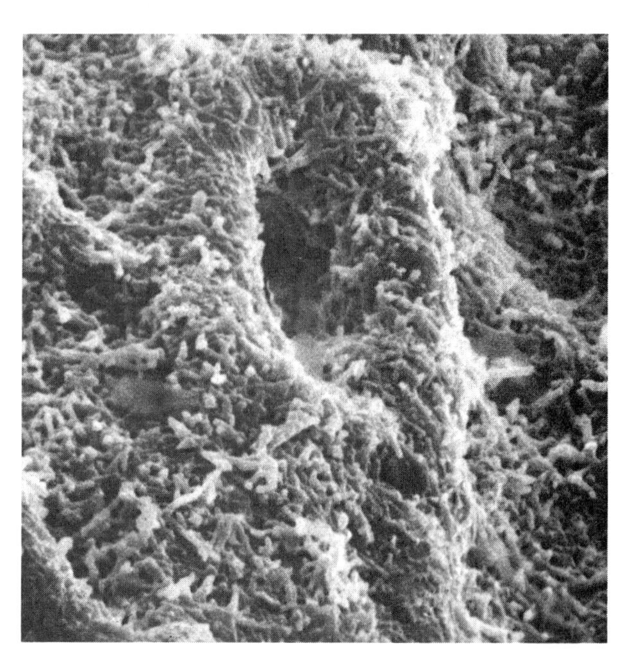

**Figure 15-5** Wax on the leaf surface of a carnation. Carnation (*Dianthus* sp.) is a common plant with a prolific layer of wax on the cuticle. The structure of wax on plant surfaces can be thin flakes, plates, rodlets, or rods. When the wax is in the form of rodlets or rods, it is visible to the naked eye as a bluish "bloom," which can easily be rubbed off the leaf. (From Troughton and Donaldson, 1972.)

## 15.2 Waxes, Cutin, and Suberin: Plant Protective Coats

The entire shoot system of an herbaceous plant is covered by a **cuticle** that slows water loss from all of its parts, including leaves, stems, flowers, fruits, and seeds (Cutler et al., 1980; Juniper and Jeffree, 1982). A scanning electron micrograph of the cuticle on a carnation leaf illustrates the waxy surface structure of the cuticle (Fig. 15-5). Without this protective cover, transpiration of most land plants would be so rapid that they would die. The cuticle also provides protection against some plant pathogens and against minor mechanical damage (Kolattukudy, 1987). It is also important in agriculture because it repels water used in various sprays containing fungicides, herbicides, insecticides, or growth regulators. Because of the hydrophobic nature of the cuticle, most spray formulations contain a detergent to reduce the surface tension of water and allow it to spread on the foliage.

Most of the cuticle is composed of a heterogeneous mixture of components collectively called **cutin**, whereas the remainder consists of overlaying waxes and of pectin polysaccharides attached to the cell wall (Fig. 15-6). Cutin is a heterogeneous polymer consisting largely of various combinations of members in two groups of fatty acids, a group with 16 carbons and one with 18 carbons (Kolattukudy, 1980a, 1980b; Holloway, 1980). Most of these fatty acids have two or more hydroxyl groups, similar to ricinoleic acid mentioned in Section 15.1. The polymeric nature of cutin arises from ester bonds uniting hydroxyl groups and carboxyl groups in the various fatty acids. Small amounts of phenolic compounds are also present in cutin, and these are thought to bind by ester linkages the fatty acids to pectins of the epidermal cell walls.

The cuticular waxes include a variety of long-chain hydrocarbons that also have little oxygen. Many waxes contain long-chain fatty acids esterified with long-chain monohydric alcohols, but they also contain free long-chain alcohols, aldehydes, and ketones ranging from 22 to 32 carbon atoms, and even true hydrocarbons containing up to 37 carbons. One long-chain primary alcohol from the cuticle with 30 carbons, *triacontanol*, is a plant growth stimulant (Section 18.3). Cutins and waxes are synthesized by the epidermis and are then somehow secreted onto the surface. Waxes accumulate in various patterns, one of which is the rodlike pattern shown in Fig. 15-5.

A frequently less-distinct protective coating over underground plant parts is generally called **suberin**. Suberin also covers cork cells formed in tree bark by the crushing action of secondary growth, and it is formed

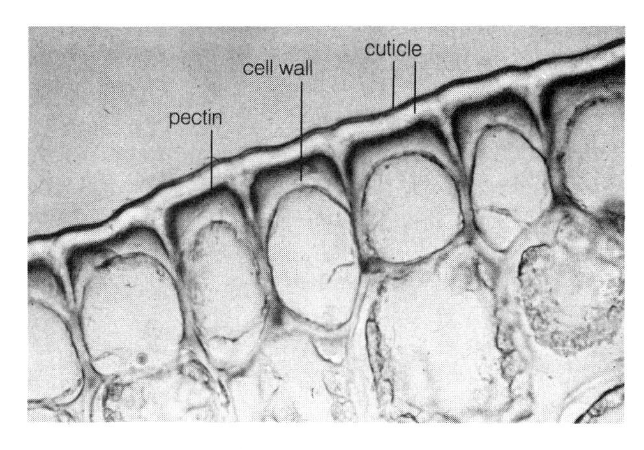


**Figure 15-6** Fine structure of the cuticle on the upper surface of a *Clivia miniata* (scarlet kafarlily) leaf. The cuticle (external dark and light regions) covers the pectin portion fused to the outer part of the cell wall. (Courtesy P. J. Holloway.)

by many kinds of cells as scar tissue after wounding (for example, after leaf abscission and on potato tubers cut for planting). Suberin also occurs in walls of noninjured root cells as a Casparian strip in endodermis and exodermis (see Section 8.3) and in bundle sheath cells of grasses. Kolattukudy (1987) concluded that "plants resort to suberization whenever physiological or developmental changes or stress factors require the erecting of a diffusion barrier." However, at the molecular level, events that induce suberization are unknown. A waxy, lipid portion (up to half of the total suberin) is a complex mixture of long-chain fatty acids, hydroxylated fatty acids, dicarboxylic acids, and long-chain alcohols. Most members of these groups have more than 16 carbon atoms (Holloway, 1983). The remainder of suberin contains phenolic compounds, of which ferulic acid (see Fig. 15-11) is a major component. As in cutin, phenolics are thought to bind the lipid fraction of suberin to the cell wall. Thus suberin is similar to cutin in having an important lipid-polyester fraction but differs both in having a much more abundant phenolic fraction and in the kinds of fatty acids present.

## 15.3 The Isoprenoid Compounds

Numerous plant products having some of the general properties of lipids form a diverse group of compounds with (or formed from) a common structural five-carbon unit. They are called **isoprenoids**, **terpenoids**, or **terpenes** (Grayson, 1988; Fraga, 1988; Croteau, 1988). The nomenclature varies with the author, as described by Loomis and Croteau (1980), but many authors use the term *terpene* for isoprenoids that lack oxygen and are pure hydrocarbons (Robinson, 1980). Included as isoprenoids are hormones such as the gibberellins and abscisic acid (see Chapters 17 and 18), farnesol (a probable stomatal regulator in sorghum), xanthoxin (a precursor of the hormone abscisic acid, Section 18.4), sterols, carotenoids, turpentine, rubber, and the phytol tail of chlorophyll.

Thousands of isoprenoids have been found in the plant kingdom, and the actual number present has yet to be estimated well. Many of these are of interest because of their commercial uses and because they illustrate the ability of plants to synthesize a vast complex of compounds not formed by animals. For most isoprenoids, especially the smaller ones, no function in the plant is presently known. Nevertheless, many influence other plant or animal species, with resulting benefits to the species containing them; such compounds (exclusive of foods) that influence another species are sometimes called **allelochemics** (Barbour et al., 1987; Whittaker and Feeny, 1971; Putnam and Tang, 1986). **Allelopathy** (Greek, *allelon*, "of one other"; *pathos*, "disease") is usually considered a special case of allelochemy involving a negative chemical interaction between different plant species. For isoprenoids and other compounds produced by plants, allelochemy against insects and other animal herbivores is much more evident than allelopathy, but several cases of allelopathy caused by isoprenoids are known (Putnam and Heisey, 1983; Putnam and Tang, 1986; Bernays, 1989; Elakovich, 1987).

Except for isoprene ($C_5H_8$) itself, the isoprenoids are dimers, trimers, or polymers of isoprene units, in which these units are usually joined in a head-to-tail fashion:

$$\text{(head)}-CH_2-\overset{\displaystyle CH_3}{\overset{|}{C}}=CH-CH_2-\text{(tail)}$$

isoprene unit

The isoprene unit is synthesized entirely from acetate of acetyl CoA by what is usually called the **mevalonic acid pathway** because mevalonate is an important intermediate. Three acetyl CoA molecules provide the five carbons for one isoprene unit, whereas the sixth carbon is lost from mevalonate pyrophosphate as $CO_2$. These reactions are described in most biochemistry books and thus will not be explained here. A description of a few of the isoprenoids is given below.

### Sterols

All sterols (steroid alcohols) are *triterpenoids* built from six isoprene units. The most abundant in green algae and higher plants (given in order of abundance) are *sitosterol* (29 C), *stigmasterol* (29 C), and *campesterol* (28 C). Figure 15-7 illustrates the structures of these and a few other sterols, including *cholesterol* (widespread in

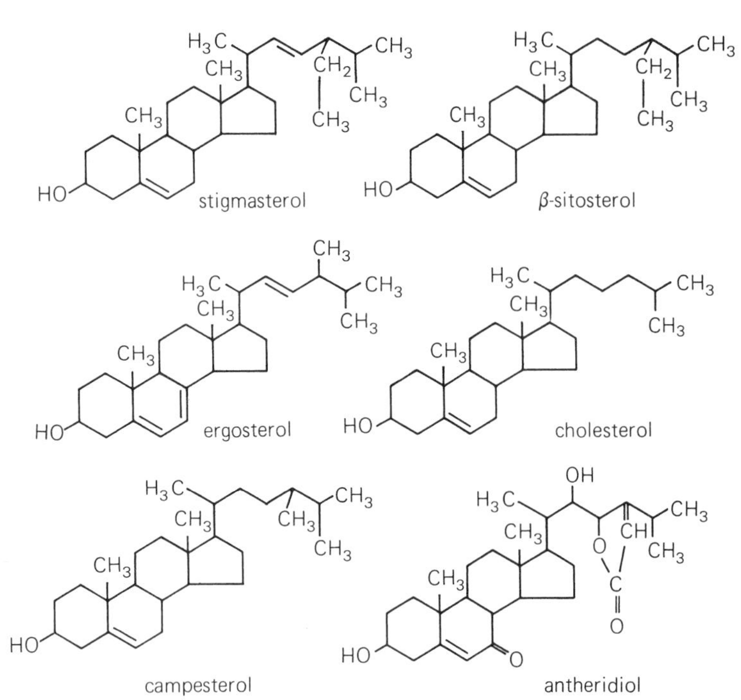

**Figure 15-7** Some plant and fungal sterols. Ergosterol probably occurs in trace amounts in some plants, but antheridiol has been found only in certain fungi.

trace amounts in plants), *ergosterol* (rare in plants but common in some fungi, converted by UV radiation of sunshine to vitamin $D_2$), and *antheridiol*, a sex attractant secreted by female strains of the aquatic fungus *Achlya bisexualis*. Altogether, more than 150 sterols are known to exist in nature. The reactions by which they are synthesized in plants were reviewed by Heftmann (1983), Benveniste (1986), Gray (1987), and Harrison (1988).

Sterols exist not only in the forms shown but also as **glycosides**, in which a sugar (usually glucose or mannose) is attached to the hydroxyl group of the sterol, and as esters, in which the hydroxyl group is attached to a fatty acid (Goad et al., 1987). Free sterols probably exist in all membranes of all organisms except bacteria, and there is little doubt that their contribution to membrane stability is one of their most important functions. None of the sterol glycosides and esters seems to exist in membranes, and their functions remain largely unknown. Besides a membrane function, certain sterols have allelochemical activity (Harborne, 1988). Relatively few examples have been well documented, even though hundreds probably exist in nature.

One example concerns **cardiac glycosides**, sterol derivatives that cause heart attacks in vertebrates but are used medicinally to strengthen and slow the heartbeat during heart failure. This example is related to coevolution of certain milkweeds (*Aesclepias* spp.), monarch butterflies, and blue jays. The milkweeds produce several bitter-tasting cardiac glycosides that protect them against herbivory by most insects and even cattle. However, monarch butterflies have adapted to these glycosides, and the glycosides that their larvae (caterpillars) ingest later cause vomiting in blue jays that eat the adult butterflies. The birds react to vomiting by rejecting other monarchs on sight alone, so considerable immunity from predation occurs among these butterflies from only one emetic experience.

Similar cardiac glycosides present in species of the genus *Digitalis* and in other unrelated species include various **digilanides** that have been used since prehistoric times as sources of arrow poisons (Robinson, 1980). These are toxic because they inhibit Na-K ATPases of heart muscle membranes. Nevertheless, if heart failure occurs because of hypertension or atherosclerosis, digitalis therapy gives a slower and stronger heart beat. In the United States several million people with heart disease routinely use *digitoxin, digoxin,* or some other digilanide from *Digitalis* (foxglove) species (Lewis and Elvin-Lewis, 1977).

Sterols are of further importance to humans because of their use as precursors for synthesis of certain synthetic animal hormones, including the female ovarian hormone progesterone. (See also the boxed essay in this chapter.) Several insect-molting hormones (**ecdysones**) exist in plants, and insects rely upon these and other sterols to form the hormones they contain (Heftmann, 1975; Sláma, 1979, 1980). Plant sterols are thus "vitamins" for many insects. Still other sterol derivatives, the **triterpenoid saponins** (sterols or sterol-like compounds attached to a short chain of one or more sugars), have many biological activities in animals. For example, some cause foaming in the intestinal tract, leading to bloat in cattle that eat young alfalfa plants. The cattle are not repelled by these saponins, however;

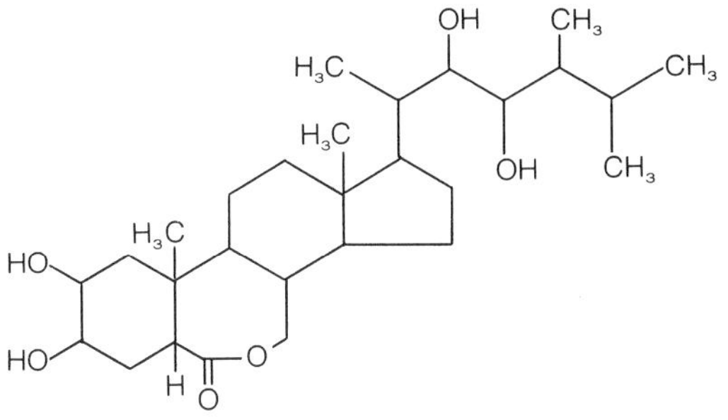

**Figure 15-8** Structure of brassinolide.

indeed, the young plants are eaten more vigorously than are older plants with much lower saponin contents. The number of triterpenoid saponins found in plants grows almost daily. A review by Mahato et al. (1988) listed 420 such compounds discovered between 1979 and 1986.

Recently, good evidence was obtained for the existence in plants of certain mammalian steroid **estrogens**, including *estrone, estriol,* and *estradiol* (Hewitt et al., 1980). Whether these and related steroids normally function within the plant as sex or growth hormones was long suspected and frequently claimed, but with little evidence. This subject is now being investigated more actively; some positive and many negative results have been obtained (Hewitt et al., 1980; Guens, 1978, 1982). A recently discovered group of steroid derivatives called **brassins** or **brassinosteroids** have distinct growth-promoting activity in some plants, especially in stems. These compounds were first isolated from bee-collected pollen grains of rape (*Brassica napus*), a mustard (Grove et al., 1979). The structure of one brassin, **brassinolide**, is shown in Figure 15-8. Interestingly, brassinolide is chemically similar to the ecdysone (insect-molting) hormones. The importance of brassins to plant physiology and their mechanism of action remain to be demonstrated, but progress is being made (Meudt, 1987; Mandava, 1988; also see Section 18.3).

## The Carotenoids

Carotenoids are a group of isoprenoids discussed in Chapter 11 in relation to their functions in photosynthesis. They are yellow, orange, or red pigments that exist in various kinds of colored plastids (**chromoplasts**) in roots, stems, leaves, flowers, and fruits of various plants (see Section 1.6). Two types of carotenoids exist: Carotenes are pure hydrocarbons, whereas xanthophylls also contain oxygen, often two or four atoms per molecule. Both types generally contain 40 carbon atoms made from eight isoprene units. Neither type is water-soluble, but both dissolve readily in alcohols, petroleum ether, acetone, and many other organic solvents.

More than 400 different carotenoids have been found in nature, although only a few are found in any given species (Spurgeon and Porter, 1980). β-Carotene, the most abundant carotenoid found in higher plants, imparts the orange color to carrot roots. Lycopene, another carotene, gives tomato fruits their red color. Lutein, a xanthophyll, is apparently present in all plants and is the predominant xanthophyll of most leaves. The structures of β-carotene, lutein, and lycopene shown in Figure 10-4 are typical of several carotenoids.

Two functions of carotenoids in leaves seem well established. As mentioned in Chapters 10 and 12, some of the carotenoids in chloroplasts participate in photosynthesis, and others prevent photooxidation of chlorophylls. Most yellow flowers contain carotenoids, especially the xanthophyll type, and it is believed that they benefit certain plants by attracting pollinating insects. As described in Figure 18-17, the xanthophyll violaxanthin also acts as the metabolic precursor of abscisic acid. The abundance of orange β-carotene in carrot roots probably results from cultivation and selection by humans, and the carotene carrots contain is both attractive and useful to us because our livers convert it to vitamin A. Recent evidence suggests that β-carotene also protects against certain cancers, probably by acting as an anti-oxidant. No function of β-carotene in carrot roots has been found.

## Miscellaneous Isoprenoids and Essential Oils

Numerous miscellaneous isoprenoid compounds are present in various amounts among certain members of the plant kingdom. In these, isoprene units are condensed into ring compounds commonly containing carbon atom numbers of 10 (the *monoterpenoids*), 15 (the *sesquiterpenoids*), 20 (the *diterpenoids*), or 30 (the *triterpenoids*). Terpenoids with 25 carbons are rarely found. Many of the terpenoids containing 10 or 15 carbons are called **essential oils** because they are volatile and contribute to the *essence* (odor) of certain species. As one example, MacLeod et al. (1988) found 71 volatile compounds in orange peels, most of which were monoterpenoids, mainly limonene. Essential oils are widely used in perfumes. Some of the volatile hydrocarbons released from plants, including isoprene itself, also contribute to smog and other forms of air pollution. Frits Went (1974) estimated that as much as 1.4 billion tons of volatile plant products, mostly hydrocarbon terpenes, are released by plants each year, especially over tropical forests. The Blue Mountains in Australia and the Smoky Mountains of Tennessee and North Carolina in the United States were probably named because of atmospheric scattering of blue light by tiny particles derived from terpenes. Still other essential oils attract insects toward flowers (aiding pollination) or to other plant parts on which insects feed or lay eggs (Metcalf, 1987).

One of the best-known essential oils is turpentine,

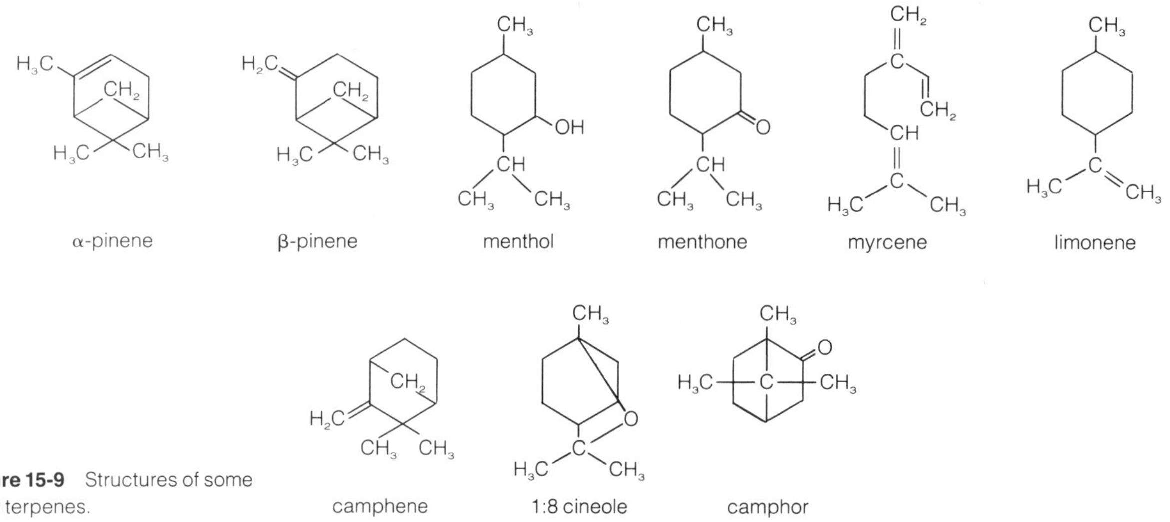

Figure 15-9 Structures of some C-10 terpenes.

present in certain specialized cells of members of the genus *Pinus*. The turpentine of some species consists largely of *n*-heptane, although monoterpenoids such as *α-pinene*, *β-pinene*, and *camphene* (Fig. 15-9) are also present. These compounds and the related *myrcene* and *limonene* (Fig. 15-9) represent important terpenoids affecting tree-killing bark beetles. Such beetles are highly destructive to coniferous forests of North America, causing millions of dollars of damage annually. Useful summaries of the complex relations between bark-beetle attack and resistance by conifers were given by Johnson and Croteau (1987) and Harborne (1988). In ponderosa pine, limonene is one of the insect repellants, whereas α-pinene acts as an attractant or aggregation pheromone. Trees that have high limonene and low α-pinene contents are rarely attacked by those pine beetles.

The essential oils sometimes contain hydroxyl groups or are chemically modified in other ways. The structures of two modified monoterpenoids, *menthol* and *menthone*, both components of mint oils, and of *1:8 cineole*, the major constituent of eucalyptus oil, are also shown in Figure 15-9. Cineole apparently performs an important function in pollination of orchids by male euglossine bees. Such bees are attracted by the orchid flowers, an important constituent of which is 1:8 cineole (Dressler, 1982).

A more complex terpenoid derivative called **glaucolide A** made from three isoprene units is representative of so-called bitter principles largely restricted to the family Asteraceae (Compositae). Bitter principles apparently repel, largely by their taste, numerous chewing insects and mammals. Glaucolide A from species in the genus *Veronia* repels various lepidopterous insects, white-tailed deer, and cottontail rabbits, for example (Mabry and Gill, 1979).

Complex mixtures of terpenes containing 10 to 30 carbon atoms make up the **resins**, which are common in coniferous trees and in several angiosperm trees of the tropics. Resins and related materials are formed in leaves by specialized epithelial cells, which line the resin ducts, and are then secreted into the ducts, where they accumulate. Resins protect trees against many kinds of insects. The ability of conifers to form additional resin ducts from xylem parenchyma when attacked by bark beetles helps protect them against damage by the insects (Johnson and Croteau, 1987).

### Rubber

Rubber is also an isoprenoid compound—the largest one of all. It contains some 3,000 to 6,000 isoprene units linked together in very long, unbranched chains. Most natural rubber is commercially obtained from the latex (milky protoplasm) of the tropical plant *Hevea brasiliensis*, a member of the family Euphorbiaceae. About one-third of this latex is pure rubber. It has been reported, however, that over 2,000 plant species form rubber in varying amounts, and various other species have been used as commercial rubber (notably *Castilla elastica*). Various *Taraxacum* (dandelion) species are among the most well-known North American species possessing this ability. Guayule (*Parthenium argentatum*), a plant common to Mexico and the southwestern United States, also produces rubber; it was studied extensively during World War II and was finally selected in 1978 by the U.S. Congress for development as a natural rubber crop.

## 15.4 Phenolic Compounds and Their Relatives

Flowering plants, ferns, mosses, liverworts, and many microorganisms contain various kinds and amounts of **phenolic compounds**. With important exceptions, the functions of most phenolics are obscure. Many presently appear simply to be by-products of metabolism, but this view likely reflects our still poor knowledge of ecological biochemistry.

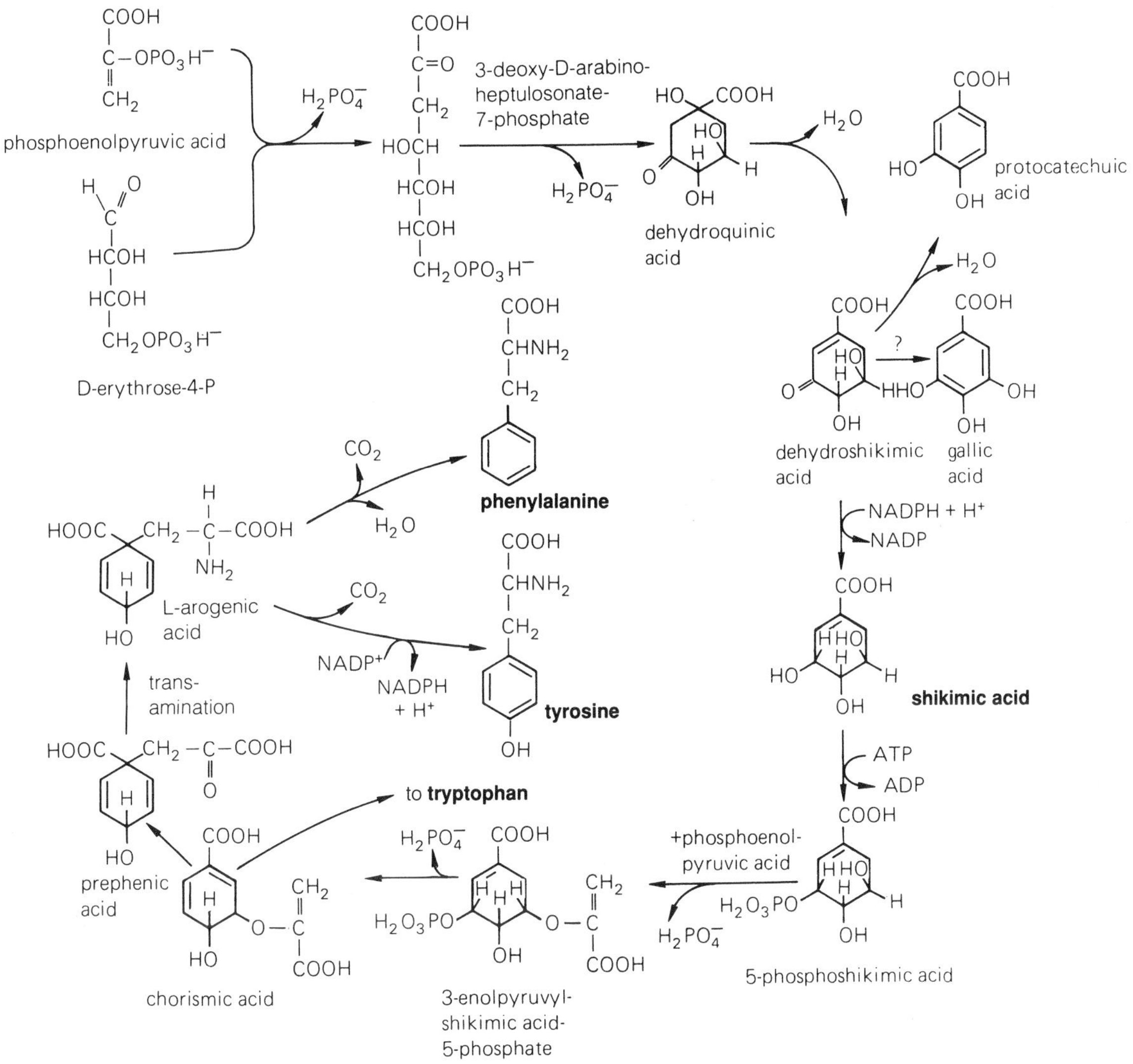

**Figure 15-10** Biosynthesis of phenylalanine and tyrosine from respiratory intermediates in the shikimic acid pathway. All of the carbon atoms seem to arise from phosphoenolpyruvate (two molecules) and from erythrose-4-phosphate (one molecule). ATP is also required.

All phenolic compounds have an aromatic ring that contains various attached substituent groups, such as hydroxyl, carboxyl, and methoxyl ($-O-CH_3$) groups, and often other nonaromatic ring structures. Phenolics differ from lipids in being more soluble in water and less soluble in nonpolar organic solvents. Some phenolics, however, are rather soluble in ether, especially when the *p*H is low enough to prevent ionization of any carboxyl and hydroxyl groups present. These properties greatly aid separation of phenolics from one another and from other compounds.

## The Aromatic Amino Acids

Phenylalanine, tyrosine, and tryptophan are aromatic amino acids formed by a route common to many phenolic compounds. Two small phosphorylated compounds are precursors of these amino acids and of many other phenolic compounds. These two compounds are PEP, from the glycolytic pathway of respiration (Chapter 13), and erythrose-4-phosphate, from the pentose-phosphate respiratory pathway (Chapter 13) and from the photosynthetic Calvin cycle (Chapter 11). These two compounds combine, producing a seven-carbon phosphorylated compound that then forms a ring structure called dehydroquinic acid, which is then converted by two reactions into a rather stable compound called *shikimic acid*. These steps, outlined in Figure 15-10, make up what is called the **shikimic acid pathway**. The shikimic acid pathway also exists in fungi and bacteria but not in animals; we require phenylalanine, tyrosine, and tryptophan in our diets because we lack this pathway. Details of the pathway are given in articles by Floss (1986), Jensen (1985, 1986), and Siehl and Conn (1988).

An interesting aspect of the shikimic acid pathway

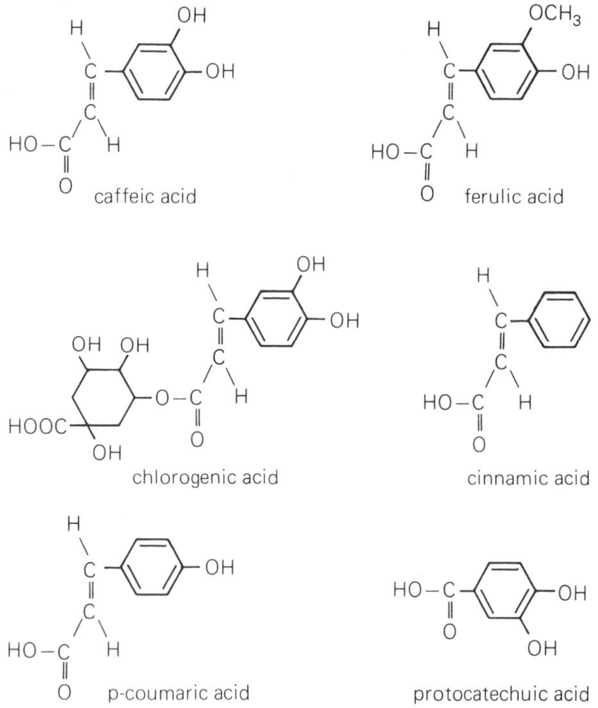

**Figure 15-11** Structures of phenolic acids often found in plants. All are shown in the *trans* form. Chlorogenic acid is an ester formed from caffeic and quinic acids.

is its inhibition by a popular herbicide called *glyphosate* (sold commercially as Roundup). Chemically, glyphosate is N-(phosphonomethyl) glycine, with the structure $HOOC-CH_2-NH-CH_2-PO_3H_2$. It blocks the shikimate pathway mainly by inhibiting the reaction leading from 5-phosphoshikimate and PEP to 3-enolpyruvyl-shikimic acid-5-phosphate (lower right of Fig. 15-10), although the first reaction of the pathway is somewhat sensitive to glyphosate (Jensen, 1986). All plants that absorb the herbicide are injured or killed (usually after one to two weeks), especially because they cannot synthesize phenylalanine, tyrosine, and tryptophan, whereas animals without the shikimate pathway are far less sensitive.

## Miscellaneous Simple Phenolics and Related Compounds

Many other phenolics also arise from the shikimic acid pathway and subsequent reactions. Among these are the acids *cinnamic, p-coumaric, caffeic, ferulic, chlorogenic* (Fig. 15-11), and *protocatechuic* and *gallic* (Fig. 15-10, upper right). The first four are derived entirely from phenylalanine and tyrosine. They are important not because they are abundant in uncombined (free) form, but because they are converted into several derivatives besides proteins. These derivatives include phytoalexins, coumarins, lignin, and various flavonoids such as the anthocyanins, all of which will be described shortly.

An important reaction in the formation of these derivatives is the conversion of phenylalanine to cinnamic acid (Reaction 15.5). This is a deamination in which ammonia is split out of phenylalanine to form cinnamic acid; the reaction is catalyzed by *phenylalanine ammonia lyase:*

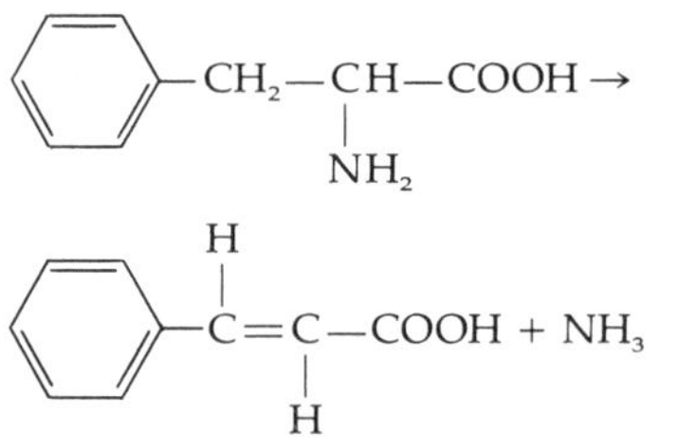

$$C_6H_5-CH_2-CH(NH_2)-COOH \rightarrow C_6H_5-CH=CH-COOH + NH_3 \qquad (15.5)$$

Subsequently, cinnamic acid is converted to p-coumaric acid by addition of one atom of oxygen from $O_2$ and a hydrogen atom from NADPH directly to the para position of cinnamic acid. A second addition of another hydroxyl group adjacent to the OH group of p-coumarate by a similar reaction forms caffeic acid. Addition of a methyl group from S-adenosyl methionine to an OH group of caffeic acid yields ferulic acid. Caffeic acid forms an ester with an alcohol group in still another acid formed in the shikimic acid pathway, quinic acid, thus producing chlorogenic acid. Formation of several such compounds was reviewed by Hahlbrock and Scheel (1989).

Protocatechuic and chlorogenic acids probably have special functions in disease resistance of certain plants. Protocatechuic acid is one of the compounds that prevent smudge in certain colored varieties of onions, a disease caused by the fungus *Colletotrichum circinans*. This acid occurs in the scales of the neck of colored onions that are resistant to the pathogen, but it is absent from susceptible white cultivars. When extracted from colored onions, it prevents spore germination and growth of the smudge fungus, and the growth of other fungi as well.

Large amounts of chlorogenic acid might similarly prevent certain diseases in resistant cultivars, but the evidence for this is weak. Chlorogenic acid is widely distributed in various parts of many plants and usually occurs in easily detectable quantities. In coffee beans the chlorogenic acid concentration is particularly high, and the soluble content of dry coffee reportedly can reach 13 percent by weight (Vickery and Vickery, 1981). A reasonable conclusion is that this acid is not very toxic to humans. It is formed in relatively large amounts in many potato tubers; its oxidation followed by a free-radical polymerization causes formation of large, uncharacterized quinones responsible for the darkening of freshly cut tubers that is well known to cooks. *Polyphenol oxidase* enzymes dependent on copper catalyze this and similar reactions, using $O_2$ as the electron acceptor (Butt and Lamb, 1981; Mayer, 1987). It is thought

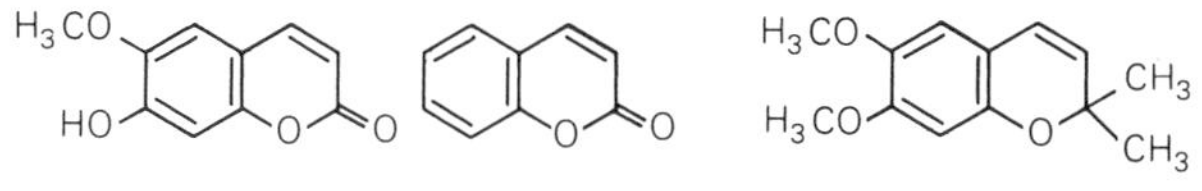

**Figure 15-12** Structures of two coumarins and of preocene 2, a plant compound that reduces levels of juvenile hormone in insects. Preocene 1,which has a similar effect, also occurs in plants; it lacks the upper methoxyl group on the benzene ring (see Bowers et al., 1976).

by some that chlorogenic acid and certain other related compounds can be readily formed and oxidized into potent fungistatic quinones by certain disease-resistant cultivars, but less readily so by susceptible ones. In this way, the infection might be well localized in the resistant plants. Ferulic acid and its derivatives certainly play a role in plant protection because they form part of the phenolic fraction of suberin.

Gallic acid is important because of its conversion to **gallotannins**, which are heterogeneous polymers containing numerous gallic acid molecules connected in various ways to one another and to glucose and other sugars. Many gallotannins greatly inhibit plant growth, and the tolerance of plants that contain them probably involves transferring them to vacuoles, where they cannot denature cytoplasmic enzymes. Gallotannins and especially other tannins are used commercially to tan leather because they cross-link proteins, denaturing them and preventing their digestion by bacteria. Gallotannins act as allelopathic agents, inhibiting growth of other species around those plants that form and release them (Rice, 1984). Other tannins are even more abundant and widespread in plants than are gallotannins, and their major function seems to be protection against attack by bacteria and fungi (Swain, 1979; also see Hemingway and Karchesy, 1989). Nevertheless, tannins almost surely also act as feeding deterrents against various herbivores, partly because of their **astringency** (ability to pucker the mouth) and partly because they inhibit both digestion and utilization of foods.

A group of compounds closely related to the phenolic acids and also derived from the shikimic acid pathway are the **coumarins**. At least 1,000 coumarins exist in nature, although only a few are usually found in any particular plant family (Murray et al., 1982). The structures of two coumarins, *scopoletin* and *coumarin* itself, are given in Figure 15-12. They are formed via the shikimic acid pathway from phenylalanine and cinnamic acid (Brown, 1981).

Coumarin is a volatile compound that is formed mainly from a nonvolatile glucose derivative upon plant senescence or injury. This is especially significant in alfalfa and sweet clover, in which coumarin causes the characteristic odor of recently mown hay. Scientists have developed certain sweet-clover strains that contain small amounts of coumarin, and others that contain it in a bound form. These strains are of economic importance because free coumarin can be converted to a toxic product, *dicumarol*, if the clover becomes spoiled during storage. Dicumarol is an anticoagulant responsible for sweet-clover disease (a hemorrhagic or bleeding disease) in ruminant animals that are fed plants that contain it.

Scopoletin is a toxic coumarin widespread in plants and often found in seed coats. It is one of several compounds suspected of preventing germination of certain seeds, causing a dormancy that exists until the chemical is leached out (for example, by a rainstorm heavy enough to provide sufficient moisture for seedling establishment). It might thus function as a natural inhibitor of seed germination. Numerous other physiological effects of coumarins are known, but clear functions for these compounds generally remain to be found.

Also given in Figure 15-12 is the structure of one of two coumarinlike compounds named **preocenes** isolated in 1976 from the plant *Ageratum houstonianum*. They cause premature metamorphosis in several insect species by decreasing the level of insect juvenile hormone, thereby causing formation of sterile adults. Decreased hormone levels also lead to reduced pheromone production by male medflies, so their sexual attractance to females is decreased (Chang and Hsu, 1981; Staal, 1986). Such compounds appear promising as insecticides that have influence only on target species.

## 15.5 Phytoalexins, Elicitors, and Plant Disease Protection

Since about 1960, various other antimicrobial compounds synthesized by plants when they are infected by certain microbes, especially fungi, have been discovered (Bailey and Mansfield, 1982; Darvill and Albersheim, 1984). These compounds were at first hypothesized to act in a way comparable to the antibodies of animals, yet they proved to have little specificity against any given microbe. They are collectively referred to as **phytoalexins** (from the Greek, *phyton*, "plant," and *alexin*, "to ward off"). In general, phytoalexins are much more toxic to fungi than to bacteria, although exceptions might exist. Compounds that act as phytoalexins include various *glyceollins* in soybean roots, *pisatin* in pea pods, *phaseollin* in bean pods, *ipomeamarone* in sweet potato roots, *orchinol* in orchid tubers, and *trifolirhizin* in red clover roots.

More than 150 phytoalexins have been identified in plants, especially dicots. Few have been found in monocots or gymnosperms, and so far none are known in nonvascular plants. Most phytoalexins are phenolic phenylpropanoids that are products of the shikimic acid pathway, although some are isoprenoid compounds and a few are polyacetylenes. It appears that nonpathogenic fungi often induce such high, toxic levels of

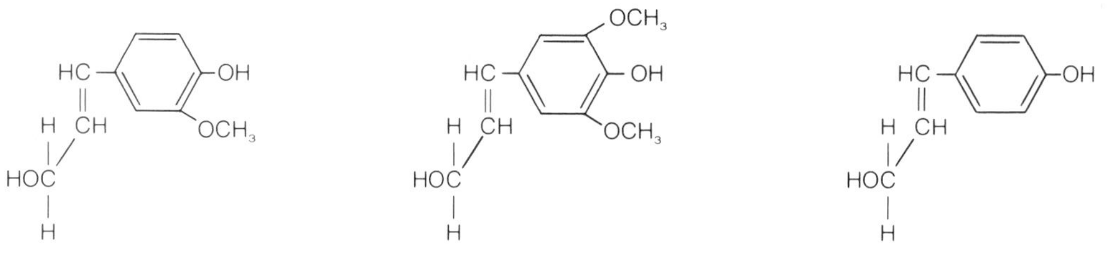

coniferyl alcohol

sinapyl alcohol

p-coumaryl alcohol

a

**Figure 15-13** (**a**) Common phenolic subunits found in lignins. (**b**) Model of partial structure of lignin, rich in coniferyl alcohol. During lignin formation, a variety of different interlocking bonds are formed by free-radical mechanisms, and bond formation depends partly on where in the joining molecules the free radicals are when collision occurs. Also, oxidation at the double bond of the three-carbon alcohol side chain in building blocks (**a**, above) causes several possible covalent bonds to form. Clearly, no lignin can be identical to another, even though all are similar.

phytoalexins in the host that their establishment is prevented, whereas pathogenic fungi are successful parasites because they either induce only nontoxic phytoalexin levels or quickly degrade the phytoalexin.

Surprisingly, several different kinds of compounds, and even viruses, can induce phytoalexin production. Compounds that cause phytoalexin production have been called **elicitors**, although known elicitors also stimulate plants to activate other defense reactions (Ebel, 1986; Boller, 1989). Some elicitors are polysaccharides produced when pathogenic fungi or bacteria attack plant cell walls (Darvill and Albersheim, 1984; Templeton and Lamb, 1988; Boller, 1989; Stone, 1989), whereas others are polysaccharides produced from degradation of fungal cell walls by plant enzymes that the fungus causes the plant to secrete. Plant-pathogen interactions are complex.

Even certain kinds of physical injury can induce phytoalexin production by plants, and so can ultraviolet radiation. A confusing thing about induction by various elicitors is that phytoalexins produced from the shikimic acid pathway and isoprenoids from the mevalonic acid pathway can be formed after adding a single elicitor. It is difficult to understand how such different metabolic pathways could be activated by a single elicitor and how quite different elicitors could activate the same kind of metabolic pathway. Apparently, such exogenous elicitors are recognized by certain proteins in membranes, which then signal the plant to produce a phytoalexin. The nature of this signal is not known, but in some cases it is clear that the signal increases transcription of mRNA molecules that code for enzymes that synthesize the phytoalexin. In the case of glyceollin production by infected soybean roots, the signal may be $Ca^{2+}$ ions (Ebel and Grisebach, 1988). In other cases certain polysaccharides produced by degradation of plant cell walls may act as signals (Darvill and Albersheim, 1984; Ryan, 1987). Even though the overall importance of phytoalexins in disease resistance is still controversial, most scientists agree that this is one of various biochemical mechanisms plants use to prevent disease.

## 15.6 Lignin

**Lignin** is a strengthening material that occurs with cellulose and other polysaccharides in certain cell walls (especially in xylem) of all higher plants. It occurs in largest amounts in wood, in which it accumulates in the middle lamella, primary walls, and secondary walls of the xylem elements. It usually occurs between the cellulose microfibrils, where it serves to resist compression forces. Resistance to tension (stretching) is primarily a function of the cellulose. The formation of lignin is considered by evolutionists to have been crucial in the adaption of plants to a terrestrial environment, because they assume that only with lignin could rigid cell walls of xylem be built to conduct sap (water and mineral salts) under tension over long distances. Lignin is considered to be the second most abundant organic compound on earth; only cellulose is more abundant. Lignin comprises 15 to 25 percent of the dry weight of many woody species (Gould, 1983). Besides the strengthening function of lignin, it also provides protection against attack by pathogens and consumption by herbivores, both insects and mammals (Swain, 1979).

Lignin is difficult to study because it is not readily soluble in most solvents. This insolubility occurs primarily because it has an unusually high molecular weight; further, in the native state it is chemically united to cellulose and other cell-wall polysaccharides by ether and probably other kinds of linkages to the hydroxyl groups of polysaccharides.

Much of what we know about lignin structure was determined by analyzing several intermediates in its synthesis. This contrasts with our knowledge of polysaccharides, proteins, and nucleic acids, the structures of which were largely determined by analyzing degradation products. In general, lignins contain three aro-

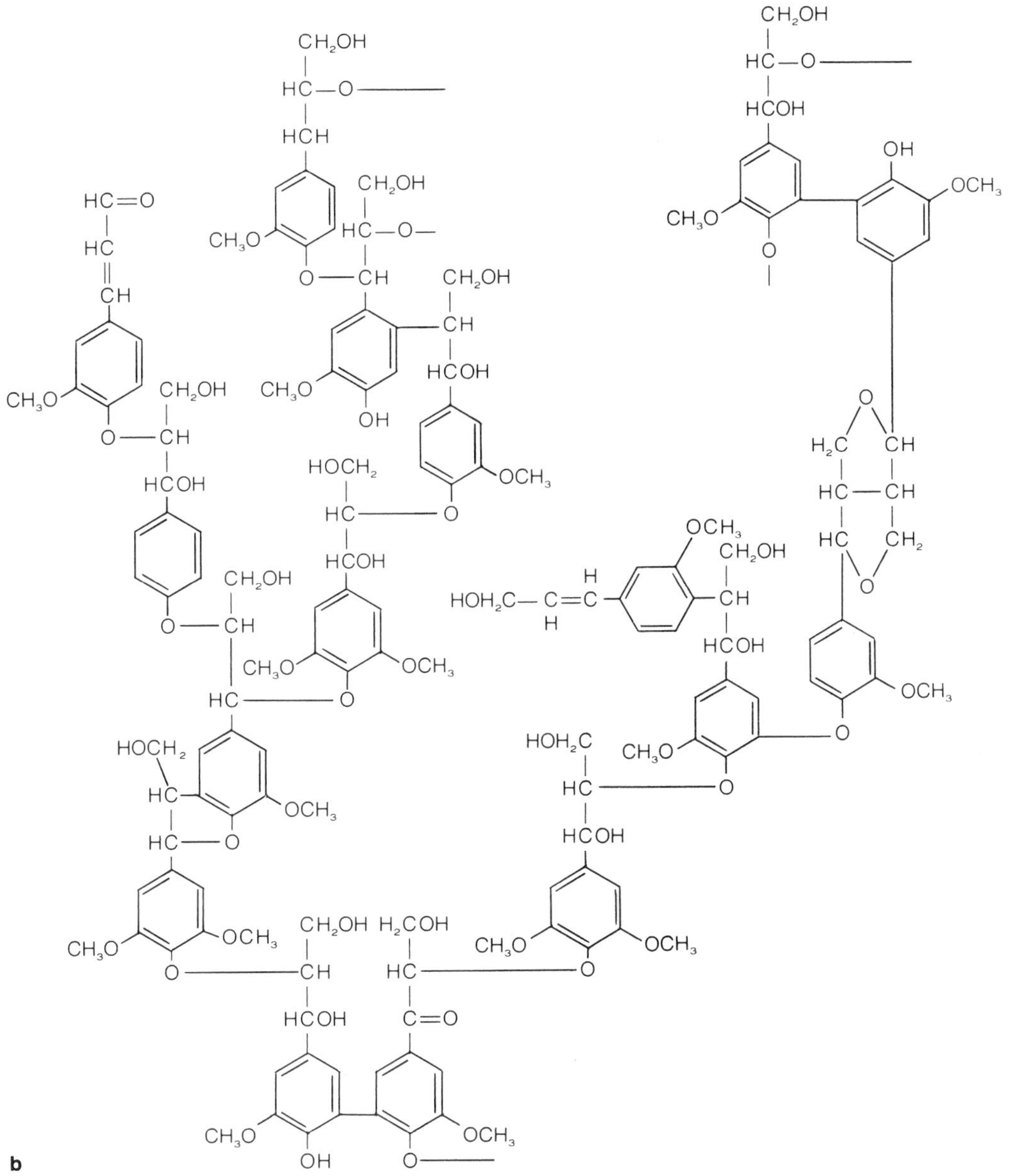

b

matic alcohols: *coniferyl alcohol,* which predominates in softwoods of conifers, *sinapyl alcohol,* and *p-coumaryl alcohol* (Fig. 15-13a). Lignins from hardwood trees and from herbaceous dicots and grasses contain less coniferyl alcohol and more of the other two. Several ways these alcohols are probably connected in lignin are depicted in Figure 15-13b.

The aromatic alcohols in lignin all arise from the shikimic acid pathway. Phenylalanine is converted to aromatic acids such as coumaric and ferulic; these are then converted to CoA esters. The esters are reduced to the aromatic alcohols by NADPH, and these alcohols are then polymerized into lignin by free-radical mechanisms. An iron-containing enzyme called **peroxidase** catalyzes two separate reactions that lead to polymerization (Mader and Amberg-Fisher, 1982). Peroxidase exists in several isozyme forms, a few of which exist in the cell walls. These isozymes apparently function first by forming $H_2O_2$ from NADH and $O_2$. Next they remove a hydrogen atom from each of two aromatic alcohols and combine the two hydrogen atoms with one $H_2O_2$ to release two $H_2O$ molecules as by-products. The remaining part of each aromatic alcohol is now a free radical, and several kinds of electronic shifts allow migration of the unpaired electron to other parts of the molecule. Many such free radicals combine spontaneously in various ways to form bonds between the alcohols, such as those proposed in Fig. 15-13, so presumably lignins always have variable structures.

## 15.7 Flavonoids

Flavonoids are 15-carbon compounds generally distributed throughout the plant kingdom (Harborne, 1988; Hahlbrock, 1981; Stafford, 1990). More than 2,000 from

plants have been identified. The basic flavonoid skeleton, shown below, is usually modified in such a way that even more double bonds are present, causing the compounds to absorb visible light and therefore giving them color. The two carbon rings at the left and right ends of the molecule are designated the A and B rings, respectively.

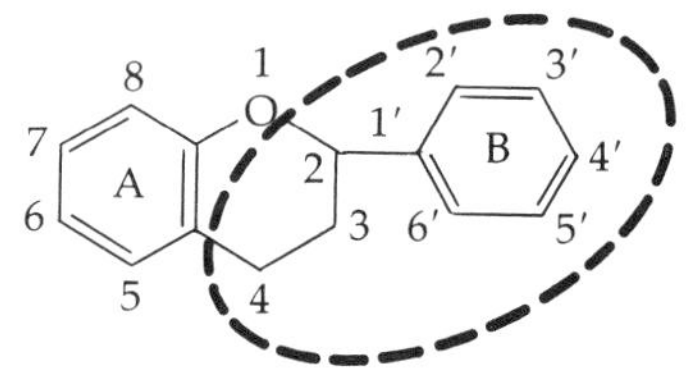

The dashed line around the B ring and the three carbons of the central ring indicate the part of flavonoids that is derived from the shikimic acid pathway. This part may be compared with cinnamic acid (Fig. 15-11), which is a precursor to it. The A ring and the oxygen of the central ring are derived entirely from acetate units provided by acetyl CoA. Hydroxyl groups are nearly always present in the flavonoids, especially attached to the B ring in the 3′ and 4′ positions (compare p-coumaric and caffeic acids of Fig. 15-11), or to the 5 and 7 positions of the A ring, or to the 3 position of the central ring. These hydroxyl groups serve as points of attachment for various sugars that increase the water solubility of flavonoids. Most flavonoids accumulate in the central vacuole, even though they are synthesized outside the vacuole.

basic anthocyanidin structure (flavylium ion)

| Position | Sugar Attached |
|---|---|
| 3 | this position is always glycosylated, commonly by glucose, galactose, rhamnose, xylose-glucose, rhamnose-glucose, or glucose-glucose |
| 5 | sometimes glycosylated; if so, by glucose |
| 7 | almost never glycosylated; if so, by glucose |

**a** hydroxylated anthocyanidins

pelargonidin (scarlet) cyanidin (crimson) delphinidin (blue-violet)

**b** methylated anthocyanidins

peonidin (rosy red) petunidin (purple) malvidin (mauve)

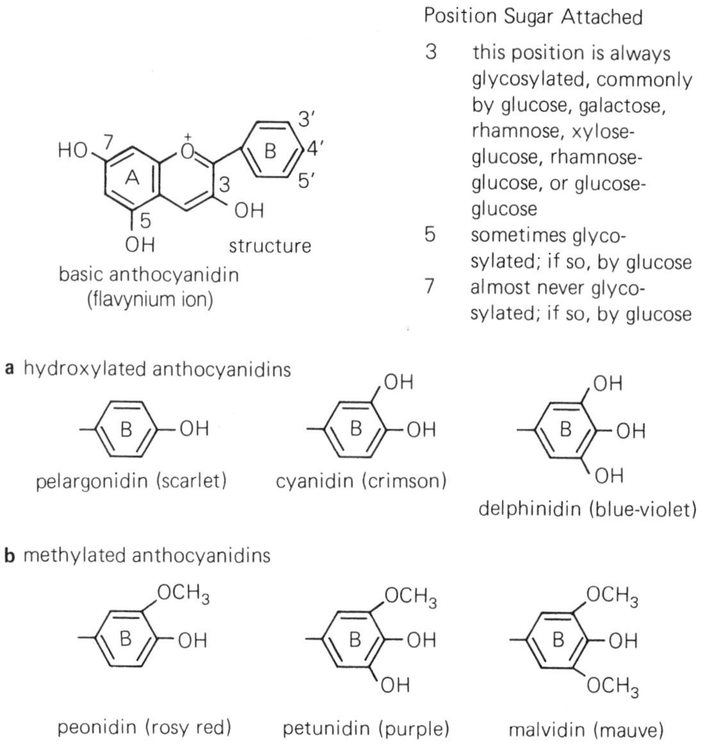

**Figure 15-14** The basic anthocyanidin ring, showing variations of B ring by hydroxylation and methylation to produce various anthocyanins. Anthocyanins are produced by attachment of sugars (glycosylation) to the 3-hydroxyl position of the anthocyanidin, and sometimes also to the 5 or 7 position.

Three groups of flavonoids are of particular interest in plant physiology. These are the **anthocyanins**, the **flavonols**, and the **flavones**. The anthocyanins (from the Greek *anthos*, "flower," and *kyanos*, "dark-blue") are colored pigments that commonly occur in red, purple, and blue flowers. They are also present in various other plant parts, such as certain fruits, stems, leaves, and even roots. Frequently, flavonoids are confined to epidermal cells. Most fruits and many flowers owe their colors to anthocyanins, although some, such as tomato fruits and several yellow flowers, are colored by carotenoids. The bright colors of autumn leaves are caused largely by anthocyanin accumulation on bright, cool days, although yellow or orange carotenoids are the predominant pigments in autumn leaves of some species.

Anthocyanins seem generally absent in the liverworts, algae, and other lower plants, although some anthocyanins and other flavonoids occur in certain mosses. They have only rarely been demonstrated in gymnosperms, although gymnosperms contain other kinds of flavonoids. Several different anthocyanins exist in higher plants, and often more than one is present in a particular flower or other organ. They are present as glycosides, usually containing one or two glucose or galactose units attached to the hydroxyl group in the central ring, or to that hydroxyl group at the 5 position of the A ring, as described in Figure 15-14. When the sugars are removed, the remaining parts of the molecules, which are still colored, are called **anthocyanidins.**

Anthocyanidins are usually named after the particular plant from which they were first obtained. The most common anthocyanidin is *cyanidin*, which was first isolated from the blue cornflower, *Centaurea cyanus*. Another, *pelagonidin*, was named after a bright red geranium of the genus *Pelargonium*. A third, *delphinidin*, obtained its name from the genus *Delphinium* (blue larkspur). These anthocyanidins differ only in the number of hydroxyl groups attached to the B ring of the basic flavonoid structure. Other important anthocyanidins include the reddish *peonidin* (present in peonies), the purple *petunidin* (in petunias), and the mauve-colored (purplish) pigment *malvidin*, first found in a member of the Malvaceae, the mallow family.

The color of anthocyanins depends first on the substituent groups present on the B ring. When methyl groups are present, as in peonidin, they cause a reddening effect. Second, the anthocyanins are often associated with flavones or flavonols, which cause them to become more blue. Third, they associate with each other, especially at high concentrations, and this can cause either a reddening or a bluing effect, depending on the anthocyanin and the *p*H of the vacuoles in which they accumulate (Hoshino et al., 1981). Most anthocyanins are reddish in acidic solution but become purple and blue as the *p*H is raised. In larkspur flowers the *p*H of epidermal cells containing delphinidin increases from 5.5 to 6.6 during aging, and the color changes from reddish purple to purplish blue (Asen et al., 1975). Because of these properties and the common presence of more than one anthocyanin, there is wide variation in the hues of flowers.

Possible functions of anthocyanins have been considered ever since their discovery. One of their useful functions in flowers is attraction of birds and bees that carry pollen from one plant to another, thus aiding pollination (Harborne, 1988). Charles Darwin long ago suggested that a fruit's beauty serves as an attractant to birds and beasts so that the fruit may be eaten and its seeds widely disseminated in the manure. Presumably, anthocyanins contribute to this beauty. Anthocyanins might also play a role in disease resistance, although evidence for this is weak. Their abundance certainly suggests some functions that have favored their evolutionary selection.

Anthocyanins and other flavonoids are of particular interest to many plant geneticists because it is possible to correlate many morphological differences among closely related species in a particular genus, for example, with the types of flavonoids they contain. The flavonoids present in related species of a genus gives taxonomists information they can use to classify and determine the lines of plant evolution (Seigler, 1981).

The flavonols and flavones are closely related to the anthocyanins, except that they differ in the central oxygen-containing ring structure, as follows:

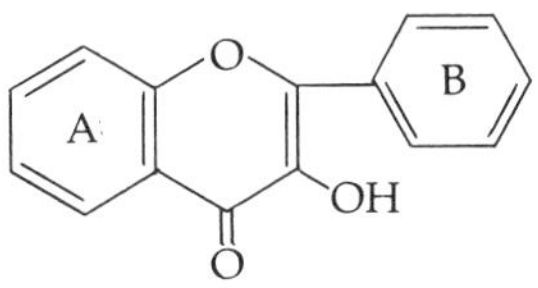

flavonols

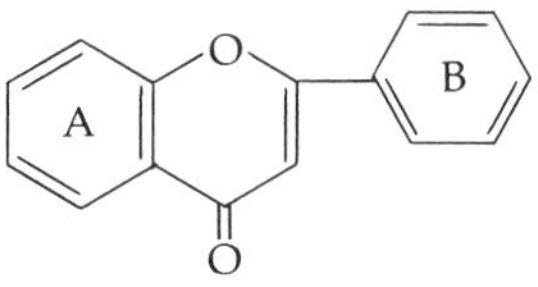

flavones

Most of the flavones and flavonols are yellowish or ivory-colored pigments, and like anthocyanins they often contribute to flower color. Even the colorless flavones and flavonols absorb ultraviolet wavelengths and therefore affect the spectrum of radiation visible to bees or other insects attracted to flowers containing them. These molecules are also widely distributed in leaves. They apparently function there as feeding deterrents and, because they absorb UV radiation, as a protection against long-wave UV rays.

Light, especially blue wavelengths, promotes formation of flavonoids (Chapter 20), and these flavonoids apparently increase the plant's resistance to long-wave UV radiation. The anthocyanins have been studied more than other flavonoids regarding the effects of light upon their biosynthesis. It has probably been known for centuries that the reddest apples are found on the sunny side of the tree; this is because anthocyanins accumulate in these fruits, and accumulation is increased by light. (Photoreceptor pigments that absorb light, causing formation of anthocyanins and other flavonoids, are described in Chapter 20). The nutritional status of a plant also affects its production of anthocyanins. A deficiency of nitrogen, phosphorous, or sulfur leads to accumulations of anthocyanins in certain plants, as mentioned in Chapter 6. Low temperatures also increase anthocyanin formation in some species, as in the coloration of certain autumn leaves on clear, sunny days accompanied by cool nights.

Certain species, especially members of the Papilionoideae subfamily of legumes, also accumulate one or more **isoflavonoids**, which differ from flavonoids in that the B ring is attached to the carbon atom of the central ring adjacent to the point of attachment in flavonoids. The functions of isoflavonoids are mostly unknown, but some act as allelochemics. For example, *rotenone*, an isoflavonoid from the root of derris (*Derris elliptica*), is a widely used insecticide. Furthermore, isoflavonoid structures resemble those of animal estrogens such as estradiol, and certain plant isoflavonoids cause infertility in female livestock, especially sheep (Shutt, 1976). Subterranean clover, in particular, accumulates especially high levels of isoflavones. These compounds cause the serious "clover disease" of sheep, first noted in western Australia in the 1960s as a decline in fertility. They are also suspected to be a factor controlling rodent populations in certain regions. Their infertility effects do not seem to deter grazing animals.

## 15.8 Betalains

The red pigment of beets is a *betacyanin*, one of a group of red and yellow **betalain** pigments that were long thought to be related to the anthocyanins, even though they contain nitrogen. Neither the red betacyanins nor the other kind of betalain pigments, the yellow *betaxanthins*, are at all structurally related to the anthocyanins, and anthocyanins and betalains do not occur together in the same plant. Betalains seem to be restricted to 10 plant families, all of which are members of the order Caryophyllales, which lacks anthocyanins. They cause colors in both flowers and fruits that range from yellow and orange to red and violet, and they also give color to vegetative organs in some cases. Like anthocyanins, their synthesis is promoted by light. Betalains also contain a sugar and a remaining colored portion. The most extensively studied member of this group is *betanin* from red beet roots, which can be hydrolyzed into glucose and *betanidin*, a reddish pigment with the following structure:

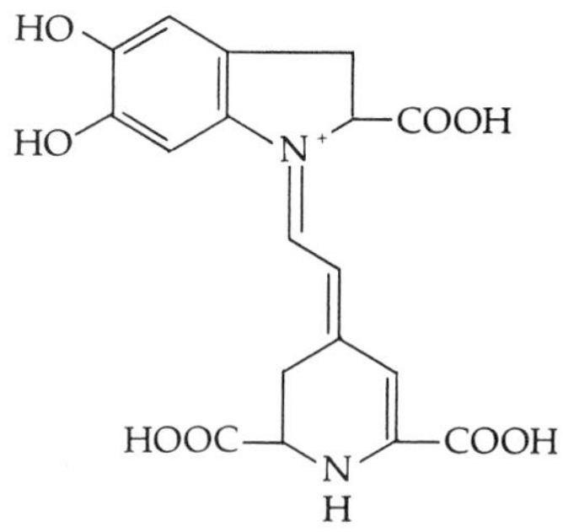

betanidin

Little is yet known of the metabolism or functions of betalains, but a role in pollination comparable to that of anthocyanins in other species seems likely (Piattelli, 1981). Protection against pathogens is another possible function (Mabry, 1980).

## 15.9 Alkaloids

Many plants contain aromatic nitrogenous compounds called **alkaloids**. Chemically, the alkaloids usually contain nitrogen in a heterocyclic ring of variable structure. This nitrogen frequently acts as a base (accepts hydrogen ions), so many alkaloids are slightly basic, as their name indicates. Most are white crystalline compounds that are only slightly water-soluble. They are of special interest because of their dramatic physiological or psychological activity in humans and other animals and because of the belief that many will prove to have important functions in plants, too.

More than 3,000 alkaloids have been found in some 4,000 species of plants, most frequently herbaceous dicots, although any given species typically contains only a few such compounds. Relatively few monocots and gymnosperms possess alkaloids. The first alkaloid to be isolated and crystallized was the drug *morphine*, isolated in 1805 from the opium poppy, *Papaver somniferum*. Other well-known alkaloids include *nicotine*, present in cultivated varieties of tobacco; *cocaine*, from leaves of *Erythroxylon coca*; *quinine*, from cuprea bark; *caffeine*, from coffee beans and tea leaves; *strychnine*, from the seeds of *Strychnos nuxvomica*; *theobromine*, from cocoa beans; *atropine*, from the poisonous black nightshade (*Atropa belladonna*); *colchicine*, from *Colchicum byzantinum*; *mescaline*, a hallucinogenic and euphoric drug from flowering heads of the cactus *Lophophora williamsii*; and *lycoctonine*, a toxic alkaloid in *Delphinium barbeyi* (larkspur). The structures of several alkaloids are shown in Figure 15-15.

theobromine caffeine cocaine

colchicine nicotine theophylline

quinine

morphine mescaline

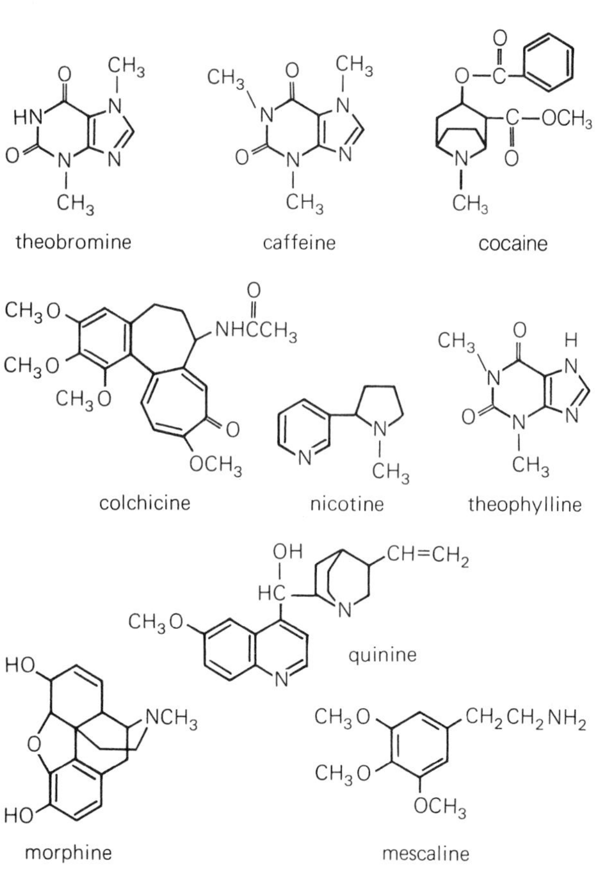

**Figure 15-15** Structures of some representative alkaloids.

Most alkaloids are probably synthesized only in plant shoots, but nicotine is produced only in the roots of tobacco. Many chemical reactions involving formation of some alkaloids are now known, although much more remains to be learned (see, for example, Herbert, 1988; Hegnauer, 1988). Synthesis of nicotine has received the most attention, initially because of its commercial importance, and now because of concern about harmful effects of smoking. *Nicotinic acid* (*niacin* or vitamin B), present in NAD and NADP molecules, is a precursor of nicotine. The nitrogen and carbon atoms of nicotinic acid, in turn, arise from a product obtained when aspartic acid and 3-phosphoglyceraldehyde are combined. Other amino acids are precursors of other alkaloids. Very few of the thousands of enzymes that must be necessary for production of the various alkaloids have been demonstrated in the plant kingdom. Furthermore, some experts have speculated that there are many thousands of alkaloids yet to be discovered in plants.

The physiological roles of alkaloids in the plants that form them are unknown, and it has been suggested that they perform no important metabolic function, being merely by-products of other more important pathways. Nevertheless, several examples are known in which they confer some protection to the plant (Robinson, 1979; Harborne, 1988). Plants containing certain alkaloids are avoided by grazing animals and leaf-feeding insects, for example. Others are used by danaid butterflies as substrates for synthesis of their courtship pheromones. Interestingly, larkspur is not avoided by cattle, even when other forage is available, and the lycoctonine in it accounts for more cattle deaths in the United States than any toxin in any other poisonous plant (Keeler, 1975).

# THREE

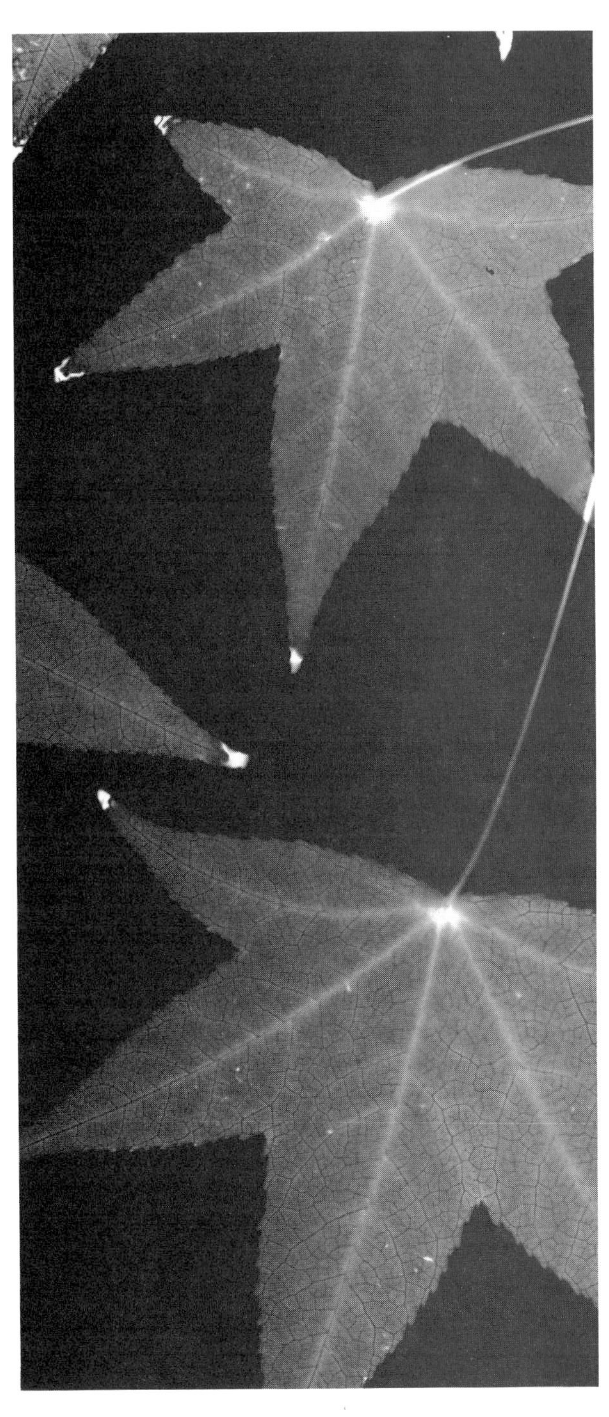

# Plant Development

# 16

# Growth and Development

Think of a favorite plant. Visualize water molecules moving into roots and up through xylem into living leaf cells, where hydrogen bonds are broken and some water molecules evaporate. Imagine $CO_2$ molecules diffusing through stomates into chloroplasts, being fixed into carbohydrates and combined with parts of other water molecules, the process being energized by ATP and NADPH arising from light-driven reactions. Think, too, of assimilates being loaded into phloem sieve tubes and moved to specific sinks, and of ions being selectively and actively absorbed or excreted, some being assimilated into organic compounds and some acting as coenzymes. True, we don't know everything that's going on in cells, but your favorite plant should certainly not seem like a static object. It is a well-organized, living thing, a machine[1] that processes matter and energy in its environment and maintains a relatively low entropy.

We are now ready to ask how the machinery came into being. We know that it began as a single cell, the zygote, which *grew* and *developed* into a multicellular organism. There was a continuous synthesis of large, complex molecules from the smaller ions and molecules that are the raw materials for growth. Cell division produced new cells, many of which became not only larger but more complex. The cells changed in different ways, producing a mature plant composed of numerous cell types. This process of cellular specialization is called **differentiation**, and the growth and differentiation of cells into tissues, organs, and organisms is often called **development**. Another useful term for the process is **morphogenesis** (Greek *morpho*, "form," and *genesis*, "origin"). By development (morphogenesis) a plant transforms itself from a fertilized egg to a mighty redwood — or a pea plant.

We know that genes govern the synthesis of enzymes, which in turn control the chemistry of cells, and that all this somehow accounts for growth and development. We do not know, however, exactly what determines which genes should be transcribed in which cells at a given time. To gain understanding of this is one of the most challenging problems for modern biologists. We do know a great deal about what happens during the growth and development of a plant. We know that chemicals called *growth substances* or *growth hormones* often play critical roles in many growth processes. Study of such substances has been an important thrust in plant physiology since early in this century. The two chapters following this introductory one are devoted to a summary of what has been learned, but much of the discussion in subsequent chapters also concerns growth substances.

It has become clear that development can be strongly modified by environment. Light, which plays an important role apart from photosynthesis, is considered in Chapters 20 and 23. Light effects are often exhibited within constraints placed by an internal timing mechanism that exists in plants and animals: the biological clock, which is the subject of Chapter 21. Plants also respond strongly to temperature changes; interesting responses, especially to low temperature, are the subjects of Chapter 22.

One of the most striking examples of morphogenesis in plants is the conversion from the vegetative to the reproductive stage. Cells must divide and differentiate in radically new ways, and hormones seem to be involved. In Chapter 23 we tell how changes in temperature and the relative lengths of day and night often modify or control flower initiation. But first, in this chapter, we summarize some general principles of growth and development.

[1]We can define a **machine** as an assemblage of parts capable of processing or directing energy and/or matter to produce a predetermined result or product. Living organisms are unique machines that, among other things, produce themselves.

## 16.1 What Is Meant By Growth?

Most commonly, **growth** means an increase in size. As multicellular organisms grow from the zygote, they increase not only in volume but in weight, cell number, amount of protoplasm, and complexity. In many studies it is important to measure growth. In theory, we could measure any one of the growth features just mentioned, but two common measurements quantify increases in volume or in mass.[2] Volume (size) increases are often approximated by measuring expansion in only one or two directions, such as length (for example, stem height), diameter (for example, of a stem), or area (for example, of a leaf). Volume measurements, as by displacement of water, can be nondestructive, so the same plant can be measured at different times. Increases in mass are often determined by harvesting the entire plant or the part of interest and weighing it rapidly before too much water evaporates from it. This gives us the **fresh mass**, which is a somewhat variable quantity because it depends on the plant's water status. A leaf, for example, often has a greater fresh mass in the morning than it does at midafternoon simply because of transpiration.

Because of problems arising from variable water contents, many people, particularly those interested in crop productivity, prefer to use the increase in **dry mass** of a plant or plant part as a measure of its growth. The dry mass is commonly obtained by drying the freshly harvested plant material for 24 to 48 hours at 70° to 80°C. The leaf that has a lower fresh mass at midafternoon will probably have a larger dry mass because it photosynthesized and absorbed mineral salts from the soil during the morning. Hence, dry mass may be a more valid estimate of what we mean by growth than is fresh mass. Of course, measurements of fresh and dry mass are usually destructive and require many samples if statistical significance is to be achieved; still, a plant grown hydroponically can be weighed at intervals for fresh mass with little effect on growth.

Sometimes dry mass does not give an adequate indication of growth. For example, when a seed, provided only with water, germinates and develops into a seedling in total darkness, the size and fresh mass increase greatly, but the dry mass decreases because of respiratory loss of $CO_2$ (Fig. 16-1). Although the total dry mass of such dark-grown seedlings is less than that of the original seed, the growing parts of the stem and root do increase in dry mass as assimilates are translocated from storage to the growing regions.

Normally, early stages in seedling development involve production of new cells by **mitosis** (nuclear division) and subsequent **cytokinesis** (cell division), but normal-appearing seedlings can be produced from seeds of some species in the absence of mitosis or cell division. When seeds of lettuce and wheat are irradiated with gamma rays from a cobalt-60 source at levels high enough to stop DNA synthesis, mitosis, and cell division, germination still occurs. Growth continues until seedlings with giant cells are produced. These seedlings, called **gamma plantlets**, can survive up to three weeks but then die, presumably because new cells are eventually necessary. Such gamma plantlets illustrate than even if we could conveniently measure the increase in cell number, this number might be a poor measure of growth. Many other examples of growth

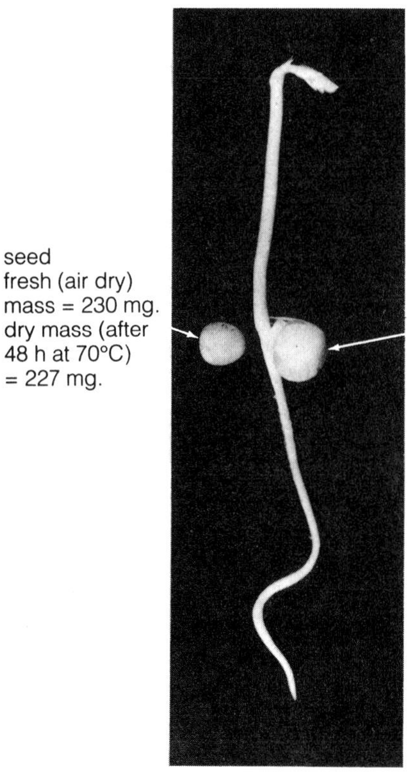


**Figure 16-1** Changes in fresh and dry mass of a pea seed as it develops into a seedling in darkness. The fresh mass increases greatly because of water uptake, but the dry mass decreases slightly because of respiration. (Photo by C. W. Ross.)

[2]An object's **weight** is not only a function of its mass but also the accelerational force being exerted upon it by gravity. Technically, weight should be expressed in newtons, the unit of force. (On earth, the weight of a 10-kg mass is about 98 newtons.) **Mass** is a fundamental quantity that does not change with the force of gravity (for example, location on earth or on the moon). It is measured by balancing it against a defined mass. A balance depends on accelerational force for its function, but the amount of force does not affect the reading. In everyday use, the term *weight* is an acceptable synonym for *mass*, but the distinction should be kept clear, and it is appropriate for plant physiologists to speak of fresh and dry *mass*, as we do here. (See Appendix A.)

without cell division are known, such as growth of certain leaves, stems, and fruits after a certain stage of development. There are also a few examples of cell division without increase in overall size, as in the maturation of the embryo sac. Nevertheless, an increase in size is the fundamental criterion of growth, even though it is not always easy to measure.

## 16.2 Patterns of Growth and Development

### Some Features of Plant Growth

Growth in plants is restricted to certain zones containing cells recently produced by cell division in a **meristem**. It is easy to confuse growth (as defined above as an increase in size) with cell division in meristems. Cell division alone does not cause increased size, but the cellular products of division do grow and cause growth. Root and shoot tips (**apices**) have meristems. Other meristematic zones are found in the vascular cambium and just above the nodes of monocots or at the bases of grass leaves. The root and shoot apical meristems are formed during embryo development while the seed forms and are called **primary meristems**. The vascular cambium and the meristematic zones of monocot nodes and grass leaves are indistinguishable until after germination; they are **secondary meristems**.

Some plant structures are determinate; others are indeterminate. A **determinate** structure grows to a certain size and then stops, eventually undergoing senescence and death. Leaves, flowers, and fruits are good examples of determinate structures, and the great majority of animals also grow in a determinate way. On the other hand, the vegetative stem and root are **indeterminate** structures. They grow by meristems that continuously replenish themselves, remaining youthful. A bristlecone pine that has been growing for 4,000 years could probably yield a cutting that would form roots at its base, producing another tree that might live for another 4,000 years. At the end of that time, another cutting might be taken, and so on, potentially forever; that is, plants can be **cloned** from individual parts. Some fruit trees have been propagated from stem sections for centuries.

Although an indeterminate meristem can be killed, it is potentially immortal. But death is the ultimate fate of determinate structures. When an indeterminate, vegetative meristem becomes reproductive (that is, begins to form a flower), it becomes determinate.

Although there are borderline cases, entire plants are in a sense either determinate or indeterminate. We use different terms, however: **Monocarpic species** (Greek *mono*, "single," and *carp*, "fruit") flower only once and then die; **polycarpic species** (*poly*, "many") flower, return to a vegetative mode of growth, and flower at least once more before dying. Most monocarpic species are **annuals** (live only one year), but there are variations on the theme. Many annuals germinate from seeds in the spring, grow during the summer and autumn, and die before winter, perpetuating themselves only as seeds. Spring wheats and ryes are commercial annuals that are planted in the spring, but seeds of *winter* wheat or rye germinate in the fall, overwinter as seedlings beneath the snow, and flower the next spring.

Typical **biennials**, such as beet (*Beta vulgaris*), carrot (*Daucus carota*), and henbane (*Hyoscyamus niger*) germinate in the spring and spend the first season as a vegetative rosette of leaves that dies back in late fall. Such a plant overwinters as a root with its shoot reduced to a compressed apical meristem surrounded by some remaining protective dead leaves (meristem plus leaves is called a **perennating bud**). During the second summer, the apical meristem forms stem cells that elongate (**bolt**) into a flowering stalk.

The century plant (*Agave americana*) may exist for a decade or more before flowering once and dying. Though a monocarpic species, it would be called a **perennial** because it lives for more than two growing seasons. It and many bamboos (*Bambusa* and other genera), which may live more than half a century before flowering once and dying, are excellent examples of the extreme monocarpic growth habit.

Polycarpic plants, perennials by definition, do not convert all their vegetative meristems to determinate reproductive ones. Woody perennials (shrubs and trees) may use only some of their axillary buds for the formation of flowers, keeping the terminal buds vegetative; alternatively, terminal buds may flower while axillary buds remain vegetative. Sometimes a single meristem forms only one flower, as in a tulip, whereas single grass or Asteraceae meristems form an inflorescence or head of flowers (for example, a sunflower). The bottle brush (*Callistemon* sp.) seems to form a terminal spike of flowers, but the apical meristem remains vegetative and continues to grow the next season, producing leaves and a woody stem. Woody perennials often become reproductive only after they are several years old. Until then, they are said to be in the **juvenile** stage. Herbaceous perennial dicots such as field bindweed (*Convolvulus arvenis*) or Canada thistle (*Cirsium arvense*) and perennial grasses die back each year in temperate climates, except for one or more perennating buds close to the soil. Some herbaceous perennial dicots form bulbs, corms, tubers, rhizomes, or other underground structures.

The seed contains a miniature plant telescoped into a tiny package, the **embryo**, which consists of embryonic root, shoot, and some primordial leaves (Dure, 1975). The gamma plantlets discussed in Section 16.1

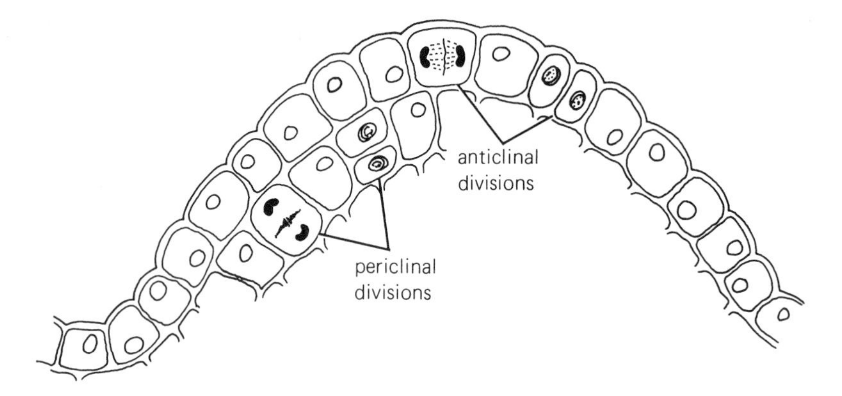


**Figure 16-2** The relation of anticlinal and periclinal divisions at the shoot apex.

can develop as far as they do because of embryo differentiation during seed formation. Normally, meristematic cells of the root and shoot apices give rise to other cells that divide to form branch roots, still more leaves, axillary buds, and stem and root tissues, including the vascular cambium. Many apical and axillary meristems eventually form flowers. In woody perennials, **lateral meristems** (**cambium**) produce **secondary xylem** and **phloem** each year, resulting in the growth in diameter of stems and roots.

## Steps in Cell Growth and Development

Although an amazing variety of forms is produced by growth and development (there are approximately 285,000 different species of flowering plants), all are accounted for by three simple (in appearance, at least) events at the cellular level. The first is **cell division**, in which one mature cell divides into two separate cells, by no means always equal to each other. The second event is **cell enlargement**, in which one or both daughter cells increase in volume. The third event is **cellular differentiation**, in which a cell, perhaps having achieved its final volume, becomes specialized in one of many possible ways. The variety of ways that cells divide, enlarge, and specialize accounts for the different tissues and organs in an individual plant and for the different kinds of plants.

To begin with, cells can divide in different planes. When the new wall between daughter cells is in a plane approximately parallel to the closest surface of the plant, the division is said to be **periclinal** (parallel to the *peri*meter; Greek *peri*, "around, surrounding"; *kline*, "bed"; denotes slope or slant, angle of incline). Alternatively, if the new wall is formed perpendicular to the closest surface, the division is **anticlinal** (Fig. 16-2). Division of a cell (cytokinesis) begins by production of a **cell plate**, which arises by the fusion of hundreds of tiny vesicles, most of which are pinched off from the ends of Golgi vesicles that contain noncellulosic polysaccharides such as pectins (see Section 1.4 and boxed essay on carbohydrate chemistry in Chapter 8). These vesicles fuse to form the pectin-rich **middle lamella**, which is bounded by membranes that were formerly part of the vesicles but now become the plasma membranes of the new daughter cells (Fig. 16-3). Subsequent formation of the new primary wall of each daughter cell also occurs, in part, by fusion of Golgi vesicles that contain other noncellulosic polysaccharides.

What guides the movement of Golgi vesicles to the equator of the cell, where the new primary dividing wall is formed during cytokinesis? Apparently, the vesicles migrate along the rodlike **microtubules** that extend toward opposite poles of the dividing cell (Section 1.6). Figure 16-3 shows numerous (barely visible) microtubules oriented with their long axes perpendicular to the cell equator. When formation of these microtubules is prevented by such antimitotic drugs as colchicine, Golgi vesicles do not move to the equator of the cell at anaphase. When the drugs are added after anaphase is nearly complete, the cell plate cannot form and cytokinesis cannot occur, so a binucleate cell is produced.

Not only does the direction of cell division have a lot to do with the formation of various structures, but the direction (or directions) of cell enlargement is also critical. Cell enlargement is largely a matter of absorption of water into the enlarging vacuole, as we shall soon see. In such elongated plant organs as stems and roots, the enlargement occurs mostly in one dimension; that is, it is in fact **elongation**. Of course, newly formed meristematic cells often enlarge in all three dimensions, but in stems and roots the enlargement soon becomes an elongation.

## Primary Wall Changes During Growth

Why do cells elongate mostly in one dimension rather than expanding equally in all directions? In Section 1.4 we explained that primary walls of growing cells consist of cellulose microfibrils embedded in an amorphous

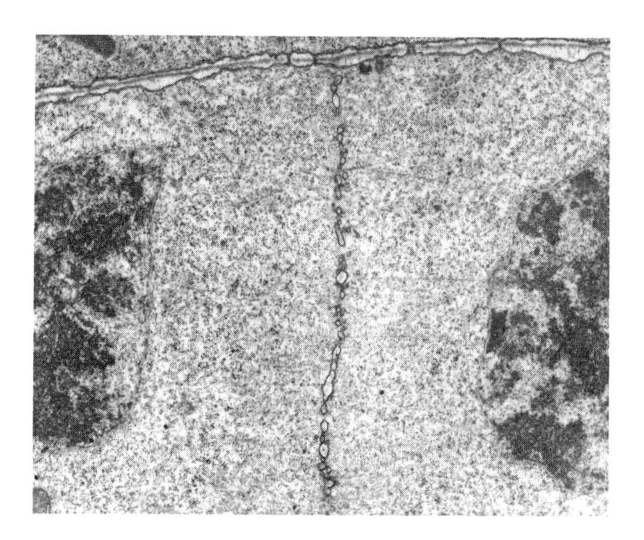

**Figure 16-3** Formation of the cell plate during cytokinesis in a cotton root tip. Pectin-rich vesicles pinched off from Golgi bodies fuse at the equator to form the new middle lamella and the two plasma membranes in contact with it. Subsequent formation of a primary wall involves noncellulosic polysaccharides that are secreted from each cell in additional Golgi vesicles and deposited into and onto the middle lamella, whereas cellulose appears to be formed in each plasma membrane without Golgi vesicle involvement. Narrow rodlike microtubules oriented perpendicular to the cell plate might function in guiding Golgi vesicles to this plate. Formation of the nuclear envelope (probably from the endoplasmic reticulum) around each daughter nucleus is nearly complete. Numerous ribosomes (tiny dots) are also visible. (Micrograph courtesy of Dan Hess.)

matrix of noncellulosic polysaccharides and some protein. Each cellulose microfibril behaves as a multistranded cable that minimizes extension in the direction of its long axis; but wall growth can occur in a direction that allows microfibrils to move apart as in an expanding spring, so growth is favored in a direction at right angles to the orientation of the microfibril axes. As growth continues, new microfibrils are deposited into the wall adjacent to the plasma membrane, so the wall retains a near-uniform thickness during growth. It is this innermost layer of microfibrils, those most recently deposited, that apparently has the most control. Apparently, when new cellulose molecules are formed during growth, existent microfibrils may be lengthened, allowing some extension parallel to their axes. (See reviews by Bacic et al., 1988; Delmer, 1987; Delmer and Stone, 1988; see also the essay on the cell wall by Nicholas C. Carpita in this chapter.)

If the orientation of new microfibrils is random, growth tends to be equal in all directions (as in fleshy fruits or leaf spongy mesophyll cells). In many young cells, however, microfibril orientation is not completely random but occurs predominantly along one axis (Fig. 16-4). Growth is then favored in a direction perpendicular to that axis, as in elongating roots, stems, and petioles. Evidence from onion root hairs suggests that the

(*Text continues on page 337.*)

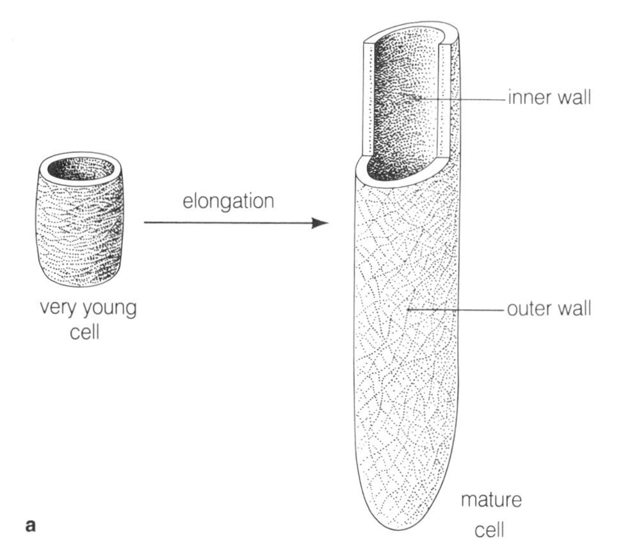


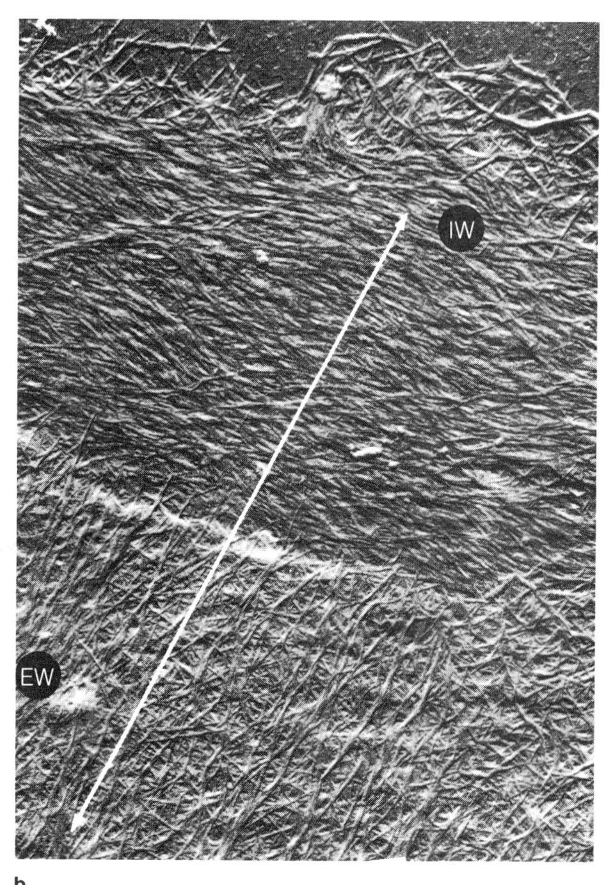


**Figure 16-4** (**a**) Changes in orientation of cellulose microfibrils during cell elongation. In the young cell, the microfibrils are oriented almost randomly, but expansion occurs longitudinally because the newly deposited microfibrils on the inner surface of the wall are oriented perpendicular to the cell's long axis. The older microfibrils on the outside of the wall are pulled into the direction of elongation during growth. (**b**) Orientation of cellulose microfibrils in the inner (younger) and outer (older) part of the primary wall. Shown is a leaf hair cell of *Juncus effusus*, a sedge. Note that the microfibrils on the interior of the wall (IW) are perpendicular to the long axis of the cell, whereas those on the exterior of the wall (EW) are parallel to the long axis of the cell, and those in between are intermediate in orientation. The direction of cell elongation is given by the long arrow. (From Jensen and Park, 1967.)

# A Special Importance of the Primary Cell Wall in Plant Development

*Nicholas C. Carpita*

*Nick Carpita was born in Indiana but grew up near the Gulf of Mexico in Clearwater, Florida. After undergraduate work at Purdue University (Indiana) and doctoral studies at Colorado State University (Ph.D. in 1977), he was a postdoctoral fellow at the Michigan State University—Department of Energy Plant Research Laboratory, where he was fascinated by the* in vivo *and* in vitro *synthesis of cellulose in cotton fibers. In 1979 he returned to Purdue University, where he is now Professor of Plant Physiology. In this essay he gives us some insight into his interests in the molecular structures and biosynthesis of cell-wall polysaccharides in cereal grasses; if space allowed, he could also tell us about his work on the molecular control of cell enlargement, cell-wall adaptation to osmotic stress, and the structure and biosynthesis of fructans.*

One of the more obvious distinctions between animal cells and plant cells is the plant cell wall, a rigid but dynamic matrix that surrounds each cell. Of course, animal cells possess extracellular matrices that function in adhesion and can direct cell shape to a certain extent, but evolution of a thick, cellulosic wall totally altered the pattern of development in plants. Plant cells, which are rarely without their walls, are fixed spatially within the plant, autonomous in some ways but also dependent on communication with neighboring cells for coordinating cell expansion and differentiation.

For structural integrity, plant cells develop turgor by accumulating solutes far in excess of the surrounding medium and using the cell wall to control volume in resisting the pending inrush of water. The wall becomes an interface through which a plant cell perceives its environment. Porosity of the wall is quite small (4 ± 2 nm; Carpita et al., 1979), but tensile strength is enormous, with breaking pressures of the primary wall of almost 1,000 MPa (Carpita, 1985). Normal cell turgor never approaches breaking pressure, yet during differentiation the primary walls are unusually pliant and slowly and discretely stretch to give rise to the myriad of shapes and sizes that plant cells possess. How does the plant cell do this? How is shape altered? How do environmental factors and hormonal signals act through modulation of gene expression to alter the biochemical and physical properties of the cell wall and make differentiation possible? I believe these are the most fundamental of the questions that remain to be explained in plant development.

I first began to appreciate the importance of the cell wall in control of cell shape when I was a graduate student. I studied how phytochrome induced lettuce seed germination, specifically how phytochrome led to alteration of the water relations and growth physics to cause the radicle to push its way out through the tough endosperm layer that mechanically controls dormancy. To better quantify the changes in water potential, embryos isolated from seeds were given promoting red light or inhibiting far-red light, and their growth rates were measured as increases in fresh mass or length. In short, phytochrome ultimately induces a lowering of the $\Psi$ that enables the radicles to grow in more concentrated solutions of PEG [polyethylene glycol; see Fig. 3.6] or in the seed, and thus they develop enough pressure to burst through the endosperm. When comparing data from length and fresh mass, however, I noted that the radicles of the red-light-treated seeds were always longer and thinner than those from far-red-light-treated seeds. The significance of this effect is unclear, but the observation stimulated my curiosity about the cell wall. The reason for this alteration in cell form had to be a change in the cell-wall chemistry that modulated the direction of expansion.

Although I never pursued this particular problem further, I wanted to delve deeper into the biochemistry of cell-wall structure and synthesis and had the good fortune to be able to work with Professor Debbie Delmer at the Michigan State University—Department of Energy (MSU—DOE) Plant Research Laboratory. There I worked specifically on cellulose biosynthesis in a beautiful model system in which unfertilized cotton ovules cultured *in vitro* produced fiber cells much like those in the real plant. I was also interested in hormone modulation of cell-wall synthesis during cell expansion of excised maize coleoptiles, and Debbie gave me the freedom to work on related problems such as these. In preliminary experiments I was amazed to find that over 80 percent of labeled L-arabinose taken up by the coleoptiles ended up in cell walls, whereas much of the labeled D-xylose remained in cytoplasmic or vacuolar pools. I must admit that I have always had a strong impulse to do an experiment first and search the literature later. The reverse chronology often saves a lot of time, but then one misses the opportunity to be amazed now and again! The salvage pathway for arabinose into UDP-arabinose and UDP-xylose had been worked out two decades before, and it was known since the mid-1920s that grasses were notably enriched in arabinoxylans. It was still intriguing to me that the salvage pathway was so efficient for arabinose and not for xylose, and what a dynamic compartment the cell wall must be. My fascination with the cell wall of grasses has occupied most of my time since I came to Purdue University over 10 years ago.

Most published models of the primary cell wall of plants are those of dicots. Walls of grasses (the family Poaceae) are very different from those of dicots and even other monocots that have been examined. It's ironic that comparative studies of the effects of light and hormones on the growth physics and water relations of tissues from grasses and dicots have revealed so many similarities because we have only recently begun to appreciate that grasses and dicots use completely different materials to give rise to properties of elasticity, yield threshold, and extensibility that are so important in growth.

Fundamentally, the dicot primary cell wall (Bacic et al., 1988) is a network of cellulose microfibrils, each of which consists of several dozen linear chains of (1-4)β-linked D-glucose, condensed to form long crystals that wrap around each cell. Although each chain may be just several thousand glucose units long (ca. 100 μm), they begin and end at different places, so the microfibrils themselves are huge, and rarely does electron microscopy reveal the beginning or the end of a microfibril. In elongating cells, the microfibrils are wrapped transversely around the longitudinal axis like tightly wound springs. Because the force generated by turgor pressure of a relatively large cell (100 μm diameter) is borne by a very thin primary wall (0.1 μm), a turgor pressure of 1.0 MPa generates at least 250 MPa of tangential and longitudinal force that the wall must resist (Carpita, 1985). The orientation of the microfibrils prevents the growing cylinders from becoming spherical, just as a "slinky" toy or tightly wound spring is difficult to pull outward. The "slinky" is quite easily stretched, however. How does the cell wall resist the enormous longitudinal force generated by turgor? Here lies the importance of the noncellulosic matrix of the primary wall. These polysaccharides are the "glue" that holds the microfibrils together (side by side) to resist this force. *The orientation of synthesis of the cellulose microfibrils establishes ultimate cell shape, whereas the dynamic interaction of the cellulose and noncellulosic polysaccharide matrix dictates the rate of cell expansion.* Much of the work in laboratories today is directed toward understanding this dynamic interaction, particularly the chemical structures of its constituents.

In dicots, the principal "glue" is xyloglucans, which are also linear chains of (1-4)β-D-glucan; but unlike cellulose, they possess numerous xylose units added at regular sites at the 0-6 position of the glucose units of the chain, fanning out like wings to make a flattened ribbon. One surface is able to bind tightly to the surface of cellulose microfibrils, and many models envision xyloglucan as a monolayer coating of the microfibrils. Considering that there are about equal amounts of xyloglucan and cellulose in the primary wall, however, a large amount of the xyloglucan must span the milieu between cellulose microfibrils. Several models suggest the existence of covalent interactions of xyloglucan with other polymers. It is equally likely, however, that the chains could self-interact noncovalently, like a chain-link fence, to tie the microfibrils together. Because this interaction is a major determinant in resistance to longitudinal force, many workers have investigated the rate of hydrolysis of xyloglucan as a possible major determinant of growth rate.

This cellulose-xyloglucan framework is also embedded in a matrix of pectic polysaccharides. These jellylike polymers are some of the most complex polymers known and are thought to perform numerous functions such as determining wall porosity, providing charged surfaces that modulate wall *p*H and ion balance, and even serving more subtle roles as recognition molecules that signal appropriate developmental responses to symbiotic organisms, pathogens, and predators. A tight interaction of the pectic substances with xyloglucan may also constitute a control of hydrolysis during cell expansion. Two fundamental constituents of pectins are **polygalacturonic acids (PGAs)**, which are helical homopolymers of (1-4)β-D-galacturonic acid, and **rhamnogalacturonans (RGs)**, which are contorted rodlike heteropolymers of (1-2)β-D-rhamnose-(1-4)α-D-galacturonic acid repeating units. The latter polymer is the reason for the complexity of pectic structure. Other polysaccharides such as **arabinans**, **galactans**, and **arabino galactans** of various configurations and sizes are attached to many of the rhamnose residues.

Pectins interact in the wall in many ways. First, the helical chains of PGAs can be condensed by crosslinking with $Ca^{2+}$ to form "junction zones," linking several chains together and forming a gel. There are stretches of PGA at the ends or within RG to link these two types of polymers together. Actually, the PGA and RG are secreted as methyl esterified polymers, and the enzyme pectin methylesterase located in the cell wall cleaves some of the methyl groups to initiate binding to $Ca^{2+}$. Some investigators believe that the activity of this enzyme may also exert some control on the activity of enzymes responsible for xyloglucan metabolism. Pectins can be crosslinked further via ester linkages with **dihydroxycinnamic acids**, such as deferulic acid, to form covalent attachments to other polymers. The size of the junction zones consisting of PGA and the size and frequency of polymer substitution on RG can constitute a fine control of wall porosity and matrix charge that, in turn, can influence developmental phenomena in ways we probably do not yet appreciate.

Once elongation is complete, the primary wall is locked into shape. A principal locking molecule is the hydroxyproline-rich glycoprotein **extensin**. Extensin is a rodlike, cation-rich protein that is thought to cross-link to itself or to the pectic substances to prevent further expansion of the cellulose microfibrils. Extensin may even have initiation sites for lignification. This is a field that is rich in research potential because of the recent discovery of several other proteins that appear during differentiation or in response to either pathogen invasion or wounding by insect predation.

(*continued*)

# A Special Importance of the Primary Cell Wall in Plant Development *(continued)*

The grass primary cell wall (Carpita, 1987) is also composed of cellulose microfibrils similar in structure to those of dicots, but that's where the similarities end. Instead of xyloglucan, the principal polymers that interlock the microfibrils are **glucuronoarabinoxylans (GAXs)**, linear chains of (1-4)β-D-xylose with wings of single arabinose units and, much less frequently, single glucuronic acid units. The degree of arabinose and glucuronic acid substitution varies markedly from GAXs whose xylose units are nearly all substituted to chains in which only 10 percent or less of the xylose units bear side-groups. The degree of substitution greatly affects their chemistry. Like cellulose, the unsubstituted (1-4) linked xylan molecules can tightly hydrogen bond to cellulose or to each other, but the side-groups greatly interfere with this binding.

In dividing and elongating cells, very highly substituted GAXs are abundant, whereas during the course of elongation and differentiation, more and more unsubstituted GAX accumulates. I believe therein lies the explanation for the salvage pathway for arabinose. The arabinosylated GAXs may be the soluble synthesized form that is converted to other GAXs through selective hydrolysis of the arabinose units. The arabinose is then absorbed and salvaged into nucleotide sugars for synthesis of new GAX. The arabinose units may serve an additional function in the cell wall. Grasses are notably poor in pectin. They contain both PGA and RG, but the highly substituted GAX is closely associated with these pectins. The cell-wall model that I suggested based on all of these observations was one of cellulose microfibrils interlaced with unsubstituted GAXs. Additional GAXs of varying degrees of substitution functionally replaced the pectic substances that predominate in the dicot cell wall, and the degree of substitution by arabinose and glucuronic acid controls porosity and surface charge.

Most remarkable in grasses is the synthesis of developmental-stage-specific polysaccharides. Although the model just depicted may hold for newly divided cells, the constituents of the wall change markedly when cells begin to expand. Some arabinans that are no longer made during cell expansion are found in the walls of dividing cells. Instead, new polymers, called **mixed-linkage β-D-Glucans (β-D-glucans)**, are synthesized along with GAX. β-D-glucans are unbranched homopolymers of glucose containing a mixture of (1-4) linear oligomers and 1, 3-β-D-glucose "kinks." The fine structure is more complicated, however. In most of the polymer, the (1-3) kinks occur at specific intervals to link approximately 10 cellotriose units to every five cellotetraose units. This produces a flattened, corkscrewlike polymer about 50 residues long. These corkscrew units are linked via special regions that contain longer stretches of the linear (1-4) linked units. Some contiguous (1-3) linked units may be spliced into these special regions to give added flexibility to the macromolecule, making it a "molecular thread." The β-D-glucan macromolecule is thought to be the microfibril "glue" of the grasses analogous to the dicot xyloglucan, and the longer stretches of (1-4) linked units are significant in its metabolism. An endo-β-D-glucanase that can hydrolyze the β-D-glucan from microfibrils to permit expansion was discovered in the cell walls of developing maize seedlings. Investigations in several laboratories are now focused on precisely how the β-D-glucans and their specific hydrolases function in cell elongation.

Just as in dicots, the grasses are locked into form during differentiation. Unlike dicots, however, they do not use the hydroxyproline-rich extensin. Grasses possess similar kinds of structural proteins, but a large proportion of the cross-linking function resides with hydroxycinnamic acids to form grass lignin, even in the primary cell wall.

If one can now appreciate the complex structure of the cell wall, the interaction of its components, and its subtle alterations during growth, imagine how such a complicated structure is synthesized. Except for cellulose, all the cell-wall polymers are synthesized in the endoplasmic reticulum and Golgi apparatus, secreted in some soluble form to the wall, and then organized into the insoluble matrix. We know very little about the synthesis, and no synthase complex has ever been purified. Although these synthases may be quite interesting and may be relevant polypeptides of developmentally expressed genes, we do not yet have a handle on getting at those genes. We do know a little about how xyloglucan is made: the import mechanisms of nucleotide sugars into the lumen of the Golgi, the coordinated use of UDP-glucose and UDP-xylose, and even possible involvement of membrane potential and *p*H gradients in driving both synthesis and secretion of the polymer-laden vesicles to the cell surface. As I write this essay, Dr. David Gibeaut, a research associate in my laboratory, has developed an ingenious new technique to isolate the Golgi apparatus from contaminating compounds from vacuole and cytosol to provide the first unequivocal demonstration of the *in-vitro* synthesis of the grass mixed-linkage β-D-glucan.

Although we have assembled a reasonable catalog of the protein and polysaccharide constituents of the plant cell wall, we still have only rudimentary visions of how the wall is organized in three-dimensional space, how it is altered specifically during growth, how such a fine physiological control of wall metabolism is achieved, and least of all, how it is synthesized. Each enzymic component of the machinery—the synthases, the structural proteins, the hydrolases, and even the enzymes of the salvage pathways—must have its own special regulation. The more we learn about these mechanisms the greater will become our understanding of how the regulatory genes that control them make development possible. Questions about how the developmental machinery works await fresh ideas and approaches.

orientation of microtubules and thus possibly microfibrils is much like a spring (a helix or spiral) coiled around the elongating cell (Lloyd, 1983).

If the pattern of cellulose microfibril deposition is so important in controlling final cell shape, what controls that orientation? In our discussion of *cytoskeleton dynamics in cell-wall formation* in Section 1.6, we noted that cellulose formation may be controlled by enzymes located on the plasmalemma: *globules* on the outer face and *rosettes* on the inner face of the membrane. Movement of these structures seems to be guided by the microtubules that often appear in close association with them (see Fig. 1-15). Evidence that microtubules are involved comes from the use of certain drugs that prevent microtubule formation. Addition of these drugs causes new cellulose microfibrils to be randomly oriented, and removal allows renewed production of transversely oriented microtubules and microfibrils. If microtubules control microfibril arrangement, we must learn what controls microtubule patterns.

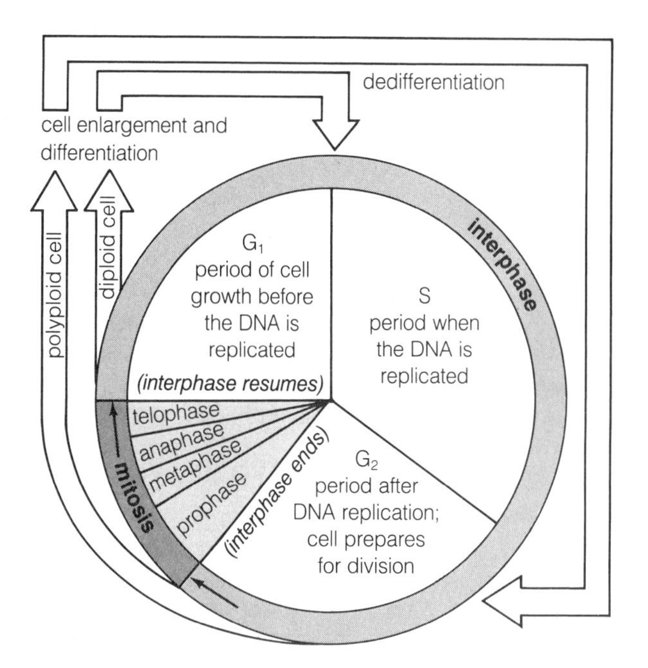


**Figure 16-5** A generalized diagram of the cell cycle. There is great variation among different cells in the length of time a cell stays in any phase. In plant cells (less frequently in animal cells), wounding or some other treatment will often cause differentiated cells to again become meristematic — to dedifferentiate. Where they reenter the cell cycle will depend upon where they left it. If the cell leaves the cycle after S but before mitosis, it will be polyploid. (Adapted from Starr and Taggart, 1981.)

## The Cell Cycle

Cell biologists, including plant physiologists, are interested in the series of repeating events called the **cell cycle**, which is reviewed in Figure 16-5. This cycle is concerned largely with the time of DNA replication in relation to nuclear division (John, 1982). After mitosis there is a period of cell growth before DNA replication (**$G_1$**), then DNA replication (**S**), then growth after replication (**$G_2$**), and finally mitosis to complete the cycle. In terms of the events we have been discussing, one of the daughter cells produced by mitosis may not continue in the cell cycle but may enlarge and differentiate. If this occurs before DNA replication, the differentiated cell will have the normal diploid number of chromosomes and amount of **chromatin** (genetic material), but in plants it is not unusual for differentiation to occur after DNA replication, so that the differentiated cell has more than the diploid quantity of chromatin. Sometimes chromosomes are further duplicated without cell division, and the differentiated cell becomes polyploid. Often these polyploid cells are larger than their diploid counterparts. As the diagram illustrates, differentiated plant cells may sometimes reenter the cell cycle by a process called **dedifferentiation**, after which they again have the ability to divide; that is, they are again meristematic (see Section 16.6).

## The Physics of Growth: Water Potentials and Yield Points

How do plant cells grow in volume? Such growth is primarily caused by an uptake of water that stretches the cell walls, but new cell-wall and membrane materials are synthesized so that the wall usually doesn't get thinner. In a few cases (for example, root hairs and pollen tubes), the wall increases in area only at the tip, but in most higher-plant cells, growth occurs throughout the lateral surfaces.

The biophysics of cell growth has been an active area of research during recent years, and there are several excellent reviews (for example, Boyer, 1988; Cleland, 1981; Cosgrove, 1986, 1987; Dale, 1988; and Taiz, 1984). Much of the current thinking goes back to a 1965 paper by James A. Lockhart, then at the University of Hawaii in Honolulu (Lockhart, 1965).

What causes a cell to absorb water and enlarge? An old hypothesis suggested that the wall and plasmalemma were extended in small stepwise increments by metabolic activities of the cell, and that water entered at each step to fill the void. The opposite, modern interpretation is that water pressure (turgor) drives growth by forcing the wall and membranes to expand. The rate of water movement into a cell is governed by two factors: the water potential gradient and the permeability of the membrane to water. Hence, the rate of cell enlargement is also proportional to these factors, to a first approximation.

Equation 3.1 in Section 3.2 introduced the basic osmotic relations:

$$\Psi = \Psi_s + \Psi_p = \Psi_s + P$$

where $\Psi$ = water potential, $\Psi_s$ = osmotic or solute potential, and $\Psi_p$ = pressure potential, which is a real pressure, $P$, called turgor pressure. The difference in water potential ($\Delta\Psi$) inside and outside a cell is:

$$\Delta\Psi = (\Psi_{se} + P_e) - (\Psi_{si} + P_i) \quad (16.1)$$

where e donotes external and i, internal.

Assuming that outside pressure is negligible and including a factor for the relative hydraulic conductivity of the membrane to water ($L$), we can write the equation for the rate of cell enlargement as follows:

$$\frac{1}{V} \cdot \frac{dV}{dt} = L(\Delta\Psi) = L(\Psi_{se} - \Psi_{si} - P_i) \quad (16.2)$$

where $V$ is cell volume, $dV$ is an incremental change in the volume of the cell, and $dt$ is an infinitesimal increment of time. Thus $dV/dt$ is the rate of cell increase in volume or the rate of growth. But growth is a geometric or logarithmic function of time (see Section 16.3, which follows). To express this, many workers divide $dV/dt$ by the cell volume to give a **relative growth rate** of the cell, which we can define as $r$. A similar form of the relative growth rate uses the natural logarithm of cell volume: $d\ln V/dt$.

Equations 16.1 and 16.2 show that there are two ways for water potential inside a cell to be more negative than water potential outside the cell, making water uptake and growth possible: *Solutes* inside the cell might *increase*, making the osmotic potential inside more negative, or *pressure* inside the cell might *decrease*. As it turns out, solute concentrations inside many growing cells remain constant within the limitations of measurement. In those cases, the driving force for growth must be a decreasing turgor pressure. Pressure in the cell is caused by the mechanical resistance of the cell wall to being stretched. If this resistance is lowered so that the wall relaxes, its stretching leads to a lowered pressure, which reduces cell water potential, leading to a larger water potential gradient ($\Delta\Psi$) and movement of water into the cell.

This principle was perceived by A. N. J. Heyn in the Netherlands in the early 1930s. Heyn (1931) oriented a stem segment horizontally, fixed it at one end, and placed a weight on the other end. The weight caused the stem to bend down. When the weight was removed, the stem returned *part way* to its original position. Heyn reasoned that the weight caused the cell walls to stretch **elastically**, as would a rubber band (accounting for the bending back), and **plastically**, as bubble gum stretches without returning to its original dimensions (accounting for the irreversible bending). In Heyn's experiments, plastic stretching was increased by auxin, a plant growth regulator (see Section 17.2). Thus Heyn introduced the concept of wall **extensibility**, which has both elastic and plastic components.

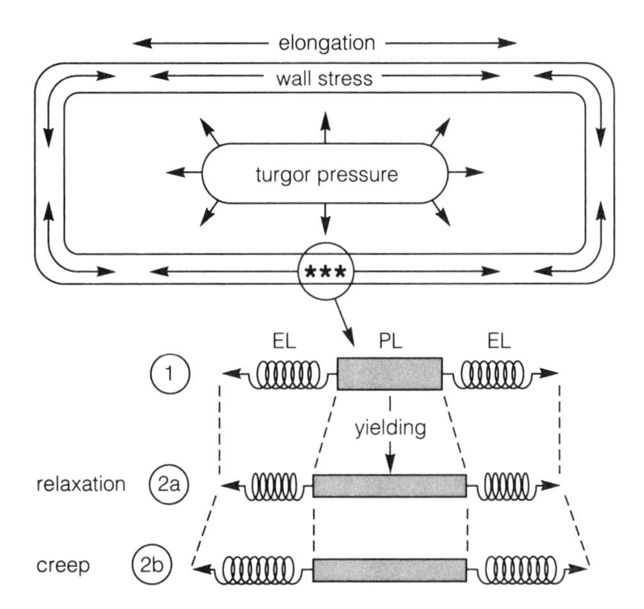


**Figure 16-6** A mechanical model of the growing cell wall. Turgor pressure exerts a force against the wall and sets up a tensile stress in the plane of the wall. Point 1 shows that both elastic (EL) and plastic (PL) elements bear the wall stress. Elastic elements are shown as springs in which extension is proportional to stress. In going from point 1 to point 2a, the plastic elements yield (relax), but the elastic elements contract (showing that wall stress has decreased), so there is no net change in wall length. At point 2b, wall stress is maintained constant (elastic elements do not contract) as the wall yields and thus allows expansion. (From Cosgrove, 1987.)

Plastic wall stretching is achieved as the cell wall is **loosened**, so the cellulose microfibrils can slide past each other more easily. This is called **shear**; it involves a breaking of bonds between adjacent microfibrils, but the exact mechanism of wall loosening (**wall yielding**) is not yet fully understood.

We can think of the plastic and elastic elements in the wall as being in series with each other, as in Figure 16-6. As the plastic elements relax, they stretch, allowing the elastic elements to shorten; this can happen only if wall stress and turgor pressure are reduced. If water enters almost instantaneously as pressure decreases in response to relaxation of the plastic elements, the elastic elements might shorten only infinitesimally, so pressure decreases only infinitesimally. In this steady-state, growth process, called **creep**, wall stress and turgor remain constant.

Relative cell growth rate ($r$) proves to be proportional to the extent that turgor pressure ($P$) exceeds a value called the **yield threshold** or **yield point** ($Y$). This empirical relation is shown in Equation 16.3 (Lockhart, 1965):

$$r = \phi(P - Y) \quad \text{for } P > Y \quad (16.3)$$

The proportionality factor $\phi$ is the **wall extensibility**. Because actual wall properties that lead to deformation (**rheological properties**) are complex and difficult to measure, the yield point ($Y$) is expressed as the minimum *pressure* necessary to cause cell growth. Turgor pressure is the cause of existing wall stress and is thus an indirect but equivalent measure of it. Together, Equations 16.2 and 16.3 show that relative cell growth rate depends upon five interrelated factors: conductivity of walls and membranes to the water that bathes the cells, the difference in osmotic (solute) potential inside and outside the cell, cell turgor pressure, and two wall-yielding properties: extensibility and yield threshold.

Technically, **wall relaxation** is the reduction in wall stress at constant cell-wall dimensions. During recent years, wall relaxation has been measured in living tissues to obtain data about the yield threshold and wall extensibility. Two approaches have been used: In the first, the growing tissue is isolated from a water supply to prevent water uptake; in the second, water uptake is prevented in a pressure chamber by increasing the external pressure on the tissue just enough to reduce growth to zero. In either approach, wall stress is monitored by measuring turgor pressure, and one convenient way to do this is with the pressure probe (Section 3.6). The techniques also permit an estimation of hydraulic conductance ($L$). These studies have validated Equations 16.2 and 16.3 by direct measurements (Cosgrove, 1985).

As wall relaxation occurs, turgor pressure drops exponentially until the yield point ($Y$) is reached at some value above zero. The rate of turgor-pressure decrease can be used to evaluate the wall extensibility ($\phi$).

Before *in vivo* wall-relaxation techniques were developed, extensibility could be measured only on dead tissues (reviewed by Cleland, 1981). The most widely used technique for measuring the mechanical properties of dead cell walls is called the **Instron technique** (Cleland, 1984) after the instrument that is used to stretch the tissues.[3] Stem sections or other samples are prepared by boiling them in alcohol to remove the proteins from the cell walls and to disrupt the protoplasts. A sample is then subjected to two successive extensions at a constant rate of **strain** (deformation), and the **stress** (force) along the walls is measured as a function of extension. The first extension involves both an elastic (reversible) and a plastic (irreversible) component, but the second extension is entirely reversible. From the slopes of the two curves, plastic and elastic extensibility can be determined.

*In vivo* stress relaxation uses the natural stresses on the wall (cell turgor), rather than an externally applied stress. Turgor applies hydraulic stress in all directions from within the cell instead of a pulling stress applied externally in only one direction. Furthermore, current, metabolically controlled aspects of wall loosening are measured, instead of merely measuring the potential extensibility that was present at the time the tissue was prepared.

[3]The instrument is used by physicists and engineers to measure the mechanical properties (stress/strain relations) of various materials.

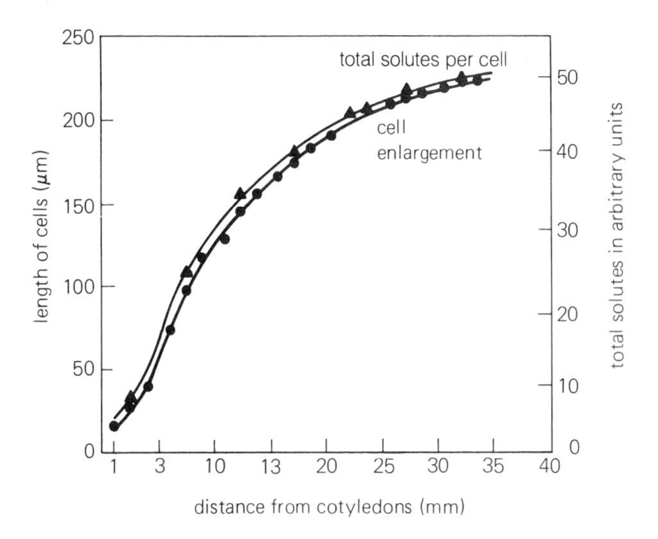


**Figure 16-7** Relation between cell length and solute content in epidermal cells of sunflower hypocotyls. The cells are at different stages of development, so the abscissa could logically be labeled *time* or *age* instead of *distance from cotyledons*. Etiolated seedlings were 90 h old and 45 to 50 mm tall. Lengths and widths of the growing epidermal cells were measured microscopically. Solute concentrations were measured from incipient plasmolysis values and then converted to arbitrary units by multiplying by the cell length. (Cell diameters remained almost constant during growth.) Plants were grown in a peat-sand mixture and were watered only with tap water. (Data from Beck, 1941.)

Interestingly, all the techniques—including the initial indirect experiments of Heyn plus some techniques not discussed here—clearly show that auxins cause markedly increased plastic extensibility of the cell walls. It is also clear that cell growth is driven by wall loosening, as just described.

In an enlarging cell, water that enters would soon dilute the solutes and thus lower $\Psi_{si}$, if solutes were not absorbed from the surroundings or synthesized in the tissue. Such solute accumulation often does accompany growth, as is illustrated in Figure 16-7. The sunflower hypocotyl cells represented in the figure expanded in length 15-fold, yet their solute concentrations remained essentially constant as the increased solutes closely matched cell-size increases. In a more recent study, also with sunflower seedlings, the accumulated solutes proved to be mostly glucose, fructose, and $K^+$ translocated from the cotyledon cells (McNeil, 1976).

What would happen in a tissue growing in pure water with no access to a solute supply such as mineral salts from a soil solution or sugars derived from storage or photosynthesis? Water uptake would dilute the existing solutes, and $\Psi_{si}$ would rise toward zero as in the Höfler diagram of Figure 3-3. Because $\Delta\Psi$ across the plasma membrane is very small (as little as 0.0003 MPa; Cosgrove, 1986), $P_i$ must decrease, and growth will eventually stop when the yield threshold is reached, unless the threshold decreases. Usually, growth stops in the absence of a solute supply, apparently because the wall either retains its rigidity or becomes even less plastic. Obviously, a plant requires water as the driving force for growth, but continued water uptake requires mineral salt absorption or sugars and other organic solutes provided by translocation or photosynthesis. This fact (along with the essential functions of mineral elements, sugars, and other organic solutes in metabolic processes) is essential to understanding how the mineral environment influences growth.

The extreme sensitivity of growth to water stress (Hsiao, 1973) occurs because the yield threshold ($Y$) is often very close to cell turgor ($P_i$; Matyssek et al., 1988). When soil dries, or solutes increase in the solution bathing the roots (lowering $\Psi_{se}$), growth stops when $P_i$ equals $Y$, which is well before $P_i$ reaches zero and before tissues wilt. Furthermore, Equation 16-3 shows that the growth rate is sensitive to the extent by which $P_i$ is raised above the turgor threshold (but see the discussion of the *mechanics of gravitropic bending* in Section 19.4).

Equations 16.2 and 16.3 imply that growth can be modified by changes in hydraulic conductance, extensibility, or yield threshold. In the alga *Nitella*, the cell apparently modifies its yield threshold to maintain a constant growth rate over a range of $\Psi_{se}$ values (Green et al., 1971). We have noted that, in higher plants, auxins can increase wall extensibility. Cytokinins and gibberellins can also increase extensibility and perhaps lower yield thresholds in certain sensitive tissues (Cleland, 1981).

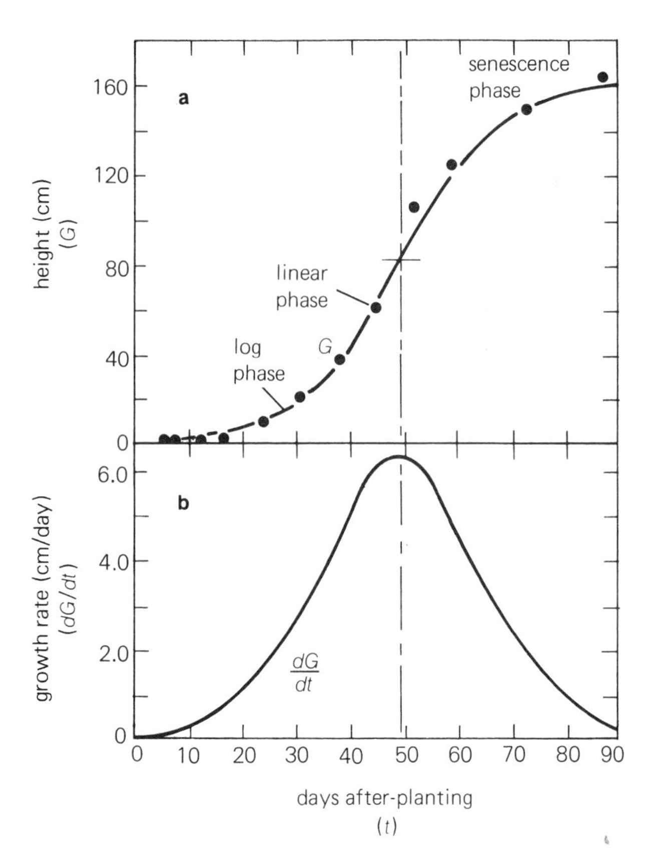


**Figure 16-8** A nearly ideal sigmoid growth curve and bell-shaped growth-rate curve for maize. The rate curve in **b** is the first derivative (the slope) of the total-growth curve in **a**. (Drawn from data of Whaley, 1961.)

## 16.3 Growth Kinetics: Growth Through Time

### Whole Organs: The S-Shaped Growth Curve

Many investigators have plotted size or weight of an organism against time, producing a growth curve. Often the curve can be fitted with a simple mathematical function, such as a straight line or a simple S-shaped curve. Although the metabolic and physical processes that produce the growth curves are too complex to be explained by simple models, simple curves are often useful in interpolating from measured data. In addition, the coefficients that must be supplied so that the equations fit the curves can be used to categorize the effects of an experimental treatment (such as an irrigation regime or application of a growth regulator) on the growth of observed plants or plant organs.

An idealized **S-shaped (sigmoid) growth curve** exhibited by numerous annual plants and individual parts of both annual and perennial plants is illustrated for maize in Figure 16-8a. The curve shows cumulative size as a function of time. Three primary phases can usually be detected: a logarithmic phase, a linear phase, and a senescence phase (Sinnott, 1960; Richards, 1969).

In the **logarithmic phase**, the size ($V$) increases exponentially with time ($t$). This means that the **growth rate** ($dV/dt$) is slow at first (Fig. 16-8b), but the rate continuously increases. The rate is proportional to the size of the organism; the larger the organism, the faster it grows. A logarithmic growth phase is also exhibited by single cells, such as the giant cell of the alga *Nitella*, and by populations of single-celled organisms, such as bacteria or yeasts, in which each product of division is capable of growth and division. Mathematicians have drawn an analogy between this logarithmic phase and the growth of money that draws compound interest. The accumulated interest also draws interest, so the total principal grows exponentially.

In the **linear phase**, increase in size continues at a constant, usually maximum rate for some time (Fig. 16-8b). (The constant growth rate is indicated by a constant *slope* in the upper, height curves and by the horizontal part of the lower, rate curves: part of the rate curve for Alaska pea, all of the curve for Swartbekkie pea in Fig. 16-9.) It is not always clear exactly why the growth rate in this phase should be constant, rather than proportional to the increasing size of the organism, but if we are measuring the growth of a single, unbranched stem, the linear phase may simply express the constant activity of its apical meristem.

The **senescence phase** is characterized by a decreasing growth rate (note drop in rate curve in Fig. 16-8b) as the plant reaches maturity and begins to senesce. We will discuss senescence later.

Although the curves of Figure 16-8 are representative of many species, growth curves of other species and organs are often different. In Figure 16-8 the linear phase is hardly detectable, so the logarithmic and senescence phases are almost continuous. More commonly, the linear phase is extended. The Swartbekkie pea (Fig. 16-9) is a rather extreme example. Its growth rate was constant at about 21 mm height increase per day for nearly two months. (Senescence phase is not shown, although it occurred later.) Alaska pea, another tall cultivar, showed a more sigmoid growth curve and bell-shaped rate curve flattened on top because of an extended linear phase.

Growth curves of apple, pear, tomato, banana, strawberry, date, cucumber, orange, avocado, melon, and pineapple fruits are sigmoid, whereas raspberry, grape, blueberry, fig, currant, olive, and all stonefruits (peach, apricot, cherry, and plum) show interesting double-sigmoid growth curves in which a first "senescence" phase (flat part of the curve) is followed by another logarithmic phase leading to a second sigmoid part of the curve (Coombe, 1976; Fig. 16-9c).

Fewer data are available for height growth of perennial species, especially trees, but sigmoid curves would probably be produced, usually with important flat portions caused by winter or dry periods (Zimmermann and Brown, 1971). For specific seasons, data for shoots of trees are readily available, and modified sigmoid curves are indeed observed. Figure 16-10 shows such curves for pines and deciduous hardwood trees. Note that there were important differences in actual height growth and in lengths of the growing period among the various species. Among the hardwoods, all except the yellow poplar essentially stopped elongating before August. Among the pines, the exotic (imported; not native) red and white species stopped elongating in late spring, whereas the native species grew taller during a longer time period. Typically, elongation is more rapid during the long days of late spring and early summer (Section 23.2), but there are exceptions.

It is common for trees to cease growing taller temporarily in late summer, when temperatures are still warm and days are relatively long (Kramer and Kozlowski, 1960). Sometimes growth resumes again before

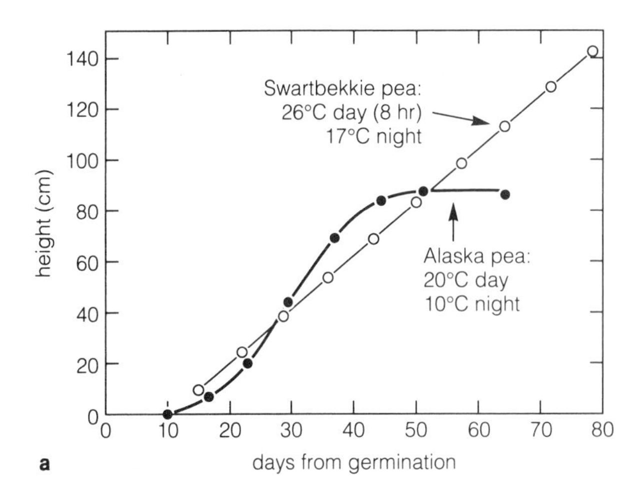


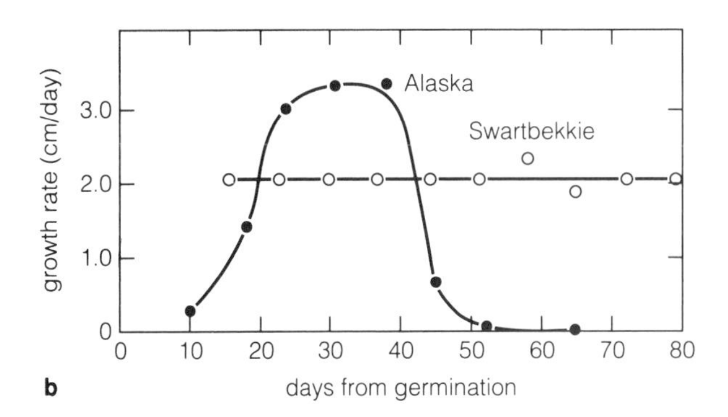


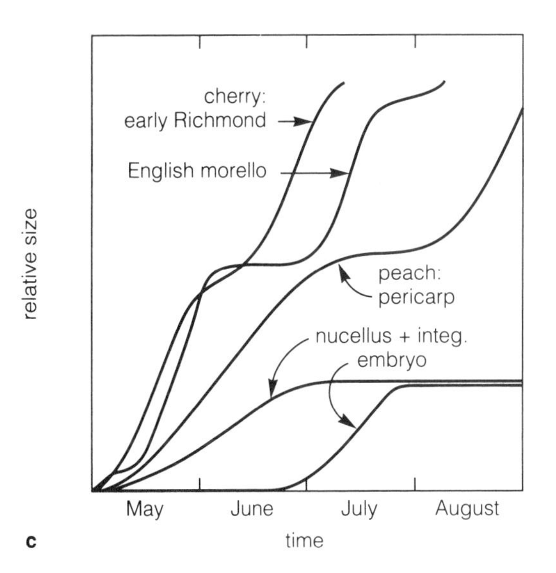


**Figure 16-9** Various growth curves that do not show a classical sigmoid shape. (**a**) Growth curves for two tall pea varieties. Note extended linear phase for Swartbekkie pea. (**b**) Growth-rate curves derived from data in **a** as in Fig. 16-8b. The bell-shaped curve for Alaska pea differs only in detail from that of Fig. 16-8b, but the bell-shape does not even appear for Swartbekkie, with its extended constant growth rate. (Data from Went, 1957.) (**c**) Growth curves for two cherries and a peach cultivar. (Data from Tukey, 1933 and 1934.)

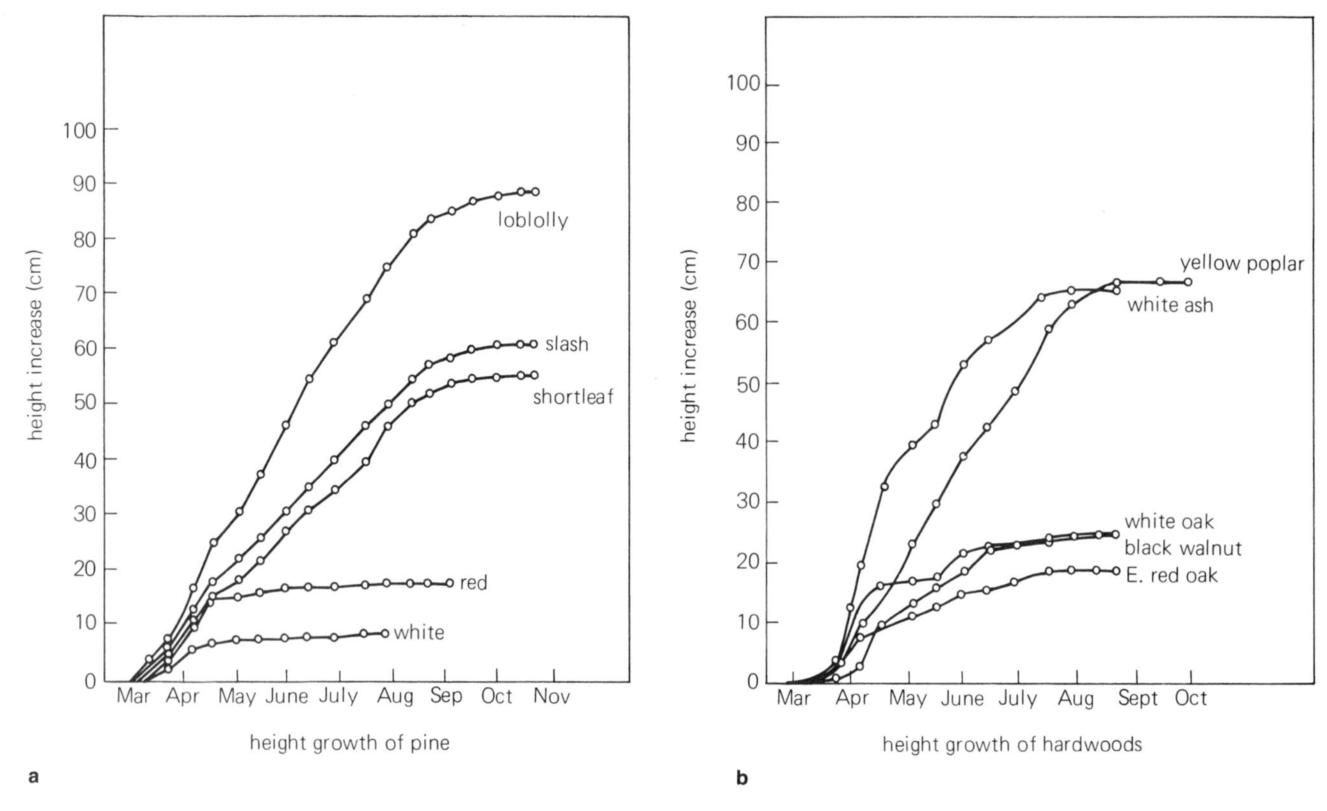


**Figure 16-10** Shoot elongation in pines (**a**) and deciduous trees (**b**) in North Carolina during one growing season (1983). All trees had been planted a few years earlier at the same location, but not all are native to the southeastern United States. Note differences in growth rates and in lengths of the periods of active growth. (Data from Kramer, 1943.)

winter dormancy, a deeper dormancy that results in part from the increasing night lengths and decreasing day lengths and in part from the low temperatures of autumn (Chapters 22 and 23). Growth in stem diameter (caused by growth of cells produced by the vascular cambium) continues, at a decreasing rate, until well after growth in height stops. Because xylem cells produced in summer have smaller diameters than those in spring, annual rings consisting of spring and summer wood form, from which estimates of age can be made for most temperate-zone tree species. In deciduous trees, photosynthesis continues until the leaves become senescent and yellow; in evergreens, until temperatures become too cold. Because of this, increases in dry weight and radial growth often continue several weeks after stem elongation ceases. Root growth can continue as long as water and nutrients are available and soil temperatures remain high enough; that is, dormancy such as that found in shoots does not occur in the roots examined to date. Because of this, orchard trees often benefit by continued root growth from late summer fertilization and irrigation.

## The Flow Analogy of Plant Growth

Because plants grow as meristems that produce new cells that then enlarge and differentiate, they leave a record of their growth history—and provide a prophesy of potential future growth. We all know that much of the history of a log can be inferred by examining the annual rings of its cross section. A narrow ring means difficult growing conditions, and a wide ring with large cells means more ideal conditions during that year. The same principle applies at the cellular level at the stem or root tip. Some phase of cell development by division, cell enlargement, and cell differentiation is occurring at every moment among the cells in a growing stem or root. Dividing cells are found in the apical meristems, elongating cells a bit farther from the tip, and differentiating cells still farther from the tip. The history of a differentiated cell can be inferred from the younger cells closer to the tip; conversely, the future of a young cell can be inferred by examining the more mature cells farther from the tip. The same is true on a larger scale for the leaves along a stem: The history of leaf production can be inferred by examining the pattern formed by the leaves or leaf scales along the stem, and future stages of leaf development of a young leaf primordium near the stem tip become quite clear from an examination of older leaves farther down the stem.

These features of stems reveal that indeterminate growth in a plant is a *flow process*. An analogy can be drawn to a waterfall. As long as the flow rate is constant, the shape of the waterfall also remains constant. Yet at

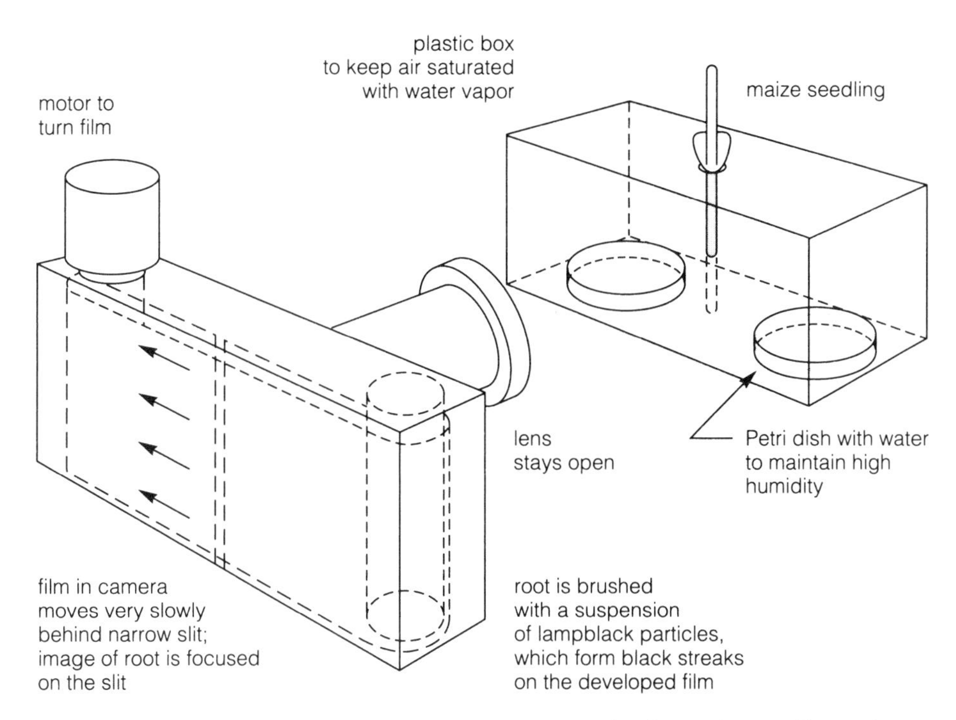


**Figure 16-11** The camera setup used to take streak photographs of a growing root. The root is brushed with a suspension of lampblack and put in a moist chamber mounted in front of the camera. The camera lens is left open, and the film is moved at a constant, slow rate past a narrow, vertical slit mounted directly in front of the film. The black spots on the growing root will appear as streaks in the resulting photograph, as in Figure 16-12.

any moment the molecules of water that combine to produce this form are not the ones that produce the form at any other moment. In a similar way, an upper stem of a pea plant, for example, appears constant from day to day, but the individual cells making up the tip and youngest leaves are continually changing—flowing, as it were—from the meristematic region of cell division toward the more mature parts of the stem. (Other examples of flowing structures with constant form include a ship's wake and a candle's flame.) Realizing that plant growth is analogous to fluid dynamics has provided some powerful mathematical tools for the analysis of plant development (Silk and Erickson, 1979; Silk, 1984).

A traditional approach to the study of growth has its origins in the extensive studies of Julius von Sachs (whom we met in our discussion of mineral nutrition and other topics), who made marks with India ink at equal intervals along a growing root tip and then examined the distances between the marks some time (often 24 hours) later (reviewed in Erickson, 1976; Erickson and Silk, 1980). Of course, the marks in the elongation zone were farther apart when remeasured, and those in the part of the root where differentiation had occurred were the same distances apart as when they were marked. This approach (which actually originated with Henri Louis du Hamel du Monceau, who inserted fine silver wires into the roots of walnut seedlings in 1758) is limited because the rates of elongation of cells between the marks do not remain constant throughout the interval between marking and measurement some hours later; they are constantly changing, as we have just seen.

What is needed is a technique to measure the change in the length of cells in very short time intervals—so short that the growth rates remain constant, for all practical purposes, during those intervals, which should be on the order of seconds. Such a technique has been developed with **streak photographs**, in which the image of a growing root tip or other organ is focused on a slit in the film plane of a camera, and then the film moves slowly by the slit while the lens remains open (Fig. 16-11). To identify points on the growing organ, one may brush it with a suspension of lampblack particles. Results of such a technique are shown in Figure 16-12. The steepness of the streaks (each one representing a particle of lampblack) represents the rate of movement of that point on the root. Because the top of the plant is fastened in place in relation to the camera, the tip of the root is moving fastest and produces the steepest streaks. Points in the differentiated zone are not moving at all, so their streaks appear horizontal.

Such figures can be analyzed in various ways. If we imagine that the root tip is standing still, we can plot

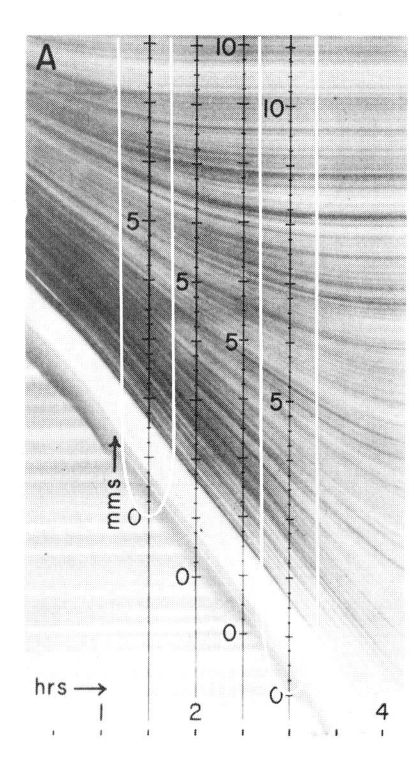


**Figure 16-12** Growth of a maize root as recorded in a streak photograph. (See Fig. 16-11.) The outlines in white suggest the positions of the root at 1.5 and 3.0 h. Near the top of the photograph, the root has stopped growing, so the streaks are horizontal. Because the root is attached above the top of the photograph, the tip is moving most rapidly; streaks made by the tip are steepest. Note curving streaks that represent portions of the root that were growing rapidly when the photograph was started (top left of photo) and then slowed in their growth as their cells began to differentiate (upper right of photograph). The superimposed scales make it possible to measure the distance of any streak from the root tip at different time intervals. Such data were used to produce the curves of Fig. 16-13. (Streak photo courtesy of Ralph O. Erickson; from Erickson and Goddard, 1951. See also Erickson and Sax, 1956 and Erickson and Silk, 1980.)

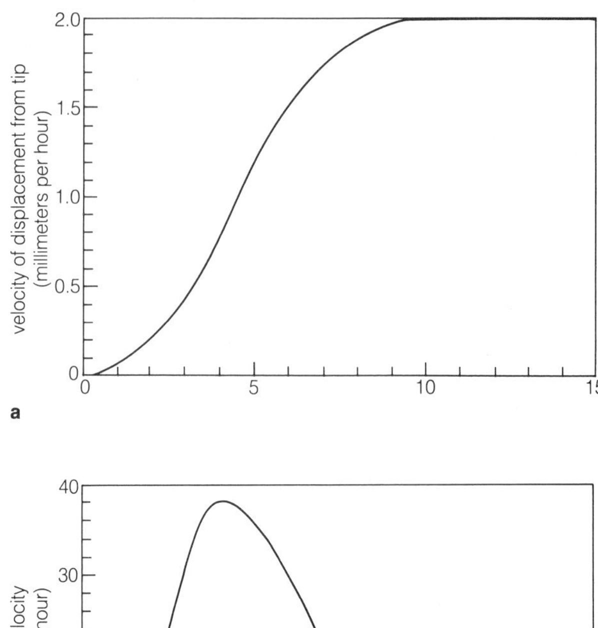


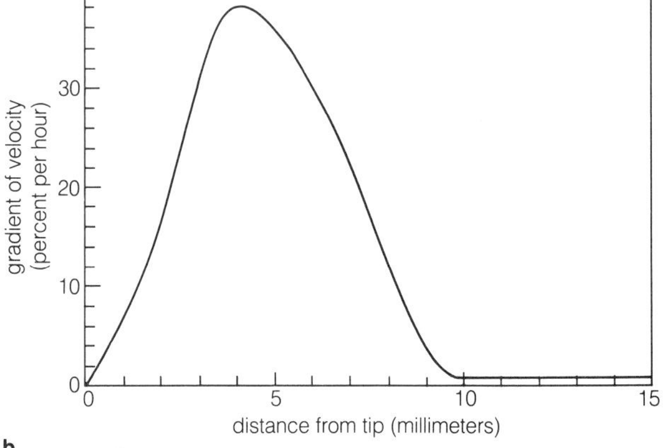


**Figure 16-13** Distribution of growth in a maize root. If the velocity of displacement from the tip of any point along the root is plotted as a function of its distance from the tip (**a**), a sigmoid curve is obtained. If the growth rate (slope of growth curve) of any point along the root is plotted as a function of distance from the tip (**b**), a bell-shaped curve is obtained. Compare with Figure 16-8. (Data from Erickson and Sax, 1956.)

the velocity of displacement of any point (any streak) as a function of distance from the tip. This velocity can be determined by measuring the distance between the tip and the streak at one time and then measuring the distance at a later time (along a vertical axis to the right of the first measurement). Velocities of displacement can also be determined by measuring the slopes of the streaks at a given time and performing appropriate mathematical adjustments. Results of such measurements produce data such as those plotted in the displacement curve in Figure 16-13a. An S-shaped curve is again evident, but in this case it represents distribution of growth along the root.

Measuring the slope of that curve gives the growth rate for any individual cell at any given distance from the tip (rate curve in Fig. 16-13b). This is similar but not identical to the bell-shaped curve of Figure 16-8b. Both curves of Figure 16-13 show that cells 10 mm from the root tip are still growing, and the rate curve shows that the maximum rate of growth occurs about 4 mm from the tip. Note that if growth of the maize root is a steady-flow process, growth rates measured for individual cells along the root provide information about the growth rate of any single cell as a function of time; that is, the scales on the abscissas of Figure 16-13 could be changed from distance to time.

One of the most fascinating results of the study of stem growth as a flow process comes from an examination of growth of the **hypocotyl** or **epicotyl**[4] **hook**

[4]The **hypocotyl** is the portion of the stem below the cotyledons and above the root; the **epicotyl** is the portion of the stem above the cotyledons. In some species (for example, lettuce, soybean, and bean) the hook forms in the hypocotyl just below the cotyledons; in other species (for example, pea), the hook forms in the epicotyl below the first true leaves, and the cotyledons remain in the ground.

observed on many dicot seedlings grown in the dark (Fig. 16-14; see discussion of the etiolation syndrome in the introduction to Chapter 20). Apparently the hook protects the seedling as it thrusts up through the soil. Light causes the hook to straighten, but the growth of the hook can be observed over intervals of many hours under dim safelights of suitable wavelengths. Photographs such as those in Figure 16-15 allow an analysis of growth, and no lampblack is needed because individual surface hairs can be recognized and used as markers.

Is the hook simply raised in elevation by elongation of cells below, or are new cells continually being produced in the meristem at the tip of the hook, enlarging and elongating as they pass through the hook and eventually differentiating below the hook? Examination of the surface hairs proves the latter to be the case. Cells flow through the hook just as water flows over a waterfall. Because cell divisions stop before the cells reach the hook, the hook must form when cells on the outside elongate more than those on the inside; the stem becomes straight below the hook (or the hook straightens) as cells elongate more rapidly on the inside than on the outside. That is, the form of the hook is determined by internal, closely coordinated factors that control the rate of cell elongation on opposite sides of the stem! The morphogenetic program in control of this phenomenon is completely unknown, and in that respect it is quite

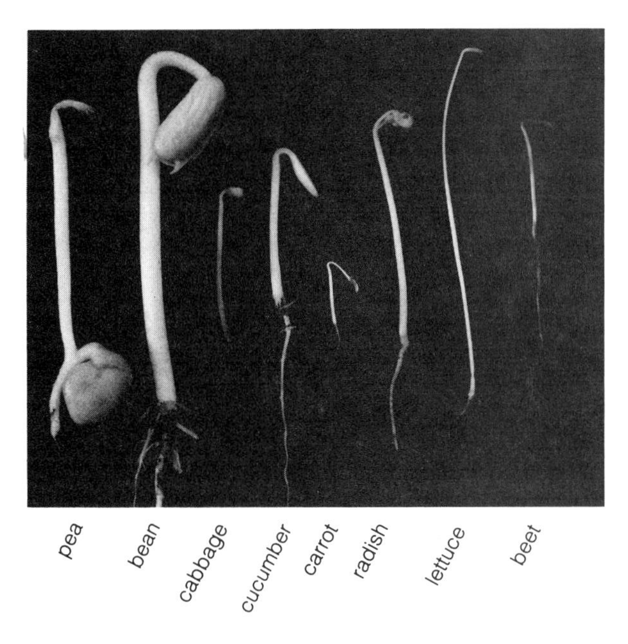


**Figure 16-14** Photograph of the seedling hooks of various species of dicots.

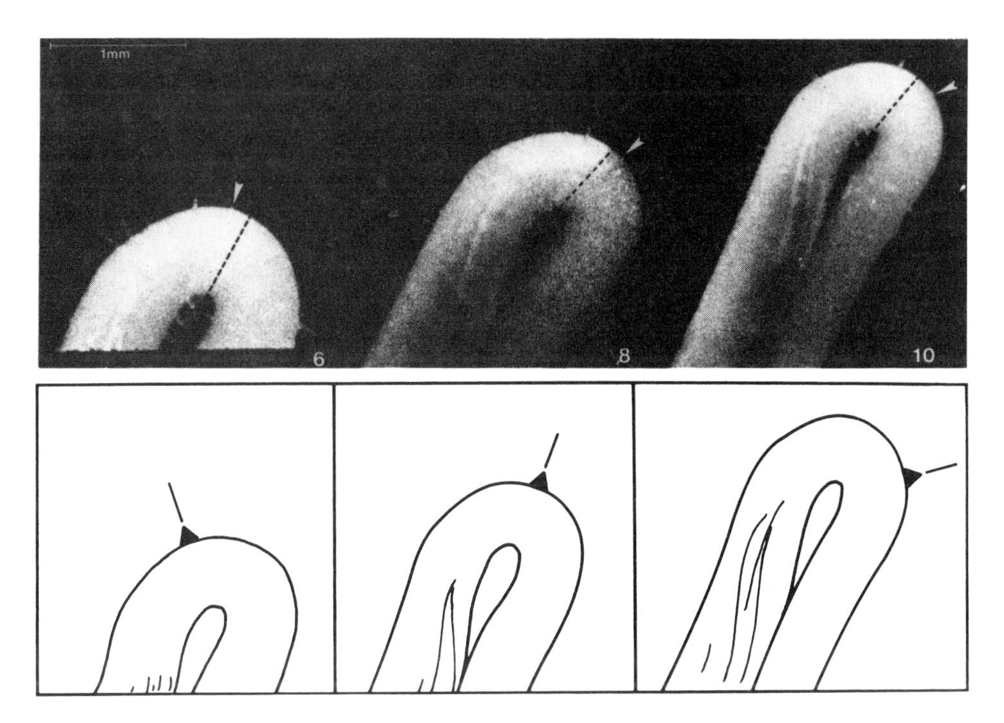


**Figure 16-15** Growth of the hypocotyl hook of a lettuce seedling photographed at 6, 8, and 10 hours after an initial observation. At bottom is a schematic drawing of the lettuce seedling during growth. The arrow (top) or black triangle (bottom) shows how a point on the hook grows through the hook with time, as discussed in the text. (Figure courtesy of Wendy Kuhn Silk; it is a composite based upon Silk and Erickson, 1978; and Silk, 1980, 1984.)

representative of our understanding of the ultimate control factors of morphogenesis in general. Whatever these control factors are, they are themselves adjusted by light of suitable irradiance and wavelength, so that the growth rates that control formation of the hook become altered in such a way that the hook straightens.

The analogy of the waterfall breaks down when we discuss the control mechanisms. Clearly the shape of the waterfall is determined by both gravity and the flow channel (the rocks or cliffs that direct where the water will flow). The flow of cells through an epicotyl hook, on the other hand, is somehow internally determined by whatever morphogenetic program is in charge of the organism's growth.

## 16.4 Plant Organs: How They Grow

Having examined several of the general principles of plant growth, it is now time to consider a few special features of the various plant organs.

### Roots

*Organization of the young root* In the great majority of species, seed germination begins with protrusion of the **radicle** (embryonic root), rather than the **epicotyl** (shoot), through the seed coat (Bewley and Black, 1978; Feldman, 1984). In some species (for example, sugar pine, *Pinus lambertiana*), cytokinesis occurs in the radicle before germination is complete. In others (maize, barley, broadbean, lima bean, and lettuce), few if any mitoses occur before radicle protrusion; elongation is caused by growth of the cells that formed when the embryo was developing on the mother plant. Continued growth of the primary root of the seedling and of branch roots derived from it requires activity of apical meristems. A typical root tip is illustrated in Figure 16-16.

The oldest cells of the root cap are in the **distal** part (that part farthest from the point of attachment to the rest of the plant; that is, the tip). In a more **proximal** position (closer to the meristem) are the young cells being formed from the apical meristem. The root cap protects the meristem as it is pushed through the soil and acts as the site of gravity perception for roots (see gravitropism in Section 19.4). Furthermore, it secretes a polysaccharide-rich slime or **mucigel** over its outer surface that lubricates the root as it slides through the soil. This requires the activity of Golgi vesicles, as shown in Figure 1-13. As the root grows, the mucigel continues to cover its surface as it matures. This mucigel harbors microorganisms and probably influences formation of mycorrhizae, root nodules, and ion uptake in unknown ways (Barlow, 1975; Foster, 1982).

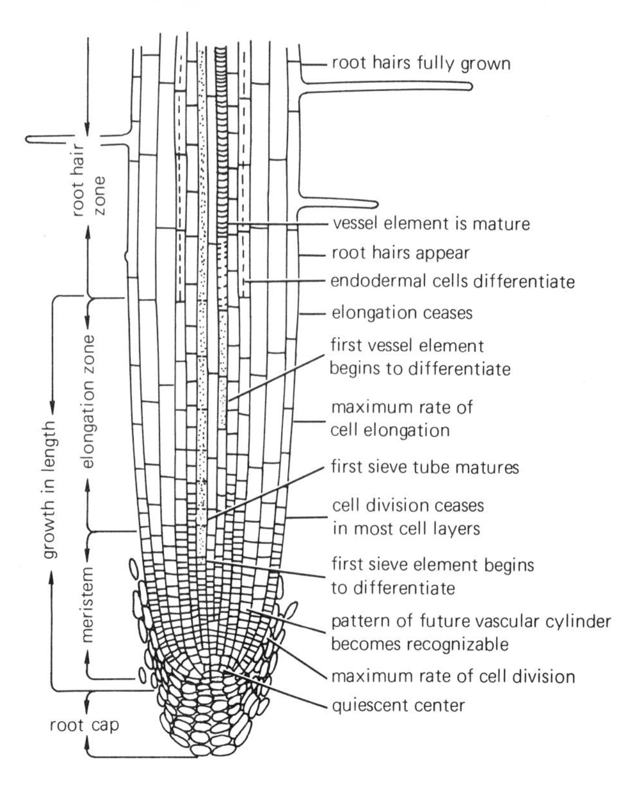


**Figure 16-16** Simplified diagram of the growing zone of a root, in longitudinal section. The number of cells in a living root is normally much greater than is shown in this diagram (compare with Fig. 5-9). In a root such as that of Figure 16-13, the point of maximum elongation is about 4 mm from the tip. (From *The Living Plant*, Second Edition, by Peter Martin Ray. Copyright © 1972, by Saunders College Publishing, a division of Holt, Rinehart & Winston, Inc. Reprinted by permission of the publisher.

Cells produced by divisions in the apical meristem develop into the epidermis, cortex, endodermis, pericycle, phloem, and xylem. Microscopists can detect where cell division is occurring (that is, where the meristem is) by observing cells in any of the mitotic stages. Because a doubling of the DNA content usually means that mitosis and cytokinesis will follow (see Fig. 16-5), another clever method to locate DNA synthesis is to provide cells with radioactive thymidine that is incorporated into DNA. Autoradiography is then used to detect DNA synthesis. Just proximal to the root cap, there is typically a small zone called the **quiescent center**, where division seldom occurs (Clowes, 1975). If the meristem or the root cap is damaged, the quiescent center becomes active and can regenerate either of these parts.

*Formation of lateral roots* The frequency and distribution of lateral root formation partly controls the overall shape of the root system and hence the zones of the soil that are explored. Lateral or branch roots generally begin development several millimeters to a few centimeters distal to the root tip. They originate in the peri-

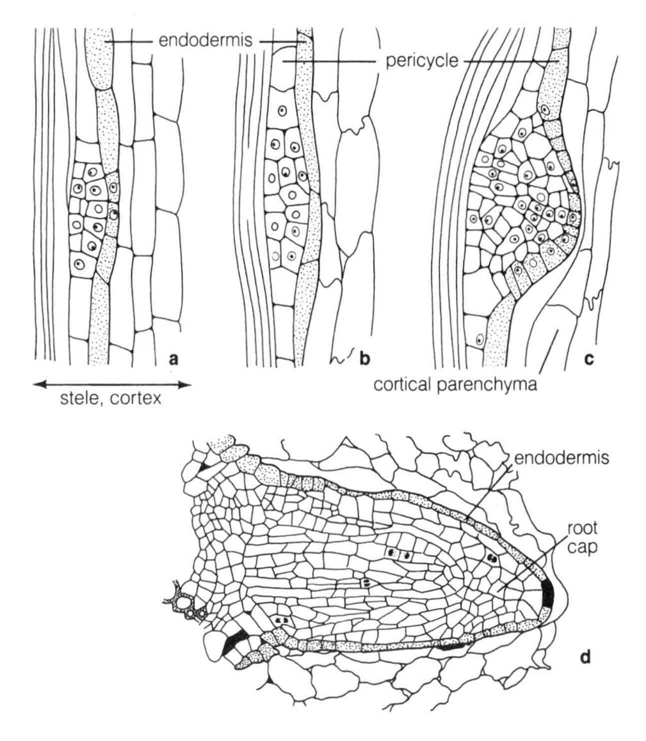


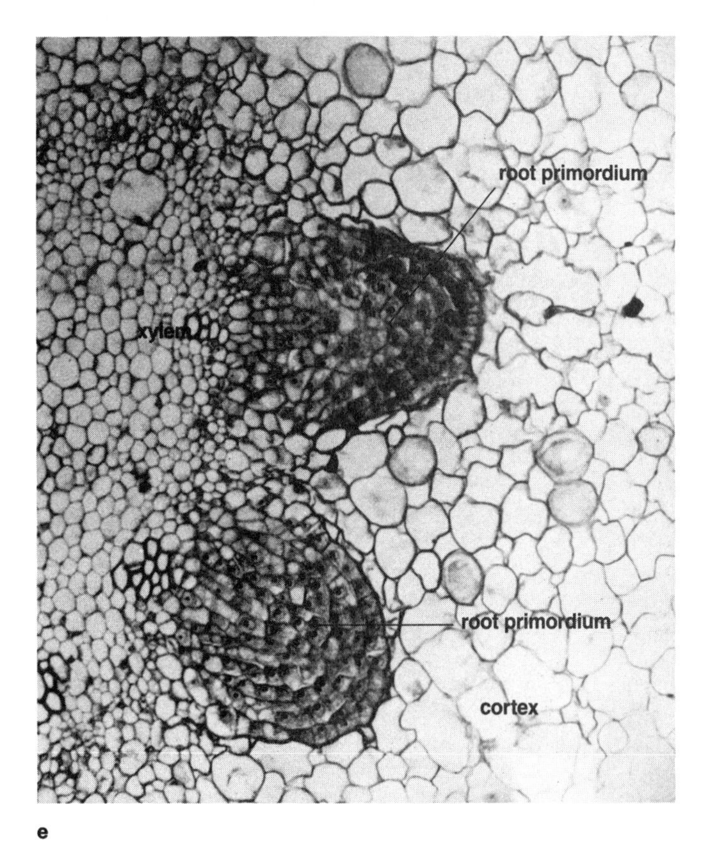


**Figure 16-17** The origin of secondary roots. Growth begins with divisions in the pericycle (**b** and **c**) that result in the establishment of a small mass of cells. These become the root primordium, which grows outwardly through the cortex. Frequently, the endodermis divides in pace with growth of the branch root, covering it, as in **d**, until it breaks out of the main root. (From Jensen and Salisbury, 1972.)

cycle, usually opposite the protoxylem points, growing outwardly through the cortex and epidermis, as illustrated in Figure 16-17. This growth probably involves secretion by the branch root of unidentified hydrolytic enzymes that digest the walls of the cortex and epidermis.

*Radial growth of roots* The roots of gymnosperms and most dicots develop a vascular cambium from procambial cells located between the primary phloem and primary xylem near or in the root-hair zone (Esau, 1977; Fahn, 1982; Mauseth, 1988). Because this cambium produces expanding new xylem cells (toward the inside) and phloem cells (toward the outside), it is indirectly responsible for most of the increased width of these roots. Most monocots do not form a vascular cambium, and the small radial enlargement they undergo is caused mainly by increases in diameter of nonmeristematic cells.

After the vascular cambium initiates secondary growth, a **cork cambium (phellogen)** arises in the pericycle. This becomes a complete cylinder that forms **cork (phellem)** toward the outside and, later, some secondary cortex (**phelloderm**) inside. The epidermis, exodermis if there is one, original cortex, and endodermis are sloughed off, leaving xylem (at the center), vascular cambium, phloem, secondary cortex, cork cambium, and finally cork cells in the mature root. Water-repellent suberin (Section 15.2) is deposited in the cork cell walls.

As the root system grows, more of it becomes suberized. For example, in mature loblolly pine (*Pinus taeda*) and yellow poplar or tulip tree (*Liriodendron tulipifera*), the surface area of unsuberized roots during the growing season is almost always less than 5 percent of the total. Apparently, suberized roots absorb water and mineral salts through lenticels, through tiny crevices formed by penetration of branch roots, and through holes left when branch roots die (Section 5.3).

## Stems

The apical meristem of the shoot forms in the embryo and is the place where new leaves, branches, and floral parts originate. The basic shoot tip structure is similar in most higher plants, both angiosperms and gymnosperms. Figure 16-18 shows photomicrographs of the terminal shoots of a representative dicot and a monocot.

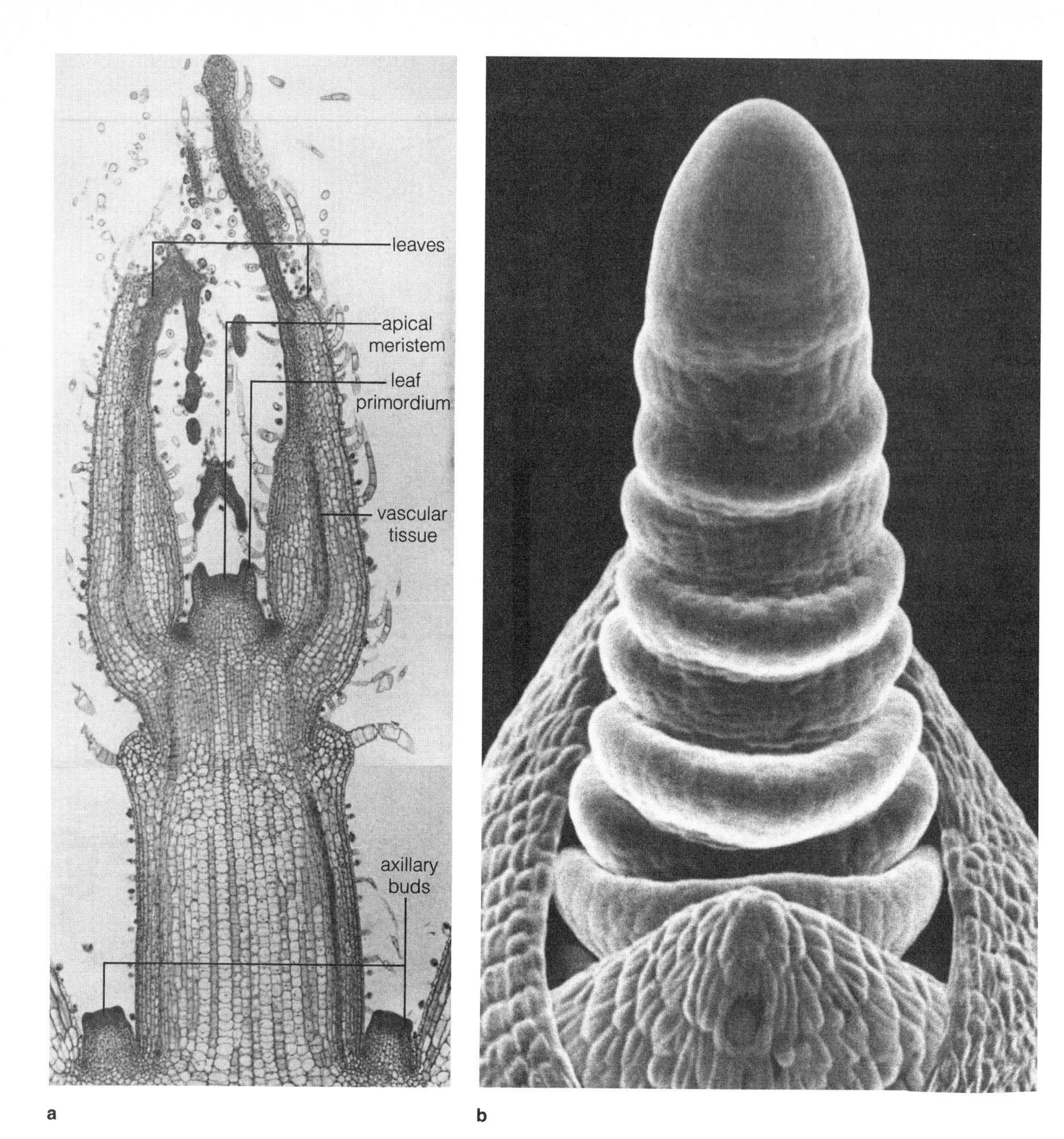


**Figure 16-18** (**a**) Longitudinal section through the apical part of the shoot of a dicot. (From Jensen and Salisbury, 1972.) (**b**) A scanning electron micrograph of the apical meristem of wheat at a late vegetative stage. Leaf primordia in grasses are formed as ridges around the shoot axis. (Micrograph courtesy of John Troughton; see Troughton and Donaldson, 1972.)

In growing stems, the region of cell division is much farther from the tip than it is in roots (Sachs, 1965). In many gymnosperms and dicots, some cells divide and elongate several centimeters below the tip. In grasses, growth also occurs far down the stem, but it is restricted to specific, repeating regions. Near the shoot tip of young monocots, the leaf primordia are very close together, and the internodes are formed later by division and growth of cells between these primordia. At first, these divisions occur throughout the length of the young internode, but later the meristematic activity becomes restricted to the region at the base of each internode and just above the node itself. These repeating meristematic regions are called **intercalary meristems** because they are intercalated (inserted) between regions of older, nondividing cells. Each internode consists of older cells at the top and younger cells near the base derived from the intercalary meristem.

## Leaves

The earliest sign of leaf development in both gymnosperms and angiosperms usually consists of divisions in one of the three outermost layers of cells near the surface of the shoot apex (see Figs. 16-2 and 16-18). Periclinal divisions followed by growth of the daughter cells cause a protuberance that is the **leaf primordium**, whereas anticlinal divisions increase the surface area of the primordium. Both kinds of divisions are important for further development of leaves and for growth in other parts of the plant.

Leaf primordia do not develop randomly around the shoot apex. Rather, each species typically has a characteristic arrangement, or **phyllotaxis**, causing opposite or alternate leaves (Richards and Schwabe, 1969). Alternate leaves are arranged in several species-specific ways much studied by mathematicians as well as plant physiologists. No one knows why a given leaf primordium develops where it does, but a recent model (Chapman and Perry, 1987) suggests that one substance diffuses from the apex and another from the ends of the vascular system; the second substance is consumed by formation of a leaf. The model based on these assumptions can account for most leaf arrangements. Traditional theories involve competition for space or various inhibitors.

The shape of the leaf primordium is produced by the magnitude and direction of its cellular divisions and expansions. The direction of expansion is controlled by the yielding properties of the cell wall, so the cell division planes, which are the planes in which the new walls are deposited, affect the primordium shape. Because cell division is accompanied by a coordinated amount of cell expansion, the primordium appears long and narrow when most of the early divisions are periclinal. When more of the divisions are anticlinal, the young organ is shorter and wider.

Subsequent leaf development is highly variable, as shown by the almost endless variety of leaf shapes. Continued outward extension occurs by both periclinal and anticlinal divisions at the primordium tip (apex or distal end). Later, often when the leaf is only a millimeter or so long, meristematic activity begins throughout its length. In grass leaves and conifer needles, this activity ceases first at the distal end and finally resides at the leaf base. An increase in width of the leaf blade in angiosperms results from meristems producing new cells along each margin of the leaf axis, but these cease activity well before the leaf matures. In grasses, the basal meristem is an intercalary one that remains potentially active for long periods, even after leaf maturity. It can be stimulated by defoliation caused, for example, by a grazing animal or a lawnmower. Growth distribution in a grass leaf base is shown in Figure 16-19.

Figure 16-20 shows a crabgrass (*Digitaria sanguinalis*) leaf with its base encircling the stem, an encirclement that results from periclinal divisions in the primordium all the way around the shoot apex (see also Fig. 16-18b). The basal meristem in a grass leaf sheath often lies immediately external to an intercalary meristem of the stem.

In dicot leaves, most cell divisions stop well before the leaf is fully grown, frequently when it is half or less of

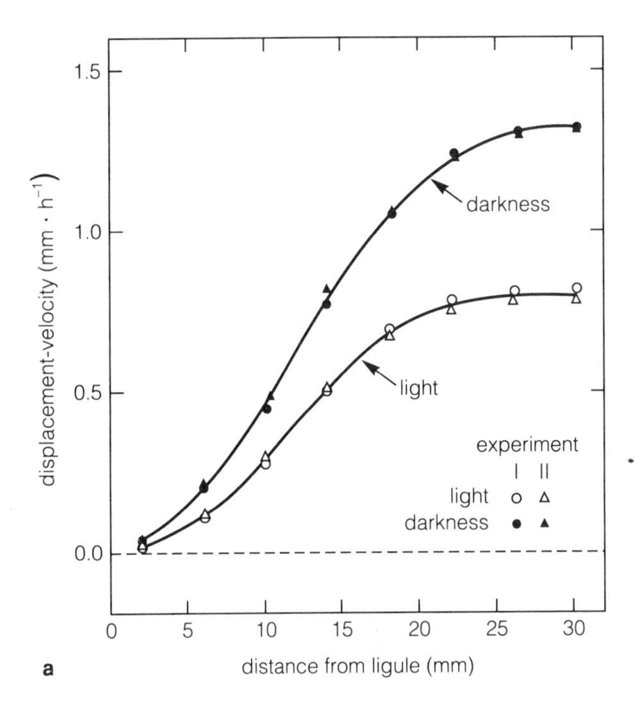


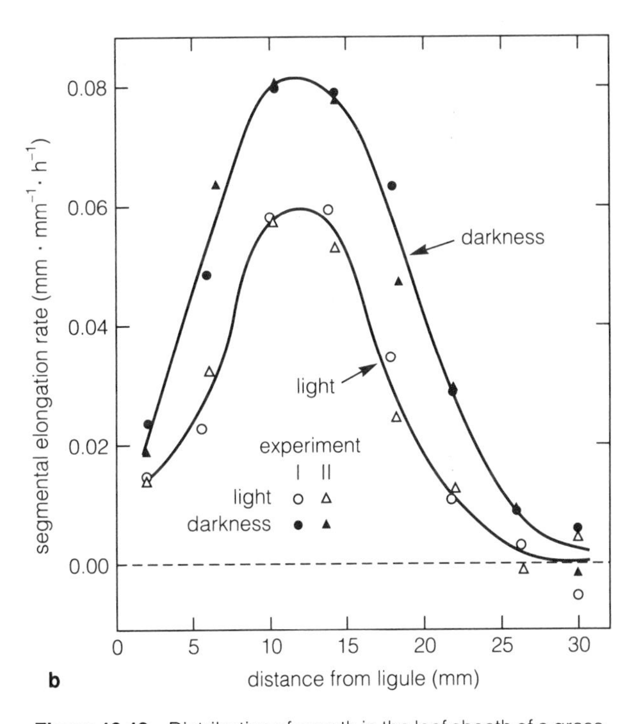


**Figure 16-19** Distribution of growth in the leaf sheath of a grass leaf, *Festuca arundinacea* (tall fescue) in light and darkness. Holes were pierced at intervals in the leaf base with a needle, and measurements were made of the distances between the holes after the leaves had elongated about 4 mm. As in Figure 16-8b, the rate curves (**b**) are derived from the slopes of the "growth curves" (displacement-velocity curves) in **a**. The elongation rate was faster in darkness than in light. (Modified from Schnyder and Nelson, 1988.)

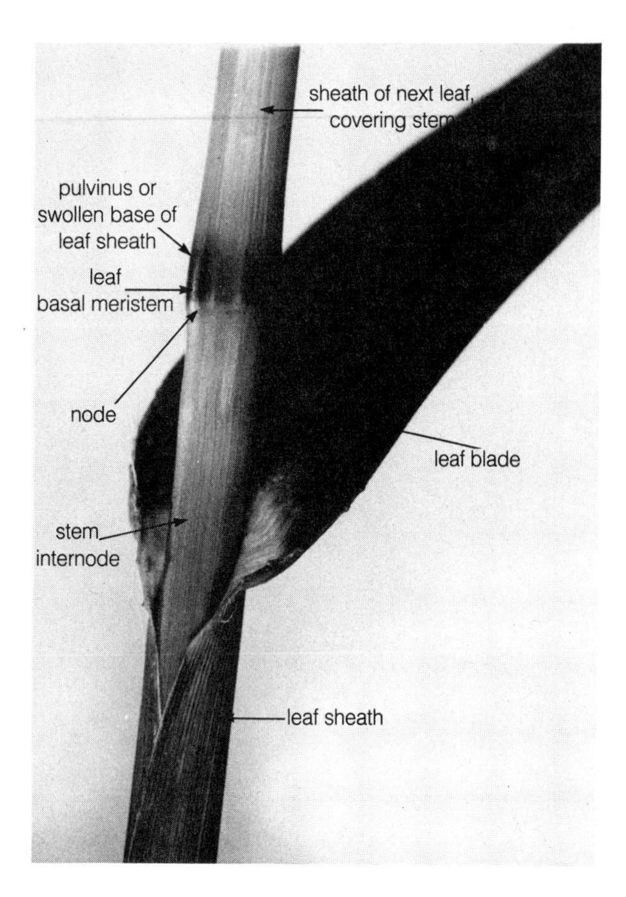


**Figure 16-20** The relation of the leaf blade and sheath to the stem in a grass (crabgrass). The cylindrical leaf sheath partly replaces the stem in providing support. (From W. W. Robbins, T. E. Weier, and C. R. Stocking, 1974, *Botany, An Introduction to Plant Biology*, John Wiley and Sons, Inc., New York. Copyright © 1974 by John Wiley and Sons, Inc. Reprinted by permission of John Wiley & Sons, Inc.)

its final size (Dale, 1988). In a bean primary leaf, cell division is complete when it has reached slightly less than one-fifth its final area, so the final 80 percent of leaf expansion is caused solely by the growth of preformed cells. This growth occurs over the entire leaf area, but not uniformly. The same is true for many other dicots. Cells in the young leaf are relatively compact. As leaf expansion occurs, mesophyll cells stop growing before epidermal cells do, so the expanding epidermis then pulls the mesophyll cells apart and causes development of an extensive intercellular space system in the mesophyll (see Fig. 4-5).

A few leaf primordia and sometimes even floral primordia are usually detectable near the shoot apex of the embryo in the seed, but most primordia (especially in perennial species) are formed after germination. In conifers and deciduous trees, the early rapid growth in spring usually involves expansion of leaf primordia formed during the previous season and extension of the internodes between these primordia; only in late summer are new primordia formed. Some of these new primordia form part of the bud that is usually dormant during the winter or during a long dry period.

## Flowers

After establishment of roots, stems, and leaves, flowers and then fruits and seeds form, perpetuating the species and completing the life cycle. Most angiosperm species produce bisexual (**perfect**) flowers containing functional female and male parts, whereas such others as spinach, cottonwoods, willows, maples, and date palms are **dioecious**, containing **imperfect** staminate (male) and pistillate (female) flowers on different individual plants. **Monoecious** species such as maize, cocklebur, squash, pumpkin, cucumber, and many hardwood trees form staminate and pistillate flowers at different positions along a single stem. The balance of male and female flowers may determine yields of such crops as cucumbers. The reproductive structures of conifers develop in unisexual **cones (strobili)**. Most conifers are monoecious, although junipers and certain others are dioecious.

**Anthesis**, the opening of flowers that makes their parts available for pollination, is sometimes a spectacular phenomenon, usually associated with full development of color and scent. Whereas many flowers remain open from anthesis until **abscission** (falling off), others such as tulips open and close at certain times of the day for several days. Opening is usually caused by faster growth of the inner compared with the outer parts of the petals, but continued opening and closing is probably a response to temporary changes in turgor pressure across the two sides. Opening and closing are influenced by temperature (see Fig. 22-2) and atmospheric vapor pressure, but the major factor is often an internal clock set by the daily dawn and/or dusk signals (see Chapter 21). For example, evening primrose flowers (*Oenothera* species) usually open in the evening about 12 hours before dawn, but they can be rephased to open in the morning by artificially reversing the light-dark cycles. The light influencing this response is absorbed by the flowers themselves.

After anthesis and pollination, the petals eventually wither, die, and abscise. In some species, withering follows anthesis rapidly. For example, in *Portulaca grandifolia* and many morning glories, including *Ipomea tricolor* and *Pharbitis nil*, opening of the flower occurs in the morning, and the corolla withers by late afternoon (Fig. 16-21). Such withering is commonly associated with extensive transport of solutes from the flowers to other plant parts, often to the ovary, with rapid water loss. There is an accelerated breakdown of protein and RNA from petals and sepals during withering, and hydrolytic enzymes such as proteases and ribonucleases are apparently activated by hormonal changes to cause such breakdown. Nitrogenous products such as amino

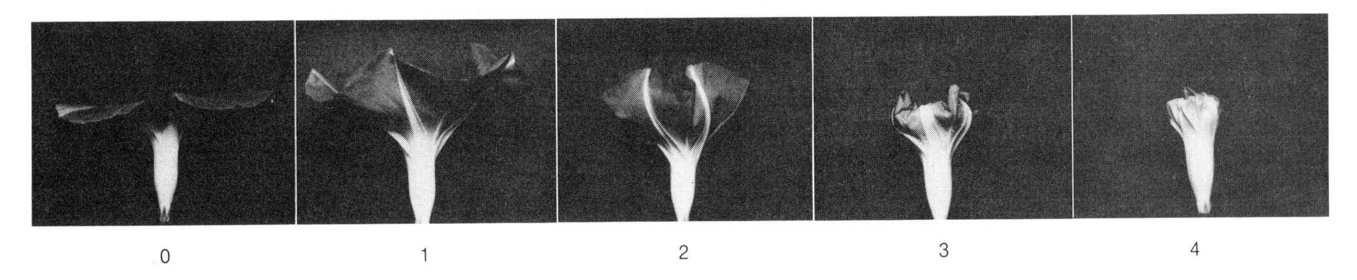


**Figure 16-21** Stages of flower fading in *Ipomea tricolor*. Stage 0 represents the fully open corolla; stages 1 to 4, progressive phases of fading. Flower opening (stage 0) begins at about 06:00, whereas fading and curling in stage 1 begins at about 13:00 of the same day. Curling is caused by turgor changes in the rib cells. The cells in the inner side of the rib lose solutes and water while the outer rib cells expand, causing curling. (Photographs courtesy of Hans Kende; from Kende and Baumgartner, 1974.)

acids and amides are then transported to seeds and other tissues in which growth is occurring, so nutrients are conserved. Although withering and color fading are common, certain rose and *Dahlia* species lose petals that are still turgid and that contain most of their original protein.

## Seeds and Fruits

*Chemical changes in growing seeds and fruits* The zygote, embryo sac, and ovule develop into the **seed**, whereas the surrounding ovary develops into the **fruit** (the **pericarp**). Numerous anatomical and chemical changes occur. Often, sucrose, glucose, and fructose accumulate in the ovules until the endosperm nuclei become surrounded by cell walls; then concentrations of these sugars decrease as they are used in cell-wall formation and starch or fat synthesis. These sugars arise largely from sucrose and other sugars transported through the phloem into the young seeds and fruits (Chapter 8). Most nitrogen in immature seeds and fruits is present in proteins, amino acids, and the amides glutamine and asparagine. The amino acids and amides decrease in concentration as storage proteins are formed in the protein bodies.

The roles of enzymes and nucleic acids in developing seeds are important to seed longevity. For a mature seed to germinate after remaining alive for long periods, it must either possess all the enzymes necessary for germination and seedling establishment or have the genetic information available to synthesize them. Some enzymes essential to germination are produced in a stable form during seed development; others are translated from stable messenger RNA, transfer RNA, and ribosomal RNA molecules synthesized during seed maturation; and still others are formed from newly transcribed RNA molecules only after the seed is planted (Spencer and Higgins, 1982). Thus, different seeds control enzyme production in various ways, and there are different mechanisms even in the same seed for control of specific enzymes. Loss of water during seed maturation is critical, leading to important but poorly understood changes in the physical and chemical properties of the cytoplasm. As a result, dry seeds respire extremely slowly and remain alive through extended drought or cold periods (Chapter 22).

Chemical composition of edible fruits and the transformation of carbohydrates during ripening have been widely studied (Hulme, 1970; Coombe, 1976; Rhodes, 1980). In apples, the concentration of starch increases to a maximum and then decreases somewhat until harvest as it is converted to sugars. In apples and pears, fructose is often the most abundant sugar, but lesser amounts of sucrose, glucose, and sugar alcohols are also present. Grapes and cherries contain about equal amounts of glucose and fructose, but sucrose is often undetectable. The hexose concentration in grapes can reach unusually high values. Concentrations of glucose and fructose in some cultivars reach 0.6 M for each sugar, giving the mature fruits an unusually negative osmotic potential and a sweet taste. During ripening of oranges, grapes, grapefruits, pineapples, and various berries, organic acids (principally malic, citric, and isocitric) decrease and sugars increase, so the fruits become sweeter. In lemons, however, the acids continue to increase during ripening, so the *p*H decreases and the fruits remain sour. Lemon fruits contain virtually no starch at any time during development, although other fruits (for example, bananas, apples, and peaches) contain much starch when they are immature. A few (for example, avocado and olive) store lipids.

Numerous other changes in fruit composition have been studied, including transformation of chloroplasts to carotenoid-rich chromoplasts, accumulation of anthocyanin pigments, and accumulation of flavoring components. The use of gas chromatography has allowed identification of hundreds of volatile substances such as aliphatic or aromatic esters, aldehydes, ketones, and alcohols that contribute to the flavor and aroma of strawberries and other fruits (Nurnsten, 1970). This provides a basis for improvement of fruit flavors by plant breeding and for development of artificial flavoring substances.

*Importance of seeds for fruit growth* Development of fruits usually depends either on germination of pollen grains on the stigma (pollination) or on pollination plus subsequent fertilization. Furthermore, extracts of pollen grains added to certain flowers will simulate natural pollination and fertilization by causing ovary growth and wilting and abscission of the petals. Developing seeds are also usually essential for normal fruit growth. If seeds are present only in one side of a young apple fruit, only that side of the fruit will develop well. Seeds are also essential for normal strawberries.

Normal production of fruits lacking seeds is called **parthenocarpic fruit development**. It is especially common among fruits that produce many immature ovules, such as bananas, melons, figs, and pineapples. Parthenocarpy can result from ovary development without pollination (citrus, banana, and pineapple), from fruit growth stimulated by pollination without fertilization (certain orchids), or from fertilization followed by abortion of the embryos (grapes, peaches, and cherries).

**Figure 16-22** Delay of senescence in soybean plants caused by daily removal of flower buds. (Photograph courtesy of A. Carl Leopold; from Leopold and Kriedemann, 1975.)

*Relations between vegetative and reproductive growth* Gardeners have long practiced the technique of removing flower buds from certain plants to maintain vegetative growth. A commercial example is the *topping* (removal of flowers and fruits) of tobacco plants, which encourages leaf production. Such an effect on soybeans is shown in Figure 16-22.

There is a competition for nutrients among vegetative and reproductive organs. Developing flowers and fruits, especially young fruits, possess a large drawing power for mineral salts, sugars, and amino acids. During the accumulation of these substances by the reproductive organs, there is often a corresponding decrease in the amounts present in the leaves. Studies with radioactive tracers show that nutrient accumulation in developing flowers, fruits, or tubers occurs largely at the expense of materials in nearby leaves. There is usually a competition among individual fruits of the same plant for nutrients. For example, fruit size decreases with increasing number of fruits allowed to form on tomato plants or apple trees.

The mechanism by which fruits can divert nutrients out of leaves and into their own tissues, sometimes against apparent concentration gradients, is not understood but is probably controlled by phloem unloading, as discussed in Section 8.4. Various hormones, especially cytokinins (see Chapter 18), might also be involved.

Actually, the situation is more complex than a simple competition for nutrients. In the cocklebur (*Xanthium strumarium*), induction of flowering by long nights causes leaf senescence just as rapidly when flower buds are removed as when they are allowed to develop normally. In other cases, perhaps some inhibitor that causes premature death is transported into the vegetative organs.

Factors that stimulate shoot growth may retard flower, tuber, and fruit development. High nitrogen fertilization causes luxuriant stem and leaf growth of tomato plants but may reduce fruit development. Similarly, excess nitrogen stimulates leaf growth but sometimes inhibits growth of potato tubers or apples and reduces sugar content but not size in sugar-beet roots. Although agriculturists often put much stock in the idea that high nitrogen reduces fruit yield, the situation is seldom that simple. There is no real evidence that high nitrogen inhibits fruit production (total biomass) per plant if other nutrients, especially phosphorus and potassium, are adequate.

Do processes interfering with vegetative growth also stimulate flower development? Sometimes they do. Heavy pruning, drought, tying branches to the ground, or various other mutilation procedures may stimulate flowering. Furthermore, such commercial growth retardants as Phosphon D, CCC, Amo-1618 (see Chapter 17) and B995 (N-dimethylamino succinamic acid) inhibit growth of stems, and this stunting is sometimes accompanied by earlier appearance of flower buds or a greater number of flowers per plant. These chemicals are used in commercial chrysanthemum production, for example, but they inhibit flowering in some other species. (These matters are discussed further in Chapter 23.)

## 16.5 Morphogenesis: Juvenility

The life cycles of many perennial species include two phases in which certain morphological and physiological characteristics are rather distinct. After germination, most annual and perennial seedlings enter a rapidly growing phase in which flowering usually cannot be induced. A characteristic morphology, especially evident in leaf shapes, is sometimes produced during this time. Plants having these characteristics are said to be in the **juvenile phase**, as opposed to the **mature** or **adult phase**.

The juvenile phase with respect to flowering varies in perennials from only one year in certain shrubs up to 40 years in beech (*Fagus sylvatica*), with values of 5 to 20 years common in trees. In Chapter 23, this reaching of maturity after which flowering can occur is described as attaining a *ripeness to respond*. Such long juvenile phases in conifers and other trees pose serious obstacles to genetic programs designed to improve their quality. Another common physiological difference between perennials in the juvenile and adult phases is the ability of stem cuttings to form adventitious roots. In the adult phase, the rooting ability is usually diminished and is sometimes lost.

The juvenile and adult morphologies of leaves are examples of **heterophylly**, which is well illustrated for an annual dicot by the bean, which always forms simple primary leaves at first and compound trifoliate leaves later (see Fig. 19-6), and by the pea, which has quite reduced, scalelike juvenile leaves. Among perennials, many junipers form needlelike juvenile leaves and scalelike adult leaves. The many species of *Acacia* and *Eucalyptus* often have juvenile-leaf forms that are strikingly different from adult forms. English ivy (*Hedera helix*), another perennial, has been studied extensively. Its juvenile growth habit is that of a creeping vine, but later it becomes shrublike and forms flowers. Its juvenile leaves are palmate with three or five lobes, whereas its adult leaves are entire and ovate. Although attainment of the adult phase is usually fairly permanent, juvenility in ivy can be induced in shoots that develop from lateral buds of mature stems by treating the leaf just above this lateral bud with gibberellic acid ($GA_3$; Chapter 17). ABA prevents this reversion caused by $GA_3$, suggesting that a balance of gibberellins and ABA might normally be involved in the transition from one state to another. On the other hand, certain gibberellins induce flowering and thereby terminate juvenility in many gymnosperm species (Section 23.10). Or is it a matter of semantics? There is evidence from tissue-culture studies (see next section) that floral tissue is itself very juvenile; what could be more juvenile than the developing embryo? Thus, although we define juvenility as an inability to form flowers, the flowers themselves may be thought of as being juvenile!

## 16.6 Morphogenesis: Totipotency

We noted earlier that in differential growth, cells in a plant become different even though their genes are identical. How do we know that all the cells in an organism have identical genes? First, events during chromosomal duplication and separation during mitosis strongly suggest this. Second, many plant cells are **totipotent**. By this we mean that a nonembryonic cell has the potential to dedifferentiate into an embryonic cell and then to develop into a complete new plant, if the environment is suitable. A root parenchyma cell, for example, may begin to divide and produce an adventitious bud and finally a mature, flowering shoot. All the genes for production of the whole plant must exist in such differentiated root cells. This could not happen if their genes had been irreversibly altered during root differentiation. Totipotency is also illustrated by development of cultured callus tissues into new plants, and partial totipotency occurs when adventitious roots develop from stem cells and when xylem and phloem are regenerated from wounded cortex cells. In fact, totipotency might be advantageous to plants mainly because it provides them with a mechanism for healing wounds and reproducing vegetatively by cloning.[5]

In each of these examples of totipotency, several cells cooperate to form primordia from which the whole plant arises. Experiments in which plants develop from single cells were pioneered by Frederick C. Steward and coworkers at Cornell University in the 1950s and are associated with his work on cytokinins (Section 18.1). Steward found that single cells broke away from pieces of callus derived from carrot-root phloem. When conditions were changed, single cells in the cell suspensions would occasionally divide to form multicellular embryoids. From these, new plants capable of producing seeds were formed (Steward, 1958; Steward et al., 1958). Cloning from single cells had been achieved.

Even after Steward's experiments, there was some question about whether single cells were totipotent because Steward's embryoids had always developed in the presence of many cells in suspension, although each plant apparently came from a single cell. Vasil and Hildebrandt (1965) answered this question by producing entire plants from isolated single cells. Nevertheless, the fact that *some* cells are totipotent does not prove that *all* cells have this property; in any tissue culture there are many cells that do *not* become embryos.

[5]A **clone** consists of *all* the organisms that have been produced asexually from an individual parent organism. Thus, a clone is normally a *group* of organisms. The popular press seems to assume that a clone is an individual, which is possible but not the original definition of the term.

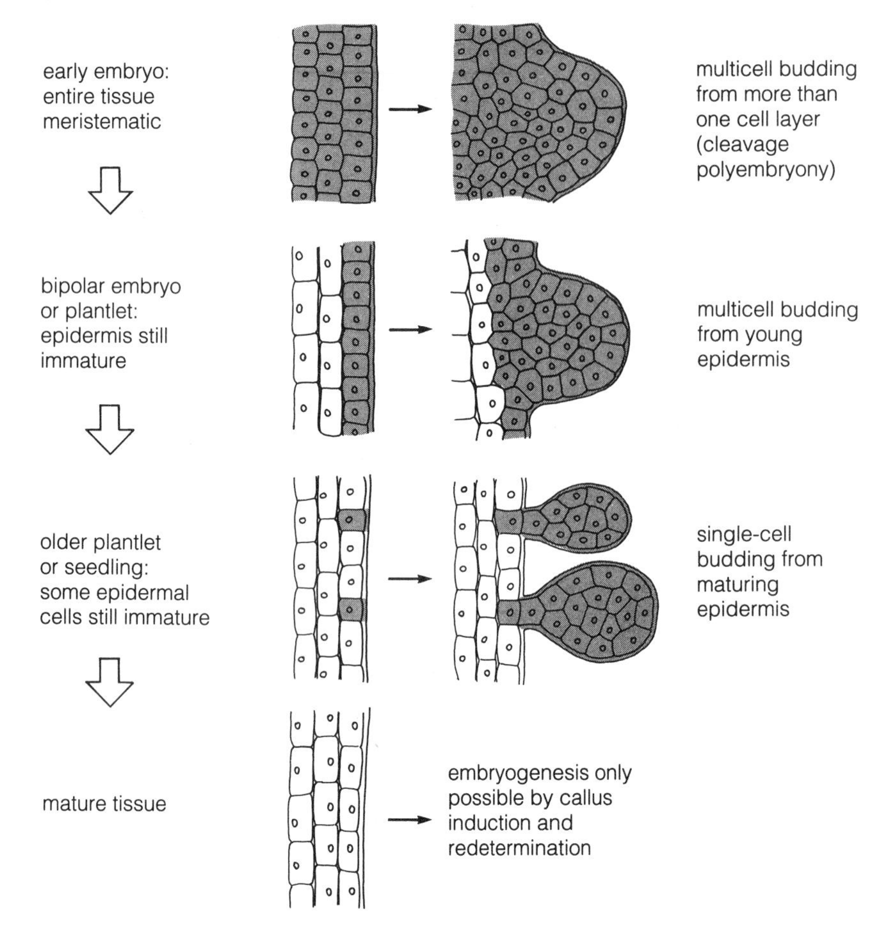


**Figure 16-23** The probable relationships of superficial or multilayer, and single-cell or multicellular initiation patterns in direct somatic embryogenesis. Shading indicates preembryonically determined cells. (From Williams and Maheswaran, 1986; used by permission.)

Even haploid pollen grains develop into callus tissues and then whole plants (Sunderland, 1970; Sangwan and Norreel, 1975). Sometimes the cells of these plants contain predominantly triploid and diploid chromosome numbers, although some of the cells are haploid. Apparently the diploid and triploid cells result from **endoreduplication** (doubling of chromosomes in mitosis, with lack of subsequent cytokinesis) or nuclear fusion, as discussed in Section 16.2.

Steward's original observations have developed into a rather extensive field of research that deals with **somatic embryogenesis**, in which **somatic** (nonreproductive) haploid or diploid cells develop into differentiated plants through characteristic embryological stages (see reviews by Shepard, 1982; Williams and Maheswaran, 1986). The process occurs naturally in many species, sometimes from cells associated with seed development, such as nucellus or synergid cells. In such a case, the embryo in the seed was not formed from a union of gametes. Such a mode of reproduction is called **apomixis**; it is quite common among the flowering plants and is only revealed by careful study because seeds appear normal. Other examples of natural somatic embryos are the miniature plantlets that form along the leaf margins of *Bryophyllum* (see Fig. 23-3).

What does it take for a cell or tissue to form a somatic embryo? This seems to depend very much on the status of the cell or tissue (Fig. 16-23). Somatic embryogenesis occurs more easily in tissues that are themselves still somewhat embryonic (juvenile, we might say). In an early embryo, for example, whole groups of cells can cooperate to form a new meristem. As the embryo becomes more mature, only groups of *epidermal* cells can form an embryo; with more maturity, only individual epidermal cells can undergo embryogenesis. When the tissues are mature, embryogenesis is only possible after a callus has been induced, and it is much more likely if the callus comes from reproductive tissues including floral buds, ovules, somatic tissues of anthers, mature or immature embryos, or cotyledons (all can be thought of as juvenile), rather than stems, shoot buds, leaf primordia, root primordia, and so on. There

**Table 16-1 Some Principles of Differentiation.**

| Principle | Example or Discussion |
|---|---|
| 1. Once synthesized by enzymes, many large molecules and other structures are arranged into fairly stable three-dimensional structures by spontaneous self-assembly. | After synthesis, polypeptides fold into the most stable structure for their watery medium. Ribosomes, microtubules, and viruses also often assemble from component parts spontaneously. Even certain cell types (for example, in sponges) can reassemble after being dissociated. |
| 2. Genes control the kinds of enzymes a cell can make, but environment determines whether or not enzymes function effectively. | Nitrate reductase appears in roots in response to nitrate. Temperature often determines enzyme activity. |
| 3. Environment sometimes determines transcription and translation of genetic information into functioning enzymes. | Chromatin (genetic material) from shoot apices of peas would not make globulin *in vitro*, but chromatin from pea cotyledons would do so, as occurs in the intact plant. The genes for globulin were **repressed** in the shoots but not in the cotyledons. |
| 4. The *position* of a cell in relation to other cells may determine how it responds in differentiation. If the new cells resemble the ones directing their differentiation, the process is called **homeogenetic induction**. | New cambial cells form from cortical cells adjacent to existing procambial cells; the process is called **redifferentiation**. This might be caused by something released from the existing cells. In some cases, plant cells that are already differentiated **dedifferentiate** before becoming differentiated as a new cell type. |
| 5. If the new cells are different from the ones causing the changes, the process is called **heterogenetic induction**. | In some dicots, leaf hairs are located only over vascular bundles, suggesting that the presence of the bundles controls differentiation of the hair cells. |
| 6. Differentiation sometimes seems to be controlled by **field effects**, in which differentiation may occur in "fields" that do not overlap. | A minimum and constant distance is maintained between differentiating stomates on a leaf, so stomatal patterns are not random. Growth substances could be involved in many of these cases. |
| 7. Tissue differentiation usually requires an initial act of cell division. After *dedifferentiation*, mitosis and cytokinesis occur; then differentiation occurs in the daughter cells. Often, the two daughter cells are not alike. | Cambial cells do not themselves differentiate into xylem or phloem cells. Epidermal cells divide to produce one large and one small cell on a young root surface; the small one is a **trichoblast** and will become a root hair (Fig. 16-24). Similar processes are involved in formation of guard cells, subsidiary cells, and sieve-tube elements and companion cells. This is emphasized by observation of gamma plantlets, in which cell division cannot occur because of radiation treatment; no root hairs, leaf hairs, or stomates are formed. |

See the second edition of this textbook for a much more detailed discussion of the material in this table.

is also evidence that, as Steward originally suggested, isolation of a cell from its neighbors (that is, breaking of plasmodesmatal connections) can often predispose it to embryogenesis. The role of growth regulators in these scenarios is not yet well understood. It is clear that auxins can induce callus formation on many stems, and specific growth regulators, particularly cytokinins, often must be added to culture media to induce embryogenesis. But it is also evident that the predisposition of the tissue to such treatments also determines their success. This is an illustration of the critical role of sensitivity of the target tissue to the growth substance, which we discuss further in later chapters.

Although much remains to be learned about the mechanisms of plant morphogenesis in general and somatic embryogenesis in particular, a number of important applications have been and will be developed. Artificial seeds provide one good example. A highly productive hybrid (for example, one of wheat or maize) might be induced to form thousands of somatic embryos, perhaps from an immature embryo taken from a maturing seed, and the new embryos might then be induced to become quiescent. Encased in some protective coating, they would be ready for planting when needed.

## 16.7 Some Principles of Differentiation

Ultimately, it is the goal of biologists to understand morphogenetic events such as those we have been discussing in this chapter by understanding what occurs at the cellular level and what control mechanisms are involved. We close this chapter with Table 16-1, which reviews some suggestions about such mechanisms. It should be quite evident as you study the table that these suggestions bear little direct relationship to the phenomenological data that we have been discussing; that is, we have a few ideas about how morphogenesis is controlled, and we have vast bodies of information about morphogenetic processes, but it is seldom possible to use the ideas about control mechanisms to understand the observed phenomena. Nevertheless, that remains the goal. Reaching it would be a crowning achievement of our time — or any time.

It should also be evident by now that we cannot proceed much further without a thorough discussion of plant hormones and growth regulators. That is the special purpose of the next two chapters and will often be the theme in the rest of this book.

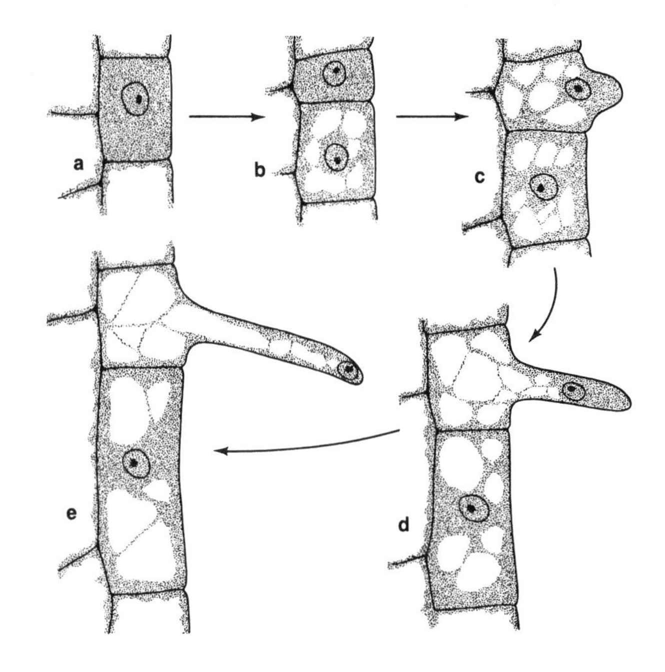


**Figure 16-24** (**a**) An unequal cell division of a young epidermal cell precedes formation of a root hair and an ordinary epidermal cell. (**b**) This division forms a trichoblast (smaller upper cell) and an atrichoblast (larger lower cell). (**c, d,** and **e**) The trichoblast develops into a root hair. (From Jensen and Salisbury, 1972.)

# 17

# Hormones and Growth Regulators: Auxins and Gibberellins

In the previous chapter we reviewed some of the growth regions of plants and introduced a few of the numerous effects of certain plant hormones on growth and development. In this and the next chapter we summarize knowledge about such hormones and related growth regulators. Subsequent chapters present more details about the roles of these compounds in specific developmental processes.

There are currently only five generally acknowledged groups of hormones, even though more will almost surely be discovered, and we review some likely candidates in the next chapter. The five groups include four *auxins*, many *gibberellins* (84 to date), several *cytokinins*, *abscisic acid*, and *ethylene*. In this chapter we will examine only the first two of these groups.

## 17.1 Concepts of Hormones and Their Action

### Hormones Defined

What is a plant hormone? Most plant physiologists accept a definition that is similar to that developed for animal hormones: *A* **plant hormone** *is an organic compound synthesized in one part of a plant and translocated to another part, where in very low concentrations it causes a physiological response*. The response in the target organ need not be promotive because processes such as growth or differentiation are sometimes inhibited by hormones, especially abscisic acid. Because the hormone must be synthesized by the plant, such inorganic ions as $K^+$ or $Ca^{2+}$ that cause important responses are not hormones. Neither are organic growth regulators synthesized only by organic chemists (for example, 2,4-D, an auxin) or synthesized only in organisms other than plants. The definition also states that a hormone must be translocated in the plant, but nothing is said about how or how far; nor does this mean that the hormone will not cause a response in the cell in which it is synthesized. (A good example involves ethylene and fruit ripening; almost surely ethylene promotes ripening of the very cells that make it, as well as that of others.) Sucrose is not considered a hormone, even though it is synthesized and translocated by plants, because it causes growth only at relatively high concentrations. Hormones are often effective at internal concentrations near 1 $\mu$M, whereas sugars, amino acids, organic acids, and other metabolites necessary for growth and development (excluding enzymes and most coenzymes) are usually present at concentrations of 1 to 50 mM.

The idea that plant development is influenced by special chemicals in plants is not new. About 100 years ago, the famous German botanist Julius von Sachs suggested that specific organ-forming substances occur in plants. He supposed that one substance caused stem growth and others leaf, root, flower, or fruit growth. No such organ-specific chemicals have ever been identified. Because of hormones' very low concentrations in plants, the first hormone to be discovered (indoleacetic acid) was not identified until the 1930s, and even then it was first purified from urine. Because it could evoke so many different responses when added exogenously to plants, it was considered by many to be the only plant hormone, a notion disproved when the numerous effects of gibberellins were discovered in the 1950s.

As more hormones were identified and their effects and endogenous concentrations studied, it became apparent that not only does each hormone affect the responses of many plant parts, but that those responses depend on the species, the plant part, its developmental stage, the hormone concentration, interactions among known hormones, and various environmental factors. Therefore, it is risky to generalize about the effects of hormones on processes of growth and development in a particular plant organ or tissue. Nevertheless, von Sachs's concept that different tissues can respond differently to different chemicals is certainly valid.

## The Concept of Differential Sensitivity to Hormones

In the early 1980s, Anthony J. Trewavas repeatedly emphasized the concept that differential sensitivity is much more important in determining the effects of a hormone than is the concentration of that hormone within plant cells (see, for example, Trewavas, 1982, 1987, and the discussion published jointly by Trewavas and Cleland, 1983). Although many researchers have argued convincingly against Trewavas's overall conclusion, his papers forced researchers to consider and, when possible, measure tissue sensitivity to hormones. Nowadays, both sensitivity and hormone concentration receive attention in many studies of hormone action (also see Firn, 1986 and Section 19.5).

We know that if plant hormones present in micromolar or submicromolar concentrations are to be active and specific, three major parts of a response system must be present. First, the hormone must be present in sufficient quantity in the proper cells. Second, the hormone must be recognized and bound tightly by each of the groups of cells that respond to the hormone (the **target cells**). Protein molecules have the necessary complex structures to recognize and select among much smaller molecules (as described for enzymes in Chapter 9), and based on knowledge about hormonal action in animals, hormone-binding proteins in the plasma membrane of plant cells are being identified. Such proteins are called **receptor proteins**. Third, the receptor protein (the configuration of which is presumably changed during hormone binding) must cause some other metabolic change that leads to **amplification** of the hormonal signal or messenger. In fact, various amplification processes might occur in sequence before response to the hormone finally occurs.

In light of such a response system, the myriad of responses of various plant parts to exogenous application of different plant hormones is no longer so puzzling. Developmental changes even in a single tissue of a given species are almost surely accompanied by a change not only in hormone concentration but also in both the frequency or availability of receptor proteins and the capacity to amplify the hormonal signal. Other plant parts or parts from a different species may respond in a different way.

## The Effects of Hormones on Gene Activity

There is now conclusive evidence (to be described in this chapter and in Chapter 18) that one thing plant hormones do is control gene activity. How genes are controlled biochemically is still largely unknown, but much of what we do know is summarized in Chapter 24. We wish to emphasize here that activation of genes represents a large amplification process because repeated transcription of DNA into messenger RNA (mRNA), followed by translation of the mRNA into enzymes with high catalytic activity at low concentrations, can lead to many copies of an important cellular product. These products then determine what an organism is composed of and therefore what it looks like (its phenotype).

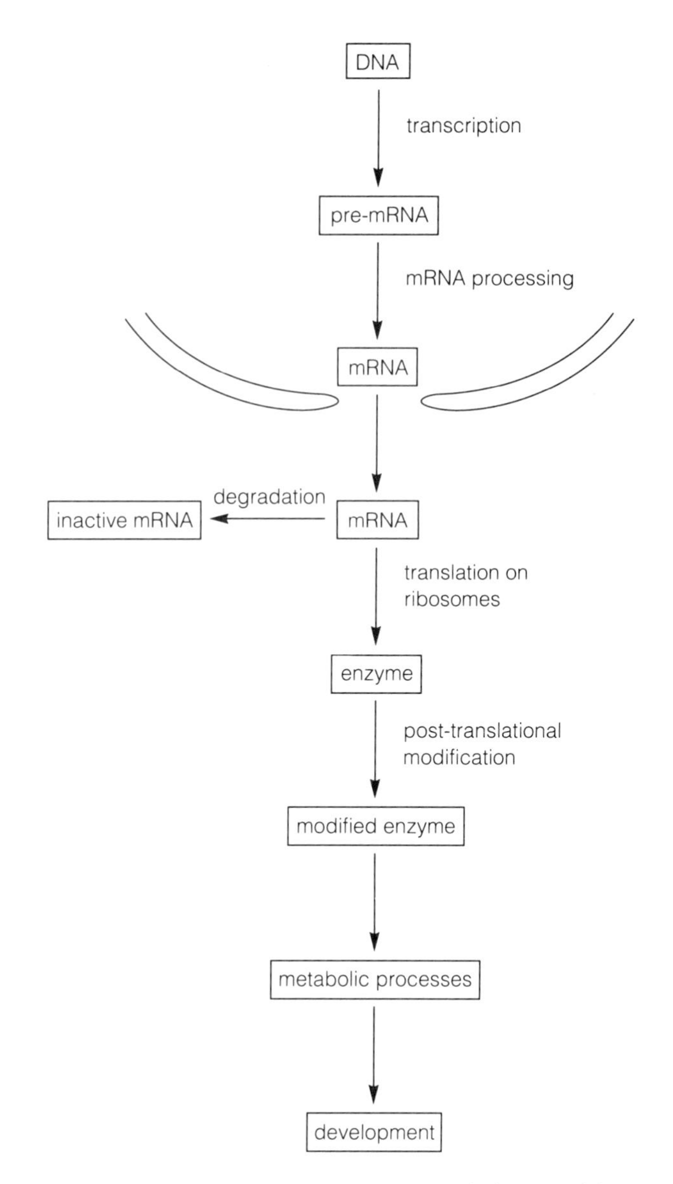


**Figure 17-1** Possible sites of hormonal control of gene activity.

There are various control points in the flow of genetic information from DNA to a molecular product. One of these, perhaps the most important, occurs at the level of transcription. Another control point, also in the nucleus, concerns processing of the mRNA, because most mRNA molecules are partially degraded and have some pieces rearranged before they ever leave the nucleus (see Appendix C). These processing steps are controlled by enzymes whose actions must be regulated, and hormones might affect such regulation. Next,

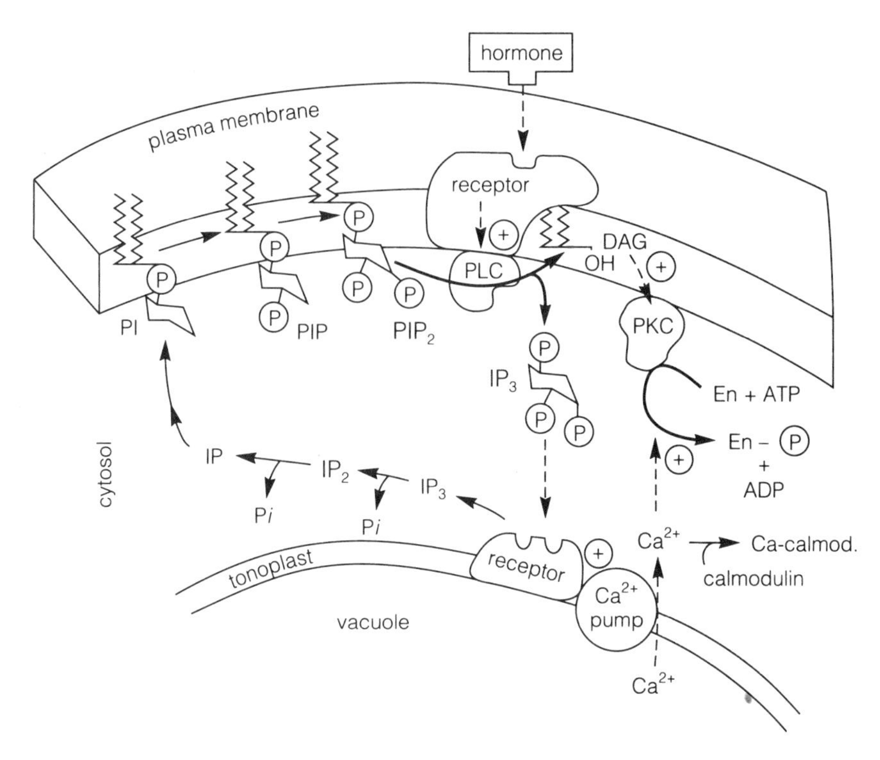


**Figure 17-2** Model for initial hormone transduction at the plasma membrane. Binding of a hormone to its receptor causes activation ( + ) of nearby phospholipase c (PLC). PLC hydrolyzes a membrane lipid, phosphatidylinositol-4,5-bisphosphate ($PIP_2$) to release inositol-1,4,5-trisphosphate ($IP_3$) and a diacylglycerol (DAG). $IP_3$ moves to the tonoplast in plant cells, where it combines with a receptor that activates ( + ) a $Ca^{2+}$ pump or transporter that moves $Ca^{2+}$ from the vacuole to the cytosol. DAG, which remains membrane-bound, activates protein kinase c (PKC). PKC is also activated by $Ca^{2+}$ released from the vacuole, so various enzymes become phosphorylated by PKC. Calcium also activates other protein kinases and other enzymes, when free or bound with calmodulin. $IP_3$ loses phosphates by hydrolysis to form $IP_2$ and IP, which is then converted back to phosphatidylinositol (PI) and other phosphoinositide lipids (PIP and $PIP_2$) in the plasma membrane. (Modified from various sources.)

mRNA leaves the nucleus, probably via a nuclear pore (see Figure 1-11), and in the cytosol it can either be translated on ribosomes or degraded by ribonucleases. If it is translated into an enzyme, post-translational modification of the enzyme can occur by processes such as phosphorylation, methylation, acetylation, glycosidation, and so on (described in Section 9.2). These processes, too, might be affected by hormones (or by light or some other environmental signal). Possible points of control are summarized in Figure 17-1.

## Sites of Hormonal Activity

Research by physiologists and biochemists has emphasized that many animal hormones, especially peptide hormones, act first not in the nucleus but in the plasma membrane, where receptor proteins exist. Furthermore, hormone reception in animals quickly causes one or two major transduction systems to be set in motion. One of these, which is unlikely to occur in plants, involves activation of an enzyme called *adenyl cyclase* that forms *cyclic AMP* from ATP. Cyclic AMP activates numerous enzymes in animals, especially protein kinases that phosphorylate enzymes and modify their activity. However, cyclic AMP itself probably is not important in plants (Bressan et al., 1976; Spiteri et al., 1989), and cyclic AMP-dependent protein kinases have long been sought but not found in plants; thus this system will not be discussed further here.

A second transduction system of animals has received far more support by plant researchers, even though its involvement as a plant hormone or environmental signal has not yet been proven. It was reviewed by Boss (1989), by Marmé (1989), by Einspahr and Thompson (1990), and in several articles in the book edited by Morré et al. (1990). We will describe it here in general terms as a potential explanation for how various plant hormones might function.

The process begins with binding of the primary hormone to a receptor protein in the plasma membrane (external surface) of a target cell (Fig. 17-2). Next, the bound hormone-receptor complex activates a nearby membrane enzyme called **phospholipase c (PLC)**. Phospholipase c then hydrolyzes one of a group of rather nonabundant membrane phospholipids called phosphoinositides. **Phosphoinositides** are phospholipids that contain inositol (for example, phosphatidylinositol, abbreviated PI in Fig. 7-11) or similar lipids in which hydroxyl groups of inositol are esterified to one or two phosphate groups (at carbon 4 or at carbons 4 and 5). Phospholipase c hydrolyzes the latter, phosphatidylinositol 4,5-bisphosphate ($PIP_2$), between the glycerol and the phosphate attached to carbon 1 of the inositol phosphate portion, so that it releases **inositol-1,4,5-trisphosphate ($IP_3$)** and **diacylglycerol (DAG)**; DAG represents glycerol now esterified only to two fatty acids.

$IP_3$ and DAG both have further activity and can cause a cascade of responses. $IP_3$ is highly water-soluble and moves, in animal cells, to the endoplasmic reticulum, where it causes release of stored $Ca^{2+}$ into the cytosol. In plant parenchyma cells with large central vacuoles, most of the cell's $Ca^{2+}$ is stored not in the endoplasmic reticulum but in the vacuole, where the concentration is often in the mM range (at least 1,000 times that of the cytosol). There is now good evidence (summarized by Memon et al., 1989) that $IP_3$ stimulates release of vacuolar $Ca^{2+}$ to the cytosol; thus the site of $Ca^{2+}$ release might differ between plant and animal cells based on their different structures and functions.

DAG is not water-soluble (because of the two fatty acids still attached), so it functions within the plasma membrane, where it is likely quite mobile. DAG activates an enzyme in the membrane called **protein kinase c (PKC)**. This enzyme uses ATP to phosphorylate certain enzymes that regulate various stages of metabolism; for some enzymes phosphorylation causes activation, and for others it causes deactivation (see, for example, Section 11.5). Regardless, the kinds of metabolic products become changed by enzyme phosphorylation, and so can the cell's behavior and growth pattern. We don't yet know how.

Increased levels of $Ca^{2+}$ in the cytosol caused by $IP_3$ also activate certain enzymes, including several protein kinases (Blowers and Trewavas, 1989; Poovaiah and Reddy, 1990; Budde and Randall, 1990). Some such kinases require activation by free $Ca^{2+}$; others require activation by Ca-calmodulin (Section 6.7). When the $Ca^{2+}$ concentration begins to rise in the cytosol, four $Ca^{2+}$ combine to form a chelate or complex with inactive calmodulin, making an active Ca-calmodulin complex. That complex itself then activates certain enzymes. So far, those enzymes known to be activated by the Ca-calmodulin complex in plants include various protein (enzyme) kinases, **$NAD^+$ kinase** (an enzyme that uses ATP to phosphorylate $NAD^+$ to $NADP^+$), and an ATPase from plasma membranes that transfers excess $Ca^{2+}$ out of the cell. Therefore, a primary hormonal stimulus finally leads to modified enzyme activity, altered metabolic processes, and eventually a physiologically and morphologically different kind of cell. Many such changes by various hormones and environmental stimuli interact to help make a different tissue, organ, or plant.

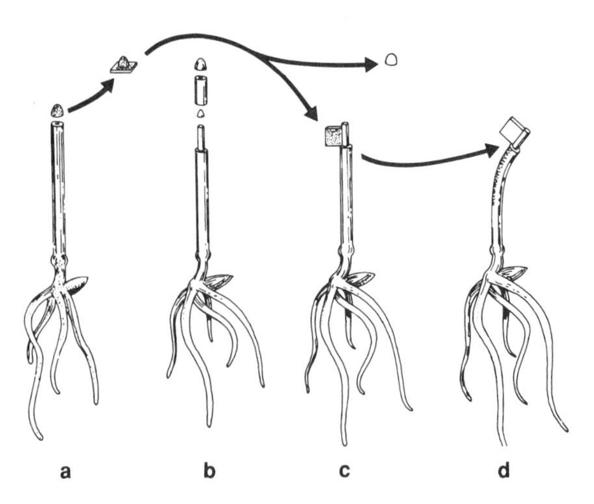


**Figure 17-3** The demonstration by Went of auxin in the *Avena* coleoptile tip. Auxin is indicated by stippling. (**a**) The tip was removed and placed on a block of gelatin. (**b**) Another seedling was prepared by removing the tip, waiting a period of time, and removing the tip again because a new "physiological tip" sometimes forms. (**c**) The leaf inside the coleoptile was pulled out, and the gelatin block containing the auxin was placed against it. (**d**) Auxin moved into the coleoptile on one side, causing it to bend. (From Salisbury and Parke, 1964.)

In trying to relate our models of Figures 17-1 and 17-2, we might wonder how any change in metabolism caused by modified enzyme activity in the cytosol could affect gene expression at any early control point in Figure 17-1 (say, through transcription, mRNA processing, translation, or mRNA stability). There are no good answers yet, but we seem to be gradually encircling an expanding problem that has considerable but finite complexity. Control over activity of certain enzymes after initial hormone reception seems to be an important key, and secondary or tertiary messengers such as $IP_3$, DAG, and $Ca^{2+}$ are often involved. Furthermore, in animals, some steroid (sex) hormones are believed to be absorbed by cells and move into the cytosol where they combine with receptor proteins; then the steroid-receptor complex moves into the nucleus and affects gene activity. These steroid hormones require no plasma-membrane receptor and no second messenger system. Do some plant hormones act the same way? If so, which ones? Do plants also have sex hormones, in the sense that such hormones promote male versus female flower formation or even the far more common case in which bisexual flowers are formed? Some tentative answers to these questions will be given in this and the following chapters.

## 17.2 The Auxins

The term **auxin** (Greek *auxein*, "to increase") was first used by Frits Went, who, as a graduate student in Holland in 1926, discovered that some unidentified compound probably caused curvature of oat coleoptiles toward light (see his essay at the end of this chapter). This curvature phenomenon, called phototropism, is described in Section 19.4. The compound Went found is relatively abundant in coleoptile tips, and Figure 17-3 indicates how he showed its existence. The critical demonstration was that a substance present in the tips could diffuse from them into a tiny block of agar. The activity of this auxin was detected by the curvature of the coleoptile caused by enhanced elongation on the side to which the agar block was applied.

Went's auxin is now known to be **indoleacetic acid (IAA**, Fig. 17-4a), and some plant physiologists still equate IAA with auxin. Nevertheless, plants contain three other compounds that are structurally similar to IAA and cause many of the same responses as IAA; these should be considered to be auxin hormones (Fig. 17-4a). One of these is **4-chloroindoleacetic acid (4-chloroIAA)**, found in young seeds of various legumes (Engvild, 1986). Another, **phenylacetic acid (PAA)**, is widespread among plants and is frequently more abundant than IAA, although it is much less active in causing typical responses of IAA (Wightman and Lighty, 1982; Leuba and LeTorneau, 1990). The third, **indolebutyric acid (IBA)**, is of more recent discovery; it was formerly thought to be only an active synthetic auxin, but it occurs in maize leaves and various dicots (Schneider et al., 1985; Epstein et al., 1989) and so is probably widespread in the plant kingdom.

Little is known about transport characteristics of 4-chloroIAA, PAA, or IBA and whether they in fact normally function as plant hormones, although this seems likely. Three additional compounds found in many plants have considerable auxin activity. They are readily oxidized to IAA *in vivo* and are probably active only after conversion. We do not yet consider them to be auxins, but rather only auxin precursors. These are *indoleacetaldehyde, indoleacetonitrile,* and *indole ethanol* (Fig. 17-4a). Each has a structure similar to IAA, but each lacks the carboxyl group.

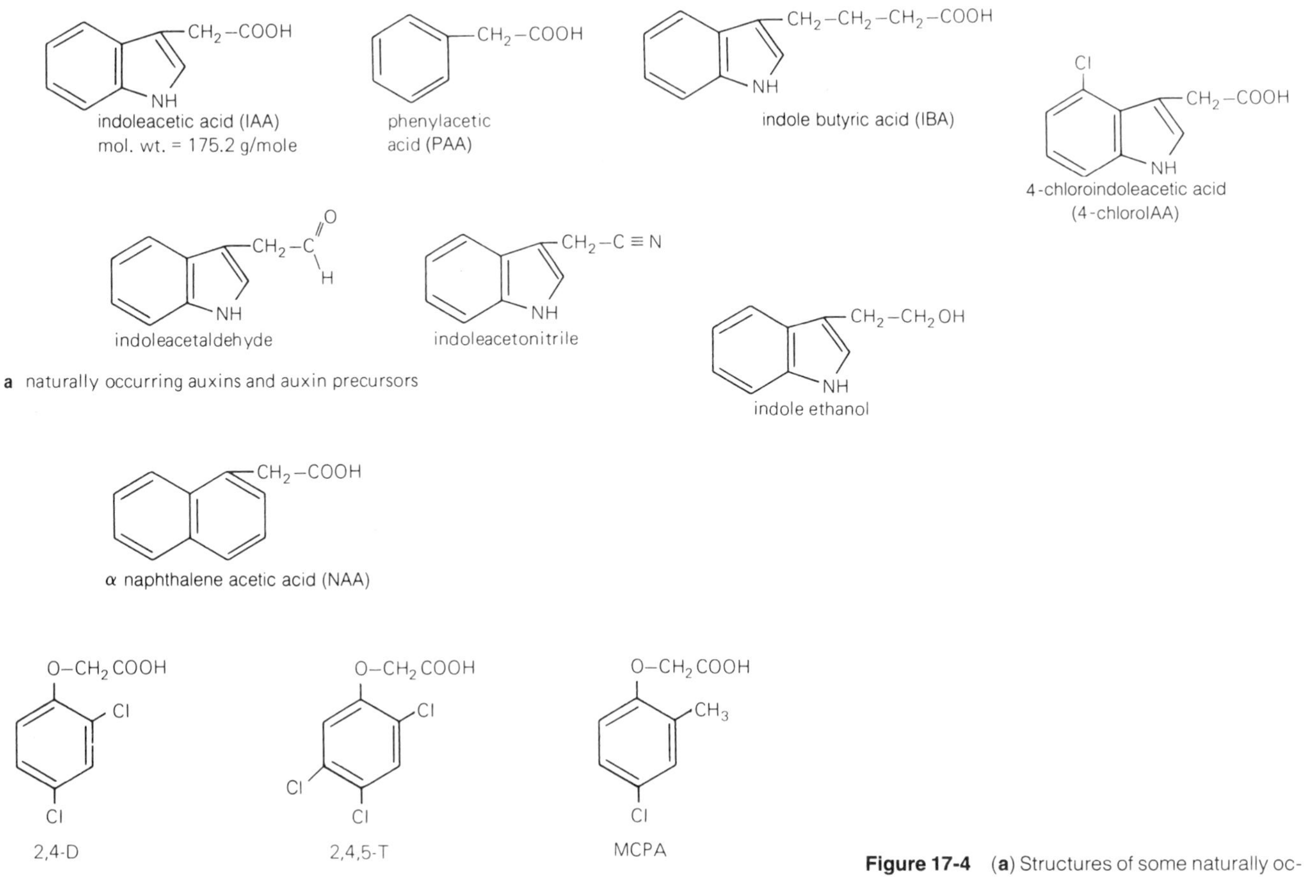

$CH_2-COOH$
indoleacetic acid (IAA)
mol. wt. = 175.2 g/mole

$CH_2-COOH$
phenylacetic acid (PAA)

$CH_2-CH_2-CH_2-COOH$
indole butyric acid (IBA)

Cl, $CH_2-COOH$
4-chloroindoleacetic acid (4-chloroIAA)

$CH_2-CHO$
indoleacetaldehyde

$CH_2-C\equiv N$
indoleacetonitrile

$CH_2-CH_2OH$
indole ethanol

**a** naturally occurring auxins and auxin precursors

$CH_2-COOH$
α naphthalene acetic acid (NAA)

$O-CH_2COOH$, Cl, Cl
2,4-D

$O-CH_2COOH$, Cl, Cl, Cl
2,4,5-T

$O-CH_2COOH$, $CH_3$, Cl
MCPA

**b** synthetic auxins

**Figure 17-4** (**a**) Structures of some naturally occurring compounds having auxin activity, and (**b**) structures of other compounds that are only synthetic auxins.

Certain compounds synthesized only by chemists also cause many physiological responses common to IAA and are generally considered to be auxins. Of these, *α-naphthalene acetic acid (NAA), 2,4-dichlorophenoxyacetic acid (2,4-D),* and *2-methyl-4-chlorophenoxyacetic acid (MCPA)* (Fig. 17-4b) are best known. Because they are not synthesized by plants, they are not hormones. They are instead classified as **plant growth regulators**, and many other kinds of compounds also fit this category. The term *auxin* has become much more encompassing since Went's discovery of IAA because so many compounds are structurally similar to IAA and cause similar responses. Nevertheless, without precisely defining an auxin, we emphasize that each of the known auxin-like compounds is similar to IAA in having a carboxyl group attached to another carbon-containing group (usually $-CH_2-$) that in turn is connected to an aromatic ring.

## Synthesis and Degradation of IAA

IAA is chemically similar to the amino acid tryptophan (although often 1,000 times more dilute) and is probably synthesized from it. Two mechanisms for synthesis are known (Fig. 17-5), both of which involve removal of the amino-acid group and the terminal carboxyl group from the side chain of tryptophan (Sembdner et al., 1980; Cohen and Bialek, 1984; Reinecke and Bandurski, 1987). The preferred pathway for most species probably involves donation of the amino group to an α-keto acid by a transamination reaction to form indolepyruvic acid and then decarboxylation of indolepyruvate to form indoleacetaldehyde. Finally, indoleacetaldehyde is oxidized to IAA. The enzymes necessary for the conversion of tryptophan to IAA are most active in young tissues, such as shoot meristems and growing leaves and fruits. In these tissues the auxin contents are also

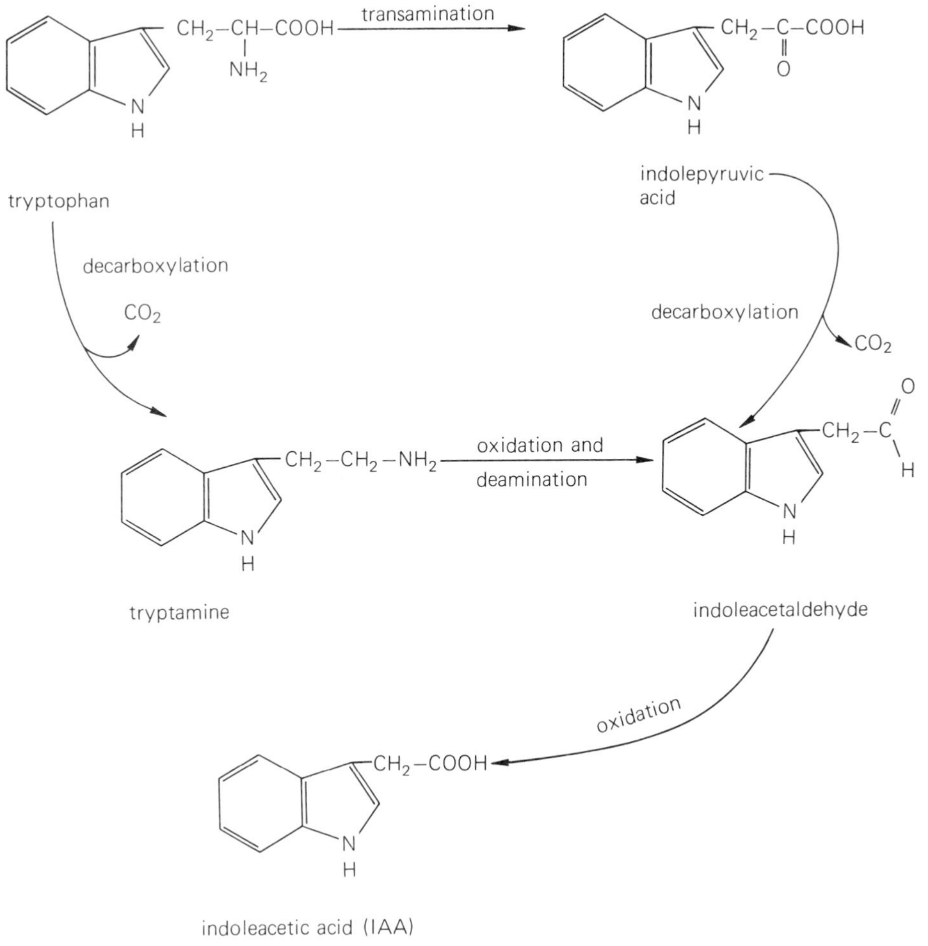

tryptophan $\xrightarrow{\text{transamination}}$ indolepyruvic acid

tryptophan $\xrightarrow{\text{decarboxylation}}$ tryptamine + $CO_2$

indolepyruvic acid $\xrightarrow{\text{decarboxylation}}$ indoleacetaldehyde + $CO_2$

tryptamine $\xrightarrow{\text{oxidation and deamination}}$ indoleacetaldehyde

indoleacetaldehyde $\xrightarrow{\text{oxidation}}$ indoleacetic acid (IAA)

**Figure 17-5** Possible mechanisms of formation of IAA in plant tissues.

highest, suggesting that IAA is synthesized there. Nevertheless, two recent reports indicate that IAA is derived not from the L-form of tryptophan but from the D-form, usually considered unnatural (McQueen-Mason and Hamilton, 1989; Tsurusaki et al., 1990). This possibility clearly needs careful investigation to determine how important it is and how generally it occurs.

It seems logical that plants should have mechanisms to control the amounts of hormones that are as potent as IAA. The rate of synthesis is one mechanism, and temporary inactivation by formation of **auxin conjugates** is another. In conjugates, also called **bound auxins**, the carboxyl group of IAA (studied most among auxins) is combined covalently with other molecules to form derivatives. Numerous IAA conjugates are known, including the peptide *indoleacetyl aspartic acid* and the esters *IAA-inositol* and *IAA-glucose* (Cohen and Bandurski, 1982; Bandurski, 1984; Caruso, 1987). In general, plants can release IAA from these conjugates by hydrolase enzymes, indicating that conjugates are storage forms of IAA. In cereal-grain seedlings, these conjugates are important forms in which IAA can be transported, especially from the endosperm of the seed through the xylem toward coleoptile tips and young leaves. This appears to be the way grass coleoptile tips get the auxin that was discovered by Went.

Other processes for IAA removal are degradative, and there are two types. The first involves oxidation by $O_2$ and loss of the carboxyl group as $CO_2$. The products are variable, but *3-methyleneoxindole* is usually a principal one. The enzyme that catalyzes this reaction is *IAA oxidase*. Several IAA oxidase isozymes exist, and all or nearly all are identical to the *peroxidases* involved in early steps of lignin formation (Section 15.6). In a study with beech and horseradish, for example, 20 peroxidase isozymes were found, and all had IAA-oxidase activity (Gove and Hoyle, 1975). Synthetic auxins are not destroyed by these oxidases, so those auxins persist in plants much longer than does IAA. Conjugated auxins are also resistant to IAA oxidases.

More recently, a second pathway for degradation of IAA has been found to occur in dicots and monocots (Reinecke and Bandurski, 1987). In this pathway the carboxyl group of IAA is not removed, but carbon 2 of the heterocyclic ring is oxidized to form *oxindole-3-acetic acid*. Also, *dioxindole-3-acetic acid* exists in various species, and it has been oxidized both at carbons 2 and 3 of the heterocyclic ring. Details of this degradative pathway are still unknown, but it might prove to be much more important than that involving IAA oxidase.

## Auxin Transport

A surprising thing about the ability of IAA to act as a hormone is the way in which it is transported from one organ or tissue to another. In contrast to movement of sugars, ions, and certain other solutes, IAA is not usually translocated through the phloem sieve tubes or through xylem, but instead primarily through parenchyma cells in contact with vascular bundles (Jacobs, 1979; Aloni, 1987a, 1987b). IAA will move through sieve tubes if applied to the surface of a leaf mature enough to export sugars, but normal transport in stems and petioles is from young leaves and then downward along the vascular bundles. Even synthetic auxins applied to plants move as IAA moves.

This transport has features that are different from those of phloem transport. First, *auxin movement is slow*, only about 1 cm $h^{-1}$ in both roots and stems, but this is still 10 times faster than diffusion would predict. Second, *auxin transport is polar*, always occurring in stems preferentially in a basipetal (base-seeking) direction, regardless of whether the base is normally down or whether the plant is turned upside down. Transport in roots is also polar but preferentially in an acropetal (apex-seeking) direction. Third, *auxin movement requires metabolic energy*, as evidenced by the ability of ATP-synthesis inhibitors or the lack of oxygen to block it. Other strong inhibitors of polar auxin transport are *2,3,5,-triiodobenzoic acid (TIBA)* and *α-naphthylthalamic acid (NPA)*, although TIBA and NPA interfere specifically with auxin transport and not with energy metabolism. Those two compounds have often been called **antiauxins**.

How can the polar transport of auxins occur? The most popular hypothesis indicates, first, that cells use plasma membrane ATPases to pump $H^+$ from the cytosol into cell walls (see Section 7.8). The lower *p*H of cell walls (about 5) keeps the carboxyl group of an auxin less dissociated than in the cytosol, where the *p*H is higher (7 to 7.5). Noncharged auxins then move from the wall into the cytosol by cotransport with $H^+$. (Cotransport or symport is explained in Section 7.10.) In fact, studies with isolated plasma-membrane vesicles indicate that this cotransport involves absorption of one $H^+$ per IAA molecule (Sabater and Rubery, 1987; Rubery, 1987; Heyn et al., 1987). Inside the cytosol the higher *p*H causes the auxin's carboxyl group to dissociate and attain a negative charge. As the concentration of charged auxin (for example, $IAA^-$) builds up in the cytosol, its outward movement is more favored thermodynamically. However, polar transport through an organ requires that the auxin move out only from the basal end of the cell opposite to that which it entered. This preferential exit at the basal end assumes that some carrier in that region of the membrane transports charged auxins out toward the cell wall, where the low *p*H again causes most of them to become noncharged. This chemiosmotic hypothesis for IAA transport is summarized in Figure 17-6.

The crucial problem with this transport hypothesis

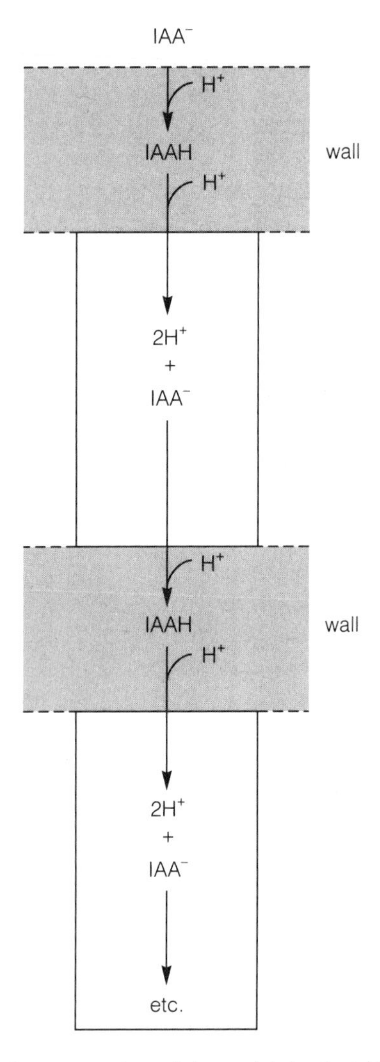


**Figure 17-6** Chemiosmotic model to explain basipetal transport of IAA in living cells. ATP-driven proton pumps in the plasma membrane (not shown) keep the wall *p*H lower than that of the cytosol. Two IAA receptor proteins are thought to exist (neither is shown). One receptor transports IAAH (undissociated IAA) into the top of the cell by cotransport with protons down their free-energy gradient; another receptor at the base of the cell transports $IAA^-$ out of the cell.

has been learning how charged auxins are transported out at the basal ends of cells, because such transport requires cells to be polarized so that they absorb at one end and secrete at the other. (We encountered a similar problem for ion transport into dead xylem cells of roots in Section 7.3; we mentioned that living pericycle or xylem parenchyma cells in roots seem to secrete into dead xylem cells only those ions that they absorbed from cells closer to the root surface.) Direct evidence for the polarized location of an auxin transporter at the basal end of pea stem cells was obtained by Jacobs and Gilbert (1983). This transporter is blocked by NPA, which probably explains how that antiauxin prevents basipetal auxin transport. TIBA apparently blocks at the same site (Goldsmith, 1982).

The mechanism of polar transport of auxins still requires more study, but polar transport down the stem from young leaves or meristematic cells of the shoot tip certainly occurs. The interesting problem of how it might be controlled was studied by Jacobs and Rubery (1988). They found that certain flavonoids abundant in plant cells (see Section 15.7), especially *quercetin*, *apigenin*, and *kaempferol*, are powerful inhibitors of the basal transporter that causes auxin efflux from cells. These compounds might act as part of an auxin-transport control system. This transport is probably important for regulation of such processes as renewed vascular-cambium activity in woody plants during spring, normal differentiation of xylem and phloem at the bases of leaves, growth of stem cells, and perhaps for inhibition of lateral bud development. It is this transport to coleoptile cells directly below the agar block in Figure 17-3 that causes the elongation that results in curvature.

## Extraction and Measurement of Auxins and Other Hormones

A major question confronting us repeatedly is whether a given hormone helps control some particular physiological process *in vivo*. In most cases control is first suspected because exogenous supply of low concentrations of the hormone (or a synthetic analog) promotes the process. A minimum requirement to determine *in vivo* involvement is to extract the hormone and relate its tissue concentration to the magnitude of the response. Notice, however, that the assumption in such studies is that the response is in fact limited by the level of endogenous hormone. But what if hormone receptors or the capacity of the cells to amplify the hormone signal are instead limiting? In that case, little relation might exist between response and hormone concentration, but almost no studies have yet been done to measure concentrations both of hormones and receptors.

Hormone levels are difficult to measure, and assays of known hormone receptors are still in their infancy. For hormone analysis we must have a method that not only is very sensitive but also highly specific, so that other cellular components do not interfere. This general problem represents one of many in which physiologists with important agricultural, horticultural, and forestry problems have had to collaborate with chemists and biochemists.

The first step is to extract the hormone with an organic solvent that will neither extract numerous contaminating compounds nor destroy the hormone we seek. Second, by partitioning the hormone in other immiscible solvents or by using various chromatography procedures, the hormone can be purified much further (reviewed by Yokota et al., 1980; Brenner, 1981; Morgan and Durham, 1983; Horgan, 1987). At this stage, the technique used most frequently historically was to

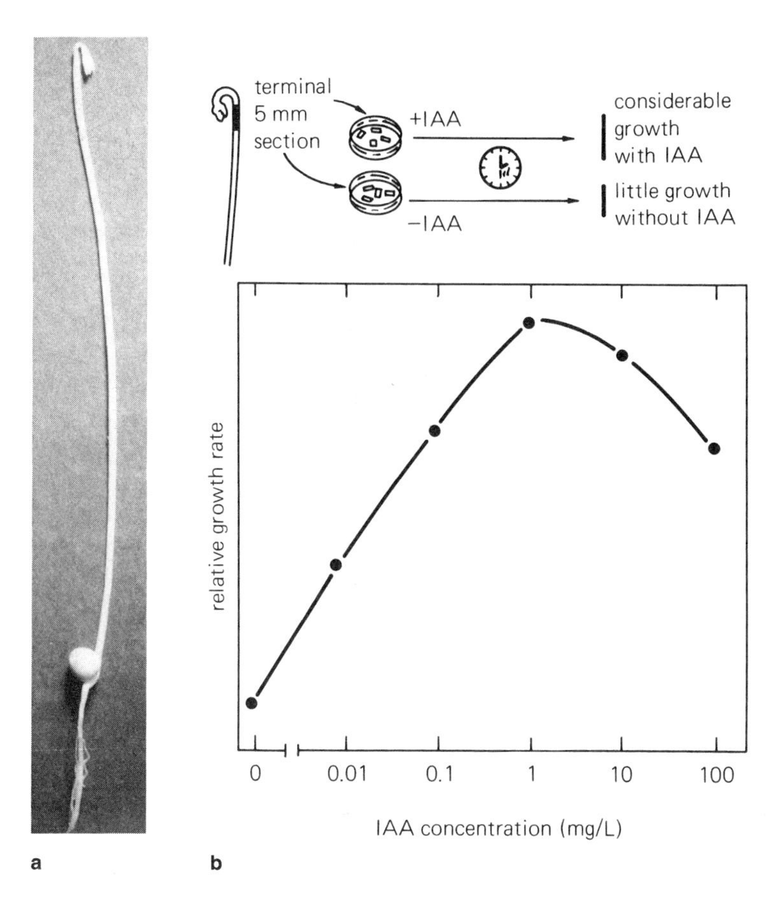


**Figure 17-7** (**a**) Week-old pea seedling grown in darkness. The third (upper) internode is used in auxin bioassays, as shown in **b**. (Photo by C. W. Ross and Nicholas Carpita.) (**b**) Top, technique used in auxin bioassay using apical sections from etiolated pea stems (epicotyls). Sections are placed in petri dishes containing sucrose and certain mineral salts. Growth is often measured 12 to 24 h later. Bottom, influence of the IAA concentration upon growth rate of pea stem sections. Note that auxin concentrations are plotted logarithmically and that an optimum concentration is reached that, when exceeded, results in less growth. (After Galston, 1964.)

measure the amount of the partially purified hormone by a biological assay, or **bioassay**. Bioassays take advantage of the great sensitivity and specificity of certain plant parts or single gene mutants that are deficient in certain hormones (for example, gibberellin-deficient dwarf maize or rice). Reviews of theory and descriptions of methods used in various bioassays are given in the book by Yopp et al., 1986.

In one example of the use of a bioassay, for years physiologists tediously but with great skill analyzed auxins using Went's coleoptile curvature test (Fig. 17-3), measuring the extent of curvature caused by the auxin diffusing from an agar block. Another easier but less sensitive and less specific bioassay for auxins was then developed. It involves cutting elongating sections from coleoptiles or from dicot stems and then growing the sections in a petri dish or other container with various amounts of the partially purified auxin sample and certain other solutes as needed. Over a certain auxin concentration range (often three orders of magnitude), elongation of the sections increases with the amount of auxin added (Fig. 17-7). This *straight-growth test* is subject to interference from inhibitory compounds such as abscisic acid and many phenolics often extracted along with the auxin. Cytokinins also inhibit elongation of stem sections, although this is seldom a bioassay problem because cytokinins are chemically different from auxins and do not contaminate reasonably well purified extracts of auxins. Under certain conditions, gibberellins have little influence on elongation of coleoptile sections, and this makes the assay relatively specific. Nevertheless, most straight-growth tests are far less specific and sensitive than modern physicochemical methods.

Nowadays, bioassays for auxins have been replaced whenever possible by the use of modern instruments of separation and quantitation, including high-performance liquid chromatography (HPLC) and gas chromatography (GC), which are then followed by the use of mass spectrometry (MS) to obtain proof of structure (reviewed by Brenner, 1981; Horgan, 1987).[1]

Another extremely sensitive detection method is **immunoassay**, in which an antihormone antibody made by animal cells is used to react with the hormone

[1]The use of GC-MS to identify plant hormones unequivocally requires a full-scan mass spectrum, together with a carefully determined retention time, both of which are compared with the authentic hormone. To identify and quantify small amounts of hormone, selected ion monitoring (SIM) following GC yields a vastly increased instrument sensitivity that is often in the picogram range.

in a cuvette assay (Weiler, 1984; Pence and Caruso, 1987; Yopp et al., 1986). Immunoassays are more rapid and often 10,000 times as sensitive as any bioassay, but they are commonly subject to negative or positive interference unless the hormone has been well purified beforehand (for example, see Cohen et al., 1987). Furthermore, one must either prepare or purchase the antibody, which frequently is not easy.

Why couldn't we just treat an intact stem or coleoptile and measure its growth response as a bioassay? This is an important question that concerns whether or not elongation of coleoptile sections (classically studied so long) and stems of dicots and conifers are usually limited by a supply from above of one of the four known auxins. For many years the hypothesis has been that enough endogenous auxin is usually supplied to stems and intact plants by basipetal transport from coleoptile tips of grasses or from young leaves above, so that exogenous auxin does not enhance growth. Nevertheless, stems of intact plants of some species elongate faster when auxins are added, at least for several hours (Hall et al., 1985; Tamini and Firn, 1985; Carrington and Esnard, 1988). These experiments show that elongating parts of stems of some species are indeed auxin-deficient but not seriously deficient in auxin receptors or other factors. Auxin bioassays such as the coleoptile curvature test and the straight-growth test probably depend on removal of the responding part from the normal auxin supply: principally, young leaves. As we note in Chapter 19, turning coleoptile sections to the horizontal position gravistimulates them and thereby changes their sensitivity to auxin.

In general, a deficiency of a hormone must be created experimentally (as by removing young leaves) to show that adding a hormone has an effect. However, for gibberellins, abscisic acid, and perhaps cytokinins, genetic mutants that are deficient in these hormones exist. These mutants are providing much information about the importance of hormones to growth and development, especially for gibberellins, as will be explained later in this chapter. No useful mutants that fail to synthesize auxins and also exhibit slow elongation of stems have yet been discovered (Reid, 1990). However, there are certain mutants that contain normal auxin levels but behave in some ways that suggest they are auxin-deficient. One of these is the *diageotropica (dgt)* recessive mutant of tomato. Plants homozygous for the mutation have diageotropic (diagravitropic) shoot systems that grow more or less horizontally. They fail to respond to added auxins either by faster production of ethylene or by faster elongation of excised hypocotyl sections (Kelly and Bradford, 1986), and they also have abnormal vascular tissue, thin stems, altered leaf morphology, and no branching of lateral roots, but apparently they do transport auxins polarly down the stem (Daniel et al., 1989). Stems but not roots of these plants lack what appears to be an important auxin receptor (probably a protein) in the endoplasmic reticulum (Hicks et al., 1989). Perhaps this lack prevents the shoot system from responding to auxins. If so, we should be able to learn the structure of the receptor protein, how it functions, and the gene that codes for it, and then we may learn what controls activity of that gene. Furthermore, we should be able to learn much more about the functions that auxins truly control in plants. More will be said of this probable receptor protein later.

IAA is also assumed to contribute to the growth of leaves, flowers, fruits, and stems of grasses and conifers, but in these cases, too, the only evidence usually comes from experiments on effects of exogenous auxins on detached organs or parts of organs. Roots have been studied more intensively and require special attention.

## Effects of Auxins on Roots and Root Formation

IAA exists in roots at concentrations similar to those in many other plant parts. As first shown in the 1930s, applied auxins will promote elongation of sections excised from roots or even of intact roots of many species, but only at extremely low concentrations ($10^{-7}$ to $10^{-13}$ M, depending on the species and age of roots). At higher concentrations (but still as low as 1 to 10 $\mu$M), elongation is almost always inhibited. The assumption is that root cells usually contain enough or almost enough auxin for normal elongation. Indeed, many excised roots grow for days or weeks *in vitro* without added auxin, indicating that any requirement they might have for this hormone is satisfied by their ability to synthesize it. The best experiments to date concerning auxin levels in roots have investigated only whether or not roots contain IAA and whether that level of IAA normally promotes root growth. Based on what we now know about the presence of four auxins in the plant kingdom, each of the root auxins must be reinvestigated with modern methods and analyses.

One of many questions about how auxins act concerns how they can inhibit root growth at micromolar concentrations. Part of this inhibition has long been assumed to be caused by ethylene because *auxins of all types stimulate many kinds of plant cells to produce ethylene*, especially when relatively large amounts of auxin are added. In most species ethylene retards elongation of both roots and stems (see Section 18.2). Nevertheless, results reported by Eliasson et al. (1989) strongly indicate that IAA can inhibit elongation of attached roots of pea seedlings but fails to affect ethylene production from the same roots soon after they are excised (Fig. 17-8). These and other results indicate that auxins inhibit growth of pea roots, at least, by an unknown mechanism that is independent of ethylene. We must await many further answers about how auxins can inhibit or, at much lower concentrations, even enhance

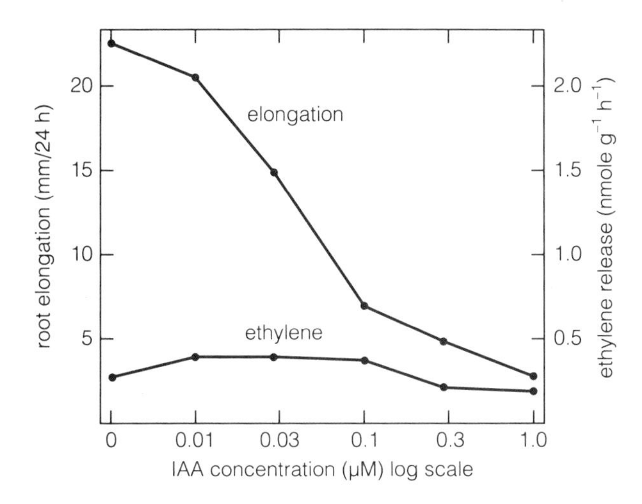


**Figure 17-8** Inhibition of pea-root elongation by IAA without enhancement of ethylene production. Seedlings with roots near 3 cm long were grown for 24 h with roots immersed in auxin concentrations as low as 0.01 μM. Then 1-cm root sections were removed and placed in sealed tubes in contact with paper wetted with IAA, and then ethylene was collected for 2 h. (From Eliasson et al., 1989.)

root elongation. Nevertheless, the ability of excised roots to grow in tissue culture for weeks or months means that such roots need not depend on any shoot-produced auxin for growth. That could mean that, when excised, roots soon adapt to form the only auxin or auxins they need. It could also mean that roots always have the ability to synthesize enough auxins for their growth.[2]

Physiologists have also investigated whether auxins affect the usual process of root formation that helps balance growth of root and shoot systems. There is good evidence that auxins from stems strongly influence root initiation. Removal of young leaves and buds, both of which are rich in auxins, inhibits the number of lateral roots formed. Substitution of auxins for these organs often restores the plant's root-forming capacity. Thus there is an important difference in exogenous auxin effects on root elongation, in which an inhibition is usually observed, and in root initiation and early development, in which promotion is observed (Wightman et al., 1980). Nevertheless, roots of several species grown in tissue culture without a shoot system form lateral roots, showing that under these conditions they either don't require an auxin or they make enough of their own.

Auxins also promote adventitious root development on stems. Many woody species (for example, apples, most willows, and Lombardi poplar) have preformed adventitious-root primordia in their stems that remain arrested for some time unless stimulated by an auxin (Haissig, 1974). These primordia are often at nodes or on the lower sides of branches between nodes. Apple burr-knots on stems contain up to 100 root primordia each. Even stems without preformed root primordia will form adventitious roots, which result from division of an outer layer of phloem.

Adventitious-root formation on stem cuttings is the basis for the common practice of asexual reproduction of many species, especially ornamentals in which it is essential to maintain genetic purity. Julius von Sachs obtained evidence in the 1880s that young leaves and active buds promote root initiation, and he suggested that a transmissible substance (a hormone) was involved. In 1935, Went and Kenneth V. Thimann showed that IAA stimulates root initiation from stem cuttings, and the first practical use of auxins developed from this demonstration. The synthetic auxin NAA (Fig. 17-4) is usually more effective than IAA, apparently because it is not destroyed by IAA oxidase or other enzymes and therefore persists longer. Indolebutyric acid (IBA) is used to cause rooting even more commonly than NAA or any other auxin. IBA is active even though it is rapidly metabolized to IBA-aspartate and at least one other conjugate with a peptide (Wiesman et al., 1989). It was suggested that conjugate formation stores the IBA, and then gradual release keeps the IBA concentration at a proper level, especially during later stages of root formation.

Commercial powders into which cut ends of stems are dipped to facilitate root production usually contain IBA or NAA mixed with inert talcum powder and often with one or more nonhelpful B vitamins (Fig. 17-9). Propagation from cut leaves can also be promoted by auxins. With some species (apples, pears, and most gymnosperms), rooting from stems is minimal with or without auxin. However, it is now known that many failures with auxins are associated with use of cuttings from mature plants. When trees or shrubs are still in the juvenile stage (usually also preflowering; see Section 16.5), they root much more easily with auxins, especially IBA.

The site of adventitious-root formation on stems in most species is the physiological basal position away from (distal to) the stem apex. Even if cut shoots are inverted in a humid atmosphere, the roots will usually form near the top, away from the original stem tips and where auxin is presumably accumulated by polar movement. In many species adventitious roots form near the

[2]Incidentally, such root tissue cultures require the addition of one or more of the B vitamins (especially thiamine, vitamin $B_1$) for successful growth. Does this mean that these vitamins, known to act as coenzymes in many metabolic reactions, are also root-growth hormones? Or is it just that excised roots lose their otherwise normal ability to synthesize them? We don't know, yet various commercial firms sell B vitamins as part of powders or solutions with claims that they promote growth or formation of roots on stem cuttings. There is little or no evidence to support such claims.

**Figure 17-9** Promoting root growth from cuttings by treatment with an auxin. (From Kormondy et al., 1977.)

bases of stems of intact plants, sometimes only as primordia, but sometimes they emerge as the prop-roots do from the nodes on maize stalks. Added auxin often causes emergence of many adventitious roots in the lower internodal stem region, as in tomato plants. Adventitious roots are not restricted to the base of stems but can form on the lower surface of stems that are placed in a horizontal position and kept moist. Higher auxin levels develop in the zone of root emergence before root development occurs. In nature this would allow weak stems to develop additional supportive roots to supplement the existing root system.

## Auxin Effects on Lateral-Bud Development

In stems of most species, the apical bud exerts an inhibitory influence (**apical dominance**) upon the lateral (axillary) buds, preventing or slowing their development. This extra production of undeveloped buds has definite survival value, for if the apical bud is damaged or removed by a grazing animal or a wind storm, a lateral bud will then grow and become the leader shoot. Apical dominance is widespread and was reviewed by Hillman (1984), by Tamas (1987), and by Martin (1987). It also occurs in bryophytes and pteridophytes and in some roots. Another dominant effect of the shoot apex is to cause branches below to grow somewhat horizontally; this horizontal growth often prevents shading of the lower branches and increases photosynthetic productivity of the whole plant.

Gardeners have long removed apical buds and young leaves to increase branching. This technique (called **pinching**) also allows branches to grow more vertically, especially the uppermost branch. In many species, continual removal of the youngest visible leaves is as effective as removal of the entire shoot apex, suggesting that a dominance factor, an inhibitor, arises in those young leaves. If an auxin is added to the cut stump after the shoot apex is discarded, lateral-bud development and vertical orientation of existing branches are again retarded in many species. This replacement of the bud or young leaves with an auxin suggests that the inhibitory compound they produce is IAA or another auxin. Although the review by Tamas (1987) strongly favors the hypothesis that an endogenous auxin is the inhibitor that normally prevents lateral-bud growth, other authors have been skeptical. The amount of IAA that must be added to a cut stump (from which the apex was removed) to prevent development of the lateral buds is often 1,000 times as great as the IAA content of the apical bud itself. Such high doses cause cell division and elongation of the cut stump, which makes it a nutrient sink that could divert nutrients from the lateral buds and prevent their growth indirectly. Studies with $^{14}C$-labeled IAA show that this hormone moves down the stem from the cut surface but does not enter lateral buds in detectable levels. Furthermore, direct application of IAA to the lateral buds does not inhibit their growth and sometimes even promotes it.

A review by Hillman (1984) emphasized the difficulty of analyzing hormone levels in buds to learn whether these levels correlate with the degree of growth inhibition; the problem is that it is technically very difficult to analyze many small buds in which hormones are so dilute. Hillman's review also explained that measurements of hormone levels in entire tissues or organs are of little significance without knowledge of hormone levels within crucial cells or cell organelles. Nevertheless, he and his colleagues (Hillman et al., 1977) had previously analyzed IAA concentrations in 1,500 to 5,000 whole lateral buds of *Phaseolus* (bean) plants in each of two treatments involving decapitated plants (with growing buds) and nondecapitated plants (with nongrowing buds). They found a higher IAA concentration in growing than in nongrowing buds 24 h after decapitation, which supports the idea that IAA is not the inhibitor that prevents outgrowth of lateral buds.

Recently, Gocal et al. (1990) also used GC-MS techniques to measure IAA concentrations in lateral buds of bean plants (as did Hillman et al., 1977) but with se-

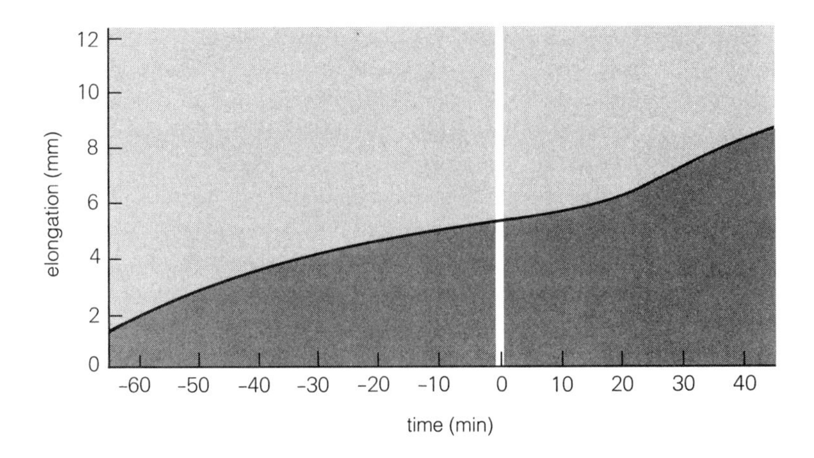


**Figure 17-10** Representation of shadowgraph record of growth of excised oat (*Avena sativa* cv. Victory) coleoptile sections. The incubation medium was changed from water to 3 mg/L IAA at the time corresponding to the vertical white line. The rapid elongation (steeper slope) at the beginning of the record is the result of tactile stimulation of the coleoptiles during experimental manipulations. (Reproduced from *The Journal of General Physiology*, 1969, 53:1–20, by copyright permission of the Rockefeller University Press, courtesy M. L. Evans and P. M. Ray.)

lected ion monitoring as the analytical indicator. This method allowed quantitation of IAA in as few as 60 lateral buds. Time studies (2 to 24 h) showed that after removal of the apical bud and its subtending shoot, growth promotion of the larger lateral bud in the axis of one of the two primary leaves was accompanied by increases in the amount and concentration of IAA within the bud. At 8 h after decapitation the concentration of IAA in the bud was nearly 10 times that in the comparable but slow-growing bud from control plants. These results are consistent with those of Hillman et al. (1977) obtained at only one time (24 h) after decapitation. The amount of IAA in lateral buds increases after decapitation, contrary to predictions that it should decrease when the apical bud is removed. The results of Gocal et al. (1990) also demonstrate the great advantage in sensitivity provided to GC-MS techniques by selected ion monitoring.

In Chapter 18 we give reasons that suggest that repressed lateral buds are deficient in cytokinins, but even that hypothesis has not yet been supported by careful analyses of cytokinin levels. Abscisic acid, ethylene, and gibberellins have also been investigated in relation to apical dominance, but little evidence that they act as transmissible inhibitors or promoters has been obtained. There are other growth regulators in plants (see Section 18.6); one of these might be important, but as yet there is little supporting evidence. Other unsatisfying hypotheses to explain apical dominance were reviewed by Phillips (1975), by Rubenstein and Nagao (1976), and by Hillman (1984).

## Possible Mechanisms of Auxin Action

Numerous researchers have repeatedly emphasized that we still do not understand how any plant hormone acts biochemically. Although this statement is true, it is relative. We do understand many biochemical and physiological processes that are controlled by hormones, although the hormonal effects initiating these processes have not yet been clarified. One of the most thoroughly investigated effects of auxins is enhanced elongation of sections from oat and maize coleoptiles and from stems of various dicots. In this and other test systems, researchers have been interested in how fast an auxin (or other hormone for other systems) could cause some detectable response, because the *earlier the response the more likely it has to do with the primary effect of the hormone*.

In a careful review of how auxins function, Cleland (1987a) emphasized that growth promotion of coleoptile or stem sections by auxins is both rapid and dramatic. It can start within 10 minutes and can continue for many hours, and during this time the growth rate can increase from 5 to 10 times (Fig. 17-10). This growth, with or without auxin, requires water absorption, which means that the cells must maintain a water potential (see Chapter 3) more negative than that of the solution that surrounds them (as explained in Section 16.2). For auxin-induced growth, the water potential is maintained not only more negative than that of the surrounding solution but also more negative than that of control sections. This occurs because cell walls of auxin-treated cells yield more easily, so the pressure potential required to force cell expansion of these cells never needs to become as high as in nontreated cells. The conclusion from much research is that auxins cause **wall loosening**, a term describing the more rapidly extensible or plastic nature of walls from cells treated with auxins.

In a review of this subject, Ray (1987) described three mechanisms considered in the last 30 years to explain wall loosening, all of which have been more or less rejected. The last and most popular mechanism requires mention here, partly because so much evidence supports it and because results of only a few (though important) experiments will likely cause its general rejection. This mechanism, which became known as the **acid-growth hypothesis**, states that auxins cause receptive cells in coleoptile or stem sections to secrete $H^+$ into their surrounding primary walls and that these $H^+$

ions then lower the *p*H so that wall loosening and fast growth occur. The low *p*H presumably acts by allowing the function of certain cell-wall–degrading enzymes that are inactive at a higher *p*H. These enzymes presumably break bonds in wall polysaccharides, allowing the walls to stretch more easily. Reviews largely supporting the acid-growth hypothesis are those by Rayle and Cleland (1979), Taiz (1984), Evans (1985), and Cleland (1987a, 1987b).

The acid-growth hypothesis was questioned seriously with respect to elongation of dicot stems when L. N. Vanderhoef and his colleagues (see the review by Vanderhoef, 1980) found that a low *p*H in cell walls of soybean-hypocotyl sections causes more rapid elongation only for an hour or two, whereas the sections grow faster in auxin for a day or two. Also, pea-stem sections elongate faster with an added auxin whether or not external salts such as KCl are provided, but only if such salts are present does auxin promote acidification of cell walls. More recently, Kutschera and Schopfer (1985) and Kutschera et al. (1987) concluded that auxins do not promote elongation of maize-coleoptile sections via wall acidification. Their results showed that even though auxins lower the wall *p*H to about 5, it requires a still lower *p*H (3.5 to 4) to increase substantially wall loosening in the absence of auxin. This is another way of partially separating the effects of auxin and low wall *p*H on growth. Nevertheless, the ability of auxins to lower the wall *p*H may contribute to growth promotion for short time periods.

Separation of growth promotion (accompanied by wall loosening) and wall acidification had already been shown for cytokinin effects on cucumber-cotyledon growth (Rayle et al., 1982; Ross and Rayle, 1982), indicating that walls can be loosened hormonally without wall acidification. Furthermore, research with both cucumber cotyledons and maize coleoptiles verified that a potent growth promoter from fungi, **fusicoccin**, can acidify cell walls enough to promote their growth. Fusicoccin is a diterpene glucoside that was identified by plant pathologists in the 1960s as the principal toxin responsible for disease symptoms caused by the fungus *Fusicoccum amygdali* on peach, almond, and prune trees (reviewed by Marré, 1979). It has the remarkable abilities to activate a plasma-membrane ATPase that transports $H^+$ from the cytosol into the wall, to lower the wall *p*H, to enhance wall loosening, and to promote cell growth. Even though fusicoccin can increase growth of coleoptiles and cotyledons as it promotes $H^+$ efflux, auxins cannot promote $H^+$ efflux enough to promote maize-coleoptile growth, nor can cytokinins promote $H^+$ efflux enough to enhance cotyledon growth. What these results mean, in effect, is that auxins and other hormones must cause wall loosening and cell expansion in some (perhaps most) species by some unknown mechanism.

As mentioned in the first section of this chapter, not all cells respond to any particular hormone. Therefore, we should ask which cells respond to auxins. For sections of coleoptiles and dicot stems, it is mainly the epidermis that elongates in response to auxins. Normally, subepidermal layers such as the hypodermis (if present), the cortex, and the pith contain cells that are under pressure and are ready to elongate. Their elongation is restricted because they are bound via continuous cell-wall polysaccharides to epidermal cells that cannot stretch as fast. The overall result is that subepidermal layers elongate just enough to keep slower-growing epidermal cell walls under slight tension. Apparently, even though epidermal cells have positive pressure potentials (are under turgor pressure), their walls are being stretched. It seems as though an internal pressure and an external stretching or tension ought to force those epidermal cells to grow unusually fast, yet their walls simply don't stretch rapidly unless more auxin is added to make them looser (reviewed by Cosgrove, 1986 and by Kutschera, 1987, 1989; also see Cosgrove and Knievel, 1987). Sections of stems or coleoptiles placed in an auxin solution respond by developing looser epidermal walls. These epidermal cells then elongate faster, and their elongation also allows elongation of connected subepidermal cells, so that the entire coleoptile or stem elongates faster.

Now that we know that the epidermis responds first to auxins, experiments with the epidermis seem especially important in terms of what auxins do and how fast they do it. This work has begun in attempts to learn whether auxins activate genes in the epidermis (Dietz et al., 1990). But even before physiologists concentrated on special cell layers of stems, much research pioneered by Joe L. Key and Thomas J. Guilfoyle at the University of Georgia showed that auxins caused rapid changes in gene activity in soybean-hypocotyl sections (Fig. 17-11). That work was soon followed by comparable results with pea-stem sections, conferring more generality to the principle that auxins can change a few gene products (proteins) as fast as they promote elongation. This research was important because it showed that auxins not only affect the kinds of proteins formed but that they also do so rapidly (that is, before or as soon as growth promotion starts). (See reviews by Key, 1987; Guilfoyle, 1986; Theologis, 1986; Guilfoyle and Hagen, 1987; Hagen, 1987; and Key, 1989.) This well-documented effect of auxins on growth promotion and gene activity needs to be related to the model depicted in Figure 17-1. Where does control lie? Much evidence indicates that the main control is at transcription, but control over mRNA stability has not yet been assessed (Key, 1989). Furthermore, it has not yet been proven that the proteins induced by auxins are directly involved in growth. Nevertheless, recent research is concentrating on all detectable early effects of any auxin, whether

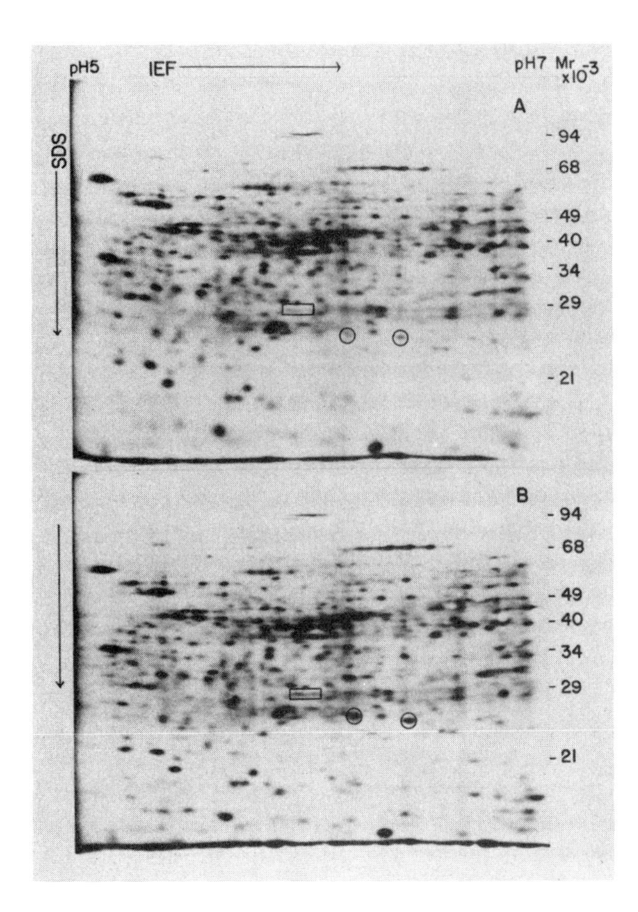


**Figure 17-11** IAA-induced increases in certain polypeptides synthesized by messenger RNAs in maize coleoptiles (A, controls). Coleoptile sections in B were exposed to 50 μM IAA for 20 min to promote elongation; then messenger RNAs from sections with or without auxin were isolated. These mRNAs were translated into proteins by a cell-free extract of wheat germ containing $^{35}$S-methionine to label all proteins formed. Proteins were boiled with a detergent (sodium dodecyl sulfate, SDS), which converts heteropolymers or homopolymers (see Chapter 9) into individual polypeptide chains. These polypeptides were then separated by two-dimensional polyacrylamide gel electrophoresis, first from left to right by isoelectric focusing (IEF), which separates on the basis of charge, and then from top to bottom, which separates on the basis of molecular weight. The molecular weights in thousands of grams per mole are listed at the right of each photograph. Radioactive polypeptides were visualized as spots or streaks by placing the gel next to a film sensitive to radioactivity (see autoradiography, Chapter 8). (Unpublished data of L. L. Zurfluh and T. J. Guilfoyle.) Three boxed or circled polypeptides become more abundant soon after auxin treatment.

naturally occurring or synthetic, especially in relation to Figure 17-2, in which hormone receptors and their actions are emphasized.

An important question is whether hormone receptors can in fact be found in plants. The term *receptor* implies that biochemical action must follow hormone binding *in vivo;* otherwise the binding might have little physiological significance. Many proteins bind nonspecifically either ionically or by van der Waals forces to small molecules, so it is essential to show for, say, a true auxin receptor, that binding occurs at low and physiologically reasonable concentrations of an auxin and that the protein won't bind to molecules with similar structures that lack auxin activity. A few binding proteins for auxins have been purified and antibodies made against them. In some cases adding the antibody in low concentrations to an isolated plant part prevents the physiological action of an auxin, strongly indicating that the binding protein that caused formation of the specific antibody is indeed a hormone receptor.

To date, the main purported auxin-receptor protein has been sequenced (indirectly, by sequencing a DNA that codes for it), and it is apparently a dimer made of two polypeptides of about 20 kDa each. This purported receptor exists largely in the ER (likely the same receptor as described earlier for the *diageotropica* tomato mutant), but it also exists near the outer surface of the plasma membrane (reviewed by Napier and Venis, 1990). Now, in terms of the model depicted in Figure 17-2, we need to know what happens next—that is, how amplification occurs. Accumulating evidence suggests that auxins act at the plasma membrane to cause changes in metabolism of inositol phospholipids and inositol phosphates, a finding that is largely consistent with that model.

## Auxins as Herbicides

Work at the Boyce Thompson Institute in New York in the 1940s established that 2,4-D has auxin activity. Subsequent work there and in England showed that 2,4-D, NAA, and certain other related compounds are effective **herbicides**, or plant killers. Four of the most widely used auxin herbicides have been 2,4-D, 2,4,5-T, MCPA (Fig. 17-4), and derivatives of picolinic acid such as *picloram* (sold under the trade name Tordon). The popularity of these herbicides is derived from their high phytotoxicity, their relatively low cost, and their property of affecting dicots much more than monocots (Klingman et al., 1982; Moreland, 1980). Because of this selectivity, they are often used to kill broadleaf dicot weeds in cereal-grain crops and lawns. For grass pastures and rangelands in which woody perennials such as sagebrush and mesquite are often a problem, 2,4,5-T was particularly effective, but the U.S. Environmental Protection Agency forced its removal from the market primarily because it contained traces of a powerful toxin (a *dioxin*). Several derivatives of benzoic acid such as *dicamba* also have auxin activity and are more effective than the others against deep-rooted perennial weeds, including field bindweed or wild morning glory (*Convolvulus arvensis*), Canada thistle (*Cirsium arvense*), and dandelion (*Taraxacum officinale*).

In spite of much research to determine how auxin herbicides kill only certain weeds, their mechanism of action is generally unknown. Part of their selectivity against broadleaf weeds results from greater absorption and translocation than in grasses, but more important factors are involved. It is sometimes stated that they cause a plant to "grow itself to death," but this phrase is misleading. Certain parts of some organs do indeed grow faster than other parts, so we see twisted and deformed leaf blades, petioles, and stems because of unequal growth. Much of this results from epinastic effects (see Fig. 18-12) that arise from the common property of all auxins to enhance ethylene production; ethylene is notorious for causing epinasty. But unequal growth that causes twisting results from inhibition of one part and promotion of the other. Overall growth of plants is definitely retarded and is eventually stopped if enough herbicide is absorbed and translocated. Modern hypotheses suggest that these compounds alter DNA transcription and RNA translation so greatly that enzymes needed for coordinated growth are not produced properly.

Elsewhere in this text we consider other possible auxin functions in relation to phototropism and gravitropism (Chapter 19) and in delaying abscission of leaves, flowers, and fruits (Chapter 18).

## 17.3 The Gibberellins

Gibberellins were first discovered in Japan in the 1930s from studies with diseased rice plants that grew excessively tall (for historical reviews, see Phinney, 1983 and Thimann, 1980). These plants often could not support themselves and eventually died from combined weakness and parasite damage. As early as the 1890s, the Japanese called this the **bakanae** ("foolish seedling") **disease**. It is caused by the fungus *Gibberella fujikuroi* (the asexual or imperfect stage is *Fusarium moniliforme*). In 1926, plant pathologists found that extracts of the fungus applied to rice caused the same symptoms as the fungus itself, demonstrating that a definite chemical substance is responsible for the disease.

In the 1930s, T. Yabuta and T. Hayashi isolated an active compound from the fungus, which they named

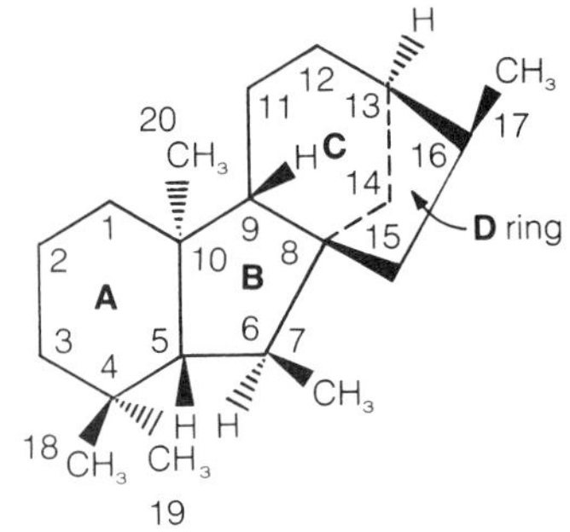

**Figure 17-12** Structure of the *ent*-gibberellane skeleton and of six active gibberellins. The gibberellins are numbered as for *ent*-gibberellane, except that the carbon 20 methyl group of *ent*-gibberellane has been oxidized and then released as $CO_2$ to form the 19-carbon gibberellins shown here. Furthermore, in these gibberellins (and some others), the carbon 19 methyl group of *ent*-gibberellane has been oxidized to a carboxyl and then used to form a lactone ring.

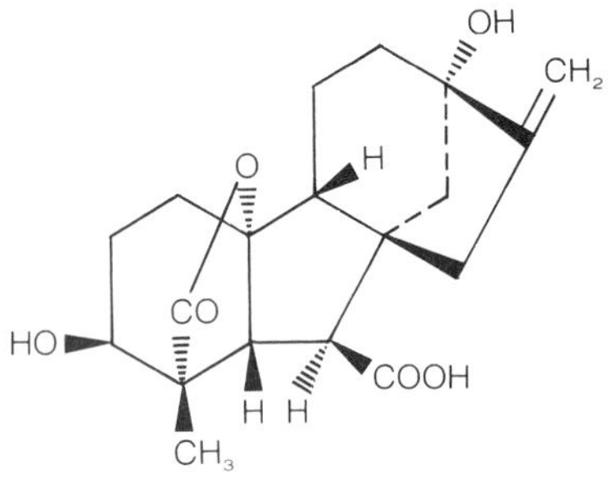

*ent*-gibberellane skeleton

$GA_1$ 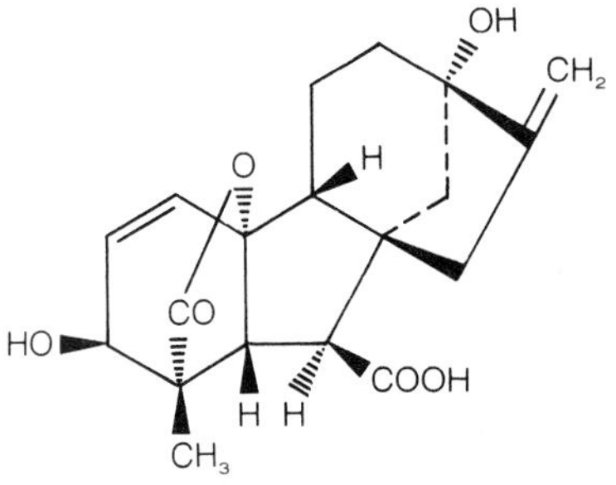 $GA_3$ 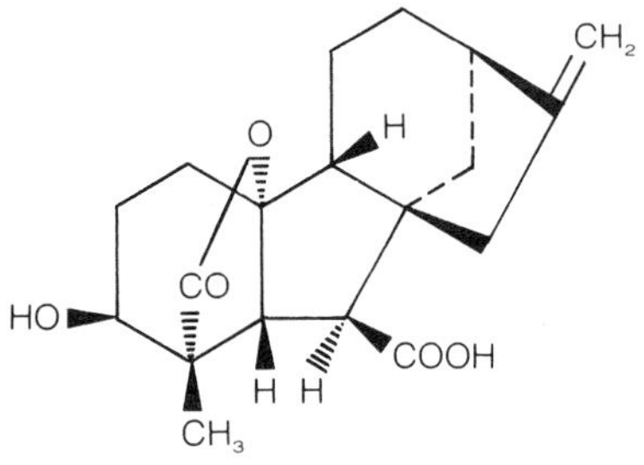 $GA_4$

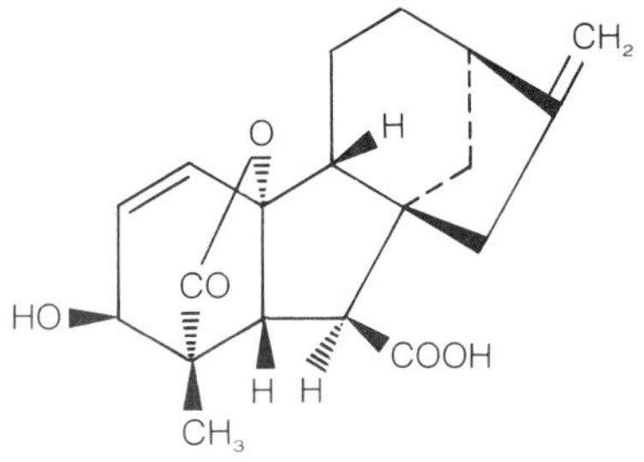

$GA_7$ 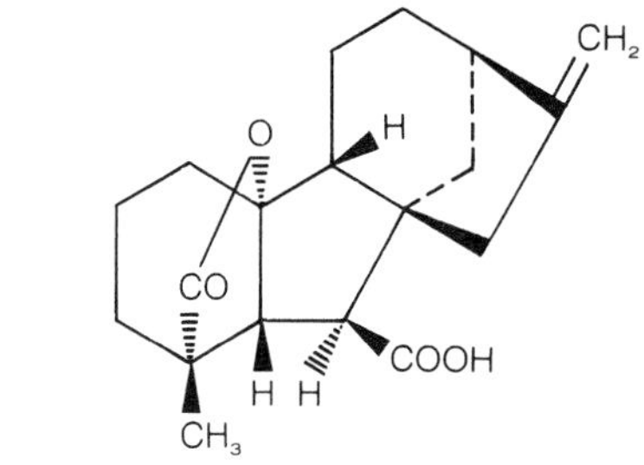 $GA_9$ 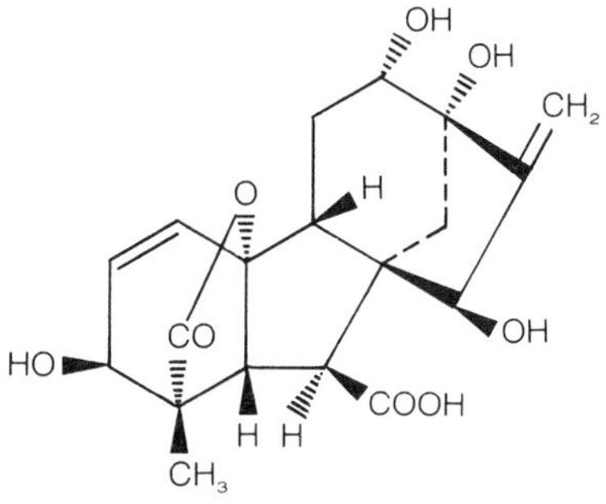 $GA_{32}$

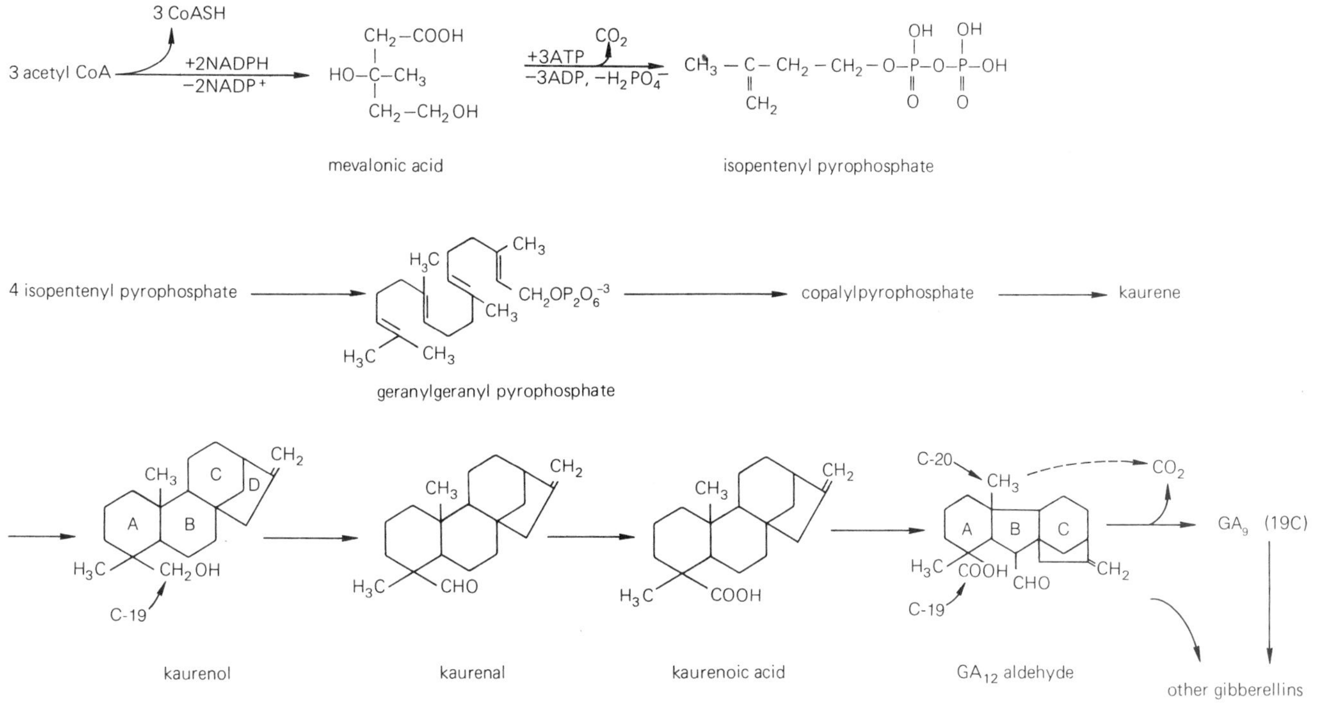

**Figure 17-13** Some reactions of gibberellin biosynthesis. Many steps indicated as single arrows actually involve more than one enzyme-catalyzed reaction, especially those before kaurene.

**gibberellin**. Thus the first gibberellin was discovered as early as the discovery of IAA; yet because of preoccupation with IAA and synthetic auxins, the lack of early contact with the Japanese, and then World War II, western scientists did not become interested in gibberellin effects until the early 1950s.

As of 1990, 84 gibberellins had been discovered in various fungi and plants (reviewed by Sponsel, 1987; by Graebe, 1987; and by Takahashi et al., 1990). Of these, 73 occur in higher plants, 25 in the *Gibberella* fungus, and 14 in both. Seeds of the cucurbit *Sechium edule* contain at least 20 gibberellins, and seeds of the common bean (*Phaseolus vulgaris*) contain at least 16, but most species may contain fewer.

All gibberellins are derivatives of the *ent-gibberellane skeleton*. The structure of this molecule with its ring-numbering system, along with structures of six active gibberellins, is shown in Figure 17-12. All gibberellins are acidic and are named **GA** (for gibberellic acid) with a different subscript to distinguish them. All gibberellins have either 19 or 20 carbon atoms grouped in either four or five ring systems. The fifth ring system (not present in *ent*-gibberellane) is the lactone ring shown attached to ring A in the gibberellins of Figure 17-12. All gibberellins have one carboxyl group attached to carbon 7, and some have an additional carboxyl attached to carbon 4, so all could be called gibberellic acids. However, $GA_3$, the first highly active and longtime commercially available gibberellin (purified from the culture medium of the fungus *G. fujikuroi*), has historically been called **gibberellic acid**. The number of hydroxyl groups on rings A, C, and D ranges from none (as in $GA_9$) to four (as in $GA_{32}$), with carbon 3 or carbon 13 or both most commonly hydroxylated.

Gibberellins exist in angiosperms, gymnosperms, ferns, and probably also in mosses, algae, and at least two fungi. Recently they were found also to exist in two species of bacteria (Bottini et al., 1989; Atzorn et al., 1988). It should be noted, however, that some of the 84 known gibberellins are probably just physiologically inactive precursors of other active ones, and still others are inactive hydroxylated products. It does not seem likely that any plant depends on all of the gibberellins it contains, but this has not been studied well enough for us to be sure. Furthermore, the 25 gibberellins in *G. fujikuroi* have no known function (although one can speculate that they might enhance hydrolysis of starch to sugars in host plants by inducing formation of amylase enzymes, thereby obtaining a sugar food source).

## Metabolism of Gibberellins

As mentioned in Section 15.3, gibberellins are isoprenoid compounds. Specifically, they are diterpenes synthesized from acetate units of acetyl coenzyme A by the mevalonic acid pathway. *Geranylgeranyl pyrophosphate* (Fig. 17-13), a 20-carbon compound, serves as the donor for all gibberellin carbon atoms. It is converted to *copalylpyrophosphate*, which has two ring systems, and the latter is then converted to *kaurene*, which has four ring systems. Conversion of kaurene farther along the pathway involves oxidations occurring in the endoplasmic reticulum, producing the intermediate compounds *kaurenol* (an alcohol), *kaurenal* (an aldehyde), and *kaurenoic acid*; each compound is oxidized successively more.

The first compound with a true gibberellane ring system is the aldehyde of $GA_{12}$, a 20-carbon molecule. From it arise both 20- and 19-carbon gibberellins, probably also in the ER. $GA_{12}$ aldehyde is formed by extrusion of one of the carbons of the B ring in kaurenoic acid (Fig. 17-13) and contraction of that ring. All plants likely use the same reactions to form $GA_{12}$ aldehyde, but from this point in the pathway different species use at least three different pathways to form different gibberellins. In all cases, however, the aldehyde group extending downward from ring B in $GA_{12}$ aldehyde is oxidized to a carboxyl group that is necessary for biological activity of all gibberellins.

In general, 19-carbon gibberellins are more active than 20-carbon gibberellins, and the carbon lost from the 20-carbon molecules is that of the methyl group attached between rings A and B of $GA_{12}$ aldehyde. It becomes oxidized to a carboxyl group, which is then released as $CO_2$. In most gibberellins the fifth (lactone) ring system is formed from the carbon 19 carboxyl group of $GA_{12}$ aldehyde to produce $GA_9$. Other important modifications of the ring systems can occur. For example, $GA_1$ (Fig. 17-12) has one hydroxyl group attached to ring A and another attached between rings C and D. As we shall describe, $GA_1$ seems especially important in causing stem elongation.

Certain commercial growth retardants that inhibit stem elongation and cause overall stunting do so in part because they inhibit gibberellin synthesis. These products include *Phosphon D, Amo-1618, CCC* or *Cycocel, ancymidol,* and *paclobutrazol*. The first two block conversion of geranylgeranyl pyrophosphate to copalylpyrophosphate (see Fig. 17-13). Phosphon D also inhibits subsequent formation of kaurene, whereas ancymidol and paclobutrazol block oxidation reactions between kaurene and kaurenoic acid. In many plants growth inhibition by each can be completely overcome by $GA_3$, which suggests that their major effects are to inhibit gibberellin synthesis. However, Phosphon D, Amo-1618, and CCC inhibit sterol synthesis in tobacco, indicating that they are not specific inhibitors of gibberellin formation. The use of plant-growth retardants, including those that act by blocking gibberellin synthesis, was reviewed by Grossmann (1990).

The commonly used $GA_3$ appears to be only slowly degraded, but during active growth most gibberellins are rapidly metabolized by hydroxylation to inactive products. Also, they can readily be converted to conjugates that are largely inactive. These conjugates might be stored or translocated before their release at the proper time and place. Known conjugates include **glucosides**, in which glucose is connected in an ether bond to one of the —OH groups or in an ester bond to a carboxyl group of the gibberellin. Another important metabolic process is conversion of highly active gibberellins to less active ones. For example, Douglas-fir shoots, which show little vegetative growth in response to most applied gibberellins, can effectively hydroxylate $GA_4$ to the much less active $GA_{34}$.

Which parts of plants synthesize gibberellins? Clearly, if we find these hormones in a plant organ, they may have been either synthesized there or translocated there. Immature seeds contain relatively high amounts of gibberellins compared with other plant parts, and cell-free extracts from seeds of some species can synthesize gibberellins. These and other results indicate that much of the high gibberellin content of seeds results from biosynthesis, not from transport. The ability of other plant parts to synthesize gibberellins is less well established because fewer direct biochemical data are available. Nevertheless, it is likely that most plant cells have some ability to synthesize gibberellins.

Young leaves are thought to be major sites for gibberellin synthesis, as is the case for auxins. This hypothesis is consistent with the fact that when the shoot tip and young leaves are excised and the cut stump is then treated with either a gibberellin or an auxin, stem elongation is promoted as compared to cut stems not given either hormone. The implication is that young leaves normally promote stem elongation because they transport both hormones to the stem. This is curious, because young leaves are translocation sinks via the phloem, not sources. For auxins we know that transport does not usually occur via the phloem but instead occurs polarly in cells attached to vascular bundles, so no problem in explaining their transport arises. But for gibberellins, transport other than by diffusion occurs through both xylem and phloem and is not polar. How gibberellins could be transported effectively from young leaves to cause stem elongation, if indeed this occurs, is unknown.

Roots also synthesize gibberellins; but exogenous gibberellins have little effect on root growth, and they inhibit adventitious-root formation. These hormones can be detected in the xylem exudates of roots and stems when these organs are excised and root pressure forces the xylem sap out. Inhibitors of gibberellin synthesis decrease amounts of gibberellins in these exudates. Repeated excision of part of the root system causes marked decreases in the concentrations of gibberellins in the shoot, suggesting either that much of the shoot's gibberellin supply arises from the roots via the xylem or that repeatedly excised roots cannot supply water and mineral nutrients in sufficient amounts to maintain the ability of the shoot to synthesize its own gibberellins.

## Gibberellin-Promoted Growth of Intact Plants

Gibberellins have the unique ability among recognized plant hormones to promote extensive growth of intact plants of many species, especially dwarfs or biennials in the rosette stage. With some exceptions (to be noted later), they generally enhance elongation of intact stems much more than that of excised stem sections, so their effects are opposite to those of auxins in this respect. An early demonstration of elongation caused by an ether-soluble substance extracted from bean seeds was made by John W. Mitchell and his colleagues (1951) (Fig. 17-14). They were not sure what caused this unusual growth promotion, but they demonstrated that IAA was not responsible. We now know that seeds of beans and many other dicots are rich sources of gibberellins and that the symptoms Mitchell and coworkers observed are identical to those caused by several gibberellins.

**Figure 17-14** Growth stimulation of *Phaseolus vulgaris* by a gibberellin-containing extract prepared from seeds of the same cultivar (Black Valentine). An ether extract of seeds was evaporated and 125 μg of the residue were mixed with lanolin and applied as a band around the first internode of the plant on the right. Plants were photographed three weeks after treatment. Plant on the left was untreated. (From Mitchell et al., 1951.)

Most dicots and some monocots respond by growing faster when treated with gibberellins, but several species in the family Pinaceae show few or no elongation responses to $GA_3$ (Pharis and Kuo, 1977). However, they do respond well to a mixture of $GA_4$ and $GA_7$ (Pharis et al., 1989). Cabbages and other species in the rosette form that have short internodes sometimes grow 2 m tall and then flower after $GA_3$ application, whereas untreated plants remain short and vegetative. Short bush beans become climbing pole beans, and dwarf genetic mutants of rice, maize, and peas phenotypically exhibit the tall characteristics of normal varieties when treated with $GA_3$. Watermelons, squash, and cucumbers elongate fastest in response to gibberellins without a carbon 13 hydroxyl group ($GA_4$, $GA_7$, $GA_9$). Dwarf meteor peas are sensitive to as little as $10^{-9}$ gram (one nanogram) of $GA_3$, so their growth has long been used as a gibberellin bioassay. Dwarf rice (cv. Tanginbou) can even respond to as little as 3.5 picograms ($3.5 \times 10^{-12}$ g) of $GA_3$ (Nishijima and Katsura, 1989). Reviews of dwarf mutants in relation to gibberellins have been written by Reid (1987), by Hedden and Lenton (1988), and by Reid (1990).

Five different dwarf mutants of maize (Fig. 17-15) grow as tall as their normal counterparts after gibberellin application. Each of the mutants contains a mutation on a different gene, and each mutation controls a different enzyme needed in the pathway of gibberellin synthesis. These plants are **gibberellin-synthesis mutants**, most of which are underproducing dwarfs. The studies of Bernard O. Phinney, J. MacMillan, and their colleagues (for example, MacMillan and Phinney, 1987) indicated that only $GA_1$ controls stem elongation in maize and that all dwarf mutants lack enzymes to convert other gibberellins to $GA_1$. Growth of hybrid cultivars of maize in which heterosis exists is not appreciably stimulated by gibberellins because these hybrids presumably contain enough $GA_1$ to allow growth (Rood et al., 1988). However, their inbred counterparts did respond to $GA_3$ by faster elongation. Much evidence now indicates that $GA_1$ is the primary gibberellin needed for elongation of dwarf peas, sweet peas, rice, tomato, rape, and some wheat cultivars. When $GA_3$ or other gibberellins promote elongation of dwarfs, they probably do

**Figure 17-15** Five recessive dwarf maize mutants that are deficient in gibberellin production. An-1 is the anther ear-1 mutant. Notations under the plants indicate the point in the pathway of gibberellin synthesis that is blocked because of the mutation in that plant. (From Phinney and Spray, 1987.)

so by first being converted to $GA_1$. Mutants that overproduce gibberellins and that have unusually long internodes have also been found. In *Brassica rapa* (syn. *campestris*), the mutant gene causes overproduction of $GA_1$ (Rood et al., 1990).

It is likely that most species require $GA_1$ for stem elongation, although the mere presence of this hormone is not sufficient in many cases. Thus, many **gibberellin-sensitivity mutants** are also known in maize, pea, and wheat (reviewed by Reid, 1990 and by Scott, 1990). These mutants seem to have adequate levels of $GA_1$ but cannot respond to it. Among several possible reasons, lack of receptor proteins is an obvious possibility that is under investigation. Some of the dwarf and semidwarf wheat cultivars respond well to fertilizers by exhibiting increased grain yields, and these cultivars are being used in plant-breeding experiments.

## Gibberellin-Promoted Germination of Dormant Seeds and Growth of Dormant Buds

The buds of evergreens and of deciduous trees and shrubs that grow in temperate zones usually become dormant in late summer or early fall (see Chapter 22). Dormant buds are relatively hardy during cold winters and drought. Seeds of many noncultivated species are also dormant when first shed and will not sprout even when exposed to adequate moisture, temperature, and oxygen. Dormancy of buds and seeds is often overcome (broken) by extended cold periods in winter, allowing growth in the spring when conditions are favorable. For some species, bud dormancy can also be overcome by the increasing day lengths that occur in late winter, and for seeds of many species dormancy is broken by brief periods of red light when they are moist (see Section 20.6).

Gibberellins overcome both kinds of seed dormancy and both kinds of bud dormancy in many species, acting as a substitute for low temperatures, long days, or red light. In seeds one gibberellin effect is to enhance cell elongation so that the radicle can push through the endosperm, seed coat, or fruit coat that restricts its growth. Buds have been less carefully investigated, and whether a stimulation of cell division in addition to elongation is necessary is not known but is likely.

## Flowering

As we describe in Chapter 23, the time at which a plant forms flowers is dependent upon several factors, including its age and certain properties of the environment. For example, the relative durations of daylight and darkness have important influences on several species. Some species flower only if the period of daylight exceeds a critical length, and others flower only if this period is shorter than some critical length. Gibberellins can substitute for the long-day requirement in some species, again showing an interaction with light. Gibberellins also overcome some species' need for an inductive cold period if they are to flower or to flower sooner (vernalization). Evidence is accumulating that indicates that some gibberellins are much more effective in enhancing flowering than others.

## Gibberellin-Stimulated Mobilization of Foods and Mineral Elements in Seed Storage Cells

Soon after a seed germinates, the young root and shoot systems begin to use the mineral nutrients, fats, starch, and proteins present in storage cells of the seed. The young seedling depends upon these food reserves before it can absorb mineral salts from the soil and before it can extend its shoot system into the light. The mineral salts are readily translocated via the phloem into and throughout the young roots and shoots, if such salts are mobile. The seedling has a problem with fats, polysaccharides, and proteins because these molecules are not translocated. How is this problem solved? We touched on this matter briefly in Chapters 8, 13, 14, and 15, when we discussed how storage polymers are converted into sucrose and into mobile amino acids or amides. Gibber-

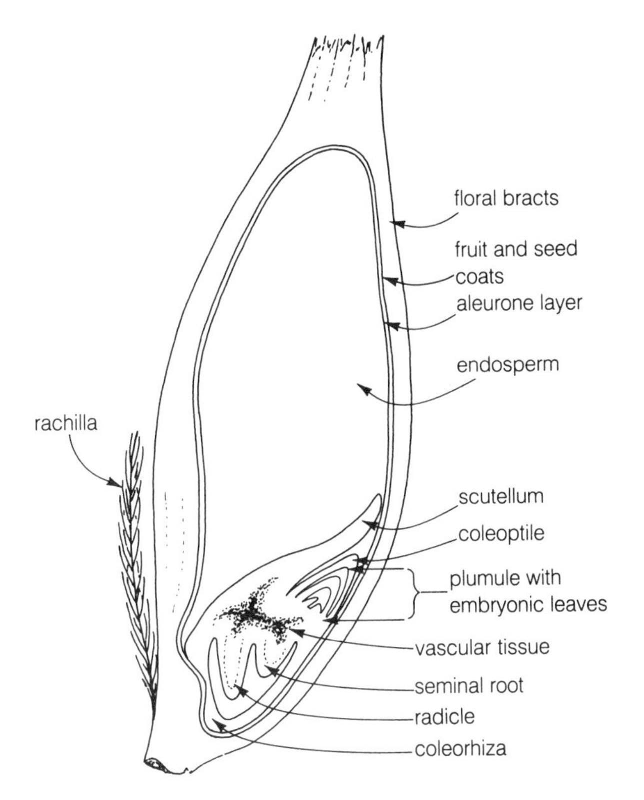


**Figure 17-16** Barley seed sectioned to illustrate major tissues. (Original drawing by Arnold Larsen, Colorado State University Seed Laboratory.)

**Figure 17-17** Gibberellin-stimulated digestion of the endosperm in barley half-seeds. The embryo half of each seed was removed before treatment with (top to bottom) 5 μL of 0.1 μM $GA_3$, 0.001 μM $GA_3$, or $H_2O$. (Courtesy J. E. Varner.)

ellins stimulate these conversions, especially in cereal grains.

The embryo (germ) of seeds of cereal grains and of other grasses is surrounded by food reserves present in the metabolically inactive cells of the **endosperm**; the endosperm in turn is surrounded by a thin living layer most commonly two to four cells thick called the **aleurone layer** (Fig. 17-16). After germination occurs, primarily in response to increased moisture, the aleurone cells provide hydrolytic enzymes that digest the starch, proteins, phytin, RNA, and certain cell-wall materials present in the endosperm cells.

One of the necessary enzymes for these digestion processes is α-amylase, which hydrolyzes starch (see Section 13.2). If the embryo is removed from a barley seed, the aleurone cells do not produce and secrete most hydrolytic enzymes, including α-amylase. This suggests that the barley embryo normally provides some hormone to the aleurone layer and that this hormone stimulates aleurone cells to manufacture these hydrolytic enzymes. This hormone, which appears to be a gibberellin, also stimulates secretion of the hydrolytic enzymes into the endosperm, where they digest food reserves and cell walls. Reserve mineral elements also become more readily available as a result of gibberellin action. Figure 17-17 illustrates endosperm degradation in barley half-seeds (from which the embryo was removed) in response to as little as $9 \times 10^{-12}$ gram (9 pg) of $GA_3$. The increase in α-amylase in aleurone layers of such half-seeds, frequently used as a gibberellin bioassay, results mainly from enhanced transcription of the gene that codes for α-amylase (reviewed by Akazawa et al., 1988 and by Fincher, 1989).

In grass seeds (including barley), gibberellins are probably synthesized in the **scutellum** (cotyledon) and perhaps also in other parts of the embryo. The kind of gibberellin synthesized likely depends upon the species, but in barley seeds $GA_1$ and $GA_3$ seem most important. Nevertheless, although the aleurone layers of barley, wheat, and wild oats (*Avena fatua*) respond to added $GA_3$ or to certain other gibberellins by synthesizing α-amylase and other hydrolytic enzymes, some cultivated oat cultivars and most maize cultivars do not. There is considerable genetic variability among cereal-grain seeds relative to gibberellin responses.

Although the aleurone layer is responsible for enzymes that digest some of the reserve foods in the endosperm, evidence has existed for about 100 years that the scutellum also secretes enzymes that cause digestion (reviewed by Akazawa and Hara-Nishimura, 1985 and by Akazawa et al., 1988). The portion of the scutellum that faces the endosperm is composed of a single layer of columnar cells whose internal structure is rich in endoplasmic reticulum and dictyosomes typical of

## Why a Biologist? Some Reflections

*Frits W. Went*

*With the discovery of auxin, Frits W. Went's fame was assured. In this essay, he tells how that happened while he was a young student working in his father's laboratory at the University of Utrecht in the Netherlands. After completing the doctoral degree, he spent five years in Java, a Dutch possession, and almost 20 years at the California Institute of Technology, where he continued hormone work and developed interests in desert ecology. He moved in 1958 to St. Louis, Missouri, and then in 1964 to the Desert Biology Laboratory at the University of Nevada, where he continued his desert studies. Professor Went died on May 1, 1990.*

Years ago I tried to discover what made a biologist become a biologist. Soon I found that there are about as many different motivations as biologists, but a few were more prevalent than others. Every human being is born with an enormous amount of intellectual curiosity. If this has not been blunted through unfortunate experiences in youth, then an inspiring teacher may guide this urge for understanding toward biological or other problems. Early acquaintance with plants or animals can direct an inquisitive mind—even without teacher direction—toward the mysteries of life: the problems of growth, form, function, environment, and heredity. The orderly mind may be attracted toward taxonomy or biophysics, whereas the mind intrigued by complexity may select ecology, and the mind trying to understand interrelationships may become a physiologist. The mechanically inclined person might attack biological problems with delicate instruments, whereas the artist might try to solve problems of shape and color in nature.

There is an equal plethora of methodological approaches to the solution of biological problems. Since Francis Bacon in the 17th century pointed out the inevitability of cause and effect, the experimental or inductive approach has taken precedence over the old Aristotelian deductive approach, which through pure reasoning forced life into a straitjacket of axioms or preconceived ideas.

Not a few bright minds have been misled by their or others' sophisticated hypotheses. Such complicated but artificial fabrications as the phlogiston theory of the 18th century and the ether theory of the 19th century have held back real understanding. It is difficult for us to say which present-day ideas may be discarded in the centuries to come, but the Second Law of Thermodynamics might well be one of them.

An important aspect of the motivation to become a biologist may be humanitarian—the urge of the person to take a positive part in the well-being of his fellow human beings. Most disciplines in biology, particularly agriculture and medicine, have an input to society, and let us not forget teaching, one of the most important motivations of all. This desire to transmit the knowledge gathered by our culture to the generations to come is quite the contrary of the attitude of a professor 100 years ago, who admonished his successor never to tell the students everything he knew!

In my own case, ever since I chose botany over chemistry or engineering as my life's endeavor I have been intrigued with the form and function of a plant and with its place and role in the environment. When walking in the country, I am forever wondering why a particular plant grows in a particular spot; why it has its particular shape; why it does not grow 100 m farther on; why only some plants grow in the desert or tropics; why a limited number of plants have become weeds; why some plants are closely similar to others that lived 200 million years ago, whereas most are recently evolved; why some trees can be tapped for sugar, whereas most cannot. And I want to look inside the plant to find how it grows and functions, why it branches the way it does, and how it responds to its environment in such a precise way.

Some of these questions have already been answered, at least partially. Yet in many cases the answers don't satisfy me; they are either not general enough, too simple, or blatantly anthropocentric. This means that for me nature is still full of interesting problems.

One of the early problems that attracted me as a student was phototropism. Many of my fellow students thought that our predecessors in my father's laboratory—Blaauw, Arisz, Bremekamp, and Koningsberger, with their doctoral theses on phototropism—had preempted this subject. But such others as Dolk, Dillewijn, and Gorter were fascinated by the unsolved problems of plant responses to environment, and we had almost nightly bull sessions. I had to fulfill my military service obligations, which left only evenings and nights free for more productive projects. We discussed the newest publications of Paal, Seubert, Nielsen, and Stark—dissecting, interpreting, or repeating them.

This was in early 1926, an exciting time, with the growth-

---

secreting cells. The evidence indicates that the scutellum is probably even more important than the aleurone layer for providing enzymes that digest endosperm reserves in several species. This seems especially true during the first two days, when little activity of the aleurone layer can be detected, although the aleurone layer contributes substantially after the seed has germinated. Interestingly, gibberellins have no significant

substance concept just around the corner. Our discussions were usually based on Paal's theory: that the stem tip normally produces a growth-promoting factor. The hottest point of debate was whether or not in phototropism this factor was destroyed by the light. To wind up the argument, I asserted that I would "prove that the growth regulator from the tip was light-stable." Consequently, I had to extract it from seedlings and then expose it to light. For this, I prepared a tiny cube of gelatin, stuck it on a needle, and placed cut stem tips on it all the way around. When I removed the tips after an hour and placed the gelatin cube on one side of the seedling stump, nothing happened at first. But in the course of the night, the stump started to curve away from the gelatin block. It had acquired the capacity of the stem tips to grow! At 3:00 A.M. on April 17, 1926, I ran home to my parents' house nearby, burst into their bedroom, and said excitedly: "Father, come and see, I've got the growth substance."

My father (who was also my major professor) sleepily turned around and said: "Fine. Repeat the experiment tomorrow (which was my day off from military service); if it is any good, it will work again, and then I can see it."

Then followed an exciting time. I lived for my nights in the laboratory. Every experiment seemed to work, and I learned a lot about the behavior of the growth substance in and outside the stem tip. Of course, I chose this subject for my thesis work. But then, with completion of my military obligation, something unexpected happened. Although I improved my whole experimental procedure, rebuilt the temperature and humidity controls of the lab room, grew much better seedlings, and worked much more cleanly, the growth substance seemed to have disappeared, for none of my test plants responded any more — until I found that bacteria lurking in the gelatin ate all of the growth substance overnight! My procedure had changed because I prepared the blocks during the day and left them overnight. When I pressed an icebox into service (refrigerators were at that time unavailable), the bacteria ceased their activity and all experiments succeeded again.

To begin with, my approach to scientific problems was rather naïve. I mentioned already that I set out early to *prove* that the growth substance was light-stable. I soon learned that experiments cannot prove anything; they can only *test* a hypothesis. Thus, my experiments to test whether or not auxin is moved along a gradient by decrement unexpectedly showed that its transport was polar. And my tests on the behavior of auxin inside the seedling in unilateral light showed that this deflected the strictly downward auxin stream laterally, thus laying a solid basis for the Cholodny-Went theory of phototropism. Further work suggested that other growth factors were involved in the action of auxin. In an unaccountable way, I was later accused of promulgating theories, whilst I was only presenting experimental facts (which were finally, although many years later, accepted as true).

After receiving my Ph.D., I was faced by a completely new challenge. The working conditions for a tropical botanist in Java — before the blessing of air conditioning — were harder. There was less equipment, which operated less well in the oppressive moist heat, so my main efforts went to ecology or, rather, applied physiology. I found tropical plants admirably suited for work on root initiation, but only after moving to California, where I could again devote full time to physiological problems, could I work out a proper test to study root formation.

Whereas all my auxin experiments were based on more or less logical sequences of deductions leading to the crucial experiments, my ecological work was mostly a set of questions asked of nature, after observations of nature had posed the problems. We constructed a phytotron, in which such environmental factors as temperature, light, humidity, wind, or rain could be controlled. With this new tool I could establish, for example, that the profuse flowering of the desert in certain years depended upon the precise germination response of seeds from desert annuals to temperature and rain and not upon a mystic "survival of the fittest" or a "struggle for existence." In the last 30 years, phytotrons have helped make ecology an experimental rather than a descriptive science; now the extreme complexity of the organism in its total environment can be reduced to experimentally manageable subunits. It is satisfying when laboratory experiments give us a better insight into the mechanism of life, but for me the greatest thrill comes when these new insights help me understand what goes on in nature, when laboratory knowledge is applicable in the field. Thus, nature not only provides the inspiration; it is also the ultimate arbiter. The laboratory is only an interlude between perceiving and understanding.

[See also F. W. Went, 1974, Reflections and speculations, *Annual Review of Plant Physiology* 25:1–26.]

effect on digestion induced by the scutellum, even though this organ is believed to produce gibberellins that activate the aleurone layer.

Gibberellins have much less dramatic effects on mobilization of food reserves in dicots and gymnosperms than in cereal grains, although in some species the presence of the embryonic axis is still essential for normal degradation of these reserves in the food-

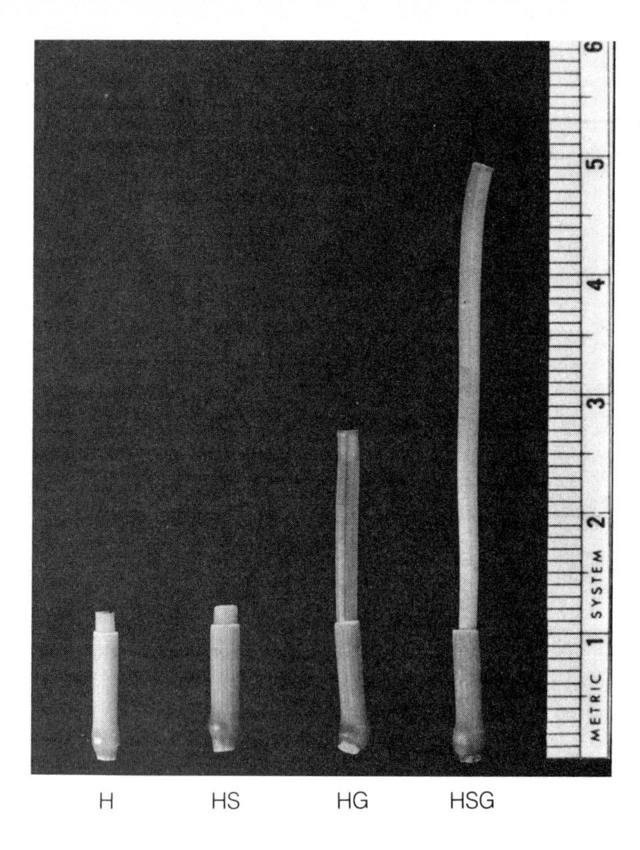


**Figure 17-18** Effect of $GA_3$ and sucrose on the growth of 1-cm oat-stem segments. The segments are shown after 60 h of treatment in Hoagland's nutrient solution (H), Hoagland's + 0.1 M sucrose (HS), Hoagland's + 30 μM $GA_3$ (HG), and Hoagland's + sucrose + $GA_3$ (HSG). A centimeter ruler indicates actual size. Elongation of the leaf sheaths did not occur, but growth of cells derived from the intercalary meristem (see Fig. 16-20) accounts for the stem elongation illustrated. (From Adams et al., 1973.)

storage cells. In castor bean (*Ricinus communis*), a dicot in which the endosperm remains well developed in the mature seed, degradation of fats does not require the presence of the embryo, even though fat breakdown is increased by added gibberellins (Mariott and Northcote, 1975). Whether this means that gibberellins are already present in sufficient amounts in the endosperm itself is not yet known. In other dicots and in gymnosperms, degradation of starch and fat is not affected by added gibberellins, but cytokinins sometimes supplant the normal role of the embryo in hastening fat breakdown.

## Other Gibberellin Effects

Gibberellins (especially $GA_4$ and $GA_7$) cause **parthenocarpic** (seedless) **fruit** development in some species, which suggests a normal function in fruit growth, and gibberellins formed in young leaves might also renew activity of the vascular cambium in woody plants. Other important effects of gibberellins are the delay of aging (senescence) in leaves and citrus fruits and effects on leaf shapes; the latter is a response especially apparent in leaves that show heterophylly or phase changes (see Chapter 16). Little was known about hormonal control of flower growth until the last several years, but gibberellins are now strongly implicated in growth of petals in some species (reviewed by Raab and Koning, 1988). To date little or no promotion of petal expansion has been attributed to other plant hormones.

## Possible Mechanisms of Gibberellin Action

The many effects of gibberellins suggest that they have more than one primary site of action. Thus far, research with hormone receptors has neither verified nor disproved that idea. Even a single effect such as enhanced stem elongation in whole plants results from at least three contributing events. *First, cell division is stimulated in the shoot apex*, especially in the more basal meristematic cells from which develop the long files of cortex and pith cells (Sachs, 1965). Careful work by Liu and Loy (1976) showed that gibberellins promote cell division because they stimulate cells in the $G_1$ phase to enter the S phase and because they also shorten the S phase. Increased cell number leads to more rapid stem growth because each of the cells can grow.

*Second, gibberellins sometimes promote cell growth because they increase hydrolysis of starch, fructans, and sucrose into glucose and fructose molecules*. These hexoses provide energy via respiration, they contribute to cell-wall formation, and they also make the cell's water potential momentarily more negative. As a result of the decrease in water potential, water then enters more rapidly, causing cell expansion and diluting the sugars. In sugarcane stems gibberellin-promoted growth results in part from increased synthesis of invertase enzymes that hydrolyze incoming sucrose into glucose and fructose (Glasziou, 1969). In dwarf peas, the activities of both invertase and amylase enzymes rise with increased growth (Broughton and McComb, 1971). The same is true for α-amylase in a dwarf maize. Less quantitative work with other species indicates that gibberellin-induced stem growth is associated with increases in amylase activity in small water plants and in certain trees, suggesting that the result is generalizable; to date, however, we have no data for conifers. Results with winter-wheat stems indicate that gibberellins promote hydrolysis of fructans by fructan hydrolase enzymes (Zhang, 1989), suggesting that these enzymes represent another kind of hydrolase induced by gibberellins.

*Third, gibberellins often increase cell-wall plasticity*. An excellent example of this occurs in internodes of oat, in which growth promotion of young cells derived from the intercalary meristem is unusually dramatic. Here no enhancement of cell division occurs. Elongation caused by $GA_3$ is 15 times as great as in the untreated sections (Fig. 17-18), provided that sucrose and mineral salts are present to provide energy and to prevent excessive dilution of the cell contents (that is, prevent a rise in the osmotic potential). A significant increase in wall plasticity occurs, and a similar phenomenon explains gibberellin-promoted growth in lettuce-hypocotyl sections and in entire hypocotyls of cucumber seedlings (Taylor and Cosgrove, 1989).

Not only is stem elongation promoted by gibberellins, but so is growth of the whole plant, including leaves and roots. We stated that application of gibberellins directly to leaves promotes their growth slightly and influences their shapes, yet direct application to roots usually has almost no effect on the roots themselves. But if the gibberellins are applied in any manner whereby they can move into the shoot apex, increased cell division and cell growth apparently lead to increased elongation of the stem and (in some species) to increased development of the young leaves. In species in which faster leaf development occurs, enhanced photosynthetic rates then increase growth of the whole plant, including the roots.

How might gibberellins loosen cell walls and also increase formation of hydrolytic enzymes leading to stem elongation? We have no evidence about wall-loosening mechanisms except that, in contrast to the initial growth burst caused by auxins, $H^+$ ions are not involved (Stuart and Jones, 1978; Jones and MacMillan, 1984; reviewed by Métraux, 1987). In oat internodes there is a lag of nearly 1 h before promotion of elongation can be detected. This delay should allow ample time for gibberellins to increase gene activation and promote formation of specific enzymes that cause physiological processes. For lettuce-hypocotyl sections, lags of less than 20 min occur. And intact dwarf peas were reported to elongate faster within 10 min after treatment with $GA_3$ (McComb and Broughton, 1972). In this case, hydrolases that attack cell-wall polysaccharides might be synthesized faster or simply become more active in gibberellin-treated cells. For both dwarf maize seedlings (whole shoots) and stems of pea seedlings, $GA_3$ was shown to induce specific changes in the kinds of proteins synthesized (Chory et al., 1987). These changes occurred before enhancement of growth by the hormone, so some of the induced proteins might be growth-promoting enzymes. This situation is similar to that of auxins described in Section 17.2.

## Commercial Uses of Gibberellins

Considering the numerous effects of gibberellins, it seems logical that they would be used in commercial applications. Major limiting factors have been their cost and their frequent promotion of fresh weights but not of dry weights, especially regarding the possible application in growth of pastures and hay crops. We still must rely on the *Gibberella* fungus to synthesize $GA_3$ at reasonable cost, even for physiological experiments. Nevertheless, $GA_3$ is used extensively in the Central and Imperial valleys of California to increase the size of Thompson seedless-grape berries and the distance between grape bunches (Fig. 17-19).

**Figure 17-19** Effects of gibberellin and girdling on growth of Thompson seedless grapes. (Courtesy J. LaMar Anderson, Utah State University, Logan.)

When applied at the right time and with the proper concentration, gibberellins cause the grape bunches to elongate so that they are less tightly packed and less susceptible to fungal infections. Usually the plants are sprayed twice, once at bloom and again at the fruit-set stage (Nickell, 1979). A mixture of $GA_4$ and $GA_7$ is now used to enhance seed production in Pinaceae seed orchards (Carlson and Crovetti, 1990), as is $GA_3$ in seed orchards of certain members of the families Taxodiaceae and Cupressaceae (Nagao et al., 1989). Gibberellins are also used by some breweries to increase the rate of malting through the GAs' enhancing effects on starch digestion. Celery plants, which are valued for the lengths and crispness of their stalks, respond favorably to gibberellins, but the poor storage qualities of such stalks limit wide use of these hormones in celery production. Gibberellins have also been sprayed on fruits and leaves of navel-orange trees (when the fruits have lost most of their green color) to prevent several rind disorders that appear during storage. Here the hormones delay senescence and maintain firmer rinds. Gibberellins are now used commercially in Hawaii to increase sugarcane growth and sugar yields. These and other potential effects of gibberellins were reviewed by Martin (1983) and by Carlson and Crovetti (1990).

# 18

# Hormones and Growth Regulators: Cytokinins, Ethylene, Abscisic Acid, and Other Compounds

The more we learn about growth and development, the more complex these processes seem to become. In the previous chapter we explained that both processes depend upon IAA and gibberellins but that these hormones generally influence different parts of the plant in different ways. In spite of the complexities, we now realize that both hormone types must be considered if we are to understand growth. In this chapter we discuss the other three kinds of hormones that are currently known (cytokinins, ethylene, and abscisic acid), emphasizing that although each has distinct effects, growth and development normally involve an interplay among all known hormones and probably others as yet undiscovered. We also mention some additional compounds that are sometimes active as growth substances.

## 18.1 The Cytokinins

About 1913, Gottlieb Haberlandt discovered in Austria that an unknown compound present in vascular tissues of various plants stimulated cell division that caused cork-cambium formation and wound healing in cut potato tubers. This discovery was apparently the first demonstration that plants contain compounds, now called **cytokinins**, that stimulate cytokinesis. In the 1940s, Johannes van Overbeek found that the milky endosperm from immature coconuts is also rich in compounds that promote cytokinesis. In the early 1950s, Folke Skoog and his colleagues, who were then interested in auxin-stimulation of plants grown in tissue cultures, found that cells in pith sections from tobacco stems divided much more rapidly when a piece of vascular tissue was placed on top of the pith, verifying Haberlandt's results.

Skoog and his coworkers tried to identify the chemical factor from the vascular tissues using growth of tobacco pith cells as a bioassay system. These cells were cultured on agar media containing known sugars, mineral salts, vitamins, amino acids, and IAA. IAA itself increased growth for a time by causing relatively enormous cells to be formed, but these cells did not divide; many were polyploids with several nuclei. In seeking substances that would promote cell division, they found a highly active adenine-like compound in yeast extracts. This led to investigations of the ability of DNA to promote cytokinesis (because DNA contains adenine) and, in 1954, to the discovery by Carlos Miller (then a student of Skoog) of a very active compound formed by partial breakdown of aged or autoclaved herring-sperm DNA. They named this compound **kinetin** (reviewed by Miller, 1961).

Although kinetin itself has not been found in plants and is not the active substance found by Haberlandt in phloem, related cytokinins are present in plants. F. C. Steward, also using tissue-culture techniques in the 1950s, found in coconut milk several cytokinins that enhance cell division in carrot root tissues. The most active of these were subsequently shown by D. S. Letham (1974) to be compounds previously given the common names **zeatin** and **zeatin riboside**. In 1964 zeatin had first been identified almost simultaneously by Letham and by Carlos Miller, both of whom used the milky endosperm of corn (*Zea mays*) as a source. Since then, other cytokinins with adenine-like structures similar to kinetin and zeatin have been identified in various parts of seed plants. None of these cytokinins is present in DNA, nor are they breakdown products of DNA, but some occur in transfer-RNA (and sometimes in ribosomal-RNA) molecules of seed plants, yeasts, bacteria, and even primates, and more than 30 exist as unbound, free cytokinins. It is one or more of the unbound cytokinins that cause the physiological responses described in this chapter, but those in transfer RNA (tRNA) probably have unknown functions.

Figure 18-1 shows the structures of the free-base form of the three most commonly detected and most

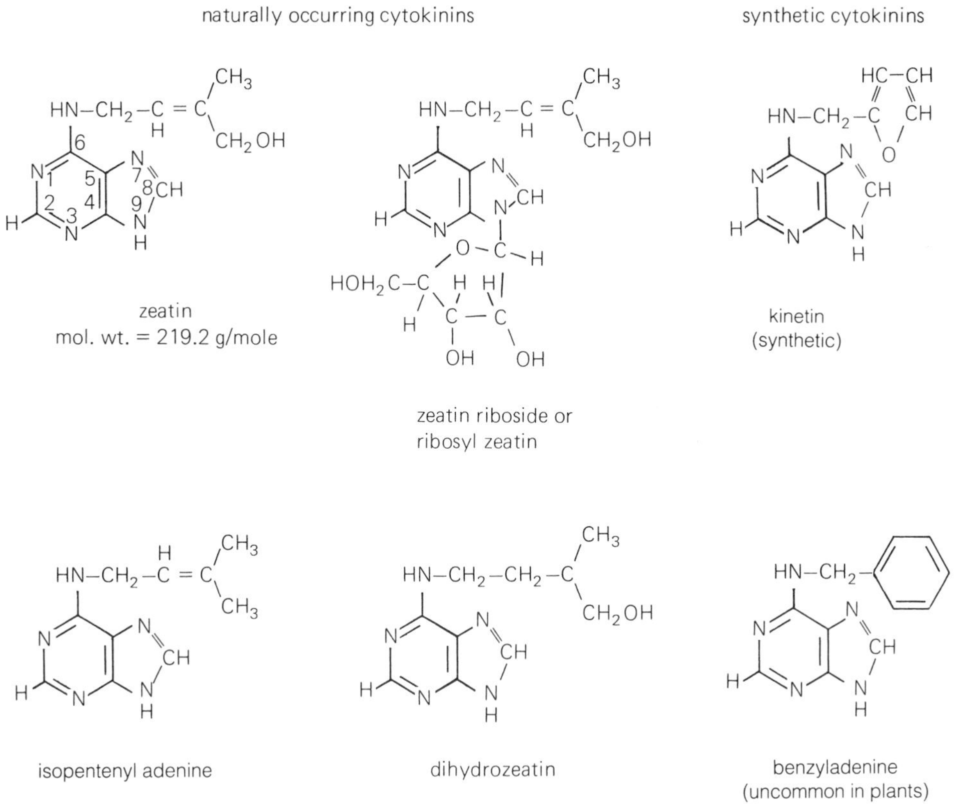

naturally occurring cytokinins

synthetic cytokinins

zeatin
mol. wt. = 219.2 g/mole

zeatin riboside or
ribosyl zeatin

kinetin
(synthetic)

isopentenyl adenine

dihydrozeatin

benzyladenine
(uncommon in plants)

**Figure 18-1** Structures of common natural and synthetic (kinetin) cytokinins. All these are adenine derivatives in which the purine ring is numbered as shown for zeatin (upper left). Zeatin and zeatin riboside can exist with groups arranged about the side-chain double bond either in the *trans* (as shown) or *cis* (with $CH_3$ and $CH_2OH$ groups interchanged) configuration. The *cis* form predominates in tRNA-bound cytokinins, but the *trans* form exists in free zeatin and zeatin riboside.

physiologically active cytokinins in various plants: zeatin, **dihydrozeatin**, and **isopentenyl adenine (IPA)**. Also shown are kinetin and another synthetic cytokinin, **benzyladenine**, both of which are highly active. Kinetin is probably not formed by plants, but there are two reports that benzyladenine or its riboside exist in plants (Ernst et al., 1983; Nandi et al., 1989). Note that all cytokinins have a side chain rich in carbon and hydrogen attached to the nitrogen protruding from the top of the purine ring. Each cytokinin can exist in the **free-base** form shown or as a **nucleoside**, in which a ribose group is attached to the nitrogen atom of position 9 (see ring-numbering system for zeatin in Fig. 18-1). An example is zeatin riboside, a relatively abundant cytokinin in many plants. Furthermore, the nucleosides can be converted to **nucleotides**, in which phosphate is esterified to the 5′-carbon of ribose, as in adenosine-5′-phosphate (AMP). In a few cases, evidence for formation of nucleoside diphosphates and triphosphates similar to ADP and ATP has also been obtained, but all these nucleotides seem to be less abundant than the free-base or nucleoside forms.

Two questions now arise: How do we define a cytokinin, and should the free bases, nucleosides, and nucleotides all be considered cytokinins? Not every expert would agree on the same definition, but a reasonable one should depend partly on early discoveries that cytokinins promote cytokinesis (cell division) in tissues grown *in vitro*, such as cultures from tobacco pith, carrot phloem, or soybean stems. In fact, R. Horgan (1984) defined them as substances that, in the presence of optimal auxin concentrations, induce cell division in the tobacco-pith or similar assay system grown on an optimally defined medium. Other authors prefer to include in the definition the facts that such compounds are adenine derivatives and that they have common and important effects in addition to promoting cytokinesis. We will describe these additional effects later, but because all of them promote cytokinesis, it seems reasonable to define **cytokinins** as *substituted adenine compounds that promote cell division in the above-mentioned tissue systems*. The question as to whether the free-base, the nucleoside, or a nucleotide form is the active form has not yet been convincingly answered. Most evidence favors the free base as the active form (Letham and Palni, 1983; Van der Krieken et al., 1990). The chemistry and biolog-

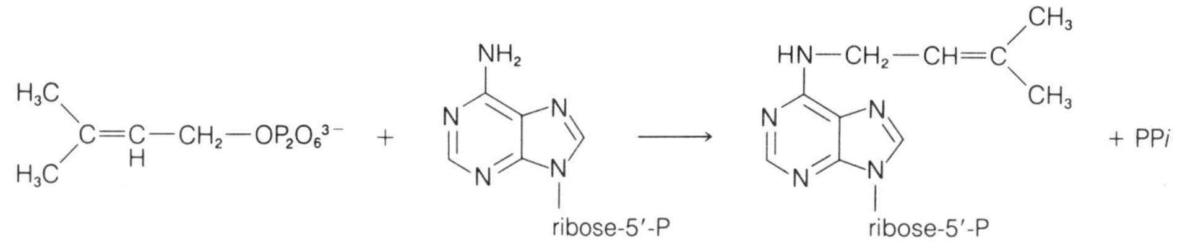

Figure 18-2 Formation of isopentenyl AMP, a precursor of isopentenyl adenine.

ical activity of more than 200 natural and synthetic cytokinins was reviewed by Matsubara (1990); that review gives us fairly good ideas of the chemical structure necessary for cytokinin activity, and in general the free bases in Figure 18-1 apparently have nearly ideal structures.

Cytokinins also exist in mosses, in brown and red algae, and apparently also in diatoms, and they sometimes promote algal growth. It is likely that they are widespread if not universal in the plant kingdom, but very little is known of their functions except in angiosperms, in some conifers, and in mosses. Certain pathogenic bacteria and fungi contain cytokinins that are believed to influence disease processes caused by these microbes, and cytokinin production by nonpathogenic fungi and bacteria is thought to influence mutualistic relations with plants, such as formation of mycorrhizae and root nodules (Greene, 1980; Ng et al., 1982; Sturtevant and Taller, 1989).

## Cytokinin Metabolism

Two important questions about cytokinin metabolism should be asked: How do plants synthesize cytokinins, and how do plants regulate the amounts of cytokinin they contain? A breakthrough in our knowledge of biosynthesis came from the demonstration by Chong-Maw Chen and D. K. Melitz (1979) that tobacco tissues contain an enzyme called **isopentenyl AMP synthase** (previously discovered in a slime mold) that forms **isopentenyl adenosine-5′-phosphate (isopentenyl AMP)** from AMP and an isomer of isopentenyl pyrophosphate. (The latter compound is a product of the mevalonate pathway and is an important precursor of sterols, gibberellins, carotenoids, and other isoprenoid compounds; see Section 15.3. ) The isomer involved is Δ-2-isopentenyl pyrophosphate, where the prefix Δ means that the molecule has a double bond between carbons 2 and 3. The reaction that occurs in tobacco tissues is shown in Figure 18-2. Note that pyrophosphate (PP*i*) is released from the isopentenyl group and that the latter is added to the amino nitrogen attached to carbon 6 of the purine ring.

The isopentenyl AMP formed in this reaction can then be converted to isopentenyl adenosine by hydrolytic removal of the phosphate group by a phosphatase enzyme, and isopentenyl adenosine can be further converted to isopentenyl adenine by hydrolytic removal of the ribose group. Furthermore, isopentenyl adenine can be oxidized to zeatin by replacement of one hydrogen by an —OH in a methyl group of the isopentenyl side chain (compare structures in Fig. 18-1). Dihydrozeatin is then formed from zeatin by reduction (with NADPH) of the double bond in the isopentenyl side chain (Martin et al., 1989). These reactions probably account for formation of the three major cytokinin bases, but other possibilities for biosynthesis exist.

Cellular levels of cytokinins are also affected by their degradation and by their conversion to presumably inactive derivatives other than nucleosides and nucleotides. Degradation occurs largely by **cytokinin oxidase**, an enzyme system that removes the five-carbon side chain and releases free adenine (or, when zeatin riboside is oxidized, free adenosine). Formation of cytokinin derivatives is more complex because many conjugates can be formed (Letham and Palni, 1983). The most common conjugates contain either glucose or alanine; those that contain glucose are called **cytokinin glucosides**.

In one kind of glucoside, carbon 1 of glucose is attached to the hydroxyl group of the side chain of either zeatin, zeatin riboside, dihydrozeatin, or dihydrozeatin riboside. In the second kind of glucoside, carbon 1 of glucose is attached to a nitrogen atom (via a C—N bond) at either position 7 or 9 of the adenine ring system in any of the three major cytokinin bases. In alanine conjugates, alanine is connected in a peptide bond to the nitrogen at position 9 of the purine ring. No function of any of these conjugates is known, but the glucosides might represent storage forms or, in some cases, special transport forms of cytokinins. According to McGaw (1987), the alanine conjugates are not likely to represent storage forms but instead are irreversibly formed products of cytokinin removal. It is unlikely that any such conjugates represent physiologically active cytokinins.

## Sites of Cytokinin Synthesis and Transport

If we knew how actively the reactions that form isopentenyl AMP, isopentenyl adenine, zeatin, and dihydrozeatin occur in various organs and tissues, we would have good biochemical information about the sites of cytokinin biosynthesis. Unfortunately, this information is not yet available, so less direct methods have been used to determine where cytokinins are formed. One method has been to find out where they are most abundant. In general, cytokinin levels are highest in young organs (seeds, fruits, and leaves) and in root tips. It seems logical that they are synthesized in these organs, but in most cases we cannot dismiss the possibility of transport from some other site. For root tips, synthesis is almost surely involved, because if roots are severed horizontally, cytokinins are exuded (by root pressure) from the xylem of the remaining lower portions for periods up to four days (Skene, 1975; Torrey, 1976). It is unlikely that these lower portions could store enough cytokinins derived from some other source to act as a rather long-term supplier for the xylem.

Evidence such as this has led to the widespread idea that root tips synthesize cytokinins and transport them through the xylem to all parts of the plant. This might explain their accumulation in young leaves, fruits, and seeds into which xylem transport occurs, but the phloem is generally a more effective supply system for such organs that have limited transpiration. Although root tips probably do represent an important cytokinin source for various plant parts, small, rootless tobacco plants effectively convert radioactive adenine into various cytokinins (Chen and Petschow, 1978). Furthermore, radioactive adenine was converted into several cytokinins not only by pea roots but also by pea stems and leaves (Chen et al., 1985). Carrot roots were also investigated, and the results indicated that it was primarily the cambial regions of the root that synthesized cytokinins (Chen et al., 1985). This observation and other studies indicate that shoots can synthesize some of the cytokinins they require.

Transport of various kinds of cytokinins certainly occurs in the xylem (Jameson et al., 1987), but sieve tubes also contain cytokinins, as evidenced by the presence of the latter in aphid honeydew. Further evidence for transport in the phloem is provided by experiments with detached dicot leaves. When a mature leaf is cut off plants of some species and kept moist, cytokinins move to the base of the petiole and accumulate there. This movement probably occurs through the phloem, not the xylem, because transpiration strongly favors xylem flow from the petiole to the leaf blade. Cytokinin accumulation in the petiole implies that mature leaf blades can supply young leaves and other young tissues with cytokinins via the phloem, provided of course that such leaves can synthesize cytokinins or receive them from roots. Nevertheless, if a radioactive cytokinin is added to the surface of a leaf, very little of that which is absorbed is transported out. These and many other results indicate that cytokinins are not readily distributed in the phloem. Almost surely, young leaves, fruits, and seeds that are transport sinks do not readily transport their cytokinins elsewhere via either xylem or phloem. Our tentative conclusion is that except for delivery from roots via the xylem, transport of cytokinins within the shoot is rather limited.

a

b

**Figure 18-3** (**a**) Callus grown from the scutellum of a rice seed. (**b**) Embryogenic callus that has formed a young shoot (S) and root (R) system. (Courtesy M. Nabors and T. Dykes.)

## Cytokinin-Promoted Cell Division and Organ Formation

We explained that a major function of cytokinins is to promote cell division. Skoog and his colleagues found that if the pith from tobacco, soybean, and other dicot stems is cut out and cultured aseptically on an agar medium with an auxin and proper nutrients, a mass of unspecialized, loosely arranged, and typically polyploid cells called a **callus** forms (see Section 16.6). Figure 18-3a illustrates the general appearance of a callus. If a cytokinin is also provided, cytokinesis is greatly promoted, as already mentioned. The amount of growth from new cells serves as a sensitive and highly specific bioassay for cytokinins and is important in how we define these compounds (reviewed by Skoog and Leonard, 1968 and by Skoog and Armstrong, 1970).

a b

**Figure 18-4** Development of (**a**) tomato and (**b**) petunia plants from a callus, illustrating totipotency. (Photographs courtesy of Murray Nabors and R. S. Sangwan.)

Skoog and his colleagues also found that if a high cytokinin-to-auxin ratio is maintained, meristematic cells are produced in the callus; these cells divide and give rise to others that develop into buds, stems, and leaves. But if the cytokinin-to-auxin ratio is lowered, root formation is favored. By choosing the proper ratio, calli from many species (especially dicots) can be made to develop into an entire new plant. The ability of calli to regenerate whole plants represents a tool to select plants with resistance to drought, salt stress, pathogens, and certain herbicides, or those with other useful characteristics.

The way in which a callus forms a new plant is variable. Frequently, with relatively high cytokinin-to-auxin ratios, only a shoot system first develops; then adventitious roots are formed spontaneously from the stems while still in the callus. (Roots can also be induced by common horticultural techniques to form from stems of young shoots taken out of the callus; see Sections 16.6 and 17.1.) This formation of shoots or of shoots and adventitious roots by the callus is called **organogenesis**. Sometimes, however, calli become embryogenic (Fig. 18-3b) and form an embryo that develops into a root and shoot; this is called **embryogenesis**. Formation of young plants from calli is shown in Figure 18-4. Both cytokinins and auxins must usually be added to the medium if embryogenesis is to occur, but little information indicates how they act as control agents.

Cytokinins and IAA are important in controlling formation and development of tumorous outgrowths (galls) on stems of many dicots and gymnosperms, a condition called **crown gall**. This disease is caused by the bacterium *Agrobacterium tumefaciens* (closely related to the nitrogen-fixing members of *Rhizobium*). The galls can be grown in sterile culture without addition of cytokinin or auxin; that is, the cells are autonomous for these hormones. *A. tumefaciens* contains several **plasmids** (small circles of DNA that can occur independently of the bacterium's own DNA molecule; see Chapter 24); one of these plasmids, called the *Ti plasmid*, contains a section of DNA that is transferred to the host plant's stem cells during infection and is responsible for the rapid and unorganized growth of the galls. This section of DNA is called *T-DNA* (the T stands for transferred).

The T-DNA contains, among other genes, one that codes for the enzyme isopentenyl AMP synthase (which acts in the reaction shown in Fig. 18-2) and two that code for enzymes that convert tryptophan to IAA. Mutation of these different genes causes changes in levels of cytokinins and IAA and in shoot morphology. If all three genes are mutated so that they are inactivated, tumors don't develop and hormone levels are low. If only the isopentenyl AMP synthase gene is inactivated, cytokinin levels decrease and galls grow slowly and form numerous roots by organogenesis. If either of the auxin biosynthetic genes is inactivated, galls grow slowly, form much less IAA, and produce leafy shoots with few or no roots. These results are entirely what would have been expected based on the cytokinin-to-auxin ratio effects first discovered by Skoog. Good reviews of crown-gall genes and hormone effects are those by Morris (1986, 1987) and by Weiler and Schroder (1987), whereas more recent papers generally supporting the above conclusions are those by Spanier et al. (1989) and by Smigocki and Owens (1989).

## Cytokinin-Delayed Senescence and Increased Nutrient-Sink Activities

When a mature but still-active leaf is cut off, it begins to lose chlorophyll, RNA, proteins, and lipids from the chloroplast membranes more rapidly than if it were still attached, even if it is provided with mineral salts and water through the cut end. This premature aging or senescence, evident by leaf yellowing, occurs especially fast if the leaves are kept in darkness. In dicot leaves, adventitious roots often form at the base of the petiole, and then senescence of the blade is greatly delayed. The roots apparently provide something to the leaf that keeps it physiologically young. This something almost surely contains a cytokinin transported through the xylem.

Two main pieces of evidence suggest that a cytokinin is involved: Many cytokinins will partially replace the need for roots in delaying senescence, and the cytokinin content of the leaf blade rises substantially when the adventitious roots form (see the review by Van Staden et al., 1988). In sunflowers, the cytokinin content of the xylem sap increases during the period of rapid growth and then decreases greatly when growth stops and flowering begins, which suggests that a reduction in cytokinin transport from roots to shoots might allow senescence to occur faster (Skene, 1975).

How cytokinins retard senescence in detached oat leaves has been investigated extensively by Kenneth V. Thimann, a pioneer in auxin research, and his colleagues at the Thimann Laboratories in Santa Cruz, California (see Thimann, 1987). When leaves of oats and many other species are cut off and floated on a solution of dilute mineral salts, they begin to senesce, characterized first by degradation of proteins into amino acids and later by loss of chlorophyll. This senescence occurs much more rapidly in darkness than in light, and cytokinins added to the solution on which the leaves are floating essentially replace the light effect by delaying senescence. Thimann (1987) suggested that cytokinins do this by maintaining the integrity of the tonoplast membrane. Otherwise, proteases from the vacuole would leak into the cytoplasm and hydrolyze both soluble proteins and proteins of chloroplast and mitochondrial membranes. Consistent with this idea, Y. Y. Leshem and his colleagues in Israel have obtained much evidence suggesting that cytokinins protect membranes against degradation (Leshem, 1988). Their results strongly indicate that cytokinins act by preventing oxidation of unsaturated fatty acids in membranes. Such prevention probably occurs because cytokinins both inhibit formation and speed breakdown of free radicles such as **superoxide** ($O_2^{\cdot -}$) and the **hydroxy radicle** ($OH^{\cdot}$) that otherwise oxidize membrane lipids (Thompson et al., 1987; Leshem, 1988).

The delay of senescence by cytokinins appears to be a natural, partially root-controlled phenomenon and is associated with other interesting phenomena. Cytokinins cause transport of many solutes from older parts of the leaf and even from older leaves into the treated zone. A dramatic illustration of this is shown in Figure 18-5. Here the oldest (primary) leaves of a bean plant were painted at four-day intervals with the synthetic cytokinin benzyladenine. Normally these leaves become senescent sooner than the trifoliate leaves above, but in this example the senescence pattern was reversed. The treated primary leaves withdrew nutrients from the adjacent trifoliate, causing it to senesce first. (Note also that the benzyladenine apparently did not move effectively from the treated leaves to the younger trifoliate leaves just above.)

Further studies with bean plants showed that two kinds of treatments can greatly delay senescence of the primary leaves and can even reverse their senescence once they have become pale green-yellow. One treatment is to cut off the leaves and stem above, and the other is to dip the primary leaves once into a solution of benzyladenine (Venkatarayappa et al., 1984). Other studies with many dicots and monocots show that if only one part of a leaf is treated, radioactive metabolites added to another part of the same leaf or to an adjacent leaf migrate through the phloem into the treated zone and accumulate there (see, for example, Gersani and Kende, 1982). The implication is that young leaves can remove nutrients from older ones partly because they are rich in cytokinins and, therefore, that cytokinins enhance the ability of young tissues to act as sinks for phloem transport. Whether or not these hormones are involved in the normal transport of mobile nutrients into twigs and larger branches of woody plants before leaf fall in autumn is an interesting question. That cytokinins in reproductive structures might have survival value by enhancing movement of sugars, amino acids, and other solutes from mature leaves into seeds, flowers, and fruits is also an interesting hypothesis.

**Figure 18-5** Senescence of a trifoliate bean leaf caused by treating the primary leaves of cuttings with the synthetic cytokinin benzyladenine (30 mg/L) at 4-day intervals. (From Leopold and Kawase, 1964.)

When certain fungi that cause rust and mildew diseases infect leaves, areas of dead and dying cells are produced. As leaves senesce, these necrotic areas are often surrounded by several green and starch-rich cells, even when the rest of the leaf has become yellow and senescent. These **green islands** are rich in cytokinins that are probably synthesized by the fungus (Greene,

1980). The cytokinins presumably help maintain food reserves for the fungus and influence the subsequent course of the disease.

The ability of cytokinins to retard senescence also applies to certain cut flowers and fresh vegetables. An excellent review of flower-petal senescence is that by Borochov and Woodson (1989). The concentration of cytokinins in rose and carnation petals decreases as aging occurs, and applied cytokinins slow this aging process. Carnations have been studied most, and for that species solutions containing dihydrozeatin or benzyladenine are most effective (Van Staden et al., 1990). For most cut flowers, however, exogenous cytokinins cannot overcome the senescence-promoting effects of ethylene produced by the flowers (see Section 18.2). The storage lives of brussel sprouts and celery can be increased by relatively inexpensive commercial cytokinins such as benzyladenine, but such a treatment is not permitted for foods sold in the United States, even though we are constantly exposed to natural cytokinins in food from plants. The influence of cytokinins and other hormones on storage of fruits and vegetables was reviewed by Ludford (1987).

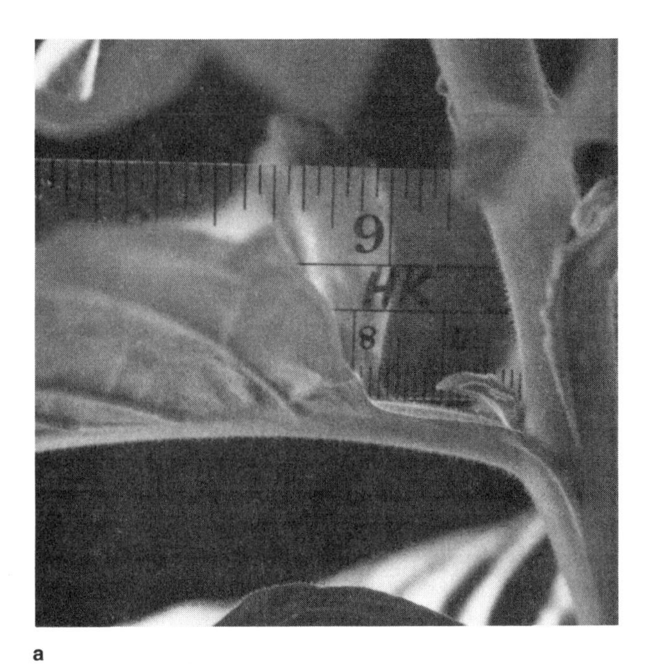


a

**Figure 18-6** Promotion of lateral-bud expression in a cytokinin-overproducing mutant of tobacco. The lateral bud at the 12th node (numbering from shoot apex) is shown for the wild-type (**a**) and the mutant (**b**). (From Medford et al., 1989.)

## Cytokinin-Promoted Lateral-Bud Development in Dicots

If a cytokinin is added to a nongrowing lateral bud dominated by the shoot apex above it (a situation termed apical dominance; see Section 17.2), the lateral bud often begins to grow. In early studies of this phenomenon, the synthetic kinetin was the main compound used, and growth of the lateral bud continued for only a few days. Prolonged elongation of the bud could be caused only by adding IAA or a gibberellin to it. Another cytokinin, benzyladenine, sometimes causes substantially more elongation than kinetin, but its effects have been studied with only a few species. Pillay and Railton (1983) showed that benzyladenine and zeatin dramatically enhance elongation of pea lateral buds for at least two weeks, whereas isopentenyl adenine and kinetin promote only short-term growth. The reason that the closely related hormones zeatin and isopentenyl adenine cause such different effects is unknown, but the authors speculated that isopentenyl adenine is only weakly active because it is slowly hydroxylated to the much more active zeatin in the buds. Results reported by King and Van Staden (1988) generally support the importance of hydroxylation. Other evidence that quiescent lateral buds cannot synthesize active cytokinins also exists, but we are still uncertain about the relative importance of cytokinins and other hormones and nutritional factors in controlling lateral-bud development.

In a recent genetic-engineering experiment, cytokinin levels were increased in whole tobacco and *Arabidopsis thaliana* plants by a novel technique (Medford et al., 1989). The general procedures of genetic-engineering experiments are explained in Chapter 24, but essentially a bacterial gene that codes for isopentenyl AMP synthase (Fig. 18-2) was inserted by bacterial infection into the genome of wounded cells of cut leaf discs. These wounded cells develop into a callus carrying the new gene, and then the callus forms plants by organogenesis. Along with this structural gene, a promoter gene that activates it was also inserted; the promoter gene is activated only at relatively high temperatures (40 to 45°C), when the plants are subjected to heat shock (for more about heat-shock genes, see Sections 24.5 and 26.6). After the transformed plants had developed at normal growth temperatures, heat shock was given for 15 min, and then normal temperatures were again provided. After four more hours, leaves of transformed and heat-shocked tobacco plants contained, relative to heat-shocked but nontransformed plants, six times as much isopentenyl AMP, 23 times as much zeatin riboside monophosphate, 46 times as much zeatin riboside, and 80 times as much zeatin. Cytokinin levels were not measured in *Arabidopsis*, but both it and tobacco showed various morphological changes. The most pronounced morphological effect of high cytokinin levels was extensive development of lateral buds (Fig. 18-6). These interesting experiments show the effects of abnormally high cytokinin levels and further support the idea that cytokinins can overcome apical dominance, but they don't allow comparison between cytokinin-deficient and normal plants; mutants that are deficient in cytokinins are needed for such comparisons.

b

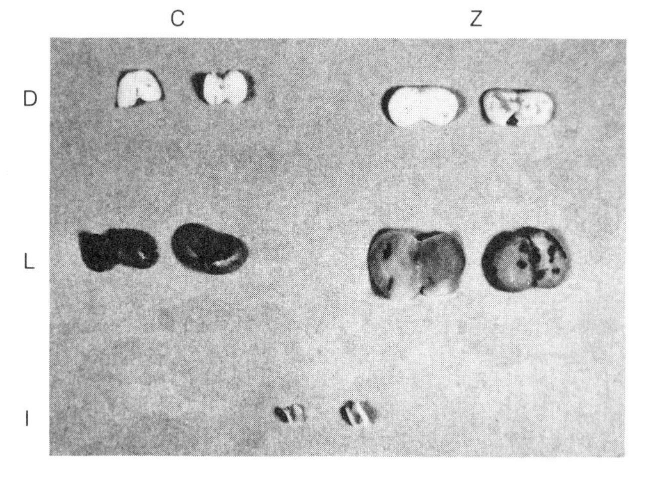


**Figure 18-7** Promotion of excised radish cotyledon enlargement by zeatin and light. Cotyledons at the bottom labeled I represent initial cotyledons excised from 2-day-old, dark-grown seedlings before growth studies. Excised cotyledons were incubated for 4 days on filter papers held in petri dishes containing 2 mM potassium phosphate (*p*H 6.4) alone (controls, C) or also with 2.5 $\mu$M zeatin (Z). Cotyledons exposed to light (L) received continuous fluorescent radiation at a level near the photosynthetic light compensation point. Cotyledons incubated 4 days in darkness (D). (Unpublished results of A. K. Huff and C. W. Ross.)

In another example of transformed tobacco, IAA-deficient plants were produced by inserting a gene that codes for an enzyme that converts IAA to an inactive conjugate with the amino acid lysine (Harry Klee, personal communication). Tobacco plants cannot readily degrade this conjugate, so their IAA becomes unavailable. Like cytokinin overproducers, they also branched excessively compared to nontransformed control plants. These results indicate that the cytokinin-to-auxin ratio is important in controlling apical dominance (lateral-bud repression); high ratios favor bud development and low ratios favor dominance.

Enhanced lateral branching also occurs in two bacterial diseases in which the pathogen synthesizes a cytokinin. One is a fasciation disease caused by *Corynebacterium fascians* that occurs in various dicots such as chrysanthemum, garden peas, and sweet peas. In **fasciations**, normally round stems become flattened and numerous lateral buds develop into branches, often forming a broomlike (witches' broom) bundle of stems. In garden peas, symptoms of this disease can be duplicated by adding a cytokinin to young plants. Highly pathogenic strains of the bacterium contain a plasmid; nonpathogenic strains lack this plasmid (Nester and Kosuge, 1981). Pathogenic strains synthesize and release into their growth media several cytokinins that almost certainly contribute to the fasciation disease.

*Corynebacterium fascians* also causes certain kinds of witches' brooms in trees, again accompanied by production of multiple lateral buds that grow into branches. Two other pathogens (*Exobasidium* spp.) that cause witches' brooms also produce cytokinins. In these cases, too, cytokinins are thought to cause the onset of disease symptoms.

## Cytokinin-Enhanced Cell Expansion in Dicot Cotyledons and in Leaves

When seeds of many dicots are germinated in darkness, the cotyledons emerge above ground but remain yellow and relatively small. If the cotyledons are exposed to light, growth increases greatly, even if the light energy provided is too low to allow photosynthesis. This is a photomorphogenetic effect controlled partly by phytochrome (as described in Chapter 20), but cytokinins are probably involved as well. If the cotyledons are excised and incubated with a cytokinin, the growth rate is doubled or tripled relative to controls that lack added hormone, whether in light or darkness. Growth is caused entirely by water uptake that drives cell expansion, because the dry weight of the tissues does not increase.

This growth promotion occurs with more than a dozen known species, including radish, sugar beet, lettuce, sunflower, cocklebur, white mustard, squash, cucumber, pumpkin, muskmelon, and fenugreek. Most of these species contain fats as the major food reserve in the cotyledons. Furthermore, the cotyledons normally emerge above ground and become photosynthetic in each species. No response has been found in species with cotyledons that remain underground after germination, or in beans, in which cotyledons emerge but do not become leafy. Figure 18-7 shows the promo-

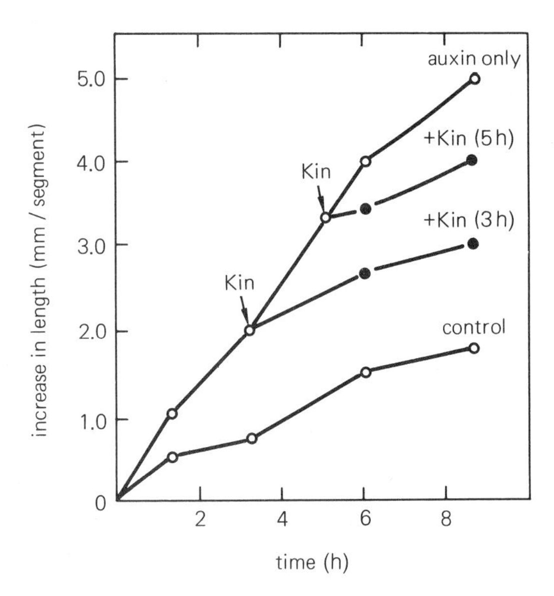


**Figure 18-8** Inhibition of auxin-induced elongation in soybean hypocotyl sections by 4 μM kinetin added at different times of incubation (arrows). (From Vanderhoef et al., 1973.)

tive effects of zeatin on radish-cotyledon enlargement in both light and darkness; it also demonstrates that light is effective in the absence of zeatin. Auxins do not promote growth of cotyledons, and gibberellins also have little effect when the cotyledons are cultured in water or in darkness; thus this response provides a useful bioassay for cytokinins (Letham, 1971; Narain and Laloraya, 1974).

Do cytokinins promote cotyledon growth only by increasing expansion of cells already present, or do these hormones promote cell division and expansion of the resulting daughter cells? All results indicate that they increase both cytokinesis and cell expansion, especially the latter. Remember, however, that cytokinesis does not increase the growth of any organ itself because it is only a division process. Therefore, overall growth requires cell expansion, and growth promotion by cytokinins involves faster cell expansion and production of larger cells.

Because cotyledons in which growth is promoted by cytokinins become photosynthetic organs, we might ask whether true leaves also require cytokinins for growth. Definite promotive effects on intact dicot leaves of some species occur after repeated applications of cytokinins, but the effects are usually small and may arise indirectly through attraction of metabolites from other organs. If discs are cut from dicot leaves with a cork borer and kept moist, cytokinins increase expansion through enhanced cell growth, again suggesting a normal function of cytokinins from some other organ, perhaps roots, in leaf growth. Further evidence that cytokinins from roots promote leaf growth comes from experiments in which some or all of the roots were removed from bean and winter rye (*Secale cereale*). Leaf growth from rootless plants soon slowed in both species, but application of a cytokinin to the leaves restored much of this growth. So far as we are aware, no studies on the effects of cytokinins on growth of conifer needles have been made.

## Effects of Cytokinins on Stems and Roots

Normal growth of stems and roots is thought to require cytokinins, but the endogenous amounts are seldom limiting. As a result, applications of exogenous cytokinins also fail to increase growth of these organs. This finding was also observed with tobacco and *Arabidopsis* plants in the genetic-engineering experiment described above in which endogenous cytokinin levels were significantly raised in transformed plants (Medford et al., 1989). Suppose, however, that we stop delivery of cytokinins (and gibberellins) from roots to shoots by removing the roots. Can we now add cytokinins and gibberellins and restore shoot growth, especially stem elongation? In sunflower and pea, growth restoration was unsuccessful, but in soybean success was obtained. Conflicting results with so few (only dicot) species justify no general conclusion. Experiments designed to provide more general answers to this question seem straightforward.

Another approach to determine the importance of cytokinins to normal growth of stems and roots is to excise sections and grow them *in vitro*, just as was done in experiments with auxins and gibberellins (Chapter 17). In such experiments the assumption is that excised sections will be depleted of cytokinins when separated from shoot tips or root tips that presumably represent hormone sources. However, nobody has ever demonstrated by actual measurements that excised sections become cytokinin-deficient. When root or stem sections are grown *in vitro* with an exogenous cytokinin, elongation is almost always retarded relative to control sections. For example, data showing the sharply antagonistic effects of an auxin and of kinetin on elongation of soybean-hypocotyl sections are plotted in Figure 18-8. Interestingly, although elongation is inhibited, stem sections usually become thicker by radial expansion of cells, so the overall fresh weight of treated sections is not much different from that of control sections.

What can we conclude from results showing only inhibition of elongation? We might conclude that elongating stems and roots do not require cytokinins. Alternatively, even though such organs might need the hormones for elongation, they may already contain sufficient amounts. In both cases we could argue that exogenous cytokinins inhibit *in vitro* growth by causing excess internal concentrations. There seems no easy way to solve this problem without measuring the inter-

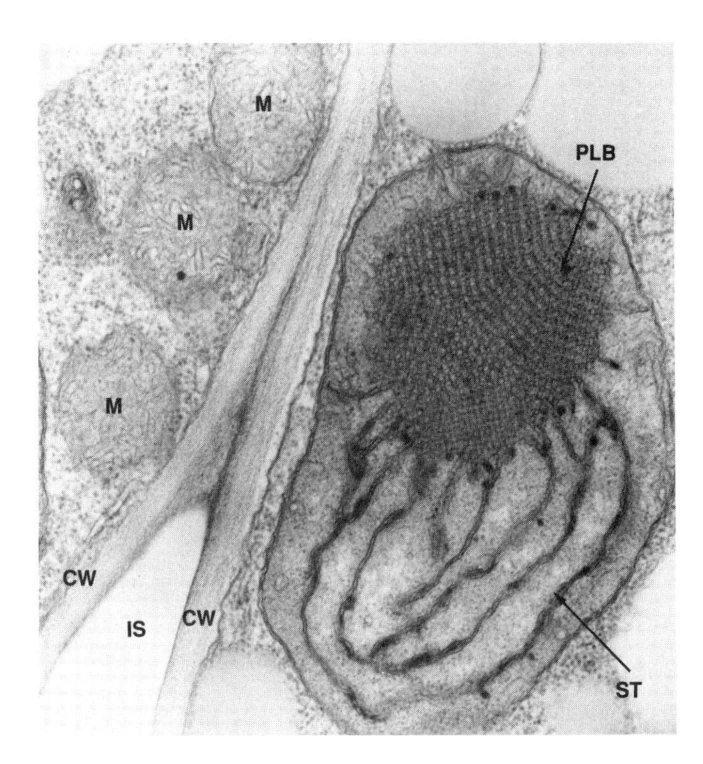


**Figure 18-9** Etioplast from a cotyledon of a dark-grown radish seedling, illustrating the prolamellar body (PLB) and stroma thylakoids (ST) radiating from it. Also shown are two adjacent cell walls (CW), an intercellular space (IS) between the walls, and mitochondria (M) in the cell at the left. (Courtesy Nicholas Carpita.)

nal concentrations of cytokinins in excised sections, especially in epidermal cells that probably limit the overall rate of elongation. (Note the relation of this problem to that of assessing the importance of auxins to root elongation described in Section 17.2.)

However, there are two known cases in which applied cytokinins do promote elongation: sections from young wheat coleoptiles (Wright, 1966) and intact watermelon hypocotyls, especially those of a dwarf cultivar (Loy, 1980). In the wheat coleoptiles, growth promotion occurs only if the tissues are still young and cell division is still occurring, but it was found that cytokinins cause growth by promoting cell elongation, not division. In the dwarf watermelons, enhanced hypocotyl elongation also occurs in response to exogenous cytokinin, primarily because the rate of cell elongation is increased; this increase results from cytokinins applied either to the shoot tip or to the roots.

In summary, exogenous cytokinins can promote cell expansion in young leaves, cotyledons, wheat coleoptiles, and watermelon hypocotyls, but much remains to be learned about the normal role of these hormones in cell expansion, especially in stems and roots. As usual, even less is known about trees in general and about conifers in particular.

## Cytokinin-Promoted Chloroplast Development and Chlorophyll Synthesis

If angiosperm seedlings are grown in darkness, we can remove a young leaf or cotyledon from a seedling and test whether adding a cytokinin to it has any effect on chloroplast development or chlorophyll synthesis. This experiment is possible because in darkness no chlorophyll is formed and chloroplast development is blocked. Young plastids are then arrested at the stage of proplastids (see Section 10.2) or, more commonly, at the stage of etioplasts. **Etioplasts** (from dark-grown or etiolated seedlings) are yellow because of the carotenoids they contain, and they have an interesting system of internal membranes closely arranged into an internal lattice called the **prolamellar body** (Fig. 18-9). Upon exposure to light, the prolamellar body gives rise to much of the thylakoid system found in normal green chloroplasts, and this development is accompanied by formation of special thylakoid proteins that become attached to chlorophylls in the two photosystems and light-harvesting complexes.

Addition of cytokinin to etiolated leaves or cotyledons several hours before they are exposed to light has two main effects: It enhances subsequent development (in the light) of etioplasts into chloroplasts, especially by promoting grana formation, and it increases the rate of chlorophyll formation. Much is known about the details of these effects (Parthier, 1979; Guern and Péaud-Lenoël, 1981; Lew and Tsuji, 1982); the principal reason for both is probably that cytokinins enhance formation of one or more proteins to which chlorophylls bind and become stabilized. We suspect that endogenous cytokinins normally increase chloroplast development in leaves in a similar way. More will be said of the ability of cytokinins to activate synthesis of a protein that binds chlorophylls *a* and *b* in relation to mechanisms of cytokinin action (see below).

## The Mechanisms of Cytokinin Action

The variability of cytokinin effects suggests that they might have different mechanisms of action in different tissues, yet the simpler view is that a common primary effect is followed by numerous secondary effects that depend on the physiological state of the target cells. As with other hormones, amplification of the initial effect must occur because cytokinins are present in such low concentrations (0.01 to 1 $\mu$M). Some promotive effect of cytokinins on RNA and enzyme formation was long suspected, partly because the effects of cytokinins are usually blocked by inhibitors of RNA or protein synthesis. No specific effects on DNA synthesis have been observed, although exogenous cytokinins often increase cell division and might normally be required for that process.

Many investigators have tried to determine

whether plants have a special receptor protein that would bind cytokinins and then lead to various physiological effects depending on the cell type. Several proteins that bind cytokinins more or less specifically have been found in various plant parts, but nearly all of them don't bind with high enough specificity or with enough affinity for active cytokinins (Napier and Venis, 1990). An interesting exception is a binding protein from barley leaves that binds zeatin with unusually high affinity and that binds other cytokinins in approximate relation to their biological activities (Romanov et al., 1988). Far more work with other species must be done before we learn whether this is a physiologically meaningful hormone-receptor protein. Meanwhile, other approaches can be used to determine how cytokinins act.

Enhanced cytokinesis is one of the most important cytokinin responses because it allows commercial micropropagation of several crops from tissue cultures. Biochemical aspects of this long-known response are being studied. Fosket and his colleagues (Fosket, 1977; Fosket et al., 1981) concluded that cytokinins promote division of cells in tissue culture by increasing the transition from $G_2$ to mitosis (see The Cell Cycle, Section 16.2) and that they do this by increasing the rate of protein synthesis. Some of these proteins could be enzymes or structural proteins needed for mitosis. Of course, protein synthesis could be increased by stimulating formation of messenger RNAs that code for those proteins, but no increase in messenger RNA production has been observed. Fosket and his colleagues concluded that cytokinins act specifically on translation. One of several pieces of evidence for that conclusion is that ribosomes in cytokinin-treated cells are frequently grouped in large protein-synthesizing polysomes, rather than in smaller polysomes or as free monoribosomes (the latter being characteristic of slowly-dividing cells). There is as yet no explanation as to how polysome formation and translation are increased by cytokinins, and no special enzyme or other protein that might lead to mitosis has been discovered in cytokinin-treated cells.

From studies of cytokinin-activated cell division in apical meristems, Houssa et al. (1990) obtained results largely consistent with those of Fosket and others. They found that benzyladenine caused a large decrease in the length of the S phase of the cell cycle (from $G_2$ to mitosis, during which DNA and cell-division proteins are synthesized). They suggested that some nuclear protein is a target for cytokinins. Presumably, that protein would enhance cell division rather directly: by controlling synthesis of DNA, for example. Remember, however, that nuclear proteins that could act as targets for the action of cytokinins or other hormones within the nucleus are synthesized in the cytosol during translation and that the nucleus does not produce its own proteins. Therefore, cytokinins might have so-called target effects in the nucleus only after first enhancing production of one or more nuclear proteins by translation in the cytosol.

Certain other cases of cytokinin action (for example, growth promotion) also appear to involve effects on translation, as evidenced by increased polysome levels, faster incorporation of radioactive amino acids into proteins, and inhibition of the physiological response by inhibitors of protein synthesis. This finding has given rise to the popular concept that whereas auxins and gibberellins mainly influence transcription in the nucleus, cytokinins mainly influence translation in the cytosol. This might not be true.

Chen et al. (1987) showed that benzyladenine changes the kinds of mRNAs formed by excised pumpkin cotyledons in which cytokinins promote cell expansion, cell division, and chlorophyll synthesis. Amounts of some mRNAs were enhanced by benzyladenine, whereas those of others were suppressed. The earliest changes were detected 1 h after addition of the cytokinin, and cytokinin action in these organs and in other plant parts is commonly observed even later, considerably later than effects of auxins or gibberellins in plant parts that respond to these hormones.

The simplest interpretation of changes in mRNA levels caused by cytokinins is that transcription of some genes is enhanced and that of other genes is repressed. (We made the same interpretation for auxins and gibberellins in Chapter 17 based on similar results.) However, we must also remember that the presence of a particular mRNA molecule depends partly on its rate of formation during transcription and partly on its rate of degradation (that is, its stability; see Fig. 17-1). Cytokinins might act only at transcription, or by influencing only mRNA stability, or both. In other studies with excised cotyledons, enhanced formation of polysomes appeared to result from faster mRNA synthesis because of a more active RNA polymerase (Ananiev et al., 1987; Ohya and Suzuki, 1988).

In at least three cases cytokinins affect the amounts of mRNA molecules that code for known proteins. Two proteins and their mRNAs are strongly **upregulated** (formed faster or degraded slower). These proteins are a chorophyll *a*/*b* binding protein (which becomes part of LHCII in thylakoids) and the small subunit protein of rubisco. When dark-grown (etiolated) leaves are exposed to cytokinins in darkness or to light without cytokinin, these two proteins and their mRNAs become much more abundant than in leaves not provided cytokinin (reviewed by Flores and Tobin, 1987 and by Cotton et al., 1990). Both mRNAs are coded by nuclear genes; this suggests action of cytokinin on transcription in the nucleus, yet Flores and Tobin (1987) obtained evidence that cytokinins act instead by increasing stability of those mRNAs, thereby allowing faster translation of their genetic messages into proteins.

The third example of cytokinin control of a known protein and its mRNA concerns the protein phytochrome. (Phytochrome, discussed in Chapter 20, is a protein-pigment complex that controls many developmental processes in the life of plants.) Formation of this protein and its mRNA is **downregulated** (formed slower or accumulated in smaller amounts) by the cytokinin zeatin and by red light absorbed by phytochrome itself (Cotton et al., 1990). Whether zeatin acts by deactivating the phytochrome gene in the nucleus or promotes degradation of phytochrome mRNA is not yet known. These results are especially interesting because they show common effects on a specific protein and its mRNA of a cytokinin and of red light absorbed by phytochrome. (As described in Chapter 20, red light and cytokinins have other physiological effects in common.) Furthermore, Bracale et al. (1988) found that light and benzyladenine caused similar changes in polypeptides and in plastid morphology during conversion of etioplasts to chloroplasts. From these many effects of cytokinins, we can summarize by saying that the evidence does not allow us to know with certainty whether cytokinins generally act on transcription, on mRNA stability, or on translation because evidence for each is available. Perhaps cytokinins affect all three processes in different species or plant parts.

Once we finally discover how cytokinins affect protein synthesis, there will still remain the problem of how newly translated enzymes or other proteins then cause cytokinesis, cell expansion, and other effects. At Colorado State University we attempted to determine how cell expansion of detached cotyledons is increased, without any knowledge of what kind of enzyme is involved. We found for both radish and cucumber cotyledons that cytokinin treatment causes increased plasticity (but not elasticity) of the cell walls; that is, the walls become loosened such that they can expand faster irreversibly under the existing turgor pressure (Thomas et al., 1981). Cytokinin-treated cotyledons grow with only about 0.15 MPa of turgor pressure compared to about 0.90 MPa for untreated cotyledons (Rayle et al., 1982). We also found that whatever the mechanism of cell-wall loosening, it is almost certainly not caused by acidification of the wall (Ross and Rayle, 1982), so the acid-growth mechanism is not applicable. As with auxins and gibberellins, cytokinins cause cells to alter their walls in some way that makes them more plastic, but the nature of this alteration and the enzyme or enzymes that cause it remain to be discovered. Whether cytokinins commonly act on the plasma membrane and by transduction lead to enhanced levels of Ca-calmodulin (see Fig. 17-2) is an unresolved subject that is undergoing active research.

## 18.2 Ethylene, a Volatile Hormone

The ability of certain gases to stimulate fruit ripening has been known for many years. Even the ancient Chinese knew that their picked fruits would ripen more quickly in a room with burning incense. In 1910, an annual report by H. H. Cousins to the Jamaican Agricultural Department mentioned that oranges should not be stored with bananas on ships because some emanation from the oranges caused the bananas to ripen prematurely. (Healthy oranges produce almost no ethylene, so it was probably fungi on infected oranges that produced the emanation.) This report was apparently the first suggestion that fruits release a gas that stimulates ripening, but it was not until 1934 that R. Gane proved that ethylene is synthesized by plants and is responsible for faster ripening.

Another historical practice implicating still a different role for ethylene was the building of bonfires near their crops by Puerto Rican pineapple growers and Filipino mango growers. These farmers apparently believed that the smoke helped to initiate and synchronize flowering. Ethylene causes these effects in both species, so it is almost surely the most active component in the smoke. Stimulation of fruit ripening is a widespread phenomenon, whereas promoted flowering appears to be restricted to mangos and most bromeliad species, including the pineapple.

Still another effect of gases was reported as early as 1864. Before the use of electric lights, streets were lighted with illuminating gas. Sometimes the gas pipes leaked, and in certain German cities this caused the leaves to fall off the shade trees. Ethylene causes senescence and abscission of leaves, so again it presumably was responsible.

The Russian physiologist Dimitry N. Neljubow (1876–1926) first established that ethylene affects plant growth. In 1901 he identified ethylene in illuminating gas and showed that it causes a **triple response** on pea seedlings: inhibited stem elongation, increased stem thickening, and a horizontal growth habit. Furthermore, leaf expansion is inhibited, and normal opening of the epicotyl hook is retarded. In this chapter we will describe these and other effects of ethylene. Books by Abeles (1973), Roberts and Tucker (1985), and Mattoo and Suttle (1990) and reviews by Lieberman (1979), Beyer et al. (1984), M. S. Reid (1987), Mattoo and Aharoni (1988), and Borochov and Woodson (1989) describe these and other ethylene effects in greater detail.

### Ethylene Synthesis

Ethylene production by various organisms can often be readily detected by gas chromatography because the molecule can be withdrawn from tissues under vacuum and because gas chromatography is so sensitive. Only a few bacteria reportedly produce ethylene, and no algae are known to synthesize it; furthermore, it generally has little influence on their growth. However, several fungal species produce it, including some that normally grow in soils. It is suspected that ethylene released by

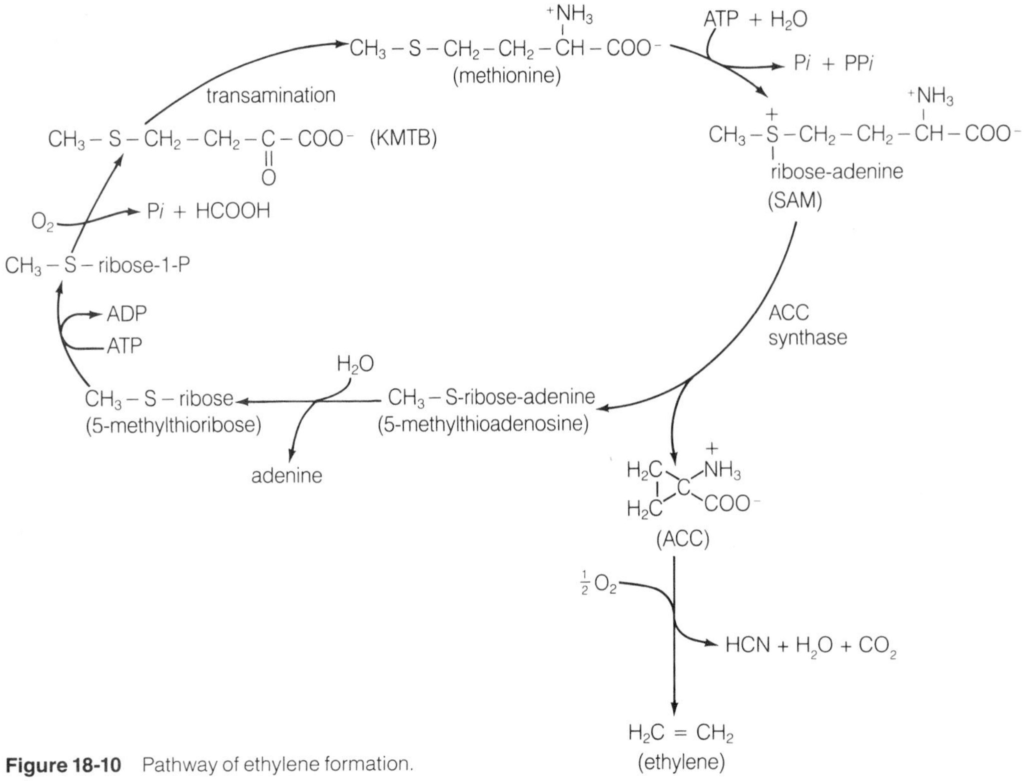

Figure 18-10 Pathway of ethylene formation.

soil fungi helps to promote germination of seeds, control growth of seedlings, and retard diseases caused by soil-borne organisms.

Essentially all parts of all seed plants produce ethylene. In seedlings, the shoot apex is an important site of production. Nodes of dicot-seedling stems produce much more ethylene than do internodes when equal tissue weights are compared. Stems produce more ethylene when they are laid horizontally (see Section 19.5). Roots release relatively small amounts, but again auxin treatment usually causes the rate of release to rise. Production in leaves generally rises slowly until the leaves become senescent and abscise. Flowers also synthesize ethylene, especially just before they fade and wither, and in most species this gas causes their senescence and abscission. The highest known rate of ethylene release was recorded from fading flowers of Vanda orchids: 3.4 mL $h^{-1}$ $kg^{-1}$ fresh weight (Beyer et al., 1984).

In many fruits, little ethylene is produced until just before the respiratory climacteric signaling the onset of ripening, when the content of this gas in the intercellular air spaces rises dramatically from almost undetectable amounts to about 0.1 to 1 $\mu$L per L. These concentrations generally stimulate ripening of those fleshy and nonfleshy fruits that display a climacteric rise in respiration (see Section 13.13), if the fruits are sufficiently developed to be susceptible to the gas (Tucker and Grierson, 1987). Sections of ripe apple or pear fruits or even apple peelings are often used as an ethylene source in laboratory demonstrations. Nonclimacteric fruits synthesize little ethylene and are not induced to ripen by it. It is clear, however, that most climacteric fruits, including nonfleshy fruits, normally ripen partly in response to the ethylene they produce. In nonclimacteric fruits such as cherries, grapes, and citrus fruits, ethylene seems to play no role in natural ripening, although it is used commercially to de-green oranges and lemons (M. S. Reid, 1987).

Interestingly, numerous mechanical and stress effects such as gentle rubbing of a stem or leaf, increased pressure, pathogenic microorganisms, viruses, insects, waterlogging, and drought increase ethylene production. Ancient Egyptian civilizations unknowingly took advantage of increased ethylene production resulting from injury by gashing immature sycamore figs to stimulate ripening. When figs only about 16 days old are gashed, they ripen within as little as four days.

Early researchers discovered that ethylene is derived from carbons 3 and 4 of the amino acid methionine. A second important advance was made in the laboratory of Shang-Fa Yang at the University of California, Davis, when it was found that an unusual amino-acid–like compound, **1-amino-cyclopropane-1-carboxylic acid (ACC)**, is involved as a close precursor of ethylene. Yang and his colleagues elucidated several other reactions of the pathway of ethylene formation (reviewed by Yang and Hoffman, 1984; Yang et al., 1985; Imaseki et al., 1988; Kende, 1989). Figure 18-10 shows this pathway; note that the sulfur atom of methionine is conserved by a cyclic, salvage process. Without this salvage, the amount of reduced sulfur might limit the amount of methionine and the rate of ethylene synthesis. Other notable features of the pathway are that ATP is essential for conversion of methionine to S-adenosylmethionine (SAM) and that $O_2$ is needed in the final conversion of ACC to ethylene. (The requirements for ATP and $O_2$ almost certainly explain why ethylene production nearly stops under severely hypoxic conditions.) Evidence also indicates that four of the carbon atoms of the ribose unit of SAM are salvaged and reappear in methionine. An intermediate, *α-keto-γ-methylthiobutyric acid* (KMTB), is important in the salvage of these carbons.

Two potent inhibitors of ethylene synthesis have been discovered, both of which are useful tools for investigating the pathway of ethylene formation and studying effects of reduced ethylene production in tissues. These compounds, *aminoethoxyvinylglycine (AVG)* and *aminooxyacetic acid (AOA)*, are well-known inhibitors of enzymes that require pyridoxal phosphate as a coenzyme. AVG and AOA block conversion of SAM to ACC, but they have no other important effect on the pathway. This and other studies with the purified enzyme show that *ACC synthase* is a pyridoxal phosphate-dependent enzyme.

The final reaction in the pathway, conversion of ACC to ethylene, is catalyzed by an oxidative enzyme called the *ethylene-forming enzyme (EFE)*. This enzyme has not been well purified, probably because it is tightly bound in or on a membrane. Studies with isolated vacuoles of *Vicia faba* showed that these organelles contain most of the ACC in cells and that they form most of the ethylene (Mayne and Kende, 1986; Kende, 1989). Presumably, therefore, EFE is on or in the tonoplast. However, work with three other species showed that both the plasma membrane and the tonoplast probably synthesize ethylene (Bouzayen et al., 1990). Interestingly, ethylene formation is accompanied on a 1 to 1 basis by formation of *hydrogen cyanide* (HCN). Plants have ways to metabolize away the HCN (in the process saving the nitrogen and carbon); otherwise the cyanide might poison cytochrome oxidase in mitochondria and inhibit respiration (see Section 13.7).

The control of ethylene synthesis has received considerable study, especially regarding the promoting effects of auxins, wounding, and drought stress, and aspects of the fruit-ripening stage. It is now accepted that the rate-limiting step in ethylene formation is usually catalyzed by ACC synthase. In stems of mung bean (*Phaseolus aureus*) and pea seedlings, IAA increases ethylene formation by a factor of several hundred. In these and other tissues, auxins induce additional formation of ACC synthase, and the enhanced formation of ACC resulting from action of that enzyme then leads to increased ethylene production. Wounding also increases

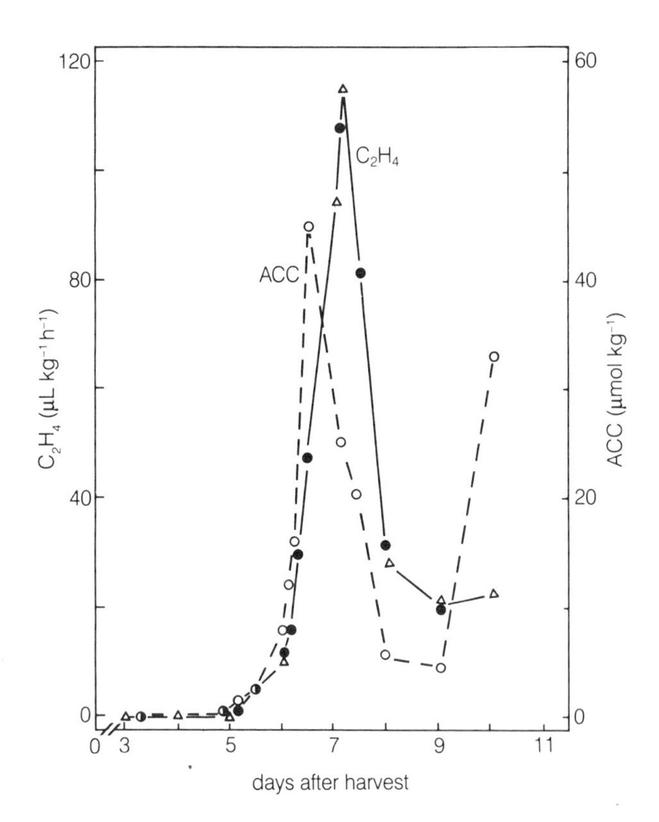


**Figure 18-11** Changes in ACC content and rate of ethylene production in ripening avocado fruits. (From Hoffman and Yang, 1980.)

ethylene production by inducing formation of ACC synthase.

In ripening climacteric fruits, formation of ACC again limits ethylene formation. In preclimacteric avocado fruits, for example, the concentration of ACC rose from near zero to over 40 μmol $kg^{-1}$ of fruit tissue just before a peak of ethylene synthesis (Fig. 18-11). Ripening soon followed. The levels of ACC and ethylene decreased greatly about two days after the peaks occurred, but the level of ACC then rose again without additional ethylene synthesis. Application of ACC to preclimacteric fruits did not cause the rise in ethylene release, indicating that the climacteric is accompanied not only by increased production of ACC from SAM but also by an increased ability to convert ACC to ethylene.

McKeon and Yang (1987) pointed out that enhanced ethylene production by ripening fruits requires development of large increases in activity of both ACC synthase and EFE and that in some fruits EFE becomes the rate-limiting enzyme. A curious autocatalytic (positive-feedback) ability of ethylene to stimulate its own formation occurs in many senescing organs, including leaves, flower petals, and over-ripened fruits. This effect results at first from ethylene-promoted activity of EFE, but then it is followed by much larger increases of ACC-synthase activity, and the increased ACC formed by that enzyme probably explains the ability of one rotten apple in a barrel to spoil the others. More importantly, diffusion of ethylene through intercellular spaces within a fruit probably coordinates ripening of quite different tissues within the fruit.

Besides oxygen, other environmental factors—light and carbon dioxide—affect ethylene synthesis in leaves. Light inhibits synthesis in photosynthetic cells, mainly by interfering with conversion of ACC to ethylene. Carbon dioxide promotes synthesis by enhancing conversion of ACC to ethylene. It seemed logical that the opposing effects of light and $CO_2$ are explained by the photosynthetic removal of $CO_2$ in daylight, and this was found to be true. Carbon dioxide not only activates EFE in leaves but induces its synthesis as well. The rather complex (or tissue-dependent) interactions of $CO_2$ and ethylene were reviewed by Sisler and Wood (1988); also see the article by Zhi-Yi and Thimann (1989).

**Figure 18-12** Severe epinasty in cocklebur. Plant on the left is an untreated control; plant on the right was dipped into a solution of 1 mM naphthalene acetic acid (an auxin) about 2 days before the photograph was taken. Epinasty in response to ethylene without added auxin is very similar. (Photo by F. B. Salisbury.)

## Ethylene Effects on Plants in Waterlogged Soils and Submerged Plants

Because of the requirement for $O_2$ to convert ACC to ethylene, we would expect waterlogged roots to produce less ethylene. This is true, but waterlogged tomato plants nevertheless exhibit ethylene toxicity symptoms (Kawase, 1981; Jackson, 1985a, 1985b). These symptoms, some of which are also characteristic of other plant species, include chlorosis of leaves, decreased stem elongation but increased stem thickness, wilting, epinasty and eventual abscission of leaves, decreased root elongation often accompanied by adventitious-root formation, and increased susceptibility to pathogens. In many species, including tomato, aerenchyma formed in the root cortex (see Fig. 13-14) increases oxygen movement to the roots from the shoots. Furthermore, transport of cytokinins and gibberellins from the roots to the shoot via the xylem is reduced. These symptoms in response to excess ethylene result from the following events.

Waterlogged soils rapidly become hypoxic (see Section 13.13) because water fills the air spaces and $O_2$ replenishment around the roots is reduced by the very slow movement of the gas through water. Ethylene synthesis is then inhibited because $O_2$ is required to convert ACC to ethylene, but the ethylene that is synthesized is trapped in the roots because its escape through water is reduced by a factor of about 10,000 compared to air (Jackson, 1985b). This ethylene then causes some of the cortical cells to synthesize **cellulase**, an enzyme that hydrolyzes cellulose and is partly responsible for degrading cell walls. These cortex cells also lose their protoplasts and disappear, forming the air-filled aerenchyma tissue.

Even before the aerenchyma develops, however, ACC accumulates and is transported in the xylem to the shoots. The well-aerated shoots then rapidly convert this ACC to ethylene, which in turn causes leaf epinasty (see Section 19.2). An extreme case of epinasty is shown in Fig. 18-12. Epinasty of petioles occurs because mature parenchyma cells on the upper side of the petiole elongate in the presence of ethylene, whereas those on the lower side do not. This physiological difference in morphologically similar cells is not understood, but it emphasizes again that only some cells are targets for a given hormone. Ethylene also retards elongation of the stem, increases its radial expansion, causes leaf senescence, and promotes formation of adventitious roots on the stems (especially in tomato).

## Ethylene Effects on Elongation of Stems and Roots

Although ethylene causes epinasty of leaves by promoting elongation of cells on the upper side, it usually inhibits elongation of stems and roots, especially in dicots. (Inhibited stem elongation in peas forms part of the previously mentioned triple response.) When elongation is inhibited, stems and roots become thicker by enhanced radial expansion of the cells (Fig. 18-13). In dicot stems, altered cell shapes are apparently caused by a more longitudinal orientation of cellulose microfibrils being deposited in the walls, preventing expansion parallel to these microfibrils but allowing expansion perpendicular to them (for a review, see Eisinger, 1983). No comparable studies of changes in shapes of cells or cellulose microfibril orientation seem to have been performed with roots, but almost certainly the changes are similar.

Root and stem thickening caused by ethylene is of survival value for dicot seedlings emerging from the soil. In these species, a hook in the epicotyl or the hypocotyl is formed in response to endogenous ethylene (see Fig. 16-14) shortly after germination; then this hook pushes up through the soil and makes a hole through which the cotyledons or young leaves can be safely drawn. If the soil is excessively compact, the hook and primary root become unusually thick, probably because ethylene is synthesized faster when the compacted cells are subjected to increased mechanical pressure and because the ethylene escapes less rapidly in compacted soils. This resulting thickness increases the strengths of both stems and roots, allowing them to push through the compacted soil. Growth is slow, however, because of retarded elongation.

In the cereal grains maize and oat, ethylene has effects on the mesocotyl (first internode; see Fig. 20-10) that are similar to effects on dicot stems: inhibited elongation and greater thickening (Camp and Wickliff, 1981). How general this is among members of the grass family is unknown, but the advantage to seedlings in compact soils should be the same as the advantage to dicots. We can also speculate that the loss of gravitropic sensitivity by stems of dicot seedlings is advantageous in compact soils because a more horizontally growing stem might more likely encounter a crack in the soil than one that grows upright.

Although retarded stem elongation is common among land plants, certain dicots and ferns that grow at least part of the time with their roots and stems underwater respond to ethylene by enhanced elongation. Among these species are *Callitriche platycarpa* (a starwort), *Ranunculus sceleratus, Nymphoides peltata*, and the water fern *Regnellidium diphyllum* (Jackson, 1985a, 1985b; Ridge, 1985). When submerged, the stems of these plants elongate rapidly so that the leaves and upper stem parts are kept at the water surface. Submergence causes ethylene accumulation in the stems, which is responsible for the rapid elongation. In some species it is the stem that elongates in response to ethylene; in others petioles elongate. In either case flowers or leaves are kept above water, thereby allowing pollination and photosynthesis in air. A similar phenomenon occurs in stems of deep-water rice (Raskin and Kende, 1985). In this species internode lengths of up to 60 cm have been recorded, and the plants grow and even reproduce with stems in several meters of water (Jackson, 1985b). The stems are hollow, with many air spaces in the cortex; both characteristics probably help the parts above water to act as a snorkel system and aerate the roots. These examples contrast with those of most species, in which ethylene inhibits stem elongation, and such examples emphasize the different responses of somewhat similar cells to the same hormone (Osborne et al., 1985).

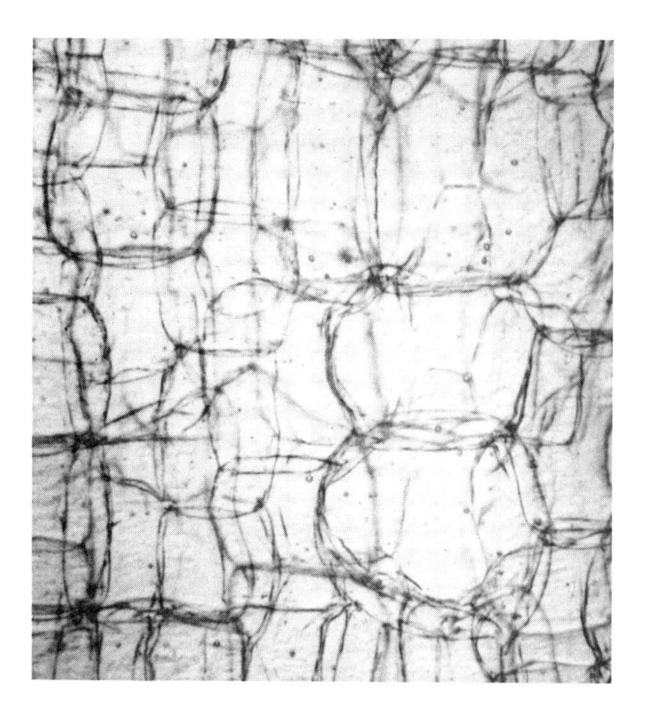

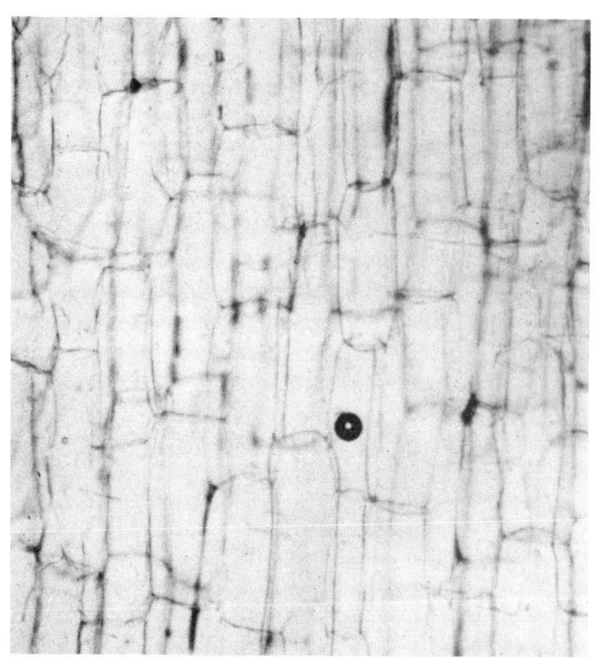

**Figure 18-13** Effects of ethylene on cell elongation and radial expansion in the upper internode of pea seedlings. Plants were grown 4 days in darkness, then treated with 0.5 $\mu$L/L ethylene (top photo) or left as controls (bottom photo). Cell sections were made 24 h after ethylene treatments began. (From Stewart et al., 1974.)

## Ethylene Effects on Flowering

The induction of flowering in mangos and bromeliads by ethylene (mentioned earlier) is unusual because in most species this gas inhibits flowering. Nevertheless, the indirect use of ethylene to promote flowering has been widely used in the pineapple industry in Hawaii. In the 1950s, fields were often sprayed with the auxin NAA, now known to stimulate ethylene synthesis in the plants. As a result, pineapple fields flower faster and, more importantly, mature fruits appear at about the same time, which allows a one-harvest mechanical operation.

An ethylene-releasing substance called **Ethrel** (trade name) or **ethephon** (common name) is commercially available. It is *2-chloroethylphosphonic acid* ($Cl-CH_2-CH_2-PO_3H_2$), which rapidly breaks down in water at neutral or alkaline *p*H to form ethylene, a $Cl^-$ ion, and $H_2PO_4^-$. Because ethephon can be translocated throughout the plant, it replaced NAA as a promoter of flowering in pineapples (Nickell, 1979), and it is used in various other aspects of horticulture such as fruit production. For example, it is sprayed on some tomato fields late in the season to cause uniformly ripened fruits and enable more efficient mechanical harvesting (M. S. Reid, 1987). It is often used as well in research as a source of ethylene.

## Additional Ethylene Effects

Ethylene causes numerous other effects on plants, many of which have not been studied much. One rather well-studied example is induction of flower senescence. As with climacteric fruits, numerous flowers undergo a climacteric rise in respiration and in ethylene production, and in these flowers the ethylene eventually causes senescence. In fact, as with fruits, there is evidence that when petal cells become sufficiently sensitive to ethylene, they respond by a burst of autocatalytic ethylene production as ACC synthase is induced (Borochov and Woodson, 1989). Soon the petals begin to wilt in response to an increase in permeability of the plasma membrane and tonoplast, which is followed by loss of solutes and then water to the cell walls and probably to the intercellular spaces. In some species pollination increases the rate of ethylene production, and ACC is one of the substances translocated from the stigma that cause ethylene release and senescence (Woltering, 1990).

Another effect of ethylene in some species is promotion of adventitious-root formation, an effect also caused (separately) by auxins (M. S. Reid, 1987; Mudge, 1988). Effects of ethylene on expression of flower sex in monoecious species also occur; good examples are cucurbits such as squash, pumpkin, and melon. Ethylene strongly promotes formation of female flowers in these plants and in others in different families (Abeles, 1973; Durand and Durand, 1984). In some species ethylene breaks seed dormancy (Taylorson, 1979), and in nature soil fungi might contribute some of this ethylene.

## Relation of Ethylene to Auxin Effects

The ability of IAA and all synthetic auxins to increase ethylene production raises the question of whether many auxin effects are really caused by ethylene instead. Indeed, ethylene appears to be responsible in some cases (see reviews by Abeles, 1973, and by Burg, 1973); included are leaf epinasty, inhibition of stem and leaf elongation, flower induction in bromeliads and mangos, inhibition of epicotyl- or hypocotyl-hook opening in dicot seedlings, and increased percentage of female flowers in dioecious plants. Also, release of auxin by germinating pollen grains promotes ethylene production in the stigma, which contributes to senescence of flowers in some species (Stead, 1985). As described in Section 18.7, abscission of leaves, flowers, and fruits involves interactions among auxins, ethylene, cytokinins, and abscisic acid. Nevertheless, growth promotion, initial stages of adventitious-root production, and many other effects of auxins appear to be independent of ethylene production. Only in certain plant parts, and only when the auxin concentration becomes relatively high, is ethylene production great enough to account for certain auxin effects.

## Antagonists of Ethylene Action

At high concentrations (5 to 10 percent), $CO_2$ inhibits many ethylene effects, perhaps by acting as a competitive inhibitor of ethylene action. In ripening fruits it interferes with the ability of ethylene to catalyze its own formation (the climacteric ethylene burst). Such interference probably results from retarded conversion of ACC to ethylene (Cheverry et al., 1988). Therefore, in this case the ability of $CO_2$ to inhibit ethylene action leads to decreased ethylene production. Because of this inhibition, $CO_2$ is often used to prevent overripening of picked fruits and some vegetables (reviewed by Knee, 1985; Weichmann, 1986; Kader et al., 1989). Such fruits are stored in an airtight room or container in which the gas composition is controlled. An ideal atmosphere for many fruits contains 5 to 10 percent $CO_2$, 1 to 3 percent $O_2$, and no ethylene. Removal of some oxygen is important because this slows ethylene synthesis. However, if too much $O_2$ is removed, glycolysis is stimulated by the Pasteur effect (see Section 13.13), causing excess sugar breakdown. Another technique useful in fruit storage is to partially evacuate the container, thereby removing $O_2$ and ethylene from the tissues into the atmosphere.

Inhibition of ethylene action by $CO_2$, although common, is not universal. One reason for this is that carbon dioxide's inhibitory property is lost as the tissue ethylene concentration approaches or exceeds 1 $\mu$L/L, a concentration that gives about half-maximal activity in nearly every response to ethylene studied (Beyer et al., 1984). For this reason and because of the high concentrations of $CO_2$ required, it seems unlikely that $CO_2$ often acts *in vivo* as an antagonist of ethylene action.

A much more effective inhibitor of ethylene action is $Ag^+$, the silver ion (Beyer, 1976). Among the ethylene effects found by Beyer to be nullified or inhibited by $Ag^+$ (added as $AgNO_3$) were the triple responses of etiolation of pea seedlings, promotion of abscission of leaves, flowers, and fruits of cotton, and induction of senescence in orchid flowers. Silver thiosulfate proved even more effective in delaying senescence of cut flowers than silver nitrate (Halevy and Mayak, 1981; M. S. Reid, 1987). An example of this effect is shown in Figure 18-14.

More recently, various synthetic, volatile olefin compounds were found to be strong inhibitors of ethylene action (Sisler and Yang, 1984; Bleecker et al., 1987; Sisler, 1990a, 1990b). Among these compounds, *trans-cycloctene* and *2,5-norbornadiene* are particularly effective. For example, ethylene-induced senescence of carnation petals can be greatly delayed by atmospheric norbornadiene, and petals partly senescent from ethylene treatment can recover if ethylene is removed and norbornadiene added (Wang and Woodson, 1989). These compounds bind to normal ethylene receptors, thereby preventing ethylene action.

**Figure 18-14** Delay of carnation-flower senescence by silver thiosulfate (STS), an inhibitor of ethylene action. Flowers were pretreated for only 10 min with 4 mM STS and were then kept in deionized water for 10 days at 20°C. (Courtesy M. S. Reid.)

## How Ethylene Acts

Many ethylene effects are accompanied by increased synthesis of enzymes, the kind of enzyme depending on the target tissue. When ethylene stimulates leaf abscission, cellulase and other cell-wall–degrading enzymes appear in the abscission layer (see Section 18.7). When fruit ripening or flower senescence occurs, several kinds of necessary enzymes are produced. When cells are injured, phenylalanine ammonia lyase (see Section 15.4) appears, which is an important enzyme in the formation of phenolic compounds thought to be involved in wound healing. When certain fungi infect cells, ethylene induces the plant to form two enzymes that degrade fungal cell walls: *β-1,3-glucanase* and *chitinase* (Boller, 1988). Many scientists conclude that ethylene is a signal for plants to activate mechanisms against fungal attack. In several cases, increases in amounts of mRNAs that code for these enzymes are caused by ethylene treatment. Almost certainly, ethylene enhances transcription of various nuclear genes, the kind depending on species, organ, tissue, and other factors.

Where ethylene acts within cells is an important but unsolved question. We do know that it binds with one or more receptor proteins and that these receptors are located in membranes. We don't yet know if the receptors are in the plasma membrane. Reviews of progress in identifying ethylene receptors in various plants are those by Napier and Venis (1990) and by Sisler (1990a, 1990b). Considerable evidence suggests that these receptors contain copper at their active sites. If a primary receptor exists in plasma membranes, we might expect the existence of a transduction system similar to that illustrated in Figure 17-2.

Several ethylene mutants have now been found after treatment of seeds with chemical mutagens such as *ethyl methanesulfonate* (Reid, 1990; Scott, 1990; Guzman and Ecker, 1990). Some are synthesis mutants and some are response mutants. All except one of the synthesis mutants are ethylene overproducers (as in pea, *Arabidopsis thaliana*, and in tomato). In general, these mutants (not exposed to ethylene) exhibit the triple response mentioned earlier when they are seedlings; when they are older they exhibit other responses similar to those of plants treated with ethylene. Three ethylene-insensitive mutants of *Arabidopsis* are known. Identification of the genes and mRNA molecules affected should help identify altered receptor proteins, assuming that the genes don't control some step in a transduction pathway after ethylene binds to a receptor.

One of the best-studied insensitive mutants of *Arabidopsis*, called *etr*, lacks several responses to ethylene that exist in wild-type plants (Bleecker et al., 1988). These absent responses include inhibition of root and hypocotyl elongation, reduced chlorophyll content of leaves, enhanced activity of peroxidase isozymes, accelerated leaf senescence, and enhanced germination of partly dormant seeds. The *etr* mutant is compared with nonmutants in Figure 18-15.

**Figure 18-15** The *etr Arabidopsis* mutant growing above wild-type seedlings in an atmosphere containing ethylene. Note differences in stem length, stem thickening, hook opening, and gravitropic response. (From A. B. Bleecker, M. A. Estelle, C. Somerville, and H. Kende, 1988. Insensitivity to ethylene conferred by a dominant mutation in *Arabidopsis thaliana*. Science 241:1086–1089. Copyright 1988 by the American Association for Advancement of Science. Photograph by Kurt Stepnitz.)

## 18.3 Triacontanol, Brassins, Salicylic Acid, and Turgorins

**Triacontanol** is a 30-carbon, saturated primary alcohol first isolated from shoots of alfalfa. It is very insoluble in water (less than $2 \times 10^{-16}$ M or $9 \times 10^{-14}$ g $L^{-1}$), yet colloidal suspensions of this compound significantly enhance growth of maize, tomato, and rice plants when sprayed on the foliage of seedlings at concentrations as low as 0.1 nanograms per liter (reviewed by Ries, 1985). Maize and rice reportedly respond by increased growth within 10 min. Little is known about the mechanism of action of triacontanol, but it is of potential importance in increasing crop yields.

The **brassins** or **brassinosteroids** are recently discovered steroid growth-promoters first isolated from pollen grains of rape plants but now known to be present in several other species as well. The nature of these compounds is described in Section 15.3. They have various effects on plant growth and act partially by increasing sensitivity to auxins.

**Salicylic acid** (2-hydroxybenzoic acid), the active ingredient in aspirin (acetyl-salicylic acid), is a plant hormone important for some known physiological responses. In Chapter 13 we mentioned production of heat and aromas in the appendix of the inflorescence in the *Arum* lilies. A cause of this production is salicylic acid produced in staminate flower primordia and translocated to the appendix (see Fig. 13-10). There it promotes activity of cyanide-resistant respiration (see Section 13.9) that leads to heat production and volatilization of compounds that attract insect pollinators (reviewed by Meeuse and Raskin, 1988).

Another effect of salicylic acid is promotion of resistance to certain plant pathogens, including tobacco mosaic virus (Malamy et al., 1990), tobacco necrosis virus, and the fungal pathogen *Colletotrichum lagenarium* (Métraux et al., 1990). In the reports mentioned, inoculation of leaves with the virus or fungal pathogen caused substantial increases in tissue (or phloem sieve tube) concentrations of salicylic acid. And this compound causes production of one or more pathogenesis-related (PR) proteins that increase disease resistance in the infected leaves and in adjacent leaves. Clearly, salicylic acid meets the criteria for a plant hormone and it almost surely has many undiscovered physiological roles.

Another group of growth regulators, the **turgorins**, are described in Section 19.2.

## 18.4 Polyamines

These compounds are polyvalent cations that contain two or more amino groups, including the amino acids lysine and arginine (see Fig. 9-2). Among the most abundant and physiologically active polyamines are **putrescine** ($NH_2(CH_2)_4NH_2$), **cadaverine** ($NH_2(CH_2)_5NH_2$), **spermidine** ($NH_2(CH_2)_3NH(CH_2)_4NH_2$), and **spermine** ($NH_2(CH_2)_3NH(CH_2)_4NH(CH_2)_3NH_2$). These compounds exist free or bound to various phenolic compounds such as coumaryl and caffeoyl groups.

In contrast to hormones, which are often present in micromolar concentrations, polyamines frequently exist in millimolar concentrations. Among several physiological effects, they promote cell division, stabilize membranes, stabilize isolated protoplasts (probably by effects on membranes), promote development of some fruits, minimize water stress of various kinds of cells, and delay senescence of detached leaves. (For a recent review, see Evans and Malmberg, 1989.)

Little is known about polyamines' primary mechanism of action, but their positively charged amino groups cause them to combine with negatively charged phosphate groups in DNA and RNA in the nucleus and in ribosomes. As a result of this combination, they often increase transcription of DNA and translation of RNA in plant and animal cells. Evans and Malmberg (1989) concluded that polyamines are not plant hormones (they are poorly translocated and too abundant), but they might be considered plant growth regulators or merely one of several kinds of metabolites needed for certain developmental processes. We expect to learn much more about their functions in both plants and animals.

## 18.5 Abscisic Acid (ABA)

The discussion above about ethylene indicates that in several ways it can be considered a stress hormone because it is produced in much higher amounts when plants are subject to various kinds of stresses. But still another hormone called **abscisic acid (ABA**; introduced in Section 4.6) often gives plant organs a signal that they are undergoing physiological stress. Among these stresses are lack of water, saline soils, cold temperatures, and frost. ABA often causes responses that help protect plants against these stresses. As we shall explain, it also helps cause normal embryogenesis and formation of seed-storage proteins, and it prevents premature germination or growth of many seeds and buds.

In 1963 abscisic acid was first identified and chemically characterized in California by Frederick T. Addicott and his coworkers, who were studying compounds responsible for abscission of cotton fruits. They named one active compound *abscisin I* and called a second (much more active) compound *abscisin II*. Abscisin II proved to be ABA. In the same year two other research groups had very likely discovered ABA as well. One group was led by Philip F. Wareing in Wales; they were studying compounds that caused dormancy of woody plants, particularly *Acer pseudoplatanus*. They named their most active compound *dormin*. The other group was led by R. F. M. Van Steveninck, first in New Zealand and then in England; they were studying a compound or compounds that accelerated abscission of flowers and fruits of the yellow lupine (*Lupinus luteus*). Because it became evident (in 1964) that dormin and the lupine compound were identical to abscisin II, physiologists agreed in 1967 to call the compound abscisic acid. ABA appears to be universal among vascular plants; it is also present in some mosses, some green algae, and some fungi, but not in bacteria.

### Chemistry, Metabolism, and Transport of Abscisic Acid

ABA (Fig. 18-16) is a 15-carbon sesquiterpenoid synthesized partly in chloroplasts and other plastids by the mevalonic acid pathway (see Section 15.3). Thus early reactions in ABA synthesis are identical to those of isoprenoids such as gibberellins, sterols, and carotenoids.

Biosynthesis of ABA in most (perhaps all) plants occurs indirectly by degradation of certain (40-carbon) carotenoids present in plastids (reviewed by Zeevaart and Creelman, 1988, and by Creelman, 1989). More recent evidence for this process comes from work by two active research groups, one led by Jan A. D. Zeevaart at Michigan State University (see Zeevaart et al., 1989 and Rock and Zeevaart, 1990) and the other by Daniel C. Walton at the State University of New York in Syracuse (see Sindhu et al., 1990). Chloroplasts in leaves contain the carotenoids from which ABA arises, whereas in roots, fruits, seed embryos, and certain other plant parts the necessary carotenoids are in other chromoplasts, leucoplasts, or proplastids. Only some of the reactions that lead from carotenoids to ABA have been well identified, but a tentative pathway is depicted in Figure 18-17. All reactions that form xanthoxin probably occur in plastids, but subsequent steps likely occur somewhere in the cytosol.

Essentially, the carotenoid *violaxanthin* with a *trans* configuration at all double bonds is probably converted by an unknown enzyme to *9-cis-violaxanthin*, which has the same *cis* configuration as ABA at carbons 2 and 3 (Fig. 18-17). Next, *9-cis*-violaxanthin is somehow oxidized with $O_2$ and is split to release an unknown compound or compounds (with a total of 25 carbons) and *xanthoxin*, a 15-carbon epoxide with a structure similar to that of ABA. Xanthoxin is converted to *ABA aldehyde* by opening of the epoxide ring and by oxidation (by either $NADP^+$ or $NAD^+$) of the ring hydroxyl group to a keto group. Finally, the aldehyde group on the side chain of ABA aldehyde is oxidized to the carboxyl group of ABA. Interestingly, this latter oxidation almost surely requires a coenzyme that contains molybdenum (Walker-Simmons et al., 1989), providing another essential function for molybdenum in plants.

Inactivation of ABA can occur in two ways. One is by attachment of glucose to the carboxyl group to form an ABA-glucose ester (Fig. 18-16). This ester appears to be restricted to the vacuole. Note that a comparable inactivation by attachment of glucose also occurs with IAA, gibberellins, and cytokinins. Another inactivation process is oxidation with $O_2$ to form *phaseic acid* and *dihydrophaseic acid* (see Fig. 18-16).

Transport of ABA occurs readily in both xylem and phloem and also in parenchyma cells outside vascular bundles. In parenchyma cells there is usually no polarity (as contrasted to the case of auxins), so movement of ABA within plants is similar to that of gibberellins.

### ABA-Induced Stomatal Closure

The importance of ABA as a stress hormone was first suggested in 1969 by S. T. C. Wright and R. W. P. Hiron at Wye College, London University. They found that the ABA content of wheat leaves rose by a factor of 40 during the first half hour of wilting. Many researchers then showed that application of ABA to leaves would cause

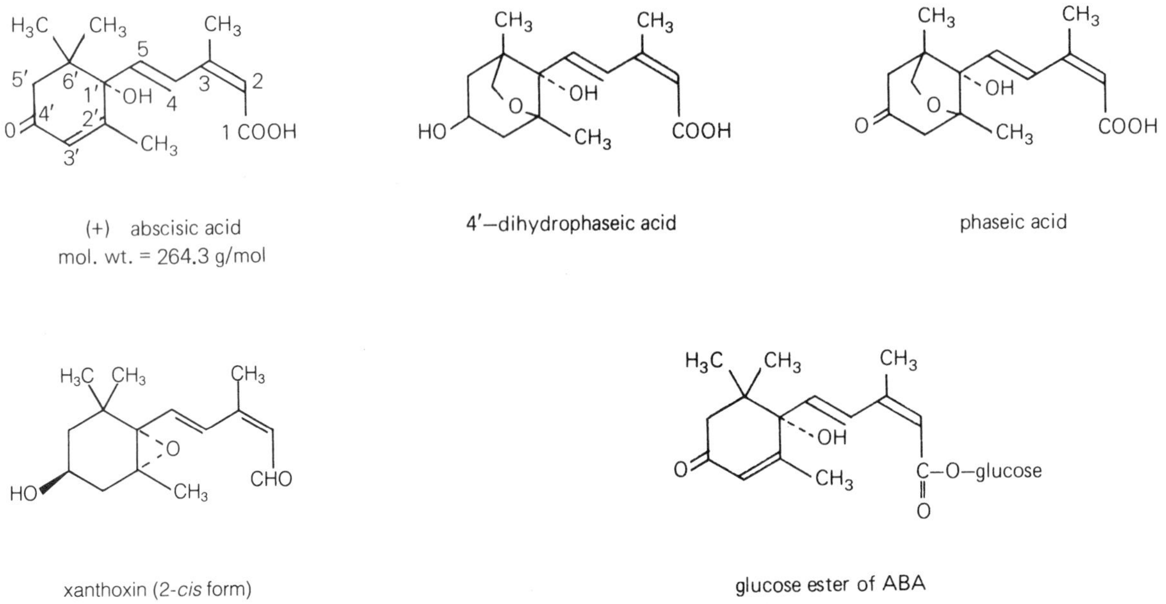

(+) abscisic acid
mol. wt. = 264.3 g/mol

4′–dihydrophaseic acid

phaseic acid

xanthoxin (2-*cis* form)

glucose ester of ABA

**Figure 18-16** Structure of abscisic acid (ABA) and some related compounds. ABA (upper left) has one asymmetric carbon atom (1′ in the ring). The form synthesized by plants is the dextrorotatory (+) product shown with the S (sinister) configuration about the asymmetric (chiral) carbon atom. But commercial ABA is a racemic (±) mixture. Both forms are biologically active.

stomatal closing in numerous species and that stomates would stay closed in light or darkness for several days, probably depending on how long it took for the plant to metabolize away the ABA (reviewed by Raschke, 1987).

The ABA content of monocot and dicot leaves rises substantially when leaves are subjected to water stress, both when leaves are detached from roots or left intact. Recently it has even been possible to measure the ABA concentration in single guard cells using separated cells and enzyme-linked immunoassay procedures (Harris and Outlaw, 1990). Water stress causes at least 20-fold increases in ABA contents, up to 8 femtograms per cell (1 fg equals $10^{-15}$ g). It was found that water-stressed roots also form more ABA and that this ABA is transported through xylem to the leaves, where it causes stomates to close (see Section 4.6).

There is now evidence that this root-supplied ABA comes largely from water-stressed tips of shallow roots and that it serves as a signal to leaves that water in the soil is becoming depleted (Davies and Mansfield, 1988; Zhang and Davies, 1989). Stomates close in response to ABA from either leaves or roots, and they are then drought-protected. Of course, because photosynthesis nearly stops, shoot growth is restricted (further reducing water loss), but growth of deeper roots can continue until they too become dry. ABA causes stomates to close by inhibition of an ATP-dependent proton pump in the plasma membrane of guard cells. This pump normally transfers protons out of guard cells, leading to rapid influx and accumulation of $K^+$ and then to osmotic water absorption and stomatal opening. However, ABA, acting in the free space on the outer surface of guard-cell plasma membranes, shuts off $K^+$ influx, so $K^+$ and water leak out, turgor is reduced, and stomates close.

How water stress actually causes ABA production in leaves has been carefully investigated. It seems that loss of turgor rather than a more-negative osmotic potential is the primary signal (reviewed by Zeevaart and Creelman, 1988). This turgor loss probably causes an unknown signal from the plasma membrane to activate certain nuclear genes that lead to enhanced ABA synthesis. Several results suggest that it is the plasma membrane that responds to decreased turgor and that it does so by transporting $Ca^{2+}$ into the cell at an increased rate (Lynch et al., 1989). Reviews by Owen (1988) and by Skriver and Mundy (1990) logically suggest that $Ca^{2+}$ and phosphoinositols then act in the signal-transduction chain (see Fig. 17-2) to activate genes needed for ABA synthesis. Furthermore, $Ca^{2+}$ and phosphoinositols seem also to be involved in the *action* of ABA when it rapidly causes stomatal closing, but in that case no gene activation is involved.

Research with several mutants that are unable to form much ABA (ABA-synthesis mutants) shows that several genes and enzymes are necessary to synthesize ABA (reviewed by Zeevaart and Creelman, 1988 and Skriver and Mundy, 1990; also see Walker-Simmons et al., 1989). The block to synthesis in a barley mutant is the inability to convert ABA aldehyde to ABA (see Fig.

18-17). In one tomato mutant (*notabilis*, or *not*), conversion of a carotenoid to a 15-carbon intermediate (perhaps xanthoxin) is blocked, whereas two others (*flacca*, or *flc* and *sitiens*, or *sit*) and the *droopy* mutant of potato cannot convert ABA aldehyde to ABA (Duckham et al., 1989). Three ABA-deficient maize mutants have blocks in carotenoid biosynthesis; they are albinos that lack protection against chlorophyll photooxidation. A *wilty* (*wil*) mutant of pea and an *Arabidopsis* mutant (*aba*) each have low levels of ABA, but affected steps in ABA synthesis have not yet been identified. The three mutants of tomato and droopy potato wilt when subjected to even mild water stress because insufficient ABA is present to cause stomatal closure. In fact, *droopy* stays partly wilted day and night; spraying leaves of these mutants with ABA prevents much of their wilting.

## ABA as a Possible Defense Against Salt and Cold Stresses

There is now considerable evidence to suggest that ABA levels increase not only when plants are stressed by an inadequate water supply, but also by saline soils, by chilling and freezing temperatures, and in some species even by high temperatures. In most (probably all) of these examples, the actual stress is deficiency of water in the protoplast. We mentioned above that water stress acts by turgor loss to activate genes that control ABA synthesis, and it is likely that other stresses also enhance ABA synthesis by effects on transcription. In many cases applied ABA can partly reduce the plant's reaction to the stress factor. For example, ABA "hardens" plants against frost damage (reviewed by Guy, 1990 and by Tanino et al., 1990) and against excess salt (reviewed by Skriver and Mundy, 1990). Figure 18-18 illustrates cold hardening by application of ABA to roots.

In one of the most careful studies of salt stress, Ray A. Bressan, Paul M. Hasegawa, and their colleagues at Purdue University examined the role of salt (NaCl) and ABA on cultured tobacco cells and whole plants (see, for example, Singh et al., 1987; Hasegawa et al., 1987; Iraki et al., 1989; Schnapp et al., 1990). Salt stress causes formation of several new proteins, especially a low-molecular-weight protein named *osmotin* that accumulates in abundance and is suspected of helping protect against the stress. Osmotin is also formed in several other species when salt-stressed. In tobacco, both salt stress and ABA induce formation of osmotin by effects on transcription. Salt is required to maintain osmotin synthesis, but in the presence of exogenous ABA and the absence of salt, high osmotin levels are transient. It would be interesting to learn just how ABA and proteins such as osmotin protect cells against salt stress, if indeed such proteins are truly protective.

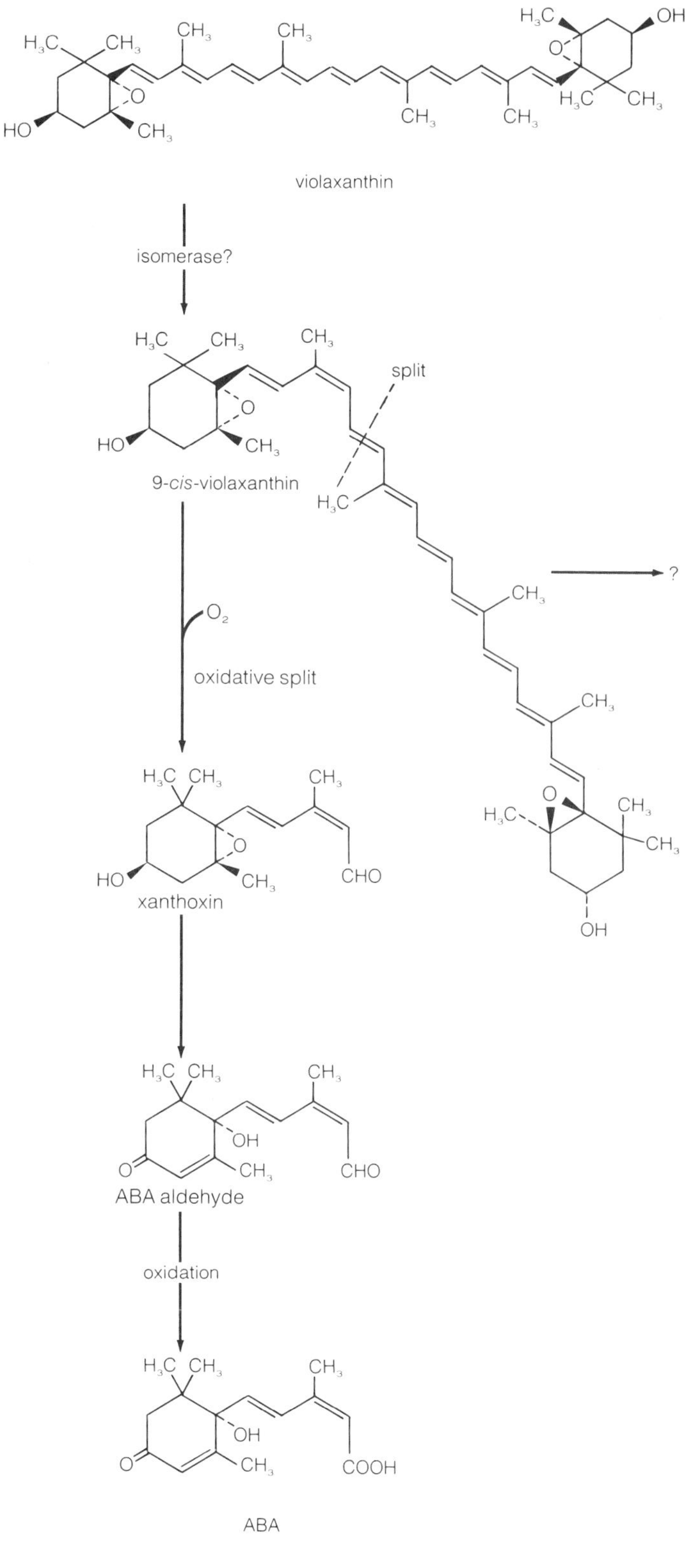

**Figure 18-17** Proposed reactions for synthesis of ABA from the carotenoid violaxanthin.

**Figure 18-18** Protection against freezing by ABA in Puma rye seedlings. Seedlings were grown 10 days at 25°C, were treated by a soil drench with $10^{-4}$ M ABA on two consecutive days, and were then exposed to cold temperatures and photographed 3 days later. Controls (extreme left) were killed at −3°C, whereas plants in soils previously treated with ABA survived at −9°C but not at −11°C. (Courtesy Grant C. Churchill and Larry V. Gusta.)

## Effects of ABA on Embryo Development in Seeds

Development of embryos after pollination has been studied extensively by removing immature embryos and growing them in tissue culture. Hormonal and genetic effects on embryo development were reviewed by Quatrano (1987), Skriver and Mundy (1990), Kermode (1990), and Bewley and Marcus (1990). Embryo development has conveniently been divided into three major stages: mitosis and cell differentiation; cell expansion and accumulation of food reserves (proteins, fats, starch, and so on); and maturation, during which the seed dries and passes into a resting or dormant state.

If embryos of many species are removed from the mother plant about halfway through development and are cultured *in vitro*, they are capable of germinating and developing into seedlings. An interesting question is what causes such embryos to fail to germinate in moist fruits on the mother plant (to exhibit **vivipary**) before they begin to desiccate and mature. ABA has been most investigated in this problem, mainly because it can inhibit germination of many mature seeds. Three approaches have been used to test ABA involvement: (1) measurement of the effects of exogenous ABA on development and growth of cultured embryos, (2) determination of endogenous levels of ABA at various times during development, and (3) study of ABA levels in viviparous seeds of maize and in ABA-synthesis mutants (maize, *Arabidopsis*) that have very low levels of ABA in all plant parts.

According to Quatrano (1987), endogenous ABA is strongly linked to initiation of the normal maturation pathway and to inhibition of precocious germination (vivipary). Furthermore, in many species exogenous ABA can cause or speed formation of special groups of seed-storage proteins in cultured embryos that either fail to synthesize these proteins or form them much more slowly. Evidence such as this indicates that normal increases in ABA levels during the early and middle stages of seed development control deposition of reserve proteins. Activation of transcription is the common cause of this effect of ABA. Whether ABA controls deposition of starch and fats in developing embryos is an interesting question that seems not to have been studied.

## Effects of ABA on Dormancy

The most common (but not universal) response of cells to ABA is growth inhibition. The early results of Wareing and his colleagues that led to the discovery of dormin (ABA) showed that levels of this compound increased considerably in leaves and buds when bud dormancy occurred in the relatively short days of late summer. They also found that direct applications of ABA to nondormant buds caused dormancy. These results suggested that ABA is a bud-dormancy hormone that is synthesized in leaves that detect day length (see Chapter 23) and translocated to buds to induce dormancy. However, much additional work with other woody plants argues strongly against such a hormonal role.

Perhaps the most convincing result is that direct application of ABA to buds slows or stops growth but does not cause development of bud scales and other characteristics of dormant buds. Other results with $^{14}C$-labeled ABA show that extremely little of it moves from leaves to buds when dormancy begins. Furthermore, short-day treatments that induce dormancy in various species cause no rise in ABA levels of buds of several species.

During the last two decades numerous studies concerning the possible importance of ABA in causing dormancy of seeds have been conducted (reviewed by Bewley and Black, 1982 and by Berrie, 1984). Exogenous ABA is a potent inhibitor of seed germination in many species. Furthermore, some studies show that ABA levels decrease in whole seeds when dormancy is broken by some environmental treatment (for example, exposure to light or cold temperatures); other studies with other species show no such decreases. One conclusion from these results might be that ABA causes seed dormancy in some species but not in others. This seems reasonable because (as described in Chapters 20 and 22) many other compounds are associated with seed dormancy, especially (in a dormancy-breaking role) gibberellins. However, it is doubtful whether analyses of whole seeds, including storage tissues, could provide the crucial information needed about changes in ABA levels in the radicle cells that grow and cause germination when dormancy is overcome (also see Section 20.4).

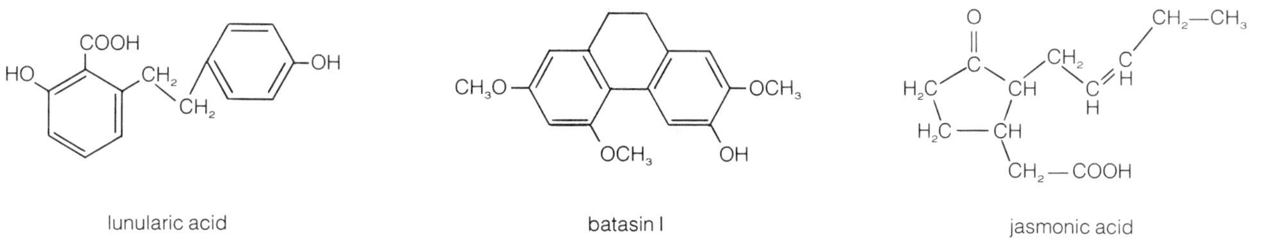

lunularic acid

batasin I

jasmonic acid

**Figure 18-19** Structures of lunularic acid, batasin I, and jasmonic acid.

### ABA and Abscission

The role of ABA in causing abscission of leaves, flowers, and fruits is controversial. Various reviewers evaluate published data in different ways. Addicott (1983) made a strong case for an important role of endogenous ABA in causing abscission, especially relative to the importance of ethylene. Milborrow (1984) concluded that exogenous ABA causes abscission but does so much less effectively than exogenous ethylene. More recently, Osborne (1989) reviewed effects of ethylene and ABA on abscission and concluded that ABA probably has no direct role in abscission; rather, it acts indirectly by causing premature senescence of cells in the organ that is shed, which in turn provokes a rise in production of ethylene. According to Osborne, ethylene, not ABA, is clearly the initiator of the actual abscission process.

### How ABA Acts

ABA seems to have three major effects, depending on the tissue involved: (1) effects on the plasma membrane of roots, (2) inhibition of protein synthesis, and (3) specific activation and deactivation of certain genes (transcription effects).

The effect on root membranes is to make them more positively charged, thereby increasing the tendency with which excised root tips stick to negatively charged glass surfaces. This effect is probably involved in the rapid loss of $K^+$ ions from guard cells (which involves inhibition of a plasma membrane ATPase) and perhaps in the ability of ABA to rapidly inhibit auxin-induced growth. Interference with synthesis of proteins and other enzymes could help explain long-term effects on growth and development, including the proposed role in seed dormancy and the inhibition of gibberellin-promoted hydrolase activity in cereal-grain seeds.

However, the ability of ABA to selectively control the transcription of certain genes, depending on cell type, represents a powerful control over developmental processes. The next few years of research should elucidate much about how ABA and other hormones and environmental factors control transcription. Progress in this field to date is summarized in Chapter 24.

## 18.6 Other Inhibitory Growth Regulators

ABA is a widespread growth regulator and is often an inhibitor, but many other compounds that usually inhibit growth have been discovered. The structures of these compounds have few similarities (Fig. 18-19). In liverworts, *lunularic acid* is present in **gemmae** (vegetative propagules about 1 mm in diameter formed in gemma cups on the upper surface of thalli of the mother plant). There is evidence, reviewed by Milborrow (1984), that lunularic acid prevents germination of these gemmae until they fall from the mother plant's thallus and the acid is leached out. Furthermore, growth of the entire thallus seems to be partly controlled by lunularic acid in response to day length. During short days the concentration of the inhibitor is low and thalli grow rapidly; during long days the reverse is true. An extensive survey of nonvascular plants by Gorham (1977) showed that lunularic acid is present in many species of lower plants but not in algae (nor in vascular plants, so far as we know).

**Batasins** are compounds in yam plants (*Discorea batatus*) that seem to cause dormancy in **bulbils** (vegetative reproductive structures) that arise from swelling of the aerial lateral buds. The structure of batasin I is shown in Figure 18-19. Batasins are concentrated in the skin of the bulbils and are absent from the core. A lengthy exposure to cold temperature (stratification or prechilling; see Section 22.1) that breaks dormancy causes batasins to disappear, whereas their amounts increase during the initial development of dormant bulbils (Hasegawa and Hashimoto, 1975). However, it is not known whether batasins are transported into or accumulate within the bud cells whose failure to grow actually causes dormancy.

**Jasmonic acid** (Fig. 18-19) and its methyl ester (*methyl jasmonate*) occur in several plant species and in the oil of jasmine (reviewed by Parthier, 1990). Jasmonates have been found in 150 families and 206 species (including fungi, mosses, and ferns), so they are perhaps ubiquitous in plants. They are formed biosynthetically from free linolenic acid (see Table 15-1), probably as a result of action of the enzyme *lipoxygenase*. These compounds have been shown to inhibit growth of certain plant parts and to strongly promote leaf senescence. Their functions remain to be proved, but a promoting role in senescence seems most likely.

## 18.7 Hormones in Senescence and Abscission

The processes of deterioration that accompany aging and that lead to death of an organ or organism are called **senescence**. Although meristems do not senesce and might potentially be immortal, all differentiated cells produced from meristems have restricted lives. Senescence therefore occurs in all nonmeristematic cells, but at different times. Many evergreen species retain their leaves for only two or three years before they die and abscise, but bristlecone pines (*Pinus aristata*) retain functional needles for up to 30 years. In deciduous trees and shrubs the leaves die each year, but again the stem and root systems remain alive for several years. In perennial grasses and herbs such as alfalfa, the aboveground system dies each year, but the crown and roots remain largely viable. In herbaceous annuals, leaf senescence progresses from old to young leaves, followed by death of stem and roots after flowering; only the seeds survive. For reviews of senescence, see books by Thimann (1980), Thomson et al. (1987), and Noodén and Leopold (1988) and general reviews by Goldthwaite (1987), Kelly and Davies (1988), Noodén and Guiamét (1989), Stoddart and Thomas (1982), and Sexton and Woolhouse (1984).

What causes senescence? One important thing to realize is that senescence is genetically programmed into each species and into organs and tissues of individual plants. Senescence of leaves is accompanied by early losses in chlorophyll, RNA, and proteins, including many enzymes. Because these and other cellular constituents are constantly being synthesized and degraded, loss could result from slower synthesis or faster breakdown or both. Slow synthesis is expected when nutrients normally arriving in an organ are diverted elsewhere, as, for example, when flowering and fruit formation occur. One theory for leaf senescence, therefore, is that flower and fruit development cause a competition for nutrients. In Section 16.4 we mentioned the competition between vegetative and reproductive organs for nutrients essential for growth, and we showed in Figure 16-22 how removal of all flowers postpones leaf senescence in soybean (an annual). As emphasized by Noodén and Guiamét (1989), a striking feature of monocarpic plants (those that flower once and then die; see Section 16.2) is the sharp shift in investment of nutrient resources (minerals and carbohydrates) away from vegetative parts and toward reproductive parts. Growth of roots and stems and production of new leaves decrease and often stop early in the reproductive phase, partly because meristematic activity slows or stops in vegetative organs.

Nutrient removal by flowers or fruits is not the entire explanation of senescence, even for soybeans, because the young flowers could not possibly divert enough nutrients to cause death of the leaves. In cocklebur plants, short-day and long-night conditions induce flowering and leaf senescence, but even if all flower buds are removed leaf senescence still occurs. Furthermore, development of male flowers on staminate spinach plants induces senescence as effectively as development of both flowers and fruits on female plants, even though staminate flowers divert far fewer nutrients than do fruits and seeds. Kelly and Davies (1988) concluded from an extensive literature review that diversion of assimilates to developing fruit is no longer accepted by most researchers as the strongest regulator of senescence. They proposed instead that development of the reproductive phase itself somehow causes diversion of nutrients to flowers and fruits, slowing of vegetative growth, and then the later stages of senescence. They emphasized that as reproductive structures become strong sinks, vegetative organs somehow also become weaker ones. In roots, loss of sink strength is accompanied by decreasing transport of mineral nutrients and cytokinins upward through the xylem. It is quite likely that a decreased cytokinin supply to leaves is partly responsible for the start of leaf senescence.

Contrary to the effects of cytokinins, ethylene and ABA promote senescence. In fruits the effect of ethylene is manifested by rapid ripening followed by abscission; in flowers the common result is wilting, fading of color, export of nutrients, withering, and then abscission; in leaves we observe loss of chlorophyll, RNA, and protein, transport of nutrients, and then abscission. The effect of ethylene in senescence and abscission seems to be much more dramatic than that of ABA. To what extent natural rises in ABA levels contribute to senescence and abscission is still uncertain, and how change from a vegetative to a reproductive state could signal gradual formation of ABA (if it in fact does) is also unknown.

What advantage is there to abscission of senescent leaves, flowers, and fruits? For fruits, the importance in perpetuation of the species is obvious because fruits contain seeds; for flowers we suspect that the reasons involve removal of a useless organ that might act as a potential portal for infection and that, in some species, would shade new leaves the next growth season. When senescence occurs, a degradation of proteins into mobile amino acids and amides commonly occurs as well. Many other large molecules (except for those in cell walls) are also degraded into smaller, more readily translocatable forms in which nutrients are conserved

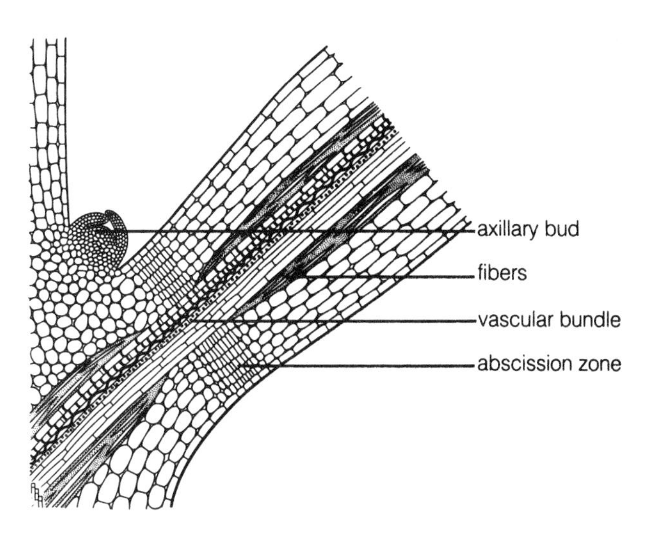


**Figure 18-20** The abscission layer. (From Addicott, 1965.)

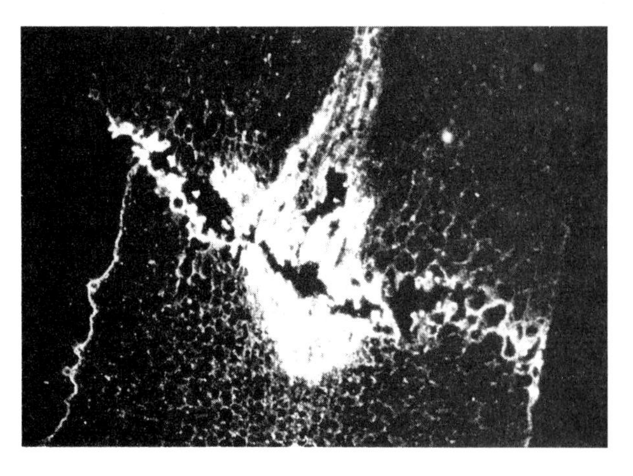

**Figure 18-21** Immunochemical localization of cellulase mRNA by antisense RNA in a bean-leaf abscission zone. For explanation of antisense RNA, see Section 24.4. (Courtesy Mark Tucker.)

by storage in other parts of the plant. This nutrient economy helps forest trees survive on unfertile soils. Leaves that abscise apparently could not withstand cold winters and would shade new leaves the next spring, so their loss, preceded by nutrient salvage, increases survival and productivity of individual perennial plants.

In most species, abscission of leaves, flowers, or fruits is preceded by formation of an **abscission zone** or **abscission layer** at the base of the organ involved (Kozlowski, 1973; Addicott, 1982; Sexton et al., 1985). In leaves this zone is formed across the petiole near its junction with the stem (Fig. 18-20). In many compound leaves, each leaflet also forms an abscission zone.

The abscission zone consists of one or more layers of thin-walled parenchyma cells resulting from anticlinal divisions across the petiole (except in the vascular bundle). In some species these cells are formed even before the leaf is mature. Just before abscission, the middle lamella between certain cells in the distal region (that region farthest from the stem) of the abscission zone is often digested. This digestion involves synthesis of polysaccharide-hydrolyzing enzymes, most importantly cellulase and pectinases, followed by their secretion from the cytoplasm into the wall. Formation of these enzymes is accompanied by a rapid rise in respiration in cells of the proximal region of the abscission zone. This rise is similar to that which occurs in climacteric fruits, and it also involves increases in polyribosomes characteristic of cells actively synthesizing proteins. Furthermore, one or more layers of these proximal cells increase in size (both length and diameter), whereas the cells of the abscission zone distal to the breaking point do not.

These wall-digestion processes, accompanied by pressures resulting from unequal growth in expanding proximal and senescent distal cells of the zone, cause a break. So long as high auxin concentrations are maintained in leaf blades, abscission is delayed. However, senescence leads to decreased auxin levels in those organs, and a buildup in ethylene concentration often begins. Ethylene, a powerful and widespread abscission-promoter of various plant organs in many plant species, acts by causing cell expansion and by inducing synthesis and secretion of the cell-wall degrading hydrolases. This action results from effects on transcription because mRNA molecules that code for these hydrolases (at least cellulase) increase in abundance after ethylene treatment (Fig. 18-21).

# 19

# The Power of Movement in Plants

*We believe that there is no structure in plants more wonderful, as far as its functions are concerned, than the tip of the radicle. If the tip be lightly pressed or burnt or cut, it transmits an influence to the upper adjoining part, causing it to bend away from the affected side; . . . If the tip perceives the air to be moister on one side than on the other, it likewise transmits an influence to the upper adjoining part, which bends toward the source of moisture. . . . [W]hen [the tip] is excited by gravitation the [adjoining] part bends towards the centre of gravity. In almost every case we can clearly perceive the final purpose or advantage of the several movements.*

*Charles and Francis Darwin,*
The Power of Movement in Plants, *1880*

The Darwins recognized over a century ago that plants, although rooted to the soil, exhibit a range of fascinating movements that often — but not always — can be interpreted as adaptations to their environments. Of course, study of these manifestations preceded the Darwins; plant movements have always elicited intense interest from people who enjoyed studying plants. Julius von Sachs (1832–1897), for example, in addition to his work on plant nutrition and other subjects, pioneered modern scientific studies of plant movements and was often cited by the Darwins. Among other things, he invented the clinostat (Section 19.5), which he used to study phototropism and gravitropism. Because there are many kinds of plants and many ways they move, the topic is a complex one. In this chapter we present an overview and an introduction to this captivating subject. (See Hart, 1990, for a general review.)

## 19.1 Some Basic Principles

Most of the movements we will discuss fall into two natural categories: the **tropisms** (Greek, *trope*, "turn"), in which the direction of the environmental stimulus determines the direction of movement, and the **nastic movements** (Greek, *nastos*, "pressed close"), which are triggered by an external stimulus (sometimes interacting with an internal timing mechanism) but in which the stimulus direction *does not* determine the direction of movement. Both nastic and tropistic movements are often the result of differential (irreversible) growth (Chapter 16), but they can also be caused by differential and *reversible* uptake of water into special cells, called **motor cells**, that collectively form a **pulvinus** (plural: **pulvini**). Stems growing away from gravity or toward a light source are examples of tropisms, and daily leaf movements or stomatal opening and closing (Chapter 4) exemplify nastic movements.

In a chapter on movements we could also discuss such diverse topics as migration of slime molds, gliding of blue-green algae, cytoplasmic streaming, orientation of chloroplasts and nuclear migration, and action of cilia and flagella. Some of these movements are examples of **taxis**, in which an organism (usually single or few-celled; rarely an organ such as the pollen tube) moves toward or away from some stimulus such as light, warmth or cold, or a chemical gradient. These phenomena are worthy of extensive discussion, but most occur in the lower plants, which we do not emphasize in this text. Hence, we will not discuss them here. (See Haupt and Feinleib, 1979.)

Because plants we do discuss don't exist in a dark vacuum, they are always being influenced by the energy and mass parameters of their environments. Often, changes in one or more of these parameters induce either tropistic or nastic responses, or both. An environmental change that induces a plant movement (or other response) is a **stimulus**. Often a stimulus **induces** in the plant a process that continues after the stimulus no longer exists in its initial form. This is the case with daily leaf movements that change their course in response to the rising or setting of the sun; similarly, a stem will continue to bend toward the light or away from gravity for some time after the light or gravitational stimulus has been removed.

Stimuli always act on some machinery that is part of the plant, and the part that receives (**perceives**, we say) the stimulus is the **receptor**. Once the stimulus has been perceived, it is changed (**transduced**) into some other form, often called a **signal**, that then leads to some **motor response** (growth or pulvinar action) that is the actual cause of the plant movement. Wolfgang Haupt in Germany and Mary Ella Feinleib in Massachusetts edited a volume (1979) about plant movements. In their introduction, they discussed the three steps in plant movements that we have just introduced and that form a framework for research on the topic:

1. **Perception** How does a plant or plant part detect the environmental stimulus that causes the response? For example, what pigment absorbs the light that causes phototropism, or what in the cells or tissues responds to gravity? Where in the plant is the perception mechanism located? And what is the mechanism? These questions have been difficult to answer for plants because such plant organs as leaves, stems, and roots, unlike eyes and ears, are not specialized to respond only to one stimulus.
2. **Transduction** How does the perception mechanism transduce its perceived stimulus to cells in the organ where movement occurs? What **signal** does it send — that is, what biochemical or biophysical changes occur in response to the environmental stimulus? This has been an especially active field of research for much of the 20th century, especially with regard to the tropisms. Researchers have looked for electrical signals (for example, action potentials; Section 19.2) and for chemical messengers (hormones).
3. **Response** What actually happens during movement? Any hypothesis put forth to explain the mechanisms of perception and transduction must account for the observed motor responses. This seems obvious enough, yet the details of each response were rather neglected for several decades. Early workers in the late 1800s and early 1900s carefully studied growth of cells on opposite sides of a bending organ, for example, and later researchers must have assumed that everything possible had already been learned. Yet many early results were either overlooked or forgotten, so we have returned to such studies, especially at the cellular level.

Two generalizations have come from work on nastic movements and tropisms conducted during the past two or three decades. Because we must necessarily limit the scope of our discussions in the following pages, these generalizations will not always be apparent. They are: *First, similar mechanisms within a plant often cause different responses*. For example, changes in cell volume caused by $K^+$ and then water movement in and out of cells accounts for such diverse responses as stomatal opening and closing and thigmonastic leaf folding in the sensitive plant; $K^+$ could also be important in some tropisms. *Second, different mechanisms can produce similar responses in different or even the same organisms*. For example, different pigment systems are apparently responsible for phototropic bending in different organisms, although the exceptional ones, which respond to *red* (instead of blue) light, are rare and won't be discussed.

## 19.2 Nastic Movements

Leaves or leaflets of compound leaves often exhibit nastic movements. An upward bending of an organ is called **hyponasty**; downward bending is **epinasty**. Often, these leaf movements are caused by pulvini at the base of the petiole, blade, or leaflet, but they also occur in many plants with no pulvini. Epinasty occurs, for example, when cells on the top of the petiole or blade, especially in the main veins, grow (elongate irreversibly) more than those on the bottom. Generally speaking, nastic movements are reversible, whether they are controlled by pulvini or by changing relative growth rates on top and bottom of an organ. If epinastic leaf bending is caused by more rapid growth of cells on top of the petiole and blade than on the bottom, for example, bending changes to hyponastic when cells on the lower side begin to grow faster.

### Nyctinasty

In many species leaf movements, from nearly horizontal during the day to nearly vertical at night, have been recognized for more than 2,000 years. As discussed in Chapter 21, typical **nyctinastic** (Greek, *nux*, "night") movements are rhythmic processes controlled by interactions between the environment and the biological clock. Here we will emphasize the motor responses involved in these movements.

It is sometimes convenient to study species with doubly compound leaves, in which each leaf bears several pinnae and each pinna bears several pairs of opposite pinnules (leaflets) attached to a single rachilla. Examples include the silk tree (*Albizzia julibrissin*; Satter and Galston, 1981), the sensitive plant (*Mimosa pudica*), and the raintree (*Samanea saman*). Such doubly compound leaves often exhibit striking **sleep movements**, as illustrated in Figure 19-1. At night, the tips of opposite leaflets of *Albizzia* fold together (*close*), rise upward, and point toward the distal end of the rachilla. *Samanea* leaflets fold down instead of up (Fig. 19-1). In either case, cells in the pulvinus that swell during opening are called **extensors**, whereas those that shrink are called **flexors**. The pulvinus of *Samanea* is a straight cylinder during the day when the leaflets are open and a curved

a

b

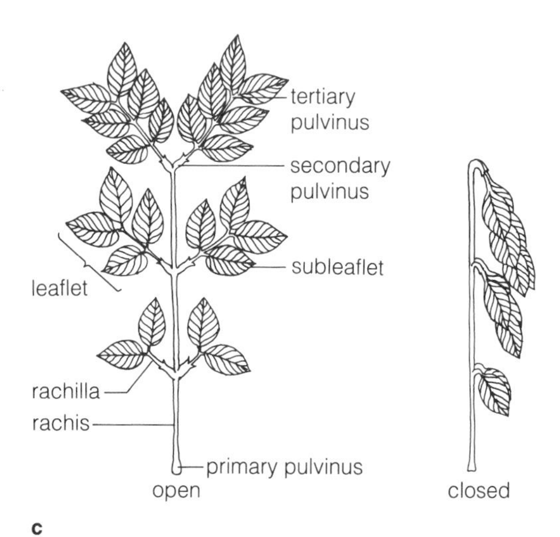


c

**Figure 19-1** *Albizzia julibrissin* leaves in the normal daytime position (**a**) and in typical night (sleep) position (**b**). (Photos courtesy of Beatrice M. Sweeney.) (**c**) *Samanea saman* leaf in the daytime and the nighttime positions. (Drawings courtesy of Ruth Satter.)

cylinder or hook at night when leaflets are closed (Fig. 19-1).

What causes water to flow from one side of a pulvinus to another? In 1955, Hideo Toriyama (see also Toriyama, 1962) observed in pulvini of the sensitive plant (*Mimosa pudica*) that $K^+$ moves out of the cells that lose water. Subsequently, Ruth L. Satter and colleagues (Satter and Galston, 1981; Satter et al., 1970, 1988) found that the $K^+$ concentration of *Albizzia* pulvini is unusually high (almost 0.5 M) and that leaflet closing is accompanied by loss of $K^+$ from ventral cells and absorption of $K^+$ by dorsal cells (although dorsal cells do not receive the $K^+$ that the ventral cells lose, which must be stored in other cells or in the pulvinus apoplast). Furthermore, changes in $Cl^-$ parallel those of $K^+$ (Schrempf et al., 1976). The same changes have been observed in *Samanea* (review in Satter et al., 1988). Hence, in *Mimosa, Albizzia*, and *Samanea*, water movement occurs in response to the osmotic driving force of ion transport, just as in stomate opening and closing (Section 4.6).

Satter's group's recent work (see Satter et al., 1988; Moran et al., 1988) has concentrated on the mechanisms of ion transport into and out of extensor and flexor cells. Several lines of evidence suggest the presence in membranes of depolarization-activated, $K^+$-selective channels that provide outward pathways for $K^+$ diffusing from shrinking cells. $Cl^-$ appears to be cotransported into motor cells via $H^+$/anion symport (see Chapter 7). $H^+$ transport appears to be the primary mechanism. In some manner, changes from light to dark or dark to light, acting via phytochrome and a blue-absorbing pigment (see Chapter 20), activate these mechanisms, possibly via effects on the phosphatidylinositol cycle described in Section 17.1 (see Satter et al., 1988).

It is interesting that light acts in opposite ways in the extensor and flexor cells. For example, the transition from white light to darkness *activates* the $H^+$ pump in flexor cells and *inactivates* the $H^+$ pump in extensor cells. Because leaves open and close on an approximately 24-h schedule under constant conditions, the biological clock must also activate these mechanisms. Furthermore, light acts not only directly on the ion-absorption mechanisms but also indirectly by resetting the clock, as we discuss in Chapter 21.

In related work, S. Watanabe and T. Sibaoka (1983; see also Sibaoka, 1969) examined responses of leaflet pairs from detached pinnae of *Mimosa*. During the daytime (from 6:00 to 16:00), closed leaflet pairs opened in response to blue or far-red light, but not to orange or red light (see the high-irradiance reaction in Chapter 20). At night they did not respond to the light but instead opened in response to applied auxins (IAA, α-naphthaleneacetic acid, or 2,4-D).

Olle Björkman and Stephen B. Powles (1981) described an interesting response of the redwood sorrel (*Oxalis oregana*). This shade plant (Section 12.3) photosynthesizes in redwood forests at light levels only

1/200th of full sunlight. Sunflecks penetrate the forest canopy and could damage this delicate species. With only a 10-second lag period after sunlight strikes the leaves, they begin to fold downward, the folding being complete in about six minutes. When shade returns, there is a 10-minute lag period, but the leaves revert to their horizontal position in about half an hour. Blue wavelengths of light are sensed by a small pulvinus where each leaflet joins the petiole.

Though somewhat less spectacular, the folding of leaflets when light levels are too high is common in many legumes. Black locust (*Robinia pseudoacacia*), in the Fabaceae, folds its leaves upward to a vertical position in intense sunshine. At night, the leaves move downward so that their lower sides face each other. Cold air also causes a downward movement.

## Hydronasty (or Hygronasty)

As in nyctonasty, **hydronasty** involves a folding or rolling up of leaves, but the latter occurs in response to water stress instead of light. These processes reduce exposure of the leaf surface to dry air, supplementing stomatal closure to reduce transpiration. The danger of photoinhibition is also minimized. Folding and rolling movements are caused by loss of turgor in thin-walled motor cells called **bulliform cells**, as illustrated for Kentucky blue grass (*Poa pratensis*) in Figure 19-2. The bulliform cells have little or no cuticle, so they lose water by transpiration faster than other epidermal cells. As their turgor pressure decreases, constant turgor in cells on the lower side of the leaf causes the folding. This is only one of several mechanisms by which plants resist drought (Chapter 26).

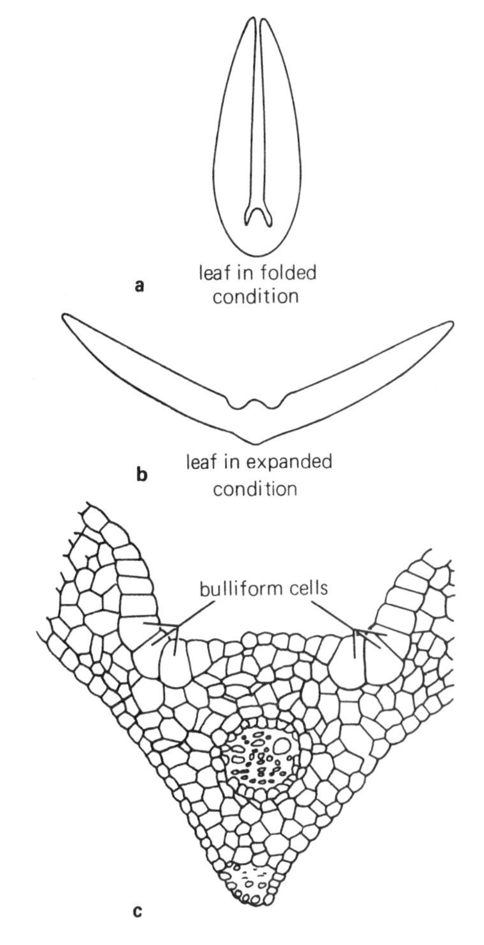


**Figure 19-2** Outline drawings of blue grass (*Poa pratensis*) leaf (**a**) in folded condition and (**b**) in expanded condition. (**c**) A detailed drawing of the midportion of the leaf showing the bulliform cells. Changes in the turgor of these cells control the folding and opening of the leaf. (From Meyer and Anderson, 1952.)

## Thigmonasty

Nastic movements resulting from touch — **thigmonasty** (Greek, *thigma*, "touch") — are widespread. They are especially striking in certain members of the subfamily Mimosoideae in the family Fabaceae (Leguminosae) (Ball, 1969). The most notable example is *Mimosa*, the sensitive plant. Upon being touched, shaken, heated, rapidly cooled, or treated with an electrical stimulus, the leaflets and leaves rapidly fold together (Fig. 19-3). When only one leaflet is stimulated, a stimulus moves throughout the plant, collapsing the other leaflets. The advantage to the plant of this response is uncertain, but one idea is that the folding leaflets startle away insects before they begin to eat the foliage. The folding is caused by water transport out of motor cells in pulvini, an occurrence associated with an efflux of $K^+$, as noted.

Transmission of signals in *Mimosa* has been investigated for many years (Roblin, 1982; Samejima and Sibaoka, 1980; Simons, 1981; Umrath and Kastberger, 1983). There is evidence for two distinct mechanisms, one *electrical* and the other *chemical*. The electrical response was first studied extensively by Jagadis Chunder Bose, in India, between 1907 and 1914 (see Sinha and Bose, 1988) and then in better experiments by A. L. Houwink (1935) in Holland. An **action potential** (Pickard, 1973) is a change in voltage that forms a characteristic peak when plotted as a function of time (Fig. 19-4). Action potentials in *Mimosa* are similar to those occurring in animal nerve cells but are much slower. In both plant and animal cells, they are caused by fluxes of specific ions across cell membranes. In *Mimosa*, they travel through parenchyma cells (connected by plasmodesmata) of the xylem and phloem at velocities up to about 2 cm $s^{-1}$, whereas action potentials travel along animal nerve cells at velocities of tens of meters per second.

The action potential will not pass through a pulvinus from one leaflet to another unless the chemical response is also elicited, in which case several leaflets may fold up. The chemical response, first reported by the Italian scientist Ubaldo Ricca (1916a, 1916b), is caused

**Figure 19-3** Response of the sensitive plant. The tip of one leaf is stimulated (in this case, by using a flame) without stimulating the rest of the plant. After 14 s, the petiole of the leaf has collapsed, and many of the leaflets have folded up as water leaves pulvinal cells at the bases of petioles and leaflets. As the stimulus travels back along the stem, other leaves collapse and their leaflets fold. This process continues in the last photograph (taken after about 1.5 min). (Photographs by F. B. Salisbury.)

by a substance that moves through the xylem vessels along with the transpiration stream. Ricca cut through a stem and then connected each cut end with a narrow, water-filled tube. When a leaf on one side of the tube was wounded, a leaf on the other side folded. The active substance, formerly called **Ricca's factor** but now identified as a turgorin (see below), can be extracted from wounded cells and applied to a cut branch, and its folding effects can then be measured. Its movement elicits electrical responses that travel ahead of it from one leaflet to another in parenchyma cells. Rapidly transported wound signals have also been observed in pea tissues (Davies and Schuster, 1981; Van Sambeek and Pickard, 1976).

## Turgorins: Hormones that Control Nastic Movements

Hermann Schildknecht (1983, 1984), an organic chemist at the University of Heidelberg in Germany, and his group have done extensive work isolating and identifying compounds (like Ricca's factor) that activate pulvini in the leaves of such plants as *Mimosa* and *Acacia karroo*, which is not sensitive to touch but exhibits nyctinasty. For a bioassay, a *Mimosa* leaf is placed in a solution with the suspected active substance, which is then transported in the transpiration stream to the pulvini, where their membranes respond, causing the leaf pinnules to fold up if the substance is active. Two so-called **periodic**

**Table 19-1 Molecular Structure of Several Turgorins along with the Minimum Concentration Required to Produce an Effect.**

Glucose — Gallic Acid (structure with substituents $R^1$, $R^2$, $R^3$, $R^4$ and COOH)

| Turgorin | Structure | | | | Minimum Conc. |
|---|---|---|---|---|---|
| | $R^1$ | $R^2$ | $R^3$ | $R^4$ | [mol/L] |
| PLMF 1 | $CH_2OSO_3H$ | OH | OH | OH | $2.33 \times 10^{-7}$ |
| PLMF 2 | $CH_2OSO_3H$ | $OSO_3H$ | OH | OH | $1.96 \times 10^{-7}$ |
| S-PLMF 2 | $CH_2OSO_3H$ | OH | OH | H | $2.42 \times 10^{-6}$ |
| M-LMF 5 | COOH | OH | OH | OH | $2.75 \times 10^{-6}$ |
| PLMF-synth | $CH_2OSO_3H$ | OH | H | H | $2.51 \times 10^{-5}$ |
| LMF-synth | COOH | OH | H | H | $1.57 \times 10^{-3}$ |

Source: Schildknecht (1986).

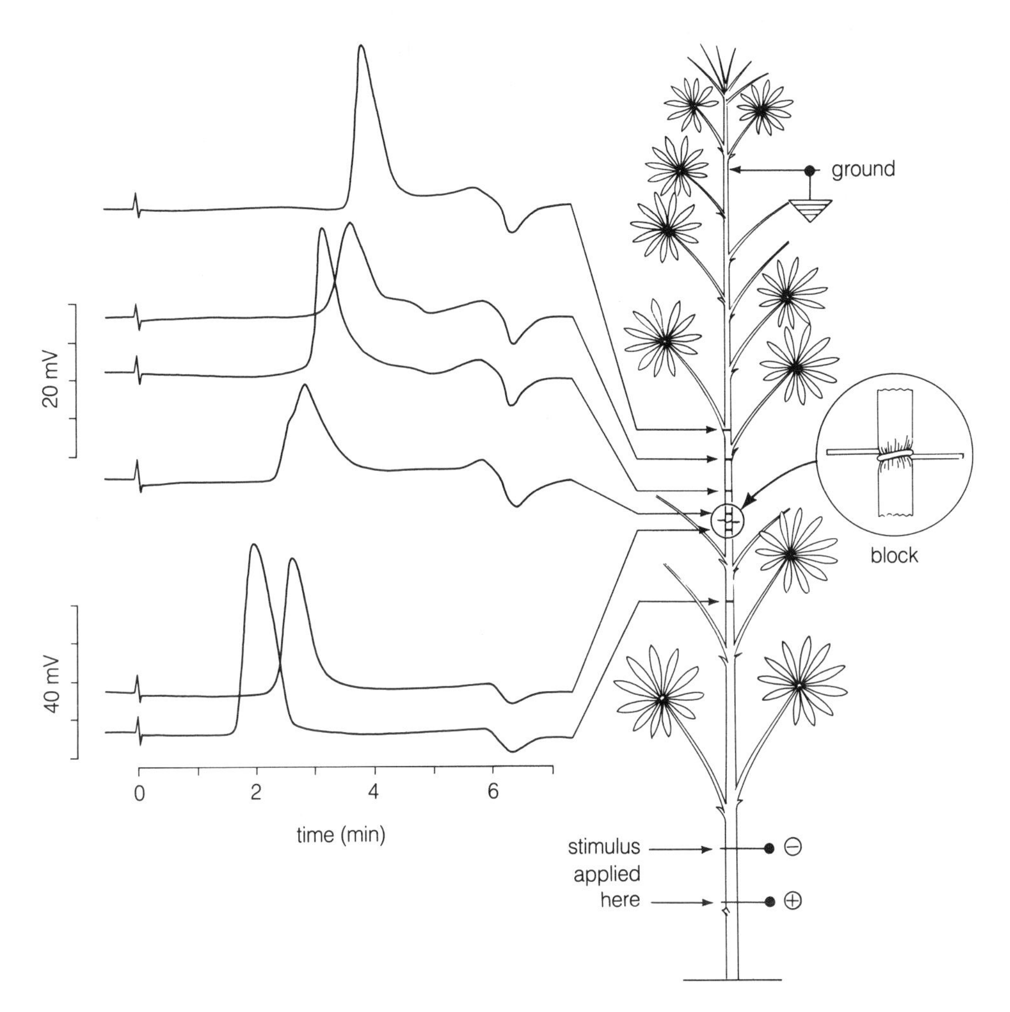


**Figure 19-4** An example of action potentials measured in *Lupinus angustifolius*. The purpose of this study was to examine movement of action potentials past a block in the stem produced by tightening a thread around the stem (detail drawing in circle). The drawing of the stem shows the location of the anode (+) and the cathode (−), the block, and the ground, which is necessary to measure the action potentials. Arrows indicate the locations of electrodes along the stem. The action-potential curves show voltages as a function of time, and the small blip to the left on each curve is an artifact caused by application of the stimulus (the direct electric current between anode and cathode). (From Zawadski and Trebacz, 1982, with permission.)

**leaf movement factors (PLMFs)** from *Acacia* proved to be the β-glucosides of gallic acid, with the glucoside bonding at its para-hydroxyl group. Several other compounds with closely related structures were also identified from the extracts of other plants (Table 19-1), and the most active were the β-D-glucoside-6-sulfate and the β-D-glucoside-3,6-disulfate of gallic acid; they were called PLMF 1 and PLMF 2. Extracts from *Mimosa* contained PLMF 1 plus another compound, PLMF 7. *Robinia* extracts contained PLMF 1.

Some *Oxalis* species react to touch or shaking, much as *Mimosa* does. Schildknecht and coworkers (see Schildknecht, 1983, 1984) isolated from *Oxalis stricta* both PLMF 1 and another compound, PLMF 3, which is a derivative of protocatechuic acid (see Figure 15-10) instead of gallic acid. Glucose-6-sulfate (Table 19-1) was again part of the molecule.

Schildknecht suggested that these compounds form a new class of plant hormones, which he called **turgorins** because they act on the turgor of pulvinus cells. As is the case with other phytohormones, they are active at low concentrations ($10^{-5}$ to $10^{-7}$ M), and at least in some cases they meet the criterion of translocation. It will be interesting to see if sensitivity to the compounds also plays a role in their action. Recently, Peter Kallas, Wolfram Meier-Augenstein, and Schildknecht (1990) have demonstrated the presence of a PLMF-1-specific receptor (presumably a protein) in the outer side of the plasma membrane of *Mimosa*.

## The Venus's-Flytrap

One of the few well-studied examples in which an action potential is obviously useful to a plant is the excitation by an insect of one or more sensory epidermal hairs[1] of the Venus's-flytrap, *Dionaea muscipula*. Action potentials move from the hairs into the bilobed leaf tissues and cause the lobes to snap shut within a half second or so (Fig. 19-5). This usually traps the insect, which is then digested by enzymes secreted from the leaf, providing a nitrogen and phosphate source for the plant. About 500 other flowering plants are **carnivorous**

[1]Here is an example in which a plant has a special organ that reacts to an environmental stimulus.

a

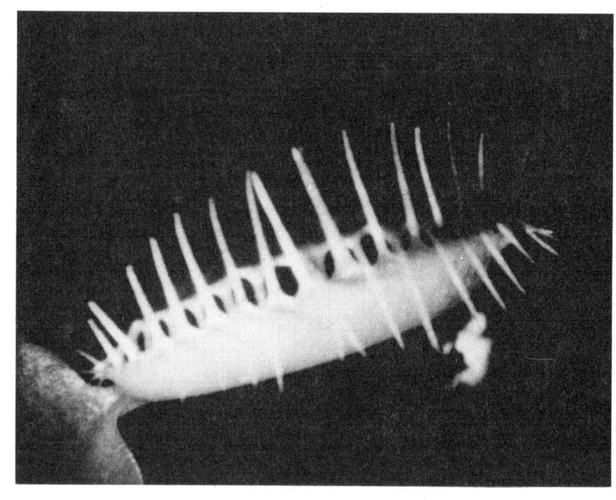

b

**Figure 19-5** Venus's-flytrap (*Dionaea muscipula*). (**a**) Open trap, "set" to capture an insect. (**b**) Closed trap. (Photographs supplied by Steven Williams.)

(flesh eaters), and the capture mechanisms are diverse and often independent of action potentials (Heslop-Harrison, 1978). In the case of the Venus's-flytrap, Stephen E. Williams and Alan B. Bennett (1982) showed that the rapid closing is another example of acid growth (Section 17.2). Hydrogen ions are rapidly pumped into the walls of cells on the outside (lower side) of each leaf trap in response to the action potentials from the trigger hairs on the inside (upper side). The protons apparently loosen the cell walls so rapidly that the tissue actually becomes flaccid, allowing cells to quickly absorb apoplastic water and causing the outside of each leaf to expand and thus the trap to snap shut. The trap gradually opens as growth of the inside surfaces overcomes the rapid growth of the outside surface during closing.

The tests of Williams and Bennett's hypothesis are a model of the scientific method in action. First, they marked the inside and the outside surfaces of the leaf (trap) with dots, measuring distances between them before and after closing. The outside expanded about 28 percent, but the inside did not change, confirming that growth of the outside closes the trap. The inside grew slowly during the 10 hours of reopening, but there was no further change on the outside, so this is a true, irreversible, rapid-growth response rather than action of a pulvinus. Second, their hypothesis suggested that infiltrating the leaves with neutral buffers should neutralize the protons pumped out of the cells and thereby prevent closing. This occurred, and infiltrating the leaves with a sorbitol solution with the same osmotic potential as the neutral buffer did not prevent closing. Third, they predicted that infiltrating leaves on the outside with acidic buffers (*p*H 3.0 to 4.5) should cause closing without stimulation of the trigger hairs. This also occurred. Fourth, because hydrogen pumping requires much ATP, they predicted that ATP levels would drop rapidly in the tissues during closing. They found a drop of about 29 percent during closing. All their tests were positive, strongly suggesting acid growth in Venus's-flytrap closing.

Nevertheless, a more recent paper of Dieter Hodick and Andreas Sievers (1989) in Germany suggests that the acid-growth hypothesis is doubtful. In their experiments, buffering the apoplast at *p*H 6 did not prevent movement of traps that had been cut several times from the margin to the midrib to facilitate buffer diffusion into the mesophyll. They obtained other data as well, which we will not discuss, that failed to support the acid-growth mechanism.

## Thigmo- and Seismomorphogenesis

In the early 1970s, Mordecai J. Jaffe (1973, 1980; Giridhar and Jaffe, 1988) began to investigate developmental effects of mechanical stimulation, especially rubbing, on plants. Most vascular plants studied responded with slower stem elongation and increased stem diameter, which produced short, stocky plants (Fig. 19-6), sometimes only 40 to 60 percent as tall as controls. Jaffe called these and similar developmental responses to mechanical stress **thigmomorphogenesis**. Clearly, they are related to thigmonasty, but rapid plant movements are not involved. Flexing stems also causes these responses, and in nature the bending effects of wind influence plant development in this way. Cary A. Mitchell at Purdue University (1977; Mitchell et al., 1975; Hammer et al., 1974) has also studied these effects extensively. He has placed plants on a shaker table (typically for only seconds to minutes, once to several times each day) such that they are mechanically shaken but not physi-

**Figure 19-6** The effect of the number of daily rubs of the stem (once up and once down, between thumb and forefinger, with moderate pressure) on the growth of young bean plants. From left to right, the number of stimuli were 0, 2, 5, 10, 20, or 30. (Courtesy of Mordecai J. Jaffe; see Jaffe, 1976.)

cally contacted. The same basic responses appear, and Mitchell terms the phenomenon **seismomorphogenesis** (Greek, *seismos*, from *seiein*, "to shake"). In either case, shorter, stronger plants are produced, and these are less easily damaged by natural mechanical stresses (especially wind) than are their taller, more slender, greenhouse-grown counterparts. The taller plants are really "unnatural" because they occur only in greenhouses—and then only when conditions are just right (typically, relatively low light levels). Plants grown outside, hardened by wind and bright sunlight, show little further response to mechanical stress.

Rubbing by farm machinery and by animals almost certainly contributes to the hardening. Spraying tomato plants with water for 10 seconds once each day reduced their growth in a greenhouse to about 60 percent of the growth of controls (Wheeler and Salisbury, 1979). The stockier plants were desirable in greenhouse culture, but unfortunately fruit yields were significantly reduced.

Salisbury (1963) found that simply measuring the lengths of cocklebur (*Xanthium strumarium*) leaves with a ruler at daily intervals slowed their growth and caused premature senescence. Neel and Harris (1971) found that young sweetgum (*Liquidambar styraciflua*) trees elongated more slowly and set winter terminal buds (a *developmental* response!) when their trunks were vibrated or shaken for only 30 seconds each day. Inhibitory effects of mechanical stress on *flowering* of a few species have also been observed. Such responses may prove to be as common and thus as important to plants as responses to light, temperature, or gravity. Certainly, they will confuse the outcome of any experiment if controls and treated plants do not receive comparable mechanical stress.

What causes these responses? Because the plants at first show no symptoms of injury but only altered growth, a change in growth-regulator patterns is suspected. The decreased stem elongation and increased stem thickening suggest that ethylene production plays a role, and indeed increased ethylene has been observed following mechanical stimulation (see Section 18.2). Auxin transport is also apparently inhibited (Mitchell, 1977), extractable GA-like activity disappears (Beyl and Mitchell, 1983), and calcium may be involved.

How a change in growth-regulator balance might come about is unknown, but Jaffe (1980) found a lower electrical resistance of bean stems within a few seconds after rubbing, followed by a slower rise back toward the normal level. This finding probably suggests a change in membrane permeability that allows ions such as $K^+$ to pass rapidly from the symplast into the apoplast following mechanical stimulation. (The increased ionic concentration would account for the greater flow of current and thus lowered resistance.) A change in membrane permeability might affect amounts of hormone at the subcellular sites at which they act and might also affect subsequent production of growth regulators by altering the availability of precursor molecules.

Janet Braam and Ronald W. Davis (1990) found that mRNA in *Arabidopsis* plants increased 10 to 100 percent 30 minutes after the plants were mechanically stressed by spraying with water or by other means. Typically, such stressed plants became less than 50 percent as tall as unstressed controls. The mRNA was produced by at least four and probably five genes that were activated by the mechanical stress. One of those genes coded for calmodulin and two others coded for two closely related proteins. Calcium levels increased in stressed cells, so it is possible that calcium-calmodulin is a secondary messenger in the gene activation; indeed, if this is true, the calmodulin may activate the gene for its own production. These studies could be an important breakthrough toward understanding molecular mechanisms of plant responses to mechanical stress.

## 19.3 Tropisms: Directional Differential Growth

You are probably familiar with plant response to the *direction* of an environmental stimulus in which unequal (differential) growth (usually cell elongation) occurs on different sides of an organ. Roots grow downward and stems upward in response to gravity (**gravitropism**[2]). Stems and leaves frequently orient themselves with respect to light rays (**phototropism**), but roots seldom exhibit phototropism. **Thigmotropism** is a response to contact with a solid object that is exhibited by climbing plants that grow around a pole or the stem of another plant (Jaffe and Galston, 1968). There are other tropisms as well.

There has been considerable interest in the tropisms for well over two centuries, and they have been studied with modern scientific techniques since at least the time of von Sachs over a century ago. It is humbling to realize that in spite of thousands of technical papers reporting studies on tropisms we remain baffled by the basic mechanisms. Although we have a vast amount of data, most of it is descriptive or phenomenological.

Yet progress has been made in learning about tropisms during recent decades, and the space agencies of the United States, Europe, Japan, China, and the Soviet Union are currently supporting research on gravitropism. This research includes both ground-based studies and experiments in the microgravity of orbiting spacecraft. Most studies are presently being conducted in the context of the three postulated steps that were discussed in Section 19.1: perception, transduction, and response.

## 19.4 Phototropism

### Coleoptiles and Stems

The Darwins (1880) described many experiments on plant tropisms. They observed that canary grass (*Phalaris canariensis*) coleoptiles would not bend toward dim light if the tips were cut off or shaded. They also studied dicots and some other monocots. It was their experiments (plus those of others) localizing phototropic sensitivity in the tips of coleoptiles that eventually led to the experiments of Frits Went and thus to the discovery of the first known auxin (Chapter 18).[3]

The Darwins found that oat (*Avena*) coleoptiles also bent toward light when their tips were covered; that is, some phototropic sensitivity occurs below the tips, but the **tip response** is about a thousand times more sensitive than the **base response**. Thus, if dim light is used, most of the response is localized in the tip, and this is evident because curvature toward the light begins at the tip and moves gradually down the coleoptile as the stimulus is transmitted from the tip to the tissues below. If higher light levels are used, however, bending begins simultaneously along the entire length of the coleoptile. **Hypocotyls** (seedling stem below the cotyledons) of such dicot seedlings as sunflower show no apical-bud response (see Dennison, 1979, 1984).

*Perception: Dose-response relations* Several important quantitative questions can be asked about the phototropic response. Of primary importance is the question of **reciprocity**: Is the response proportional to the duration of the exposure, to its **energy** or **photon flux** (energy or photons per unit area per unit time = **irradiance** or **fluence rate**; often loosely called *intensity*), or to the total dose; that is, the product of exposure dura-

[2]Formerly **geotropism**, which means "tropistic response to the earth." Because the response is actually to a gravitational (or other accelerational) force, *gravitropism* is a preferable term, just as several decades ago *phototropism* (the general response to the direction of light) replaced *heliotropism* (response to the sun—a phenomenon of considerable importance, as we shall see).

[3]The discovery of auxins also depended upon experiments with roots done by the Polish plant physiologist Ciesielski (1872), who was cited by the Darwins.

tion times flux (called **fluence**; see Appendix B)? If the response is to total dose, then flux and duration bear a *reciprocal* relationship to each other,[4] so the **law of reciprocity** is said to hold.

Figure 19-7 shows results of experiments by Zimmermann and Briggs (1963), who irradiated oat (*Avena*) coleoptiles with three different photon fluxes of blue light for different exposure times.[5] Closely similar results of work by Brigitte Steyer (1967), who examined 12 dicot species as well as four monocots, are also shown in the figure. Her results were further extended by Baskin and Iino (1987) with four more dicots. Results are plotted as degrees of phototropic curvature as a function of fluence; they are **dose-response curves** (or **fluence-response curves**). In the figure from Zimmermann and Briggs, the first, ascending part of the curve is the same for all three photon fluxes (irradiances), so reciprocity holds over this low-fluence range. This part of the curve is called **first-positive curvature**. At higher irradiances (part of the curve labeled C), meaning shorter exposure times, first-positive curvature is followed by decreasing curvature with increasing fluence, producing a bell-shaped curve. In a few monocots such as oat (*Avena*) coleoptiles (none of the dicots tested by Steyer or by Baskin and Iino), this trend continues until the organs bend *away* from the light, a response that is called **first-negative curvature**. With further increase in fluence, a **second-positive curvature** and even a **third-positive curvature** become evident (Du Buy and Nuernbergk, 1934). At the intermediate irradiances used by Zimmermann and Briggs (labeled B), the decending part of the curve is greatly reduced, but second-positive curvature is still apparent. At the lowest light levels (longest exposure times, labeled A), only a shoulder in the curve suggests second-positive curvature. Clearly, reciprocity holds only for the first-positive part of the dose-response curves. Why is that?

Blaauw and Blaauw-Jansen (1970a and 1970b) obtained dose-response curves for at least 24 different irradiance levels. Their results confirm those of Zimmermann and Briggs. The second-positive curvature occurs at the same *time* at all irradiance levels (about 40 minutes after the beginning of irradiance). Thus, as the irradiance levels decrease, first-positive curvature is delayed (reciprocity holds), but second-positive curvature comes at the same time regardless of irradiance level, so eventually the two come together, eliminating first-negative curvature.

Thought and a few experiments suggest that light really has two effects in phototropism. First, it acts as a trigger for the bending response, which is what we have been emphasizing. Second, it decreases the sensitivity of the organ to subsequent light. This effect is nondirectional and is referred to as a **tonic effect**. For example, if a coleoptile is exposed to one second of 0.03 W $m^{-2}$ of unilateral blue light, positive curvature occurs—but only if the coleoptile has previously been in the dark. Much less positive curvature occurs if the coleoptile has already been exposed to 10 seconds of the same unilateral irradiance. In this case, the negative area of the curve is approached. Actually, the tonic effect is easy to observe even when the light comes from above the coleoptile instead of from one side (Meyer, 1969). When two exposures are given, the tonic effect caused by the first exposure gradually decays away, so after about 20 to 25 minutes the exposure to the second irradiance is the same as if the first exposure had not been given. Recovery has occurred. Hence, Blaauw and Blaauw-Jansen concluded that the curvatures beyond the first-positive response are not really independent phenomena but are the result of the desensitization of the first-positive curvature system.

With *Avena* coleoptiles, *red* light given just prior to blue light shifts the region of first-positive and first-negative curvatures to tenfold higher irradiances. The second-positive curvature was shifted to threefold lower irradiances by pretreatment with red light (Zimmermann and Briggs, 1963; Curry, 1969). Thus, red light changes the sensitivity of the tissue to the blue light that actually causes the bending. Using brief exposures, Zimmermann and Briggs were able to reverse the red effect (maximum at 660 nm) with subsequent exposures to far-red light (about 730 nm). This is a test for the phytochrome pigment system discussed in Chapter 20, so phytochrome (activated by red light) plays a role in determining sensitivity of coleoptiles to the blue light that causes bending (the topic of the next section).

Dark-grown (*etiolated*; see Chapter 20) stems and other tissues conduct light much as the fiber-optic light pipes used in modern communication systems (Mandoli and Briggs, 1982, 1983, 1984). A coleoptile tip that has just penetrated the soil surface will conduct ("pipe") light down to the primary leaf, mesocotyl, and roots. The relative distribution within the tissue may change so that there is more light on the "shaded" side than on the "lighted" side. As the light is transmitted through the tissue, some wavelengths are absorbed more than others, so its spectral composition changes. Further spectral changes occur as the etiolated tissue becomes green in response to the light. These responses could complicate phototropism and other light responses.

[4]Flux × time = fluence ($J\ s^{-1}\ m^{-2} \cdot s = J\ m^{-2}$), hence: flux = fluence/time.

[5]In the early part of this century, A. H. Blaauw in the Netherlands performed a number of similar experiments in the field of phototropism, as we note later. In one, he reported that *Avena* coleoptiles would curve in response to only about one-thousandth of the light from a full moon—but it took 43 hours of exposure to obtain this response! With the highest light levels available to him, he got the same response with an exposure time of only 0.08 seconds.

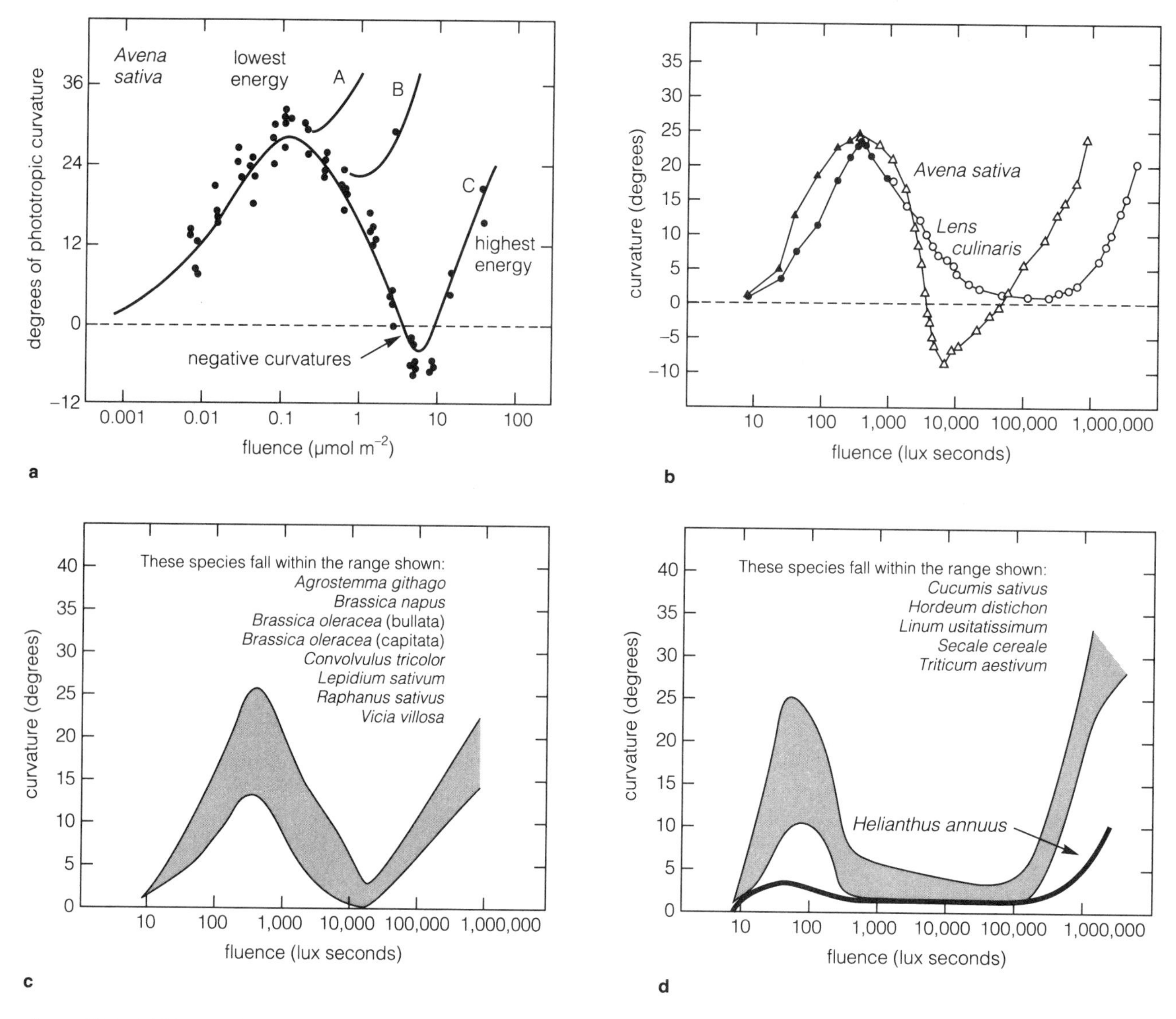


**Figure 19-7** (**a**) Phototropic response of oat (*Avena*) coleoptiles as caused by increasing fluences of unilateral blue light at 436.8 nm following red light. Note first-positive, decreasing, and second-positive curvatures. Light energy was held constant, and exposure times were varied to give the total doses shown along the abscissa. Irradiances were 0.014, 0.14, and 1.4 μmol $m^{-2}$ $s^{-1}$ for lines labeled A, B, and C. Points are actual data (to show scatter) but apply only to curve C. (From Zimmermann and Briggs, 1963.) (**b**) Similar dose-response curves obtained by Brigitte Steyer (1967) with oats (*Avena sativa*) and lentils (*Lens culinaris*), a dicot. First-negative curvature is apparent only for *Avena*. (**c**) and (**d**) Figures that summarize results with the other 12 dicots and three monocots tested by Steyer. None showed negative curvature, and *Helianthus* exhibited virtually no first-positive curvature, although second-positive curvature is evident. (Steyer used white light measured in lux; one lux second of white light is approximately 0.2 μmol $m^{-1}$. Common names of species are: *Agrostemma* = common corncockle; *Brassica napus* = winter rape; *Brassica oleracea* = cabbage; *Convolvulus* = bindweed; *Lepidium* = pepperwort; *Raphanus* = radish; *Vicia* = vetch; *Cucumis* = cucumber; *Hordeum* = barley; *Linum* = flax; *Secale* = rye; *Triticum* = wheat; and *Helianthus* = sunflower. Barley, rye, and wheat are monocots.)

*Perception: Action spectra and the photoreceptor pigment* To identify the pigment responsible for any photochemical process, an essential step is to compare the action spectrum for the process with the absorption spectra of pigments suspected to be involved (Section 10.3). Over 80 years ago, A. H. Blaauw (1909) in Holland found that blue light was most effective in causing phototropic curvature. Since then, increasingly detailed action spectra have been measured, and plant physiologists have suggested that one or both of two common yellow pigments, carotenoids and flavins, might absorb the light that causes phototropism. (Some yellow pig-

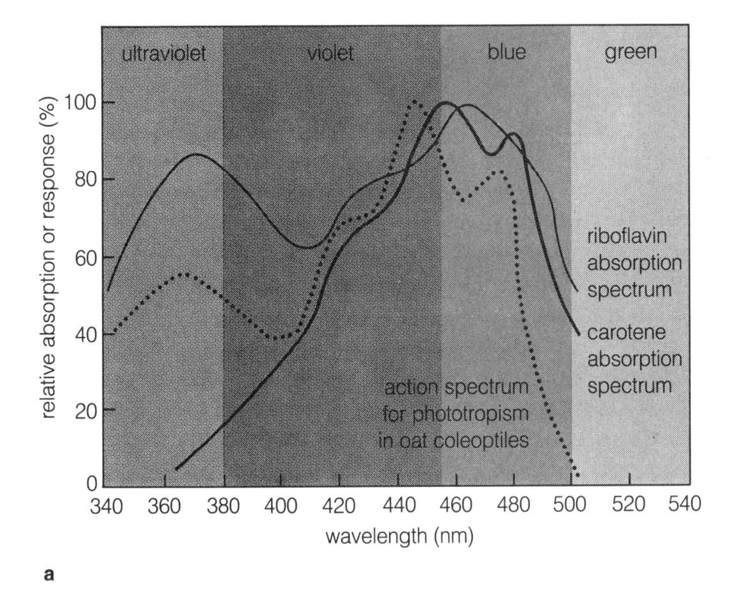


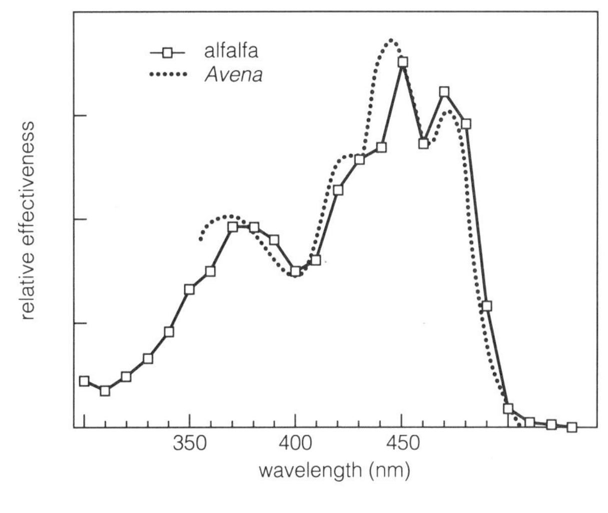


**Figure 19-8** (**a**) An action spectrum for phototropism compared with the absorption spectra of riboflavin and carotene. (Data assembled from various sources. See especially Thimann and Curry, 1960; Dennison, 1979.) (**b**) Action spectra for phototropism in alfalfa (*Medicago sativa*) hypocotyls compared with an action spectrum for oat coleoptiles (*Avena sativa*). The two action spectra are essentially identical. (From Baskin and Iino, 1987; used by permission.)

ments absorb blue and sometimes ultraviolet wavelengths, and the remaining wavelengths combine to produce the sensation of yellow in the human eye.) Figure 19-8 shows the absorption spectra for β-carotene and riboflavin and a representative phototropism action spectrum for *Avena* coleoptiles. The figure also includes a comparison of the *Avena* (monocot) action spectrum with one obtained by Baskin and Iino (1987) for both the ascending and descending arms of the bell-shaped, dose-response curve (as shown in Fig. 19-7) of alfalfa (*Medicago sativa*, a dicot). It is clear that the action spectra for the monocot and the dicot are identical, suggesting that the same blue-absorbing pigment functions in phototropism in both plant groups. The pigment (pigments?) has been called **cryptochrome** (Schopfer, 1984). We will discuss it further in Chapter 20.

As you can see from examining Figure 19-8a, it is not an easy matter to distinguish between the two possible pigment systems by comparing available absorption and action spectra. The strong peak in the ultraviolet (360 to 380 nm) favors riboflavin as the absorbing pigment, but the two peaks in the blue-violet part of the spectrum (about 450 and 480 nm) favor carotene. Nevertheless, accumulating evidence has suggested that a flavin pigment is the primary photoreceptor in phototropism. For example, certain phototropically active fungi (for example, *Phycomyces*) have action spectra for various responses that are almost identical to those for phototropism in higher plants, and a flavin attached to a protein (**flavoprotein**) appears to be the only pigment involved (Muñoz and Butler, 1975). Upon light absorption, the flavoprotein becomes oxidized by reducing a *b*-type cytochrome in the plasmalemma (Brain et al., 1977). Furthermore, certain higher-plant mutants that are exceptionally low in carotene nonetheless respond phototropically, and certain herbicides that block the formation of carotenoid pigments do not eliminate the phototropic response.

There are some interesting complications in our attempts to determine the photoreceptor pigment. For one thing, an inactive screening pigment can cause peaks in the action spectrum (see an example and a review in Vierstra and Poff, 1981); that is, although the light absorbed by the pigment may not itself be transduced to a phototropic response, the presence of the pigment accentuates the steepness of the light gradient across the coleoptile. Of course, it is the plant response to the light *gradient* that leads to phototropic bending. Wavelengths that are absorbed by the screening pigment will cause a steeper gradient and thus will appear as peaks in the action spectrum, even if the energy absorbed by that pigment has no further effect on phototropic bending.

*Transduction in phototropism* As early as 1909, Blaauw (see also Blaauw, 1918) proposed that light acted in phototropism rather directly by inhibiting growth on the irradiated side of a stem or coleoptile. But in 1926, N. Cholodny theorized and Frits Went demonstrated that auxin apparently migrates from the irradiated to the shaded side of a unilaterally irradiated coleoptile tip (Section 17.2). Figure 19-9 shows results of a similar experiment reported by Briggs (1963). In that experiment, the amount of auxin transported into agar blocks

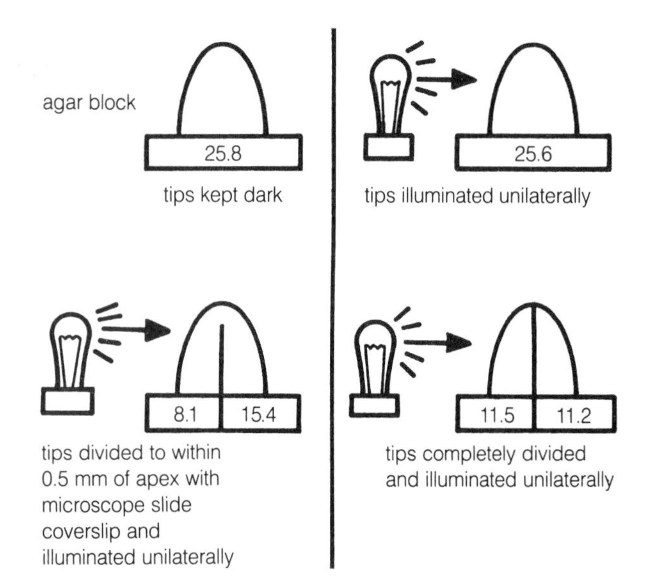


**Figure 19-9** Experiments showing that unilateral illumination of maize-coleoptile tips leads to a transport of auxin from the illuminated to the shaded side of the tips and apparently does not cause a destruction of auxin. Numbers on the agar blocks represent the degrees of curvature caused by application of the blocks unilaterally to decapitated oat-coleoptile stumps. In the partially split tips, part of the auxin was transported laterally above the dividing barrier, but in the completely split tips this was not possible. (From Briggs, 1963.)

at the base of a coleoptile tip is indicated by the degrees of curvature elicited when the agar blocks were placed on decapitated coleoptiles, as in the standard *Avena* curvature test for auxin (see Fig. 17-3). Tips exposed to light exported as much auxin as those kept in the dark, suggesting that auxin destruction by light did not occur, as Blaauw's model might suggest. When the agar and part of the coleoptile were divided with a thin piece of mica and the tip exposed to light from one side, the amount of curvature was twice as great for the block under the shaded side as for that under the irradiated side. Division of the entire tip with mica prevented lateral auxin transport in the tip, so auxin amounts collected from both sides were the same.

The **Cholodny-Went model** suggests, then, that light from one side in some way causes a transport of auxin toward the shaded side, accounting for the basic transduction mechanism in phototropism. The model suggests many tests, and it has guided experimentation for almost seven decades. Many experiments have supported it, yet in the late 1970s it was again called into question. Richard D. Firn and John Digby (1980), for example, presented criteria by which the Cholodny-Went hypothesis might be tested, and they concluded that the criteria had not been met. Their suggestions stimulated considerable research. In an excellent review, Briggs and Baskin (1988) summarized Firn and Digby's proposed tests of the Cholodny-Went model by listing the following four criteria: (1) In a phototropically curving organ, acceleration of growth on the shaded side should accompany retardation of growth on the irradiated side; (2) the development of a lateral auxin gradient must accompany or precede the appearance of differential growth; (3) it must be shown that auxin is indeed a factor limiting growth in the responsive organ; and (4) it must be shown that the auxin differential established is sufficient to account for the growth differential that is observed.

We will discuss the first criterion in the next section. Several studies on transport of $^{14}$C-IAA showed that the direction and rate of transport in coleoptiles could, indeed, be influenced by light (for example, Pickard and Thimann, 1964; Shen-Miller et al., 1969), although results were often controversial. Iino and Briggs (1984) showed that when curvature began in the apical regions of the maize coleoptile and moved downward, the rate of movement was consistent with known rates of auxin transport. Baskin et al. (1985) made similar measurements of coleoptiles to which auxin had been applied to one side near the tip and obtained a rate of basipetal movement of growth stimulation similar to that found in photostimulation. In further work, Baskin et al. (1986) applied auxin to coleoptile tips and reported that auxin was indeed limiting for growth, that there was an approximately linear relation between auxin concentration and growth rate over a range that spanned the rates occurring in phototropic bending, and that an auxin gradient established at the coleoptile tip was well sustained during its basipetal transport. In view of these facts, Briggs and Baskin (1988) concluded that, in a monocot coleoptile (maize), evidence for the Cholodny-Went model is quite strong.

Nevertheless, there is good evidence that a different mechanism acts in at least some dicots. Franssen and Bruinsma (1981) found no auxin gradient across a phototropically stimulated sunflower hypocotyl, but they did demonstrate a gradient of an inhibitor, xanthoxin, across the hypocotyl, with 60 to 70 percent of the inhibitor on the lighted side. Hasegawa et al. (1989) isolated and characterized three inhibitors that formed a similar gradient across phototropically stimulated radish hypocotyls. Koji Hasegawa, Masako Sakoda, and Johan Bruinsma (1989) critically tested the Cholodny-Went model as it applies to the *Avena* coleoptile, the organ that led to the model's formulation in the first place. They repeated Went's experiment (as shown for maize in Fig. 19-9) and obtained the same results when the agar blocks were placed on properly prepared coleoptiles: The block from the dark side caused the most bending. But when they analyzed the blocks with a physiochemical assay, they found equal amounts of IAA in both blocks, and there was about 2.5 to 7 times as much as was indicated by the curvature test. This finding suggested that the blocks contained an inhibitor along with the auxin, and they found two as yet unidentified inhibitors in the blocks, with more of each in the block from the illuminated side. In spite of the evidence in favor of Cholodny-Went summarized above, these results strongly support an inhibitor model as suggested by Blaauw, even for grass coleoptiles.

*The growth response* An important test of the Cholodny-Went and Blaauw hypotheses asks what actually happens during phototropic bending of a coleoptile or hypocotyl. As noted in the first criterion above, the Cholodny-Went hypothesis suggests that growth on the light and shaded sides of the bending organ should be compensatory: as much growth reduction on the light side as promotion on the shaded side. The Blaauw hypothesis suggests inhibition on the lighted side with little change (or some inhibition) on the shaded side. Manual measurements of growth were made a century ago; more recent measurements used photographic techniques. Marks (for example, minute glass beads) are placed along both sides of coleoptiles or hypocotyls illuminated from one side, and photographs are taken at intervals during phototropic bending. Distances between the markers are then carefully measured on projected images of the photographic negatives, plotting growth rates on the two sides as a function of time during photostimulation. The experiment sounds simple enough, but there are mine fields to trap the unwary.

Franssen et al. (1982) reported that when high light levels were used to elicit the basal response, growth on the irradiated side stopped almost instantaneously upon the beginning of irradiation, whereas growth of the shaded side continued, typically at about the same rate as before the beginning of unilateral irradiation. This clearly supports Blaauw, but as it turns out, the Cholodny-Went hypothesis could also be correct.

It has long been known that red light (acting through phytochrome; see Chapter 20) and blue light can inhibit growth of stems and coleoptiles, as we noted in our discussion of the tonic effect. But what if this treatment is applied to the entire stem or coleoptile (for example, from above) to saturate the general inhibitory response, and *then* the organ is irradiated unilaterally with phototropically active light? Several workers take such an approach (for example, Baskin, 1986; Baskin et al., 1985; Rich et al., 1987). Results of work by Iino and Briggs (1984) are shown in Figure 19-10. Under these conditions, growth on the lighted and the shaded sides is completely compensatory, as predicted by Cholodny-Went. The conclusion seems to be that under some conditions and in some species there is a general inhibition of growth by light, as Blaauw suggested, but that superimposed on this, or under some conditions, instead of this, there is compensatory growth possibly caused by transport of auxin (or an inhibitor?), as suggested by the Cholodny-Went model.

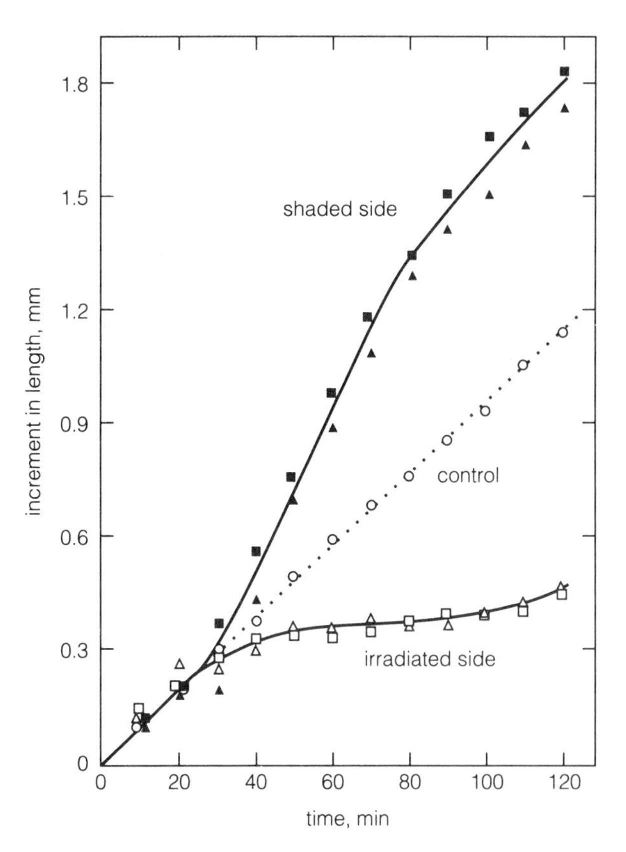


**Figure 19-10** Time courses of growth increments on shaded and irradiated sides of maize (*Zea mays*). After coleoptiles were marked 15 mm from the tip, whole shoots (squares) or just tips (triangles) were unilaterally irradiated for 30 s with blue light (fluence = 5.0 $\mu$mol m$^{-2}$). Control plants (circles) were treated the same way as irradiated plants but without phototropic induction. Plants were grown under red light (0.15 $\mu$mol m$^{-2}$ s$^{-1}$). Each point represents the mean of 10 plants. (From Iino and Briggs, 1984; used by permission.)

## Leaf Mosaics

Leaves also respond phototropically to light. If, for example, half of a cocklebur leaf blade in the light is covered with aluminum foil (simulating natural shading), two responses occur. One is an elongation of the petiole on the covered side, so it bends and displaces the leaf toward the irradiated side of the blade. The other is an upward bending (hyponasty) of the shaded side of the leaf. Such responses in a number of leaves would presumably move them from the shade into the light whenever this is physically possible. The result is that leaves often barely overlap; instead, they form patterns called **leaf mosaics**. Such mosaics appear in leaf canopies of many trees (observed by standing under them and looking upward), as well as in ivy climbing the walls of a building (Fig. 19-11) or growing in a house in which light comes predominantly from a single window throughout the day. Are leaf mosaics examples of true tropisms? Sometimes it is difficult to decide in such borderline cases.

**Figure 19-11** A typical leaf mosaic exhibited by Boston ivy (*Parthenocissus tricuspidata*) growing on the side of a building. Note how almost every leaf is exposed to light. Leaf mosaics often develop in house plants that depend on light from one direction (as from a window).

Perhaps when one part of a leaf blade is shaded by another, the shaded part transports more auxin to its side of the petiole than the brightly illuminated side transports to its corresponding petiole part. Or perhaps the irradiated side transmits an inhibitor. These possibilities remain to be tested. There is opportunity to study how plants maximize the amount of photosynthetically active radiation to which they are exposed (or minimize it when irradiances are high enough to be damaging), as in the following example.

## Solar Tracking

Many plants are capable of **solar tracking**, in which the flat blade of the leaf remains nearly at right angles to the sun throughout the day, maximizing the light harvested by the leaf. The phenomenon was studied by the Darwins (1880) and was neglected until H. C. Yin performed some important basic studies (1938). James Ehleringer and I. Forseth (1980) documented the importance of solar tracking in the desert and in other natural ecosystems, and C. M. Wainwright (1977) studied a desert lupine (*Lupinus arizonicus*). Similar responses have been observed in leaves of such species as cotton, soybean, beans, alfalfa, and various wild members of the Malvaceae, such as *Malva neglecta* and *Lavatera cretica*. Most of our understanding of the physiological mechanisms involved in solar tracking, however, depends upon studies by Amnon Schwartz and Dov Koller (1978, 1980) at the Hebrew University of Jerusalem (others are noted by Schwartz et al., 1987).

Solar tracking is a true tropism because the orientation of the leaves is determined by the direction of the sun's rays, but it is neither *positive* nor *negative*, as is stem phototropism; it is **diaphototropism (diaheliotropism)** in which the organ orients at right angles to the light rays. Leaf orientation is controlled by motor cells in a pulvinus where the blade joins the petiole. Movement of water in and out of these motor cells is completely reversible and is almost certainly controlled by osmotic solutes, including $K^+$.

The pattern of solar tracking is illustrated in Figure 19-12. The leaf **lamina** (blade) follows the sun throughout its daily course across the sky, much as a radio telescope tracks a moving satellite. At sunset, the laminae are almost vertical, facing the point on the western horizon where the sun is setting. Within an hour or two the laminae assume a "resting" position at right angles to the petiole, and an hour or two before sunrise they start moving again, turning to face the point on the eastern horizon where the sun came up the day before and will rise again on the following day. Thus, these leaf movements respond to the direction of the sun's rays not only during the daylight hours, but also during the following night. Moreover, if plants are left in total darkness for several days, they continue to reorient their leaves each 24 hours to the point where the sun (or artificial laboratory light) first appeared the last time it

reappears as soon as the new starch grains are able to *settle*; they usually remain suspended in the cytoplasm for a few hours after they first appear. (Note also that coleoptiles of maize mutants with smaller amyloplasts than the wild type are less sensitive to gravity.)

Just when everyone thought the issue was settled, Timothy Caspar, Chris Somerville, and Barbara G. Pickard (1985) presented a poster at the annual meeting of the American Society of Plant Physiologists at Brown University, on which they reported studies with a mutant of *Arabidopsis* that could not synthesize starch but that still responded to gravity. This was serious evidence against a statolith theory of graviperception, but Fred Sack and John Kiss (1988) had a poster at the Reno, Nevada, meeting on which they reported that the gravitropic response was significantly slower in the starchless mutant. They suggested that starchless plastids of the mutant did settle in the statocytes, but more slowly than did amyloplasts of the wild type. Hence, *starch* must not be necessary for gravity perception, but it might *aid* perception when present, and settling of plastids may well be necessary. (Companion papers from the two research groups have now been published in a single issue of *Planta*: see Caspar and Pickard, 1989; Kiss et al., 1989.)

Could other cell organelles settle and cause the gravitropic response? We have just seen that starchless plastids of *Arabidopsis* can settle and probably lead to gravitropic bending. To settle, a statolith must have a density significantly greater than the medium in which it is suspended. Starch has a density of about 1.3, which is greater than that of the cytosol (which is close to 1.0). Densities of other organelles in the cell are often very close to that of the cytosol, so they would not be expected to settle. This is true of the nucleus, for example. There is evidence that some organelles besides amyloplasts or starchless plastids do settle, but the rates do not correlate with presentation times (Shen-Miller and Hinchman, 1974). Even if an organelle is dense enough, it will not settle if it is too small; it will remain in suspension because of its Brownian movement. This is true of ribosomes, for example. Thus amyloplasts or starchless plastids remain the most likely candidates for the role of statoliths.

*Transduction: The role of an inhibitor* It is clear that the root cap sends an inhibitor to the lower side of the root, slowing growth there so that the root bends downward. Figure 19-13 illustrates several experiments that demonstrate this. Various treatments designed to block inhibitor movement from the root cap into one side of the root cause bending toward the side that does receive the inhibitor. (See Wilkins, 1975; Wilkins and Wain, 1974.)

Near the end of the 1970s it seemed clear that the inhibitor from root caps was ABA (Wilkins, 1979), but now this is very much in doubt (Wilkins, 1984). Michael L. Evans, Randy Moore, and Karl-Heinz Hasenstein (1986) told the story in an excellent review. Although ABA inhibits root growth, it does so only at concentrations significantly higher than those thought to occur naturally, and roots grown in the presence of a compound that inhibits ABA synthesis continue to respond to gravity. Furthermore, the roots of a maize mutant known to be unable to synthesize ABA also respond to gravity. Roots will even curve down when they are immersed in a high concentration of ABA that would surely overwhelm any natural ABA gradient within the root.

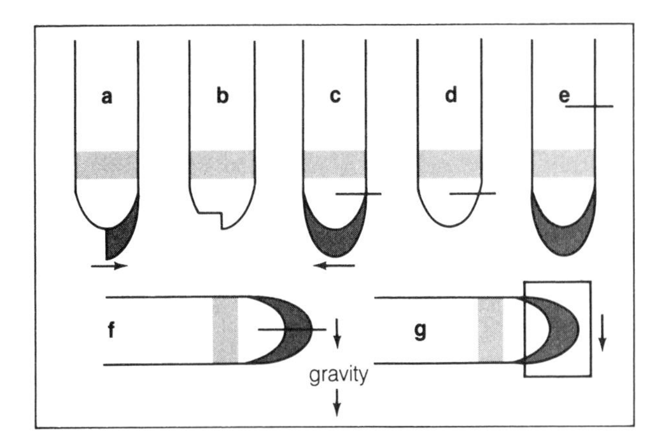


**Figure 19-13** Diagrammatic representation of various treatments applied to the roots of maize (*Zea mays*) and peas (*Pisum sativum*). The experiments suggest that the lower side of the root cap produces an inhibitor of root growth that is transmitted to the growing cells (shaded). Arrows show the direction of curvature of the root after treatment. (**a**) The vertical tip bends toward the remaining portion of a root cap, but (**b**) removal of cap and a portion of the meristem does not cause bending. (**c**) Insertion of a horizontal barrier between the cap and the growing zone causes bending away from the side in which the barrier was inserted, but (**d**) such a barrier does not cause bending in the absence of the root cap or (**e**) when the barrier is above the growing zone. Insertion of a horizontal barrier in a horizontal root (**f**) nearly prevents bending (suggested by the short arrow), but a vertical barrier (**g**) has little effect on bending. (From Shaw and Wilkins, 1973. Used by permission.)

Now it appears that IAA is the inhibitor. IAA, which is present in roots, inhibits root growth at concentrations 100 to 1,000 times lower than those of ABA. If the root cap is removed and a small block of agar containing IAA is added to one side, the root bends toward that side. Radioactive IAA applied uniformly to a horizontal root moves to the lower side, and auxin-transport inhibitors overcome the root's ability to respond to gravity (but also often inhibit root growth, weakening this evidence). Thus IAA could be the effective inhibitor in root gravitropism, although other substances cannot be completely ruled out (Feldman, 1981; Jackson and Barlow, 1981; Suzuki et al., 1979).

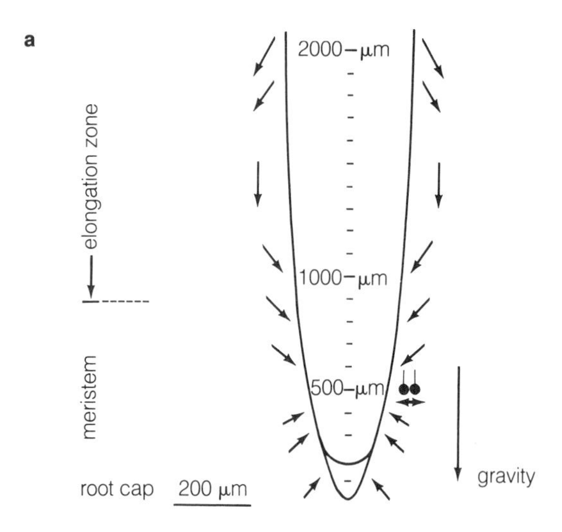


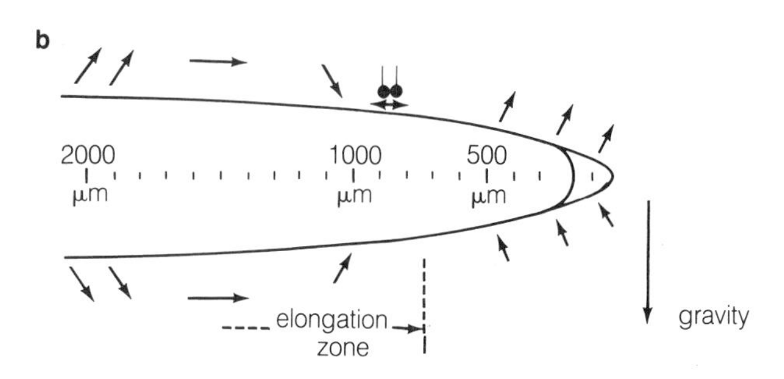


**Figure 19-14** (**a**) The qualitative current pattern measured with a vibrating electrode (symbolized by two heavy dots and vertical lines) around a vertical root. Arrows indicate the direction of current flow—that is, the direction of movement of positive ions. The large arrow indicates the direction of the gravity vector. (**b**) A hypothetical pattern of current flow of a 24-h-old *Lepidium* root that has been tilted to the horizontal for 20 min. The current pattern is based on a possible interpretation of the measurements of acropetal and basipetal currents. Note the reversal in flow direction on top near the tip. (From Behrens et al., 1982.)

Acid growth (Section 17.2) has been implicated in root gravitropism, which is suggestive of a *promotive* auxin action (Mulkey et al., 1981). Roots embedded in agar containing a *p*H indicator (bromcresol purple) and allowed to respond to gravity become more acidic on the *upper* side, where most growth is taking place (Mulkey and Evans, 1981), and less acidic on the lower side, where growth is reduced. Auxin on the *lower* side may occur at a supraoptimal concentration that inhibits rather than promotes acid growth, and high levels of auxin have been shown to inhibit acid efflux from roots (Evans et al., 1980).

*Calcium and electric currents: A model* Calcium appears to take part in gravitropism. Radioactive or nonradioactive $Ca^{2+}$ moves to the bottom of gravistimulated roots, and when a strong ligand of $Ca^{2+}$ (such as EDTA: *ethylene-diaminetetraacetic acid*) is applied to roots and other tissues, gravitropic bending can be completely prevented (Lee et al., 1983). Although the ligand could be doing something else in addition to chelating $Ca^{2+}$, it is significant that excess added $Ca^{2+}$ reverses the ligand effect and that roots treated with the ligand continue to grow at their normal rate, at least for several hours. As with auxin, if $Ca^{2+}$ is prevented from moving in the root, gravitropism is inhibited. Furthermore, if $Ca^{2+}$ is added in an agar block to one side of the root cap, the root will curve toward that side, in some cases making a 360° loop! One more piece of indirect evidence for $Ca^{2+}$ participation in root gravitropism concerns electric currents that have been measured in gravireponding roots (Behrens et al., 1982, 1985; Björkman and Leopold, 1987a, 1987b). These currents could be caused by a flow of $H^+$ ions that could be a counterflow to movement of $Ca^{2+}$.

Figure 19-14 summarizes some observations of roots made at the University of Bonn in Germany with minute vibrating electrodes. Such techniques allow the measurement of current flow around test roots and clearly indicate that such flow changes during gravitropic stimulation. Björkman and Leopold (1987a, 1987b) measured current shifts in maize roots with a 3- to 4-min lag period corresponding to the presentation time in those organs. Tanada and Vinten-Johansen (1980) measured changes in electrical potential on the surface of soybean hypocotyls within 1 min of horizontal placement.

How can settling statoliths act within statocytes to lead to further steps in the response chain? Rosemary G. White and Fred D. Sack (1990) noted that amyloplasts of maize and barley (*Hordeum vulgare*) are closely associated with microfilament bundles. They suggested that such bundles could link settling statoliths with plasma membranes and other cellular organelles such as the endoplasmic reticulum, pulling on them as they settle and thus triggering the gravitropic response. Several other workers have reported that statoliths sometimes fall onto the ER. Evans et al. (1986) noted that the ER is known to be rich in $Ca^{2+}$, and that some $Ca^{2+}$ might be released from the ER when it is contacted or pulled upon by amyloplasts (Sievers et al., 1984). This released $Ca^{2+}$ might then activate **calmodulin**, a small protein known to be a powerful activator of many enzymes important to cellular function, not only in plants but also in animals and in some microorganisms. The calmodulin in turn might activate both calcium and auxin pumps in cell membranes, accounting for the movement of $Ca^{2+}$ and auxin toward the bottom of a gravistimulated root. Auxin normally flows through the center of the root (the stele) toward the root cap. If the root is vertical, the auxin is slowly redistributed symmetrically to the root cortex; if the root is gravistimulated, the above mechanism might account for the movement of greater amounts of auxin into the lower

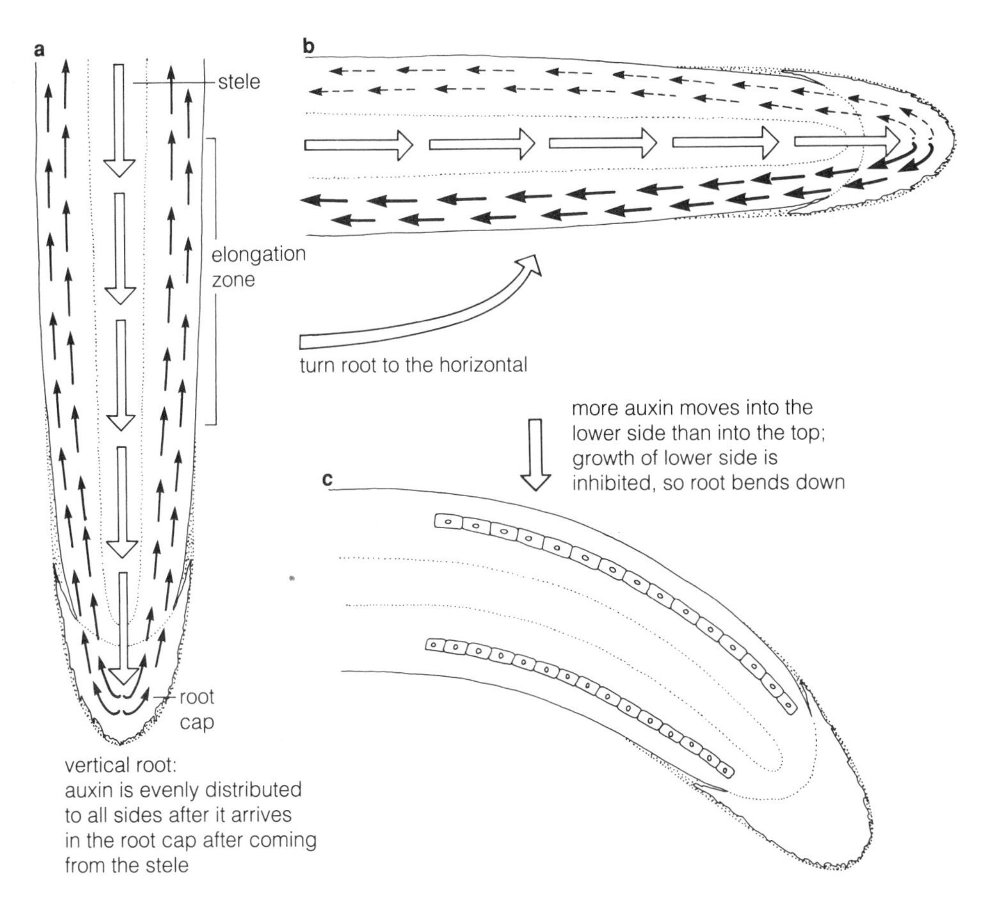


**Figure 19-15** Patterns of auxin flow in a root as affected by gravistimulation, according to the Evans, Moore, and Hasenstein model. In a vertical root (**a**), auxin travels from the elongating zone toward the root cap through the stele. After the auxin enters the cap, some moves laterally and then symmetrically back into the elongation zone. When the root is turned to the horizontal (**b**), the pattern of auxin flow becomes asymmetric, with more flowing into bottom than into top tissues. Accumulating $Ca^{2+}$ in some way increases the rate of auxin flow back into the elongating tissue, where the auxin inhibits growth and thus leads to downward bending (**c**).

cortex tissues. All this is quite theoretical, but accumulating evidence supports the hypothesis. Figure 19-15 illustrates the proposed model.

One other important detail complicates our understanding of the gravitropic response of roots. In certain cultivars of maize (and also in radish and other species less studied), the nature of the gravitropic response depends on light. Roots of seedlings growing in the dark are diagravitropic; that is, they tend to grow horizontally. When exposed to light (as when the root grows to within a few millimeters of the surface), these roots become positively orthogravitropic and bend downward. Red wavelengths are most effective, and the phytochrome system (Chapter 20) has been implicated.

## Stems and Coleoptiles

Measurements of growth of top and bottom surfaces of horizontal coleoptiles, hypocotyls, and stems give results strikingly similar to those observed with high irradiances in phototropic bending: The top surface ceases growth almost immediately while growth of the bottom surface speeds up or sometimes continues at about the same rate or even slower (see, for example, MacDonald et al., 1983). Most plant physiologists have assumed that stem bending occurs because auxin moves to the lower tissues, promoting their growth (again the Cholodny-Went model, supported by work of Dolk, 1930, who performed experiments comparable to those of Fig. 19-9). But there are complications. Any proposed mechanism must account for the observed response, so we will examine this more closely before considering the models.

*The response: The mechanics of stem bending* As noted by Ann Bateson and Francis Darwin in 1888, if a growing stem is turned on its side and restrained so that it cannot bend upward, when the restraint is released some hours to days later it does bend, often to more than 90° and within 1 to 10 seconds (see Wheeler and Salisbury, 1981; Mueller et al., 1984). Cells on the bottom of such a restrained stem elongate, but not as much as if the stem were not restrained. Because the stem is held straight and cells are not dividing, top cells are stretched. During restraint, the diameters of bottom cells increase, whereas those of top cells decrease; upon release from restraint, top cells shorten and become thicker, while those on the bottom elongate somewhat and become thinner. The result is the rapid bend (Sliwinski and Salisbury, 1984).

In a horizontal stem, cell growth on top does not merely virtually stop; those cells will not grow even if they are stretched by elongation of bottom tissues. This situation seems opposite to that described in Section 16.2, where we consider the physics of cell growth. There we concluded that cell growth occurs because wall loosening decreases pressure within the cells, which in turn makes the water potential more negative, so that water enters the cells osmotically, driving cell expansion. In a restrained horizontal stem, however, pressure in bottom tissues increases while growth continues, and growth stops in top cells in spite of decreasing pressure. Top cells return nearly to their original dimensions when the stem is released from restraint and bending occurs (Mueller et al., 1984; Sliwinski and Salisbury, 1984). Pressure and tension refer to the tissue as a whole, however, so a pressure difference between the protoplasts of individual cells and their surrounding apoplast could still account for growth or lack of it. Perhaps the yield threshold of cell walls in upper tissues is exceptionally high compared with that in lower tissues.

It has long been known that the stem epidermis and cortical cells play a critical role in stem growth (Diehl et al., 1939; von Sachs, 1882; Thimann and Schneider, 1938). This is illustrated by splitting a typical stem longitudinally (as children split dandelion peduncles); the two halves bend outward some 30 to 50°. Bending occurs because the outer tissues are under tension with respect to the inner tissues. A stem grows as pith cells take up water and expand against the resistance produced by the other cells in the stem, which grow with less force.

*Perception* In stems, the sites of gravity perception and response are the same. Even after much stem above the rapidly growing region has been removed, that region will continue to bend upward in response to gravity when the stem is turned on its side. This is true for coleoptiles, hypocotyls, and mature stems. The Darwins (1880; pp. 511–512 in 1896 edition) removed coleoptile tips from *Phalaris* and noted that the organs "bowed themselves upwards as effectually as the unmutilated specimens in the same pots, showing that sensitiveness to gravitation is not confined to their tips." (Some authors have incorrectly considered the stem or coleoptile tip to be the site of gravity perception in stems, as the root cap is the site of perception in roots.)

As with roots, amyloplasts are thought to be the statoliths of stems. In many species of angiosperms, the amyloplasts are confined to one or two layers of cells, called the **starch sheath**, just outside the vascular bundles. The starch sheath generally forms the inner layer of the cortex, which consists of several layers of parenchyma cells and often a layer of collenchyma cells just below the epidermis (Fig. 19-16). Because of its anatomical relationship to the vascular tissues, the starch sheath may be thought of as an endodermis, although there are usually no Casparian strips around starch-sheath cells (Esau, 1977; Mauseth, 1988). As with roots, the evidence that amyloplasts in stems play the role of statoliths is strongly suggestive but not conclusive. Amyloplasts can be seen to settle in the starch-sheath cells of stems, and all gravitropically responsive stems so far studied do have amyloplasts. (The distribution is somewhat different in coleoptiles, amyloplasts occurring internal to the vascular tissues instead of outside them.)

*Transduction* The Cholodny-Went hypothesis as it applies to gravitropism in stems and coleoptiles has been questioned by several workers (see, for example, Digby and Firn, 1976; Firn and Digby, 1980). Considering the rapid cessation of growth on the top of a stem laid on its side compared with the often normal growth rates on the bottom, it is difficult to imagine that changes in auxin concentration could occur rapidly enough or reach magnitudes capable of accounting for these differences. Growth rates on the top and bottom of a horizontal stem can differ by a factor of 10 or more, especially when parts of the upper surface do not grow at all or even shrink significantly. If auxin concentration is in complete control, upper cells would have to be almost completely depleted of auxin within a short time after the stem was turned on its side. Many workers have found only minimal gradients in auxin concentration across gravitropically stimulated stems. Sometimes, with good modern techniques (see Weiler, 1984), no gradients (Mertens and Weiler, 1983) or auxin transport (Phillips and Hartung, 1976) could be detected, especially in dicots. Again, we might validly ask: If gradients do occur, are they a result of gravitropic bending, rather than its cause?

Even though the Cholodny-Went hypothesis is being questioned by some workers, others are coming to its defense, both by reviewing many experiments that have supported the hypothesis over the years and providing new experimental data that also seem to support the hypothesis (for example, Pickard, 1985). MacDonald and Hart (1987) have suggested a modified and elegant Cholodny-Went model. They proposed that dicot stem tissues differ in their sensitivity to auxin; the epidermis is thought to be more auxin-responsive and subepidermal tissue less responsive. Auxin transport during gravitropism might be limited in distance, with

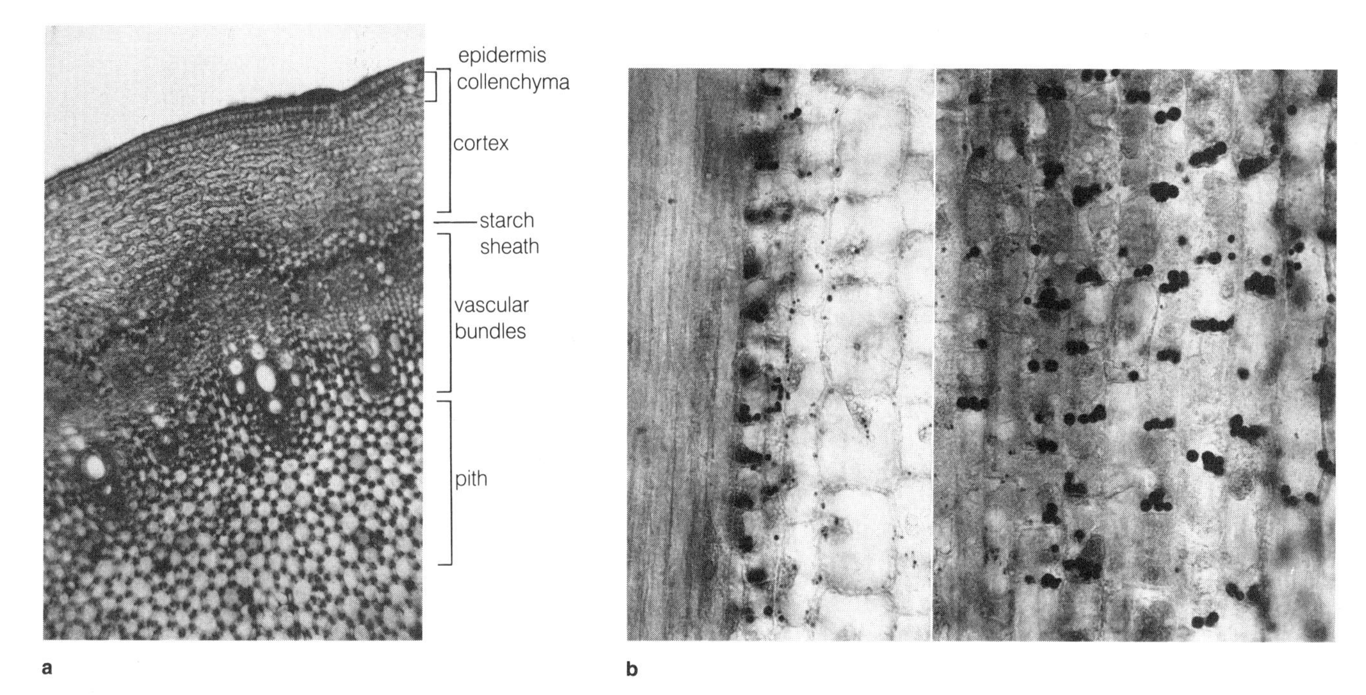


**Figure 19-16** (**a**) Free-hand cross section of a castor bean (*Ricinus*) stem, which has a well-defined starch sheath just outside the vascular tissues. The starch was stained with iodine solution ($I_2 \cdot KI$). (From Salisbury et al., 1982.) (**b**) Sections of the region of a pea stem that contains sedimented amyloplasts (stained with $I_2 \cdot KI$) in the starch sheath. Both micrographs are longitudinal sections, but the one on the right is in the plane of the sheath and shows only sheath cells. In the section on the left, the starch sheath is shown between larger vacuolated cortical cells and narrower and more elongate vascular cells. 275 × (Photographs courtesy of Fred Sack.)

auxin moving out of responsive epidermal cells on top of the stem into cortex below and from nonresponsive cortex in the bottom into epidermis. In such a case, it might be technically very difficult to measure auxin gradients; they might not appear at all when a horizontal stem is split into upper and lower halves and auxin is measured in each half stem.

*Transduction: Auxin and the gravitropic memory* Even if auxin gradients do not always account for gravitropic bending, it is clear that bending does not occur when insufficient auxin is present. This was shown by Leo Brauner and Achim Hager (1958) at the University of Munich in Germany. They removed the tips of sunflower hypocotyls (presumably a source of auxin); and then, after four days, they gravistimulated the hypocotyls by turning them to the horizontal position. Bending did not occur, but if the hypocotyls were returned to the vertical and the cut stump supplied with an auxin (IAA) solution, the hypocotyls bent in the predicted direction. Brauner and Hager called this phenomenon the **gravitropic memory** (or *Mneme*).

In other memory experiments, Brauner and Hager gravistimulated intact hypocotyls in the cold (4°C) for 30 min to 5 h. No bending occurred. When the hypocotyls were turned to the vertical and warmed to 20°C, they bent in the predicted direction. The degree of bending was a function of the logarithm of the time they were gravistimulated in the cold.

Gravitropic bending of decapitated mature stems of castorbean, cocklebur, and tomato was also accentuated by treatment with 1 percent IAA in lanolin, even though the high concentration of auxin, applied in an indiscriminate way to the cut stump of the stem, would probably swamp out the auxin gradients required by the Cholodny-Went model (Sliwinski and Salisbury, 1984).

We noted that Mulkey and Evans (1982) eliminated gravitropic bending of maize roots by treatment with several inhibitors of auxin transport. Wright and Rayle (1983) also inhibited gravitropic bending in sunflower hypocotyls with auxin-transport inhibitors, but again growth was inhibited. Bandurski et al. (1984) measured an IAA asymmetry in horizontal maize mesocotyls within 15 min after seedlings were placed on their sides. The asymmetry was small, however; 56 to 57 percent of the IAA was in the mesocotyl lower halves and 43 to 44 percent in the top halves. Harrison and Pickard (1989) observed larger differences (up to 3.5 times as much on the bottom as on the top during the main phase of curvature) in hypocotyls of tomato (*Lycopersicon esculentum*), and significant asymmetries were observed within 5 to 10 min. They concluded that these data provide strong support for a Cholodny-Went mechanism, but some of their data also support a changing-sensitivity hypothesis, as discussed below.

*Transduction: Substances besides auxin* There have been studies with growth regulators besides auxin. Gibberellins (GAs), for example, occur in higher concentrations on the bottom of gravitropically stimulated stems (reviewed by Wilkins, 1979). Sometimes, significant gradients are not observed until after bending has occurred, but such gradients were established in leaf-sheath pulvini of cereal grasses (see below) during bending (Pharis et al., 1981).

In some experiments ethylene appeared to play a positive role in gravitropic stem bending (Wheeler et al., 1986, plus references cited therein; Balatti and Willemöes, 1989). Four inhibitors of ethylene action or synthesis ($Ag^+$, $CO_2$, $Co^{2+}$, and aminoethoxyvinylglycine, abbreviated AVG) all reduced the rate of gravitropic bending in cocklebur, tomato, and castor bean. Some of the rate of bending could be restored by surrounding the stems with low concentrations of ethylene. But ethylene seems to play no role in the graviresponse of dandelion peduncles (Clifford et al., 1983), cereal leaf-sheath pulvini (Kaufman et al., 1985), or tomato hypocotyls (Harrison and Pickard, 1986). What is the ethylene doing when it does promote gravitropic bending? It would be logical to think that the ethylene might in some way contribute to the inhibition of growth on the top of a stem placed horizontally. This would be in keeping with its known inhibitory effect on elongation of vertical stems (Section 18.2). Yet when ethylene is measured in the tissues, it is found to increase at about the same rate as gravitropic bending, but in the *bottom* tissues instead of the top tissues (Wheeler et al., 1986). Adding ethylene to maize roots enhances curvature ($>90°$) and prolongs auxin asymmetry (personal communication from Michael Evans; manuscript in preparation).

Clifford et al. (1982) attached a thread to the top of a vertical dandelion peduncle and used the thread to pull the peduncle to one side with a 2-gram weight and a pulley system. After the stress was removed, the peduncle bent away from the force, much as it bends when turned to its side and allowed to respond to gravity. The peduncle was displaced slightly from the vertical by the stress, but displacing unstressed peduncles by the same amount did not lead to bending. The stress response might suggest that settling amyloplasts cause their gravieffects by applying mechanical stress within the statocytes.

Calcium ions ($Ca^{2+}$) apparently play an important role in shoot as well as root gravitropism. For one thing, $Ca^{2+}$ concentrations have been observed to be higher on the tops of horizontal stems (Slocum and Roux, 1983), and $Ca^{2+}$ is known to inhibit cell elongation (perhaps by overcoming auxin effects).

*Transduction: The possible role of changing sensitivity to auxin* It should be evident by now that there is much to learn about transduction in gravitropic stem bending. Much more seems to be involved than a simple transport of auxin, but what are the alternatives? One alternative mechanism was noted in Chapter 17. Gravistimulation could lead to a change in tissue sensitivity to auxin. If tissues in the lower side of a horizontal stem became more sensitive to auxin than those in the upper side, the lower side would respond more to the auxin present in the stem, leading to more growth and upward bending whether or not auxin concentrations in the stem changed. Most plant physiologists who have worked on stem gravitropism have failed to consider this possibility, but Brauner and his colleagues tested the hypothesis for over a decade. This began, perhaps, with the paper (Brauner and Hager, 1958) on the gravitropic memory. One explanation they suggested was that sensitivity to auxin changed during gravistimulation of decapitated or cold sunflower hypocotyls, so their tissues responded differently to the auxin that was presented after they were turned to the vertical.

Brauner and his colleagues found several pieces of evidence suggesting that sensitivity to auxin did change in response to gravistimulation[7] (see reviews in their final paper, Brauner and Diemer, 1971, and in Salisbury et al., 1988). For example, they applied auxin in agar to induce different auxin gradients across sunflower hypocotyls and then measured bending and tissue auxin concentration. A given measured auxin gradient in vertical hypocotyls produced a given bending, but the same measured auxin gradient produced much more bending in horizontal hypocotyls (providing that highest concentrations were in bottom tissues). A change in tissue sensitivity to auxin seems the most likely explanation for these results.

What is sensitivity? The dictionary defines **sensitivity** as the capacity of an organism or physical system (for example, a microphone or a photocell) to respond to a stimulus. In this sense, sensitivity is synonymous with responsiveness. It is measured and expressed quantitatively by varying the stimulus and observing the response; that is, by obtaining a dose-response curve. When stem sections are immersed in solutions with a wide range of auxin concentrations, increasing elongation follows increasing auxin concentration, usually plotted on a logarithmic scale, until a maximum response is obtained, after which further increases in auxin lead to decreasing responses (see Fig. 17.7).

Except for the supersaturation part of this response

[7]They also found that sensitivity to auxin decreased in decapitated hypocotyls and that this decreased sensitivity was more important in their gravitropic-memory experiments than was depletion of auxin (Brauner and Böck, 1963).

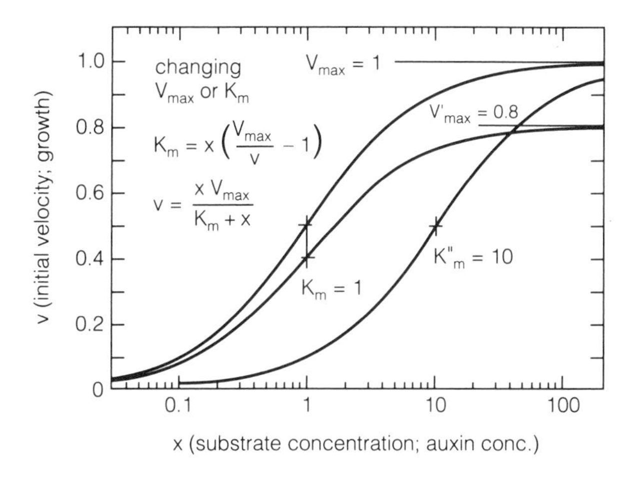


**Figure 19-17** Michaelis-Menten equations in which initial velocity (or, roughly, growth rate or initial growth rate) is plotted as a logarithmic function of substrate (auxin) concentration, showing a change in $V_{max}$ without changing $K_m$ or a change in $K_m$ without changing $V_{max}$. (From Salisbury et al., 1988.)

curve, it is analogous to the classical Michaelis-Menten curves obtained when reaction rate of an enzymatically controlled reaction is plotted as a function of substrate concentration (Fig. 19-17; see also Fig. 9-10b). The maximum reaction rate is called $V_{max}$, and the concentration that produces a rate that is half of $V_{max}$ is called the Michaelis constant, $K_m$. In an *in vitro* enzymatic reaction, $V_{max}$ is determined by the concentration of enzyme. Because auxin must bind to something (for example, a protein) to carry out its promotion of growth, we can think that the near-equivalent of $V_{max}$ in an auxin dose-response curve might be determined by the quantity of auxin binding sites in the tissue. Different levels of growth produced by optimal auxin concentrations suggest different levels of $V_{max}$ sensitivity; high levels of $V_{max}$ indicate high levels of sensitivity. $K_m$ indicates the binding strength of the enzyme for its substrate—or auxin for its binding sites. A smaller value for $K_m$ means that a lower concentration will produce a given effect; hence, a *smaller* $K_m$ means a *greater* $K_m$ sensitivity of the system. Other factors, including auxin penetration into the tissue, can influence the shape of the dose-response curves, but it is nevertheless helpful to consider $V_{max}$ and $K_m$ sensitivity.

In a series of experiments, hypocotyl sections from sunflower or soybean were turned to the horizontal and immersed in a wide range of buffered auxin concentrations (0, $10^{-8}$ to $10^{-2}$ M IAA). Vertical control sections were also used. The sections were photographed at 30-min intervals, and the projected images (negatives) were analyzed with a digitizer-computer system (Salisbury et al., 1988; Rorabaugh and Salisbury, 1989). Figure 19-18a shows bending as a function of time after the beginning of gravistimulation, and Figure 19-18b shows bending and growth of upper, lower, and vertical surfaces as a function of auxin concentration 4 h after immersion (beginning of gravistimulation). As auxin concentration increased past about $10^{-7}$ M IAA, stem bending began to decrease, and at $10^{-4}$ M IAA the sections were actually bending down instead of up. The

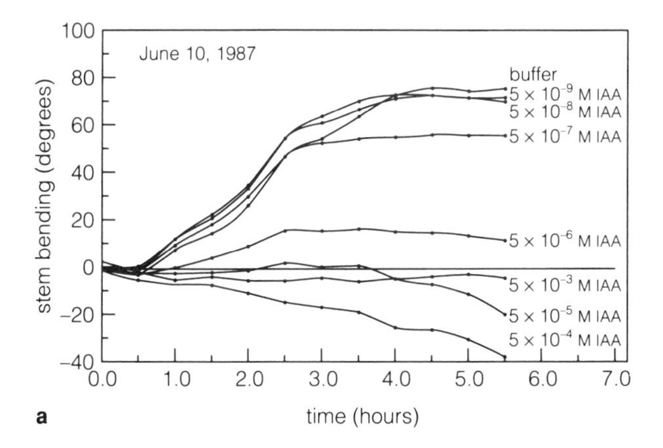


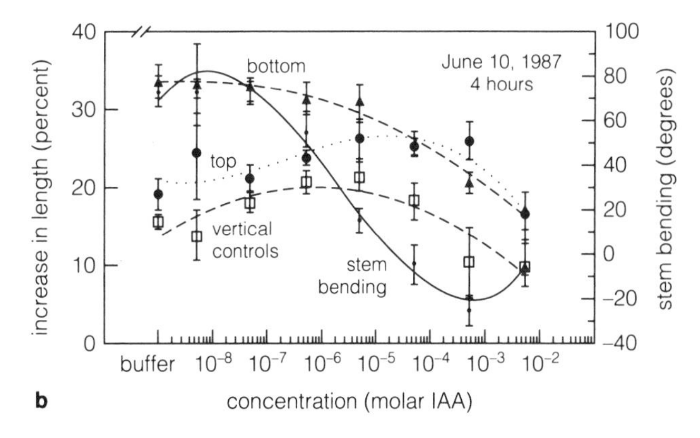


**Figure 19-18** (**a**) Gravitropic bending as a function of time for sunflower hypocotyl sections immersed in buffer with no auxin or in auxin solutions with a range of concentrations from $10^{-8}$ to $10^{-2}$ M IAA. Note that increasing the auxin concentrations reduces rate of upward bending and finally causes downward bending. (**b**) Summary graph showing stem bending and growth of upper and lower and vertical surfaces of sunflower hypocotyl sections as a function of the concentrations of buffer or auxin solutions in which they were immersed. Bending curve is the 4-h points of **a**, and all surface measurements were at 4 h. In all experiments of this type, in buffer or low auxin solutions, lower surfaces grow much more than upper surfaces, but this is reversed at the higher auxin concentrations; this observation accounts for the bending curve and suggests that $K_m$ sensitivity of lower surfaces is much greater than that of upper surfaces. Maximum growth of the lower surface ($V_{max}$) is always greater than that of the upper surface, and maximum growth of vertical controls is always lower than that of either upper or lower surfaces. (From Salisbury et al., 1988.)

dose-response curves for the upper and lower surfaces explain this and are very different from each other. Growth of the lower surface was highest in buffer and low auxin concentrations and then decreased with increasing auxin levels. Growth of the upper surface, on the other hand, was low at low auxin concentrations but increased to a maximum and then decreased at the highest auxin levels, producing the usual bell-shaped curve.

These results, by the definition given above, show that sensitivities of upper and lower surfaces to *applied* auxin were strongly influenced by gravistimulation. $V_{max}$ sensitivity was highest for lower surfaces, and $V_{max}$ sensitivities of *both* horizontal surfaces were higher than those of vertical surfaces. $K_m$ sensitivity of the lower surface is somewhere off the scale to the left of the graph, which implies a very high $K_m$ sensitivity. $K_m$ sensitivity of the upper surface is somewhere along the left arm of the bell-shaped curve; it is much lower (that is, $K_m$ has a much larger value for upper surfaces). $K_m$ sensitivity of vertical surfaces may be intermediate between that of upper and lower surfaces, but it varies from experiment to experiment. If the binding-site explanation for these sensitivities is correct, then gravistimulation increases the *number* of auxin binding sites in lower compared with upper tissues, and in both lower and upper tissues compared with vertical tissues. And binding sites in lower tissues also have a much greater affinity for auxin.

Do these results with applied auxin accurately tell what happens in the tissues during the normal course of gravistimulation and response? This remains to be seen, but the evidence is compelling enough to strongly suggest that changing sensitivity to auxin plays a crucial role in gravitropism of some plants, especially such dicots as sunflower and soybean. This could be in addition to or instead of an auxin-transport mechanism. How can settling statoliths account for changing sensitivity to auxin? Could the statocytes be polarized so that movement of statoliths *away* from the stem surface (that is, on top) causes *reduced* sensitivity and movement *toward* the surface causes *increased* sensitivity to auxin? Research will be needed to find out.

*Clinostat experiments* In 1873 Julius von Sachs (see von Sachs, 1882) suggested that weightlessness might be simulated for a plant by rotating the plant slowly about a horizontal axis. If the perception time exceeds the rotation time, then the vectors of gravitational stimulation would add algebraically to zero, as though the plant being so rotated were truly weightless. Sachs suggested the name **clinostat** for the apparatus that is used to rotate plants as described (usually about their longitudinal stem/root axis, but tumbling end over end should theoretically be the same and does indeed usually produce the same effects). Around the turn of the century, hundreds of clinostat experiments were done in attempts to assess the importance of gravity (and unilateral light) for plants. Now that space exploration provides orbiting laboratories, it is possible to compare the early results with those obtained by allowing plants to grow in true microgravity[8] conditions.

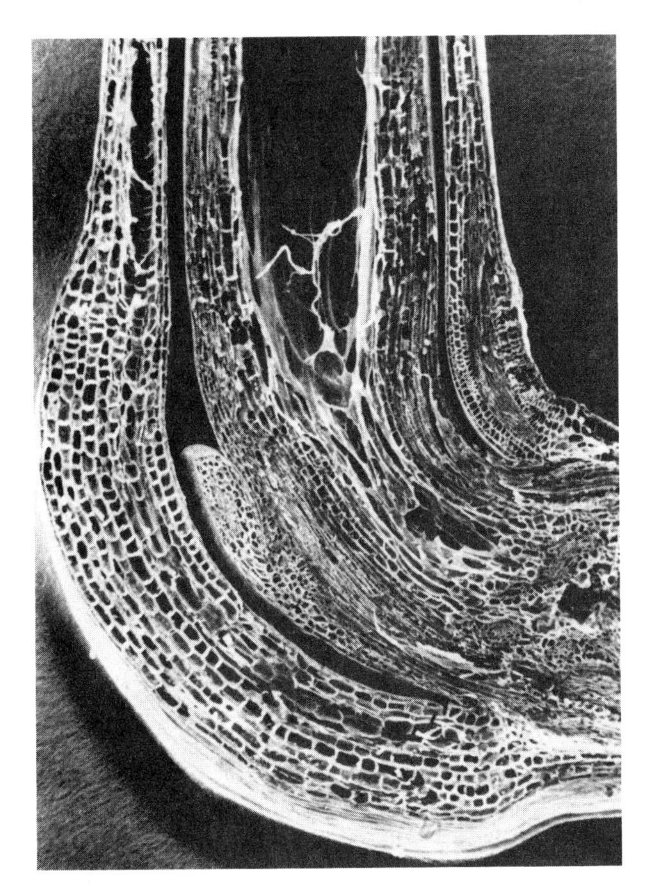

**Figure 19-19** The gravitropic response in a grass (*Muhlenbergia*) pulvinus as shown with the scanning electron microscope. Note the large cells on the bottom (convex side) compared with the much smaller cells on top (concave side). An axillary bud is seen within the leaf sheath on the elongating side. 30×. (Micrograph courtesy of Peter B. Kaufman; see Dayanandan et al., 1976.)

The most noticeable response of plants on a clinostat is leaf epinasty (downward bending; Section 19.2). It was suggested that this and other clinostat responses might be caused not by the simulated weightlessness of clinostating, but rather by the mechanical stresses imposed by leaf flopping and other strains (Tibbitts and Hertzberg, 1978). Of course, a plant on a clinostat is not really weightless. As it rotates, its leaves and other or-

[8]An orbiting object is nominally *weightless* because it is in *free fall*, falling toward the earth at a rate that exactly compensates its tendency (because of its inertia) to move in a straight line. Thus it would have no weight on a spring balance that was also in orbit. But objects inside an orbiting spacecraft are subject to slight accelerational forces caused by adjustments in the orbit or activities of the astronauts. These forces are on the order of one thousandth of the accelerational force caused by gravity on the earth's surface (that is, about 0.001 × g). Workers in the field speak of **microgravity**, but *milli*gravity would be more correct.

gans move in response to gravity. Perhaps these mechanical stresses lead to production of ethylene and leaf epinasty. (Ethylene is known to cause leaf epinasty.) Two kinds of experiments suggest that this simple explanation for clinostat-caused epinasty is not correct after all (Salisbury and Wheeler, 1981). For one thing, when vertical plants experience mechanical stresses in several ways comparable to the stresses produced on a clinostat, leaf epinasty does not appear. Second, when gravity is compensated for by the simple procedure of inverting plants every 5 to 30 minutes so that they are upside-down half the time, epinasty does appear, just as on a clinostat. The inversions can be done very carefully in an attempt to reduce mechanical stresses, and control plants can be inverted at the same time but immediately returned to the vertical, so that they receive essentially the same mechanical disturbance but are not upside-down half the time; such controls show no epinasty. These experiments suggest that plants grown in an orbiting satellite should display epinasty, and experiments in the space-shuttle laboratory, as well as a few earlier experiments in orbiting laboratories, confirm this prediction.

*False pulvini of grasses* In grass stems, the gravity-sensing tissue is located near the nodal region. In the Panicoid grasses, such as maize and sorghum, the base of the internode is swollen and responds to gravity if it is not fully mature. In some cases, the base of the leaf sheath can also sense gravity (Gould, 1968). In the Festucoid grasses, such as oats, wheat, and barley, only the leaf-sheath base possesses this specialized tissue. The gravity-sensing tissue is often called a pulvinus, but it is actually a **false pulvinus** because it acts by differential *growth* and not by reversible changes in cell volume, as do the true pulvini that control nyctinastic leaf movements in some dicots (Section 19.1). The grass false pulvinus is a highly specialized organ of meristematic and maturing cells that both detects and responds to gravity (Dayanandan et al., 1976, 1977). As the cells mature, they become parenchyma and collenchyma with small amounts of vascular tissue. Response to gravity consists of differential cell elongation, the lowermost cells elongating while the uppermost do not grow at all (Fig. 19-19). Continued differential growth of the pulvinus can bring the shoot to a near vertical position, usually after 48 to 72 hours (Fig. 19-20).

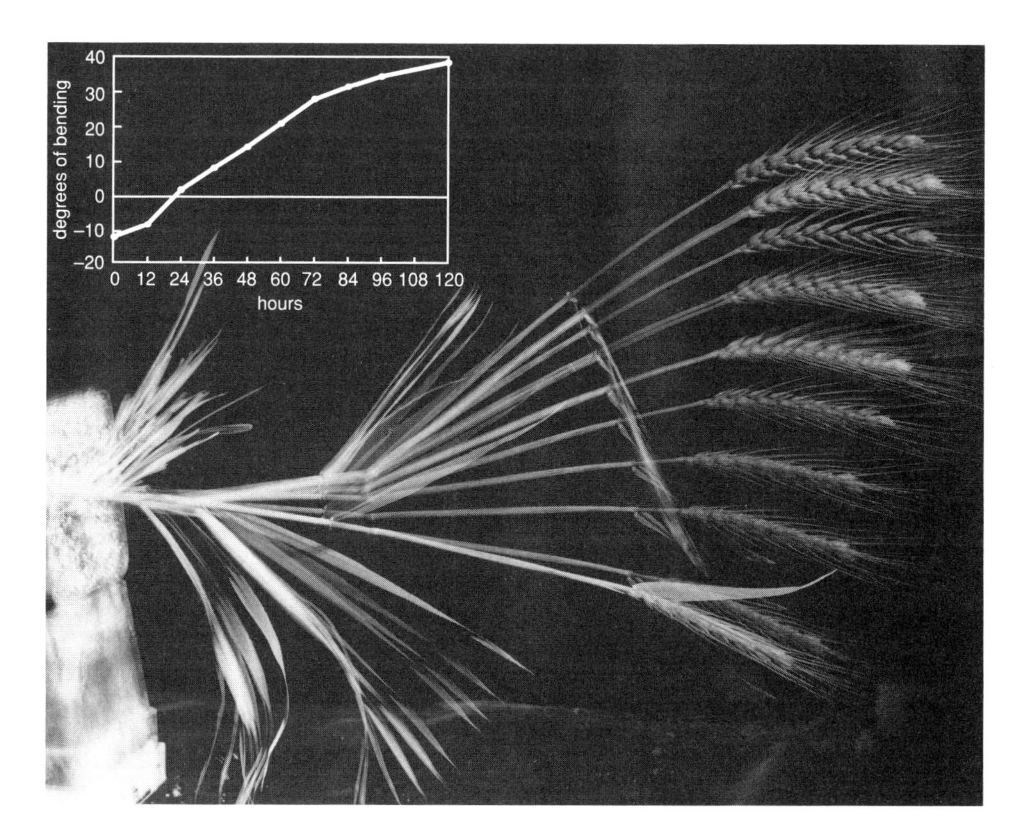


**Figure 19-20** A multiple exposure showing gravitropism in a wheat stem. The plant was turned on its side for the first exposure (after removing a number of other stems from the pot), and the other exposures were made at about 12-h intervals. The course of bending is shown by the graph in the upper left corner; it was unusually slow, perhaps because the nodes were too mature. Note that only two nodes responded; the node just inside the rim of the pot was mature and did not respond. (Photograph by Linda Gillespie and F. B. Salisbury. Three times over the weekend, the two photographers misunderstood their assignments; each came in and made an exposure but at slightly different times so they didn't meet—as you can see by the double images.)

## Studying the Gravitropic Responses of Cereal Grasses

*Peter B. Kaufman*

*Peter B. Kaufman is a Professor of Botany at the University of Michigan in Ann Arbor. He is a dynamic lecturer who has participated in writing general botany textbooks for students in the more applied botanical sciences. As he travels to scientific meetings, he always carries a camera to take pictures for use in his teaching and writing. As an expert in the field of plant gravitropism in cereal grasses, he often collaborates with scientists from all over the world. Here he tells of his current interests, updated for this edition of our textbook.*

We embarked on studies of gravitropic responses in grass shoots in about 1978 in response to the need for basic research on how cereal grass shoots recover from lodging (falling over due to the action of wind and/or rain)—or in more technical terms, the negative gravitropic curvature response in the shoots. Such studies are of importance for agriculture because lodging in cereal grasses is responsible for serious losses in grain yield. Any way by which we can understand how lodging can be prevented is of immense benefit to agriculture. There is another reason why we are exploring this interesting response in cereal grains. We intend to grow such grains as rice and wheat in NASA space vehicles such as the Space Shuttle, Space Lab, and Space Station, where gravity influences are almost absent. We want to grow such plants in space for food (grain production) for occupants of the space vehicles. Further, we want to use grain crops to enrich the cabin atmosphere with oxygen and moisture and to recycle organic wastes. [See the essay on *Limiting Factors and Maximum Yields* on page 560.] Finally, we want to grow cereals in space to understand how they grow and how we can control their growth in a near-weightless environment. One of the major questions is: How do cereal grass shoots grow when their gravity-perception mechanisms are not stimulated? And, if they tend to grow in all directions, how can we direct their growth to obtain normal shoot growth and acceptable yields?

Here on earth, we have been vigorously pursuing three basic questions concerning the mechanisms by which cereal grass shoots grow upwards when they are lodged (gravistimulated). They are (1) How is gravity perceived in the shoots? (2) How are hormones involved in the transduction process? and (3) What is the physiological and metabolic basis of the differential growth response that takes place in the grass pulvini?

It was most exciting for us, with the help of Casey Lu, P. Dayanandan, Il Song, Tom Brock, and C. I. Franklin, when we found that starch grains in amyloplasts are the graviperceptive organelles in the cells near vascular bundles in the leaf-sheath false pulvini. They cascade to the bottoms of the cells (statocytes) within 2 minutes, starting within 15 seconds. If the stem segments with pulvini are placed in the dark for five days, the starch grains disappear and the pulvini will not curve upward when gravistimulated. However, if after dark treatment we feed them with sucrose, the pulvini reform starch and are once again competent to respond to gravity.

Once these starch statoliths fall, how is the gravity stimulus transduced? We can think of several possibilities: (1) They act as pressure probes on the plasma membrane to open more ion and/or hormone channels; (2) they serve as such information carriers as $Ca^{2+}$ or hormone-deconjugating enzymes; and (3) they may provide substrate for energy production, cell-wall synthesis, or osmotic-pressure maintenance during growth.

---

Starch statoliths sediment within 2 min of horizontal placement, and a curvature response is initiated within 15 to 30 min of gravity stimulation. If barley nodes are treated with α-amylase or placed in the dark for five days, the starch disappears and no graviresponse occurs; reconstitution of the starch by treatment with 0.1 M sucrose restores the upward-bending response in these false pulvini (Kaufman et al., 1986; Song et al., 1988).

As with stems and roots, there are many complications and few positive answers. For example, IAA, GA, and ethylene are all known to accumulate more in the lower half of the pulvinus following gravistimulation, but the ethylene first appears about 5.5 h after bending begins (see the personal essay of Peter Kaufman above).

Brock and Kaufman (1988) reported evidence for changes in tissue sensitivity to auxin in the false pulvini of oats (*Avena sativa*). The experiments were basically similar to those described for sunflower and soybean.

*Other organs* Stamens, flower peduncles, various fruits, leaves, and other organs are also known to be gravitropically sensitive, although relatively little study has been devoted to these organs. In experiments in which stems are restrained to the horizontal as described above, leaf blades that are not restrained assume a more or less horizontal position. It appears that

During the transduction process, we find that both IAA and GAs become asymmetrically distributed in the pulvinus. Typical top/bottom hormone ratios are 1:1.5 to 1:2.0. Tom Brock in our lab found that gravistimulated pulvini become more sensitive (responsive) to exogenously added IAA than do those in vertical orientations. Further, they only bend in response to $GA_3$ treatment in the horizontal (gravistimulated) position.

How can we account for the top/bottom asymmetries in endogenous-free IAA and free $GA_3$ that develop during the graviresponse? Because neither IAA nor $GA_3$ are transported downward to any significant extent in gravistimulated pulvini, they must be increasingly released from bound form or increasingly synthesized from top to bottom of the pulvinus. We are currently investigating both possibilities for auxin and native GAs in cereal grasses, especially the kinetics of change in amounts of free IAA and its conjugates and of free GAs and their glucosyl-ester conjugates during upward bending. We are studying the GAs in collaboration with Dick Pharis [see his personal essay in Chapter 23] at the University of Calgary and have carried out our initial studies on free IAA and its conjugates with Bob Bandurski and Jerry Cohen at Michigan State University and the USDA at Beltsville, respectively. Both are actively helping us with the auxin extraction and purification procedures for identifying the free IAA and its conjugates by GC/MS (gas chromatography/mass spectrometry).

What about the last part of the gravitropic response in cereal grass shoots, namely, the differential growth response? We know that we are dealing with growth that occurs as a result of differential cell elongation. No cell division is involved in graviresponding cereal-grass pulvini. What is the mechanism by which differential cell elongation occurs? We know from recent studies of P. Dayanandan that cellulose synthesis is required; "Daya" also found that both RNA and protein synthesis are necessary. Carrying this a step further, we began to look at the kinds of proteins synthesized in graviresponding pulvini compared with those from upright pulvini. Il Song developed superb methods in our lab for extracting salt- and alkali-soluble proteins and for separating them by SDS/PAGE (sodium dodecyl sulfate/polyacrylamide gel electrophoresis) and isoelectric focusing. What we find so far is that early (within 2 hours) after barley shoots are first gravistimulated, at least five new proteins are made in the lower halves of the graviresponding pulvini where cell elongation is greatest. We think that one of these is glucan synthase and another is invertase. Top candidates for the early, newly synthesized proteins are wall-loosening enzymes such as endo-β-glucanase. In collaboration with David Rayle at San Diego State University and Nick Carpita and David Gibeaut at Purdue University, we found that differential top/bottom wall loosening and β-D-glucan content (mixed β-1,3- and β-1,4-linked glucan) change markedly (increase more in the lower halves) in response to gravistimulation. Thus, important steps in the growth-response mechanism involve both cell-wall loosening and cell-wall synthesis. With the help of Casey Lu, Donhern Kim, and "Raja" Karuppiah, we are now focusing on cell-wall hydrolases/acid growth as part of the cell-wall loosening process, sucrose hydrolysis by invertase, glucan synthase-mediated cellulose synthesis, and β-D-glucan synthesis as part of the cell-wall synthesis step. We are currently making monoclonal antibodies to invertase and glucan synthase to do tissue-specific enzyme localization, cloning and sequencing the genes for both enzymes, and determining how IAA and GAs cause upregulation of their expression.

much of a plant is sensitive to gravity, including those leaves, lateral roots, rhizomes, stolons, and branches that grow at some orientation other than the vertical (plagiotropic growth). The angle will depend on the plant's species and age, the organ involved, and sometimes on various environmental factors.

## 19.6 Other Tropisms and Related Phenomena

Several other tropisms have been described over the years, but most have not been studied in detail. Various workers (including the Darwins, 1880) have reported that plants will grow toward water (**hydrotropism**), certain chemicals (**chemotropism**), electric currents (**electrotropism**), and so on. Many plant organs, especially tendrils, respond to touch (**thigmotropism**). Tendrils bend toward the point of contact, wrapping around a support. Thigmotropism was reviewed by Jaffe and Galston (1968).

Mordecai J. Jaffe, H. Takahashi, and Ronald L. Biro (1985) studied a mutant of pea (*Ageotropum*), the roots of which responded neither to gravity nor to light but did respond to a humidity (moisture) gradient (**hygrotropism**). The roots often grew up out of the soil, but if they encountered relative humidity in the atmosphere that was below 80 to 85 percent, they turned and grew back into the soil. Removal of the root cap eliminated most of the hygrotropic response, although the rate of root growth was not affected, suggesting that the hygrotropic response was localized in the root cap, as is the gravitropic response.

## Circumnutation: Nastic or Tropistic?

Again it was the Darwins (1880) who called our attention to the fact that a stem tip appears to trace a more or less regular ellipse as the stem grows, completing a single cycle in various times that depend on species (for example, less than 2 h for a sunflower seedling). The Darwins called this phenomenon **circumnutation**. Its utility for climbing plants is obvious because the stem is likely to encounter some kind of support as it circumnutates. Any function it may serve in nonclimbing stems is not yet evident, but the Darwins suggested that virtually all plant movements that we have been discussing were modifications of circumnutation.

They also suggested that circumnutation occurs in response to some internal, rhythmical control; that is, that circumnutation is endogenously controlled, is a nastic movement. More recently, Johnsson (1971, 1979) provided a mathematical model based on the assumption that circumnutation is really a matter of gravitropic overshoot, in which case it would be a tropistic response. As a stem reaches the vertical in response to gravity, it may continue to grow more on what was the lower side, passing the vertical before an opposite gravitropic response takes over. This overshooting would lead to an oscillation. The mathematical model shows that it is not difficult to convert a back-and-forth waving to an elliptical or even circular motion.

If the overshoot model is correct, circumnutation should not occur on a clinostat, nor in an orbiting space laboratory. Although the effect could be secondary, many species do stop circumnutating on a clinostat (Brown and Chapman, 1988); others do not. In a space experiment, sunflower seedlings continued to nutate for many hours in microgravity (less than $10^{-3}$ g; Brown and Chapman, 1984). Other ways have been found to separate gravitropism from nutation. For example, when laid on their sides some stems continue circumnutations during upward bending (Britz and Galston, 1982, 1983), an observation that is incompatible with an overshoot theory (Heathcote and Aston, 1970). The weight of evidence seems to be against an overshoot mechanism, but such a mechanism could function in some species and not others.

## Reaction Wood

As a plagiotropic tree branch grows, it might be expected to bend downward because of its increased weight and distance from the trunk. Such a phenomenon is sometimes observed, but it is resisted by formation of *reaction wood* (Scurfield, 1973; Wilson and Archer, 1977). **Reaction wood** is the increased xylem produced on either the upper or lower side of a branch by more rapid division of the vascular cambium on that side. In conifer (softwood) limbs, reaction wood forms on the lower side and by expansion *pushes* such limbs more upright, maintaining a more constant angle. Tracheid walls become abnormally thick and contain more lignin and less cellulose than usual. Such wood is called **compression wood**. In angiosperm (hardwood) trees, reaction wood forms on top and *contracts* to pull the branch toward the trunk by tension. Called **tension wood**, it contains more fibers and fewer vessel elements than normal wood. The fibers and fiber tracheids attain secondary walls that are thicker than in normal wood because they form an additional cellulose-rich layer in the secondary wall. In contrast to conifer compression wood, tension wood in hardwoods does not become unusually rich in lignin. Reaction wood is reviewed in a three-volume, encyclopedic treatise by Tore E. Timell (1986), in which some 8,100 references are cited!

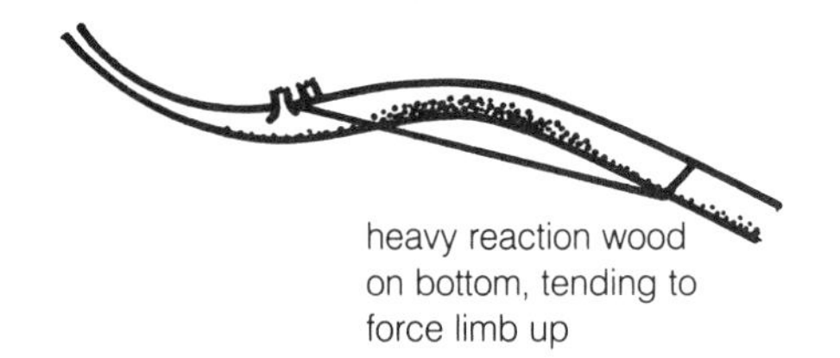


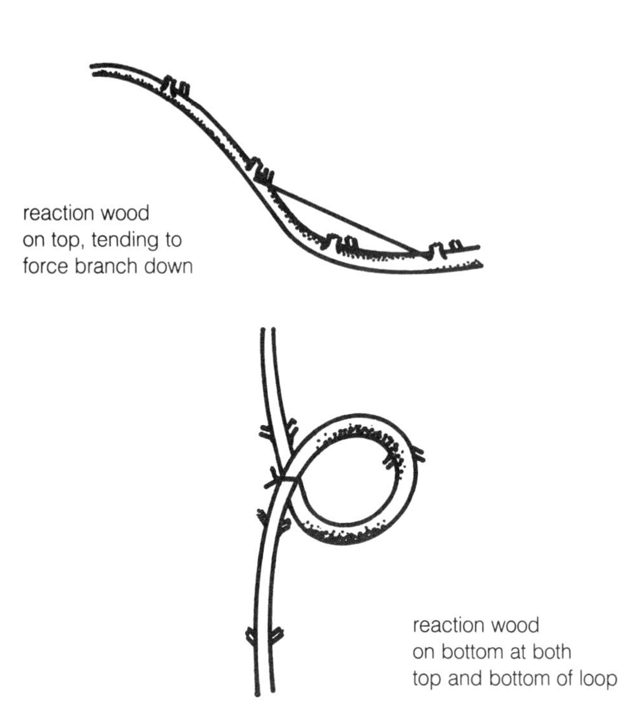


**Figure 19-21** Summary of some experiments that cause formation of reaction wood in conifers (indicated by shading).

What is reaction wood a reaction to? It could be a gravitropic response (a mechanism to maintain plagiotropic growth), or a response to tensions and pressures resulting from bending, or both. If a cable is tied to a pine limb, causing the end to bend down (Fig. 19-21), reaction wood forms on the lower side. Either explanation might account for this effect. If the cable causes an upward bend, reaction wood forms on the upper side. One might imagine that either pressures (compression) on cells of the concave side or tensions on the convex side are more likely to account for the positions of reaction wood than is the direction of the gravitational force. But if a young leader of a pine tree is wrapped into a complete loop, reaction wood forms on the lower side of both horizontal parts of the loop. Because the lower side of the upper part is under pressure and the lower side of the bottom part is under tension, the second hypothesis is inadequate. On balance, gravity appears to be more important than other factors in the formation of reaction wood, but other factors do seem to be involved (Timell, 1986).

Redistribution of IAA or other hormones, especially ethylene, might explain these results, but the cause of hormone movement may prove complex. When young plants of Arizona cypress (*Cupressus arizonica*) were treated with ethylene, branches bent upward (hyponasty). Other treatments that cause an increase in endogenous ethylene (for example, decapitation, $GA_3$, and certain levels of IAA) also induced branch hyponasty. Furthermore, when mercuric perchlorate was used to remove ethylene from the air around the plants, branches grew downward. Thus, ethylene could be involved in plagiotropic growth of branches, or the ethylene could reflect changes in auxin or other compounds (Blake et al., 1980).

Clearly, much has been learned since the Darwins investigated the power of movement in plants over a century ago. But it is equally clear that much remains to be learned. Few mechanisms are really understood.

# 20

# Photomorphogenesis

Light is an important environmental factor that controls plant growth and development. A principal reason for this, of course, is that light causes photosynthesis. Furthermore, light influences development by causing phototropism (see Section 19.4). Numerous other effects of light that are quite independent of photosynthesis also occur; most of these effects control the appearance of the plant—that is, its structural development or morphogenesis (origin of form). The control of morphogenesis by light is called **photomorphogenesis**.

For light to control plant development, the plant must first absorb the light. There are four kinds of photoreceptors known to affect photomorphogenesis in plants:

1. **phytochrome**, which absorbs most strongly red and far-red light and about which we know the most. Phytochrome also absorbs blue light. At least two major kinds of phytochrome are known.

2. **cryptochrome**, a group of similar, unidentified pigments that absorb blue light and long-wave ultraviolet wavelengths (the UV-A region, about 320 to 400 nm). Cryptochrome was named because of its special importance in cryptogams (nonflowering plants).

3. **UV-B photoreceptor**, one or more unidentified compounds (technically not pigments) that absorb ultraviolet radiation between about 280 and 320 nm.

4. **protochlorophyllide** ***a***, a pigment that absorbs red and blue light and becomes reduced to chlorophyll *a*.

In this chapter we describe some effects of each of these photoreceptors, emphasizing phytochrome because most is known about it and because it seems to be the most important photoreceptor in vascular plants. Phytochrome and other photoreceptors control morphogenic processes beginning with seed germination and seedling development and culminating with formation of new flowers and seeds.

As implied from the above statement that photomorphogenesis is controlled at many stages in a plant's life cycle, individual processes are highly specific for a particular plant part at a particular developmental stage. (In Chapters 17 and 18 we emphasized the same phenomenon for regulation of developmental processes by hormones.) Light itself carries no morphogenic information, and it is also unlikely that the kind of photoreceptor is a specific information carrier. Rather, the *competence to respond or sensitivity of the cells is the crucial factor*. Hans Mohr, a leading researcher in plant photomorphogenesis for many years, has emphasized repeatedly that photomorphogenesis has two important stages: *pattern specification*, in which cells and tissues develop and become competent to react to light, and *pattern realization*, during which time the light-dependent process occurs (Mohr, 1983). Another important aspect of photomorphogenesis is the *need for an amplification system*, as emphasized for hormone action in Figure 17-2. The number of molecules in a plant that become altered by light may be several thousand or several million times the number of photons that cause the response. In many (but far from all) cases, gene activation represents part of the amplification process.

Some photomorphogenic effects of light can easily be noted by comparing seedlings grown in light with those grown in darkness (Fig. 20-1). Large seeds with abundant food reserves eliminate the need for photosynthesis for several days. Dark-grown seedlings are **etiolated** (French, *etioler*, "to grow pale or weak"). Several differences caused by light are apparent:

1. Chlorophyll production is promoted by light.

2. Leaf expansion is promoted by light, but less so in the monocot (maize) than the dicot (bean).

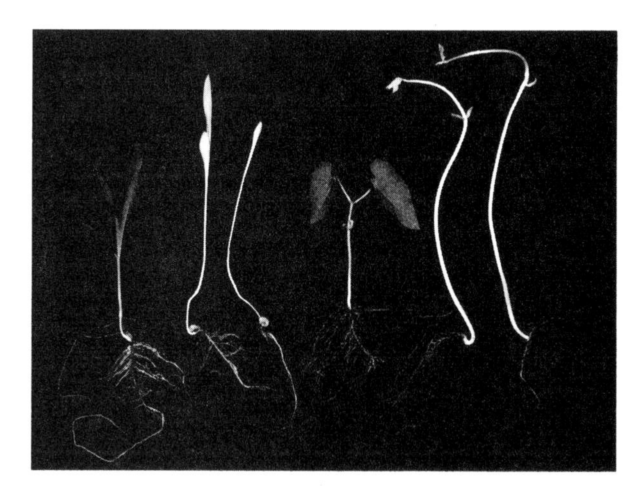

**Figure 20-1** Effects of light on seedling development in a monocot (maize) and a dicot (bean). The plant at the left of each group was germinated and grown in a greenhouse, whereas the other representatives of each were grown in continuous darkness for 8 days. (Photograph by Frank B. Salisbury.)

3. Stem elongation is inhibited by light in both species. (The maize stem is short and not visible in Figure 20-1 because it is surrounded by leaf sheaths that extend nearly to ground level in such young plants.)
4. Root development is promoted by light in both species.

All these differences seem advantageous for a seedling if it is to extend its stem through the soil and if its leaves are to reach the light. More of the food reserves in the endosperm (maize) or cotyledons (bean) are used to extend the stem upward in darkness than in light, and less food is used to develop leaves and roots and to form chlorophyll, all of which are less important for a dark-grown plant. Besides these light effects, many others are essential to monocots, dicots, gymnosperms, and many lower plants.

## 20.1 Discovery of Phytochrome

The discovery and isolation of phytochrome and the demonstration of its importance as a pigment that controls photomorphogenic responses is one of the most brilliant and important of all physiological accomplishments. Most of the research leading to phytochrome detection and isolation was accomplished at the U.S. Department of Agriculture Research Station in Beltsville, Maryland, between 1945 and 1960. The history of phytochrome discovery was summarized by one of its pioneers, Harry A. Borthwick (1972), by Briggs (1976), and more briefly by Furuya (1987a). Sterling B. Hendricks, another pioneer, described some aspects of the discovery of phytochrome in his essay in this chapter.

An important observation had already been made at Beltsville by W. W. Garner and H. A. Allard in about 1920. They found that the relative durations of light and dark periods control flowering in certain plants (see Chapters 21 and 23). Then, in 1938, it was discovered by others that the cocklebur, which requires nights that are longer than some critical minimum length to flower (that is, it is a short-day plant), is prevented from flowering by a brief interruption of its dark period with light. Red light proved much more effective than other wavelengths, not only for interrupting the long nights that otherwise induce cockleburs and Biloxi soybeans to flower, but also for promoting expansion of pea leaves. Red light interrupting a dark period was also the most effective color of light in promoting flowering of Wintex barley and other long-day plants that require nights that are shorter and days that are longer than some critical length.

Borthwick and Hendricks then collaborated with experts who were familiar with seed dormancy in many species. They constructed a large spectrograph that, with a bright light source, could separate various colors of light over areas so great that even potted plants could be lined up and exposed to different wavelengths. They obtained an action spectrum with a peak in the red wavelengths for promotion of germination of Grand Rapids lettuce seeds, only 5 to 20 percent of which will usually sprout in darkness. It had already been shown in the 1930s that red light promoted germination of such seeds but that blue or far-red light inhibited germination to rates below even those obtained in darkness. **Far-red light** *includes those wavelengths just longer than the red, covering approximately the range from 700 to 800 nm.* (Those wavelengths longer than about 760 nm are invisible to humans and technically are near infrared, as shown in Appendix B, Figure B-2. Visible far-red wavelengths appear dark red to us.) The Beltsville group then made a remarkable discovery.

When seeds were exposed to far-red light just after a promotive treatment with red light, promotion was nullified; but if red light was given after far-red light, germination was enhanced. By repeatedly alternating brief red and far-red light treatments, they found that the color of light last applied determined whether the seeds germinated or not, with red light promoting and far-red light nullifying that promotion (Fig. 20-2). Furthermore, inhibition of flowering in short-day plants by interruption of a long night with red light was largely overcome by immediately following red light with far-red light.

By that time investigators realized that a blue pigment was present that absorbed red light (called $\mathbf{P_r}$); but its concentration was too low to give color to etiolated

## The Discovery of Phytochrome

### *Sterling B. Hendricks*

*The discovery of a new process in the biological world is always exciting, and when it proves to be an important one, it may be epochal. This was true for the discovery of phytochrome. In 1970, we asked Sterling B. Hendricks to tell us of the discovery. Sterling Hendricks died on January 4, 1981.*

In 1945, Harry A. Borthwick, Marion W. Parker, and I set out to find out something about how plants recognize day lengths. Our method was to note changes in flowering induced by breaking long nights with periods of light of various wavelengths and intensities—or, more exactly, to measure action spectra. A low energy of 660-nm (red) light proved most effective in preventing flowering of the short-day soybean and cocklebur and in inducing flowering of the long-day barley and henbane. These oppositely responding plants had the same action spectrum, in all details, for flowering change. The near infrared, 700 to 900 nm, appeared to be ineffective.

Instead of elaborating the finding about flowering, we (with Frits W. Went) next chose to measure action spectra for inhibition of etiolation (the elongation of stems and restriction of leaf expansion that occur in the dark). Suppression of stem lengthening of barley and enhancement of pea-leaf size by light gave the flowering action spectrum. It was exciting to know that such diverse displays were related in initial cause.

A need for light in germination of some seeds had been known for about a century. Our first measurements (in 1952 with Eben H. Toole and Vivian K. Toole) were on lettuce seeds, which Lewis H. Flint and E. D. McAllister had studied 17 years earlier. Again, 660-nm light was most effective for promoting germination. Seeds placed in the 700- to 800-nm region of the spectrum, however, germinated less than the 20 percent value of those in darkness (controls). Flint and McAllister had found germination to be suppressed in this region to 50 percent germination when seeds were initially potentiated by light. The differences in experimentation had no immediate import to us.

But one day in 1952, during an action spectrum measurement, the thought came that the 700- to 800-nm region had not been correctly tested in flowering controls. If plants had been first exposed to 660-nm light to potentiate the flowering change, instead of being taken directly from darkness, the 700- to 800-nm region might change them back to their dark condition. The 700- to 800-nm region, in truth, might cause a response similar to darkness rather than being without effect, as we had earlier supposed. This region, when tested with lettuce seeds, did indeed suppress potentiated germination from a high to a low value, with maximum effectiveness near 730 nm. The reversal of the potentiated response with short-day plants for flowering also was well enough borne out to support a generalization.

The photoreversibility was the touchstone for deeper understanding of the equivalence of initial action in diverse phenomena. It also led to eventual isolation of the pigment *phytochrome*. The day in 1952, though, held a further nuance that is not so widely appreciated, nor was it quickly grasped by us. We had done quite a different experiment than had Flint and McAllister. The lettuce seeds in our experiments were exposed for short times to low total energy of moderate intensity, whereas they used high total energy given by continuous exposure at low intensity. The photoreversibility was thus quickly and convincingly evident to us but not to them—as was its implication of determination at the molecular level by quick change in form of a light-absorbing pigment.

In an earlier paragraph, the reversibility of flowering was expressed as ". . . well enough borne out. . . ." The reversal, in truth, was rather poor. The more we tried to enhance it by increasing the energy through long exposure, the poorer it became. In fact, 730-nm radiation at low intensity for an hour inhibited cocklebur flowering as effectively as did low-energy exposure at 660 nm. This was recognition of what is now called the "high-irradiance reaction." Flint and McAllister were likely dealing with this reaction in lettuce seeds rather than with the simple reversibility of phytochrome. The high-energy reaction, which has many facets in nature, is also thought to arise in part from phytochrome, but in a more complex way than by photoreversibility alone. Finally, our initial objective—determining how plants recognize day length (or night length)—is somewhat a will-o'-the-wisp. Phytochrome action is involved, but what the action might be and how it ties in with endogenous biological rhythms is still under vigorous debate. [This remains true in 1991.]

maize seedlings, in which it was first detected by absorption changes with a spectrophotometer. Researchers also decided that the pigment could be converted by red light to a different form (called $\mathbf{P}_{fr}$) that absorbed far-red light (a form that eventually proved to be olive-green in color) and that the blue pigment could be regenerated with far-red light. The olive form produced by red light was deduced to be the active form, whereas the blue form seemed to be inactive. These ideas were based on physiological and spectrophotometric studies with seeds or etiolated plants, and they needed to be verified by extracting the pigment and studying it *in*

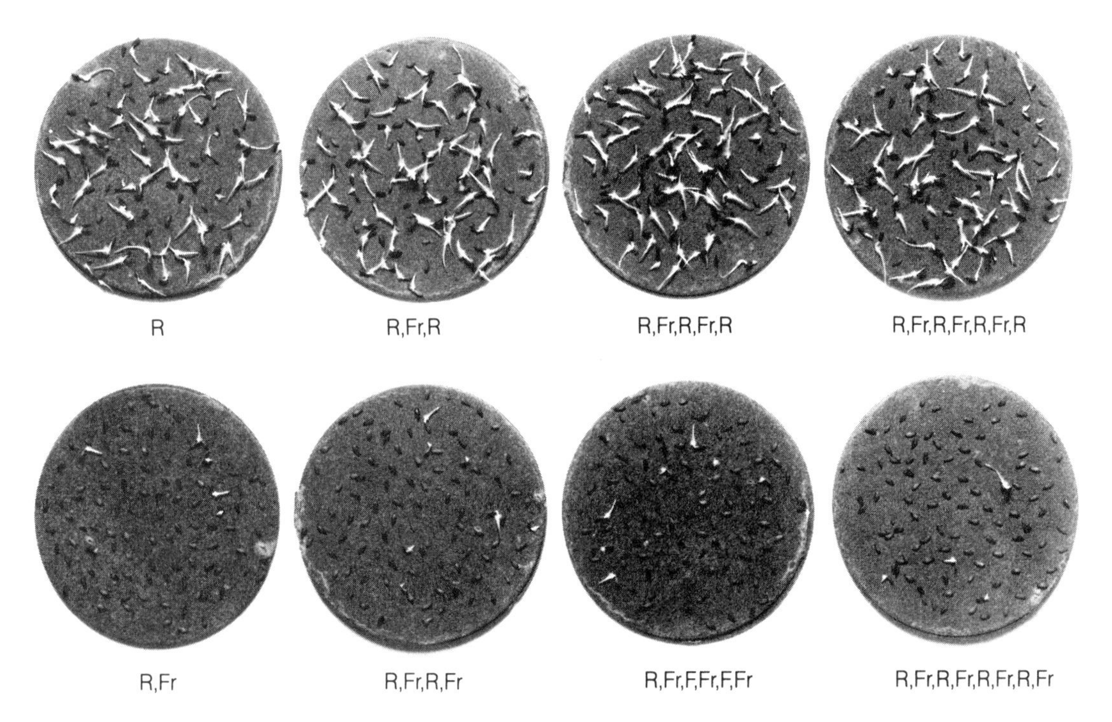


**Figure 20-2** Reversal of lettuce-seed germination with red (R) and far-red (Fr) light. Red exposures were for 1 min and far-red for 4 min. If the last exposure is to red light, seeds germinate; if it is to far-red light, they remain dormant. Temperature during the half hour required to complete the treatments was 7°C; at all other times it was 19°C. (Courtesy Harry Borthwick.)

*vitro*. (This has been the scientific approach for all biological pigments, including rhodopsin for vision, chlorophylls and carotenoids in photosynthesis, and cytochromes in respiration.)

In the early 1960s, H. W. Siegelman and other protein chemists purified phytochrome from homogenates of cereal-grain seedlings by using column chromatography and other techniques routinely used to purify proteins. They demonstrated that isolated phytochrome changes color reversibly upon exposure to either red or far-red light. Essentially all the early deductions based only upon physiological experiments with whole plants had been verified. Even the absorption spectrum of each form of phytochrome was measured. These investigators' conclusions are shown by the photoreversible scheme below:

$$P_r \underset{\text{far-red light}}{\overset{\text{red light}}{\rightleftharpoons}} P_{fr}$$

## 20.2 Physical and Chemical Properties of Phytochrome

Nearly all studies of phytochrome have been made with pigment purified from etiolated seedlings. There are two reasons for this. First, seedlings grown in total darkness contain 10 to 100 times as much total phytochrome as seedlings grown in light. Second, they have no chlorophyll that absorbs blue and red light and interferes with spectrophotometric studies of phytochrome. Only recently has phytochrome from green plants been studied much, and we return to its properties later.

Absorption spectra of highly purified phytochrome molecules from etiolated angiosperms have maxima in red wavelengths at about 666 nm for the red-absorbing bright blue form ($P_r$) and at about 730 nm for the far-red-absorbing olive form ($P_{fr}$; Vierstra and Quail, 1983, 1986). Figure 20-3a shows an example of these absorption spectra. The absorption spectrum for $P_{fr}$ has a shoulder in the red region (near 666 nm) that is caused by $P_r$, not $P_{fr}$; $P_r$ is present because it is impossible to convert more than about 85 percent of the $P_r$ to $P_{fr}$ in a phytochrome sample. More effective conversion cannot occur because part of the $P_{fr}$ is converted back to $P_r$ when $P_{fr}$ itself absorbs red light. Figures 20-3b and 20-3c illustrate action spectra in the red and far-red regions for various physiological responses. *Similarities of the absorption spectra of phytochrome and the action spectra of plant responses is one important evidence suggesting that phytochrome is the pigment causing such responses. A second evidence is that responses caused by red light are almost always nullified by an immediate subsequent exposure to far-red light. A third evidence is that only low irradiance levels of either red or far-red light that are capable of interconverting phytochrome from one form to another cause these responses.*

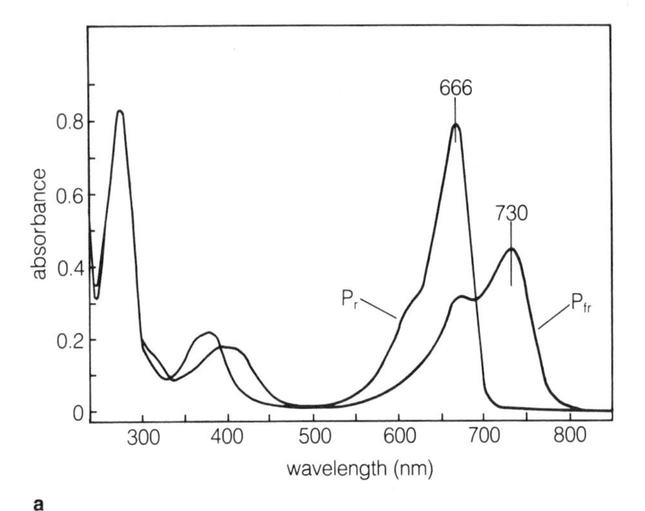


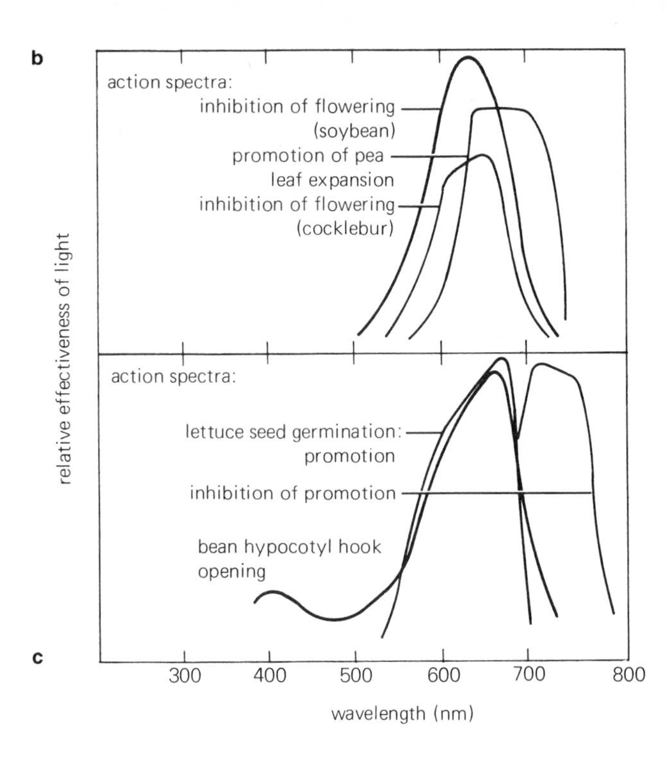


**Figure 20-3** A comparison of the absorption spectra of both forms of phytochrome with action spectra for various physiological processes. (Absorption spectra are from Vierstra and Quail, 1983. Action spectra in **b** were redrawn from data of Parker et al., 1949. Action spectra for promotion of bean hypocotyl hook opening in **c** were redrawn from Withrow et al., 1957, and spectra for promotion and subsequent inhibition of Grand Rapids lettuce were redrawn from Borthwick et al., 1954.)

In leaves and stems, light absorbed by chlorophyll alters both the action spectra for phytochrome responses caused by $P_{fr}$ and the amount of light (absorbed by $P_r$) required for the response. Action-spectra peaks are shifted toward shorter wavelengths near 630 nm (at which wavelengths chlorophyll absorbs less), and much more energy is needed for response (Jose and Schäfer, 1978; also see action-spectra peaks for inhibition of flowering of soybean and cocklebur in Fig. 20-3). Both $P_r$ and $P_{fr}$ absorb violet and blue light, but low irradiance levels of these wavelengths are much less effective than red or far-red light for the physiological processes we have described so far. Because neither $P_r$ nor $P_{fr}$ absorbs green light effectively and because our eyes are especially sensitive to green, safelights with filters that transmit only low-intensity green light are used in physiological experiments in which phytochrome participates. In general, green safelights and green filters are highly useful in phytochrome studies, but they must provide low irradiances and they must be tested to be sure that they cause no response. In some especially sensitive responses to be mentioned later, not even a green safelight is safe.

Chemically, phytochrome is a homodimer of two identical polypeptides, each with a molecular weight of about 120 kDa (reviewed by Vierstra and Quail, 1986). Each polypeptide has a prosthetic group called a **chromophore** that is attached via a sulfur atom in a cysteine residue of the polypeptide. This chromophore is an open-chain tetrapyrrole similar to the photosynthetic phycobilin pigment of red algae and cyanobacteria. It is the chromophore, not the protein, that absorbs the light that causes phytochrome responses. When $P_r$ is converted to $P_{fr}$ by red light, there is apparently a *cis-trans* isomerization in the chromophore (Fig. 20-4). Alteration of the chromophore in phytochrome then causes several unidentified subtle changes in the structure of the phytochrome protein (Song and Yamazaki, 1987; Hansjörg et al., 1989). The changes in protein structure are somehow responsible for the physiological activity of $P_{fr}$ and the inactivity of $P_r$.

During the mid-1980s reports about chemical, spectral, and immunological properties of phytochrome from green plants began to appear (reviewed by Furuya, 1989). It is now clear that two major types of phytochrome exist: **type 1** and **type 2** (Furuya, 1989; Tokuhisha and Quail, 1989; Cordonnier, 1989). Type 1 predominates in etiolated seedlings, and type 2 predominates in green plants and seeds (at least oat seeds). Type 2 from green oat seedlings is slightly smaller (118 kDa) than that from etiolated oats (124 kDa), but it too exists as a dimer *in vivo* and has spectral properties similar but not identical to those of etiolated type 1. One important spectral difference in types 1 and 2 of oats is that the $P_r$ form of type 2 has an absorption maximum near 654 nm, as opposed to the 666 nm peak of type 1 $P_r$ (Tokuhisha and Quail, 1989). If type 2 $P_r$ absorbed maximally at 654 nm in all species, that would help explain the differences in action spectra of dark-grown and green plants plotted in Figure 20-3; however, type 2 $P_r$ from green pea seedlings has an absorption peak at 667 nm, the same as type 1 from dark-grown seed-

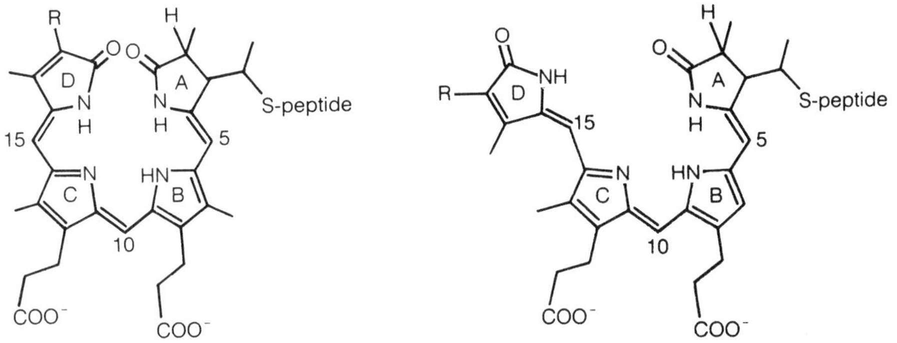

Figure 20-4 Postulated structure of the chromophore of $P_r$ (left) and of $P_{fr}$ (right). (From Rüdiger, 1987. Also see Rüdiger, 1986.)

lings. Type 2 $P_{fr}$ formed from both green pea and oat seedlings has an absorption maximum *in vitro* at 724 nm, as opposed to 730 nm for type 1 phytochrome from various plants.

Based on immunological reactions, the proteins of phytochrome types 1 and 2 are quite different, even though some important similarities exist (Cordonnier, 1989). However, even within a single green plant more than one type 2 form occurs. For example, Sharrock and Quail (1989) obtained evidence for four or five different genes in *Arabidopsis thaliana;* three of these (one type 1 and two type 2) were active *in vivo,* so they could be cloned (see Chapter 24), allowing isolation of enough DNA for chemical analysis. These three genes were found to have nucleotide sequences that differ significantly, indicating that each gene controls formation of a different phytochrome protein. The difference between the type 2 phytochromes is as great as that between either of them and type 1 (roughly 50 percent homology in amino-acid sequences predicted by the genetic code in all comparisons). The structures of these genes and the phytochromes they code for are now being studied intensively. Meanwhile, it is clear that whenever we speak about phytochrome, we are usually speaking of a group of related proteins that are attached to a similar if not identical chromophore. The functions of separate members in this group are likely to differ; otherwise plants would probably not produce more than one kind.

## 20.3 Distribution of Phytochrome Among Species, Tissues, and Cells

What we have said so far applies to phytochrome of angiosperms. Does phytochrome exist in other kinds of plants? It does in gymnosperms, liverworts, mosses, ferns, and some green algae, and recently it was reported to exist in certain red and brown algae (López-Figueroa et al., 1989). It is probably absent in bacteria and fungi.

Little is known about chemical properties of phytochrome in species other than angiosperms. In a green alga, in several pine species, and in the ancient gymnosperm *Ginko biloba, in-vivo* absorption peaks for $P_r$ and $P_{fr}$ occur at slightly shorter wavelengths than in angiosperms. However, even angiosperms display some variability in these peaks, partly because chlorophyll and other neighboring molecules influence the *in-vivo* absorption spectra of phytochrome. It is possible that the chromophore of all phytochromes is identical, but this has not been determined. The association of phytochrome with other molecules in a cell may influence the *in-vivo* absorption spectrum, but purified type 1 and 2 phytochromes differ, as already mentioned. These differences occur because either the chromophores are chemically distinct or the different proteins associated with a common chromophore cause changes in its absorption properties. At this time it is not possible to explain different absorption by phytochromes in gymnosperms and lower plants compared to angiosperms.

Phytochrome is present in most organs of all plants investigated, including roots. *In all plants phytochrome is present and synthesized entirely as $P_r$; apparently no $P_{fr}$ can be synthesized in darkness.* Phytochrome contents of plant parts have long been measured *in vivo* by the decrease in absorbance of the tissues when $P_{fr}$ is produced from $P_r$ by exposure of the tissues to red light. In the early 1970s, Lee H. Pratt and Richard A. Coleman developed an immunological technique for phytochrome determination that allows its identification in specific cells and subcellular organelles. Pratt and Coleman's original technique is nearly 1,000 times more sensitive than spectrophotometric methods and is also applicable to green tissues. This basic method has now been improved and made more sensitive and specific (Pratt, 1986; Pratt et al., 1986; McCurdy and Pratt, 1986). (See Dr. Pratt's essay in this chapter concerning the importance of immunological techniques to research on phytochrome and other aspects of plant physiology.) Both light microscopy and electron microscopy can be used in immunological assays.

# Antibodies and the Study of Phytochrome

*Lee H. Pratt*

*Dr. Pratt is now a Professor of Botany at the University of Georgia. In 1982 he received the prestigious Charles A. Shull award from the American Society of Plant Physiologists for some of the work he describes in this essay, which he has updated for this edition of* Plant Physiology.

**Antibodies**, also known as **immunoglobulins**, are proteins synthesized by an animal in response to exposure to a foreign agent. They are only one component of an animal's immune system, the study of which is referred to as **immunology**. Substances to which antibodies can bind are known as **antigens**. A natural function of antibodies is to destroy antigens such as bacteria and viruses. A goal of this essay is to demonstrate with reference to phytochrome that even though antibodies are not made by plants, they represent an invaluable research tool to the plant physiologist.

Following training as a graduate student in the laboratories of Winslow Briggs and Norman Bishop, I accepted in 1967 a postdoctorate position with Warren Butler, who had been part of the group responsible for the isolation of phytochrome. It was in his laboratory that I was first exposed to the notion that antibodies could be a powerful research tool. Warren Butler had already prepared antibodies to phytochrome by injecting the purified pigment into rabbits and subsequently harvesting the antibodies from blood taken from the rabbits. Because antigens can be virtually anything, including proteins, carbohydrates, plant hormones, or even nucleic acids, because antibodies can be readily labeled with such tags as fluorescent dyes, enzymes, or electron-dense metals, and because antibodies are highly specific with respect to the antigen to which they bind, antibodies can be used as antigen-specific stains. When used in this way with suitably prepared tissue samples, antibodies permit the detection of virtually any substance with the convenience of light microscopy or the resolution of electron microscopy. This particular immunochemical application is referred to as **immunocytochemistry**. Butler was especially interested in applying this technique to learn how phytochrome is distributed throughout a plant and where it is located within a cell.

When I established my own laboratory at Vanderbilt University in 1969, I began independently to explore possible applications of immunochemical methods. Having been trained as a botanist, however, my lack of background in immunology prevented me from seeing the vast potential of antibodies as a research tool. The best use of antibodies that I could come up with in my first grant proposal was to suggest that I should make antibodies to phytochrome because "cross reactivity between phytochromes isolated from different plant sources and their antibodies should give a relative measure of the extent of homogeneity between phytochromes from different taxonomic groups. . . ." Although eventually we did get around to pursuing this objective, once we began to make antibodies to phytochrome we found even more interesting things to do with them.

One of the more interesting things was the immunocytochemical visualization of phytochrome, as suggested originally by Warren Butler. With input from Ludwig Sternberger, a pioneer in this area, with the participation of Richard Coleman, initially a technician and subsequently a graduate student in my laboratory, and with our newly produced antibodies to phytochrome, we had almost immediate success. For the first time, the distribution of phytochrome throughout entire seedlings could be observed on a cell-by-cell basis. Although the terminology was not then in vogue, we were in fact assessing the expression of phytochrome genes at the protein level. A remarkable observation was that for two otherwise comparable, adjacent cells, only one might be accumulating detectable levels of phytochrome, indicating that the spatial regulation of gene expression is exquisitely precise.

With subsequent adaptation of electron-dense probes, it was also possible to determine where within the cell phytochrome was found, thereby testing, for example, the hypotheses that phytochrome might be in the nucleus interacting directly with the genome of the cell, or that it might be interacting with membranes. Apart from rare exceptions, we found no phytochrome in the nucleus. With the participation of John Mackenzie, Jr., however, we demonstrated in 1975 that $P_{fr}$ had a different distribution in the cell as compared to $P_r$, which is indicative of a possible association of $P_{fr}$ with membrane. Nevertheless, more recent information obtained with gold-labeled antibodies by David McCurdy in my laboratory, and also by Volker Speth, Veit Otto, and Eberhard Schäfer in Freiburg, indicates that this different distribution of $P_{fr}$ does not involve an association with membranes. Thus the data to date fail to support either hypothesis. Related immunocytochemical work done in my laboratory by Mary Jane Saunders, a postdoctoral investigator, and Michele Cope, a graduate student, are consistent with this conclusion.

Once we began to appreciate the tremendous potential of immunocytochemistry as a research tool, I decided in 1974 to sit in on a course in immunology at Vanderbilt's School of Medicine. This experience led to several other uses of antibodies. Not only did Robert Hunt learn to use them for affinity purification of phytochrome, a procedure that is many-fold more rapid and more powerful than other approaches to protein purification, but he also began to use antibodies for the detection of phytochrome in light-grown, photosynthetically competent plant tissues. Although the chlorophyll and exceedingly low abundance of phytochrome in such tissues prevents direct spectrophotometric detection of phytochrome, immunochemical assays are not perturbed by chlorophyll and, as Robert Hunt also learned, are up to several thousand times more sensitive. Subsequently, with the help of Susan Cundiff, Harry Stone, and Maury Boeshore, all of whom were graduate students working in my laboratory, and of Marie-Michèle Cordonnier, a postdoctoral investigator, we also used antibodies to follow phytochrome's appearance in developing seedlings, to characterize its fate during the course of the so-called phytochrome destruction reaction, to search for possible changes in its molecular properties *in vivo* as a function of its form, and to initiate comparative studies of phytochrome from different sources, as had been proposed originally as justification for making the antibodies. As was true for the work mentioned above, the resulting information could not have been obtained in most instances without the use of antibodies.

The immunochemical work mentioned so far relied solely on the use of antibodies recovered from serum of rabbits injected with phytochrome. Because phytochrome is a large protein, it is a correspondingly complex antigen, possessing perhaps 10 or so **epitopes**, each of which is that small portion of an antigen against which an antibody is made. And because an animal typically makes about 5 or 10 different antibodies to each epitope, the set of heterogeneous antibodies directed to phytochrome in the serum of a rabbit would be expected to number between 50 and 100. Not only is this set of so-called **polyclonal antibodies** a complex mixture of immunoglobulins, but there is the inherent possibility that it is contaminated by unwanted immunoglobulins, some of which might yield spurious data. Clearly, the ability to select a single antibody from such a pool of immunoglobulins would not only eliminate the possibility of spurious results arising from the presence of unwanted antibodies; it would also provide an immunoglobulin that would identify only a single epitope.

Fortunately, about 15 years ago a procedure to obtain such chemically pure antibodies, known as **monoclonal antibodies**, was developed. In collaboration with Marie-Michèle Cordonnier, who had by then taken a position at the University of Geneva, and with the technical assistance of Cecile Smith, we took advantage of this procedure to begin raising monoclonal antibodies directed to phytochrome. Although their production is complex, it is simple in principle. From the spleen of a mouse injected with an antigen such as phytochrome, one isolates cells that synthesize and secrete antibody. These cells are fused with tumor-derived **myeloma** cells, yielding what are called **hybridomas**. Hybridomas not only retain the ability to synthesize and secrete antibody; they also possess the immortality of the myeloma. A key step is then to isolate a single hybridoma cell that produces an antibody of interest and grow it in large quantities as a monoclonal cell line, which in turn secretes a single monoclonal antibody. Because a hybridoma can be preserved in liquid nitrogen, the antibody it produces can be available in whatever quantity one wants over an unlimited period of time.

Of the applications that we have made of monoclonal antibodies, two stand out. In one case we have relied upon their epitope specificity to probe discrete regions of the phytochrome molecule, thereby permitting us to learn something about its structure/function relationships. This is possible because a given monoclonal antibody interacts predominantly with only about 6 out of 1,130 amino acids in the phytochrome polypeptide. In the other case we have used the antibodies to differentiate among different phytochromes within the same plant. Even though these phytochromes are all clearly related, it is possible to discriminate among them because appropriate monoclonal antibodies can be sensitive to only a few amino-acid changes within the molecule.

Together with Marie-Michèle Cordonnier (first at the University of Geneva and subsequently at CIBA-GEIGY Biotechnology in North Carolina), Sandy Stewart at CIBA-GEIGY, and Yukio Shimazaki, a postdoctoral investigator in my laboratory, we used monoclonal antibodies to help elucidate the relationship between the different phytochromes that are most abundant in dark- and light-grown plants. This work expanded upon the discovery in 1983 by James Tokuhisha in Peter Quail's laboratory that these two phytochromes are different. Subsequently, we developed monoclonal antibodies directed specifically to phytochrome from light-grown oat plants. With the participation of Yu-Chie Wang, a graduate student in my laboratory, and Marie-Michèle Cordonnier and Sandy Stewart, we discovered that phytochrome from green oats is itself heterogeneous; that is, the use of monoclonal antibodies led to the discovery that there are in a single plant not just two, but at least three different photoreversible chromoproteins, each of which possesses the characteristics required to be called phytochrome.

Thus, the application of antibodies to the study of phytochrome has led to a number of critical advances, many of which could not have been made in any other way. Perhaps the most exciting of these advances is the discovery that a single plant contains at least three phytochromes, which leads in turn to several as-yet-unanswered questions. In particular, what is the total number of phytochromes in a single organism? Is this number greater than three? In addition, does each of these phytochromes have a different biological function, or can any one of them take the place of the others? For those of us who have worked with phytochrome for many years, it is inherently difficult to use the term *phytochrome* in the plural sense (that is, *phytochromes*), and yet that is what we must now do.

Of course, not all of our work has required the use of antibodies, nor have antibodies been responsible for all, or even most, advances in knowledge of photomorphogenesis. Nevertheless, because all immunochemical methods can at least in principle be applied to the study of any antigen, and because virtually any substance found in a plant can be an antigen, it should be evident that antibodies are an irreplaceable and invaluable research tool for any plant physiologist. That this is the case is perhaps best indicated by the exponential growth in the number of articles that describe application of immunochemical methods in, for example, the journal *Plant Physiology*. When we began in 1969, such articles were rare, whereas today they are commonplace.

Pratt and Coleman's results with dark-grown plants show that root-cap cells of grass seedlings contain high amounts of type 1 phytochrome, consistent with absorbance by the cap of light that enhances gravitropic sensitivity in certain grasses and other plants (see Section 19.5). Phytochrome distribution in grass shoots is variable, but oat, rye, rice, and barley seedlings all have high concentrations in the apical regions of the coleoptile, near the shoot apex, and (except for oat) in the growing leaf bases. Subcellularly, phytochrome exists in the nucleus and throughout the cytosol, but it does not seem to be in any organelle or membrane or in the vacuole (Warmbrodt et al., 1989). Unfortunately, knowing the location of phytochrome within cells has told us essentially nothing about how it acts. Furthermore, mainly because type 2 phytochromes in green plants are proving to be immunologically distinct from those of type 1, the immunocytochemical methods pioneered by Pratt are just now being successfully applied to locating phytochrome in green plants.

Recently, the formation of type 1 and 2 phytochromes in germinating seeds of oat and pea was studied. Both spectrophotometric and immunological methods were used when possible. Some interesting observations have emerged so far (reviewed by Colbert, 1988, 1990; Thomas et al., 1989; Furuya, 1989). Seeds of both species contain mainly type 2 phytochromes. When germination and development occur in either light or darkness, the amount of type 2 phytochromes increases gradually in late stages of imbibition, when synthesis of other proteins also begins. In 3-day-old oat seedlings, the total amount only approximately triples compared to that in the seed. In pea (studied only through 12 h of imbibition), type 2 phytochromes increase about the same amount in light or darkness. Later, the total amount of type 2 phytochromes increases even more in both species as the plant grows in light.

The amount of type 1 phytochrome in each species shows developmental changes similar to those of type 2 phytochromes in light, so that older light-grown seedlings contain roughly half type 1 and half type 2 phytochromes (Nagatani et al., 1989, 1990). However, when seeds germinate and seedlings develop in darkness, the amount of type 1 phytochrome increases by roughly 100 times; thus dark-grown seedlings contain far more *total* phytochrome than do light-grown seedlings, as mentioned earlier. This is true for all species investigated. This abundance of type 1 phytochrome probably allows these seedlings to intercept even very weak light and develop into normal green plants. When seedlings receive light, one of their responses is to lose most of their type 1 phytochrome. This occurs for three main reasons: they stop making more mRNA needed to synthesize that phytochrome, type-1-phytochrome mRNA appears to be an unstable mRNA (is rapidly hydrolyzed), and most of the type-1-phytochrome protein is rapidly degraded.

We see from the above that in darkness a gene that codes for type 1 phytochrome becomes strongly activated, but in light it is deactivated. An interesting fact is that the light that causes this deactivation is red, and it is absorbed by phytochrome itself. $P_{fr}$ then down-regulates formation of $P_r$. Whether type 1 or a type 2 $P_{fr}$ down-regulates type 1 is generally unknown, but in oat seedlings the evidence indicates that a type 2 is responsible (Thomas et al., 1989). The essay by Dr. James Colbert in this chapter describes some of his research concerning phytochrome's regulation of its own synthesis.

The ratio of $P_{fr}$ to the total amount of phytochrome of both forms is denoted by $\phi$:

$$\phi = \frac{P_{fr}}{P_r + P_{fr}} = \frac{P_{fr}}{P_{total}}$$

Red light of 667 nm converts about 86 percent of phytochrome into $P_{fr}$, so $\phi$ is 0.86. Far-red light above 720 nm removes nearly all $P_{fr}$, so $\phi$ approaches zero. Even very low irradiances with red and far-red light are adequate to establish the photoequilibrium between $P_r$ and $P_{fr}$ because both forms absorb these wavelengths so effectively. In sunlight or under incandescent light used in growth chambers, the irradiance of far-red photons is only slightly less than that of red photons (see Appendix Fig. B-3 or Fig. 20-8). For sunlight, the ratio of red photons to far-red photons is about 1.1 to 1.2. Nevertheless, $P_r$ absorbs red light more effectively than $P_{fr}$ absorbs far-red light; furthermore, $P_r$ is converted more efficiently to $P_{fr}$ than $P_{fr}$ is converted back to $P_r$. Therefore, sunlight acts primarily as a red source that forms more $P_{fr}$ than $P_r$. The $\phi$ value in sunlight is about 0.6.

In etiolated plants given a brief red-light treatment, some of the $P_{fr}$ that is formed disappears gradually, even in darkness. Two processes account for this. The first process is **destruction**, because after an interval of time in darkness it is no longer possible to regenerate as much $P_r$ in tissues by an exposure to far-red light; thus the total amount of detectable phytochrome is less than before. (The mechanism of this destruction will be described later.) The second process is **dark reversion** to $P_r$, which usually requires a few hours. (It should be emphasized that destruction and reversion also occur in the light but are not directly caused by light.) Reversion occurs in most dicots and gymnosperms but has not been detected in monocots or in any of the 10 dicot families often classified as part of the order Caryophyllales. Because of destruction and reversion, we must modify our idea that light simply sets up a photostationary reversible state between $P_r$ and $P_{fr}$. For type 1 phytochrome and dark-grown seedlings, reversion and destruction (when either is applicable) must be added, as shown in Figure 20-5. However, type 2 phytochromes

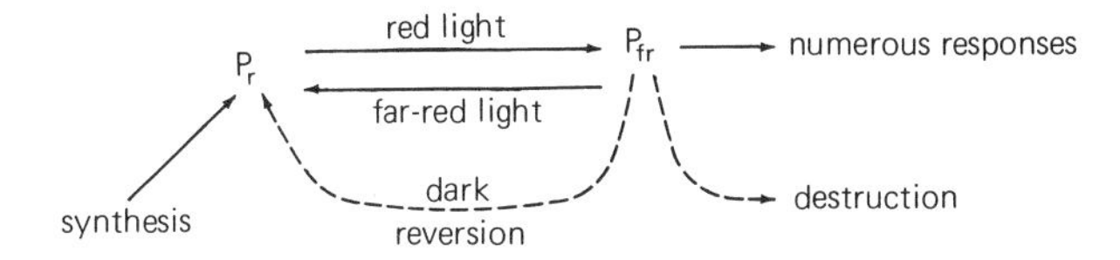


**Figure 20-5** A summary of some transformations of phytochrome. Dashed lines indicating dark reversion and destruction do not seem to occur with type 2 $P_{fr}$ molecules.

of green plants appear to be far more stable and do not disappear in darkness by these reversion or destruction processes. Furthermore, perhaps because seeds largely have type 2 phytochrome, there is no evidence for dark destruction in them, even when they are moist and phytochrome can undergo phototransformations.

Degradation of type 1 $P_{fr}$ is now known to occur by a highly selective process in which $P_{fr}$ is first attached to a small protein called ubiquitin (Jabben et al., 1989; Pollmann and Wettern, 1989). **Ubiquitin** has been found in all eukaryotes examined, but not in prokaryotes; it is a small protein of only 76 amino acids, and its amino-acid sequence is highly conserved in various organisms. Attachment of ubiquitin to $P_{fr}$ or to other proteins targets them for degradation, and to carry out degradation ATP and three enzymes are required. Proteases then recognize and hydrolyze the targeted proteins and release free ubiquitin.

## 20.4 Cryptochrome, the Blue/UV-A Photoreceptor

The effects of blue light on plants were discovered in 1864 by Julius von Sachs, who observed that phototropism (see Section 19.4) is caused only by blue (and violet) wavelengths. Since then, many effects of blue light in many kinds of plants and fungi have been discovered. A review by Senger and Schmidt (1986) showed action spectra for 16 effects of blue, violet, and near-UV radiation (also see Senger and Lipson, 1987). The near-UV range consists of wavelengths just shorter than our eyes can detect; it is commonly called **UV-A radiation** and extends from 400 nm down to 320 nm. (Reviews of UV effects on plants are those by Caldwell, 1981; Wellmann, 1983; and Coohill, 1989). Therefore, the responses we are concerned with here are essentially in the region of 320 to 500 nm.

A typical action spectrum for a cryptochrome response (phototropism) is depicted in Figure 19-8. However, action spectra and the fluence (total photons absorbed) required to cause a response vary with the organism and the response. Senger and Schmidt (1986) concluded that several somewhat different blue/UV-A photoreceptors are involved. We will call all of these **cryptochrome**; none has yet been identified, which currently makes the *crypto* part of the word especially appropriate. As described in Section 19.4, cryptochrome is likely a flavoprotein (a protein with riboflavin attached). It probably exists associated with a cytochrome protein in or tightly bound to the plasma membrane (reviewed by Galland and Senger, 1988 and in two volumes on blue-light effects edited by Senger, 1987).

We describe several effects of light absorbed by cryptochrome in subsequent parts of this chapter; sometimes the activated pigment acts independently, and sometimes it reinforces the effects of $P_{fr}$ or of the UV-B receptor. It should be noted that even though cryptochrome absorbs UV-A radiation, the largest peak in action spectra is usually in the blue-violet region near 450 nm. Also, because far more photons of blue and violet light normally strike plants as compared to UV photons, most photoresponses caused by cryptochrome probably result from absorbance of blue and violet wavelengths, hereafter simply called blue.

## 20.5 Dose-Response Relations in Photomorphogenesis

In the late 1950s, when properties of phytochrome were gradually being discovered, it became apparent that action spectra for certain processes in dark-grown seedlings were quite different depending on whether light was applied for a short period of time (usually less than 5 min) or was applied for several hours.

In short exposures, for example, light-promoted formation of the purple pigment anthocyanin in white mustard (*Sinapis alba*) seedlings exhibits a peak near 660 nm, where $P_r$ absorbs most effectively. However, when exposures of several hours are given, there is a large peak near 725 nm in the far-red region and a smaller peak in the blue region. Also, when etiolated lettuce seedlings are given extended periods of light, hypocotyl elongation is inhibited (as for bean in Fig. 20-1), and again the action spectrum has a distinct peak in the far-red region near 720 nm. Other peaks occur in the blue and UV-A region, but here (and frequently in other examples) there is no action in the red region where $P_r$ absorbs (Fig. 20-6). Plant responses that require relatively high irradiance levels and that have action spectra that are atypical of commonly observed phytochrome responses came to be known as **high-irradiance reactions** (or **responses; HIR**). *Whereas most phytochrome responses are saturated by energies of red light equal to as little as 200 J $m^{-2}$* (which is less than one percent of the energy in all visible wavelengths provided by 1 min of full sunlight), *HIR require at least 100 times more energy*. (Note: For red light, 200 J $m^{-2}$ corresponds to a fluence of about $10^{-3}$ moles of photons per square meter.) Nevertheless, as emphasized by Smith

## Phytochrome Genes and Their Expression: Working in the Dark

*James T. Colbert*

*Dr. Jim Colbert went to graduate school at the University of Wisconsin with the intention of studying plant anatomy. After completing a Master's degree investigating the vascular anatomy of sugarcane, the excitement of the emerging field of plant molecular biology enticed him to change his research direction and study the molecular biology of phytochrome. In this essay he tells us about his involvement in the isolation of the first phytochrome clones and in subsequent studies on regulation of phytochrome gene expression. He is now an Assistant Professor of Botany at Iowa State University, where in addition to continuing his research on phytochrome he has taught courses in general botany, plant physiology, and plant molecular biology.*

The importance of phytochrome in regulating plant development has been known for many years. However, until recently, relatively little was known about the genes that encode the phytochrome protein molecule. In the summer of 1981 I had the good fortune to become a graduate student in the laboratory of Dr. Peter Quail at the University of Wisconsin. Peter wanted to characterize phytochrome by isolating phytochrome genes and determining their nucleotide sequence as an aid in understanding the mechanism of phytochrome action. As part of my graduate program I worked on the isolation of phytochrome **cDNA clones**. We hoped that such clones would provide both a starting point for the molecular characterization of phytochrome and clues about how phytochrome regulates plant development. Together with a post-doctoral scientist, Howard Hershey, and another graduate student, James Lissemore, I spent the next two years in an effort to produce and identify phytochrome cDNA clones.

We chose to work with dark-grown (etiolated) oat seedlings because these seedlings were known to possess high amounts of phytochrome protein. We learned that phytochrome mRNA was also abundant in etiolated oat seedlings. This observation resulted in my spending a great deal of time working in the dark, and it led to preparation of our **cDNA library** [see Chapter 24] from etiolated oat seedlings. To screen our cDNA library for phytochrome clones, we first selected cDNA clones that encoded mRNA species that were more prevalent in etiolated seedlings than in light-treated seedlings. Subsequently, we used **in-vitro translation** with radioactive amino acids to learn which cDNAs corresponded to mRNAs that coded for phytochrome. My major responsibilities included performing the *in-vitro* translations, immunoprecipitating the *in-vitro* synthesized polypeptides, running electrophoresis of the immunoprecipitated polypeptides on **polyacrylamide gels**, and analyzing the gels by **fluorography**. Developing the X-ray film that showed the results of the fluorographic analysis was exciting because each time you knew that there was a chance of detecting a phytochrome clone. The day I developed the first fluorograph showing that one of our cDNA clones encoded phytochrome, I actually ran out of the darkroom and down the hall to find Peter and Howard. It was an exciting moment that is clearly etched in my memory; we had discovered the first phytochrome clone. As it turned out, that first cDNA clone was quite small and was itself not very useful for molecular characterization of phytochrome, but it allowed us to isolate additional phytochrome clones much more easily.

After obtaining phytochrome clones, we began to investigate the regulation of phytochrome gene expression. We were interested in the regulation of phytochrome gene expression because it seemed likely that a gene that regulated many aspects of plant development would itself be carefully regulated. We were able to demonstrate that the amount of phytochrome mRNA in etiolated oat seedlings was

---

and Whitelam (1990), the term HIR is somewhat a misnomer because the fluence rate (irradiance level) required to cause these responses is very much less than that provided by sunlight.

In a review of HIR, Mancinelli (1980) concluded that, depending on the species and the response being studied, HIR usually have three general kinds of action spectra. In one kind there is a peak in a single spectral region (usually blue/UV-A). Examples include promotion of anthocyanin synthesis in dark-grown sorghum seedlings, unrolling of leaves of rice seedlings, coiling in tendrils of pea seedlings, and phototropisms. In a second kind of action spectrum there are peaks in two spectral regions (usually blue/UV-A and red). This kind of response is seen in seedlings that have either been grown continuously in light or for a while in darkness followed by greening and de-etiolation in light. In the third general type of action spectrum, three spectral regions show activity (blue/UV-A, red, and far-red); this response is characteristic of many in etiolated seed-

dramatically decreased by red light. The red-light-induced down-regulation of phytochrome mRNA abundance was reversible by exposure of the seedlings to far-red light, so we concluded that, at least in oat seedlings, phytochrome regulates the abundance of its own mRNA. Subsequently, James Lissemore, Alan Christensen (a postdoctoral scientist), and I demonstrated that production of $P_{fr}$ by red light results in a decrease in transcription of the phytochrome genes. The decrease in transcription is remarkably rapid, being detectable within 5 min after exposure of etiolated oat seedlings to red light. I found it very intriguing (and still do) that phytochrome — a photoreceptor that regulates numerous events during plant development, including the expression of other genes — also regulates the expression of its own genes.

Genes coding for phytochrome have now been cloned from several plant species, including zucchini, pea, rice, maize, and *Arabidopsis*. Unfortunately, determination of the nucleotide sequences of these phytochrome genes has not allowed elucidation of the function of phytochrome. Nevertheless, knowing both the nucleotide sequences and the derived amino-acid sequences has allowed determination of those regions of the phytochrome molecule that are highly conserved through evolution and are, therefore, likely to be important to the function of phytochrome. For a protein that plays a pivotal role in regulating plant development, the similarity of overall sequence between dicot phytochromes and monocot phytochromes has proved to be surprisingly small. Of course, some regions (for example, the chromophore-binding region) exhibit quite similar sequences. Another contribution from attempts to characterize phytochrome at the molecular level has been the discovery, most notably in work done by Robert Sharrock and Peter Quail with *Arabidopsis*, that several distinct phytochrome genes are present in at least some plant species. Whether the phytochrome molecules produced by these distinct genes perform special functions within the plant remains to be determined.

Down-regulation of phytochrome mRNA abundance by red light occurs in most of the plant species that have been investigated. Our current research interest is to understand how the abundance of phytochrome mRNA rapidly decreases in red-light-treated etiolated oat seedlings. Although transcription of the oat phytochrome genes is known to be decreased by red light, the abundance of phytochrome mRNA would not be expected to decrease rapidly unless phytochrome mRNA is either inherently unstable or is destabilized as a result of the production of $P_{fr}$. Evidence from a wide range of organisms has demonstrated that some mRNA species are much more, or much less, stable than the average mRNA. Clearly, the relative stability of a particular mRNA species has an important role in determining the amount of that mRNA in the cell.

Our current evidence supports the idea that oat phytochrome mRNA is inherently unstable. It appears likely that the cellular machinery responsible for the selective degradation of mRNA molecules recognizes phytochrome mRNA as an mRNA species that should be rapidly degraded. Little is known about why some mRNA species are much less stable than others, but it seems likely that information specifying the relative stability of particular mRNA species would be carried in the nucleotide sequence of the mRNA molecules. We would like to learn what part(s) of the phytochrome mRNA molecule result in the relatively rapid degradation of this mRNA. We hope that our research will lead not only to increased understanding of how phytochrome gene expression is regulated but will also "shed some light" on the more general question of how plant cells regulate mRNA stability. [Authors' note: Several of the research efforts described here by Dr. Colbert are published in his two reviews listed in the Reference Section of this text.]

lings. The HIR action spectrum for lettuce in Figure 20-6 seems somewhat unusual because although it shows peaks in only two general regions, red light is not included but far-red light is. Nevertheless, it is rather common for far-red light rather than red light to act in dark-grown seedlings. When seedlings are exposed to light and turn green (and now contain far less unstable type 1 phytochrome), they usually lose most or all of their sensitivity to far-red light for HIR (although not for other photomorphogenic processes). In addition to their requirement for high irradiance levels, HIR are further characterized by having no red/far-red reversibility and no reciprocity between time and irradiance level. (Reciprocity is explained in Section 19.4.)

Hans Mohr (1986) concluded that for several responses in etiolated angiosperm seedlings, activation of cryptochrome (usually to cause an HIR) is necessary for seedlings if they are to become competent to respond to red light acting through phytochrome; that is, cryptochrome allows $P_{fr}$ to be expressed fully. In a care-

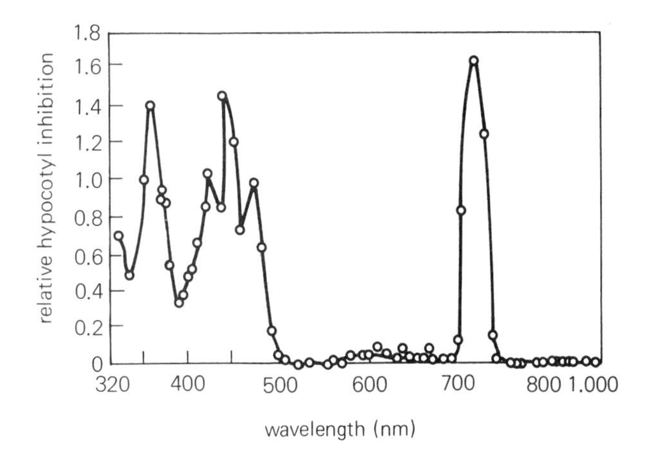


**Figure 20-6** Action spectrum for inhibition of hypocotyl elongation in etiolated lettuce seedlings. Data are expressed relative to inhibition by blue light at 447 nm, designated 1.0. Light was applied continuously to the entire seedling for 18 h, starting 54 h after planting the seeds. Hypocotyl elongation was measured at the end of the light treatment. (From Hartmann, 1967.)

ful review of various pigments involved in photomorphogenesis, Mancinelli (1989) arrived at a similar conclusion. What this seems to mean for plants in nature is that both cryptochrome and phytochrome frequently cooperate to cause photomorphogenesis. Mohr (1986) suggested that recognition by a plant of a large portion of the solar spectrum using both cryptochrome and phytochrome is advantageous. Even though this may be true, in many cases we still do not understand to what extent phytochrome itself absorbs the blue and UV-A wavelengths that cause HIR. Furthermore, for etiolated seedlings the frequent lack of action by red light in HIR but action in the far-red region near 720 nm at first seems difficult to explain as a phytochrome response.

For anthocyanin production in white mustard, Hartmann (1967) demonstrated that the peak near 720 nm does indeed result from $P_{fr}$ action. For HIR to occur, $P_{fr}$ must be present for relatively long times, usually many hours, but it need be present only in relatively low amounts (a few percent of total phytochrome). This amount of $P_{fr}$ cannot persist in etiolated seedlings treated continuously with red light because the $P_{fr}$ formed by red light is destroyed and because it reverts slowly to $P_r$. Furthermore, as described earlier, red light effectively shuts down the supply of type 1 $P_r$, the predominant form in etiolated seedlings, so in continuous red light the $P_{fr}$ reverts, is destroyed, and is also formed less rapidly. However, with continuous far-red light there is always a low amount of $P_{fr}$ present because $P_r$ absorbs some far-red light and is converted to $P_{fr}$. This $P_{fr}$ persists and functions, and thus we see responses to far-red light by HIR in etiolated seedlings. In light-grown seedlings, however, red light is more effective than far-red light in HIR, probably because type 2 $P_{fr}$ is more stable and because red light forms much more of it than does far-red light (Kronenberg and Kendrick, 1986; Smith and Whitelam, 1990). The importance of HIR in seeds and light-grown plants, which is generally less well understood, will be described in subsequent sections.

Plants grown in darkness respond not only to low and high fluence rates, but also in what have become known as **very low fluence responses (VLFR)**. One of these (inhibition of elongation of the mesocotyl or first internode in oat seedlings; morphology described in Section 20.7) was discovered by Blaauw et al. (1968), but it was research by Dina Mandoli in Winslow R. Briggs's laboratory at the Carnegie Institution of Washington in Stanford, California, that emphasized the quantitative aspects of two VLFR and how widespread these effects probably are (see Mandoli and Briggs, 1981). They observed that many scientists who were studying phytochrome responses used green safelights when they watered and handled the plants. For many responses this presents no problem, but both the mesocotyl and the coleoptile of oats are sensitive even to green safelights. Mandoli germinated and grew oats in darkness for 4.3 days, then illuminated by automation for various times with red, green, or far-red light. Growth was measured 1 d after light treatment began. All wavelengths inhibited elongation of the mesocotyl and promoted that of the coleoptile, but red light was by far the most effective for both responses. Wide fluence-response curves for red light are plotted in Figure 20-7.

Curves for the two responses are almost mirror images of each other, and they exhibit two distinct phases or regions of effect (separated by a plateau) as the fluence is increased logarithmically over 11 orders of magnitude. The most sensitive region shown for both responses has a threshold total-fluence requirement of about $10^{-10}$ moles of photons per square meter of surface area exposed (equivalent to the visible photons in about one second of full moonlight); this region is saturated by about 3,000 times that fluence. These could be the most sensitive plant photoresponses known. These VLFR cannot be nullified by far-red light because far-red light has no effect at very low fluences, but at much higher fluences it causes the same response; this response similarity probably occurs because far-red light forms some $P_{fr}$, as mentioned before. The second phase for response of each organ requires a fluence that is about 10,000 times higher. These **low-fluence responses (LFR)** represent rather typical phytochrome responses in terms of photon requirements, and they are nullified by far-red light.

When Mandoli and Briggs (1981) increased fluence higher than needed to saturate the LFR, another plateau occurred in which no further inhibition took place; that is, no HIR was evident. However, a HIR was observed

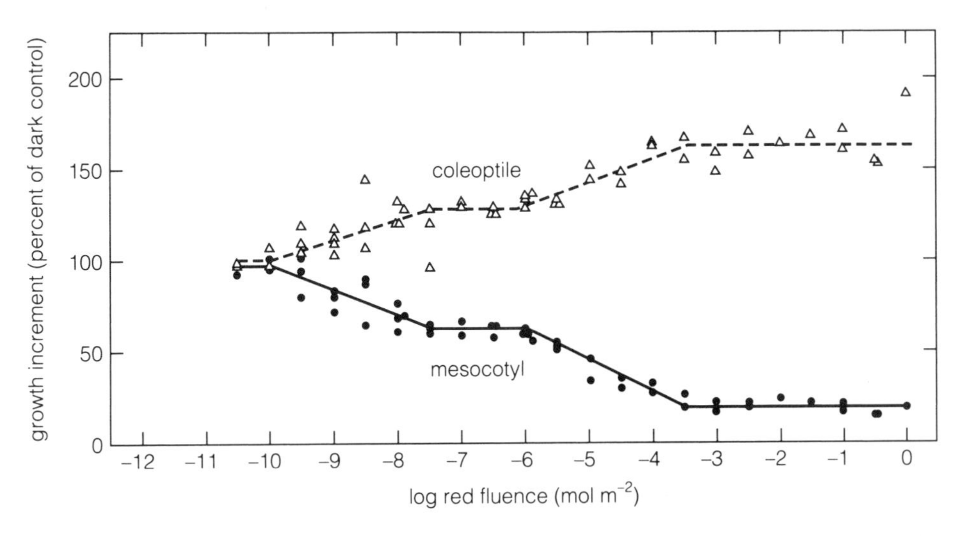


**Figure 20-7** Effects of various fluences of red light on elongation of the coleoptile and mesocotyl of dark-grown oat seedlings. (From Mandoli and Briggs, 1981.)

by Schäfer et al. (1982) in experiments in which red or far-red light was provided continuously during a 24-h period, even though the total fluence was not greater than that in the last plateau region in Figure 20-7. (We noted earlier that many HIR require long exposure periods.)

All of these results are important because they show three quantitatively distinct sensitivity regions and because they explain why certain other VLFR cannot be nullified with far-red light. Furthermore, these results force researchers to look for more VLFR, being careful to avoid green safelights that might saturate these responses before the experiment essentially even starts. As Mandoli reportedly quipped, "reagent-grade darkness" is required! With a knowledge of these properties of phytochrome and cryptochrome, let us consider some other physiological processes they control, beginning with seed germination.

## 20.6 The Role of Light in Seed Germination

### Some Examples of Light-Dependent Germination

The importance of light for germination of certain seeds has probably been recognized for hundreds of years, but the first comprehensive study was described by Kinzel in 1907 (Rollin, 1972). Kinzel reported that among 964 species, 672 showed enhanced germination in light. In a recent study of temperate herbaceous annuals and perennials (all of them noncultivated species), Baskin and Baskin (1988) observed that among 142 species, germination of 107 was promoted by light, 32 species showed no response, and only three showed light inhibition.

Seeds of most species that respond to light are undomesticated and rich in fat but so small that their seedlings might not reach light at the soil surface before their food reserves were used up. Most of our cultivated seeds do not require light, no doubt because of human selection against a light requirement. Seeds of some wild species even show inhibited germination in light, as observed both by Kinzel and the Baskins, sometimes because of the blue light but especially because of the far-red light present (Frankland and Taylorson, 1983). Kinzel found 258 of the 964 species he studied to be inhibited by sunlight relative to darkness. The far-red wavelengths of sunlight are nearly always those that are most inhibitory, presumably because they decrease the amount of $P_{fr}$ in the seed to a level below that already present and needed for germination. Although blue light is also inhibitory sometimes, we are uncertain whether it is absorbed by phytochrome or by cryptochrome. Little research has been performed on this subject, so we concentrate here on what appear to be effects involving phytochrome.

Seeds that require light for germination are said to be **photodormant**. As discussed in Section 22.3, we use the term **dormancy** as a general description of seeds or buds that fail to grow when exposed to adequate moisture and air and a favorable temperature for growth. Seeds that normally germinate in darkness but are inhibited by light are considered to become dormant after exposure to the light; essentially this classification is used in the excellent books on physiology of seeds by Bewley and Black (1982, 1985). However, other authors define as dormant a living but dry pea seed that needs only water and air to germinate at room temperature; clearly there is considerable confusion and disagreement in terminology concerning dormancy in seeds, buds, tubers, and corms.

Lang et al. (1987) listed 54 terms that have been used to describe dormancy, the most common of which are *rest* and *quiescence*. Many more terms involve the word *dormancy* preceded by an adjective such as primary, secondary, innate, and so on. Lang et al. argued reasonably for unity in terminology and suggested use of three new terms only (*endodormancy*, *ecodormancy*, and *paradormancy*). Although this terminology might be useful for some cases of dormancy, their examples for seed dormancy (Table 4 of Lang et al., 1987) are often unsatisfactory. To use their terminology, we need to understand much more about the causes of dormancy and how it is broken. This need for more information certainly applies to photodormancy, a topic we consider next.

## Interactions of Light and Temperature in Photodormant Seeds

A further aspect of effects of light on germination is an interaction with temperature. Temperature has almost no effect on photochemical interconversions of $P_r$ and $P_{fr}$, but the chemical reactions controlled by $P_{fr}$ and those influencing its destruction are quite temperature-sensitive. An example of crucial temperature control occurs in seeds of Grand Rapids lettuce (*Lactuca sativa*) and peppergrass (*Lepidium virginianum*). Light usually promotes their germination (Fig. 20-2), but extended exposure to 35°C after a single light treatment, or exposure to that temperature in continuous light, keeps them dormant. Similarly, seeds of the Great Lakes cultivar of lettuce usually do not require light to germinate; but if they are soaked at 35°C, they become photodormant and then germinate only in light at that or a lower temperature. On the other hand, cool temperatures of 10 to 15°C allow Grand Rapids lettuce (at least some seed lots) to germinate well even without a light treatment. Still another example is provided by Kentucky bluegrass (*Poa pratensis*), in which temperatures alternating between 15°C and 25°C substitute for light in causing germination.

Evidence shows that temperature responses such as these are sometimes caused by effects of temperature on the amounts of $P_{fr}$ in the seeds. High temperatures decrease the $P_{fr}$ level in some species by increasing its rate of reversion to $P_r$, although in general other factors have not been evaluated. Destruction of $P_{fr}$ in seeds has been generally ignored because such destruction is uncommon or typically absent (perhaps because of stable type 2 phytochrome, as mentioned in Section 20.3). It is likely that many effects of temperature have nothing to do with changes in amounts of $P_{fr}$ in the seeds (for example, when Grand Rapids lettuce germinates in darkness at 10 or 15°C). There are two other possibilities: One is that certain temperatures increase sensitivity and allow even minute amounts of $P_{fr}$ (formed and trapped by desiccation during seed maturation) to act; another is that certain temperatures simply cause the same biochemical reactions as those caused by $P_{fr}$. We don't yet understand these biochemical reactions, but evidence for involvement with gibberellins is described in a later section of this chapter.

Both photodormant and nondormant seeds usually imbibe water and swell, unless their seed coats prevent water uptake. Even dead seeds can swell! But only nondormant seeds continue to absorb water and grow after imbibition is complete (after their colloids are fully hydrated). Dormancy is broken by light only when seeds are partially or fully imbibed. The time required for imbibition varies from as little as an hour to almost two weeks, depending on how permeable the seed or fruit coats are to water and how large the seed is. Only then is $P_r$ sufficiently hydrated to be transformed to $P_{fr}$. In seeds that survive many years in the soil, $P_r$ is stable and only awaits the proper combination of moisture, light, and temperature to become $P_{fr}$ and cause germination. When Grand Rapids lettuce seeds are imbibed, exposed to light to form $P_{fr}$, and then immediately dehydrated, they will germinate in darkness upon remoistening for up to a year. This shows that $P_{fr}$ is stable in dry seeds for long periods. That $P_{fr}$ is stable in dry seeds is also shown by the requirement of lettuce seeds for a red-light treatment after they are dried under far-red light. An implication of stable $P_r$ and $P_{fr}$ is that whether a seed requires light to germinate depends on how much $P_{fr}$ was produced in it during ripening on the mother plant.

The amount of chlorophyll that covers the embryo as the seed ripens is especially important in determining whether or not seeds of a given species will be photodormant (Cresswell and Grime, 1981). In general, embryos that are covered during ripening by maternal tissues that contain high amounts of chlorophyll require light to germinate, whereas those that are covered by maternal tissues with little or no chlorophyll do not. The apparent reason for this is that chlorophyll absorbs the red wavelengths and prevents $P_{fr}$ formation in the ripening embryos, so the mature seeds (from which most chlorophyll has disappeared) then require red wavelengths to promote germination.

The relative lengths of night and day during ripening also affect photodormancy in some species (for example, lettuce, *Chenopodium album*, *Portulaca oleraceae*, and *Carrichtera annua*); long days usually favor photodormancy, whereas short days usually favor nondormant seeds (Bewley and Black, 1982). In *Chenopodium*, long days just before or during seed development increase the red-light requirement of the seeds by increasing seed-coat thickness, but in other species the reason for day-length effects is unknown.

## Ecological Aspects of Photodormancy in Seeds

What possible ecological benefit is light sensitivity to seeds lying in litter near the soil surface? Answers to such questions frequently involve speculation, as do those given here. For buried seeds whose germination is promoted by light, germination when they are partly uncovered more nearly assures that the seedlings will be able to photosynthesize, grow, and perpetuate the species. A light requirement for buried seeds might distribute germination over several years and thereby help perpetuate the species because only a fraction of the seeds in the soil might be disturbed and exposed to light in a given season. An unfavorable growing year might otherwise destroy most of the plants. For seeds whose germination is inhibited by light, germination is prevented until they are well-covered by litter, when they would more likely have sufficient water to grow. Koller (1969) described two light-inhibited species that inhabit coarse, sandy soils of the Negev desert in Israel, in which species germination is prevented unless the seeds are well buried, because moisture is more plentiful below ground than at the soil surface.

Another idea is that phytochrome provides seeds with a clue about whether they are covered by a canopy of other plants or instead exist in an open area. This possibility has been studied extensively by physiological ecologists. The idea developed from two facts. First, far-red light usually inhibits germination of light-requiring seeds (and even of seeds that germinate moderately well in darkness). In nature this is a HIR. Second, leaves in a canopy transmit considerably more far-red light than red light. Most of the blue, red, and some of the green wavelengths are removed by leaves through photosynthesis and reflectance, but most of the far-red light passes through to seeds below and converts their active $P_{fr}$ to inactive $P_r$. Figure 20-8 illustrates the spectral distribution of radiation above and below the canopy in a sugar-beet field. The lowest curve for radiation filtered by leaves (ground level, within rows) shows a small peak of transmission in the green region (540 nm) and a much larger one in the far-red region. Under such a canopy, no more than 10 percent of the total phytochrome would exist as $P_{fr}$ (that is, $\phi = 0.10$). If seeds ripen on a plant under a canopy transmitting such a high ratio of far-red light to red light, they are likely to require direct sunlight to form additional $P_{fr}$ before they can germinate. In a forest, many seeds requiring relatively high amounts of $P_{fr}$ might never sprout until a fire, death of old trees, or timber removal eliminates the canopy. Seeds in soils of evergreen forests seem especially affected in this respect, but seeds in deciduous forests are probably also affected in some climates, depending on when new leaves develop relative to the temperature and light requirements of seeds below. The light-filtering effect of plant canopies is also important to agriculture because germination of many weed seeds is promoted by direct sunlight but inhibited by light filtered from crop plants that have developed above them. Hundreds of species have been investigated in this respect (reviewed by Bewley and Black, 1982; Frankland and Taylorson, 1983).

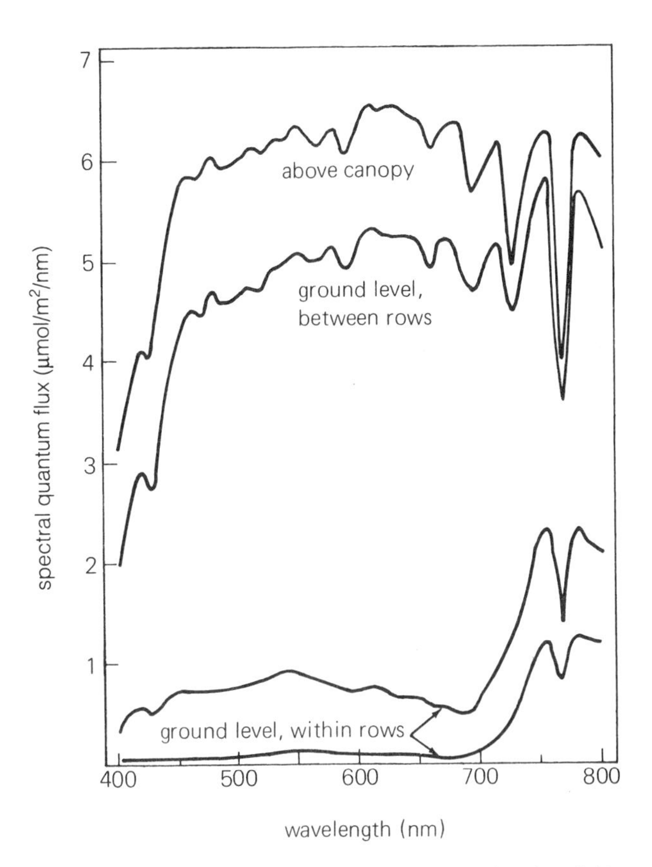


**Figure 20-8** Influence of shading upon wavelengths of sunlight present in various portions of a sugar-beet field. Within the rows (bottom two curves) there is much less attenuation of far-red light than of the other wavelengths, so shaded plants contain a higher proportion of phytochrome in the $P_r$ form than do unshaded plants. (From Holmes and Smith, 1975.)

Ecologically, the light sensitivity of seeds from species that live in shaded conditions is different from that of pioneer species that live in more open areas (Grime, 1979, 1981; Bewley and Black, 1982). As you might predict, seeds of species that live in shaded conditions are less inhibited by the far-red light transmitted by plant canopies than are those that invade more open areas.

## Is Phytochrome the Only Pigment Active in Germination?

Although both promotive and inhibitory effects of both far-red and blue radiation have often been attributed to phytochrome, several results are difficult to interpret in this way. In tansy (*Phacelia tanacetifolia*), germination is suppressed not only by extended periods of far-red and blue light but also by extended red light, all acting through the HIR system. If only $P_{fr}$ is important in allowing germination, why should wavelengths that cause widely different $\phi$ values all be inhibitory? The answers are largely unknown.

We do not yet understand how HIR caused by far-red or blue light inhibits germination. Most seed researchers have ignored blue-light effects, partly because blue light is preferentially absorbed by cryptochrome and very little is known about that pigment, partly because each form of phytochrome absorbs blue light to some extent (Fig. 20-3a), and partly because seed coats and fruit coats absorb blue wavelengths much more than red or far-red wavelengths and so much blue light never reaches the embryo. All these factors make it difficult to evaluate to what extent cryptochrome and phytochrome participate in mediating blue-light effects on germination of seeds of any species. Nevertheless, there is evidence that cryptochrome does contribute to promotion of germination at low irradiance levels of blue light (Small et al., 1979).

High-irradiance reactions caused by far-red light have been studied much more than blue-light effects. All result from absorption by phytochrome, even though the action-spectrum peak is usually near 720 nm rather than 730 nm, where $P_{fr}$ absorbs maximally (Frankland and Taylorson, 1983; Frankland, 1986; Cone and Kendrick, 1986). There are three important reasons why these effects cannot be explained simply by removal of $P_{fr}$ needed for germination. First, the action-spectrum peak is 10 nm shorter than the 730 nm peak of absorption by $P_{fr}$. Second, high irradiances that give widely different $\phi$ ratios (and $P_{fr}$ levels) are all inhibitory; for tansy, even red light is inhibitory, as mentioned. And third, high irradiances are inhibitory even after a previous treatment with low-irradiance red light that promotes germination (by causing $P_{fr}$ formation) has completed its action; that is, formation of $P_{fr}$ by red light occurs early in the germination process and high-irradiance effects are noted much later, as late as just before the radicle protrudes through the seed coat. We do not understand these effects. Clearly, there is much more to learn about phytochrome before we can understand the role of other pigments in germination.

## The Nature of Photodormancy

We now ignore inhibitory effects of HIR and return to promotion of germination by low-irradiance effects (LFR). Given that $P_{fr}$ causes photodormant seeds such as lettuce to germinate, why do they not sprout in its absence? To answer this it is essential to identify that part of the seed in which $P_{fr}$ must be formed.

This problem was approached in lettuce seeds by separately covering the cotyledons and the hypocotyl-radicle tissues with aluminum foil before exposure to red or far-red light (Ikuma and Thimann, 1959). Both red and far-red light were much more effective when they were absorbed by the hypocotyl-radicle region than by the cotyledons, indicating that $P_{fr}$ formation is essential in the cells that actually grow and cause germination. Furthermore, $P_{fr}$, which itself cannot move from cell to cell, seemingly causes no movement of a germination promoter from cotyledons to the radicle. (Absorption of light by the cotyledons enhanced germination somewhat, but this absorption probably resulted in the scattering of light to the hypocotyl-radicle region.) Light is easily scattered over many cell distances in plant parts, including lettuce seeds, as shown by Widell and Vogelmann (1988) with fiber-optic probes and as reviewed by Vogelmann (1989). From similar experiments with *Citrullus colocynthus* and from laser-millibeam experiments with *Cucurbita pepo* (pumpkin), it was also concluded that the radicle-hypocotyl region is the light-sensitive region of seeds (reviewed by Bewley and Black, 1982).

If we detach the lettuce embryo from the surrounding endosperm, seed coat, and fruit coat, the radicle itself now elongates in darkness or after a brief exposure to either red or far-red light. However, radicles from embryos exposed to red light begin to elongate sooner and at a faster rate than those given darkness (Fig. 20-9). Those given far-red light elongate at a rate essentially the same as those given darkness. This growth-promoting effect of red light is nullified by a short treatment with far-red light after exposure to red light, as expected from germination results with whole seeds. Furthermore, if naked lettuce embryos from imbibed seeds are removed under a green safelight and placed in solutions with negative water potentials (such as polyethylene glycol), those embryos subsequently given red light will absorb water and grow in a solution having a more-negative water potential than those kept in the dark or given far-red light (Nabors and Lang, 1971). Our conclusion is that $P_{fr}$ increases the growth potential of the radicle cells, presumably those in the elongating region, by decreasing their water potential so that they more easily absorb water from soils and germinate.

These facts suggest that germination of light-requiring lettuce seeds fails in darkness because the radicle cannot grow with sufficient force to break through the layers that surround it. Of these layers, the lettuce radicle is restricted almost entirely by the tough endosperm, even though it is only two or three cell layers thick. The endosperm is also the restrictive layer in *Phacelia tanacetifolia*, *Datura ferox*, tomato, and various *Syringa* (lilac) species. For lettuce and *Datura ferox*, both increased thrust of the radicle and weakening of the endosperm barrier near the radicle tip are important (Carpita et al., 1979; Tao and Khan, 1979; Psaras, 1984; Sánchez et al., 1986, 1990). For other seeds, we might reasonably expect $P_{fr}$ to increase germination either by increasing radicle thrust or by weakening surrounding barriers to its growth, or both. Photodormancy and seed dormancy in general is less a mystery when we

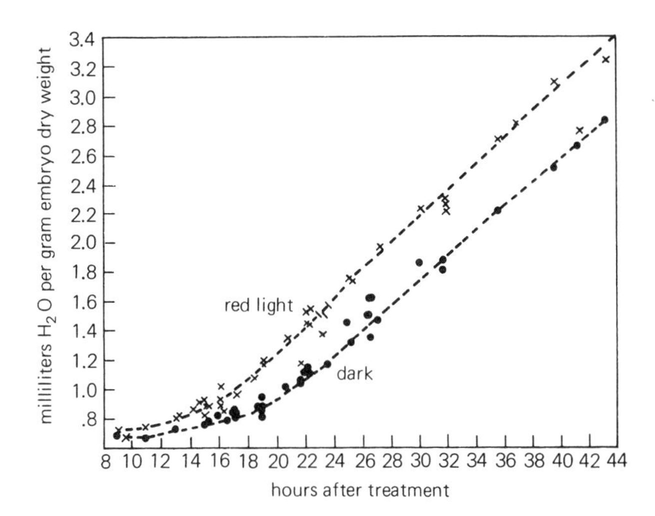


**Figure 20-9** Stimulation of growth by red light in naked embryos of lettuce seeds. Seeds were soaked in distilled water for 3.5 h; then some were given a 10-min treatment with red light. Endosperm, seed coat, and fruit coat were then removed under a green safelight, and growth (monitored by measuring water uptake) was measured at the time periods indicated. Each point represents the response of one embryo. (From Nabors and Lang, 1971.)

consider germination to be a struggle between the growth potential of the radicle and the growth-restrictive mechanical effects of surrounding layers. In some cases, the external restriction is great; in others it is of little consequence, and only a small increase in the radicle's growth potential caused by $P_{fr}$ or some other factor is enough to cause germination. Nevertheless, even when the restriction is small, removal of $P_{fr}$ by far-red light usually reduces germination. This has been demonstrated not only for native seeds under a forest canopy but also for cultivated crops such as pumpkin and maize.

## Effects of Hormones on Photodormancy

In many seeds that are photodormant or dormant for other reasons, applied gibberellins substitute for the light or other environmental requirement; and for a few species such as lettuce, cytokinins also substitute for light or partially replace it. In highly dormant celery cultivars, the requirement for light is not effectively replaced by either gibberellic acid or a cytokinin, but application of both hormones breaks dormancy in darkness (Thomas, 1989). Auxins usually do not promote germination of photodormant or nondormant seeds and are instead either innocuous at low concentrations or inhibitory at high concentrations. The role of ethylene is less clear: It cannot break photodormancy, but it can partially overcome other kinds of seed dormancy in cocklebur, *Amaranthus retroflexus*, and in certain peanut and clover cultivars. Ethylene can also partially overcome dormancy caused by high temperatures in lettuce and can overcome certain photodormancy problems in cocklebur (Esashi et al., 1983). As mentioned in Chapter 18, applied abscisic acid almost always retards germination, unless covering layers prevent the embryo from absorbing it. There seems little doubt that in many species ABA contributes to development of dormant seeds and prevents vivipary, as described in Section 18.5.

Collectively, these results suggest that $P_{fr}$ might break photodormancy by causing synthesis of a gibberellin or a cytokinin or by destroying an inhibitor such as ABA. The evidence concerning this supposition is controversial (Bewley and Black, 1982; De Greef and Frédéricq, 1983; Carpita and Nabors, 1981), but no one has yet measured hormone changes that occur specifically in the radicle or hypocotyl cells responsible for germination. This seems essential to understanding relations among light, growth promoters, and growth inhibitors in photodormancy and in other kinds of seed dormancy discussed in Chapter 22. Analyses of whole seeds for hormone levels might not be helpful to understand hormonal aspects of dormancy because the whole seed is so large relative to the tissues that control germination. Nevertheless, the literature continues to contain reports of hormone analyses of whole seeds.

During the 1980s direct evidence for the importance of gibberellins in overcoming photodormancy became available from studies of mutants. As mentioned in Chapter 17, dwarf mutants of maize, garden pea, sweet pea, rice, wheat, tomato, and bean are known, some of which have blocks in gibberellin synthesis and some blocks in responses to gibberellins (reviewed by Reid, 1987, 1990; Hedden and Lenton, 1988; and Scott, 1990). No reduced germination was reported for any of them, but they were selected for dwarfism, not poor germination, and they are cultivated species in which dormancy is unusual. Researchers in the Netherlands found dwarf mutants of tomato and *Arabidopsis thaliana* that also fail to germinate without applied gibberellin; these are gibberellin-deficient mutants (Groot et al., 1987, 1988; reviewed by Karssen et al., 1989). In tomato, wild-type seeds with normal gibberellin contents germinate well in darkness without added gibberellin, and no photodormancy exists in the cultivar these researchers studied. With *Arabidopsis*, wild-type seeds will not germinate in darkness, but red light acting via $P_{fr}$ overcomes their dormancy. Dormancy can also be overcome in the wild-type with a gibberellin, especially a mixture of $GA_4$ and $GA_7$ that is sometimes as effective as $GA_3$ for breaking seed dormancy at concentrations that are lower by 1,000 times. In one of the well-studied gibberellin-deficient *Arabidopsis* mutants, no germination occurred in water even with light, but high germination in light occurred with 1 $\mu$M $GA_{4+7}$ or in darkness with 100 $\mu$M $GA_{4+7}$. Therefore, applied gibberellin overcomes the genetic block and the light requirement. The conclusions from this and other work with these mutants are that normal wild-type tomato and *Arabidopsis* seeds must have gibberellins to germinate and that a requirement for light in *Arabidopsis* exists largely because light induces formation of one or more gibberellins.

Other research by the Netherlands group indicates that in tomato, gibberellins probably induce weakening of the tough endosperm near the radicle tip. As in other seeds with a tough endosperm, the endosperm of tomato is rich in **galactomannans** (cell-wall polysaccharides), and $GA_{4+7}$ induces an increase in activity of three enzymes that hydrolyze those galactomannans. This hydrolysis presumably makes the endosperm weak enough that the radicle can grow through it and cause germination. If this is how gibberellins function in tomato, their action on the endosperm is similar to their action combined with that of red light in breaking dormancy of seeds of *Datura ferox* (Sánchez et al., 1986, 1990). As we emphasized in Chapter 17, a common action of gibberellins in various plant parts is to induce activity of certain kinds of hydrolytic enzymes.

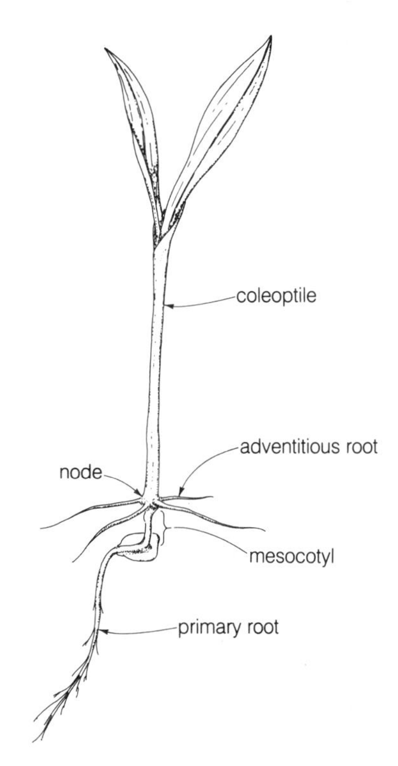


**Figure 20-10** Some morphological characteristics of a week-old maize seedling grown in light. The coleoptile has stopped elongating, and two leaves have broken through it and have largely unrolled. The shoot apex is at the node where adventitious (prop) roots originate. The mesocotyl is the first internode formed above the seed-storage tissues and the scutellum (cotyledon) in the seed.

## 20.7 The Role of Light in Seedling Establishment and Later Vegetative Growth

Once germination is accomplished, further plant development still remains subject to control by light. We introduced some of these controls in Section 20.1 and Figure 20-1. We now evaluate the roles and actions of phytochrome, cryptochrome, and the UV-A receptor in plant development.

### Development of Poaceae (Grass) Seedlings

After a grass seed or cereal-grain seed germinates, its coleoptile elongates until the tip breaks through the soil. Between the scutellum (see Fig. 17-17) and the base of the coleoptile is an internode called the **mesocotyl** (first internode; see Fig. 20-10) that in most grass species elongates greatly after germination of deeply planted seeds. (In wheat, rye, barley, and bamboos, the mesocotyl is detectable in embryos but does not elongate in darkness or light; reviewed by Hoshikawa, 1969.) Elongation of the mesocotyl, coleoptile, and leaves enclosed by the coleoptile is necessary to carry leaves into the light and to establish near the soil surface the adventitious roots produced at the node just above the mesocotyl (Fig. 20-10). Elongation of the mesocotyl has received attention for more than 40 years. As mentioned earlier, all results show that mesocotyl elongation is extremely sensitive to light.

Elongation of the coleoptile must equal or exceed that of the leaves it encloses as they grow upward together; otherwise the leaves would grow out of the coleoptile and probably be broken off in the soil. Growth rates of these two organs are coordinated until they reach the soil surface and are exposed to light. After exposure to light, the leaves become green and photosynthetic, and they break through the coleoptile tip. Leaf emergence occurs because light promotes leaf elongation and decreases the extent to which coleoptiles can elongate (even though it speeds their elongation when they are a few days younger; see Thomson, 1954; Schopfer et al., 1982; Smith and Jackson, 1987). The light promotion of leaf growth, the speeding of young coleoptile elongation, and the inhibition of final coleoptile length are phytochrome responses to sunlight.

In the maize seedling grown in light from a seed planted near the soil surface (shown in Fig. 20-10), the mesocotyl had elongated very little, and the first two leaves had emerged from the coleoptile. Each of these leaves was rolled up inside the coleoptile, but when exposed to light they began to unroll (flatten out). Rolling was still evident only at the point of departure from the broken coleoptile.

Substantial unrolling of grass leaves is controlled by a typical phytochrome response in which low fluences of red light are promotive and subsequent far-red light nullifies the red-light effect (Fig. 20-11). Low fluences of far-red light are without effect, and low fluences of blue light are only slightly promotive, except in rice. Nevertheless, wheat and barley leaves unroll even further if they are exposed up to 9 h with dim red light, so both a LFR and a HIR to red light occur (Virgin, 1989). Unrolling is caused by more rapid growth (likely as a result of wall loosening) of cells on the concave (later to be uppermost) side, as opposed to the convex side. Applied gibberellins and, in some species, cytokinins replace the need for light (De Greef and Frédéricq, 1983).

These results suggest that $P_{fr}$ causes rolled leaves to form gibberellins or cytokinins that then cause unrolling. This hypothesis might be correct for gibberellins because $P_{fr}$ promotes gibberellin production and release from young plastids in rolled wheat and barley leaves before the leaves unroll. No studies showing light effects on cytokinin contents of unrolling leaves are available, so for now it seems safest to conclude that light might induce leaf unrolling by causing production of gibberellins in concave cells. Alternatively, the concave cells might become more sensitive to the hormone levels they already contain when exposed to light.

## Development of Dicot Seedlings

In dicots, underground and food-rich cotyledons either remain underground by **hypogeal development**, as in pea, or emerge above ground **epigeally**, as in beans, radish, and lettuce. In either case, a hook formed near the stem apex pushes up through the soil and pulls with it the fragile young leaves or cotyledons (see Figs. 16-14 and 18-15). (In Figure 20-1 this hook in bean had moved, as it does in seedlings that develop epigeally, into the epicotyl [stem section above cotyledons] and had opened somewhat, perhaps by slight light exposure during watering.) As mentioned in Section 18.2, this hook forms as a result of unequal growth on the two sides of the hypocotyl or epicotyl in response to ethylene soon after germination. As the hook emerges from the soil, red light acting through $P_{fr}$ promotes opening of the hook. Hook opening apparently results from inhibition by light of ethylene synthesis in the hook. Differential growth that results from faster elongation of cells on the lower (concave) side than on the upper (convex) side causes hook opening (see Section 18.3). Accompanying this opening, light increases leaf-blade expansion, petiole elongation, chlorophyll formation, and chloroplast development, as also occurs in grass leaves (Fig. 20-1).

Most of the light promotion of leaf growth, at least in dicots, is caused by a HIR (Dale, 1988). A good example is provided by the primary leaves of bean. Plants grown 10 d under dim red light have slightly larger leaves and substantially more cells than those kept in darkness. When they are transferred to white light, cell expansion and leaf growth increase greatly. In this case blue light acting through an HIR system causes cell expansion by enhancing acidification of epidermal cell walls, thus loosening them so that the whole leaf expands faster under the existing turgor pressure (Van Volkenburgh, 1987). Light effects on chlorophyll formation and chloroplast development result first from a triggering action of $P_{fr}$ that causes production of **delta-aminolevulinic acid (ALA)** from glutamic acid (reviewed by Kasemir, 1983; Hoober, 1987; Beale, 1990).

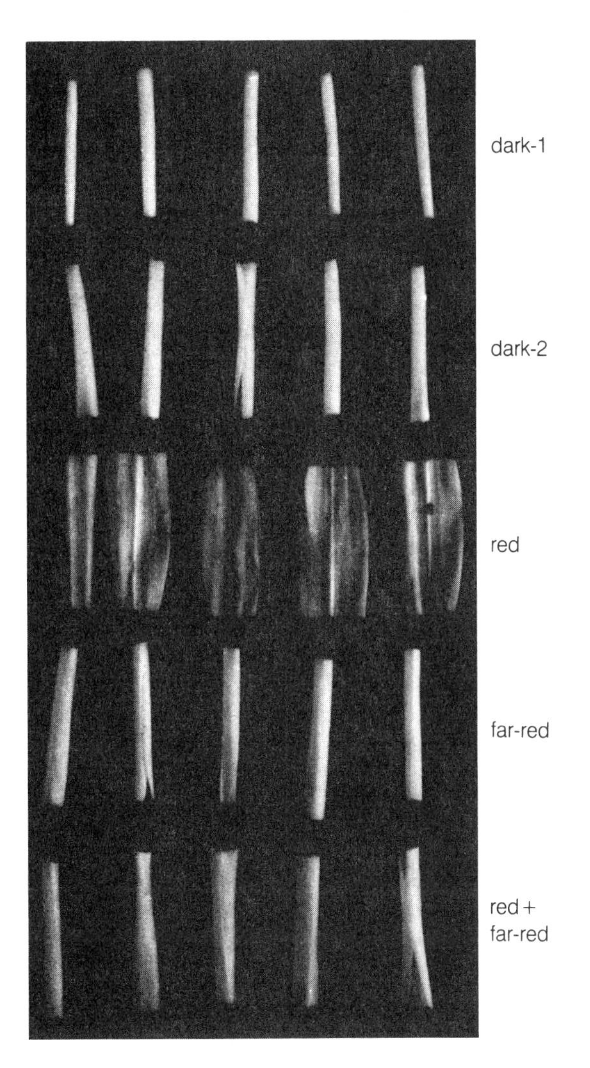


**Figure 20-11** Effect of pretreatment with red and far-red light on unrolling of leaf sections from etiolated maize seedlings. Red light promotes opening, whereas subsequent far-red treatment nullifies the red-light effect. (From Klein et al., 1963.)

ALA is the metabolic precursor that is converted into each of the four pyrrole rings of chlorophyll. Nevertheless, ALA is not converted all the way to chlorophyll without higher irradiances of red or blue light. Instead, the metabolic pathway stops when a compound often called **protochlorophyll** is formed. More accurately, protochlorophyll is **protochlorophyllide *a***, which differs from chlorophyll *a* (Fig. 10-4) only by the absence

**Figure 20-12** Effects of continuous white light (which contains blue light) and of continuous red light on growth and development of pea seedlings. Plant on the left was grown in a growth chamber with white light, that in the center under dim red light (approximately 1 $\mu$mol m$^{-2}$ s$^{-1}$), and that on the right in total darkness. (Photograph courtesy Marta J. Laskowski, Timothy W. Short, and W. R. Briggs.)

**Figure 20-13** Growth of *Chenopodium album* after 21 days under two different red/far-red ratios. Both plants were grown to the three-leaf stage under identical conditions; the one on the right was then provided light enriched in far-red. The estimated $\phi$ values in the two plants were 0.71 (left) and 0.38 (right). Each plant received the same amount of photosynthetically active radiation (400–700 nm). (From Morgan and Smith, 1976.)

of a phytol tail and two hydrogen atoms. Protochlorophyllide *a* is rapidly reduced to chlorophyllide *a* in red or blue light because protochlorophyllide *a*, like chlorophylls, absorbs those photons effectively. Addition of the phytol tail, an isoprenoid formed from the mevalonic acid pathway (see Chapter 15), completes formation of chlorophyll *a*; some of the chlorophyll *a* is then converted to chlorophyll *b*. Chloroplast development depends strongly on chlorophyll formation and, therefore, on both of these light effects. All of these responses lead, within a few hours, to photosynthesis in grass leaves as they break through the coleoptile and in cotyledons or young leaves of dicots as they break through the soil. Cotyledons of conifer seedlings somehow form chlorophyll and become photosynthetic even in darkness, but their needles require light for these processes.

As photosynthesis begins in leaves and cotyledons, shorter and thicker stems are produced. Of course, a dark-grown seedling cannot elongate after its food supplies are exhausted; but while carbohydrates or fats are still plentiful, light still inhibits stem elongation. This inhibition of stem elongation was apparently first recorded by Julius von Sachs in 1852, who observed that stems of many species do not grow as fast during daylight as they do at night. We now realize that blue, red, and far-red light all contribute to this phenomenon and that cryptochrome and phytochrome both participate.

Figure 20-12 shows effects of continuous white and red light on development of pea seedlings (Laskowski and Briggs, 1989). Studies of hypocotyl elongation in dark-grown cucumber, radish, and a few other seedlings show that inhibition by red light acts through $P_{fr}$ formation and that blue light acts on cryptochrome. Several interesting differences in responses to red and blue light were observed (Cosgrove, 1986, 1988). First, inhibition by red light occurs because of $P_{fr}$ formation in the cotyledons, not in the hypocotyl itself, whereas blue light acts directly on the hypocotyl. Second, elongation slows within 30 s after exposure to blue light, whereas the red-light effect requires at least 15 min. Third, hypocotyls of plants exposed to blue light begin to elongate again in darkness much faster than those exposed to red light. Fourth, responses to blue light are often HIR, whereas those to red light are typically LFR. Finally, in lettuce and in *Sinapis alba*, responses to blue light but not to red light are gradually lost as the seedlings de-etiolate.

### Photomorphogenic Effects in Vegetative Growth

In many well-established and growing plants, other photomorphogenic processes occur. If green and photosynthetic dicots are exposed to red light (acting through phytochrome) or to blue light (acting through cryptochrome), stem elongation is inhibited. If plants grow under a leaf canopy, where the light received is primarily far-red light, $P_{fr}$ is removed from their leaves and their stems become considerably elongated (Fig. 20-13). The same is true of conifers studied so far. Branching of stems is simultaneously retarded in many species under a canopy, so plants use more energy to raise the stem apex toward the canopy top than when unshaded.

In agricultural crops planted in rows, plants in exposed outer rows are often shorter and more highly-branched than are those within the field because of this effect. A similar phenomenon is often seen with plants on greenhouse benches. Also, in thick stands of lodgepole pine, such as are abundant in Yellowstone National Park in Wyoming and in many other mountainous areas of the western United States, the results of retarded branching and the death of branches in reduced light are forests of trees having long, straight trunks that provide relatively knot-free timber. This principle is now used by foresters to select distances between transplanted seedlings in reforestation work. Plants that do not elongate in response to increased (relative to red) far-red radiation are those that normally grow in the shade of others (for example, on the forest floor). These shade plants seem to be adapted to an environment in which it is impossible to elevate their leaves above the overhead canopy by stem elongation (Morgan, 1981). Smith (1986) and Smith and Whitelam (1987, 1990) concluded from such studies that an important function of phytochrome in light-grown plants is to act as a shade-avoidance mechanism.

More recent studies show that plants respond to adjacent plants by far-red reflection signals even before one shades the other (Ballaré et al., 1987, 1990; Casal and Smith, 1989; Smith et al., 1990). It was noted that stems of various dicots show increased rates of elongation even when they grow near another plant. The large reflection of far-red light by leaves and stems is detected by neighboring plants and causes enhanced stem elongation in them. Apparently, this far-red light removes $P_{fr}$ from the absorbing plants and allows them to modify their growth patterns in anticipation of being shaded.

**Figure 20-14** Promotion of tillering in wheat plants grown at low densities. Plants were grown at densities of 1,000 (left) or 300 (right) plants per square meter. The latter density approximates typical field plantings. (Courtesy Bruce Bugbee.)

Branching at the bases of grass stems (called **tillering**) is also controlled in part by the phytochrome system. Tiller (branch) formation is retarded in closely spaced cereal-grain crops or in thick pastures because the light transmitted to the stem bases is rich in far-red wavelengths, causing low $\phi$ values there. Increasing the $\phi$ value with light rich in red wavelengths promotes tillering, just as it promotes branching in dicots (Deregibus et al., 1983; Casal et al., 1985; Kasperbauer and Karlen, 1986). Crowding effects on tillering in wheat are shown in Figure 20-14.

## 20.8 Photoperiodic Effects of Light

In many species, responses to light, especially light absorbed by phytochrome, are influenced by the time of day in which light is received. The effects of light in

**Figure 20-15** Growth of Douglas fir (*Pseudotsuga menziesii*) after 12 months on photoperiods of 12 h (left), 12 h plus a 1-h interruption near the middle of the dark period (middle), and 20 h (right). (From Downs, 1962.)

interrupting the normal dark period or in prolonging the normal period of daylight are referred to as **photoperiodic effects** (Vince-Prue, 1975, 1989). These responses mainly concern bud dormancy of perennial plants and production of flowers and seeds by perennials and (especially) nonperennials (see Chapter 23). Figure 20-15 illustrates the importance of day length in controlling bud dormancy (and therefore overall growth) of Douglas fir, a conifer. In general, long days promote elongation of stems of most species, and short days lead to the changes associated with autumn (for example, dormancy and frost-hardiness of buds).

## 20.9 Light-Enhanced Synthesis of Anthocyanins and Other Flavonoids

Most plants form anthocyanin pigments and other flavonoids in specialized cells in one or more of their organs, and this process is frequently promoted by light. A simple example is the faster development of the red color resulting from anthocyanin in apple fruits on the sunny side (as opposed to the shady side) of a tree. Production of flavonoids requires sugars as a source of the phosphoenolpyruvate and erythrose-4-phosphate (see Sections 11.2 and 13.10) that provide carbon atoms needed for the flavonoid B ring and as a source of acetate units needed for the flavonoid A ring (see Section 15.7). Sugars, especially sucrose, can arise from degradation of starch or fat in storage organs during seedling development or from photosynthesis in chlorophyll-containing cells. It is no surprise, therefore, that anthocyanin synthesis is increased by light acting photosynthetically in leaves or green apple fruit skins; but light promotes synthesis of these pigments in organs that photosynthesize little or not at all, including autumn leaves, flower petals, and etiolated seedlings, showing that at least one other pigment participates.

Action spectra for anthocyanin production in several species are shown in Figure 20-16. In general, maximum responses occur in the red, far-red, and blue regions, whereas green light (approximately 550 nm) is almost without effect. Peaks in the yellow, orange, red, and far-red regions in various species vary considerably both in wavelengths and heights. Blue light is effective in nearly all species, and in sorghum red and far-red light are ineffective. A detailed action spectrum in the blue region for synthesis of the aromatic-acid precursors of the B ring in *Cucumus sativus* hypocotyls (Smith, 1972) is similar to that of phototropism shown in Figure 19-8 and to the inhibition of hypocotyl elongation in lettuce seedlings by blue light shown in Figure 20-6, suggesting that effective blue wavelengths are absorbed primarily by cryptochrome. Red and far-red wavelengths act via phytochrome and independently of photosynthesis in etiolated seedlings, but in green apple skins photosynthesis also contributes. High irradiance levels characteristic of the HIR system are required for these red- and far-red-light effects, and both phytochrome and cryptochrome are probable photoreceptors in most species. Mancinelli (1985, 1989) described numerous species and the probable photoreceptors involved in formation of anthocyanins.

Numerous attempts have been made to determine the site or sites of light action in biochemical pathways leading to both the A and B rings of flavonoids. The accumulation of flavonoids in many leaves during senescence in autumn suggests a relation among protein hydrolysis, phenylalanine appearance, and the use of phenylalanine in ring-B formation. Because phenylalanine can be used in various metabolic pathways, control by light of the first step in its conversion to ring B was suspected. This step requires the enzyme phenylalanine ammonia lyase (Chapter 15, Reaction 15.5), and light promotes its activity in various organs of many plants. Nevertheless, several other flavonoid-synthesizing enzymes not mentioned in our book also exhibit increased activity after light treatment, indicating that production of both rings occurs more rapidly in light.

Another receptor that absorbs light that promotes flavonoid synthesis is called the **UV-B receptor**. Several such receptors are thought to exist, and they might be DNA (reviewed by Beggs et al., 1986). In leaf epidermal cells of parsley seedlings and in parsley cell-suspension cultures, UV-B is highly effective; the same is true for

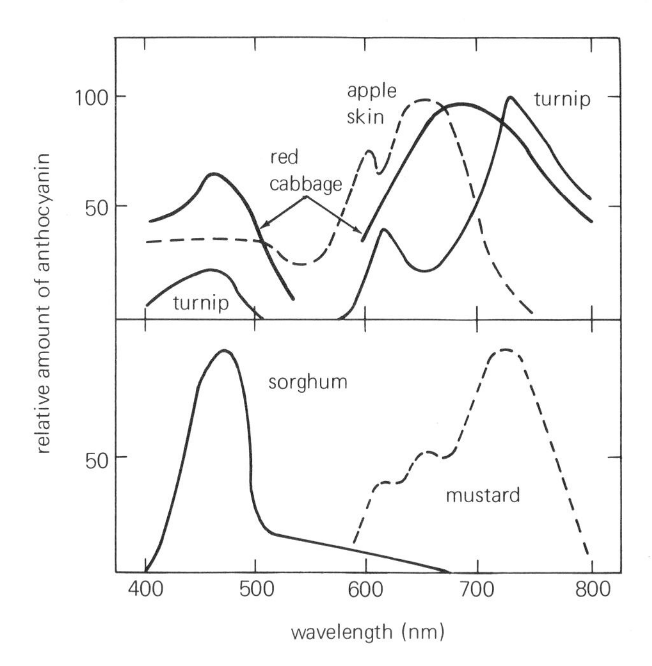


**Figure 20-16** Action spectra for anthocyanin formation in various species after prolonged irradiance. The apple fruits contained chlorophyll, but the turnip, cabbage, sorghum, and mustard seedlings were probably chlorophyll-free when irradiation began. (Data redrawn from various sources. Apple skin data are from Siegelman and Hendricks, 1958; red cabbage and turnip, Siegelman and Hendricks, 1957; sorghum, Downs and Siegelman, 1963; mustard, Mohr, 1957.)

maize coleoptiles. The peak of action is about 300 nm, and several other species show this response to UV-B. In most of these species there appears to be coaction between phytochrome and either cryptochrome or the UV-B receptor.

In a series of excellent studies, Klaus Hahlbrock and his colleagues in Germany tracked down the effect on flavonoid formation in parsley cell cultures of a white-light source that was rich in blue and UV wavelengths. Irradiation first caused a coordinated increase in activity of enzymes that convert phenylalanine to *p-coumaryl coenzyme A*, a precursor of the flavonoid B ring. Next there was a rise in activity of **chalcone synthase**, the key enzyme that forms the basic flavonoidlike structure by condensing p-coumaryl CoA (for the B ring) with three acetate groups of *malonyl CoA* molecules (for the A ring). Studies indicated that light causes these enzymes to be synthesized more rapidly by increasing amounts of the mRNA molecules that code for them. Time-course studies first showed enhanced synthesis of enzymes that convert phenylalanine to p-coumaryl CoA, then a peak in activity of chalcone-synthase transcription, next a peak in chalcone-synthase activity, and finally a rise in amounts of flavonoids (Beggs et al., 1986; Hahlbrock and Scheel, 1989). Mechanisms by which light activates genes that control formation of these flavonoids and of anthocyanins are now being studied intensively. The ability of UV radiation to cause formation of UV-absorbing flavonoids appears to be a way plants protect themselves against UV radiation.

Like the flavonoids, lignins are also formed from the shikimic acid pathway with participation of phenylalanine ammonia lyase. In seedlings or in immature parts of older plants undergoing xylem differentiation or formation of xylem from the vascular cambium, lignin formation in xylem cell walls is promoted by light. This process is partly responsible for the greater stiffness of seedlings grown in light as opposed to darkness.

## 20.10 Effects of Light on Chloroplast Arrangements

When irradiance levels are high, chloroplasts are usually aligned along radial walls of the cells, becoming shaded by each other against light damage. In weak light and often in darkness, chloroplasts are separated into two groups distributed along the walls nearest to and farthest from the light source, thereby maximizing light absorption. This movement of plastids, which depends upon the direction of light as well as its irradiance level, is an example of **phototaxis** (movement of an entire organism or organelle in response to light; see Haupt, 1986 and Section 19.1).

In mosses and angiosperms, phototactic responses to both low and high irradiances are maximal under blue wavelengths, and phytochrome does not participate (Inoue and Shibata, 1973; Seitz, 1987). Action spectra suggest that cryptochrome is again involved. In some green algae, including *Mougeotia* and *Mesotaenium*, and in the fern *Adiantum*, however, both phytochrome and cryptochrome absorb low-irradiance light responsible for movement of the chloroplast or chloroplasts to regions of the cells in which light absorption is increased. Also, in *Mougeotia*, the effective phytochrome molecules are located near the plasma membrane. In this species each cell contains one long, ribbon-shaped chloroplast that does not actually migrate but simply rotates so that it faces light of low or moderate irradiance, whereas only its narrow edge faces light of high irradiance.

In all species the chloroplast itself does not absorb the light causing its phototaxis; instead, light absorbed elsewhere in the cell causes chloroplast movements through effects on cytoplasmic streaming, which effects result from interactions between microfilaments and microtubules. Ecologically, chloroplast movements seem important mainly to increase light absorption at low irradiances and to decrease absorption when irradiances are so high that they might cause solarization (see Section 12.3) or other photodestructive effects.

## 20.11 How Photoreceptors Cause Photomorphogenesis

We do not yet understand how photoreceptors cause photomorphogenesis. Nevertheless, two main kinds of effects that differ in their rapidity seem to exist. One effect can be considered a fast membrane-permeability effect; the other can be considered a slower effect on gene expression. Research on control of gene expression is now proceeding especially rapidly, yet such control might often depend on earlier light-induced effects on changes in membrane permeability that influence fluxes of ions (especially $Ca^{2+}$) across the plasma membrane, tonoplast, ER, or nuclear envelope. In this section we summarize some examples of each main kind of effect with citations (especially of reviews) that mention past progress and future research questions. Keep in mind that primary effects of phytochrome and cryptochrome might act on some part of the receptor-transduction system for possible hormone action depicted in Figure 17-2.

Early research by the discoverers of phytochrome and by others suggested that $P_{fr}$ acts primarily by changing permeability of membranes and that photomorphogenic responses somehow result from those changes (reviewed by Roux, 1986 and by Kendrick and Bossen, 1987). Subsequent research verified that in some cells of some species there are rapid (seconds to minutes) responses, not only to $P_{fr}$ but in other cases to blue-activated cryptochrome as well. One of the most rapid effects of $P_{fr}$ was discovered to occur in excised barley root tips, and it also occurs in mung bean root tips; this phenomenon is called the **Tanada effect** (Tanada, 1968).

When excised barley root tips are swirled gently in a glass beaker with IAA, ATP, ascorbic acid, and certain mineral salts, a brief red-light treatment causes them to begin within seconds to stick to the walls of the beaker. In this case, beaker walls had a slight negative charge because they had been washed in dilute phosphate. This means that red-light-treated root tips attained a slight positive surface charge, perhaps because of $H^+$ transport from cytosol to cell walls by an ATPase in the plasma membrane (see Section 7.8). A brief far-red-light treatment rapidly decreases the positive charge of the root tips and causes their release from the beaker walls. The phytochrome for this response is in the root cap.

Another rapid effect of $P_{fr}$ described in Section 19.2 involves promotion of nyctinastic or sleep movements of certain legumes. In this case $P_{fr}$ also acts on membranes; this is known to be true because the electropotential across plasma membranes becomes perturbed and potassium ions are then rapidly transported from extensor cells to flexor cells of leaf pulvini (see Fig. 19-1). Research by Moysset and Simon (1989) concerning *Albizzia* nyctinastic movements indicates that $Ca^{2+}$ is involved in the transduction process resulting from $P_{fr}$ formation. The effect on rotation of *Mougeotia* chloroplasts caused by $P_{fr}$ is rapid and is complete within 15 min. Changes in membrane permeability again seem to be involved, especially a $P_{fr}$-induced increase in cytosolic $Ca^{2+}$ (Haupt, 1986, 1987; Lew et al., 1990). The fast (less than 30-s lag time) for blue-light-induced suppression of hypocotyl elongation in dark-grown cucumber seedlings also involves an effect on membrane permeability (Spalding and Cosgrove, 1988).

We mentioned earlier that cryptochrome is probably a flavoprotein present in the plasma membrane, and so it is not surprising that unusually rapid changes in membrane permeability could be caused by its activation by light. However, there is no biochemical evidence that phytochrome is part of any membrane as either an integral or a peripheral protein. In fact, comparison of its amino-acid sequence with that of known integral proteins indicates that it is not an integral protein (Vierstra and Quail, 1986). Nevertheless, we could more easily explain its effects on membrane permeability if it were to quickly bind to the plasma membrane.

In contrast to the concept proposed by workers in Beltsville, Maryland, that the primary effect of phytochrome is on membrane permeability, Hans Mohr's research with *Sinapis alba* seedlings led to his suggestion (as early as 1966) that phytochrome controls activation and deactivation of specific genes, depending on the developmental state of and kinds of cells involved. Much research since then demonstrates that phytochrome, cryptochrome, and UV-B receptors can indeed control gene activity in the sense that each can regulate the amount of enzyme produced, in contrast to post-translational effects on enzyme activity. We have mentioned some of these examples in this chapter. Others are described in recent reviews (see Link, 1988; Nagy et al., 1988; Thompson, 1988; Thompson et al., 1988; Moses and Chua, 1988; Marrs and Kaufman, 1989; Okamuro and Goldberg, 1989; Watson, 1989; Simpson and Herrera-Estrella, 1990). These reviews also emphasize that there are several cases in which light seems to act primarily by controlling stability of certain mRNA molecules instead of their synthesis.

Clearly, light can control photomorphogenesis by several mechanisms. It is possible that phytochrome present near the plasma membrane causes changes that increase $Ca^{2+}$ influx (or sometimes efflux) and that this signal is used in a transduction process to alter gene activity. There is now much evidence that $Ca^{2+}$ fluxes are altered by $P_{fr}$ and that Ca-calmodulin is often part of the transduction process (reviewed by Marmé, 1989 and by Morse et al., 1990; also see Moysset and Simon, 1989). Even more recent evidence suggests that $P_{fr}$

causes activation of the Ca-calmodulin-dependent enzyme $NAD^+$ kinase in apical buds of both etiolated pea seedlings and light-sensitive lettuce seeds (Kansara et al., 1989; Zhang et al., 1990). Furthermore, accumulating evidence suggests that both $P_{fr}$ and activated cryptochrome lead to rapid increases in phosphorylation of certain proteins (Park and Chae, 1989; Short and Briggs, 1990). It is known that free $Ca^{2+}$ or Ca-calmodulin can activate different protein kinases in plants (Blowers and Trewavas, 1989), so light-stimulated absorption of $Ca^{2+}$ across the plasma membrane or its release from an internal compartment such as the ER or vacuole is probably essential. As mentioned in earlier chapters, changes in activity of enzymes by their phosphorylation and dephosphorylation is another mechanism that controls cellular activities. It is becoming obvious that light can cause photomorphogenesis by diverse biochemical processes, but clarification of the role of light-induced changes in cytosolic $Ca^{2+}$ concentrations will probably lead to a common theme for some of them.

# 21

# The Biological Clock: Rhythms of Life

Change is the only thing in its environment that an organism can count on. Almost nothing is really constant. In a study of the alpine tundra on the north end of Rocky Mountain National Park in Colorado (Salisbury et al., 1968), several kinds of environmental changes appeared: Wind velocities changed significantly in less than a second. Temperatures, light levels, and humidities sometimes changed radically in time intervals from minutes to perhaps 5 or 6 hours. All these and other changes were superimposed on a daily (**diurnal**[1]) cycle. Weather cycles typically lasted several days. In one summer, for example, heavy storms were separated by intervals of from 10 to 22 d, with an average of about 13. Exceptionally clear days were separated by intervals of 5 to 14 d, with an average of about 10. These erratic weather cycles were always superimposed on the **annual** cycle of the seasons.

Furthermore, **tidal (lunar)** cycles influence the tidal-zone environment and could be important in other habitats as well. Weather trends may be related to the 11-year sunspot cycle, and long-term climatic changes, such as those that produced the ice ages, occur over periods of centuries to millennia.

It would be to an organism's advantage to anticipate and adjust to these environmental changes. Three of them — those related to the mechanics of the solar system — are regular enough so that this should be possible: the daily, tidal, and annual cycles. For an organism to anticipate and prepare for these regular changes in its environment, it needs a clock and various associated mechanisms. The timing system should have at least two broad sets of characteristics.

*First*, it should be accurate. It should not be unduly influenced by capricious factors of an organism's environment, those that cannot accurately be predicted: temperature, light levels during the day (which vary because of clouds and shading), wind velocity, moisture, and so on. Even if a biological clock were not highly sensitive to these factors, which would be surprising and impressive, it would be even more impressive if such a clock could run with the accuracy achieved by our own mechanical and electronic clock systems. Yet without such accuracy, it might soon get out of phase with the environment and, therefore, be of no benefit to the organism.

But there is an alternative to inherent accuracy: The biological clock might be regularly reset or synchronized by some dependable feature of the organism's environment. It wouldn't matter much if the clock were to gain or lose an hour or two a day, if it were reset each day at sunrise or sunset or both. Presumably, the clock might have one status appropriate for daytime and another for night; resetting at dawn and dusk might keep the clock's status in synchronization with the environment. Incidentally, it is conceivable that a "clock" might simply be driven by changing features in the environment. In this case, the organism would only track environmental changes without actually keeping time or anticipating future environmental events.

*Second*, there must be coupling mechanisms that allow the organism to take advantage of the clock's timekeeping. It is reasonable, for example, that a plant might conserve energy if it could direct and concentrate its available resources to the photosynthetic mechanism during the day and to other metabolic mechanisms during the night. Because nights are virtually always cooler than days, organisms might adjust their temperature optima for critical metabolic processes accordingly.

We have indeed found many such manifestations of biological time measurement. It appears that virtually all eukaryotic and at least some prokaryotic organisms have biological clocks. Sometimes it is easy to recognize clock control of metabolism and activity. There are also examples of highly sophisticated clock responses that we might not have expected.

[1]The term *diurnal* has often been used in the literature synonymously with *daily*, but it also implies the daytime or the light span, just as *nocturnal* implies night.

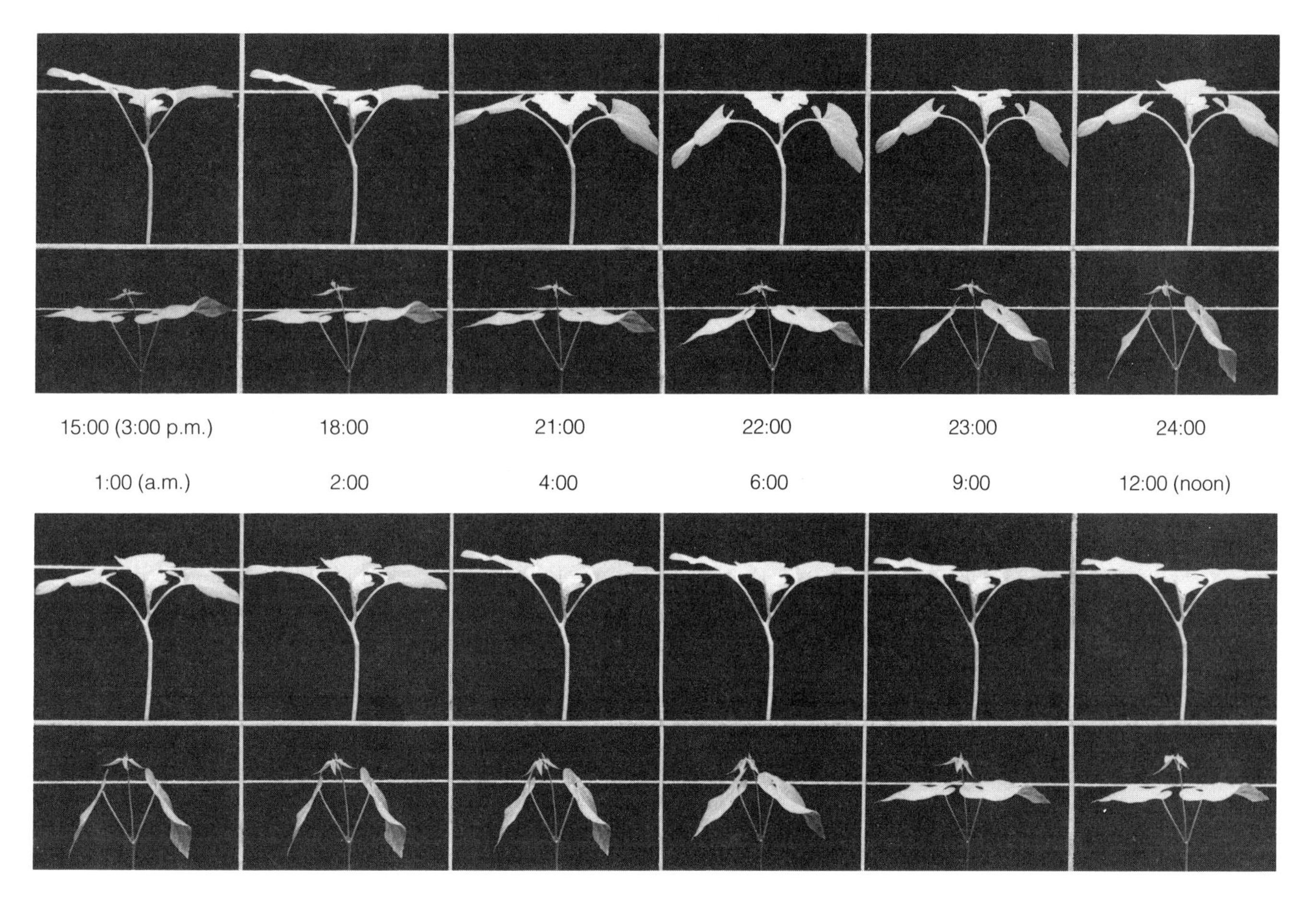


**Figure 21-1** Leaf movement in cocklebur (top row) and bean (second row). Plants were photographed at hourly intervals from noon to noon. Twelve photographs were selected for the figure. The bean leaves drop more sharply and somewhat later than the cocklebur leaves. Note that the young cocklebur leaves bend up instead of down during part of the night. (Photographs by F. B. Salisbury.)

Confronted with these observations, we immediately ask: What is the mechanism of these clocks? How do they work? The observed phenomena imply a few things about their nature, but at the moment little is known about clock mechanisms. Plants prove to be good subjects for the study of mechanisms because it is easy to make observations at the cellular and biochemical levels, where timing is apparently located. (Several reviews have been written, including those by Bünning, 1973, 1977; Hillman, 1976; Koukkari et al., 1987a and 1987b; Koukkari and Warde, 1985; Luce, 1971; Moore-Ede et al., 1982; Satter and Galston, 1981; and Sweeney, 1983, 1987.)

## 21.1 Endogenous or Exogenous?

As illustrated for two plants in Figure 21-1, the leaves of many species exhibit one position during the daytime (typically nearly horizontal) and another position in the middle of the night (typically nearly vertical). This observation was made at least as early as 400 B.C.E. by Androsthenes, who was the historian of Alexander the Great. These are nastic movements (Chapter 19).

Over two and a half centuries ago the French astronomer Jean Jacques d'Ortour de Mairan (1729) was perceptive enough to recognize a fundamental problem in relation to this cycle of "sleep movements" in plants. He wondered whether the movements were driven by changes in the environment (the daily light-dark cycle), or whether they might be controlled by some time-measuring system within the plant. If leaves moved only in response to *external* changes, timing was **exogenous**; if in response to an *internal* clock, timing was **endogenous**. Using a "sensitive" plant (probably *Mimosa pudica*), de Mairan observed the movements even after the plants had been placed in deep shade. Because these motions did not require intense sunlight during part of the 24-h cycle, he suggested that the movements were endogenously controlled. He extended an invitation to botanists and physicists to pursue this research, noting that scientific progress can be slow.

It was indeed slow. It was 30 years before his experiment was confirmed, and it was 250 years before an

# Potato Cellars, Trains, and Dreams: Discovering the Biological Clock

## *Erwin Bünning*

*As we've seen in this chapter, it was Erwin Bünning as a postdoctoral fellow, along with several colleagues, who discovered the free-running, circadian nature of the biological clock in plants, a discovery that made endogenous timing seem to be the only acceptable explanation. In 1963–64, one of us [F. B. S.] spent a sabbatical year with Bünning in Tübingen, where he was Director of the Botanical Institute, and had the opportunity to hear him tell of his early work. This is Salisbury's translation of a letter from Bünning, sent in response to a request that he record some of his experiences. The letter is dated June 2, 1970. Professor Bünning passed away on October 4, 1990.*

The story went something like this. At the Institute for the Physical Basis of Medicine in Frankfurt, the biophysicist Professor Dessauer (an X-ray specialist) became interested in the effects of the ionic content of the air upon humans. Those were the years when people began to be interested in atmospheric electricity, cosmic rays, and so on. Naturally, humans could not be used as experimental objects, and so in 1928 Dessauer searched for botanists to work on plants. One whom he found was Kurt Stern, who lived in Frankfurt; the other was me, who had just finished my doctoral work in Berlin. So we began in August of 1928 to contemplate the problem. In the process we came upon the work of Rose Stoppel, who had been studying the diurnally periodic movements of *Phaseolus* (common bean) leaves. In the process she had found, as had several other authors, that under "constant" conditions in the darkroom, most leaves reached the maximum extent of their sinking (their maximum night position) at the same time, namely, between 3:00 and 4:00 A.M. Her conclusion: Some unknown factor synchronized the movement. Could this be atmospheric ions? We had Rose Stoppel visit us from Hamburg for two or three weeks so that we could become familiar with her techniques. In our group we always called her *die Stoppelrose* ("stubble rose"), and the name was most appropriate. She was energetic and persistent, so persistent that she just died this January in her 96th year. Her results also appeared in our experiments: The night position usually occurred at the indicated time. Then we investigated the effects of air that had been enriched with ions or air from which all the ions had been removed. The result: Nothing changed—atmospheric ions are not "factor x."

We then decided that our research facilities at the Institute were insufficient. Hence, after *die Stoppelrose* had left, Stern and I moved to his potato cellar, where with the help of a thermostat, we obtained rather constant temperature. Contrary to the practice of Stoppel, who turned on a red "safe" light to water her plants, we went into the cellar just once a day with a very weak flashlight and felt around with our fingers for the pots and recording apparatus so that we could water the plants and so that we could see if everything was in order with the recorders. The flashlight was weakened with a

endogenous clock was generally recognized, although a few other early workers, including Augustin de Candolle, Charles Darwin, and Julius von Sachs, were also interested in these movements and published preliminary studies relating to them. The early investigator who probably devoted the most time and effort to this topic was Wilhelm Pfeffer (see Pfeffer, 1915), who from 1875 to 1915 wrote many papers about leaf movements of the common bean plant (*Phaseolus vulgaris*). (We mention Pfeffer's work on osmosis in Chapter 2.) Much of his extensive work is still of interest (Bünning, 1977). In spite of his early skepticism, he became convinced that such an endogenous clock must exist. Ironically, he was unable to provide experimental data that were sufficiently convincing to convince his contemporaries. During Pfeffer's time, zoologists were also observing and reporting rhythmic behavior in animals.

The real breakthrough came in the 1920s. In Hamburg, Germany, Rose Stoppel continued the researches of Pfeffer on leaf movements in bean plants. Using Pfeffer's method, she fastened the stem and petiole to bamboo sticks and attached the leaf blade with a thread to a lever contacting a moving drum coated with lamp black. The lever traced a record of the leaf movements. Stoppel observed that when leaf movements were measured in a darkroom at constant temperatures, the maximum vertical position was observed at the same time each day. A **rhythm** is the repeated occurrence of some function, and the **period** of the rhythm is the time between recurrences of a recognizable point on the cycle (for example, the maximum vertical position). Because the rhythm's period was almost exactly 24 h, Stoppel reasoned (as we have in the introduction) that a biological clock could probably not be this accurate; some factor in the environment must be resetting the clock on a daily schedule. Because the plants were in the dark and

dark red filter so that one could see only for a few centimeters' distance. In those days it was the dogma of all botanical textbooks that red light had absolutely no influence upon plant movements or upon photomorphogenesis. We did one other thing differently from Stoppel. Since Kurt Stern's house was a long way from our laboratory, we didn't make our daily control visit in the morning, rather only in the afternoon. The result: Most of the maximal night positions no longer appeared between 3:00 and 4:00 A.M., but rather between 10:00 and 12:00 A.M. Hence we concluded: The dogma is false. Red light must synchronize the movements so that a night position always appears about 16 h after the light's action. That was "factor X." When we eliminated this hardly visible red light, we found that the leaf movement period was no longer exactly 24 h but 25.4 h. [This circadian feature of the clock was the key to understanding its endogenous nature—Ed.]

So that was about how that story went. Naturally, I could also tell you the story about how I came upon the significance of the endogenous rhythms for photoperiodism. That was about in 1934. Of course, I had already often asked myself how such an endogenous rhythm might ever have any selection value (in evolution), and I had already expressed the opinion in 1932 in a publication (Jahrbuch der Wissenschaftlichen Botanik 77: 283–320) that some interaction between the internal plant rhythms and the external environmental rhythms must be of significance for plant development. But there were coincidences in the story. As a young scientist, one naturally had to allow himself to be seen by the power-wielding people of his field, so that he might receive invitations for promotion. Hence, I traveled in 1934 from Jena to Königsberg in Berlin to introduce myself to the great Professor Kurt Noack. We discussed this and that. He mentioned that discoveries were being made that were so remarkable one simply couldn't believe in them. Such a one, for example, was photoperiodism. He, as a specialist in the field of photosynthesis, must certainly know that it makes no difference whatsoever what program is followed in giving the plant the necessary quantities of light. Then, as I was riding back on the train, the idea came to me—aha, for the plant it does make a difference at which time light is applied, if not exactly in photosynthesis, nevertheless for its development!

I could present a third story, one from very recent years. I have long felt that the daily leaf-movement rhythms had no selection value in themselves. As you know, I have recently changed my mind. The movements could indeed be important in avoiding a disturbance of photoperiodic time measurement by moonlight. Before I had begun to test the idea experimentally or even to think about it, it came very simply one night in a dream. The dream (apparently as a memory of one of my visits to the tropics): tropical midnight, the full moon high above on the zenith, in front of me a field of soybeans, the leaves, however, not sunken in the night position, rather broadly horizontal in the day position. My thoughts (in the dream): How shall these plants know that this is not a long day? They had better hide themselves from the moon if they want to flower.

at constant temperature, this factor could not be daylight or temperature, so Stoppel called it *factor X*.

Two young botanists in Frankfurt, Erwin Bünning and Kurt Stern, were looking for a research problem involving such subtle physical factors as the ionic content of the atmosphere. Could such a factor time the leaf movements in Stoppel's experiments? They found that atmospheric ions had no effect on the rhythmic movements, but they identified Stoppel's factor X as the red light she used while watering her plants. When the light was eliminated, Stoppel's prediction held true: The maximum vertical position of the leaves came about an hour and a half later each day, so the leaf-movement cycle was soon out of phase with day and night outside the darkroom (Bünning and Stern, 1930; Bünning et al., 1930). Bünning tells this story in his personal essay in this chapter.

The **free-running period** is the period of a rhythm that continues under constant environmental conditions. Because the free-running period was greater than 24 h, there seemed to be no alternative to an endogenous clock. The rhythms were not simply tracking the normal day and night cycle; they drifted out of phase with outside day and night. Normally, they were reset or **entrained** to the natural cycle, probably by dawn or dusk or both, but the free-running clock betrayed its inaccuracy and thus displayed its endogenous nature.

In the 1950s, Franz Halberg (see Halberg et al., 1959) at the University of Minnesota suggested that rhythms with a free-running period of approximately but not exactly 24 h should be called **circadian**. He coined the term from the Latin *circa*, which means "approximately," and from *dies*, or "day."

Pfeffer, and even A. P. de Candolle in 1825 (see Moore-Ede et al., 1982; Sweeney, 1987), had observed rhythms with circadian free-running periods, but they had not attached the proper significance to their observations. Yet Antonia Kleinhoonte, working independently in Delft, Holland, arrived at exactly the same conclusions as those of Bünning, Stern, and Stoppel. Results of her and their experiments were published between 1928 and 1932 (see, for example, Kleinhoonte, 1932). Stern later emigrated to America; neither he nor Kleinhoonte continued to work on the biological clock, but Bünning continued his studies well into his formal retirement.

## 21.2 Circadian Rhythms

Several fields of modern knowledge clearly qualify as *biology* rather than *botany* or *zoology*. Examples are genetics, the pathways of cellular respiration, and the cell theory. Biological clocks and various rhythms occur in virtually all eukaryotes that have been studied carefully, including protistans, fungi, plants, and animals. Although many attempts to discover rhythms in prokaryotes (monerans) had negative results, or only rhythms with very short periods were detected, clear rhythms in photosynthesis and nitrogen fixation have been reported in the blue-green alga *Synechococcus* (Grobbelaar et al., 1986; Mitsui et al., 1986, 1987). Nitrogen fixation is inhibited by oxygen, which is produced in photosynthesis, so the rhythm ensures that the two processes will go on at different times of the day, even in continuous light.

Several rhythms have been studied in single-celled organisms. Phototaxis in the green alga *Euglena* and a mating reaction in *Paramecium* are good examples. There has been much study of the biological clock in *Gonyaulax polyedra*, a marine dinoflagellate. Beatrice Sweeney (see her personal essay in this chapter) and J. Woodland Hastings, then located at the Scripps Institution of Oceanography at La Jolla, California, were the first to work intensively with this organism, which has rhythms for three important variables. Most spectacular is a rhythm of **bioluminescence** observable when a suspension of *Gonyaulax* cells is tapped or otherwise jarred, causing them to emit light. The amount of light they emit follows a circadian rhythm, with its peak normally occurring near midnight. Johnson et al. (1984) showed that the bioluminescence rhythm matches a rhythm in the activity of luciferase, the enzyme that causes the light emission. There is also a rhythm in cell division, with the maximum occurring near dawn. The third rhythm is in photosynthesis. A maximum quantity of $CO_2$ is fixed (standard test conditions) near noon; that is, the photosynthetic mechanism is adjusted by the clock to anticipate the environment, just as we speculated in our introduction. Rhythms, including bioluminescence, have also been observed in related organisms, but one highly bioluminescent dinoflagellate has no bioluminescence rhythm!

Note that we are dealing here with a population of organisms rather than with individuals, as in Bünning's original studies. Such studies with populations are quite common. But even in such a population, the individual organisms are keeping their own time, as we learned from experiments with *Gonyaulax* in which cultures on different schedules were mixed with each other: Their individual rhythms continued after being mixed.

There are several known fungal circadian rhythms. One is a rhythm in formation of conidia (asexual spores) in *Neurospora crassa*. A series of dark bands, one per day, appears in the mycelia growing on agar from one end to the other of a long culture tube. This rhythm is of interest because several *Neurospora* clock mutants are known (see below). Another fungal example is a rhythm of spore discharge in *Pilobolus*. Yet true circadian rhythms in fungi are rare, although there are many noncircadian rhythms, often driven by environmental cycles (Lysek, 1978; Piskorz-Binczycka et al., 1989).

Many rhythmic phenomena besides leaf movements have been observed in higher plants (Koukkari and Warde, 1985). These include petal movements, growth rates of various organs, concentrations of pigments and hormones, stomatal opening and closing, discharge of fragrance from flowers, times of cell division, metabolic activities (for example, photosynthesis and respiration), and even the volume of the nucleus. Many plants exhibit a rhythm in sensitivity to such environmental factors as light and temperature. Some species flower or grow well only when temperatures during the part of the cycle that normally comes at night (**subjective night**) are lower than temperatures during **subjective day**. Or light given during subjective night may inhibit some responses. Thus the clock adjusts a plant's metabolism to coincide with its cycling environment.

Several insects exhibit rhythms that are convenient for study (Saunders, 1976). The time of day at which adult *Drosophila* emerge from the pupae (**eclosion**) follows a circadian rhythm and is an intensively studied example, as is the activity of the cockroach. **Activity** or **running cycles** have also been studied in birds and rodents. These rhythms are useful because they often continue under constant environmental conditions for months or even years, and they are relatively easy to study by equipping cages with automatic devices that continuously record the organism's activity.

Rhythms with obvious outward manifestations (leaf and petal movements, animal activities, and others) could be less important than the internal metabolic changes controlled by the clock. We have mentioned several metabolic rhythms of plants, and comparable

## Women in Science

### *Beatrice M. Sweeney*

*There is much concern about the role of women in our society. It has often seemed to me [F. B. S.], whether fact or only an illusion, that an unusually high percentage of those who have contributed to our understanding of the biological clock have been women. I was discussing this idea with Beatrice M. Sweeney during a symposium on biological clocks several years ago. She agreed to write a few personal thoughts about these matters for use in our textbook. She was located in the Department of Biological Sciences, University of California at Santa Barbara. She died in 1989.*

Dear Frank:

Until you remarked upon it at the Christmas meeting at Sacramento, I hadn't noticed the unusual number of women in the field of biological clocks. I am not accustomed to thinking about the sex of my scientific colleagues. In this respect, I suppose I am a truly liberated woman. I think I owe the fact that I am not conscious of whether or not scientists are women to my good fortune in receiving my training in Dr. Kenneth Thimann's laboratory at Harvard, where even long ago all graduate students were on an equal footing. We all regarded ourselves as superior, and so our chances of turning out that way were much increased. This brings me back to the subject of women in science, since I often think women have a difficult time believing that they really can do first-class research in a field usurped exclusively for so long by men, at least in the top prestige bracket. In rhythm work, women have, I believe, found relief from their feelings of inferiority, because they recognize in this field a comfortable familiarity. To get up in the middle of the night, perhaps several times, is not strange to them. What matters that it is a series of tubes full of *Gonyaulax* and not a baby that demands their attention? I notice that the predilection of women for chronobiology does not seem to be declining. The traditions set by such women as Rose Stoppel, Antonia Kleinhoonte, Marguerite Webb, and Janet Harker are being ably perpetuated by Audrey Barnett, Ruth Halaban, Marlene Karakashian, Laura Murray, Ruth Satter, and Therese van den Driessche, and I'm sure you can name others. Of course, I still regard myself as active and expect any day to crack the problem of the basic circadian oscillator.

I would like to say something personal to the young ladies who are perhaps considering a future as scientists, now that the sole production of children is no longer in the interest of world well-being. My work in science has been the very stuff of life to me, endlessly frustrating and rewarding. I should like to relate to these young women what fun it is to work, rather than be a tourist, in strange parts of the world. Imagine the delight of traveling northward inside the Great Barrier Reef on a 60-foot boat usually devoted to the hunting of alligators and to have available a microscope with which to see the details of the strange animals and plants, the familiar yet unfamiliar phytoplankton, the algae growing within the giant clams and coral. I could mention also the jungles of New Guinea at night, ringing with the stridulations of locusts at multiple high frequencies, flashing with fireflies. Or the beaches of Jamaica, white and shining and fringed with the bending fronds of the coconut palms like long eyelashes. The knowledge of flora and fauna acquired in scientific study immeasurably increases the pleasure of viewing an unfamiliar part of the world.

But most of all I'd like to say that the day-to-day research and teaching of an academic profession provides an endlessly varying and interesting way of life, to my tastes infinitely more satisfying than cooking, dusting and shopping, even than bringing up children. Perhaps my four children have benefited rather than suffered from the fact that I have other interests and satisfactions than themselves. At least it is interesting to see that three of the four are in some way pursuing science as a career.

---

cycles (for example, potassium levels in the blood, urine excretion, and body temperature) have also been documented in animals (Moore-Ede et al., 1982). These often help an organism adjust to its environment. The adjustments can be subtle; an organism may be far more sensitive to toxic chemicals, herbicides (Koukkari and Warde, 1985), or ionizing radiation (for example, X-rays) during a part of its circadian cycle. Perhaps animals conserve energy during the inactive part of their cycle by lowering resistance to factors not likely to be encountered. *In any case, anyone doing biological experiments must be aware of the profound effects of the physiological clock on virtually all life functions.* The time when a treatment is given is often decisive.

Circadian rhythms are also common in humans, as we will see in the final section of this chapter. In general, the same principles apply to humans as apply to plants and other organisms.

**Table 21-1 Types of Rhythms, Period Ranges, and Examples of Vascular Plants, Fungi, Algae, and Some Other Organisms Displaying These Rhythms.[a]**

| Domain or Type | Period | Examples: Process | Examples: Approximate Period | Examples: Organism |
|---|---|---|---|---|
| Ultradian[b] | <20 h | Glycolysis | 1.8 min | *Saccharomyces carlsbergensis* |
| | | Protoplasmic streaming | 2–2.5 min | *Physarum polycephalum* |
| | | Leaf movements | 3 min | *Desmodium gyrans* |
| | | Sap flux | 20 min | *Gossypium areysianum* |
| | | Auxin transport | 25 min | *Zea mays* |
| | | Transpiration | 30 min | *Avena sativa* |
| | | Leaf movements | 36 min | *Gossypium hirsutum* |
| | | Enzyme activity | 1–5 h | *Pisum sativum* |
| | | Leaf cell viscosity | 2–3 h | *Helodea densa* |
| | | Enzyme activity | 12–15 h | *Chenopodium rubrum* |
| **Circadian**[c] | 20–28 h | Conidia formation (fungus) | ca. 24 h | *Neurospora crassa* |
| | | Coleoptile growth | 24 h | *Avena sativa* |
| | | Leaf movement | 22.67 h (continuous darkness) | *Coleus blumei* x *C. frederici* |
| | | Leaf movement | 23.06 h (continuous dim light) | *Coleus blumei* x *C. frederici* |
| | | Leaf movement | 24.79 h (continuous bright light) | *Coleus blumei* x *C. frederici* |
| Infradian | >28 h | Gametangia formation | 4 d | *Derbesia tenuissima* |
| | | Growth pattern (fungi) | 4 d | *Aspergillus ochraceus*, *Colletotrichum lindemuthianum*[1] |
| **Tidal** | 12.4 or 24.8 h | Leaves sand only when covered with water | 12.4 h | *Synchelidium* (an amphipod)[2] |
| | | Migration up out of sand at low tide | 24.8 h | *Hantzschia virgata* (a diatom)[3] |
| **Semilunar** | 14.8 d (ca. 15 d) | Egg release | 16 d | *Dictyota dichotoma* |
| | | Spawning | 14.8 d | *Leuresthes tenuis* (grunion)[4] |
| | | Mating, dark of the moon | 14.8 d | *Eunice* (a polycheate worm)[5] |
| Menstrual | ca. 28 d | Menstruation | ca. 28 d | Human females |
| **Lunar (circalunar) (rare)** | 29.6 d (ca. 30 d) | Population size | 29.6 d | Various zooplankton[6] |
| **Circannual (circannian)** | 12 ± 2 months | Seed germination | 1 yr | *Solanum acaule* |
| | | Fat accumulation and hibernation | 1 yr | Golden-mantled ground squirrel[7] |
| Infraannual | Several years | Flowering | 30–40 yr | Various bamboos[8,9] |

[a]Modified from Koukkari and Warde (1985), where references are given for their examples, with additions from Sweeney (1987); superscript numbers here refer to the following references: 1: Jerebzoff, 1965; 2: Enright, 1963; 3: Palmer and Round, 1967; 4: Walker, 1952; 5: Hauenschild et al., 1968; 6: Fryer, 1986; 7: Pengelley and Asmundson, 1969; 8: Janzen, 1976; 9: Sweeney, 1987.

[b]According to another scheme (Reinberg, 1971), the period range of ultradian rhythms is 0.5 to 20 h, and anything with a period shorter than 0.5 h is referred to as a *high-frequency rhythm*. Koukkari et al. (1987a) have also singled out ultradian oscillations in the period range of 30 to 240 min as a special group that appears to be ubiquitous in plants, animals, and microorganisms. Most of the ultradian rhythms exhibit considerable variability and change of pattern over time.

[c]Rhythms that match natural cycles are shown in **boldface**. Most of these appear to be endogenously controlled and are temperature-compensated. Those not shown in boldface do not match natural cycles and are usually sensitive to temperature.

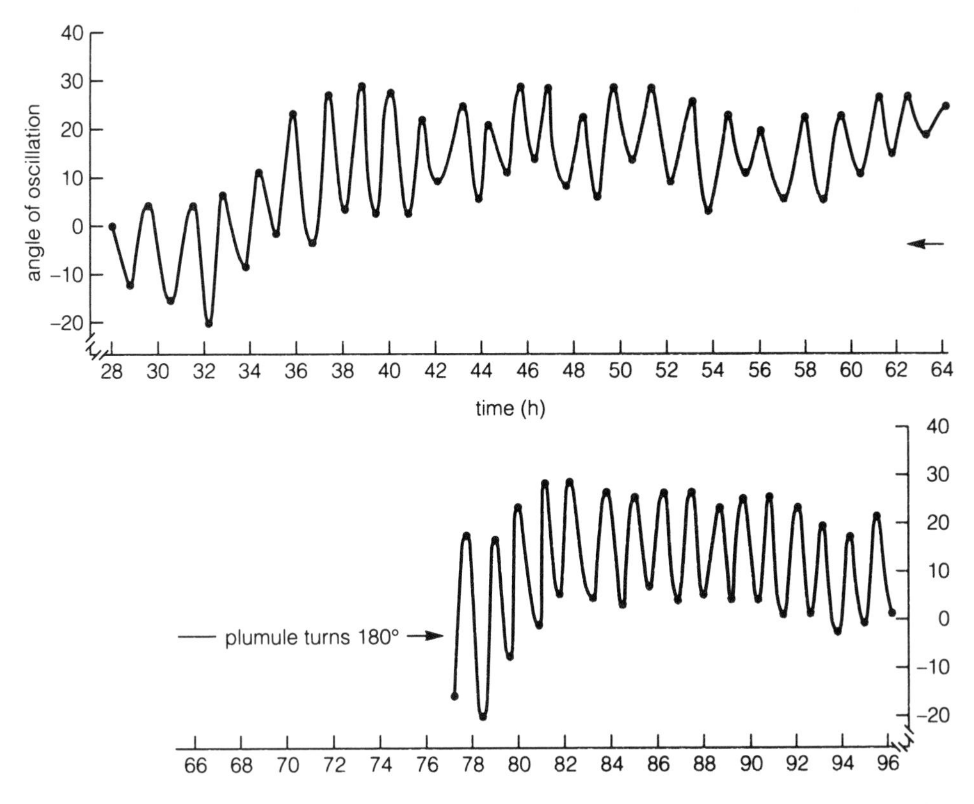


**Figure 21-2** Cycles of circumnutation in the plumule of *Pisum sativum* cv. Alaska germinated in darkness and transferred to constant red light of low irradiance (330 mW $m^{-2}$) before hour 28. The period was 77 min at 26°C. (After Sweeney, 1987; from Galston et al., 1964.)

## 21.3 The Spectrum of Biological Rhythms

Although much study remains to be done, many biologists are intrigued with the frequency spectrum of rhythms (Table 21-1). There are many rhythms and several ways to classify them, each classification providing some insights. Perhaps the most logical classification is based on period length (or **frequency**, which is the number of cycles per unit time interval and thus the reciprocal of the period). Table 21-1 lists examples of **ultradian** (period less than 20 h), circadian (20 to 28 h), and **infradian** (longer than 28 h) rhythms. The rhythms can also be classified according to whether they match some natural cycle of the environment (printed in boldface in Table 21-1) or do not (not in boldface). As we discuss below, rhythms that match environmental cycles (daily, tidal, and so on) typically share a common feature: insensitivity of the period to temperature changes. The periods of rhythms that do not match environmental periodicities are usually quite different at different temperatures.

Except for the tidal rhythms with a 12.4-h periodicity, which are difficult to place in Table 21-1, most of the ultradian rhythms do not match environmental cycles. Circumnutation movements of the shoot (Section 19.6) provide a good example of an ultradian rhythm in plants. Galston et al. (1964) observed circumnutation in *Pisum sativum* (pea, cultivar Alaska) stems with a period of 77 min at 26°C, and the period was highly temperature-sensitive, with $Q_{10} = 2$ (Fig. 21-2). Rapid leaf and root movements with periods of 4 to 5 min, 15 min, and so on, exist in many plants. One of the shortest ultradian cycles involves beating of flagella and cilia, with frequencies of about 200 beats per second. It has been suggested that the ultradian rhythms (for example, glycolysis in Table 21-1) were the raw material upon which evolution acted to produce the periods that match environmental cycles. This may be so. It has also been suggested that circadian and other temperature-compensated rhythms are the summation of ultradian cycles (just as a quartz clock sums the ultrahigh frequency vibrations of its quartz crystal to produce seconds, minutes, and so on); this idea seems much less likely because most ultradian cycles are so highly sensitive to temperature.

A few known infradian rhythms do not match environmental cycles, and again many of these are temperature-sensitive. Fungal growth, gametangia formation, and sporulation provide good examples (Table 21-1), but conidia formation in *Neurospora crassa* is a typical circadian rhythm. Because of the 28-d period, it has been suggested that the human menstrual cycle might be a circalunar rhythm, but the cycle is not synchronized by either the phases of the moon or the tides. On the other hand, the cycle shortens as menopause approaches and can be synchronized in one or more women (for example, in a dormitory) by odors given off from another woman.

A few cycles of several years are known; we will call them **infraannual**. Flowering of various bamboo species with a period of 30 to 40 yr or longer provides an excellent example. A bamboo species from mainland China, *Phyllostachys bambusoides*, has shown a seeding cycle of 120 yr, with synchronized flowering of widely separated members of the species (Janzen, 1976). For obvious reasons, no one has undertaken the proper tests to see if such long cycles are truly endogenously controlled, but specimens from a single culture have been taken to widely separated parts of the world, such as the gardens at Kew and the tropical rain forests of Jamaica, where they continue to flower in synchrony with each other, suggesting that control is endogenous (Janzen, 1976; Young and Haun, 1961). Such observations have been reported for many species, including the *P. bambusoides* mentioned above.

The tides are caused by the action of gravitational forces from the sun and the moon on the earth, and thus they are highly regular. The period from one full moon to another has been measured to at least eight significant figures; it is 29.530589 d long! It is easy to understand the tides if we begin by considering only the moon and imagining a "model earth" uniformly covered with water (Fig. 21-3). The earth and the moon revolve about their mutual center of gravity, which lies just within the larger earth's surface on the side closest to the moon, and this creates a centrifugal force at all points on the earth; this force directed away from the moon. On the side nearest the moon, the moon's gravity pulls the earth's water toward the moon and against the centrifugal force. Gravity is stronger than the opposing centrifugal force, so water rises toward the moon to produce a high tide. On the side away from the moon, the centrifugal force of the earth's revolution about the earth-moon center of gravity greatly exceeds the moon's gravitational pull, so the water again bulges away from the earth. Between these opposite sides, the moon's gravity and the centrifugal force are about equal, and the water drops below mean sea level.

Because the earth rotates in relation to the sun once every 24 h, the moon passes above any point on the earth's surface about once each 24 h, but the moon is revolving about the earth-moon center of gravity in the same direction the earth is rotating, so the moon appears to rise each day about 50 min later than it did the day before. That is, the moon passes overhead once each 24.8 h, so, on our model earth there is a high tide (a **flood tide**) twice each day, once when the moon is overhead and once when it is on the opposite side of the earth. At the times between flood tides, there is a low tide (an **ebb tide**). Flood tides are separated from each other by about 12.4 h, as are ebb tides.

Now consider the effect of the sun. Its gravitational pull on the earth is much smaller than that of the moon, but the same principles apply, with the earth-sun center of gravity being close to the center of the sun. When the sun is on the same side of the earth as the moon (at the time of the **new moon**), its gravitational pull combines with that of the moon to produce especially high flood tides and low ebb tides, collectively called **spring tides**. When the sun is on the opposite side of the earth from the moon (at the time of the **full moon**), its gravity combines with the centrifugal force of the earth-moon system to also produce high flood tides and low ebb tides, also called spring tides. When the moon-earth-sun system forms a right angle (moon in its first or third quarter), the sun's effect on the tides goes against that of the moon, but the moon's effect is greater than that of the sun, so there are still flood and ebb tides, but they are respectively lower and higher than at the spring tides; these are the **neap tides**. Spring tides are separated by about 14.8 d (half the lunar cycle), as are neap tides.

There are many complications. For one thing, the tide lags behind the moon's passage by several hours. There are also effects of coastal shape, sea bottom configuration, wind, atmospheric pressure (causing atmospheric and land tides much smaller than ocean tides), the elliptical orbits of the moon and the earth, and a few other factors. Thus the actual tides vary greatly from place to place. A few places (for example, the Gulf of Mexico) have only one flood and one ebb tide each day. The two daily tides we have been describing are most common in the Atlantic Ocean. In the Pacific, there are two tides each day, but one flood tide may be so low that it can hardly be distinguished from the following ebb tide; or one of the ebb tides may be almost as high as a flood tide. In any case, depending on location, organisms living in the tidal zone are exposed to daily changes in water level, and these changes are superimposed on the lunar cycle of spring and neap tides. If the organisms have cycles adapted to these tides, they would be expected to have periods of 12.4 or 24.8 h, or 14.8 d—and possibly 29.5 d.

Many examples of such **tidal** and **semilunar** cycles are known, but true **lunar (circalunar)** cycles are rare. Some insects emerge and mate in large numbers just after the full moon, and some marine organisms swarm and spawn at the same time, but a rhythmic crash in the population of zooplankton at the full moon in Cahora Bassaz Reservoir in Mozambique was caused by fish that feed during the intense moonlight (reviewed by

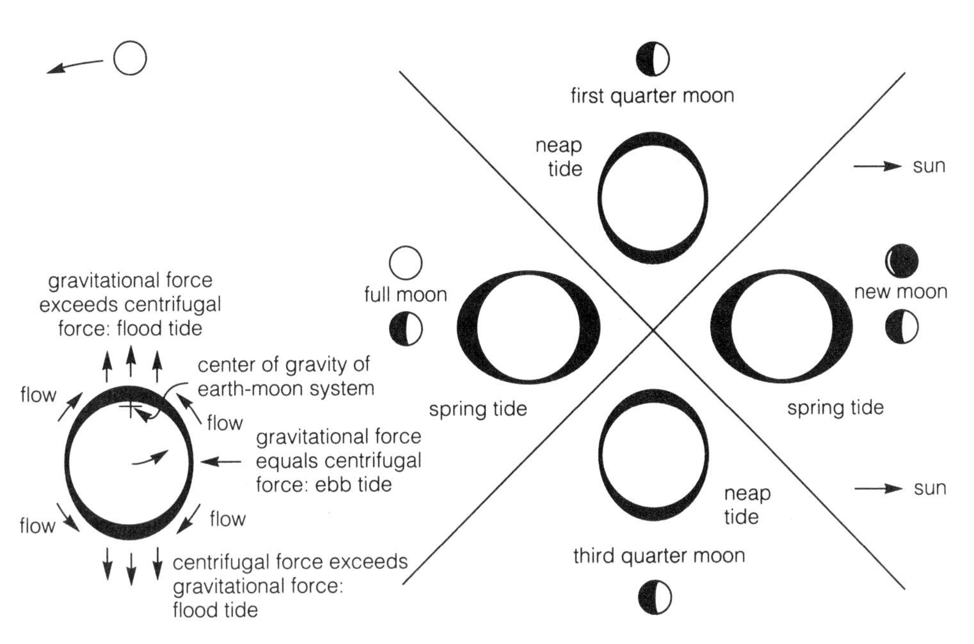


**Figure 21-3** The origin of the tides. In the earth-moon system, the earth and the moon are revolving around their common center of gravity, and this produces a centrifugal force on the earth in a direction away from the moon. At the same time, the moon's gravity is pulling on the earth. As the figure shows, **flood tides** occur on the side of the earth facing the moon, where the moon's gravity predominates, and on the side away from the moon, where the centrifugal force predominates. Between are **ebb tides**, which occur where the moon's gravity and the centrifugal force are about equal. In the earth-moon-sun system, the sun and the earth have the same relationship as the moon and the earth, but the moon's influence is greater than that of the sun. When the force from the sun and the moon combine at the new moon and at the full moon, flood tides are especially high and ebb tides especially low; these are the **spring tides**. When the forces from the moon and the sun oppose each other at the first and the third quarter moons, the flood tides are lower and the ebb tides higher; these are the **neap tides**.

Fryer, 1986). Studies of such rhythms under constant laboratory conditions have not been carried out.

Tidal rhythms in some phytoplankton that migrate out of the sand at low tide proved to be circadian rhythms entrained to the tides by light penetrating the murky water (reviewed by Sweeney, 1987), but examples are known (especially among invertebrate animals) of organisms that follow tidal cycles of the coast where they were collected after being brought into the laboratory and held in constant conditions. The diatom in Table 21-1 is one good example, and activity, color, and metabolism of fiddler crabs and other tidal-zone animals provide others. True tidal rhythms are not rare (Palmer, 1975, 1990).

**Semilunar rhythms** are known but are not common. For example, the grunion (*Leuresthes tenuis*), a small fish living off the coast of Southern California, spawns from late February to early September during three to four nights at the new and the full moons (spring tides) and as the tides are descending. Females lay their eggs in the sand, where they are fertilized by the males and remain until the next spring tide. If the timing were not exactly right, the eggs would be washed out of the sand and could not survive (Walker, 1952).

**Circannual rhythms** are also known. Germination of many seeds appears to be best at certain times during the year, even though the seeds have been stored under conditions of constant temperature, light, and moisture. Spruyt et al. (1983) documented an annual rhythm in the sensitivity of dark-grown bean seedlings to irradiation with red light. Opening of the epicotyl hook (see Chapters 16 and 20 and Fig. 16-14) was maximal between February and June (twice as sensitive to light as in December), although seeds were germinated under identical conditions. Pigment synthesis alternated in sensitivity with hook opening. Entrance into and termination of hibernation of ground squirrels follow a rhythm of about a year, and their amount of daily wheel-running activity also follows such a cycle.

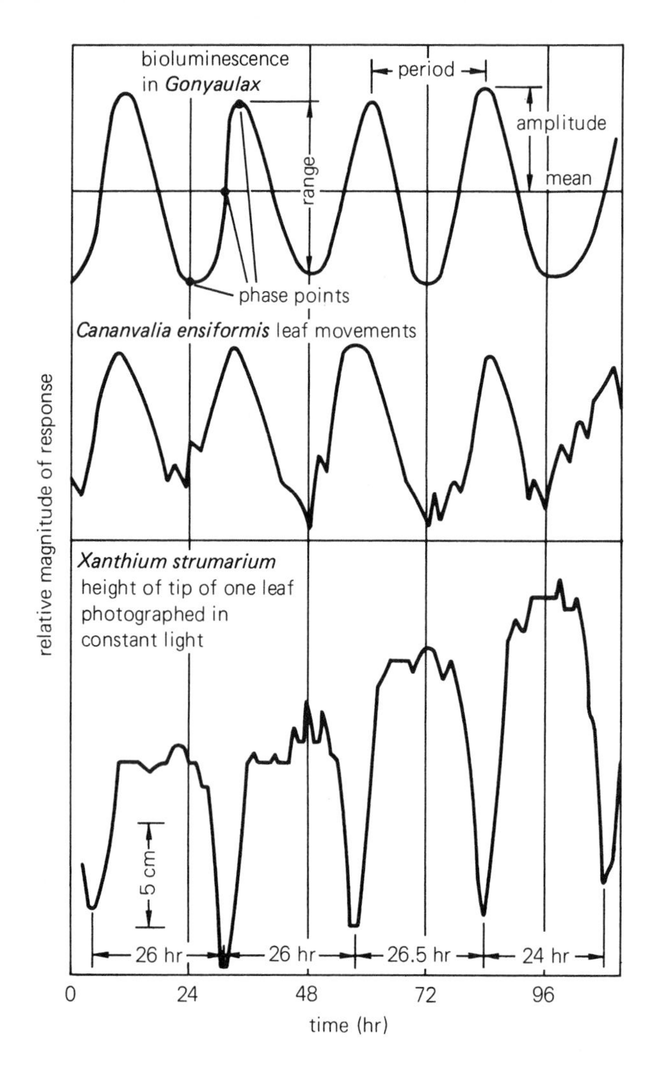


**Figure 21-4** Some representative data for various circadian rhythms. Top: bioluminescence intensity in *Gonyaulax* measured for plants kept under constant conditions of dim light. Some characteristics of circadian cycles are indicated. Middle: leaf movements of *Cananvalia ensiformis* recorded on a kymograph; high points on the graph indicate low leaf positions. Light and dark conditions are indicated by the bar. Note gradual shift of the peak during darkness as the cycles progress. Bottom: leaf movement of cocklebur (*Xanthium strumarium*) recorded by time-lapse photography. High points indicate high leaf positions. Period lengths between the troughs are indicated. Note increase in absolute height of the leaves, particularly at the peaks but also at the troughs. This is largely caused by growth of the stem during the course of the experiment, but the increase in range of leaf movement is also apparent. All light was from fluorescent lamps.

## 21.4 Basic Concepts and Terminology

In discussions of biological rhythms (and we will now concentrate on circadian rhythms), it is helpful to use the terminology applied to physical oscillating systems, although this terminology is sometimes modified somewhat (Halberg et al., 1977; Koukkari et al., 1987a, 1987b). Consider three rhythm characteristics (Fig. 21-4): First, the **period** is the time between comparable points on the repeating cycles, as we have seen. The term **phase** is used in a specialized sense as any point on a cycle that is recognizable by its relationship to the rest of the cycle. It may be specified as the time after an initial reference time, such as midnight of the first cycle, or it may be given in degrees with the assumption that one cycle (one period) equals 360°. If the cycle is described by a sinusoidal function (approximately fits a sine or cosine curve), then the maximum phase on the cycle is called the **acrophase**. Hence, the period is the time between acrophases. Botanists in particular have used the term *phase* to mean a recognizable *portion* of a cycle—for example, the part that normally occurs during the light span, the so-called photophile phase.

Second, the **range** is the difference between the maximum and the minimum values, and the **amplitude** equals half the range, or the extent to which the observed response varies from the **mean** (Fig. 21-4). Third is the **pattern** of the cycle. Many rhythms follow a sinusoidal curve (as in the bioluminescence rhythms of Fig. 21-4), but there are many variations. A sharp maximum might be accompanied by a broad minimum, for example, or the slope of the curve approaching the maximum might be steep, whereas that approaching the minimum might be less so.

When plants or animals are exposed to an environment that fluctuates according to some period and the rhythms exhibit the same period, they are said to be **entrained** to the environment, as opposed to free-running. As we shall discuss later, this entrainment to the environment can be brought about by several factors, particularly an oscillating light and dark environment, separated by **dawn** (lights on) and **dusk** (lights off). Such an entraining cycle is called a **synchronizer** or **Zeitgeber**, a German word meaning "time giver." The term **entrainment** is used when the Zeitgeber is a fluctuating environment with several regular cycles. If an environmental stimulus is given only once (for example, a single flash of light) and the phase of the rhythm is shifted in response to it, then the rhythm is said to have been **phase-shifted** or **rephased**.

## 21.5 Rhythm Responses to Environment

Many investigators have expended much effort in obtaining data relating to the biological clock, especially as it is exhibited by circadian rhythms. Far too much detail has accumulated for discussion here. We can only consider a few effects of light, temperature, and applied chemicals.

## Light

The work of Bünning, Stern, and Stoppel showed that entraining factors might be as subtle as a weak red light; hence, light was of obvious interest as a possible Zeitgeber. It was readily apparent that the rhythms could be entrained to some light-dark cycle other than a 24-h one. The rhythms could be entrained by shorter cycles of 20 to 21 h (in rare cases, even 10 to 16 h) or longer cycles of 28 to 38 h. The cell-division rhythm of *Gonyaulax*, for example, can be entrained to a 14-h cycle (7:7, light:dark), but it immediately returns to a 24-h period when returned to constant conditions.

Another approach was to allow a rhythm to become strongly established by a cycling environment and then to let it run free in continuous darkness. A brief interruption of light was then given at various times during the free-running rhythm. Typically, when the flash of light was given during subjective day, there was virtually no effect upon the rhythm; that is, if light comes during the phases typical of day in a natural cycling environment, the following phases of the cycle are not much influenced. When the light interruption comes during early subjective night, however, the rhythm is typically *delayed* (that is, the next peak comes later than otherwise expected). It is as though the flash of light were acting as *dusk*, but by coming later, it caused a delay. As the light flash is given later and later during subjective night, the extent of the delay increases until a point is reached at which the flash of light suddenly results in an *advance* of the rhythm, rather than in a delay (that is, the next peak comes earlier than expected). The flash of light is then acting as *dawn* rather than as dusk (Fig. 21-5).

By carefully studying curves such as those in Figure 21-5, one can account for the phenomenon of entrainment; that is, taking into account the advancing effects of dawn (lights on) and the retarding effects of dusk (lights out) allows us to predict the phases of a rhythm in relation to both the normal 24-h cycle and to cycles that are other than 24 h in length. Colin S. Pittendrigh (see, for example, Pittendrigh, 1967), a zoologist, and his coworkers pioneered this approach.

In entrainment by light, some photoreceptor pigment is absorbing the light and is thereby changed in such a way that it can lead to an advance or a delay of the clock. It would be interesting to understand the photobiochemical mechanism of this response. A first step in understanding is identification of the photoreceptor pigment, which is approached by determining the action spectrum for the response, as discussed in Chapters 10, 19, and 20. Light of carefully controlled spectral quality is given at various times to test its effectiveness on phase shifting or entrainment.

Action spectra have been determined for the various rhythms of *Gonyaulax*, for a mating rhythm of *Paramecium*, for the conidiation rhythm of *Neurospora*, for

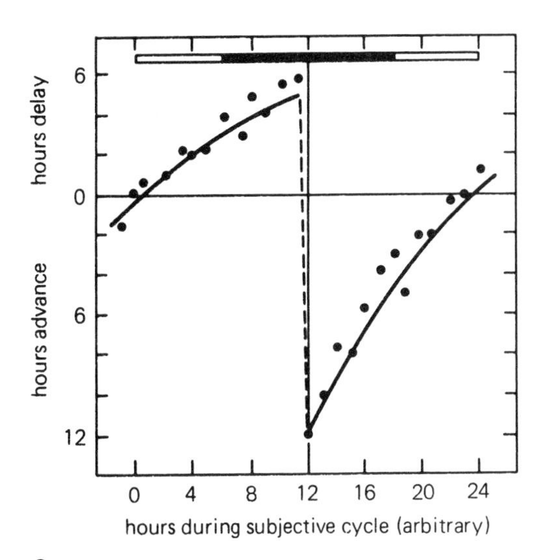


**Figure 21-5** (**a**) Phase shift in petal movements of *Kalanchoe blossfeldiana* following 2-h exposures to orange light given at various times during an extended period of continuous darkness. Bar at the top indicates subjective status of the rhythm; that is, dark part of the bar indicates petal closure or subjective night. As light interruption approaches the middle of subjective night, there is an increasing delay. Following the middle of subjective night, there is an advance, which decreases as subjective day is approached. (Data after Zimmer, 1962.) (**b**) Phase-response curves for five circadian rhythms in five different organisms, including the experiment of Zimmer, shown in **a**. Notice how similar they are. Time points are plotted in the middle of the light exposure. (From Sweeney, 1987.)

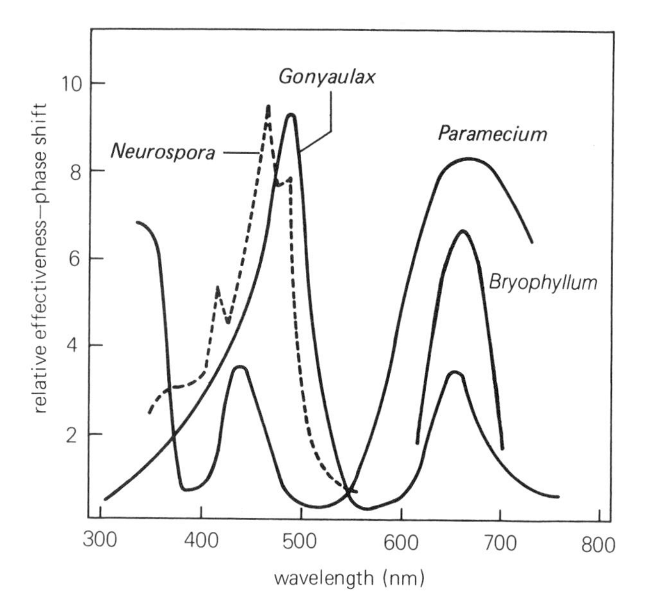


**Figure 21-6** Approximate action spectra for phase shifting in the rhythms of *Neurospora*, *Gonyaulax*, *Paramecium*, and *Bryophyllum*. (From Ehret, 1960; Muñoz and Butler, 1975; Sweeney, 1987; and Wilkins and Harris, 1975.)

$CO_2$ evolution in *Bryophyllum* (Fig. 21-6), and for eclosion of *Drosophila* (fruit fly) and diapause in *Pectinophora* (the pink bollworm). The action spectra are different in the several organisms, as though various available photoreceptors have been pressed into the service of the circadian clock (Sweeney, 1987), but most have strong responses in the blue part of the spectrum. The response of *Gonyaulax* to red may be due to chlorophyll, but the reason for the even stronger red response of *Paramecium* is not clear. Victor Muñoz and Warren L. Butler (1975) in La Jolla, California, isolated a flavoprotein-cytochrome *b* complex from *Neurospora* with an absorption spectrum that closely matches the action spectrum of Figure 21-6 (see discussion in Section 19.4). They suggested that this or a similar pigment may couple light and the clock in the many organisms, both plant and animal, that respond to blue light.

Lars Lörcher (1957), working in Bünning's laboratory in Tübingen, Germany, found that the rhythms of dark-grown bean plants were most effectively established by red light and that this establishment was reversed by an immediate exposure to far-red light. Furthermore, the leaf-movement rhythms often continued for several days when the plants were maintained at a constant temperature and under continuous light, provided that the light was rich in red wavelengths but contained none of the far-red part of the spectrum. When far-red light was present, the rhythms damped out more quickly. These observations implicated the phytochrome system, but Lörcher found that other wavelengths were also effective in entrainment, providing the plants had been grown in the light rather than in the dark.

There are other examples in which the phytochrome system has been implicated. Philip J. C. Harris and Malcolm B. Wilkins (1978a, 1978b) in Glasgow, Scotland, reset the $CO_2$-evolution rhythm in *Bryophyllum* (a succulent with CAM) leaves held in the dark only with red light (600 to 700 nm; Fig. 21-6). They were unable to reverse this red rephasing with far-red light, but far-red light applied simultaneously with or immediately after red light abolished the rhythm completely. Esther Simon, Ruth L. Satter, and Arthur Galston (1976), on the other hand, could reset the leaflet-movement rhythm of excised *Samanea* (a semitropical leguminous tree; see Fig. 19-1) pulvini with only 5 min of red light, and the red effect was completely cancelled by subsequent exposure to far-red light. But there were additional complications. When longer irradiation times were used, blue and far-red light could also reset the *Samanea* rhythms, but the resetting curves were qualitatively different (reviewed by Satter and Galston, 1981, and Gorton and Satter, 1983). Clearly, phytochrome can interact with the biological clock in some plants at least (but not in fungi, animals, or perhaps protistans), but it is equally clear that there are complications that remain to be understood.

For a long time zoologists simply assumed that the photoreceptor was the eye of the animal with which they were working. But Michael Menaker (1965) at the University of Texas in Austin showed that the activity rhythm of blind sparrows (both eyes removed) could be entrained by light signals. It is the pineal gland of the brain that actually responds to the light. Enough light, especially red light, penetrates the skull to be effective. This gland, especially prominent in birds, has been known since antiquity as the "third eye"! Menaker and his coworkers (see Zimmerman and Menaker, 1979) have studied the pineal responses to light and their effects on the clock. The activity clocks, at least, seem to be located in the pineal gland; effects are transmitted via hormones rather than nerve impulses.

## Temperature

Bünning had investigated the question of temperature effects on the clock in 1931. He reported that the effect of temperature on the response was unexpectedly small, but the leaf-movement period of his bean plants was more sensitive to temperature than that of some animals studied subsequently. It was Pittendrigh (1954) who realized that the clock would be of little value to an organism if the rate at which it ran were strongly dependent upon temperature, as are most metabolic functions. He had heard of the eclosion rhythm of *Drosophila* pupae, discovered by H. Kalmus in Germany, so he studied this rhythm at several temperatures and found its period to be nearly constant over a wide temperature range (shortened slightly at higher temperatures). Thus temperature insensitivity of the biological clock was discovered—or at least appreciated.

Most of the ultradian rhythms are sensitive to temperature, but some are not. Respiration and total cell protein in a soil amoeba (*Acanthamoeba castellanii*) oscillated with a period averaging 76 min, for example, and the period remained the same at several temperatures from 20 to 30°C (Lloyd et al., 1982).

Frank Brown and H. Marguerite Webb at Northwestern University in Chicago reported in 1948 that a color change observable in the fiddler crab had a period of *exactly* 24 h that was virtually independent of temperature. But Brown and Webb did not deduce a temperature-insensitive clock; rather, they considered their results to be evidence *against* an endogenous clock. Indeed, they returned to Stoppel's concept of a factor X in the environment that was responsible for actual time measurement. Brown (see Brown, 1983) championed this idea for three decades before his death. He and his coworkers observed numerous biological responses to subtle environmental factors such as geomagnetic fields (see, for example, Brown and Chow, 1973a, 1973b; Blakemore and Frankel, 1981). These responses and factors are of considerable interest in themselves, but few workers think that they can account for biological time measurement. Just as Pfeffer, nearly a century ago, was unable to convince his colleagues that the clock was endogenous (it took the experiments of Bünning, Stern, Stoppel, and Kleinhoonte to do that), so was Brown unable to convince most of us that the clock is exogenous. Most workers now believe that Brown and Webb's experiments with fiddler crabs were just an especially outstanding demonstration of the clock's "temperature independence." But "most workers" could still be wrong. Few workers (including the present authors) have given Brown's ideas an honest, in-depth appraisal.

We are faced with somewhat of a paradox in our discussion of temperature effects on the biological clock. Changes in temperature of only 2.5°C or less can synchronize the rhythms (act as a Zeitgeber) in *Neurospora* and other organisms, and temperature can also influence the amplitude of the response. Such effects must surely be important in nature. Still, the period of a free-running rhythm is relatively temperature-insensitive. So some aspects of the clocks are sensitive, and others insensitive, to temperature.

Temperature effects on the period vary in different organisms. In *Gonyaulax* the $Q_{10}$ for this effect is slightly less than 1 (as temperature increases, the free-running period becomes slightly longer), whereas it is equivalent to about 1.3 for bean-leaf movements. $Q_{10}$ values that very closely approach 1.00 have been reported. For example, the value is approximately 1.02 for biological time measurement that controls flowering of cocklebur (Salisbury, 1963).

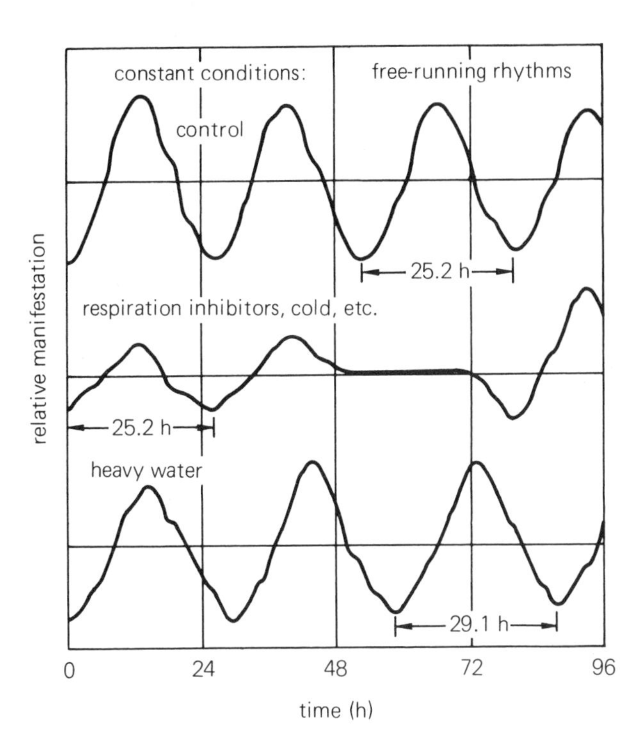


**Figure 21-7** Effects of various chemicals and other factors on free-running rhythms. Most compounds that inhibit metabolism reduce the amplitude but do not affect the period (compare center record with control above). Heavy water increases the length of the period (29.1 h compared with 25.2 h for eclosion in *Drosophila*). Curves are schematic and do not represent actual data.

## Applied Chemicals

If we could find some chemical that would clearly inhibit or promote biological time measurement, we could test its mode of action and perhaps gain insight into the operation of the biological clock. Many chemicals have been studied, but we must be careful in interpreting the results. A given chemical may, and often does, influence the amplitude of the rhythm—the "hands on the clock"—without influencing its mechanism as indicated by the period (Fig. 21-7). There are some instances in which the amplitude can be completely damped out; yet when the inhibitor is removed, the cycle proves to be in phase with that of control organisms having the same period and to which no inhibitor was applied. This is one aspect of the general problem of **masking**. As Sweeney (1987) said: "Only the uninhibited use inhibitors."

In the early 1960s, Bünning (see his 1973 book) and his coworkers reported that several chemicals (colchicine, ether, ethyl alcohol, urethane) applied to bean plants appeared to act directly on the clock itself, rather than on its manifestations. In these cases the period was lengthened somewhat, and the amplitude was unaffected or only slightly decreased. It was suggested that the feature the compounds had in common was the ability to influence membranes, and that therefore the clock may be associated with cellular membranes.

Heavy water (deuterium oxide) caused a considerable increase in the phototaxis period of *Euglena* and also in the bean-leaf movements, and it slowed the measurement of time in flowering controlled by photoperiod (see Section 21.7 and Chapter 23). Although heavy water could influence many processes within the plant, membranes are among the structures affected. Lithium ions ($Li^+$) lengthen the period and otherwise affect the rhythms in several plants and animals (reviewed by Engelmann and Schrempf, 1980), again suggesting involvement of membranes. Valinomycin, a $K^+$ ionophore, also caused phase shifts in bean leaf movements (Bünning and Moser, 1972) and in *Gonyaulax* (Sweeney, 1987), but some other $Ca^{2+}$ and $K^+$ ionophores were ineffective with *Gonyaulax*. Neither actinomycin, which inhibits DNA translation, nor chloramphenicol, which inhibits translation on 70S ribosomes, caused phase shifts of the *Gonyaulax* bioluminescence rhythm, but anisomycin and cycloheximide, which inhibit translation on 80S ribosomes, did reset the rhythm. It was also shown that RNAs extracted at different times are different. All of this suggests that the clock might involve $K^+$ transport and protein synthesis (translation) on 80S ribosomes, both of which occur in the cytoplasm.

## 21.6 Clock Mechanisms

What and where is the clock, and how does it work? Could there be more than one clock or clock mechanism in the same or in different organisms?

We know more about the location and number of clocks than we do about how they work. So far, there is no evidence that the clock has different mechanisms in different organisms, but much study remains to be done. The several rhythms in *Gonyaulax* all keep the same period for many days under constant conditions and respond with the same phase shift to a stimulus such as a light pulse, strongly suggesting that there is only one clock in each *Gonyaulax* cell (reviewed by Sweeney, 1987), but this is not the case for multiple rhythms in multicellular organisms. Sleep cycles can get out of phase with cycles of potassium excretion in humans, for example. Multiple clocks must occur.

Most of the evidence for location of the clock in membranes was reviewed in Section 21.5, under "Applied Chemicals." For awhile it was thought that the nucleus might be involved; a circadian rhythm in volume of the nucleus has been reported in several organisms (reviewed by Bünning, 1973). The large, single-celled alga *Acetabularia* has its nucleus located in its holdfast, where it can easily be removed. Such enucleated cells exhibit a photosynthetic rhythm that persists for at least a month, as shown in five different laboratories (Bünning, 1973; Sweeney, 1987). Even small fragments of the *Acetabularia* cell show such rhythms. Yet *Acetabularia* is not exactly the most typical of cells.

In searching for clock mechanisms, researchers have used not only inhibitors but also clock mutants in various organisms, especially *Neurospora crassa*, *Drosophila*, and *Chlamydomonas* (a single-celled, motile, green alga). A few conclusions are worthy of our attention.

Approximately 12 *Neurospora* clock mutants for conidial banding have been found (see reviews by Feldman, 1982, 1983). Their periods range from 16.5 to 29 h. Seven of the mutants map to a single gene locus called *frq* (frequency), which apparently plays a key role in clock organization. At 25°C, the periods of the *frq* mutants all differ from that of the wild type by some multiple of 2.5 h. Furthermore, the *frq* cycles are restricted to a 7-h part of the wild-type cycle, which is shortened in *frq-1* to 2 h and lengthened in *frq-7* to 14.5 h. The short-period *frq* mutants retain temperature compensation, but the long-period mutants do not.

The situation in *Drosophila melanogaster* is especially interesting. Three mutants are known, one that shortens both the eclosion *and* the activity rhythms, one that lengthens these rhythms, and one that produces short, noisy rhythms with multiple periodicities (Dowse et al., 1987; Konopka and Benzer, 1971). All three mutations occur at the same chromosomal locus. These same mutated genes also influence the ultradian rhythm of the male courtship song and in the same direction; for example, the shortened circadian rhythms occur in the same flies as shortened ultradian rhythms (Kyriacou and Hall, 1980). Jackson et al. (1986) have identified and sequenced a biologically active segment of DNA that restores rhythmicity when it is transduced into flies carrying the mutations. The protein for which the DNA codes appears to be a proteoglycan (a protein-polysaccharide). Such efforts should eventually illuminate the biochemical mechanisms of the controlling clock.

In the meantime, several workers have attempted to construct hypotheses or models to explain the clock mechanism. Such a model must explain the long periods (compared with periods of known chemical oscillating systems), the resetting effects of temperature and especially light, temperature compensation, the various chemical effects, and the apparent key role of membranes.

Temperature compensation of the clock poses an interesting problem because the $Q_{10}$ for most biochemical reactions is appreciably greater than 1.0. Reactions involving hydrolysis of ATP, for example, often have a $Q_{10}$ of about 2.0. How, one might ask, if the living organism is fundamentally a biochemical system, can we account for temperature insensitivity?

We can envision certain feedback systems. The products of one reaction or function might inhibit the velocity of an earlier one. Such feedback inhibition of other processes is well known in living organisms. As temperature increases, the velocity of the first reaction (the time-measuring reaction) might increase, but the amount of inhibitory product would then increase at a proportional rate, so the reaction would maintain a near-constant rate over a wide range of temperatures. The $Q_{10}$ value of less than 1 observed in *Gonyaulax* could be accounted for by such a scheme, assuming that with increasing temperature the inhibitory product were produced somewhat faster than the time-measuring reaction is accelerated. It is difficult to imagine any other way to account for this observation. Such feedback (possibly over-compensating, as in *Gonyaulax*) could be invoked in a **temperature-compensated system** of time measurement based on biochemical reactions. Furthermore, some chemical reactions catalyzed by enzymes that have temperature coefficients close to 1 are now known.

As we noted above, there are several evidences for the possible participation of membranes in the basic clock mechanism (reviewed by Engelmann and Schrempf, 1980). A special *slow* diffusion across a membrane (normally a fast process) with a gradual buildup of some key substance on one side could account for the long period. Indeed, the concentrations of some ions, especially $K^+$ in certain plant parts, have been found to oscillate in a circadian manner, and such ions phase-shift certain plant rhythms when applied externally. Circadian rhythms in permeability and transport properties of membranes have been reported. We can imagine how light and perhaps temperature might influence membranes and thus the clock. We have seen that phytochrome apparently plays a role in rephasing plant clocks and that it is thought to interact with membranes.

With these and other ideas in mind, several clock mechanisms involving membranes have been proposed (reviewed by Engelmann and Schrempf, 1980). Beatrice Sweeney published a model in 1972, followed by modifications in 1974 and 1976 (see Sweeney, 1976). Some *substance X*, possibly $K^+$, is transported into an organelle (thylakoids within chloroplasts in Sweeney's model). When the concentration inside reaches some critical value, active transport is stopped, and substance *X* leaks out until a critical low concentration is achieved, reinitiating active transport and thus another cycle. The transport phase requires energy, but the leakage phase does not, matching evidence for a tension/relaxation system reported by several authors (for example, Bünning, 1960). At higher temperatures, leakage is accelerated more than transport, slowing the tension phase but speeding the relaxation phase and thus accounting

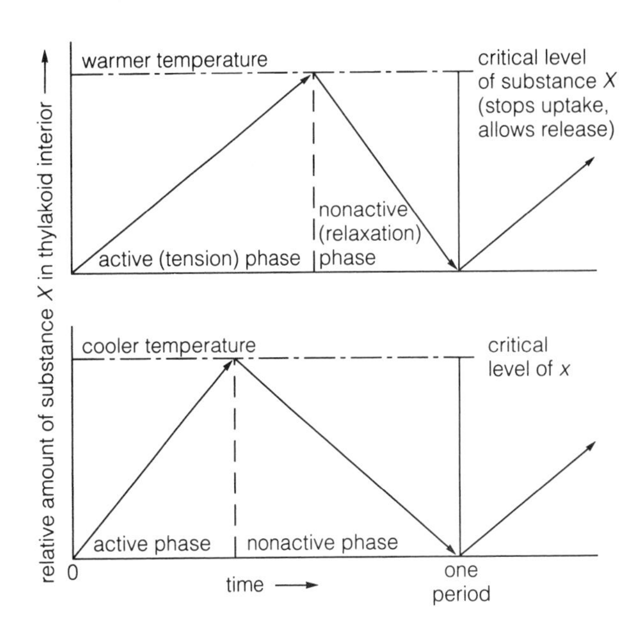


**Figure 21-8** Temperature compensation of the circadian period according to Sweeney's model. The model postulates that timekeeping depends on the concentration of some substance *X* ($K^+$?) in the interior of thylakoid double membranes in chloroplasts of *Gonyaulax* and perhaps other organisms. In the active phase, there is both an uptake and a release of substance *X* into and from the interior of the thylakoids. At higher temperatures (upper part), substance *X* is released faster compared with lower temperatures (lower part), thus leading to a reduced net rate of uptake during the active phase. Therefore, the active phase must be longer for the concentration of *X* to reach the threshold that switches to the passive part, but the passive part is faster at the higher temperatures, accounting for a constant period (temperature compensation). (See Engelmann and Schrempf, 1980, for a review of this and other membrane models.)

for temperature compensation (Fig. 21-8). Light shifts the phase by accelerating the leakage rate through the membranes. This causes a delay during tension but an advance during relaxation.

Other membrane models are similar to Sweeney's in their basic concepts, but some include protein synthesis and other features to account for various observations. An example based on the flagellate *Euglena* involves $NAD^+$, the mitochondrial $Ca^{2+}$-transport system, $Ca^{2+}$, calmodulin, $NAD^+$ kinase, and $NADP^+$ phosphatase (Goto et al., 1985). No one thinks that any one model will prove entirely correct, but at least a start has been made.

## 21.7 Photoperiodism

Even before the work of Bünning, Stern, Stoppel, and Kleinhoonte, the biological clock had been clearly demonstrated as a component of living plants, although the discovery was not widely recognized by those working on rhythms until perhaps the 1950s. Wightman W. Garner and Henry A. Allard were two scientists working in the United States Department of Agriculture research laboratories at Beltsville, Maryland in the 1910s. They had made two observations that they could not explain. Maryland Mammoth tobacco (a new hybrid) grew at that latitude to a height of 3 to 5 m during the summer months but never flowered, although it flowered profusely when only about 1 m tall after being transplanted to the winter greenhouse. They had also noticed that all the individuals of a given cultivar of soybean would flower at about the same time in the summer regardless of when they had been planted in the spring; that is, large plants sown in early spring bloomed at the same time as the much smaller plants sown in the summer. Garner and Allard suspected that some factor of the environment was responsible for flowering of these two species.

They tested various environmental factors that might differ between the summer fields and the winter greenhouses: light levels, temperatures, soil moisture, and soil nutrient conditions. But no combination of these factors resulted in flowering of the tobacco plants. Garner and Allard realized that day length varied throughout the season and was a function of latitude, so they tested what must have seemed like the remote possibility[2] that this could control flowering. It did. When the days were shorter than some maximum length, the tobacco plants began to bloom. When the days were shorter than some other maximum length, the soybean plants also began to bloom.

Garner and Allard found that some plants (for example, spring barley and spinach) flower in response to days that are *longer* than some critical length. These they called **long-day plants**, and those such as tobacco and soybean, which flower when the day length is less than some maximum, **short-day plants**. Some species they studied showed no response to day length; these they called **day-neutral plants** (also known as **day-length-indifferent plants**). Garner and Allard's results were published in 1920, and they named the discovery of flowering in response to the relative lengths of day and night **photoperiodism**, after a suggestion from A. O. Cook, a colleague at Beltsville.

Subsequent work indicated a critical role for the dark period (Chapter 23), but it is now clear that plants are capable of measuring the length of the light or the dark period or both. Here was a clear-cut demonstration of biological time measurement. Completely comparable responses were subsequently discovered in animals, including effects on insect life cycles, fur color (for example, in the arctic hare), breeding times (gonad size), and migration of many birds and other animals. So Garner and Allard's discovery proved to be a fundamental principle of biology.

## 21.8 Photoperiod-Rhythm Interactions

Garner and Allard (1931) also did some experiments, now mostly forgotten, in which they subjected plants to cycles of light and darkness that did not add up to 24 h. The results were complex and not always easy to explain, but it seems clear that when the total of light and dark deviates much from 24 h, growth and other responses are inhibited. Apparently, the light-dark cycle must agree at least approximately with the circadian cycles of the plant. When the cycle deviates too much from 24 h, the rhythmic change in light and darkness can no longer entrain the rhythms, as we have already noted.

Tomato plants exhibit some interesting rhythm interactions. Although short days slightly promote flowering, plants will usually flower over a wide range of day lengths, from very short to about 18 h long; that is, they are nearly day-neutral. But stem height is strongly influenced by day length; plants held under 16-h days have stems about twice as long as those held under 8-h days (see Fig. 23-2). Furthermore, many tomato cultivars actually die when the days account for more than 18 h of a 24-h cycle. This is true even when the days are extended by low light levels, so plants kept under long days are not photosynthesizing much more than plants kept under short days.

With sensitive tomato plants, it is even possible to observe a slight but significant inhibition of growth when light is given in the middle of the dark period. To see this, Highkin and Hanson (1954) gave tomato plants either 16 h of continuous light plus 8 h of darkness or 14 h of light with another 2 h in the middle of the dark period. Tomato plants are also adapted to the normal alternation in temperature between day and night; they produce more flowers when night temperatures are lower than day temperatures. This phenomenon was called **thermoperiodism** by Frits Went (discussed in Chapter 22).

Keun Chang Yoo and Shunpei Uemoto (1976) at Kyushu University, Japan, found an interesting relationship between a circannual rhythm and a flowering response to photoperiod. Radish (*Raphanus sativus*) seeds were stored at 25°C, and each month some were germinated and grown in continuous light. An annual rhythm in flowering appeared, as shown in Figure 21-9. Plants did not flower during autumn and winter, even with 60 d of continuous light, but for two years during

[2]We'll see in Chapter 23 (box essay, p. 510) that others had discovered the phenomenon, but their thinking was not yet crystallized, and Garner and Allard were probably unaware of the work of the others when they initiated their own experiments.

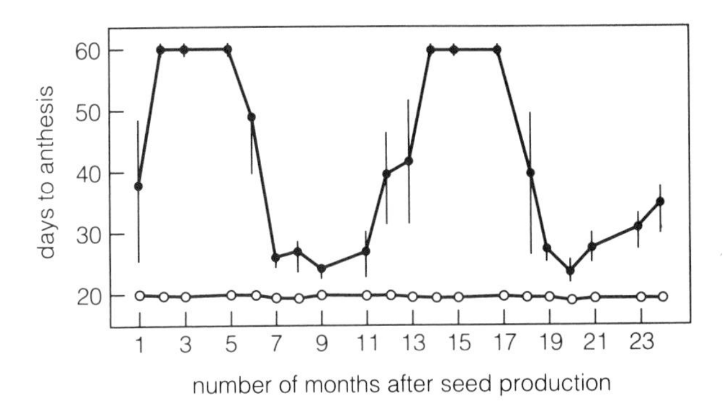


**Figure 21-9** Flowering of radish (*Raphanus sativus* cv. Wase-shijunichi), measured as days to anthesis, as affected by the period of seed storage. Flower promotion is indicated by fewer days to anthesis. The vertical lines indicate the period from the first to the last anthesis. Storage began in July, and plants that were germinated two months later did not flower, even after 60 d in continuous light. Seven to 10 months later, plants flowered in about 25 d, but 14 to 17 months later, plants again would not flower after two months of continuous light, clearly demonstrating an annual rhythm of response. Cold treatment (10 d at 5°C) abolished the annual rhythm, such that all plants flowered after 20 d. (From Yoo and Uemoto, 1976.)

spring and summer, plants flowered after about 25 d of continuous light. If plants were treated with cold temperatures (Chapter 22), the annual rhythm of flowering was abolished, and plants flowered in about 20 d.

An important question concerns whether the clock that measures the lengths of day and night in photoperiodism is the same as the clock that controls the circadian rhythms. We will save that discussion for Chapter 23 (Photoperiodism), but here we can state that the photoperiodism and circadian clocks have many features in common but can be separated from each other in various ways, and thus they appear to be different.

## 21.9 How the Clocks Are Used

In the introduction we deduced the likely existence of a biological clock based on the ways it (or they) might be used in a cycling environment. Rhythms of activity in animals, for example, allow a species to occupy a niche not only in space but also in time; a nocturnal animal and a diurnal animal might use the same space but at different times. Plant rhythms in metabolism adjust the plant to prevailing light and temperature conditions. Rhythms in flower opening must be closely coupled to the time memory of honeybees (noted below). The phenomenon of photoperiodism confers upon the organism possessing it the ability to occupy a particular niche in *seasonal* as contrasted to *daily* time. Several species that flower in response to photoperiodism may do so at different times and in sequence throughout the season, providing a rather constant source of nectar for insect pollinators. Given a time during the season with minimal flowers but high availability of pollinators, there would be a selective advantage to any species able to flower during that time.

### Leaf Movements

But do leaf movements perform any valuable function in nature? Pfeffer settled this question in his mind by tying the leaves to a bamboo framework so that movements could not occur. Because his plants exhibited no apparent ill effects, he concluded that the movements were some by-product of the evolutionary process and of no selective value to the plant.

Charles Darwin suggested that leaf positions might play a role in heat transfer between a plant and its environment (Section 4.9). A horizontal leaf is in a good position to receive sunlight during the day, but at night more heat might radiate from such a leaf into space, making it more likely to freeze on cold nights. *Vertical* leaves in a plant community, however, might radiate more to each other, keeping each other warm! Darwin and his son Francis (1880) tested this idea experimentally and obtained positive (although not very striking) results. Similar, more recent experiments (Enright, 1982; Schwintzer, 1971) confirmed the basic observation, but temperature differences between vertical and horizontal leaves were small, less than 1°C. On the other hand, Alan P. Smith (1974) studied *Espeletia schultzii* in the high Andes of Venezuela at an elevation of 3,600 m. *Espeletia* is a large rosette species (0.5 m diameter, 0.25 m height) with pubescent leaves that close around the single apical bud at night and open during the day; the plant is typical of several others at alpine elevations in the tropics. Leaves of some plants were held open with wires, others were held closed, leaves of some plants were removed, and control plants were left intact. The bud-core temperatures of plants with open or no leaves dropped below freezing on a cold, clear night, and young leaves wilted and died, showing that nyctinasty in this species is indeed adaptive. The rosette form of the leaves also forms a parabolic reflector and radiator that warms the bud during the day, presumably promoting its metabolism and development.

Transpiration can be greatly reduced in leaves that fold up at night (for example, *Albizia, Mimosa, Samanea*). When some species are fully folded, no stomates are exposed; all are fully protected (Mauseth, 1988; Wilkinson, 1971).

Bünning suggested (see his personal essay in this chapter) that a horizontal leaf would be in a better position to absorb the light of the full moon (at its zenith at midnight). Because such absorption might upset the photoperiodism response, a vertical leaf position at night could be a protective device to ensure successful time measurement in photoperiodism. But this conclusion requires more study, as we discuss in Chapter 23.

### Time Memory

While Bünning, Stern, and Stoppel were discovering the endogenous nature of the biological clock in plants, Ingeborg Behling in Germany (1929) was making an important discovery about the clock in honeybees. She found that honeybees were easily trained to feed at a certain time during the day. As noted, this must be an adaptation to plant rhythms of flower opening and nectar production. It is as though the clock in the honeybee can have a "rider" attached to it, indicating the time of day and informing the honeybee 24 h later that it is time to feed. It remains to be seen whether this is the same clock that controls circadian rhythms.

Humans have a comparable time-measuring system. Our time memory is most frequently manifested by waking up at a predetermined time. This is particularly impressive because the person must translate a learned concept (clock time) into some form that will "adjust the rider" on his or her biological clock. Human time memory can often be most impressively demonstrated under hypnosis, especially with posthypnotic suggestion.

### Celestial Navigation

In spite of the extremely intriguing discoveries made during the 1920s and 1930s, only a small minority of biologists showed any great interest in the biological clock until the early 1950s. At that time American botanists became interested in the work of Bünning because it seemed to bear a direct relationship to photoperiodism, a topic in which there was considerable interest. Indeed, Bünning had proposed a theory to correlate photoperiodism and the circadian rhythms (see his personal essay in this chapter). Zoologists, particularly Pittendrigh at Princeton University, also began to take notice of Bünning's discovery and of other work going on in Europe. The discovery of temperature compensation was especially stimulating.

Then Gustav Kramer, K. von Frisch, and others working primarily in Germany found that certain birds and other animals could tell direction on the earth's surface by the position of the sun in the sky. Because this position changes, the organism must be able to correct for the time of day, apparently by the use of some kind of clock. Subsequent work in a planetarium even implicated the stars. Until then, some of the manifestations of the clock (for example, the leaf movements) did not seem to be of much value to the organism. In the case of **celestial navigation**, however, the clock must be used, so this spectacular discovery (along with temperature compensation) caught the attention of biologists all over the world and led to a surge of interest in biological clocks.

## 21.10 Some Important Implications of the Biological Clock

Rhythms have important implications for our modern life. We'll mention a few; you should be able to think of others.

### Agriculture

Rhythmic responses of plants to chemicals such as herbicides (see examples in Koukkari and Warde, 1985) provide a good example of an agricultural implication. Little work has been done, but the mechanisms of response could vary and might often be indirect. Leaf position, for example, might determine the amount of herbicide that falls on a leaf from above. Because many crop yields depend on how much the crop flowers (seed and fruit crops) or doesn't flower (such vegetative crops as sugarcane and spinach), and because photoperiodism often influences flowering, the photoperiodism clock is of considerable importance in agriculture. With a seed crop such as the short-day soybean, yields of certain cultivars may be maximal over a range of latitudes (that is, day lengths) as narrow as 80 km. If the plants are too far south, late summer days are too short, and flowering occurs before the plants have had a chance to develop a sufficient number of leaves; too far north, and plants flower too late in the season and may be damaged by frost. Some plants (for example, soybeans) are most sensitive to cold stress at certain times during the day (Couderchet and Koukkari, 1987; King et al., 1982). Berthold Schwemmle (1960) summarized his own experiments (as well as those of several other workers) in which alternating temperature during the daily cycle could be substituted for alternating light, with high temperature (for example, 35°C) acting like light. We will return to these matters in the next two chapters.

### The Physiological Clock in Humans

We have noted the existence of a few circadian cycles in humans (reviews: Luce, 1971; Moore-Ede et al., 1982; Thompson and Harsha, 1984). There are rhythms in sleep (for example, waking just before the alarm); in alertness and speed in calculation; in hormone levels, heart rate, body temperature, excretion of urine, sensitivity to drugs, births and deaths (both mostly at night); and in many other phenomena. (There is also much pseudoscience associated with human cycles; see box essay on page 483.)

Not all human rhythms have been studied under

## Biorhythms and Other Pseudosciences

There is talk about biorhythms. The notion is that human behavior is controlled by three cycles, each of which is initiated at the moment of birth: a physical cycle of 23 d, a sensitivity (emotional) cycle of 28 d, and an intellectual cycle of 33 d (Mackenzie, 1973; Thommen, 1973). The first half of each cycle is supposed to be the time when one is most positive in the attribute of that cycle; during the second half, one is supposed to be negative. The days on which there is a crossover from plus to minus or minus to plus are *critical days*, and if the critical points for two or three cycles fall on the same day (which happens about six times a year for two cycles and once for three), you had better watch out!

The concept was developed from about 1897 to 1932 by certain medical doctors and others in Vienna, Berlin, Innsbruck, Philadelphia, and elsewhere. It is usually presented as a "scientific" doctrine. Some companies such as the Ohmi Railway Company in Japan have calculated the cycles for their employees, warning them of critical days. The accident rate is reported to have dropped by more than 50 percent!

Yet there is little objective evidence to support the hypothesis (Rodgers et al., 1974). Most evidence is anecdotal: Such and such movie star is reported to have had a serious accident on a triple-critical day. Many adherents to the doctrine swear by it. But, of course, their results might well be examples of the self-fulfilling prophecy. After you have plotted your charts for several months in advance, you will be expecting good and bad days, and subconsciously or otherwise you may adjust your life to meet these expectations. If you keep a careful diary, you might test the theory by plotting your charts for a *past* interval covered by the diary and looking to see whether anything special happened on the good or bad days—but to be objective, you must also note special things that happened on the other days as well.

It should be clear from the discussions in this chapter that the basic premise on which the concept of biorhythms rests has no foundation in scientific observation. Rhythms clearly exist in organisms, including humans, but they have three features at total variance with those of the biorhythm hypothesis: They are typically *circa*, approximating but almost always varying from an exact period length, unless they are continually entrained to a cycling environment; they often vary from individual to individual within a species; and they are relatively easy to shift by various environmental factors. Their periods are plastic, not rigid. The rhythms discussed in this chapter could never maintain exact periods of 23, 28, or 33 d from the time of birth—the same for everyone—throughout the proverbial three score years and ten of an individual life.

Speaking of matters pseudoscientific, we find that the botanical sciences have spawned their share. There are those who suggest that talking to your plants, praying over them, or singing to them makes them grow better (and perhaps it does, if you thereby increase the $CO_2$ concentration around them or take better care of them). There have been several papers purporting that music makes plants grow better. There are even special records on the market; their makers claim that they provide the best music for plants. (This needs much work, but it is possible that sound waves might vibrate the cellular organelles and influence plant growth—but classical music and not rock music?!?)

Several years ago, the media reported "experiments" in which a polygraph (lie detector) that was attached to a plant registered wild responses when bad things happened, such as another plant being "murdered" in the same room or brine shrimp being dunked in boiling water. Do plants really have feelings and emotions, responding to "thought waves" from an experimenter?

If so, it surely remains to be demonstrated. The "experiments" apparently worked once, but no one has been able to repeat them consistently. And consistent, objective verification is what we must demand in science (Galston, 1974). Truly, progress often depends on startling and unexpected discoveries, but it is as common for such claimed discoveries to be mistaken interpretations or to be due to poorly designed experiments as for them to be advances in knowledge. We are entitled, even obligated, to test all claims by insisting upon verification by objective observers who thoroughly understand the role of controls in an experiment and who understand all the factors that might influence the outcome. For example, the polygraph responses observed when "murder" was perpetrated near the plant attached to the machine were at about the same level as the "noise" to be expected if the polygraph were attached to an inanimate object. Could the results have been due to coincidence? Of course, and it is likely so.

constant conditions, but there have been some impressive studies. Jürgen Aschoff and Rütger Wever (1981) and Wever (1979) told how they used underground bunkers in Andechs, near Munich, as special living quarters in which volunteers (often students studying for final, comprehensive examinations) could stay several weeks in constant conditions. Body temperature, sleeping and activity, and urine volume and chemistry of 147 volunteers were recorded automatically. Most subjects showed circadian periods of about 25 h for these parameters (some were as long as 27 h), and most of the principles learned with other organisms applied to humans, although humans are more easily entrained by such social and intellectual cues as the presence of other people or a wrist watch (Aschoff et al., 1975; Sulzman, 1983).

It was discovered with humans and confirmed with squirrel monkeys that rhythms with different periods could exist in a single individual and that the different cycles were controlled by at least two separate clocks (Sulzman, 1983). For example, activity and urinary calcium excretion both had a period of 33 h, whereas body temperature, urine volume, and urinary potassium excretion all had a free-running period of 25 h in the individual being studied.

Such studies have several implications for modern life. For example, some human disorders such as wintertime depression can be cured or helped by bright light (reported as 2,500 lux or more, which is still only about 3 percent of full sunlight) given at the appropriate time of day (Moore-Ede et al., 1982; Czeisler et al., 1986). The bright light is required each day to reset the human-activity and other clocks in some individuals.

Jet air travel across time zones has a strong effect upon the internal timing system of the passengers. It may take several days for the clock in a tourist — or a diplomat — to adjust to a new time zone. North-south travel of comparable distance but within the same time zone is much less fatiguing, showing that it is our clocks that are affected. For most people, travel from east to west is easier than going in the other direction. Because our free-running periods are usually *longer* than 24 h, it is easier to delay our cycles than to shorten them. It also helps to expose ourselves to bright light: Take a walk in the sun if possible. A benzodiazepine drug (triazolam or Halcion), which is used to treat insomnia, was effective in resetting the clock in the golden hamster (Turek and Losee-Olson, 1986).

Knowledge of the biological clock is applied in many hospitals where drugs can be administered in the lowest doses at times when patients are most sensitive to them. Such knowledge should also be, but seldom is, applied to shift workers, medical interns (who often must work 36-h shifts), pilots, and others. More recognition should be given to the difficulties of staying alert when one's biological clock says it is time to sleep. It may well be significant that the accidents at Three-Mile Island and Chernobyl happened between 2 and 4 in the morning!

# 22

# Growth Responses to Temperature

Plant growth is notoriously sensitive to temperature. Often a change of a few degrees leads to a significant change in growth rate. Each species or variety has, at any given stage in its life cycle and in any given set of study conditions, a **minimum temperature** below which it will not grow, an **optimum temperature** (or range of temperatures) at which it grows at a maximum rate, and a **maximum temperature** above which it will not grow and may even die. Figure 22-1 shows curves for growth rate as a function of temperature. The growth of various species is typically adapted to the temperatures of their natural environments. Alpine and arctic species have low minima, optima, and maxima; tropical species have much higher **cardinal temperatures**. Plants close to the minimum or maximum temperatures are often under stress, the topic of Chapter 26.

Often different tissues within the same plant have different cardinal temperatures. A classical and easily demonstrable example of this is the difference in optimum growth temperatures for the upper and lower **tepal** surfaces (collective term for petals and sepals that are similar in appearance, especially in members of the lily family) of the tulip or crocus flower. Studies in Germany, dating from those of Julius von Sachs in 1863, have demonstrated that low temperatures (3 to 7°C) are optimal for growth of the lower surface of tepal tissues, causing tulip or crocus flowers to close, whereas higher temperatures (10 to 17°C) are optimal for growth of the upper surface of tepal tissues, causing flowers to open (Fig. 22-2). An abrupt change in temperature of only 0.2 to 1°C often results in rapid growth and the opening or closing of tulip or crocus flowers, although the optimal growth temperatures for the two sides of the tepals are about 10°C apart. This temperature-induced movement (a *nastic* movement; Chapter 19) of the tepals caused by growth is termed **thermonasty**.

Temperatures influence more than tissue growth, however. Often, specific temperature regimens initiate critical steps in the life cycle: seed germination, flower initiation, and induction or breaking of dormancy in perennial plants. And these developmental responses are often influenced by environmental factors in addition to temperature, including irradiance, photoperiod, and moisture. These interactions are diverse and complex, so the topics of this chapter sometimes stray somewhat from its central theme, which is plant response to temperature, especially low temperature.

## 22.1 The Temperature-Enzyme Dilemma

In our speculations about plant growth response to temperature, we often postulate the occurrence of enzyme reactions that are influenced by two opposing factors: With increases in temperature, increased kinetic energy of the reacting molecules leads to an increased rate of reaction, but increasing temperature also

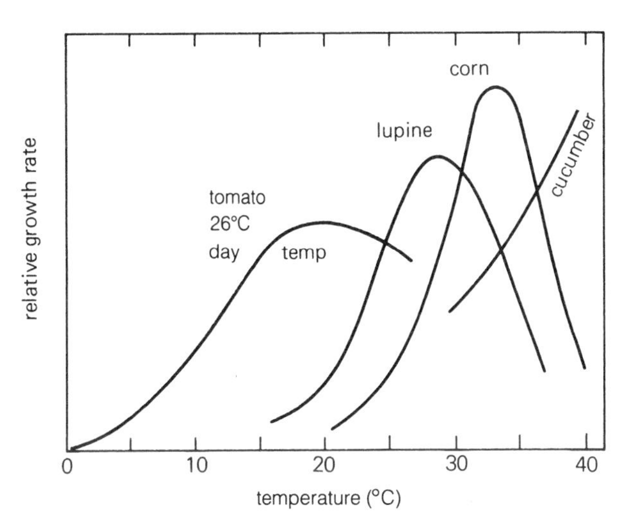


**Figure 22-1** Plant growth as a function of temperature for four species. With tomato, day temperature was constant and night temperature varied.

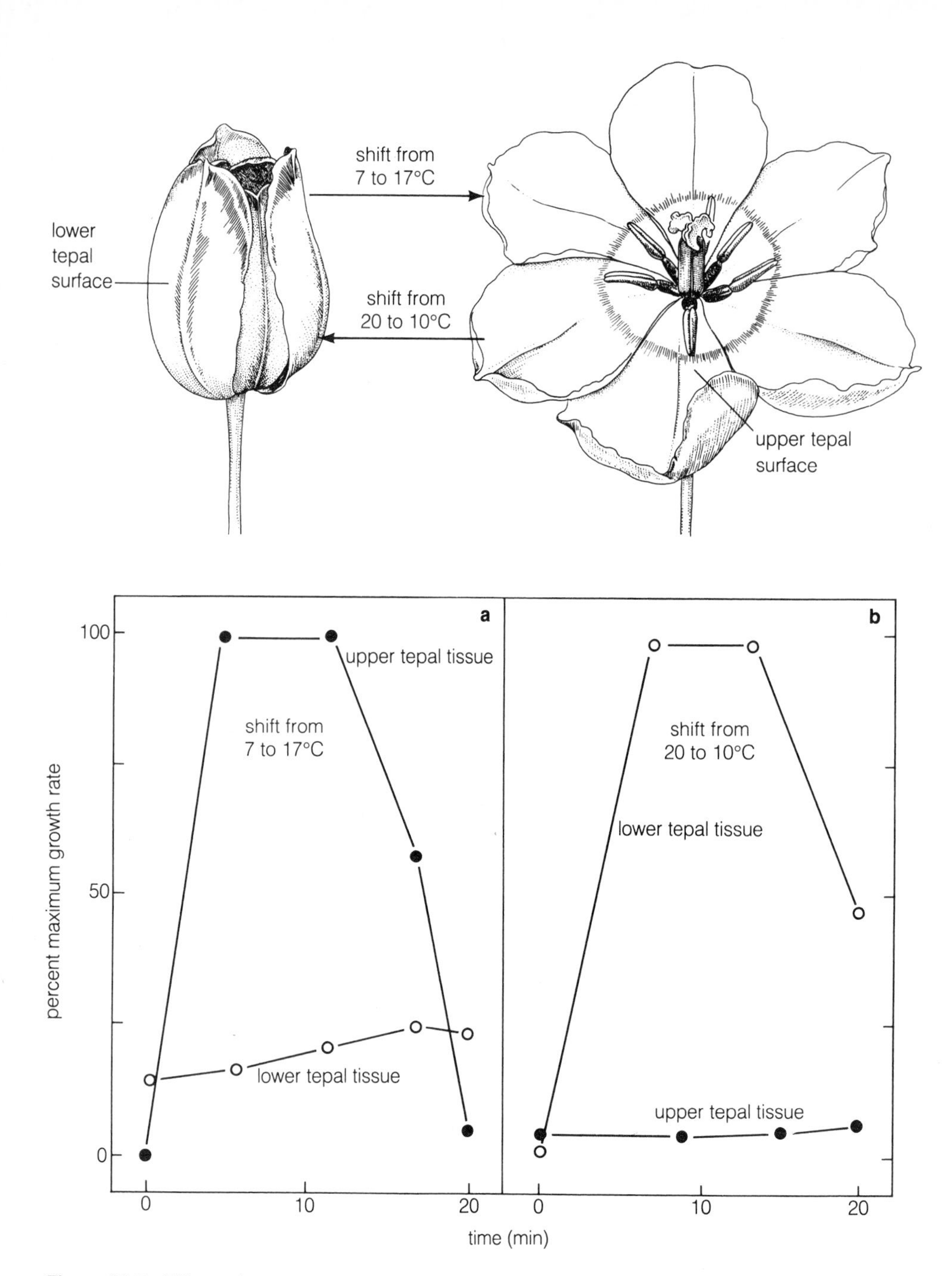


**Figure 22-2** Effects of changing temperature on opening and closing of tulip flowers. (**a**) As temperature increases, growth of upper tepal tissue speeds up for a short time, whereas growth of lower tissue remains about constant, causing flower opening. (**b**) When temperature decreases, growth of lower tepal tissue speeds up and growth of upper tissue remains the same, accounting for flower closing. (Figure prepared by Stanley H. Duke based on data of Wood, 1953.)

causes an increased rate of enzyme denaturation. Subtracting the destruction curve from the reaction curve produces an asymmetrical curve (Fig. 22-3) with its own minimum, optimum, and maximum (cardinal) temperatures. The curve applies to respiration, photosynthesis, and many other plant responses in addition to growth.

In our discussion of $Q_{10}$ values in Section 2.3, we saw that equal small increments of temperature cause many chemical reactions to increase in rate by some equal multiplication of the rate at the lower value. Thus if $Q_{10} = 2$, then the reaction rate approximately doubles for any interval of 10°C. From data for several reactions that follow this rule, Svante Arrhenius (Swedish chem-

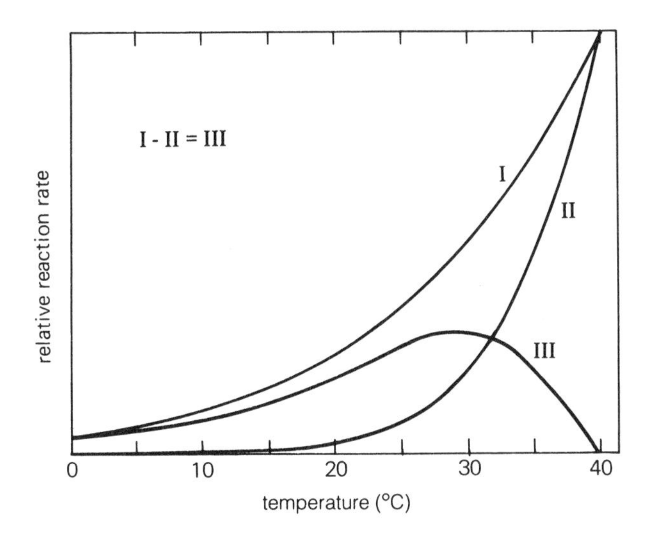


**Figure 22-3** Enzyme activity and temperature. I. Rate of reaction with a $Q_{10}$ of 2, which is typical of many chemical reactions, including those (at lower temperatures) controlled by enzymes. II. A reaction with a $Q_{10}$ of 6, which is typical of denaturation of protein. III. The expected curve for enzyme denaturation (II) subtracted from the curve for enzymatically controlled reaction rate (I). This curve is typical of the temperature responses of enzymatically controlled reactions over the entire temperature range of such reactions.

ist, 1859–1927, 1903 Nobel Prize-winner in chemistry) derived in 1889 the following relation, called the **Arrhenius equation**:

$$\frac{d \ln k}{dT} = \frac{A}{T^2}; \quad \text{or} \quad \ln \frac{k_2}{k_1} = A \frac{T_2 - T_1}{T_1 T_2} \tag{22.1}$$

where $k_1$ and $k_2$ are reaction rates at temperatures $T_1$ and $T_2$, respectively, and $A$ is a constant. From these equations, Arrhenius derived a somewhat more complex equation, which has the following form (see texts on physical chemistry for the more detailed equation):

$$\log k = a - b\frac{1}{T} \tag{22.2}$$

where log $k$ is the logarithm of the reaction rate, $T$ is the absolute temperature in kelvins, and $a$ and $b$ are constants. The equation is that of a straight line when log $k$ is plotted as a function of the reciprocal of the kelvin temperature ($1/T$) in the **Arrhenius plot**. This has been done for countless physiological processes or enzymatic reactions, as in the examples of Figure 22-4. The slope

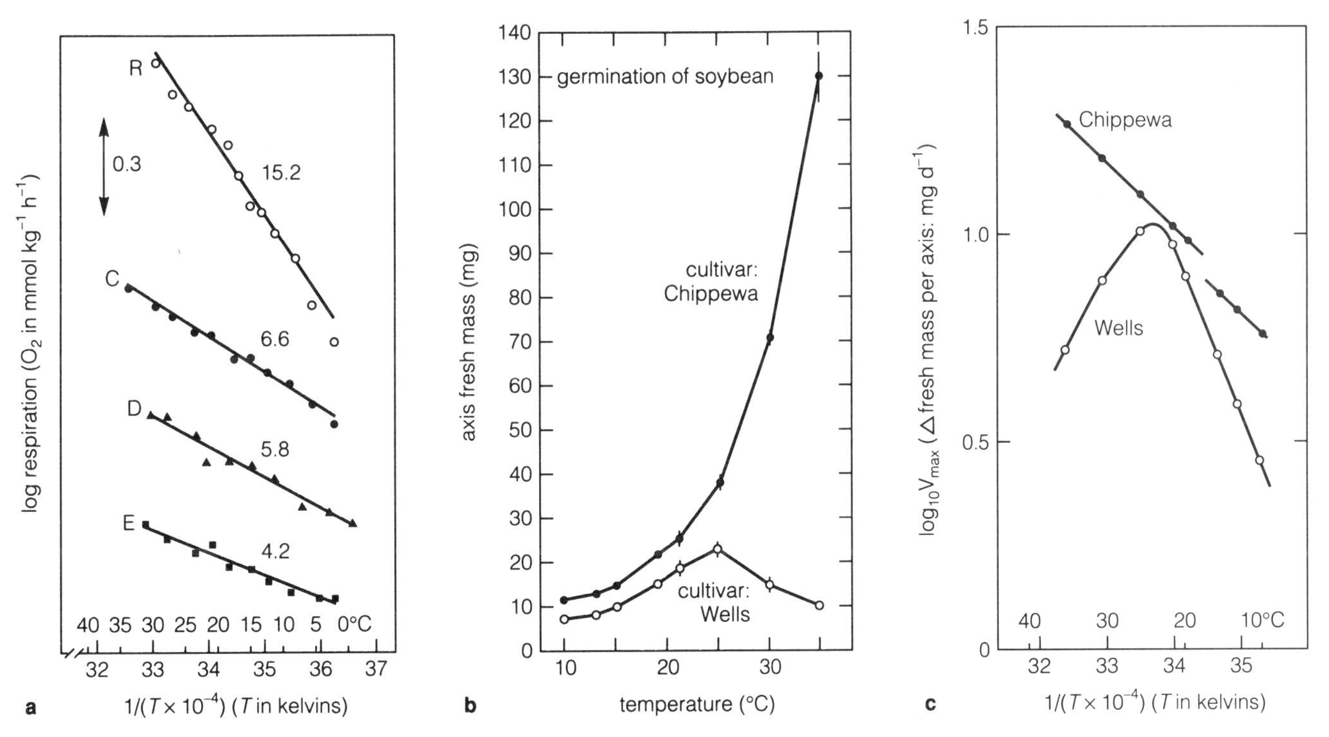


**Figure 22-4** The Arrhenius plot as a way to analyze effects of temperature on various processes including metabolic reactions, other reactions, germination, or other plant or animal responses). (**a**) Simple Arrhenius plots of root respiration for *Ranunculus* (R), *Carex* (C), *Dupontia* (D), and *Eriophorum* (E). Numerals on the curves refer to apparent activation energies in kJ $mol^{-1}$. In such a plot, the ordinate is the logarithm of the response (in this case, respiration), and the abscissa is the reciprocal of the absolute temperature. (Temperatures in degrees Celsius are also shown; note that in such a reciprocal plot, low temperatures appear to the right instead of the left.) (From Earnshaw, 1981; used by permission.) (**b**) Effect of germination temperature on fresh mass of the seedling axis of two cultivars of soybeans. (Vertical lines represent standard deviations.) (**c**) Arrhenius plots of the data shown in **b**. *Changes*, Δ, in fresh mass per day (growth *rates*) are shown in **c**. (From Henson et al., 1980; used by permission.)

of the line is given by the constant $b$, and it is possible to derive the **energy of activation** ($E_a$) from this slope.[1] This is the minimal energy required for the process being measured to occur.

For a given process occurring within an organism, the Arrhenius plot is normally linear within the range of temperatures in which the organism is capable of living for an extended period. Any bend, break, or inflection in an Arrhenius plot denotes a change in sensitivity to temperature of the process being measured. A sharp drop indicates that protein denaturation exists at the upper temperature limit of the plot, and a discontinuity (break) or inflection (change in the slope) may exist at the lower temperature limit of the plot. Such an inflection or discontinuity in an Arrhenius plot indicates that the process being studied is sensitive to low temperature. An increase in the slope of the plot below the temperature of inflection indicates that the energy of activation ($E_a$) has increased and has become more limiting to the maximal velocity ($V_{max}$, the maximal rate of the process under given conditions) than at temperatures above the inflection temperature. This situation is commonly observed in physiological processes and enzymatic reactions of plants sensitive to chilling (see Section 26.5). Arrhenius plots of different cultivars within the same species can be quite different for the same process, such as growth following germination plotted in Figure 22-4b.

Once we've become familiar with positive responses as the temperature increases from the minimum to the optimum, it could come as a surprise to learn that certain processes are promoted as temperature *decreases* toward the freezing point. In **vernalization**, exposing certain plants to low temperatures for a few weeks results in the ability to form flowers, usually after plants are returned to normal temperatures. Low temperatures in the autumn often cause or contribute to the development of dormancy in many seeds, buds, or underground organs, and the low temperatures of winter may contribute to the breaking of dormancy in these same organs. The interesting paradox of the low-temperature response is that low temperature first causes dormancy to develop in plants, but then further low temperatures cause a breaking of the dormancy. If we oversimplify dormancy induction by thinking of it as a simple *negative* slowing down of plant processes at decreased temperatures, then we must surely consider the breaking of dormancy in an opposite, *positive* sense. This is the fascinating dilemma of the low-temperature response.

[1]In equation 22.2, the slope $b = E_a/2.303R$, where $R$ is the gas constant. Thus, $E_a = 2.303R\,b$. The constant $a$ equals log $A$, where $A$ is called the frequency factor and equals the collision frequency of molecules in a reaction.

We will consider five different positive responses to low temperatures. The first, vernalization, has been studied extensively, so it provides a good starting place. The second, also studied extensively, is the breaking of seed dormancy by exposure of moist seeds to low temperatures. This treatment has often been referred to as **stratification**, but the term **prechilling** is now more popular because it is more descriptive.[2] We will have more to say about seed dormancy and germination than their responses to temperature. The third process is closely related: the breaking of winter dormancy in the buds of perennial woody plants. The fourth process has been studied less: the low-temperature induction of development of underground storage organs such as tubers, corms, and bulbs. The fifth process has been studied still less: the effects of low temperatures on the vegetative form and growth of certain plants.

In each of these five processes, we are concerned mostly with **delayed effects** (sometimes called **inductive effects**) upon some developmental plant process. Such effects, in which the response appears some time after completion of the stimulus, are also observed in response to other environmental factors, such as day length (photoperiodism). In fact, low-temperature and day-length effects are frequently interrelated in the five plant responses.

Ultimately, the low-temperature response could involve activation of specific genes, a switching of the morphogenetic program in response to low temperatures. Could the genes respond directly to low temperature? If not, then what in the cell actually responds, transducing the low-temperature signal to a physiological change? Are the transducers located in the cytoplasm or the nuclei of the cells in which the program is readjusted? Or in other cells? There are few answers to these questions, but they could guide future research.

## 22.2 Vernalization

The process of vernalization (but not the term) was described in at least 11 publications in the United States during the mid-19th and early 20th centuries (for example, in the *New American Farm Book* in 1849), but it was completely overlooked by "establishment" science until 1910 and 1918, when J. Gustav Gassner (see his 1918 paper) in Germany described the vernalization of cereals. Much of the early work on plant development took place in Europe; the United States and Canada were apparently preoccupied with subduing the frontier.

[2]It is easy to confuse vernalization (an effect on flowering) with stratification or prechilling (an effect on germination) because both processes can occur when moist seeds are exposed to low temperature.

In the 1920s, the term *vernalization* was coined by Trofim Denisovich Lysenko, who under Stalin was allowed to exercise absolute political control over Soviet genetic science, decreeing in the process that Soviet geneticists should accept the dogma of the inheritance of acquired characteristics (see Caspari and Marshak, 1965). Vernalization, from Latin, translates into English as "springization," the implication being that winter cultivars were converted to spring or summer cultivars by cold treatment. We now realize, although apparently Lysenko did not, that the genetic makeup is not changed by the low-temperature treatment. The cold supplied artificially by the experimenter simply substitutes for the natural cold of winter, as for fall-planted winter cereals such as wheat or rye.

The term *vernalization* has been widely misused. Any plant response to cold has sometimes been referred to as vernalization, as has any promotion of flowering by any treatment (even day length). We shall restrict the term **vernalization** to *low-temperature promotion of flowering*.

## The Response Types

There are numerous vernalization responses, depending not only on species but frequently on varieties and cultivars within species. In classifying the response types, there are several factors to consider. To begin with, we may differentiate *delayed* from *nondelayed* responses. Most plants that have been studied respond after a delay, although a few (for example, Brussels sprouts) form flowers during the cold treatment itself.

An appropriate way to classify the response types is according to the *age* at which the plant is sensitive to cold. The **winter annuals**, especially cereal grasses, were studied during the 1930s and 1940s, particularly in the Soviet Union and by Frederick G. Gregory and O. Nora Purvis (see Purvis, 1961) at Imperial College in London. Such plants respond to low temperatures as seedlings or even as seeds, providing that sufficient oxygen and moisture are present. Petkus rye (*Secale cereale*) seeds are normally planted in the fall, when they usually germinate, spending the winter as small seedlings. Or moist seeds may be exposed to low temperatures in a cold chamber for a few weeks. Plants subsequently form flowers at normal temperatures in approximately seven weeks after growth begins in the spring. Without the cold treatment, 14 to 18 weeks are required to form flowers, but ultimately flowers do appear. Because the cold requirement is a **quantitative** or **facultative** one (low temperatures result in *faster* flowering) but not a **qualitative** or **absolute** one (in which flowering *absolutely depends* upon cold), we have another basis for classification. Most winter annuals are delayed and quantitative in their response, although some (for example, Lancer wheat) have an absolute cold requirement.

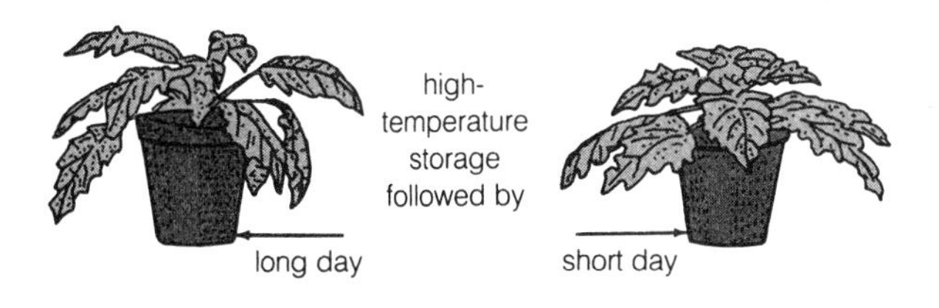


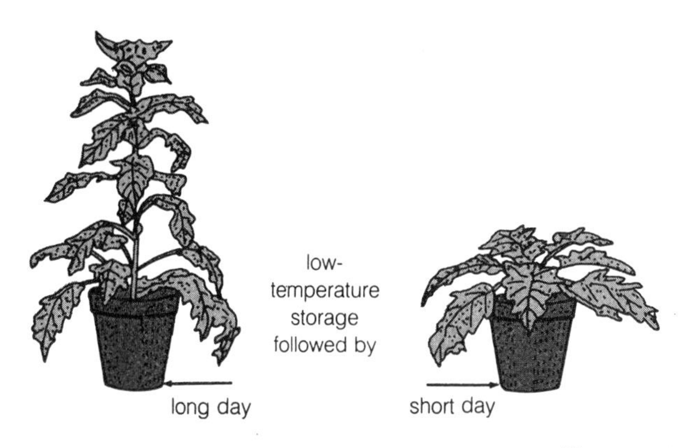


**Figure 22-5** Bolting (flowering) response of henbane (*Hyoscyamus niger*), a typical rosette species, to storage at high or low temperature followed by long- or short-day treatment. Only cold followed by long days induces flowering (bolting).

With Petkus rye there are two interesting complications. Short-day treatments will substitute to a certain extent for low temperature, and flowering of previously vernalized, growing plants is strongly promoted by long days. All winter annuals so far studied are promoted not only by cold but also by the subsequent long days of late spring and early summer.

The **biennials** live two growing seasons, then flower and die (Section 16.2). Examples include several cultivars of beets, cabbages, kales, Brussels sprouts, carrots, celery, and foxglove. They germinate in the spring, forming vegetative plants that are typically a rosette (Fig. 22-5, lower right). The leaves often die back in the autumn, but their dead bases protect the crown with its apical meristem. With the coming of the second spring, new leaves form, and there is a rapid elongation of a flowering shoot, a process called **bolting**. Exposure to the winter cold between the two growing seasons induces flowering. Most biennials must experience several days to several weeks of temperatures slightly above the freezing point to subsequently flower; they have an absolute cold requirement, as contrasted to the facultative winter annuals. Sugar-beet plants may be kept vegetative for several years by not being exposed to cold (Fig. 22-6). Flowering of many biennials is also promoted by long days following the cold, and some may absolutely require this treatment (for example, the European henbane *Hyoscyamus niger*, Fig. 22-5). Other biennials are day-neutral following vernalization.

**Figure 22-6** A 41-month sugar-beet plant kept vegetative by never being exposed to low temperature. (Photograph courtesy of Albert Ulrich; see Ulrich, 1955. The technician at the Earhart Plant Research Laboratory at the California Institute of Technology is Helene Fox.)

Many species of cold-requiring plants do not fall readily into the categories of winter annuals or biennials. Flowering of several perennial grasses, for example, is promoted by cold. Some of these have a subsequent short-day requirement for flowering. The chrysanthemum is a short-day perennial that has been studied extensively because of its photoperiodism response. Its cold requirement, which must be met once before it can respond to the short days, was overlooked because plants are propagated vegetatively, and cuttings carry the vernalization effect with them. Certain woody perennials have a low-temperature requirement for flowering (Chouard, 1960), and several annual garden vegetables will flower somewhat earlier in the season if they are exposed to a short vernalization treatment (Thompson, 1953).

To summarize: Many different species are induced to flower by cold periods, in some plants there is a quantitative and in others a qualitative response, and flowering of many species also requires or is promoted by suitable day length. These responses to environment prepare a plant for the annual climatic cycle. We are not dealing with an endogenous timer as in the previous chapter but with a complex system in which a plant responds to one season by becoming prepared for the next.

## Location of the Low-Temperature Response

It is the bud, presumably the meristem, that normally responds to cold by becoming vernalized. Only if buds are cooled will the plants subsequently flower. Embryos or even isolated meristems from rye seeds have also been vernalized. In another approach, various parts of a vernalized plant have been grafted onto an unvernalized plant. If a vernalized meristem is so transplanted, it will ultimately flower, but if a meristem from an unvernalized plant is grafted onto a vernalized plant after removal of the vernalized meristem, the growing transplanted meristem remains vegetative (see the extensive review by Lang, 1965b).

S. J. Wellensiek (1964) in the Netherlands has suggested that vernalization requires dividing cells. Several studies support his conclusion, although some seeds respond even at temperatures a few degrees below the freezing point, at which temperature cell division seems unlikely and microscopic investigation has failed to reveal it. That cell division or DNA replication in nondividing cells is indeed necessary would be a significant finding. We noted in Table 16-1 that DNA must replicate before cellular differentiation; perhaps only when the DNA is temporarily separated from the chromosomal proteins can gene activation or inactivation occur.

## Physiological Experiments

Sometimes in plant physiology we are unable to study biochemical or biophysical events directly but must instead take a more indirect approach. When whole plants are manipulated in various ways, and when results are observed and deductions are made based on our understanding of plant function at the molecular level, we have conducted **physiological experiments**; they typically produce descriptions of **phenomenology**. This chapter includes many examples.

One physiological investigation on vernalization determined optimum temperatures for the process (Fig. 22-7). Vernalization proceeds at a maximum rate over a fairly wide range of cool temperatures (depending somewhat upon the species), and vernalization occurs even at a few degrees below freezing. Usually the lower limit is set by the formation of ice crystals within the tissues. There is typically a fairly broad optimum (about 0 to 10°C in Fig. 22-7), and some effects have been observed at temperatures as high as 18 to 22°C in some species. Another type of physiological study has determined the most effective vernalization times. Minimum lengths for any observable effect vary from 4 days to 8 weeks, depending on the species. Saturation times vary from 3 weeks for winter wheat to 3 months for henbane.

If immediately following vernalization treatment a plant is exposed to high temperatures, it often does not flower. This reversal is referred to as **devernalization**. To be strikingly effective, devernalizing temperatures must be about 30°C or higher with winter rye, and they

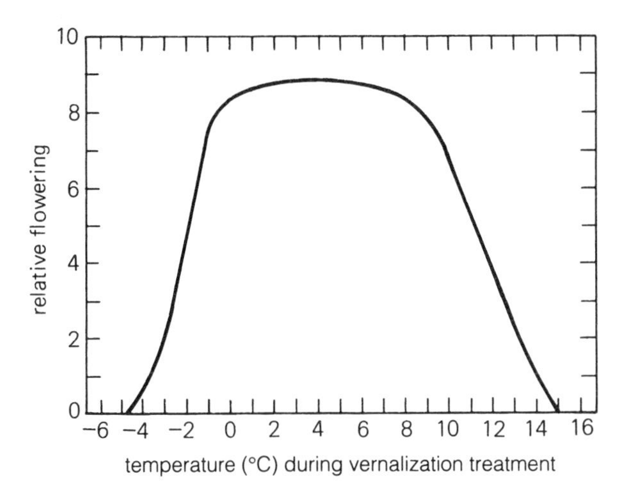


**Figure 22-7** Final relative flowering response as a function of temperature during vernalization. The data represent response of moist Petkus rye seeds to a 6-week period of treatment. (From Salisbury, 1963; see also Purvis, 1961.)

must be applied for a few days and within 4 or 5 days (somewhat longer in other species) after the low temperature. Actually, some devernalization can be observed when plants are exposed to any temperatures higher than those that will cause vernalization. In winter rye, 15°C is the neutral temperature; any temperature below this speeds flowering, and any higher temperature delays it. Anaerobic conditions experienced just after vernalization also cause devernalization, even at neutral temperatures. After devernalization, most species can be revernalized with another cold treatment.

## Vernalin and Gibberellins

If the apical meristem itself responds to low temperatures, a translocated flowering stimulus or hormone does not seem likely. And in most cases, the vernalization effect is not translocated from one meristem to another, either within the same plant or when a vernalized bud or plant is grafted to a nonvernalized plant. There are exceptions, however, as reported as early as 1937 by Georg Melchers in Germany (reviewed by Lang, 1965b). The work has since been extended somewhat in the Soviet Union. When Melchers grafted a vernalized henbane plant to a vegetative receptor plant that had never experienced low temperature, it flowered. Dissimilar response types will also transmit the flowering stimulus across a graft union; cold-requiring plants can be induced to flower without the cold period by being grafted to a noncold-requiring variety, for example. The reverse, though less clear-cut, will also occur. It must be emphasized, however, that transmission is limited to a few species. In most cases, transmission across a graft union fails.

If the experiment is to succeed, a living graft union must be formed between the two plants; conditions favoring transport of carbohydrates also favor transport of the stimulus. If the receptor is defoliated or darkened, for example, while photosynthesizing leaves are left on the donor, the receptor must obtain its nutrients from the donor, which favors movement of the vernalization stimulus across the graft union.

Melchers postulated the existence of a hypothetical vernalization stimulus, which he called **vernalin**. The logical thing to do would be to isolate and identify it. Many futile attempts have been made, but results with gibberellins show that their properties are similar to those expected for vernalin. Anton Lang (1957) found that gibberellins applied to certain biennials induced them to flower without a low-temperature treatment (Fig. 22-8). Others (for example, Purvis, 1961) induced winter annuals by treating seeds with gibberellins. It was then shown that natural gibberellins build up within several cold-requiring species during exposure to low temperature. Gibberellins clearly seem to be involved in vernalization.

Is gibberellin equivalent to vernalin? For various reasons, plant physiologists have been reluctant to answer in the affirmative. When gibberellins (primarily

**Figure 22-8** Anton Lang's induction of flowering in the carrot by application of $GA_3$ (center) or by vernalization (right). The control plant (left) and the center plant were held at temperatures above 17°C, whereas the plant at the right was given 8 weeks of cold treatment. The center plant was treated with 10 μg of $GA_3$ each day for the 8 weeks. (Photograph courtesy of Anton Lang; see Lang, 1957.)

only $GA_3$) are applied to a cold-requiring rosette plant, for example, the first observable response is the elongation of a vegetative shoot, followed by the development of flower buds on this shoot. When plants are induced to flower by exposure to cold, however, flower buds are apparent as soon as the shoot begins to bolt. If the flowering response to applied gibberellins cannot be equated with the natural induction of flowering, several questions come to mind: Could gibberellins induce changes within the plant that in turn lead to flowering or even to production of vernalin? Could several different molecules influence the morphogenetic program in essentially the same way?

Mikhail Chailakhyan (1968) in the Soviet Union suggested that there are two substances involved in flower formation, one a gibberellin or gibberellinlike material and the other a substance he called **anthesin**. Plants that require low temperatures or long days or both might lack sufficient gibberellins until they have been exposed to the inducing environment, whereas short-day plants might contain sufficient gibberellins but lack anthesin. Chailakhyan agreed that cold-requiring plants produce Melchers's vernalin in response to cold, but the vernalin was then converted during long days to gibberellins, at least in those plants requiring long days following the cold.

An elegant experiment performed many years ago by Melchers (1937) supports the two-substance point of view. A noninduced short-day plant (Maryland Mammoth tobacco) was grafted to a noninduced cold-requiring plant (henbane), causing the latter to flower. Apparently, each contained one of the essential substances for the flowering process but had to obtain the other from the plant to which it was grafted, a phenomenon that succeeded for the henbane but not for the tobacco.

### Vernalization and the Induced State

Flower development typically follows the cold treatment by days or weeks. How permanent is this **induced state**, the vernalized condition of the plant before flowering? One henbane variety requires low temperatures followed by long days for flowering. After vernalization, flowering can be postponed by providing only short days (see Fig. 22-5). No loss of the vernalization stimulus appeared in such plants after 190 days, even though all the original leaves exposed to the cold had died. Only after 300 days was there any loss of the vernalized condition. In many other species, the induced state appears to be highly stable. Certain cereal seeds, for example, can be moistened (to 40 percent water, which is too little for germination), vernalized, and then dried out and maintained for months to years without loss of the vernalized condition. We noted that chrysanthemums propagated by cuttings retain their initial vernalized condition. Yet the induced state is far less permanent in several other species.

## 22.3 Dormancy

### The Phenomenon

Only a few plants function actively close to the freezing point. How, then, do plants live where temperatures remain close to or below freezing for several weeks or months each year? Most commonly, such plants become dormant or quiescent, in which conditions they remain alive but exhibit little metabolic activity. Leaves and buds of many evergreens show this reduced activity during the winter, and deciduous perennials lose their leaves and form special inactive buds. Seeds of most species in cold regions are dormant or quiescent during the winter. Certain changes occur in the cells of such seeds that allow them to resist subfreezing temperatures (see Chapter 26).

It seems appropriate that temperature itself might play a regulatory role in the survival of plants in cold regions. The dormant or quiescent condition of evergreen leaves and buds often develops in response to low temperatures, the effects typically being accentuated by short days. Then, subsequent growth in the spring is frequently dependent upon prolonged exposure of dormant buds and seeds to the cold of winter. The buds or seeds accumulate or sum the periods of exposure to cold. Thus, they measure the length of the winter and anticipate spring, when it is safe to resume growth and lose hardiness.

As we've seen so often in our discussions, the situation becomes complex. Plants commonly respond to multiple environmental cues. Germination of seeds, for example, is influenced not only by temperatures, but also (depending always upon species) by light, breaking of the seed coat to permit penetration of the radicle and perhaps entry of oxygen and/or water, removal of chemical inhibitors, and maturity of the embryo.

### Concepts and Terminology of Dormancy

Seed physiologists usually define **germination** as those events that commence with imbibition and terminate when the **radicle** (embryonic root; or in some seeds, the cotyledons/hypocotyl) elongates or emerges through the seed coat (Bewley and Black, 1982, 1984; Mayer, 1974). A seed may remain **viable** (alive) but unable to germinate or grow for several reasons. These can be roughly classified into *external* or *internal* conditions. An internal situation that is easy to understand is an embryo that has not reached a morphological maturity capable of germination (for example, in certain members of the Orchidaceae, Orobanchaceae, or the genus *Ranunculus*). Only time will allow this maturity to develop.

Germination of seeds of wild plants is often limited in this or some other internal way, but seeds of many domestic plants may be limited only by lack of moisture and/or warm temperature.

To distinguish between these two different situations, seed physiologists have used two terms: **Quiescence** is the condition of a seed when it is unable to germinate only because suitable external conditions are not available (for example, the seed is simply too dry or too cold), and **dormancy** is the condition of a seed when it fails to germinate because of internal conditions, even though external conditions (for example, temperature, moisture, and atmosphere) are suitable.

There are problems with this terminology. Dormant seeds are frequently induced to germinate by some specific *change* in the environment, such as light or a period of low temperature. Where do we draw the line on conditions that are considered "suitable external conditions"? Furthermore, in one sense it is *always* the internal conditions that are limiting. (If water is limiting, it is lack of water in the cells of the embryo inside the seed.) In another sense, external conditions only allow germination by influencing the internal ones. We can at least be more precise, stating conditions rather than depending upon the word *suitable*. Thus we could define dormancy as the condition of a seed when it fails to germinate even though (1) ample external *moisture* is available, (2) the seed is exposed to *atmospheric conditions* typical of those found in well-aerated soil or at the earth's surface, and (3) *temperature* is within the range usually associated with physiological activity (say, 10 to 30°C). (A seed physiologist will define conditions still more precisely.) Accordingly, quiescence is the condition of a seed when it fails to germinate *unless* the foregoing conditions are available (Jann and Amen, 1977). Furthermore, the concept of dormancy carries with it the idea of *induction*; in virtually every case, germination does not occur during the treatment that breaks dormancy but only later.

As mentioned briefly in Section 20.6, Lang et al. (1987; see also Salisbury, 1986) suggested that we use the term **ecodormancy** in the sense that *dormancy* is defined here and **endodormancy** in the sense of *quiescence*. They also suggested **paradormancy** as a form of dormancy that is controlled by plant parts other than the one being considered (for example, dormancy of a bud controlled by nearby leaves). If this suggestion is adopted, it might clear up some of the confusion that has surrounded the dormancy terminology.[3]

[3]Researchers who study fruit trees (**pomologists**) use a different terminology (Samish, 1954). The concept of dormancy as defined in this text is called *rest* by them, whereas the term *dormancy* is used by them in exactly the same sense as *quiescence*. You should be on guard as you study the literature, especially for the term *dormancy*, which may be used in either sense. To avoid confusion, in this text *dormancy* is never used in the sense of *quiescence*.

One other term has been widely used in studies in this field. **Afterripening** is used by some authors in reference to any changes that go on within the seed (or the bud) during the breakdown of dormancy. Other authors (for example, Leopold and Kriedemann, 1975) have used the term in a more restricted sense, limiting it to maturation changes that occur in the embryo during storage. We can think of afterripening as the dormancy-breaking changes that occur in the seed with time but in the absence of any other special treatment (Bewley and Black, 1984).

## 22.4 Seed Longevity and Germination

It is an impressive idea that a living organism can go into a sort of suspended animation—can remain alive but not grow for a long period of time, only to begin active growth when conditions are finally suitable. There were reports of the successful germination of Emmer wheat from the ancient silos at Fayum (stored about 6,400 years before the present, or B.T.P.) or from the tomb of Tutankhamen at Thebes (4000 to 5000 B.T.P.), for example, but examination of such seeds has shown them to be not only dead but without any high molecular-weight, nucleic-acid components (see Osborne, 1980). Bewley and Black (1982, 1984) suggested that at least one of these reports (from 1843; reproduced in their discussion) must have been a hoax. Nevertheless, the life span of some seeds is indeed great, in some cases exceeding the human life span (Mayer and Poljakoff-Mayber, 1989).

Table 22-1 lists the longevities of several seeds. *Mimosa glomerata* had a probable longevity of 221 years, but a rather more typical life span for seeds is from 10 to 50 years. Viable seeds of a lupine (*Lupinus arcticus*) were found in a lemming burrow (along with the remains of a lemming) buried deep in permanently frozen silt of the Pleistocene Age in the central Yukon (Porsild et al., 1967). Material surrounding the seeds was dated by radio-carbon techniques at 14,000 B.T.P., but there is no proof that the seeds themselves were that old (Bewley and Black, 1982). Viable lotus (*Nelumbo nucifera*) seeds found in peat in drained lakes of the Pulantien basin of South Manchuria have been estimated to be anywhere from very young (some results of radio-carbon dating) to 1,000 or 2,000 years old (archeological dating of the peat). Similar lotus seeds were found in a boat in a lake near Tokyo, and the *boat* was radio-carbon dated at 3000 B.T.P., which again says nothing about the seeds. But Priestly and Posthumus (1982) have radio-carbon dated portions of a viable lotus seed from Pulantien at about 466 B.T.P. at the time of germination. Of viable seeds dated by circumstantial evidence only, the ages of *Canna compacta* are probably most credible (Lerman and

**Table 22-1 Some Representative Life Spans for Seeds.**

| | Viability (%)[a] | | Age at Test | Storage Conditions |
|---|---|---|---|---|
| | Initial | Final | | |
| Sugar maple (*Acer saccharinum*) | | – | <1 week | |
| English elm (*Ulmus campestris*) | | – | ca. 6 months | |
| American elm (*Ulmus americana*) | 70 | 28 | 10 months | dry storage |
| *Heavea, Boehea, Thea*, sugarcane, etc. | 85 | – | <1 y | |
| Wild oats (*Avena fatua*) | 56 | 9 | 1 y | buried 20 cm in soil |
| Alfalfa (*Medicago sativa*) | 50 | 1 | 6 y | buried 20 cm in soil |
| Yellow foxtail (*Setaria lutescens*) | 57 | 4 | 10 y | buried 20 cm in soil |
| Cocklebur (*Xanthium strumarium*) | 91 | 15 | 16 y | buried 20 cm in soil |
| Canada thistle (*Cirsium arvense*) | 90 | 1 | 21 y | buried 20 cm in soil |
| Kentucky bluegrass (*Poa pratensis*) | 89 | 1 | 30 y | buried 20 cm in soil |
| Red clover (*Trifolium pratense*) | | 1 | 30 y | buried 20 cm in soil |
| Tobacco (*Nicotiana tabacum*) | | 13 | 30 y | buried 20 cm in soil |
| Button clover (*Medicago orbicularis*) | | – | 78 y | herbarium |
| Clover (*Trifolium striatum*) | | – | 90 y | herbarium |
| Big trifoil (*Lotus uliginosus*) | | 1 | 100 y | dry storage |
| Red clover (*Trifolium pratense*) | | 1 | 100 y | dry storage |
| Locoweed (*Astragalus massiliensis*) | | – | 100–150 y | herbarium |
| Sensitive plant (*Mimosa glomerata*) | | – | 221 y | herbarium |
| Indian lotus (*Nelumbo nucifera*) | | – | 1,040 y | peat bog |
| Arctic lupine (*Lupinus arcticus*) | | – | 10,000 y ? | frozen silt, lemming burrows; seeds could be modern |

[a]A dash in the "Final" column means that some seeds were viable at the time of the test, but percentage viability was not reported.

Source: From various sources. See summaries in Altman and Dittmer, 1962 and in Mayer and Poljakoff-Mayber, 1989.

Cigliano, 1971). Apparently, the seeds had been inserted into young growing fruits of a walnut species, such that at maturity the fruits had healed, producing rattles. The seeds were entirely enclosed within the hardened shells. Shell material was carbon dated at 620 ± 60 years B.T.P.

Storage conditions always influence seed viability. Increased moisture usually results in a more rapid loss of viability, but a few seeds can live for long intervals submerged in water (for example, *Juncus* sp. for seven years or more). Many domestic seeds, such as those of pea, soybean, and bean, remain viable longer when their moisture content is reduced and they are stored at low temperature. Storage in jars or in the air at moderate to high temperatures usually results in dehydration and severe cellular rupture when the seeds are hydrated. Cellular rupture injures the embryo and releases nutrients that are good substrates for pathogens. Normal oxygen levels are generally detrimental to seed life spans. Viability is usually lost most rapidly when seeds are stored in humid air at temperatures of 35°C or warmer. Some loss can be due to internal pathogens. Some seeds remain alive longer when buried in soil than when stored in jars on a laboratory shelf, perhaps because of differences in light, $O_2$, $CO_2$, moisture, and ethylene.

A few seeds have an unusually short life span. Seeds of *Acer saccharinum*, *Zizana aquatica*, *Salix japonica*, and *S. pierotti* lose their viability within a week if kept in air. Seeds of several other species remain viable only a few months to less than a year. Such seeds are said to be **recalcitrant**. Often, they die when only a little moisture is lost, or they cannot withstand cool temperatures (for example, seeds of tropical tree crops). This is a serious problem for long-term seed storage in liquid nitrogen with the goal of genetic conservation (as at the National Seed Laboratory in Fort Collins, Colorado).

How do long-lived seeds remain viable so long?

While a seed remains alive, it retains its stored foodstuffs within its cells; as soon as it dies, some of these begin to leak out. Dormant but viable seeds can remain intact on wet filter paper for months; as soon as they die, they are overgrown by bacteria and fungal hyphae, which live on the food that leaks out. There is evidence that viable seeds produce antibiotics that prevent attack by pathogens. But what maintains the integrity of the membranes? No one knows.

What happens during germination? Although it is an oversimplification, seed physiologists speak of four stages: (1) hydration or imbibition, during which water penetrates into the embryo and hydrates proteins and other colloids, (2) the formation or activation of enzymes, leading to increased metabolic activity, (3) elongation of radicle cells, followed by emergence of the radicle from the seed coat (germination proper), and (4) subsequent growth of the seedling. The covering layers around the embryo—the endosperm, the seed coat, and the fruit coat—can interfere with penetration of water and oxygen or both, and they can prevent emergence of the radicle by acting as a mechanical barrier (Section 20.6). In other seeds, they apparently prevent leaching of inhibitors out of the embryos or contain inhibitors themselves. What are the causes of dormancy, what ecological advantages do dormancy mechanisms confer, and how are various forms of dormancy broken to allow germination?

## 22.5 Seed Dormancy

### Impaction and Scarification

One of the easiest examples of dormancy to understand is the presence of a hard seed coat that prevents absorption of oxygen or water. Such a hard seed coat is common in members of the family Fabaceae (Leguminosae), although it does not occur in beans or peas, which points out that dormancy is uncommon in domesticated species. In a few species, water and oxygen are unable to penetrate certain seeds because entry is blocked by a corklike filling (the **strophiolar plug**) in a small opening (**strophiolar cleft**) in the seed coat. Vigorous shaking of the seeds sometimes dislodges this plug, allowing germination. The treatment is called **impaction**, and it has been applied to seeds of *Melilotus alba* (sweet clover), *Trigonella arabica*, and *Crotallaria egyptica*.

*Albizzia lophantha* is a small, leguminous, understory tree in southwest Western Australia (Dell, 1980). Most seeds germinate only in ash beds after fire; less than 5 percent germinate without heat. As it turns out, entry of water into the seed is prevented by a small strophiolar plug until this plug pops out when the seed is heated. Thus the distribution of this plant is controlled by fire through the presence of a strophiolar plug.

Breaking the seed coat barrier is called **scarification**. Knives, files, and sandpaper have been used. In nature, the abrasion may be by microbial action, passage of the seed through the digestive tract of a bird or other animal, exposure to alternating temperatures, or movement by water across sand or rocks. In the laboratory and in agriculture (when needed), alcohol or other fat solvents (which dissolve away the waxy materials that sometimes block water entry) or concentrated acids may be used. The seeds of cotton and many tropical tree legumes, for example, may be soaked for a few minutes to an hour in concentrated sulfuric acid and then washed to remove the acid, after which germination is greatly improved.

Scarification is of considerable ecological importance. The time required for scarification to be completed by some natural means may protect against premature germination in the autumn or during unseasonal warm periods in winter. Scarification in the digestive tracts of birds or other animals leads to germination after the seeds are more widely dispersed. Seeds washed down a gully in the desert are not only scarified but often end up in a spot where there is more water. Dean Vest (1972) demonstrated an interesting symbiotic and mutualistic relationship between a fungus and the seeds of shadscale (*Atriplex confertifolia*) growing in the deserts of the Great Basin. The fungus grew on the seed coats, scarifying them so that germination could occur. Fungal growth occurred only when temperatures and moisture conditions were suitable during early spring, the most likely time for survival of the seedlings.

As noted in relation to *Albizzia*, fire is another important natural means of scarification. Several seeds, particularly in conditions such as those of the chaparral vegetation of Mediterranean climates (for example, Southern California), are effectively scarified by the fires that are so common there. The result is a relatively rapid recovery of the area following the fires. Additionally, the fires remove the leaf canopy that normally absorbs red light and that leaves the spectrum enriched in far-red light, thus inhibiting seed germination (see Section 20.7).

### Osmotic and Chemical Inhibitors

What prevents seeds in a ripe tomato from germinating inside the fruit? Temperature is usually ideal, and there is ample moisture and oxygen. If the seeds are removed from the fruit, dried, and planted, they germinate rapidly, indicating that they are mature enough for germination. Indeed, they even germinate if taken directly out of the fruit and floated on water. In the fruit, the osmotic potential of the juice is too negative to permit germination (Bewley and Black, 1984). Specific inhibitors might also be present, just as ABA in the developing endosperm of alfalfa seeds inhibits germination of the embryo. Other fruits may filter out wavelengths of light that are necessary for germination. (Most of us have noticed a germinating seed inside an orange, so **vivipary**, as discussed in Section 18.5, does occur.)

Often chemical inhibitors are also present in the seeds, and often these must be leached out before germination can occur. In nature, when enough rain falls to leach inhibitors from the seed, the ground will be adequately wet for survival of the new seedling (Went, 1957). This is especially important in the desert, where moisture is more limiting than other factors such as temperature. Vest (1972) found that shadscale seeds contained enough sodium chloride to inhibit them osmotically (see also Koller, 1957). Usually the inhibitor is more complex than table salt (Evenari, 1957; Ketring, 1973), and inhibitors include representatives from a wide assortment of organic classes. Some are cyanide-releasing complexes (especially in rosaceous seeds), whereas others are ammonia-releasing substances. Mustard oils are common in the Brassicaceae (Cruciferae). Other important organic compounds include organic acids, unsaturated lactones (especially coumarins, parasorbic acid, and protoanemonin), aldehydes, essential oils, alkaloids, and phenolic compounds. ABA is often present in dormant seeds, but in many if not all cases it disappears long before dormancy is broken (Bewley and Black, 1984; Walton, 1977). Thus, ABA may be a potent inhibitor of germination when it is present, but there must be more to a seed's dormancy.

Germination inhibitors occur not only in seeds but also in leaves, roots, and other plant parts. When leached out or released during decay of litter, they may inhibit the germination of seeds or root development in the vicinity of the parent plant. Substances produced by one plant that harm another are called **allelopathics** (Chapter 15). (Of course, allelopathics do not produce dormancy in the usual sense.) Actually, some compounds produced by other organisms act as germination *promoters*. For example, nitrate is a commonly used germination promoter in seed-physiology laboratories and is produced by decay of virtually any plant or animal residue.

Before leaving the topic, we should note that many known compounds that are not natural products may strongly influence germination one way or the other. These include many of the growth regulators currently of commercial importance (for example, Dalapon and others). Thiourea has been used in the laboratory as a germination promoter, and nitrate and nitrite are frequently used to stimulate germination of many weed seeds, especially graminaceous species.

## Prechilling

Many seeds, particularly those of rosaceous species such as the stone fruits (peach, plum, cherry), many

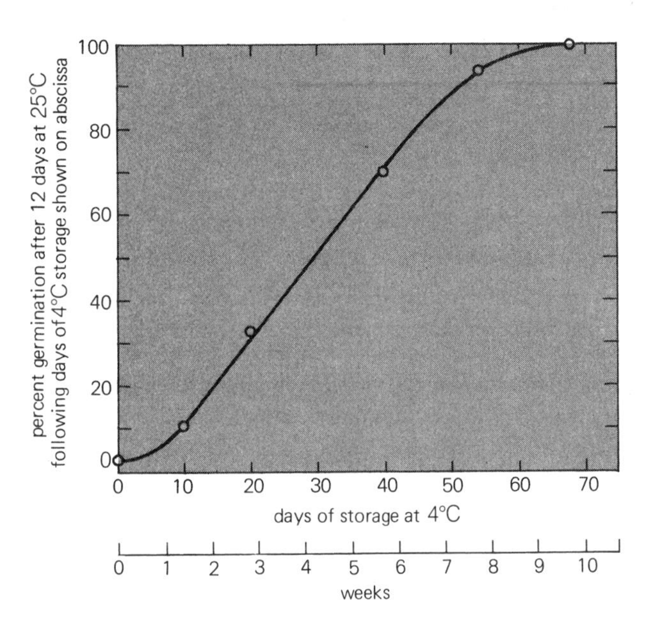


**Figure 22-9** Germination of apple seeds as a function of storage time at 4°C. (Data from Villiers, 1972.)

other deciduous trees, several conifers, and several herbaceous *Polygonum* species will not germinate until they have been exposed for weeks to months to low temperatures and oxygen under moist conditions (Fig. 22-9). Crocker and Barton (1953) listed 62 such species, and numerous others have been found since. Rarely, moist seeds respond to *high* temperatures, and several seeds respond best when daily temperatures *alternate* between high and low. The practice of layering the seeds during winter in flats containing moist sand and peat is called **stratification**. Because seeds in the flats must be cooled before they will germinate, a more popular and descriptive term than stratification nowadays is **prechilling**. Prechilling in seed laboratories and for physiological experiments is routinely performed in incubators or growth chambers. In nature, the low-temperature requirement protects seeds from precocious germination in the fall or during an unseasonal warm period in winter.

What chemical changes go on within the seed during prechilling, allowing it subsequently to germinate when conditions are right? Most seeds, including those that require cold, are rich in fats and proteins but have little starch (Nikolaeva, 1969; Lang, 1965a), and during the cold treatment the embryo of some species grows extensively by transfer of carbon and nitrogen compounds from food-storage cells. Sugars accumulate, and these might be required as sources of energy and to attract water osmotically, later causing germination. Even in cold-requiring seeds such as European ash (*Fraxinus excelsior*), in which the embryo is already fully developed before stratification, a massive degradation of fat occurs in the embryo itself during the cold. The protein content rises, and starch then appears.

Perhaps inhibitors disappear during prechilling, and/or growth promoters such as gibberellins or cytokinins accumulate (Khan, 1977). Auxins have little effect on germination, but in many cases gibberellins will substitute for all or part of the cold treatment, just as they often do in vernalization. Perhaps they accumulate during stratification in amounts that overcome dormancy, but most of the data obtained so far argue against this interpretation (Bewley and Black, 1982, 1984). Cytokinin effects are usually less dramatic and are much less widespread. In a sense, the far-red-absorbing form of phytochrome ($P_{fr}$) is a growth regulator that is required for germination of many seeds, as discussed in Chapter 20.

Both inhibitor disappearance and hormone accumulation in whole seeds have been observed, but there are numerous contradictions. In Section 20.6 we discussed similar studies with light-requiring seeds. We concluded that the radicles themselves should be analyzed because changes in the rest of the seed could mask the important changes in relatively small radicles. I. Arias and coworkers (1976) measured gibberellins in the embryonic axis and in the food-storage cotyledon cells of the hazel tree (*Corylus avellana*), a species in which gibberellins fully overcome the prechilling requirement. Although during chilling there was little accumulation of gibberellin in either part, chilling allowed the embryonic axis but not the much larger cotyledons to synthesize much gibberellin when returned to a germination temperature of 20°C. The GA concentration became 300 times as great in the axis as in the cotyledons. Similar studies are needed with other seeds, especially now that such modern analysis methods as GC-MS and selected ion monitoring allow sensitive analysis of hormones and other compounds in small plant parts (see Section 17.2).

The molecular basis for the breaking of any kind of dormancy in seeds remains to be discovered, partially because some reports seem to contradict each other. For example, compounds that inhibit respiration, such as nitrite, cyanide, azide, malonate, thiourea, and dithiothreitol, can often break seed dormancy. On the other hand, Roberts and Smith (1977) and others have shown that elevated levels of oxygen, which should promote respiration, can induce germination of certain dormant seeds.

Where does the dormancy mechanism lie? Consider three possibilities: that the seed coat contains a chemical that inhibits elongation of the radicle; that the seed coat or endosperm acts as a mechanical barrier to elongation; and/or that the radicle itself lacks the ability to grow until chilled. Isolated, prechilled embryos from many seeds will subsequently grow when placed in warmer temperatures, but nonprechilled isolated embryos will not. In these cases, cold temperatures must act directly on the embryo. Consistent with this, growing embryos in prechilled walnut seeds can exert mechanical pressure at least 1.0 MPa greater than nonchilled embryos that are incapable of breaking the shells. In various species of lilac, including *Syringa vulgaris*, prechilling has no effect on the mechanical resistance of the endosperm or on its inhibitor content, but radicles of chilled embryos will elongate in a solution with a water potential about 0.5 MPa more negative than that in which nonchilled radicles will grow (Junttila, 1973). Thus there is good evidence that the embryo itself responds to the cold, but there is little direct evidence that inhibitors in seed coats are affected—although they are often present.

Prechilling of seeds sometimes has a strong delayed effect on growth in addition to its dormancy-breaking action. If the embryos of peach seedlings are excised from their cotyledons, they germinate without prechilling, but the seedlings are frequently stunted and abnormal. When excised embryos are treated with low temperature, they grow into normal seedlings. Thus it is prechilling and not the presence of the cotyledons that ensures their normality. Because stunted plants often lose their dwarf habit when sprayed with gibberellins, the accumulation of gibberellins or other hormones during prechilling could account for these results, or prechilling could increase the potential to synthesize gibberellins.

### Light

In Sections 20.1 and 20.6, we mentioned that light controls germination of many seeds, and we discussed some of the complications in this response. Clearly there are several environmental cues, often interacting in intricate ways, that control the germination process.

## 22.6 Bud Dormancy

In temperate regions, seed and bud dormancy have much in common, and in buds inducing dormancy is as critical as breaking dormancy. Bud dormancy almost always develops before fall color and the senescence of leaves. Buds of many trees stop growing in midsummer, sometimes exhibiting a little growth again in late summer before going into deep dormancy in autumn. Flower buds that will grow the next season typically form on fruit trees in midsummer. The leaves remain green and photosynthetically active until early autumn, when leaf senescence occurs in response to short, bright, cool days. As chlorophyll is lost, the yellow and orange carotenoid pigments become apparent, and anthocyanins (primarily cyanidin glycoside) are synthesized. Fruits such as apples often mature during this time. Frost hardiness also develops in response to the low temperatures and short days of autumn.

Bud dormancy is induced in many species by low temperatures, but there is also a response to day length, especially if temperatures remain high. With several deciduous trees studied at Beltsville, Maryland (Downs and Borthwick, 1956), short-day treatment resulted in formation of a dormant terminal bud and cessation of internode elongation and leaf expansion, but often the leaves were retained. Long nights, each interrupted by an interval of light, had the same effect as long days. The buds of birch (*Betula pubescens*) detect the day length directly, but in other species leaves usually detect the photoperiod, although dormancy occurs in the buds (Wareing, 1956). Perhaps this correlative phenomenon, like others, is caused by a growth regulator, which could be abscisic acid (Section 18.5).

There are always interactions. In the Beltsville study with deciduous trees, the short-day induction of dormancy was observed at temperatures between 21°C and 27°C, but at temperatures between 15°C and 21°C there was little stem growth during either long days or short days; low temperature prevailed over day length.

Different genetic races within a species, called **ecotypes** (see Section 25.4), often have quite different dormancy responses. For example, Thomas O. Perry and Henry Hellmers (1973) found that a northern (Massachusetts) race of red maple (*Acer rubrum*) developed winter dormancy in response to short days and cold temperatures in growth chambers, but a southern race from Florida did not. Ole M. Heide (1974) studied Norway spruce (*Picea abies*). Trees from Austria (47° latitude) stopped elongating at day lengths of 15 h or less, but trees from northern Norway (64° latitude) stopped when day lengths were 21 hours or less. Both stopped growing well before killing frosts. Temperature had little influence, but trees from high elevations stopped elongating at day lengths that were longer than those required to halt the growth of trees at the same latitude but at low elevations. Heide also found that roots did not respond to photoperiods applied to the tops. With few exceptions, roots continue to grow as long as nutrients and water are available, until soil temperatures become too cold (Kramer and Kozlowski, 1979). Clearly, such trees are well adapted to the environments in which they naturally occur. The Florida maples, for example, are restricted to warm, southern climates because they cannot enter dormancy soon enough in the fall.

Withholding water frequently accelerates development of dormancy, as does the restriction of mineral nutrients, particularly nitrogen. This is probably important for species that enter dormancy before the high temperatures and drought occurring in the tropics or in dry climates. Situations are also known in which dormancy develops in response to *changing* day length (and even to changing soil temperature).

Partial bud dormancy precedes true dormancy, and it can easily be reversed by moderate temperatures and long days (or continuous light). Gradually, however, attempts to induce active growth fail, and then the plant has reached the true dormancy that requires special treatments to overcome (Vegis, 1964).

Morphology is important in dormancy phenomena. A dormant bud typically has greatly shortened internodes and specially modified leaves called **bud scales**. These scales prevent desiccation, briefly insulate against heat loss, and restrict movement of oxygen to the meristems below. They may also respond to ambient light and/or perform other functions. In a sense, bud scales are analogous to the seed coat.

The hormonal factors involved in dormancy are not known, but in trees abscisic acid has been implicated in the response (Walton, 1980). In the mid-1960s, one group of researchers reported that they were able to induce the formation of resting buds in a number of tree species by feeding ABA through the leaves, but no one has been able to reproduce their results. Phillips et al. (1980) listed numerous examples of conflicting data on whether ABA accumulates in dormant tissues. Because of these conflicting data, it is now impossible to conclude that ABA normally causes dormancy.

Dormancy is also overcome by specific temperatures, day lengths, or both. The temperature effect was studied as early as 1880 (see Leopold and Kriedemann, 1975), but the day-length effect has been recognized only since the late 1950s. Because leaves respond to day length in the induction of dormancy and in flowering, it seemed reasonable that leaves are the only organs that respond to day length. But it is now known that dormancy is broken by long days in several leafless trees: beech, birch, larch, yellow poplar, sweetgum, and red oak, for example. Except for beech, these species also respond to cold periods. In other species, cold must be followed by long days. Even in midwinter, certain deciduous species will respond to long-day treatment (particularly continuous light).

A midsummer dormancy occurs in some species (especially evergreens), during which the stems cease to elongate for a period of time. This is typically broken by exposure to more long days.

What organ responds to the long days that overcome dormancy? Apparently the bud scales themselves respond, or enough light penetrates to bring about the response within the primordial leaf tissues inside the bud. Probably both the short-day induction and the long-day breaking of dormancy are phytochrome responses, but the case is not clear-cut. In some studies of short-day induction, red light is most effective in the night interruption, and its effect is reversed somewhat by a subsequent exposure to far-red light, but this reversal has failed in several other studies.

Dormancy in many buds can be broken by exposure to low temperatures. Days to months may be required at temperatures below 10°C. With fruit trees, 5 to 7°C is more effective than 0°C. Considerable work has been done with fruit trees to determine the minimum cold period required to break dormancy because this period determines how far south they can be grown in the northern hemisphere. Apples, for example, may require 1,000 to 1,400 h at about 7°C. Headway has been made in selecting peach cultivars with a shorter chilling requirement than normal, for example, which allows them to be cultivated where the winters are warmer. Incidentally, high temperatures following cold will reinduce dormancy in apple trees, a situation closely analogous to devernalization.

The effects of chilling upon the breaking of dormancy are not translocated within the plant but are localized within the individual buds. A dormant lilac bush, for example, may be placed with one branch protruding outside through a small hole in the greenhouse wall. The branch exposed to the low temperatures of winter will leaf out in early spring, but the rest of the bush inside the greenhouse remains dormant.

Several chemical treatments of the bud will break dormancy. For instance, 2-chloroethanol ($ClCH_2CH_2OH$), often called ethylene chlorohydrin, has been used with success for many years. Applied in vapor form, it breaks dormancy of fruit trees. Another simple but often effective treatment is immersion of the plant part in a warm water bath (40 to 55°C). Often a short exposure (15 s) is effective. Applied gibberellins break bud dormancy in many deciduous plants, just as they break dormancy of many cold-requiring seeds and induce flowering of many cold-requiring plants.

## 22.7 Underground Storage Organs

In many cases, temperature conditions will induce the formation of such underground storage organs as bulbs, corms, and tubers. In some species, dormancy is also broken or subsequent growth influenced by storage temperatures. In other species day length also influences formation of the organs.

### The Potato

Potato tubers develop over a wide range of temperatures and day lengths from swellings at the tips of underground stems called **stolons**,[4] which are derived from nodes at the base of the stem in the soil. Physiologists (Vreugdenhil and Struik, 1989) outlined the following four steps in tuber formation: (1) Stolon induction and initiation, (2) stolon growth (elongation and branching), (3) cessation of longitudinal growth of the stolon, and (4) tuber induction and initiation, which results in radial growth of the stolon tip to form a tuber. These steps can be separated experimentally because they are affected somewhat differently by different environmental conditions and by different hormonal treatments.

Stolon initiation can occur even before the leafy shoot has emerged, so it does not depend on signals from the shoot. It occurs over a wide range of temperatures and day lengths, but development of the stolons into tubers usually (depending on the cultivar) requires more specific conditions. Apparently, for stolon initiation it is important for levels of gibberellins to be high and for levels of cytokinins not to be too high. Long days favor stolon elongation, but short days result in cessation of stolon growth (Chapman, 1958). Short days also result in a lowering of gibberellins in the plant, and that may be what causes the stolons to stop elongating. It is possible to restrict stolon elongation without radial growth of the stolon (which forms the tubers), but normally these two processes go hand in hand. Ethylene stops stolon elongation (for example, in response to mechanical resistance in the soil), but ethylene also stops tuber formation (Mingo-Castel et al., 1976). When conditions are favorable, growth of tubers is initiated. This is more than a response to lowered gibberellins and ethylene, which are both negative influences; there is also excellent evidence for a positive **tuber-inducing substance** that forms in the leaves of some cultivars in response to short days. All the expected features of photoperiodism are present, including a *critical night* and an inhibitory effect of a light interruption given during the dark period (see Chapter 23 and Chapman, 1958). There are significant differences among cultivars, but in one study tuber formation did not require short days but proceeded at any day length (a day-neutral response) when the night temperature was below 20°C. Tuberization was optimal at night temperatures of about 12°C. Such an interaction between photoperiodism and temperature is common, as it is in vernalization and dormancy.

In sensitive cultivars on long days, no tubers will form at any soil temperatures unless the *shoots* are exposed to low temperatures. Hence, the leaves must detect both the photoperiod and the temperature and must transmit the tuber-inducing substance to the stolons. There have been many attempts to isolate this substance, and recently Yasunori Koda and coworkers (1988) have isolated a highly active material from potato leaves. It will induce tubers *in vitro* (single-node stem segments) at concentrations of $3 \times 10^{-8}$ M, in the range of active concentrations for auxin and other growth reg-

[4]**Stolons** are usually defined as aboveground, horizontal stems, as in the strawberry. Underground horizontal stems are **rhizomes**. Potato "stolons" (the term used by physiologists who work with potatoes) are usually underground, but they can be aerial; in darkness, even aboveground potato buds develop into stolons.

ulators. It proved to be a complex molecule similar to jasmonic acid.[5]

Because it is an underground stem, the potato tuber exhibits stem characteristics. Its eyes are the axillary buds, and they remain inactive in response to the presence of the apical bud. When the potato is cut up to produce seed pieces, this apical dominance is lost, and the axillary buds grow if dormancy has been broken. There are practical reasons both to prolong and to break tuber dormancy. The longer tubers can be stored during winter and spring in the dormant condition, the higher their price when sold. In potato "seed" certification, however, it is desirable to break dormancy prematurely to test for pathogens in sample tubers. The time normally required to break dormancy is somewhat shorter when the tubers are stored at about 20°C than at lower temperatures, but there is no clear-cut temperature effect. Certainly there is no cold requirement.

It is possible to break dormancy in potato tubers by the chemical treatments that are effective in breaking bud dormancy of aboveground stems (2-chloroethanol, gibberellins, hot water, and so on). Thiourea also causes sprouting but can result in as many as eight sprouts from a single eye, rather than the usual single sprout. Dormancy can also be induced or prolonged by spraying such growth regulators as maleic hydrazide or chloropropham on the foliage before harvest or on the tubers after harvest. Storage temperature is also important. Tubers sprout somewhat prematurely at high temperatures, and at low temperatures (≈0 to 4°C) starch turns to sugar. If a single storage temperature must be used, the ideal compromise seems to be about 10°C. Modern potato-processing facilities, however, store the tubers at a much lower temperature (about 2°C), and when workers are ready to slice and fry the tubers to make chips, they move them to a higher temperature storage area for several days so that the sugar will be converted back to starch. If this is not done, the sugar caramelizes during frying to produce a dark brown or even black color that is undesirable in the final potato chips.

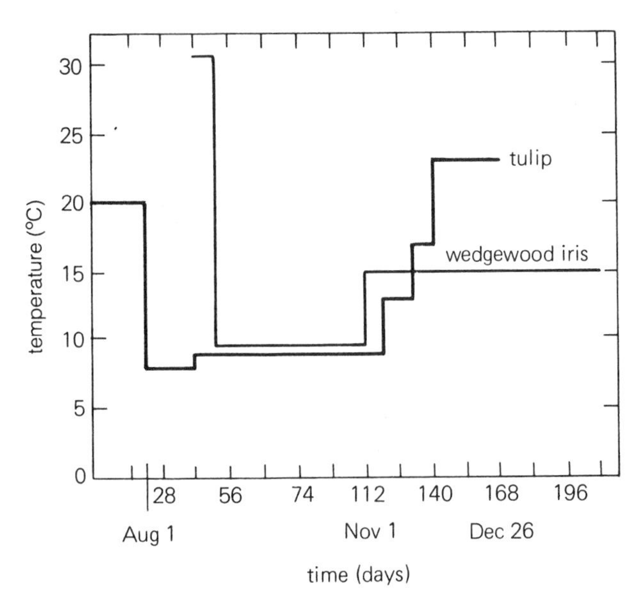


**Figure 22-10** Temperature treatment for early flowering of *Tulipa gesneriana* W. Copeland and of *Iris xiphium* Imperator. With the tulip, flower initiation begins and is well under way during the 20°C treatment. Moving to storage rooms at 8 and 9°C provides an acceleration in blooming, so flowers are produced at Christmas. Continuous 9°C gives equal earliness, but quality is poor unless the 20°C treatment is given first. The bulbs are planted in a controlled-temperature greenhouse about midway during the low-temperature treatment. The temperature is first raised when the leaf tips are visible, then again when they are 3 cm long, and finally when they are 6 cm long. With iris, the short period at high temperature is essential to flowering, although actual initiation of flower primordia does not occur until the bulbs have been moved from low temperature to 15°C, at which time the sprouts are about 6 cm long. Again the 9°C treatment is to ensure earliness. At temperatures much above 15°C during the last part of the treatment, abnormal flowers are sometimes produced. Low light levels will also result in "blasted" flowers at this time, especially if the temperatures are not right. If extremely high temperatures (38°C) are used during the first flower-induction period, flower parts are increased or decreased, or tetramerous, pentamerous, or dimerous flowers result. (Data from Annie M. Hartsema, 1961; figure from Salisbury, 1963.)

## Bulbs and Corms

There has been little investigation of how bulbs, corms, and rhizomes are induced to form, but much work has been done in Holland, supported largely by the Dutch bulb industry, to determine the optimum storage conditions (primarily storage temperature as a function of time) that will result in the formation of leaves, flowers, and stems at desirable times and with the desirable properties. The approach was to observe the morphology of the bulb carefully in the field during a normal season, and then to repeat these observations with bulbs stored under accurately controlled temperatures. The goal was to shorten the time to flowering, a process called **forcing**. This work has been going on since the 1920s (see Hartsema, 1961; Rees, 1972).

Here are a few generalizations: Bulbs must reach a critical size, which often requires two or three years, before they begin to respond to storage temperatures by forming flower primordia. In some cases (for example, tulip), leaf primordia are formed before the flowers, but sometimes leaf and flower formation are nearly simultaneous. Specific temperatures are often required for flower initiation or subsequent stem elongation. The pattern of change and the optimum temperatures usually match the climate where the bulbs are native.

[5]The tuber-inducing substance was identified as 3-oxo-2(5-?-D-glucopyranosoloxy-2-*cis*-pentenyl)-cyclopentane-1-acetic acid (Vreugdenhil and Struik, 1989).

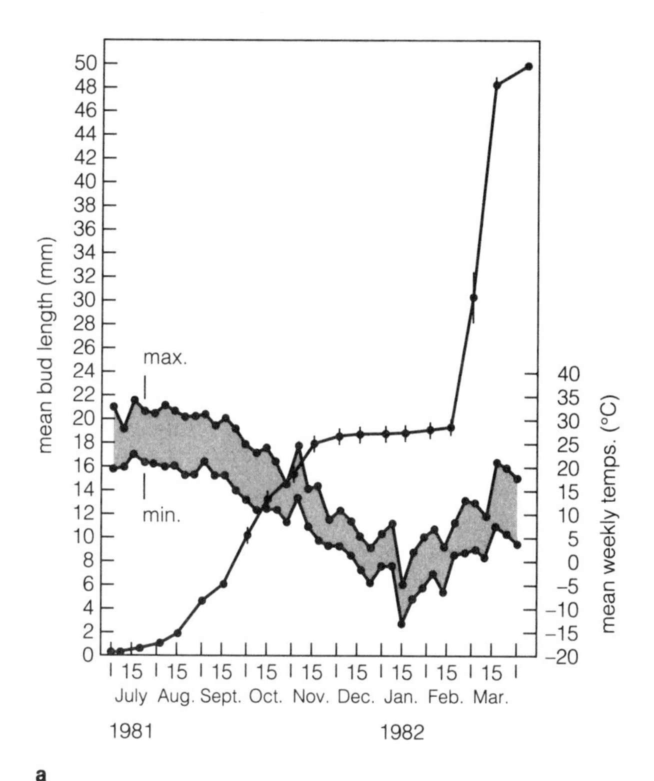


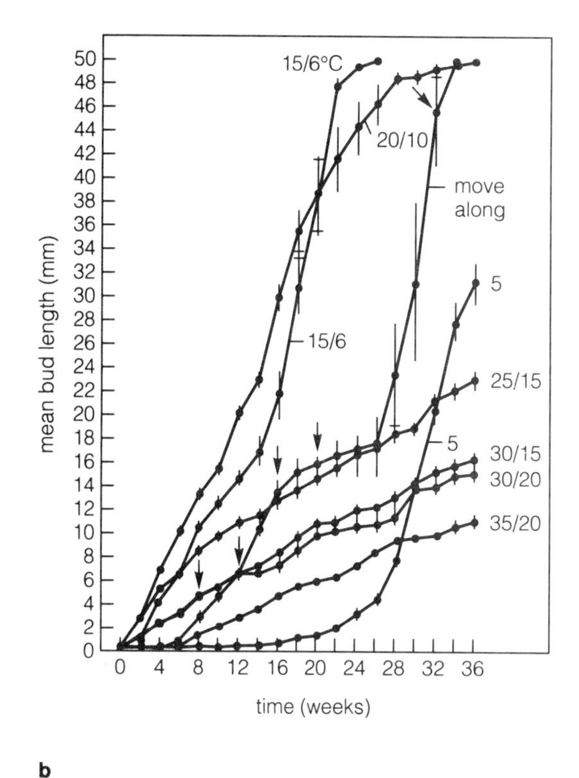


**Figure 22-11** (**a**) The length of *Pediomelum subacaule* buds during the year, compared with maximum and minimum temperatures. Plants were kept in moist vermiculite in an open, nontemperature-controlled greenhouse. (**b**) Length of *Pediomelum* buds, also in moist vermiculite, held at controlled day/night temperatures as labeled on the figure. The curve labeled "move along" was subjected to simulated seasonal temperatures in the field. Note that plants held at intermediate temperatures (especially 15/6°C and 20/10°C) grew throughout the experiment, whereas plants held at warmer temperatures never grew as fast; plants held at 5°C failed to grow at first but grew rapidly after about 24 weeks. (From Baskin and Baskin, 1990; by permission.)

There are several patterns: In some species, flower primordia form before the bulbs can be harvested. This allows little control of flower formation during storage, so study of these has been limited. In others, flower primordia form during the storage period after harvest in the summer but before replanting in the fall, making control easier. Figure 22-10 shows a storage-temperature regimen designed to cause rapid flowering of tulips in time for Christmas. Note that temperatures that induce flowering are relatively high compared to those effective in vernalization of seeds and whole plants. Nevertheless, the response is similar.

In most bulbous irises (Fig. 22-10), the actual flower primordia appear during the low temperatures of winter (9 to 13°C optimum), but a high-temperature (20 to 30°C) pretreatment is essential if flower formation is to occur at all. This is a true example of induction similar to vernalization, but the response is to high rather than low temperatures. In each example, the plants are adapted so that their flowering, vegetative growth, and dormancy are nicely synchronized with seasonal changes in temperature.

Jerry M. and Carol C. Baskin (1990) studied a small plant, *Pediomelum subacaule*, that grows in cedar glades of Tennessee, Georgia, and Alabama. The plant is a perennial that emerges in early spring, flowers, and becomes dormant in late June and early July, when the shoot and absorbing roots die, leaving a small shoot bud at the top of a tuberous storage root, about 50 mm below the soil surface. Plants don't grow during the dry summer. Elongation of the bud occurs in autumn and late winter but not during the coldest part of midwinter (Fig. 22-11a). The Baskins subjected dormant roots, buried 50 mm below the surface of moist vermiculite, to several combinations of day and night temperatures, as shown in Figure 22-11b. One set of temperatures ("move along" in Fig. 22-11b) approximated field temperatures. Shoot elongation, measured at intervals by temporarily removing the plants from the vermiculite, was very slow at the highest temperatures but continued throughout the year at intermediate (cool) temperatures. Plants held at the lowest temperature (5°C) did not grow much for about 20 weeks but finally grew quite rapidly. Plants subjected to simulated field temperatures grew much as plants in the field do. These results showed that the plants were never truly dormant but only quiescent because they would grow at any time if temperatures were right and moisture was available. But the temperature optimum for growth decreased with time, so rapid growth occurred in autumn and late winter. These physiological changes controlling plant response to temperature guaranteed that plants would emerge in early spring when ample soil moisture was available, become dormant during the dry season, and repeat the sequence the next year. Apparently quiescence was caused by drought (came much later in wet years) rather than long days, as was the case with *Anemone coronaria* (Kadman-Zahavi et al., 1984).

## 22.8 Thermoperiodism

The temperature discussion so far has mostly been concerned with the annual temperature cycle, but Frits Went (1957) described **thermoperiodism**, a phenomenon in which growth and/or development is promoted by alternating day and night temperatures. We noted that potato tubers form in response to low night temperatures; fruit set on tomato plants is also promoted by low night temperatures. Stem elongation and flower initiation are also thermoperiodic responses in some species. An original implication of the thermoperiodism concept was that plant productivity was higher under a thermoperiodic environment. For some species, including certain tomato cultivars, this is true, but fluctuating day and night temperatures are not essential for optimum growth of numerous other species. Cocklebur, sugar beet, wheat, oats, bean, and pea grow as well at an *optimum* constant temperature as they do when day and night temperatures vary. An experimenter must be careful to compare various thermoperiodic regimens with the optimum constant temperature rather than some other temperature (Friend and Helson, 1976).

Some plants grow better when the environment fluctuates on a 24-h cycle, presumably to coincide with the phases of their circadian clock. Thus, some species grow poorly when both light and temperature are constant. Varying temperature on a 24-h cycle prevents or reduces the injury caused to tomato plants by continuous light and temperature, if light levels are high enough. Indeed, many thermoperiodic responses interact with the light environment, typically via photoperiodism and probably via balances in the phytochrome system.

One of the most spectacular examples of thermoperiodism reported by Went (1957) involved *Laothenia charysostoma* (formerly *Baeria*), a small annual composite commonly seen during spring in mountain valleys and foothills and occasionally in the western portions of the Mojave Desert of California. It is extremely sensitive to night temperature. Grown under short-day conditions in Went's experiments, plants survived only

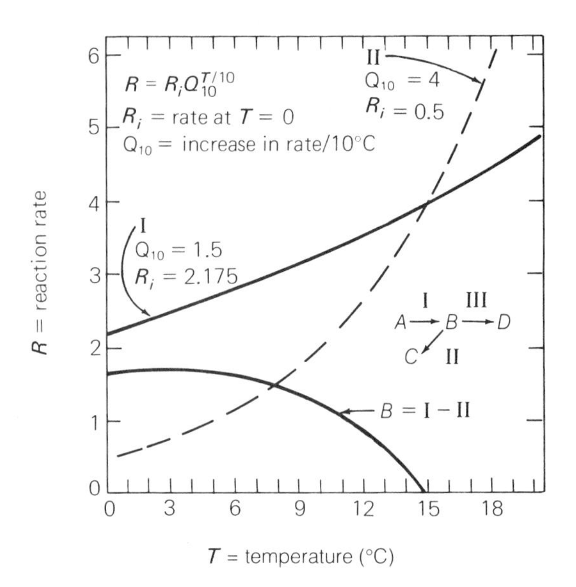


**Figure 22-12** Sample curves showing hypothetical reaction rates as a function of temperatures for reactions with $Q_{10}$ values of 1.5 or 4.0. If the reaction with $Q_{10}$ = 1.5 is considered to be reaction I in the reactions shown in the circle (and discussed in the text) and if the reaction with $Q_{10}$ = 4.0 is considered to be reaction II, then the hypothetical product *B* will be proportional to curve II minus curve I, as shown (curve *B*). Compare the shape of curve *B* with those of the curves in Figures 22-1, 22-3, and 22-7. (From Salisbury, 1963.)

two months when the night temperature was 20°C. At lower temperatures, they grew for at least 100 days. They died rapidly at night temperatures of 26°C. Many species do not grow particularly well at night temperatures this high, but how can we account for death at this temperature? The *Laothenia* plants flourish when the day temperature is well above 26°C, providing only that the night temperature is low enough. Went reported that other plants native to California acted similarly in his experiments.

Perhaps, as pointed out earlier, different tissues within the same plant have different cardinal temperatures. For proper growth and development of the entire plant, the temperature range during the day should include near-optimal temperatures for the growth of all necessary tissues. Normally, soil temperatures are different from air temperatures, so plants may have different cardinal temperatures for roots and shoots. By maintaining roots and shoots at the same temperature, optimal growth and development may not be obtained.

## 22.9 Mechanisms of the Low-Temperature Response

How can we understand positive plant responses to low temperature? We might be dealing with some kind of hormonal or metabolic block. Such a block could be a chemical inhibitor or the lack of some necessary substance within the plant, or both. An inhibitor could disappear, or a growth regulator could arise at low temperatures, influencing flowering, germination, subsequent seedling growth, and so on. Gibberellins and ABA often seem to play roles. Are the mechanisms the same in the several responses we have described? Surely the diversity is great enough that we might not expect a common mechanism, but in many cases there are striking similarities.

Remember the paradox introduced at the beginning of the chapter: If low temperatures reduce the rate of chemical reactions, how can we account for increased production of some growth promoter or increased destruction of an inhibitor at low temperatures compared to high ones? In the 1940s, Melchers and Lang, and Purvis and Gregory, simultaneously and independently suggested a model (Fig. 22-12) that is not unlike that of Figure 22-3. There might be two hypothetical interacting reactions, one (I) with a fairly low temperature coefficient or $Q_{10}$, the other (II) with a higher $Q_{10}$. Products of reaction I are acted upon by reaction II. If the rate of reaction I exceeds that of II, then the product ($B$) of reaction I will accumulate; if the reverse is true, the product ($C$) of reaction II will accumulate. Even if the $Q_{10}$ of reaction I is relatively low but the reaction progresses at low temperatures more rapidly than reaction II, then we can explain the accumulation of $B$ at low temperatures. With increasing temperature, the rate of reaction II increases much more rapidly than the rate of reaction I, so at some critical temperature, $B$ will be used as fast as it is produced and hence will not accumulate. Reaction II would be devernalization, and the fact that devernalization fails after two or three days at the neutral temperature might indicate a third reaction (III) that converts $B$ to $D$, a stable end product. Of course, the model is naive speculation because many other factors could play roles: enzyme synthesis, enzyme activation, membrane permeability changes, phase changes, transport of nutrients, and so on. In half a century no one has actually found such a mechanism in organisms, yet the principle has a certain logic and might still turn out to be valid.

The compensated feedback systems discussed in relation to temperature independence in the biological clock (Section 21.6) could provide for an overall reaction with a negative temperature coefficient. The *product* of one reaction might inhibit the *rate* of another. Or, at low temperatures a substance might accumulate because another compound inhibiting its production might not. Again, different temperature coefficients would be required. Because gibberellins increase in some seeds and buds as dormancy is broken, they may be equivalent to $B$ or $D$ in Figure 22-12. Or gibberellins might leak out of a storage compartment when membranes become much more permeable at low temperatures (Arias et al., 1976). In a few species, cytokinins or ethylene could play this role.

What if we are dealing with destruction of an inhibitor at low temperatures rather than the synthesis of a promoter? We have only to reverse the roles of the two hypothetical reactions in the model. The destruction (or conversion) reaction must have a fairly rapid rate at low temperatures and a low $Q_{10}$. The synthesis of the inhibitor, on the other hand, must be low at low temperatures but must have a high $Q_{10}$.

# 23

# Photoperiodism

The synchronization of organisms with seasonal time is a truly spectacular manifestation. Often, this synchronization is concerned with reproduction: It is appropriate and adaptive for young animals to be born at specific times of year, for all members of a given angiosperm species to flower at the same time (ensuring an opportunity for cross-pollination), and for mosses, ferns, conifers, and some algae to form reproductive structures in a given season. Many other plant responses, such as stem elongation, leaf growth, dormancy, formation of storage organs, leaf fall, and development of frost resistance, also occur seasonally. Frequently, these seasonal responses are synchronized by **photoperiodism** (introduced in Chapter 21). Much of what we see happening in the natural world is happening because plants and animals are able to detect the lengths of day or night or both.

## 23.1 Detecting Seasonal Time by Measuring Day Length

In a nonmountainous region on the equator, sunrise and sunset occur at the same time each day throughout the year, so the lengths of day and night remain constant. Exactly at the poles, the sun remains above the horizon for six months each year and below for the other six months. Again, the day and the night are about equal; each is six months long! As one travels from the equator toward the poles, the days become longer in summer and shorter in winter (Fig. 23-1a). This is because the equator is tipped 23.5° to the plane of the ecliptic (the earth's orbit around the sun), so during winter the pole is tipped 23.5° away from the sun and during summer 23.5° toward the sun.

The rate at which day length changes varies during the year (Fig. 23-1b). Near the times of the summer and winter solstices, when days are longest and shortest, there is little change from day to day; during spring and autumn, the rate of change is much more rapid, as days become longer during spring, shorter during fall. Thus, an organism might detect the season by measuring day and night lengths and how much they change, but because the absolute day lengths at any time of year depend so strongly upon latitude, organisms must be "calibrated" to their location.

Study of the photoperiodic responses of organisms could contribute important information to our understanding of natural ecosystems, but of the approximately 300,000 species of plants, only a few hundred have been grown with different artificial photoperiods. Not many surprising facts have appeared from this work. As one might expect, plants that grow at latitudes far from the equator respond in various ways (mostly flowering has been studied) to longer days than do plants growing closer to the equator. It was surprising, however, to learn that such tropical plants as *Kalanchoe blossfeldiana* respond to day length, detecting the small changes that occur 5 to 20° from the equator (about 1 min day$^{-1}$ at 20° latitude in March or September).

It was also interesting to learn that different ecotypes (see Section 25.4) within a single species often have different responses to day length. In three representative studies, specimens of two short-day plants, lambsquarters (*Chenopodium rubrum*; Cumming, 1969) and cocklebur (*Xanthium strumarium*; McMillan, 1974), and one long-day plant, alpine sorrel (*Oxyria digyna*; Mooney and Billings, 1961), were collected at various latitudes throughout North America. Alpine sorrel plants were also collected in the arctic and cocklebur plants from all over the world. In each case, the day length that induced flowering was longer for individuals collected farther north. Often, no morphological differences could be detected among individuals having a greatly different flowering response. Charles Olmsted (1944) even found that sideoats grama (*Bouteloua curtipendula*) had short-day strains (*ecotypes*) at the southern end of its range and long-day strains at the northern

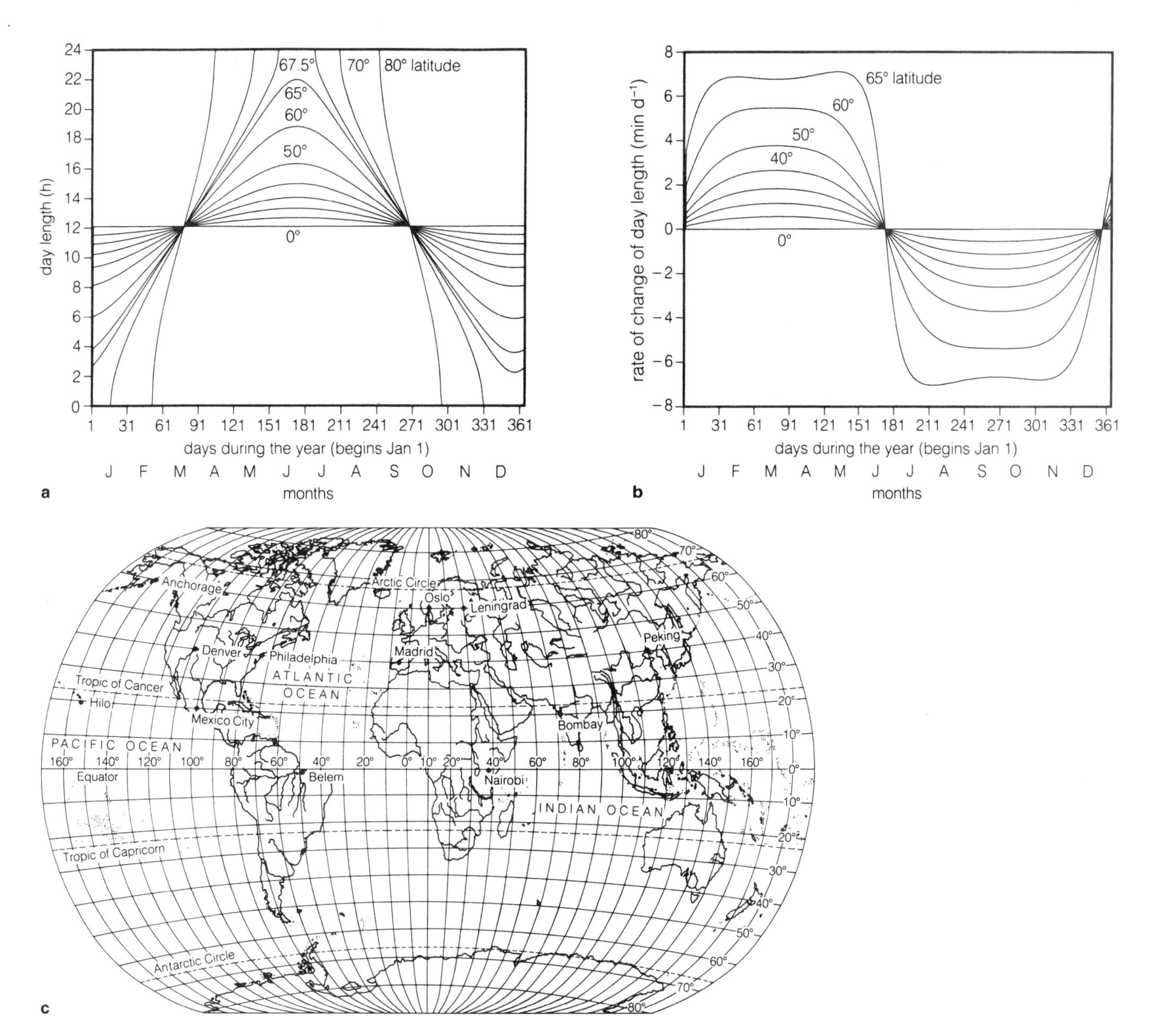


**Figure 23-1** (**a**) Day length as a function of time during the year for various northern latitudes shown on the map (**c**). Day lengths are calculated as the time from the moment the upper part of the sun (rather than its center) first touches the eastern astronomical horizon (that is, the horizon as it would appear on the open ocean) in the morning until the upper part of the sun just touches the western astronomical horizon in the evening. Thus the day length at the equator defined this way is slightly longer than 12 h. The Arctic Circle, 66.5° latitude, is where the center of the sun just touches the astronomical horizon at midnight on the summer solstice, but part of the sun is visible above the horizon at midnight for a few days before and after the summer solstice, as you can see from the day-length curve for 66.5° latitude. (**b**) The rate of change of day lengths as a function of time during the year for several latitudes. Note that the rate of change is quite steady during much of the year (that is, the curves are relatively flat), especially at far northern latitudes. (The curves were generated and drawn by computer; programs were written by Michael J. Salisbury, based on equations and explanations supplied by Roger L. Mansfield of Astronomical Data Service, 3922 Leisure Lane, Colorado Springs, CO 80917.)

end. Cultivars and varieties of many other species (for example, cotton, soybean, rice, wheat, chrysanthemum, and certain native grasses) have also been compared, although not always from a wide latitudinal range, and the same diversity has become apparent. Much of this work has been done in Scandinavia (see, for example, Bjørnseth, 1981; Hay and Heide, 1983; Junttila and Heide, 1981). In all these cases the plants flower at some appropriate time. For example, the various cocklebur species in temperate zones flower 6 to 8

**Table 23-1 Representative Species Having Various Flowering Responses to Photoperiodic Treatment.[a]**

I. Known species that flower in response to a single inductive cycle

| SHORT-DAY PLANTS (SDP) | Approximate Critical Night[b] | LONG-DAY PLANTS (LDP) | Approximate Critical Day[b] |
|---|---|---|---|
| *Chenopodium polyspermum* Goosefoot | | *Anagallis arvensis* Scarlet pimpernel | 12–12.5 |
| *Chenopodium rubrum* Red goosefoot | | *Anethum graveolens* Dill | 11 |
| *Lemna paucicostata* Duckweed | | *Anthriscus cerefolium* Salidcherril | |
| *Lemna perpusilla* Strain 6746 Duckweed | 12 | *Brassica campestris* Bird rape | |
| *Oryza sativa* cv Zuiho Rice | 12 | *Lemna gibba* Swollen duckweed | |
| *Pharbitis nil* cv Violet Japanese morning glory | 9–10 | *Lolium temulentum* Darnel ryegrass | 14–16 |
| *Wolffia microscopia* Duckweed | | *Sinapis alba* White mustard | ca. 14 |
| *Xanthium strumarium* Cocklebur | 8.3 | *Spinacia oleracea* Spinach | 13 |

II. Some species that require several cycles for induction[c]

SHORT-DAY PLANTS

1. SDP (qualitative or absolute)

   *Cattleya trianae* Orchid
   *Chrysanthemum morifolium* Chrysanthemum cultivar
   *Cosmos sulphureus* cv Yellow Cosmos
   *Glycine max* Soybean
   *Kalanchoe blossfeldiana* Kalanchoe
   *Perilla crispa* Purple common perilla
   *Zea mays* Maize or corn

   SDP at high temperature; quantitative SDP at low temperature

   *Fragaria* × *ananassa* Pine strawberry

   SDP at high temperature; day-neutral at low temperature

   *Pharbitis nil* Japanese morning glory
   *Nicotiana tabacum* Maryland Mammoth tobacco

   SDP at low temperature; day-neutral at high temperature

   *Cosmos sulphureus* cv Orange flare Cosmos

   SDP at high temperature; LDP at low temperature

   *Euphorbia pulcherrima* Poinsettia
   *Ipomea purpurea* cv Heavenly blue Morning glory

2. Quantitative SDP

   *Cannabis sativa* cv Kentucky Hemp or marijuana
   *Chrysanthemum morifolium* Chrysanthemum cultivar
   *Datura stramonium* (older plants are day-neutral) Jimsonweed datura
   *Gossypium hirsutum* Upland cotton
   *Helianthus annuus* Sunflower
   *Saccharum spontaneum* Sugarcane

   Quantitative SDP; require or are accelerated by low-temperature vernalization

   *Allium cepa* Onion
   *Chrysanthemum morifolium* Chrysanthemum cultivar

LONG-DAY PLANTS

3. LDP (qualitative or absolute)

   *Agropyron smithii* Bluestem wheatgrass
   *Arabidopsis thaliana* Mouse ear cress
   *Avena sativa*, spring strains Oats
   *Chrysanthemum maximum* Pyrenees chrysanthemum
   *Dianthus superbus* Lilac pink carnation
   *Fuchsia hybrida* cv Lord Byron Fuchsia
   *Hibiscus syriacus* Hibiscus
   *Hyoscyamus niger*, annual strain Black henbane
   *Nicotiana sylvestris* Tobacco
   *Raphanus sativus* Radish
   *Rudbeckia hirta* Black-eyed Susan
   *Sedum spectabile* Showy stonecrop

   LDP; require or accelerated by low-temperature vernalization

   *Arabidopsis thaliana*, biennial strains Mouse ear cress
   *Avena sativa*, winter strains Oats
   *Beta saccharifera* Sugar beet
   *Bromus inermis* Smooth bromegrass
   *Hordeum vulgare* Winter barley
   *Hyoscyamus niger*, biennial strain Black henbane
   *Lolium temulentum* Darnel ryegrass
   *Triticum aestivum* Winter wheat

   LDP at low temperature; quantitative LDP at high temperature

   *Beta vulgaris* Common beet

   LDP at high temperature; day-neutral at low temperature

   *Cichorium intybus* Chicory

   LDP at low temperature; day-neutral at high temperature

   *Delphinium cultorum* Florists larkspur
   *Rudbeckia bicolor* Pinewoods coneflower

   LDP; low-temperature vernalization will substitute (at least partly) for the LD requirement

   *Spinacia oleracea* cv Nobel Spinach
   *Silene armeria* Sweetwilliam silene

4. Quantitative LDP

   *Hordeum vulgare* Spring barley
   *Lolium temulentum* cv Ba 3081 Darnel ryegrass
   *Nicotiana tabacum* cv Havana A Tobacco
   *Secale cereale* Winter rye
   *Triticum aestivum* Spring wheat

   Quantitative LDP; require or accelerated by low-temperature vernalization

   *Digitalis purpurea* Foxglove
   *Pisum sativum* Late flowering garden pea
   *Secale cereale* Winter rye

   Quantitative LDP at high temperature; day-neutral at low temperature

   *Lactuca sativa* Lettuce
   *Petunia hybrida* Petunia

DUAL DAY-LENGTH PLANTS

5. Long-short-day plants

   *Aloe bulbilifera* Aloe
   *Kalanchoe laxiflora* Kalanchoe
   *Cestrum nocturnum* (at 23°C, day-neutral at >24°C) Night-blooming jasmine

6. Short-long-day plants

   *Trifolium repens* White clover

   Short-long-day plants; require or accelerated by low-temperature vernalization
   *Dactylis glomerata* Orchard grass
   *Poa pratensis* Kentucky bluegrass
   (in these plants, SD is required for induction and LD for development of inflorescence)

   Short-long-day plants; low temperature substitutes for SD effect and, after low temperature, plants respond as LDP
   *Campanula medium* Canterbury bells

INTERMEDIATE-DAY PLANTS

7. Plants flower when days are neither too short nor too long

   *Chenopodium album* Lambsquarters, goosefoot
   *Coleus hybrida* cv Autumn Coleus
   *Saccharum spontaneum* Sugarcane

AMBIPHOTOPERIODIC PLANTS

8. Plants quantitatively inhibited by intermediate day lengths

   *Chenopodium rubrum* ecotype 62° 46′ N at 25°C (responds as quantitative intermediate-day plant at 15 to 20°C and as a quantitative LDP at 30°C) Goosefoot
   *Madia elegans* Tarweed
   *Setaria verticillata* Hooked bristlegrass

DAY-NEUTRAL PLANTS

9. Day-neutral plants: These are the plants with least response to day length for flowering. They flower at about the same time under all day lengths but can be promoted by high or low temperature or by a temperature alternation.

   *Cucumis sativus* Cucumber
   *Fragaria-vesca semperflorens* European alpine strawberry
   *Gomphrina globosa* Globe amaranth
   *Gossypium hirsutum* Upland cotton
   *Helianthus annuus* Sunflower
   *Helianthus tuberosus* Jerusalem artichoke
   *Lunaria annua* Dollar plant
   *Nicotiana tabacum* Tobacco
   *Oryza sativa* Rice
   *Phaseolus vulgaris* Kidneybean
   *Pisum sativum* Garden pea
   *Zea mays* Maize or corn

   Day-neutral plants; require or accelerated by low-temperature vernalization

   *Allium cepa* Onion
   *Daucus carota* Wild carrot
   *Geum* sp. Avens
   *Lunaria annua* Dollar plant

[a]Mostly from Vince-Prue, 1975 and Salisbury, 1963b.
[b]Critical night or day often depends on conditions (for example, temperature), age of the plant, number of inductive cycles, and cultivar. Hence, some are not shown, and those that are shown are only representative.
[c]Note that single species often appear in several categories, indicating variabilities of varieties and cultivars within species. To conserve space, the lists have been greatly abbreviated.

weeks before the average killing frost in autumn, allowing time for seed ripening. As you might imagine, there are many implications of photoperiodism for agriculture (Vince-Prue and Cockshull, 1981), including control of flowering in many ornamentals and field crops (for example, sugar cane, in which flowering reduces sugar yield).

Why hasn't the role of photoperiodism in ecology been studied more intensively? Partially, perhaps, because the importance of day length in the life of a plant calls no attention to itself until the plant is moved to another latitude or to artificial conditions of light and temperature. Plants must be well adapted to the day lengths at the latitudes where they exist, or they could not exist there, so ecologists are not likely to notice the day-length response. Plant physiologists are concerned with such things, but so far they have faced the challenges of understanding the *mechanism* of photoperiodism rather than its ecological significance.

## 23.2 Some General Principles of Photoperiodism

Since the days of Tournois, Klebs, and Garner and Allard (see the box entitled "Some Early History" on page 510), well over a thousand papers describing studies on photoperiodism have been published. The most striking initial impression to be gained from this vast body of facts might be that there are no broad generalities, no

**Figure 23-2** Some representative day-neutral (tomato), short-day (lambsquarters, Japanese morning glory, and cocklebur), and long-day (henbane, radish, muskmelon, petunia, barley, and spinach) plants. Note the strong effects of day length (accentuated by the extension of day length with incandescent light, which is rich in far-red wavelengths) on vegetative form of all plants, especially tomato, in which both short-day and long-day specimens are flowering. In nearly all cases, plants exposed to long days have longer stems. (Photographs by F. B. Salisbury.)

sweeping laws to help us understand the photoperiodism response. Each species and often each cultivar or variety within a species seems to have its own features of response; probably no two respond exactly alike.

Such a situation certainly poses a challenge for a plant physiology student and for the authors of a plant physiology text! But things are not quite as hopeless as they may appear. Though every rule seems to have its exception, some generalizations can nonetheless be made: Principles of photoperiodism apply whether the process being controlled is the initiation of a soybean flower or of a female pine cone or the development of a potato tuber. In the next ten sections, we present ten generalizations. We will present experimental data based on several species, and we will note a few exceptions. In your reading, don't worry about remembering all the details — experts in the field have difficulty doing that — but let those details help you understand the generalization presented at the end of each section.

We will document only a few points with specific references to the literature. Several reviews include detailed references (see, for example, Atherton, 1987; Bernier, 1988; Bernier et al., 1981; Evans, 1969a, 1975; Halevy, 1985; Hillman, 1979; Salisbury, 1981b, 1982, 1989; Schwabe, 1971; Vince-Prue, 1975, 1989; Vince-Prue et al., 1984; and Zeevaart, 1976a, 1976b).

## 23.3 Photoperiod During a Plant's Life Cycle

Photoperiodism is a widespread phenomenon in nature. In their first paper, Garner and Allard (1920) suggested that bird migrations might be controlled by photoperiod, and soon photoperiodism in birds was demonstrated (Rowan, 1925). Since then, many animal responses to photoperiod have been documented, including several developmental changes in insects, fur (pelage) changes in mammals, and promotion of reproduction in insects, reptiles, birds, and mammals. Virtually every aspect of plant growth and development is influenced by photoperiod (Table 23-1; Vince-Prue, 1975).

### Seed Germination

To begin with, the germination of certain seeds depends on the photoperiod applied to the parent plant.

Mature seeds of certain species are also influenced in their germination by photoperiod. There are both long-day and short-day seeds for germination, and this has been shown to be a true photoperiodic effect by obtaining the long-day response with an interruption of a long dark period (that is, interruption of the night for plants on short-day cycles). Birch seeds, for example, germinated only on long days or when the long dark period was interrupted with light (Black and Wareing, 1955).

## Some Features of the Vegetative Shoot

Garner and Allard (1923) suggested that stem elongation in response to long days is probably the most widespread photoperiodic phenomenon. Hundreds of papers published since then support this observation. Stem elongation in response to long days has been observed in conifers, in which the response is often very strong (see Fig. 20-15), and also in both monocots and dicots among the flowering plants (angiosperms). In the few possible exceptions, long days lead to flowering, and flowering may terminate stem elongation. But much more commonly, plants that flower in response to long days do so by the rapid stem elongation we have called bolting. One way to avoid the complication of flowering is to observe effects of day length on stem elongation in plants that are day-neutral for flowering. Again, the elongation in response to long days is often highly apparent (see, for example, the two tomato plants in Fig. 23-2). Related effects are a suppression of branching by long days (promotion by short days) in several species and effects on **tillering** (formation at the crown of separate flowering stems) in grasses. Many temperate-zone grasses, such as barley, tiller more under short days, but rice (tropics and subtropics) tillers more under long days.

## Some Early History

Lloyd T. Evans

Julien Tournois

Martin Bopp

Georg Klebs

William J. van der Woude and Donald T. Krizek

Wightman W. Garner

William J. van der Woude and Donald T. Krizek

Henry A. Allard

If day length plays such a decisive role, why was photoperiodism not discovered sooner? To be sure, A. Henfrey had suggested in 1852 that day length might influence plant distribution, but the measurement of time by plants must have seemed unlikely to 19th-century botanists. Even the discoverers seemed to resist the ideas generated by their own data. Probably the first to realize the role of day length was Julien Tournois, who studied the flowering and sexuality of hops (*Houblon japonais*) and hemp (*Cannabis sativa*) in Paris in 1910. He noticed an extremely early flowering of his greenhouse plants in the winter, but at first he convinced himself that they were flowering in response to the decreased *quantity* of light rather than to its duration. In his third paper (Tournois, 1914) he finally grasped the point: "Precocious flowering in young plants of hemp and hops occurs when, from germination, they are exposed to very short periods of daily illumination." And: "Precocious flowering is not so much caused by shortening of the days as by lengthening of the nights." He had planned further experiments but was killed at the front in World War I.

Across the lines—in Heidelberg, Germany—Georg Klebs (1918) had probably also discovered the role of day length. He made plants of house leek (*Sempervivum funkii*) flower by exposing them to several days of continuous illumination. Klebs was convinced that nutrition controlled reproduction in plants, but in this case he felt that the additional light, which caused his plants to flower, was acting catalytically and not as a nutritional factor. But, it could be argued, because *Sempervivum* required long days to flower, the key factor was the additional photosynthesis provided by the added light. Tournois's hops and hemp, however, flowered with *less* light; so he tried lower intensities extended over a longer time to see if they would provide the same response as short durations. They did not, so the time factor seemed to be controlling. Garner and Allard (1920; see Section 21.7) followed the same line of reasoning, separating in their experiments the effects of light quantity from those of light duration.

Incidentally, nutritional factors often do play at least a quantitative role in plant reproduction. Effects of the **carbohydrate/nitrogen** ratio on flowering were being studied during and before World War I, particularly by Klebs (1904, 1910, 1918), who proposed that the ratio controlled flowering (see also Fischer, 1916). In some agricultural species, especially perennials (for example, apples and also tomato, which grows as a perennial in the tropics), too much nitrogen accelerates vegetative growth at the expense of flowering and fruiting. (Sugar yields from sugar beets are also decreased by too much nitrogen near the end of the growing season.) On the other hand, most annual field crops (for example, wheat and other cereals, maize) have greatly increased yields of fruits (seeds) in response to heavy nitrogen fertilization.

E. J. Kraus and H. R. Kraybill (1918) published a paper on "Vegetation and Reproduction with Special Reference to the Tomato," which is most often cited by American authors as the source of the carbohydrate/nitrogen hypothesis. J. Scott Cameron and Frank G. Dennis, Jr., (1986) have called our attention to the fact that Kraus and Kraybill neither provided evidence for the hypothesis in their paper nor strongly supported it in their discussion, although there is a supporting statement in their summary. In any case, application of the knowledge that too much nitrogen can reduce yields of some crops may have had as much effect in agriculture as our understanding of photoperiodism.

In spite of these many examples, long-day-induced stem elongation may not always be a true photoperiodic phenomenon. As we saw in Section 20.6, stems of many plants elongate in response to enriched far-red light compared with red wavelengths (that is, low $P_{fr}/P_{total}$), sometimes when the light treatment is applied only for a few minutes before the dark period. These responses can be independent of photoperiodism, in which an interruption of the dark period under short days gives the same response as long days, but photoperiodism often plays a role (as in Fig. 20-15).

Many features of leaves are strongly influenced by day length. Long days, for example, often promote leaf expansion and stomatal density and decrease leaf succulence, organic acids, and chlorophyll. Anthocyanin may be increased or decreased by long days, depending on species.

## Roots and Storage Organs

Rooting of cuttings has been promoted by long days, both when applied to the cutting itself and when applied to the mother plant from which the cutting was taken. Considerable study has been devoted to the formation of underground storage organs. Potato tubers are induced by short days as discussed in Chapter 22, in which the strong interaction between day length and temperature was emphasized. So-called root tubers (true roots) of cassava (*Manihot esculenta*, a tropical crop) are also promoted by short days, as are tubers and root storage organs such as those of dahlia, radish, and many other species, but bulb formation in onions is a long-day response.

## Vegetative Reproduction

Various kinds of vegetative reproduction are also influenced by photoperiod. For example, long days typically cause strawberry plants to form runners and *Bryophyllum* to form foliar plantlets on its leaf margins (Fig. 23-3). Timothy grass (*Phleum pratense*) formed such **viviparous plantlets** in 12- to 14- but not 16-h photoperiods (Junttila, 1985).

**Figure 23-3** Foliar plantlets that are produced on the leaf margins of *Bryophyllum* under long days. (Photograph by F. B. Salisbury.)

## Sexual Reproduction

In a few, not very well documented reports, reproductive organs of bryophytes formed in response to long days; other species responded to short days. Reproduction in conifers is also influenced by day length, sometimes promoted by short days and sometimes by long days. Photoperiod effects on flower induction in angiosperms is the most studied photoperiod effect. As we shall see, the situation is highly complex, with hundreds of known examples of plants in which flowering is promoted by short days, by long days, and by complex combinations of day length and temperature.

Once the flower has formed in response to day length, its further development often is also strongly influenced by photoperiod (Vince-Prue, 1975). Most typically, the same day lengths that led to the production of flowers also lead to an increased rate of floral development. But some plants have a photoperiodic requirement for initiation but are day-neutral for floral development (*Kalanchoe blossfeldiana*, *Fuchsia hybrida*) or the opposite (*Bougainvillea*, *Phaseolus vulgaris*). A strawberry hybrid (*Fragaria* x *ananassa*) is even SD[1] for initiation and LD for development, and *Callistephus chinensis* is LD for initiation and SD for floral development. Most perennial, temperate-region grasses have a dual induction requirement for flowering: primary induction of floral primordia caused by short days and/or low temperature, and secondary induction required to complete floral development and seed formation caused by long days and higher temperatures. Heide (1989) found that two grasses from the High Arctic (*Poa alpina* and *Poa alpigena*) retained these responses even though they had never experienced short days except when they were frozen. If induction was marginal, flowers took on vegetative characteristics that could lead to vegetative reproduction (called *vivipary* by Heide).

Sex expression in many species is often strongly influenced by photoperiod, but no simple relationship exists: Either femaleness or maleness can be promoted by either short days or long days, depending on species and cultivars. Different cultivars of cucumber (*Cucumis sativus*) provide good examples. It now seems clear that gibberellins produced in the leaves promote maleness and cytokinins produced in roots promote femaleness, especially in hemp and spinach (Chailakhyan and Khrianin, 1987), and this phenomenon can be influenced by photoperiod.

Many studies have investigated effects of photoperiod and other factors on seed filling, and thus yield, of agricultural plants. In a few cases, photoperiod seemed clearly to influence seed development (for example, soybeans), but the situation is usually complicated by a host of other factors. For one thing, the amount of plant

[1]We will use the following abbreviations: LD(s) = long day(s), SD(s) = short day(s), LDP(s) = long-day plant(s), and SDP(s) = short-day plant(s).

available to produce seeds is often strongly influenced by photoperiod, and this in turn influences yield. Often there is a delicate balance among the various factors. Northern cultivars of soybeans may be so sensitive to photoperiod that they give maximum yields only within a band of latitude about 80 km wide (Hamner, 1969; see also Board and Settimi, 1988). Cultivars for more southern latitudes are successful over a wider range.

### The Autumn Syndrome

As might be expected, temperate-zone plants are often influenced by the short days of autumn. Typically, the response is also strongly modified by temperature (Chapter 22). Short days often promote leaf abscission, reduced stem elongation, reduced chlorophyll production, increased formation of other pigments, dormancy, and development of frost hardiness. Annual plants typically senesce and die at the end of the growing season, often long before autumn arrives. Sometimes, this senescence is promoted by the same photoperiod that stimulates flowering, flower development, and seed filling.

### Spring Rejuvenation in Woody Perennials

We saw in Chapter 22 that the buds of woody plants often break dormancy in spring in response to the low temperatures of winter, and often this phenomenon is also promoted by long days. In a few cases (for example, birch), the long days may promote bud break even in the absence of a previous low-temperature treatment.

The conclusion of this section: *Many aspects of a plant's life cycle are influenced by photoperiod; long days almost always promote stem elongation, and short days applied to species of temperate regions induce the autumn syndrome; virtually all other plant responses may be promoted by either short days or long days, or plants may be day-neutral in response, always depending on species, cultivars, and varieties.*

## 23.4 The Response Types

Most studies of photoperiodism have emphasized the flowering process, as we will for the remainder of this chapter. Figure 23-4 summarizes some possible effects of day length in each 24-h cycle on relative flowering. In a truly day-neutral plant, which is probably rare, flowering is independent of day length, so a horizontal line appears in the figure. Flowering of long-day plants is promoted by LDs, so their curves (solid lines) slope upward to the right. SDPs take longer to flower on LDs, so their curves (dashed lines) slope downward to the right. As in vernalization (Section 22.2), there is both a **facultative** and an **absolute response** to photoperiod. Curves representing plants with an absolute day-length requirement cross the abscissa at the day length called the **critical day**. It is also possible to speak of the **critical night**: that night length that must be exceeded for flowering of SDPs or inhibition of flowering of LDPs. Actually, the facultative or the absolute response of most if not all species depends on plant age, history, growing conditions, and perhaps other factors (Bernier, 1988).

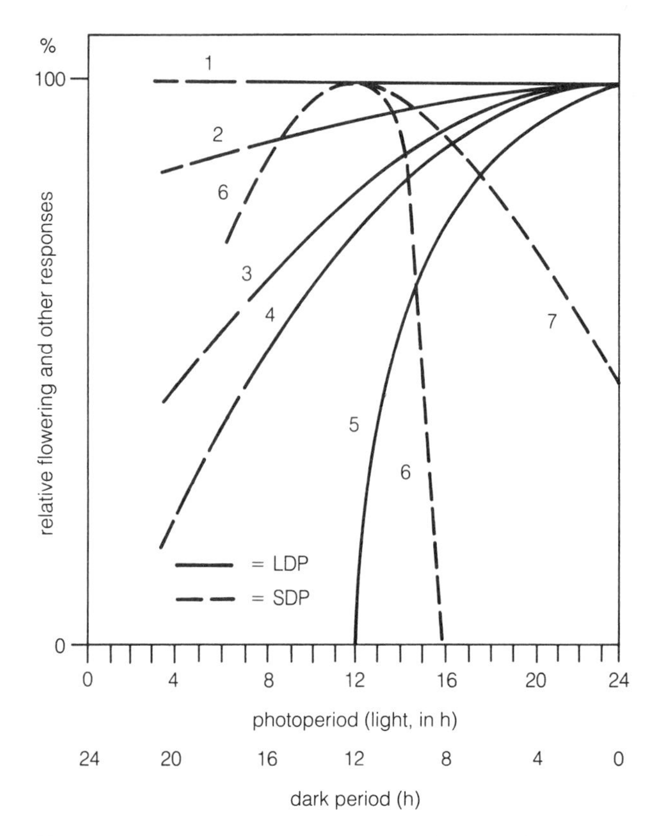


**Figure 23-4** Diagram illustrating flowering (and other) responses to various day lengths. Flowering can be measured in various ways, such as counting the number of flowers on each plant, classifying the size of the buds according to a series of arbitrary stages (see Fig. 23-6), or taking the inverse of the number of days until the first flower appears. Curve 1: A truly day-neutral plant, flowering about the same at all day lengths; such a response is probably very rare. Curves 2, 3, and 4: Plants quantitatively promoted in their flowering (or other responses) by increasing day lengths. The three curves represent three species that are promoted to different degrees. Curve 5: Qualitative or absolute long-day plant such as henbane; this example flowers only when days are longer than 12 h. Curve 6: Qualitative short-day plant such as cocklebur; this example flowers only when days are shorter than 15.7 h and nights are longer than 8.3 h. Note that cocklebur also fails to flower if days are shorter than about 5 h but flowers as days get longer (a typical long-day response). (With many species, there is little or no flowering when days are unusually short.) Curve 7: Quantitative short-day plant, which flowers on any day length but better under short days. Note that other species, not shown here, have different critical day and night lengths, not just the 12-h and 15.7-h days shown for henbane and cocklebur.

One other complication is apparent. SDPs often will not flower if the days are *too* short. Note that this minimum light requirement in SDPs is represented by a curve that slopes upward to the right in Figure 23-4. Such a curve expresses the LD response, so SDPs act as LDPs when the days are extremely short!

This brief day-length requirement is for more than just photosynthesis, although photosynthesis is obviously required to produce the plant, or the seed from which the plant came, in the first place. Flowering of the short-day succulent *Kalanchoe blossfeldiana* can be induced with only one second of red light each day, and some SDPs that can be kept indefinitely in the dark by being fed sucrose nevertheless show a minimum red-light requirement for the photoperiodism response. *Thus the photoperiodism response seems to require some minimal amount of $P_{fr}$ each day* (Vince-Prue, 1983, 1989). This is illustrated by the following experiment: When the day was shorter than 5 h, far-red light given at the end of the day (reducing the $P_{fr}$ to a low level) inhibited induction of the SDP *Pharbitis nil* (Japanese morning glory). In that case, $P_{fr}$ had to be present well into the dark period for flowering to occur, but if the day was longer than 5 h, end-of-day far-red light did not inhibit flowering, showing that $P_{fr}$ had apparently accomplished whatever it did during the light period. Thus this requirement for a minimal amount of $P_{fr}$ does not seem to be related to time measurement or to the specific events that occur during light or darkness.

Ireland and Schwabe (1982a, 1982b) established that the SDPs *Xanthium* and *Kalanchoe* require $CO_2$ during the light period for photoperiodic induction, but only certain photosynthesis inhibitors duplicated the inhibition caused by lack of $CO_2$. They concluded that photoperiodic induction in *Xanthium* required some product of $CO_2$ fixation and that a step in the photosystem II electron-transport pathway inhibited by DCMU (diuron) might be crucial.

The SD and LD responses are completely opposite. Curve 6 in Figure 23-4 could represent cocklebur (*Xanthium strumarium*), a classical SDP; curve 5 could represent henbane (*Hyoscyamus niger*), a classical LDP. *Both* flower when days are about 12.5 to 15.7 h long (nights 11.5 to 8.3 h).

Sometimes the critical day or night can be determined for a population of plants within rather narrow limits (for example, within 5 to 10 min). In other cases, the limits are far less exact and may spread out over one or more hours. With cocklebur, the greater the number of inductive cycles, the more exact is the critical day or night.

Still other ways in which species differ is in their ripeness to respond (see below) and in the number of SD or LD cycles required to induce flowering. Species that require only a single inductive cycle for flowering (Table 23-1) have been widely used because one can perform experimental treatments (for example, application of a chemical) at various times in relation to the single cycle without the complications of effects on previous or subsequent cycles.

Representative LD, SD, and day-neutral plants with some critical day lengths are shown in Figure 23-2 and listed in Table 23-1. The three response types observed by Garner and Allard form the foundation for any classification of photoperiod response types. Since their time, a few other categories have been discovered. To produce flowers, a few species (item 5, Table 23-1) require LDs followed by SDs, as occurs in late summer and fall. When maintained either under continuous LDs or continuous SDs, they remain vegetative. The counterparts to these **long-short-day plants**, the **short-long-day plants** (item 6), require SDs followed by LDs, as in spring. There are a few species, at least (item 7), that flower only on **intermediate day lengths** and remain vegetative when days are either too short or too long. They have counterparts that remain vegetative on intermediate day length, flowering only on longer or shorter days (item 8).

There are several interesting interactions between photoperiod and temperature. We have seen that a vernalization requirement is often followed by a requirement for LDs (Section 22.2), as occurs during the LDs of late spring following winter. In other cases, a plant may exhibit a given response type at one temperature but not at another. It may, for example, have a qualitative or quantitative SD response at temperatures above, say, 20°C but be essentially day-neutral at cooler temperatures. Two examples of plants that are absolute SDPs at high temperature and absolute LDPs at low temperature have been reported: poinsettia (*Euphorbia pulcherrima*) and morning glory (*Ipomea purpurea*). They are day-neutral only at the intermediate temperature. *Silene armeria*, a LDP, is induced on SDs at either high (32°C) or low (5°C) temperatures, but only to low levels of flowering. Specific day-length requirements at certain temperatures prove to be quite common.

Vernalization and a given photoperiodic treatment are sometimes interchangeable. For example, in a variety of Canterbury bells (*Campunula medium*), vernalization is fully replaced by SDs, but LDs are required following either treatment. Indeed, many short-long-day plants flower after a cold treatment followed by long days (Evans, 1987). We noted above that most perennial, temperate-region grasses have a dual induction requirement for flowering, with short days and/or low temperature followed by long days and higher temperatures required to complete induction (Heide, 1989). Still other complications and interactions are known. For example, *Pharbitis nil* (probably should be called *Ipomea nil*) has become a prototype SDP, yet it can be induced to low levels of flowering under LDs by low temperatures, high light levels, treatment with growth retardants, removal of roots, and low nutrient levels. Bernier (1988) suggested that there are alternate pathways to flowering in all species. Surely such a diversity of response types is of ecological importance.

Day-length sensitivity can be governed by a single gene, as in Maryland Mammoth tobacco and a mutant of *Arabidopsis*, or several genes may be involved, as in sorghum and wheat (review by Bernier, 1988). Response to day length may be either dominant or recessive, again depending on species. Both single/multiple and dominant/recessive genes can govern vernalization as well as photoperiod. In the case of wheat, genes controlling day length and cold sensitivity proved to be independent of each other, probably controlling different component processes of flowering. The genes that control the flowering response often influence other aspects of plant growth as well. Examples are the flowering genes in wheat, which also influence stem height and tillering.

Molecular genetics (Chapter 24) holds considerable promise as a means to unravel the steps in the flowering process (Law and Scarth, 1984). The first goal is to identify genes that control flowering (and several have been identified). Then, the genes might be cloned and the proteins they code for synthesized and studied. Eventually, it should be possible to understand the chemical reactions that are controlled by those proteins/enzymes.

There undoubtedly are numerous photoperiod-response types with respect to plant responses other than flowering. Most remain to be studied, but examples are known, such as that of tossa jute (*Corchorus olitorius*), in which stem elongation in response to LDs occurred only above 24°C; below that temperature the plant was day-neutral with respect to stem elongation (Bose, 1974).

Our concluding generalization is simply stated: *Although the basic response types are short-day, long-day, and day-neutral plants, there is a wide diversity of photoperiodic response types and much plasticity of response.*

## 23.5 Ripeness to Respond (Competence)

Only a few plants respond to photoperiod when they are small seedlings. *Pharbitis* (Japanese morning glory) responds to SDs in the cotyledonary stage, and some species of goosefoot or lambsquarters (*Chenopodium* spp.; Cumming, 1959) respond and flower as minute seedlings. In laboratory studies they can be grown on filter paper in a Petri dish. Most species, such as the cocklebur, must attain a somewhat larger size; cotyledons do not respond. Henbane must be 10 to 30 days old before it will respond to LDs. Certain monocarpic bamboo species and several polycarpic trees will not flower until they are 5 to 40 or more years old, but it is not known whether or not they then respond to photoperiod (see Section 16.2). Klebs called the condition a plant must achieve before it will flower in response to the environment *Blühreife*, which translates as **ripeness to flower**; but a more descriptive term is **ripeness to respond**. In many species the number of required photoperiodic cycles decreases as the plant gets older; that is, ripeness to respond increases with age. Often the plant finally flowers independently of the photoperiod; it becomes day-neutral. On the other hand, leaves of the scarlet pimpernel (*Anagallis arvensis*) are most sensitive to LDs when the plant is a seedling; sensitivity actually declines in leaves that are most recently produced, so the plant becomes more difficult to induce as it gets older.

Individual leaves must also reach a ripeness to respond. In some species, the leaf is maximally sensitive when it is first mature (fully expanded), but it is the half-expanded cocklebur leaf, the one growing most rapidly, that is most sensitive; leaves less than 10 mm long will not respond.

The only way to know whether induction has occurred is to observe the change at the meristem from vegetative to reproductive growth—that is, to observe the formation of flowers. This meristematic change is called **evocation** (Evans, 1969b). Ripeness to respond might depend on the status of the leaves (that is, on their ability to respond to photoperiod as just discussed) or on the ability of the meristem to undergo evocation. If evocation can occur, the meristem is said to be **competent** (it has achieved a condition called **competence**). A test for competence is to graft a meristem onto a flowering plant of the same species. If it flowers, it was competent; if not, it may not have achieved competence. In general, young woody meristems are not competent, but herbaceous meristems of any age are. As usual, however, there are many exceptions (Bernier, 1988).

The concept of ripeness to respond is identical to that of *juvenility*, defined as the condition of a plant before it is mature enough to flower (Section 16.5). Another related concept is that of **minimum leaf number**, which is the minimum number of leaves the plant produces from seedling to earliest flower under the most ideal conditions for flowering.

The conclusion of this section: *Before a plant can flower in response to its environment (particularly day length and temperature), the leaves that detect the environmental change (meristems in vernalization) must reach a condition called ripeness to respond, and meristems must be competent to respond to the stimulus from the leaves. There is a great diversity among species and plant organs in the age at which they achieve these conditions.*

## 23.6 Phytochrome and the Role of the Dark Period

In the 1930s, Karl C. Hamner and James Bonner (1938) at the University of Chicago studied photoperiodic induction of cocklebur. They asked which was more important, day or night? In one experimental approach, days and nights were varied to give cycles that did not equal 24 hours. The critical night, but not the critical day, remained constant, indicating the importance of the dark period. In another approach, days were interrupted with darkness or nights with light. Interruption of the day with darkness had little or no effect, but interruption of the night with light inhibited flowering of SDPs and (in later experiments) promoted flowering of LDPs (Fig. 23-5). This was the discovery of the **night break phenomenon**.

Once it was known that a period of light during the dark period would nullify the effect of darkness, several possibilities for experimentation immediately became apparent. Researchers could ask: Which is more important in the night break, the level of light used (its irradiance) or the total quantity of light energy (fluence, as calculated by multiplying irradiance by the time interval over which it is applied)? Within rough limits, the total quantity of energy proves to be the determining factor (that is, *reciprocity* applies; see Section 19.4). Researchers could then ask: How dark is dark? Light applied even at very low irradiances during the entire dark period is effective (especially in inhibiting flowering of SDPs). In cocklebur, for example, it is effective at 3 to 10 times the irradiance from the full moon, or roughly 0.00001 to 0.0003 that of sunlight.

When during the dark period is light most effective? Usually at some constant time after the beginning of an inductive dark period for SDPs or an inhibitory dark period for LDPs (Fig. 23-5). This most effective time is often equivalent to the critical night.

Which wavelengths of light are most effective? In the early 1940s it became apparent that red light was considerably more effective than other wavelengths. Action spectra for inhibition in SDPs and promotion in LDPs are typical of those for other phytochrome responses (see Fig. 20-3). Thus in the early 1950s, immediately after far-red reversibility was discovered in germination of lettuce seed, cocklebur plants were irradiated in the middle of a long inductive dark period with red light followed by far-red light. If the far-red light followed the red irradiation immediately, plants flowered; if about 30 min were allowed to elapse between the red and the far-red exposures, the far-red light no longer nullified the effects of the red light. Apparently, $P_{fr}$ completes its inhibitory act within 30 min in cocklebur leaves.

Let's state a preliminary conclusion: *The dark period plays an important role in the photoperiodic response because a night break inhibits flowering of short-day plants and promotes flowering of long-day plants. Phytochrome apparently detects the light, and its effectiveness depends upon the time of irradiation.*

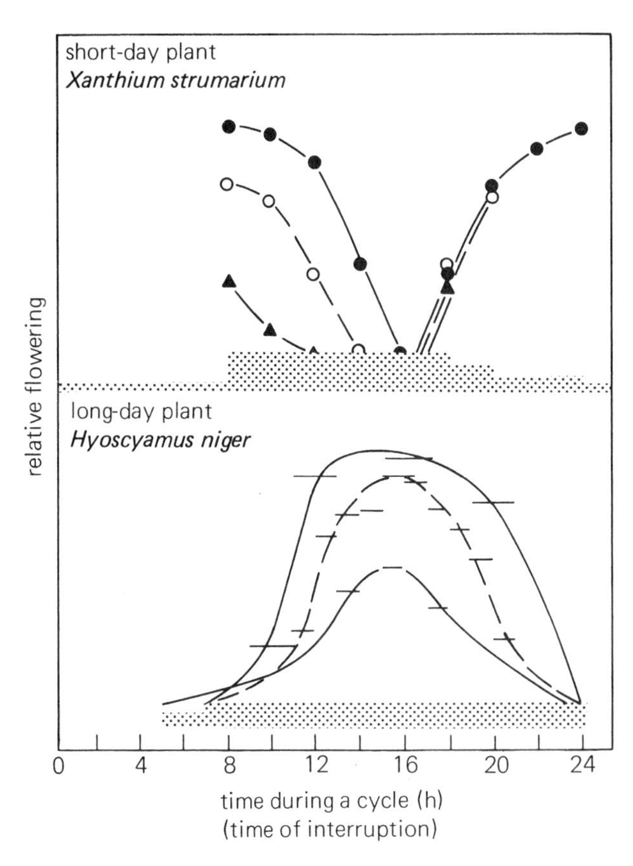


**Figure 23-5** Effects of a light interruption given at various times during dark periods of various lengths (shaded bars) on subsequent flowering of an SDP and an LDP. Interruptions for *Xanthium* (cocklebur) were 60 s; with *Hyoscyamus* (henbane), durations of interruptions are indicated by lengths of data lines. With SDPs, a night break inhibits flowering; with LDPs, such a break promotes flowering. With the cocklebur, a night break 8 h after the beginning of darkness inhibited flowering completely regardless of how long the dark period was. (Data for *Xanthium* from Salisbury and Bonner, 1956; those for *Hyoscyamus* from Claes and Lang, 1947.)

Now we shall examine some complications. LDPs are often less sensitive and somewhat more quantitative in their response to a night break than are SDPs. Using four photoflood lamps, for example, flowering is completely inhibited in cocklebur plants by a few seconds of light given about 8 h after the beginning of an inductive dark period. With many LDPs, however, flowering continues to be promoted as the duration of a night break (using comparable high levels of light) increases from seconds to hours. Furthermore, whereas red light is most effective in a night break with SDPs or a brief night break with LDPs, a mixture of red and far-red wavelengths can be more effective with LDPs when it is applied as a long night break, as a day-length extension, or during the continuous light that best induces flowering in most LDPs.

We noted above that far-red light given at the beginning of a dark period could inhibit induction in *Pharbitis* if the day was shorter than 5 h, and from that (and many other experiments not discussed) it has been concluded that SDPs require some level of $P_{fr}$ at some time during the day or night. Yet red light (which produces $P_{fr}$) is clearly most effective at *inhibiting* induction of SDPs when it is given during the dark period. About 80 percent of the total phytochrome is in the $P_{fr}$ form at the end of a day that consists of fluorescent or red light (60 percent following sunlight), but after a few hours of darkness red light (which produces $P_{fr}$) is highly effective at inhibiting induction of SDPs. Hence, we must conclude that the $P_{fr}$ initially present at the beginning of darkness soon disappears. $P_{fr}$ *produced by red light at the proper time during darkness* is highly effective at inhibiting induction (see Fig. 23-5), whereas the positive role of $P_{fr}$ does not seem to be related to specific events in the induction cycle.

Based on these observations, it has been postulated that phytochrome must exist in *two* forms in SDPs. One of them is necessary for induction of SDPs and is highly stable in darkness (although we will not review the evidence for stability here; see Vince-Prue, 1989). The other is highly labile in darkness but inhibits induction of SDPs when it is produced at the proper time. Interestingly enough, as we noted in Chapter 20, the use of antibodies raised against the phytochrome apoprotein has revealed that there are at least two distinct types of phytochrome present in plants (Furuya, 1989; Jordan et al., 1986). It remains to be seen whether these two forms of phytochrome will turn out to be the two forms that have been postulated as described above, but some workers (for example, Rombach, 1986) suggest that this might be likely.

It would help considerably if we could measure the various forms of phytochrome in leaves at different times during photoperiodic induction. So far, it has been difficult to make these measurements in green tissues. One approach is to treat plants with an herbicide (called Norflurazon or Zorial) that inhibits chlorophyll and carotenoid formation in white light (Jabben and Deitzer, 1979). The nearly colorless seedlings were supplied with sucrose through their roots to keep them alive. Flowering was induced in the LD Wintex barley by LDs, and several phytochrome responses were shown to be the same in these colorless plants as in normal dark- and light-grown controls. Phytochrome was measured *in vivo* (in intact shoot tips) with a spectrophotometer before and during destruction caused by light. Levels were identical in etiolated seedlings with or without herbicide treatment.

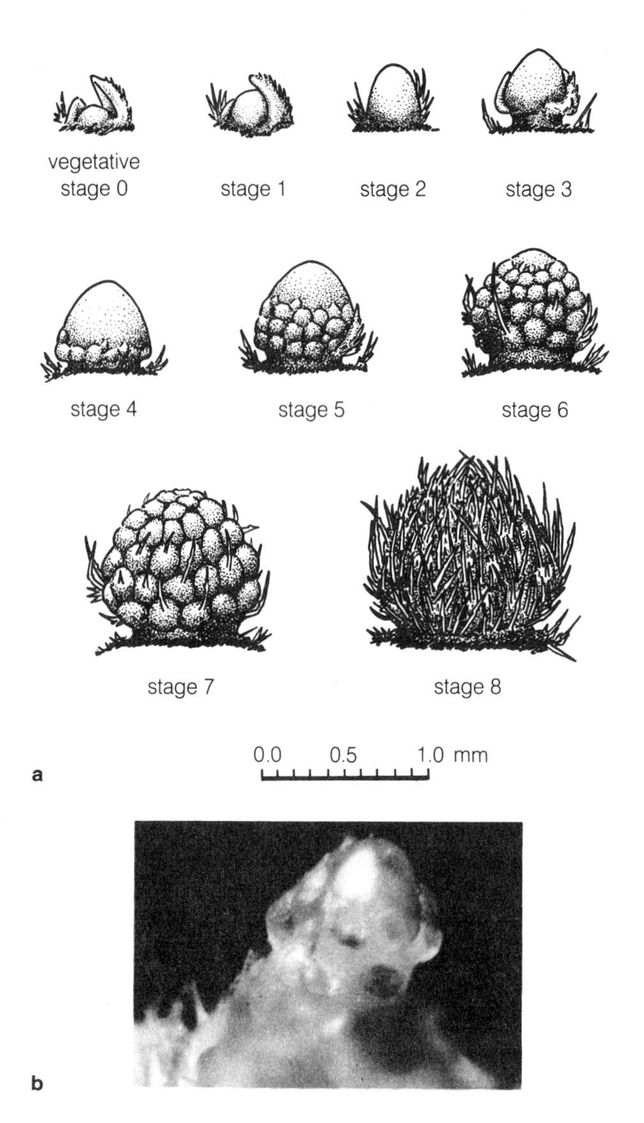


**Figure 23-6** (**a**) Drawings of the developing terminal (staminate) inflorescence primordium of cocklebur, illustrating the system of floral stages devised by Salisbury (1955). (**b**) Photograph through a dissecting microscope of a cocklebur inflorescence primordium at stage 3. (Photo by F. B. Salisbury.)

In Chapter 20 we noted that the concentration of $P_{fr}$ may drop (and it is apparently the *concentration* that is important in photoperiodism) either by *destruction* or by *reversion* to $P_r$. In the case of reversion, there is no loss in total phytochrome. Reversion can be quite rapid: 19 min at 22°C for conversion of half of the $P_{fr}$ in *Pharbitis* (Rombach, 1986), for example. This could account for the sensitivity of several species of SDPs to red light as soon as an hour after the beginning of darkness. As yet, however, reversion has not been shown to occur in grasses or members of the order Caryophyllales (also called Centrospermae).

high enough to influence flowering in the middle of a dark period, even though sensitivity to light at that time increases by about an order of magnitude compared with dusk. Furthermore, the full moon is low in the sky at temperate latitudes[2] (for example, low in the southern sky in the northern hemisphere), so its rays do not strike the plant from directly above but rather at a low angle. Nevertheless, some experiments have shown a slight photoperiod response to moonlight (von Gaertner and Braunroth, 1935; Kadman-Zahavi and Peiper, 1987).

Our conclusion: *Plants respond sharply to the light changes at dawn and dusk, "ignoring" the changes in irradiance during both day and night, although there might be slight responses to moonlight.*

## 23.9 The Florigen Concept: Flowering Hormones and Inhibitors

Not long after photoperiodism was discovered, workers around the world wondered which part of the plant detected day length. It was soon apparent that the leaf responded. Using a SDP, for example, an experimenter could enclose the leaf for 16 h in a black paper envelope while leaving the rest of the plant under LDs or continuous light. Such a treatment soon induced flowering. A similar experiment with an LDP prevented flowering. Covering the bud but not the leaves under LDs did not lead to flowering in SDPs but did in LDPs.[3]

If the leaf detects the photoperiod but the bud becomes the flower, there must be some stimulus transmitted from the leaf to the bud. In the 1930s, Mikhail Chailakhyan in the Soviet Union (see his 1968 review) grafted induced plants to noninduced plants held under noninducing day lengths and observed that the flowering stimulus would cross a graft union, causing the noninduced plant to flower. He suggested that the stimulus was a chemical substance, a hormone as opposed to some electrical or nervous stimulus. Chailakhyan named the hypothetical stimulus **florigen** (Latin, *flora*, "flower" and Greek, *genno*, "to beget").

The grafting studies have provided two important bits of information: Florigen moves only through a living tissue union between the two graft partners and probably only through phloem tissue. Furthermore, florigen frequently seems to move with the assimilate stream. If the receptor partner is defoliated or held under low irradiance levels, movement of assimilates into

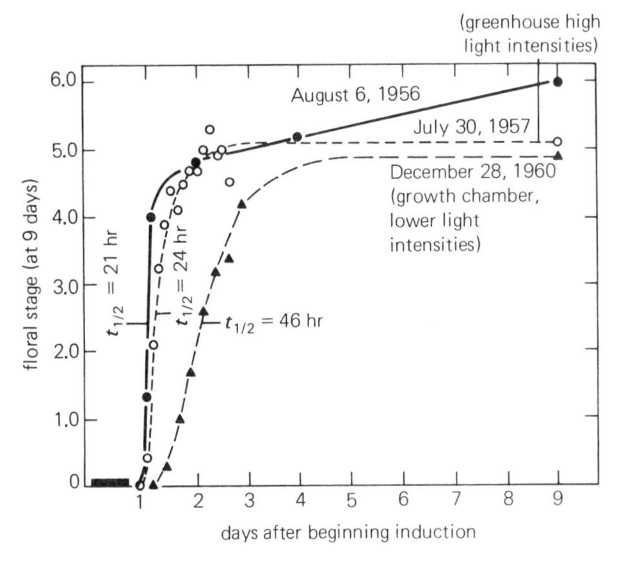


**Figure 23-15** Three translocation curves obtained by defoliation of *Xanthium* plants at various times following a 16-h inductive dark period. Floral stages (determined on the ninth day) are shown as a function of the times when leaves were cut off. (Actually, plants were defoliated to a single, highly sensitive leaf before the long dark period; this leaf was then removed after the inductive dark period at the times shown.) Numbers on the abscissa represent noon of the indicated day; the bar on the abscissa suggests the inductive dark period. Approximate times after beginning of the dark period when half the stimulus was out of the leaf are indicated by $t_{1/2}$. Dates refer to the day plants were subjected to the dark treatment. Plants represented by the $t_{1/2}$ = 46 h curve were kept under about 400 μmol $m^{-2}$ $s^{-1}$ of fluorescent light in growth chambers (23°C); other plants were in a greenhouse. (From Salisbury, 1963a.)

the receptor is promoted, and so is movement of florigen. On the other hand, in some species florigen is exported from extremely young leaves that would be expected to be importing assimilates, so florigen might move by other mechanisms in addition to the assimilate stream.

Many different varieties or species representing different response types have been grafted to each other (see Lang, 1965; Vince-Prue, 1975; Zeevaart, 1982). Florigen produced by one response type will often induce flowering in another type; an induced SDP grafted to an LDP in SDs will induce the LDP to flower, for example. Apparently, although florigen is produced in response to widely different environmental conditions, it is the same compound, or at least physiologically equivalent, in many if not all angiosperms.

A third experiment that is difficult to interpret in any way but by the florigen concept was first performed in the early 1950s. Plants from several species that require only a single inductive cycle were defoliated at various times after that cycle, and the level of flowering some time later was plotted as a function of the time of defoliation. Results with cocklebur are shown in Figure 23-15; the SD Japanese morning glory and the LD pe-

[2] The sun is higher in summer, which causes the higher summer temperatures, but the full moon is on the opposite side of the earth from the sun; hence, it is low.

[3] Actually, green stems of some species will respond to photoperiod if they are given a sufficient number of inductive cycles.

rennial ryegrass have also been studied. Apparently, when the leaves are cut off immediately following the inductive long dark period (SDPs) or long light period (LDPs), plants remain vegetative because the hormone has not yet been exported from the leaves. Export is complete some hours later, however, because defoliated plants subsequently flower almost as well as plants from which leaves were not removed.

There is also strong evidence for inhibitory substances or processes. Indeed, both promoters and inhibitors must influence flowering. If a plant is induced by exposing only one leaf to suitable photoperiodic conditions, flowering is frequently inhibited by the noninduced leaves. The presence of leaves on a receptor plant in a graft experiment also reduces receptor flowering. These inhibitory effects are often caused by influences on the translocation of assimilates. For example, a long-day leaf growing between an induced short-day leaf and the bud may be exporting assimilates directly to the bud and thereby effectively blocking movement of assimilates — and florigen — from the induced leaf to the bud. Many inhibitor experiments can be understood in this way, and often the explanations are supported by tracer experiments.

Some effects cannot be explained on the basis of translocation, however. If the whole explanation were photosynthesis and transport of assimilates, then light levels should be important. Yet a brief night interruption at low irradiance can produce a typical long-day inhibitory effect in a single leaf on a short-day plant (Gibby and Salisbury, 1971; King and Zeevaart, 1973). In some species, flowering seems to be repressed under noninductive conditions strictly by inhibitors coming from the leaves. Such plants will flower when defoliated. Examples are the LD henbane and various SD cultivars of strawberry (*Fragaria* sp.). In some species (for example, *Xanthium* and *Pharbitis*) promoters may be dominant, although modified by inhibitors; in others there seems to be a true balance (darnel ryegrass); in still others (henbane and strawberry cultivars) inhibitors may be dominant but modified by promoters.

Chailakhyan and I. A. Frolova in Moscow, cooperating with Anton Lang of Michigan State University (Lang et al., 1977), grafted plants of both SD and LD cultivars of tobacco onto a day-neutral (DN) cultivar and then grew them under various light conditions. When exposed to SDs, the SD graft partners caused the DNPs to flower earlier, and under LDs, LD partners also promoted earlier flowering of DN partners. These results again confirm the presence of a positive-acting flowering hormone. When the SD graft partners were maintained under LDs, flowering in the DN species was slightly or not at all retarded, but under SDs, the LD tobacco completely prevented flowering of the DN partner. Here, in one experiment, is clear-cut evidence both for flowering promoters produced under favorable day lengths in both a LDP and a SDP and for an inhibitor produced under unfavorable day lengths in a LDP (and possibly in a SDP).

We need to isolate and identify florigen(s) and flower inhibitor(s). So far, this has not been successful, partially because of the lack of a reliable bioassay. Numerous solvents and application techniques have been tried to obtain extracts from induced plants and apply the extracts or fractions to noninduced plants. Sporadic successes have been reported, but it has never been possible to produce completely reproducible results. Cleland (1978) and Salisbury (1982) have recounted the stories of these efforts. (A few relevant papers include Cleland and Ajami, 1974; Carr, 1967; Hodson and Hamner, 1970; and Lincoln et al., 1966.)

Chailakhyan and Lozhnikova (1985; review in Bernier, 1988) reported that ethanolic extracts from leaves of SD and LD tobaccos, both grown in SDs, consistently caused flower formation in seedlings of the SDP *Chenopodium* kept in LDs, and extracts from the two tobaccos grown in LDs elicited flowering in the LDP *Rudbeckia* kept in SDs. So far, the meaning of these experiments is not clear.

Failure to isolate florigen has led to skepticism about its existence. For example, Sachs (1978) and Sachs and Hackett (1983; see also Bernier, 1988; Bernier et al., 1981, Vol. II) questioned the florigen concept and suggested that photoperiodic induction causes a diversion of nutrients (sucrose and so on) within the plant, leading to floral initiation. Nutrients directed toward the bud might cause evocation. Inhibitory effects might be caused by a diversion to sinks other than meristems that can become flowers. Sachs and Hackett noted, for example, that high irradiances (presumably leading to high rates of photosynthesis) can sometimes override and replace photoperiodic signals. Flowering was correlated with soluble-solid percentages in *Bougainvillea* (Ramina et al., 1979) and was promoted by added sucrose in the quantitative LDP *Brassica campestris* grown in sterile culture (Friend et al., 1984). A $GA_{4/7}$ mixture strongly promoted "flowering" in *Pinus radiata* and caused a significant reallocation of dry matter within the terminal buds to developing long-shoot primordia (potential seed-cone buds; Ross et al., 1984). Thus, it seems clear that transport patterns can change during photoperiodic induction as Sachs and Hackett emphasized, but it is not clear that this *causes* flowering rather than being a result of it. The defoliation and grafting experiments outlined previously are especially difficult to explain by a nutrient-diversion hypothesis.

The lack of a florigen bioassay is a serious problem. When extracts are applied to plants to see if they have florigenic activity, we wonder whether the active material is able to penetrate the plants and move to the bud. Perhaps its effects are nullified by the presence of inhibitors produced in the test plants in noninductive conditions. Some workers have tried to overcome these problems by applying these extracts directly to the bud itself. Again, however, successes have been only rather minor.

So far, few attempts have been made to extract and isolate floral inhibitors. Furthermore, an inhibitory effect might be a process rather than a substance. For example, LD leaves on an SDP might in some way absorb and destroy promoting substances produced in SD leaves.

Because attempts to extract promoters and inhibitors have been so disappointing, other more indirect approaches have been taken. Experimenters have added various antimetabolites to LDPs and SDPs, seeking those that apparently inhibit the flowering process in a specific way. For example, a compound may be effective only when applied during florigen synthesis. Again, results have not been promising. It appears that respiration and protein and nucleic acid synthesis are essential for a leaf to synthesize florigen, but then these processes are involved in virtually all aspects of the life of a plant. Before we are carried further afield, let us state the conclusion for this section: *There is much circumstantial evidence that flower initiation is controlled or strongly influenced by hormones: one or more positively acting florigens and one or more negatively acting inhibitors. These substances remain to be identified.*

## 23.10 Responses to Applied Plant Hormones and Growth Regulators

Because plant hormones and growth regulators can influence virtually every aspect of plant growth and development, it is logical to investigate their effects on flowering. Many compounds that will induce or inhibit flowering in one species or another when applied at appropriate concentrations are now known. (Sometimes, a compound will inhibit at one concentration and promote at another.) There is an important potential for practical application of this knowledge because the induction of flowers often plays such an important role in agriculture. Work with hormones and growth regulators could also lead to better understanding of the flowering process. Again, however, there are about as many exceptions as rules, and the possibility that induction might change the sensitivity or responsiveness of the plant to growth regulators (Chapters 17 and 18) has hardly been examined. Let's summarize a few tentative observations.

### Auxins and Ethylene

In many species, auxins inhibit flowering. In SDPs, inhibition occurs before translocation of florigen from the leaf is complete, after which there may be marginal promotive effects. Promotions have also been observed in LDPs held under days just too short for induction. Auxin concentrations required to inhibit flowering usually produce severe epinasties and other responses, and measured plant auxin levels seldom correlate with flowering in any meaningful way. Hence, although endogenous auxins may influence flowering, they probably do not control it, at least not in all species.

Auxins clearly cause flowering in some bromeliads, including the pineapple. In pineapple and cocklebur, applied auxins cause production of ethylene, which itself influences flowering in the same way as auxins (inhibition in SDPs, promotion in bromeliads). In the bromeliads, IAA is relatively ineffective because it is apparently broken down by the plant's enzymes. Thus synthetic auxins such as NAA or 2,4-D (see Figure 17-4) must be used. In *Guzmania*, another bromeliad, ethylene or ACC (ethylene precursor; Chapter 18) treatment or shaking the plants for 15 s, which presumably produces ethylene (Chapter 19), causes flowering (De Proft et al., 1985). Treatment with AVG (which inhibits ethylene synthesis; Figure 18-10) prevents ethylene release and flowering, but not in plants treated with ACC (which releases ethylene in spite of AVG). These results suggest that ethylene is the only controlling factor of flowering in *Guzmania* and perhaps other bromeliads. Juvenile plants do not respond to these ethylene treatments, suggesting, perhaps, that such plants are not yet sensitive to ethylene as far as flowering is concerned.

### Gibberellins

Shortly after the gibberellins (GAs) became available, it was found that $GA_3$ (then the most available GA) could substitute for the cold requirement of several species that require vernalization and also for the LD requirement of several LDPs (see Fig. 22-8). There were some important exceptions among LDPs (for example, *Scrofularia hyecium* and *Melandrium* sp.), and gibberellins normally failed to replace SDs in SDPs, although again there were a few exceptions (for example, cosmos and rice). $GA_3$ promoted flowering in the SDP *Pharbitis nil* when it was applied early in an inductive dark period but not when it was applied later (Ogawa, 1981), a result opposite to that obtained with cocklebur (Greulach and Haesloop, 1958; Salisbury, 1959). In many species gibberellins increased in stems and leaves under LDs, so it seemed reasonable to assume that gibberellins might account for at least part of the flowering requirement (a florigen complex?) in LDPs and also for the common stem elongation under LDs.

The situation is not simple, however. In LDPs, for example, LDs increase both the rate of synthesis and the rate of destruction of GAs, and many studies found no correlation between extractable GAs and flowering.

# Gibberellins, a Fascinating and Highly Diverse Class of Plant Hormones

*Richard P. Pharis*

*Dick Pharis is a Canadian plant physiologist who got his early training in the United States but escaped in 1975 to the Canadian Rockies. His interests in gibberellins have led him into research on the physiology of flowering in conifers and herbaceous and woody angiosperms and also into the physiological basis for hybrid vigor and inherently superior growth of trees. He is now a Professor of Botany at the University of Calgary.*

I approached plant hormone physiology in a rather circuitous way. As a teenager I grew up spending virtually every weekend and most of the summer backpacking and fishing in the Cascade or Olympic mountains of Washington State. These wilderness travels stimulated an interest in forestry, and I took my Bachelor of Science degree in forestry at the University of Washington. There I was stimulated by contact with David Scott, Dick Walker, and Daniel Stuntz. A master's degree with Frank Woods in ecological physiology (silvics) at Duke University followed, and work over the next two years in my doctoral program with two exceptional plant physiologists, Paul Kramer and Aubrey Naylor, told me that plant physiology was to become my first love.

A first position as a plant physiologist with the United States Department of Agriculture (Forest Service) in Roseburg, Oregon, taught me that field ecophysiologists didn't solve problems; they just delineated and described them [but see Chapter 25!]. Budget cuts delayed laboratory construction, so I accepted an offer from Henry Hellmers and James Bonner at the California Institute of Technology to work on the photosynthetic efficiency of conifers. It was there that I "fell in love" with gibberellins!

My two years, 1963 and 1964, at Caltech were still "exciting times." (The team would be dismantled a few years later.) Seminars, chats over coffee, and wine and cheese on Fridays with James's molecular-biology group gave me a unique glimpse into the infant field of plant molecular biology. Just as fascinating were the interactions with other faculty and Fellows in the controlled-environment laboratory, the phytotron. Anton Lang, Bernard O. Phinney, Jan Zeevaart, Erich Heftmann, Harry Highkin, Fred Ruddat, and many others provided me with an unparalleled learning experience in gibberellin physiology. Gibberellins and flowering were the order of the day.

In 1965 I had an opportunity most young scientists can just dream about: an appointment as assistant professor at a young university only 1.5 h from Banff National Park. Those early days at the University of Calgary were unbelievably good, complete with small numbers of bright, interested students, good funding for equipment and research grants, and department heads and deans who gave me, and the plant physiology colleagues who arrived after me, both moral and fiscal support.

But I digress.

Gibberellins — there are now 84 or more of them of quite diverse structure (see Fig. 17-12). The GAs that are biologically active are presumed to be those that have 19 carbon atoms, possess a lactone ring, and have a carboxyl group at carbon 7, but methyl esters of $GA_9$ and $GA_{73}$ are potent antheridiogens in *Lygodium*. Gibberellins with zero, one, two, three, and four hydroxyls are known, and GAs with five or more hydroxyls may well exist. Many hydroxyls occur in $\alpha$ and $\beta$ form, and epoxide and keto forms occur as well.

This is an extremely complex array of structures for a phytohormone class. From the pioneering work of Bernie Phinney, Jake MacMillan, James Reid, and their colleagues, we can now assign a specific physiological function (shoot elongation) to $GA_1$ and $GA_3$, which are carbon-3$\beta$ and carbon-13 dihydroxylated. But, we might ask, do only these two GAs function causally in plant growth and development? Over the past 20 years we have worked with several flowering systems in which differential efficacy for different GAs is the rule. Here are some of our results:

For the first few years in Calgary I continued my work with GA-induced "flowering" (actually, cone formation) in the Cupressaceae and Taxodiaceae (in which $GA_3$ and many other biologically active GAs induce flowering) and attempted to extend these successes to the Pinaceae. Unfortunately, in typical Murphy's law fashion, the more commercially important pines, spruces, firs, and larches do not generally flower in response to $GA_3$. The question of why not, and ways to promote their precocious and abundant flowering, has been studied intensively in my lab since 1968, primarily with colleague Stephen Ross of the British Columbia Forest Service Research Division. Carefully timed stress treatments that would promote flowering in young Pinaceae conifer seedlings were used and endogenous GAs examined, initially by bioassay, with identifications and precise quantifications by GC-MS-SIM coming much later. The bioassay trends told us that the stress treatments (mainly water stress and high temperature) that promoted flowering in nature yielded a large increase in endogenous, less polar (for example, with only one or no hydroxyl groups) GA-like substances while reducing more polar GA-like substances. This trend led us to test, in 1972, the application of $GA_4$, $GA_7$, and $GA_9$, yielding our first success: A $GA_{4/7}$ mixture, and especially $GA_{4/7} + GA_9$, stimulated flowering in seedlings of Douglas fir. Now, some 17 years later, tree breeders and seed-orchard managers throughout the world are using $GA_{4/7}$ to cost-effectively produce seed from genetically superior Pinaceae, conifer-families genotypes.

What then triggers conifers (of all three families) in nature to flower, both precociously and more abundantly? Environmental stress, most notably drought, probably blocks biosynthetic conversion of less polar, native-conifer GAs ($GA_4$, $GA_7$, and $GA_9$) into the more polar $GA_1$ and/or $GA_3$. A buildup of these less polar GAs in lateral primordia or apical meristems could then cause the differentiation or initiation of next year's conebud. In a wet, cool summer, GA conversion is probably not hindered, and sexual differentiation would not occur but growth would be stimulated.

I also maintained a parallel research program that examined the possible hormonal basis for inherently superior growth. In it we used hybrid genotypes of maize and known parent crosses (F1) of superior conifer genotypes. Stewart Rood (University of Lethbridge) and I began the maize work in 1980, and it was also a collaborative venture among physiologists, breeders, and chemists. It showed, using precise and accurate methods (GC-MS-SIM with use of [$^2H_2$]-labeled GAs as quantitative internal standards), that heterosis in maize is causally related to the concentration of GAs, most notably $GA_1$ and its $C_{20}$ precursor, $GA_{19}$, in the shoot cylinder that contains the apical meristem (Rood et al., 1988, and references cited therein). In essence, parental genotypes of maize showing inbreeding depression are GA-deficient, and their F1 hybrids, which show heterosis, have near optimal amounts of endogenous GAs.

The work on F1 conifer families began in 1982 and is still going on. However, it now appears that we can successfully test for inherent growth potential at ages 2 to 6 months in F1 crosses of commercially important conifers by growing the seedling progeny in near-optimal (phytotron) conditions. Intriguingly, as with maize, endogenous GA levels (in this case, $GA_9$ in needles) are significantly correlated with inherent rapidity of growth. Thus, one may now be able to screen for inherently fast-growing families by age 6 months and then promote their flowering (with $GA_{4/7}$ plus a stress treatment) by approximately age 2 to 4 years, thereby obtaining F1 seed by the age of about 4 to 6 years. Such a rapid turnover of generations in forest tree breeding was unheard of just a few years ago.

For most physiologists, work with GAs in conifers would be classed as an esoteric adventure. The "more important" question would be: Are GAs causal factors in the flowering of photoperiodically sensitive plants? By the late 1970s, many physiologists had concluded that although GAs were likely effectors of stem bolting in LDPs, they were not causal factors in floral initiation.

In 1982 I began work that was intended to reexamine the possible role of GAs as florigenic substances in photoperiodically sensitive plants. This work has been a continuing collaborative effort with physiologists Lloyd Evans and Rod King at the CSIRO Division of Plant Industry in Canberra, Australia, and an organic chemist, Lewis Mander, from the Research School of Chemistry, Australian National University. We have used the LDP, *Lolium temulentum*, strain Ceres, which will flower in response to one LD of a 16-h to 24-h duration. If the apex is dissected out at about day 21 (prior to rapid bolting) after the floral-inductive treatment (either LD or GA application), a comparison of floral stage with stem elongation can be obtained. Under noninductive SD there is no flowering and virtually no stem elongation. For a 24-h inductive LD, there is 100 percent flowering, with only a minimal elongation of the stem at day 21. Applied $GA_3$ under SD can mimic the LD response insofar as inflorescence initiation, but $GA_3$ also causes a large amount of stem elongation, even by day 21.

We have found that inductive LD increases the level of endogenous GAs and GA-like substances several fold in the apices and leaves within just a few hours to days. The most consistent increases in apex GA-like substances occurred in a chromatographic region where polyhydroxylated GAs, such as $GA_{32}$, are known to elute. These early results have led to an extensive program of organic synthesis of GAs by Lew Mander's group. Testing these GAs of widely differing structure has been accomplished systematically on *Lolium* (Evans et al., 1990). Although a number of native GAs in *Lolium temulentum* have been identified by GC-MS, none of the putative polyhydroxylated GA-like substances has, as yet, been characterized.

From this systematic testing of GAs in *Lolium*, we can draw some general conclusions about GA structure/function activity. First, the GA must have a carboxyl group at carbon 7 and a double bond in ring A. A florigenic GA must also be hydroxylated, and the point of hydroxylation is critical in determining whether the GA will be highly florigenic and/or whether it will promote stem elongation. GAs hydroxylated at carbons 13, 15β, and 12β are highly florigenic and give only minimal increases in stem elongation, but hydroxylation at carbon 3β tends to reduce florigenic activity while significantly promoting stem elongation. Thus, $GA_1$ (ring A dihydro, C-3β, C-13 dihydroxylated), is a very poor florigen in *Lolium* but an excellent elongator of the stem. Conversely, $GA_{32}$ ($\Delta^{1,2}$, C-3β, C-12α, C-13, C-15β-hydroxylated) is almost a thousand times more effective in flower induction than $GA_1$.

The *Lolium* story is still incomplete, and our work on SDPs and woody angiosperms (where GAs are usually inhibitory to flowering) is just beginning. For me, the last two decades in plant hormone physiology have been an exciting period. Advances in methodology (HPLC separation; GC-MC identification; quantitation by bioassay originally, now by use of GC-MS-SIM with stable isotope-labeled internal standards) have allowed imaginative physiological probes to be explored in a definitive fashion.

Whatever successes have come about could not have been gained without the enthusiastic long-term collaboration of many colleagues. Research on gibberellins is now moving very quickly toward molecular aspects. Even so, the chemist, biochemist, and physiologist are essential players on the new team, as will be, at the applications stage, scientists in agronomy, horticulture, and forestry.

Zeevaart (1976a) summarized "conclusive evidence" that flower formation and stem elongation are separate processes, with GA promoting stem elongation but not flower formation (in *Silene*, for example). Although GAs were reduced to levels below those that could be detected, flowering occurred. Flower formation and stem elongation proved to be under control of two separate but closely linked genes (Wellensiek, 1973).

Recently, the confusion has been somewhat reduced as investigators have examined effects on flowering of several different GAs (over 84 now being identified). In the LDP *Lolium temulentum*, for example, some GAs were as much as 1,000 times as effective in inducing flowering as were others, and specific molecular structures were clearly related to effectiveness (Evans et al., 1990; Pharis et al., 1987a). King et al. (1987) described a similar situation with *Pharbitis nil*: Some GAs were much more effective at promoting flowering than others, and (as implied above) promotion occurred when the GAs were applied 11 to 17 h before a single dark period; 24 h later the same dosages were inhibitory.

GAs seem to be particularly important in the formation of cones in conifers (reviewed by Pharis et al., 1987b). Y. Kato (for example, Kato et al., 1958) in Japan, Richard Pharis (see, for example, Pharis and Kuo, 1977) in Calgary, Canada, and others have pioneered in this work. Most conifers require several years before they attain ripeness to respond, but by spraying with $GA_3$, Pharis induced the formation of male strobili on Arizona cyprus (*Cupressus arizonica*) when plants were only 55 days old. He and his coworkers were unable to induce cone formation with $GA_3$ in other members of the Pinaceae, but they now achieve this by applying the less polar GAs ($GA_4$, $GA_5$, $GA_7$, and $GA_9$; especially the mixture $GA_{4/7}$). Because the breeding of conifers is important to the lumber industry, these observations are significant. Breeding times are reduced from years to months. Richard P. Pharis expands on the role of GAs in flowering and cone formation in his personal essay on page 526.

### Cytokinins

Cytokinins have been observed to promote flower formation in a few species. A combination of a cytokinin (benzyladenine) and $GA_5$ has induced flowering in one SD cultivar of chrysanthemum. In another cultivar, benzyladenine could substitute for the latter part of photoinduction. Zeatin applied to root-forming cuttings of the LDP *Anagallis arvensis* inhibited or slightly promoted flowering, depending on its concentration and the root formation stage (Bismuth and Miginiac, 1984). In most LDPs and SDPs, cytokinins do not affect flowering.

### Abscisic Acid

When certain SDPs (for example, Japanese morning glory and *Chenopodium rubrum*) are already slightly induced and then treated with abscisic acid (ABA), flowering is promoted; ABA inhibits flowering in a few LDPs (for example, *Spinacia*). But no conclusions can be drawn because ABA inhibits flowering in other SD and LD species and is completely innocuous in still others. ABA concentrations are generally, but not always, higher in shoots under LDs than under SDs.

### Sterols

A substance called TDEAP (tris-[2-diethylaminoethyl]-phosphate trihydrochloride), which inhibits synthesis of cholesterol in animals, inhibits flowering of cocklebur and Japanese morning glory. Does inhibition by TDEAP imply that florigen is related to cholesterol; that is, that it is a sterol of some kind? Not according to results obtained so far. Application of various sterols does not overcome the effect of TDEAP, nor could any significant changes in sterol fractions be detected during induction. Sterol biosynthesis can be inhibited by compounds other than TDEAP without influencing flowering one way or the other. Furthermore, TDEAP has been shown to block photosynthesis and export of assimilates (and florigen?) from the leaf (Zeevaart, 1979). Thus, evidence that florigen is a sterol is at best flimsy.

### $\gamma$-Tocopherol

Battle et al. (1976, 1977) measured $\gamma$-tocopherol (a form of vitamin E) in leaves of cocklebur plants at various times during an inductive dark period and following 20-min light interruptions given at various times during such a dark period. After 10 to 12 h of continuous darkness, the $\gamma$-tocopherol had increased in the leaves to levels five times those at the beginning of the dark period. Red-light interruptions up to and including 8 h after the beginning of darkness maintained the level of $\gamma$-tocopherol at about that measured at the beginning, and the effect was completely reversed by far-red light. When the light interruption was given 9 h after the beginning of darkness, however, the $\gamma$-tocopherol level was almost as high as in dark controls. The point at which both floral induction and the level of $\gamma$-tocopherol could no longer be reversed by light was reached at the same time. Of the many correlative measurements of specific substances, this is one of the most interesting, but to our knowledge, the work has not been pursued any further.

### Other Substances

Several other substances have been reported to influence flowering. For example, there has been considerable work on polyamines (see Bernier, 1988, for references). The prostaglandins provide another example. Prostaglandins are hormonelike, 20-carbon carboxylic acids found in animals. They have various effects, presumably mediated at their location in membranes. There are limited reports that they also occur in plants. E. G. Groenewald and J. H. Visser (1974, 1978) found that various inhibitors of prostaglandin synthesis, especially aspirin, inhibited flowering of *Pharbitis* (SDP) to various degrees and that certain prostaglandins applied to excised shoot apices of *Pharbitis* held under short days promoted flower development. Groenewald et al. (1983) found that the concentration of prostaglandin $F_{2\alpha}$ in *Pharbitis* was about 20 times higher under short days than under long days, and Janistyn (1982) found prostaglandin $F_{2\alpha}$ in flowering but not vegetative plants of *Kalanchoe blossfeldiana*. Even though this evidence for participation of prostaglandins in flowering is circumstantial, it is nevertheless interesting.

To state our growth regulator conclusion: *The flowering response is often influenced by applied plant growth regulators and other compounds, but the few patterns that can be tentatively discerned have many exceptions. Several compounds will cause flower formation, but there is no convincing evidence that florigen is or is not one or more of the well-known plant hormones—although a role for gibberellins seems increasingly likely.*

## 23.11 The Induced State

The flowering stimulus acts quite differently in different species (see review by Zeevaart, 1976b). In cocklebur, young leaves that are allowed to grow out under LDs after a plant has been induced by SDs can be grafted to vegetative receptors, causing them to flower. The young leaves apparently become induced by the older leaves. As a matter of fact, as many as five vegetative cocklebur plants have been grafted in series to an induced plant at the end of the chain, with flowers forming on all plants. The LDP *Silene armeria* and the LSDP *Bryophyllum daigremontianum* are also capable of such **indirect induction**.

Perilla, on the other hand, acts quite differently. An excised leaf can be induced by SDs and can then be grafted to a series of vegetative receptors for several months, inducing each to flower. But none of the leaves on the receptor plants becomes induced. Thus the flowering condition in perilla is not as "contagious" as it is in cocklebur. We can conclude that *plants exhibit at least two forms of the induced state, one in which induced leaves can induce other leaves, and the other in which induced leaves cannot induce other leaves but can themselves be induced when isolated from the plant.*

## 23.12 Floral Development

Much could be said about the anatomy and physiology of floral development. We do not discuss the topic in detail because this chapter is devoted to photoperiodism, with its emphasis on time measurement and detection of light and dark. Nevertheless, we note that evocation at the apical meristems changes the course of meristematic development from the vegetative to the reproductive mode and results in many interesting physiological changes. In many species, arrival of the flowering stimulus leads to an immediate increase in mitotic activity, and nuclear size often increases, as does the size of the nucleolus. Frequently there is a buildup in the number of ribosomes and mitochondria and in the amount of RNA in the apical cells. The buildup in RNA may occur *before* the stimulus arrives. Does this mean that some other stimulus precedes the main one? Or that florigen at a concentration that is too low to cause flowering arrives sooner, causing the observed changes in RNA and other factors? (For reviews of cellular changes that accompany conversion of the bud to the floral state, see Bernier, 1988; Jacqmard et al., 1976; Havelange, 1980; and Kinet et al., 1985.)

We might ask if evocation comes about as a cascade of events originally initiated by some single stimulus. Such a view has been inherent in the florigen concept. But there is evidence for partial evocation in which, for example, changes in irradiance might lead to certain changes at the meristem, whereas changes in cytokinins lead to meristematic events that are quite different. Bernier (1988) concluded that evocation is a process "consisting of a limited number of sequences of changes, each sequence being controlled independently of the others." Again, the picture is much more complicated than we might have imagined a few years ago.

Our final conclusion: *Different changes at the meristems can be caused by different stimuli and at different times. Although the stimuli remain to be identified, it is nevertheless clear that some or all of them are formed in the leaves in response to photoperiod and other environmental factors.*

## 23.13 Where Do We Go from Here?

Since the first edition of this book was published in 1969, much has been written about photoperiodism and the flowering process. During the 1970s, activity in this field decreased noticeably, or at least shifted toward empirical studies of flowering in different species. We have examined some important studies conducted during the 1980s, however, especially those relating to phytochrome and time measurement. Molecular-genetic approaches are being developed. The most important problems were outlined in 1983 during the symposium held in Littlehampton, England (Vince-Prue et al., 1984), and then at the 12th Annual Riverside Symposium in Plant Physiology in 1989 (Lord and Bernier, 1989). Perhaps a major accomplishment achieved during the past decade was the realization, with increasing vividness, that the problems are exceedingly complex. Differences among species are extensive and significant, much more so than we might have imagined a few decades ago. Studies on reproduction biochemistry have told us virtually nothing. Except within broad limits, there is no way to predict how a given growth regulator will influence flowering of a species not yet tested.

Nevertheless, resolving the unanswered questions about the flowering process as it might be controlled by photoperiodism or other environmental stimuli might help us understand the processes of plant development, and the puzzle of understanding development, plant or animal, remains one of the most significant unsolved problems in modern biology. Just what is time measurement in photoperiodism? What is induction? What is the floral stimulus? How does it act in evocation? Someday we may know.

# 24

# Molecular Genetics and the Plant Physiologist

***Guest Authors: Ray A. Bressan and Avtar K. Handa***

Avtar K. Handa

Ray A. Bressan

*In 1972 Ray Bressan left his home in Illinois to work on a Ph.D. with Cleon Ross at Colorado State University. After graduation he accepted a position as a postdoctoral fellow in the laboratory of Philip Filner at the Department of Energy/Plant Research Laboratory at Michigan State University. Avtar Handa, who had just graduated from the Tata Institute of Fundamental Research in Bombay, India, was also a new postdoctoral fellow in Filner's laboratory. Both Ray and Avtar—and Nick Amrhein working with Phil Filner—had just finished a fruitless search for cyclic AMP in plant tissues, so they had much in common. Ray and Avtar both left the Plant Research Laboratory on the same day to go to Purdue University in Indiana, Ray as a member of the Horticulture Department and Avtar as a postdoctoral fellow in the Botany Department. Soon Avtar also joined the Horticulture Department, where Ray is now Professor of Plant Physiology and Avtar is Professor of Plant Molecular Biology.*

*When the two friends arrived at Purdue, the science of molecular genetics was just beginning to expand with explosive force. As a postdoctoral fellow, Avtar applied gene-cloning techniques to study seed-storage proteins with Brian Larkins, and Ray has also been active in applying the new techniques to traditional problems in plant physiology. In this chapter—which could easily be expanded into a book—Ray and Avtar summarize the techniques of molecular genetics and point out how they can be applied in plant physiology. We hope this chapter will provide a well-organized review for those who have already studied the topic and a tantalizing introduction for those who have not.*

All the biochemical processes that determine form and function (phenotype) of plants are the result of information encoded within the DNA sequence of the genome and of the interaction of that information with the environment. This information is converted into biochemical activity and the resulting macromolecular structure of the plant through biosynthesis of specific enzymes or proteins via transcription and translation (see Appendix C for review). In other words, all of the morphology and physiology of plants is based on metabolic processes, and in turn, all of the metabolic processes are the result of the conversion of genetic information into the enzymes and proteins that control metabolism. In one sense, then, understanding plant physiology could result from knowing the genetic information upon which it is based, along with understanding how the conversion, or expression, of that genetic information into enzymes or proteins is controlled.

From the traditional plant physiologist's point of view, this approach is essentially working backwards. Physiologists usually first try to characterize the physiological process as some type of response or characteristic of the plant. They then attempt to base this response or characteristic on a metabolic process controlled by specific enzymes or proteins. More recently, however, many physiologists are learning to characterize the genetic bases of the metabolic processes that control physiological phenomena.

It has long been recognized that once we discovered the basis of genetic control of physiological processes, our understanding of these processes would increase dramatically. Within recent years our ability to study the regulation of expression of specific genes has indeed increased tremendously.

The ability to isolate and produce unlimited replications (cloning) of specific nucleotide sequences (genes) is rapidly allowing plant physiologists to study physiological phenomena by identifying the genetic regulatory mechanisms that control physiology. Here we describe these and other procedures, their general usefulness to plant physiologists, and some important advances that have already been made by their use — all with the rapid advances of molecular-genetic techniques that have become tools for plant physiologists in mind (Murphy and Thompson, 1988).

Historically, molecular gene cloning became possible as the result of much work on nucleic-acid biochemistry and biology, especially in the field of microbiology because microorganisms serve as the basic tools of gene cloning. The detailed knowledge of the infection and replication cycles of viruses and other extrachromosomal DNA (**plasmids**) harbored by bacterial cells played an indispensable role in gene cloning. The discovery in the 1970s of **endonuclease class II** (**restriction enzymes**; see Smith and Wilcox, 1970; Kelly and Smith, 1970) that can cut DNA molecules at specific locations in the nucleotide sequence, and in the 1960s of **ligases** that can join DNA fragments together, formed the basis for the "cutting" and "splicing" of DNA necessary to clone specific genes or parts of genes. The early work of Marmur and Doty (1962) and Britten and Kohne (1968), which established procedures for **denaturation** (breaking base-pair bonds) and **renaturation** (reforming base-pair bonds) of DNA and RNA, formed the basis of the rapid and simple nucleic acid **hybridization** (hydrogen bonding of the complementary bases on two separate strands of DNA) detection methods that are now so widely used. Then, in the early 1970s, H. W. Boyer, S. N. Cohen, and P. Berg (see Cohen et al., 1973; Jackson et al., 1972) were the first to work out the detailed procedures for producing in bacterial hosts unlimited copies of specific genes (Grunstein and Hogness, 1975). With the availability of unlimited quantities of DNA molecules of specific sequence, these DNAs could be used along with the hybridization procedures as specific **DNA probes** (segments of DNA that can be used to detect complementary strands by hybridization) to study the factors controlling gene expression.

Plant physiologists have begun to use cloned genes to study how gene expression controls many physiological phenomena, including water use, assimilation of carbon and other basic nutrients, expression of tolerance to biotic and abiotic stresses, flowering, seed germination, and many other processes. For example, why do leaves have fully developed, functional chloroplasts and roots do not? The genes necessary to form functional chloroplasts are present in root cells, but obviously they are not expressed. What controls this expression? We will now briefly describe the strategies and procedures for obtaining and using cloned genes for studying gene expression in plants. As in other fields of science, much of the learning involved in mastering the discipline is concerned with learning a special language; hence, we define each new term as we go along.

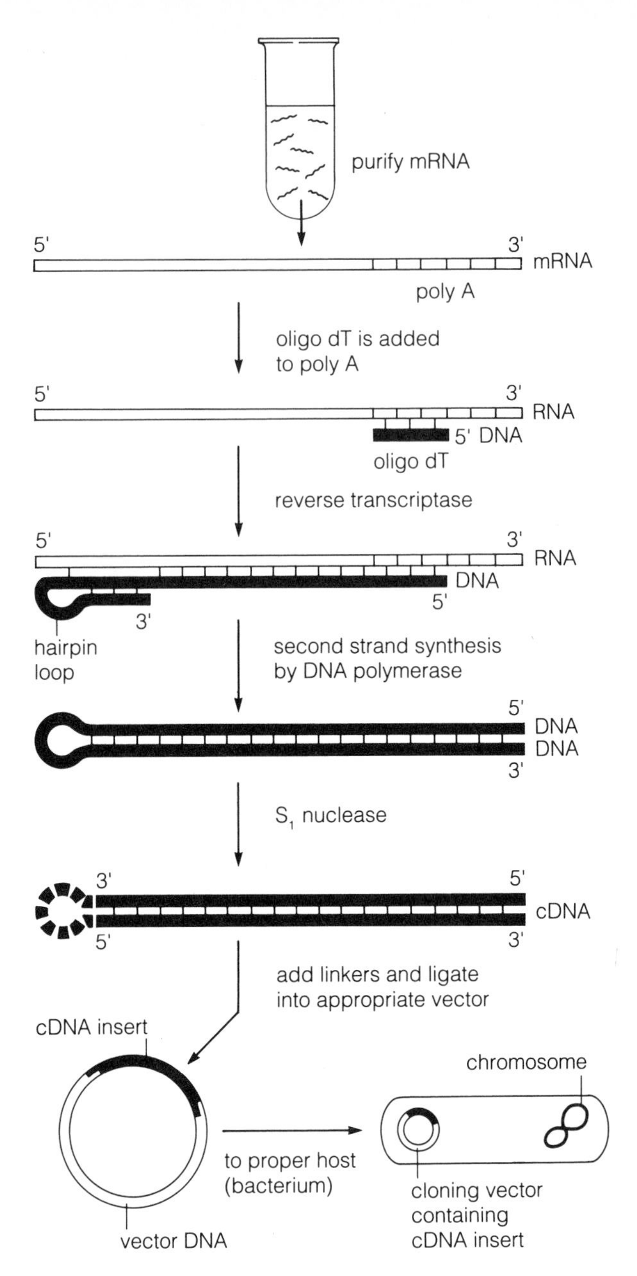


**Figure 24-1** Construction of a cDNA library from poly (A)$^+$ mRNA. Shown is synthesis of both cDNA strands, insertion into a vector, and transfer of the vector containing insert DNA to a suitable bacterial host, where it will be multiplied (cloned).

## 24.1 Gene Cloning

### Construction of a cDNA Library

To clone a gene (**gene cloning**) generally means to cause a bacterium to replicate relatively large amounts of the gene such that the gene can be easily recovered in sufficient quantity to perform experiments (Fig. 24-1). Gene cloning begins by construction of a cDNA library. (**cDNA** refers to copy-DNA or to complementary DNA because the DNA in the gene is copied from messenger

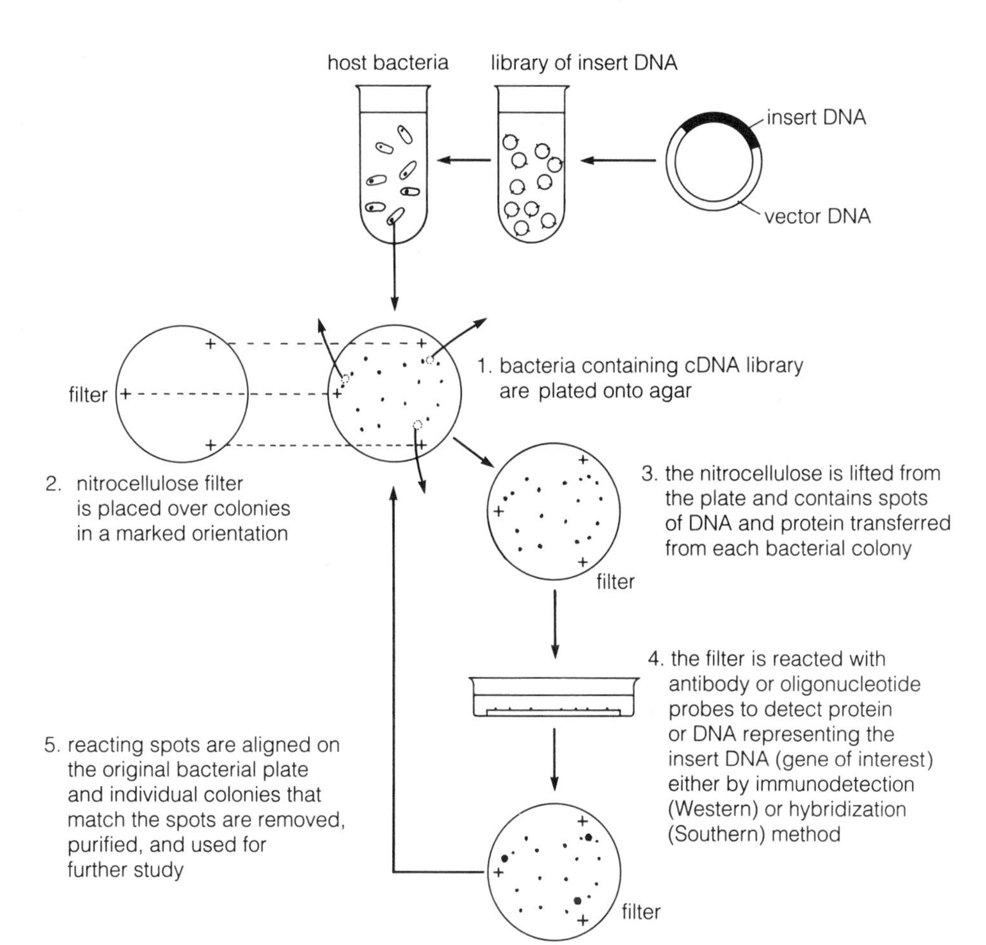


**Figure 24-2** Screening the cDNA library with either an antibody or DNA probe. Bacterial cells containing specific insert DNA are isolated in this way.

RNA that is complementary in base pairs to the gene being copied or cloned.) Cloning is initiated by isolation of all of the messenger RNA being formed from genes currently active in the cells or tissues. Isolation of mRNA is made rather easy because the mRNAs that code for structural and enzymatic proteins contain a polyadenine sequence on the 3′ end that is thought to function in the stability of the mRNA. These **poly(A)$^+$ RNAs** can thus be separated from the more abundant ribosomal and transfer RNAs that lack this **poly A tail.** These polyadenylated mRNAs are then converted to **complementary DNAs (cDNA)** using the enzymes *reverse transcriptase* and *DNA polymerase*. The cDNAs are then made into a library by their **insertion** (joining or **ligating**) into an appropriate **cloning vector** (a segment of DNA that can be transferred to and replicated in living bacterial cells). The cloning vector is usually a bacterial plasmid or virus (**phage**) present in the bacterium. The **library** then consists of this large number of vector DNA molecules containing different inserts that have the nucleotide sequences of the plant mRNAs that were isolated originally. Plasmids or phages containing the cDNA inserts are capable of being absorbed by a host bacterium, where they can self-replicate as the bacterial cells multiply, greatly enlarging the library. As a rule, only one plasmid with one cDNA insert infects a single bacterial cell, so each bacterial cell in the culture of infected cells probably contains a specific cDNA that is different from those in other cells.

Cloning a gene requires the isolation of an individual clone of bacteria that contains, on a cloning vector, a unique cDNA insert representing the corresponding unique mRNA from which it was made during construction of the cDNA library. This isolation involves some type of screening process to distinguish one particular cDNA insert from all the others; that is, to choose one cDNA from the library. The type of screening procedure used depends on the type of gene to be isolated. In most of the screening procedures that we will describe, a first step is to plate out the bacterial culture by diluting it and spreading the suspension onto an agar plate such that each bacterium is at some distance from all the others. Each bacterium will grow into a colony, and the challenge is to find the colony with the desired cDNA.

## Isolation of Specific Genes from Libraries

*Antibody screening* One of the simplest procedures for screening the cDNA clones involves the use of antibodies (Fig. 24-2). This technique, of course, first requires purification of the protein encoded by the gene of interest so that antibodies against the protein can be made in a suitable animal.

Screening the cDNA library with the antibody requires that the cDNA library be constructed using an appropriate **expression vector**. Such vectors allow expression (transcription) of the cDNA insert because the vector insertion site is next to a segment of vector DNA referred to as the promoter. **Promoters** are sections of DNA that allow binding of RNA polymerase to the DNA chain to be transcribed. Because of the activity of the promoter, RNA polymerase will then transcribe the cDNA insert into relatively large amounts of mRNA. This mRNA is then translated into the protein encoded by the cDNA insert by translation on the bacterial host ribosomes. Thus any bacterial colony with cells that harbor the cDNA encoding this protein will produce large amounts of the protein.

Specific detection of this protein with antibody is then possible because of the specificity of antibody-antigen recognition (Capra and Edmonson, 1977). Screening is accomplished by plating onto agar a large number of bacterial cells that contain the vector with different cDNA inserts. Transfer of the proteins encoded by the insert DNA to a nitrocellulose filter is achieved by first overlaying the nitrocellulose onto the bacterial colonies that are growing on the agar surface. Later the nitrocellulose is lifted from the agar surface, but only after marking its orientation so that visible spots on the filter will correspond to the locations of specific bacterial colonies or bacteriophages containing insert DNA that can later be picked off of the plate. Protein that is antigenic to the antibody used for screening is then visualized on the nitrocellulose using a Western blotting procedure (see Section 24.2). The result is that one can isolate a bacterial strain that has the cDNA of interest and is capable of expressing it as a protein.

*Oligonucleotide screening* Another technique widely used to isolate cDNA clones involves the use of an oligonucleotide probe. An **oligonucleotide probe** is a short piece of DNA (usually 10 to 50 nucleotides) that has a sequence similar to that of the desired cDNA insert. Oligonucleotides are chemically synthesized, and their sequence is usually decided by deducing the triplet base codons needed to code for a specific amino acid. The amino-acid sequence of a protein of interest is first obtained, and then the oligonucleotide is made from this protein-sequence information. Screening the library is performed essentially as with antibodies (see Fig. 24-2) but with some important differences. Expression vectors are not needed because DNA, not protein, is detected. Detection is accomplished by hybridization of $^{32}P$-labeled oligonucleotide to the complementary insert DNA.

*Differential screening* Attempts to isolate genes involved in a physiological process of interest may not always require prior identification and purification of a protein; **differential** (or +, −) **screening procedures** can be used. These procedures utilize differential expression of genes, usually before and after some inductive signal. Auxin-induced genes are a good example (Walker and Key, 1982; Hagen et al., 1984). Many genes that are potentially involved in the physiological responses to auxin are likely to be induced to produce mRNA after auxin treatment. Even though the proteins encoded by the induced genes remain unknown, the genes themselves can be isolated by first preparing a cDNA library from mRNA produced after auxin (or some other inductive) treatment. This library will contain cDNAs representing mRNAs that are both induced and noninduced (present before induction) by auxin. Then two sets of radiolabeled cDNAs are made: one from mRNA from noninduced tissues and one from induced tissues. Each of these cDNA populations is used as a probe to screen separately the same cDNA library by the procedures described for use of oligonucleotides (see above). Bacterial colonies with inserts representing mRNAs similar in sequence to both cDNA probes will react positively to both. Any colony with an insert that represents an induced gene will react positively only to the cDNA probes made from induced mRNA. Confirmation that the cloned cDNA represents an induced gene is accomplished by using the cloned cDNA as a probe in Northern blots (see Section 24.2) to quantify expression of the gene at the RNA level.

*Heterologous-probe screening* Specific cDNAs may be isolated by using what are known as **heterologous probes**, which are usually cDNA clones that have been previously obtained from another species. If there is sufficient sequence identity of the genes between species, such clones will allow successful screening of a cDNA library (as with oligonucleotide probes) made using mRNA from another species.

*Transposon tagging* **Transposons** are discrete DNA sequences that are mobile (McClintock, 1948; Fedoroff, 1989); that is, they have the ability, with the aid of an enzyme (*transposase*), to be cut out of one position in the genome and rejoined at another position in the genome. If transposons are rejoined to the genomic DNA sequence within the sequence of a functional gene, disfunction of that gene will result because its correct sequence will be disrupted. This genetic mutation will then be "tagged" because the mutated gene will contain the transposon DNA sequence inserted within its sequence. Because many transposon sequences are cloned, radiolabeled probes can be made from them, and by using a library made from genomic DNA containing the transposon-induced mutation, such probes can be used to identify bacteria with vectors containing the mutated gene with its inserted transposon se-

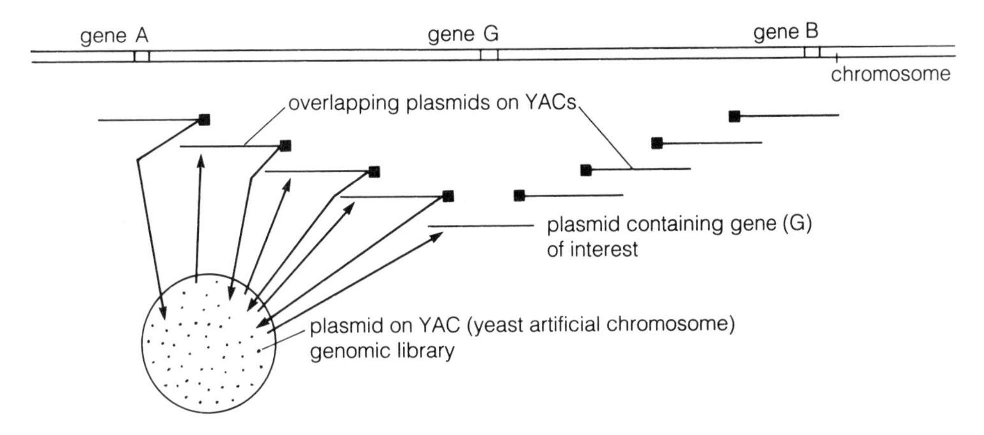


**Figure 24-3** Chromosome walking from gene A or B to gene G. Cosmids containing insert DNA that hybridized to the end of the preceding insert are progessively isolated until gene G is reached.

quence. Such an identified bacterial clone will contain, of course, an insert of the mutated version of the gene of interest. However, the wild-type gene may be obtained by constructing a probe from DNA sequences that are next to (flanking) the transposon DNA and that represent the sequence of the original gene before transposon insertion. Such a probe will specifically detect from a library made from wild-type genomic DNA the colonies containing the wild-type gene without a transposon insert.

*Gene cloning by chromosome walking* Genetic maps, which have been constructed based by phenotypic markers (traits), define the relative distance between two genes on a chromosome. It has been demonstrated recently that **restriction fragment length patterns** or **polymorphisms (RFLP)** of DNA can be used as genetic markers (Tanksley et al., 1989).

If a slight change in the DNA sequence from different plants (cultivars, species, and so on) results in a change in the position of a restriction enzyme site, a different restriction fragment length pattern will result. Because it is possible to clone and identify thousands of unique restriction fragments from a genome of a particular organism, the use of RFLPs has made it possible to obtain genetic markers covering an entire genome. The DNA in a chromosome is like a long continuous thread along which all genes are connected to each other in a linear manner; thus RFLP marker-based genetic maps can be used to clone genes that can be linked to a particular RFLP. This is accomplished by **walking** from the site of a known RFLP, which is co-inherited with the genetic trait (gene) of interest, toward the exact locus (position along the DNA) of the gene of interest.

Genome walking is performed by first creating a genomic library of the plant of interest. (**Genomic libraries** are made from the total DNA of the organism rather than from mRNAs.) The organism's DNA is cut into pieces, usually 10 to 30 kb (kilobases) in length, with an appropriate restriction enzyme. The pieces are then ligated into a suitable vector that can be harbored in a bacterial host. The library is probed with a cloned restriction fragment to which the gene of interest is linked (determined by co-inheritance), and **recombinant clones** (vector plus insert DNA) containing at least part of probe DNA and some adjacent DNA sequences are selected (Fig. 24-3). DNA inserts from these clones are digested with restriction enzymes, and a map of the order of the DNA segments is constructed. DNA fragments representing the extremities of the inserts are isolated and used as hybridizing probes to select additional clones from the library. Each new clone will contain DNA inserts that are sequences continued from the previous insert. This walking cycle is repeated several times until the desired position (based on a linkage map) is reached.

If the fragment representing the extreme end of a DNA insert from a clone contains a **repetitive DNA** sequence, a fragment next to the extreme end fragment is used as a probe to select the next sequence of genomic DNA (Britten and Kohne, 1968) because identical repetitive DNA sequences exist at many places in the genome, and a repetitive DNA probe would not necessarily identify a sequence adjacent to the end of the previous fragment. Genomic libraries for chromosome walking are generally constructed in lambda-phage replacement vectors that allow about 20 kb walking in each step. This technique is more suitable to small genomes, such as that of *Arabidopsis thaliana*, because relatively few recombinant phages (containing inserts) can contain the entire genomic DNA sequence. For plants containing larger genomes, including most of the important crop species such as tomato, onions, or cereals, chromosome walking becomes tedious and time-consuming, but it can be used to isolate genes for which the only information known is the phenotypes they control. This technique has been used successfully to isolate several genes linked to human diseases (Watkins, 1988).

*Complementation or shotgun cloning* Genes that cause phenotypes for which mutant or variant forms are available may be isolated with a procedure in which the mutation is complemented by the transfer (transformation) of a cloned gene. A genomic library, which consists of inserts within transformation vectors that represent the active wild-type genome, is used to transform tissues from mutant plants. Many transformed plants are then produced and screened (or selected, if the phenotype allows) for plants that revert to the wild-type phenotype. A reverted phenotype would be expected in a plant originating from a cell that received a wild-type version of the mutated gene from the library. This gene could then be isolated from a library made from DNA from the transformed revertant plant using the vector DNA as a probe to screen the library.

The major drawback of this procedure is the need to produce a sufficient number of transformed plants to represent all of the clones in the library. Genomic DNA inserts are usually around 10 to 20 kb long. Most plant genomes contain 1 to 6 million kb of DNA. Therefore, it would require anywhere from 50,000 to 600,000 transformed plants to represent a complete genomic library. Screening this many plants would be prohibitively tedious. However, some species such as *Arabidopsis thaliana* have much smaller genomes (only about 70,000 kb), which requires screening only 3,500 to 7,000 transformed plants. If the gene of interest is suitable, plant-gene libraries can be used to complement bacterial mutants, of which many more transformants can be screened for reversion.

## Characterization of Cloned Genes

Several methods have been developed to establish or prove the identity of a cloned gene. Complementation of a mutant phenotype with the cloned DNA sequence is considered a proof for the identity of a gene involved in determining that particular phenotype. However, as just noted, identification of a gene using this method requires analysis of a large number of transformed plants, and in most cases it has not yet become practical to use this procedure. Another method that has been used by several researchers is based on immunoprecipitation of the *in-vitro*-translated radiolabeled product of a mRNA species that is selected by hybridization of the complementary mRNA to the cloned DNA sequences.

Generally, the cloned DNA sequence is immobilized on a membrane filter and hybridized to either total RNA or the poly(A)$^+$ RNA from the tissue of interest. After hybridization, the filter is washed to remove the nonspecifically adsorbed RNA species, and the RNA species that is hybridized to the cloned DNA is then released by denaturation of the RNA-DNA hybrid. The mRNA is then translated *in vitro* using a radiolabeled amino acid and ribosomes, enzymes, transfer RNA, and other factors from either wheat germ or rabbit reticulocytes. The radiolabeled polypeptide products are immunoprecipitated by specific antibodies raised against the specific protein for which the cloned DNA is suspected to encode and are separated using electrophoresis on a denaturing polyacrylamide gel. If the size of the immunoprecipitated polypeptide from the hybrid-released mRNA is identical to that of the immunoprecipitated polypeptide from total poly(A)$^+$ RNA from a tissue that is known to make that particular protein, this is considered positive evidence for having cloned the gene of that particular protein. This method, however, does not unequivocally establish the identity of the cloned gene; this can be done only by comparison of the DNA and amino-acid sequences.

Two methods for sequencing DNA molecules were published in 1977. Maxam and Gilbert (1977) developed a chemical method in the United States; Sanger et al. (1977) developed an enzymatic method in the United Kingdom. Both of these procedures use high-resolution denaturing polyacrylamide gel electrophoresis capable of resolving radiolabeled, single-stranded oligodeoxynucleotides that differ in length by only one nucleotide and range from a few bases up to 500 bases in length.

The enzymatic method (also known as **dideoxy sequencing**) developed by Sanger et al. (1977) is based on the ability of DNA polymerases from *Escherichia coli* or other organisms to elongate DNA chains by adding new deoxynucleotides at the 3′-end of the chain, resulting in synthesis of a complementary copy of a single-stranded DNA molecule from a DNA template. DNA polymerase elongates (but cannot initiate) DNA chains, so a "primer" DNA chain is first annealed to the single-stranded DNA molecule (template) to be sequenced. Sanger et al. (1977) used incorporation of a dideoxynucleotide derivative of either G (guanine), A (adenine), T (thymine), or C (cytosine) to specifically terminate DNA-chain elongation at the elongating end in either a G, A, T, or C base. This termination occurs because the dideoxynucleotide derivatives lack a 3′-hydroxyl group to continue the phosphate backbone of the DNA polymer. Thus whenever incorporation of a dideoxynucleotide occurs at the 3′-end of the chain, elongation is terminated.

In practice, a set of four independent reactions of DNA-chain elongation is carried out in the presence of a small amount of dideoxynucleotide derivative of either G, A, T, or C, plus a radiolabeled deoxynucleotide (usually deoxyATP). Under the appropriate concentrations of deoxynucleotides and dideoxy thymidine triphosphate, there is termination of elongation of some of the chains during the synthesis of the complementary DNA strand at every position along the template where there is A, the base complementary to T. When the reaction mixture contains dideoxy C, there is termination of

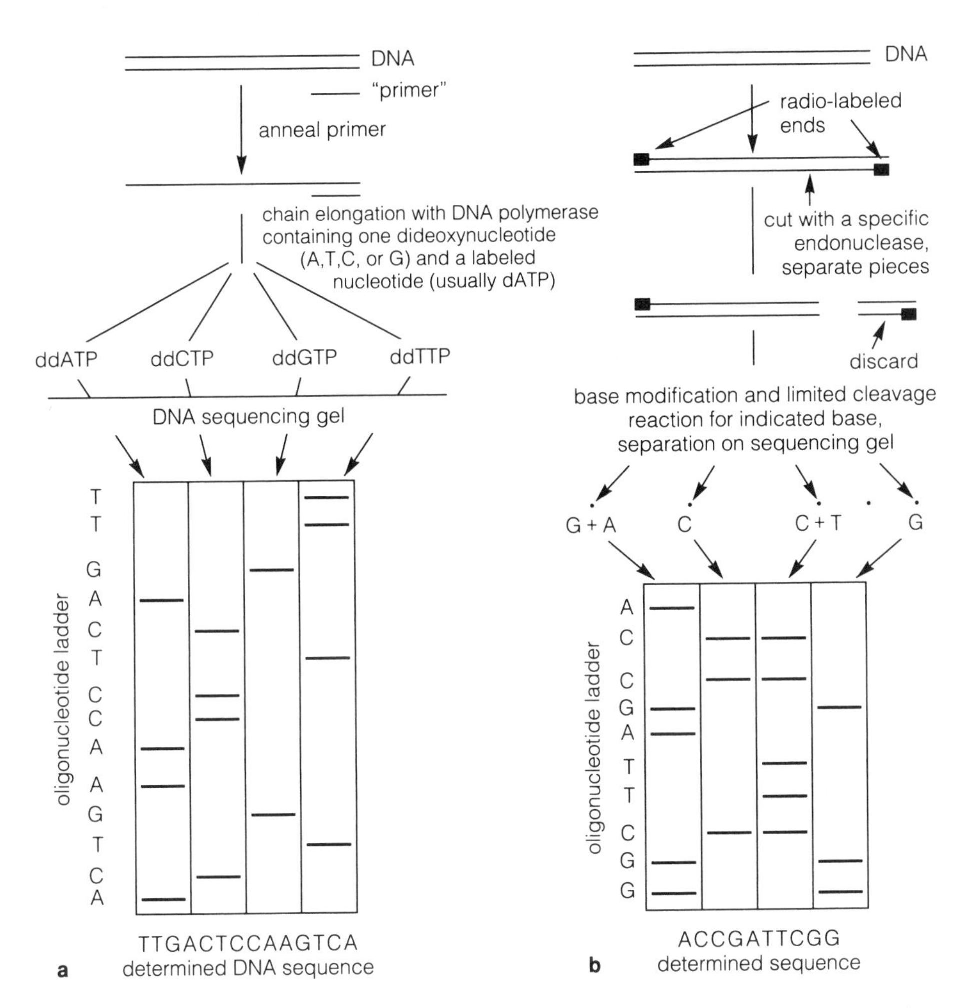


**Figure 24-4** Dideoxy (**a**) and chemical (**b**) DNA-sequencing strategies. For dideoxy-sequencing strategy, the single-stranded DNA (obtained by denaturing of double-stranded DNA) is annealed to an oligonucleotide that serves as a "primer" for DNA polymerase. Chain elongations are allowed to occur in four separate reaction mixtures in the presence of four deoxynucleotide triphosphates, including a radiolabeled deoxynucleotide triphosphate (usually dCTP) and one of the four dideoxynucleoside triphosphates. The four reactions are terminated and products are separated on a sequencing gel. In the chemical sequencing strategy, the double-stranded DNA is end-labeled using 5′-polynucleotide kinase, cut with a specific endonuclease, and separated into two separate fragments on an agarose gel. One of the labeled fragments (marked with a square in the figure) is subjected to base modification and limited cleavage reaction (see text) and separated on a sequencing gel. The DNA-sequencing polyacrylamide gel is fixed, dried, and autoradiographed on X-ray film. The DNA sequence is determined by reading the oligonucleotide ladder in the four lanes.

some of the chains at every position where there is a G in the complementary strand, and so on for the other two cases. The oligonucleotides from these four reaction mixtures are each separated independently using high-resolution denaturing polyacrylamide gel electrophoresis, which results in identification of the relative lengths of oligonucleotides ending in either G, A, T, or C. Upon visualization using autoradiography, the resulting "**ladder**" is read directly for the DNA sequence of the DNA molecule (Fig. 24-4). Very large DNA molecules can be sequenced with this method either by using restriction enzymes to construct smaller derivatives of the larger molecules or by using chemically synthesized oligonucleotide "primers" to start the chain elongation at different positions along the DNA template molecules.

The **chemical DNA sequencing method** developed by Maxam and Gilbert (1977) is based on random cleavage of DNA at every one or two bases using base-specific chemical reagents. First, the oligodeoxynucleotide molecule to be sequenced is radiolabeled with $^{32}P$ either at the 3′- or 5′-end using the enzyme *polynucleotide kinase;* then it is subjected to base modification to cleave the DNA strand specifically at either G, A, T, or C. This results in strands that differ in length by one nucleotide. In the first reaction, dimethyl sulfate is used to methylate nitrogen 7 of guanine; this methylation opens the ring between carbon 8 and nitrogen 9. Next, the methylated product is displaced from the DNA chain by piperidine; this represents G-specific cleavage. In the second reaction, cleavage at A and G is obtained by first treating the oligodeoxynucleotide with formic acid, which weakens adenine and guanine glycosidic bonds (those between deoxyribose and either of the bases) by protonating the purine ring; this treatment is followed by cleavage with piperidine. In the third reaction, the rings of thymine and cytosine are opened up by hydrazine treatment, and the modified thymine and cytosine molecules are cleaved with piperidine. In the presence of NaCl, hydrazine splits only the cytosine ring, resulting in cleavage at sites containing cytosine, and this constitutes the fourth reaction. After the chemical reactions are completed, products from each of the four reactions are separated independently using a sequencing gel, as described for Sanger's method, and the DNA sequence is read directly from the resulting autoradiogram. The major advantage of this method is that it overcomes problems associated with DNA-polymerase synthesis of DNA from a template strand (for example, secondary structure within the template DNA molecule can prematurely terminate elongation of the oligodeoxynucleotide chain).

## 24.2 Analysis of Gene Expression in Plants

Several methods have been developed to study the expression of a gene at both the transcriptional and translational levels by detecting the products of either transcription (mRNA) or translation (protein). These methods are also very useful in determining the expression of a gene that has been transferred into plant cells (**transformation**). However, the majority of genes that have been transferred into plant cells from other plants or from other organisms are expressed only in certain cell types at specific stages of development. Thus, it becomes important to regenerate plants from the transformed cells before studying the expression of an introduced gene. Rapid progress is being made in the use of plant tissue culture to regenerate plants from cultured plant cells or protoplasts. In the following few paragraphs we describe some of the techniques that have proven useful in analyzing gene expression in plants during their growth and development or in response to environmental stimuli.

### Immunological Methods to Quantify Expression of a Gene at the Protein Level (Western Blotting)

The inherent specificity of an antibody for an antigen has allowed development of specific assays for many proteins (Capra and Edmonson, 1977). The development of advanced protein-purification techniques has helped us obtain highly purified protein preparations that can be used either to determine amino-acid sequences or to raise specific antibodies directly. If some of the sequence of a gene (or of the protein that it encodes) is known, synthetic peptides can be made using this sequence information; amino acids can be chemically connected in a specific sequence. The proper amino-acid sequence can be deduced from the nucleotide sequence of the gene using the genetic code. Such prepared peptides have been used to raise antibodies that can be used to detect the proteins encoded by the gene.

Once antibodies are produced, sensitive quantification of a specific protein can be accomplished by **Western blotting** (Burnette, 1981). First, the plant tissues (leaves, stems, roots, cultured cells, and so on) are homogenized in an extraction buffer to extract total proteins. The extracted proteins are separated using one- or two-dimensional electrophoresis on polyacrylamide gels. Then the separated proteins are transferred to a suitable "membrane" (commonly a sheet of nitrocellulose) using electrophoretic transfer. These "membranes" are then treated with either radiolabeled or enzyme-linked antibodies that either produce radioactivity that can be detected on photographic film or convert a noncolored substrate to a colored product so that the polypeptides of interest can be visualized on the membrane. Densitometric scanning can be used to quantify the amount of the protein of interest.

### RNA-DNA Hybridization to Quantify Expression of a Gene at the RNA Level (Northern Blotting)

In the **Northern blot** procedure (Thomas, 1980), either total RNAs or poly(A)$^+$ RNAs are isolated from plant tissue and separated (based on their sizes) by electrophoresis in an agarose gel in the presence of a strong denaturing agent such as formaldehyde or methyl mercury. A sheet of nitrocellulose or another suitable membrane that retains RNA molecules is laid on the gel, and an appropriate buffer is allowed to flow through the gel. This deposits (blots) RNAs onto the membrane (Fig. 24-5). Alternatively, the separated RNAs are transferred to the membrane using electrophoresis. The RNAs are fixed (bound) there either by baking for 1 to 2 h at 80°C or by exposure to ultraviolet light. A radiolabeled DNA probe, specific for (complementary to) the gene under study, is then hybridized to the RNA on the filter to identify the complementary mRNA species. Using appropriate standards, the quantities of mRNA species of interest can be determined easily. In addition to quantification of particular mRNA molecules, this method also allows determination of their sizes.

A modification of this technique, referred to as **slot-blot analysis**, is useful for rapidly determining levels of

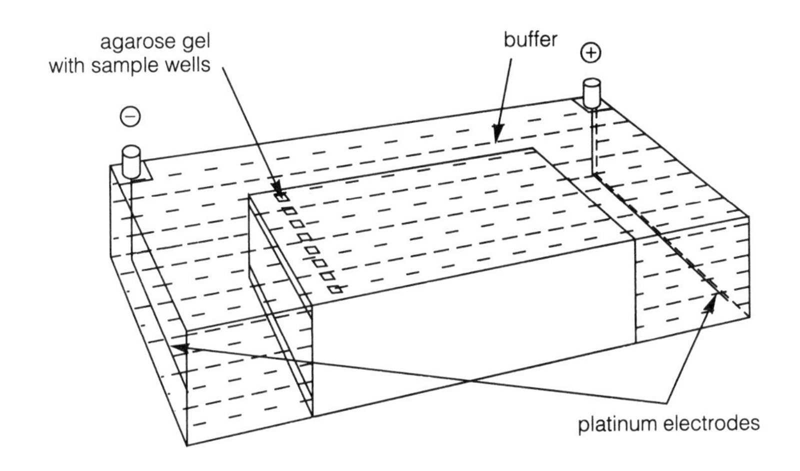


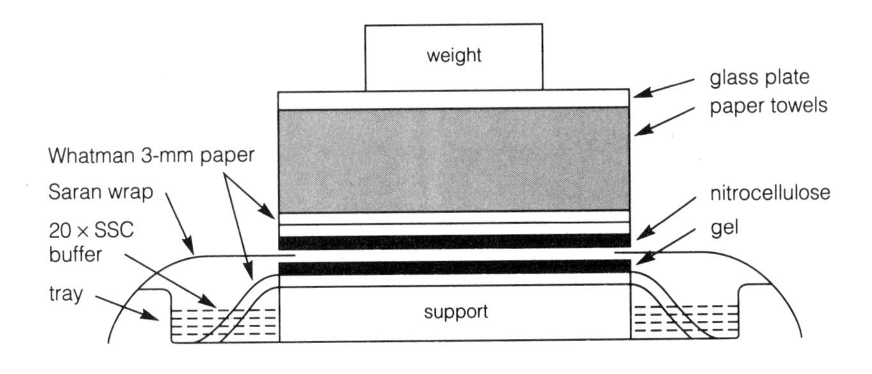


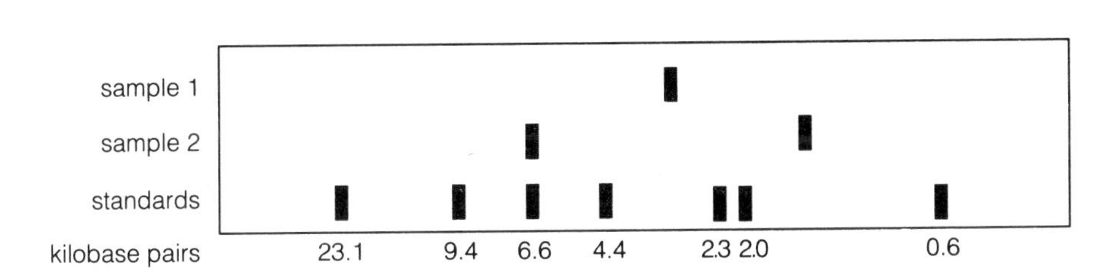


**Figure 24-5** Some techniques to separate DNA and RNA and to carry out hybridization of nucleic acids. (**a**) Electrophoretic separation of DNA or RNA molecules; (**b**) transfer of RNA or DNA to nitrocellulose membrane for Southern or Northern blotting. Usually DNA or RNA standards also are separated on a lane during electrophoresis, allowing determination of the size of DNA or RNA fragments in the unknown sample. In **c** is a diagrammatic autoradiogram showing bands containing nucleic acids that hybridized to radioactive DNA or RNA probe.

selected mRNA species. In this method total RNAs or poly(A)$^+$ RNAs are not separated on an agarose gel but are instead blotted directly onto a nitrocellulose filter using a commercial slot-blot device. After fixing or binding to the membrane filter as described above, the blotted membranes are processed as explained above.

## *In-Situ* Localization of Protein or mRNA Encoded by a Gene of Interest

Spatial localization of particular mRNAs or proteins at cellular and tissue levels allows us to determine what type of cells express a particular gene. This information is extremely important for producing transgenic plants. Localization of a mRNA species is accomplished by RNA-DNA hybridization, whereas the protein of interest is localized using specific antibodies. To localize a mRNA species, a complementary DNA strand is radiolabeled, usually with $^{32}$P or $^{35}$S, and hybridized to a tissue-section preparation (Cox et al., 1984). After hybridization, the mRNA species of interest is localized with autoradiography by placing the sectioned tissue onto a sheet of X-ray film. Proteins are generally localized using specific antibodies that are either radiolabeled, coupled with an enzyme, or linked with an electron-dense heavy metal such as gold. The light microscope or electron microscope is then used to localize proteins within a cell.

## Use of Reporter Genes to Study Expression of Genes

It is generally not possible to determine the regulation of expression of a particular gene during plant growth and development because the specific products of most genes have not yet been identified or are difficult to measure without an antibody. To overcome this difficulty, genes encoding proteins or enzymes that can be readily assayed have been used as **reporter genes** to study mechanisms that regulate expression of various genes. In this technique the DNA sequences that encode the protein of a gene of interest are replaced with DNA sequences that encode a readily assayed reporter gene, leaving the original *cis* (flanking) DNA sequences intact. These **chimeric constructs** (pieces of DNA of different origins joined together) are then introduced into plants by one of several transformation procedures described below. The factors and circumstances (for example, light or a hormone) that affect expression of the original gene product by the *cis*-acting regulatory DNA sequences can then be determined by assaying the product of the reporter gene in the resulting transgenic plants (see Section 24.4). ***Cis*-acting** refers to a DNA sequence that functions at a specific position adjacent to the amino-acid coding sequence of a gene. This **gene fusion technique** (joining together parts of different genes) has been successful in the study of both tissue- and cell-specific gene expression during growth and development. Moreover, reporter genes have been used to examine the expression of plant genes altered by mutation.

To avoid interference from endogenous enzyme activity, reporter genes are selected that encode enzymes that are normally not formed by plants. To date, several reporter genes, mostly those encoding bacterial enzymes, have been used to study expression of plant genes. *Chloramphenicol acetyl transferase* (CAT), *neomycin phosphotransferase* (NPT), and *beta-glucuronidase* (GUS) have been used extensively in plants. Beta-glucuronidase seems to be most promising, especially considering the ease with which this enzyme can be quantified (Jefferson et al., 1987). It is also useful for histochemical analysis, which helps to determine the location of gene expression at the cellular level. It is expected that several more reporter-gene systems will be developed in the near future, which will improve our ability to study gene expression in plants.

## Role of Genome Organization and Gene Amplification in Gene Expression (Southern Blotting)

Genomes of higher plants contain amounts of DNA ranging between $10^7$ and $10^{12}$ nucleotide pairs. Based on DNA-DNA hybridization kinetics to ascertain the degree to which these DNA sequences are repeated at various locations within the genome, plant genomic DNA, like DNA of other eukaryotic organisms, is divided into three categories: highly repetitive DNA, middle repetitive DNA, and unique or low copy number DNA (Britten and Kohne, 1968). However, unlike mammalian cells, in which essentially all DNA is present in the nucleus and mitochondria, DNA in plants is present in three cellular compartments: the nucleus, mitochondria, and plastids (including chloroplasts).

The relative position of most DNA is fixed in a linear order on the chromosomes, and thus DNA is inherited unchanged from parents to offspring. Examples have begun to accumulate, however, that show that the copy number of some DNA sequences and their relative positions on the chromosomes can change rapidly under certain environmental and growth conditions, especially when exposed to tissue culture and various stresses. Mechanisms underlying these changes in plant genomes are not yet understood, but it seems that transposable elements and gene amplification have a significant role in these changes. **Transposable elements**, sometimes called **controlling elements**, are genetic entities consisting of small DNA sequences capable of jumping from one position to another in the plant genome during development.

**Gene amplification**, a process that allows selective duplication of a DNA sequence, plays an important role in adaptability of eukaryotic organisms to various stresses. It has also been suggested that an abundant class of repetitive DNA, called "**selfish DNA**," is capable of amplification, transposition, or both and can then cause changes in genome organization. Selfish DNA is so named because it is thought to have evolved with no purpose other than to promote its own replication and persistence. Selective gene amplification and transposition are rapidly being recognized as mechanisms that can cause a novel phenotype by altering the function of a particular gene. These mechanisms are thus of great interest to plant biologists.

Structural properties of genes in eukaryotic organisms, including plants, are rather complex. Instead of a continuous DNA sequence that directly encodes mRNA for a protein, most genes are interrupted by additional DNA sequences that do not encode any part of the amino-acid sequence. The DNA sequences that encode for amino acids are called **exons**, whereas those that interrupt coding regions are called **introns**. Most genes in plants are interrupted by several introns. After transcription of a gene, the intron sequences are precisely excised from the newly transcribed mRNA, resulting in mature mRNA. Enzymes capable of excising introns have been demonstrated to occur in various eukaryotic cells.

The organization within the genome of genomic sequences, including those of particular genes, has been extensively studied using a technique known as

**Southern blotting.**[1] In this technique, DNA sequences (from genomic DNA or from cloned DNA sequences) restricted with an appropriate restriction endonuclease are size-fractionated by electrophoresis in agarose or acrylamide gels. DNA fragments in these gels are visualized by UV light after staining with ethidium bromide and are then photographed. After denaturing, DNA fragments are transferred from the gel to a nitrocellulose (or another suitable) membrane. To transfer DNA sequences, a piece of nitrocellulose is placed on a gel such that a flow of an appropriate buffer through the gel elutes DNA fragments and deposits (blots) them onto the membrane. (Alternatively, the separated DNA fragments can be transferred to the membrane by electrophoresis.) After transfer of the DNA molecules, the membrane is baked for 1 to 2 h at 80°C or is exposed to UV light to bind DNA molecules to the membrane. Labeled DNA probes that represent specific sequences of a gene under investigation are then hybridized to DNA that is bound to the filter and visualized by autoradiography. Mismatching between the DNA sequences bound to the membrane and those used as probes can be investigated by removing unmatched sequences with a single strand-specific endonuclease such as S1 nuclease. Information obtained using this technique can define the size and number of copies of a particular gene in the genome and ascertain whether variant forms exist. This technique is routinely used for examining restriction fragment length polymorphisms and for chromosome walking.

## 24.3 Genetic Modification of Plants Using Recombinant DNA Technology

The ability to transform plant cells with foreign DNA has proved to be a valuable tool of molecular genetics. **Totipotency** (the ability to regenerate a plant from a single somatic cell; see Chapter 16) has permitted studies of the regulation of expression of foreign genes during plant growth and development. These **transgenic plants** (that contain one or more genes transferred from a different kind of organism) have been used to analyze and define DNA sequences such as **promoter** and **enhancer** elements that control developmental and tissue-specific expression and the rate of transcription of genes. Moreover, transgenic plants have played an important role in our understanding of how environments (for example, light, temperature, moisture, nutrition, and so on) and plant growth regulators affect expression of specific genes. Transformation techniques have also made it possible to transfer genes across sexual barriers and to introduce genes from one species to another. This is the essence of genetic engineering, and it would not be an overstatement to say that genetically engineered crop plants will help to alleviate chemical pollution and food problems in the 21st century. Two examples of transformed plants are shown in Figure 24-6.

### Development of Selectable Markers

Although **recombinant DNA techniques (gene cloning)** made it possible to isolate useful genes from plants, several important requirements needed to be met to transform plant cells and obtain transgenic plants. Because the transformation process is inherently inefficient, not all cells will incorporate the foreign gene into its genome. Therefore a **selectable marker** procedure to select the transformed cells from among the nontransformed ones was needed. Selectable markers currently being used normally encode for resistance to an antibiotic or to a chemical that is toxic to plant cells. Once these selectable markers are introduced (by transformation) into plant cells, the transformed cells can easily be selected because only the cells containing the selectable marker gene can survive in the selection medium containing the toxic agent. Genes for resistance to *kanamycin, neomycin, hygromycin, chloramphenicol, aminoethyl cysteine,* and *methotrexate* have been used as selectable markers to select transformed plants. In addition, *opine synthase, beta-galactosidase,* and *beta-glucuronidase* have been used in screening procedures to identify transformed cells.

### Transformation Vectors

*Agrobacterium Ti-based vectors* The discovery that part of the crown gall plasmid DNA (the **T or transfer DNA**; see Section 18.1) of *Agrobacterium tumefaciens* is naturally transferred and **integrated** (joined chemically into another DNA strand) into plant-cell chromosomes provided the first dependable and predictable method to deliver foreign DNA (not normally present in the recipient genome) into plants. Extensive understanding of the **tumor-inducing (Ti) plasmid** at the molecular level resulted in the removal of genes responsible for the formation of tumors (Fraley et al., 1986). These "disarmed" Ti plasmids were used to develop vectors that can deliver DNA and incorporate it into the genome of plant cells. More recently, binary vector-containing strains of *Agrobacterium* have been developed, in which the genes required for infection by *A. tumefaciens* and

[1] Southern blotting is named after E. M. Southern, the scientist who developed this technique (see Southern, 1975). The other blotting techniques, Northern and Western, were subsequently given the names of directions in recognition of Southern's important contribution — and perhaps to show that scientists have some sense of humor.

a b

**Figure 24-6** Examples of transformed plants. (**a**) Tobacco plants infected with alfalfa mosaic virus. The diseased plant on the left was nontransformed, whereas the transgenic one on the right contains the gene for the alfalfa mosaic virus coat protein. Resistance to virus in the transformed plant is somehow conferred by expression of the virus coat protein. (From Tumer et al., 1987.) (**b**) Soybean plants sprayed with glyphosate, an herbicide that blocks an enzyme of the shikimic acid pathway called EPSP synthase [see Fig. 15-10 and Section 15.4]. The plant on the right was transformed before herbicide treatment with the gene that codes for EPSP synthase, so that plant contained high amounts of the enzyme and was able to catalyze synthesis of EPSP (5′-enolpyruvyl shikimate-3′-phosphate) to maintain formation of aromatic acids and other essential products of the shikimic acid pathway. (Courtesy Monsanto Agricultural Co.)

mobilization of the T-DNA are contained on one plasmid that cannot integrate its DNA into the plant genome (Bevan, 1984). In these bacterial strains the T-DNA sequences required for integration of foreign DNAs into plant chromosomes are present on a second (hence binary) plasmid that is used to carry the foreign DNA sequence but does not contain tumor-inducing genes (Fig. 24-7). The availability of binary vectors containing selectable marker genes has greatly facilitated cloning and the introduction of foreign DNA into plant cells. Several transformation techniques have been developed to introduce DNA cloned on an *Agrobacterium*-based vector into plant cells, including (1) inoculation of intact or decapitated plants after wounding, (2) *in-vitro* inoculation of meristematic or embryonic explants, (3) cocultivation of bacteria with growing protoplast-derived cells, and (4) transformation using leaf discs or other tissues from which plants can be regenerated.

*Virus-based vectors* Several mammalian DNA viruses have been used to develop vectors for stable transformation of mammalian cells. Therefore, the potential of plant DNA viruses to deliver foreign DNA into plant cells is being investigated. Cauliflower mosaic virus (CaMV) and geminivirus have received the most attention in this regard. The entire nucleotide sequence of CaMV DNA and some of the sequence of geminiviruses has been determined. However, the mechanisms by which these viruses infect and multiply in plants are rather complex, and much work remains to be accomplished before any of these viruses can be used to develop useful vectors for plant transformation.

*Transformation by naked DNA* Compared with transformation of bacteria and fungi with naked DNA (DNA without attached protein), the introduction of naked DNA into plant cells is a more recent development. Initial studies showed that treatment of plant protoplasts with poly-1-ornithine or polyethylene glycol containing $Ca^{2+}$ facilitates absorption of DNA by isolated plant protoplasts. It was further demonstrated that some DNA absorbed by plant protoplasts results in stable integration and expression of marker genes in the genome of transformed plants. Methods based on **electroporation** of naked DNA increase the frequency of transformed cells. In this technique, plant protoplasts and the DNA are placed in a chamber, and short electrical pulses are provided with a generator. The electrical pulses cause temporary openings in the membranes of plant protoplasts and subsequent uptake of DNA. Foreign DNA transferred to protoplasts by electroporation becomes integrated into the genome and is expressed, depending upon the regulatory sequences present in the recombinant DNA molecule.

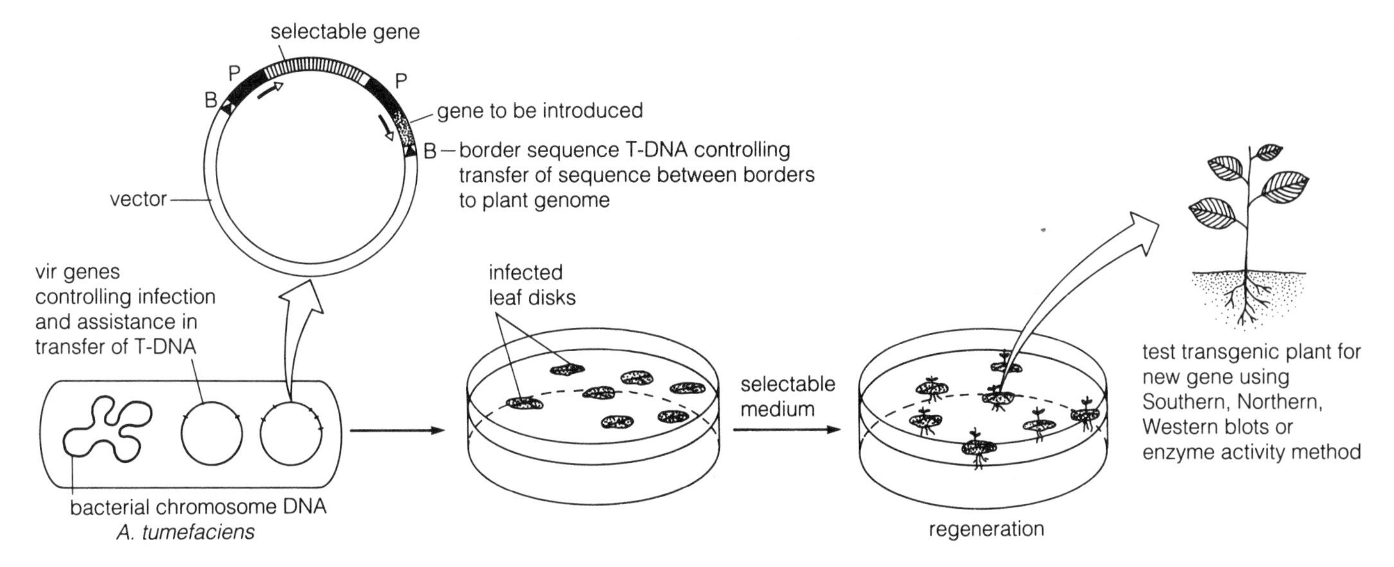


**Figure 24-7** Transformation of plant leaf discs to produce regenerated plants containing a transferred foreign gene present within the T-DNA borders on the vector.

*Transformation of plant cells using a high-velocity microprojectile* *Agrobacterium* has a relatively limited host range and does not effectively infect monocots. This problem, coupled with the need to use protoplasts and subsequent plant regeneration for electroporation or direct DNA absorption, resulted in development of a new technology based on bombardment of cells with particles coated with naked DNA. To transform plant cells by this method, intact or explanted meristematic tissue is bombarded with high-speed **microprojectiles** usually made of gold or tungsten, which are coated with the DNA molecules to be introduced into the plant cells (Klein et al., 1988). These microprojectiles are propelled into the cell by an electric discharge or a cartridge blast. Once inside the cell, the microprojectiles release some of the DNA, which then undergoes stable integration into the host genome. The transformed cells within the treated meristems are then selected based on the expression of a selectable marker, or they are grown directly into plants. Transgenic plants can then be screened out of the population. It appears to be possible to transform virtually any species this way and to achieve integration of foreign DNA into the mitochondrial and chloroplast genomes as well.

## 24.4 Mechanisms Controlling Expression of Genes

Because accumulation of mRNA, protein, and finally enzyme activity in a cell is a result of the net synthesis, degradation, and functional activity of each protein species, regulatory mechanisms that modulate the ultimate expression of a gene encompass several controls, both post-transcriptionally and post-translationally. It is becoming clear that a large number of auxiliary transcription factors are required to regulate gene expression in eukaryotic cells at the transcriptional level. Molecular biology has helped us devise techniques that can be used to understand the basis of various regulatory mechanisms that control gene expression during plant development. In the following paragraphs, we explain some of these techniques and the regulatory mechanisms that have been identified by using them.

### Identification of *cis*-Acting Elements

It has been shown that information required to direct cell- or tissue-specific expression of a gene is present in DNA sequences that are adjacent to the DNA sequences that encode a particular protein. Identification and characterization of these controlling DNA segments, which are also known as ***cis*-acting elements**, has provided a wealth of information concerning gene expression in eukaryotic cells (Mitchell and Tjian, 1989). Generally, the *cis*-acting elements are located within a few hundred base pairs of the 5′- or 3′-ends of the protein-encoding sequences (Fig. 24-8). To determine the exact location of DNA sequences that regulate expression of a gene in a particular cell type during development, both the 5′- and 3′-flanking regions of genes are progressively deleted with the help of specific restriction enzymes. The ability of these sequences with deleted flanking regions to continue to provide cell- or tissue-specific expression of a reporter gene is then determined using transgenic plants. Based on the levels of expression of the reporter gene from these **deletion constructs** (*construct* refers to a DNA sequence modified by the scientist) in the transgenic plant, the exact se-

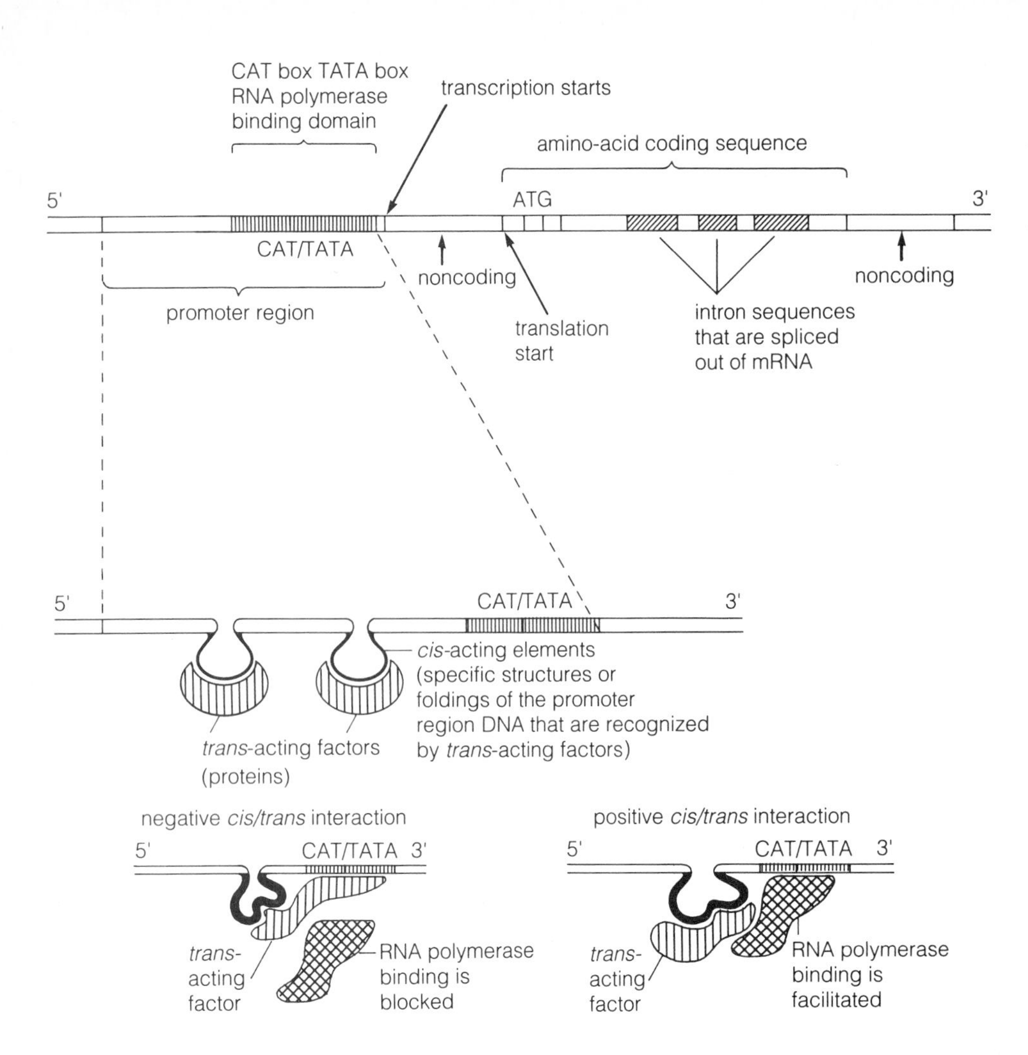


**Figure 24-8** Model of interaction between *cis*-promoter elements and a *trans*-acting factor. Interactions that affect the transcriptase both negatively and positively are illustrated. Specific nature of these interactions is presently conjecture.

quences required for the developmental-specific expression of a gene can be determined (Fig. 24-9). This technique was first used by Nam Hai Chua's group at Rockefeller University in New York to determine the DNA sequences required for the light-regulated expression of plant genes (see Moses and Chua, 1988; Kuhlemeier et al., 1987). Since then, this technique has been used by several researchers.

## Characterization of *trans*-Acting Elements Responsible for Expression of a Gene

Biochemical and genetic analyses led to the discovery that multiple ***trans*-acting factors** (commonly proteins) recognize and bind to the ***cis*-acting** elements located within the promoter region of eukaryotic genes; there the factors control synthesis of mRNA by RNA polymerase II (Fig. 24-8). The presence of *trans*-acting elements is usually demonstrated by retardation of electrophoretic mobility of *cis*-acting DNA sequences of a gene by the nuclear extract from cells actively expressing that gene (VarShavsky, 1987). This technique is based on the concept that the functionally important domains of the *cis*-acting DNA sequences are specifically recognized by the *trans*-acting proteins. Thus binding of the *trans*-acting protein will increase the size of any DNA fragment containing its corresponding *cis*-acting element, thereby decreasing its electrophoretic mobility. Recently, several *trans*-acting proteins have been identified and purified from eukaryotic cells (Mitchell and Tjian, 1989). Many *trans*-acting proteins have common or similar sequence domains that apparently account for their ability to interact with DNA in the double-stranded conformation.

## Molecular Techniques to Alter Expression of Specific Genes

*Complementary RNA (antisense) sequences* Naturally occurring and experimentally induced genetic variants are valuable assets for determining the role of a particular gene product in controlling physiological processes. Obtaining such mutants is difficult, however. Recently, RNA that is complementary to a specific mRNA has proven to be quite useful as a tool to prevent translation of that mRNA (Van der Krol et al., 1988; Smith et al., 1988). The **complementary RNA** (also named the **antisense RNA**) hybridizes, or forms a duplex with, the mRNA through hydrogen-bond base pairing; this inhibits translation of that specific mRNA (Fig. 24-9). Although the mechanism by which translation is inhibited has not yet been elucidated, it has been suggested that such duplex mRNA is either rapidly degraded, that its processing in the nucleus is impaired, or that its attachment to ribosomes and subsequent translation is blocked. Production of antisense RNA for the gene for chalcone synthase (see Section 20.9) in petunia and tobacco resulted in a new flower phenotype with altered pigmentation (Van der Krol et al., 1988). The altered flower pigmentation in the transgenic plants was caused by reduced steady-state levels of mRNA (about 1 percent of the wild-type expression level) for chalcone synthase. In another study, introduction of a gene of tomato that expresses antisense RNA for *polygalacturonase* (an enzyme that hydrolyzes pectins in cell walls) resulted in over 95 percent inhibition of the total polygalacturonase activity in fruits of the transgenic plants (Smith et al., 1988). In most plant studies, the CaMV 35-S promoter, which when present at the 5′-end of a gene allows a high level of mRNA synthesis, has been used to drive expression of the antisense RNA genes. The use of *cis*-acting regulatory sequences along with antisense sequences will eventually allow construction of developmentally or environmentally conditioned mutants.

Antisense RNA methodology may eventually have wide application both in basic and applied plant biology. We may be able to use antisense methodology to reduce expression of deleterious genes and thereby enhance overall crop productivity or remove from plants chemicals that are hazardous to human health, such as nicotine from tobacco, caffeine from coffee or tea, or numerous carcinogenic substances from various plants.

*Exchanging or altering* cis-*acting elements (promoters)* Generally, DNA sequences encoding a polypeptide of interest can be joined to *cis*-acting regulatory DNA sequences from other genes, and the chimeric construct is introduced into plant cells by one of several of the transformation methods described earlier. In addition to plant promoters, several promoters have been characterized from *Agrobacterium* and plant viruses. In general, these promoters are constitutively expressed in plants. The list of these promoters is growing rapidly, and soon it will become possible to modulate expression of a particular gene to a desired level in specific cells or tissues at specific times during development by using appropriate *cis*-acting sequences. This technology will have far-reaching implications for crop productivity. For example, using CaMV 35-S promoters, Roger Beachy's research group at Washington University in St. Louis obtained transgenic plants that express high levels of various virus coat proteins (Powell et al., 1986). Expression of these coat proteins provided the plants immunity against the viruses (as in Figure 24-6a). Similar approaches may help plants grow under stressful conditions, once the gene products required for stress resistance are characterized. Moreover, the ability to control expression of a particular gene in a cell- or tissue-specific manner will help determine the role of various genes in controlling plant growth and development.

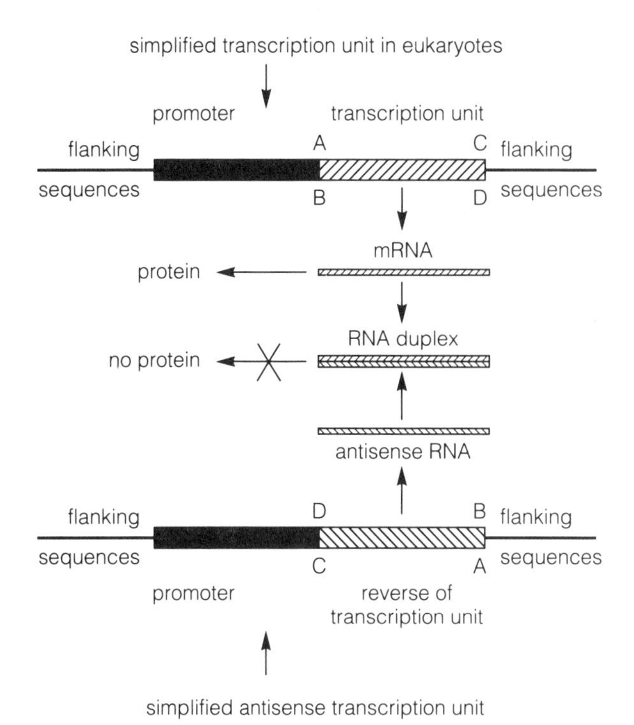


**Figure 24-9** Schematic representation of the genetic structure of an antisense-RNA transcription unit and one of the possible mechanisms (formation of RNA duplex) by which antisense RNA inhibits production of protein from mRNA. Several other mechanisms have been proposed by which antisense RNA inhibits production of a protein from mRNA, including degradation of RNA duplex by ribonuclease H, blockage of processing of mRNA to mature form, and inhibition of transport of mRNA to cytoplasm.

## 24.5 Examples of Isolated Genes that Affect Physiological Processes

Because we have emphasized that there is an underlying genetic basis for all physiological phenomena, we present several examples of how control of physiological processes have become better understood through the use of molecular-genetic approaches.

## Seed-Storage-Protein Genes

The first plant genes isolated and studied were those encoding seed storage proteins. The high abundance of these proteins and the abundance of their mRNAs greatly facilitated isolation of cDNAs complementary to their mRNAs from cDNA libraries.

The specific accumulation of these proteins only in tissues of the developing ovule provided an opportunity to examine how gene structure might control the particular part of the plant in which a gene would be expressed. For example, the gene encoding the bean seed storage protein *phaseolin* was introduced into tobacco plants using a gene-transfer system. The phaseolin protein, which is not normally made by tobacco plants, was synthesized at high levels in the tobacco seeds but not in the leaves. The information necessary to cause the specific synthesis of the phaseolin protein in seeds but not other plant parts was carried on the transferred phaseolin gene. More interestingly, even though the seed storage proteins of tobacco are made in both the embryo and endosperm, the phaseolin protein was synthesized only in the embryo, just as it is in the bean plant. Not only is the tissue specificity of this gene manifested in a foreign host; so also is the temporal specificity of its expression. The phaseolin protein begins to accumulate in bean seeds 15 days after anthesis. The transfer of the phaseolin gene to tobacco resulted in phaseolin accumulation 15 days following anthesis, even though tobacco seed proteins begin to accumulate much earlier (at 9 days post-anthesis; see Kuhlemeier et al., 1987).

Isolation and study of seed-storage-protein genes have revealed other interesting features of some plant genes (Higgins, 1984). Many seed-storage-protein genes, including those encoding the maize storage proteins, the *zeins*, exist as **multigene families** represented in the genome by many copies, 5 to 25 copies in the case of zeins. The gene copies (the gene family) that encode these proteins within the genome often differ slightly in their sequences. The members of a gene family are usually clustered together within the genome, and it has been speculated that gene families arise as the result of gene duplication followed by sequence divergence through accumulation of mutations. In several cases some members of gene families are not expressed; that is, no functional mRNA is produced from that gene and it is referred to as a **pseudogene**. The function of gene families is still uncertain, but it has been suggested that extra copies of a given gene may facilitate its evolution by providing greater tolerance of deleterious mutations.

## Photomorphogenesis and the Quest for Light-Regulated Genes

The necessity of light for plants is obvious because plants derive their organic nutrients from photosynthesis. Apart from the central role of light in the nutrition of plants, however, is the effect of light on various other physiological processes. We have seen in previous chapters how light affects such processes as seed germination, flowering, and senescence.

Exactly how does light control these activities? In the introduction of this chapter we described a central framework by which various physiological processes of plants can be controlled by differential activities of genes. Until recently, precise mechanisms by which gene activity is controlled were not known. After we gained the ability to clone and sequence specific genes, researchers isolated several genes whose activities were suspected to be controlled by light. Many scientists reasoned that important clues to the question of how light could control a physiological process through an effect on gene expression might lie in the specific sequence or structure of the genes themselves.

We must keep in mind that some responses of plants to light probably do not involve induction of altered expression of genes (see Section 20.11). Some responses such as stomatal opening (see Section 4.4) and leaf movements (see Section 19.2) are very rapid and are more likely the result of the effect of light on a molecular process controlled by genes that are already expressed. However, as described in Section 20.11, light affects other processes by controlling activity of specific genes.

The effect of light on activities of several plant enzymes has been known for many years (reviewed by Kuhlemeier et al., 1987). Also, by using *in-vitro* translation assays, investigators found that unique mRNAs appeared in plant tissues treated with light. Therefore, light must somehow affect the direct expression of specific genes coding for these proteins. Within recent years the expression of several genes encoding many proteins, including the large and small subunit of rubisco, chlorophyll *a*/*b* binding protein, the 32 kDa electron-acceptor protein of photosystem II, PEP carboxylase, 3-phosphoglyceraldehyde dehydrogenase, phytochrome, and pyruvate,phosphate dikinase, are regulated in certain species by light. Useful reviews are by Kuhlemeier et al. (1987), Link (1988), Nagy et al. (1988), Thompson (1988), Okamuro and Goldberg (1989), Watson (1989), Gilmartin et al. (1990), and Simpson and Herrera-Estrella (1990). The most carefully studied genes that have activities controlled by light are those coding for the small subunit of rubisco and the chlorophyll *a*/*b* binding protein. These genes are under the control of phytochrome (see Chapter 20). Little is known of how the $P_{fr}$ form of phytochrome transduces the light signal to activate these genes, but some information about the unique structure of these light-regulated genes has allowed us to speculate about the possible molecular mechanism between $P_{fr}$ and gene expression.

The structure of the gene RbcS, which encodes the small subunit of rubisco, has been carefully studied.

Specific DNA sequences that cause the gene to be light-induced have been identified by use of an assay system that allows us to determine the effect of removing specific portions of the RbcS gene on its ability to be light-induced. Such studies revealed that a sequence of DNA **upstream** (opposite to the direction of transcription) from the amino-acid-coding sequences was necessary for the light-regulated transcription of the gene. This DNA sequence has been called the **light-regulated element (LRE)** and is an example of a *cis*-acting genetic regulatory element (Moses and Chua, 1988). Addition of the LRE sequence to other non-light-regulated genes caused them to become light-regulated. Surprisingly, it was also found that the LRE controls tissue-specific expression of the RbcS gene. The RbcS gene is specifically expressed in leaves but not roots, and this specificity of expression can be conveyed to another non-tissue-specific gene when the LRE was added to it (Kuhlemeier et al., 1987).

## Hormonally Regulated Genes

As powerful regulators of plant physiological processes, plant hormones are chemical signals that can control the phenotype of the plant. As we have already stated, the information within the genome is the basis for the phenotype, and interactions between the genome and the environment bring about all changes in the phenotype. The activity of plant hormones represents an important mechanism by which the environment interacts with the genome to control phenotype.

*Environmentally altered hormone levels* The environment (light, temperature, nutrition, and so on) often affects the amount and types of hormones made by various plant tissues (see Chapters 17, 18, and 20). Therefore, environmental regulation of genes controlling hormone biosynthesis may represent a powerful mechanism for controlling plant development. The specific combination of hormones and even the particular relative amounts of hormone combinations have long been suspected to be important in controlling plant morphogenesis. A classic example is the demonstration by Skoog and Miller (1957) that the ratio of auxin to cytokinin controls formation of roots or shoots by undifferentiated plant tissues (see Section 18.1). The introduction into donor plants of genes obtained from *Agrobacterium tumefaciens* that can alter the levels of auxin and cytokinin demonstrated conclusively that the ratio of these hormones does indeed control root and shoot formation (Klee et al., 1987; Medford et al., 1989). Transfer of a genetic element that contains a functional gene for synthesis of auxin (but not of cytokinin) resulted in high auxin-to-cytokinin ratios and root formation. Transfer of an element containing a functional gene for cytokinin (but not for auxin) synthesis produced tissues with low auxin-to-cytokinin ratios, and shoots developed from these tissues.

*Control of hormone receptivity by developmentally regulated genes* We should keep in mind that the control of gene expression by various plant hormones is likely a multistep process. Plant hormones most likely act as more general switching signals that are activated by binding to a hormone receptor present in only certain cells or tissues (see Section 17.1). The hormone receptor complex or a second messenger might then act as a *trans*-acting DNA binding complex that activates particular sets of genes, which themselves may even produce specific *trans*-acting signals. Therefore, the effect of a particular hormone on gene expression will differ depending on the presence of receptors in the tissue being treated. This specific responsiveness of different tissues greatly increases the specificity of action of the hormone.

In addition, many physiological responses are a direct effect of environmental or hormonal signals that do not involve any induced change in gene expression but result instead from interaction of the environment or hormone with a specific receptor that is biochemically active and does not interact with DNA. However, the gene encoding such a receptor is usually developmentally controlled as well and is active only in certain tissues at a certain time, just as are receptors that interact with DNA and control gene expression. A likely example of a hormone response involving a biochemically active receptor is the specific appearance of ABA receptors in guard cells, which allow ABA to control stomatal aperture.

Examples of genes whose expression is controlled by gibberellin, IAA, ABA, and ethylene have been found in tissues of several plant species. The actions of many of these substances were described in Chapters 17 and 18.

## Temperature-Regulated Genes

Temperature has obvious and profound effects on plant growth and development. Molecular responses of plant tissues to elevated temperatures have been the subject of intense study in the last decade (see Chapter 26). Plants and other organisms ranging from bacteria to humans respond to increased temperature by synthesizing a group of specific proteins referred to as **heat-shock proteins (hsps)**. Plant tissues produce two major classes of hsps: high-molecular-weight hsps from 70 to 95 kDa and low-molecular-weight hsps from 15 to 30 kDa. If the temperature shift is large enough, synthesis of hsps is dramatically increased, whereas synthesis of most other proteins is dramatically decreased. The temperature shift required to induce hsp synthesis varies depending on the plant species.

The regulation of synthesis of many hsps involves

controlling transcription of the hsp gene into mRNA, although control can also be at translation. A heat-shock-induced promoter has been used to convey temperature inducibility to a reporter gene.

There has been much speculation and some controversy over the role of hsp genes in tolerance to high temperatures. The ubiquity of these genes in diverse organisms and the highly conserved sequence of some of the hsp genes suggests that these genes have some important fundamental function. Some evidence that hsp genes are important for **thermoprotection** has emerged. The results of a number of experiments first performed by Lin et al. (1984) and Key et al. (1985) showed that tissues that have been heat shocked previously and have synthesized hsps can survive a subsequent heat treatment that is otherwise lethal. More recent experiments provided evidence that continued synthesis of most hsps by some plant tissues is not required to tolerate elevated temperatures. However, hsps may play an important protective role in the transition or adjustment period after exposure to high temperature or may possibly prevent injury during brief cycles of high temperature.

Some hsps have been identified as proteases. Ubiquitin is a well-studied heat-induced protein that targets other proteins such as phytochrome for proteolysis (see Section 20.3).

## Developmentally Regulated Genes

Development of an organism occurs through both space and time. Therefore, we may think of developmentally regulated genes as those regulated over time or among spatially separated tissues. Many genes that are expressed only during the development of specific tissues have been identified (reviewed by Goldberg et al., 1989). These include the seed-protein genes; early- and late-expressed genes during embryogenesis; embryo-specific, germination-specific, and cotyledon-, root-, leaf-, petal-, and anther-specific genes. Many genes are expressed in young plants but not older plants, and vice versa.

The technique of ***in-situ* hybridization**, in which a labeled RNA or DNA sequence is used as a probe to detect the presence of complementary sequences in thin tissue sections, allows the identification of very specific locations of mRNAs that are encoded by specific genes. Expression of some genes can be shown to occur in only certain cells of plant tissues, such as the epidermis or cells surrounding the vascular tissue of a leaf or cotyledon.

Intense interest and activity is now being focused on the precise regulatory mechanisms that control these developmental features of plant genes. We are certain that the rapid progress that is being made will allow a dramatic expansion of our level of understanding of the role of gene expression in the physiology of plants. Already, hundreds of plant genes have been cloned, and the entire DNA sequence encoding at least 30 known plant enzymes has been determined. We encourage students of plant physiology to become more familiar with this developing technology, which inevitably will affect all aspects of plant physiological studies in the future.

# FOUR

# Environmental Physiology

# 25

# Topics in Environmental Physiology

It's high summer in the Rockies. You are on your way to the West Coast when you and your traveling companions decide that you will take an extra day to detour over the Continental Divide through Rocky Mountain National Park. You are afraid that your subcompact car might begin to boil over, but to your great relief you make it without mishap to the high point on the Trail Ridge Road; the sign says 12,183 feet above sea level! You pull into a parking stall, turn off the engine, and set out for a short tramp on the tundra.

Those little alpine plants are really something, you think to yourself. How can they look so healthy when they are covered with snow nine or ten months of the year? They must have to photosynthesize at rapid rates even at low temperatures to succeed in a place like this. How do their genes differ from those of the prairie plants that you saw yesterday?

Just last spring you completed the undergraduate course in plant physiology. Your copy of Salisbury and Ross (which you prudently decided not to turn in at the student book exchange!) lies on your pile of gear back in the car. Remembering the book's general outline, you begin to formulate questions and observations in your mind: Is there something about the water relations of these plants that allows them to withstand frost? The cells in their leaves must be turgid because of osmosis. The soil seems to have ample moisture; would transpiration cool these plants even below the chilly air temperature? How do they respond to the low temperatures up here? Is there anything special about the way they absorb minerals from this cold soil? Is there anything unusual about translocation of assimilates in these plants? Probably not.

Is there something special about their enzymes that allows them to function optimally at low temperatures? Photosynthesis must be highly efficient in these plants, and the low temperatures at night probably reduce dark respiration. Do they perhaps utilize C-4 photosynthesis? Do the ones growing almost on the bare rock have some means of nitrogen fixation—mycorrhizae, associated bacteria, or something? The flowers are brightly colored. Are their pigments produced through some special biochemistry? Could the colors be intensified some way by the bright light and intense ultraviolet radiation? Are the plants being "sunburned"?

Being dormant so much of the year could mean that these plants have unusual timing mechanisms. Do they grow slightly while under the snow, so that they are ready to go when summer finally comes? What causes them to flower? Do they go dormant in the fall in response to the shorter days? What hormonal mechanisms if any mediate these responses to the environment? Do the leaves fold up at night to resist radiant loss of heat to the cold sky above? Surely auxins, gibberellins, cytokinins, ethylene, and inhibitors play important roles in making these plants what they are, where they are. How do they do it?

By asking questions like these, you are beginning to think like an environmental physiologist. And, of course, you don't have to be at the top of Trail Ridge Road to have such thoughts. Perhaps your walk was in the Sonoran Desert near Tucson, Arizona, or in the relic tall-grass prairie of Nebraska. Maybe you were checking on the yield of a wheat field in Kansas. Were you lucky enough to visit the steaming rain forests of Brazil? Or if not, then the near-tropics of the Florida Everglades? Maybe your questions developed as you walked along the Appalachian Trail, or in the Great Smoky Mountains of Tennessee or North Carolina. Similar questions could occur to you on a farm in California, in Central Park in New York City, or as you putter in your backyard garden.

In all these situations and thousands more, you can ask related questions about plants in their environments. Indeed, you can develop these questions into a science. Though the questions may be similar, the sciences that develop from them may have different names. If you are a dyed-in-the-wool physiologist, you

## The Challenge of a New Field: Plant Physiological Ecology

*Park S. Nobel*

*There are many physiological ecologists who began their careers as traditional ecologists, but there are few who began as physicist-engineer-physiologists. Park Nobel is one of those rare ones who came from physics to ecology. Indeed, he is now a recognized leader in this field and the author of comprehensive texts on biophysical plant physiology and ecology. Here he tells the story of his changing and developing research interests.*

A sabbatical year spent in Canberra, Australia in 1973–1974 completely changed my career. I had left for Australia as a laboratory scientist studying the ion and water relations of chloroplasts, especially the use of irreversible thermodynamics to interpret osmotic responses. Based on my desire to learn more about environmental matters, the Guggenheim Foundation had funded a project to study the chloroplasts of guard cells. However, the available techniques required far more chloroplasts than could readily be isolated from guard cells. So, capitalizing on a relatively unused wind tunnel, the year was spent developing equations describing the boundary layers adjacent to cylindrical and spherical objects, which challenged my undergraduate training in engineering (bachelor's in engineering physics from Cornell University) and graduate training in physics (master's from California Institute of Technology). After a Ph.D. in biophysics from the University of California, Berkeley in 1965, I had steadily moved from the physical sciences that I enjoyed studying to the biological sciences, where I felt that discoveries loomed and that each person could make significant contributions.

Upon returning to Los Angeles, I wanted to continue research on boundary layers of cylinders and spheres and thought immediately of the magnificent succulent plants that had caught my attention during trips to the desert. Two other professional changes took place. First, the equipment in my chloroplast laboratory matched the request of a newly hired faculty member, so I agreed to exchange the contents of my laboratory for cash, thus burning my cellular bridges but gaining finances to begin in a new field. Secondly, I joined the Laboratory of Biomedical and Environmental Sciences to study the "biophysics of desert plants" such as agaves and cacti, beginning initially with water relations and later returning to boundary-layer considerations.

In 1974 an undergraduate (Larry Zaragoza) and a graduate student (Bill Smith, now at the University of Wyoming) started working in my laboratory on an intriguing anatomical problem based on my course in plant physiological ecology. Specifically, just as ground area is appropriate for discussing productivity by crops and leaf area is appropriate for discussing $CO_2$ uptake by leaves, mesophyll cell-wall area seemed appropriate for discussing cellular aspects of photosynthesis. Recognition of the importance of mesophyll cell surface area per unit leaf area, or $A^{mes}/A$, allowed a separation of the biochemical effects on photosynthesis occurring within mesophyll cells from effects due to anatomy. For example, sun leaves differed from shade leaves in that the former had a much higher $A^{mes}/A$, which leads to a much greater area within the leaves for $CO_2$ entry into the mesophyll cells. This led to a more quantitative understanding of the higher photosynthetic capacity per unit leaf area of sun leaves compared to the often fourfold thinner shade leaves on the same plant. A postdoctoral fellow in my laboratory, David Longstreth (now at Louisiana State University) and a technician, Terry Hartsock, helped elucidate the influence of salinity and nutrients on $A^{mes}/A$. We showed that light primarily influenced

might be quite happy with the term **environmental physiology**. If your field is agriculture, you may call your science **crop ecology** or even **crop physiology**. Traditionally, if you work with vegetables (for "kitchen gardens") or with ornamentals, you call your science **horticulture**, but you may still be asking questions that concern environmental physiology. If your interest is in field crops (cereals, forage crops, root crops, and others), you may call your science **agronomy**. If you are interested in how plants grow in their natural environments, then your field is **physiological ecology**, which also has its applied aspects, such as **forestry** and **range management**. Indeed, most students in agriculture and in forestry and range management are required to take a course in plant physiology, mainly for the environmental aspects of the subject.

## 25.1 The Problems of Environmental Physiology

The specific questions that guide research depend upon the specific field of endeavor. In agriculture, for example, most research is guided by the economics of obtain-

$A^{mes}/A$; salinity, water stress, and temperature influenced both anatomy and cellular photosynthetic properties; and nutrients affected mainly the cellular properties.

With regard to desert succulents, the obvious place to begin was with water relations. This ranged from analyzing the extension of the growing season caused by the storage of water within the stem of the barrel cactus *Ferocactus acanthodes* to the crucial shifting of water from the succulent leaves to the inflorescence of the common monocarpic century plant *Agave deserti*. The latter species in its native habitat in the Sonoran Desert was found to have the highest annual water-use efficiency ever reported for any plant: 40 g $CO_2$ fixed per kg water vapor transpired.

Once the importance of the water status was beginning to be understood, the complicated responses of agaves and cacti to temperature were investigated, often with the use of computer models. This ranged from seeing what day/night air temperatures led to maximal nocturnal $CO_2$ uptake by these CAM plants to studying the influence of morphology on distributional limits. For the latter a simulation model was developed with Don Lewis that included the conventional energy-budget terms popularized by David Gates, Klaus Raschke, and others, plus additional heat-conduction and storage terms required for describing the massive organs of succulent plants. [See Section 4.8.] This allowed studies on the influence of spines and apical pubescence in protecting the meristem of cacti from freezing, which can extend their northern and high-elevation distributional limits. The freezing tolerances of cacti from North and South America were compared. Analysis of the high-temperature tolerance proved equally exciting, as many species of agaves and cacti could survive exposure to 65°C, an extremely high temperature for the survival of vascular plants. [See Section 26.6.]

As another application of modeling, we investigated the relationships between the morphology of desert succulents and light interception. Beginning with a field determination that $CO_2$ uptake by *F. acanthodes* was limited by the photosynthetic photon flux—even during clear days in the high-radiation environment of a desert—and incorporating the chance observation that the terminal cladodes (flattened stems) of platyopuntias usually tend to face east-west, we have shown that other morphological features of cacti are adaptations to maximize light interception at times of the year most favorable for growth. Thus modeling had again been combined with morphology to provide insight into plant physiology.

More recently, the influence of water status, temperature, and photosynthetic photon flux on net $CO_2$ uptake over 24-h periods has been determined for various species of agaves and cacti. An index has been created for each of these environmental factors, ranging from zero when field values of that factor cause net $CO_2$ uptake to be zero (such as during extended drought or extreme temperatures), up to a maximum of unity when no limitation occurs. The product of the three indices is the Environmental Productivity Index, which estimates the overall influence of the environment on net $CO_2$ uptake and hence on plant growth and productivity. A component nutrient index has been proposed to incorporate edaphic factors. Also, the Environmental Productivity Index has been predicted on a worldwide basis, including simulating the effects of the doubled atmospheric $CO_2$ levels expected during the next century.

To close, I offer a quote from Pasteur that summarizes my own philosophy of plant studies: "Chance favors only the prepared mind." Studying the many fields that impinge on plant physiology and ecology, such as calculus, chemistry, physics, and even areas of engineering, may seem remote from a direct study of plants, but such fields can provide the tools to enhance overall understanding of plant responses to the environment.

ing maximum yields of a specifically harvested part and the highest possible quality for the lowest cost and energy inputs. Environmental physiology always plays a role in such research.

Studies in physiological ecology apply the methods of physiology to the problems of ecology. Traditionally these problems have centered on the questions of plant and animal distribution. The ever-present assumption is that organisms occur where they do in nature because they are adapted to their environments. Plants or their seeds that occur in deserts can withstand drought and high temperatures, for example, whereas the deciduous trees of temperate forests grow only when there is ample moisture and moderate temperatures. Hence, studies in physiological ecology may attempt first to measure the microenvironments of plants or animals in the field and then simulate these environments to study the same organisms in the laboratory. But ecology has many interesting problems besides those of distribution.

Near the end of the 1970s, four eminent physiological ecologists were given the assignment of editing four volumes on physiological ecology for the *Encyclopedia of Plant Physiology* (New Series). (See the essay of Park S.

**Table 25-1 Some Topics of Physiological Plant Ecology.**[a]

| Topic | Some References in This Text |
|---|---|
| 1. Responses to the physical environment | |
| Radiation: Measurement of parameters | Appendix B |
| Responses to irradiance (quantum flux) | |
| Photosynthesis (PAR and PPF) | Chapters 10–12 |
| Other responses | Chapter 20 |
| Spectral distribution (radiation quality) | Appendix B |
| Phytochrome responses | Chapter 20 |
| Other nonphotosynthetic responses (blue light, etc.) | Chapters 19, 20 |
| Photoperiod and biological-clock mechanisms | Chapters 21, 23 |
| Responses to ultraviolet and ionizing radiation | |
| Temperature: Normal and extreme | Chapter 22 and many others |
| Wind as an ecological factor | |
| Energy exchange between a plant and its environment | Chapter 4 |
| Fire as an ecological factor | |
| The soil environment | |
| Aquatic environments | |
| 2. Water in the soil-plant-atmosphere continuum (Schulze et al., 1987; stress physiology: Osmond et al., 1987) | Chapters 2–5, 26 |
| Water in tissues and cells | Chapters 3–5 |
| Water uptake, storage, and transport | Chapters 3–5, 7, 8 |
| Water loss through stomates and cuticle | Chapter 4 |
| Plant responses to flooding | Chapters 13, 26 |
| Seed and spore germination | Chapters 20, 22 |
| 3. Photosynthesis | Chapters 10–12 |
| Ecological significance of different $CO_2$-fixation pathways (Pearcy et al., 1987) | Chapter 12 |
| Modeling of photosynthetic response to environment | |
| Water use and photosynthesis | Chapter 12 |
| Plant life forms and their carbon, water, and nutrient relations | |
| 4. Responses to the chemical and biological environment (individual organisms) | |
| Features of soil chemistry (limestone vs. silicate rocks, soil *p*H, essential or toxic ions) | |
| The physiology of salt-tolerant plants (halophytes), osmoregulation (Osmond et al., 1987) | Chapters 6, 7, 26 |
| Ecology of nitrogen nutrition (including $N_2$ fixation) | Chapter 14 |
| Plant-"plant" interactions | |
| Mutualism (mycorrhizae, lichens) | Chapter 7 |
| Parasitism (viral, bacterial, fungal) | Chapter 7 |
| Competition | |
| Allelopathy (Bazzaz et al., 1987) | Chapter 15 |
| Plant-animal interactions | |
| Pollination, fruit and seed ecology | |
| Herbivory (Bazzaz et al., 1987) | Chapter 15 |
| Carnivorous plants | Chapter 19 |
| Mutualism | Chapter 15 |
| Parasitism | |
| Competition | |
| 5. Ecosystem processes (populations forming a great variety of communities) | |
| Mineral cycling or transfer | |
| Energy flow through ecosystems | |
| Productivity (photosynthesis again) | Chapter 12 |
| Human influences | |
| Biocides and growth regulators | |
| Pollution: Atmosphere, water, soil | |
| Agriculture: Controlled or artificial ecosystems | |
| Responses to multiple environmental factors (Chapin et al., 1987) | Chapter 25 |

[a]This table is based on the tables of contents in the four volumes edited by Lange et al. (1981–1983) and on a special issue of BioScience devoted to plant physiological ecology (see Mooney et al., 1987, and references cited in the table). Often several chapter titles were concerned with a single topic (suggesting intense interest in that field); in the table such titles are combined under a single topic. Indeed, many titles were modified to fit the format of the table.

Nobel, one of those four eminent physiological ecologists.) After considering work that was going on all over the world, they commissioned various authors to write review articles relating to these topics. Although the emphasis was on natural ecosystems, they did not hesitate to commission many authors who were interested in agricultural and other applied problems. In 1987, another summary of the field appeared in a special issue of BioScience (see Mooney et al., 1987). Table 25-1 summarizes the topics of environmental physiology based on the tables of contents of the four volumes and the special issue of BioScience.

To begin with, environmental physiologists study plant responses to the physical environment, as reflected by the first large category in the table. Furthermore, much of the emphasis in studies of environmental physiology has been on water in the soil-plant-atmosphere continuum and on environmental effects on photosynthesis. In addition, a number of other interesting interactions between individual plants and their chemical and biological environments are being investigated. During recent years, ecologists have become increasingly interested in the ways individual plants and animals interact with each other. Parasitism, herbivory, and related topics provide examples. The next level of study concerns interactions of plant and animal populations with each other and with the physical environment. Perhaps these topics are being emphasized more by ecologists than by physiologists. Mineral cycling and energy flow through ecosystems, and ecosystem productivity as well, have been especially active areas of research, but human influences have also received much attention—especially in the news media! (See Barbour et al., 1987 for a good general discussion of plant ecology.)

In the remainder of this chapter, we consider some principles of plant responses to the environment, including the question of the nature of the environment, and then we examine two examples of environmental physiology: the role of genetics and responses to radiation.

## 25.2 What Is the Environment?

Dictionaries define **environment** as the circumstances, objects, or conditions by which one is *surrounded*.[1] Should environment include *everything* that surrounds an organism? Are a cricket's chirps or low-energy radio waves coming from a distant planet part of a plant's environment? In the broadest sense, yes. But *if* they have absolutely no effect on the plant, it seems unreasonable to think of them as part of the plant's **operational environment**, which is the complex of climatic, **edaphic** (soil), and biotic factors that act on an organism or an ecological community and ultimately determine its form and survival.

Such a definition helps somewhat, but it may not always be easy to know for sure whether an environmental factor is part of the operational environment. The cricket's chirp is clearly part of the operational environment of another cricket, but as yet we have no reason to believe that it is part of a plant's operational environment. How about the radio waves? We know of no way that such waves could act upon either crickets or plants, but it is possible that such an action remains to be discovered.

George G. Spomer (1973) at the University of Idaho has defined the operational environment in a way that provides further insight. Applying concepts of thermodynamics (Chapter 2), he points out that an environmental factor interacts with an organism only when the factor heats or does work on the organism or the organism heats or does work on the factor. If the radio waves pass through the plant unchanged, for example, then the plant is also unchanged, and there could be no interaction; the radio waves would not be part of the plant's operational environment. The cricket's chirp does work upon the hearing apparatus of another cricket, so it is part of the second cricket's operational environment. According to Spomer, such an interaction involves a direct transfer of mass (matter) or energy across the boundary between an organism and its environment.

In the diary of his first trip into the Sierra written over a century ago (see Muir, 1976), John Muir said: "When we try to pick out anything by itself, we find it hitched to everything else in the universe." This is a statement of the **holistic concept**, which suggests that everything in the universe interacts with everything else. At some highly theoretical level this might be true (although no one can test the idea), but Spomer's analysis of interactions by energy or mass transfer helps to place the holistic concept in a proper perspective. The transfer/interactions (if they occur at all) might be so infinitesimal that they have no practical consequences. Sound waves from the crickets' chirp act on the plant as well as the other crickets' ears, but the energy transfer is probably too small to have any significant effect on the plant's metabolism or other activities—especially compared with energy inputs from other sources. The plant has no way to extract *information* from the cricket's chirp, although another cricket does.

Environmental factors that fit Spomer's definition of operational factors include light, heat, water, electrical potentials, various gases, mineral elements, and organic substances. These factors can be directly transferred across the boundary between the organism and its environment. Temperature, *p*H, electrical potentials, gravitational forces, partial pressures of gases, concentrations, and water potential are not operational factors because they are not themselves transferred across boundaries. Instead, they indicate a *potential* for transfer. If the *temperature* inside an organism is lower than the temperature outside, the difference indicates the potential for transfer of *heat* across the boundary. Different *p*H values indicate a potential for transfer of hydrogen ions, partial gas pressures indicate a potential for gas transfer, concentrations indicate potentials for transfer of dissolved substances, water potentials indicate potentials for transfer of water, and so on.

[1]If this is the case, is there such a thing as the *internal environment* of a cell? No, especially if we are careful in our use of the language. It is correct to speak of internal conditions, and any cellular organelle such as a chloroplast has its own environment, but we should not speak of "internal environment."

**Table 25-2 Environmental Factors That Are Operational for Many Plants.**

| Factor | Units | Potential | |
|---|---|---|---|
| | | Measure | Units |
| Energy factors | | | |
| Radiation (including light) | $J\ m^{-2}$<br>cal[a] $m^{-2}$ | Radiation level (irradiance) compared with absorptivity of pigment | $W\ m^{-2}$<br>photons: mol $m^{-2}\ s^{-1}$<br>cal[a] $m^{-2}$ min[a]$^{-1}$ |
| Heat | $J\ kg^{-1}$<br>cal[a] $kg^{-1}$ | Temperature | Kelvin<br>°C[a] |
| Gravity[b] | | | |
| Mass factors | | | |
| Gases | kg or mol | Pressure or partial pressure | Pa ($N\ m^{-2}$)<br>bars[a]<br>(0.1 MPa = 1.0 bar) |
| Liquids and solutions | | Density (concentration) | $kg\ m^{-3}$ |
| Water | kg or mol | Water potential | pascals ($N\ m^{-2}$) |
| Solution (Solute in water) | kg or mol | Chemical potential, concentration | mol $m^{-3}$<br>$kg\ m^{-3}$<br>mol L[a]$^{-1}$<br>(%[a] or ppm[a]) |
| Hydrogen ions in water | kg or mol | *p*H ($H^+$ potential) (concentration) | *p*H units<br>(as above) |
| Solids | kg or mol | Concentration (seldom used) | kg or mol $m^{-3}$ |

[a]These are not SI units; they should usually be avoided, but plant physiologists still use liters (L), minutes (min), and degrees Celsius (°C); see Appendix A.
[b]Gravity is an energy factor in the environment but does not lend itself to this kind of analysis.

The environment—that is, the universe—may be too complex to describe, but if we limit ourselves to factors that are probably part of the operational environment of plants, the task is somewhat simpler. Table 25-2 lists the major energy and mass factors and the potentials generally used to describe their levels.

## 25.3 Some Principles of Plant Response to Environment

Although environmental physiology has blossomed only during the past three or four decades, some good studies go back over 100 years. Justus Liebig proposed a basic premise of the science in 1840, for example, as we'll see in a moment. This section also summarizes a few other generalizations that guide the research of environmental physiologists.

### Saturation and Limiting Factors

Perhaps the most fundamental principle of plant responses to environment—and the one most frequently encountered in this text—is that of **saturation**. Organisms respond to virtually any environmental parameter according to a common pattern: As a parameter increases, it reaches a **threshold** above which it begins to have an effect, after which the response increases until the system becomes **saturated** by the parameter. Then, as the parameter's level or concentration continues to increase, response remains constant or begins to decrease if at such high levels the parameter becomes toxic or inhibitory. Figure 25-1 shows the expected pattern. Looking back through this text, we can find figures that illustrate much of the phenomenon for temperature (Fig. 22-1), photosynthesis (Figs. 12-4, 12-8, and 12-9), mineral nutrition (Figs. 6-5 and 6-6), enzyme action (Fig. 9-10), and transport of ions across membranes

(Fig. 7-15). We discussed the curves in terms of Michaelis-Menten enzyme kinetics (Section 9.5) and applied this concept to our discussion of changing sensitivity to auxin in the gravitropic response of dicot stems (Section 19.5, Fig. 19-17). The saturation level approached as the factor increases is called $V_{max}$ in the Michaelis-Menten discussion, and it suggests the maximum availability of reaction sites for the reaction (for example, auxin binding sites). The factor level (for example, concentration) that produces one half of $V_{max}$ is called the Michaelis constant, $K_m$, and it indicates the affinity of the reaction sites for the factor (for example, auxin). Thus dose-response curves of the saturation type give us insights into what is going on between the factor and whatever it interacts with in the organism; we can talk about $V_{max}$ and $K_m$ sensitivity changes.

Figure 25-1 shows the basic dose-response curve with three phases or zones: deficiency, tolerance, and toxicity. This was emphasized by Victor E. Shelford (1913) in what he called the **law of tolerance**. As long as addition of a factor leads to an increased response, we can say the factor is **deficient**. If increasing the factor does not change the response, the factor is present in the zone of **tolerance**. The lowest level of the factor to give maximum response is the **optimum**. When addition of the factor causes decreased response, it is present in the zone of **toxicity** or **inhibition**. Between the optimum and toxicity, we speak of **luxury consumption**. Of course, a nonessential element will have no effect until it becomes toxic, and this toxicity can sometimes be lethal, whether the element or other factor is essential or not. (See the discussion in Berry and Wallace, 1989.)

It is easy to understand these curves between the minimum and the optimum and to understand the concept of saturation: The organism simply utilizes the factor being considered until its capacity for this utilization is used up or saturated. But what about the toxicity or inhibition present in so many of these examples? Explanations for toxicity vary, depending on the phenomenon being considered. When growth is inhibited by high temperature, we have suggested enzyme denaturation as the explanation (see Fig. 22-3), but this is not always satisfying. Recall *Laothenia chrysostoma* (Section 22.8), which dies within 30 days when *night* temperatures are 26°C; surely enzymes are not being denatured at these temperatures because *day* temperatures of 26°C or above are not harmful. The causes are no doubt complex, involving perhaps the production of an inhibitor during warm nights (implying a phytochrome interaction?). Superoptimal concentrations of mineral nutrients might become toxic because they begin to interact with systems in the organism other than those that were responding on the ascending part of the curves.

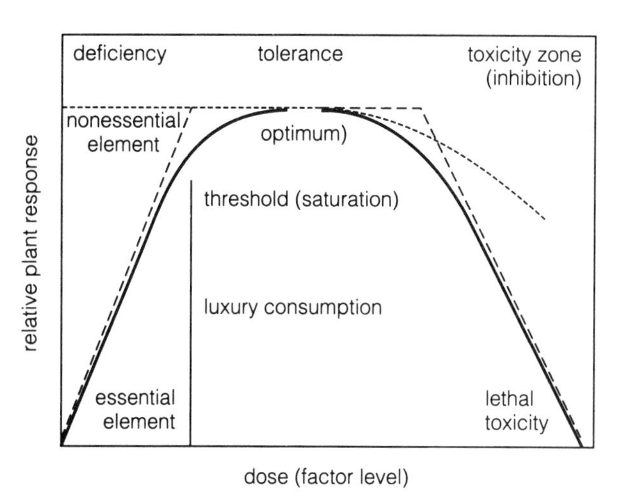


**Figure 25-1** A generalized dose-response curve showing an organism's response to an environmental parameter. The curve is three-phased, with zones of deficiency, tolerance, and toxicity or inhibition. Note that a nonessential factor can be toxic at high levels. Minimum, optimum, and maximum (only optimum is shown) are called cardinal points. Typically, there is a wide zone of tolerance; because additional amounts of the parameter in this zone do not induce additional plant response (typically, yield), we speak of luxury consumption.

Justus Liebig's (1841) book, published in Germany in 1840 and translated as *Organic Chemistry in Its Applications to Agriculture and Physiology*, had an immense impact on thought about plants. It almost became a best seller. In the book Liebig formulated his **law of the minimum**, which in retrospect can be derived from and understood on the basis of saturation curves. The law states: "The growth of a plant is dependent upon the amount of 'foodstuff' presented to it in minimum quantities." This is the deficiency zone of the dose-response curve (see Fig. 25-1). F. F. Blackman (1905) in England discussed the principle (without reference to Liebig) and proposed the term **limiting factor** for that "foodstuff presented . . . in minimum quantities."

Figure 25-2 illustrates Liebig's and Blackman's concept with a simple experiment in mineral nutrition involving two levels of phosphorus in nutrient solutions with a wide range of nitrogen concentrations. The threshold for nitrogen is extremely low, but below that level plants do not grow at all (that is, they stay the same size they were when transplanted). *As nitrogen increases, plants respond the same at both phosphorus levels until phosphorus becomes limiting at the nitrogen saturation level.* A higher concentration of phosphorus leads to a higher saturation level for nitrogen. Such ideal curves are called **Blackman curves**.

The practical implications of Liebig's law were and continue to be obvious and important. In agriculture (Liebig was probably the first important agricultural

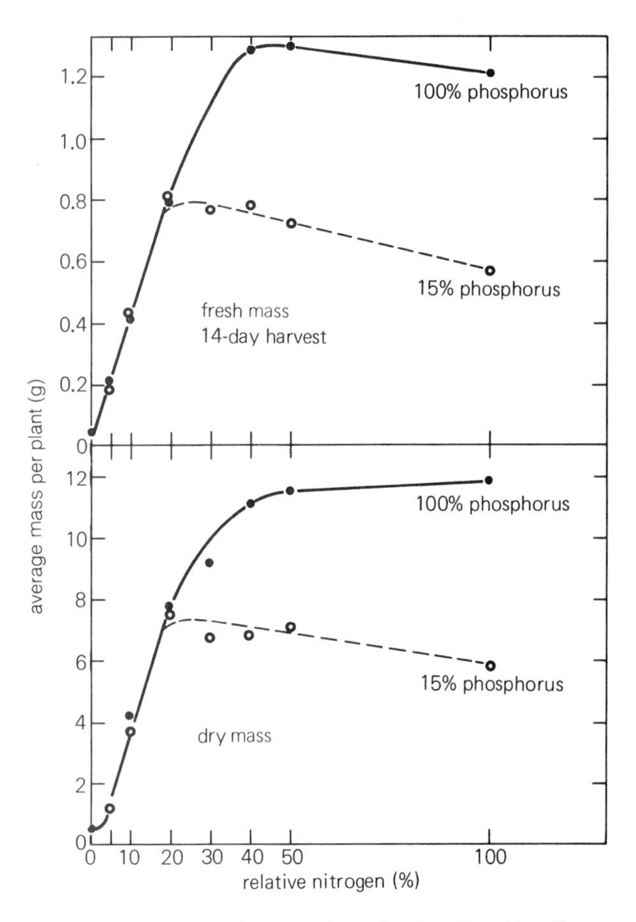


**Figure 25-2** Results of an experiment in mineral nutrition. Tomato plants were grown in vermiculite and watered with nutrient solutions containing various concentrations of nitrate ($NO_3^-$) and one of two concentrations of dihydrogen phosphate ($H_2PO_4^-$), as shown. Fresh and dry mass, plotted as a function of nitrate concentration, produces typical "Blackman curves," nicely illustrating Liebig's law. Nitrate in the 100-percent nitrogen solution was 15 mM, and phosphorus in the 100-percent solution was 1.0 mM (0.15 mM in the 15-percent solution). Plants were watered with the nutrient solutions on Mondays, Wednesdays, and Fridays and with distilled water on the other days. Curves represent fresh and dry weights of 14-day-old plants; note the slight differences in the shapes of the curves. Data are means of six plants. (See Salisbury, 1975, for more details.)

chemist), the challenge is to discover the limiting factor and to supply sufficient amounts of it. If plant yield is limited by insufficient amounts of nitrogen, then more nitrogen is applied. When enough nitrogen has been applied, then perhaps phosphorus becomes limiting and needs to be applied. This approach has had spectacular success since 1840; much more food can now be produced on a hectare of land than before. Of course, the limiting factor might be one or more of many things besides mineral nutrients: water, damage by pests (diseases, insects), competition from weeds, $CO_2$ concentration (especially in greenhouses), or the plant's genes (hence the success of breeding programs), to name the important examples.

Few attempts have been made to discover and remedy *all* environmental factors that might be limiting, but the exploration of space has provided an impetus to do just that. If, for example, we were to establish a permanent colony on the moon or Mars, we would have to grow food for the inhabitants, and because *all* environmental factors would have to be controlled with equipment made incredibly expensive by the costs of transporting it to the moon or to Mars, it becomes essential to learn how to grow and prepare food with the highest possible efficiency and to recycle plant nutrients without a buildup of potentially toxic materials. The United States National Aeronautics and Space Administration has funded research on such a bioregenerative system, called **CELSS** (Controlled Ecological Life-Support System), as described in the nearby box essay.

Liebig's law may be applied in physiological ecology studies because a plant's geographic distribution may be limited on one of its borders by the factor presented to it in the "least" amount. This is always a relative matter because highly disparate quantities of the different elements and environmental parameters are required by plants (see Tables 6-1 and 6-3 for examples), and toxicities could also be limiting.

## Interaction of Factors

Unfortunately, things are not as simple as Liebig's law might imply. Under carefully controlled conditions such as those represented in Figure 25-2, everything may work out as the law predicts. In the real world, things seldom work out so well; the curves are not identical in the ascending parts where only one factor is supposed to be limiting. Figure 25-3 shows a more typical response; the experiment was the same as that illustrated in Figure 25-2, but the plants were somewhat older.

There are several ways to explain a failure of Liebig's law. Most probably all boil down to a single idea: The extent to which Liebig's law might function (that is, the extent to which Blackman curves might be obtained in multiple-factor studies) depends upon the extent to which the factors under consideration enter into reactions within the organism. Say, for example, that we plot photosynthesis as a function of irradiance at two $CO_2$ levels. If we could study a single chloroplast, then the saturating irradiance might be quite distinct. But in the real world, bottom cells in a leaf or lower leaves on the plant will not get as much light as upper cells or leaves; they will not all photosynthesize to the same extent. Diffusion of $CO_2$ into the leaf will also not be uniform. As a result, there will not be a sharp break in the curve (nor a distinct saturation irradiance), and both light and $CO_2$ can limit photosynthesis at the same time.

This idea can be understood a bit easier, perhaps, with equilibrium constants. Think of a reaction with

wheat. Examination of our wheat plants showed that many heads were not filled with wheat grains. Pollenation and grain development were not proceeding normally. Our high yields were obtained because of the otherwise ideal conditions and the high density of planting made possible by these ideal conditions.

As it turned out, we had chosen too high a temperature for normal pollination and grain development. When we lowered the temperature to 20°C (and to 17°C for a 4-h dark period), our life cycle was extended from the previous 59 days to 89 days, but the harvest index reached 45 percent and the yield of edible wheat nearly tripled to 60 g $m^{-2}$ $d^{-1}$—*five times* the world record in the field! Such yields would allow a CELSS farm of only 13 $m^2$ to supply a single human being! Of course, a real CELSS farm would be considerably larger than this to provide a safety factor, to allow for other less efficient crops ("man cannot live on bread alone"), and perhaps to allow the use of lower irradiances, which prove to be photosynthetically more efficient (see Box Figure).

Have we finally reached the maximum genetic yield potential for wheat? And how can we tell if we have? It is possible to approach tentative answers to these questions by considering that photosynthesis always sets an upper limit on yield. No green plant that depends on light for growth can produce more chemical-bond energy than it receives as radiant energy. Furthermore, biochemical studies and the second law of thermodynamics tell us that photosynthesis can never be 100 percent efficient (Chapter 11). The biochemical considerations suggest a maximum theoretical efficiency of around 30 percent. In addition, there are other limitations. Some of the fixed energy will be required for growth of the plant and to maintain the living cells already in existence. Furthermore, not all the light energy that falls on the crop will be absorbed and used in photosynthesis (although, with our dense wheat plantings, absorption is 95 to 98 percent). Taking all these things into consideration (Bugbee and Salisbury, 1988, 1989), it appears that we could never hope for a conversion of light energy to chemical-bond energy greater than 15 to 20 percent, and 13 percent may be an even more realistic figure. As the Box Figure shows, almost 11 percent of the light energy, applied at the lowest level (400 μmol $m^{-2}$ $s^{-1}$) to the wheat plants over their entire life cycle, was converted to chemical-bond energy in the harvested biomass. And we measured (by gas exchange) even higher conversion efficiencies during the maximum growth period of the crop (lower during establishment of the canopy and maturing of the leaves and heads). In short, we estimate whether we are approaching maximum yield by comparing our productivities with the theoretical limits set by photosynthesis. If they approach those limits, only light must be limiting; if not, something else is limiting.

These experiments have demonstrated that maximum yields for an ecosystem such as a crop of wheat in a growth chamber can greatly exceed those obtained under the very best of field conditions. Does this mean that wheat plants in the field are functioning at only a fifth or less of their genetic potential? Probably not. Actually, individual wheat plants in the field could be yielding somewhat better than the plants in our growth chambers. Planting densities in the field are on the order of 200 to 500 plants $m^{-2}$, whereas the experiment shown in the Box Figure used 2,000 plants $m^{-2}$, and we have had densities as high as 6,000 $m^{-2}$ in some experiments. The incredible yields achieved in our chambers are caused not only because of the ideal water and nutrient conditions, the elevated $CO_2$ (about 1,200 μmol $mol^{-1}$), and the near perfect temperature, humidity, and wind speed, but especially because these conditions combined with high light levels (sunlight equivalent for 20 hours a day; 2.5 times as much as could be received anywhere on earth) allow the plants to be crammed together in the dense plantings essential for such yields. A leaf in full light in our chambers may yield more than its counterpart in the field because of these ideal conditions and because the life cycle has been shortened (from about 120 days in the field), but each *plant* is not yielding five times its counterpart in the field. Plants in the field could do better with more ideal light, $CO_2$, nutrient, and water conditions, but they do surprisingly well as it is. (More details in Salisbury, 1991.)

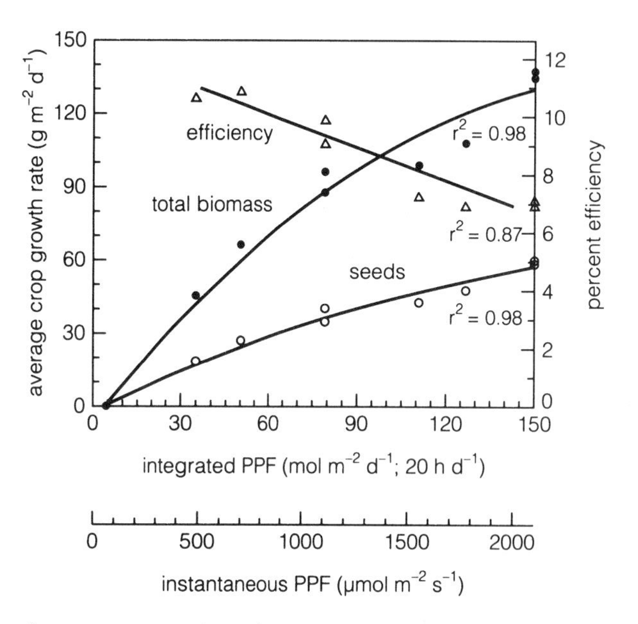


Average crop growth rate (total biomass and biomass of seed, which is edible grain) and percent efficiency (chemical-bond energy of total biomass as percent of input light energy) as a function of irradiance applied to wheat plants in a controlled environment. Irradiance is shown as instantaneous photosynthetic photon flux (PPF: flux of photons that are effective in photosynthesis; 2,000 μmol $m^{-2}$ $s^{-1}$ = full sunlight) and PPF integrated over the 20-h day. (See Bugbee and Salisbury, 1988.)

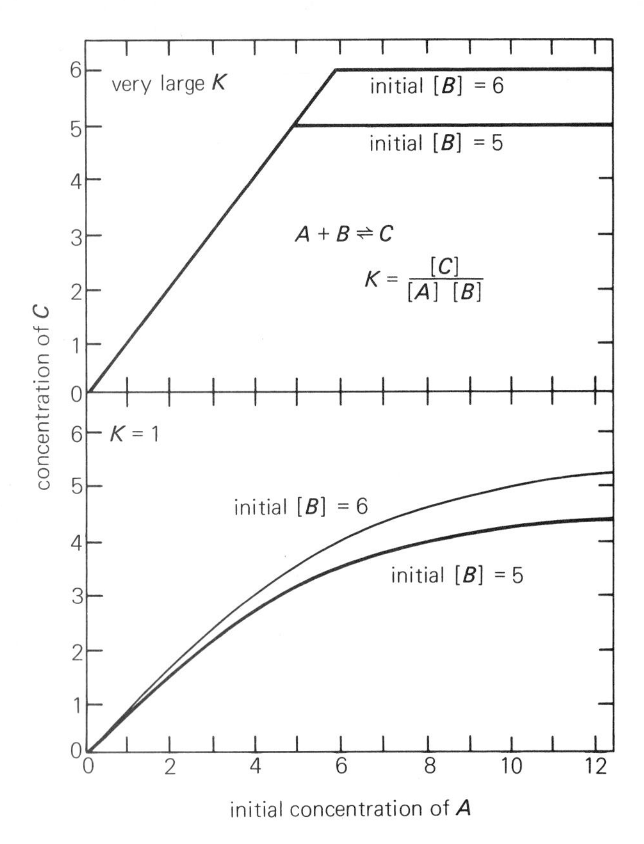


**Figure 25-4** An illustration of the principle of limiting factors as it might be observed in a chemical equilibrium reaction. Concentration of a hypothetical product ($C$) is plotted as a function of initial concentration of one reactant ($A$) in the presence of two initial concentrations of the other reactant ($B$). (Top) With a very large equilibrium constant, virtually all available $A$ enters into the formation of $C$ until $B$ becomes limiting at its two concentrations. This is the ideal limiting factor response ("Blackman curves"). (Bottom) If $K$ is only 1, then even at low concentrations, both $A$ and $B$ limit the amount of product formed.

There are many complications. In one form of interaction, the minimal or toxic level of one factor (for a given organism) may depend on levels of one or more other factors. Sodium or potassium ions, for example, may be quite toxic to plants at low concentrations when they are supplied to the roots in solution by themselves (along with a suitable anion, such as chloride). Addition of small amounts of calcium ions raises the minimum toxic concentration of the sodium or potassium ions to much higher levels.[2] Another example is the enhancing effect of one drug on the response of an animal to another drug. If two factors interact in such a way that the response to them given together is greater than the sum of responses to each given alone, we speak of **synergism**.

[2]Are we dealing with calcium deficiency instead of sodium or potassium toxicity? No, because other ions supplied by themselves are not toxic.

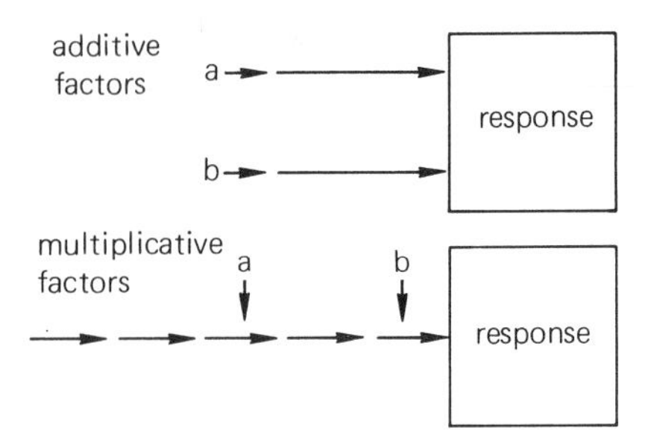


**Figure 25-5** An illustration of additive and multiplicative factors according to the analysis of Mohr (1972).

Some powerful mathematical tools have been developed to help us understand factor interactions in nature or in controlled experiments. One of these is **regression analysis**, which is a highly valuable tool in such situations as field observations, in which data must be taken as they come. When an experiment can be carefully designed in advance (for example, when treatments can be set up according to statistical criteria; that is, randomly), **Fisher's analysis of variance** is used widely and appropriately. (See statistics textbooks for descriptions of these methods.)

Such studies indicate whether or not two environmental factors interact. If, under the specified conditions, they both influence a given response but do not interact, they may be either additive or multiplicative in their effects (Fig. 25-5). When they are **additive** they act upon different causal sequences that lead to the response. Say that a compound is made in two different compartments in the cell; one factor may influence one compartment and another factor the other compartment. Stem growth in the white mustard plant, for example, can be influenced oppositely by gibberellic acid and by red light (phytochrome: $P_{fr}$). The two responses are purely additive, as shown in Figure 25-6. **Multiplicative responses** are more common. The two factors act on different steps in the same causal sequence (Fig. 25-5), so the effect of one is always some fraction of the effect of the other. Rate of stem growth in the white mustard plant as influenced by red light is determined by the concentration of ions or sucrose in the growth medium (Fig. 25-7), for example. Analysis of variance shows when factors add or multiply in controlling a response; that is, when they do *not* interact. Any other result in an analysis of variance indicates an interaction of factors, and there are many kinds of interaction (see Lockhart, 1965; Mohr, 1972).

## The Kinds of Plant Responses to Environment

In addition to the quantitative manner in which an organism responds to environment, there are other ways of classifying responses. These types of responses are

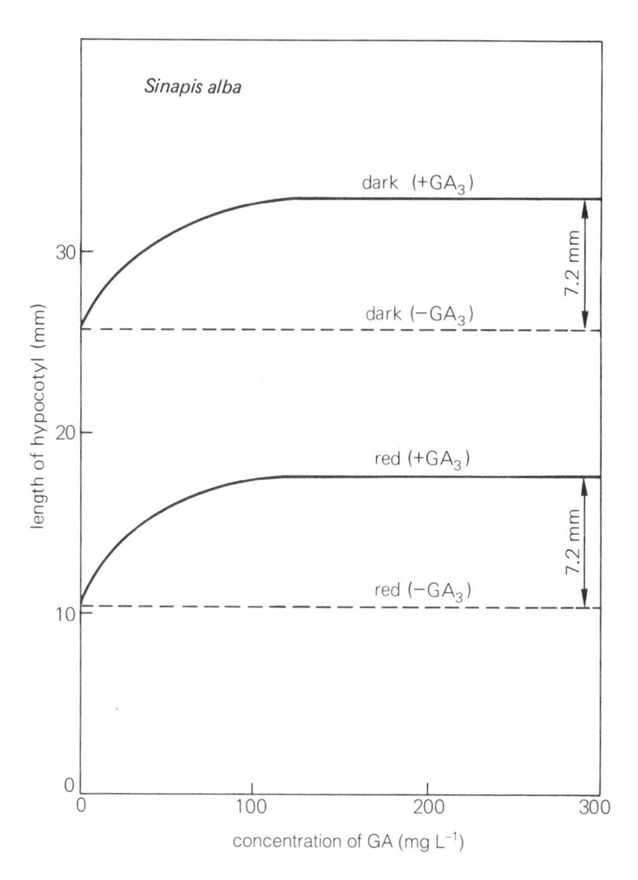


**Figure 25-6** An empirical example of "numerically additive behavior" in response to two factors. Hypocotyl lengthening in the mustard seedling was investigated under the control of light (continuous standard red light) and exogenously applied gibberellic acid ($GA_3$). It is apparent that the dose-response curve for exogenously applied $GA_3$ is the same with and without light. Note that $GA_3$ promotes lengthening of the hypocotyl, whereas red light inhibits lengthening. Hypocotyl length was measured 72 h after sowing. (After Mohr, 1972.)

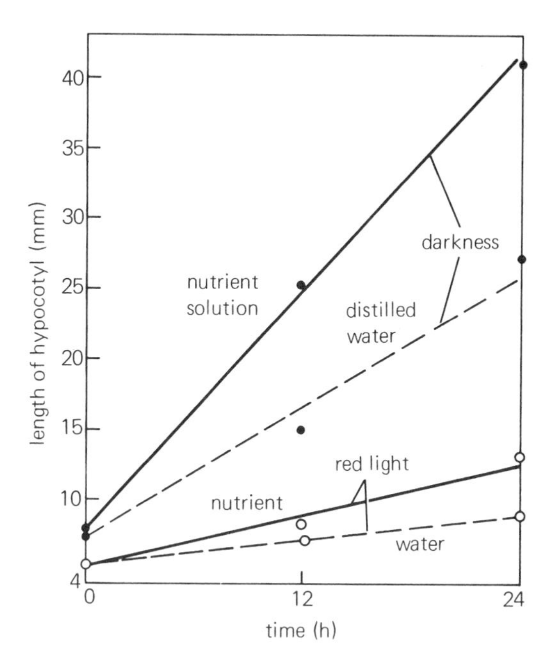


**Figure 25-7** An illustration of multiplicative effects. Growth of the hypocotyl of white mustard (*Sinapis alba*) is plotted as a function of time. Red light inhibits compared with darkness, and distilled water inhibits compared with a complete nutrient solution (Knop's solution). The *percentage* of inhibition is constant at all times. (From Schopfer, 1969.)

summarized in Table 25-3, the ideas of which have been modified several times since they were originally presented by Anton Lang at the Annual Meeting of the American Institute of Biological Sciences held at Purdue University in 1961.

The table illustrates three contrasting sets of response types—and some other ideas as well. First, we can contrast immediate responses (photosynthesis in response to irradiance) with those that are delayed for some time (many seconds to many days) after the beginning of the environmental stimulus that initiates them. This distinction is clearly related to the second set of response types, which is the contrast between responses provided by the energy input from the environment (for example, photosynthesis or enzymatically controlled reactions) and those responses that are *triggered* by some environmental change but use the energy provided by the plant's metabolism to produce the response, rather than energy derived from the change in the environment that causes the response. In the latter situation, we speak of an **amplification** of the environmental parameter. For example, the energy required to cause a stem to bend toward a unilateral light source must be much greater than the energy provided by the relatively few photons that initiate the phototropic bending (which may be applied for only a brief instant; bending takes much longer). In the third set of responses, a few plant responses (germination, perhaps) are triggered by some environmental change (that is, they are switched on or off by the change). One good example is the snapping shut of the Venus's-flytrap, a carnivorous plant with trigger hairs on the inner surfaces of two opposing special leaves. When these hairs are touched (usually by an insect), acid may be secreted into the walls of cells on the outside of the leaves causing them to grow rapidly and causing the trap to snap shut (Section 19.2).

More commonly, plant responses are not only initiated by an environmental change, but the extent of the change determines the extent of the response; that is, the response is **modulated** by the environmental change—even though the response may be considerably delayed.

**Homeostasis** is a particularly interesting organismal response to environment in which some feature of the organism's internal condition is maintained relatively constant or is allowed to vary only within re-

**Table 25-3 Types of Plant Responses to Environment.**

| Type of Response | Characteristics | Examples |
|---|---|---|
| 1. Direct (nondelayed) | As environment changes, plant response changes immediately (or almost). | Photosynthesis (light level), transpiration (heat load), enzyme-controlled reactions (temperature). |
| 2. Triggered or on-off | Environmental factor crosses a threshold, response begins even if factor returns to original level; often response is delayed. Sometimes amplification occurs. | Germination in response to such factors as low temperature or red light. (Other examples are rare: sensitive plant, carnivorous plants, etc.) |
| 3. Modulated (quantitative), delayed responses | Level of response determined by level (potential) of environmental factor; often both amplification and delay; often interact with biological clock. | Phototropism, gravitropism, many phytochrome responses, vernalization, photoperiodic induction (many responses such as flowering, stem elongation, tuber formation), rhythm setting, and plant growth in response to germination temperature (e.g., peach seedlings). (Many examples.) |
| 4. Homeostasis | Maintenance of (nearly) constant internal conditions in spite of changes in the environment; usually (always?) achieved through negative feedback. | Bird and mammal body temperature and blood chemistry, interactions between stomatal aperture and photosynthesis, internal concentrations of growth regulators. |
| 5. Conditioning effects | Gradual changes in the plant in response to continued exposure to some environmental condition. | Development of frost or drought resistance (low water potential). |
| 6. Carryover effects | Effects of growth conditions carried over two or more generations. | Experiments with inbred pea plants (see text); other, less well documented examples. |

stricted limits in spite of much wider changes, often in the same environmental factor, outside the organism. By far the best and most studied examples are among animals (for example, body temperature in mammals and birds), but the phenomenon is also recognized in plants. For example, $Ca^{2+}$ and phosphate levels are maintained within relatively close limits within the cytoplasm, primarily by movement through the tonoplast in and out of the vacuole (see Boller and Wiemken, 1986). If these ions become too concentrated in the cytosol, they can damage important enzymes there (see Section 6.7), but they also contribute in a positive way to cell metabolism. ($Ca^{2+}$ levels in the cytosol are typically in the micromolar range; in the vacuole, concentrations are often about 10 mM.)

**Carryover effects** could be of considerable ecological significance, although they have been studied carefully by only a very few workers. In the early 1950s, Harry Highkin, then at the California Institute of Technology, discovered that genetically pure, inbred pea plants grew poorly when the day and night temperatures were equal and held constant (10, 17, or 20°C; see Highkin and Lang, 1966). When the pea plants were grown for several generations under these adverse conditions, each generation (up to about the fifth) grew more poorly than the previous one. When Highkin reversed the situation (germinated seeds from stunted plants under optimum conditions), it required at least three generations to reach the maximum level of growth. Such a carryover of environmental effects from one generation to the next seems quite contrary to most of our concepts of genetics. (The environment does not change nucleotide sequences in the genes.) Nevertheless, the phenomenon is real, and it has since been confirmed by other workers. Seed suppliers and others say they have been aware of the phenomenon for years. Highkin and the others were careful to demonstrate that the effects were not caused by genetic selection. Apparently, the developing embryo (or perhaps the stored food material in the cotyledons) is in some way conditioned by the environment and by the parent plant so that the effect carries over through a number of succeeding generations.

## 25.4 Ecotypes: The Role of Genetics

We assume that the distribution of a given species is determined by its genetics, but what if there is genetic diversity within a species? Is it possible that the dwarf *Potentilla glandulosa* plants of the high Sierra Nevada, for example, have a genetic composition that allows them to get along at relatively low temperatures, whereas the larger *Potentilla glandulosa* plants found at lower elevations have a genetic composition that allows them to do well only at higher temperatures? When we think about the principles of evolutionary gene-pool change, we might certainly expect such situations. The reproductive processes are relatively slow, which makes the rate of gene flow within the gene pool relatively slow, but climatic pressures on the population differ depending upon location. Hence, we might expect the genetic composition of a population to vary throughout the range of that population.

We can imagine two possible explanations for the dwarfed versus large *Potentillas*: First, their genetic compositions might be alike, but their different appearances might be caused by the different climates to which they are exposed; and second, the differences could be caused by actual genetic diversity. How do we distinguish between these two possible explanations? Obviously, the thing to do is to bring the different plants together and grow them either in a **uniform garden** or in a controlled-environment facility. In the 1920s, Göte Turesson (1922) in Sweden developed the uniform-garden approach to a high level of precision. Jens Clausen, William Hiesey, and David Keck (1940s) of the Carnegie Laboratory at Stanford University in California and others followed suit. As it turns out, both environment and genetics are important.

Environmental effects on plant morphology (that is, appearance) and physiology are common. Turesson called plants with similar genetic makeups that exhibit differences caused by varied natural environments **ecophenes**. This is usually not emphasized in discussions like this one because the genetic differences we are about to discuss are so obviously important. Nevertheless, we should realize that the environment can and does produce many different ecophenes from any uniform genetic stock. Numerous effects of temperature, light, nutrients, and other factors on plant growth and development have been emphasized in this and several other chapters.

Turesson and others also found genetic differences in representatives taken from the different areas of a species's distribution. Turesson called these different genetic representatives of the population **ecotypes**. When the *Potentillas* from the Sierra Nevada, the Coast Range, and other locations were brought together in a uniform garden, they continued to exhibit striking and significant morphological differences (Fig. 25-8). Many species have now been studied, and it seems obvious that different environments will exert different selection pressures, resulting in different genetic compositions that are directly correlated with geography.

As might be expected, selection also works on the physiological responses to environment (Billings, 1970; see also Tieszen et al., 1981). For example, photoperiodic ecotypes have been demonstrated in several species (Section 23.4). Alpine sorrel (*Oxyria digyna*) plants collected from several locations in the arctic flowered only in response to days longer than 20 h, whereas those collected from the southern Rockies flowered in response to days longer than 15 h, for example. The arctic plants also reached peak photosynthesis rates at lower temperatures than their southern counterparts, but the alpine plants that grew at high elevations and relatively lower $CO_2$ pressures were more efficient in utilizing $CO_2$. Any competent taxonomist would classify all the alpine-sorrel specimens as the same species, but careful observation revealed a number of morphological differences between the northern and the southern representatives, as well as the physiological differences noted.

**Figure 25-8** Photograph of three *Potentilla glandulosa* specimens grown in a uniform garden at Mather, California, and collected on June 5 to 18, 1935. Plants were originally collected from three locations — coastal, mid-Sierran (Mather), and alpine stations — in California 5 to 13 years previously. (From Clausen et al., 1940.)

Many other examples could be cited. In one study, Olle Björkman (1968) at the Carnegie Laboratories in Stanford, California, studied the enzyme that fixes $CO_2$ in Calvin-cycle photosynthesis (ribulose-1,5-bisphosphate carboxylase: rubisco; see Chapter 11). First he examined two ecotypes of goldenrod (*Solidago virgaurea*): one that normally grows in the sun and one that grows in the shade. He found that the sun ecotype had several times as much enzyme as the shade ecotype, even when the shade ecotype was grown in the sun. This finding correlated closely with photosynthesis rates of the two ecotypes. Björkman also studied different species, some collected from the deep shade of a California coastal-redwood forest and others collected from several sunny locations, all close to Stanford. Again, the enzyme had a higher activity per unit leaf area in sun plants than in shade plants.

The carryover effects discussed in the last subsection could be complications in uniform-garden and other studies. Apparent differences in the first generations could be caused by carryover rather than by genetics. Whether this might be the case when plants are transplanted rather than raised from seed has not been studied, but most workers in the field feel that the complications are minor (but see Clements et al., 1950).

## 25.5 Plant Adaptations to the Radiation Environment

There are several ways that radiation (the visible portion of which is light) varies in nature. Almost all variations are of potential importance to plants. As an example of the research activity in environmental physiology—or physiological ecology—we shall review the radiation environment and plant responses to it (Smith, 1983).

### The Radiation Environment

The basic parameters of the natural radiant-energy environment are controlled by the astronomical characteristics of the earth—a nearly spherical, rotating planet with an atmosphere, its equator tilted 23.5° to the plane of its orbit around the sun. Latitude north and south of the equator, the daily rotation, and the seasons resulting from the tilted equator determine the sun's elevation above the horizon at any point on the earth's surface at any given time of day and on any day of the year. These factors also determine day length (see Fig. 23-1). The sun's elevation above the horizon determines, first, the length of the atmospheric pathway through which the sun's rays must travel to reach a point on earth and, second, the area of horizontal surface that will be irradiated by a given cross-sectional area of the sun's rays. The larger the surface irradiated by a given cross-sectional area (that is, the farther one is from the equator), the cooler the climate is likely to be. Day length is a function of the sun's elevation when it is at the zenith and the angle the sun's daily path makes with the horizon; the more acute the angle, the shorter the day is in winter (down to a 0-h day) and the longer it is in summer (up to a 24-h day).

These astronomically determined characteristics of the radiant-energy environment can be modified at any time and place on the earth's surface by other factors, including weather (clouds), atmospheric composition (for example, natural or human-caused pollution), shading by topography, shading by vegetation, reflection, and such human-controlled factors as shade from buildings and glass through which the radiation must pass. We might expect plants to have become adapted to virtually all the natural variations. Consider the following four aspects of the radiation environment and how they can vary.

*Irradiance or photon flux*[3] In northern (or southern) latitudes, because of both the lower solar elevation and the longer atmospheric pathway, irradiances at noon and per unit area are greatly reduced compared with those in equatorial regions. Yet total daily photon flux in the north is often greater in summer than that at equatorial latitudes because of the north's long summer days (Fig. 25-9). Of course, weather and other factors often modify both instantaneous and integrated photon fluxes.

*Spectral composition: Quality* At sunrise or sunset at all latitudes (when the sun's elevation is less than 10°), the total light environment (that part of the spectrum to which the human eye is sensitive) becomes enriched in blue (380 to 500 nm) and far-red (700 to 795 nm; Fig. 25-10) wavelengths. Far-red light also increases relative to orange-red (595 to 700 nm; see Fig. 23-12). Blue wavelengths are enriched because much of the irradiation comes from the blue sky (atmospheric scattering); far-red light is enriched because the longer the *direct* atmospheric path through which the sun's rays pass, the more the short wavelengths (blue) are preferentially removed by scattering.

Ultraviolet radiation is greatly reduced by stratospheric ozone. In the shade, blue light is enriched relative to longer wavelengths because most light comes from the blue sky, but perhaps the most profound quality change is caused by leaf shading (Section 20.7). This phenomenon occurs because chlorophyll absorbs much of the red light but the leaves transmit the far-red light, so leaf shade is greatly enriched in far-red light. Clouds do not change the spectral composition profoundly, although blue light is somewhat enriched and scattering of all wavelengths is increased. Natural (for example, volcanic) and human-caused pollution can influence quality in various ways, mostly by reducing blue wavelengths. Light passing through snow can be influenced in various ways, depending on the condition of the snow (Marchand, 1987; Richardsen and Salisbury, 1977).

[3]As discussed in Appendix B, **irradiance** refers to radiant energy striking a unit surface area in a unit time. **Fluence** is energy per unit area (not cosine-corrected) without the time factor. Plant physiologists and others have often used the term *intensity* to express the idea of irradiance, but physicists reserve **intensity** for the energy being emitted by a light *source*. Irradiance can be expressed in energy units as watts per square meter ($W\ m^{-2}$), which is the same as joules per second per square meter ($J\ s^{-1}\ m^{-2}$), or as the number of moles of photons per square meter per second (usually *micro*moles: $\mu mol\ m^{-2}\ s^{-1}$), with the wavelength range being specified in either case. When using energy terms, we should speak of **photosynthetically active radiation (PAR)**; on a photon basis, we speak of **photosynthetic photon flux (PPF)**. (It is not necessary to add the term *density* because it is redundant to *flux*.)

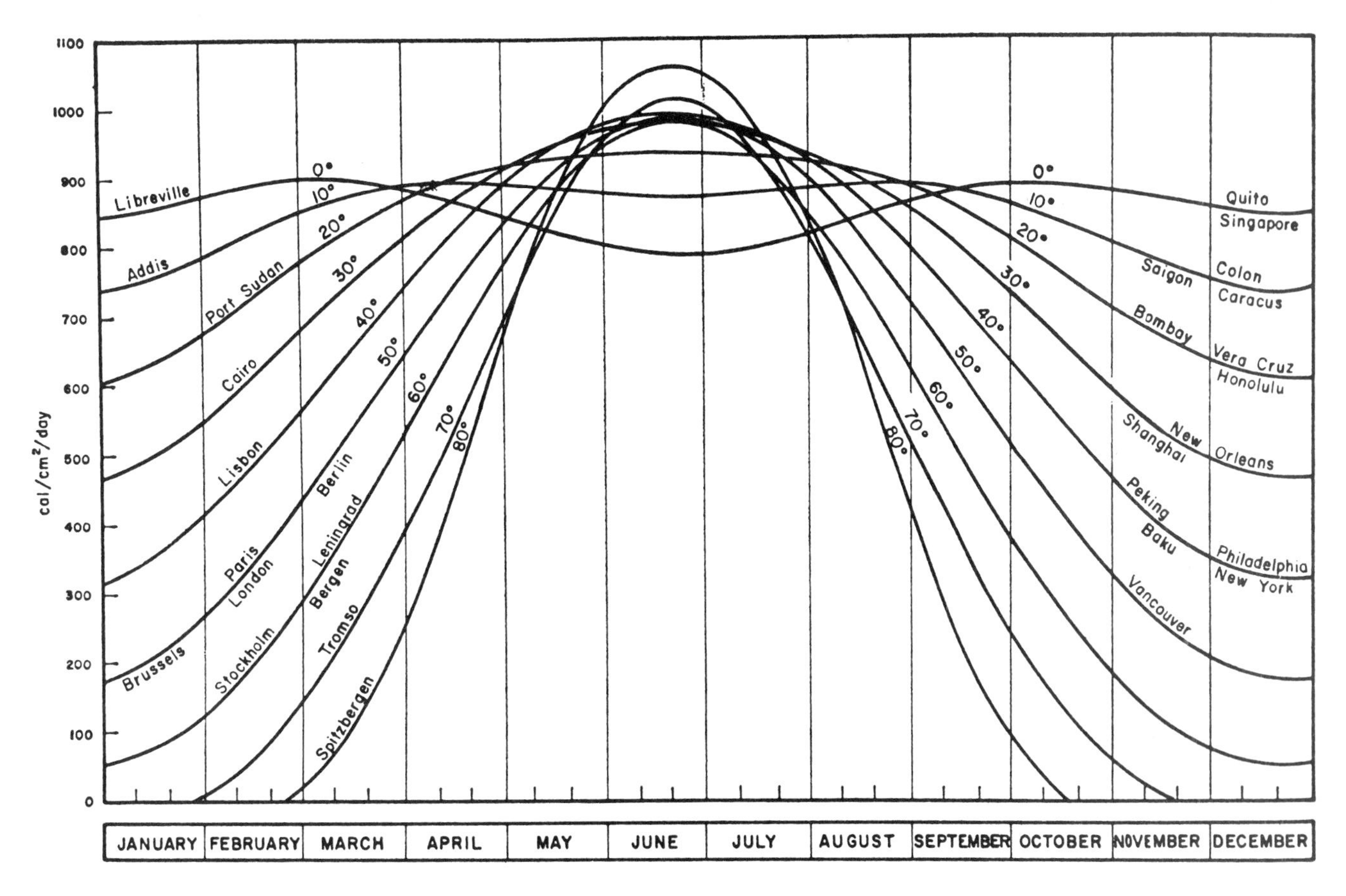


**Figure 25-9** Daily totals of the undepleted solar radiation received on a horizontal surface for different geographical latitudes as a function of the time of year, based on the solar constant value of 1.353 kJ $m^{-2} s^{-1}$. (Note that 1,000 cal $cm^{-2} d^{-1}$ = 41.84 MJ $m^{-2} d^{-1}$; from Gates, 1962.)

Moonlight (reflected sunlight) has slightly less blue light than sunlight but is otherwise similar. Its irradiance is about six orders of magnitude lower than that of sunlight (see Fig. 23-12). Starlight has an irradiance about 2.5 orders below that of moonlight but with a spectrum similar to that of sunlight.

*Duration or diurnal cycling of light* We have already noted the crucial effects of latitude upon day length, but it is important to note that the modifying factors of clouds, topographic and leaf shade, reflections, and so on influence day length proportionally much less than they influence other aspects of the radiation environment. This influence will depend upon the sensitivity with which a given plant can detect changes in light level (Section 23.8), but plants are apparently highly sensitive to the slight changes in irradiance that occur rapidly just before sunrise and just after sunset. Thus the length of day detected by a plant may be changed only slightly by clouds and shade (say, 7 to 10 minutes out of 12 hours, which is a little over 1 percent). Effects on irradiance and spectrum can be much greater than this.

*Direction* The direction of the sun's rays is a function of the sun's elevation and position. By obscuring the sun as a "point" source of radiation, clouds and other shade destroy the directionality of the incoming radiation to a greater or lesser degree.

## Plant Responses to Radiant Energy

In the following paragraphs we will briefly review the categories of plant response to radiant energy (see Table 25-4), most of which have already been discussed in various chapters in this book. Photosynthesis is a good example of a topic that is especially important to modern plant physiological ecology; hence we emphasize it in the following discussion.

*Photosynthesis: Carbon gain and carbon allocation* "All flesh is grass," said Isaiah (40:6), the Hebrew prophet, some 25 centuries ago. Today we might say that all grass and hence all flesh is sunlight, air, and water. Photosynthesis forms the foundation for virtually all life, and it is the primary metabolic process of any ecosystem. It seems clear that understanding the photosynthesis of the various species in a plant com-

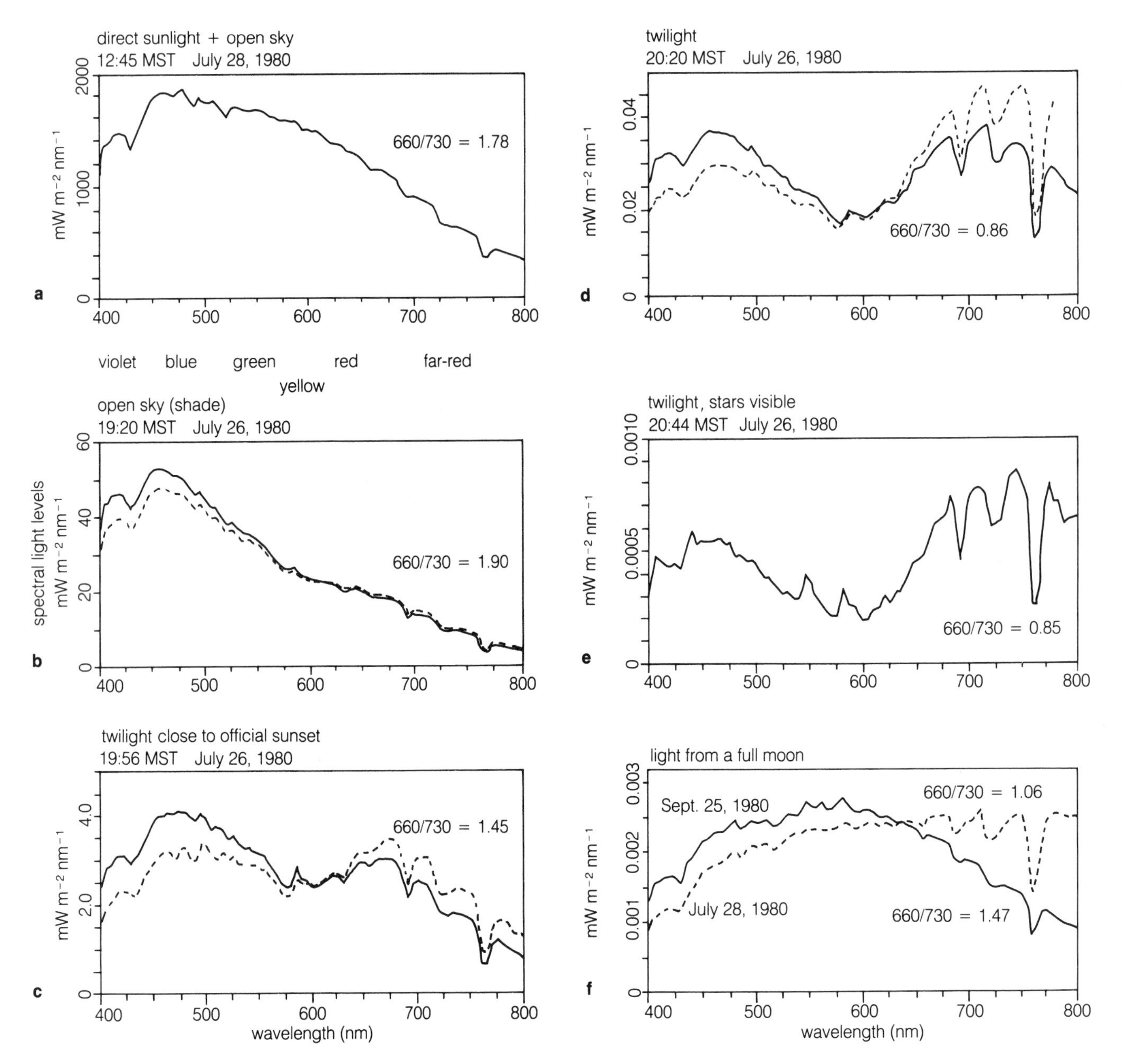


**Figure 25-10** Spectral distributions of natural light energies, including sunlight (**a**), skylight measured at four times before and during twilight (**b** to **e**), and light from the full moon (**f**). Note the greatly different ordinate scales for the various curves. The energy levels at the end of the twilight measurement (**e**) are an order of magnitude lower than light from the full moon. Curves in **b**, **c**, and **d** were made while light levels were changing rapidly, each scan requiring 10 min; hence, they were "corrected" by lowering the long-wavelength (red) end where the scans began by an amount proportional to the scans taken 12 min later and by raising the short-wavelength (blue) end in the same way. That is, the curves were "rotated" (by computer) around their center points, down on the red end and up on the blue end. Dashed lines are original data; solid lines are after computer rotation. Dashed line for moonlight (**f**) was obtained with a fiber-optics probe aimed directly at the full moon; solid line includes skylight. (From Salisbury, 1982.)

**Table 25-4 Adaptations to the Radiant-Energy Environment.**

| Primary Plant Responses | Irradiance L = low M = medium H = high | Spectrum UV = ultraviolet B = blue G = green R = red FR = far-red | Duration/cycle P = photoperiodism C = circadian clock | Direction important? |
|---|---|---|---|---|
| | Environmental Characteristics | | | |
| 1. Photosynthesis, leaf development | H | B–R, G | C | no |
| 2. Seed germination, bud break | L, H | R/FR, B | P | no |
| 3. Etiolation syndrome | L | R/FR | | no |
| 4. Stem elongation, apical dominance | H | R/FR | P | no |
| 5. Stem orientation | L | B–UV | | yes |
| 6. Leaf orientation: | | | | |
| circadian | M | R/FR | C | no |
| solar tracking | H | B | C | yes |
| 7. Reproduction and storage organs | L, H | R/FR | P | no |
| 8. Dormancy | L, H | R/FR | P | no |
| 9. Damage by | | | | |
| ultraviolet radiation | H | UV | | no |
| photoreversal | H | B | | no |

munity will take us a long way toward understanding the community (Pearcy et al., 1987).

**Leaf photosynthetic capacity** (photosynthetic rate when all environmental factors are optimal) has been studied for many species and has been found to vary nearly 100-fold, with the capacity being highest for plants found in environments that are rich in resources (Mooney and Gulmon, 1979). Highest photosynthetic capacities are found among annuals and grasses that grow in deserts, where light is not limiting and other resources are intermittently abundant, but evergreen desert shrubs that must endure periods of drought have very low photosynthetic capacities (Ehleringer and Mooney, 1984). The photosynthetic capacity of many leaves is highly plastic, depending strongly on available resources. Recall the photosynthetic properties of sun and shade plants (see Figs. 12-4 and 12-6, reviewed by Björkman, 1981). Shade plants have an extremely low level of dark respiration, so the compensating light level is also extremely low, and positive net assimilation is achieved at light levels much below those required by sun plants to reach compensation. But shade plants are saturated at low levels at which sun plants are just beginning to photosynthesize at moderately high rates. Shade plants never achieve such high rates as sun plants and can be damaged (**photoinhibited**) by light levels that are not even saturating for sun plants.

When a sun plant is grown at low irradiances (say, 1/20 of full sunlight; see Fig. 12-6), it acclimates so that its photosynthesis curve approaches but does not quite reach that of shade plants. Individual leaves on a single plant can also develop either as sun leaves or as shade leaves. The ability to acclimate to low irradiances has been studied mostly with C-3 species, but many C-4 plants also adjust to low irradiances. On the other hand, shade plants apparently do not become acclimated to high light levels. Grown under such conditions, plants may actually become less photosynthetically capable at all irradiances.

Temperatures can also affect photosynthesis. Temperatures near the soil, for example, can be as much as 10°C higher than air temperatures higher in the canopy.

There are also internal controls on photosynthetic capacity. Removing such carbon sinks as developing fruits typically reduces leaf photosynthetic capacity, but if some leaves are removed, photosynthetic capacity of the remaining ones often increases. Experiments of this type have usually been done with crop plants, but the generalizations probably apply to wild species as well. It appears that leaf photosynthetic capacity is determined primarily by the availability of photosynthetic enzymes, particularly rubisco. There is evidence to suggest that stomatal conductance adjusts to this capacity rather than being a strong component of the capacity itself.

The discovery of C-4 and CAM photosynthesis has had strong implications for environmental physiology. C-4 species are especially well-adapted to warm and dry environments with high irradiances because C-3 plants experience greater $O_2$ inhibition and thus photorespiration as temperatures increase. The $CO_2$ pump of C-4 plants allows for higher photosynthetic rates when $CO_2$ levels are low within the leaf, so stomatal conductances can also be low, leading to the high water-use efficiency. Also because of the pump (and because rubisco is catalytically superior in some C-4 plants; Seemann et al., 1984), a C-4 leaf can get along with less $CO_2$ than a C-3 leaf, so nitrogen-use efficiency can also be higher (more photosynthesis for a given amount of nitrogen in the plant). Thus C-4 plants are prevalent in nitrogen-poor tropical grasslands and savannas. In the central Great Plains of North America, C-3 grasses are active during the spring, but C-4 species are active during summer. Desert summer annuals are almost exclusively C-4 and winter annuals C-3 (Mulroy and Rundel, 1977). Tropical-forest understories do not have enough light and may be too cool to offer much advantage for C-4 photosynthesis. Although C-3 photosynthesis is favored only in the cooler northern forests, there are many successful C-3 plants in warm environments.

CAM species have the highest water-use efficiencies of any plants, so it is not surprising that desert succulents use this photosynthesis mode in which stomates close during the day but open at night, allowing absorption of $CO_2$ and its fixation into organic acids. Nevertheless, CAM species are found in many dry microhabitats such as rock outcroppings in temperate humid environments or on the branches of trees in the tropical or temperate rain forests. (These plants are called **epiphytes**: plants growing on other plants—for example, orchids and bromeliads.) CAM is also used in some unusual ways. *Isoetes howellii* is an aquatic, nonflowering plant (order: Lycopsida) in vernal pools in California, where $CO_2$ can be depleted during the day by photosynthesis but builds up at night as organisms respire. CAM allows *Isoetes* to take up the $CO_2$ at night. Early in the morning the plants shift into C-3 photosynthesis, but as $CO_2$ becomes depleted, they become dependent on the internal release of $CO_2$ from the organic acids formed at night by CAM (Keeley and Busch, 1984). An *Isoetes* species that grows in Peruvian bogs above 4,000 m elevation has no stomates but instead takes up $CO_2$ through its roots; $CO_2$ is fixed in the stems at least partially by CAM (Keeley et al., 1984; Raven et al., 1988). One big advantage for CAM plants might be that they can internally recycle $CO_2$ under drought conditions when the stomates stay closed both day and night. When the rains come, they can respond with increased $CO_2$ uptake almost immediately following water uptake. As described in Chapter 11, some CAM plants then shift to a C-3 photosynthetic mode. This may be more important than the high water-use efficiency.

Respiration plays an important role in carbon accumulation during plant growth (reviewed by Amthor, 1989). It is a difficult role to determine, however, because it is not easy to know how much respiration goes on when plants are in the light. Usually, it is assumed that dark respiration stays the same during the light, but we have seen (Chapter 11) that there is good evidence that this is not so. In any case, it is clear that some portion of the energy captured in photosynthesis is used for growth (that is, to synthesize new molecules) and for maintenance of living cells. This portion may be on the order of 30 to 40 percent of the energy captured in photosynthesis. There is ecological significance in the way plants differ in this percentage. Some plants use much more energy than others, for example, in synthesizing protective secondary compounds such as tannins or alkaloids, or such structural compounds as lignin (Chapter 15).

If a leaf is to benefit a plant, its carbon gain must exceed the carbon costs of its construction, maintenance, and protection. When the leaf is young it is a liability for the plant, using more fixed carbon than it produces. When it is fully expanded it typically is profitable to the plant, fixing more carbon than it uses. With senescence comes a decline in carbon fixation, and the leaf again becomes a liability as nitrogen and other minerals are mobilized and exported before the leaf falls. Evergreen leaves usually have a higher cost of formation per unit of dry weight than deciduous leaves, and they become productive much more slowly when conditions are harsh (for example, drought, shade). Nevertheless, leaf formation is the essence of plant growth. The process is closely analogous to money in the bank growing by compound interest, as we have seen (Chapter 16). As a leaf becomes profitable, it can contribute to the production of other leaves that will also eventually become profitable.

*Photosynthesis: Leaf morphology and canopy architecture* Leaf area and leaf morphology are strongly influenced by light levels during development. Compared with shade leaves, sun leaves have less area per leaf, are thicker (often have more layers of palisade mesophyll consisting of longer cells—see Fig. 12-5), weigh more per unit leaf area, are more densely distributed on the stem with shorter petioles (shade each other more), and have more chlorophyll per unit dry weight. Epidermis, spongy mesophyll, and vascular systems are also more developed in sun leaves. Because neutral-density screens that produce shade but do not change the spectrum appreciably have influenced leaf morphology, the enriched far-red light of leaf shade must not be essential, but careful studies might show that it contributes to the production of shade leaves.

Structure of the leaf canopy determines how much light will be absorbed by individual leaves in a plant community, as has long been recognized for agricultural systems. The most important trait for high productivity is rapid closure of the canopy at the beginning of the growing season. Until the canopy closes, not all of the available light can be absorbed. After closure, nearly vertical leaves, as in grasses, allow more light to penetrate into the canopy, and when light rays are more or less parallel to the leaves, irradiance per unit leaf area is not high enough to saturate photosynthesis. This is another reason why such grasses as wheat and rice can produce such high yields (see essay on CELSS on page 560). Upper leaves on plants with more or less horizontal leaves may be light-saturated and may shade the lower leaves so that they do not get enough light. On the other hand, they also shade shorter competitors.

Ideally, a plant should have a **leaf area index** (ratio of leaf surface area to ground surface area; see Chapter 12) that allows for optimum photosynthesis of the plant as a whole. If the index is too low, not enough light will be absorbed; if too high, lower leaves will not get enough light and will thus be a liability. Spruce (*Picea excelsa*) needles have a low photosynthetic capacity, but the leaf area index is so high that the tree as a whole is more productive than a deciduous beech (*Fagus sylvatica*) with leaves that have twice the photosynthetic capacity. The longer growing season for the spruce also contributes to its high productivity, but not as much as the high leaf area index (Schulze et al., 1977).

*Photosynthetic utilization of sunflecks by understory plants* Most laboratory studies of photosynthesis use a steady light source applied to a leaf (or a suspension of algae) held in a rigid position in a transparent chamber. This is necessary to obtain quantitative data such as those we discussed in Chapters 10 through 12. But such steady-state conditions are not typical of leaves in natural or agricultural environments. As leaves flutter in the breeze and clouds pass rapidly overhead, irradiance reaching the chloroplasts changes rapidly. Even algae are often subjected to changing irradiance as waves reflect and refract the incoming light. An interesting situation in which such changing light levels can be extreme is the forest floor, where sunflecks penetrate the upper canopy, irradiating a leaf (or some portion of it) for anywhere from a fraction of a second to tens of minutes. How important for the photosynthesis of understory plants are these brief moments of high photon flux?

Robert W. Pearcy (1988, 1990) and his coworkers have studied the role of sunflecks, mostly in tropical forests. Their investigations have led to some interesting insights into the physiological ecology of such forests and to an increased understanding of photosynthesis itself. They also provide an excellent example of leading-edge research in physiological ecology during the 1980s.

The light received from sunflecks varies greatly, but in one study, the diffuse irradiance on the forest floor equaled 10 to 20 $\mu$mol m$^{-2}$ s$^{-1}$, whereas light from about 120 sunflecks was almost always over 50 $\mu$mol m$^{-2}$ s$^{-1}$ and sometimes reached 1,200 $\mu$mol m$^{-2}$ s$^{-1}$, accounting for 12 to 65 percent of the light energy reaching the forest floor. Many sunflecks did not reach maximum irradiance levels because *part* of the sun was obscured by leaves in the canopy above (a **penumbral** effect). Usually the sunflecks came in clusters, often within 1 min of each other, followed by periods with no sunflecks.

Measuring photosynthesis during sunflecks is not easy because highly sensitive equipment is required. Among other instruments, Pearcy has used a photosynthesis chamber that clamps onto a leaf, covering about 22 cm$^2$ of leaf area but having a volume of only 4.5 cm$^3$. Applying proper corrections for the time it takes for gas to pass through the system, virtually instantaneous changes in $CO_2$ uptake can be monitored with a sensitive infrared gas analyzer. Such measurements have shown that some 30 to 60 percent of the photosynthesis of understory plants in a natural environment is typically caused by the sunflecks. But to study the phenomenon in detail, a researcher cannot depend on natural sunflecks. Rather, a lamp is used to produce *light*flecks, allowing precise control of the duration and irradiance received by the leaf in the photosynthesis chamber, which may be attached to a leaf in the forest instead of the laboratory.

Photosynthesis in a leaf does not increase instantly as irradiance increases, nor does it drop back to low levels when irradiance decreases. This is well illustrated in Figure 25-11, which shows photosynthesis ($CO_2$ uptake; assimilation = $A$), stomatal conductance ($g$), and calculated internal $CO_2$ partial pressure ($P_i$) as a function of time, during which irradiance was increased from 6 to 520 $\mu$mol m$^{-2}$ s$^{-1}$. What accounts for these lags in getting photosynthesis started and in shutting it down?

The lag in starting photosynthesis after irradiance increases is called an **induction effect**. Two factors apparently account for it. One is a change in stomatal conductance (Fig. 25-11). Stomates open in response to increasing irradiance, as we saw in Chapter 4. But this effect is not sufficient to account for the lag, as indicated by the observation that internal $CO_2$ partial pressure can remain relatively high during the early phases of the induction effect. There is now good evidence (summarized by Pearcy, 1988) that rubisco must be light-

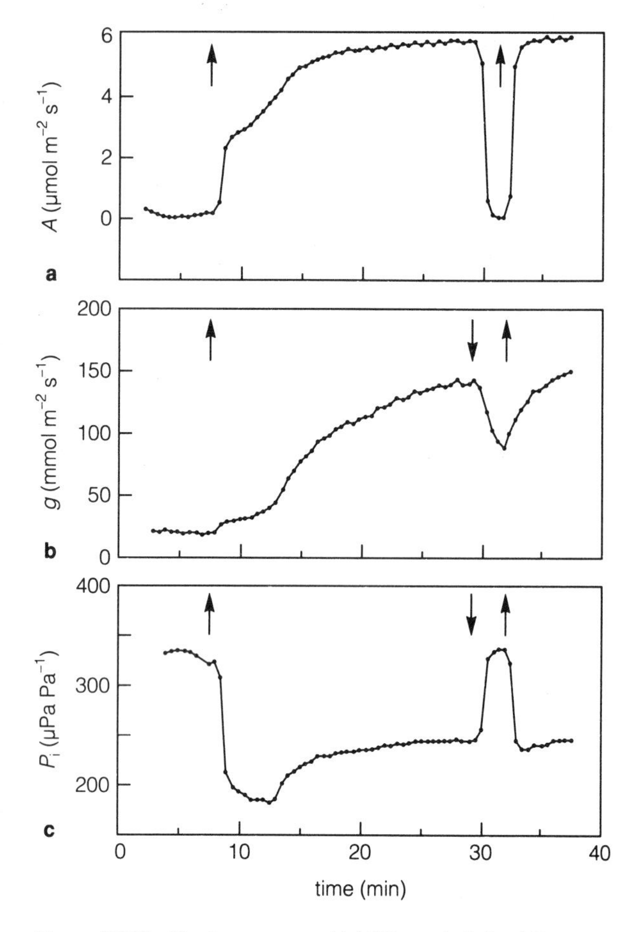


**Figure 25-11** The time course of (**a**) $CO_2$ assimilation ($A$), (**b**) stomatal conductance to water vapor ($g$), and (**c**) partial pressure of $CO_2$ ($P_i$) within a leaf, all as a function of time and in response to changes in photon irradiance, as indicated by the arrows. The low PPF was 6 μmol m$^{-2}$ s$^{-1}$, and the high PPF (the lightfleck) was 520 μmol m$^{-2}$ s$^{-1}$. The leaf was on a sapling of *Argyrodendron peralatum* in the understory of the Curtain Fig Forest, Australia. (From Pearcy, 1988; used by permission.)

activated before photosynthesis can proceed at maximum rates, and that this, along with stomatal changes, accounts for induction. In any case, after induction has occurred, photosynthesis begins much more rapidly for subsequent sunflecks. Figure 25-12 shows response of leaves to a lightfleck before and after induction. The leaf that was already induced showed much more photosynthesis than the other leaf, but the other leaf continued to fix $CO_2$ at a relatively high level *after* the lightfleck. This post-irradiation $CO_2$ fixation can be highly significant, resulting in even more fixed carbon than would be predicted based on steady-state photosynthesis rates at the low and high irradiation levels. The post-irradiation $CO_2$ fixation depends on metabolite pools (perhaps ribulose-1,5-bisphosphate) that are built up during the lightfleck and utilized after the light has dropped back to low levels. With induced leaves, the effect is evident as a tailing off of the curve (Fig. 25-12), but proportionally it is much less than for noninduced leaves.

*Other photosynthetic adaptations* Still other photosynthetic adaptations to the light environment have been studied. Although terrestrial plants and green algae best use blue and red wavelengths, for example, various brown, red, and blue-green algae (cyanobacteria) also use the green wavelengths that penetrate ocean waters most deeply. The photosynthetic efficiency of certain algae and higher plants is known to be under the control of circadian clocks (Chapter 21), so highest rates are measured (under standard conditions) during subjective day and lowest rates during subjective night.

*Seed germination and spring bud break in deciduous plants* As discussed in Chapter 20, seed germination can be promoted or inhibited by light, depending on species and other factors such as temperature, and germination can often be inhibited strongly by leaf shade or by artificial light enriched in far-red wavelengths (Morgan and Smith, 1981). This is also true for many species that are light-insensitive when tested with white light. Indeed, several hundred species have now been studied, and on the order of 80 percent or more respond to the balance in red/far-red light, presumably via the phytochrome system. Although it is true that many species normally germinate in spring before canopy leaves have appeared, it also seems clear that many other species germinate only after the canopy has been removed as described in Section 20.6.

Some seeds that respond to photoperiod are known (Vince-Prue, 1975; see Section 23.3). A few species produce seeds whose germination is promoted by short days, but promotion by long days is more common (for example, rice; Bhargava, 1975). It is also known that the photoperiod to which mother plants are exposed while their seeds are developing may strongly influence subsequent seed germination. Frequently, as discussed in Section 20.6, short days applied to the mother plant increase germinability of seeds compared with those that develop under long days (for example, *Amaranthus retroflexus*, Kigel et al., 1977; *Chenopodium polyspermum*, Pourrat and Jacques, 1975), although a few exceptions have been reported (germination is best when mother plants are exposed to long days: for example, lettuce, Koller, 1962; more examples are in Mayer and Poljakoff-Mayber, 1989). We currently fail to understand how day length influences the ability of seeds to germinate.

Spectral effects have not been reported, but a few species are known in which buds detect the lengthening days of spring and respond by becoming active (for example, *Betula pubescens* and *Liquidambar styraciflua*; Downs and Borthwick, 1956); the response is often modified by temperature (for example, degree of winter chilling).

*The etiolation syndrome* Dark-grown seedlings exhibit a constellation of characteristics that are apparently adaptations to germination at some depth in the soil (Section 20.7; Mohr, 1972). Internodes elongate, leaves do not expand, chlorophyll does not develop, and roots do not grow rapidly. Many dicotyledons have a hook at the tip of the stem that pushes up through the soil (see Fig. 16-14), and grasses have a coleoptile that protects the emerging leaf. When the seedling reaches the light, internode elongation slows, leaves expand, chlorophyll develops, roots grow more rapidly, and the hook straightens or the coleoptile is penetrated by the leaf. The entire character of the developing seedling changes (see Fig. 20-5). Phytochrome is in control of many features because red light is effective and its effects are reversed by far-red light.

*Stem elongation and apical dominance* Leaf shade—and even light reflected from nearby plants—exerts strong control over stem extension and lateral bud growth in many de-etiolated plants (Section 20.7; Ballaré et al., 1990; Morgan and Smith, 1981); there are close similarities to etiolation. This is best shown by providing a constant amount of photosynthetically active light to controls and supplementing this for treated plants with additional far-red light. Species that normally grow in open areas sometimes increase their growth rate as much as 400 percent in response to such irradiation. Axillary buds remain dormant. Sometimes this could allow the plants to overtop the canopy and reach full sunlight while conserving energy by not branching. Species that normally grow under a forest canopy usually do not respond to the increased far-red light, which also seems adaptive because there is seldom any chance for them to overtop their canopy.

Stem elongation is also promoted in many species by long days (Section 23.3). To distinguish this from the effect of far-red light, extension of the day should be with light low in far-red wavelengths (for example, fluorescent light), or a night break should be given to see whether that promotes elongation. Leaf number and area are also promoted by LDs in many species. These responses to day length are probably adaptive because they produce taller, leafier plants during the time of year when growing conditions are likely to be most suitable—given that ample moisture is available.

*Stem and leaf orientation: Phototropism* Phototropic bending of stems toward a light source is a response to low irradiances of blue or ultraviolet radiation, probably mediated by a flavin pigment (Section 19.4). Such a response might allow a plant to grow laterally around an overhead obstruction. This could be especially important for seedlings growing through the obstructions on a forest floor, for example, or for plants growing inside but near the mouths of caves or under riverbanks. Most of the time in most ecosystems, however, plant stems apparently do not respond phototropically. Stems and branches, rather, bear a constant relation to the direction of the gravitational force.

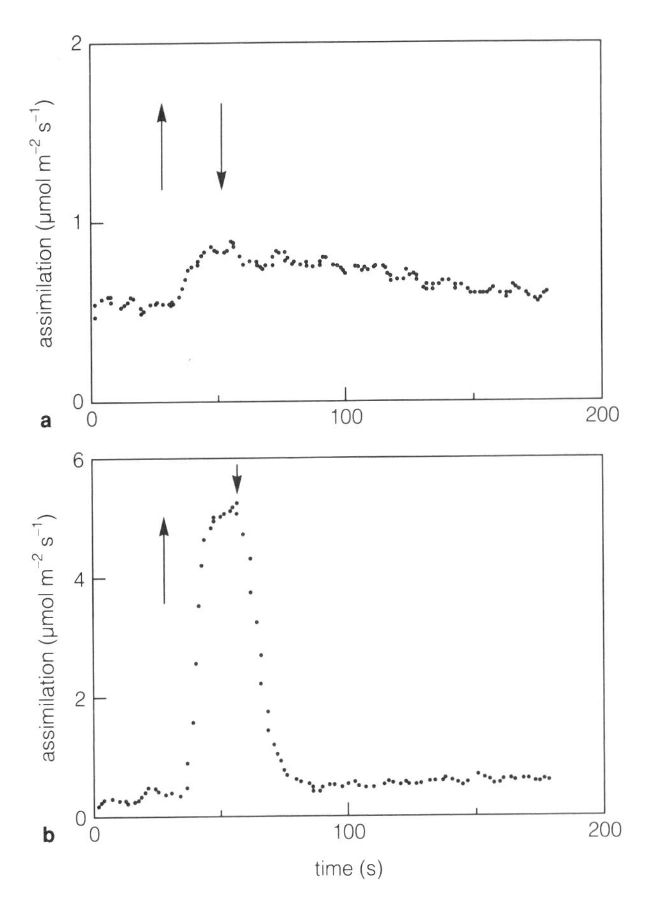


**Figure 25-12** Response of assimilation (photosynthesis) of an *Alocasia macrorrhiza* leaf to 20-s lightflecks (indicated by the arrows) either (**a**) before light induction of the leaf or (**b**) after full induction. The lightflecks had an irradiance of 500 μmol $m^{-2}$ $s^{-1}$. (From Pearcy, 1988; used by permission. See also Chazdon and Pearcy, 1986.)

Leaves and flowers of many species do exhibit solar tracking, a strong phototropism (Section 19.4). Leaf blades of many plants (for example, cotton, soybean, beans, and alfalfa) follow the sun during the day, maintaining their blades at right angles to the sun's rays, much as a radio telescope tracks a moving satellite. This maximizes the irradiance on the leaf during the day.

*Circadian leaf movements* Study of the circadian sleep movements exhibited by many plants has provided another vast body of literature (Chapter 21).

Whether the movements are adaptive has long been argued, and proposed answers are not completely convincing. For example, although a horizontal daytime position is rather well suited for photosynthesis, the nearly vertical night position might also effectively lower irradiance from moonlight — if the full moon were almost directly overhead, which is only the case at midnight in the tropics and subtropics and in winter in temperate zones (Salisbury, 1981a, 1981b, 1982). In any case, light is effective in synchronizing the internal plant rhythms with the external cycling environment. Both dawn and dusk are often critical, and phytochrome has been implicated.

*Reproduction and storage organs* There are many plant responses to photoperiod, but the most studied is flower induction in angiosperms. Hundreds of species have now been studied, although only a few in any detail (Chapter 23). From an ecological standpoint, the most striking discovery is surely the diversity of response types. There are species, varieties, or cultivars that respond to SDs, LDs, intermediate days, LDs followed by SDs, SDs followed by LDs, and all of these combined with various temperature interactions. Some plants have an absolute requirement for a given day length; others may only be promoted in their flowering by some day length. Different latitudinal ecotypes or cultivars within a species often have different day-length requirements.

A photoperiod requirement for flowering assures that the plant will flower at an appropriate season (for example, after sufficient vegetative growth has been achieved); this will be nearly the same time each year, within the limits of temperature modification of the photoperiodism response. Furthermore, all the plants in a population will flower at essentially the same time, facilitating cross pollination (and often providing beautiful **aspect dominance** in an ecosystem).

Responses to twilight, discussed in Chapter 23, have strong ecological implications. In addition, light levels rather than photoperiod influence flowering of a few species, but study of them has been minimal.

Development of storage organs is often influenced by photoperiod, and this could also be of ecological significance. Responses are similar to those described for flower induction (Chapter 23).

*Dormancy: The autumn syndrome* In many plants, especially deciduous woody perennials, short days lead to leaf senescence and abscission, inhibition of stem elongation, terminal (resting) bud formation, frost hardiness, and other developmental changes characteristic of winter dormancy (Chapters 22 and 23). In a sense, this is the ecological opposite of increased stem elongation and leaf growth in response to long days, although that response is as often observed in herbaceous annuals as in woody perennials. Typically, the development of dormancy is also promoted by low temperatures.

### Damage Caused by Ultraviolet Irradiation

Damage to plants by ultraviolet radiation might be ecologically significant and could become more so if the stratospheric ozone layer is reduced (Caldwell, 1981). However, there is a wide diversity in plant resistance to UV radiation, and UV damage involving DNA can be reversed by high irradiances with blue light (called **photoreactivation**). Barnes et al. (1988) studied potential effects of increased ultraviolet (UV-B) radiation on competition between wheat (*Triticum aestivum*) and wild oats (*Avena fatua*) and found that wheat had the advantage, especially when precipitation was high. They could not account for these results by effects on photosynthesis (Beyschlag et al., 1988), but they concluded that plant morphology was altered, allowing the wheat to overtop the wild oats to a greater degree under UV-B treatment than under natural conditions.

Clearly, radiation, especially light, influences plants in a multitude of ways. The more we learn of these matters, especially with reference to individual species, the more we will know about environmental physiology.

# 26

# Stress Physiology

An important branch of environmental physiology is concerned with how plants and animals respond to environmental conditions that deviate significantly from those that are optimal for the organism in question—or, in a broader sense, for organisms in general. As a division of physiological ecology, this field, called **stress physiology**, can contribute to our understanding of what limits plant distribution. Most research in the field, however, is concerned with how adverse environmental conditions limit agricultural yields. One of the first challenges encountered is how to define the word *stress*.

## 26.1 What Is Stress?

In 1972, Jacob Levitt (see Levitt, 1972, 1980) proposed a definition of biological stress derived from physical science. *Physical stress* is any force applied to an object (for example, a steel bar); *strain* is the change in the object's dimensions (for example, bending) caused by the stress. Levitt suggested that **biological stress** is any change in environmental conditions that might reduce or adversely change a plant's growth or development (its normal functions); **biological strain** is the reduced or changed function.

Recall our discussion of limiting factors and the law of tolerance (Section 25.3). When environmental conditions are such that the plant is responding maximally to some factor (is at or close to the optimum part of the curve in Fig. 25-1), it is not being stressed by that factor. Any change in environmental conditions that results in plant response that is less than the optimum might be considered stressful. Of course, such a concept is sometimes easier to discuss in a theoretical way than it is to apply. Consider a plant suddenly subjected to reduced light levels. Because photosynthesis is immediately reduced, the lower light levels would be the *stress* and the diminished photosynthesis the *strain*. Stem elongation would probably be promoted, too; so, were *high* light levels a stress for stem elongation? Probably we would conclude that the promoted stem elongation rates actually constituted the strain because they led to taller stems with less mechanical strength—but that could be an advantage if the leaves were thereby carried above shading competitors into higher light levels. It's all a question of what is "best" or "normal" for a given plant, and the answer to this question can be highly subjective, depending upon circumstances and judgments. Most studies in stress physiology have been concerned with conditions that are much more obviously stressful; for example, conditions that limit yield.

Levitt defined **elastic biological strain** as those changes in an organism's function that return to the optimal level when conditions are again optimum (that is, when biological stress has been removed). If the functions do not return to normal, the organism is said to exhibit **plastic biological strain**. The analogy with physical objects is clear: An elastic deformation strain in a steel bar, for example, disappears when the stress is removed; a plastic deformation does not (that is, the bar remains bent).

Plant physiologists have emphasized such plastic strains as those caused by the stresses of frost, high temperature, limited water, or high salt concentrations. Elastic strains in plants, such as reduced photosynthesis in response to low light (it returns to normal with the return of high light levels), have been less studied by stress physiologists, although they must be extremely common and have been emphasized in studies of stress in animals.

Levitt (1972, 1980) distinguished between avoidance and tolerance (hardiness) to any given stress factor. In **avoidance**, the organism responds by somehow reducing the impact of the stress factor. For example, a plant in the desert might avoid the dry soil by extending its roots down to the water table. If the plant develops **tolerance**, on the other hand, it simply tolerates or en-

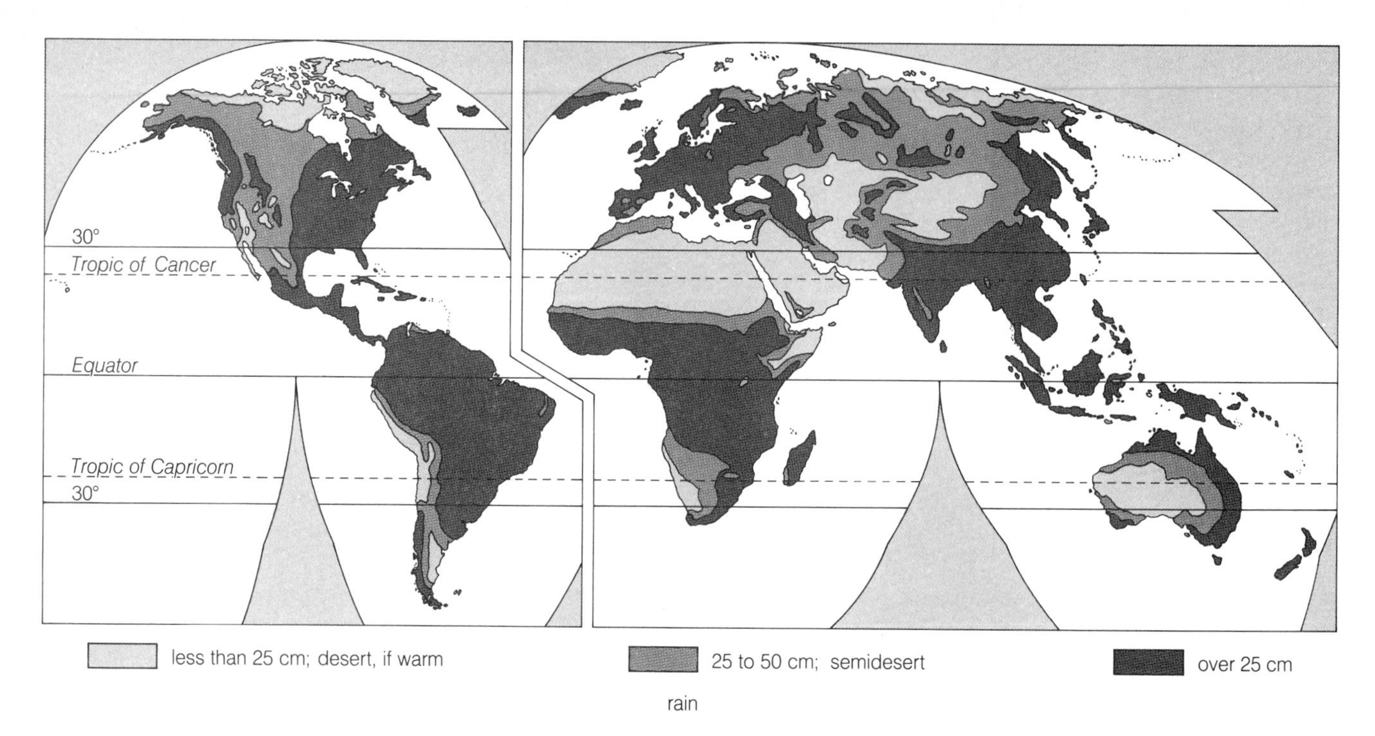


**Figure 26-1** Map of the world showing areas of extremely low (less than 250 mm) and low (between 250 and 500 mm) precipitation (usually as rain). Horse-latitude deserts usually occur between north or south latitudes of about 20 to 30°. By far the most striking example is the Sahara Desert of North Africa, extending to the deserts of the Arabian peninsula and the Near East, but the horse-latitude deserts of Mexico, South America, South Africa, and especially Australia are also evident. Rain-shadow deserts are important in North America (the Great Basin northward through Canada to the Arctic), where they are caused by the Sierra Nevada, the Cascades, the Rocky Mountains, and other ranges. Rain-shadow deserts are also important in Central Asia (the Gobi Desert and others), where they are caused by the Himalayas and other mountain ranges. Note the low precipitation in the Arctic — in some areas as low as the Sahara and other extreme deserts. Because of the low temperatures in polar regions, evaporation is greatly reduced compared with warm areas such as the subtropics; hence more water is available for plant growth. Yet moisture as well as low temperatures can limit plant growth in the low-rainfall polar regions. (From Jensen and Salisbury, 1984.)

dures the adverse environment. Creosote bush is a good example of a desert plant that is drought-tolerant. It simply dries out but survives anyway; it tolerates or endures the dryness of its protoplasm.

For the most part, Levitt's definitions are based on concepts that have been developing for well over a century. The terms *avoidance* and *tolerance* were often used early in the 20th century (for example, by Shantz, 1927), but the derivation of the concepts of *stress* and *strain* from their physical counterparts has not been widely accepted by stress physiologists, and the terms have sometimes been criticized (for example, by Kramer, 1980). The problem is that the term *stress* has often been used in the sense of biological *strain* as defined by Levitt. We introduce the terms because they emphasize the difference between the cause (the stress) and the effect (the strain).

Walter Larcher (1987) at the University of Innsbruck in Austria noted that we can keep this distinction clearly in mind if we use certain modifiers for the term *stress*: **stress factor** = Levitt's stress and **stress response** = biological strain. Larcher pointed out that Levitt's concept works best when we are dealing with individual stress factors, although stress responses are typically caused by more than one stress factor (Larcher et al., 1990). Hot summer weather, for example, may produce stress factors of high light levels (photodestruction of chlorophyll), low humidity, dry soil, and high temperatures. Furthermore, stress responses are typically complex, are exhibited by various parts of the plant, and may involve such *stress hormones* as abscisic acid (ABA) and ethylene, which are distributed throughout the plant.

In the 1930s, Hans Selye (see Selye, 1936, 1950) developed a concept of stress in human medicine. In his lexicon, *stress* was a syndrome of reactions within an organism in response to one or more stress factors — and any environmental factor could act as an agent of stress when it deviated from its optimum level for the organism, as we have noted already. Larcher (1987) suggested that we should speak of a **state of stress** when we use the term in Selye's sense. Selye's work emphasized the dynamic nature of stress responses in animals, and in many cases the stages of development of a state of stress can be applied to plants as well. When the stress factor is first experienced, there is the **alarm reaction**, in which the function of interest deviates markedly from the norm. Then comes the **resistance stage** (or **restitution phase**), in which the organism adapts to the stress factor and the function often returns toward its normal state (but may not completely achieve it). Finally, if the stress factor increases or continues for a long time, the **stage of exhaustion** may be reached, in which the function may again strongly deviate from the norm; this can eventually lead to death.

One more difficulty with the stress concept needs to be mentioned. Many plants found in what seem to be the most stressful conditions on earth—hot deserts, salty soils, or high mountaintops—often appear healthy and to be species not found anywhere else. If they are flourishing and apparently can't survive under other "less stressful" conditions, is it valid to think of them as being stressed? Actually, many such plants grow better under less stressful conditions if they have the chance. It has been suggested that they normally do not occur in more moderate situations because they cannot compete with the plants already growing there (Barbour et al., 1987). They are stressed in their native habitats in the sense that energy must be expended to overcome the harmful effects of the stress factors (Larcher, 1987)—for example, to pump salts into the vacuole from the cytoplasm, where they could denature enzymes. For that matter, as we saw in the essay on page 560, studies with canopies of wheat in controlled environments suggest that crops even in the most productive fields are stressed; it is possible to create environments in which they will yield more (but not *much* more on an individual-plant basis).

## 26.2 Stressful Environments

Recalling our discussion of the operational environment (Section 25.2; Spomer, 1973), we realize that an environmental stress means that some potential in the environment differs from the potential within the organism in such a way that there is a driving force for transfer of energy or matter into or out of the organism that could lead to a stress response. Low outside water potentials, for example, provide a driving force for loss of water; low temperatures can lead to loss of heat. But what are the environmental limits for the existence of life on our planet? We might look for answers by examining those environments in which productivities are lowest and thus the stresses might be the greatest.

**Figure 26-2** The desert east of Phoenix, Arizona, on the northern edge of the horse latitudes. A giant saguaro (pronounced sah-WAH-ro) cactus dominates the picture, with a much smaller cholla (CHAW-yuh) at its base and to the left. Numerous desert shrubs dominate the vegetation; the Superstition Mountains are in the background. The photograph was taken in mid-July. (Saguaro is *Cereus giganteus*; cholla is an *Opuntia* species, a common desert genus with many species. Photograph by F. B. Salisbury.)

### Deserts and Other Dry Areas

A desert (Fig. 26-1) is an area of low rainfall—an area of **drought**—with less than about 200 to 400 mm of precipitation per year, depending upon temperatures, potential for evaporation, season of precipitation, and other factors. Deserts often have sparse but fascinating vegetation (Fig. 26-2). The most extensive deserts occur in the so-called **horse latitudes**, which range from approximately 20 to 30° north and south of the equator (30 to 35° over oceans). In these regions, air that has ascended in other latitudes descends and is thereby compressed and warmed, forming a zone of high pressure. Warm air holds more moisture than cold air, so precipitation does not occur, and the descending air does not produce many surface winds.[1]

[1]Reportedly, the horse latitudes received their name because horses died in the becalmed ships and had to be thrown overboard.

North of the northern horse latitudes (and on the southern tip of South America, south of the southern horse latitudes) deserts also occur. Global air movements in the northern and southern temperate zones are predominately from west to east (the **westerlies**). Storm systems moving this way in the northern hemisphere rotate in a counterclockwise direction (clockwise in the southern hemisphere), so a storm center is preceded by south winds and followed by north or west winds. As storms approach a mountain range, the rising air expands and cools and can then hold less moisture, resulting in precipitation on the western slopes, which are typically covered with lush forests. On the eastern slopes, the air descends, compresses, warms, and can hold more moisture. Areas east of the mountains have low precipitation and are called **rain-shadow deserts** because they occur in the "rain shadows" of mountains. The descending warm and dry winds on the eastern slopes of the Rockies are called **chinooks** or snow eaters. North and east of the Alps, such winds are called the **Föhn**. The deserts of the Great Basin occur in the rain shadow east of the Sierra, but the plains east of the Rocky Mountains are not rain-shadow deserts because they receive moisture moving north from the Gulf of Mexico.

Deserts in the horse latitudes are typically hot and dry all year, whereas most rain-shadow deserts are cold during winter. Because air above deserts is usually dry, it absorbs relatively little incoming sunlight or outgoing long-wave thermal radiation. Thus deserts are hot during the daytime and relatively cold at night. Air warmed in desert valleys rises, whereas air that cools at higher elevations flows down canyons and gullies, especially during the night. Hence, wind is common and can lead to dune formation, although the great sand dunes associated with deserts, as portrayed by moviemakers and others, are not as common as we are led to believe (except in the Sahara Desert). Perhaps so-called **desert pavement** is more common. This consists of a surface layer of small stones, finer material having been eroded away, mostly by wind.

Desert soils are often salty because the low rainfall does not leach away the salts as they form by weathering of soil particles and rock. The actual status of a desert soil depends considerably upon the time during the year when rain does fall. Mediterranean climate zones, for example, usually occur just to the north of the northern horse latitudes or to the south of the southern horse latitudes. The horse latitudes follow the sun and shift away from the equator during summer, so Mediterranean climates experience summer drought that can last from six to nine months. Yet they may have considerable rainfall during the winter months when the horse latitudes have shifted back toward the equator. Technically, their abundant winter moisture keeps these areas from qualifying as true deserts. Because of the high precipitation during part of the year, these soils are usually less salty than other desert soils — but at the same time, they have some of the characteristics of desert soils (for example, relatively high *p*H). Summer showers in the Arizona desert (moisture from the Gulf of Mexico, mostly) lead to a unique and relatively lush desert vegetation, but temperatures are high, as is evaporation, and total precipitation is low enough that these areas qualify as true deserts.

Actually, the uncertainties of world climatic patterns mean that some regions in the temperate zone that are normally blessed with ample rainfall for productive agriculture may experience droughts that extend for several weeks to months and/or reduced precipitation that may last for several years; examples are the dust bowls of the southern plains in the 1930s, the widespread 1988 summer drought in much of the United States, and the serious drought in Europe during the summer of 1983. Thus **water stress** (water potential negative enough to damage plants) occurs in many parts of the world besides deserts.

## Tundras and Other Cold Areas

**Tundras** are areas on the earth's surface where temperatures are too low to permit the growth of trees (Fig. 26-3). Such areas occur on the tops of mountains (**alpine tundra**) and in the far north (**arctic tundra**). (Antarctic tundras are very limited in extent.) The polar tundras occur because the sun is relatively low in the sky even in summer (and actually below the horizon for days to months in winter poleward of the arctic and antarctic circles), so the slanting solar rays must follow a long pathway through the atmosphere and then strike horizontal surfaces at such acute angles that the energy of a given cross section of solar radiation is spread over a relatively large horizontal area. These combined effects (long atmospheric pathway and low solar angle) become increasingly important as one moves from the equator toward the poles and result in colder temperatures occurring at any given elevation. Stated another way, a given cold temperature occurs at lower elevations as the poles are approached. The **upper tree limit (tree line)**, for example, occurs at lower and lower elevations when one moves from the equator (where it is typically 3,500 to 4,500 m) until it reaches sea level some distance above the arctic circle. Alexander von Humboldt described this phenomenon in 1817; the principle is called **Humboldt's law**.

Why are temperatures low in alpine tundras? If a volume of gas expands or contracts without exchanging heat with its surroundings, the expansion or contraction is said to be **adiabatic**. As air rises, it expands because pressures are lower at higher elevations. If the expansion is adiabatic, the temperature of the air will decrease because there will be less heat per unit volume

**Figure 26-3** Alpine tundra high above tree line (note trees in distance at center right) on the northern border of Rocky Mountain National Park, Colorado, U.S.A., in the Mummy Range. The elevation is about 3,420 m (11,200 ft) above sea level, and the area was one of several sites for a physiological-ecology study of tundra plants. Because of different slopes, snow accumulation areas, substrates, and other features, several distinct vegetation types can be recognized in the tundra, but all are characterized by small plants, often with colorful flowers. (Photograph by F. B. Salisbury.)

of air. Dry air in the earth's atmosphere cools about 1°C for each 100 m of vertical rise. Thus, if dry air at 30°C on the valley floor is raised by global winds to a mountain top 1,500 m (4,921 ft) above, it will cool to about 15°C unless it is warmed or cooled by the mountainside or sunshine on the way up. This rate of cooling is an expression of the **adiabatic gradient**, or **adiabatic lapse rate**, for dry air in the earth's atmosphere (Fig. 26-4). Adiabatic cooling during expansion is the basic reason that higher air is nearly always cooler than lower air.

There is another important reason why alpine tundras are cool. At night any surface usually radiates more heat into the sky than radiates to it from the atmosphere above, so the surface cools. This is especially true for alpine surfaces because the atmosphere above is colder and thinner than it is above lower surfaces. So tundras cool faster at night than do lower areas. The cooled surfaces cool the adjacent air, which contracts, becomes more dense, and often flows down canyons or off slopes, replacing warmer air that is rising in the valleys.

During the day, high gusty winds are characteristic of alpine tundras, partly because of rapid temperature fluctuations but also because the mountain peaks deflect the global air movements that always occur high in the atmosphere. These winds increase evaporation from plants and soil, but humidity is usually high, which reduces evaporation. Virtually everything fluctuates rapidly in the alpine environment: wind velocity, temperature, light level (under partially cloudy skies, which are common), and humidity (see introduction to Chapter 21). The winter winds mean that much of the alpine tundra is blown free of snow, whereas deep snow drifts form in more protected areas. Some plants must thus endure the frigid temperatures of exposed areas while other plants avoid the winter extremes under a thick insulating blanket of snow.

High mountains near the equator have alpine tundras that do not experience winter (snow cover) but that can experience night temperatures as low as −6 to −11°C at any time during the year (Körner and Larcher, 1988; Sakai and Larcher, 1987). Actively growing plants living there must avoid or resist freezing the same as do plants in more temperate alpine tundras. These tropical tundras are characterized by a few succulent rosette plants, often with stems several meters tall (for example, species of *Dendrosenecio* and *Lobelia*).

Alpine and arctic tundras share similar average low temperatures but otherwise differ considerably in such factors as radiation flux, which is much lower in the arctic but is extended over longer days in summer. Light levels increase in alpine tundras because of a thinner layer of scattering atmosphere. Even when the sky is overcast, the diffused light measurable within clouds shrouding the tundra is much brighter than the light below clouds in the valleys. Highest irradiances are observed when the sky is partly cloudy, so that direct

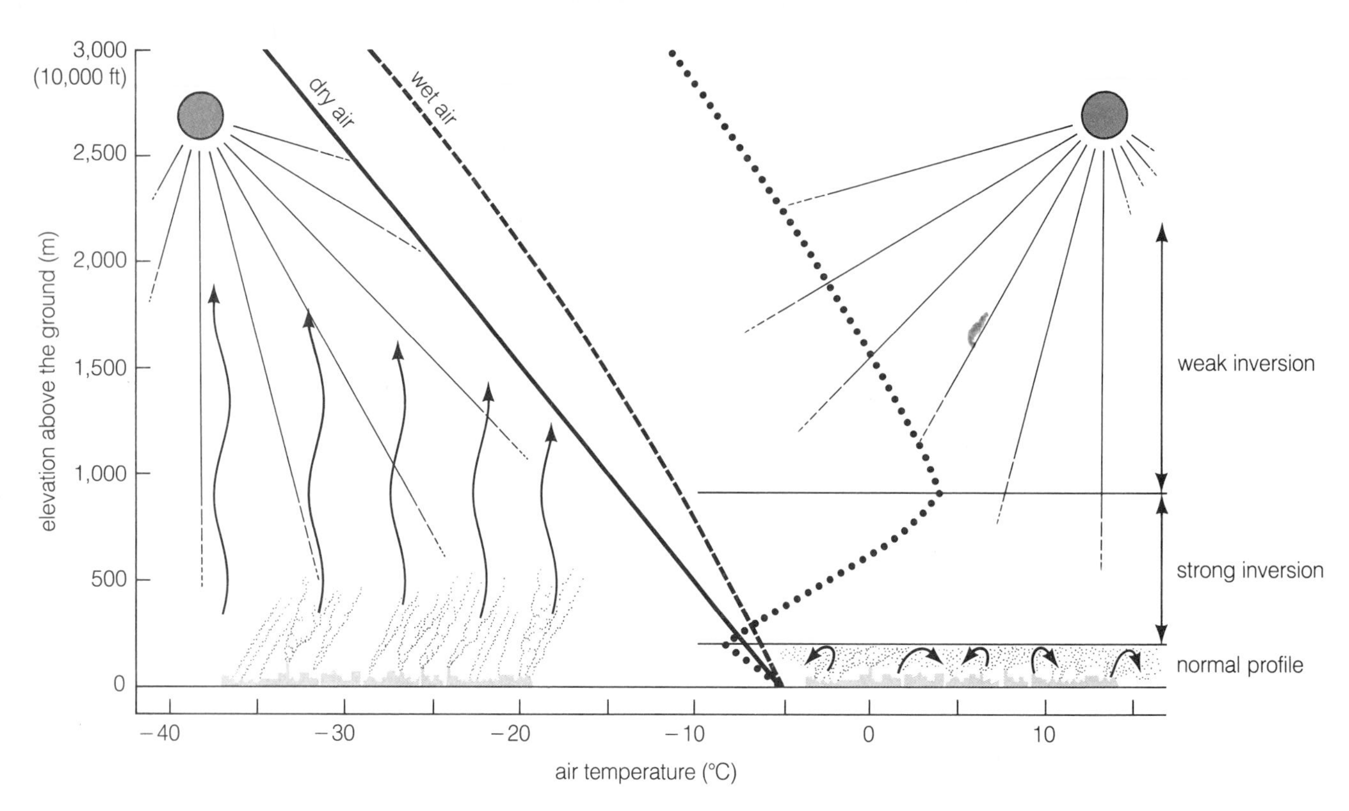


**Figure 26-4** The adiabatic lapse rate and atmospheric inversion. The two dividing lines that angle to the left between the two parts of the figure show temperature profiles (temperature as a function of elevation) that are examples of the adiabatic lapse rates for dry and wet air. These show the temperature of a volume of air that is allowed to expand adiabatically (without exchanging heat with its surroundings) and thus to cool as it rises through the elevations shown. If the temperature gradient is steeper than the adiabatic lapse rate, as is usually the case, air warmed by contact with the warm ground rises by convection (left side of figure). If the gradient is less steep than the lapse rate, air does not rise by convection (right side). The dotted line shows an actual temperature profile measured with a sounding balloon on a cold December 31, 1963, above Salt Lake City, Utah. Below about 200 m, the temperature gradient was steeper than the lapse rate, and air rose—but only to the bottom of the layer of air in which the profile was actually reversed from its normal condition. The air was warmer instead of colder with increasing elevation up to about 900 m. This was an unusually strong inversion—and a collector of atmospheric pollutants. The inversion continued even above 900 m because the temperature gradient was still not as steep as the lapse rate, even for wet air. (From Jensen and Salisbury, 1984.)

rays of the sun and reflected rays from the clouds both fall on the same area. Irradiances may be as high as 1,500 W $m^{-2}$ (solar constant = 1,350 W $m^{-2}$). Ultraviolet radiation can be high and perhaps important for organisms in alpine tundras, but it is very low in the arctic. The fact that tundra vegetation is so similar in alpine zones and in the arctic (often the same or closely related species but separate ecotypes; Section 25.4) testifies to the importance of low temperature.

What is special about the plants of the alpine or arctic tundras? Of course they are able to survive freezing, but why are they invariably small compared with their counterparts of warmer climates? Körner and Larcher (1988) noted that, contrary to widespread beliefs, the photosynthetic capacity of plants from cold regions is not essentially different from that of plants in temperate regions, when comparable life forms are considered. Prevailing leaf temperatures seldom limit seasonal photosynthetic carbon gain. Körner and Larcher suggested that cold affects tundra plants in two ways: indirectly, via the briefness of the growing season, and directly, via effects of reduced temperature on the developmental processes of growth, particularly cell division. Ecologists have largely overlooked the role of developmental processes in natural ecosystems, but tundra plants may well provide a logical example of the importance of such effects.

Having emphasized the low temperatures of tundras, we should note that plant leaves are not themselves always cold. Warmed by the sun, they may reach temperatures 20°C above air temperature (see, for example, Salisbury and Spomer, 1964; reviewed by Körner and Larcher, 1988). But temperatures are almost always low at night, and the growing season is very short. Daily temperature fluctuations are often 25°C and may reach 50°C (Körner and Larcher, 1988).

We use the tundras as examples of regions on earth where low temperatures are especially important, but freezing temperatures are also important every year in temperate zones, and frost can be damaging to native and especially crop plants even in Mediterranean and subtropical climates. Even in the tropics, unusually low temperatures (although they may be as much as 10°C above freezing) can cause chilling injury to such sensitive plants as bananas.

### Other Stressful Environments

There is a great variety of stressful environments on the earth's surface in addition to those we have described (Crawford, 1989). Some are of limited scale; others may be nearly worldwide in distribution. Flooding, for example, can produce a stressful condition quite opposite to the extremely negative water potentials of deserts, but damage results from the exclusion of oxygen rather than from the high water potentials. High temperatures, often close to or even above the boiling point, occur in hot springs, near volcanoes, in piles of decaying organic matter, and in extremely hot deserts, such as in Death Valley, California or Saudi Arabia. Living organisms often occur in these situations, as we shall see. We have already discussed (Chapters 12 and 25) stresses caused by low light levels in forests and at depths in bodies of water.

There are many spots on earth where soils are either highly acidic or highly alkaline, or where they are deficient in several nutrients (as are sand and some volcanic materials) or specific nutrients (especially nitrogen). The open oceans are deficient in many nutrients because these are carried to the ocean depths in the dead bodies of various organisms, especially **plankton** (single-celled plants and other organisms) as they settle to the bottom. Thus low nutrient concentrations (potentials) present special problems over the nearly two-thirds of the earth's surface covered with water; they are **nutrient deserts**.

The extreme pressures and total darkness that exist at great depths in the ocean, where organisms nonetheless live, surely seem stressful. Yet the organisms are alive and seem to be doing well (often existing on the "rain" of organic matter from above). The concept of stress that comes readily to our minds is a function of the environments to which we are ourselves adapted.

## 26.3 Water Stress: Drought, Cold, and Salt

Although there are many complications, a creosote bush growing in the desert, a white mangrove growing in a coastal forest with its roots in salt water, and a white spruce living in the north woods are all stressed, at least at certain times during the year, by a common factor: negative water potentials (water stress). We will begin our survey of this subject with the plants of the desert.

### The Xerophytes

We can classify plants according to their response to available water: **Hydrophytes**[2] grow where water is always available, as in a pond or marsh; **mesophytes** grow where water availability is intermediate; and **xerophytes** grow where water is scarce most of the time. Solutes strongly influence water potential and can have specific toxicities, so ecologists further classify plants that are sensitive to relatively high salt concentrations as **glycophytes** and those that are able to grow in the presence of high salt concentrations as **halophytes**. (Such organisms that are not plants are called **halophiles**.)

All the plants of the deserts are called xerophytes, but different species survive the drought in various ways. Figure 26-5 expands the concepts of avoidance and tolerance introduced above. There are several forms of avoidance, but tolerance is always a matter of endurance. Homer LeRoy Shantz (1927) used four terms in classifying xerophytes. The terms are descriptive and are shown in Figure 26-5: *escape, resist, avoid,* and *endure*. Xerophytes in the desert are actually exposed to a wide range of water potentials. Plants such as the palms that grow at an oasis, where their roots reach the water table, or other plants such as mesquite (*Prosopis glandulosa*) and alfalfa (*Medicago sativa*) that have roots that extend as much as 7 to 10 m down to the water table, never experience extremely negative water potentials. They are *water spenders*. They certainly avoid the drought. Of course, such plants must be able to use the available soil water while they are extending their roots to the water table. (In the Judean Hills of Israel, certain trees have extended their roots down along tight fissures in solid limestone to depths of 30 m or more, dissolving the rock as they go.)

The so-called **desert ephemerals** are annual plants that *escape the drought* by existing only as dormant seeds during the dry season. When enough rain falls to wet the soil to a considerable depth, these seeds often germinate, perhaps in response to the leaching away of germination inhibitors (Chapter 22). Many of these plants grow to maturity and set at least one seed per plant before all soil moisture has been exhausted. They are eminently well suited to dry regions and thus are xerophytes in the true sense of the word, yet their active and metabolizing protoplasm is never exposed to extremely negative water potentials and is not drought-hardy. As with each group of xerophytes outlined here,

[2]The names are from Greek: *-phyte*, "plant"; *hydro*, "water"; *meso*, "middle"; *xero*, "dry"; *glyco*, "sweet"; and *halo*, "salt."

the ephemerals form a class that consists of many species, each with its own special characteristics and ways of responding to different amounts of water or nutrients (Ludwig et al., 1989).

Succulent species such as the cacti, century plant (*Agave americana*), and various other crassulacean-acid-metabolism (CAM) plants (Section 11.6) are *water collectors*; they resist the drought by storing water in their succulent tissues. Enough water is stored, and its rate of loss is so extremely low (because of an exceptionally thick cuticle and stomatal closure during the daytime), that they can exist for long periods without added moisture. MacDougal and Spaulding (1910) reported that a stem of *Ibervillea sonorae* (a desert succulent from Mexico in the cucumber family, the Cucurbitaceae) stored "dry" in a museum used stored water to form new growth every summer for eight consecutive summers, decreasing in weight only from 7.5 to 3.5 kg! Because their protoplasm is not subjected to extremely negative water potentials, succulents are drought avoiders and are not truly drought-tolerant. The water potential in their tissues is often about $-1.0$ MPa. Some of the succulents, especially the cacti, have extensive shallow root systems that absorb (collect) surface moisture after a storm, storing it in their succulent tissues.

Some species that are subjected to periodic drought can switch from CAM, which conserves water because stomates are closed during the day, to C-3 photosynthesis when water becomes available (see Section 11.6). *Clusia rosea* is a tree that begins as an *epiphyte* (seeds germinate on branches of other trees) in the rain forest. During the time between rain storms, the young plants may be seriously water-stressed because their roots are not in the soil. Schmitt et al. (1988) found that plants would switch from CAM to C-3–photosynthesis quickly in response to changing moisture and irradiance levels.

Many nonsucculent desert plants have other adaptations that reduce water loss; they are *water savers*. For example, it is common for desert shrubs and other plants to have small leaf blades. This condition increases heat transfer by convection, lowering leaf temperature and thus reducing transpiration (Sections 4.8 and 4.9). Such leaves will still be as warm as the air temperature, but air temperatures are seldom fatal. Other adaptations that reduce transpiration include sunken stomates, shedding of leaves during dry periods, and heavy pubescence on leaf surfaces (Ehleringer et al., 1976). It is also important that such plants increase root resistance to prevent water loss to dry soil (Schulte and Nobel, 1989). Although these modifications may indeed reduce the loss of water, they never completely prevent it and are by themselves insufficient protection against extreme drought.

As water evaporates from plants, salts in the protoplasm could reach levels that could damage crucial enzymes. An important adaptation found in many organisms subjected to water and other stresses is the accumulation of certain organic compounds such as sucrose, amino acids (especially proline), and several others that lower the osmotic potential and thus the water potential in cells without limiting enzyme function. As water stress increases, such compounds appear in the cells of many xerophytes (see Section 26.4); the resulting drop in osmotic potential is called **osmotic adjustment** or **osmoregulation** (Morgan, 1984).

Perhaps most impressive among the xerophytes are plants that simply endure the drought. They lose large quantities of water, so their protoplasm is subjected to extremely negative water potentials, yet they are not killed. Such **euxerophytes** (true xerophytes) exhibit *dehydration tolerance* or **hardiness** rather than mere avoidance. Plants that only avoid drought are of great interest to ecologists, but they do not challenge our physiological understanding to the degree that euxerophytes do. Incidentally, many characteristics of the drought avoiders, such as small leaves and sunken stomates, also occur in the drought endurers. Yet in the euxerophytes the ultimate weapon against drought is the ability to endure it — to be drought-tolerant.

The ability of some euxerophytes to endure drought is phenomenal. Water content of the creosote bush (*Larrea divaricata*), a desert shrub of both North and South America, drops to as little as 30 percent of the final fresh weight before the leaves die. With most plants, levels below 50 to 75 percent are lethal. Some of the most spectacular euxerophytes (called **poikilohydric** by some ecologists) are mosses and ferns — plants that we normally associate with wet environments. Their ability to dry out and then become metabolically active immediately upon rehydration apparently depends upon special features not common to other plants. Examples are *Selaginella lepidophylla* (the resurrection plant), certain grasses, and *Polypodium* (a fern).

Much work on desert xerophytes has been done in the Negev Desert of Israel. Researchers (Evenari et al., 1975) there have studied algae, lichens, and mosses that can tolerate extreme and prolonged desiccation, as well as extreme cold and heat when they are dry. They can take up water directly and instantaneously from dew, rain, or even a moist atmosphere (some when relative humidity is as low as 80 percent), and such water absorption leads to an instantaneous switching on of metabolic activity. (Air at 80 percent RH and 20°C has a water potential of $-30$ MPa.)

Israeli workers have identified among higher plants most of the features just discussed, plus a few others. An important adaptation, for example, is that of **heteroblasty**, the ability of a single plant to produce morphologically and physiologically different seeds. Such seeds have different germination requirements, so only a few seeds in a given crop germinate at any given

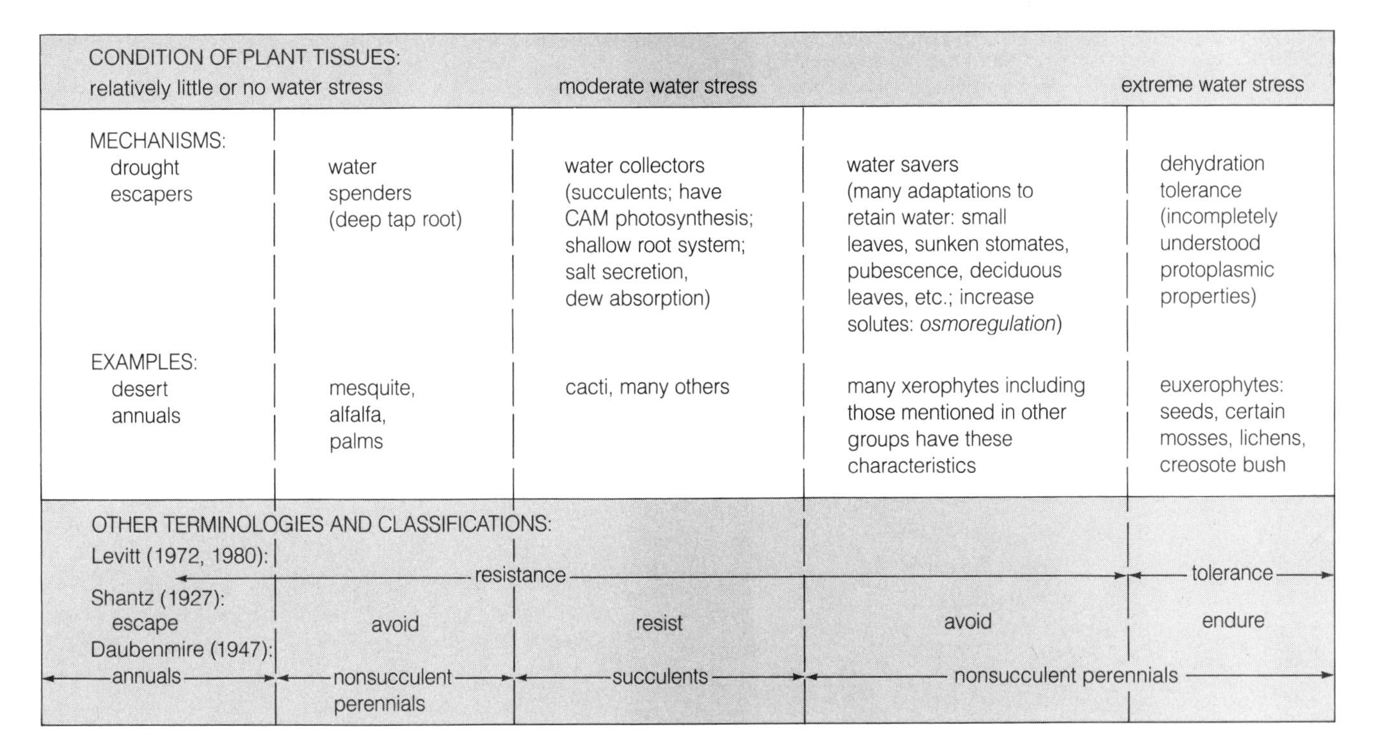


**Figure 26-5** An approach to a classification of plant responses to water stress. Some other approaches are also shown.

time. The risk inherent in seedling survival is thus distributed over a variety of environmental conditions and sometimes over several years. These workers have also studied stomatal regulatory mechanisms in desert plants. In the most interesting examples, when water stress was low in the leaf, stomates opened when the temperature was raised; when water stress was higher and the temperature was again raised, stomates closed.

In the Negev Desert, plants often utilize dew. Other plants may also use dew, but the extent to which this contributes to plant growth remains somewhat controversial (Rundel, 1982). Later we'll consider an interesting example of salt secretion on leaves: The salt absorbs moisture from the air, and this water is then absorbed into the leaves.

Incidentally, insect physiologists (Edney, 1975) reported that certain insects absorb water from an atmosphere with a relative humidity as low as 50 percent—and thus a water potential of almost $-100$ MPa! Liquid within the insect's body might have a water potential of only $-1.0$ to $-2.0$ MPa and would be in equilibrium with an atmosphere of about 99 percent relative humidity. How absorption occurs remains a mystery.

There are numerous other adaptations of desert plants that cannot be discussed here for lack of space. For example, the **water-use efficiency** (the ratio of dry-matter production to water consumption) increases as soil-water availability decreases (Ehleringer and Cooper, 1988), and allelopathics (Section 15.3) are often produced by desert plants, restricting the germination or growth of competing plants, which reduces competition for water.

## Water Stress in Mesophytes

Although desert xerophytes are studied by physiological ecologists, much work on water stress has been done by foresters and agriculturists (for example, Mussell and Staples, 1979). Water can limit crop growth and productivity virtually anywhere, either because of unexpected dry periods or normally low rainfall that makes regular irrigation necessary. When one must irrigate to obtain a crop, how much and how often should water be applied to the field? The answer, which could influence the expenditure of large sums of money, may be strongly influenced by research on stress physiology. Here we are seldom concerned with the severe stresses endured by desert plants; rather, we are interested in the extent to which withholding relatively small amounts of water might influence crop yield.

Extensive research has continued for well over a century. Thousands of papers on plant responses to drought have now been published. Many reviews and volumes that report on symposia have been published since 1980 (see, for example, Bewley and Krochko, 1982; Bradford and Hsiao, 1982; Greenway and Munns, 1980;

**Table 26-1 Generalized Sensitivity to Water Stress of Plant Processes or Parameters.**[a]

| Process or Parameter Affected | Sensitivity to Stress: Very Sensitive → Relatively Insensitive; Tissue $\Psi$ Required to Affect Process[b] (0 MPa, −1.0 MPa, −2.0 MPa) | Remarks |
|---|---|---|
| Cell growth | ———— - - - | Fast-growing tissue |
| Wall synthesis | ———— | Fast-growing tissue |
| Protein synthesis | ———— | Etiolated leaves |
| Protochlorophyll formation | ———— | |
| Nitrate reductase level | ———— | |
| ABA accumulation | - - - ———— | |
| Cytokinin level | ————— | |
| Stomatal opening | - - - - ———————— - - - - - - | Depends on species |
| $CO_2$ assimilation | - - - - ———————— - - - - - - | Depends on species |
| Respiration | - - - ———— | |
| Proline accumulation | - - - ————— | |
| Sugar accumulation | ————— | |

[a]Length of the horizontal lines represents the range of stress levels within which a process first becomes affected. Dashed lines signify deductions based on more tenuous data.
[b]With $\Psi$ of well-watered plants under mild evaporative demand as the reference point.
Source: From Hsiao, 1973.

Hanson and Hitz, 1982; Kramer, 1983; Levitt, 1980; Marchand, 1987; Morgan, 1984; Schulze, 1986; Staples and Toenniessen, 1984; Tranquillini, 1982; and Turner and Kramer, 1980).

Theodore Hsiao (1973; Bradford and Hsiao, 1982) has been especially active in this field. Table 26-1 (his 1973 summary table, which remains valid) outlines the sequence of events that occurs when water stress develops rather gradually as water is withheld from a plant growing in a substantial volume of soil. It is important to realize that the later events are almost undoubtedly indirect responses to one or more of the early events rather than to water stress itself.

Cellular growth appears to be the most sensitive response to water stress (Fig. 26-6). Decreasing the external water potential ($\Psi$) by only $-0.1$ MPa (sometimes less) results in a perceptible decrease in cellular growth (irreversible cell enlargement) and thus root and shoot growth (Neumann et al., 1988; Sakurai and Kuraishi, 1988). Hsiao suggested that this sensitivity is responsible for the common observation that many plants grow mainly at night when water stress is lowest. (But temperature, photoinhibition, and endogenous rhythms could also be involved.) The inhibition of cell expansion is usually followed closely by a reduction in cell-wall synthesis. Protein synthesis may be almost equally sensitive to water stress. These responses are observed only in tissues that are normally growing rapidly (synthesizing cell-wall polysaccharides and protein as well as expanding). It has long been observed that cell-wall synthesis depends upon cell growth (Section 16.2). The effects on protein synthesis are apparently controlled at the translational level, the level of ribosome activity.

At slightly more negative water potentials, protochlorophyll formation is inhibited, although this observation is based on only a few studies. Many studies indicate that activities of certain enzymes, especially nitrate reductase, phenylalanine ammonia lyase (PAL), and a few others, decrease quite sharply as water stress increases. A few enzymes, such as α-amylase and ribonuclease, show increased activities. It was thought that such hydrolytic enzymes might break down starches and other materials to make the osmotic potential more negative, thereby resisting the drought (osmotic adjustment), but careful studies don't always support this idea. Nitrogen fixation and reduction also decrease with water stress, a finding that is consistent with the observed drop in nitrate-reductase activity. At levels of stress that cause observable changes in enzyme activities, cell division is also inhibited. And stomates begin to close, leading to a reduction in transpiration and photosynthesis.

There has been much controversy about whether the commonly observed drop in photosynthesis in response to only moderate stress is caused by stomatal closure or more directly by the water stress itself (see, for example, Kaiser, 1987). Calculations of the internal $CO_2$ concentration seemed to show that it remained high under moderate water stress, suggesting that photosynthesis itself was inhibited. But because the calculations were subject to several errors, this conclusion was questioned. Nevertheless, when the errors are taken into account, evidence remains suggesting that high transpiration rates can lead to decreased photosynthesis (Bunce, 1988). Measurements of responses of several photosynthetic enzymes to water stress also suggest a direct effect (Vu and Yelenosky, 1988). Final conclusions will depend on future research, and species differences will probably be important.

At about the level of stress that elicits effects on enzymes, abscisic acid (ABA) begins to increase markedly (at least 40-fold; see Sections 4.6 and 18.5) in leaf tissues and, to a lesser extent, in other tissues, including roots (reviewed by Bradford and Hsiao, 1982; Salisbury and Marinos, 1985; Walton, 1980). This leads to stomatal closure and reduced transpiration, as we discussed in Section 4.6. In addition, ABA inhibits shoot growth, further conserving water, and root growth appears (in some studies) to be promoted, which could increase the water supply. There is also evidence that suitably low concentrations of ABA increase the rate of water conductance through roots, which would reduce the water stress in the shoots. Most of these adaptations involving ABA are best observed in mesophytes; xerophytes often have other adaptations (Kriedemann and Loveys, 1974).

There is evidence that ABA normally plays a role in the resistance of mesophytes to water stress. Most studies have been done with drought-sensitive and drought-resistant cultivars of crop plants (see, for example, Quarrie, 1980). Often, resistant cultivars have higher levels of ABA when they are exposed to stress, and sensitive cultivars can be phenotypically converted to resistant types by application of ABA. But there always seem to be exceptions. A yellow lupine (*Lupinus luteus*), for example, is remarkably insensitive to applied ABA.

As a plant responds to water stress, what causes the increased ABA production in its tissues? Evidence suggests that lowered cell turgor is the trigger (reviewed by Bradford and Hsiao, 1982), but how turgor might control ABA synthesis remains unknown.

It is interesting that ABA increases in leaves in response to several kinds of stress, including nutrient deficiency or toxicity, salinity, chilling, and waterlogging. Reduced cell turgor and water potential are not involved in all of these. It may well be that ABA is a kind of universal stress hormone, its production controlled or triggered by several mechanisms. In all cases it seems to reduce growth and metabolism and thus conserve resources, which will then be available during recovery if and when the stress is removed.

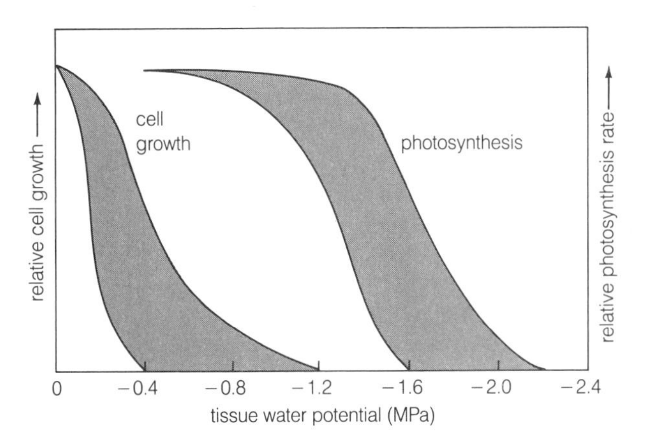


**Figure 26-6** Cell growth and photosynthesis as a function of decreasing tissue water potentials. Shaded areas include ranges of response as observed with several species in different experiments. Cell growth (leaf enlargement, for example) is much more sensitive to decreasing water potential than is photosynthesis. (See Boyer, 1970; Acevedo et al., 1971.)

Ethylene also seems to play a role in stress reactions. It appears in response to various stress factors, including excess water (Jackson, 1985), plant pathogens, air pollution, root pruning, transplanting, handling (Section 19.2), and perhaps drought (Tietz and Tietz, 1982).

Although stresses are relatively mild at $\Psi = -0.3$ to $-0.8$ MPa, interactions and indirect responses begin to be the rule. Cytokinins decrease in leaves of some species at about these levels. At slightly more negative water potentials, the amino acid proline begins to increase sharply, sometimes building up to levels of 1 percent of tissue dry weight. Increases of 10- to 100-fold are common. Depending on species, other amino acids and amides, especially betaine, also accumulate when the stress is prolonged. Proline arises *de novo* from glutamic acid and ultimately probably from carbohydrates. These compounds contribute to osmotic adjustment (discussed below).

At higher levels of stress ($\Psi = -1.0$ to $-2.0$ MPa), respiration, translocation of assimilates, and $CO_2$ assimilation drop to levels near zero. Hydrolytic-enzyme activity increases considerably, and ion transport can be slowed. Actually, in many species respiration often increases, not really dropping off until water stresses of $-5.0$ MPa are reached.

Plants usually recover if watered when stresses

are −1.0 to −2.0 MPa, meaning that, in spite of the severity of the water stress, the stress response was elastic—or at least somewhat elastic, because growth and photosynthesis in young leaves frequently do not reach the original rates for several days, and old leaves are often shed. Clearly, because growth is especially sensitive to water stress, yields can be noticeably decreased with even moderate drought. Cells are smaller and leaves develop less during water stress, resulting in reduced area for photosynthesis. Furthermore, plants may be especially sensitive even to moderate drought during certain stages, such as tassel formation in maize. Ultimately then, in the sense of final yield, stress responses are really plastic, even with moderate water stress.

## Salt Stress

A common and important stress factor in deserts is the presence of high salt concentrations in the soil (reviewed by Flowers et al., 1977). Soil salinity also restricts growth in many temperate regions besides deserts (Greenway and Munns, 1980). Millions of acres have gone out of production as salt from irrigation water accumulates in the soil. A plant faces two problems in such areas, one of obtaining water from a soil of negative osmotic potential and another of dealing with the high concentrations of potentially toxic sodium, carbonate, and chloride ions. Some crop plants (for example, beets, tomatoes, rye) are much more salt-tolerant than others (for example, onions, peas), and many crops have cultivars that are relatively salt-tolerant.

In the study of salt tolerance, the **euhalophytes** (*true* halophytes that tolerate or endure high levels of salt) are particularly interesting. Several such species grow best where salt levels in the soil are high, as in deserts or in soils saturated with brackish waters on the sea coasts or close to the shores of extremely salty waters such as the Great Salt Lake, where the salt content may be saturated at levels as high as 26 percent by weight.[3] They also grow in nonsalty soils. *Allenrolfea* (iodinebush), *Salicornia* (pickleweed or samphire), and *Limonium* (sea lavender, marsh rosemary) are representative genera. Species of *Atriplex* (shadscale) and *Sarcobatus* (black greasewood) grow in somewhat less salty soils, and certain bacteria (archaebacteria?) and blue-green algae (cyanobacteria) live in the waters of the Great Salt Lake. In general, prokaryotes and archaebacteria are more resistant to environmental stresses than are eukaryotes.

Barbour (1970; see also Barbour et al., 1987) reviewed the literature that suggests that *no* angiosperms are **obligate halophytes**, plants that cannot grow unless the soil is salty. All halophytes studied so far have sometimes been found growing naturally in nonsalty soils and will grow well when planted in nonsalty soils. Normally, they are not abundant in nonsalty soils because they cannot compete with the glycophytes that normally grow there. As we discuss below, however, members of the genus *Halobacterium* (prokaryotes, again) accumulate large amounts of salt into their cells and cannot survive except in salty environments.

In terrestrial halophytes, the osmotic potential of leaf cell sap is invariably highly negative. Sap from tissues of actively growing *Atriplex* species, having no special cold hardiness, for example, freeze only when temperatures drop below −14°C, implying that their osmotic potentials are as low as about −17.0 MPa. This contrasts with a normal −1.0 to −3.0 MPa in most plants. In some cases, the xylem sap does not have a highly negative osmotic potential but may be almost pure water. To obtain water from the surrounding soil, the water potential within the xylem sap must then be greatly lowered by tension. This was demonstrated by Scholander and his coworkers for mangrove trees (see Fig. 5-17).

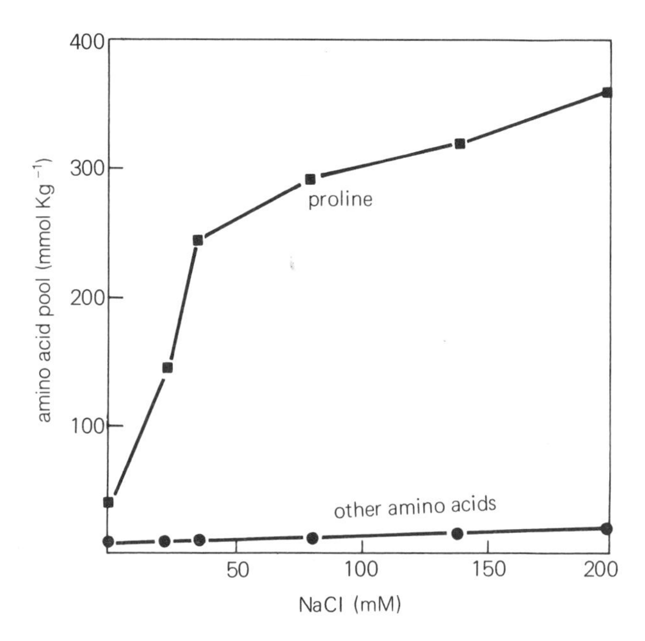


**Figure 26-7** Amino acid and proline accumulation in *Triglochin maritima* grown at different salinities. Cuttings of *T. maritima* were grown in a nonsaline medium for two weeks before being transferred to saline media. Shoot tissues were harvested for analysis after 10 days of saline treatment. The extent of proline accumulation in this specialized halophyte is extreme compared with most plants, but accumulation of proline or other compatible solutes as drought or salt stress increases is common. Values in stressed mesophyte leaf tissues seldom exceed 200 mmol $kg^{-1}$ of dry mass. (Note that values in this figure are for fresh mass; from Stewart and Lee, 1974.)

[3]During the early 1980s, several wet years caused the volume of the Great Salt Lake to quadruple; its salt content dropped to about 7 percent.

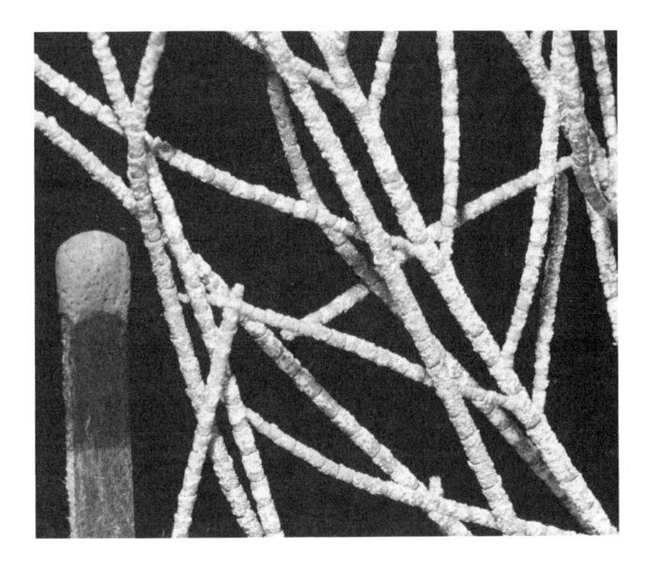

**Figure 26-8** Tamarisk (*Tamarix pentandra*) leaves collected at Barstow in the Mojave Desert, California, showing heavy incrustations of salt. The paper match indicates the scale. (Photograph by F. B. Salisbury.)

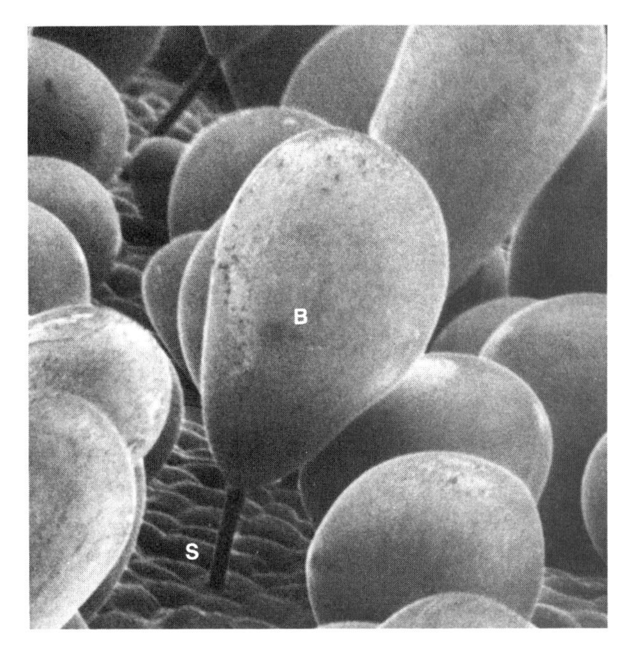


**Figure 26-9** A scanning electron micrograph of salt bladders on the leaf of saltbush (*Atriplex spongiosa*). This species of saltbush is indigenous to Australia and is salt-tolerant because it has developed a mechanism to control the $Na^+$ and $Cl^-$ concentrations of its tissues by accumulating salt in epidermal bladders on the surface of the aerial parts of the plant. Salt from the leaf tissues is transferred through the small stalk cell (S) and into the balloonlike bladder cell (B). As the leaf ages, the salt concentration in the cell increases, and eventually the cell bursts or falls off the leaf, releasing the salt outside the leaf. 450 ×. (From Troughton and Donaldson, 1972.)

Some halophytes are referred to as **salt accumulators**. In these species (for example, *Atriplex triangularis*; Ungar, 1977) the osmotic potential continues to become more negative throughout the growing season as salt is absorbed. Even in these plants, however, the soil solution is not taken directly into the plant. Based upon quantities of water transpired by the plant, it is easy to calculate that if the complete soil solution were absorbed, the plant would contain 10 to 100 times as much salt as is actually observed. Instead, water moves into the plant osmotically and not simply in bulk flow. The endodermal layer in the roots probably constitutes the osmotic barrier.

Halophytes in which the salt concentration within the plant does not increase during the growing season are known as **salt regulators**. Salt-tolerant derivatives of wheat, for example, limit the accumulation of $Na^+$ and $Cl^-$ ions under salt stress compared with sensitive cultivars (Schachtman et al., 1989). Mangroves provide another spectacular example, excluding nearly 100 percent of the salt (Ball, 1988).

Often salt does enter the plant, but because the leaves swell by absorbing water, concentrations do not increase very much, if at all. This leads to the development of **succulence** (a high volume/surface ratio), a common morphological feature of halophytes. Ice plant (*Mesembryanthemum crystallinum*) is a good example (see Flowers et al., 1977). Rapid growth is another mechanism that dilutes the salt. In these cases, and when salt is excluded by the roots as in mangroves, organic compounds without the toxic effects of salt build up in the tissues, maintaining osmotic balance with the soil solution. Proline is a common example (Fig. 26-7), but other amino acids and such other compounds as galactosyl glycerol and organic acids also occur (Hellebust, 1976). As we shall see in a subsequent section, these compounds function in osmotic adjustment.

Sometimes excess salt is exuded on the surface of the leaves, helping to maintain a constant salt concentration within the tissue (Fig. 26-8). In certain halophytes there are readily observable salt glands on the leaves, sometimes consisting of only two cells (Fig. 26-9). Although $Na^+$ ions are essential for some salt-tolerant species, it is probable that sodium is transported out of the cytosol by counter transport with incoming $H^+$ (see Section 7.10). This moves much of the ion out of the cytoplasm of both root and leaf cells, inwardly to the central vacuoles and outwardly to the extracellular spaces.

*Nolana mollis*, a dominant, succulent shrub of the Atacamba Desert of northern Chile, grows where rainfall is less than 25 mm $y^{-1}$, although high fog and a relative humidity around 80 percent are common. The plant is almost always wet to the touch. Mooney et al. (1980) found that salt glands on the leaves secrete salt (mostly NaCl) that absorbs water hygroscopically from the atmosphere. If the leaves are washed with distilled

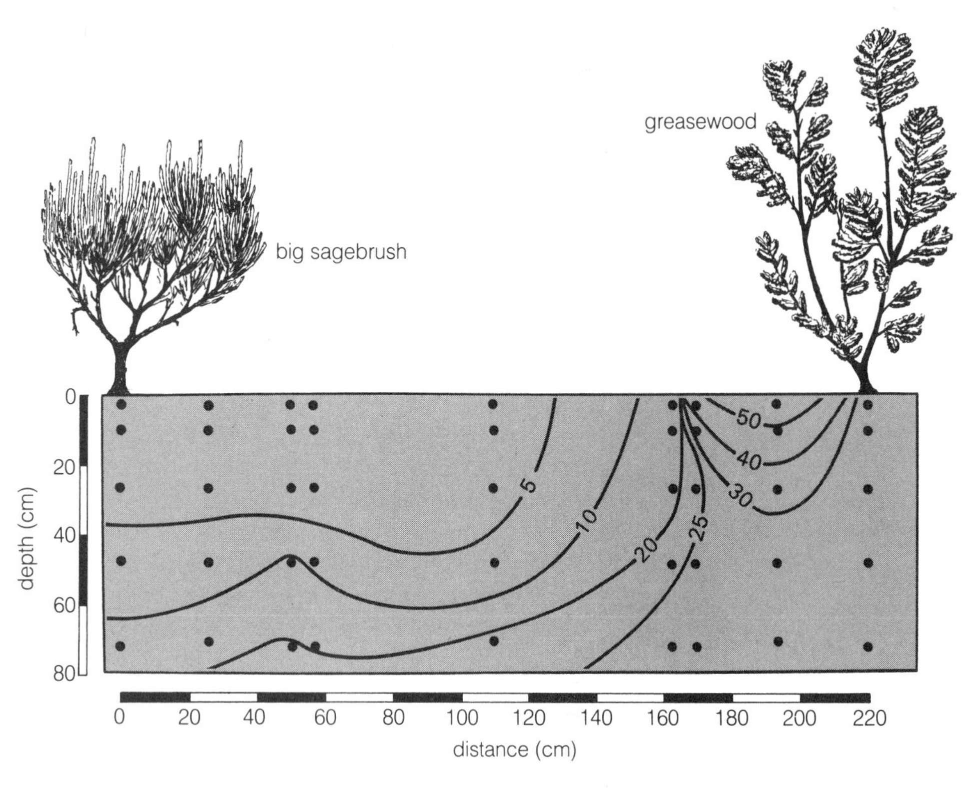


**Figure 26-10** Salt concentrations in the soil (numbers in lines represent percent exchangeable sodium) as a function of position under a sagebrush plant (*Artemisia tridentata*) and under a greasewood plant (*Sarcobatus vermiculatus*) in the Escalante Desert of southern Utah. Note high sodium under greasewood. Dots are sampling points. (Data from Fireman and Hayward, 1952.)

water and blotted dry, they remain dry until they have had a chance to secrete more salt. If filter paper is soaked in solution collected from the leaves (or in concentrated NaCl solution) and then dried in an oven, it also absorbs water from the moist atmosphere and becomes wet to the touch. Can the plant absorb the water on its leaves? Mooney and his coworkers suggest two pathways of absorption: directly into the leaves or through the roots after the salty solution has dripped onto the ground. Either pathway would require the expenditure of metabolic energy by mechanisms that are not known to exist in plants, although they do exist (but are not understood) in insects and arachnids. The researchers calculate that ample respiratory energy is available, if a mechanism exists for its use.

Actually, large quantities of both organic and inorganic materials are leached from the leaves of many plants, both halophytes and glycophytes. Some of the leaching brought about by washing the leaves is caused by removal of materials within the tissues as well as washing off materials that have been exuded at the surface. In any case, materials that are washed from the leaves to the soil, or that fall with the leaves, are recycled back to the plant and to other plants. Species that absorb large amounts of salt and then lose it may considerably increase the salinity of the surface soil. Thus, as in other environments, desert plants sometimes profoundly influence the soil upon which they grow. Fireman and Hayward (1952) found, for example, that greasewood (*Sarcobatus vermiculatus*) in the Escalante Desert of Utah brought salts from depths, depositing them on the surface (Fig. 26-10), probably as leaves fell and decayed. The result was high salt concentrations beneath the greasewood plants, especially when compared with soil beneath big sagebrush (*Artemesia tridentata*) plants, which did not redistribute salt in this manner. Clearly, physiological differences between these two species are of considerable ecological significance.

Another potential problem for plants growing on saline soils is obtaining enough potassium. This problem exists because sodium ions compete with the uptake of $K^+$ by a low-affinity mechanism (Chapter 7), and $K^+$ is commonly present in such soils in much lower concentrations than is $Na^+$. The presence of $Ca^{2+}$ appears to be crucial. If sufficient calcium is present, a high-affinity uptake system having preference for transport of $K^+$ can operate well, and the plants can then obtain sufficient potassium and restrict sodium (LaHaye and Epstein, 1969). It is possible that calcium fertilization of some saline soils low in $Ca^{2+}$ might increase their agricultural productivity. A favorable effect of $Ca^{2+}$ on soil structure could also be important. Gypsum ($CaSO_4$) is sometimes used, providing both $Ca^{2+}$ and some acidity, which helps in leaching out the $Na^+$. Elemental sulfur is also sometimes applied; it becomes oxidized to produce sulfuric acid, which aids in $Na^+$ leaching. Sulfuric acid itself has been applied with some success.

There has been considerable interest during recent years in the mechanisms by which $Ca^{2+}$ overcomes the harmful effects of $Na^+$ (Cramer et al., 1985, 1986, 1988). For example, growth of maize roots was highly sensitive to NaCl (75 mM) in the nutrient solution, but this could be completely overcome by addition of $Ca^{2+}$ (10 mM), *providing* that the $Ca^{2+}$ was supplied *before* the sodium. This and other studies support the proposal that $Ca^{2+}$ protects membranes from adverse effects of $Na^+$, thereby maintaining membrane integrity and minimizing leakage of cytosolic $K^+$.

A related line of research examines new polypeptides (proteins) that appear in response to salt stress (Hurkman et al., 1988; Ramagopal, 1987a, 1987b; Rouxel et al., 1989; see the discussion of heat-stress proteins in Section 26.6 below and the discussion of salt-stress proteins in Section 18.5). These proteins might be responsible for the membrane responses. A low-molecular-weight protein called *osmotin* seems to be especially important. Based on the observation that high $Na^+$ concentrations produced long and narrow cortical cells of cotton roots, Kurth et al. (1986) suggested that the $Na^+/Ca^{2+}$ ratio might influence the deposition of microtubules in the cytoskeleton and thus the deposition of microfibrils in the wall. Iraki et al. (1989) studied the cell walls of isolated tobacco (*Nicotiana tabacum*) cells adapted to grow in strong salt solutions (for example, 0.428 M NaCl). In such solutions, cells were only one-fifth to one-eighth the volume of unadapted cells. The walls of the adapted cells were much weaker than normal walls because carbon was diverted from wall synthesis to formation of molecules important in osmotic adjustment (see Section 26.4 below).

In the context of studies on the role of membranes in plant stress responses, we note that many workers are currently attempting to understand the mechanisms of ion transport across membranes and that these mechanisms typically involve proton pumps ($H^+$-ATPases) that establish $H^+$ gradients across the membrane, which gradients can be used to drive solute uptake into plant cells through various secondary transport mechanisms. Calcium often plays a crucial role in the models that are developed (see Chapter 7 and such references as Butcher and Evans, 1987 and Giannini and Briskin, 1989). Blumwald and Poole (1987) showed that increased $Na^+$ in the growth medium of salt-tolerant cells from sugar beet induced a doubling in activity of $Na^+/H^+$ antiport activity at the tonoplast.

### The Lower Temperature Limits for Survival and Growth

There is apparently no lower temperature limit for survival of spores, seeds, and even lichens and certain mosses in the dry condition. Such test objects have been held within a fraction of a degree of absolute zero for several hours with no apparent damage. Even active tissue may survive these low temperatures if it is experimentally cooled so rapidly that intracellular water freezes into extremely small crystals that don't damage the cytoplasm. But the lower temperature limits for survival under more normal circumstances depend strongly on the species and the extent to which the tissues have been hardened against frost—as we shall see in Section 26.4. Actively growing plants often can survive only a few degrees below 0°C, whereas many hardened plants can survive to about −40°C, and a few acclimated plants (willows and conifers, for examples) seem to have no low-temperature limits for survival (Sakai and Larcher, 1987).

What are the low-temperature limits for active growth as compared with survival? Several higher plants are able to grow and even flower under the snow, where the temperature is close to 0°C or sometimes below (Richardson and Salisbury, 1977). Such plants include native species such as the snow buttercup (*Ranunculus adoneus*), which forms flowers under snowbanks in the high mountains, as well as winter cereals (for example, winter wheat and winter rye) and ornamentals (crocuses, snowdrops, tulips, hyacinths, and daffodils—the last three not necessarily actually growing under snow). These are called **geophytes** or **spring ephemerals**. They grow slowly during the winter and thus often have a significant head start when the snow melts. The native species, especially, then grow rapidly and flower before later species overtop them or before trees above them leaf out.

It has been reported that lichens can photosynthesize at −20 to −40°C. Kappen (1989) reviewed these reports and questioned the data because the lichens might have been warmed to above the nearby air temperatures by radiation and may not have been totally frozen. Kappen's own measurements, carried out in Wilkes Land, Antarctica, detected vigorous photosynthesis at −10°C when the thalli had been sprayed with water before freezing. Frozen lichens covered with snow also exhibited active photosynthesis. Certain bacteria can grow at temperatures on the order of −22°C. Snow algae grow at the freezing point of pure water and below (Aragno, 1981). Actively growing giant rosette plants (*Dendrosenecio* sp. and *Lobelia telekii*) above tree line on equatorial mountains can experience night temperatures as low as −13°C at any time. Bodner and Beck (1987) found that such plants could photosynthesize when their cell water was supercooled to −8°C. When cells were dehydrated by freezing, photosynthesis stopped, but it resumed immediately after thawing, a feature not seen in most flowering plants. Similar results were found with *Rhododendron ferrugineum* in the Austrian Alps in autumn (Larcher and Nagele, 1985). The lower limits for active growth of organisms have not been determined.[4]

## Frost and Freezing Injury

Although productivity of world ecosystems is probably limited more by water than by any other environmental factor, low temperature is perhaps most limiting to plant distribution (Parker, 1963). To grow even in subtropical regions subject occasionally to freezing or even to near-freezing temperatures, plants must be capable of some acclimation to low temperatures. Plants that grow in polar regions must tolerate extremely low temperatures; only a few species can achieve this. Frost and chilling damage in crop plants is an important hazard nearly everywhere.

We might expect that death of a plant results from water expansion upon freezing and subsequent disruption of cell walls and other anatomical features. Careful examination during the early decades of the 19th century showed, however, that plants actually contract rather than expand upon freezing. This is because ice crystals grow into the extracellular air spaces. Furthermore, although it must occur, ice is almost never observed within the living cells of tissues that have frozen naturally. (Ice is, however, observed in the dead xylem cells of trees in winter; see Section 5.5.) Nor is damage to cells other than collapse observed. There is no rupture of cell walls or even of cell membranes, although there is ample evidence that membranes are damaged during thawing (Steponkus, 1984; for a recent scanning-electron-microscope study, see Pearce, 1988). With rapid tissue cooling in the laboratory (for example, 0.3 to 5°C $min^{-1}$), however, ice forms within the cells, and cellular components are damaged.

Such rapid cooling does occur in nature. Acclimated American arborvitae (*Thuja occidentalis*) tissues were capable of withstanding temperatures to −85°C when cooled slowly, but southwest-facing foliage was injured when the temperature dropped 10°C per minute from 2 to −8°C at sunset. Such changes duplicated in the laboratory also injured plants. Injury symptoms could not be duplicated by any form of desiccation, and it was concluded that the winter burn was caused by the rapid temperature drop (White and Weiser, 1964). Nevertheless, Sakai and Larcher (1987) questioned the conclusion that intracellular freezing is the cause of **sunscald** on south- and southwest-facing tree trunks. They suggested that the damage typically occurs in early or late winter before hardiness has had a chance to develop or after hardiness has begun to disappear.

Typically, ice crystals begin to form in the extracellular spaces, and water from within the cells diffuses out and condenses on the growing ice masses, which may become several thousand times as large as an individual cell. The cell acts as an osmotic system, with the osmotic concentration inside increasing as water diffuses out through the plasmalemma, dehydrating the cell. When these ice crystals melt in frost-hardy plants, the water goes back into the cells and they resume their metabolism. In nonacclimated plants, damage to membranes and other cellular components may have occurred, so metabolism cannot be resumed and the water does not reenter the cells completely. In the next section, we summarize what is known about the nature of frost damage and how plants can become resistant to it.

## Hardening (Acclimation)

We have noted that plants can become **acclimated** to various stress factors by developing tolerance (becoming hardy) against the stress factor that induced the change and often against other stress factors as well. For example, plants exposed to low water potentials, high light levels, and such other factors as high-phosphorus and low-nitrogen fertilization become drought-tolerant (hardy) compared with plants of the same species not treated this way. Such acclimation to drought is of considerable importance to agriculture. It is a good example of a conditioning effect.

Actively growing plants, especially herbaceous species, are damaged or killed by temperatures of only −1 to −5°C, but many of these plants can be acclimated to survive winter temperatures of −25°C or lower. In regions where air temperatures drop below this, many plants have underground meristems protected from extreme air temperatures by soil or snow (Fig. 26-11). They avoid or escape cold rather than being extremely cold-hardy. Most species that survive freezing temperatures tolerate some ice formation in their tissues. Generally, hardier plants survive with more of their water frozen than do less hardy plants. But there are apparently several mechanisms of hardiness (see excellent discussions in Burke et al., 1976; Levitt, 1980; Sakai & Larcher, 1987; and Steponkus, 1981, 1984).

In practical terms, minor increases in hardiness could have a major impact on world food production. Winter wheats and winter rye yield 25 to 40 percent

[4]One of us (Salisbury) attended a NASA-supported Conference on Environmental Extremes held in San Diego, California, on February 10–11, 1966. There it was stated by experts in the field that certain bacteria grow at −22°C (as we have noted), that molds have exhibited active growth in cold-storage lockers at −38°C, and that spores were formed by these organisms at −47°C! We are currently unable to document these claims, and they should be viewed with suspicion (but see Allen, 1965).

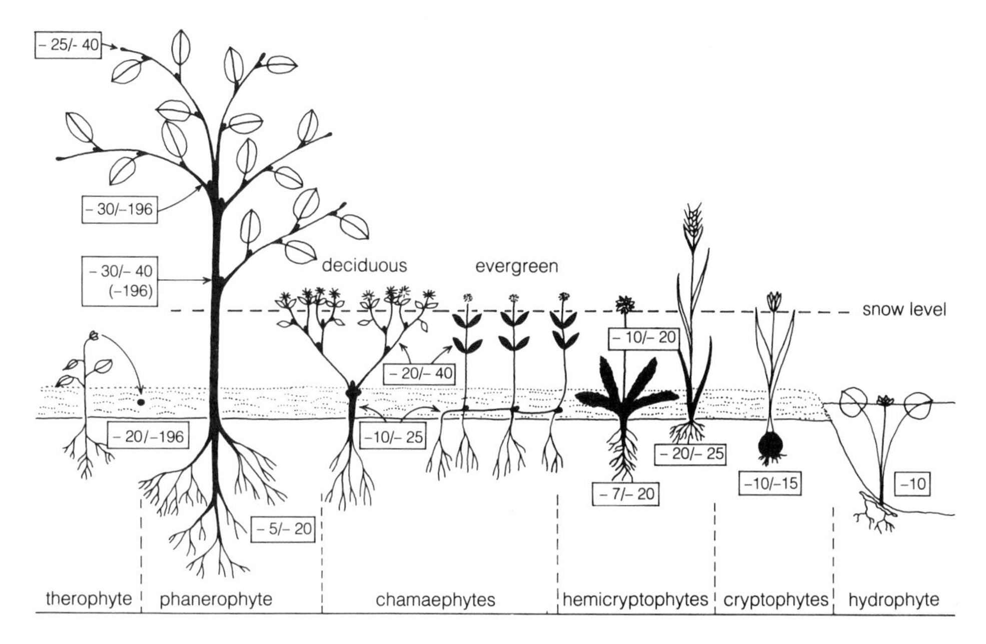


**Figure 26-11** Typical temperature ranges of frost resistance for **life forms** according to Raunkiaer (1910). The overwintering parts are shown in black. (From Larcher, 1983; used by permission.)

more than comparable spring cultivars, for example, because they make better use of spring rains. If the winter wheats and rye could be made 2°C more cold-hardy, they could replace much of the large areas of spring wheats and rye in North America and the Soviet Union.

Frost hardiness typically develops during exposure to relatively low temperatures (for example, 5°C) for several days. Temperatures down to −3 to −10°C are sometimes required for maximum acclimation (Larcher and Bauer, 1981; Weiser, 1970). Short days also promote acclimation in several species, and there are indications that a stimulus may move from leaf tissue to the stems. The development of frost hardiness is a metabolic process requiring an energy source. Apparently this can be provided by light and photosynthesis. Factors that promote more rapid growth inhibit acclimation: high nitrogen in the soil, pruning, irrigation, and so on. In general, nongrowing or slowly growing plants are more resistant to several environmental extremes, including air pollution. Water-stressed plants are also more resistant to air pollution, partially because their stomates are closed.

Salt hardiness can be increased somewhat by exposure to saline conditions. Salt hardening is minimal, however, compared with drought or cold acclimation. Clearly, hardening against drought, freezing, and high salt concentrations is often a matter of hardening against water stress, but as we shall see in the next section, there are many complications.

## 26.4 Mechanisms of Plant Response to Water and Related Stresses

How does a euxerophyte differ from an ordinary mesophyte? Or how does a halophyte differ from a glycophyte? What is special about plants that can survive extremely low temperatures? Many different proposals have been presented to account for tolerance and acclimation, and a few of these may be common to the three kinds of stress (all related to water stress) that we are discussing here. For example, protoplasmic viscosity usually increases with high water stress, often to the point at which the protoplasm becomes brittle. Euxerophytes maintain protoplasmic plasticity much better at a given water stress than do mesophytes, and the hydrolytic activity (breakdown of starch, protein, and so on) is also less noticeable in euxerophytes. To a certain extent, these features also appear in plants capable of surviving extremely low temperatures, but halophytes often approach the problem in somewhat different ways. Here are some possibilities.

### Proposals About Responses of Mesophytes to Mild Water Stress

How are mesophytes damaged by mild water stress? At least five possibilities have been proposed. First, it is known that water activity (indicating its ability to enter

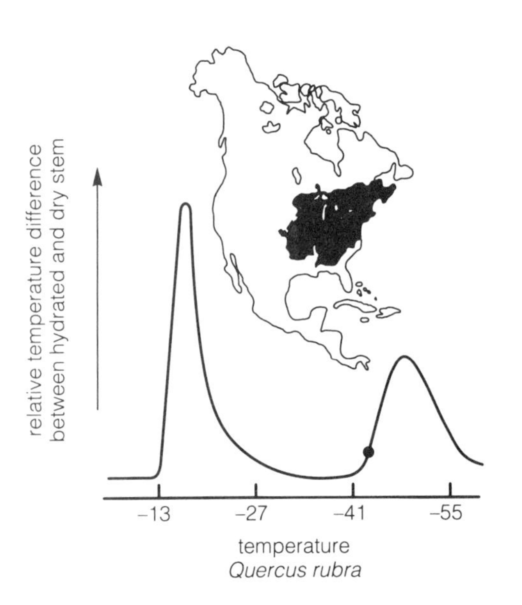


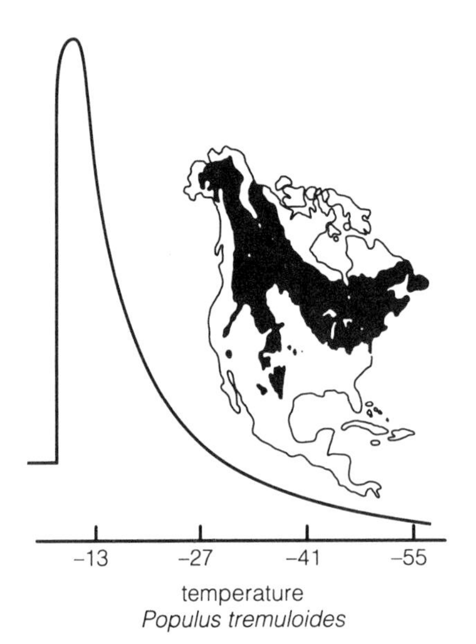


**Figure 26-12** Differential thermal analysis (see also Fig. 26-13) of winter-hardy xylem and the natural ranges of northern red oak (*Quercus rubra*; deep supercools) and quaking aspen (*Populus tremuloides*; does not deep supercool). On the thermal-analysis curves (see Fig. 26-14 for explanation of such curves), the peaks to the left show the heat of fusion released when extracellular water freezes; the peak to the right in red oak shows the heat of fusion released when the supercooled sap inside of xylem parenchyma cells freezes. Aspen has no such peak and can withstand temperatures to −60°C. (From George and Burke, 1984; used by permission.)

into chemical reactions) is a function of water potential and is thus lowered by water stress. Nevertheless, water activity is closely related to water concentration, and as we have seen (Section 2.7), water concentration changes only slightly as water potential changes considerably. At the maximum stress levels of interest to agriculturists ($\Psi = -1.0$ to $-2.0$ MPa), water activity is lowered only slightly—probably not enough to be of any real consequence in chemical reactions. Second, solutes increase in concentration as water is lost. This could be important, but it is probably not very important under mild water stress simply because the concentration changes amount to only a few percent. Third, water stress might result in special changes in membranes. Such effects have indeed been demonstrated, but because comparable effects can be caused by other factors without noticeable plant response, it does not seem likely that this is an important aspect of plant response to water stress. Fourth, water stress might upset the hydration of macromolecules—the "ice" structure of the water molecules surrounding enzymes, nucleic acids, and so on. If this water of hydration is upset, function would also be influenced. Levitt (1962) has suggested that dehydration of key enzymes would cause disulfide bonds within proteins to break and reform, sometimes reforming between adjacent molecules, leading to enzyme denaturation when molecules are rehydrated. But again, it has been calculated that mild water stress would not have much influence on the structure of water of hydration, which involves only a small percentage of the water in a cell anyway. Amazingly enough, studies have shown that considerable water can be lost from a cell before enzyme function is noticeably influenced. But there could be exceptional enzymes that have not been studied.

Fifth, even the mildest water stress may profoundly change the turgor pressure within plant cells. Pressure changes of this magnitude ($P = 0.1$ to $1.0$ MPa) probably have little effect upon most enzyme activities (judging by observed responses as well as thermodynamic principles), but such changes could be the stimulus to which some special response mechanism in the cell reacts in transducing water stress to the observed cellular responses. In the large-celled marine green alga *Valonia*, ion uptake decreases with slight increases in cellular turgor pressure. Such responses have not been observed in most higher-plant cells, although red-beet tissue responds this way to decreased turgor. The observation may serve as a model for what might be taking place (see discussion in Hellebust, 1976). We have already noted that ABA is apparently produced in response to decreasing leaf-cell turgor (Sections 4.6 and 26.3), and we have emphasized that cell expansion, meaning plant growth in general, is highly sensitive to water stress, probably via decreasing cell turgor (see Fig. 26-6).

## Frost Damage and Frost Acclimation

Several good summaries review the status of our knowledge of frost effects (for example, Krause et al., 1988; Li, 1984; Marchand, 1987; Sakai and Larcher, 1987). Frost resistance is based either on tolerance to extracellular ice formation and thus severe cell dehydration (as in

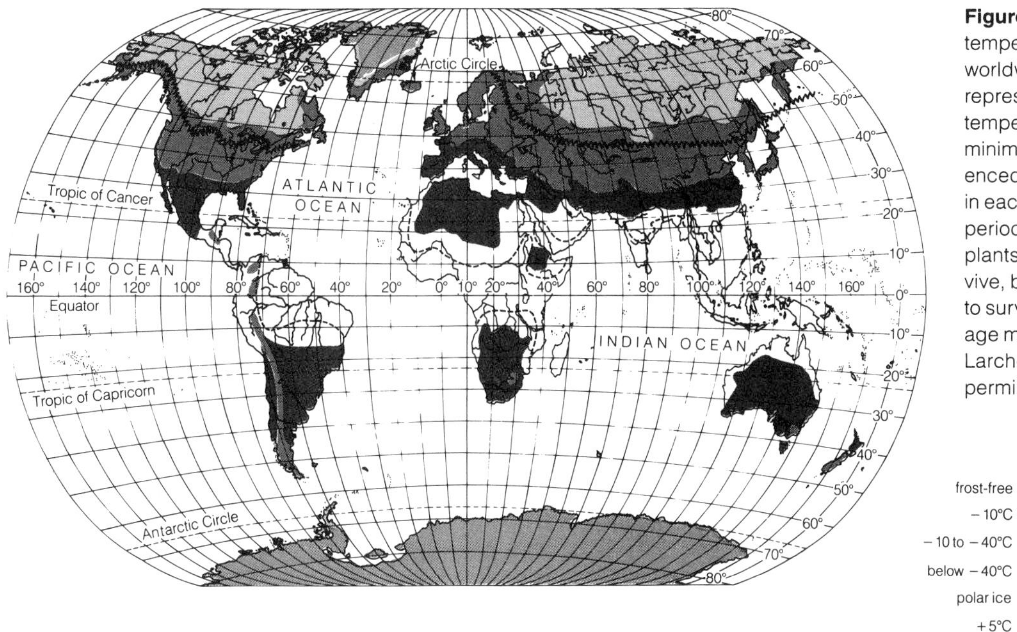


**Figure 26-13** Tentative map of low-temperature thresholds limiting worldwide plant distribution. The lines represent average annual minimum temperatures rather than the lowest minimum temperatures ever experienced (or experienced once or twice in each century). When such unusual periods of cold occur, at least a few plants in the population usually survive, but populations cannot continue to survive if they cannot tolerate average minimum temperatures. (From Larcher and Bauer, 1981; used by permission.)

drought tolerance) or on frost avoidance, especially supercooling. We will discuss the dehydration effect first.

Although several significant differences in protoplasm have been observed between hardy and sensitive plants, most recent studies have emphasized either the compatible solutes (**cryoprotectants** in this context) that we discuss below or alterations in the various membrane systems in the cell, along with the patterns of proteins that are associated with these membranes (Gilmour et al., 1988; Guy and Haskell, 1987). The changed protein patterns also occur in response to other stresses, as we note below in the section called "Other Stress Proteins." There are also changes in growth regulators (especially ABA; Lalk and Dörffling, 1985), and Quader et al. (1989) observed reversible changes in elements of the cytoskeleton in response to cold stress.

There have been many suggestions about membrane changes in response to cold (Steponkus, 1984). For example, a low sterol/phospholipid ratio in the plasmalemma might stabilize the lamellar bilayer during freeze dehydration, but the evidence is contradictory (see Krause et al., 1988 for review). It is clear that in most species, freezing stress leads to an inhibition of photosynthesis after thawing and that the photosynthetic reactions of thylakoid membranes are temporarily impaired. Particularly, photosystem II is inhibited. Nevertheless, it appears that $CO_2$ assimilation is more sensitive to freezing stress than is activity of the thylakoids. Krause et al. (1988) reported that the light-regulated enzymes of the carbon reduction cycle appear to be the first to be affected. These enzymes reside in the chloroplast stroma.

## Deep Supercooling and Ice Nucleation

Water in xylem tissues of most deciduous fruit trees and forest species, and in many dormant buds (see, for example, Ashworth, 1984), does not freeze until temperatures drop as low as about −40°C (lowest on record is −47°C). This phenomenon is called **deep supercooling** (Burke et al., 1976). Ice nucleation is relatively improbable in small water volumes about the size of a plant cell. Water can be supercooled to −38°C, and the solute-containing water in acclimated xylem cells apparently acts like water droplets in that the xylem cells are isolated from one another by air spaces, dry cell walls, and the plasmalemma. As in similarly acclimated herbaceous plants, ice forms in the bark and buds when temperatures are only a few degrees below the freezing point of pure water, but the crystals form in the spaces between cells, and tissues are not damaged by this. Xylem tissue in most hardwoods is too rigid and too impervious to water to permit the formation of such crystals, however. When freezing does occur, xylem ray parenchyma cells are killed, the wood becomes dark and discolored, and vessels are filled with gummy occlusions (Section 5.6). Wood-rotting organisms often invade such injured trees, which frequently die. Such species do not grow where winter temperatures drop below about −40°C (Figs. 26-12 and 26-13). This limit is set by the supercooling process; it proves to be the weak link in plants' survival.

Extremely hardy woody plants (for example, birches, alders, quaking aspen, willows) native to the boreal forests of the Northern Hemisphere do not undergo deep supercooling. Extracellular freezing oc-

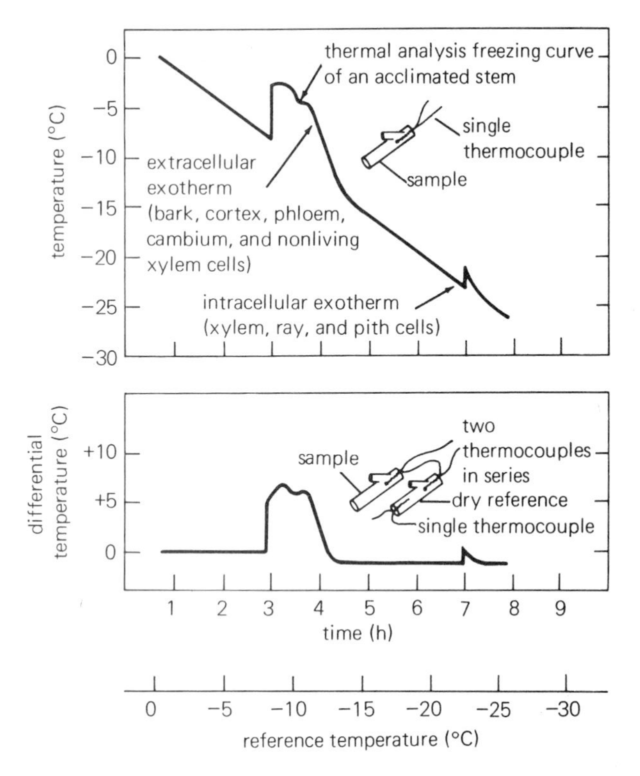


**Figure 26-14** Thermal analysis and differential thermal analysis as methods to observe deep supercooling in stem samples. In both experiments the sample is cooled over a period of time, and the sample temperature (top) or the differential temperature between the sample and a dry reference (bottom; see also Fig. 26-12) is monitored. The peaks or *exotherms* show release of the heat of fusion and thus indicate freezing points. The first peak (left, warmest temperature) indicates freezing of extracellular water; the second peak (right, coldest temperature) presumably indicates freezing of water in cells. Cell freezing typically leads to the death of the cells. (From Burke et al., 1976.)

curs, and the cells are extremely tolerant of the dehydration of their protoplasm. As ice masses form, they draw water from the cells until all but the water of hydration (bound water) is removed. Such dormant, winter-hardy, woody plants (typically softwoods) can readily survive the −196°C of liquid nitrogen. These same plants, when actively growing, may be killed by −3°C! Thus cold acclimation is much more spectacular than drought or salt hardening.

Deep supercooling is observed by measuring stem temperatures with a small thermocouple device while temperature is lowered (Fig. 26-14). When the xylem ray cells finally freeze, the released heat of fusion causes a sharp temperature rise, as when osmotic potentials are determined cryoscopically (see Fig. 3-10).

Both tissue rigidity and submicroscopic anatomical features appear to be important in xylem supercooling (George and Burke, 1984). Membranes may play a role in the deep supercooling of living tissues by keeping ice nuclei from reaching the supercooled water. But water in wood with no living cells can also supercool, although not quite as much as in living wood. This suggests that the microstructure of cell walls can also separate supercooled water from ice crystals nearby. Apparently, as water leaves the cells to condense on the crystals, tension builds within the cells, lowering the water potential so that its vapor pressure remains in equilibrium with the ice. This model needs more testing, but it is supported by current evidence.

Plants with no appreciable level of hardiness may escape light frost damage by supercooling to −2 to −10°C. When they do freeze it is because of various things that provide nuclei for ice crystals. Sometimes these occur within the tissues, as has been shown for wood of *Prunus* spp. (Gross et al., 1988); other times they are on the surface of the leaves. Certain species of bacteria (for example, *Pseudomonas syringae* and *Erwinia herbicola*) have been found, for example, that initiate ice formation at relatively warm temperatures. Spraying frost-sensitive plants in the field with suspensions of these bacteria causes the plants to be killed by light frosts, whereas spraying with other bacterial species protected the plants by facilitating supercooling (Anderson et al., 1982; Lindow, 1983).

## Compatible Solutes: Osmotic Adjustment (Osmoregulation)

We have noted in several places that certain substances such as proline, betaine, and various carbohydrates build up in cells that are subject to drought or high salt concentrations. This is called **osmotic adjustment** or **osmoregulation** (Flowers et al., 1977; Jefferies, 1981; Morgan, 1984; Turner and Jones, 1980).

Consider the situation with high salt concentrations. Because there are excessive dissolved salts in the soil solution (or sea water in the case of mangroves and other plants that grow in similar situations), the osmotic potential is negative enough to cause water to diffuse out of the tissues into the surrounding solutions—unless the water potential in the tissues is at least as negative. Actually, if the tissues are to absorb water and survive, their water potential must be more negative than that of the surrounding solution. One way to overcome this problem would be for the cells simply to accumulate salt to the same or higher concentrations as those outside the plant. Why doesn't this happen? (We noted that those plants that do accumulate salt usually dilute it by becoming succulent or accumulate the salt into their vacuoles, where it could still dehydrate the cytoplasm osmotically.)

In all eukaryotes so far studied, salts such as NaCl denature the enzymes and thus cannot be tolerated in the cytoplasm itself (although this effect may be somewhat less drastic than has been assumed; see Cheeseman, 1988). When they occur in halophyte cells, they

**Table 26-2 Examples of Some Organisms and the Compatible Solutes That Increase in Their Cells During Osmotic Adjustment.**

| Organism | Compatible Solute |
|---|---|
| BACTERIA | |
| Various halophiles and nonhalophiles (e.g., *Klebsiella, Salmonella, Streptococcus*) | amino acids (glutamate, proline, etc.) |
| *Halobacterium salinarium* (halophile; an Archaebacterium; no osmotic adjustment) | NaCl |
| FUNGI | |
| *Chaetomium globosum* (a terrestrial form) | polyhydric alcohols (mannitol, arabitol, glycerol) |
| *Saccharomyces rouxii* (an osmophilic form) | arabitol |
| MICROALGAE | |
| *Chlorella pyrenoidosa* (freshwater) | sucrose |
| *Dunaliella* spp. (marine and halophilic) | amino acids, glycerol |
| *Scenedesmus obliquus* (freshwater) | carbohydrate (sucrose + raffinose, glucose, fructose) |
| ANGIOSPERMS | |
| glycophytes: *Chloris gayana, Hordeum vulgare* (barley) | betaine and proline |
| halophytes: *Aster tripolium, Mesembryanthemum nodiflorum, Salicornia fruticosa, Triglochin maritima* | proline |
| halophytes: *Atriplex spongiosa, Spartina townsendii, Suaeda monoica* | betaine |

Source: Mostly from Flowers et al., 1977 and Yancey et al., 1982.

are on the order of 10 times more concentrated in the vacuole than in the cytosol. Apparently, many plants that tolerate the various kinds of water stress do so by synthesizing in their cytoplasm compounds that can exist at high concentrations without denaturing the enzymes essential for the metabolic processes of life. The organic compounds that can be tolerated have been referred to as **compatible solutes** (or **compatible osmotica**). Paul H. Yancey and his coworkers (1982) pointed out that the number of compatible solutes discovered among the five kingdoms of living organisms is relatively limited. Apparently only a few compounds can exist at relatively high concentrations in the cytoplasm without damaging the enzymes there. Table 26-2 lists a few species that exhibit different degrees of resistance to water stress (produced by drought, cold, or salt), along with the compatible solutes found in their cells. Figure 26-15 shows the chemical structures of a few of these compounds.[5]

The degree of osmotic adjustment is a function of the degree of outside water stress caused by salt in the surrounding medium (see Fig. 26-10), by drying soil, and by frost hardening. The regulation or adjustment occurs in halophytes, xerophytes, and mesophytes. How does the regulation occur? Current research is concerned with this question, but complete answers remain to be determined (LeRudulier et al., 1984). We have already emphasized the role of cell turgor in synthesis of ABA and in controlling the rate of cell growth as it is influenced by water stress. Most workers in the field suspect that changes in cell turgor also activate and control the degree of compatible-solute synthesis. In the marine algae, for example, turgor pressure remains constant over a wide range of external salinities, indicating that it must be regulated and suggesting that cells are sensitive to changes in turgor. Perhaps the physical properties of the plasmalemma change as the force with which it is pressed against the wall changes. Felix D. Guerrero and John E. Mullet (1988) observed rapid changes in leaf translatable RNAs in response to a reduction in turgor of pea leaves. There is also the interesting observation that levels of putrescine, a polyamine, increased to over 60 times normal in oat-leaf cells within 6 h in response to osmotic stress (sorbitol and other osmotica dissolved in the media on which leaf segments were floated; Flores and Galston, 1982). Yet observed maximum concentrations of putrescine are still far too low for the compound to be acting as a compatible solute. Could the change be part of the transduction between cell turgor and the synthesis of such solutes? Future research is required to find out.

Membranes of organisms that produce compatible solutes must not only prevent the external salt from

[5]Roles in addition to that of compatible solute have been suggested for some of the compounds in Table 26-2. Proline, for example, might also serve in nitrogen storage and transport.

**polyols**

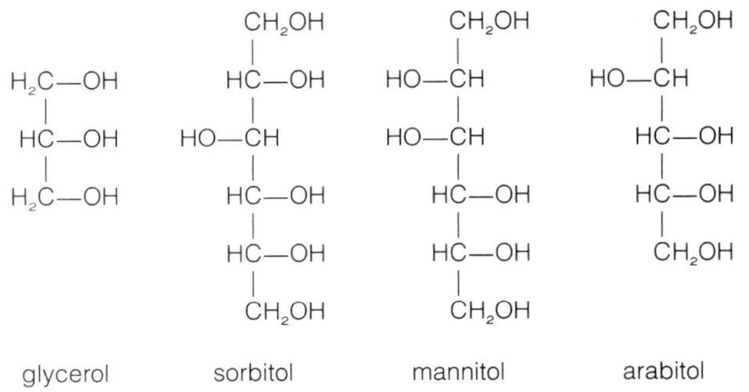

glycerol sorbitol mannitol arabitol

**amino acids**

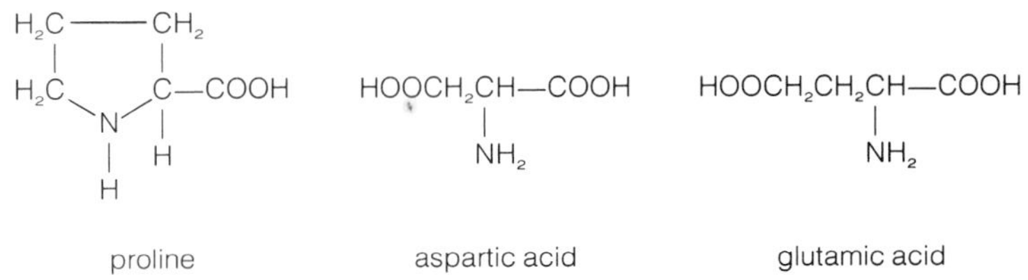

proline aspartic acid glutamic acid

**methylated quaternary ammonium compounds**

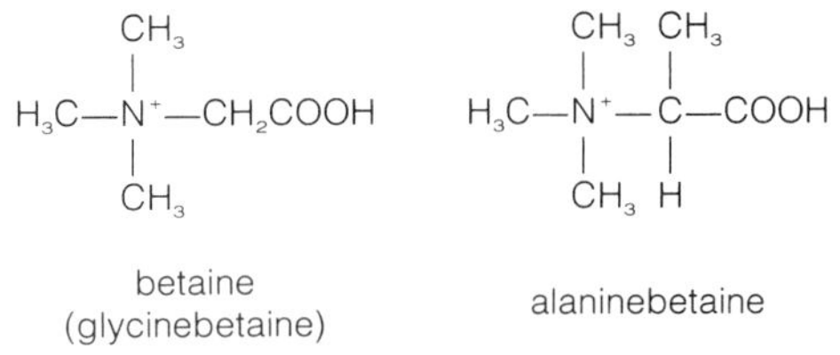

betaine (glycinebetaine) alaninebetaine

**Figure 26-15** Molecular structures of some compatible solutes found in stressed plants and other organisms. (See also Table 26-2.)

entering the cell but must also prevent the compatible osmoticum from leaking out. Gimmler et al. (1989) studied isolated plasma membranes of the extremely salt-tolerant unicellular green alga *Dunaliella parva*, which synthesizes large amounts of glycerol as a compatible solute. The membrane is not only highly impermeable to glycerol, but it has an effective ion-pumping capacity that actively exports the salt that leaks in. The plasma-membrane ATPases responsible for this pumping require unusually high concentrations of divalent cations (up to 100 mM $Mg^{2+}$ or $Ca^{2+}$). Furthermore, fairly high concentrations (up to 800 mM) of NaCl or $NaNO_3$ (but not $Na_2SO_4$) were required for maximum activity of these enzymes, and the ATPases were extremely resistant to salt. About 2.5 M NaCl (about a 15 percent solution) was required for half-maximal inhibition of activity. Thus, the compatible solute (glycerol) in the cytosol of these cells allows the enzymes there to be more or less like their counterparts in glycophytes, but the enzymes in the plasma membrane, which are exposed to high salt concentrations, have the necessary adaptations.

There are a few organisms that seem to be the exception to the rule about compatible solutes; actually, they illustrate the importance of the rule. Certain species of *Halobacterium* (Table 26-2) accumulate large quantities of sodium chloride in their cells. These interesting organisms are considered by many bacteriologists to belong to a sixth kingdom, the Archaebacteria (or a third kingdom, if organisms are divided into Prokaryotes, Eukaryotes, and Archaebacteria). Some of these organisms are unable to grow and reproduce unless they are in highly concentrated salt solutions such as the Dead Sea or the Great Salt Lake. Whereas most enzymes of eukaryotic halophytes are identical (or nearly so) to their counterparts in glycophytic eukaryotes, the enzymes of halobacteria have been extensively modified. They maintain their metabolic activities only when they exist in strong salt solutions, being denatured by more dilute solutions. Thus, in an evolutionary sense it seems that the halobacteria have followed the extremely difficult pathway of modifying hundreds to thousands of enzymes so that they can tolerate high salt concentrations, whereas the other halophytes and salt-tolerant eukaryotes have taken the genetically much simpler approach of producing compatible solutes that do not harm cytoplasmic enzymes but provide a water-potential balance between the cytoplasm and the surrounding environment of high water stress (Yancey et al., 1982).

Although we have emphasized salt tolerance in our discussion of compatible solutes, it is important to note that these compounds also appear in many plants during frost acclimation, where they are called **cryoprotectants**. Indeed, increases in sugars and polyhydric alcohols during hardening were observed many years ago, although the correlation with hardiness was not always perfect (see the reviews in Krause et al., 1988; Sakai and Larcher, 1987). Free amino acids and such other substances as glycinebetaine and polyamines have also been observed. (Note that the term *osmoregulation* is usually not used with reference to cryoprotectants.)

## 26.5 Chilling Injury

Tropical or subtropical plants grown in southern regions of temperate North America are sometimes damaged by frost or even by temperatures slightly above freezing. Such crops include citrus, cotton, maize, rice, sorghum, soybean, sugarcane, and sweet potato. Certain tropical fruits such as bananas are damaged even by a few hours below 13°C. (Never put bananas in the refrigerator!) Timing is crucial: Rice plants exposed to temperatures below 16°C at the time of pollen-mother cell division will not produce a crop. It has been estimated that worldwide rice production would decrease 40 percent if worldwide mean temperature dropped only 0.5 to 1.0°C.

As in our discussion of water-stress effects, we are faced with the question of mild chilling effects (such as those that damage bananas), as contrasted to the death caused by severe freezing. Many mechanisms have been proposed to account for **chilling injury**. Because chilling disrupts all the metabolic and physiological processes in plants, it seemed almost futile to look for a single key reaction that might be responsible. Nevertheless, it is possible that just such a key response has been identified (Graham and Patterson, 1982; Lyons, 1973). As the temperature is lowered in chilling-sensitive plants, lipids in cellular membranes solidify (crystallize) at a critical temperature that is determined by the ratio of saturated to unsaturated fatty acids. This critical temperature for a phase transition from liquid to crystalline often proves to be equivalent to the temperature that causes chilling damage. Development of tolerance to chilling temperatures in chilling-sensitive plants apparently involves changes in this ratio. An increase in the proportion of unsaturated fatty acids or in the quantity of sterols causes the membrane to remain functional at lower temperatures.

The lipid model suggests that the membrane normally exists in a liquid-crystalline condition. In this state its enzymes have their optimal activity, and its permeability is thus under control. Below the critical temperature, the membrane exists in a solid-gel state. This change in state should bring about a contraction resulting in cracks or channels that lead to increased permeability (Yoshida et al., 1979). This would lead to the observed upset of solute balances (leakage from chill-damaged cells or organelles of ions and other solutes; also upset of proton transport; see DuPont and Mudd, 1985). Enzyme activities would also be upset, leading to imbalances with nonmembrane-bound enzyme systems. Thus, metabolites such as those produced in glycolysis would be expected to accumulate because they could not be acted upon by the mitochondrial enzyme systems. Such accumulation of glycolysis intermediates has indeed been observed.

It has been suggested that little ATP would be formed because of the importance of membranes in its formation and because of these imbalances between the mitochondria and the glycolytic systems. Similar events would probably take place between the chloroplast and the cytoplasm around it. Kasamo (1988) studied ATPases in the plasmalemma and tonoplast of chilling-sensitive and chilling-insensitive cultured cells of rice (*Oryza sativa*). Clearly, the activities of the ATPases were dependent upon the status of the membranes as influenced by temperature. Neuner and Larcher (1990) used *in vivo* chlorophyll fluorescence techniques to detect differences in chilling susceptibility of two soybean cultivars; again, photosynthetic membranes were involved (see also Larcher and Neuner, 1989, and Larcher et al., 1990).

If the temperature is raised soon enough, the membranes return to the liquid-crystalline state (because this phase transition is completely reversible), and the cell recovers. If metabolite buildup and solute leakage are allowed to occur to any great extent, however, cells are injured or killed.

Some cultivars are more sensitive to chilling than are others, although their fatty-acid ratios appear to be the same. These differences could be caused by different sensitivities to the accumulated metabolites rather than to the initial effects on the membranes. Chilling effects can often be avoided if tissues are exposed to high temperatures for brief intervals between chilling periods, provided that initial chilling was not too prolonged. In terms of the hypothesis given, this would allow the metabolism of accumulating metabolites, and so toxic levels are not allowed to develop.

In summary, there is much support for the lipid-membrane (with associated proteins) hypothesis for chilling injury, and the basic idea might even be extended to gain at least a partial understanding of frost hardening (Harvey et al., 1982; Steponkus, 1984). But there are still problems to be solved, mostly lack of correlation between lipid compositions and chilling sensitivity in some studies (summarized by Graham and Patterson, 1982).

## 26.6 High-Temperature Stress

Elevated temperatures typically accompany drought conditions and are an important environmental stress factor in themselves. This is especially true for euxerophytes that are hardly cooled by transpiration.

### The Upper-Temperature Limits for Survival

Biologists have long been fascinated by growth of organisms at high temperatures (see, for example, Argano, 1981; Brock, 1978; Kappen, 1981; and Steponkus, 1981). Plants typically die when exposed to temperatures of 44 to 50°C, but some can tolerate higher temperatures. Stem tissues near the soil line of plants in the desert, for example, may reach levels considerably above this, and *Tidestromia oblongifolia* photosynthesizes optimally in Death Valley, California, at these temperatures (see Fig. 12-4). *Stipa* spp., *Carex humulis*, and *Bothriochloa ischaemum* also can survive temperatures of 65 to 70°C (Larcher et al., 1989).

A few bacteria (*Thermotoga* sp.) can grow in thermal springs, such as those in Yellowstone National Park, at temperatures up to 90°C (Pool, 1990). But the Archaebacteria live at the highest temperatures (Pool, 1990). Members of the genus *Pyrodictium* were found growing at temperatures up to 110°C in fields of hot, bubbling, sulfurous mud (**solffatara fields**) in a hydrothermal system off Isola Vulcano, Italy, and most recently, Archaebacteria of the genus *Methanopyrus* were found growing at that temperature in sediment samples taken by the research submersible *Alvin* at the Guaymas Basin hot vents in the Gulf of California, at depths of 2,000 to 2,010 m (Huber et al., 1989). In the laboratory, the organisms grew best at 98°C but did not grow at 80°C or 115°C. They also grew best in 0.2 to 5 percent NaCl[6] (optimum, about 1.5 percent). Growth was strongly inhibited by traces of oxygen. At present, 110°C is the accepted upper temperature limit for growth of organisms.[7]

Such extreme conditions lead to denaturation of enzymes and unfolding of nucleic acids in most organisms. So how do the Archaebacteria manage? No one knows, but it has been suggested that special lipids occur in their membranes, that their proteins have some special structure (although their amino acids are the same as in other organisms), and that their DNA is protected by special histonelike proteins. When the proteins are added to DNA *in vitro*, it can withstand temperatures 30°C higher than usual (Pool, 1990).

Eukaryotic organisms have not been found growing at temperatures above 56 to 60°C (temperatures of hot springs that support green algae), and the upper limit for animals seems to be about 45 to 51°C. Photosynthesis apparently does not occur even in blue-green algae (prokaryotes) at temperatures above 70 to 72°C. Several dry spores and seeds of higher plants will survive temperatures well above 100°C, but these do not actively grow at such temperatures. In general, dry and dormant structures withstand various stresses well.

The deleterious effects of high temperatures on higher plants occur primarily in photosynthetic functions, and the thylakoid membranes, particularly the photosystem II complexes located on these membranes, are apparently the most heat-sensitive part of the photosynthetic mechanism (Santarius and Weis, 1988; Weis and Berry, 1988). Long-term acclimations can be superimposed on a more rapid adaptive adjustment to high temperatures that occurs within a few hours. In addition to the heat effects on primary photochemical reactions, there is evidence that rubisco and other enzymes of carbon metabolism are adversely affected. Energy dissipated by photorespiration can exceed that consumed by $CO_2$ assimilation. Light and other factors can

[6]Some Archaebacteria are known to live in saturated salt solutions (about 36 percent), in highly alkaline (*p*H 11.5) or acidic (*p*H of 1 or less) waters, and at the extreme pressures of ocean depths (1,300 to 1,400 atm; Pool, 1990).

[7]As we noted in the third edition of this text, J. A. Baross and J. W. Deming (1983) found Archaebacteria in hot water rising from sulphide-encrusted vents located along tectonic rifts and ridges of the deep ocean floor. Some communities occurred at temperatures exceeding 360°C! They claimed that they grew these bacteria at 250°C in a titanium syringe pressurized to 265 atmospheres (26.85 MPa) and containing enriched sea water. Unfortunately, the work was not duplicated by other researchers, and Baross and Deming have lost their cultures and are now unable to extend the work. They do not retract their claim, however.

cause an increase in tolerance to heat, but plants' acclimation to high temperatures is minimal (a few degrees) compared with acclimation to drought or to freezing temperatures. For example, soybean seedlings exposed for 2 h to 40°C are subsequently able to survive an otherwise lethal 2-h exposure to 45°C. This small acclimation could nevertheless be significant because high-temperature extremes may exceed normal high temperatures by only a few degrees. Many of the changes that appear during acclimation to heat stress are reversible, but if the stress is too great, irreversible changes can occur, and these can lead to death.

Plants that are hardy to high temperatures exhibit high levels of water of hydration and high protoplasmic viscosity, characteristics that are also exhibited by euxerophytes. High-temperature-adapted plants also are able to synthesize at high rates when temperatures become elevated, allowing synthetic rates to equal breakdown rates and thereby avoiding ammonia poisoning. These features were observed long ago, and some of them may be functions of the heat-shock proteins discussed in the next paragraphs.

### Heat-Shock Proteins

During the past two decades there has been considerable interest in the so-called heat-shock response (see, for example, Key et al., 1985; Kimpel and Key, 1985; Lindquist and Craig, 1988; Ougham and Howarth, 1988; Sachs and Ho, 1986; and Section 24.5). Organisms ranging from bacteria to humans respond to high temperatures by synthesizing a new set of proteins, the **heat-shock proteins** (**HSPs** or **hsps**); typically, synthesis of most of the normal proteins is repressed. The HSPs are studied by electrophoresis on SDS gels and by other methods. Some of these HSPs have relatively high molecular weights (for example, a 70-kilodalton HSP is common in many organisms, including plants), but a group of 20 to 30 low-molecular-weight HSPs (15 to 27 kDa) may be unique to plants. The HSPs appear rapidly, often becoming a substantial portion of the total proteins within 30 min after an abrupt shift from, say, 28°C to 41°C. Their synthesis continues during the next 3 to 4 h, but after 8 h the pattern of synthesis is essentially the same as it was at the initial low temperature. The HSPs also appear when the increase in temperature is more gradual, as might occur under natural conditions. Three or four hours after return to the normal temperature, HSPs are no longer produced, but many of the HSPs are still present, indicating that they are quite stable. Heat-shock mRNAs have also been studied. The kinetics of their appearance and disappearance matches that of the HSPs in the expected way.

It is becoming apparent that the HSPs play a role in heat tolerance, perhaps by protecting essential enzymes and nucleic acids from heat denaturation. Without such protection, nucleic acids might be cleaved by specific metal ions that leak into the cytoplasm from outside (or from the vacuole) as membranes become more permeable at the higher temperatures (Burke and Orzech, 1988).

### Other Stress Proteins

Special proteins appear in response to stresses other than high temperatures (Sachs and Ho, 1986). Soybean seedlings treated with 50 to 75 μM arsenite for a few hours developed tolerance to a subsequent heat treatment and produced a pattern of proteins very similar to the HSPs of this species (Key et al., 1985). Water stress induces changes in protein and mRNA patterns, but these do not necessarily match the patterns of HSPs (Bray, 1988; Bensen et al., 1988; Ramagopal, 1987a, 1987b; and Ranieri et al., 1989). Bhagwat and Apte (1989) studied proteins induced in a nitrogen-fixing cyanobacterium (*Anabaena* sp.) by heat shock, salinity, and osmotic stress. They found 15 new polypeptides, four of which were unique to heat shock. Four others were induced by all three stresses. Michalowski et al. (1989) studied the time course for induction of mRNA in response to salt stress in the ice plant as the plant shifted to CAM metabolism. Special proteins formed in response to heavy metals and to ultraviolet light have also been noted (Sachs and Ho, 1986).

There have been several reports of **cold-acclimation proteins (CAPs)**; for example, Gilmour et al. (1988) found a pattern of new polypeptides (four 47-kDa, a 160-kDa, and others) that appeared in cold-hardened plants of *Arabidopsis thaliana*. The significance of the CAPs remains to be discovered, but Hincha et al. (1989) showed, in accordance with earlier work, that on a molecular basis these proteins are several orders of magnitude more efficient than sucrose in protecting thylakoid membranes from freezing damage.

## 26.7 Acidic Soils

Plants are found growing on soils in a *p*H range of at least 3 to 9, and the extremes provide another stress to which some species are adapted. Cranberries, for example, grow on acid bogs, whereas certain desert species normally grow only on soils with high *p*H. In general, we know far too little about why some plants are native to low-*p*H soils and others are native to soils with higher *p*H values. Certainly one of the reasons is competition. If we use hydroponic techniques to study the growth of various species that apparently prefer different *p*H levels, we usually find that they do reasonably well over a wide *p*H range. But in nature, even a slight advantage of one species over another can eventually lead to elimination of the less well-adapted one.

Soil factors closely correlated with $pH$ are probably more important than the concentration of $H^+$ ions *per se*. For example, high rainfall leads to leaching of calcium and formation of acidic soils, so calcium is usually low in acidic soils and abundant in soils of high $pH$ (calcareous soils). Moderate concentrations of this element favor development of root nodules on many legumes (Chapter 14), so nitrogen-fixing legumes will grow better on soils rich in calcium than on most acidic soils. The less-abundant calcium in acidic soils may also limit plant growth simply because $H^+$ is much more toxic to roots in the absence of calcium. One of the beneficial effects of liming acid soils no doubt derives from this fact. (**Liming** involves addition of calcium in various forms, often in mixtures: CaO, which is lime or burned lime; $Ca(OH)_2$, which is water-slaked or hydrated lime; or $CaCO_3$, which is limestone, dolomite, or air-slaked lime.)

The $pH$ also strongly influences the solubility of certain elements in the soil and the rate at which they are absorbed by plants. Iron, zinc, copper, and manganese are less soluble in alkaline soils than in acidic soils because they precipitate as hydroxides at high $pH$. Iron-deficiency chlorosis is thus common on soils in the western United States, which are often alkaline. Phosphate, absorbed largely as the monovalent $H_2PO_4^-$ ion, is more readily absorbed from nutrient solutions having $pH$ values of 5.5 to 6.5 than at lower or higher $pH$ values. In soils of high $pH$, more of the phosphate is present as the less readily absorbed divalent $HPO_4^{2-}$ ion. Furthermore, much of this is usually present as insoluble calcium phosphates. In soils of low $pH$, in which $H_2PO_4^-$ should predominate, the frequent high concentrations of aluminum ions cause its precipitation as aluminum phosphate.

The relatively high concentrations of available aluminum in many acidic soils (those below about $pH$ 4.7) can inhibit growth of some species, not only because of detrimental effects on phosphate availability but apparently also by inhibiting absorption of iron and by direct toxic effects on plant metabolism. Some species (for example, azaleas) not only tolerate these high aluminum concentrations but thrive on such soils. Still other species tolerate amounts of various heavy metals that are toxic to most plants. An example is bentgrass (*Agrostis tenuis*) that is grown in Wales and Scotland on mine tailings having unusually large amounts of lead, zinc, copper, and nickel. This grass does not exclude such toxic metals but somehow accumulates them without being injured appreciably. We don't understand the tolerance mechanism, although it has been suggested that specific chelating agents (for example, in root cell walls) form strong complexes with the metal ions and prevent their reaction with sensitive protoplasmic constituents such as enzymes. Secretion of these metals into the vacuoles would also decrease their toxic effects. (See box essay in Chapter 6 on tolerance to heavy metals.)

## 26.8 Other Stresses

Although we have examined the most important environmental stress factors, there are still others that might be discussed. For example, certain species of plants not only survive but flourish on soils derived from serpentine rock (Kruckeberg, 1954; Whittaker, 1954) or highly acidic material derived from rock that has been strongly modified by percolation of hot water in ancient hot springs (Salisbury, 1964, 1985). The serpentine soils are very deficient in calcium and apparently have toxic amounts of other elements. Somehow serpentine species have adjusted to these stresses. Material derived from hydrothermally altered rock has most of its phosphate tied up in forms unavailable to most plants. There is little available nitrogen, whereas iron, aluminum, and calcium occur in superabundant amounts. Such conditions are not only stressful but fatal to many species; yet, again, a few species have adjusted to these stress factors. Often, plants from both serpentine and hydrothermally altered material can also grow well on more normal soils. Much needs to be learned about such situations.

If space permitted we could discuss at some length ultraviolet light as a stress factor. We have mentioned the topic in other chapters, including the final paragraphs of the previous chapter.

We are becoming acutely aware of the importance of atmospheric pollutants as stress factors. Because of the implications for agriculture, and for the health and productivity of the world's forests and other ecosystems as well, these stress factors have been the objects of much research in recent years. Often they form valid and important topics of plant physiology that we will arbitrarily not discuss, mostly because of space limitations.

In spite of the broad spectrum of environmental stress factors, it is proper to end this chapter and our textbook by returning to the importance of water in the life of most plants. Water stress is either of primary importance or is a contributing factor to reduced growth and yields of plants growing on much of the earth's surface.

# APPENDIX A

## The Système Internationale: The Use of SI Units in Plant Physiology

At the Eleventh General Conference on Weights and Measures, which met in October 1960 in Paris, the metric system of units was given the name *Système Internationale d'Unites* (International System of Units), with the abbreviation SI (Système Internationale) in all languages. The system was designed to simplify the metric system then in use and to unify the application of units in all the sciences and other human endeavors. At subsequent meetings held every three or four years, further refinements were made. For example, the Fourteenth General Conference (1971) adopted the **mole (mol)** as an SI base unit for amount of substance and the **pascal (Pa)** as a derived unit of pressure equal to 1 newton per square meter.

Plant physiologists and other scientists have attempted to apply the system. In some cases this has meant giving up long-accepted and familiar units; in other cases the new units seemed unreasonable and even illogical, although with the passage of time, several SI units that at first seemed unacceptable seem much less so now. SI units are being applied by an increasing number of plant physiologists. Indeed, studies have shown that greater familiarity with the system leads to more frequent use.

A goal of this appendix is to reflect the current use of units in the science of plant physiology rather than to push for the strict usage of SI units beyond their current application in the field. Our criterion for the use of SI units in this text is their use in the most current literature of plant physiology. As it turns out, this means that nearly all the SI units are used.

In the physical sciences, a fundamental measure of a quantity, or one of a set of such measures, is called a **dimension**. For example, space has the dimension of length ($l$), which when multiplied by itself gives area ($l^2$) and when multiplied again gives volume ($l^3$). Other dimensions used to measure physical quantities are mass ($m$), time ($t$), and temperature ($T$). These and combinations of these can express the dimensions of any physical quantity.

If the dimensions of physical quantities are to be communicated in numerical terms, it is essential that the communicants accept the same **units**. For example, the basic unit of length in the metric system as developed during the French Revolution (1791 to 1795) was the **meter**, which was defined as equivalent to the length of a bar preserved in Sèvres, France; in 1960 it was defined as being the length equal to 1 650 763.73 wavelengths in vacuum of the radiation corresponding to the transition between the levels $2p_{10}$ and $5d_5$ of the krypton-86 atom. In 1983, again in response to advancing technology, the meter was redefined as the distance light travels in a vacuum during a time interval of 1/299 792 458 of a second.

The International System of Units recognizes seven base units, each with its own name and symbol, which are the same (with slight spelling differences) in all languages. The seven units are the meter (length), kilogram (mass), second (time), ampere (electric current), kelvin (temperature), candela (luminous intensity), and mole (amount of substance). These units are shown with their symbols in Table A-1 on page 602.

Note that a quantity of substance can be expressed either in terms of its mass or the number of particles of which it is composed: "The **mole** is the amount of substance of a system that contains as many elementary entities as there are atoms in 0.012 kilogram of carbon 12. When the mole is used, the elementary entities must be specified and may be atoms, molecules, ions, electrons, other particles, or specified groups of such particles" (see Goldman and Bell, 1986). Plant physiologists and others include photons among the particles that can be expressed in moles. Note that 1 mole of a substance contains **Avogadro's number** of particles (now defined as the number of atoms in 0.012 kg of carbon 12 $\approx 6.022045 \times 10^{23}$ particles).

The **ampere** is defined as the current required to produce, in vacuum, a force of $2 \times 10^{-7}$ newtons per meter of length between two parallel conductors of infinite length and 1 meter apart. Because force (the newton) is defined in terms of length, mass, and time (see Table A-2), current could also be defined in those terms. Luminous intensity (the candela) is defined in terms of the light intensity perceived by the human eye as compared with the intensity of freezing platinum. It is of value to engineers who are concerned with artificial lighting for human beings, but other measures of radiation can be derived from power (the watt) per unit area or the number (moles) of photons per unit area per unit time. The unit for power also combines length, mass, and time. Thus, although the International System of Units recognizes seven base units, only the four mentioned in the first paragraph are truly basic in that they are not derived from any other units — and temperature could be derived from the other three. The radian (plane angle) and the steradian (solid angle) are *supplementary units* in the International System of Units, but in physics they are considered to be units derived from base units.

Table A-2 shows some common SI units that are derived from the base units and are of value to plant physiologists. Table A-3 lists the prefixes that are preferred in the International System of Units, along with four other prefixes that are part of the system and that were commonly used in the metric system but that are nonpreferred; they should be avoided when the others can conveniently be used. Table A-4 on page 604 summarizes most of the style conventions that govern the use of SI units. Table A-5 on page 605 lists discarded metric units with their acceptable SI equivalents, although several of these units are still widely used by plant physiologists and by others. Table A-6 on page 606 briefly discusses some of these exceptions and others.

**Table A-1 The Seven Base Units.**

| Quantity | Unit | Symbol |
|---|---|---|
| Length ($l$) | meter | m |
| Mass (not weight) ($m$) | kilogram[a] | kg |
| Time ($t$) | second | s |
| Electric current ($I$) | ampere | A |
| Thermodynamic temperature ($T$) | kelvin | K (not °K) |
| Luminous intensity ($I$) | candela[b] | cd |
| Amount of substance ($n$, $Q$) | mole[c] | mol |

[a]Note that, for historical reasons, the gram is *not* the SI base unit for mass. The kilogram is the only base unit with a prefix. Note further that **weight** is technically a measure of the *force* produced by gravity, whereas the kilogram (base unit) is a unit of **mass**. Thus it is technically incorrect to use the word *weight* in conjunction with the unit kilogram; the proper unit for weight is the newton. (On earth, the weight of a 10-kg mass is about 98 newtons.) *Although in many technical fields and in everyday use the term "weight" is considered an acceptable synonym for "mass,"* plant physiologists should use the term "mass" whenever it is appropriate.

Mass is a fundamental quantity that does not change with the force of gravity (for example, with location). The weight of objects, on the other hand, is about 1 percent less at the equator than at the poles and is 82 percent less on the moon.

A balance *balances* the mass of an unknown object against a defined mass; hence, a balance measures true mass. All balances depend upon an accelerational force for their function, but the magnitude of the accelerational force does not affect the reading. Unfortunately, the magnitude of accelerational force does affect the measurement of mass on electronic "balances" because they are really scales that measure weight. This is usually not a serious problem because the force of gravity is constant for a given location, and electronic balances and spring scales are calibrated with a standard set of objects of known mass.

All objects with a mass also have a volume and thus displace some air, which has a density of 1.205 kg $m^{-3}$ (1 atm, dry air, 20°C). A correction for this volume displacement would be necessary in some situations (for example, measuring the mass of a helium balloon!), but most plant tissues have a density similar to that of water (1,000 kg $m^{-3}$), so the correction is only about 0.1 percent.

[b]As a unit of luminous intensity, the candela is based on the sensitivity of the human eye; we know of no application in plant physiology. The lux (lx) is a measure of illuminance based on the candela (1 lx = 1 cd · sr $m^{-2}$); it has been widely used in plant physiology but should be avoided.

[c]The mole should always be used to report the amount of a pure substance, and in such cases the type of substance must be specified. To report the amount of a mixture or of an unknown substance, mass must be used.

**Table A-2 Derived Units of Interest to Plant Physiologists.**

| Quantity | Unit Name | Symbol | Definition |
|---|---|---|---|
| Area ($A$) | square meter | $m^2$ | m·m |
| Volume ($V$) | cubic meter | $m^3$ | m·m·m |
| Speed or velocity[a] ($v$) | meters per second | $m\ s^{-1}$ | $m\ s^{-1}$ |
| Force ($F$) | newton | N | $kg{\cdot}m\ s^{-2}$ |
| Energy ($E$), work ($W$), heat | joule | J | $N{\cdot}m\ (m^2{\cdot}kg\ s^{-2})$ |
| Power | watt | W | $J\ s^{-1}\ (m^2{\cdot}kg\ s^{-3})$ |
| Pressure ($P$) | pascal | Pa | $N\ m^{-2}\ (kg\ s^{-2}\ m^{-1})$ |
| Frequency ($\nu$) | hertz | Hz | cycle $s^{-1}$ |
| Electric charge ($Q$) | coulomb | C | A·s |
| Electric potential ($\psi$)[b] | volt | V | $W\ A^{-1}\ (J\ A^{-1}\ s^{-1};\ J\ C^{-1})$ |
| Electric resistance ($R$) | ohm | $\Omega$ | $V\ A^{-1}$ |
| Electric conductance ($G$) | siemens | S | $A\ V^{-1}\ (\Omega^{-1})$ |
| Electric capacitance ($C$) | farad | F | $C\ V^{-1}$ |
| Concentration | moles per cubic meter | $mol\ m^{-3}$ | $mol\ m^{-3}$ |
| Irradiance (energy) | watts per square meter | $W\ m^{-2}$ | $J\ s^{-1}\ m^{-2}$ |
| Irradiance (moles of photons) | moles per square meter second | $mol\ m^{-2}\ s^{-1}$ | $mol\ m^{-2}\ s^{-1}$ |
| Spectral irradiance (moles of photons) | moles per square meter second nanometer | $mol\ m^{-2}\ s^{-1}\ nm^{-1}$ | $mol\ m^{-2}\ s^{-1}\ nm^{-1}$ |
| Magnetic field strength | amperes per meter | $A\ m^{-1}$ | $A\ m^{-1}$ |
| Activity (of radioactive source) | becquerel | Bq | $s^{-1}$ |

[a]Technically, velocity is a vector quantity requiring specification of a magnitude (speed) and a direction, but magnitude is most important in plant physiology.
[b]Note that the symbol for electric potential (lowercase psi: $\psi$) is easily confused with the symbol for water potential (uppercase psi: $\Psi$).

**Table A-3 Preferred and Nonpreferred Prefixes[a] (multiples and submultiples).**

| Preferred | | | | | | Nonpreferred | | |
|---|---|---|---|---|---|---|---|---|
| kilo | k | ($10^3$) | milli | m | ($10^{-3}$) | hecto | h | ($10^2$) |
| mega | M | ($10^6$) | micro | μ | ($10^{-6}$) | deka | da | (10) |
| giga | G | ($10^9$) | nano | n | ($10^{-9}$) | centi | c | ($10^{-2}$) |
| tera | T | ($10^{12}$) | pico | p | ($10^{-12}$) | deci | d | ($10^{-1}$) |
| peta | P | ($10^{15}$) | femto | f | ($10^{-15}$) | | | |
| exa | E | ($10^{18}$) | atto | a | ($10^{-18}$) | | | |

[a]Pronunciation of the first syllable of every prefix is accented to assure that the prefix will retain its identity.

**Table A-4 Abbreviated Summary of SI Style Conventions (Rules).**

NAMES OF UNITS AND PREFIXES

1. Unit names begin in lowercase, except at the beginning of a sentence or in titles or headings in which all main words are capitalized. (But note that use of the "degrees Celsius" is an exception; use of "degrees centigrade" is obsolete.)
2. Apply only one prefix to a unit name (for example, nm, not mμm). The prefix and unit name are joined without a hyphen or space between. In three cases, the final vowel of the prefix is dropped: megohm, kilohm, and hectare.
3. If a compound unit involving division is spelled out, the word *per* is used (not a slash or solidus, except in tables in which space may be limited). Only one *per* is permitted in a written unit name.
4. If a compound unit involving multiplication is spelled out, the use of a hyphen is usually unnecessary, but it can be used for clarity (for example, newton meter or newton-meter).
5. Plurals of unit names are formed by adding an "s," except that hertz, lux, and siemens remain unchanged, and henry becomes henries.
6. Names of units are plural for numerical values greater than 1, equal to 0, or less than −1. All other values take the singular form of the unit name. Examples: 100 meters, 1.1 meters, 0 degrees Celsius, −4 degrees Celsius, 0.5 meter, −0.2 degree Celsius, −1 degree Celsius, 0.5 liter.

SYMBOLS FOR UNITS

7. Symbols are used when units are used in conjunction with numerals.
8. Symbols are never made plural (that is, by addition of "s").
9. A symbol is not followed by a period except at the end of a sentence.
10. Symbols for units named after individuals[a] have the first letter capitalized, but (except for Celsius) the name of the unit is written in lowercase (see rule 1), whereas most other symbols are not capitalized. A recent exception is the liter, which is best symbolized with a capital L to avoid confusion with the numeral one (1); a script $\ell$ has also been used.
11. Symbols for prefixes greater than kilo are capitalized; kilo and all others are lowercase. It is important to follow this rule, because some letters for prefixes are the same as some symbols (or another prefix): G for giga and g for gram; K for kelvin and k for kilo; M for mega and m for milli and for meter; N for newton and n for nano; and T for tera and t for metric ton.
12. Use numerical superscripts ($^2$ and $^3$) to indicate squares and cubes; do not use sq., cu., or c.
13. Exponents also apply to the prefix attached to a unit name; the multiple or submultiple unit is treated as a single entity. Thus $\mu m^3$ is the same as $10^{-18}$ $m^3$.
14. Never begin a sentence with a symbol (and preferably not with a numeral).
15. Compound symbols formed by multiplication may contain a product dot (·) to indicate multiplication; international rules say that this may be replaced with a period or a space. In the United States, the product dot is recommended. In this book, we have used the product dot only when symbols are in the numerator; we have used a space between symbols in the denominator (that is, symbols with a negative exponent). (Although not an official SI rule, one should avoid inserting nonsymbol words between symbols, except perhaps the first time for clarity: micromoles of $CO_2$ per mole of air; thereafter, simply μmol $mol^{-1}$.)
16. Do not mix symbols and spelled-out unit names (for example, W per square meter), and *never* mix SI units or their accepted relatives (for example, liter, minute, hour, day, plane angle in degrees) with units of another system such as the English system (for example, miles per liter, kg $ft^{-3}$, or the quantity of fat in a food given as grams per ounce).
17. Unit symbols are printed in Roman type (upright letters); italic letters are reserved for quantity symbols (usually variables), such as *A* for area, *m* for mass, *t* for time, and $\Psi$ for water potential. For typewriting or longhand, underlining may be used as a substitute for italics. The Greek mu, μ, the prefix symbol for micro, should be written in Roman type (not in italics, as is often done).

NUMERALS

18. A space (or hyphen, see next entry) is left between the last digit of a numeral and its unit symbol and between unit symbols when more than one is used. Exceptions are the degree symbol for angles or latitudes (for example, 30° north) and the degree Celsius (°C), which is an inseparable symbol. Temperatures may be written 20°C or 20 °C, depending on editorial preference, but not 20° C. It is also incorrect to use 12° to 25°C (that is, to use ° without C); 12 to 25°C is correct.
19. When a quantity is used in an adjectival sense, it is preferable to use a hyphen in lieu of a space between the number and the unit name or symbol (except ° and °C). For example: a 500-W lamp, a 10-mL sample, and so on. This avoids confusion about adjectives.
20. The period is used as a decimal marker, although some countries (for example, Germany and Great Britain) use a comma or a raised period.
21. To avoid confusion (because some countries use a comma as a decimal marker), a space is used instead of a comma to group numerals into three-digit groups; this rule may be followed to the right as well as to the left of the decimal marker. Omission of the space is preferred when there are only four digits, unless the numeral is in a column with others that have more than four digits. (Because it is common in the United States, we have used the comma in this text, except in Appendices A and B.)
22. Decimal fractions are preferred to common fractions.
23. Decimal values less than 1 have a zero to the left of the decimal.

[a]Individuals after whom units are named include: Antoine Henri Becquerel (France, 1852–1908), Anders Celsius (Sweden, 1701–1744), Charles Augustin de Coulomb (France, 1736–1806), Michael Faraday (England, 1791–1867), Heinrich Rudolf Hertz (Germany, 1857–1894), James Prescott Joule (England, 1818–1889), Lord William Thomson Kelvin (Scotland, 1824–1907), Sir Isaac Newton (England, 1643–1727), Georg Simon Ohm (Germany, 1787–1854), Blaise Pascal (France, 1623–1662), Sir William Siemens (Germany, Great Britain, 1823–1883), Count Allessandro Giuseppe Antonio Anastasio Volta (Italy, 1745–1827), and James Watt (Scotland, England, 1736–1819).

**Table A-4 (continued)**

24. Multiples and submultiples are generally selected so that the numeral coefficient has a value between 0.1 and 1000. For comparison, however, especially in tables, similar quantities should use the same unit, even if the values fall outside this range. Exceptions occur when the differences are extreme (for example, 1500 m of 2-mm wire).

25. With numerals, do not substitute the product dot (·) for a multiplication sign ( × ). (For example, use 2 × 2, not 2·2.)

THE DENOMINATOR

26. For a compound unit that is a quotient, use "per" to form the name (for example, meters per second) and a slash (/; solidus) to form the symbol, with no space before or after the slash (for example, m/s). Compound units may also be written with negative exponents (for example, m $s^{-1}$), as we have done in this text.

27. The denominator cannot be a multiple or submultiple of an SI unit (for example, μN $m^{-2}$ is acceptable, but N $\mu m^{-2}$ is not). (But see item 5 in Table A-6.) Because the kilogram is an SI base unit, it may (should) be used in denominators. (This rule is often broken by plant physiologists and others, but it should be followed whenever possible; for example, it is just as easy to use mmol $kg^{-1}$ as nmol $mg^{-1}$.)

28. Use of two or more "pers" or slashes in the same expression is not recommended because they are ambiguous (see above); negative superscripts avoid this problem: J $K^{-1}$ $mol^{-1}$ (not J/K/mol); J/K mol is acceptable because all symbols to the right of the slash belong to the denominator.

**Table A-5 Some Discarded Metric Units.**

| Discarded Metric Unit | Acceptable SI Unit |
|---|---|
| micron (μ) | micrometer (μm) |
| millimicron (mμ) | nanometer (nm) |
| angstrom (Å) | 0.1 nanometer (nm) |
| bar (bar) | 0.1 megapascal (MPa); 100 kilopascal (kPa) |
| calorie (cal) | 4.1842 joule (J) |
| degree centigrade (°C) | degree Celsius (°C) |
| liter (L or l or liter)[a] | 0.001 $meter^3$ |
| hectare (ha) | 10 000 $m^2$ or 0.01 $km^2$ |
| einstein (E) | mole of photons or quanta (mol) |
| parts per million (ppm) | mg $kg^{-1}$<br>μmol $mol^{-1}$ (e.g., $CO_2$ in air)<br>(Use kg for mixed substances and mol for pure substances and gases.)<br>1000 $mm^3$ $m^{-3}$ (volume; e.g., liquids) |
| parts per billion (ppb) | μg $kg^{-1}$<br>nmol $mol^{-1}$<br>$mm^3$ $m^{-3}$ (volume; e.g., liquids) |

[a]The liter is particularly difficult to discard because there is no preferred SI unit for volume between the cubic millimeter ($mm^3$) and the cubic meter ($m^3$; because the centimeter and the decimeter are nonpreferred — see Table A-3). (Note that 1 000 000 000 $mm^3$ = 1 $m^3$, but 1 000 000 $mm^3$ = 1 L = 0.001 $m^3$.)

**Table A-6 Exceptions or Special Cases for the Plant Sciences.**

1. Days (d) can be used when integration over a longer period of time is important (for example, measurements of growth: kg $d^{-1}$); some special applications may justify the use of minutes (min) and hours (h), but the second (s) is the base unit in SI.
2. The liter is still widely used (abbreviation was l but L is now officially recommended; some journals allow ℓ or spelled out *liter* to avoid confusion with the numeral 1, but L is preferred).
3. Molar concentrations are used (moles $liter^{-1}$ = M); use millimolar (mM) instead of $\times 10^{-3}$ M unless concentrations are being compared over a range that exceeds three orders of magnitude. Some plant physiologists now use moles $meter^{-3}$ (1 mol $m^{-3}$ = 1 mM), which is the SI unit of concentration.
4. The hectare (square hectometer; equals 10 000 $m^2$) is widely used in agriculture, but the $m^2$ is preferable.
5. Sometimes it is impossible to avoid using units with prefixes in denominators (as W $m^{-2}$ $nm^{-1}$ when speaking of light energy per nanometer of wavelength). In some cases it may be preferable to write out information for further clarity. For example, a strict editor would insist that a temperature gradient of 1 K $mm^{-1}$ be written as 1000 K $m^{-1}$. It would be better to state: ". . . a temperature gradient of 1 K over a distance of 1 mm was measured."
6. The candela (one of the seven base units: luminous intensity) is not used in plant science because it is based on the sensitivity of the human eye. The lux (lx) has been widely used in plant physiology; because it is based on the candela (1 lx = 1 cd · sr $m^{-2}$), it should not be used. Radiant energy measurements should be accompanied by a description of the source and, when possible, its spectral characteristics. (Incidentally, 1 lux = 0.0929 foot-candles and 1 ft-c = 10.76391 lx. The foot-candle is the English unit of illuminance; it has also been widely used by plant physiologists.)
7. The centimeter is widely used, but the millimeter and meter are preferable (see Table A-3).
8. The angstrom (Å) continues to be used in measurements of atomic dimensions, but the nanometer (1 nm = 10 Å) is preferred.
9. The bar (bar or mbar) is also widely used as a unit of pressure, especially in meteorology, but multiples of the pascal (MPa or kPa) are preferred in plant physiology (as in this text and in most current publications).
10. The calorie (cal) and kilocalorie (kcal or Cal) are widely used but should be replaced with the joule (J) and kilojoule (kJ). The calorie is defined as being exactly equal to 4.18400 joules. (Several values for the calorie have been used; the value given here is called the "thermochemical calorie" and is the one accepted by the U.S. Bureau of Standards, now the National Institute of Standards and Technology.)
11. Many plant physiologists and others use the **dalton (Da)** as the *unit of atomic mass* (1 Da = 1 g $mol^{-1}$; for example, "the molecular weight of sucrose is 342.30 Da, while that of rubisco is over 500 kDa"). While the dalton is not an SI unit, it is convenient and we have used it throughout this text.
12. It is common among plant physiologists, biochemists, and others to describe the accelerational forces produced during centrifugation, or those experienced in an orbiting satellite or on the surface of some other plant or satellite, as multiples of the average gravitational force at the earth's surface. There is no SI unit to express this multiple of earth's gravity and a lower-case g is the symbol for the gram. An appropriate symbol is **xg** (times gravity): "The mixture was centrifuged for 30 min at 20 000 xg. Accelerational forces in the satellite are $10^{-3}$ xg."

# Important References for Application of SI Units


Anonymous. 1979. Metric Units of Measure and Style Guide. U.S. Metric Association, 10245 Andasol Avenue, Northridge, CA 91103. [A convenient 14-page guide to the correct use of SI units.]

Anonymous. 1985. Radiation quantities and units. ASAE Engineering Practice: ASAE EP402. American Society of Agricultural Engineers, 2950 Niles Road, St. Joseph, MI 49085-9659.

Anonymous. 1988. Use of SI (metric) units. ASAE Engineering Practice: ASAE EP285.7. American Society of Agricultural Engineers, 2950 Niles Road, St. Joseph, MI 49085-9659.

Anonymous. 1989. Guidelines for measuring and reporting environmental parameters for plant experiments in growth chambers. ASAE Engineering Practice: ASAE EP411.1. American Society of Agricultural Engineers, 2950 Niles Road, St. Joseph, MI 49085-9659.

Anonymous. [no date] Standard Practice for Use of the International System of Units. ASTM E380-89. American Society for Testing and Materials, 1916 Race Street, Philadelphia, PA 19103. [Perhaps the best guide on SI units in the United States.]

Boching, P. M. 1983. Author's Guide to Publication in Plant Physiology Journals. Desert Research Institute Pub. No. 5020. Reno, NV.

Buxton, D. R. and D. A. Fuccillo. 1985. Letter to the editor. Agronomy Journal 77:512–514. [This letter includes a summary of a survey of 97 journals; 77 percent either required or encouraged the use of SI units.]

Campbell, G. S. and Jan van Schilfgaarde. 1981. Use of SI units in soil physics. J. Agron. Educ. 10:73–74.

Council of Biology Editors. 1983. CBE Style Manual, Fifth Edition. Council of Biology Editors, Bethesda, MD.

Downs, R. J. 1988. Rules for using the international system of units. HortScience 23:811–812.

Goldman, D. T. and R. J. Bell. (Eds.). 1986. The International System of Units (SI). National Bureau of Standards Special Publication 330. U. S. Department of Commerce/National Bureau of Standards. [This is the United States edition of the English translation of the fifth edition of "Le Système Internationale d'Unites (SI)", the definitive publication of the International Bureau of Weights and Measures. There is also a British version with slight differences, as in the spelling of "metre," "litre," and "deca." The U. S. version is for sale by the Superintendent of Documents, U. S. Government Printing Office, Washington, DC 20402.]

Incoll, L. D., S. P. Long, and M. R. Ashmore. 1977. SI units in publications in plant science. Current Advances in Plant Sciences 28:331–343.

Metric Practice Advisory Group. 1985. Metric Editorial Guide, Fourth Edition. American National Metric Council, 1010 Vermont Ave. N.W., Suite 320, Washington, DC 20005.

Monteith, J. L. 1984. Consistency and convenience in the choice of units for agricultural science. Expl. Agric. 20:105–117.

Savage, M. J. 1979. Use of the international system of units in the plant sciences. HortScience 14:493–495.

Thien, S. J. and J. D. Oster. 1981. The international system of units and its particular application in soil chemistry. J. Agron. Educ. 10:62–70.

Vorst, J. J., L. W. Schweitzer, and V. L. Lechtenberg. 1981. International system of units (SI): Application to crop science. J. Agron. Educ. 10:70–72.

Weast, Robert C. (Ed.). 1990 (and new editions each year). CRC Handbook of Chemistry and Physics. CRC Press, Boca Raton, FL.

# APPENDIX B

## Radiant Energy: Some Definitions

Plants are strongly influenced by the radiant energy in their environment. Hence, a plant physiologist must understand the nature of radiant energy and how it interacts with plants. The principles of radiation and how radiation interacts with matter are taught in physics and physical chemistry classes. Thus, the following information is presented with two purposes in mind: to provide a brief review of the topic and to provide an accessible reference source for ideas or terms once understood but no longer remembered clearly. The format is a series of definitions, which could have been listed alphabetically. But assuming that you will want to review the entire topic in a logical fashion, we have arranged the definitions so that one builds upon another. Individual terms can be found by scanning the list or checking in the index.

### B.1 Basic Concepts and Terms

**Radiant energy (radiation)** A form of energy that is emitted or propagated through space or some material medium. It is said to be *electromagnetic* and is propagated in the form of pulsations or waves. Certain concepts and equations appropriately describe the wave nature of radiant energy, but this energy also behaves like a stream of particles. These particles without rest mass can also be described by certain equations and by reference to certain manifestations. The equations even relate the wave idea to the particle concept, but we do not yet fully understand radiant energy. The term is sometimes extended (perhaps incorrectly) to include streams of subatomic or atomic particles that do have mass, such as electrons, positrons, or the atomic nuclei that make up primary cosmic rays. Radiant energy that we can see is *light*.

**The wave nature of radiant energy** Several phenomena, including diffraction, interference, and polarization (mentioned below) suggest that radiant energy is propagated in the form of waves. Because familiar waves (for example, water or sound waves) are propagated through a medium, it was postulated that radiant energy is also propagated through a medium, called the *ether*. Careful experiments designed around the turn of the century failed to prove the existence of the ether, and the concept has been rejected. Nevertheless, the wave nature of radiant energy continues to be apparent, even though it apparently needs no medium for its propagation.

**Frequency** ($\nu$ = Greek nu, sometimes $F$) The number of wave crests (peaks in energy) passing a given point in a given interval of time. Frequency is usually expressed in terms of energy crests (vibrations or waves) per second ($s^{-1}$).[1] Green light has a frequency of about $6 \times 10^{14}$ pulsations $s^{-1}$; radio waves have frequencies between about $10^4$ and $10^{11}$ $s^{-1}$.

**Velocity** ($c$) The distance traveled by a peak of radiant energy in some specified interval of time. The velocity of all forms of radiant energy is the same in a vacuum and is equal to $3.00 \times 10^8$ m $s^{-1}$ (300 000 km $s^{-1}$ or 186 000 miles $s^{-1}$). Velocity is virtually identical to this value in air but is slower in media such as water ($2.25 \times 10^8$ m $s^{-1}$) or crown glass ($1.98 \times 10^8$ m $s^{-1}$).

**Wavelength** ($\lambda$ = Greek lambda) The distance between waves or crests of energy in electromagnetic radiation. The wavelength is equal to the velocity divided by the frequency: $\lambda = c/\nu$. Likewise, the frequency is

[1]There are three ways of writing units when some occupy the position of a denominator; all three are equivalent: m per s, m/s, and m $s^{-1}$. The last is widely used, partially because it makes it easier to cancel units in equations such as those that follow (for example, see *Mole of quanta*, below).

equal to the velocity divided by the wavelength: $\nu = c/\lambda$. Wavelengths of radiant energy vary from much shorter than the diameter of an atom to several kilometers in length (see *The electromagnetic spectrum*, below). Green light has a wavelength of about 500 nm or $5 \times 10^{-7}$ m; radio waves have wavelengths between about $10^{-3}$ and $10^{4}$ m.

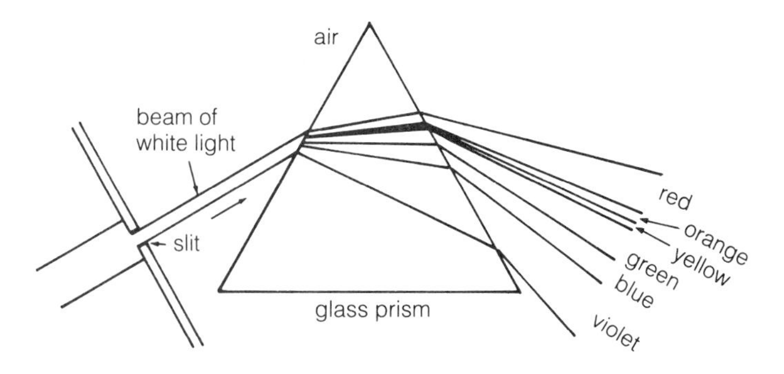


**Figure B-1** White light is dispersed into its component colors by refraction when it is passed through a glass prism.

## B.2 Wave Phenomena

**Refraction** The change in direction (bending) that takes place when a ray of radiant energy passes from one medium into another in which its velocity is different. Refraction at the surface between glass and air makes it possible to form glass into lenses. Light is refracted within leaves as it passes from air into a cell wall or the cytoplasm; it may be refracted several times within a leaf. Because different wavelengths are refracted to different degrees, wavelengths are separated into a **spectrum** when they pass through a prism (Fig. B-1).

**Diffraction and interference** Diffraction includes those phenomena produced by the spreading of waves around and past obstacles that are similar in size to the wavelength. Interference phenomena are caused by reinforcement when energy crests (waves) are superimposed upon each other (are in phase) or by the opposite effect, which occurs when waves are out of phase, canceling or damping each other. Thus, as waves are diffracted, they may reinforce or cancel each other by interference, producing a rainbow effect (separating the various wavelengths). Plant physiologists often use two devices that operate according to these principles. **Diffraction gratings**, which consist of fine lines ruled very close together on a transparent surface, separate a mixture of wavelengths into a spectrum similar to that produced by a prism. **Interference filters** have a thin layer of a reflective medium on a glass surface; the layer is of such thickness that one wavelength (or multiples thereof) is strongly reinforced by passing through the filter, whereas other wavelengths are canceled.

**Polarization** Light waves normally vibrate in many directions at right angles to the direction of propagation. When light is polarized, the wave is made to vibrate in more or less one direction; it vibrates in a plane, so it is said to be *plane polarized*. Light becomes polarized when it is passed through certain substances or is reflected. Many molecules important in plants and other living things, when in solution, will rotate the plane of polarization of a beam of polarized light. These **optically active molecules** typically contain at least one asymmetric carbon atom (an atom with four different groups attached to it).

## B.3 Particle Phenomena

**The particulate nature of radiant energy** Radiant energy exists in units that cannot be further subdivided. In the photoelectric effect, for example, one electron may be ejected from a surface upon the absorption of one *particle* or *packet* of radiant energy. The particulate nature of radiant energy is described by certain equations, many of which include terms for the frequency or the wavelength — terms derived from the wave concepts of radiant energy.

**Quantum or photon** Terms often used interchangeably for the particles of energy in electromagnetic radiation. (The *quantum* is the unit quantity of energy in the quantum theory, whereas the *photon* is a quantum in the electromagnetic field at a specific wavelength or frequency.)

**Photon energy ($E$)** The energy ($E$) of a quantum or photon is equivalent to the frequency ($\nu$) times Planck's constant ($h$): $E = h\nu$. Thus the energy of a photon is directly proportional to the frequency of the radiation; higher frequencies have more-energetic photons. Because the frequency is equal to the velocity divided by the wavelength, the energy of a photon will be inversely proportional to the wavelength: $E = hc/\lambda$. Longer wavelengths (lower frequencies) have less-energetic photons. These equations are useful in calculating the energy relations of photosynthesis and other plant processes that depend on light energy.

**Planck's constant ($h$)** A universal constant of nature relating the energy of a photon to the frequency of the oscillator that emitted it (that is, the frequency of the radiant energy). Its dimensions are energy times time per quantum or photon. It is equal to $6.6255 \times 10^{-34}$ J·s photon$^{-1}$, $1.58 \times 10^{-34}$ cal·s photon$^{-1}$, or $6.6255 \times 10^{-27}$ erg·s photon$^{-1}$.

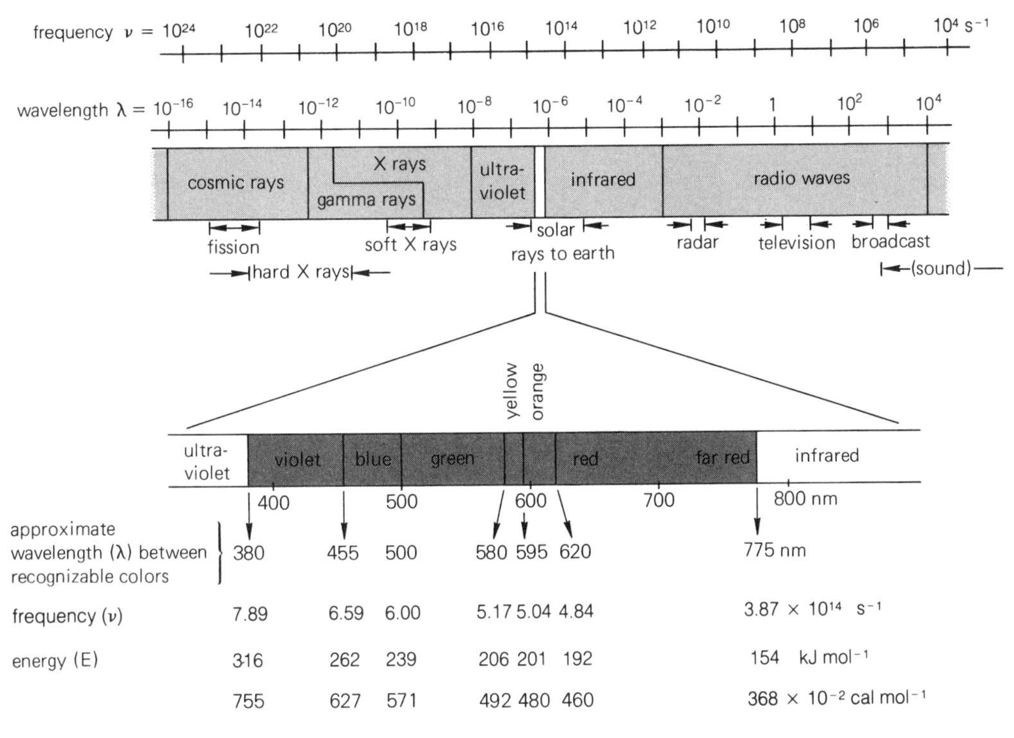


**Figure B-2** The electromagnetic spectrum, using both frequency ($\nu$) and wavelength ($\lambda$) in m. Most of the spectrum is shown, and the visible portion is expanded to depict the region that appears to the human eye to have various colors.

**Quantum yield** ($\phi$ = Greek phi) An expression of efficiency when the absorption of a photon by a molecule results in some photochemical reaction. The quantum yield ($\phi$) is equal to the ratio of the number of molecules reacted ($M$) to the number of photons absorbed ($q$): $\phi = M/q$. Quantum yields for photosynthesis and for interconversion of the two forms of phytochrome (see Chapters 10 and 20) are widely studied values.

**Mole of quanta** (formerly, **einstein**) A number of quanta or photons equal to Avogadro's number: $6.02 \times 10^{23}$ mol$^{-1}$. Because the einstein is not an SI unit, the *mole of quanta* is becoming increasingly common; see *Photosynthetic photon flux (PPF)*, below. The energy ($E$) in a mole of red light ($\lambda$ = 660 nm or $6.6 \times 10^{-7}$ m, $\nu = 4.545 \times 10^{14}$ s$^{-1}$) can be calculated as follows:

$$E = \frac{hc}{\lambda} = \frac{(6.6255 \times 10^{-34}\,\text{J·s photon}^{-1})\,(3.0 \times 10^{8}\,\text{m s}^{-1})}{(6.6 \times 10^{-7}\,\text{m})}$$

or

$$E = h\nu = (6.6255 \times 10^{-34}\,\text{J·s photon}^{-1}) \times (4.545 \times 10^{14}\,\text{s}^{-1})$$

$$E = 3.01 \times 10^{-19}\,\text{J photon}^{-1}$$

$$E\,\text{mol}^{-1} = (3.01 \times 10^{-19}\,\text{J photon}^{-1}) \times (6.02 \times 10^{23}\,\text{photon mol}^{-1})$$
$$= 181\,000\,\text{J mol}^{-1}\ (43\,000\,\text{cal mol}^{-1})$$
$$= 181\,\text{kJ mol}^{-1}$$

Blue light ($\lambda = 4.50 \times 10^{-7}$ m, $\nu = 6.67 \times 10^{14}$ s$^{-1}$) has a frequency 6.67/4.545 times that of red light, so its energy per photon is 1.467 times that of red light = $4.42 \times 10^{-19}$ J photon$^{-1}$ or 266 kJ mol$^{-1}$ (63 100 cal mol$^{-1}$).

## B.4 The Spectrum and Light Sources

**The electromagnetic spectrum** The known distribution of electromagnetic energies arranged according to either wavelengths, frequencies, or photon energies (Fig. B-2). At one end of the spectrum is radiant energy of extremely short wavelengths and, consequently, of extremely high frequencies and energetic photons. At this end of the spectrum are cosmic rays. (Primary cosmic rays are atomic nuclei—about 87 percent are protons—and thus are not photons. They have quantum energies, however, so they can be placed on the

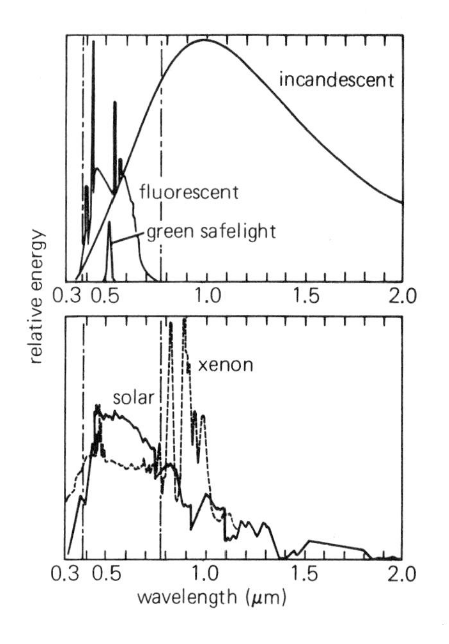


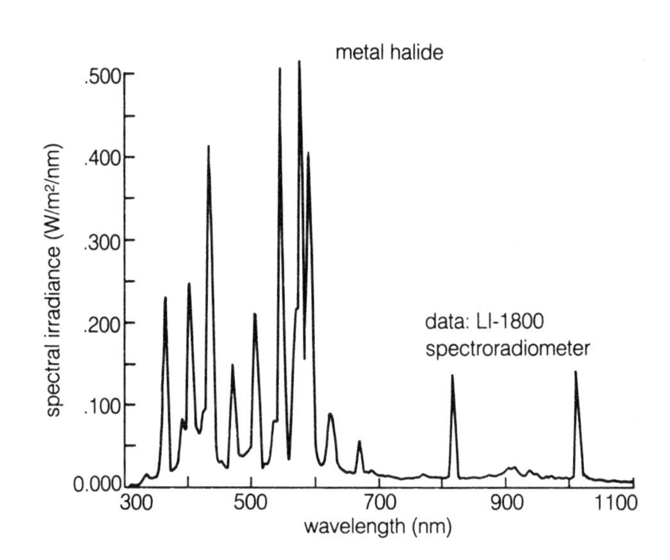


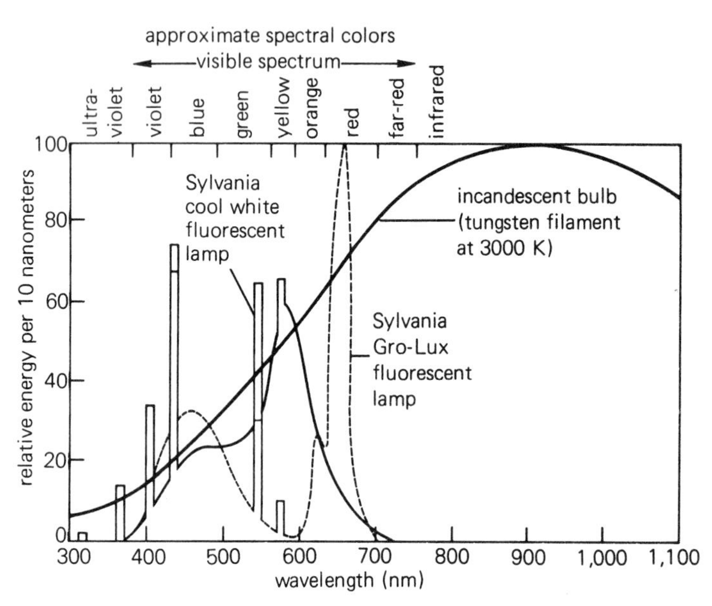


**Figure B-3** Emission spectra for several light sources. Note incandescent lamp peak at about 1.0 μm, mercury emission lines in fluorescent lamp spectra, infrared peaks (0.85 to 1.05 μm) from the xenon lamp, and solar peak in the middle portion of the visible spectrum (between the dashed vertical lines).

spectrum. Secondary cosmic rays do include highly energetic photons.) Slightly longer wavelengths (lower frequencies, less-energetic photons) are gamma rays, which overlap broadly with the X-ray part of the spectrum. Ultraviolet radiation has wavelengths slightly shorter than those in the visible or light part of the spectrum, and infrared radiation has wavelengths longer than those of visible light; radio waves are longer still. The entire spectrum extends over at least 20 orders of magnitude, with the visible portion being a part of only one order of magnitude.

**Light** The visible portion of the spectrum. The term is sometimes incorrectly used to include the ultraviolet and infrared portions of the spectrum as well.

**Color** The appearance of objects as determined by the response of the eye to the wavelengths of light coming from these objects. Short wavelengths produce the sensation we call violet or blue; the longest wavelengths produce the sensation of red. Colors of things are due to *pigments*, which absorb some or all of these wavelengths; for example, a pigment is blue if it absorbs all wavelengths *except* blue. A black pigment absorbs all visible wavelengths. (Is black a color?)

**Light sources** Because plant physiologists deal continually with responses of plants to light, it is important to know something about the spectral characteristics of potential sources of light to which the plants are exposed (as in, for example, special growth chambers).

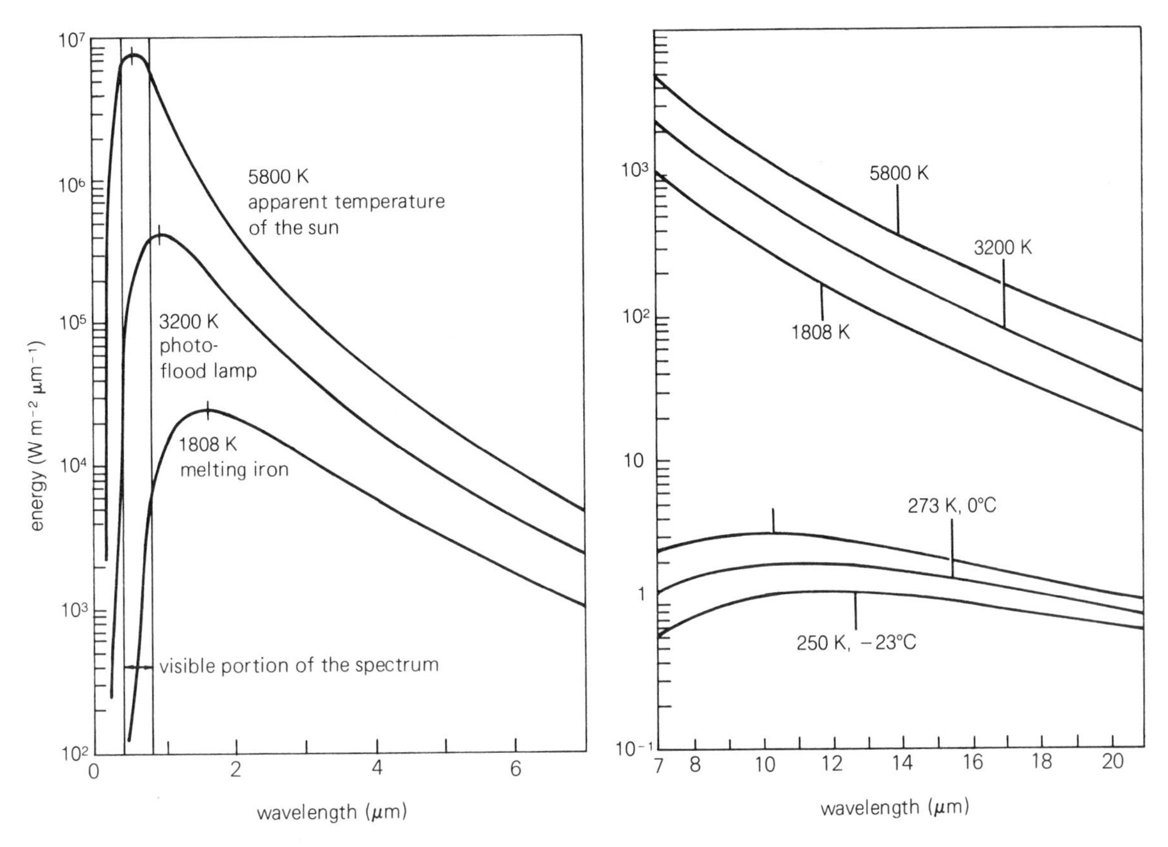


**Figure B-4** Black-body emission spectra compared over a wide range of energy emissions and wavelengths. The spectra would apply for any perfect black-body radiator. Note shifts in the peaks toward longer wavelengths (Wien's law), flattening of the curves, and decreases in total energy (Stefan-Boltzmann law) as temperatures decrease.

Spectral distributions for several light sources are shown in Fig. B-3.

**Wien's law** An equation that relates spectral output (spectral quality) to temperature of the radiating object. The peak of emitted radiant energy ($\lambda_{max}$) shifts toward shorter wavelengths with increasing temperature. This peak multiplied by the absolute temperature ($T$) of the source is equal to a constant, Wien's displacement constant ($w$ = 2897 μm·K): $\lambda_{max}\ T = w$. Wien's law is illustrated for a wide range of temperatures in Figure B-4. For example, we can apply Wien's law to an object at room temperature (25°C = 298 K): $w/T = \lambda_{max}$, 2897 μm·K / 298 K = 9.7 μm. Thus objects at room temperatures emit maximally in the far-infrared part of the spectrum ($\lambda_{max}$ = approximately 10 μm or 10 000 nm); at temperatures approximating those of an incandescent filament, the peak of emission is in the near-infrared ($\lambda_{max}$ = about 1 μm); and at the temperature of the sun or other stars, the emission peak is in the visible part of the spectrum ($\lambda_{max}$ = about 0.50 μm).

**Color temperature** As expressed by Wien's law, the emission peak of objects is a function of absolute temperature. Thus the spectral distribution of the light emitted by incandescent sources (such as an incandescent filament or the surface of a star) can be indicated in terms of absolute temperature. Epsilon Orionis and Sirius, with surface temperatures of 28 000 K and 13 600 K, respectively, are blue-white stars; the sun (5 800 K) has an emission peak in the green-yellow part of the spectrum; and Betelgeuse (3 600 K) is a red star. Color temperatures are used in photography; sensitivity of film must be balanced to the color temperature of the light source, for example.

**Emissivity** The curves shown in Figure B-4 are for perfect black-body radiators. Actually, such an ideal is seldom achieved. In practice, just as objects fail to absorb some wavelengths, they also fail to emit some wavelengths; the curve for emission is the same as the curve for absorption. Thus, because a leaf has an absorptivity coefficient of about 0.98 in the far-infrared part of the spectrum, it also has an emissivity of about 0.98, with most of the radiant energy being emitted in that part of the spectrum at normal temperatures.

Incidentally, the atmosphere is far from a perfect black body, even in the far-infrared, as indicated by the solar spectrum in Figure B-3. The "dips" in the solar spectrum represent absorption—and thus also emission—bands or lines caused by various atmospheric constituents, especially water and carbon dioxide. This phenomenon must be taken into account in calculating the thermal radiation coming from the atmosphere and the thermal radiation absorbed by the atmosphere after being emitted by objects on the ground.

**Incandescent light sources** Light sources such as the sun, the hot filament in an incandescent lamp, or the **plasma** (a "gas" consisting of charged particles) of an arc lamp that emit radiant energy in the visible spectrum because of their high temperatures. Large portions of the energy from these sources are in the infrared part of the spectrum. The hotter the incandescent source, the more the peak of the spectrum is shifted toward blue wavelengths (see *Wien's law*, above). Spectra from incandescent sources are continuous rather than consisting of lines.

**Fluorescent lamps** Lamps, often used in plant growth chambers, that produce light by *fluorescence* (see *Fluorescence and phosphorescence*, below). The spectrum from such lamps consists of individual lines from the mercury emission superimposed on a continuous spectrum from the phosphor. Such light is usually especially rich in blue wavelengths but can be enriched with red wavelengths. Special fluorescent tubes have been developed for plant growth. They are rich in blue and red wavelengths (absorbed by chlorophyll), but experiments with these lamps give mixed results; often plants grow just as well under ordinary fluorescent lamps.

**Sodium and mercury vapor lamps** Lamps in which an electric current passing through hot vapor causes light emission at specific wavelengths: orange for the sodium vapor lamps and blue and green for the mercury vapor lamps. These lamps are called **high intensity discharge (HID) lamps**. In one highly efficient version, they contain metal salts of halide elements such as fluorine, chlorine, bromine, or iodine; these **metal halide lamps** produce a much broader spectrum than do the mercury and sodium vapor lamps, although the spectrum consists mostly of lines (is not continuous). By suitable combinations of the halides, an intensely white light can be produced.

In recent years there have been many studies in which plants have been grown with the various lamps. There is considerable difference in response depending on species; but, surprisingly, some species (for example, wheat) grow extremely well under HID lamps—even the low-pressure sodium vapor lamps. The light from these lamps consists of the double-line emission from the excited sodium atom centered closely on 589 nm, which in no way resembles the absorption spectrum of the leaf. Nevertheless, several species grow to maturity and produce seed and fruits when illuminated only with light from low-pressure sodium vapor lamps (Cathy and Campbell, 1980).[2]

High-pressure sodium vapor lamps, like the low-pressure sodium vapor lamps, emit predominately orange light but produce a considerably broader spectrum than do the low-pressure lamps with their single wavelength. In recent years, a number of plant-growth laboratories have experimented with different combinations of these lamps and have found that many species grow well under a mixture of high-pressure sodium and metal-halide lamps or under each of these lamps alone. These lamps are much more efficient than incandescent lamps and thus require much less power to operate at a given irradiance level, although they are expensive to install and to replace.

## B.5 Radiation Quantities

**Irradiance** **Radiant energy flux** is the radiant energy falling upon a surface in an interval of time (for example, $J\ s^{-1}$). **Irradiance** is the *radiant energy flux* received on a unit plane surface (for example, $J\ s^{-1}\ m^{-2}$). Because $1\ W = 1\ J\ s^{-1}$, $J\ s^{-1}\ m^{-2} = W\ m^{-2}$. In the past, ergs $m^{-2}\ s^{-1}$ have been used for low irradiances, but ergs are not SI units. Calories (cal = 4.18400 J)[3] have also been used, but use of the calorie is now discouraged. Irradiance can also be expressed on a photon basis as mol $m^{-2}\ s^{-1}$ (einsteins $m^{-2}\ s^{-1}$ have been used; $\mu$mol $m^{-2}\ s^{-1}$ is commonly used for the irradiances encountered in plant studies). Irradiance, the correct radiometric term, is often incorrectly called *intensity* by plant physiologists. Use of the term **intensity** should be restricted to the light emission of the *source* (that is, we say that the sun or a lamp has such and such an *intensity*).

In the literature of plant physiology, irradiance has also been given in units of *illumination* rather than total energy or photons. Such units (for example, 10.76 **lux** = 1.0 **foot-candle, ft-c**) are defined in terms of the sensitivity of the human eye. Plants, however, respond to the spectrum in ways quite different from the way the human eye responds, so such a measurement has no value unless exact information is given about the light source. Because plants respond to some wavelengths of light more than others (Chapters 10 and

[2]Cathy, Henry M. and Lowell E. Campbell. 1980. Light and lighting systems for horticultural plants. Horticultural Reviews 2: 491–537.

[3]The calorie defined by the U. S. National Bureau of Standards equals exactly 4.18400 joules. The calorie as defined in slightly different ways has slightly different values.

20), even a measurement given in energy or photon units has little if any value when the spectral distribution is not also given or implied by describing the source.

**Fluence and fluence rate** These terms, from the science of photochemistry, are coming into wide use in plant physiology. **Fluence** is the total energy or photons falling on a unit surface area summed over some arbitrary time interval (for example, J $m^{-2}$ or mol $m^{-2}$ during so many hours). Such a value may be important in describing some plant responses to the total *dose* of radiation, as in the curves that show phototropic bending as a function of the total light energy to which the plants were exposed (that is, in plotting a *dose-response curve* as in Fig. 19-7). **Fluence rate** is the fluence per unit time, and the *second* is used because it is a base SI unit ($\mu$mol $m^{-2}$ $s^{-1}$ or W $m^{-2}$).

There has been some controversy as to whether fluence and fluence rate should be measured as the energy falling on a surface from a single direction (unidirectional) or as the three-dimensional integration of energy falling on a point from all directions (for example, on a spherical receiver), so the geometry of the receiver should be specified (that is, whether it is spherical or planar). Fluence rate is therefore equivalent to *irradiance* only for unidirectional measurements.

**Photosynthetically active radiation (PAR)** Ecologists and others frequently report their measurements of irradiance in energy units for wavelengths from 400 to 700 nm, the wavelengths most active in photosynthesis. Appropriate SI units for PAR are watts per square meter (W $m^{-2}$). Sometimes photon measurements are referred to as PAR, but it would be best to reserve this designation for energy units.

**Photosynthetic photon flux (PPF)** In a photochemical process such as photosynthesis, the end product depends upon the number of quanta absorbed rather than the total light energy absorbed. A single red photon has the same effect in photosynthesis as a single blue photon, for example, although the blue photon has more energy. Hence, in the recent literature it has become common to refer to the number of photons per unit area per unit time. Instruments that respond only to light between the wavelengths of 400 to 700 nm are often used; these may be suitably filtered and calibrated to read directly in $\mu$mol $m^{-2}$ $s^{-1}$. During recent years and continuing until the present, not all plant physiologists have been aware of these conventions, so both W $m^{-2}$ and $\mu$mol $m^{-2}$ $s^{-1}$ have been referred to as PAR. Note that the source is required to equate PAR to PPF and that PPF always refers to $\mu$mol $m^{-2}$ $s^{-1}$ in the wavelength region from 400 to 700 nm. (PPF used to be called "photosynthetic photon flux *density*" or PPFD, but because the term *flux* includes the concept of *density*, *density* has been dropped.)

## B.6 Mechanisms of Absorption and Emission

**Pigment** Any substance that absorbs light (see *Color*, above). If all the visible spectrum is absorbed, the substance appears black to the human eye; if all wavelengths except those in the green part of the spectrum are absorbed (that is, green wavelengths are transmitted or reflected), the substance appears green.

**Ground state and excited state (energy level)** When atoms, ions, or molecules absorb radiant energy, they are raised to a higher **energy level** (see Fig. 10-5 and Section 10.3). Before absorbing any energy, they are said to be in the **ground state**; after, they are in the **excited state**. The actual change in energy content may result from changes in the vibration or rotation of the atom, ion, or molecule or from changes in the electronic configuration of the atoms involved. Typically, an electron is moved a greater distance from the nucleus as energy is absorbed; it is then said to have moved to a higher energy level. Due to the wave motions of the electrons orbiting the nuclei of atoms, there is not a continuous series of energy levels; rather, electrons can exist only at certain discrete distances (energy levels) from the nucleus. To move an electron to the next higher level requires, then, a discrete amount of energy. A pigment absorbs only those photons that have exactly the quantity of energy required to bring about the change in electronic configuration. Actually, there is a range of photon energies that can be absorbed in any given case because some translational, vibrational, or rotational energy changes can occur in addition to the electronic changes.

**Fluorescence and phosphorescence** An atom, ion, or molecule in an excited state may lose its excitation energy in any of three ways: First, it may be immediately lost as heat; that is, totally converted to translational, vibrational, or rotational energy. Second, it may be partially lost as heat, with the remainder emitted as visible light of any wavelength longer (a photon of lower energy) than that absorbed. If this occurs within $10^{-9}$ to $10^{-5}$ s after absorption of the original photon, it is called **fluorescence**. If the delay is longer than that ($10^{-4}$ to 10 s or more), it is called **phosphorescence**. Third, the energy may be used to cause a chemical reaction, such as in photosynthesis.

**Black body** A surface that absorbs all the radiation falling upon it. Usually the term is used in reference to some portion of the spectrum under consideration. One may speak of a black body with reference to visible light and/or infrared radiation, for example. Carbon black or black velvet provide surfaces that approximate a black body. A still more perfect approach to true black-body conditions would be a small opening in the surface of a large hollow sphere lined with carbon black. Obviously, only a minute portion of the radiation entering this opening would ever leave through the opening.

## B.7 Quantifying Absorption, Transmission, and Reflection

**Emissivity (absorptivity) coefficient** A decimal fraction expressing the portion of impinging radiation that is absorbed. A leaf, for example, has an absorptivity coefficient of about 0.98 in the far-infrared portion of the spectrum. (See *Emissivity* above.)

**Transmission** Passage of radiant energy through a substance without being absorbed.

**Reflection** When radiant energy changes direction (may "bounce back") after encountering a substance—but is not absorbed. If the radiation goes in all directions, that is called **scattering**.

**Transmittance ($T$) or percent transmission** The fraction of light transmitted by a substance. It is expressed as a decimal fraction, $T = I/I_o$, or a percentage ($100 \times I/I_o$), where $I_o$ = irradiance of *incident* radiant energy (radiation that impinges on the substance) and $I$ = irradiance of *transmitted* radiant energy. Reflection may also be expressed as a decimal fraction or a percentage.

**Absorbance ($A$)** Formerly called *optical density*. It is the logarithm of the reciprocal of transmittance ($T$):

$$A = \log \frac{1}{T} = -\log T = \log \frac{I_o}{I}$$

Absorbance is often proportional to the concentration of a pigment in a transparent (no reflection or scattering) solution, according to the following laws:

**Beer's law** Each molecule of a dissolved pigment absorbs the same fraction of light incident upon it. Thus, in a nonabsorbing medium, the light absorbed should be proportional to dissolved pigment concentration. The law often holds for dilute solutions but fails as the light-absorbing properties of pigment molecules change at higher concentrations.

**Lambert's law** Each layer of equal thickness absorbs an equal fraction of the light that traverses it. This idea can be combined with Beer's law as the **Beer-Lambert law:** *The fraction of incident radiation absorbed is proportional to the number of absorbing molecules in its path.*

**Extinction coefficient** ($\epsilon$ = Greek epsilon) The Beer-Lambert law can be stated mathematically as follows:

$$A = \log \frac{I_o}{I} = \epsilon cl$$

where

$\epsilon$ = **extinction coefficient**

$c$ = concentration of the pigment solute

$l$ = length of the path of light (for example, through a special quartz cell) in centimeters ($A$, $I_o$, and $I$ were defined in *Transmittance* and *Absorbance*, above)

The extinction coefficient is a constant for a given pigment in dilute solution and can be determined by solving the equation just given:

$$\epsilon = \frac{A}{cl}$$

If the concentration of solute is in moles liter$^{-1}$ (M), $\epsilon$ is called the **molar extinction coefficient** (with units of L mol$^{-1}$ cm$^{-1}$). If concentration is known only in grams liter$^{-1}$, $\epsilon$ is the **specific extinction coefficient** (usually with the symbol $a_s$). The extinction coefficient is a characteristic of a given absorbing molecule in a given solvent with light of a specified wavelength. It is independent of concentration only when the Beer-Lambert law holds. The more intensely colored a pigment is at a given concentration, the larger its extinction coefficient.

## B.8 Thermal Radiation

**Stefan-Boltzmann law** All objects with temperatures above absolute zero emit radiant energy (**thermal radiation**). The quantity ($Q$) emitted is a function of the fourth power of the absolute temperature ($T$) of the emitting surface, according to the Stefan-Boltzmann law:

$$Q = e\delta T^4$$

where

$Q$ = the quantity of energy radiated (in Joules or calories, using $\delta$ as below)

**Table B-1 Radiation Emitted from Black-Body Surfaces at Various Temperatures.**

| °C | K | $T^4$ | $Q = \text{J m}^{-2}\,\text{s}^{-1}$ ($\text{W m}^{-2}$) | % of $Q$ at 0°C |
|---|---|---|---|---|
| 0 | 273 | $5.55 \times 10^9$ | 315 | 100 |
| 20 | 293 | $7.37 \times 10^9$ | 418 | 133 |
| 30 | 303 | $8.43 \times 10^9$ | 478 | 152 |
| 5 477[a] | 5 750[a] | $1.09 \times 10^{15}$ | $6.20 \times 10^7$ | 19 700 000 |

[a]Average surface temperature of the sun.

$e$ = the emissivity (about 0.98 for leaves at growing temperatures); see *Emissivity*, above

$\delta$ = the **Stefan-Boltzmann constant** ($5.670 \times 10^{-8}$ $\text{W m}^{-2}\,\text{K}^{-4}$, or $8.132 \times 10^{-11}$ $\text{cal cm}^{-2}\,\text{min}^{-1}\,\text{K}^{-4}$)

$T$ = the absolute temperature in K (°C + 273)

The fourth power of absolute temperature in the expression means that the emission of radiant energy will increase greatly as temperature increases. Although the normal range of temperatures encountered by plants is narrow on the absolute-temperature scale, the fourth power function means that energy radiated by bodies in this narrow range nevertheless varies considerably (Table B-1).

**Net radiation** The difference between the radiation absorbed by an object and that emitted by it. A leaf, for example, emits radiation according to the Stefan-Boltzmann law. At normal temperatures, most of this radiation is in the far-infrared portion of the spectrum. Such emission leads to cooling. If the leaf is being illuminated by sunlight, however, it absorbs a portion of the sunlight (according to its absorptivity coefficient), which warms the leaf. Whether the leaf increases or decreases in temperature will depend upon whether more or less radiation is being absorbed or emitted—and also upon other mechanisms (convection, transpiration, and so on) that add heat to or remove heat from the leaf.

# APPENDIX C

## Gene Replication and Protein Synthesis: Terms and Concepts

Many times in this text we have spoken of "transcription," "translation," and related processes, each time assuming that our readers have some background knowledge of contemporary molecular biology. This field is discussed in textbooks of basic botany and biology, subjects you have surely studied by now. Nevertheless, this appendix is presented as a brief review and reference source to refresh your memory. Read it from beginning to end if you feel that you need a review, or scan the boldfaced terms if you want to use this appendix as a glossary.

### C.1 The Central Dogma of Molecular Biology

The so-called **central dogma** of molecular biology is concerned with protein synthesis and the transfer of **information** (the sequences of nucleotides in nucleic acids and amino acids in proteins), first from cell generation to cell generation and then from **genes** (the carriers of heredity; actually, the carriers of sequence information) to the proteins, including the **enzymes** (which control metabolic activity and thus life). **Proteins** consist of **amino acids** arranged in specified sequences, and these sequences determine the biological activity of the proteins. The **nucleic acids**, including **DNA (deoxyribonucleic acid)** and **RNA (ribonucleic acid)**, are chains of **nucleotides**, each of which consists of a nitrogen base (related to the alkaloids, page 326, and the cytokinins, pages 382 to 393) attached to a five-carbon sugar in the ring form (**ribose** or **deoxyribose**), which in turn is attached to a phosphate group. (A nucleotide, adenosine monophosphate, abbreviated AMP is shown as part of the $NADP^+$ molecule on page 220.) DNA occurs in the nucleus (and in mitochondria and plastids) and is the genetic material. Messenger-RNA (mRNA) is formed in the nucleus, copying the sequence of nucleotides from DNA. This mRNA then moves out of the nucleus into the cytoplasm, where other proteins and RNA molecules, partially in ribosomes, translate the nucleotide sequence into an amino-acid sequence. This **translation** is protein synthesis.

### C.2 The Double Helix

In DNA there are always as many molecules of **adenine** (a **purine** nucleotide) as there are molecules of **thymine** (a **pyrimidine**), and there are as many molecules of **guanine** (a purine) as there are molecules of **cytosine** plus **5-methylcytosine** (pyrimidines). The nucleotides are connected (through their phosphate and deoxyribose molecules) to form long chains, and each DNA molecule consists of two such chains that are wound together like two parallel banisters of a spiral staircase to form a **double helix**. The structures of the nucleotides are such that an adenine in one chain is always paired with a thymine in the opposite chain, and a guanine is always paired with a cytosine or 5-methylcytosine; that is, a purine is always paired with a specific pyrimidine in the opposite chain. This arrangement of paired nucleotides is called **complementary bonding**.

This structure of paired purines and pyrimidines in opposite helical (spiral) chains of DNA immediately suggests how information (the sequence of nucleotides in the chains) can be transmitted from one cellular generation to the next. In the presence of suitable enzymes (especially **DNA polymerases**), the hydrogen bonds holding the two chains of the helix together are broken so that the chains can separate from each other (probably only a small portion at a time), and free nucleotides (each with three phosphates as in ATP) can pair with their complementary bases (nucleotides) in each of the two chains. The free nucleotides are then attached to each other to form complementary chains with the exact nucleotide sequence that existed in the chains that they are replacing. Thus two double helices identical to the original are formed. As the nucleotides attach to each other to form the new chains, two phosphate groups released as pyrophosphates are immediately hydrolized to form phosphate, making the process irreversible.

## C.3 Transcription: Copying DNA to Make RNA

**Transcription** is the copying process in which the sequence of nucleotides in DNA specifies a complementary sequence of nucleotides in an RNA molecule. This process also requires specific enzymes. Only one of the two strands of DNA in a gene carries the coded directions for synthesizing a protein; it is sometimes called the **sense strand**. In transcription, the two DNA chains unwind and separate, and the sense strand serves as a **template** for RNA synthesis. The RNA chain is assembled on the DNA template when free nucleotide bases link to it according to the rules for complementary pairing. The nucleotides in RNA are slightly different from those in DNA: Uracil (a pyrimidine) occurs in RNA instead of thymine. When the nucleoside triphosphates are paired with their complementary bases in the DNA sense chain, **RNA polymerase** links them together, again releasing pyrophosphate, which is hydrolyzed to form phosphate. The RNA that is synthesized in this transcription process is called **messenger RNA (mRNA)** because it carries the message of nucleotide sequence from the nucleus, where transcription occurs, to the cytoplasm, where protein synthesis occurs.

The segment of DNA that is transcribed is a **gene**. The DNA of a **chromosome** is continuous (each chromosome contains thousands of connected genes), so there are specific initiator and terminator nucleotide sequences to mark the beginning and end of a gene. These sequences act as punctuation marks so that only a single gene segment unwinds and becomes available for transcription at any one time.

It is now known that DNA often contains segments of a single nucleotide pair or sequence repeated over and over again, but cytoplasmic mRNA does not contain comparable sequences. This is partially because large amounts of DNA are never transcribed and because many precursor mRNAs contain sequences that are removed to form the mature mRNA in a process known as **RNA splicing (editing)**.

About 5 percent of the RNA made in the nucleus is mRNA, but about 10 to 15 percent is transfer RNA (see below), and 70 to 80 percent is **ribosomal RNA (rRNA)**. Each **ribosome** is made of two subunits, each with its characteristic RNAs and proteins. (In bacteria, one subunit has one large RNA of about 1,500 nucleotides plus about 20 different proteins; the other subunit has one RNA with about 100 nucleotides plus about 35 proteins. Eukaryotic ribosomes are similar but larger.) Ribosomal RNA precursors are synthesized in the **nucleolus** and are assembled there with protein imported from the cytoplasm to form ribosome subunits. These subunits are transported to the cytoplasm for assembly into functional ribosomes. The rRNA is transcribed from **chromatin** (nucleoprotein that forms chromosomes) in a region called the **nucleolar organizer**.

## C.4 Translation: Protein Synthesis in the Cytoplasm

In protein synthesis, the nucleotide sequence consisting of four kinds of nucleotides in mRNA undergoes a **translation** to the amino-acid sequence of protein (20 kinds of amino acids). To begin with, free amino acids (**AA**) enter into a two-step process called **amino-acid activation**. In the first step, an amino acid interacts with ATP in an enzyme-catalyzed reaction that removes pyrophosphate (immediately hydrolyzed to phosphate) from the ATP and attaches the amino acid to AMP. The AA-AMP complex remains attached to the enzyme that catalyzed the reaction. In the second step, the amino acid is transferred from AMP to an appropriate molecule of **transfer RNA (tRNA)**, forming an **aminoacyl-tRNA complex** that is released from the enzyme that formed it. Much of the original energy of the ATP is transferred to the aminoacyl-tRNA complex; thus it is an "activation" of the amino acid. There is a specific enzyme that activates each amino acid, and there is also a specific tRNA or set of tRNAs that becomes attached to each amino acid.

## C.5 The Genetic Code

Part of each tRNA molecule consists of three nucleotides that form an **anticodon**. This sequence is positioned in the tRNA molecule so that it can form complementary bonds with a three-nucleotide sequence, called a **codon**, in the mRNA strand. This positioning makes it possible for the tRNA molecules, each carrying its specific amino acid, to line up along the mRNA strand so that the amino acids will be in the proper sequence as specified by the sequence of nucleotides.

Although only 20 amino acids are coded for and take part in protein synthesis, at least 60 tRNAs occur in the cytoplasm of eukaryotes. Thus a given amino acid can be carried to the mRNA at the ribosomes (where protein synthesis takes place) by more than one kind of tRNA. Note that the actual act of translation involves the linking of specific amino acids, first to specific tRNA molecules and then to each other to form protein. The amino-acid sequence is determined by the nucleotide sequence in mRNA. Translation is accomplished by each specific enzyme that recognizes a given amino acid and its appropriate tRNA, which has the anticodon sequence of nucleotides that is appropriate for that amino acid, and then by the enzymes that link the amino acids to form protein.

The sequences of nucleotides in the codons and anticodons form the **genetic code**. There are 64 ($4 \times 4 \times 4$) possible three-nucleotide sequences based on the four kinds of nucleotides in nucleic acids, but there are only 20 amino acids that form protein at the ribosomes. Hence the genetic code is **redundant** or **degenerate**, meaning that more than one codon codes for a given amino acid. The genetic code has been broken, which means that we know which amino acids are specified by which codons (see basic textbooks for tables of the genetic code).

## C.6 The Steps of Protein Synthesis

Having established the ground work, let us consider the actual steps in protein synthesis. The mRNA is transcribed in the nucleus, copying the sequence of nucleotides from DNA. We can think of the nucleotide sequences in both DNA and mRNA as consisting of three-nucleotide sequences called codons, but nothing special separates one codon from another.

After transcription in the nucleus, the mRNA enters the cytoplasm and becomes attached to a ribosome. In the surrounding medium are tRNA molecules, each kind with a specific amino acid attached to it. Among these tRNA molecules, at least one will have an anticodon that **recognizes** (has complementary nucleotides) the first codon in the mRNA chain of nucleotides. There will also be tRNAs with anticodons to recognize the next codon in the RNA chain, and so on. The two amino acids attached to the first two tRNA molecules to bond with the mRNA chain are brought into close proximity, and a peptide bond forms between them, catalyzed by an enzyme that is part of the ribosome itself. Formation of the bond involves separating the amino acid from the first tRNA and transferring it to the amino acid carried by the second tRNA molecule to become attached to the mRNA strand at the ribosome. The dipeptide formed in this way is then attached (transferred) through a peptide bond to the amino acid attached to the third tRNA molecule, and so on, until the chain of amino acids that constitutes the protein has been formed.

Thus proteins form as the ribosome moves along the mRNA chain, exposing codons along the chain that can be recognized by the anticodons of tRNA molecules with their amino acids. Peptide bonds form between amino acids that are brought together in this way. The energy to form the peptide bonds comes from the energy that is released when each amino acid is separated from its tRNA. Note that a single strand of mRNA can interact with several ribosomes at a time, forming a **polyribosome (polysome)**. In addition to the components of protein synthesis that we have been discussing, the process also requires the presence of enzymes that allow protein synthesis to begin, continue, and terminate with release of the finished polypeptide from the ribosome.

Special codons are involved in the initiation and termination of protein synthesis. In bacteria, one of the two codons AUG (adenine-uracil-guanine) or GUG always occurs as the first word in the coded message for protein synthesis. These codons have been called **initiator codons**. Unless one of them is present as the first word in the message, protein synthesis cannot proceed. Because they code for methionine and valine, these amino acids often initiate polypeptide chains in bacteria—but sometimes they are removed (edited) after the chain forms. Likewise, one of the three **terminator codons** (UAA, UAG, or UGA) must be the last word in the message, or protein synthesis cannot be successfully completed. These terminator codons specify no amino acids and evidently have no corresponding tRNA molecules with complementary anticodon sites. As the ribosome reaches one of these terminator codons, no aminoacyl-tRNA binds to the mRNA. In some manner the terminator induces release of the completed polypeptide from the ribosome.

The steps in protein synthesis have been worked out almost entirely with cell-free systems isolated from bacteria, especially *Escherichia coli*. Although it is not known how completely some of the details of the mechanism apply to eukaryotic cells, the evidence is good that the basic process outlined here takes place in much the same way in higher organisms. The genetic code, for example, is known to be the same in all eukaryotic organisms so far studied; that is, it appears to be universally conserved.

The final and critical question concerns the regulation of protein synthesis. Why are some genes "turned on," being transcribed to form mRNA, while the majority of genes are inactive? This is really the central question of development and most answers still lie in the future. We do, however, know much about what regulates protein synthesis in certain cells. In bacteria, the **operon hypothesis** was developed to help explain the regulation of gene activity. Although much evidence now supports this model, we will not discuss it here. Similar schemes are developing to help us understand gene regulation in eukaryotic organisms and these were reviewed in Chapter 24.

The conclusion of this discussion, and indeed the conclusion of this text, is that plants and all other organisms are marvelously complex, intricate constructions with an assemblage of functions that collectively interact with each other and with their environment to produce the phenomenon we call *life*. Perhaps most marvelous of all is that our brain cells, with their DNA, mRNA, proteins/enzymes, and a host of other molecules, in some incredible way allow us to contemplate it all!

# References*

## *Chapter 1 Plant Physiology and Plant Cells*


Albersheim, Peter. 1975. The walls of growing plant cells. Scientific American 232(4):81–95.
Allen, Nina Strömgren and Douglas T. Brown. 1988. Dynamics of the endoplasmic reticulum in living onion epidermal cells in relation to microtubules, microfilaments, and intracellular particle movement. Cell Motility and the Cytoskeleton 10: 153–163.
Avers, Charlotte J. 1981. Cell Biology, Second Edition. Van Nostrand Reinhold, New York.
Avers, Charlotte J. 1982. Basic Cell Biology, Second Edition. PWS Pubs. (Willard Grant Press), Boston.
Bacic, Antony, Philip J. Harris, and Bruce A. Stone. 1988. Structure and function of plant cell walls. Pages 297–371 *in* Jack Preiss (ed.), The Biochemistry of Plants, Vol. 14. Academic Press, San Diego.
Berns, M. W. 1984. Cells, Second Edition. Modern Biology. Saunders, New York.
Boller, Thomas and Andres Wiemken. 1986. Dynamics of vacuolar compartmentation. Ann. Rev. Plant Physiol. 37:137–164.
Bracegirdle, Brian and Patricia H. Miles. 1973. An Atlas of Plant Structure, Vol. 1 & 2. Heinemann Educational Books, London.
Burgess, Jeremy. 1985. An Introduction to Plant Cell Development. Cambridge University Press, Cambridge, New York, London.
Capaldi, Roderick A. 1974. A dynamic model of cell membranes. Scientific American 230(3):26–33.
Carpita, Nicholas C. 1982. Limiting diameters of pores and the surface structure of plant cell walls. Science 218:813–814.
Carpita, Nicholas, Dario Sabularse, David Montezinos, and Deborah P. Delmer. 1979. Determination of pore size of cell walls of living plant cells. Science 205:1144–1147.
Clarkson, David T. (Ed.). 1973. Ion Transport and Cell Structure in Plants. Wiley, New York.
Clegg, James S. 1983. What is the cytosol? Trends in Biochemical Sciences 8:436–437.
de Duve, Christian. 1983. Microbodies in the living cell. Scientific American 248(5):74–84.
Delmer, Deborah P. 1987. Cellulose biosynthesis. Ann. Rev. Plant Physiol. 38:259–290.
Delmer, D. P. and Bruce A. Stone. 1988. Biosynthesis of plant cell walls. Pages 373–420 *in* Jack Preiss (ed.), The Biochemistry of Plants, Vol. 14. Academic Press, San Diego.
Dustin, Pierre. 1980. Microtubules. Scientific American 243(2):66–76.
Fawcett, Don W. 1982. The Cell, Second Edition. Saunders, Philadelphia.
Frei, Eva and R. D. Preston. 1961. Cell wall organization and wall growth in the filamentous green algae *Cladophora* and *Chaetomorpha*. I. The basic structure and its formation. Proceedings of the Royal Society, Series B, 154:70–94.
Fry, Stephen C. 1986. Cross-linking of matrix polymers in the growing cell walls of angiosperms. Ann. Rev. Plant Physiol. 37:165–186.
Fry, S. C. 1988. The Growing Plant Cell Wall: Chemical and Metabolic Analysis. Longman, Harlow, Essex, UK.
Grivell, Leslie A. 1983. Mitochondrial DNA. Scientific American 248(3):78–89.
Gunning, B. E. S. and R. L. Overall. 1983. Plasmodesmata and cell-to-cell transport in plants. BioScience 33(4):260–265.
Hall, J. L. and T. J. Flowers. 1982. Plant Cell Structure and Metabolism, Second Edition. Longman Group Limited, London.
Harris, N. 1986. Organization of the endomembrane system. Ann. Rev. Plant Physiol. 37:73–92.
Herth, Werner. 1985. Plasma-membrane rosettes involved in localized wall thickening during xylem vessel formation of *Lepidium sativum* L. Planta 164:12–21.
Herth, W. and G. Weber. 1984. Occurrence of the putative cellulose-synthesizing "rosettes" in the plasma membrane of *Glycine max* suspension culture cells. Naturwissenschaften 71:153–154.
Holtzman, E. and A. B. Novikoff. 1984. Cells and Organelles, Third Edition. Saunders, Philadelphia.
Jensen, W. A. and F. B. Salisbury. 1984. Botany, Second Edition. Wadsworth, Belmont, Calif.
Lazarides, E. and Jean Paul Revel. 1979. The molecular basis of cell movement. Scientific American 240(5):100–113.
Ledbetter, Myron C. 1965. Fine structure of the cytoplasm in relation to the plant cell wall. Journal of Agriculture & Food Chemistry 13:405–407.
Ledbetter, Myron C. and Keith R. Porter. 1970. Introduction to the Fine Structure of Plant Cells. Springer-Verlag, New York.
Lloyd, Clive W. 1982. The Cytoskeleton in Plant Growth and Development. Academic Press, London and New York.
Lloyd, Clive W. 1987. The plant cytoskeleton: The impact of fluorescence microscopy. Ann. Rev. Plant Physiol. 38:119–139.
Mazia, Daniel. 1974. The cell cycle. Scientific American 230(1):54–64.
Morré, D. James, Daniel D. Jones, and H. H. Mollenhauer. 1967. Golgi apparatus mediated polysaccharide secretion by outer root cap cells of *Zea mays*. 1. Kinetics and secretory pathway. Planta (Berl.) 74:286–301.
Morré, D. James and H. H. Mollenhauer. 1974. The endomembrane concept: A functional integration of the endoplasmic reticulum and golgi apparatus. Pages 84–137 *in* A. W. Robards (ed.), Dynamic Aspects of Plant Ultrastructure. McGraw-Hill, London, New York, Toronto.
Palevitz, B. A. 1982. The stomatal complex as a model of cytoskeletal participation in cell differentiation. Pages 345–376 *in* Clive W. Lloyd (ed.), The Cytoskeleton in Plant Growth and Development. Academic Press, New York.
Porter, Keith R. and Jonathan B. Tucker. 1981. The ground substance of the living cell. Scientific American 244(3):56–67.
Powell, Andrew W., Geoffrey W. Peace, Antoni R. Slabas, and Clive W. Lloyd. 1982. The detergent-resistant cytoskeleton of higher plant protoplasts contains nucleus-associated fibrillar bundles in addition to microtubules. Journal of Cell Science 56:319–335.
Robards, A. W. (Ed.). 1974. Dynamic Aspects of Plant Ultrastructure. McGraw-Hill, London, New York, Toronto.
Roberts, Lorin W. 1976. Cytodifferentiation in Plants. Cambridge University Press, Cambridge, New York, London.
Rudolph, U. and E. Schnepf. 1988. Investigations of the turnover of the putative cellulose-synthesizing particle "rosettes" within the plasma membrane of *Funaria hygrometrica* protonema cells. I. Effects of Monensin and Cytochalasin B. Protoplasma 143:63–73.
Satir, B. H. (Ed.). 1983. Modern Cell Biology, Vol. 1. Alan R. Liss, New York.


*The lists for early chapters contain many general references that are not cited in the text (a common practice in many beginning textbooks); lists for later chapters contain almost exclusively references that are cited by name and date in the text (as occurs in advanced texts, technical papers, and reviews).

Schnepf, E. 1986. Cellular polarity. Ann. Rev. Plant Physiol. 37:23–47.
Schopf, J. William. 1978. The evolution of the earliest cells. Scientific American 239(3):111–138.
Schwartz, Lazar M. and Miguel M. Azar. (Eds.). 1981. Advanced Cell Biology. Van Nostrand Reinhold, New York.
Seagull, R. W. 1989. The plant cytoskeleton. Critical Reviews in Plant Science 8:131–167.
Seagull, R. W., M. M. Falconer, and C. A. Weerdenburg. 1987. Microfilaments: Dynamic arrays in higher plant cells. Journal of Cell Biology 104: 995–1004.
Shulman, R. G. 1983. NMR spectroscopy of living cells. Scientific American 248(1):89–93.
Smith, H. 1977. The Molecular Biology of Plant Cells. Botanical Monographs, Vol. 14. University of California Press, Berkeley, Los Angeles.
Troughton, John and Lesley A. Donaldson. 1972. Probing Plant Structure. A. H. & A. W. Reed Ltd., Auckland, New Zealand.
Troughton, J. H. and F. B. Sampson. 1973. Plants. A Scanning Electron Microscope Survey. John Wiley & Sons Australasia Pty Ltd., New York.
Valentine, James W. 1978. The evolution of multicellular plants and animals. Scientific American 239(3):141–160.
Wiebe, Herman H. 1978a. The significance of plant vacuoles. BioScience 28:327–331.
Wiebe, Herman H. 1978b. What is a plant? The significance of vacuoles and cell walls. Pages 389–392 *in* Frank B. Salisbury and Cleon W. Ross, Plant Physiology, Second Edition. Wadsworth, Belmont, Calif.
Woese, Carl R. 1981. Archaebacteria. Scientific American 224(6):98–122.
Wolfe, S. L. 1983. Introduction to Cell Biology. Wadsworth, Belmont, Calif.

## *Chapter 2 Diffusion, Thermodynamics, and Water Potential*

Campbell, Gaylon S. 1977. An Introduction to Environmental Biophysics. Springer-Verlag, New York, Berlin, Heidelberg.
Chang, Raymond. 1981. Physical Chemistry with Application to Biological Systems, Second Edition. Macmillan, New York.
Eisenberg, David and Donald M. Crothers. 1979. Physical Chemistry with Application to the Life Sciences. Benjamin/Cummings, Menlo Park, Calif.
Fast, J. D. 1962. Entropy. McGraw-Hill, New York, Toronto, London.
Fay, James A. 1965. Molecular Thermodynamics. Addison-Wesley, Reading, Mass.
Hatsopoulos, George N. and Joseph H. Keenon. 1965. Principles of General Thermodynamics. Wiley, New York.
Klein, J. J. 1970. Maxwell, his demon and the second law of thermodynamics. American Scientist 58(1):89.
Kramer, Paul J. 1983. Water Relations of Plants. Academic Press, Santa Clara, Calif.
Kramer, Paul J. 1988. Changing concepts regarding plant water relations. Plant, Cell and Environment 11:565–568.
Ling, G. N. 1967. Effects of temperature on the state of water in the living cell. Pages 5–14 *in* A. H. Rose (ed.), Thermobiology. Academic Press, New York.
Meyer, B. S. 1938. The water relations of plant cells. Bot. Rev. 4:531–547.
Milburn, John A. 1979. Water Flow in Plants. Longman Group Limited, London. (Published in the U.S. by Longman, Inc., New York.)
Moore, Walter J. 1972. Physical Chemistry, Fourth Edition. Prentice-Hall, Englewood Cliffs, N. J.
Morowitz, H. J. 1970. Entropy for Biologists. Academic Press, New York.
Nobel, Park S. 1983. Biophysical Plant Physiology and Ecology. Freeman, San Francisco.
Noy-Meir, I. and B. Z. Ginzburg. 1967. An analysis of the water potential isotherm in plant tissue. I. The theory. Aust. J. Biol. Sci. 20:695–721.
Passioura, J. B. 1982. Water in the soil-plant-atmosphere continuum. Pages 5–33 *in* O. L. Lange, P. S. Nobel, C. B. Osmond, and H. Ziegler (eds.), Encyclopedia of Plant Physiology, New Series, Vol. 12B, Physiological Plant Ecology. II. Springer-Verlag, New York, Berlin, Heidelberg.
Passioura, J. B. 1988. Response to Dr. P. J. Kramer's article, "Changing concepts regarding plant water relations," Volume 11, Number 7, pp. 565–568. Plant, Cell and Environment 11:569–571.
Renner, O. 1915. Theoretisches und Experimentelles zur Kohäsions Theorie der Wasserbewegung. (Theoretical and experimental contributions to the cohesion theory of water movement.) Jb wiss. Bot. 56:617–667.
Schulze, E-D., E. Steudle, T. Gollan, and U. Schurr. 1988. Response to Dr. P. J. Kramer's article, "Changing concepts regarding plant water relations," Volume 11, Number 7, pp. 565–568. Plant, Cell and Environment 11:573–576.
Sinclair, T. R. and M. M. Ludlow. 1985. Who taught plants thermodynamics? The unfulfilled potential of plant water potential. Australian Journal of Plant Physiology 12:213–217.
Slatyer, Ralph O. 1967. Plant-Water Relationships. Academic Press, London, New York.
Slatyer, R. O. and S. A. Taylor. 1960. Terminology in plant-soil-water relations. Nature 187:922–924.
Spanner, D. C. 1964. Introduction to Thermodynamics. Academic Press, London, New York. (Written by a plant physiologist.)
Taylor, S. A. and R. O. Slatyer. 1962. Proposals for a unified terminology in studies of plant-soil-water relations. Arid Zone Res. 16:339–349.
Tyree, M. T. and P. G. Jarvis. 1982. Water in tissues and cells. Pages 35–77 *in* O. L. Lange, P. S. Nobel, C. B. Osmond, and H. Ziegler (eds.), Encyclopedia of Plant Physiology, New Series, Vol. 12B, Physiological Plant Ecology. Springer-Verlag, New York, Berlin, Heidelberg.
Van Wylen, Gordon J. and Richard E. Sonntag. 1976. Fundamentals of Classical Thermodynamics, Second Edition (revised printing). Wiley, New York.

## *Chapter 3 Osmosis*

Campbell, Gaylon S. and G. S. Harris. 1975. Effect of soil water potential on soil moisture absorption, transpiration rate, plant water potential and growth for *Artemisia tridentata*. US/IBP Desert Biome Res. Mem. 75-44. Utah State University, Logan.
Chardakov, V. S. 1948. New field method for the determination of the suction pressure of plants. Dokl. Akad. Nauk SSSR 60:169–172.
Cline, Richard G., and Gaylon S. Campbell. 1976. Seasonal and diurnal water relations of selected forest species. Ecology 57:367–373.
Grant, R. F., M. J. Savage, and J. D. Lea. 1981. Comparison of hydraulic press, thermocouple psychrometer, and pressure chamber for the measurement of total and osmotic leaf water potential in soybeans. South African Journal of Science 77:398–400.
Green, P. B. and R. W. Stanton. 1967. Turgor pressure: Direct manometric measurement in single cells of *Nitella*. Science 155:1675–1676.
Harris, J. A. 1934. The physico-chemical properties of plant saps in relation to phytogeography. Data on native vegetation in its natural environment. University of Minnesota Press, Minneapolis.
Hellkvist, J., G. P. Richards, and P. G. Jarvis. 1974. Vertical gradients of water potential and tissue water relations in sitka spruce trees measured with the pressure chamber. Journal of Applied Ecology 11:637–667.
Höfler, K. 1920. Ein Schema für die osmotische Leistung der Pflanzenzelle. (A diagram for the osmotic performance of plant cells.) Berichte der Deutschen Botanischen Gesellschaft 38:288–298.
Hüsken, Dieter, Ernst Steudle, and Ulrich Zimmermann. 1978. Pressure probe technique for measuring water relations of cells in higher plants. Plant Physiology 61:158–163.
Jensen, William A. and Frank B. Salisbury. 1984. Botany, Second Edition. Wadsworth, Belmont, Calif.
Knipling, E. B. and P. J. Kramer. 1967. Comparison of the dye method with the thermocouple psychrometer for measuring leaf water potentials. Plant Physiology 42:1315–1320.
Kramer, Paul J. 1983. Water Relations of Plants. Academic Press, Santa Clara, Calif.
Kramer, Paul J., Edward B. Knipling, and Lee N. Miller. 1966. Terminology of cell-water relations. Science 153:889–890.
Lavenda, Bernard H. 1985. Brownian motion. Scientific American 252(2):70–84.
Lockhart, James A. 1959. A new method for the determination of osmotic pressure. American Journal of Botany 46:704–708.
Markhart, Albert H. III., Nasser Sionit, and James N. Siedow. 1980. Cell wall dilution: An explanation of apparent negative turgor potentials. Canadian Journal of Botany 59:1722–1725.
McClendon, John H. 1982. Water relations curves for plant cells: Toward a realistic Höfler diagram for textbooks. What's New in Plant Physiology 13(5):17–20.
Meyer, B. S., and D. B. Anderson. 1952. Plant Physiology, Second Edition, Van Nostrand, Princeton, N.J.
Millar, B. D. 1982. Accuracy and usefulness of psychrometer and pressure chamber for evaluating water potentials of *Pinus radiata* needles. Aust. Jour. Plant Physiol. 9:499–507.
Nonami, Hiroshi, John S. Boyer, and Ernst Steudle. 1987. Pressure probe and isopiestic psychrometer measure similar turgor. Plant Physiology 83: 592–595.
O'Leary, J. W. 1970. Can there be a positive water potential in plants? BioScience 20:858–859.
Ortega, Joseph K. E. 1990. Governing equations for plant cell growth. Physiologia Plantarum 79: 116–121.
Passioura, J. B. 1980. The meaning of matric potential. Journal of Experimental Botany 31(123): 1161–1169.
Passioura, J. B. 1982. Water in the soil-plant-atmosphere continuum. Pages 5–33 *in* O. L. Lange, P. S. Nobel, C. B. Osmond, and H. Ziegler (eds.), Encyclopedia of Plant Physiology, New Series, Vol. 12B, Physiological Plant Ecology II. Springer-Verlag, New York, Berlin, Heidelberg.
Pfeffer, Wilhelm Friedrich Phillip. 1877. Osmotische Untersuchungen. Studien zur Zell Mechanik. (Investigations of Osmosis. Studies of Cell Mechanics.) W. Engelmann Pub., Leipzig.
Pierce, Margaret and Klaus Raschke. 1980. Correlation between loss of turgor and accumulation of abscisic acid in detached leaves. Planta 148: 174–182.
Pisek, Arthur. 1956. Der Wasserhaushalt der Meso und Hygrophyten. (The water relations of meso and hydrophytes.) Encyclopedia of Plant Physiology 3:825–853.
Ray, Peter M. 1960. On the theory of osmotic water movement. Plant Physiology 35:783–795.
Rayle, David L., Cleon Ross, and Nina Robinson. 1982. Estimation of osmotic parameters accompanying zeatin-induced growth of detached cucumber cotyledons. Plant Physiology 70: 1634–1636.
Scholander, P. F., H. T. Hammel, E. D. Bradstreet, and E. A. Hemmingsen. 1965. Sap pressure in vascular plants. Science 148:339–346.
Shackel, Kenneth A. 1987. Direct measurement of turgor and osmotic potential in individual epidermal cells. Plant Physiology 83:719–722.
Slatyer, Ralph O. 1967. Plant-Water Relationships. Academic Press, London, New York.
Slatyer, R. O. and S. A. Taylor. 1960. Terminology in plant-soil-water relations. Nature 187:922–924.

Spanner, D. C. 1951. The Peltier effect and its use in the measurement of suction pressure. Journal of Experimental Botany 11:145–168.
Taylor, S. A. and Slatyer, R. O. 1962. Proposals for a unified terminology in studies of plant-soil-water relations. Arid Zone Res. 16:339–349.
Tyree, M. T. and J. Dainty. 1973. The water relations of hemlock (*Tsuga canadensis*). II. The kinetics of water exchange between the symplast and apoplast. Canadian Journal of Botany 51:1481–1489.
Tyree, M. T. and H. T. Hammel. 1972. The measurement of the turgor pressure and the water relations of plants by the pressure-bomb technique. Journal of Experimental Botany 23(74): 267–282.
Tyree, M. T. and P. G. Jarvis. 1982. Water in tissues and cells. Pages 35–78 *in* Lange, O. L., P. S. Nobel, C. B. Osmond and H. Ziegler (eds.), Encyclopedia of Plant Physiology, New Series, Vol. 12B, Physiological Plant Ecology II. Springer-Verlag, Berlin, Heidelberg, New York.
Waring, R. H. and B. D. Cleary. 1967. Plant moisture stress: Evaluation by pressure bomb. Science 155:1248–1254.
Wenkert, William. 1980. Measurement of tissue osmotic pressure. Plant Physiology 65:614–617.
West, D. W. and D. F. Gaff. 1971. An error in the calibration of xylem-water potential against leaf-water potential. Journal of Experimental Botany 22(71):342–346.
Wiebe, Herman H. 1966. Matric potential of several plant tissues and biocolloids. Plant Physiology 41:1439–1442.

### *Chapter 4 The Photosynthesis-Transpiration Compromise*

Aylor, Donald E., Jean-Yves Parlange, and A. D. Krikorian. 1973. Stomatal mechanics. American Journal of Botany 60:163–171.
Baker, J. M. and C. H. M. van Bavel. 1987. Measurement of mass flow of water in the stems of herbaceous plants. Plant, Cell and Environment 10: 777–782.
Biggins, J. (Ed.). 1987. Progress in Photosynthesis Research. Vol. IV. Martinus Nijhoff Publishers, Dordrecht, Netherlands.
Brown, H. T. and F. Escombe. 1900. Static diffusion of gases and liquids in relation to the assimilation of carbon and translocation in plants. Phil. Trans. Roy. Soc. (London) B. 193:223–291.
Cermak, J., J. Kucera, and M. Penka. 1976. Improvement of the method of sap flow rate determination in full-grown trees based on heat balance with direct electric heating of xylem. Biologia Plantarum 18:105–110.
Cowan, I. R., J. A. Raven, W. Hartung, and G. D. Farquhar. 1982. A possible role for abscisic acid in coupling stomatal conductance and photosynthetic carbon metabolism in leaves. Aust. J. Plant Physiol. 9:489–498.
Dacey, John W. H. 1980. Internal wind in water lilies: An adaptation for life in anaerobic sediments. Science 210:1017–1019.
Dacey, John W. H. 1981. Pressurized ventilation in the yellow water lily. Ecology 62:1137–1147.
Dacey, J. W. H. 1987. Knudsen-transitional flow and gas pressurization in *Nelumbo*. Plant Physiology 85:199–203.
Dacey, John W. H. and Michael J. Klug. 1982a. Tracer studies of gas circulation in *Nuphar*: $^{18}O$ and $^{14}CO_2$ transport. Physiologia Plantarum 56:361–366.
Dacey, John W. H. and Michael J. Klug. 1982b. Ventilation by floating leaves in *Nuphar*. American Journal of Botany 69:999–1003.
Davidoff, B. and R. J. Hanks. 1988. Sugar beet production as influenced by limited irrigation. Irrigation Science 10:1–17.
Davies, W. J., J. Metcalfe, T. A. Lodge, and A. R. da Costa. 1986. Plant growth substances and the regulation of growth under drought. Australian Journal of Plant Physiology 13:105–125.
Drake, B. G., K. Raschke, and Frank B. Salisbury. 1970. Temperatures and transpiration resistances of *Xanthium* leaves as affected by air temperatures, humidity, and wind speed. Plant Physiology 46:324–330.
Edwards, Mary and Hans Meidner. 1979. Direct measurements of turgor pressure potentials, IV. Naturally occurring pressures in guard cells and their relation to solute and matric potentials in the epidermis. Journal of Experimental Botany 30(17):829–837.
Esau, Katherine. 1965. Plant Anatomy. Wiley, New York.
Gates, David M. 1968. Transpiration and leaf temperature. Annual Review of Plant Physiology 19: 211–238.
Gates, David M. 1971. The flow of energy in the biosphere. Scientific American 225(3):88–100.
Gates, David M., Harry J. Keegan, John C. Schleter, and Victor R. Weidner. 1965. Spectral properties of plants. Applied Optics 4:11–20.
Gotow, Kiyoshi, Scott Taylor, and Eduardo Zeiger. 1988. Photosynthetic carbon fixation in guard cell protoplasts of *Vicia faba* L. Plant Physiology 86:700–705.
Hanks, R. J. (Ed.). 1982. Predicting crop production as related to drought stress under irrigation. Utah Agriculture Experiment Station Research Report 65:367.
Hanks, R. J. 1983. Yield and water-use relationships: An overview. Pages 393–411 *in* Limitations to Efficient Water Use in Crop Production (Chapt. 9A). ASA-CSSA-SSSA, 677 South Segoe Road, Madison, Wisconsin.
Harris, Michael J., William H. Outlaw, Jr., Rüdiger Mertens, and Elmar W. Weiler. 1988. Water-stress-induced changes in the abscisic acid content of guard cells and other cells of *Vicia faba* L. leaves as determined by enzyme-amplified immunoassay. Proceedings of the National Academy of Science USA 85:2584–2588.
Howard, R. A. 1969. The ecology of an elfin forest in Puerto Rico. I. Studies of stem growth and form and of leaf structure. J. Arnold Arbor. 50:225–267.
Huber, B. 1932. Beobachtung und Messung pflanzlicher Saftströme. (Observation and measurement of plant sap streams.) Berichte Deutsche Botanische Gesellschaft 50:89–109.
Humble, G. D. and K. Raschke. 1971. Stomatal opening quantitatively related to potassium transport: Evidence from electron probe analysis. Plant Physiology 48:447–453.
Janac, J., J. Catsky, and P. G. Jarvis. 1971. Infra-red gas analyzers and other physical analyzers. Pages 111–193 *in* Z. Sestak, J. Catsky, and P. G. Jarvis (eds.), Photosynthetic Plant Production: Manual of Methods. Dr. W. Junk (publ.), The Hague.
Jensen, William A. and Frank B. Salisbury. 1984. Botany, Second Edition. Wadsworth, Belmont, Calif.
Kramer, Paul J. 1988. Changing concepts regarding plant water relations. Plant, Cell and Environment 11:565–568.
Long, S. P. 1982. Measurement of photosynthetic gas exchange. Pages 25–34 *in* J. Coombs and D. O. Hall (eds.), Techniques in Bioproductivity and Photosynthesis. Pergamon Press, Oxford.
Mauseth, James D. 1988. Plant Anatomy. Benjamin/Cummings, Menlo Park, Calif.
Mellor, Robert S., Frank B. Salisbury, and Klaus Raschke. 1964. Leaf temperatures in controlled environments. Planta 61:56–72.
Meyer, Bernard S. and Donald B. Anderson. 1939. Plant Physiology. Van Nostrand, Toronto, New York, London.
Monteith, John Lennox. 1973. Principles of Environmental Physics. American Elsevier, New York.
Mott, Keith A. 1988. Do stomata respond to $CO_2$ concentrations other than intercellular? Plant Physiol. 86:200–203.
Nobel, Park S. 1980. Leaf anatomy and water use efficiency. Pages 43–55 *in* Neil C. Turner and Paul J. Kramer (eds.), Adaptation of Plants to Water and High Temperature Stress. Wiley, New York, Chichester, Brisbane, Toronto.
Nobel, Park S. 1983. Biophysical Plant Physiology and Ecology. Freeman, San Francisco.
Outlaw, William H., Jr. 1983. Current concepts on the role of potassium in stomatal movements. Physiologia Plantarum 59:302–311.
Outlaw, W. H., Jr. 1989. Critical examination of the quantitative evidence for and against photosynthetic $CO_2$ fixation by guard cells. Physiologia Plantarum 77:275–281.
Passioura, J. B. 1988. Response to Dr. P. J. Kramer's article, "Changing concepts regarding plant water relations," Volume 11, Number 7, pp. 565–568. Plant, Cell and Environment 11:569–571.
Penny, M. G. and D. J. F. Bowling. 1974. A study of potassium gradients in the epidermis of intact leaves of *Commelina communis* L. in relation to stomatal opening. Planta 119:17–25.
Penny, M. G. and D. J. F. Bowling. 1975. Direct determination of *p*H in the stomatal complex of *Commelina*. Planta 122:209–212.
Permadasa, M. A. 1981. Photocontrol of stomatal movements. Biology Review 56:551–588.
Pierce, Margaret and Klaus Raschke. 1980. Correlation between loss of turgor and accumulation of abscisic acid in detached leaves. Planta 148: 174–182.
Raschke, K. 1975. Stomatal action. Ann. Rev. of Plant Physiol. 26:309–340.
Raschke, K. 1976. How stomata resolve the dilemma of opposing priorities. Phil. Trans. R. Soc. Lond. B. 273:551–560.
Raschke, K. and G. D. Humble. 1973. No uptake of anions required by opening stomata of *Vicia faba*: Guard cells release hydrogen ions. Planta (Berlin) 115:47–57.
Reckmann, Udo, Renate Scheibe, and Klaus Raschke. 1990. Rubisco activity in guard cells compared with the solute requirement for stomatal opening. Plant Physiology 92:246–253.
Sakuratani, T. 1984. Improvement of the probe for measuring water flow rate in intact plants with the stem heat balance method. Journal of Agricultural Meteorology 37:9–17.
Salisbury, Frank B. 1979. Temperature. Pages 75–116 *in* T. W. Tibbitts and T. T. Kozlowski (eds.), Controlled Environment Guidelines for Plant Research. Academic Press, New York.
Salisbury, Frank B. and George G. Spomer. 1964. Leaf temperatures of alpine plants in the field. Planta 60:497–505.
Sayre, J. D. 1926. Physiology of stomata of *Rumex patienta*. Ohio Jour. Sci. 26:233–266.
Schulze, E-D. 1986. Carbon dioxide and water vapor exchange in response to drought in the atmosphere and in the soil. Annual Review of Plant Physiology 37:247–274.
Schulze, E-D., E. Steudle, T. Gollan, and U. Schurr. 1988. Response to Dr. P. J. Kramer's article, "Changing concepts regarding plant water relations," Volume 11, Number 7, pp. 565–568. Plant, Cell and Environment 11:573–576.
Sharkey, Thomas D. and Klaus Raschke. 1981a. Separation and measurement of direct and indirect effects of light on stomata. Plant Physiology 68:33–40.
Sharkey, Thomas D. and Klaus Raschke. 1981b. Effect of light quality on stomatal opening in leaves of *Xanthium strumarium* L. Plant Physiology 68:1170–1174.
Sheriff, D. W. and E. McGruddy. 1976. Changes in leaf viscous flow resistance following excision, measured with a new porometer. Journal of Experimental Botany 27:1371–1375.
Sinclair, T. R., C. B. Tanner, and J. M. Bennett. 1984. Water-use efficiency in crop production. Bio-Science 34(1):36–40.
Tarczynski, Mitchell C., William H. Outlaw, Jr., Norbert Arold, Volker Neuhoff, and Rüdiger Hampp. 1989. Electrophoretic assay for ribulose-1,5-bisphosphate carboxylase/oxygenase in guard cells and other leaf cells of *Vicia faba* L. Plant Physiology 89:1088–1093.

Thimann, Kenneth V. and Z-Y. Tan. 1988. The dependence of stomatal closure on protein synthesis. Plant Physiology 86:341–343.
Tibbitts, Theodore W. 1979. Humidity and plants. BioScience 29:358–363.
Troughton, John and Lesley A. Donaldson. 1972. Probing Plant Structure. McGraw-Hill, New York, St. Louis, San Francisco.
Vaughn, Kevin C. 1987. Two immunological approaches to the detection of ribulose-1,5-bisphosphate carboxylase in guard cell chloroplasts. Plant Physiology 84:188–196.
von Caemmerer, Suzanne and Graham Farquhar. 1981. Some relationships between the biochemistry of photosynthesis and the gas exchange of leaves. Planta 153:376–387.
Wilkinson, H. P. 1979. The plant surface (mainly leaf). Pages 97–117 *in* C. R. Metcalfe and L. Chalk (eds.), Anatomy of the Dicotyledons, Vol. 1. Clarendon Press, Oxford.
Willmer, C. M., J. C. Rutter, and H. Meidner. 1983. Potassium involvement in stomatal movements of *Paphiopedilum*. Journal of Experimental Botany 34(142):507–513.
Woodward, F. I. 1987. Stomatal numbers are sensitive to increases in $CO_2$ from pre-industrial levels. Nature 327:617–618.
Zeiger, Eduardo. 1983. The biology of stomatal guard cells. Ann. Rev. Plant Physiol. 34:441–475.
Zeiger, E., G. O. Farquhar, and I. R. Cowan. (Eds.). 1987. Stomatal Function. Stanford University Press, Stanford, Calif.
Zeiger, E. and Peter K. Hepler. 1977. Light and stomatal function: Blue light stimulates swelling of guard cell protoplasts. Science 196:887–889.
Zemel, Esther and Shimon Gepstein. 1985. Immunological evidence for the presence of ribulose bisphosphate carboxylase in guard cell chloroplasts. Plant Physiology 78:586–590.
Zhang, J. and W. J. Davies. 1990. Changes in the concentration of ABA in xylem sap as a function of changing soil water status can account for changes in leaf conductance and growth. Plant, Cell and Environment 13:277–285.
Ziegenspeck, H. 1955. Die Farbenmikrophotographie, ein Hilfsmittel zum objectiven Nachweis submikroskopischer Strukturelemente. Die Radiomicellierung und Filierung der Schliesszellen von *Ophioderma pendulum*. (Color photomicrography, an aid to the objective study of submicroscopic structural elements. Radial-micellation and filiation of guard cells of *Ophioderma pendulum*.) Photographie und Wissenschaft 4:19–22.

## *Chapter 5 The Ascent of Sap*

Apfel, R. E. 1972. The tensile strength of liquids. Scientific American 227(6):58–71.
Askenasy, E. 1897. Beiträge zur Erklärung des Saftsteigens. (Contributions to a clarification of the ascent of sap.) Naturhisto.-Med. Ver. Heidelberg 5:429–448.
Begg, J. E. and N. C. Turner. 1970. Water potential gradients in field tobacco. Plant Physiology 46:343–346.
Briggs, Lyman J. 1950. Limiting negative pressure of water. Journal of Applied Physics 21:721–722.
Carpita, Nicholas C. 1982. Limiting diameters of pores and the surface structure of plant cell walls. Science 218:813–814.
Comstock, G. G. and W. A. Côté, Jr. 1968. Factors affecting permeability and pit aspiration in coniferous sapwood. Wood Science and Technology 2:279–291.
Cooke, R. and I. D. Kuntz. 1974. The properties of water in biological systems. Annual Review of Biophysics and Bioengineering 3:95–126.
Crafts, A. S. and T. C. Broyer. 1938. Migration of salts and water into xylem of the roots of higher plants. American Journal of Botany 25:529–535.
Daum, C. R. 1967. A method for determining water transport in trees. Ecology 48:425–431.
Dixon, Henry H. 1914. Transpiration and the Ascent of Sap in Plants. Macmillan, London.
Dixon, Henry H. and J. Joly. 1894. Notes. Annals of Botany Lond. 8:468–470.
Dixon, Henry H. and J. Joly. 1895. On the ascent of sap. Roy. Soc. London Phil. Trans., B 186:563–576.
Esau, Katherine. 1967. Anatomy of Seed Plants, Second Edition. Wiley, New York.
Greenridge, K. N. H. 1958. Rates and patterns of moisture movement in trees. Pages 43–69 *in* K. V. Thimann (ed.), The Physiology of Forest Trees. Ronald, New York.
Hammel, H. T. 1967. Freezing of xylem sap without cavitation. Plant Physiology, 42:55–66.
Hartig, Th. 1878. Anatomie und Physiologie der Holzpflanzen. (Anatomy and Physiology of Woody Plants.) Springer, Berlin.
Hayward, A. T. J. 1971. Negative pressure in liquids: Can it be harnessed to serve man? American Scientist 49:434–443.
Heine, R. W. and D. J. Farr. 1973. Comparison of heat-pulse and radioisotope tracer method for determining sap-flow in stem segments of poplar. Journal of Experimental Botany 24:649–654.
Huber, B. and E. Schmidt. 1936. Weitere thermoelektrische Untersuchungen über den Transpirationsstrom der Bäume. (Further thermoelectric investigations on the transpiration stream in trees.) Tharandt Forst Jb 87:369–412.
Jensen, William A. and Frank B. Salisbury. 1984. Botany, Second Edition. Wadsworth, Belmont, Calif.
Legge, N. J. 1980. Aspects of transpiration in mountain ash *Eucalyptus regnans* F. Muell. Ph.D. Thesis, LaTrobe University, Melbourne, Australia.
Lybeck, B. R. 1959. Winter freezing in relation to the rise of sap in tall trees. Plant Physiology 34: 482–486.
MacKay, J. F. G. and P. E. Weatherby. 1973. The effects of transverse cuts through the stems of transpiring woody plants on water transport and stress in the leaves. Journal of Experimental Botany 24:15–28.
Mauseth, James D. 1988. Plant Anatomy. Benjamin/Cummings, Menlo Park, Calif.
McFarlan, Donald, Norris D. McWhirter, David A. Boehm, Cyd Smith, Jim Benagh, Gene Jones, and Robert Obojski. (Eds.). 1990. Guinness Book of World Records. Bantam Books, New York, Toronto, London, Sydney, Auckland. See pages 108–111.
Milburn, J. A. and R. P. C. Johnson. 1966. The conduction of sap. II. Detection of vibrations produced by sap cavitation in *Ricinus* xylem. Planta 69:43–52.
Münch, E. 1930. Die Stoffbewegungen in der Pflanze. (Translocation in Plants.) Fischer, Jena.
Nobel, Park S. 1983. Biophysical Plant Physiology and Ecology. Freeman, San Francisco.
Oertli, J. J. 1971. The stability of water under tension in the xylem. Zeitschrift für Pflanzenphysiol. 65:195–209.
Peterson, Carol A. 1988. Exodermal Casparian bands: Their significance for ion uptake by roots. Physiologia Plantarum 72:204–208.
Peterson, Carol A., Mary E. Emanuel, and G. B. Humphreys. 1981. Pathway of movement of apoplastic fluorescent dye tracers through the endodermis at the site of secondary root formation in corn (*Zea mays*) and broad bean (*Vicia faba*). Canadian Journal of Botany 59:618–625.
Pickard, William F. 1981. The ascent of sap in plants. Progress in Biophysics and Molecular Biology 37:181–229.
Plumb, R. C. and W. B. Bridgman. 1972. Ascent of sap in trees. Science 176:1129–1131.
Preston, R. D. 1952. Movement of water in higher plants. Pages 257–321 *in* A. Frey-Wyssling (ed.), Deformation and Flow in Biological Systems. Elsevier North-Holland, Amsterdam.
Rein, H. 1928. Die Thermo-Stromuhr. Ein Verfahren welches mit etwa ± 10 Prozent Genauigkeit die unblutige langdauerde Messung der mittleren Durchflußmengen an gleichzeitigen Gefäßen gestattet. (The thermostream clock. A procedure that allows an unbloody, long-duration measurement of the average flow quantities in simultaneous vessels with an accuracy of ± 10 percent.) Z. Biol. 87:394–418.
Renner, O. 1911. Experimentelle Beiträge zur Zenntnis der Wasserbewegung. (Experimental contributions to the knowledge of water movement.) Flora 103:171.
Renner, O. 1915. Theoretisches und Experimentelles zur Kohäsions Theorie der Wasserbewegung. (Theoretical and experimental contributions to the cohesions theory of water movement.) Jahrbuch für Wissenschaftliche Botanik 56: 617–667.
Scholander, Per F. 1968. How mangroves desalinate sea water. Physiology Plantarum 21:251–268.
Scholander, Per F., H. T. Hammel, E. D. Bradstreet, and E. A. Hemmingsen. 1965. Sap pressure in vascular plants. Science 148:339–346.
Scholander, P. F., E. Hemmingsen, and W. Garey. 1961. Cohesive lift of sap in the rattan vine. Science 134:1835–1838.
Schulze, E-D., J. Cermak, R. Matyssek, M. Penka, R. Zimmermann, F. Vasicek, W. Gries, and J. Jucera. 1985. Canopy transpiration and water fluxes in the xylem of the trunk of *Larix* and *Picea* trees — a comparison of xylem flow, porometer and cuvette measurements. Oecologia (Berlin) 66:475–483.
Sperry, John S. and Melvin T. Tyree. 1988. Mechanism of water stress-induced xylem embolism. Plant Physiology 88:581–587.
Spomer, G. G. 1968. Sensors monitor tension in transpiration streams of trees. Science 161: 484–485.
Stone, J. F. and G. A. Shirazi. 1975. On the heat-pulse method for the measurement of apparent sap velocity in stems. Planta 122:166–177.
Strasburger, E. 1893. Über das Saftsteigen. (About the ascent of sap.) Fischer, Jena.
Sucoff, E. 1969. Freezing of conifer xylem and the cohesion-tension theory. Physiologia Plantarum 22:424–431.
Troughton, John and Lesley A. Donaldson. 1972. Probing Plant Structure. McGraw-Hill, New York.
Tyree, Melvin T. and Michael A. Dixon. 1986. Water stress induced cavitation and embolism in some woody plants. Physiologia Plantarum 66:397–405.
Tyree, Melvin T. and John S. Sperry. 1988. Do woody plants operate near the point of catastrophic xylem dysfunction caused by dynamic water stress? Plant Physiology 88:574–580.
Tyree, M. T. and J. S. Sperry. 1989. Vulnerability of xylem to cavitation and embolism. Annual Review of Plant Physiology 40:19–38.
Weiser, Russell L. and Stephen J. Wallner. 1988. Freezing woody plant stems produces acoustic emissions. Journal of American Society for Horticultural Science 113:636–639.
Wiebe, Herman H. 1981. Measuring water potential (activity) from free water to oven dryness. Plant Physiology 68:1218–1221.
Wiebe, H. H., R. W. Brown, T. W. Daniel, and E. Campbell. 1970. Water potential measurements in trees. BioScience 20:225–226.
Zimmermann, M. H. 1983. Xylem Structure and the Ascent of Sap. Springer-Verlag, Berlin, Heidelberg, New York.

## *Chapter 6 Mineral Nutrition*

Agren, G. I. and T. Ingestad. 1987. Root:shoot ratio as a balance between nitrogen productivity and photosynthesis. Plant, Cell and Environment 10:579–586.
Ahmed, S. and H. J. Evans. 1960. Cobalt: A micronutrient element for the growth of soybean plants under symbiotic conditions. Soil Science 90: 205–210.
Alexander, G. V. and L. T. McAnulty. 1983. Multielement analysis of plant-related tissues and fluids by optical emission spectrometry. Journal of Plant Nutrition 3:51–59.

Allan, E. F. and A. J. Trewavas. 1987. The role of calcium in metabolic control. Pages 117–149 *in* D. D. Davies (ed.), The Biochemistry of Plants, Vol. 12, Physiology of Metabolism. Academic Press, New York.

Anderson, J. W. and A. R. Scarf. 1983. Selenium and plant metabolism. Pages 241–275 *in* D. A. Robb and W. S. Pierpont (eds.), Metals and Micronutrients: Uptake and Utilization by Plants. Academic Press, New York.

Arnon, D. I. and P. R. Stout. 1939. The essentiality of certain elements in minute quantity for plants with special reference to copper. Plant Physiology 14:371–375.

Asher, C. J. and D. G. Edwards. 1983. Modern solution culture techniques. Pages 94–119 *in* A. Läuchli and R. L. Bieleski (eds.), Encyclopedia of Plant Physiology, New Series, Vol. 15B, Inorganic Plant Nutrition. Springer-Verlag, Berlin.

Baker, A. J. M. 1981. Accumulators and excluders—strategies in the response of plants to heavy metals. Journal of Plant Nutrition 3:643–654.

Bhandal, I. S. and C. P. Malik. 1988. Potassium estimation, uptake, and its role in the physiology and metabolism of flowering plants. International Review of Cytology 110:205–254.

Bienfait, H. F. 1988. Mechanisms in Fe-efficiency reactions of higher plants. Journal of Plant Nutrition 11:605–632.

Bollard, E. G. 1983. Involvement of unusual elements in plant growth and nutrition. Pages 695–744 *in* A. Läuchli and R. L. Bieleski (eds.), Encyclopedia of Plant Physiology, New Series, Vol. 15A, Inorganic Plant Nutrition. Springer-Verlag, Berlin.

Bould, C., E. J. Hewitt, and P. Needham. 1984. Diagnosis of Mineral Disorders in Plants, Vol. 1, Principles. Chemical Publishing Co., New York.

Bouma, D. 1983. Diagnosis of mineral deficiencies using plant tests. Pages 120–146 *in* A. Läuchli and R. L. Bieleski (eds.), Encyclopedia of Plant Physiology, New Series, Vol. 15A, Inorganic Plant Nutrition. Springer-Verlag, Berlin.

Bowen, G. D. and E. K. S. Nambiar. (Eds.). 1984. Nutrition of Plantation Forests. Academic Press, New York.

Brown, J. C. and V. D. Jolley. 1988. Strategy I and strategy II mechanisms affecting iron availability to plants may be established too narrow or limited. Journal of Plant Nutrition 11:1077–1098.

Brown, P. H., R. M. Welch, and E. E. Cary. 1987. Nickel: A micronutrient essential for higher plants. Plant Physiology 85:801–803.

Brown, T. A. and A. Shrift. 1982. Selenium: Toxicity and tolerance in higher plants. Biological Reviews 57:59–84.

Brownell, P. F. 1979. Sodium as an essential micronutrient element for plants and its possible role in metabolism. Advances in Botanical Research 7:117–224.

Brownell, P. F. and C. J. Crossland. 1972. The requirement for sodium as a micronutrient by species having the $C_4$ dicarboxylic photosynthetic pathway. Plant Physiology 49:794–797.

Buchala, A. J. and F. Pythoud. 1988. Vitamin D and related compounds as plant growth substances. Physiologia Plantarum 74:391–396.

Chaney, R. L. 1988. Recent progress and needed research in plant Fe nutrition. Journal of Plant Nutrition 11:1589–1603.

Chapin, F. S., III. 1987. Adaptations and physiological responses of wild plants to nutrient stress. Pages 15–25 *in* H. W. Gabelman and B. C. Loughman (eds.), Genetic Aspects of Plant Mineral Nutrition. Martinus Nijhoff, Boston.

Chapin, F. S., III. 1988. Ecological aspects of plant mineral nutrition. Advances in Plant Nutrition 3:161–191.

Cheeseman, J. M. 1988. Mechanisms of salinity tolerance in plants. Plant Physiology 76:490–497.

Cheung, W. Y. 1982. Calmodulin. Scientific American 246(6):62–70.

Clark, R. B. 1982. Nutrient solution growth of sorghum and corn in mineral nutrition studies. Journal of Plant Nutrition 5:1039–1057.

Clarkson, D. T. 1985. Factors affecting mineral nutrient acquisition by plants. Annual Review of Plant Physiology 36:77–116.

Clarkson, D. T. and J. B. Hanson. 1980. The mineral nutrition of higher plants. Annual Review of Plant Physiology 31:239–298.

Cooper, A. 1979. The ABC of NFT (nutrient film technique). Grower Books, London.

Dalton, D. A., S. A. Russell, and H. J. Evans. 1988. Nickel as a micronutrient element for plants. BioFactors 1(1):11–16.

Donald, C. M. and J. A. Prescott. 1975. Trace elements in Australian crop and pasture production, 1924–1974. Pages 7–37 *in* D. J. D. Nichols and A. R. Egan (eds.), Trace Elements in Soil-Plant-Animal Systems. Academic Press, New York.

Dugger, W. M. 1983. Boron in plant metabolism. Pages 626–650 *in* A. Läuchli and R. L. Bieleski (eds.), Encyclopedia of Plant Physiology, New Series, Vol. 15B, Inorganic Plant Nutrition. Springer-Verlag, Berlin.

Duke, S. H. and H. M. Reisenauer. 1986. Roles and requirements of sulfur in plant nutrition. Sulfur in Agriculture, Agronomy Monograph No. 27, Madison, Wisc.

Engvild, K. C. 1986. Chlorine-containing natural compounds in higher plants. Phytochemistry 25:781–791.

Epstein, E. 1972. Mineral Nutrition of Plants: Principles and Perspectives. Wiley, New York.

Eskew, D. L., R. M. Welch, and W. A. Norvell. 1984. Nickel in higher plants. Plant Physiology 76: 691–693.

Evans, H. J. and A. Nason. 1953. Pyridine nucleotide-nitrate reductase from extracts of higher plants. Plant Physiology 28:233–254.

Ferguson, I. B. and B. K. Drobak. 1988. Calcium and the regulation of plant growth and senescence. HortScience 23:262–266.

Flowers, T. J. 1988. Chloride as a nutrient and as an osmoticum. Advances in Plant Nutrition 3:55–57.

Gabelman, H. W. and B. C. Loughman. (Eds.). 1987. Genetic Aspects of Plant Mineral Nutrition. Martinus Nijhoff, Boston.

Gauch, H. G. 1972. Inorganic Plant Nutrition. Dowden, Hutchinson and Ross, New York.

Gekeler, W., E. Grill, E. Winnacker, and M. H. Zenk. 1989. Survey of the plant kingdom for the ability to bind heavy metals through phytochelatins. Zeitschrift für Naturforschung 44c:361–369.

Gilroy, S., D. P. Blowers, and A. J. Trewavas. 1987. Calcium: A regulation system emerges in plant cells. Development 100:181–184.

Glass, A. D. M. 1989. Plant Nutrition. An Introduction to Current Concepts. Jones and Bartlett, Boston.

Graves, C. J. 1983. The nutrient film technique. Horticultural Reviews 5:1–44.

Grundon, N. J. 1987. Hungry Crops: A Guide to Nutrient Deficiencies in Field Crops. Queensland Department of Primary Industries, Brisbane, Australia.

Hanson, J. B. 1984. The function of calcium in plant nutrition. Advances in Plant Nutrition 1:149–208.

Hasegawa, P. M., R. A. Brassen, and A. K. Handa. 1987. Cellular mechanisms of salinity tolerance. HortScience 21:1317–1324.

Hepler P. K. and R. O. Wayne. 1985. Calcium and plant development. Annual Review of Plant Physiology 36:397–439.

Hewitt, E. J. 1966. Sand and Water Culture Methods Used in the Study of Plant Nutrition, Second Edition. Technical Communication No. 22 (Revised). Commonwealth Agricultural Bureau, Farnham Royal, Bucks, England.

Hewitt, E. J. and T. A. Smith. 1975. Plant Mineral Nutrition. Wiley, New York.

Hoagland, D. R. and D. I. Arnon. 1950. The water culture method of growing plants without soil. California Agriculture Experiment Station Circular 347.

Jensen, M. H. and W. L. Collins. 1985. Hydroponic vegetable production. Horticultural Reviews 7:483–558.

Johnston, M., C. P. L. Grof, and P. F. Brownell. 1984. Responses to ambient $CO_2$ concentrations by sodium-deficient $C_4$ plants. Australian Journal of Plant Physiology 11:137–141.

Joiner, J. N., R. T. Poole, and C. A. Conover. 1983. Nutrition and fertilization of ornamental greenhouse crops. Horticultural Reviews 5: 317–403.

Jones, J. B., Jr. 1982. Hydroponics: Its history and use in plant nutrition studies. Journal of Plant Nutrition 5:1003–1030.

Jones, J. B., Jr. 1983. A Guide for the Hydroponic and Soilless Culture Grower. Timber Press, Portland, Ore.

Kaufman, P. B., P. Dayanandan, and C. I. Franklin. 1985. Structure and function of silica bodies in the epidermal system of grass shoots. Annals of Botany 55:487–507.

Kinzel, H. 1989. Calcium in the vacuoles and cell walls of plant tissue. Flora 182:99–125.

Kirkby, E. A. and D. J. Pilbeam. 1984. Calcium as a plant nutrient. Plant, Cell and Environment 7:397–405.

Korcak, R. F. 1987. Iron deficiency chlorosis. Horticultural Reviews 9:135–186.

Latshaw, W. L. and E. C. Miller. 1924. Elemental composition of the corn plant. Journal of Agricultural Research 27:845–861.

Läuchli, A. and R. L. Bieleski. (Eds.). 1983a. Encyclopedia of Plant Physiology, New Series, Vol. 15A, Inorganic Plant Nutrition. Springer-Verlag, Berlin.

Läuchli, A. and R. L. Bieleski. (Eds.). 1983b. Encyclopedia of Plant Physiology, New Series, Vol. 15B, Inorganic Plant Nutrition. Springer-Verlag, Berlin.

Leonard, R. T. and P. K. Hepler. (Eds.). 1990. Calcium in Plant Growth and Development. American Society of Plant Physiologists, Rockville, Md.

Lewis, J. and B. E. F. Reimann. 1969. Silicon and plant growth. Annual Review of Plant Physiology 20:289–304.

Lindsay, W. L. 1974. Role of chelation in micronutrient availability. Pages 507–524 *in* E. W. Carson (ed.), The Plant Root and Its Environment. University of Virginia Press, Charlottesville.

Lindsay, W. L. 1979. Chemical Equilibria in Soils. Wiley, New York.

Loneragan, J. F. 1968. Nutrient requirements of plants. Nature 220:1307–1308.

Loneragan, J. F. and K. Snowball. 1969. Calcium requirements of plants. Australian Journal of Agricultural Research 20:465–478.

Longnecker, N. 1988. Iron nutrition of plants. ISI Atlas of Science 1:143–150.

Lovatt, C. J. 1985. Evolution of the xylem resulted in a requirement for boron in the apical meristems of vascular plants. New Phytologist 99:509–522.

Lovatt, C. J. and W. M. Dugger. 1984. Biochemistry of boron. Pages 389–421 *in* E. Frieden (ed.), Biochemistry of the Essential Ultra-Trace Elements, Plenum, New York.

Marmé, D. 1989. The role of calcium and calmodulin in signal transduction. Pages 57–80 *in* W. F. Boss and D. J. Morré (eds.), Second Messengers in Plant Growth and Development. Alan R. Liss, New York.

Marschner, H. 1986. Mineral Nutrition of Higher Plants. Academic Press, London.

Marschner, H., V. Romheld, and M. Kissel. 1986. Different strategies in higher plants in mobilization and uptake of iron. Journal of Plant Nutrition 9:695–713.

McMurtrey, J. E., Jr. 1938. Distinctive plant symptoms caused by deficiency of any one of the chemical elements essential for normal development. Botanical Review 4:183–203.

Mendel, R. R. and A. J. Muller. 1976. A common genetic determinant of xanthine dehydrogenase and nitrate reductase in *Nicotiana tabacum*. Biochemie und Physiologie der Pflanzen 170: 538–541.

Mengel, K. and G. Geurtzen. 1988. Relationship between iron chlorosis and alkalinity in *Zea mays*. Physiologia Plantarum 72:460–465.
Mengel, K. and E. A. Kirkby. 1987. Principles of Plant Nutrition, Fourth Edition. International Potash Institute, Worblaufen-Bern, Switzerland.
Merry, R. H. 1987. Tolerance of plants to heavy metals. Pages 165–171 *in* H. W. Gabelman and B. C. Loughman (eds.), Genetic Aspects of Plant Mineral Nutrition. Martinus Nijhoff, Boston.
Mertz, W. 1981. The essential trace elements. Science 213:1332–1338.
Miller, W. J. and M. N. Neathery. 1977. Newly recognized trace mineral elements and their role in animal nutrition. BioScience 27:674–679.
Miyake, Y. and E. Takahashi. 1985. Effects of silicon on the growth of soybean plants in a solution culture. Soil Science and Plant Nutrition 31: 625–636.
Moraghan, J. T. 1985. Plant tissue testing for micronutrient deficiencies and toxicities. Fertilizer Research 7:201–219.
Mullins, G. L., L. E. Sommers, and T. L. Housley. 1986. Metal speciation in xylem and phloem exudates. Plant and Soil 96:377–391.
Nabors, M. W. 1983. Increasing the salt and drought tolerance of crop plants. Pages 165–184 *in* D. D. Randall (ed.), Current Topics in Plant Biochemistry and Physiology, Vol. 2. University of Missouri Press, Columbia.
Neilands, J. B. and S. A. Leong. 1986. Siderophores in relation to plant growth and disease. Annual Review of Plant Physiology 37:187–208.
Oertli, J. J. 1987. Exogenous application of vitamins as regulators for growth and development of plants — a review. Z. Pflanzenernähr. Bodenk 150:375–391.
Pérez-Vicente, R., M. Pineda, and J. Cardenas. 1988. Isolation and characterization of xanthine dehydrogenase from *Chlamydomonas reinhardtii*. Physiologia Plantarum 72:101–107.
Pilbeam, D. J. and E. A. Kirkby. 1983. The physiological role of boron in plants. Journal of Plant Nutrition 6:563–582.
*Plant and Soil*, Vol. 72, 1983. This entire issue contains articles largely related to genetic variability in mineral nutrition.
Poovaiah, B. W. and A. S. N. Reddy. 1987. Calcium messenger system in plants. Critical Reviews in Plant Science 6:47–103.
Quispel, A. 1983. Dinitrogen-fixing symbioses with legumes, non-legume angiosperms and associative symbioses. Pages 286–329 *in* A. Läuchli and R. L. Bieleski (eds.), Encyclopedia of Plant Physiology, New Series, Vol. 15A, Inorganic Plant Nutrition. Springer-Verlag, Berlin.
Rauser, W. E. 1990. Phytochelatins. Annual Review of Biochemistry 59:61–86.
Raven, J. A. 1980. Short- and long-distance transport of boric acid in plants. New Phytologist 84: 231–249.
Raven, J. A. 1988. Requisition of nitrogen by the shoots of land plants: Its occurrence and implications for acid-base regulation. New Phytologist 109:1–20.
Rees, T. A. V. and I. A. Bekheet. 1982. The role of nickel in urea assimilation by algae. Planta 156:385–387.
Reisenauer, H. M. 1966. Mineral nutrients in soil solution. Pages 507–508 *in* P. L. Altman and D. S. Dittmer (eds.), Environmental Biology. Federation of American Societies for Experimental Biology, Bethesda, Md.
Resh, H. M. 1989. Hydroponic Food Production. Woodridge Press, Santa Barbara, Calif.
Robb, D. A. and W. S. Pierpont. (Eds.). 1983. Metals and Micronutrients. Uptake and Utilization by Plants. Academic Press, New York.
Roberts, D. M., T. J. Lukas, and D. M. Watterson. 1986. Structure, function, and mechanism of action of calmodulin. Critical Reviews in Plant Sciences 4:311–339.
Robinson, J. B. D. 1987. Diagnosis of Mineral Disorders in Plants, Vol. 3, Glasshouse Crops. Chemical Publishing Co., New York.
Romheld, V. 1987. Different strategies for iron acquisition in higher plants. Physiologia Plantarum 70:231–234.
Rovira, A. D., G. D. Bowen, and R. C. Foster. 1983. The significance of rhizosphere microflora and mycorrhizas in plant nutrition. Pages 61–93 *in* A. Läuchli and R. L. Bieleski (eds.), Encyclopedia of Plant Physiology, New Series, Vol. 15A, Inorganic Plant Nutrition. Springer-Verlag, Berlin.
Ryan, D. F. and F. H. Bormann. 1982. Nutrient resorption in northern hardwood forests. BioScience 32:29–32.
Samiullah, S., A. Ansari, and M. M. R. K. Afridi. 1988. B-vitamins in relation to crop productivity. Indian Review of Life Science 8:51–74.
Sanchez-Alonso, F. and M. Lachica. 1987. Seasonal trends in the elemental content of sweet cherry leaves. Communications in Soil Science and Plant Analysis 18:17–29.
Sandman, G. and P. Boger. 1983. The enzymological function of heavy metals and their role in electron transfer processes of plants. Pages 563–596 *in* A. Läuchli and R. L. Bieleski (eds.), Encyclopedia of Plant Physiology, New Series, Vol. 15B, Inorganic Plant Nutrition. Springer-Verlag, Berlin.
Sangster, A. G. and M. J. Hodson. 1986. Silica in higher plants. Pages 90–111 *in* D. Evered and M. O'Connor (eds.), Silicon Biochemistry, Ciba Foundation Symposium 121. Wiley, Chichester, England.
Scaife, A. and M. Turner. 1984. Diagnosis of Mineral Disorders in Plants, Vol. 2, Vegetables. Chemical Publishing Co., New York.
Seckback, J. 1982. Ferreting out the secrets of plant ferretin: A review. Journal of Plant Nutrition 5:369–394.
Shear, C. B. and M. Faust. 1980. Nutritional ranges in deciduous tree fruits and nuts. Horticultural Reviews 2:142–163.
Shelp, B. J. 1988. Boron mobility and nutrition in broccoli (*Brassica oleracea* var. *italica*). Annals of Botany 61:83–91.
Soltanpour, P. N., J. B. Jones, Jr., and S. M. Workman. 1982. Optical emission spectrometry. Pages 29–65 *in* A. Page (ed.), Methods of Soil Analysis, Part 2, Chemical and Microbiological Properties (Agronomy Monograph No. 9). Madison, Wisc.
Stadtman, T. C. 1990. Selenium biochemistry. Annual Review of Biochemistry 59:111–127.
Steffens, J. C. 1990. The heavy metal-binding peptides of plants. Annual Review of Plant Physiology and Plant Molecular Biology 41:553–575.
Stout, P. R. 1961. Proceedings of the Ninth Annual California Fertilizer Conference, pp. 21–23.
Sugiura, Y. and K. Nomoto. 1984. Phytosiderophores. Structural Bonding 58:107–135.
Swietlik, D. and M. Faust. 1984. Foliar nutrition of fruit crops. Horticultural Reviews 6:287–355.
Taylor, G. J. 1987. Exclusion of metals from the symplasm: A possible mechanism of metal tolerance in higher plants. Journal of Plant Nutrition 10:1213–1222.
Terry, N. 1977. Photosynthesis, growth, and the role of chloride. Plant Physiology 60:69–75.
Tinker, B. and A. Läuchli. (Eds.). 1988. Advances in Plant Nutrition, Vol. 3. Praeger, New York.
Titus, J. S. and S. M. Kang. 1982. Nitrogen metabolism, translocation, and recycling in apple trees. Horticultural Reviews 4:204–246.
Tomsett, A. B. and D. A. Thurman. 1988. Molecular biology of metal tolerances of plants. Plant, Cell and Environment 11:383–394.
Trewavas, A. J. (Ed.). 1986. Molecular and Cellular Aspects of Calcium in Plant Development. Plenum, New York.
Uren, N. C. 1981. Chemical reduction of an insoluble higher oxide of manganese by plant roots. Journal of Plant Nutrition 4:65–71.
Vallee, B. L. 1976. Zinc biochemistry: A perspective. Trends in Biochemical Sciences 1:88–91.
Walker, C. D., R. D. Graham, J. T. Madison, E. E. Cary, and R. M. Welch. 1985. Effects of Ni deficiency on some nitrogen metabolites in cowpeas (*Vigna unguiculata* L. Walp). Plant Physiology 79:474–479.
Walker-Simmons, M., D. A. Kudrna, and R. L. Warner. 1989. Reduced accumulation of ABA during water stress in a molybdenum cofactor mutant of barley. Plant Physiology 90:728–733.
Walworth, J. L. and M. E. Sumner. 1988. Foliar diagnosis: A review. Advances in Plant Nutrition 3:193–241.
Welch, R. M. 1981. The biological significance of nickel. Journal of Plant Nutrition 3:345–356.
Welch, R. M. 1986. Effects of nutrient deficiencies on seeds' production and quality. Advances in Plant Nutrition 1:205–247.
Werner, D. and R. Roth. 1983. Silica metabolism. Pages 682–694 *in* A. Läuchli and R. L. Bieleski (eds.), Encyclopedia of Plant Physiology, New Series, Vol. 15B, Inorganic Plant Nutrition. Springer-Verlag, Berlin.
White, M. C., A. M. Decker, and R. L. Chaney. 1981. Metal complexation in xylem fluid. I. Chemical composition of tomato and soybean stem exudate. Plant Physiology 67:292–309.
Wilcox, G. E. 1982. The future of hydroponics as a research and plant production method. Journal of Plant Nutrition 5:1031–1038.
Wild, A., L. P. Jones, and J. H. Macduff. 1987. Uptake of mineral nutrients and crop growth: The use of flowing nutrient solutions. Advances in Agronomy 41:171–219.
Williams, K., F. Percival, J. Merino, and H. A. Mooney. 1987. Estimation of tissue construction cost from heat of combustion and organic nitrogen content. Plant, Cell and Environment 10:725–734.
Williamson, R. E. 1984. Calcium and the plant cytoskeleton. Plant, Cell and Environment 7: 431–440.
Wilson, S. B. and D. T. D. Nicholas. 1967. A cobalt requirement for nonnodulated legumes and for wheat. Phytochemistry 6:1057–1066.
Wolf, B. 1982. A comprehensive system of leaf analyses and its use for diagnosing crop nutrient status. Communications in Soil Science and Plant Analysis 13:1035–1059.
Woolhouse, H. W. 1983. Toxicity and tolerance in the responses of plants to metals. Pages 245–300 *in* O. L. Lange, P. S. Nobel, C. B. Osmond, and H. Ziegler (eds.), Encyclopedia of Plant Physiology, New Series, Vol. 12C, Physiological Plant Ecology III, Responses to the Chemical and Biological Environment. Springer-Verlag, Berlin.

## *Chapter 7 Absorption of Mineral Salts*

Asher, C. J. and P. G. Ozanne. 1967. Growth and potassium content of plants in solution cultures maintained at constant potassium concentrations. Soil Science 103:155–161.
Ashford, A. E. A., W. G. Allaway, C. A. Peterson, and J. W. G. Cairney. 1989. Nutrient transfer and the fungus-root interface. Australian Journal of Plant Physiology 16:85–97.
Baker, D. A. and J. L. Hall. (Eds.). 1987. Solute Transport in Plant Cells and Tissues. Longmans, Harlow, England.
Bhandal, I. S. and C. P. Malik. 1988. Potassium estimation, uptake, and its role in the physiology and metabolism of flowering plants. International Review of Cytology 110:205–254.
Bienfait, F. and U. Lüttge. 1988. On the function of two systems that can transfer electrons across the plasma membrane. Plant Physiology and Biochemistry 26:665–671.
Blount, R. W. and B. H. Levedahl. 1960. Active sodium and chloride transport in the single-celled marine alga *Halicystis ovalis*. Acta Physiologia Scandinavia 49:1–9.
Blumwald, E. 1987. Tonoplast vesicles as a tool in the study of ion transport at the plant vacuole. Physiologia Plantarum 69:731–734.

Blumwald, E. and R. J. Poole. 1986. Kinetics of $Ca^{2+}/H^{+}$ antiport in isolated tonoplast vesicles from storage tissue of *Beta vulgaris* L. Plant Physiology 80:727–731.

Boller, T. 1989. Primary signals and second messengers in the reaction of plants to pathogens. Pages 227–255 *in* D. J. Morré and W. Boss (eds.), Second Messengers in Plant Growth and Development. Alan R. Liss, New York.

Briskin, D. P. and W. R. Thornley. 1985. Plasma membrane ATPase of sugarbeet. Phytochemistry 24:2797–2802.

Bugbee, B. and F. B. Salisbury. 1988. Exploring the limits of crop productivity: Photosynthetic efficiency of wheat in high irradiance environments. Plant Physiology 88:869–878.

Bush, D. and H. Sze. 1986. Calcium transport in tonoplast and endoplasmic reticulum vesicles isolated from cultured carrot cells. Plant Physiology 80:549–555.

Caldwell, M. M. 1987. Competition between root systems in natural communities. Pages 167–185 *in* P. J. Gregory, J. V. Lake, and K. A. Rose (eds.), Root Development and Function — Effects of the Physical Environment. Cambridge University Press, Cambridge.

Chabre, M. 1987. The G protein connection: Is it in the membrane or the cytoplasm? Trends in Biochemical Sciences 12:213–215.

Clarkson, D. T. 1974. Ion Transport and Cell Structure in Plants. McGraw-Hill, New York.

Clarkson, D. T. 1984. Calcium transport between tissues and its distribution in the plant. Plant, Cell and Environment 7:449–456.

Clarkson, D. T. 1985. Factors affecting mineral nutrient acquisition by plants. Annual Review of Plant Physiology 36:77–115.

Clarkson, D. T. and J. B. Hanson. 1980. The mineral nutrition of higher plants. Annual Review of Plant Physiology 31:239–298.

Cooper, H. D. and D. T. Clarkson. 1989. Cycling of amino-nitrogen and other nutrients between shoots and roots in cereals — a possible mechanism integrating shoot and root in the regulation of nutrient uptake. Journal of Experimental Botany 40:753–762.

Cram, W. J. 1988. Transport of nutrient ions across cell membranes *in vivo*. Pages 1–53 *in* B. Tinker and A. Lauchli (eds.), Advances in Plant Nutrition, Vol. 3. Praeger, New York.

Crane, F. L. and R. Barr. 1989. Plasma membrane oxidoreductases. CRC Critical Reviews in Plant Sciences 8:273–307.

Dainty, J. 1962. Ion transport and electrical potentials in plant cells. Annual Review of Plant Physiology 13:379–402.

della-Cioppa, G., G. M. Kishore, R. N. Beachy, and R. T. Fraley. 1987. Protein trafficking in plant cells. Plant Physiology 84:965–968.

Dingwall, C. and R. A. Laskey. 1986. Protein import into the cell nucleus. Annual Review of Cell Biology 2:367–390.

Dittmer, H. J. 1949. Root hair variation in plant species. American Journal of Botany 36:152–155.

Dixon, R. A. and C. J. Lamb. 1990. Molecular communication in interactions between plants and microbial pathogens. Annual Review of Plant Physiology and Plant Molecular Biology 41: 339–367.

Drew, M. C. 1975. Comparison of the effects of a localized supply of phosphate, nitrate, ammonium and potassium on the growth of the seminal root system, and the shoot, in barley. New Phytologist 75:479–490.

Drew, M. C. 1987. Function of root tissues in nutrient and water transport. Pages 71–101 *in* J. Gregory, J. V. Lake, and D. A. Rose (eds.), Root Development and Function — Effects of the Physical Environment. Cambridge University Press, Cambridge.

Drew, M. C. 1988. Effects of flooding and oxygen deficiency on plant mineral nutrition. Pages 115–159 *in* B. Tinker and A. Lauchli (eds.), Advances in Plant Nutrition, Vol. 3. Praeger, New York.

Dunlop, J. 1989. Phosphate and membrane electropotentials in *Trifolium repens* L. Journal of Experimental Botany 40:803–807.

Engelbrecht, S. and W. Junge. 1990. Subunit δ of $H^{+}$-ATPases: At the interface between proton flow and ATP synthesis. Biochimica et Biophysica Acta 1015:379–390.

Epstein, E. 1972. Mineral Nutrition of Plants: Principles and Perspectives. Wiley, New York.

Esau, K. 1977. Plant Anatomy, Third Edition. Wiley, New York.

Feldman, L. J. 1988. The habits of roots. What's up down under? BioScience 38:612–618.

Gastal, F. and B. Saugier. 1989. Relationships between nitrogen uptake and carbon assimilation in whole plants of tall fescue. Plant, Cell and Environment 12:407–418.

Giannini, J. L., J. Ruiz-Cristin, and D. P. Briskin. 1987. Calcium transport in sealed vesicles from red beet (*Beta vulgaris* L.) storage tissue. Plant Physiology 85:1137–1142.

Giaquinta, R. T. 1983. Phloem loading of sucrose. Annual Review of Plant Physiology 34:347–387.

Glass, A. D. M. 1989. Plant Nutrition: An Introduction to Current Concepts. Jones and Bartlett, Boston.

Gräf, P. and E. W. Weiler. 1989. ATP-driven $Ca^{2+}$ transport in sealed plasma membrane vesicles prepared by aqueous two-phase partitioning from leaves of *Commelina communis*. Physiologia Plantarum 75:469–478.

Granato, T. C. and C. D. Raper, Jr. 1989. Proliferation of maize (*Zea mays* L.) roots in response to localized supply of nitrate. Journal of Experimental Botany 40:263–275.

Gregory, P. J., J. V. Lake, and D. A. Rose. (Eds.). 1987. Root Development and Function — Effects of the Physical Environment. Society for Experimental Biology Seminar Series 30, Cambridge University Press, Cambridge.

Grimes, H. D. and R. W. Breidenbach. 1987. Plant plasma membrane proteins. Plant Physiology 85:1048–1054.

Gunning, B. E. S. and R. L. Overall. 1983. Plasmodesmata and cell-to-cell transport in plants. BioScience 33:260–265.

Hadley, G. 1988. Mycorrhizas and Plant Growth and Development. Kluwer Academic Publications, Boston.

Hatch, D. J., M. J. Hopper, and M. S. Dhanoa. 1986. Measurement of ammonium ions in flowing solution culture and diurnal variation in uptake by *Lolium perenne* L. Journal of Experimental Botany 37:589–596.

Hedrich, R. and J. I. Schroeder. 1989. The physiology of ion channels and electrogenic pumps in higher plants. Annual Review of Plant Physiology and Plant Molecular Biology 40:539–569.

Hedrich, R., J. I. Schroeder, and J. M. Fernandez. 1987. Patch-clamp studies on higher plant cells: A perspective. Trends in Biochemical Sciences 12: 49–52.

Heldt, H. W. and U. I. Flügge. 1987. Subcellular transport of metabolites in plant cells. Pages 49–85 *in* D. D. Davies (ed.), The Biochemistry of Plants, Vol. 12. Academic Press, New York.

Hepler, P. K. and R. O. Wayne. 1985. Calcium and plant development. Annual Review of Plant Physiology 36:397–439.

Higinbotham, N., B. Etherton, and R. J. Foster. 1967. Mineral ion contents and cell transmembrane electropotentials of pea and oat seedling tissue. Plant Physiology 42:37–46.

Hodgkin, T., G. D. Lyon, and H. G. Dickinson. 1988. Recognition in flowering plants: A comparison of the *Brassica* self-incompatibility system and plant pathogen interactions. New Phytologist 110: 557–569.

Hurt, E. C. 1987. Unravelling the role of ATP in post-translational protein translocation. Trends in Biochemical Sciences 12:369–370.

Ingestad, T. and G. I. Ågren. 1988. Nutrient uptake and allocation at steady-state nutrition. Physiologia Plantarum 72:450–459.

Itoh, S. and S. A. Barber. 1983. Phosphorus uptake by six plant species as related to root hairs. Agronomy Journal 75:457–461.

Jones, R. L. and D. G. Robinson. 1989. Protein secretion in plants. New Phytologist 11:567–597.

Kasai, M. and S. Muto. 1990. $Ca^{2+}$ pump and $Ca^{2+}/H^{+}$ antiporter in plasma membrane vesicles isolated by aqueous two-phase partitioning from corn leaves. Journal of Membrane Biology 114: 133–142.

Keegstra, K., L. J. Olsen, and S. M. Theg. 1989. Chloroplastic precursors and their transport across the envelope membranes. Annual Review of Plant Physiology and Plant Molecular Biology 40:471–501.

Klepper, B. 1987. Origin, branching and distribution of root systems. Pages 103–124 *in* P. J. Gregory, J. V. Lake, and D. A. Rose (eds.), Root Development and Function — Effects of the Physical Environment. Cambridge University Press, Cambridge.

Kochian, L. V. and W. J. Lucas. 1982. Potassium transport in corn roots. I. Resolution of kinetics into a saturable and linear component. Plant Physiology 70:1723–1731.

Kochian, L. V. and W. J. Lucas. 1988. Potassium transport in roots. Advances in Botanical Research 15:93–177.

Kramer, P. J. and T. T. Kozlowski. 1979. Physiology of Woody Plants. Academic Press, New York.

Kristen, U. 1989. Structural botany. I. General and molecular cytology: The plasma membrane and the tonoplast. Progress in Botany 50:1–13.

LaFayette, P. R., R. W. Breidenbach, and R. L. Travis. 1987. Glycosylated polypeptides of soybean root endomembranes. Protoplasma 136:125–135.

Lemoine, R., S. Delrot, O. Gallet, and C. Larsson. 1989. The sucrose carrier of the plant plasma membrane. Biochimica et Biophysica Acta 978: 65–71.

Levin, S. A., H. A. Mooney, and C. Field. 1989. The dependence of plant root:shoot ratios on internal nitrogen concentration. Annals of Botany 64: 71–75.

Loughman, B. C., O. Gasparikova, and J. Kolek. (Eds.). 1989. Structural and Functional Aspects of Transport in Roots. Kluwer Academic Publishers, Hingham, Mass.

Lüttge, U. and D. T. Clarkson. 1989. Mineral nutrition: Potassium. Progress in Botany 50:51–73.

Marx, D. H. and N. C. Schenck. 1983. Potential of mycorrhizal symbiosis in agricultural and forest productivity. Pages 334–347 *in* T. Kommedahl and P. H. Williams (eds.), Challenging Problems in Plant Health. American Phytopathological Society, St. Paul.

Mitchell, P. 1985. The correlation of chemical and osmotic forces in biochemistry. Journal of Biochemistry 97:1–18.

Nandi, S. K., R. C. Pant, and P. Nissen. 1987. Multiphasic uptake of phosphate by corn roots. Plant, Cell and Environment 10:463–474.

Nelson, N. 1988. Structure, function, and evolution of proton-ATPases. Plant Physiology 86:1–3.

Nelson, N. and L. Taiz. 1989. The evolution of $H^{+}$-ATPases. Trends in Biochemical Sciences 14: 113–116.

Nissen, P. 1986. Nutrient uptake by plants: Effect of external ion concentration. Acta Horticulturae 178:21–28.

Nissen, P. 1991. Multiphasic uptake mechanisms in plants. International Review of Cytology 126: 89–134.

Nissen, P. and O. Nissen. 1983. Validity of the multiphasic concept of ion absorption in roots. Physiologia Plantarum 57:47–56.

Nye, P. H. and P. B. Tinker. 1977. Solute Movements in the Root-Soil System. Blackwell, Oxford, England.

Pantoja, O., J. Dainty, and E. Blumwald. 1989. Ion channels in vacuoles from halophytes and glycophytes. Federation of European Biochemical Societies 255:92–96.

Pedersen, P. L. and E. Carafoli. 1987. Ion motive ATPases. Trends in Biochemical Sciences 12: 186–189.
Peterson, C. A. 1988. Exodermal Casparian bands: Their significance for ion uptake by roots. Physiologia Plantarum 72:204–208.
Poole, R. J. 1988. Plasma membrane and tonoplast. Pages 83–105 *in* D. A. Balzer and J. L. Hall (eds.), Solute Transport in Plant Cells and Tissues. Pittman, London.
Rea, P. A. and D. Sanders. 1987. Tonoplast energization: Two $H^+$ pumps, one membrane. Physiologia Plantarum 71:131–141.
Reinhold, L. and A. Kaplan. 1984. Membrane transport of sugars and amino acids. Annual Review of Plant Physiology 35:45–83.
Reisenauer, H. M. 1966. Mineral nutrients in soil solution. Pages 507–508 *in* P. L. Altman and D. S. Dittmer (eds.), Environmental Biology. Federation of American Society for Experimental Biology, Bethesda, Md.
Reynolds, E. R. C. 1987. Development of the root crown in some conifers. Plant and Soil 98: 397–405.
Robards, A. W. 1975. Plasmodesmata. Annual Review of Plant Physiology 26:13–29.
Robards, A. W. and W. J. Lucas. 1990. Plasmodesmata. Annual Review of Plant Physiology and Plant Molecular Biology 41:369–419.
Robinson, D. G. and S. Hillmer. 1990. Endocytosis in plants. Physiologia Plantarum 79:96–104.
Rochester, C. P., P. Kjellbom, and C. Larsson. 1987. Lipid composition of plasma membranes from barley leaves and roots, spinach leaves, and cauliflower inflorescences. Physiologia Plantarum 71:257–263.
Safir, E. (Ed.). 1987. Ecophysiology of VA Mycorrhizal Plants. CRC Press, Boca Raton, Fla.
Satter, R. L. and N. Moran. 1988. Ionic channels in plant cell membranes. Physiologia Plantarum 72:816–820.
Schroeder, J. I. and R. Hedrich. 1989. Involvement of ion channels and active transport in osmoregulation and signaling of higher plants' cells. Trends in Biochemical Sciences 14:187–192.
Serrano, R. 1987. Structure and function of proton translocating ATPase in plasma membranes of plants and fungi. Biochimica et Biophysica Acta 947:1–28.
Serrano, R. 1989. Structure and function of plasma membrane ATPase. Annual Review of Plant Physiology and Plant Molecular Biology 40:61–94.
Shishkoff, N. 1987. Distribution of the dimorphic hypodermis of roots in angiosperm families. Annals of Botany 60:1–15.
Smith, S. E. and V. Gianinazzi-Pearson. 1988. Physiological interactions between symbionts in vesicular-arbuscular mycorrhizal plants. Annual Review of Plant Physiology and Plant Molecular Biology 39:221–244.
Starr, C. and R. Taggart. 1989. Biology: The Unity and Diversity of Life, Fifth Edition. Wadsworth, Belmont, Calif.
Stein, W. D. 1986. Transport and Diffusion Across Cell Membranes. Academic Press, New York.
Stone, B. A. 1989. Cell walls in plant-microorganism associates. Australian Journal of Plant Physiology 16:5–17.
Sussman, M. R. and J. F. Harper. 1989. Molecular biology of the plasma membrane of higher plants. The Plant Cell 1:953–960.
Sutton, R. F. 1980. Root system morphogenesis. New Zealand Journal of Forestry Science 10:264–292.
Sze, H. 1985. $H^+$-translocating ATPases: Advances using membrane vesicles. Annual Review of Plant Physiology 36:175–208.
Tamasi, J. 1986. Root Location of Fruit Trees and Its Agrotechnical Consequences. H. Stillman, Boca Raton, Fla.
Tester, M. 1989. Plant ion channels: Whole-cell and single-channel studies. New Phytologist 114: 305–340.
Tinker, B. and A. Läuchli. (Eds.). 1988. Advances in Plant Nutrition, Vol. 3. Praeger, New York.
Wiebe, H. 1978. The significance of plant vacuoles. BioScience 28:327–331.
Wild, A., L. H. P. Jones, and J. H. Macduff. 1987. Uptake of mineral nutrients and crop growth: The use of flowing nutrient solutions. Advances in Agronomy 41:171–219.


## *Chapter 8 Transport in the Phloem*


Aloni, Beny, Jaleh Daie, and Roger Wyse. 1986. Enhancement of $^{14}C$-sucrose export from source leaves of *Vicia faba* by $Ga_3$. Pages 491–493 *in* James Cronshaw, William J. Lucas, and Robert T. Giaquinta (eds.), Phloem Transport. Alan R. Liss, New York.
Arnold, W. N. 1968. The selection of sucrose as the translocate of higher plants. Journal of Theoretical Biology 21:13–20.
Baker, D. A. and J. A. Milburn. (Eds.). 1990. Transport of Photoassimilates. Longman Scientific and Technical, Harlow & John Wiley & Sons, New York.
Barnabas, A. D., V. Butler, and T. D. Steinke. 1986. Phloem structure and transport pathways in the leaves of a seagrass. Pages 177–180 *in* James Cronshaw, William J. Lucas, and Robert T. Giaquinta (eds.), Phloem Transport. Alan R. Liss, New York.
Biddulph, O. and R. Cory. 1957. An analysis of translocation in the phloem of the bean plant using THO, $^{32}P$, and $^{14}CO_2$. Plant Physiol. 32: 608–619.
Blechschmidt-Schneider, S. 1986. The effect of cold-inhibited phloem translocation on photosynthesis and carbohydrate status of source leaves. Pages 487–489 *in* James Cronshaw, William J. Lucas, and Robert T. Giaquinta (eds.), Phloem Transport. Alan R. Liss, New York.
Bodson, M. and G. Bernier. 1985. Is flowering controlled by the assimilate level? Physiol. Veg. 23(4): 491–501.
Botha, C. E. J., R. F. Evert, and R. D. Walmsley. 1975. Observations of the penetration of the phloem in leaves of *Nerium oleander* (Linn.) by stylets of the aphid, *Aphis meril* (B. de F.). Protoplasma 86: 309–319.
Chatterton, N. Jerry, Philip A. Harrison, and Jesse H. Bennett. 1986. Environmental effects on sucrose and fructan concentrations in leaves of *Agropyron* spp. Pages 471–476 *in* James Cronshaw, William J. Lucas, and Robert T. Giaquinta (eds.), Phloem Transport. Alan R. Liss, New York.
Christy, A. Lawrence and Donald B. Fisher. 1978. Kinetics of $^{14}C$-photosynthate translocation in morning glory vines. Plant Physiol. 61:283–290.
Courdeau, Pascale, Jean-Louis Bonnemain, and Serge Delrot. 1986. The effect of auxin and of abscisic acid on the distribution of nutrients in the stem of the broadbean. Pages 597–598 *in* James Cronshaw, William J. Lucas, and Robert T. Giaquinta (eds.). Phloem Transport. Alan R. Liss, New York.
Crafts, A. S. 1961. Translocation in Plants. Holt, Rinehart & Winston, New York.
Crafts, A. S. and O. Lorenz. 1944. Fruit growth and food transport in cucurbits. Plant Physiol. 19: 131–138.
Cronshaw, J. 1981. Phloem structure and function. Ann. Rev. Plant Physiol. 32:465–484.
Cronshaw, James, William J. Lucas, and Robert T. Giaquinta. (Eds.). 1986. Phloem Transport. Alan R. Liss, New York.
De Vries, H. 1885. Über die Bedeutung der Circulation und der Rotation des Protoplasma für das Stofftransport in der Pflanze. (On the significance of the circulation and rotation of protoplasm for translocation in plants.) Botanische Zeitung 43:1–26.
Delrot, Serge. 1987. Phloem loading: Apoplastic or symplastic? Plant Physiol. Biochem. 25(5): 667–676.
Dewey, Steven A. and Arnold P. Appleby. 1983. A comparison between glyphosate and assimilate translocation patterns in tall morningglory (*Ipomoea purpurea*). Weed Science 31:308–314.
Esau, K. 1977. Anatomy of Seed Plants, Second Edition. Wiley, New York.
Eschrich, W. 1967. Beidirektionelle Translokation in Siebröhren. (Bidirectional translocation in sieve tubes.) Planta 73:37–49.
Eschrich, Walter. 1986. Mechanisms of phloem unloading. Pages 225–230 *in* James Cronshaw, William J. Lucas, and Robert T. Giaquinta (eds.), Phloem Transport. Alan R. Liss, New York.
Fahn, A. 1982. Plant Anatomy, Third Edition. Pergamon Press, Oxford, New York.
Fensom, D. S. 1972. A theory of translocation in phloem of *Heracleum* by contractile protein microfibrillar material. Canadian Journal of Botany 50:479–497.
Field, R. J. and A. J. Peel. 1971. The movement of growth regulators and herbicides into the sieve elements of willow. New Phytologist 70:997–1003.
Fisher, D. B. 1975. Structure of functional soybean sieve elements. Plant Physiol. 56:555–569.
Fondy, Bernadette R. and Donald R. Geiger. 1981. Regulation of export by integration of sink and source activity. What's New in Plant Physiology 12(9):33–36.
Foyer, Christine H. 1987. The basis for source-sink interaction in leaves. Plant Physiol. Biochem. 25(5):649–657.
Fritz, E., R. F. Evert, and W. Heyser. 1983. Microautoradiographic studies of phloem loading and transport in the leaf of *Zea mays* L. Planta 159: 193–206.
Geiger, Donald R. 1986. Processes affecting carbon allocation and partitioning among sinks. Pages 375–388 *in* James Cronshaw, William J. Lucas, and Robert T. Giaquinta (eds.), Phloem Transport. Alan R. Liss, New York.
Geiger, Donald R. and Bernadette R. Fondy. 1980. Phloem loading and unloading: Pathways and mechanisms. What's New in Plant Physiology 11(7):25–28.
Geiger, D. R., R. T. Giaquinta, S. A. Sovonick, and R. J. Fellows. 1973. Solute distribution in sugar beet leaves in relation to phloem loading and translocation. Plant Physiol. 52:585–589.
Geiger, D. R. and Wen-Jang Shieh. 1988. Analysing partitioning of recently fixed and of reserve carbon in reproductive *Phaseolus vulgaris* L. plants. Plant, Cell and Environment 11:777–783.
Geiger, D. R. and S. A. Sovonick. 1975. Effects of temperature, anoxia and other metabolic inhibitors on translocation. Volume 1, pages 480–504 *in* M. H. Zimmermann and J. A. Milburn (volume eds.), Transport in Plants. I. Phloem Transport. *In* A. Pirson and M. H. Zimmermann (series eds.), Encyclopedia of Plant Physiology, New Series. Springer-Verlag, Berlin.
Geiger, D. R., S. A. Sovonick, T. L. Shock, and R. J. Fellows. 1974. Role of free space in translocation in sugar beet. Plant Physiol. 54:892–898.
Giaquinta, R. 1976. Evidence for phloem loading from the apoplast: Chemical modification of membrane sulfhydryl groups. Plant Physiol. 57:872–875.
Giaquinta, R. T. 1983. Phloem loading of sucrose. Ann. Rev. of Plant Physiol. 34:347–387.
Giaquinta, R. T. and D. R. Geiger. 1972. Mechanisms of inhibition of translocation by localized chilling. Plant Physiol. 51:372–377.
Gifford, Roger M. 1986. Partitioning of photoassimilate in the development of crop yield. Pages 535–549 *in* James Cronshaw, William J. Lucas, and Robert T. Giaquinta (eds.), Phloem Transport. Alan R. Liss, New York.
Gifford, R. M. and L. T. Evans. 1981. Photosynthesis, carbon partitioning, and yield. Ann. Rev. Plant Physiol. 32:485–509.
Gifford, Roger M. and John H. Thorne. 1986. Phloem unloading in soybean seed coats: Dynamics and stability of efflux into attached "empty ovules." Plant Physiol. 80:464–469.
Gunning, Brian E. S. 1977. Transfer cells and their roles in transport of solutes in plants. Scientific Progress Oxf. 64:539–568.

Gunning, B. E. S., J. S. Pate, F. R. Minchin, and I. Marks. 1974. Quantitative aspects of transfer cell structure in relation to vein loading in leaves and solute transport in legume nodules. Pages 87–126 *in* M. A. Sleigh and D. H. Jennings (eds.), Transport at the Cellular Level. Cambridge University Press, London.

Hammel, H. T. 1968. Measurement of turgor pressure and its gradient in the phloem of oak. Plant Physiol. 43:1042–1048.

Hendrix, John E., J. C. Linden, D. H. Smith, C. W. Ross, and I. K. Park. 1986. Relationship of pre-anthesis fructan metabolism to grain numbers in winter wheat (*Triticum aestivum* L.). Aust. J. Plant Physiol. 13:391–398.

Heyser, W. 1980. Phloem loading in the maize leaf. Ber. Dtsch. Bot. Ges. 93:221–228.

Ho, Lim C. 1988. Metabolism and compartmentation of imported sugars in sink organs in relation to sink strength. Annual Review of Plant Physiology and Plant Molecular Biology 39:355–378.

Ho, L. C. and A. J. Peel. 1969. Investigation of bidirectional movement of tracers in sieve tubes of *Salix viminalis* L. Annals of Botany 33:833–844.

Housley, T. L. and D. B. Fisher. 1977. Estimation of osmotic gradients in soybean sieve tubes by quantitative microautoradiography. Qualified support for the Münch hypothesis. Plant Physiol. 59:701–706.

Hull, Roger. 1989. The movement of viruses in plants. Annual Review of Phytopathology 27: 213–240.

Jensen, William A. and Frank B. Salisbury. 1971. Botany: An Ecological Approach. Wadsworth, Belmont, Calif.

Kennedy, J. S. and T. E. Mittler. 1953. A method for obtaining phloem sap via the mouth-parts of aphids. Nature 171:528.

Kleier, Daniel A. 1988. Phloem mobility of xenobiotics. I. Mathematical model unifying the weak acid and intermediate permeability theories. Plant Physiol. 86:803–810.

Lang, Alexander, M. R. Thorpe, and W. R. N. Edwards. 1986. Plant water potential and translocation. Pages 193–194 *in* James Cronshaw, William J. Lucas, and Robert T. Giaquinta (eds.), Phloem Transport. Alan R. Liss, New York.

Läuchli, Andre. 1972. Translocation of inorganic solutes. Annual Review of Plant Physiol. 23: 197–218.

Lemoine, Remi, Jaleh Daie, and Roger Wyse. 1988. Evidence for the presence of a sucrose carrier in immature sugar beet tap roots. Plant Physiol. 86:575–580.

Lucas, W. J. and M. A. Madore. 1988. Recent advances in sugar transport. Pages 35–84 *in* Jack Preiss (ed.), The Biochemistry of Plants, Vol. 14. Academic Press, New York.

Madore, Monica A. and William J. Lucas. 1987. Control of photoassimilate movement in source-leaf tissues of *Ipomoea tricolor Cav.* Planta 171: 197–204.

Madore, Monica A., John W. Oross, and William J. Lucas. 1986. Symplastic transport in *Ipomoea tricolor* source leaves. Plant Physiol. 82:432–442.

Mauseth, James D. 1988. Plant Anatomy. Benjamin/ Cummings, Menlo Park, Calif.

McReady, C. C. 1966. Translocation of growth regulators. Annual Review of Plant Physiol. 17: 283–294.

Minchin, P. E. H. and J. H. Troughton. 1980. Quantitative interpretation of phloem translocation data. Plant Physiol. 31:191–215.

Münch, E. 1927. Versuche über den Saftkreislauf. (Experiments on the circulation of sap.) Ber. Deutsch. Bot. Ges. 45:340–356.

Münch, E. 1930. Die Stoffbewegungen in der Pflanze. (Translocation in Plants.) Fischer, Jena.

Oliveira, Cristina M. and C. Austen Priestley. 1988. Carbohydrate reserves in deciduous fruit trees. Horticultural Reviews 10:403–430.

Oparka, K. J. 1986. Phloem unloading in the potato tuber. Pathways and sites of ATPase. Protoplasma 131:201–210.

Passioura, J. B. and A. E. Ashford. 1974. Rapid translocation in the phloem of wheat roots. Aust. J. Plant Physiol. 1:521–527.

Pate, John S. 1975. Exchange of solutes between phloem and xylem and circulation in the whole plant. Volume 1, pages 451–473 *in* M. H. Zimmermann and J. A. Milburn (volume eds.), Transport in Plants. I. Phloem Transport. *In* A. Pirson and M. H. Zimmermann (series eds.), Encyclopedia of Plant Physiology, New Series. Springer-Verlag, Berlin.

Pate, J. S. 1980. Transport and partitioning of nitrogenous solutes. Annual Review of Plant Physiol. 31:313–340.

Pate, J. S. 1986. Xylem-to-phloem transfer—vital component of the nitrogen-partitioning system of a nodulated legume. Pages 445–462 *in* James Cronshaw, William J. Lucas, and Robert T. Giaquinta (eds.), Phloem Transport. Alan R. Liss, New York.

Pate, J. S., D. B. Layzell, and D. L. McNeil. 1979. Modeling the transport and utilization of C and N in a nodulated legume. Plant Physiol. 63:730–738.

Pate, J. S., P. J. Sharkey, and C. A. Atkins. 1977. Nutrition of a developing legume fruit. Functional economy in terms of carbon, nitrogen, and water. Plant Physiol. 59:506–510.

Patrick, John W. 1979. An assessment of auxin-promoted transport in decapitated stems and whole shoots of *Phaseolus vulgaris* L. Planta 146:107–112.

Patrick, J. W. 1983. Photosynthate unloading from seed coats of *Phaseolus vulgaris* L. General characteristics and facilitated transfer. Z. Pflanzenphysiol. 111:9–18.

Patrick, J. W. 1990. Sieve element unloading: Cellular pathway, mechanism and control. Physiologia Plantarum 78:298–308.

Peel, A. J. 1974. Transport of Nutrients in Plants. Wiley, New York.

Peretó, Juli G. and José P. Beltran. 1987. Hormone directed sucrose transport during fruit set induced by gibberellins in *Pisum sativum*. Physiol. Plantarum 69:356–360.

Peterson, C. A. and H. B. Currier. 1969. An investigation of bidirectional translocation in the phloem. Physiologia Plantarum 22:1238–1250.

Porter, Gregory A., Daniel P. Knievel, and Jack C. Shannon. 1985. Sugar efflux from maize (*Zea mays* L.) pedicel tissue. Plant Physiol. 77:524–531.

Reid, M. S. and R. L. Bieleski. 1974. Sugar changes during fruit ripening—whither sorbitol? Mechanisms of regulation of plant growth. Pages 823–830 *in* R. L. Bieleski, A. R. Ferguson, and M. M. Cresswell (eds.), Bulletin 12, The Royal Society of New Zealand, Wellington.

Roeckl, B. 1949. Nachweise eines Konzentrations-hubs zwischen Palisadenzellen und Siebröhren. (Proof for a concentration buildup between palisade cells and sieve tubes.) Planta 36:530–550.

Rogers, S. and A. J. Peel. 1975. Some evidence for the existence of turgor pressure gradients in the sieve tubes of willow. Planta 126:259–267.

Sauter, J. J., W. Iten, and M. H. Zimmermann. 1973. Studies of the release of sugar into the vessels of sugar maple (*Acer saccharum*). Canadian Journal of Botany 51:1–8.

Schmalstig, J. Gougler and Donald R. Geiger. 1985. Phloem unloading in developing leaves of sugar beet. I. Evidence for pathway through the symplast. Plant Physiol. 79:237–241.

Schmalstig, J. Gougler and Donald R. Geiger. 1987. Phloem unloading in developing leaves of sugar beet. II. Termination of phloem unloading. Plant Physiol. 83:49–52.

Shannon, Jack C., Gregory A. Porter, and Daniel P. Knievel. 1986. Phloem unloading and transfer of sugars into developing corn endosperm. Pages 265–277 *in* James Cronshaw, William J. Lucas, and Robert T. Giaquinta (eds.), Phloem Transport. Alan R. Liss, New York.

Sij, J. W. and C. A. Swanson. 1973. Effects of petiole anoxia on phloem transport in squash. Plant Physiol. 51:368–371.

Stitt, Mark. 1986. Regulation of photosynthetic sucrose synthesis: Integration, adaptation, and limits. Pages 331–347 *in* James Cronshaw, William J. Lucas, and Robert T. Giaquinta (eds.), Phloem Transport. Alan R. Liss, New York.

Thorne, John H. 1985. Phloem unloading of C and N assimilates in developing seeds. Annual Review of Plant Physiology 36:317–343.

Thorne, John H. 1986. Sieve tube unloading. Pages 211–224 *in* James Cronshaw, William J. Lucas, and Robert T. Giaquinta (eds.), Phloem Transport. Alan R. Liss, New York.

Thorne, John H. and Ross M. Rainbird. 1983. An *in vivo* technique for the study of phloem unloading in seed coats of developing soybean seeds. Plant Physiol. 72:268–271.

Troughton, J. H., J. Moorby, and B. G. Currie. 1974. Investigations of carbon transport in plants. I. The use of carbon-11 to estimate various parameters of the translocation process. J. Exp. Botany 25:684–694.

Turgeon, Robert. 1989. The sink-source transition in leaves. Ann. Rev. Plant Physiol. Plant Mol. Biol. 40:119–138.

Turgeon, Robert and Larry E. Wimmers. 1988. Different patterns of vein loading of exogenous [$^{14}$C] sucrose in leaves of *Pisum sativum* and *Coleus blumei*. Plant Physiol. 87:179–182.

Van Bel, Aart J. E. 1987. The apoplast concept of phloem loading has no universal validity. Plant Physiol. Biochem. 25(5):677–686.

Wardlaw, Ian F. and Lydia Eckhardt. 1987. Assimilate movement in *Lolium* and *Sorghum* leaves. IV. Photosynthetic responses to reduced translocation and leaf storage. Australian Journal of Plant Physiology 14:573–591.

Warmbrodt, Robert D. 1986. Solute concentrations in the phloem and associated vascular and ground tissues of the root of *Hordeum vulgare* L. Pages 435–444 *in* James Cronshaw, William J. Lucas, and Robert T. Giaquinta (eds.), Phloem Transport. Alan R. Liss, New York.

Watson, B. T. 1975. The influence of low temperature on the rate of translocation in the phloem of *Salix viminalis* L. Ann. Bot. 39:889–900.

Wiebe, H. H. 1962. Physiological response of plants to drought. Utah Science 23:70–71.

Wolswinkel, Pieter. 1985a. Phloem unloading and turgor-sensitive transport: Factors involved in sink control of assimilate partitioning. Physiol. Plant. 65:331–339.

Wolswinkel, P. 1985b. Effect of inhibitors on solute efflux from seed-coat halves and cotyledons of *Pisum sativum* L., after uptake from a bathing medium. The difference between sucrose and amino acids. J. Plant Physiol. 120:419–429.

Wolswinkel, P. and A. Ammerlaan. 1983. Phloem unloading in developing seeds of *Vicia faba* L. Planta 158:205–215.

Wright, J. P. and D. B. Fisher. 1980. Direct measurement of sieve tube turgor pressure using severed aphid stylets. Plant Physiol. 65:1133–1135.

Wyse, Roger E. 1986. Sinks as determinants of assimilate partitioning: Possible sites for regulation. Pages 197–209 *in* James Cronshaw, William J. Lucas, and Robert T. Giaquinta (eds.), Phloem Transport. Alan R. Liss, New York.

Ziegler, H. 1975. Nature of transported substances. Volume 1, pages 59–100 *in* M. H. Zimmermann and J. A. Milburn (volume eds.), Transport in Plants. I. Phloem Transport. *In* A. Pirson and M. H. Zimmermann (series eds.), Encyclopedia of Plant Physiology, New Series. Springer-Verlag, Berlin.

Zimmermann, M. H. 1961. Movement of organic substances in trees. Science 133:73–79.

Zimmermann, M. H. and H. Ziegler. 1975. Appendix III: List of sugars and sugar alcohols in sieve-tube exudates. Volume 1, pages 480–504 *in* M. H. Zimmermann and J. A. Milburn (volume eds.), Transport in Plants. I. Phloem Transport. *In* A. Pirson and M. H. Zimmermann (series eds.), Encyclopedia of Plant Physiology, New Series. Springer-Verlag, Berlin.

## *Chapter 9 Enzymes, Proteins, and Amino Acids*


Anfinsen, C. B. 1959. The Molecular Basis of Evolution. Wiley, New York.

Anfinsen, C. B. 1972. The formation and stabilization of protein structure. Biochemistry Journal 128:737–749.

Barondes, S. H. 1988. Bifunctional properties of lectins: Lectins redefined. Trends in Biochemical Sciences 13:480–482.

Bennett, W. S., Jr. and T. A. Steitz. 1980. Structure of a complex between yeast hexokinase A and glucose. II. Detailed comparisons of conformation and active site configuration with native hexokinase B monomer and dimer. Journal of Molecular Biology 140:211–230.

Budde, R. J. A. and R. Chollet. 1988. Regulation of enzyme activity in plants by reversible phosphorylation. Physiologia Plantarum 72: 435–439.

Burke, J. J., J. R. Mahan, and J. A. Hatfield. 1988. Crop-specific kinetic windows in relation to wheat and cotton biomass production. Agronomy Journal 80:553–556.

Burke, J. J., J. N. Siedow, and D. E. Moreland. 1982. Succinate dehydrogenase. A partial purification from mung bean hypocotyls and soybean cotyledons. Plant Physiology 70:1577–1581.

Cech, T. R. and B. L. Bass. 1986. Biological catalysis by RNA. Annual Review of Biochemistry 55: 599–629.

Dickerson, R. E. and I. Geis. 1969. The Structure and Action of Proteins. Benjamin, Menlo Park, Calif.

Doll, H. 1984. Nutritional aspects of cereal proteins and approaches to overcome their deficiencies. Philosophical Transactions of the Royal Society of London, Series B 304:373–380.

Duke, S. H., L. E. Schrader, and M. G. Miller. 1977. Low temperature effects on soybean (*Glycine max* L. Merr. ev. Wells) mitochondrial respiration and several dehydrogenases during imbibition and germination. Plant Physiology 60:716–722.

Ellis, R. J. 1990. The molecular chaperone concept. Seminars in Cell Biology 1:1–9.

Ellis, R. J. and S. M. Hemmingsen. 1989. Molecular chaperones: Proteins essential for the biogenesis of some macromolecular structures. Trends in Biochemical Sciences 14:339–342.

Engel, P. C. 1977. Enzyme Kinetics. Wiley, New York.

Gannon, M. N. 1986. Where are the asymptotes of Michaelis-Menten? A simple method to visualize and determine the maximum response. Trends in Biochemical Sciences 11:509–510.

Gontero, B., M. L. Cardenas, and J. Ricard. 1988. A functional five-enzyme complex of chloroplasts involved in the Calvin cycle. European Journal of Biochemistry 173:437–443.

Harpstead, D. D. 1971. High lysine corn. Scientific American 225(2):34–42.

Huber, S. C., T. Sugiyama, and T. Akazawa. 1986. Light modulation of maize leaf phosphoenolpyruvate carboxylase. Plant Physiology 82: 550–554.

Johnson, V. A. and C. L. Lay. 1974. Genetic improvement of plant protein. Journal of Agricultural and Food Chemistry 22:558–566.

Koshland, D. E., Jr. 1973. Protein shape and biological control. Scientific American 229:52–64.

Larkins, B. A. 1981. Seed storage proteins: Characterization and biosynthesis. Pages 449–489 *in* A. Marcus (ed.), The Biochemistry of Plants, Vol. 6, Proteins and Nucleic Acids. Academic Press, New York.

Minami, Y., S. Wakabayashi, S. Imoto, Y. Ohta, and H. Matsubara. 1985. Ferrodoxin from a liverwort, *Marchantia polymorpha*. Purification and amino acid sequence. Journal of Biochemistry 98: 649–655.

Naqui, A. 1986. Where are the asymptotes of Michaelis-Menten? Trends in Biochemical Sciences 11:64–65.

Orr, M. L. and B. K. Watt. 1957. Amino acid content of foods. United States Department of Agriculture Home Economics Research Report 41:1–82.

Patterson, B. D. and D. Graham. 1987. Temperature and metabolism. Pages 153–199 *in* D. D. Davies (ed.), The Biochemistry of Plants, Vol. 12, Physiology of Metabolism. Academic Press, New York.

Paulson, J. C. 1989. Glycoproteins: What are the sugar chains for? Trends in Biochemical Sciences 14:272–276.

Payne, P. I. and A. P. Rhodes. 1982. Cereal storage proteins: Structure and role in agriculture and food technology. Pages 346–369 *in* D. B. Boulter and B. Parthier (eds.), Structure, Physiology, and Biochemistry of Proteins. Encyclopedia of Plant Physiology, New Series, Vol. 14A, Springer-Verlag, Berlin.

Payne, P. I., L. M. Holt, E. A. Jackson, and C. N. Law. 1984. Wheat storage proteins: Their genetics and their potential for manipulation by plant breeding. Philosophical Transactions of the Royal Society of London, Series B 304:359–371.

Ranjeva, A. and A. Boudet. 1987. Phosphorylation of proteins in plants: Regulatory effects and potential involvement in stimulus/response coupling. Annual Review of Plant Physiology 38:73–93.

Ricard, J. 1980. Enzyme flexibility as a molecular basis for metabolic control. Pages 31–80 *in* D. D. Davies (ed.), The Biochemistry of Plants, Vol. 2, Metabolism and Respiration. Academic Press, New York.

Ricard, J. 1987. Enzyme regulation. Pages 69–105 *in* D. D. Davies (ed.), The Biochemistry of Plants, Vol. 11, Biochemistry of Metabolism. Academic Press, New York.

Ross, C. W. 1981. Biosynthesis of nucleotides. Pages 169–205 *in* A. Marcus (ed.), The Biochemistry of Plants, Vol. 6, Proteins and Nucleic Acids. Academic Press, New York.

Schmitter, J-M., J-P. Jacquot, F. de Lamotte-Guery, C. Beauvallet, S. Dutka, P. Gadal, and P. Decottignies. 1988. Purification, properties and complete amino acid sequence of the ferredoxin from a green alga, *Chlamydomonas reinhardtii*. European Journal of Biochemistry 172:405–412.

Sharon, N. and H. Lis. 1987. A century of lectin research (1888–1988). Trends in Biochemical Sciences 12:488–491.

Srere, P. A. 1987. Complexes of sequential metabolic enzymes. Annual Review of Biochemistry 56: 89–124.

Ting, I. P., I. Führ, R. Curry, and W. C. Zschoche. 1975. Malate dehydrogenase isozymes in plants: Preparation, properties, and biological significance. Pages 369–383 *in* C. L. Markert (ed.), Isozymes. II. Physiological Function. Academic Press, New York.

Wolfe, S. L. 1981. Biology of the Cell, Second Edition. Wadsworth, Belmont, Calif.


## *Chapter 10 Chloroplasts and Light*


Anderson, J. M. 1986. Photoregulation of the composition, function, and structure of thylakoid membranes. Annual Review of Plant Physiology 37:93–136.

Anderson, J. M. and B. Andersson. 1988. The dynamic photosynthetic membrane and regulation of solar energy conversion. Trends in Biochemical Sciences 13:351–355.

Andréasson, L.-E. and T. Vänngård. 1988. Electron transport in photosystems I and II. Annual Review of Plant Physiology and Plant Molecular Biology 39:379–411.

Arnon, D. I. 1984. The discovery of photosynthetic phosphorylation. Trends in Biochemical Sciences 9:258–262.

Arnon, D. I. 1988. The discovery of ferredoxin: The photosynthetic path. Trends in Biochemical Sciences 13:30–33.

Barber, J. 1987a. Photosynthetic reaction centres: A common link. Trends in Biochemical Sciences 12:321–326.

Barber, J. 1987b. Composition, organization, and dynamics of the thylakoid membrane in relation to its function. Pages 75–130 *in* M. D. Hatch and N. K. Boardman (eds.), The Biochemistry of Plants, Vol. 10, Photosynthesis. Academic Press, New York.

Chitnis, P. R. and J. P. Thornber. 1988. The major light-harvesting complex of photosystem II: Aspects of its molecular and cell biology. Photosynthesis Research 16:41–63.

Clark, J. B. and G. R. Lister. 1975. Photosynthetic action spectra of trees. I: Comparative photosynthetic action spectra of one deciduous and four coniferous tree species as related to photorespiration and pigment complements. Plant Physiology 55:401–406.

Ehleringer, J. R. and O. Björkman. 1977. Quantum yields for $CO_2$ uptake among $C_3$ and $C_4$ plants: Dependence on temperature, $CO_2$, and $O_2$ concentrations. Plant Physiology 59:86–90.

Ehleringer, J. R. and R. W. Pearcy. 1983. Variation in quantum yield for $CO_2$ uptake among $C_3$ and $C_4$ plants. Plant Physiology 73:555–559.

Evans, J. R. 1987. The dependence of quantum yield on wavelength and growth irradiance. Australian Journal of Plant Physiology 14:69–79.

Evans, M. C. W. and G. Bredenkamp. 1990. The structure and function of the photosystem I reaction centre. Physiologia Plantarum 79: 415–420.

Gest, H. 1988. Sun-beams, cucumbers, and purple bacteria. Photosynthesis Research 19:287–308.

Ghanotakis, D. F. and C. F. Yocum. 1990. Photosystem II and the oxygen-evolving complex. Annual Review of Plant Physiology and Plant Molecular Biology 41:255–276.

Glazer, A. N. and A. Melis. 1987. Photochemical reaction centers: Structure, organization, and function. Annual Review of Plant Physiology 38:11–45.

Govindjee and W. J. Coleman. 1990. How plants make oxygen. Scientific American 262(2):50–58.

Hatch, M. D. and N. K. Boardman. (Eds.). 1981. The Biochemistry of Plants, Vol. 8, Photosynthesis. Academic Press, New York.

Hill, R. and F. Bendall. 1960. Function of the two cytochrome components in chloroplasts: A working hypothesis. Nature 186:136–137.

Hope, A. B. and D. B. Matthews. 1988. Electron and proton transfers around the *b/f* complex in chloroplasts: Modelling the constraints on Q-cycle activity. Australian Journal of Plant Physiology 15:567–583.

Inada, K. 1976. Action spectra for photosynthesis in higher plants. Plant and Cell Physiology 17: 355–365.

Irrgang, K. D., E. J. Boekema, J. Vater, and G. Renger. 1988. Structural determination of the photosystem II core complex from spinach. European Journal of Biochemistry 178:209–217.

Jagendorf, A. T. 1967. The chemiosmotic hypothesis of photophosphorylation. Pages 69–78 *in* A. San Pietro, F. A. Greer, and T. J. Army (eds.), Harvesting the Sun, Photosynthesis in Plant Life. Academic Press, New York.

Kamen, M. D. 1989. Onward into a fabulous half-century. Photosynthesis Research 21:139–144.

Kirk, J. T. O. and R. A. E. Tilney-Bassett. 1978. The Plastids, Second Edition. Elsevier, Amsterdam.

Knaff, D. B. 1990. The cytochrome $bc_1$ complex of photosynthetic bacteria. Trends in Biochemical Sciences 15:289–291.

Lagoutte, B. and Mathis, P. 1989. The photosystem I reaction center: Structure and photochemistry. Photochemistry and Photobiology 49:833–844.

Malkin, R. 1988. Structure-function studies of photosynthetic cytochrome *b*-$c_1$ and $b_6$-*f* complexes. ISI Atlas of Science: Biochemistry 10: 57–64.

Marder, J. B. and J. Barber. 1989. The molecular anatomy and function of thylakoid proteins. Plant, Cell and Environment 12:595–614.

Mattoo, A. K., J. B. Marder, and M. Edelman. 1989. Dynamics of the photosystem II reaction center. Cell 56:241–246.

Mazur, B. J. and S. C. Falco. 1989. The development of herbicide resistant crops. Annual Review of Plant Physiology and Plant Molecular Biology 40:441–470.
McCree, K. J. 1972. The action spectrum, absorptance and quantum yield of photosynthesis in crop plants. Agricultural Meteorology 9:191–216.
Melis, A. and J. S. Brown. 1980. Stoichiometry of system I and system II reaction centers and of plastoquinone in different photosynthetic membranes. Proceedings of the National Academy of Sciences USA 77:4712–4716.
Mitchell, P. 1966. Chemical coupling in oxidative and photosynthetic phosphorylation. Biological Reviews 41:445–502.
Mitchell, P. 1985. The correlation of chemical and osmotic forces in biochemistry. Journal of Biochemistry 97:1–18.
Murphy, T. M. and W. F. Thompson. 1988. Molecular Plant Development. Prentice-Hall, Englewood Cliffs, N. J.
O'Keefe, D. P. 1988. Structure and function of the chloroplast *bf* complex. Photosynthesis Research 17:189–216.
Ort, D. R. 1986. Energy transduction in oxygenic photosynthesis: An overview of structure and mechanism. Pages 143–196 *in* L. A. Staehlin and C. J. Arntzen (eds.), Photosynthesis III, Photosynthetic Membranes and Light Harvesting Systems. Encyclopedia of Plant Physiology, New Series, Vol. 19. Springer-Verlag, Berlin.
Osborne, B. A. and M. K. Garrett. 1983. Quantum yields for $CO_2$ uptake in some diploid and tetraploid plant species. Plant, Cell and Environment 6:135–144.
Osborne, B. A. and R. J. Geider. 1987. The minimum photon requirements for photosynthesis. New Phytologist 106:631–644.
Pirt, S. J. 1986. The thermodynamic efficiency (quantum demand) and dynamics of photosynthetic growth. Tansley Review No. 4. New Phytologist 102:3–37.
Possingham, J. V. 1980. Plastid replication and development in the life cycle of higher plants. Annual Review of Plant Physiology 31:113–129.
Pschorn, R., W. Rukle, and A. Wild. 1988. Structure and function of $NADP^+$-oxidoreductase. Photosynthesis Research 17:217–229.
Reilly, P. and N. Nelson. 1988. Photosystem I complex. Photosynthesis Research 19:73–84.
Renger, G. 1988. The photosynthetic oxygen evolving complex: Functional mechanism and structural organization. ISI Atlas of Science: Biochemistry 10:41–47.
Rutherford, A. W. 1989. Photosystem II, the water-splitting enzyme. Trends in Biochemical Sciences 14:227–242.
Scheller, H. V. and B. L. Møller. 1990. Photosystem I polypeptides. Physiologia Plantarum 78:484–494.
Scheller, H. V., I. Svendsen, and B. L. Møller. 1989. Subunit composition of photosystem I and identification of center X as a (4Fe-4S) cluster. Journal of Biological Chemistry 264:6929–6934.
Schulz, A., F. Wengenmayer, and H. M. Goodman. 1990. Genetic engineering of herbicide resistance in higher plants. CRC Critical Reviews in Plant Sciences 9:1–15.
Siefermann-Harms, D. 1985. Carotenoids in photosynthesis. I. Location in photosynthetic membranes and light-harvesting function. Biochimica et Biophysica Acta 811:325–355.
Siefermann-Harms, D. 1987. The light-harvesting and protective functions of carotenoids in photosynthetic membranes. Physiologia Plantarum 69:561–568.
Smeekens, S., P. Weisbeek, and C. Robinson. 1990. Protein transport into and within chloroplasts. Trends in Biochemical Sciences 15:73–76.
Staehlin, L. A. and C. J. Arntzen. (Eds.). 1986. Photosynthesis III, Photosynthetic Membranes and Light Harvesting Systems. Encyclopedia of Plant Physiology, New Series, Vol. 19. Springer-Verlag, Berlin.
Steinback, K. E., S. Bonitz, C. J. Arntzen, and L. Bogorad. (Eds.). 1985. Molecular Biology of the Photosynthetic Apparatus. Cold Spring Harbor Laboratory, Cold Spring, New York.
Stemler, A. and R. Radmer. 1975. Source of photosynthetic oxygen in bicarbonate-stimulated Hill reaction. Science 190:457–458.
Trebst, A. and M. Avron. (Eds.). 1977. Plastoquinones in photosynthesis. Philosophical Transactions of the Royal Society of London, Series B 284:591–599.
Wellburn, A. R. 1987. Plastids. International Review of Cytology, Supplement 17:149–210.
Zscheile, F. P. and C. L. Comar. 1941. Influence of preparative procedure on the purity of chlorophyll components as shown by absorption spectra. Botanical Gazette 102:463–481.
Zscheile, F. P., J. W. White, Jr., B. W. Beadle, and J. R. Roach. 1942. The preparation and absorption spectra of five pure carotenoid pigments. Plant Physiology 17:331–346.

## *Chapter 11 Carbon Dioxide Fixation and Carbohydrate Synthesis*

Andreo, C. S., D. H. Gonzalez, and A. A. Iglesias. 1987. Higher plant phospho*enol*pyruvate carboxylase. Federation of European Biochemical Societies 213:1–8.
Andrews, T. J. and G. H. Lorimer. 1987. Rubisco: Structure, mechanisms, and prospects for improvement. Pages 131–218 *in* M. D. Hatch and N. K. Boardman (eds.), Photosynthesis. The Biochemistry of Plants, Vol. 10. Academic Press, New York.
ap Rees, T. 1987. Compartmentation of plant metabolism. Pages 87–115 *in* D. D. Davies (ed.), Physiology of Metabolism. The Biochemistry of Plants, Vol. 12. Academic Press, New York.
Avigad, G. 1982. Sucrose and other disaccharides. Pages 217–347 *in* F. A. Loewus and W. Tanner (eds.), Encyclopedia of Plant Physiology, New Series, Vol. 13A, Plant Carbohydrates I. Intracellular Carbohydrates. Springer-Verlag, Berlin.
Balsamo, R. A. and E. G. Uribe. 1988. Leaf anatomy and ultrastructure of the Crassulacean-acid-metabolism plant *Kalanchoe daigremontiana*. Planta 173:183–189.
Bancal, P., C. A. Henson, J. P. Gaudillère, and N. C. Carpita. 1991. Fructan chemical structure and sensitivity to exohydrolase. Carbohydrate Research. (In press.)
Banks, U. and D. D. Muir. 1980. Structure and chemistry of the starch granule. Pages 321–369 in J. Preiss (ed.), Carbohydrates: Structure and Function. The Biochemistry of Plants, Vol. 3. Academic Press, New York.
Bassham, J. A. 1965. Photosynthesis: The path of carbon. Pages 875–902 *in* J. Bonner and J. E. Varner (eds.), Plant Biochemistry, Second Edition. Academic Press, New York.
Bassham, J. A. 1979. The reductive pentose phosphate cycle and its regulation. Pages 9–30 *in* M. Gibbs and E. Latzko (eds.), Encyclopedia of Plant Physiology, New Series, Vol. 6, Photosynthesis II. Springer-Verlag, Berlin.
Beck, E. and P. Ziegler. 1989. Biosynthesis and degradation of starch in higher plants. Annual Review of Plant Physiology and Plant Molecular Biology 40:95–117.
Berry, J. A., G. H. Lorimer, J. Pierce, J. R. Seemann, J. Meek, and S. Freas. 1987. Isolation, identification, and synthesis of 2-carboxyarabinitol-1-phosphate, a diurnal regulator of ribulose-bisphosphate carboxylase activity. Proceedings of the National Academy of Sciences USA 84: 734–738.
Black, C. C., W. H. Campbell, T. M. Chen, and P. Dettrich. 1973. The monocotyledons: Their evolution and comparative biology. III. Pathways of carbon metabolism related to net carbon dioxide assimilation by monocotyledons. Quarterly Review of Biology 48:299–313.
Brown, R. H. and P. W. Hattersley. 1989. Leaf anatomy of $C_3$-$C_4$ species as related to evolution of $C_4$ photosynthesis. Plant Physiology 91: 1543–1550.
Buchanan, B. B. 1984. The ferredoxin/thioredoxin system: A key element in the regulatory function of light in photosynthesis. BioScience 34:378–383.
Burnell, J. N. and M. D. Hatch. 1985. Light-dark modulation of leaf pyruvate, P*i* dikinase. Trends in Biochemical Sciences 10:288–290.
Calvin, M. 1989. Forty years of photosynthesis and related activities. Photosynthesis Research 21:3016.
Campbell, W. H. and C. C. Black. 1982. Cellular aspects of $C_4$ photosynthesis: Mechanism of activation and inactivation of extracted pyruvate, inorganic phosphate dikinase in relation to dark/light regulation. Archives of Biochemistry and Biophysics 210:82–89.
Carpita, N., T. L. Housley, and J. E. Hendrix. 1991. New features of plant-fructan structure revealed by methylation analysis and carbon-13 N.M.R. spectroscopy. Carbohydrate Research (in press).
Carpita, N. C., J. Kanabus, and T. L. Housley. 1989. Linkage structure of fructans and fructan oligomers from *Triticum aestivum* and *Festuca arundinacea* leaves. Journal of Plant Physiology 134:162–168.
Chatterton, N. J., P. A. Harrison, J. H. Bennett, and K. H. Asay. 1989. Carbohydrate partitioning in 185 accessions of gramineae grown under warm and cool temperatures. Journal of Plant Physiology 134:169–179.
Crawford, N. A., M. Droux, N. S. Kosower, and B. B. Buchanan. 1989. Evidence for function of the ferredoxin/thioredoxin system in the reductive activation of target enzymes of isolated intact chloroplasts. Archives of Biochemistry and Biophysics 271:223–239.
Davies, D. D. (Ed.). 1987. Physiology of Metabolism. The Biochemistry of Plants, Vol. 12. Academic Press, New York.
Dey, P. M. and R. A. Dixon (Eds.). 1985. Biochemistry of Storage Carbohydrates in Green Plants. Academic Press, New York.
Downton, W. J. S. 1975. The occurrence of $C_4$ photosynthesis among plants. Photosynthetica 9: 96–105.
Edwards, G. E. and M. S. B. Ku. 1987. Biochemistry of $C_3$-$C_4$ intermediates. Pages 275–325 *in* M. D. Hatch and N. K. Boardman (eds.), Photosynthesis. The Biochemistry of Plants, Vol. 10. Academic Press, New York.
Edwards, G. E. and M. S. B. Ku. 1990. Regulation of the $C_4$ pathway of photosynthesis. Pages 175–190 *in* I. Zelitch (ed.), Perspectives in Biochemical and Genetic Regulation of Photosynthesis. Alan R. Liss, New York.
Edwards, G. E., H. Nakamoto, J. N. Burnell, and M. D. Hatch. 1985. Pyruvate, P*i* dikinase and NADP-malate dehydrogenase in $C_4$ photosynthesis: Properties and mechanism of light/dark regulation. Annual Review of Plant Physiology 36:255–286.
Ellis, T. J. 1979. The most abundant protein in the world. Trends in Biochemical Sciences 4:241–244.
Ford, D. M., P. P. Jablonski, A. H. Mohamed, and L. E. Anderson. 1987. Protein modulase appears to be a complex of ferredoxin, ferredoxin/thioredoxin reductase, and thioredoxin. Plant Physiology 83:628–632.
Forrester, M. L., G. Krotkov, and C. D. Nelson. 1966. Effect of oxygen on photosynthesis, photorespiration, and respiration in detached leaves. I. Soybean. Plant Physiology 41:422–427.
Frederick, S. E. and E. H. Newcomb. 1971. Ultrastructure and distribution of microbodies in leaves of grasses with and without $CO_2$-photorespiration. Planta 96:152–174.
Gerbaud, A. and M. Andre. 1987. An evaluation of the recycling in measurements of photorespiration. Plant Physiology 83:933–937.
Gibson, A. C. 1982. The anatomy of succulence. Pages 1–17 *in* I. P. Ting and M. Gibbs (eds.),

Crassulacean Acid Metabolism. Waverly Press, Baltimore.
Gutteridge, S. 1990. Limitations of the primary events of $CO_2$ fixation in photosynthetic organisms: The structure and mechanism of rubisco. Biochimica et Biophysica Acta 1015:1–14.
Gutteridge, S., M. A. J. Parry, S. Burton, A. J. Keys, A. Mudd, J. Feeney, J. C. Servaites, and J. Pierce. 1986. A nocturnal inhibitor of carboxylation in leaves. Nature 324:274–276.
Hanson, K. R. 1990. Regulation of starch and sucrose synthesis in tobacco species. Pages 69–84 *in* I. Zelitch (ed.), Perspectives in Biochemical and Genetic Regulation of Photosynthesis. Alan R. Liss, New York.
Hatch, M. D. 1977. $C_4$ pathway photosynthesis: Mechanism and physiological function. Trends in Biochemical Sciences 2:199–201.
Hatch, M. D. and N. K. Boardman (Eds.). 1987. Photosynthesis. The Biochemistry of Plants, Vol. 10. Academic Press, New York.
Hawker, J. S. 1985. Sucrose. Pages 48–51 *in* P. M. Dey and R. A. Dixon (eds.), Biochemistry of Storage Carbohydrates in Green Plants. Academic Press, New York.
Heldt, H. W. and U. I. Flügge. 1987. Subcellular transport of metabolites in plant cells. Pages 49–85 *in* D. D. Davies (ed.), Physiology of Metabolism. The Biochemistry of Plants, Vol. 12. Academic Press, New York.
Heldt, H. W., U. I. Flügge, S. Borchert, G. Bruckner, and J. Ohnishi. 1990. Phosphate translocators in plastids. Pages 39–54 *in* I. Zelitch (ed.), Perspectives in Biochemical and Genetic Regulation of Photosynthesis. Alan R. Liss, New York.
Hendrix, J. E., J. C. Linden, D. H. Smith, C. W. Ross, and I. K. Park. 1986. Relationship of pre-anthesis fructan metabolism to grain number in winter wheat (*Triticum aestivum* L.). Australian Journal of Plant Physiology 13:391–398.
Hendry, G. 1987. The ecological significance of fructan in a contemporary flora. New Phytologist 106:201–216.
Hesla, B. I., L. L. Tiezen, and S. K. Imbamba. 1982. A systematic survey of $C_3$ and $C_4$ photosynthesis in the *Cyperaceae* of Kenya, East Africa. Photosynthetica 16:196–205.
Holmgren, A. 1985. Thioredoxin. Annual Review of Biochemistry 54:237–271.
Huang, A. H. C., R. N. Trelease, and T. S. Moore, Jr. 1983. Plant Peroxisomes. Academic Press, New York.
Huber, S. C., J. A. Huber, and K. R. Hanson. 1990. Regulation of the partitioning of products of photosynthesis. Pages 85–101 *in* I. Zelitch (ed.), Perspectives in Biochemical and Genetic Regulation of Photosynthesis. Alan R. Liss, New York.
Husic, D. W., H. D. Husic, and N. E. Tolbert. 1987. The oxidative photosynthetic carbon cycle or $C_2$ cycle. CRC Critical Reviews in Plant Sciences 5:45–100.
Jenner, C. F. 1982. Storage of starch. Pages 700–747 *in* F. A. Loewus and W. Tanner (eds.), Encyclopedia of Plant Physiology, New Series, Vol. 13A, Plant Carbohydrates I. Intracellular Carbohydrates. Springer-Verlag, Berlin.
Kainuma, K. 1988. Structure and chemistry of the starch granule. Pages 141–180 *in* J. Preiss (ed.), Carbohydrates. The Biochemistry of Plants, Vol. 14. Academic Press, New York.
Keeley, J. E. 1988. Photosynthesis in quillworts, or why are some aquatic plants similar to cacti? Plants Today, July-Aug. 127–132.
Keeley, J. E. and B. A. Morton. 1982. Distribution of diurnal acid metabolism in submerged aquatic plants outside the genus *Isoetes*. Photosynthetica 16:546–553.
Kelly, G. J., J. A. M. Holtum, and E. Latzko. 1989. Photosynthesis. Carbon metabolism: New regulators of $CO_2$ fixation, the new importance of pyrophosphate, and the old problem of oxygen involvement revisited. Progress in Botany 50: 74–101.
Keys, A. J. 1990. Biochemistry of ribulose bisphosphate carboxylase. Pages 207–224 *in* I. Zelitch (ed.), Perspectives in Biochemical and Genetic Regulation of Photosynthesis. Alan R. Liss, New York.
Kluge, M. and I. P. Ting. 1978. Crassulacean Acid Metabolism: Analysis of an Ecological Adaptation. Springer-Verlag, Berlin.
Kortschak, H. P., C. E. Hartt, and G. O. Burr. 1965. Carbon dioxide fixation in sugarcane leaves. Plant Physiology 40:209–213.
Krenzer, E. G., Jr., D. N. Moss, and R. K. Crookston. 1975. Carbon dioxide compensation points of flowering plants. Plant Physiology 56:194–206.
Laetsch, W. M. 1974. The C-4 syndrome: A structural analysis. Annual Review of Plant Physiology 25:27–52.
Lea, P. J. and R. D. Blackwell. 1990. Genetic regulation of the photorespiratory pathway. Pages 301–318 *in* I. Zelitch (ed.), Perspectives in Biochemical and Genetic Regulation of Photosynthesis. Alan R. Liss, New York.
Loewus, F. A. and W. Tanner. (Eds.). 1982. Encyclopedia of Plant Physiology, New Series, Vol. 13A, Plant Carbohydrates I. Intracellular Carbohydrates. Springer-Verlag, Berlin.
Lüttge, U. 1987. Carbon dioxide and water demand: Crassulacean acid metabolism (CAM), a versatile ecological adaptation exemplifying the need for integration in ecophysiological work. New Phytologist 106:593–629.
Manners, D. J. 1985. Starch. Pages 149–203 *in* P. M. Dey and R. A. Dixon (eds.), Biochemistry of Storage Carbohydrates in Green Plants. Academic Press, New York.
Martin, C. E. and L. Kirchner. 1987. Lack of photosynthetic pathway flexibility in the CAM plant *Agave virginica* L. (Agavaceae). Photosynthetica 21:273–280.
Meier, H. and J. S. G. Reid. 1982. Reserve polysaccharides other than starch in higher plants. Pages 418–471 *in* F. A. Loewus and W. Tanner (eds.), Encyclopedia of Plant Physiology, New Series, Vol. 13A, Plant Carbohydrates I. Intracellular Carbohydrates. Springer-Verlag, Berlin.
Monson, R. K. and B. D. Moore. 1989. On the significance of $C_3$-$C_4$ intermediate photosynthesis to the evolution of $C_4$ photosynthesis. Plant, Cell and Environment 12:689–699.
Neales, T. F., A. A. Patterson, and V. J. Hartney. 1968. Physiological adaptation to drought in the carbon assimilation and water loss of xerophytes. Nature 219:469–472.
Ogren, W. L. 1984. Photorespiration: Pathways, regulation, and modification. Annual Review of Plant Physiology 35:415–442.
Oliver, D. J. and Y. Kim. 1990. Biochemistry and developmental biology of the C-2 cycle. Pages 253–269 *in* I. Zelitch (ed.), Perspectives in Biochemical and Genetic Regulation of Photosynthesis. Alan R. Liss, New York.
Pearcy, R. W. 1983. The light environment and growth of $C_3$ and $C_4$ tree species in the understory of a Hawaiian forest. Oecologia 58:19–25.
Pierce, J. 1988. Prospects for manipulating the substrate specificity of ribulose bisphosphate carboxylase/oxygenase. Physiologia Plantarum 72:690–698.
Pollock, C. J. and N. J. Chatterton. 1988. Fructans. Pages 109–140 *in* J. Preiss (ed.), Carbohydrates. The Biochemistry of Plants, Vol. 14. Academic Press, New York.
Pontis, H. G. and E. del Campillo. 1985. Fructans. Pages 205–228 *in* P. M. Dey and R. A. Dixon (eds.), Biochemistry of Storage Carbohydrates in Green Plants. Academic Press, New York.
Portis, A. R., Jr. 1990. Rubisco activase. Biochimica et Biophysica Acta 1015:15–28.
Preiss, J. (Ed.). 1980. Carbohydrates: Structure and Function. The Biochemistry of Plants, Vol. 3. Academic Press, New York.
Preiss, J. 1982a. Regulation of the biosynthesis and degradation of starch. Annual Review of Plant Physiology 33:431–454.
Preiss, J. 1982b. Biosynthesis of starch and its regulations. Pages 397–417 *in* F. A. Loewus and W. Tanner (eds.), Encyclopedia of Plant Physiology, New Series, Vol. 13A, Plant Carbohydrates I. Intracellular Carbohydrates. Springer-Verlag, Berlin.
Preiss, J. 1984. Starch, sucrose biosynthesis and partition of carbon in plants are regulated by ortho-phosphate and triose-phosphates. Trends in Biochemical Sciences 9:24–27.
Preiss, J. (Ed.). 1988. Carbohydrates. The Biochemistry of Plants, Vol. 14. Academic Press, New York.
Raven, J. A., L. L. Handley, J. J. MacFarlane, S. McInroy, L. McKenzie, J. H. Richard, and G. Samuelsson. 1988. Tansley Review No. 13: The role of $CO_2$ uptake by roots and CAM in acquisition of inorganic C by plants of the isoetid life-form: A review, with new data on *Eriocaulon decangulare* L. New Phytologist 108:125–248.
Salvucci, M. E. 1989. Regulation of rubisco activity in vivo. Physiologia Plantarum 77:164–171.
Salvucci, M. E., A. R. Portis, Jr., and W. L. Ogren. 1985. A soluble chloroplast protein catalyzes ribulosebisphosphate carboxylase/oxygenase activation in vivo. Photosynthesis Research 7: 193–201.
Scheibe, R. 1987. $NADP^+$-malate dehydrogenase in $C_3$-plants: Regulation and role of a light-activated enzyme. Physiologia Plantarum 71:393–400.
Seemann, J. R., J. A. Berry, S. M. Freas, and M. A. Krump. 1985. Regulation of ribulose bisphosphate carboxylase activity in vivo by a light-modulated inhibitor of catalysis. Proceedings of the National Academy of Sciences USA 82:8024–8028.
Servaites, J. C. 1990. Inhibition of ribulose 1,5-bisphosphate carboxylase/oxygenase by 2-carboxyarabinitol-1-phosphate. Plant Physiology 92:867–870.
Sharkey, T. D. 1988. Estimating the rate of photorespiration in leaves. Physiologia Plantarum 73: 147–152.
Sharkey, T. D. 1989. Evaluating the role of rubisco regulation in photosynthesis of $C_3$ plants. Philosophical Transactions of the Royal Society of London, Series B 323:435–448.
Shiomi, N. 1989. Properties of fructosyltransferases involved in the synthesis of fructan in liliaceous plants. Journal of Plant Physiology 134:151–155.
Somerville, C. R., A. R. Portis, Jr., and W. L. Ogren. 1982. A mutant of *Arabidopsis thaliana* which lacks activation of RuBP carboxylase *in vivo*. Plant Physiology 70:381–387.
Stiborová, M. 1988. Phosphoenolpyruvate carboxylase: The key enzyme of $C_4$-photosynthesis. Photosynthetica 22:240–263.
Stitt, M. 1990. Fructose-2,6-bisphosphate as a regulatory molecule in plants. Annual Review of Plant Physiology and Plant Molecular Biology 41:153–185.
Stitt, M. and W. P. Quick. 1989. Photosynthetic carbon partitioning: Its regulation and possibilities for manipulation. Physiologia Plantarum 77:633–641.
Stitt, M., S. C. Huber, and P. Kerr. 1987. Control of photosynthetic sucrose synthesis. Pages 327–409 *in* M. D. Hatch and N. R. Boardman (eds.), The Biochemistry of Plants, Vol. 10, Photosynthesis. Academic Press, New York.
Ting, I. P. 1985. Crassulacean acid metabolism. Annual Review of Plant Physiology 36:595–622.
Ting, I. P. and M. Gibbs. (Eds.). 1982. Crassulacean Acid Metabolism. Waverly Press, Baltimore.
Wagner, J. and W. Larcher. 1981. Dependence of $CO_2$ gas exchange and acid metabolism of alpine CAM plant *Sempervivum montanumon* on temperature and light. Oecologia 50:88–93.
Walker, J. L. and D. J. Oliver. 1986. Glycine decarboxylase multienzyme complex. Purification and partial characterization from pea leaf mitochondria. Journal of Biological Chemistry 261: 2214–2221.
Waller, S. S. and J. K. Lewis. 1979. Occurrence of $C_3$ and $C_4$ photosynthetic pathways in North American grasses. Journal of Range Management 32:12–28.

Winter, K. and J. H. Troughton. 1978. Photosynthetic pathways in plants of coastal and inland habitats of Israel and the Sinai. Flora 167: 1–34.

Woodrow, I. E. and J. A. Berry. 1988. Enzymatic regulation of photosynthetic $CO_2$ fixation in $C_3$ plants. Annual Review of Plant Physiology and Plant Molecular Biology 39:533–594.

## *Chapter 12 Photosynthesis: Environmental and Agricultural Aspects*

Allen, E. J. and R. K. Scott. 1980. An analysis of growth of the potato crop. Journal of Agricultural Science, Cambridge 94:583–606.

Austin, R. B., J. Bingham, R. D. Blackwell, L. T. Evans, M. A. Ford, C. L. Morgan, and M. Taylor. 1980. Genetic improvements in winter wheat yields since 1900 and associated physiological changes. Journal of Agricultural Science, Cambridge 94:675–689.

Azcón-Bieto, J. 1983. Inhibition of photosynthesis by carbohydrates in wheat leaves. Plant Physiology 73:681–686.

Baker, D. N. 1965. Effects of certain environmental factors on net assimilation in cotton. Crop Science 5:53–56.

Baker, D. N. and R. B. Musgrave. 1964. Photosynthesis under field conditions. V. Further plant chamber studies on the effects of light on corn (*Zea mays* L.). Crop Science 4:127–131.

Bauer, A. and P. Martha. 1981. The $CO_2$ compensation point of C-3 plants — a re-examination. I. Interspecific variability. Zeitschrift für Pflanzenphysiologie 103:445–450.

Berner, Robert A. and Antonio C. Lasaga. 1989. Modeling the geochemical carbon cycle. Scientific American 260(3):74–81.

Berry, J. A. 1975. Adaptation of photosynthetic processes to stress. Science 188:644–650.

Berry, J. A. and O.Björkman. 1980. Photosynthetic response and adaptation to temperature in higher plants. Annual Review of Plant Physiology 31:491–543.

Björkman, O. 1980. The response of photosynthesis to temperature. Pages 273–301 *in* J. Grace, E. D. Ford, and P. G. Jarvis (eds.), Plants and Their Atmospheric Environment. Blackwell Scientific Publications, Oxford.

Björkman, O. 1981. Responses to different quantum flux densities. Pages 57–107 *in* O. L. Lange, P. S. Nobel, C. B. Osmond, and H. Ziegler (eds.), Encyclopedia of Plant Physiology, New Series, Vol. 12A, Physiological Plant Ecology I. Springer-Verlag, Berlin.

Björkman, O. and P. Holmgren. 1963. Adaptability of the photosynthetic apparatus to light intensity in ecotypes from exposed and shaded habitats. Physiologia Plantarum 16:889–914.

Black, C. C. 1973. Photosynthetic carbon fixation in relation to net $CO_2$ uptake. Annual Review of Plant Physiology 24:253–286.

Boardman, N. K. 1977. Comparative photosynthesis of sun and shade plants. Annual Review of Plant Physiology 28:355–377.

Brown, R. H. 1978. A difference in N use efficiency in $C_3$ and $C_4$ plants and its implications in adaptation and evolution. Crop Science 18: 92–98.

Bugbee, B. G. and F. B. Salisbury. 1988. Exploring the limits of crop productivity. I. Photosynthetic efficiency of wheat in high irradiance environments. Plant Physiology 88:869–878.

Cardwell, V. B. 1982. Fifty years of Minnesota corn production: Sources of yield increase. Agronomy Journal 74:984–990.

Chang, J-H. 1981. Corn yield in relation to photoperiod, night temperature, and solar radiation. Agricultural Meteorology 24:253–262.

Clark, William C. (Ed.). 1982. Carbon Dioxide Review, 1982. Oxford University Press, New York.

Corré, N. C. 1983. Growth and morphogenesis of sun and shade plants. I. The influence of light intensity. Acta Botanica Neerlandica 32:49–62.

Demmig, Barbara and Olle Björkman. 1987. Comparison of the effect of excessive light on chlorophyll fluorescence (77 K) and photon yield of $O_2$ evolution in leaves of higher plants. Planta 171:171–184.

Edwards, G. E., S. B. Ku, and J. G. Foster. 1983. Physiological constraints to maximum yield potential. Pages 105–109 *in* T. Kommedahl and P. H. Williams (eds.), Challenging Problems in Plant Health. The American Phytopathological Society, St. Paul, Minn.

Ehleringer, J. and R. W. Pearcy. 1983. Variation in quantum yield for $CO_2$ uptake among $C_3$ and $C_4$ plants. Plant Physiology 73:555–559.

Emerson, R. and C. M. Lewis. 1941. Carbon dioxide exchange and the measurement of the quantum yield of photosynthesis. American Journal of Botany 28:789–804.

Enoch, H. Z. and Bruce A. Kimball. (Eds.). 1986. Carbon dioxide enrichment of greenhouse crops. Vol. 1: Status and $CO_2$ Sources. Vol. II: Physiology, Yield, and Economics. CRC Press, Boca Raton, Fla.

Evans, J. R. 1983. Nitrogen and photosynthesis in the flag leaf of wheat (*Triticum aestivum* L.). Plant Physiology 72:297–302.

Evans, L. T. 1980. The natural history of crop yield. American Scientist 68:388–397.

Gaastra, P. 1959. Photosynthesis of crop plants as influenced by light, carbon dioxide, temperature and stomatal diffusion resistance. Mededelinger van de Landbouwhogeschool Te Wagenigen 59: 1–68.

Gifford, R. M. and L. T. Evans. 1981. Photosynthesis, carbon partitioning, and yield. Annual Review of Plant Physiology 32:485–509.

Gifford, R. M., J. H. Thorne, W. D. Hitz, and R. T. Giaquinta. 1984. Crop productivity and photoassimilate partitioning. Science 225:801–808.

Good, N. E. and D. H. Bell. 1980. Photosynthesis, plant productivity, and crop yield. Pages 3–51 *in* P. S. Carlson (ed.), The Biology of Crop Productivity. Academic Press, New York.

Govindjee. (Ed.). 1983. Photosynthesis, Vol. II: Development, Carbon Metabolism and Plant Productivity. Academic Press, New York.

Hadley, N. F. and S. R. Szarek. 1981. Productivity of desert ecosystems. BioScience 331:747–753.

Hall, N. P. and A. J. Keys. 1983. Temperature dependence of the enzymic carboxylation and oxygenation of ribulose-1,5-bisphosphate in relation to effects of temperature on photosynthesis. Plant Physiology 72:945–948.

Hanover, J. W. 1980. Control of tree growth. BioScience 30:756–762.

Hanson, H. C. 1917. Leaf structure as related to environment. American Journal of Botany 4: 533–560.

Hargrove, T. R. and V. L. Cabanilla. 1979. The impact of semidwarf varieties on Asian rice-breeding programs. BioScience 29:731–735.

Herold, A. 1980. Regulation of photosynthesis by sink activity — the missing link. New Phytologist 86:131–144.

Hesketh, J. D. 1963. Limitations to photosynthesis responsible for differences among species. Crop Science 3:493–496.

Hicklenton, P. R. and P. A. Jolliffe. 1980. Alterations in the physiology of $CO_2$ exchange in tomato plants grown in $CO_2$-enriched atmosphere. Canadian Journal of Botany 58:2181–2189.

Ho, Lim C. 1988. Metabolism and compartmentation of imported sugars in sink organs in relation to sink strength. Annual Review of Plant Physiology and Plant Molecular Biology 39:355–378.

Jensen, William A. and Frank B. Salisbury. 1984. Botany, Second Edition. Wadsworth, Belmont, Calif.

Johnson, C. B. (Ed.). 1981. Physiological Processes Limiting Plant Productivity. Butterworths, London.

Körner, Christian and M. Diemer. 1987. *In situ* photosynthetic responses to light, temperature and carbon dioxide in herbaceous plants from low and high altitude. Functional Ecology 1:179–194.

Larcher, W. 1969. The effect of environmental and physiological variables on the carbon dioxide gas exchange of trees. Phytosynthetica 3:167–198.

Leopold, A. C. and P. E. Kriedemann. 1975. Plant Growth and Development. McGraw-Hill, New York.

Long, S. P. 1983. $C_4$ photosynthesis at low temperatures. (Commissioned view.) Plant, Cell and Environment 6:345–363.

Marzola, D. L. and D. P. Bartholomew. 1979. Photosynthetic pathway and biomass energy production. Science 205:555–559.

McClendon, J. H. and G. G. McMillen. 1982. The control of leaf morphology and the tolerance of shade by woody plants. Botanical Gazette 143: 79–83.

McCree, K. J. 1981. Photosynthetically active radiation. Pages 41–55 *in* O. L. Lange, P. S. Nobel, C. B. Osmond, and H. Ziegler (eds.), Encyclopedia of Plant Physiology, New Series, Vol. 12A, Physiological Plant Ecology I. Springer-Verlag, Berlin.

Monteith, John L. 1978. Reassessment of maximum growth rates for $C_3$ and $C_4$ crops. Experimental Agriculture 14:1–5.

Moss, D. N. and L. H. Smith. 1972. A simple classroom demonstration of differences in photosynthetic capacity among species. Journal of Agronomic Education 1:16–17.

Oquist, G. 1983. Effects of low temperature on photosynthesis. Plant, Cell and Environment 6:281–300.

Osborne, B. A. and M. K. Garrett. 1983. Quantum yields for $CO_2$ uptake in some diploid and tetraploid plant species. Plant, Cell and Environment 6:135–144.

Osmond, C. B., O. Björkman, and D. J. Anderson. (Eds.). 1980. Physiological Processes in Plant Ecology. Springer-Verlag, Berlin.

Pearcy, R. W. and J. Ehleringer. 1984. Comparative ecophysiology of $C_3$ and $C_4$ plants. (A review.) Plant, Cell and Environment 7:1–13.

Pearcy, Robert W., Olle Björkman, Martyn M. Caldwell, John E. Keeley, Russel K. Monson, and Boyd R. Strain. 1987. Carbon gain by plants in natural environments. BioScience 37:21–29.

Post, Wilfred M., Tsung-Hung Peng, William R. Emanuel, Anthony W. King, Virginia H. Dale, and Donald L. DeAngelis. 1990. The global carbon cycle. American Scientist 78:310–326.

Puckridge, D. W. 1968. Photosynthesis of wheat under field conditions. I. The interaction of photosynthetic organs. Australian Journal of Agricultural Research 19:711–719.

Radmer, R. and B. Kok. 1977. Photosynthesis: Limited yields, unlimited dreams. BioScience 27:599–605.

Revelle, R. 1982. Carbon dioxide and world climate. Scientific American 247(2):35–43.

Rycroft, M. J. 1982. Analysing atmospheric carbon dioxide levels. Nature 295:190–191.

Šesták, Z. 1981. Leaf ontogeny and photosynthesis. Pages 147–158 *in* C. B. Johnson (ed.), Physiological Processes Limiting Plant Productivity. Butterworths, London.

Somerville, C. R. and W. L. Ogren. 1982. Genetic modification of photorespiration. Trends in Biochemical Sciences 7:171–174.

Stuiver, M. 1978. Atmospheric carbon dioxide and carbon reservoir changes. Science 199:253–258.

Tans, Pieter P., Inez Y. Fung, and Taro Takahashi. 1990. Observational constraints on the global atmospheric $CO_2$ budget. Science 247:1431–1438.

Thomas, M. D. and G. R. Hill. 1949. Photosynthesis under field conditions. Pages 19–52 *in* J. Franck and W. E. Loomis (eds.), Photosynthesis in Plants. Iowa State University Press, Ames.

Wardlaw, I. F. 1980. Translocation and source-sink relationships. Pages 297–399 *in* P. S. Carlson (ed.), The Biology of Crop Productivity, Academic Press, New York.

Waterman, Lee S., Donald W. Nelson, Walter D. Komhyr, Tom B. Harris, Kurk W. Thoning, and Pieter P. Tans. 1989. Atmospheric $CO_2$ measurements at Cape Matatula, American Samoa, 1976–1987. Journal of Geophysical Research 94: 14817-14829.

Whittaker, R. H. 1975. Communities and Ecosystems, Second Edition. Macmillan, New York.

Wittwer, S. H. 1980. The shape of things to come. Pages 413–459 *in* P. S. Carlson (ed.), The Biology of Crop Productivity. Academic Press, New York.

Zelitch, I. 1982. The close relationship between net photosynthesis and crop yield. BioScience 32:796–802.

## *Chapter 13 Respiration*

Al-Ani, A., F. Bruzau, P. Raymond, V. Saint-Ges, J. M. Leblanc, and A. Pradet. 1985. Germination, respiration, and adenylate energy charge of seeds at various oxygen partial pressures. Plant Physiology 79:885–890.

Amthor, J. S. 1989. Respiration and Crop Productivity. Springer-Verlag, New York.

ap Rees, T. 1985. The organization of glycolysis and the oxidative pentose phosphate pathway in plants. Pages 390–417 *in* R. Douce and D. A. Day (eds.), Encyclopedia of Plant Physiology, New Series, Vol. 18, Higher Plant Cell Respiration. Springer-Verlag, Berlin.

ap Rees, T. 1987. Compartmentation of plant metabolism. Pages 87–115 *in* D. D. Davies (ed.), The Biochemistry of Plants, Vol. 12, Physiology of Metabolism. Academic Press, New York.

ap Rees, T. 1988. Hexose phosphate metabolism by nonphotosynthetic tissues of higher plants. Pages 1–33 *in* J. Preiss (ed.), The Biochemistry of Plants, Vol. 14. Academic Press, New York.

ap Rees, T. and J. E. Dancer. 1987. Fructose-2,6-bisphosphate and plant respiration. Planta 175:204–208.

Atkinson, D. E. 1968. The energy charge of the adenylate pool as a regulatory parameter. Interaction with feedback modifiers. Biochemistry 7: 4030–4034.

Atkinson, D. E. 1977. Cellular Energy Metabolism and Its Regulation. Academic Press, New York.

Avigad, G. 1982. Sucrose and other disaccharides. Pages 217–347 *in* F. A. Loewus and W. Tanner (eds.), Encyclopedia of Plant Physiology, New Series, Vol. 13A, Plant Carbohydrates I. Intracellular Carbohydrates. Springer-Verlag, Berlin.

Barclay, A. M. and R. M. M. Crawford. 1982. Plant growth and survival under strict anaerobiosis. Journal of Experimental Botany 33:541–549.

Beck, E. and P. Ziegler. 1989. Biosynthesis and degradation of starch in higher plants. Annual Review of Plant Physiology and Plant Molecular Biology 40:95–117.

Beevers, H. 1960. Respiratory Metabolism in Plants. Harper & Row, New York.

Biale, J. B. 1964. Growth, maturation, and senescence in fruits. Science 146:880–888.

Biale, J. B. and R. E. Young. 1981. Respiration and ripening of fruits — retrospect and prospect. Pages 1–39 *in* J. Friend and M. J. C. Rhodes (eds.), Advances in the Biochemistry of Fruits and Vegetables. Academic Press, London.

Black, C. C., L. Mustardy, S. S. Sung, P. P. Kormanik, D-P. Xu, and N. Paz. 1987. Regulation and roles for alternative pathways of hexose metabolism in plants. Physiologia Plantarum 69:387–394.

Bradford, K. J. and S. F. Yang. 1981. Physiological responses of plants to waterlogging. HortScience 16:25–30.

Brady, C. J. 1987. Fruit ripening. Annual Review of Plant Physiology 38:155–178.

Bryce, J. H., J. Azcon-Bieto, J. T. Wiskich, and D. A. Day. 1990. Adenylate control of respiration in plants: The contribution of rotenone-insensitive electron transport to ADP-limited oxygen consumption by soybean mitochondria. Physiologia Plantarum 78:105–111.

Budde, R. J. A. and R. Chollet. 1988. Regulation of enzyme activity in plants by reversible phosphorylation. Physiologia Plantarum 72: 435–439.

Budde, R. J. A. and D. D. Randall. 1990. Protein kinases in higher plants. Pages 351–367 *in* D. J. Moore, W. F. Boss, and F. A. Loewus (eds.), Inositol Metabolism in Plants, Wiley-Liss, New York.

Budde, R. J. A., T. K. Fang, and D. D. Randall. 1988. Regulation of the phosphorylation of mitochondrial pyruvate dehydrogenase complex *in situ*. Effects of respiratory substrates and calcium. Plant Physiology 88:1031–1036.

Cammack, R. 1987. $FADH_2$ as a "product" of the citric acid cycle. Trends in Biochemical Sciences 12:377.

Campbell, R. and M. C. Drew. 1983. Electron microscopy of gas space (aerenchyma) formation in adventitious roots of *Zea mays* L. subjected to oxygen shortage. Planta 157:350–357.

Carnal, N. W. and C. C. Black. 1983. Phosphofructokinase activities in photosynthetic organisms. The occurrence of pyrophosphate-dependent 6-phosphofructokinase in plants. Plant Physiology 71:150–155.

Chatterton, N. J., W. R. Thornley, P. A. Harrison, and J. H. Bennett. 1989. Fructosyltransferase and invertase activities in leaf extracts of six temperate grasses grown in warm and cool temperatures. Journal of Plant Physiology 135:301–305.

Cobb, B. G. and R. A. Kennedy. 1987. Distribution of alcohol dehydrogenase in roots and shoots of rice (*Oryza sativa*) and *Echinochloa* seedlings. Plant, Cell and Environment 10:633–638.

Copeland, L. and J. F. Turner. 1987. The regulation of glycolysis and the pentose phosphate pathway. Pages 107–128 *in* D. D. Davies (ed.), The Biochemistry of Plants, Vol. 11, Biochemistry of Metabolism. Academic Press, New York.

Cori, C. F. 1983. Embden and the glycolytic pathway. Trends in Biochemical Sciences 8:257–259.

Crawford, R. M. M. 1982. Physiological responses to flooding. Pages 453–477 *in* O. L. Lange, P. S. Nobel, C. B. Osmond, and H. Ziegler (eds.), Encyclopedia of Plant Physiology, New Series, Vol. 12B. Physiological Plant Ecology II. Water Relations and Carbon Assimilation. Springer-Verlag, Berlin.

Davies, D. D. (Ed.). 1980a. Metabolism and Respiration. The Biochemistry of Plants, Vol. 2, Metabolism and Respiration. Academic Press, New York.

Davies, D. D. 1980b. Anaerobic metabolism and the production of organic acids. Pages 581–611 *in* D. D. Davies (ed.), The Biochemistry of Plants, Vol. 2, Metabolism and Respiration. Academic Press, New York.

Davies, D. D. 1987a. Introduction: A history of the biochemistry of plant respiration. Pages 1–38 *in* D. D. Davies (ed.), The Biochemistry of Plants, Vol. 11. Academic Press, New York.

Davies, D. D. (Ed.). 1987b. Biochemistry of Metabolism. The Biochemistry of Plants, Vol. 11. Academic Press, New York.

Davies, D. D. (Ed.). 1987c. Physiology of Metabolism. The Biochemistry of Plants, Vol. 12. Academic Press, New York.

Day, D. A. and H. Lambers. 1983. The regulation of glycolysis and electron transport in roots. Physiologia Plantarum 58:155–160.

Dennis, D. T. and M. F. Greyson. 1987. Fructose-6-phosphate metabolism in plants. Physiologia Plantarum 69:395–404.

de Visser, R. 1987. On the integration of plant growth and respiration. Pages 331–340 *in* A. L. Moore and R. B. Beechey (eds.), Plant Mitochondria. Plenum Press, New York.

Douce, R. 1985. Mitochondria in Higher Plants. Academic Press, Orlando, Fla.

Douce, R. and D. A. Day. (Eds.). 1985. Higher Plant Cell Respiration. Encyclopedia of Plant Physiology, Vol. 18, New Series. Springer-Verlag, Berlin.

Douce, R. and M. Neuburger. 1989. The uniqueness of plant mitochondria. Annual Review of Plant Physiology and Plant Molecular Biology 40: 371–414.

Douce, R., R. Brouquisse, and E-P. Journet. 1987. Electron transfer and oxidative phosphorylation in plant mitochondria. Pages 177–211 *in* D. D. Davies (ed.), The Biochemistry of Plants, Vol. 11. Academic Press, New York.

Drew, M. C. 1979. Plant responses to anaerobic conditions in soil and solution culture. Current Advances in Plant Science 36:1–14.

Drew, M. C. 1988. Effects of flooding and oxygen deficiency on plant mineral nutrition. Advances in Plant Nutrition 3:115–159.

Dry, I. B. and J. T. Wiskich. 1985. Characteristics of glycine and malate oxidation by pea leaf mitochondria: Evidence of differential access to NAD and respiratory chains. Australian Journal of Plant Physiology 12:329–339.

Dry, I. B., J. H. Bruce, and J. T. Wiskich. 1987. Regulation of mitochondrial respiration. Pages 213–252 *in* D. D. Davies (ed.), The Biochemistry of Plants, Vol. 11. Academic Press, New York.

Elthon, T. E. and L. McIntosh. 1987. Identification of the alternative terminal oxidase of higher plant mitochondria. Proceedings of the National Academy of Science USA 84:8399–8403.

Elthon, T. E., R. L. Nickels, and L. McIntosh. 1989. Mitochondrial events during development of thermogenesis in *Sauromatum guttatum* (Schott). Planta 180:82–89.

Ericinska, M. and D. V. Wilson. 1982. Topical review. Regulation of cellular energy metabolism. Journal of Membrane Biology 70:1–14.

Farrar, J. F. 1985. The respiratory source of $CO_2$. Plant, Cell and Environment 8:427–438.

Geider, R. J. and B. A. Osborne. 1989. Respiration and microalgal growth: A review of the quantitative relationship between dark respiration and growth. New Phytologist 112:327–341.

Gill, C. J. 1970. The flooding tolerance of woody plants — a review. Forest Abstracts 31:671–678.

Goodwin, T. W. and E. I. Mercer. 1983. Introduction to Plant Biochemistry, Second Edition. Pergamon Press, Oxford.

Harris, N. and M. J. Chrispeels. 1980. The endoplasmic reticulum of mung-bean cotyledons: Quantitative morphology of cisternal and tubular ER during seedling growth. Planta 148:293–303.

Heldt, H. W. and U. I. Flügge. 1987. Subcellular transport of metabolites in plant cells. Pages 49–85 *in* D. D. Davies (ed.), The Biochemistry of Plants, Vol. 12. Academic Press, New York.

Hendry, G. 1987. The ecological significance of fructan in a contemporary flora. New Phytologist 106:201–216.

Henry, M. F. and E. J. Nyns. 1975. Cyanide-insensitive respiration. An alternative mitochondrial pathway. Sub-Cellular Biochemistry 4:1–65.

Huber, S. C. 1986. Fructose-2,6-bisphosphate as a regulatory metabolite in plants. Annual Review of Plant Physiology 37:165–186.

Jackson, M. B. 1985. Ethylene and responses of plants to soil waterlogging and submergence. Annual Review of Plant Physiology 36:145–174.

James, W. O. 1963. Plant Physiology, Sixth Edition. Oxford University Press, London.

Jenner, C. F. 1982. Storage of starch. Pages 700–747 *in* F. A. Loewus and W. Tanner (eds.), Encyclopedia of Plant Physiology, New Series, Vol. 13A, Plant Carbohydrates I: Intracellular Carbohydrates. Springer-Verlag, Berlin.

Jensen, W. A. and F. B. Salisbury. 1974. Botany: An Ecological Approach. Wadsworth, Belmont, Calif.

Johnson, I. R. 1990. Plant respiration in relation to growth, maintenance, ion uptake and nitrogen assimilation. Plant, Cell and Environment 13: 319–328.

Joly, C. A. and R. M. M. Crawford. 1982. Variation in tolerance and metabolic responses to flooding in some tropical trees. Journal of Experimental Botany 135:799–809.

Kawase, M. 1981. Anatomical and morphological adaptation of plants to waterlogging. HortScience 16:30–34.

Kennedy, R. A., S. C. H. Barrett, D. Vander Zee, and M. E. Rumpho. 1980. Germination and seedling growth under anaerobic conditions in *Echinochloa crus-galli* (barnyard grass). Plant, Cell and Environment 3:243–248.

Kidd, F., C. West, and G. E. Briggs. 1921. A quantitative analysis of the growth of *Helianthus annuus*. Part I. The respiration of the plant and of its parts throughout the life cycle. Proceedings of the Royal Society of London, Series B 92:368–384.

Kozlowski, T. T. 1984. Flooding and Plant Growth. Academic Press, New York.

Lambers, H. 1985. Respiration in intact plants and tissues: Its regulation and dependence on environmental factors, metabolism and invaded organisms. Pages 418–473 *in* R. Douce and D. A. Day (eds.), Encyclopedia of Plant Physiology, Vol. 18, New Series. Springer-Verlag, Berlin.

Lambers, H., R. K. Szaniawski, and R. de Visser. 1983. Respiration for growth, maintenance and ion uptake. An evaluation of concepts, methods, values and their significance. Physiologia Plantarum 58:556–563.

Lance, C., M. Chauveau, and P. Dizengremel. 1985. The cyanide-resistant pathway of plant mitochondria. Pages 202–247 *in* R. Douce and D. A. Day (eds.), Encyclopedia of Plant Physiology, Vol. 18, New Series. Springer-Verlag, Berlin.

Lipmann, F. 1975. Reminiscences of Embden's formulation of the Embden-Meyerhof cycle. Molecular and Cellular Biochemistry 6:171–175.

Malone, C., D. E. Koeppe, and R. J. Miller. 1974. Corn mitochondria swelling and contraction — an alternate interpretation. Plant Physiology 53: 918–927.

Mannella, C. A. 1985. The outer membrane of plant mitochondria. Pages 106–133 *in* R. Douce and D. A. Day (eds.), Encyclopedia of Plant Physiology, Vol. 18, New Series. Springer-Verlag, Berlin.

Manners, D. J. 1985. Starch. Pages 149–203 *in* P. M. Dey and R. A. Dixon (eds.), Biochemistry of Storage Carbohydrates in Green Plants. Academic Press, London.

Meeuse, B. J. D. and I. Raskin. 1988. Sexual reproduction in the arum lily family, with emphasis on thermogenicity. Sexual Plant Reproduction 1:3–15.

Meier, H. and J. S. G. Reid. 1982. Reserve polysaccharides other than starch in higher plants. Pages 418–471 *in* F. A. Loewus and W. Tanner (eds.), Encyclopedia of Plant Physiology, New Series, Vol. 13A, Plant Carbohydrates I: Intracellular Carbohydrates. Springer-Verlag, Berlin.

Miernyk, J. A., B. J. Rapp, N. R. David, and D. D. Randall. 1987. Higher plant mitochondrial pyruvate dehydrogenase complexes. Pages 189–197 *in* A. L. Moore and R. B. Beechey (eds.), Plant Mitochondria. Plenum Press, New York.

Mitchell, P. 1985. The correlation of chemical and osmotic forces in biochemistry. Journal of Biochemistry 97:1–18.

Møller, I. M. 1986. NADH dehydrogenases in plant mitochondria. Physiologia Plantarum 67:517–520.

Møller, I. M., A. Berczi, L. H. W. van der Plas, and H. Lambers. 1988. Measurement of the activity and capacity of alternative pathway in intact plant tissues: Identification of problems and possible solutions. Physiologia Plantarum 72:642–649.

Moore, A. L. and R. B. Beechey. (Eds.). 1987. Plant Mitochondria. Plenum Press, New York.

Moore, A. L. and P. R. Rich. 1985. Organization of the respiratory chain and oxidative phosphorylation. Pages 134–172 *in* R. Douce and D. A. Day (eds.), Encyclopedia of Plant Physiology, Vol. 18, New Series. Springer-Verlag, Berlin.

Morré, D. J., W. F. Boss, and F. A. Loewus. (Eds.). 1990. Inositol Metabolism in Plants. Wiley-Liss, New York.

Morrell, S., H. Greenway, and D. D. Davies. 1990. Regulation of pyruvate decarboxylase *in vitro* and *in vivo*. Journal of Experimental Botany 41: 131–139.

Nakamoto, H. and P. S. Young. 1990. Light activation of pyruvate, orthophosphate dikinase in maize mesophyll chloroplasts: A role for adenylate energy charge. Plant Cell Physiology 31:106.

Newton, K. J. 1988. Plant mitochondrial genomes: Organization, expression, and variation. Annual Review of Plant Physiology and Plant Molecular Biology 39:503–532.

O'Brien, T. P. and M. E. McCully. 1969. Plant Structure and Development. A Pictorial and Physiological Approach. Macmillan, London.

Pfanner, N. and W. Neupert. 1990. The mitochondrial protein import apparatus. Annual Review of Biochemistry 59:331–353.

Philipson, J. J. and M. P. Coutts. 1980. The tolerance of tree roots to waterlogging. IV. Oxygen transport in woody roots of sitka spruce and lodgepole pine. New Phytologist 85:489–494.

Pollock, C. J. and N. J. Chatterton. 1988. Fructans. Pages 109–140 *in* J. Preiss (ed.), The Biochemistry of Plants, Vol. 14. Academic Press, New York.

Pontis, H. G. and E. del Campillo. 1985. Fructans. Pages 205–228 *in* P. M. Dey and R. A. Dixon (eds.), Biochemistry of Storage Carbohydrates in Green Plants. Academic Press, New York.

Pradet, A. and P. Raymond. 1983. Adenine nucleotide ratios and adenylate energy charge in energy metabolism. Annual Review of Plant Physiology 34:199–244.

Preiss, J. (Ed.). 1988. Carbohydrates. The Biochemistry of Plants, Vol. 14. Academic Press, New York.

Prince, R. C. 1988. The proton pump of cytochrome oxidase. Trends in Biochemical Sciences 13: 159–160.

Raskin, I. and H. Kende. 1985. Mechanism of aeration in rice. Science 228:327–329.

Raymond, P., X. Gidrol, C. Salon, and A. Pradet. 1987. Control involving adenine and pyridine nucleotides. Pages 129–176 *in* D. D. Davies (ed.), The Biochemistry of Plants, Vol. 11. Academic Press, New York.

Romani, R. J., B. M. Hess, and C. A. Leslie. 1989. Salicylic acid inhibition of ethylene production by apple discs and other plant tissues. Journal of Plant Growth Regulation 8:63–69.

Sabularse, D. C. and R. L. Anderson. 1981. D-fructose-2,6-bisphosphate: A naturally occurring activator for inorganic pyrophosphate: D-fructose-6-phosphate phosphotransferase in plants. Biochemical and Biophysical Research Communications 103:848–854.

Senior, A. E. 1988. ATP synthesis by oxidative phosphorylation. Physiological Reviews 68: 177–231.

Siedow, J. N. 1990. Regulation of the cyanide-resistant respiratory pathway. Pages 355–366 *in* I. Zelitch (ed.), Perspectives in Biochemical and Genetic Regulation of Photosynthesis. Alan R. Liss, New York.

Siedow, J. N. and D. A. Berthold. 1986. The alternative oxidase: A cyanide-resistant respiratory pathway in higher plants. Physiologia Plantarum 66:569–573.

Siedow, J. N. and M. E. Musgrave. 1987. The significance of cyanide-resistant respiration to carbohydrate metabolism in higher plants. Pages 351–359 *in* A. L. Moore and R. B. Beechey (eds.), Plant Mitochondria. Plenum Press, New York.

Smith, H. (Ed.). 1977. The Molecular Biology of Plant Cells. University of California Press, Berkeley.

Steup, M. 1988. Starch degradation. Pages 255–296 *in* J. Preiss (ed.), The Biochemistry of Plants, Vol. 14. Academic Press, New York.

Steup, M., H. Robenek, and M. Melkonian. 1983. In-vitro degradation of starch granules isolated from spinach chloroplasts. Planta 158:428–436.

Stitt, M. 1990. Fructose-2,6-bisphosphate as a regulatory molecule in plants. Annual Review of Plant Physiology and Plant Molecular Biology 41:153–185.

Stitt, M. and M. Steup. 1985. Starch and sucrose degradation. Pages 347–390 *in* R. Douce and D. A. Day (eds.), Encyclopedia of Plant Physiology, Vol. 18, New Series. Springer-Verlag, Berlin.

Stommel, J. R. and P. W. Simon. 1990. Multiple forms of invertase from *Daucus carota* cell cultures. Phytochemistry 29:2087–2089.

Sung, S-J. S., D-P. Xu, C. M. Galloway, and C. C. Black, Jr. 1988. A reassessment of glycolysis and gluconeogenesis in higher plants. Physiologia Plantarum 72:650–654.

Tucker, G. A. and D. Grierson. 1987. Fruit ripening. Pages 265–318 *in* D. D. Davies (ed.), The Biochemistry of Plants, Vol. 11. Academic Press, New York.

Turner, T. F. and D. H. Turner. 1980. The regulation of glycolysis and the pentose phosphate pathway. Pages 279–316 *in* D. D. Davies (ed.), The Biochemistry of Plants, Vol. 2, Metabolism and Respiration. Academic Press, New York.

Weichmann, J. 1986. The effect of controlled-atmosphere storage on the sensory and nutritional quality of fruits and vegetables. Horticultural Reviews 8:101–127.

Wikström, M. 1984. Pumping of protons from the mitochondrial matrix by cytochrome oxidase. Nature 308:558–560.

Williams, J. H. H. and J. F. Farrar. 1990. Control of barley root respiration. Physiologia Plantarum 79:259–266.

Yamaya, T. and H. Matsumoto. 1985. Influence of $NH_4^+$ on the oxygen uptake of mitochondria isolated from corn and pea shoots. Soil Science and Plant Nutrition 31:513–520.


## *Chapter 14 Assimilation of Nitrogen and Sulfur*


Allen, O. N. and E. K. Allen. 1981. The Leguminosae. A Source Book of Characteristics, Uses and Nodulation. University of Wisconsin Press, Madison.

Andrews, M. 1986. The partitioning of nitrate assimilation between root and shoot of higher plants. Plant, Cell and Environment 9:511–519.

Appleby, C. A. 1984. Leghemoglobin and *Rhizobium* respiration. Annual Review of Plant Physiology 35:443–478.

Appleby, C. A., D. Bogusz, E. S. Dennis, and W. J. Peacock. 1988. A role for haemoglobin in all plant roots? Plant, Cell and Environment 11:359–367.

Aslam, M. and R. C. Huffaker. 1984. Dependency of nitrate reduction on soluble carbohydrates in primary leaves of barley under aerobic conditions. Plant Physiology 75:623–628.

Aslam, M. and R. C. Huffaker. 1989. Role of nitrate and nitrite in the induction of nitrite reductase in leaves of barley seedlings. Plant Physiology 91:1152–1156.

Becana, M. and C. Rodríguez-Barrueco. 1989. Protective mechanisms of nitrogenase against oxygen excess and partially-reduced oxygen intermediates. Physiologia Plantarum 75: 429–438.

Beevers, L. 1976. Nitrogen Metabolism in Plants. American Elsevier, New York.

Bell, E. A. 1980. Non-protein amino acids in plants. Pages 403–423 *in* E. A. Bell and B. V. Charlwood (eds.), Encyclopedia of Plant Physiology, New Series, Vol. 8, Secondary Plant Products. Springer-Verlag, Berlin.

Bell, E. A. 1981. The non-protein amino acids occurring in plants. Progress in Phytochemistry 7: 171–196.

Bergersen, F. J. and D. J. Goodchild. 1973. Aeration pathways in soybean root nodules. Australian Journal of Biological Science 26:729–740.

Blevins, D. G. 1989. An overview of nitrogen metabolism in higher plants. Pages 1–41 *in* J. E. Poulton, J. T. Romero, and E. E. Conn (eds.), Plant Nitrogen Metabolism. Plenum, New York.

Block, E. 1985. The chemistry of garlic and onions. Scientific American 252(3):114–119.

Bloom, A. J. 1988. Ammonium and nitrate as nitrogen sources for plant growth. ISI Atlas of Science: Animal and Plant Sciences 55–59.

Boddey, R. M. 1987. Methods for quantification of nitrogen fixation associated with Gramineae. CRC Critical Reviews in Plant Sciences 6:209–266.

Boddey, R. M. and J. Dobereiner. 1988. Nitrogen fixation associated with grasses and cereals: Recent results and perspectives for future research. Plant and Soil 108:53–62.

Bonas, U., K. Schmitz, and H. Bergmann. 1982. Phloem transport of sulfur in *Ricinus*. Planta 155:82–88.

Bond, G. 1963. *In* P. S. Nutman and B. Mosse (eds.), Symbiotic Associations, Thirteenth Symposium of the Society for General Microbiology. Cambridge University Press, Cambridge.

Bothe, H. 1987. Metabolism of inorganic nitrogen compounds. Progress in Botany 49:103–116.

Bowsher, C. G., D. P. Hucklesby, and M. J. Emes. 1989. Nitrite reduction and carbohydrate metabolism in plastids purified from roots of *Pisum sativum* L. Planta 177:359–366.

Bray, C. M. 1983. Nitrogen Metabolism in Plants. Longman, London.

Brunold, C. and M. Suter. 1989. Localizations of enzymes of assimilatory sulfate reduction in pea roots. Planta 179:228–234.

Campbell, W. H. 1988. Higher plant nitrate reductase: Arriving at a molecular view. Pages 1–15 *in* D. H. Randall, D. G. Blevins, and W. H. Campbell (eds.), Current Topics in Plant Biochemistry and Physiology, Vol. 7. University of Missouri, Columbia.

Côté, B., C. S. Vogel, and J. O. Dawson. 1989. Autumnal changes in tissue nitrogen of autumn olive, black alder and eastern cottonwood. Plant and Soil 118:23–32.

Curl, E. A. and B. Truelove. 1986. The Rhizosphere. Springer-Verlag, Berlin.

Dakora, F. D. and C. A. Atkins. 1989. Diffusion of oxygen in relation to structure and function in legume root nodules. Australian Journal of Plant Physiology 16:131–140.

Dalling, M. J. (Ed.). 1986. Plant Proteolytic Enzymes. Vols. I and II. CRC Press, Boca Raton, Fla.

Davies, D. D. 1986. The fine control of cytosolic *p*H. Physiologia Plantarum 67:702–706.

Dawson, J. O. 1986. Actinorhizal plants: Their use in forestry and agriculture. Outlook on Agriculture 15:202–207.

Day, D. A., G. D. Price, and M. K. Udvardi. 1989. Membrane interface of the *Bradyrhizobium japonicum-Glycine max* symbiosis: Peribacteroid units from soybean nodules. Australian Journal of Plant Physiology 16:69–84.

Dixon, R. O. D. and C. T. Wheeler. 1986. Nitrogen Fixation in Plants. Chapman and Hall, New York.

Djordjevie, M. A., D. W. Gabriel, and B. G. Rolfe. 1987. Rhizobium — the refined parasite of legumes. Annual Review of Phytopathology 25:145–168.

Downie, J. A. and A. W. B. Johnston. 1988. Nodulation of legumes by *Rhizobium*. Plant, Cell and Environment 11:403–412.

Duxbury, J. M., D. R. Bouldin, R. E. Terry, and R. L. Tate III. 1982. Emissions of nitrous oxide from soils. Nature 298:462–464.

Eisbrenner, G. and H. J. Evans. 1983. Aspects of hydrogen metabolism in nitrogen-fixing legumes and other plant-microbe associations. Annual Review of Plant Physiology 34:105–136.

Emmerling, A. 1880. Ladw. Versuchsshtat 24:113.

Farquhar, G. D., P. M. Firth, R. Wetselaar, and B. Weir. 1980. On the gaseous exchange of ammonia between leaves and the environment: Determination of the ammonia compensation point. Plant Physiology 66:710–714.

Giovanelli, J. 1980. Aminotransferases in higher plants. Pages 329–358 *in* B. J. Miflin (ed.), The Biochemistry of Plants, Vol. 5. Academic Press, New York.

Giovanelli, J., S. H. Mudd, and A. H. Datko. 1980. Sulfur amino acids in plants. Pages 454–506 *in* B. J. Miflin (ed.), The Biochemistry of Plants, Vol. 5. Academic Press, New York.

Givan, C. V. 1979. Metabolic detoxification of ammonia in tissues of higher plants. Phytochemistry 18:375–382.

Givan, C. V., K. W. Joy, and L. A. Kleczkowski. 1988. A decade of photorespiratory nitrogen cycling. Trends in Biochemical Sciences 13:433–437.

Granstedt, R. C. and R. C. Huffaker. 1982. Identification of the leaf vacuole as a major nitrate storage pool. Plant Physiology 70:410–413.

Grundon, N. J. and C. J. Asher. 1988. Volatile losses of sulfur from intact plants. Journal of Plant Nutrition 11:563–576.

Haaker, H. 1988. Biochemistry and physiology of nitrogen fixation. BioEssays 9:112–117.

Haaker, H. and J. Klugkist. 1987. The bioenergetics of electron transport to nitrogenase. FEMS Microbiology Reviews 46:57–71.

Haynes, R. J. and K. M. Goh. 1978. Ammonium and nitrate nutrition of plants. Biological Reviews 53:465–510.

Huber, T. A. and J. G. Streeter. 1985. Purification and properties of asparagine synthetase from soybean root nodules. Plant Science 42:9–17.

Huffaker, R. C. 1982. Biochemistry and physiology of leaf proteins. Pages 370–400 *in* D. C. Boulter and B. Parthier (eds.), Encyclopedia of Plant Physiology, New Series, Vol. 14A, Nucleic Acids and Proteins, Part 1. Springer-Verlag, Berlin.

Joy, K. W. 1988. Ammonia, glutamine, and asparagine: A carbon-nitrogen interface. Canadian Journal of Botany 66:2103–2109.

Kato, T. 1986. Nitrogen metabolism and utilization in citrus. Horticultural Reviews 8:181–216.

Keys, A. J., Bird, I. F., and M. J. Cornelius. 1978. Photorespiratory nitrogen cycle. Nature 275: 741–743.

Kondorosi, E. and A. Kondorosi. 1986. Nodule induction on plant roots by *Rhizobium*. Trends in Biochemical Sciences 11:296–298.

Lea, P. J. and K. W. Joy. 1983. Amino acid interconversion in germinating seeds. Recent Advances in Phytochemistry 17:12–35.

Le Gal, M. F. and L. Rey. 1986. The reserve proteins in the cells of mature cotyledons of *Lupinus albus* var. Lucky. I. Quantitative ultrastructural study of protein bodies. Protoplasma 130:120–127.

Lima, E., R. M. Boddey, and J. Dobereiner. 1987. Quantification of biological nitrogen fixation associated with sugar cane using a $^{15}N$ aided nitrogen balance. Soil Biol. Biochem. 19:165–170.

Lott, J. N. A. 1980. Protein bodies. Pages 589–623 *in* N. E. Tolbert (ed.), The Biochemistry of Plants, Vol. 1, The Plant Cell. Academic Press, New York.

Maga, J. A. 1982. Phytate: Its chemistry, occurrence, food interactions, nutritional significance, and methods of analysis. Journal of Agricultural and Food Chemistry 30:1–9.

Martinez, E., D. Romero, and R. Palacios. 1990. The *Rhizobium* genome. CRC Critical Reviews in Plant Sciences 9:59–93.

Maxwell, C. A., U. A. Hartwig, C. M. Joseph, and D. A. Phillips. 1989. A chalcone and two related flavonoids released from alfalfa roots induce nod genes of *Rhizobium meliloti*. Plant Physiology 91:842–847.

Miflin, B. J. (Ed.). 1980. The Biochemistry of Plants: Amino Acids and Derivatives, Vol. 5. Academic Press, New York.

Millard, P. 1988. The accumulation and storage of nitrogen by herbaceous plants. Plant, Cell and Environment 11:1–8.

Millard, P. and C. M. Thomson. 1989. The effect of the autumn senescence of leaves on the internal cycling of nitrogen for the spring growth of apple trees. Journal of Experimental Botany. 40: 1285–1289.

Neves, M. C. P. and M. Hungria. 1987. The physiology of nitrogen fixation in tropical grain legumes. CRC Critical Reviews in Plant Sciences 6:267–321.

Nielsen, S. S. 1988. Degradation of bean proteins by endogenous and exogenous proteases — a review. Cereal Chemistry 65:435–442.

Oaks, A. and B. Hirel. 1985. Nitrogen metabolism in roots. Annual Review of Plant Physiology 36: 345–366.

Oberleas, D. 1983. Phytate content in cereals and legumes and methods of determination. Cereal Foods World 28:352–357.

Pate, J. S. 1973. Uptake, assimilation and transport of nitrogen compounds by plants. Soil Biology and Biochemistry 5:109–119.

Pate, J. S. 1980. Transport and partitioning of nitrogenous solutes. Annual Review of Plant Physiology 31:313–340.

Pate, J. S. 1989. Synthesis, transport, and utilization of products of symbiotic nitrogen fixation. Pages 65–115 *in* J. E. Poulton, J. T. Romero, and E. E. Conn (eds.), Plant Nitrogen Metabolism. Plenum, New York.

Pernollet, J. 1978. Protein bodies of seeds: Ultrastructure, biochemistry, biosynthesis and degradation. Phytochemistry 17:1473–1480.

Peters, G. A. 1978. Blue-green algae and algal associations. BioScience 28:580–585.

Peters, G. A. and J. C. Meeks. 1989. The azolla-anabaena symbiosis: Basic biology. Annual Review of Plant Physiology and Plant Molecular Biology 40:193–210.

Powell, R. and F. Gannon. 1988. The leghaemoglobins. BioEssays 9:117–118.

Quispel, A. 1988. Bacteria-plant interactions in symbiotic nitrogen fixation. Physiologia Plantarum 74:783–790.

Raboy, V. 1990. Biochemistry and genetics of phytic acid synthesis. Pages 55–76 *in* D. J. Morré, W. F. Boss, and F. A. Loewus (eds.), Inositol Metabolism in Plants. Wiley-Liss, New York.

Rajasekhar, V. K. and R. Oelmüller. 1987. Regulation of induction of nitrate reductase and nitrite reductase in higher plants. Physiologia Plantarum 71:517–521.

Rajasekhar, V. K., G. Gowri, and W. H. Campbell. 1988. Phytochrome-mediated light regulation of nitrate reductase expression in squash cotyledons. Plant Physiology 88:242–244.

Raven, J. A. 1988. Acquisition of nitrogen by the shoots of land plants: Its occurrence and implications for acid-base regulation. New Phytologist 109:1–20.

Raven, J. A. and F. A. Smith. 1976. Nitrogen assimilation and transport in vascular land plants in relation to intracellular *p*H regulation. New Phytologist 72:415–431.

Rennenberg, H. 1984. The fate of excess sulfur in higher plants. Annual Review of Plant Physiology 35:121–153.

Rhodes, D., D. G. Brunk, and J. R. Magalhaes. 1989. Assimilation of ammonia by glutamate dehydrogenase? Pages 191–226 *in* J. E. Poulton, J. T. Romeo, and E. E. Conn (eds.), Recent Advances in Phytochemistry, Vol. 23, Plant Nitrogen Metabolism. Plenum, New York.

Rice, E. L. 1974. Allelopathy. Academic Press, New York.

Robertson, J. G. and K. J. F. Farnden. 1980. Ultrastructure and metabolism of the developing legume root nodule. Pages 65–115 *in* B. J. Miflin (ed.), The Biochemistry of Plants, Vol. 5. Academic Press, New York.

Rolfe, B. G. and P. M. Gresshoff. 1988. Genetic analysis of legume nodule initiation. Annual Review of Plant Physiology and Plant Molecular Biology 39:297–319.

Runge, M. 1983. Physiology and ecology of nitrogen nutrition. Pages 163–200 *in* O. L. Lange, P. S. Nobel, C. B. Osmond, and H. Ziegler (eds.), Physiological Plant Ecology III, Responses to the Chemical and Biological Environment, Encyclopedia of Plant Physiology, New Series, Vol. 12C. Springer-Verlag, Berlin.

Schank, S. C., R. L. Smith, and R. C. Littell. 1983. Establishment of associative $N_2$-fixing systems. Soil and Crop Science Society of Florida Proceedings 43:113–117.

Schiff, J. A. 1983. Reduction and other metabolic reactions of sulfate. Pages 401–421 *in* A. Läuchli and R. L. Bieleski (eds.), Encyclopedia of Plant Physiology, New Series, Vol. 15A. Springer-Verlag, New York.

Schmidt, A. 1979. Photosynthetic assimilation of sulfur compounds. Pages 481–496 *in* M. Gibbs and E. Latzko (eds.), Encyclopedia of Plant Physiology, New Series, Vol. 6. Springer-Verlag, New York.
Schubert, K. R. 1986. Products of biological nitrogen fixation in higher plants: Synthesis, transport, and metabolism. Annual Review of Plant Physiology 37:539–574.
Schwenn, J. D. 1989. Sulphate assimilation in higher plants: A thioredoxin-dependent PAPS-reductase from spinach leaves. Zeitschrift für Naturforschung 44c:504–508.
Sheehy, J. E. 1987. Photosynthesis and nitrogen fixation in legume plants. CRC Critical Reviews in Plant Sciences 5:121–158.
Sieciechowicz, K. A., K. W. Joy, and R. J. Ireland. 1988. The metabolism of asparagine in plants. Phytochemistry 27:663–671.
Simpson, R. J., H. Hamberg, and M. J. Dalling. 1982. Translocation of nitrogen in a vegetative wheat plant (*Triticum aestivum*). Physiologia Plantarum 56:11–17.
Solomonson, L. P. and M. J. Barber. 1990. Assimilatory nitrate reductase: Functional properties and regulation. Annual Review of Plant Physiology and Plant Molecular Biology 41:225–253.
Stam, H., H. Stouthamer, and H. W. van Verseveld. 1987. Hydrogen metabolism and energy costs of nitrogen fixation. FEMS Microbiology Reviews 46:73–92.
Stewart, W. D. 1982. Nitrogen fixation — its current relevance and future potential. Israel J. Botany 31:5–34.
Streeter, J. 1988. Inhibition of legume nodule formation and $N_2$ fixation by nitrate. CRC Critical Reviews in Plant Sciences 7:1–23.
Ta, T. C. and M. A. Faris. 1987. Species variation in the fixation and transfer of nitrogen from legumes to associated grasses. Plant and Soil 98:265–274.
Ta, T. C. and K. W. Joy. 1985. Transamination, deamidation and the utilisation of asparagine amino nitrogen in pea leaves. Canadian Journal of Botany 63:881–884.
Titus, J. S. and S. Kang. 1982. Nitrogen metabolism, translocation, and recycling in apple trees. Horticultural Reviews 4:204–246.
Tjepkema, J. D., C. R. Schwintzer, and D. R. Benson. 1986. Physiology of actinorhizal nodules. Annual Review of Plant Physiology 37:209–275.
Van Berkum, P. and B. B. Bohlool. 1980. Evaluation of nitrogen fixation by bacteria in association with roots of tropical grasses. Microbiological Reviews 44:491–517.
Van Beusichem, M. L., J. A. Nelemans, and M. G. J. Hinnen. 1987. Nitrogen cycling in plant species differing in shoot/root reduction of nitrate. Journal of Plant Nutrition 10:1723–1731.
Vance, C. P. and G. H. Heichel. 1981. Nitrate assimilation during vegetative regrowth of alfalfa. Plant Physiology 68:1052–1056.
Walker, J. L. and D. J. Oliver. 1986. Glycine decarboxylase multienzyme complex. Purification and partial characterization from pea leaf mitochondria. Journal of Biological Chemistry 261: 2214–2221.
Walsh, K. B., M. E. McCully, and M. J. Canny. 1989. Vascular transport and soybean nodule function: Nodule xylem is a blind alley, not a throughway. Plant, Cell and Environment 12:395–405.
Weber, E. and D. Neumann. 1980. Protein bodies, storage organelles in plant seeds. Biochemie und Physiologie der Pflanzen 175:279–306.
Wetselaar, R. and G. D. Farquhar. 1980. Nitrogen losses from tops of plants. Advances in Agronomy 33:263–302.
Wetzel, S., C. Demmers, and J. S. Greenwood. 1989. Seasonally fluctuating bark proteins are a potential form of nitrogen storage in three temperate hardwoods. Planta 178:275–281.
Winkler, R. G., D. G. Blevins, J. C. Polacco, and D. D. Randall. 1988. Ureide catabolism in nitrogen fixing legumes. Trends in Biochemical Sciences 13:97–100.
Yamaya, T. and A. Oaks. 1987. Synthesis of glutamate by mitochondria — an anaplerotic function for glutamate dehydrogenase. Physiologia Plantarum 70:749–756.

## *Chapter 15 Lipids and Other Natural Products*

Asen, S., R. N. Stewart, and K. E. Norris. 1975. Anthocyanin, flavonol copigments, and *p*H responsible for larkspur flower color. Phytochemistry 14:2677–2682.
Bailey, J. A. and J. W. Mansfield. (Eds.). 1982. Phytoalexins. Wiley, New York.
Barbour, M. G., J. H. Burk, and W. D. Pitts. 1987. Terrestrial Plant Ecology, Second Edition. Benjamin/Cummings, Menlo Park, Calif.
Beale, M. H. and J. Macmillan. 1988. The biosynthesis of $C_5$-$C_{20}$ terpenoid compounds. Natural Product Reports 5:247–264.
Beevers, H. 1979. Microbodies in higher plants. Annual Review of Plant Physiology 30:159–193.
Benveniste, P. 1986. Sterol biosynthesis. Annual Review of Plant Physiology 37:279–305.
Bernays, E. A. (Ed.). 1989. Insect-Plant Interactions. CRC Press, Boca Raton, Fla.
Boller, T. 1989. Primary signals and second messengers in the reaction of plants to pathogens. Pages 227–255 *in* W. F. Boss and D. J. Morré (eds.), Second Messengers in Plant Growth and Development. Alan R. Liss, New York.
Bowers, W. S., T. Ohta, J. S. Cleere, and P. A. Marsella. 1976. Discovery of insect anti-juvenile hormones in plants. Science 193:542–547.
Brown, S. A. 1981. Coumarins. Pages 269–300 *in* E. E. Conn (ed.), The Biochemistry of Plants, Vol. 7. Secondary Plant Products. Academic Press, New York.
Butt, V. S. and C. J. Lamb. 1981. Oxygenases and the metabolism of plant products. Pages 627–665 *in* P. K. Stumpf and E. E. Conn (eds.), The Biochemistry of Plants, Vol. 7. Secondary Plant Products. Academic Press, New York.
Chang, F. and C. L. Hsu. 1981. Preocene II affects sex attractancy in medfly. BioScience 31:676–677.
Ching, T. M. 1970. Glyoxysomes in megagameteophyte of germinating ponderosa pine seeds. Plant Physiology 70:475–482.
Conn, E. E. (Ed.). 1981. The Biochemistry of Plants, Vol. 7. Secondary Plant Products. Academic Press, New York.
Croteau, R. B. 1988. Metabolism of plant monoterpenes. ISI Atlas of Science: Biochemistry 1: 182–187.
Cutler, D. F., K. L. Alvin, and C. E. Price. 1980. The Plant Cuticle. Academic Press, New York.
Darvill, A. G. and P. Albersheim. 1984. Phytoalexins and their elicitors — a defense against microbial infection in plants. Annual Review of Plant Physiology 35:243–275.
Dressler, R. L. 1982. Biology of the orchid bees (euglossini). Annual Review of Ecology and Systematics 13:373–394.
Earle, F. R. and Q. Jones. 1962. Analyses of seed samples from 113 plant families. Economic Botany 16:221–250.
Ebel, J. 1986. Phytoalexin synthesis: The biochemical analysis of the induction process. Annual Review of Phytopathology 24:235–264.
Elakovich, S. D. 1987. Sesquiterpenes as phytoalexins and allelopathic agents. Pages 93–108 *in* G. Fuller and W. D. Nes (eds.), Ecology and Metabolism of Plant Lipids. American Chemical Society, Washington, D.C.
Floss, H. G. 1986. The shikimate pathway: An overview. Pages 13–55 *in* E. E. Conn (ed.), Recent Advances in Phytochemistry, Vol. 20. The Shikimic Pathway. Plenum Press, New York and London.
Fraga, B. M. 1988. Natural sesquiterpenoids. Natural Product Reports 5:497–521.
Fuller, G. and W. D. Nes. (Eds.). 1987. Ecology and Metabolism of Plant Lipids. American Chemical Society, Washington, D.C.
Gerhardt, B. 1986. Basic metabolic function of the higher plant peroxisome. Physiologie Vegetale 24:397–410.
Givan, C. V. 1983. The source of acetyl coenzyme A in chloroplasts of higher plants. Physiologia Plantarum 57:311–316.
Goad, L. J., J. Zimowski, R. P. Everdshed, and V. L. Male. 1987. The sterol esters of higher plants. Pages 95–102 *in* P. K. Stumpf, J. B. Mudd, and W. D. Nes (eds.), The Metabolism, Structure and Function of Plant Lipids. Plenum Press, New York and London.
Gould, J. M. 1983. Probing the structure and dynamics of lignin *in situ*. What's New in Plant Physiology 14:5–8.
Gray, J. C. 1987. Control of isoprenoid biosynthesis in higher plants. Advances in Botanical Research 14:25–91.
Grayson, D. H. 1988. Monoterpenoids. Natural Product Reports 5:419–464.
Grove, M. D., G.F. Spencer, W. K. Rohwedder, M. Nagabhushanam, J. F. Worley, J. D. Warthen, G.L. Steffens, J. L. Flippen-Anderson, and J. C. Cook. 1979. Brassinolide, a plant growth-promoting steroid isolated from *Brassica napus* pollen. Nature 281:216-217.
Guens, J. M. C. 1978. Steroid hormones and plant growth development. Phytochemistry 17:1–14.
Gurr, M. I. 1980. The biosynthesis of tri-acylglycerols. Pages 205–248 *in* P. K. Stumpf (ed.), The Biochemistry of Plants, Vol. 4, Lipids, Structures and Functions. Academic Press, New York.
Hahlbrock, K. 1981. Flavonoids. Pages 425–456 *in* E. E. Conn (ed.), The Biochemistry of Plants, Vol. 7. Secondary Plant Products. Academic Press, New York.
Hahlbrock, K. and D. Scheel. 1989. Physiology and molecular biology of phenylpropanoid metabolism. Annual Review of Plant Physiology and Plant Molecular Biology 40:347–369.
Harborne, J. B. 1988. Introduction to Ecological Biochemistry, Third Edition. Academic Press, New York.
Harborne, J. B. 1989. Recent advances in chemical ecology. Natural Product Reports 6:85–109.
Harrison, D. M. 1988. The biosynthesis of triterpenoids, steroids, and carotenoids. Natural Product Reports 5:387–415.
Harwood, J. 1980. Fatty acid metabolism. Annual Review of Plant Physiology and Plant Molecular Biology 39:101–148.
Harwood, J. 1989. Lipid metabolism in plants. CRC Critical Reviews in Plant Sciences 8:1–43.
Heemskerk, J. W. M. and J. F. G. M. Wintermans. 1987. Role of the chloroplast in the leaf acyl-lipid synthesis. Physiologia Plantarum 70:558–568.
Heftmann, E. 1975. Functions of steroids in plants. Phytochemistry 14:891–901.
Heftmann, E. 1983. Biogenesis of steroids in the Solanaceae. Phytochemistry 22:1843–1860.
Hegnauer, R. 1988. Biochemistry, distribution and taxonomic relevance of higher plant alkaloids. Phytochemistry 27:2423–2427.
Hemingway, R. W. and J. J. Karchesy. (Eds.). 1989. Chemistry and Significance of Condensed Tannins. Plenum, New York.
Herbert, R. B. 1988. The biosynthesis of plant alkaloids and nitrogenous microbial metabolites. Natural Product Reports 5:523–540.
Hewitt, S., J. R. Hillman, and B. A. Knights. 1980. Steroidal oestrogens and plant growth and development. New Phytologist 85:329–350.
Holloway, P. J. 1980. Structure and histochemistry of plant cuticular membranes: An overview. Pages 1–32 *in* D. F. Cutler, K. L. Alvin, and C. E. Price (eds.), The Plant Cuticle. Academic Press, New York.
Holloway, P. J. 1983. Some variations in the composition of suberin from the cork layers of higher plants. Phytochemistry 22:495–502.
Hoshino, T., U. Matsumoto, and T. Goto. 1981. Self-association of some anthocyanins in neutral aqueous solution. Phytochemistry 20:1971–1976.
Huang, A. H. C. 1987. Lipases. Pages 91–119 in P. K. Stumpf (ed.), Lipids, Structures and Functions. The Biochemistry of Plants, Vol. 9. Academic Press, New York.

Huang, A. H., R. N. Trelease, and T. S. Moore, Jr. (Eds.). 1983. Microbodies in Plants. Academic Press, New York.
Huber, S. C. 1986. Fructose-2,6-bisphosphate as a regulatory metabolite in plants. Annual Review of Plant Physiology 37:233–246.
Jensen, R. A. 1985. The shikimate/arogenate pathway: Link between carbohydrate metabolism and secondary metabolism. Physiologia Plantarum 66:164–168.
Jensen, R. A. 1986. Tyrosine and phenylalanine biosynthesis: Relationship between alternate pathways, regulation and subcellular location. Pages 57–81 *in* E. E. Conn (ed.), Recent Advances in Plant Phytochemistry, Vol. 20. The Shikimic Acid Pathway. Plenum Press, New York and London.
Johnson, M. A. and R. Croteau. 1987. Biochemistry of conifer resistance to bark beetles and their fungal symbionts. Pages 76–92 *in* G. Fuller and W. D. Nes (eds.), Ecology and Metabolism of Plant Lipids. American Chemical Society, Washington, D.C.
Juniper, B. E. and C. E. Jeffree. 1982. Plant Surfaces. Edward Arnold, London.
Keeler, R. F. 1975. Toxins and teratogens of higher plants. Lloydia 38:56–86.
Kolattukudy, P. E. 1980a. Cutin, suberin, and waxes. Pages 571–645 *in* P. K. Stumpf (ed.), Lipids, Structures and Functions. The Biochemistry of Plants, Vol. 4. Academic Press, New York.
Kolattukudy, P. E. 1980b. Biopolyester membranes of plants: Cutin and suberin. Annual Review of Plant Physiology 32:539–567.
Kolattukudy, P. E. 1987. Lipid-derived defensive polymers and waxes and their role in plant-microbe interaction. Pages 291–314 *in* P. K. Stumpf (ed.), The Metabolism, Structure and Function of Plant Lipids. Plenum Press, New York and London.
Lewis, W. H. and M. P. F. Elvin-Lewis. 1977. Medical Botany: Plants Affecting Man's Health. Wiley, New York.
Loomis, W. D. and R. Croteau. 1980. Biochemistry of terpenoids. Pages 363–418 *in* P. K. Stumpf (ed.), Lipids, Structures and Functions. The Biochemistry of Plants, Vol. 4. Academic Press, New York.
Mabry, T. J. 1980. Betalins. Pages 513–533 *in* E. A. Bell and B. V. Charlwood (eds.), Secondary Plant Products, Vol. 8, Encyclopedia of Plant Physiology, New Series. Springer-Verlag, Berlin.
Mabry, T. J. and J. E. Gill. 1979. Sesquiterpene lactones and other terpenoids. Pages 502–538 *in* G. A. Rosenthal and D. H. Janzen (eds.), Herbivores: Their Interaction with Secondary Plant Metabolites. Academic Press, New York.
MacLeod, A. J., G. MacLeod, and G. Subramanian. 1988. Volatile aroma constituents of orange. Phytochemistry 27:2185–2188.
Mader, M. and V. Amberg-Fisher. 1982. Role of peroxidase in lignification of tobacco cells. I. Oxidation of nicotinimide adenine dinucleotide and formation of hydrogen peroxide by cell wall peroxidases. Plant Physiology 70:1128–1131.
Mahato, S. B., S. K. Sarkar, and G. Poddar. 1988. Triterpenoid saponins. Phytochemistry 27: 3037–3067.
Mandava, N. B. 1988. Plant growth-promoting brassinosteroids. Annual Review of Plant Physiology and Plant Molecular Biology 39:23–52.
Mayer, A. M. 1987. Polyphenol oxidases in plants—recent progress. Phytochemistry 26:11–20.
Metcalf, R. L. 1987. Plant volatiles as insect attractants. CRC Critical Reviews in Plant Sciences 5:251–301.
Mettler, I. J. and H. Beevers. 1980. Oxidation of NADH in glyoxysomes by a malate-aspartate shuttle. Plant Physiology 66:555–560.
Meudt, W. J. 1987. Chemical and biological aspects of brassinolide. Pages 53–75 *in* G. Fuller and W. D. Nes (eds.), Ecology and Metabolism of Plant Lipids. American Chemical Society, Washington, D.C.
Moore, T. S., Jr. 1984. Biochemistry and biosynthesis of plant acyl lipids. Pages 83–91 *in* P. A. Siegenthaler and W. Eichenberger (eds.), Structure, Function and Metabolism of Plant Lipids. Elsevier, Amsterdam.
Mudd, J. B. 1980. Phospholipid biosynthesis. Pages 249–282 *in* P. K. Stumpf (ed.), Lipids, Structures and Functions. The Biochemistry of Plants, Vol. 4. Academic Press, New York.
Murray, R. D. H., J. Mendez, and S. A. Brown. 1982. The Natural Coumarins. Wiley, New York.
Nes, W. D. 1987. Multiple roles for plant sterols. Pages 3–9 *in* P. K. Stumpf, J. B. Mudd, and W. D. Nes (eds.), The Metabolism, Structure and Function of Plant Lipids. Plenum, New York and London.
Piattelli, M. 1981. The betalains: Structure, biosynthesis and chemical taxonomy. Pages 557–626 *in* E. E. Conn (ed.), The Biochemistry of Plants, Vol. 7. Secondary Plant Products. Plenum, New York and London.
Putnam, A. R. and R. M. Heisey. 1983. Allelopathy: Chemical interactions between plants. What's New in Plant Physiology 14:21–24.
Putnam, A. R. and C. S. Tang. (Eds.). 1986. The Science of Allelopathy. Wiley, New York.
Rice, E. L. 1984. Allelopathy, Second Edition. Academic Press, New York.
Robinson, T. 1979. The evolutionary ecology of alkaloids. Pages 413–448 in G. A. Rosenthal and D. H. Janzen (eds.), Herbivores: Their Interaction with Secondary Plant Metabolites. Academic Press, New York.
Robinson, T. 1980. The Organic Constituents of Higher Plants, Fourth Edition. Cordus Press, North Amherst, Mass.
Ryan, C. A. 1987. Oligosaccharide signalling in plants. Annual Review of Cell Biology 3:295–317.
Schutte, H. R. 1987. Secondary plant substances. Aspects of steroid biosynthesis. Progress in Botany 49:117–136.
Scriber, J. M. and M. P. Ayres. 1988. Leaf chemistry as a defense against insects. ISI Atlas of Science: Animal and Plant Science.
Seigler, D. S. 1981. Secondary metabolites and plant systematics. Pages 139–176 *in* E. E. Conn (ed.), The Biochemistry of Plants, Vol. 7. Secondary Plant Products. Plenum, New York and London.
Shutt, D. A. 1976. The effects of plant oestrogens on animal reproduction. Endeavour 35:110–113.
Siehl, D. L. and E. E. Conn. 1988. Kinetic and regulatory properties of arogenate dehydratase in seedlings of *Sorghum bicolor* (L.) Moench. Archives of Biochemistry and Biophysics 260: 822–829.
Sinclair, T. R. and C. T. de Wit. 1975. Photosynthesis and nitrogen requirements for seed production by various crops. Science 189:565–567.
Sláma, K. 1979. Insect hormones and antihormones in plants. Pages 683–700 *in* G. A. Rosenthal and D. H. Janzen (eds.), Herbivores: Their Interaction with Secondary Plant Metabolites. Academic Press, New York.
Sláma, K. 1980. Animal hormones and antihormones in plants. Biochemie Physiologie Pflanzen 175:177–193.
Smith, C. M. 1989. Plant Resistance to Insects. A Fundamental Approach. Wiley, Somerset, N. J.
Sorokin, H. 1967. The spherosomes and the reserve fat in plant cells. American Journal of Botany 54:1008–1016.
Spurgeon, S. L. and J. W. Porter. 1980. Biochemistry of terpenoids. Pages 419–483 *in* P. K. Stumpf (ed.), The Biochemistry of Plants, Vol. 4, Lipids, Structures and Functions. Academic Press, New York.
Staal, G. B. 1986. Anti-juvenile hormone agents. Annual Review of Entomology 31:391–429.
Stafford, H. A. 1990. Flavonoid Metabolism. CRC Press, Boca Raton, Fla.
Stich, K. and R. Ebermann. 1984. Investigation of hydrogen peroxide formation in plants. Phytochemistry 23:2719–2722.
Stone, B. A. 1989. Cell walls in plant-microorganism associations. Australian Journal of Plant Physiology 16:5–17.
Street, H. E. and H. Öpik. 1970. The Physiology of Flowering Plants: Their Growth and Development. American Elsevier, New York.
Stumpf, P. K. (Ed.). 1980. Lipids, Structures and Functions. The Biochemistry of Plants, Vol. 4. Academic Press, New York.
Stumpf, P. K. 1987. The biosynthesis of saturated fatty acids. Pages 121–135 *in* P. K. Stumpf (ed.), Lipids, Structures and Functions. The Biochemistry of Plants, Vol. 9. Academic Press, New York.
Stumpf, P. K., J. B. Mudd, and W. D. Nes. (Eds.). 1987. The Metabolism, Structure, and Function of Plant Lipids. Plenum, New York and London.
Swain, T. 1979. Tannins and lignins. Pages 657–682 *in* G. A. Rosenthal and D. H. Janzen (eds.), Herbivores: Their Interaction with Secondary Metabolites. Academic Press, New York.
Templeton, M. D. and C. J. Lamb. 1988. Elicitors and defence gene activation. Plant, Cell and Environment 11:395–401.
Tolbert, N. E. 1981. Metabolic pathways in peroxisomes and glyoxysomes. Annual Review of Biochemistry 50:133–157.
Trelease, R. N. 1984. Biogenesis of glyoxysomes. Annual Review of Plant Physiology 35:321–347.
Troughton, J. and L. A. Donaldson. 1972. Probing Plant Structure. McGraw-Hill, New York.
Vickery, B. and M. L. Vickery. 1981. Secondary Plant Metabolism. University Park Press, Baltimore.
Waller, G. R. (Ed.). 1987. Allelochemicals: Role in Agriculture and Forestry. American Chemical Society, Washington, D. C.
Wanner, G., H. Formanek, and R. R. Theimer. 1981. The ontogeny of lipid bodies (spherosomes) in plant cells. Planta 151:109–123.
Weast, R. C. (Ed.). 1988. Handbook of Chemistry and Physics, First Student Edition. CRC Press, Boca Raton, Fla.
Went, F. 1974. Reflections and speculations. Annual Review of Plant Physiology 25:1–26.
Whittaker, R. H. and P. P. Feeny. 1971. Allelochemicals: Chemical interactions between species. Science 171:757–770.
Yatsu, L. Y., T. J. Jacks, and T. P. Hensarling. 1971. Isolation of spherosomes (oleosomes) from onion, cabbage, and cottonseed tissue. Plant Physiology 48:675–682.

### *Chapter 16 Growth and Development*

Bacic, Antony, Philip J. Harris, and Bruce A. Stone. 1988. Structure and function of plant cell walls. Pages 297–371 *in* Jack Preiss (ed.), The Biochemistry of Plants, Vol. 14. Academic Press, San Diego.
Barlow, P. W. 1975. The root cap. Pages 21–54 *in* J. G. Torrey and D. T. Clarkson (eds.), The Development and Function of Roots. Academic Press, New York.
Beck, William A. 1941. Production of solutes in growing epidermal cells. Plant Physiology 16: 637–641.
Berlyn, G. P. 1972. Seed germination and morphogenesis. Pages 223–312 *in* T. T. Kozlowski (ed.), Seed Biology, Vol. 1. Academic Press, New York.
Bewley, J. Derek and M. Black. 1978. Physiology and Biochemistry of Seeds in Relation to Germination. Vol. 1, Development, Germination, and Growth. Springer-Verlag, Berlin, New York.
Boyer, John S. 1988. Cell enlargement and growth-induced water potentials. Physiologia Plantarum 73:311–316.
Carpita, N. C. 1985. Tensile strength of cell walls of living cells. Plant Physiology 79:485–488.
Carpita, N. C. 1987. The biochemistry of the "growing" plant cell wall. Pages 28–45 *in* D. J. Cosgrove and D. P. Knievel (eds.), Physiology of Cell Expansion During Plant Growth. American Society of Plant Physiology, Rockville, Md.

Carpita, Nicholas C., D. Sabularse, D. Montezinos, and Deborah P. Delmer. 1979. Determination of the pore size of cell walls of living plant cells. Science 205:1144–1147.
Chapman, J. M. and R. Perry. 1987. A diffusion model of phyllotaxis. Annals of Botany 60: 377–389.
Cleland, R. 1971. Cell wall extension. Annual Review of Plant Physiology 22:197–222.
Cleland, R. E. 1981. II. Wall extensibility: Hormones and wall extension. Pages 255–273 *in* W. Tanner and F. A. Loewus (eds.), Encyclopedia of Plant Physiology, New Series, Vol. 13B. Springer-Verlag, New York.
Cleland, Robert E. 1984. The Instron technique as a measure of immediate-past wall extensibility. Planta 160:514–520.
Clowes, F. A. L. 1975. The quiescent center. Pages 3–19 *in* J. G. Torrey and D. T. Clarkston (eds.), The Development and Function of Roots. Academic Press, New York.
Coombe, B. G. 1976. The development of fleshy fruits. Annual Review of Plant Physiology 26: 207–228.
Cosgrove, D. J. 1985. Cell wall yield properties of growing tissue. Evaluation by *in vivo* stress relaxation. Plant Physiol. 78:347–356.
Cosgrove, Daniel. 1986. Biophysical control of plant cell growth. Ann. Rev. Plant Physiol. 37:377–405.
Cosgrove, Daniel J. 1987. Wall relaxation and the driving forces for cell expansive growth. Plant Physiol. 84:561–564.
Dale, J. E. 1988. The control of leaf expansion. Ann. Rev. Plant Physiol. 39:267–295.
Delmer, Deborah P. 1987. Cellulose biosynthesis. Ann. Rev. Plant Physiol. 38:259–290.
Delmer, Deborah P. and Bruce A. Stone. 1988. Pages 373–420 *in* Jack Preiss (ed.), The Biochemistry of Plants, Vol. 14. Academic Press, San Diego.
Dure, L. S. III. 1975. Seed formation. Ann. Rev. Plant Physiol. 26:259–278.
Erickson, Ralph O. 1976. Modeling of plant growth. Ann. Rev. Plant Physiol. 27:407–434.
Erickson, Ralph O. and David R. Goddard. 1951. An analysis of root growth in cellular and biochemical terms. Tenth Growth Symposium. Growth 17 (Supp.):89–116.
Erickson, Ralph O. and Katherine B. Sax. 1956. Elemental growth rate of the primary root of *Zea mays*. Proc. Amer. Phil. Soc. 100:487–498.
Erickson, Ralph O. and Wendy K. Silk. 1980. The kinematics of plant growth. Scientific American 242(5):134–151.
Esau, K. 1977. Anatomy of Seed Plants, Second Edition. Wiley, New York.
Fahn, A. 1982. Plant Anatomy, Third Edition. Pergamon Press, Oxford, New York.
Feldman, Lewis J. 1984. Regulation of root development. Ann. Rev. Plant Physiol. 35:223–242.
Foster, R. C. 1982. The fine structure of epidermal cell mucilages of roots. New Phytologist 91: 727–740.
Green, P. B., R. O. Erickson, and J. Buggy. 1971. Metabolic and physical control of cell elongation rate. *In vivo* studies in *Nitella*. Plant Physiology 47:423–430.
Heyn, A. N. J. 1931. Der Mechanismus der Zellstreckung. (The mechanism of cell elongation.) Rec. Trav. Bot. Neerl. 28:113–244.
Hsiao, T. C. 1973. Plant responses to water stress. Ann. Rev. Plant Physiol. 24:519–570.
Huber, D. J. and D. J. Nevins. 1981. Partial purification of endo- and exo-β-D-glucanase enzymes from *Zea mays* seedlings and their involvement in cell wall autohydrolysis. Planta 151:206–214.
Hulme, A. C. 1970. The Biochemistry of Fruits and Their Products. Vol. 1. Academic Press, New York.
Jensen, William A. and Roderic B. Park. 1967. Cell Ultrastructure. Wadsworth, Belmont, Calif.
Jensen, William A. and Frank B. Salisbury. 1972. Botany: An Ecological Approach. Wadsworth, Belmont, Calif. (See also Botany, Second Edition, 1984.)
John, P. C. L. (Ed.). 1982. The Cell Cycle. Society for Experimental Biology Seminar Series, 10. Cambridge University Press, New York.
Kende, Hans and Bruno Baumgartner. 1974. Regulation of aging in flowers of *Ipomoea tricolor* by ethylene. Planta 116:279–289.
Kramer, Paul J. 1943. Amount and duration of growth of various species of tree seedlings. Plant Physiology 18:239–251.
Kramer, Paul J. and Theodore T. Kozlowski. 1960. Physiology of Trees. McGraw-Hill, New York.
Leopold, A. C. and P. E. Kriedemann. 1975. Plant Growth and Development. McGraw-Hill, New York.
Lloyd, Clive W. 1983. Helical microtubular arrays in onion root hairs. Nature 305:311–313.
Lockhart, James. 1965. An analysis of irreversible plant cell elongation. J. Theoret. Biol. 8:264–275.
Matyssek, Rainer, Sachio Maruyama, and John S. Boyer. 1988. Rapid wall relaxation in elongating tissues. Plant Physiol. 86:1163–1167.
Mauseth, James D. 1988. Plant Anatomy. Benjamin/Cummings, Menlo Park, Calif.
McNeil, D. L. 1976. The basis of osmotic pressure maintenance during expansion growth in *Helianthus annuus* hypocotyls. Australian Journal of Plant Physiology 3:311–324.
Nurnsten, H. E. 1970. Volatile compounds: The aroma of fruits. Pages 239–268 *in* A. C. Hulme (ed.), The Biochemistry of Fruits and Their Products, Vol. I. Academic Press, New York.
Ray, Peter Martin. 1972. *The Living Plant*, Second Edition. Holt, Rinehart & Winston, New York.
Rhodes, M. J. C. 1980. The maturation and ripening of fruits. Pages 157–206 *in* Kenneth V. Thimann (ed.), Senescence in Plants. CRC Press, Boca Raton, Fla.
Richards, F. C. 1969. The quantitative analysis of growth. Pages 2–76 *in* F. C. Stewart (ed.), Plant Physiology, Vol. 5A, Analysis of Growth: Behavior of Plants and Their Organs. Academic Press, New York.
Richards, F. C. and W. W. Schwabe. 1969. Pages 79–116 *in* F. C. Stewart (ed.), Plant Physiology, Vol. 5A, Analysis of Growth: Behavior of Plants and Their Organs. Academic Press, New York.
Robbins, W. W., T. E. Weier, and C. R. Stocking. 1974. Botany, An Introduction to Plant Biology. Wiley, New York.
Sachs, R. M. 1965. Stem elongation. Ann. Rev. Plant Physiol. 16:73–96.
Sangwan, R. S. and B. Norreel. 1975. Induction of plants from pollen grains of *Petunia* cultured *in vitro*. Nature 257:222–224.
Schnyder, Hans and Curtis J. Nelson. 1988. Diurnal growth of tall fescue leaf blades. Plant Physiol. 86:1070–1076.
Shepard, James F. 1982. The regeneration of potato plants from leaf-cell protoplasts. Scientific American 246(5):154–166.
Silk, Wendy Kuhn. 1980. Growth rate patterns which produce curvature and implications for the physiology of the blue light response. Pages 643–655 *in* H. Senger (ed.), The Blue Light Syndrome. Springer-Verlag, Berlin, Heidelberg.
Silk, Wendy Kuhn. 1984. Quantitative descriptions of development. Ann. Rev. Plant Physiol. 35: 479–518.
Silk, Wendy Kuhn and Ralph O. Erickson. 1978. Kinematics of hypocotyl curvature. American Journal of Botany 65:310–319.
Silk, Wendy Kuhn and Ralph O. Erickson. 1979. Kinematics of plant growth. J. Theor. Biol. 76: 481–501.
Sinnott, E. W. 1960. Plant Morphogenesis. McGraw-Hill, New York.
Spencer, D. and T. J. V. Higgins. 1982. Seed maturation and deposition of storage proteins. Pages 306–336 *in* H. Smith and D. Grierson (eds.), The Molecular Biology of Plant Development. University of California Press, Berkeley and Los Angeles.
Starr, Cecie and Ralph Taggart. 1981. Biology: The Unity and Diversity of Life, Second Edition. Wadsworth, Belmont, Calif.
Steward, F. C. 1958. Growth and organized development of cultured cells. III. Interpretations of the growth from free cell to carrot plant. American Journal of Botany 45:709–713.
Steward, F. C., Marion O. Mapes, and Kathryn Mears. 1958. Growth and organized development of cultured cells. II. Organization in cultures grown from freely suspended cells. American Journal of Botany 45:705–708.
Sunderland, N. 1970. Pollen plants and their significance. New Scientist 47:142–144.
Taiz, Lincoln. 1984. Plant cell expansion: Regulation of cell wall mechanical properties. Ann. Rev. Plant Physiol. 35:585–622.
Troughton, John and Lesley A. Donaldson. 1972. Probing Plant Structure. McGraw-Hill, New York.
Tukey, H. B. 1933. Embryo abortion in early-ripening varieties of *Prunus avium*. Botan. Gaz. 94:433–468.
Tukey, H. B. 1934. Growth of the embryo, seed, and pericarp of the sour cherry (*Prunus cerasus*) in relation to season of fruit ripening. Proceedings of the American Society for Horticultural Science 31:125–144.
Vasil, V. and A. C. Hildebrandt. 1965. Differentiation of tobacco plants from single, isolated cells in microculture. Science 150:889–892.
Went, Frits W. 1957. The Experimental Control of Plant Growth. Ronald, New York.
Whaley, W. Gordon. 1961. Growth as a general process. Pages 71–112 *in* W. Ruhland (ed.), Encyclopedia of Plant Physiology. Vol. 14: Growth and Growth Substances. Springer-Verlag, Berlin, New York.
Williams, E. G. and G. Maheswaran. 1986. Somatic embryogenesis: Factors influencing coordinated behaviour of cells as an embryogenic group. Annals of Botany 57:443–462.
Woodward, J. R., G. B. Fincher, and B. A. Stone. 1983. Water soluble (1-3),(1-4)-β-D-glucans from barley (*Hordeum vulgare*) endosperm. II. Fine structure. Carbohydrate Polymers 3:207–225.
Zimmermann, M. H. and C. L. Brown. 1971. Trees—Structure and Function. Springer-Verlag, Berlin.

### *Chapter 17 Hormones and Growth Regulators: Auxins and Gibberellins*

Adams, P. A., P. B. Kaufman, and H. Ikuma. 1973. Effects of gibberellic acid and sucrose on the growth of oat (*Avena*) stem segments. Plant Physiology 51:1102–1108.
Akazawa, T. and I. Hara-Nishimura. 1985. Topographic aspects of biosynthesis, extracellular secretion, and intracellular storage of proteins in plant cells. Annual Review of Plant Physiology 36:441–472.
Akazawa, T., T. Mitsui, and M. Hawashi. 1988. Recent progress in alpha-amylase biosynthesis. Pages 465–492 *in* J. Preiss (ed.), The Biochemistry of Plants, Vol. 14, Carbohydrates. Academic Press, San Diego.
Aloni, R. 1987a. The induction of vascular tissues by auxin. Pages 363–374 *in* P. J. Davies (ed.), Plant Hormones and Their Role in Plant Growth and Development. Martinus Nijhoff Publishers, Boston.
Aloni, R. 1987b. Differentiation of vascular tissues. Annual Review of Plant Physiology 38:179–204.
Atzorn, R., A. Crozier, C. T. Wheeler, and G. Sandberg. 1988. Production of gibberellins and indole-3-acetic acid by *Rhizobium phaseoli* in relation to nodulation of *Phaseolus vulgaris* roots. Planta 175:532–538.
Bandurski, R. S. 1984. Metabolism of indole-3-acetic acid. Pages 183–200 *in* A. Crozier and J. R. Hillman (eds.), The Biosynthesis and Metabolism of Plant Hormones. Cambridge University Press, Cambridge.
Blowers, D. P. and A. J. Trewavas. 1989. Second messengers: Their existence and relationship to protein kinases. Pages 1–28 *in* W. F. Boss and D. J. Morré (eds.), Second Messengers in Plant Growth and Development. Alan R. Liss, New York.

Boss, W. F. 1989. Phosphoinositide metabolism: Its relation to signal transduction in plants. Pages 29–56 *in* W. F. Boss and D. J. Morré (eds.), Second Messengers in Plant Growth and Development. Alan R. Liss, New York.

Bottini, R., M. Fulchieri, D. Pearce, and R. P. Pharis. 1989. Identification of gibberellins $A_1$, $A_3$, and iso-$A_3$ in cultures of *Azospirillum lipoferum*. Plant Physiology 90:45–47.

Brenner, M. L. 1981. Modern methods for plant growth substance analysis. Annual Review of Plant Physiology 32:511–538.

Bressan, R. A., C. W. Ross, and J. Vandepeute. 1976. Attempts to detect cyclic adenosine-3′5′-monophosphate in higher plants by three assay methods. Plant Physiology 57:29–37.

Broughton, W. J. and A. J. McComb. 1971. Changes in the pattern of enzyme development in gibberellin-treated pea internodes. Annals of Botany 35:213–228.

Budde, R. J. A. and D. D. Randall. 1990. Protein kinases in higher plants. Pages 351–367 *in* D. J. Morré, W. F. Boss, and F. A. Loewus (eds.), Inositol Metabolism in Plants. Wiley-Liss, New York.

Carlson, R. D. and A. J. Crovetti. 1990. Commercial uses of gibberellins and cytokinins and new areas of applied research. Pages 604–610 *in* R. P. Pharis and S. W. Rood (eds.), Plant Growth Substances 1988. Springer-Verlag, Heidelberg.

Carrington, C. M. S. and J. Esnard. 1988. The elongation response of watermelon hypocotyls to indole-3-acetic acid: A comparative study of excised segments and intact plants. Journal of Experimental Botany 39:441–450.

Caruso, J. L. 1987. The auxin conjugates. HortScience 22:1201–1207.

Chadwick, A. V. and S. P. Burg. 1967. An explanation of the inhibition of root growth caused by indole-3-acetic acid. Plant Physiology 42:415–420.

Chadwick, C. M. and D. R. Garrod. (Eds.). 1986. Hormones, Receptors and Cellular Interactions in Plants. Cambridge University Press, Cambridge.

Chory, J., D. F. Voytas, N. E. Olszewski, and F. M. Ausubel. 1987. Gibberellin-induced changes in the populations of translatable mRNAs and accumulated polypeptides in dwarfs of maize and pea. Plant Physiology 83:15–23.

Cleland, R. E. 1987a. Auxin and cell elongation. Pages 132–148 *in* P. J. Davies (ed.), Plant Hormones and Their Role in Plant Growth and Development. Martinus Nijhoff Publishers, Boston.

Cleland, R. E. 1987b. The mechanism of wall loosening and wall extension. Pages 18–27 *in* D. J. Cosgrove and D. P. Knievel (eds.), Physiology of Cell Expansion During Plant Growth. American Society of Plant Physiologists, Rockville, Md.

Cohen, J. D. and R. S. Bandurski. 1982. Chemistry and physiology of the bound auxins. Annual Review of Plant Physiology 33:403–454.

Cohen, J. D. and K. Bialek. 1984. The biosynthesis of indole-3-acetic acid in higher plants. Pages 165–181 *in* A. Crozier and J. R. Hillman (eds.), The Biosynthesis and Metabolism of Plant Hormones, Cambridge University Press, Cambridge.

Cohen, J. D., M. G. Bausher, K. Bialek, J. G. Buta, G. F. W. Gocal, L. M. Janzen, R. P. Pharis, A. N. Reed, and J. P. Slovin. 1987. Comparison of a commercial ELISA assay for indole-3-acetic acid at several stages of purification and analysis by gas chromatography-selected ion monitoring-mass spectrometry using a $^{13}C_6$-labeled internal standard. Plant Physiology 84:982–986.

Cosgrove, D. 1986. Biophysical control of plant cell growth. Annual Review of Plant Physiology 37:377–405.

Cosgrove, D. J. and D. P. Knievel. (Eds.). 1987. Physiology of Cell Expansion During Plant Growth. American Society of Plant Physiologists, Rockville, Md.

Daniel, S. G., D. L. Rayle, and R. E. Cleland. 1989. Auxin physiology of the tomato mutant *diageotropica*. Plant Physiology 91:804–807.

Davies, P. J. (Ed.). 1987. Plant Hormones and Their Role in Plant Growth and Development. Martinus Nijhoff Publishers, Boston.

Dietz, A., U. Kutschera, and P. M. Ray. 1990. Auxin enhancement of mRNAs in epidermis and internal tissues in the pea stem and its significance for control of elongation. Plant Physiology 93:432–438.

Einspahr, K. J. and G. A. Thompson, Jr. 1990. Transmembrane signaling via phosphatidylinositol 4,5-bisphosphate hydrolysis in plants. Plant Physiology 93:361–366.

Eliasson, L., G. Bertell, and E. Bolander. 1989. Inhibitory action of auxin on root elongation not mediated by ethylene. Plant Physiology 91: 310–314.

Engvild, K. C. 1986. Chlorine-containing natural compounds in higher plants. Phytochemistry 25:781–791.

Epstein, E., K-H. Chen, and J. D. Cohen. 1989. Identification of indole-3-butyric acid as an endogenous constituent of maize kernels and leaves. Plant Growth Regulation 8:215–223.

Evans, M. L. 1985. The action of auxin on plant cell elongation. CRC Critical Reviews in Plant Sciences 2:317–365.

Fincher, G. B. 1989. Molecular and cellular biology associated with endosperm mobilization in germinating cereal grains. Annual Review of Plant Physiology and Plant Molecular Biology 40:305–346.

Firn, R. D. 1986. Growth substance sensitivity: The need for clearer ideas, precise terms, and purposeful experiments. Physiologia Plantarum 67:267–272.

Galston, A. W. 1964. The Life of the Green Plant. Prentice-Hall, Englewood Cliffs, N. J.

Glasziou, K. T. 1969. Control of enzyme formation and inactivation in plants. Annual Review of Plant Physiology 20:63–88.

Gocal, G. F. W., R. P. Pharis, E. C. Yeung, and D. Pearce. 1990. Changes after decapitation in concentrations of indole-3-acetic acid and abscisic acid in the larger axillary bud of *Phaseolus vulgaris* L. cv. Tender Green. Plant Physiology 94: (in press).

Goldsmith, M. H. M. 1977. The polar transport of auxin. Annual Review of Plant Physiology 28:439–478.

Goldsmith, M. H. M. 1982. A saturable site responsible for polar transport of indole-3-acetic acid in sections of maize coleoptiles. Planta 155:68–75.

Gove, J. P. and M. C. Hoyle. 1975. The isozymic similarity of indoleacetic acid oxidase to peroxidase in birch and horseradish. Plant Physiology 56:684–687.

Graebe, J. E. 1987. Gibberellin biosynthesis and control. Annual Review of Plant Physiology 38:419–466.

Grossmann, K. 1990. Plant growth retardants as tools in physiological research. Physiologia Plantarum 78:640–648.

Guilfoyle, T. 1986. Auxin regulated gene expression in higher plants. CRC Critical Reviews in Plant Sciences 4:247–277.

Guilfoyle, T. and G. Hagen. 1987. Transcriptional regulation of auxin responsive genes. Pages 85–95 *in* J. E. Fox and M. Jacobs (eds.), Molecular Biology of Plant Growth Control. Alan R. Liss, New York.

Hagen, G. 1987. The control of gene expression by auxin. Pages 149–163 *in* P. J. Davies (ed.), Plant Hormones and Their Role in Plant Growth and Development, Martinus Nijhoff Publishers, Boston.

Haissig, B. E. 1974. Origins of adventitious roots. New Zealand Journal of Forestry Science 4: 229–310.

Hall, J. L., D. A. Brummell, and J. Gillespie. 1985. Does auxin stimulate the elongation of intact plant stems? New Phytologist 100:341–345.

Hedden, P. and J: R. Lenton. 1988. Genetic and chemical approaches to the metabolic regulation and mode of action of gibberellins in plants. Pages 175–204 *in* Beltsville Symposia in Agricultural Resources 12, Biomechanisms Regulating Growth and Development. Kluwer Academic Publishers, Boston.

Heyn, A., S. Hoffmann, and R. Hertel. 1987. *In-vitro* auxin transport in membrane vesicles from maize coleoptiles. Planta 172:285–287.

Hicks, G. R., D. L. Rayle, and T. L. Lomax. 1989. The *diageotropica* mutant of tomato lacks high specific activity auxin binding sites. Science 245:52–53.

Hillman, J. R. 1984. Apical dominance. Pages 127–148 *in* M. B. Wilkins (ed.), Advanced Plant Physiology. Pitman, London.

Hillman, J. R., V. B. Math, and G. C. Medlow. 1977. Apical dominance and the levels of indole acetic acid in *Phaseolus* lateral buds. Planta 134:191–193.

Horgan, R. 1987. Hormone analysis: Instrumental methods of plant hormone analysis. Pages 222–239 *in* P. J. Davies (ed.), Plant Hormones and Their Role in Plant Growth and Development. Martinus Nijhoff Publishers, Boston.

Jacobs, M. and S. F. Gilbert. 1983. Basal localization of the presumptive auxin transport carrier in pea stem cells. Science 220:1297–1300.

Jacobs, M. and P. H. Rubery. 1988. Naturally occurring auxin transport regulators. Science 241:346–349.

Jacobs, W. P. 1979. Plant Hormones and Plant Development. Cambridge University Press, Cambridge.

Jones, R. L. and J. MacMillan. 1984. Gibberellins. Pages 21–52 *in* M. B. Wilkins (ed.), Advanced Plant Physiology. Pitman, London.

Kelly, M. O. and K. J. Bradford. 1986. Insensitivity of the *diageotropica* tomato mutant to auxin. Plant Physiology 82:713–717.

Key, J. L. 1987. Auxin-regulated gene expression: A historical perspective and current status. Pages 1–21 *in* J. E. Fox and M. Jacobs (eds.), Molecular Biology of Plant Growth Control. Alan R. Liss, New York.

Key, J. L. 1989. Modulation of gene expression by auxin. BioEssays 11:52–58.

Klingman, G., F. M. Ashton, and L. J. Noordhoff. 1982. Weed Science. Principles and Practices, Second Edition. Wiley, Somerset, N. J.

Kormondy, E. J., T. F. Sherman, F. B. Salisbury, N. T. Spratt, Jr., and G. McCain. 1977. Biology. The Integrity of Organisms. Wadsworth, Belmont, Calif.

Kutschera, U. 1987. Cooperation between outer and inner tissues in auxin-mediated plant organ growth. Pages 215–226 *in* D. J. Cosgrove and D. P. Knievel (eds.), Physiology of Cell Expansion During Plant Growth. American Society of Plant Physiologists, Rockville, Md.

Kutschera, U. 1989. Tissue stresses in growing plant organs. Physiologia Plantarum 77:157–163.

Kutschera, U. and P. Schopfer. 1985. Evidence against the acid-growth theory of auxin action. Planta 163:483–493.

Kutschera, U., R. Bergfeld, and P. Schopfer. 1987. Cooperation of epidermis and inner tissues in auxin-mediated growth of maize coleoptiles. Planta 170:168–180.

Leuba, V. and D. LeTourneau. 1990. Auxin activity of phenylacetic acid in tissue culture. Journal of Plant Growth Regulation 9:71–76.

Liu, P. B. W. and J. B. Loy. 1976. Action of gibberellic acid on cell proliferation in the subapical shoot meristem of watermelon seedlings. American Journal of Botany 63:700–704.

MacMillan, J. and B. O. Phinney. 1987. Biochemical genetics and the regulation of stem elongation by gibberellins. Pages 156–171 *in* D. J. Cosgrove and D. P. Knievel (eds.), Physiology of Cell Expansion During Plant Growth. American Society of Plant Physiologists, Rockville, Md.

Mariott, K. M. and D. H. Northcote. 1975. The induction of enzyme activity in the endosperm of germinating castor-bean seeds. The Biochemical Journal 152:65–70.

Marmé, D. 1989. The role of calcium and calmodulin in signal transduction. Pages 57–80 *in* W. F. Boss and D. J. Morré (eds.), Second Messengers in Plant Growth and Development. Alan R. Liss, New York.

Marré, E. 1979. Fusicoccin: A tool in plant physiology. Annual Review of Plant Physiology 30: 273–312.

Martin, G. C. 1983. Commercial uses of gibberellins. Pages 395–444 *in* A. Crozier (ed.), The Biochemistry and Physiology of Gibberellins, Vol. 2. Praeger, New York.

Martin, G. C. 1987. Apical dominance. HortScience 22:824–833.

McComb, A. J. and W. J. Broughton. 1972. Metabolic changes in internodes of dwarf pea plants treated with gibberellic acid. Pages 407–413 *in* D. J. Carr (ed.), Plant Growth Substances 1970. Springer-Verlag, Berlin.

McQueen-Mason, S. J. and R. H. Hamilton. 1989. The biosynthesis of indole-3-acetic acid from D-tryptophan in Alaska pea plastids. Plant and Cell Physiology 30:999–1005.

Memon, A. R., M. Rincon, and W. F. Boss. 1989. Inositol trisphosphate metabolism in carrot (*Daucus carolta* L.) cells. Plant Physiology 91: 477–480.

Métraux, J-P. 1987. Gibberellins and plant cell elongation. Pages 296–317 *in* P. J. Davies (ed.), Plant Hormones and Their Role in Plant Growth and Development. Martinus Nijhoff Publishers, Boston.

Mitchell, J. W., D. P. Skags, and W. P. Anderson. 1951. Plant growth-stimulating hormones in immature bean seeds. Science 114:159–161.

Moore, T. C. 1989. Biochemistry and Physiology of Plant Hormones. Springer-Verlag, Berlin.

Moreland, D. E. 1980. Mechanisms of action of herbicides. Annual Review of Plant Physiology 31:597–638.

Morgan, P. W. and J. I. Durham. 1983. Strategies for extracting, purifying, and assaying auxins from plant tissues. Botanical Gazette 144:20–31.

Morré, D. J., W. F. Boss, and F. A. Loewus. (Eds.). 1990. Inositol Metabolism in Plants. Wiley-Liss, New York.

Nagao, A., S. Sasaki, and R. P. Pharis. 1989. *Chamaecyparis*. Pages 170–188 *in* A. H. Halevy (ed.), CRC Handbook of Flowering, Vol. VI. CRC Press, Boca Raton, Fla.

Napier, R. M. and M. A. Venis. 1990. Receptors for plant growth regulators: Recent advances. Journal of Plant Growth Regulation 9:113–126.

Nickell, L. G. 1979. Controlling biological behavior of plants with synthetic plant growth regulating chemicals. Pages 263–279 *in* N. B. Mandava (ed.), Plant Growth Substances. American Chemical Society, Washington, D.C.

Nishijima, T. and N. Katsura. 1989. A modified micro-drop bioassay using dwarf rice for detection of femtomol quantities of gibberellins. Plant and Cell Physiology 30:623–627.

Pence, V. C. and J. L. Caruso. 1987. Immunoassay methods of plant hormone analysis. Pages 240–256 *in* P. J. Davies (ed.), Plant Hormones and Their Role in Plant Growth and Development. Martinus Nijhoff Publishers, Boston.

Pharis, R. P. and C. G. Kuo. 1977. Physiology of gibberellins in conifers. Canadian Journal of Forest Research 7:299–325.

Pharis, R. P. and S. B. Rood. (Eds.). 1990. Plant Growth Substances 1988. Springer-Verlag, Heidelberg.

Pharis, R. P., L. T. Evans, R. W. King, and L. N. Mander. 1989. Gibberellins and flowering in higher plants: Differing structures yield highly specific effects. Pages 29–41 *in* E. Lord and G. Bernier (eds.), Plant Reproduction: From Floral Induction to Pollination. The American Society of Plant Physiologists Symposium Series, Vol. 1.

Pharis, R. P., J. E. Webber, and S. D. Ross. 1987. The promotion of flowering in forest trees by gibberellin $A_{4/7}$ and cultural treatments: A review of the possible mechanisms. Forest Ecology and Management 19:65–84.

Phillips, I. D. J. 1975. Apical dominance. Annual Review of Plant Physiology 26:341–367.

Phinney, B. O. 1983. The history of the gibberellins. Pages 19–52 *in* A. Crozier (ed.), The Biochemistry and Physiology of the Gibberellins, Vol. 1. Praeger, New York.

Phinney, B. O. and C. R. Spray. 1987. Diterpenes—the gibberellin biosynthetic pathway in *Zea mays*. Pages 19–27 *in* P. K. Stumpf, J. B. Mudd, and W. D. Nes (eds.), The Metabolism, Structure and Function of Plant Lipids. Plenum, New York.

Poovaiah, B. W. and A. S. N. Reddy. 1990. Turnover of inositol phospholipids and calcium-dependent protein phosphorylation in signal transduction. Pages 335–349 *in* D. J. Morré, W. F. Boss, and F. A. Loewus (eds.), Inositol Metabolism in Plants. Wiley-Liss, New York.

Raab, M. M. and R. E. Koning. 1988. How is floral expansion regulated? BioScience 38:670–674.

Ray, P. M. 1987. Principles of plant cell growth. Pages 1–17 *in* D. J. Cosgrove and D. P. Knievel (eds.), Physiology of Cell Expansion During Plant Growth. American Society of Plant Physiologists, Rockville, Md.

Rayle, D. L. and R. Cleland. 1979. Control of plant cell enlargement by hydrogen ions. Current Topics in Developmental Biology 11:187–214.

Rayle, D. L., C. W. Ross, and N. Robinson. 1982. Estimation of osmotic parameters accompanying zeatin-induced growth of detached cucumber cotyledons. Plant Physiology 70:1634–1636.

Reid, J. B. 1987. The genetic control of growth via hormones. Pages 318–340 *in* P. J. Davies (ed.), Plant Hormones and Their Role in Plant Growth and Development. Martinus Nijhoff Publishers, Boston.

Reid, J. B. 1990. Phytohormone mutants in plant research. Journal of Plant Growth Regulation 9:97–111.

Reinecke, D. M. and R. S. Bandurski. 1987. Auxin biosynthesis and metabolism. Pages 24–42 *in* P. J. Davies (ed.), Plant Hormones and Their Role in Plant Growth and Development. Martinus Nijhoff Publishers, Boston.

Rood, S. B., R. I. Buzzell, L. N. Mander, D. Pearce, and R. P. Pharis. 1988. Gibberellins: A phytohormal basis for heterosis in maize. Science 241:1216–1218.

Rood, S. B., P. H. Williams, D. Pearce, N. Murofushi, L. N. Mander, and R. P. Pharis. 1990. A mutant gene that increases gibberellin production in *Brassica*. Plant Physiology 93:1168–1174.

Ross, C. W. and D. L. Rayle. 1982. Evaluation of $H^+$ secretion relative to zeatin-induced growth of detached cucumber cotyledons. Plant Physiology 70:1470–1474.

Rubenstein, B. and M. A. Nagao. 1976. Lateral bud outgrowth and its control by the apex. The Botanical Review 42:83–113.

Rubery, P. H. 1987. Auxin transport. Pages 341–362 *in* P. J. Davies (ed.), Plant Hormones and Their Role in Plant Growth and Development. Martinus Nijhoff Publishers, Boston.

Sabater, M. and P. H. Rubery. 1987. Auxin carriers in *Cucurbita* vesicles. Planta 171:507–513.

Sachs, R. M. 1965. Stem elongation. Annual Review of Plant Physiology 16:73–96.

Salisbury, F. B. and R. V. Parke. 1964. Vascular Plants: Form and Function. Wadsworth, Belmont, Calif.

Schneider, E. A., C. W. Kazakoff, and F. Wightman. 1985. Gas chromatography-mass spectrometry evidence for several endogenous auxins in pea seedling organs. Planta 165:232–241.

Scott, I. M. 1990. Plant hormone response mutants. Physiologia Plantarum 78:147–152.

Sembdner, G., D. Gross, H. W. Liebisch, and G. Schneider. 1980. Biosynthesis and metabolism of plant hormones. Pages 281–444 *in* J. MacMillan (ed.), Hormonal Regulation of Development. I. Molecular Aspects of Plant Hormones. Encyclopedia of Plant Physiology, Vol. 9. Springer-Verlag, Berlin.

Spiteri, A., O. M. Viratelle, P. Raymond, M. Rancillac, J. Labouesse, and A. Pradet. 1989. Artefactual origins of cyclic AMP in higher plant tissues. Plant Physiology 91:624–628.

Sponsel, V. M. 1987. Gibberellin biosynthesis and metabolism. Pages 43–75 *in* P. J. Davies (ed.), Plant Hormones and Their Role in Plant Growth and Development. Martinus Nijhoff Publishers, Boston.

Stuart, D. A. and R. L. Jones. 1978. The role of acidification in gibberellic acid- and fusicoccin-induced elongation growth of lettuce hypocotyl sections. Planta 142:135–145.

Taiz, L. 1984. Plant cell expansion: Regulation of cell wall mechanical properties. Annual Review of Plant Physiology 35:585–657.

Takahashi, N., B. O. Phinney, and J. MacMillan. (Eds.). 1990. Gibberellins. Springer-Verlag, Berlin.

Tamas, I. A. 1987. Hormonal regulation of apical dominance. Pages 393–410 *in* P. J. Davies (ed.), Plant Hormones and Their Role in Plant Growth and Development. Martinus Nijhoff Publishers, Boston.

Tamini, S. and R. D. Firn. 1985. The basipetal auxin transport system and the control of cell elongation in hypocotyls. Journal of Experimental Botany 36:955–962.

Taylor, A. and D. J. Cosgrove. 1989. Gibberellic acid stimulation of cucumber hypocotyl elongation. Effects on growth, turgor, osmotic pressure, and cell wall properties. Plant Physiology 90: 1335–1340.

Theologis, A. 1986. Rapid gene regulation by auxin. Annual Review of Plant Physiology 37:407–438.

Thimann, K. V. 1980. The development of plant hormone research in the last 60 years. Pages 15–33 *in* F. Skoog (ed.), Plant Growth Substances 1979. Springer-Verlag, Berlin.

Trewavas, A. J. 1982. Growth substance sensitivity: The limiting factor in plant development. Physiologia Plantarum 55:60–72.

Trewavas, A. J. 1987. Sensitivity and sensory adaptation in growth substance responses. Pages 19–38 *in* G. V. Hoad, M. B. Jackson, J. R. Lenton, and R. K. Atkin (eds.), Hormone Action in Plant Development. Butterworths, London.

Trewavas, A. J. and R. E. Cleland. 1983. Is plant development regulated by changes in the concentration of growth substances or by changes in the sensitivity to growth substances? Trends in Biochemical Sciences 8:354–357.

Tsurusaki, K-I., S. Watanabe, N. Sakurai, and S. Kuraishi. 1990. Conversion of D-tryptophan to indole-3-acetic acid in coleoptiles of a normal and a semi-dwarf barley (*Hordeum vulgare*) strain. Physiologia Plantarum 79:221–225.

Vanderhoef, L. N. 1980. Auxin-regulated elongation: A summary hypothesis. Pages 90–96 *in* F. Skoog (ed.), Plant Growth Substances 1979. Springer-Verlag, Berlin.

Weiler, E. W. 1984. Immunoassay of plant growth regulators. Annual Review of Plant Physiology 35:85–95.

Wiesman, Z., J. Riov, and E. Epstein. 1989. Characterization and rooting ability of indole-3-butyric acid conjugates formed during rooting of mung bean cuttings. Plant Physiology 91:1080–1084.

Wightman, F. and D. L. Lighty. 1982. Identification of phenylacetic acid as a natural auxin in the shoots of higher plants. Physiologia Plantarum 55:17–24.

Wightman, F., E. A. Schneider, and K. V. Thimann. 1980. Hormonal factors controlling the initiation and development of lateral roots. II. Effects of exogenous growth factors on lateral root formation in pea roots. Physiologia Plantarum 49:304–314.

Wilkins, M. B. (Ed.). 1984. Advanced Plant Physiology. Pitman, London.

Yokota, T., N. Murofushi, and N. Takahashi. 1980. Extraction, purification, and identification. Pages 113–201 *in* J. MacMillan (ed.), Hormonal Regulation of Development. I. Molecular Aspects. Encyclopedia of Plant Physiology, New Series, Vol. 9. Springer-Verlag, Berlin.

Yopp, J. H., L. H. Aung, and G. L. Steffens. (Eds.). 1986. Bioassays and Other Special Techniques for Plant Hormones and Plant Growth Regulators. Plant Growth Regulator Society of America. Beltsville, Md.

Zhang, Y. 1989. The Influence of $GA_3$ on Fructosyl Carbohydrates in Wheat Plants. M.S. Thesis, Colorado State University.

## *Chapter 18 Hormones and Growth Regulators: Cytokinins, Ethylene, Abscisic Acid, and Other Compounds*

Abeles, F. B. 1973. Ethylene in Plant Biology. Academic Press, New York.

Addicott, F. T. 1965. Physiology of abscission. Pages 1094–1126 *in* W. Ruhland (ed.), Encyclopedia of Plant Physiology, Vol. 15, Part 2. Springer-Verlag, Berlin.

Addicott, F. T. (Ed.). 1982. Abscission. University of California Press, Berkeley.

Addicott, F. T. (Ed.). 1983. Abscisic Acid. Praeger, New York.

Ananiev, E. D., L. K. Karagyozov, and E. N. Karanov. 1987. Effect of cytokinins on ribosomal RNA gene expression in excised cotyledons of *Cucurbita pepo* L. Planta 170:370–378.

Berrie, A. M. M. 1984. Germination and dormancy. Pages 440–468 *in* M. B. Wilkins (ed.), Advanced Plant Physiology. Pitman, London.

Bewley, J. D. and M. Black. 1982. Physiology and Biochemistry of Seeds, Vol. 2. Springer-Verlag, Berlin.

Bewley, J. D. and A. Marcus. 1990. Gene expression in seed development and germination. Progress in Nucleic Acid Research and Molecular Biology 38:165–193.

Beyer, E. M., Jr. 1976. A potent inhibitor of ethylene action in plants. Plant Physiology 58:268–271.

Beyer, E. M., Jr., P. W. Morgan, and S. F. Yang. 1984. Ethylene. Pages 111–126 *in* M. B. Wilkins (ed.), Advanced Plant Physiology. Pitman, London.

Blankenship, S. M. and E. C. Sisler. 1989. 2,5-norbornadiene retards apple softening. HortScience 24(2):313–314.

Blazich, F. A. 1988. Chemicals and formulations used to promote adventitious rooting. Pages 132–149 *in* T. D. Davis, B. E. Haissig, and N. Sankhla (eds.), Adventitious Root Formation in Cuttings. Dioscordes Press, Portland, Ore.

Bleecker, A. B., S. Rose-John, and H. Kende. 1987. An evaluation of 2,5-norbornadiene as a reversible inhibitor of ethylene action in deepwater rice. Plant Physiology 84:395–398.

Bleecker, A. B., M. A. Estelle, C. Somerville, and H. Kende. 1988. Insensitivity to ethylene conferred by a dominant mutation in *Arabidopsis thaliana*. Science 241:1086–1089.

Boller, T. 1988. Ethylene and the regulation of antifungal hydrolases in plants. Oxford Survey of Plant Molecular and Cell Biology 5:145–174.

Borochov, A. and W. R. Woodson. 1989. Physiology and biochemistry of flower petal senescence. Horticultural Reviews 11:15–43.

Bouzayen, M., A. Latche, and J-C. Pech. 1990. Subcellular localization of the sites of conversion of 1-aminocyclopropane-1-carboxylic acid into ethylene in plant cells. Planta 180:175–180.

Bracale, M., G. P. Longo, G. Rossi, and C. P. Longo. 1988. Early changes in morphology and polypeptide pattern of plastids from watermelon cotyledons induced by benzyladenine or light are very similar. Physiologia Plantarum 72:94–100.

Burg, S. P. 1973. Ethylene in plant growth. Proceedings of the National Academy of Sciences USA 70:591–597.

Camp, P. J. and J. L. Wickliff. 1981. Light or ethylene treatments induce transverse cell enlargement in etiolated maize mesocotyls. Plant Physiology 67:125–128.

Chen, C-M. 1987. Characterization of cytokinins and related compounds by HPLC. Pages 23–38 *in* H. F. Linskens and J. F. Jackson (eds.), Modern Methods of Plant Analysis, New Series, Vol. 5, High Performance Liquid Chromatography in Plant Sciences. Springer-Verlag, Berlin.

Chen, C-M. and D. K. Melitz. 1979. Cytokinin biosynthesis in a cell-free system from cytokinin-autotrophic tobacco tissue cultures. FEBS Letters 107:15–20.

Chen, C-M. and B. Petschow. 1978. Cytokinin biosynthesis in cultured rootless tobacco plants. Plant Physiology 62:861–865.

Chen, C-M., J. R. Ertl, S. M. Leisner, and C-C. Chang. 1985. Localization of cytokinin biosynthetic sites in pea plants and carrot roots. Plant Physiology 78:510–513.

Chen, C-M., J. Ertl, M-S. Yang, and C-C. Chang. 1987. Cytokinin-induced changes in the population of translatable mRNA excised pumpkin cotyledons. Plant Science 52:169–174.

Cheverry, J. L., M. O. Sy, J. Pouliqueen, and P. Marcellin. 1988. Regulation by $CO_2$ of 1-aminocyclopropane-1-carboxylic acid conversion to ethylene in climacteric fruits. Physiologia Plantarum 72:535–540.

Cotton, J. L. S., C. W. Ross, D. H. Byrne, and J. T. Colbert. 1990. Down-regulation of phytochrome mRNA abundance by red light and benzyladenine in etiolated cucumber cotyledons. Plant Molecular Biology 14:707–714.

Creelman, R. A. 1989. Abscisic acid physiology and biosynthesis in higher plants. Physiologia Plantarum 75:131–136.

Davies, P. J. (Ed.). 1987. Plant Hormones and Their Role in Plant Growth and Development. Martinus Nijhoff, Boston.

Davies, W. J. and T. A. Mansfield. 1988. Abscisic acid and drought resistance in plants. ISI Atlas of Science: Animal and Plant Sciences 263–269.

Duckham, S. C., I. B. Taylor, R. S. T. Linforth, R. J. Al-Naieb, B. A. Marples, and W. R. Bowman. 1989. The metabolism of *cis* ABA-aldehyde by the wilty mutants of potato, pea and *Arabidopsis thaliana*. Journal of Experimental Botany 40: 901–905.

Durand, R. and B. Durand. 1984. Sexual differentiation in higher plants. Physiologia Plantarum 60:267–274.

Eisinger, W. 1983. Regulation of pea internode expansion by ethylene. Annual Review of Plant Physiology 34:225–240.

Ernst, D., W. Schäfer, and D. Oesterhelt. 1983. Isolation and identification of a new, naturally occurring cytokinin (6-benzylaminopurine-riboside) from an anise cell culture (*Pimpinella anisum* L.). Planta 159:222–225.

Evans, M. L. and P. M. Ray. 1969. Timing of the auxin response in coleoptiles and its implications regarding auxin action. Journal of General Physiology 53:1–20.

Evans, P. T. and R. L. Malmberg. 1989. Do polyamines have roles in plant development? Annual Review of Plant Physiology and Plant Molecular Biology 40:235–269.

Flores, H. E., C. M. Protacio, and M. W. Signs. 1989. Primary and secondary metabolism of polyamines in plants. Pages 329–393 *in* J. E. Poulton, J. T. Romero, and E. E. Conn (eds.), Plant Nitrogen Metabolism. Plenum, New York.

Flores, S. and E. M. Tobin. 1987. Benzyladenine regulation of the expression of two nuclear genes for chloroplast proteins. Pages 123–132 *in* J. E. Fox and M. Jacobs (eds.), Molecular Biology of Plant Growth Control. Alan R. Liss, New York.

Fosket, D. E. 1977. The regulation of the plant cell cycle by cytokinin. Pages 62–91 *in* T. L. Rost and E. M. Gifford, Jr. (eds.), Mechanisms and Control of Cell Division. Dowden, Hutchinson, and Ross, Stroudsburg, Pa.

Fosket, D. E., L. C. Morejohn, and K. E. Westerling. 1981. Control of growth by cytokinin: An examination of tubulin synthesis during cytokinin-induced growth in cultured cells of Paul's scarlet rose. Pages 193–211 *in* J. Guern and C. Péaud-Lenoël (eds.), Metabolism and Molecular Activities of Cytokinins. Springer-Verlag, Berlin.

Fujino, D. W., D. W. Burger, and K. J. Bradford. 1989. Ineffectiveness of ethylene biosynthetic and action inhibitors in phenotypically reverting the *Epinastic* mutant of tomato (*Lycopersicon esculentum* Mill.). Journal of Plant Growth Regulation 8:53–61.

Galston, A. W. and R. Kaur-Sawhney. 1987. Polyamines as endogenous growth regulators. Pages 280–295 *in* P. J. Davies (ed.), Plant Hormones and Their Role in Plant Growth and Development. Martinus Nijhoff, Boston.

Gersani, M. and H. Kende. 1982. Studies on cytokinin-stimulated translocation in isolated bean leaves. Journal of Plant Growth Regulation 1:161–171.

Goldthwaite, J. J. 1987. Hormones in plant senescence. Pages 553–573 *in* P. J. Davies (ed.), Plant Hormones and the Role in Plant Growth and Development. Martinus Nijhoff, Boston.

Gorham, J. 1977. Lunularic acid and related compounds in liverworts, algae, and *Hydrangea*. Phytochemistry 16:249–253.

Greene, E. M. 1980. Cytokinin production by microorganisms. The Botanical Review 46:25–74.

Guern, J. and C. Péaud-Lenoël. (Eds.). 1981. Metabolism and Molecular Activities of Cytokinins. Springer-Verlag, Berlin.

Guy, C. L. 1990. Cold acclimation and freezing stress tolerance: Role of protein metabolism. Annual Review of Plant Physiology and Plant Molecular Biology 41:187–233.

Guzman, P. and J. R. Ecker. 1990. Exploiting the triple response of *Arabidopsis* to identify ethylene-related mutants. The Plant Cell 2:513–523.

Halevy, A. H. and S. Mayak. 1981. Senescence and postharvest physiology of cut flowers — Part 2. Horticultural Reviews 3:59–143.

Harris, M. J. and W. H. Outlaw, Jr. 1990. Histochemical technique: A low-volume, enzyme-amplified immunoassay with sub-fmol sensitivity. Application to measurement of abscisic acid in stomatal guard cells. Physiologia Plantarum 78:495–500.

Hasegawa, K. and T. Hashimoto. 1975. Variation in abscisic acid and batasin content of yam bulbils — effects of stratification and light exposure. Journal of Experimental Botany 26:757–764.

Hasegawa, P. M., R. A. Bressan, and A. K. Handa. 1987. Cellular mechanism of salinity tolerance. HortScience 21:1317–1324.

Hoffman, N. E. and S. F. Yang. 1980. Changes of 1-aminocyclopropane-1-carboxylic acid content in ripening fruits in relation to their ethylene production rates. Journal of the American Society for Horticultural Science 105:492–495.

Horgan, R. 1984. Cytokinins. Pages 53–75 *in* M. B. Wilkins (ed.), Advanced Plant Physiology. Pitman, London.

Houssa, C., A. Jacqmard, and G. Bernier. 1990. Activation of replicon origins as a possible target for cytokinins in shoot meristems of *Sinapis*. Planta 181:324–326.

Huff, A. K. and C. W. Ross. 1975. Promotion of radish cotyledon enlargement and reducing sugar content by zeatin and red light. Plant Physiology 56:429–433.

Imaseki, H., N. Nakajima, and I. Todaka. 1988. Biosynthesis of ethylene and its regulation in plants. Pages 205–227 *in* Beltsville Symposium in Agriculture Research 12. Biomechanisms Regulating Growth and Development. Kluwer Academic Publishers, Boston.

Iraki, N. M., R. A. Bressan, P. M. Hasegawa, and N. C. Carpita. 1989. Alteration of the physical and chemical structure of the primary cell wall of growth-limited plant cells adapted to osmotic stress. Plant Physiology 91:39–47.

Jackson, M. B. 1985a. Ethylene and responses of plants to soil waterlogging and submergence. Annual Review of Plant Physiology 36:145–174.

Jackson, M. B. 1985b. Ethylene and the response of plants to excess water in their environment — a review. Pages 241–265 *in* J. A. Roberts and G. A. Tucker (eds.), Ethylene and Plant Development. Butterworths, London.

Jameson, P. E., D. S. Letham, R. Zhang, C. W. Parker, and J. Badenoch-Jones. 1987. Cytokinin translocation and metabolism in lupin species. I. Zeatin riboside introduced into the xylem at the base of *Lupinus angustifolius* stems. Australian Journal of Plant Physiology 14:695–718.
Kader, A. A., D. Zagory, and E. L. Kerbel. 1989. Modified atmosphere packaging of fruits and vegetables. CRC Critical Reviews in Food Science and Nutrition 28:1–30.
Kawase, M. 1981. Anatomical and morphological adaptation of plants to waterlogging. HortScience 16:30–34.
Kelly, M. O. and P. J. Davies. 1988. The control of whole plant senescence. CRC Critical Reviews in Plant Sciences 7:139–173.
Kende, H. 1989. Enzymes of ethylene biosynthesis. Plant Physiology 91:1–4.
Kermode, A. R. 1990. Regulatory mechanisms involved in the transition from seed development to germination. CRC Critical Reviews in Plant Sciences 9:155–195.
King, R. A. and J. Van Staden. 1988. Differential responses of buds along the shoot of *Pisum sativum* to isopentenyladenine and zeatin application. Plant Physiology and Biochemistry 26:253–259.
Knee, M. 1985. Evaluating the practical significance of ethylene in fruit storage. Pages 297–315 *in* J. A. Roberts and G. A. Tucker (eds.), Ethylene and Plant Development. Butterworths, London.
Kozlowski,T. T. (Ed.). 1973. The Shedding of Plant Parts. Academic Press, New York.
Leopold, A. C. and M. Kawase. 1964. Benzyladenine effects on bean leaf growth and senescence. American Journal of Botany 51:294–298.
Leshem, Y. Y. 1988. Plant senescence processes and free radicals. Free Radical Biology and Medicine 5:39–49.
Letham, D. S. 1971. Regulators of cell division in plant tissues. XII. A cytokinin bioassay using excised radish cotyledons. Physiologia Plantarum 25:391–396.
Letham, D. S. 1974. Regulators of cell division in plant tissues. XX. The cytokinins of coconut milk. Physiologia Plantarum 32:66–70.
Letham, D. S. and L. M. S. Palni. 1983. The biosynthesis and metabolism of cytokinins. Annual Review of Plant Physiology 34:163–197.
Lew, R. and H. Tsuji. 1982. Effect of benzyladenine treatment duration on delta-aminolevulinic acid accumulation in the dark, chlorophyll lag phase abolition, and long-term chlorophyll production in excised cotyledons of dark-grown cucumber seedlings. Plant Physiology 69:663–667.
Lieberman, M. 1979. Biosynthesis and action of ethylene. Annual Review of Plant Physiology 30:533–589.
Loy, J. B. 1980. Promotion of hypocotyl elongation in watermelon seedlings by 6-benzyladenine. Journal of Experimental Botany 31:743–750.
Ludford, P. M. 1987. Postharvest hormone changes in vegetables and fruit. Pages 574–592 *in* P. J. Davies (ed.), Plant Hormones and Their Role in Plant Growth and Development. Martinus Nijhoff, Boston.
Lynch, J., V. S. Polito, and A. Läuchli. 1989. Salinity stress increases cytoplasmic Ca activity in maize root protoplasts. Plant Physiology 90:1271–1274.
Malamy, J., J. P. Carr, D. F. Klessig, and I. Raskin. 1990. Salicylic acid: A likely endogenous signal in the resistance response of tobacco to viral infection. Science 250:1002–1004.
Martin, R. C., M. C. Mok, G. Shaw, and D. W. S. Mok. 1989. An enzyme mediating the conversion of zeatin to dihydrozeatin in *Phaseolus* embryos. Plant Physiology 90:1630–1635.
Matsubara, S. 1990. Structure-activity relationships of cytokinins. Plant Sciences 9:17–57.
Mattoo, A. K. and N. Aharoni. 1988. Ethylene and plant senescence. Pages 241–280 *in* L. D. Noodén and A. C. Leopold (eds.), Senescence and Aging in Plants. Academic Press, New York.
Mattoo, A. K. and J. C. Suttle. (Eds.). 1990. The Plant Hormone Ethylene. CRC Press, Boca Raton, Fla.
Mayne, R. G. and H. Kende. 1986. Ethylene biosynthesis in isolated vacuoles of *Vicia faba* L. — requirement for membrane integrity. Planta 167:159–165.
McGaw, B. A. 1987. Cytokinin biosynthesis and metabolism. Pages 76–93 *in* P. J. Davies (ed.), Plant Hormones and Their Role in Plant Growth and Development. Martinus Nijhoff, Boston.
McKeon, T. A. and S. F. Yang. 1987. Biosynthesis and metabolism of ethylene. Pages 94–112 *in* P. J. Davies (ed.), Plant Hormones and Their Role in Plant Growth and Development. Martinus Nijhoff, Boston.
Medford, J. I., R. Horgan, Z. El-Sawi, and H. J. Klee. 1989. Alterations of endogenous cytokinins in transgenic plants using a chimeric isopentenyl transferase gene. The Plant Cell 1:403–413.
Meeuse, B. A. D. and I. Raskin. 1988. Sexual reproduction in the arum lily family, with emphasis on thermogenicity. Sexual Plant Reproduction 1:3–5.
Métraux, J. P., H. Signer, J. Ryals, E. Ward, M. Wyss-Benz, J. Gaudin, K. Raschdorf, E. Schmid, W. Blum, and B. Inverardi. 1990. Increase in salicylic acid at the onset of systemic acquired resistance in cucumber. Science 250:1004–1006.
Milborrow, B. V. 1984. Inhibitors. Pages 76–110 *in* M. B. Wilkins (ed.), Advanced Plant Physiology. Pitman, London.
Miller, C. O. 1961. Kinetin and related compounds in plant growth. Annual Review of Plant Physiology 12:395–408.
Moore, T. C. 1989. Biochemistry and Physiology of Plant Hormones. Springer-Verlag, Berlin.
Morris, R. O. 1986. Genes specifying auxin and cytokinin biosynthesis in phytopathogens. Annual Review of Plant Physiology 37:509–538.
Morris, R. O. 1987. Genes specifying auxin and cytokinin biosynthesis in prokaryotes. Pages 636–655 *in* P. J. Davies (ed.), Plant Hormones and Their Role in Plant Growth and Development. Martinus Nijhoff, Boston.
Mudge, K. W. 1988. Effect of ethylene on rooting. Pages 150–161 *in* T. D. Davis, B. E. Haissig, and N. Sankhla (eds.), Adventitious Root Formation in Cuttings. Dioscorides Press, Portland, Ore.
Nandi, S. K., L. M. S. Palni, D. S. Letham, and O. C. Wong. 1989. Identification of cytokinins in primary crown gall tumors of tomato. Plant, Cell and Environment 12:273–283.
Napier, R. M. and M. A. Venis. 1990. Receptors for plant growth regulators: Recent advances. Journal of Plant Growth Regulation 9:113–126.
Narain, A. and M. M. Laloraya. 1974. Cucumber cotyledon expansion as a bioassay for cytokinins. Zeitschrift für Pflanzenphysiologie 71:313–322.
Nester, E. W. and T. Kosuge. 1981. Plasmids specifying plant hyperplasias. Annual Review of Microbiology 35:531–565.
Ng, P. P., A. L. Cole, P. E. Jameson, and J. A. Mcwha. 1982. Cytokinin production by ectomycorrhizal fungi. New Phytologist 91:57–62.
Nickell, L. G. 1979. Controlling biological behavior of plants with synthetic plant growth regulating chemicals. Pages 263–279 *in* N. B. Mandava (ed.), Plant Growth Substances. American Chemical Society, Washington, D. C.
Noodén, L. D. and J. J. Guiamét. 1989. Regulation of assimilation and senescence by the fruit in monocarpic plants. Physiologia Plantarum 77:267–274.
Noodén, L. D. and A. C. Leopold. (Eds.). 1988. Senescence and Aging in Plants. Academic Press, New York.
Ohya, T. and H. Suzuki. 1988. Cytokinin-promoted polyribosome formation in excised cucumber cotyledons. Journal of Plant Physiology 133: 295–298.
Osborne, D. J. 1989. Abscission. CRC Critical Reviews in Plant Sciences 8:103–129.
Osborne, D. J., M. T. McManus, and J. Webb. 1985. Target cells for ethylene action. Pages 197–212 *in* J. A. Roberts and G. A. Tucker (eds.), Ethylene and Plant Development. Butterworths, London.
Owen, J. H. 1988. Role of abscisic acid in a $Ca^{2+}$ second messenger system. Physiologia Plantarum 72:637–641.
Parthier, B. 1979. The role of phytohormones (cytokinins) in chloroplast development (a review). Biochemie und Physiologie der Pflanzen 174: 173–214.
Parthier, B. 1989. Hormone-induced alterations in plant gene expression.Biochemie und Physiologie der Pflanzen 185:289–314.
Parthier, B. 1990. Jasmonates: Hormonal regulators or stress factors in leaf senescence? Journal of Plant Growth Regulation 9:57–63.
Pillay, I. and I. D. Railton. 1983. Complete release of axillary buds from apical dominance in intact, light-grown seedlings of *Pisum sativum* L. following a single application of cytokinin. Plant Physiology 71:972–974.
Powell, L. E. 1987. The hormonal control of bud and seed dormancy in woody plants. Pages 539–552 *in* P. J. Davies (ed.), Plant Hormones and Their Role in Plant Growth and Development. Martinus Nijhoff, Boston.
Quatrano, R. S. 1987. The role of hormones during seed development. Pages 494–514 *in* P. J. Davies (ed.), Plant Hormones and Their Role in Plant Growth and Development. Martinus Nijhoff, Boston.
Raschke, K. 1987. Action of abscisic acid on guard cells. Pages 253–270 *in* E. Zeiger, G. D. Farquhar, and I. R. Cowan (eds.), Stomatal Function. Stanford University Press, Stanford, Calif.
Raskin, I. and H. Kende. 1985. Mechanism of aeration in rice. Science 228:327–329.
Rayle, D. L., C. W. Ross, and N. Robinson. 1982. Estimation of osmotic parameters accompanying zeatin-induced growth of detached cucumber cotyledons. Plant Physiology 70:1634–1636.
Reid, J. B. 1987. The genetic control of growth via hormones. Pages 318–340 *in* P. J. Davies (ed.), Plant Hormones and Their Role in Plant Growth and Development. Martinus Nijhoff, Boston.
Reid, J. B. 1990. Phytohormone mutants in plant research. Journal of Plant Growth Regulation 9:97–111.
Reid, M. S. 1987. Ethylene in plant growth, development, and senescence. Pages 257–279 *in* P. J. Davies (ed.), Plant Hormones and Their Role in Plant Growth and Development. Martinus Nijhoff, Boston.
Rennenberg, H. 1984. The fate of excess sulfur in higher plants. Annual Review of Plant Physiology 35:121–154.
Ridge, I. 1985. Ethylene and petiole development in amphibious plants. Pages 229–239 *in* J. A. Roberts and G. A. Tucker (eds.), Ethylene and Plant Development. Butterworths, London.
Ries, S. K. 1985. Regulation of plant growth with triacontanol. CRC Critical Reviews in Plant Sciences 2:239–285.
Roberts, J. A. and G. A. Tucker. (Eds.). 1985. Ethylene and Plant Development. Butterworths, London.
Rock, C. D. and J. A. D. Zeevaart. 1990. Abscisic (ABA)-aldehyde is a precursor to, and 1',4'-trans-ABA-diol a catabolite of, ABA in apple. Plant Physiology 93:915–923.
Romanov, G. A., V. Y. Taran, L. Chvojka, and O. N. Kulaeva. 1988. Receptor-like cytokinin binding protein(s) from barley leaves. Journal of Plant Growth Regulation 7:1–17.
Ross, C. W. and D. L. Rayle. 1982. Evaluation of $H^+$ secretion relative to zeatin-induced growth of detached cucumber cotyledons. Plant Physiology 70:1470–1474.
Schnapp, S. R., R. A. Bressan, and P. M. Hasegawa. 1990. Carbon use efficiency and cell expansion of NaCl-adapted tobacco cells. Plant Physiology 93:384–388.
Scott, I. M. 1990. Plant hormone response mutants. Physiologia Plantarum 78:147–152.
Sexton, R. and H. W. Woolhouse. 1984. Senescence and abscission. Pages 469–497 *in* M. B. Wilkins (ed.), Advanced Plant Physiology. Pitman, London.

Sexton, R., M. L. Durbin, L. N. Lewis, and W. W. Thomson. 1981. The immunocytochemical localization of 9.5 cellulase in abscission zones of bean (*Phaseolus vulgaris* cv. red kidney). Protoplasma 109:335–347.

Sexton, R., L. N. Lewis, and P. Kelly. 1985. Ethylene and abscission. Pages 173–196 *in* J. A. Roberts and G. A. Tucker (eds.), Ethylene and Plant Development. Butterworths, London.

Sindhu, R. K., D. H. Griffin, and D. C. Walton. 1990. Abscisic aldehyde is an intermediate in the enzymatic conversion of xanthoxin to abscisic acid in *Phaseolus vulgaris* L. leaves. Plant Physiology. 93:689–694.

Singh, N. K., C. A. Bracker, P. M. Hasegawa, A. K. Handa, S. Buckel, M. A. Hermodson, E. Pfankoch, F. Regnier, and R. A. Bressan. 1987. Characterization of osmotin. A thaumatin-like protein associated with osmotic adaptation in plant cells. Plant Physiology 85:529–536.

Sisler, E. C. 1990a. Ethylene-binding components in plants. *In* A. K. Mattoo and J. C. Suttle (eds.), The Plant Hormone Ethylene. CRC Press, Boca Raton, Fla.

Sisler, E. C. 1990b. Ethylene-binding receptors—Is there more than one? Pages 193–200 *in* S. Rood and R. P. Pharis (eds.), Plant Growth Regulators 1988. Springer-Verlag, Berlin.

Sisler, E. C. and C. Wood. 1988. Interaction of ethylene and $CO_2$. Physiologia Plantarum 73: 440–444.

Sisler, E. C. and S. F. Yang. 1984. Ethylene, the gaseous plant hormone. BioScience 34:234–238.

Skene, K. G. M. 1975. Cytokinin production by roots as a factor in the control of plant growth. Pages 365–396 *in* J. G. Torrey and D. T. Clarkson (eds.), The Development and Function of Roots. Academic Press, New York.

Skoog, F. and D. J. Armstrong. 1970. Cytokinins. Annual Review of Plant Physiology 21:359–384.

Skoog, F. and N. J. Leonard. 1968. Sources and structure: Activity relationships of cytokinins. Pages 1–18 *in* F. Wightman and G. Setterfield (eds.), Biochemistry and Physiology of Plant Growth Substances. Runge Press, Ottawa, Canada.

Skriver, K. and J. Mundy. 1990. Gene expression in response to abscisic acid and osmotic stress. The Plant Cell 2:503–512.

Smart C., J. Longland, and A. Trewavas. 1987. The turion: A biological probe for the molecular action of abscisic acid. Pages 345–359 *in* J. E. Fox and M. Jacobs (eds.), Molecular Biology of Plant Growth Control. Alan R. Liss, New York.

Smigocki, A. C. and L. D. Owens. 1989. Cytokinin-to-auxin ratios and morphology of shoots and tissues transformed by a chimeric isopentenyl transferase gene. Plant Physiology 91:808–811.

Smith, T. A. 1985. Polyamines. Annual Review of Plant Physiology 36:117–144.

Spanier, K., J. Schell, and P. H. Schreier. 1989. A functional analysis of T-DNA gene *6b*: The fine tuning of cytokinin effects on shoot development. Molecular and General Genetics 219:209–216.

Stead, A. D. 1985. The relationship between pollination, ethylene production and flower senescence. Pages 71–81 *in* J. A. Roberts and G. A. Tucker (eds.), Ethylene and Plant Development. Butterworths, London.

Stewart, R. N., M. Lieberman, and A. T. Kunishi. 1974. Effects of ethylene and gibberellic acid on cellular growth and development in apical and subapical regions of etiolated pea seedlings. Plant Physiology 54:1–3.

Stoddart, J. L. and H. Thomas. 1982. Leaf senescence. Pages 592–636 *in* D. Boulter and B. Parthier (eds.), Encyclopedia of Plant Physiology, New Series, Vol. 14A, Nucleic Acids and Proteins in Plants I. Springer-Verlag, Berlin.

Sturtevant, D. B. and B. J. Taller. 1989. Cytokinin production by *Bradyrhizobium japonicum*. Plant Physiology 89:1247–1252.

Tan, Z-Y. and K. V. Thimann. 1989. The roles of carbon dioxide and abscisic acid in the production of ethylene. Physiologia Plantarum 75:13–19.

Tanino, K., C. J. Weiser, L. H. Fuchigami, and T. T. H. Chen. 1990. Water content during abscisic acid induced freezing tolerance in bromegrass cells. Plant Physiology 93:460–464.

Taylorson, R. B. 1979. Response of weed seeds to ethylene and related hydrocarbons. Weed Science 27:7–10.

Thimann, K. V. (Ed.). 1980. Senescence in Plants. CRC Press, Boca Raton, Fla.

Thimann, K. V. 1987. Plant senescence: A proposed integration of the constituent processes. Pages 1–19 *in* W. W. Thomson, E. A. Nothnagel, and R. C. Huffaker (eds.), Plant Senescence: Its Biochemistry and Physiology. American Society of Plant Physiologists, Rockville, Md.

Thomas, J., C. W. Ross, C. J. Chastain, N. Koomanoff, J. E. Hendrix, and E. van Volkenburgh. 1981. Cytokinin-induced wall extensibility in excised cotyledons of radish and cucumber. Plant Physiology 68:107–110.

Thompson, J. E., R. L. Legge, and T. F. Barber. 1987. The role of free radicles in senescence and wounding. New Phytologist 105:317–344.

Thomson, W. W., E. A. Nothnagel, and R. C. Huffaker. (Eds.). 1987. Plant Senescence: Its Biochemistry and Physiology. American Society of Plant Physiologists, Rockville, Md.

Torrey, J. G. 1976. Root hormones and plant growth. Annual Review of Plant Physiology 27:435–459.

Tseng, M. J. and P. H. Li. 1990. Alterations of gene expression in potato (*Solanum commersonii*) during cold acclimation. Physiologia Plantarum 78: 538–547.

Tucker, G. A. and D. Grierson. 1987. Fruit ripening. Pages 265–318 *in* D. D. Davies (ed.), The Biochemistry of Plants, Vol. 12. Physiology of Metabolism. Academic Press, New York.

Vanderhoef, L. N., C. Stahl, N. Siegel, and R. Zeigler. 1973. The inhibition by cytokinin of auxin-promoted elongation in excised soybean hypocotyl. Physiologia Plantarum 29:22–27.

Van der Krieken, W. M., A. F. Croes, M. J. M. Smulders, and G. J. Wullems. 1990. Cytokinins and flower bud formation in vitro in tobacco. Plant Physiology 92:565–569.

Van Staden, J., A. D. Bayley, S. J. Upfold, and F. E. Drewes. 1990. Cytokinins in cut carnation flowers. VIII. Uptake, transport and metabolism of benzyladenine and the effect of benzyladenine derivatives on flower longevity. Journal of Plant Physiology 135:703–707.

Van Staden, J., E. L. Cook, and L. D. Noodén. 1988. Pages 281–328 *in* L. D. Noodén and A. C. Leopold (eds.), Senescence and Aging in Plants. Academic Press, New York.

Venkatarayappa, T., R. A. Fletcher, and J. E. Thompson. 1984. Retardation and reversal of senescence in bean leaves by benzyladenine and decapitation. Plant and Cell Physiology 25: 407–418.

Walker-Simmons, M., D. A. Kudrna, and R. L. Warner. 1989. Reduced accumulation of ABA during water stress in a molybdenum cofactor mutant of barley. Plant Physiology 90:728–733.

Walton, D. C. 1987. Abscisic acid biosynthesis and metabolism. Pages 113–131 *in* P. J. Davies (ed.), Plant Hormones and Their Role in Plant Growth and Development. Martinus Nijhoff, Boston.

Wang, H. and W. R. Woodson. 1989. Reversible inhibition of ethylene action and interruption of petal senescence in carnation flowers by norbornadiene. Plant Physiology 89:434–438.

Weichmann, J. 1986. The effect of controlled-atmosphere storage on the sensory and nutritional quality of fruits and vegetables. Horticultural Reviews 8:101–127.

Weiler, E. W. and J. Schroder. 1987. Hormone genes and crown gall disease. Trends in Biochemical Sciences 12:271–275.

Woltering, E. J. 1990. Interorgan translocation of 1-aminocyclopropane-1-carboxylic acid and ethylene coordinates senescence in emasculated *Cymbium* flowers. Plant Physiology 92:837–845.

Wright, S. T. C. 1966. Growth and cellular differentiation in the wheat coleoptile (*Triticum vulgare*). II. Factors influencing the growth response to gibberellic acid, kinetin, and indole-3-acetic acid. Journal of Experimental Botany 17:165–176.

Yang, S. F. and N. E. Hoffman. 1984. Ethylene biosynthesis and its regulation in higher plants. Annual Review of Plant Physiology 35:155–189.

Yang, S. F., Y. Liu, L. Su, G. D. Peiser, N. E. Hoffman, and T. McKeon. 1985. Metabolism of 1-aminocyclopropane-1-carboxylic acid. Pages 9–21 *in* J. A. Roberts and G. A. Tucker (eds.), Ethylene and Plant Development. Butterworths, London.

Yopp, J. H., L. H. Aung, and G. L. Steffens. (Eds.). 1986. Bioassays and Other Special Techniques for Plant Hormones and Plant Growth Regulators. Plant Growth Regulator Society of America, Beltsville, Md.

Zeevaart, J. A. D. and R. A. Creelman. 1988. Metabolism and physiology of abscisic acid. Annual Review of Plant Physiology and Plant Molecular Biology. 39:439–473.

Zeevaart, J. A. D., T. G. Heath, and D. A. Gage. 1989. Evidence for a universal pathway of abscisic acid biosynthesis in higher plants from $^{18}O$ incorporation patterns. Plant Physiology 91: 1594–1601.

Zhang, J. and W. J. Davies. 1989. Abscisic acid produced in dehydrating roots may enable the plant to measure the water status of the soil. Plant, Cell and Environment 12:73–81.

Zhang, J. and W. J. Davies. 1990. Changes in the concentration of ABA in xylem sap as a function of changing soil water status can account for changes in leaf conductance and growth. Plant, Cell and Environment 13:277–285.

Zhi-Yi, T. and K. V. Thimann. 1989. The roles of carbon dioxide and abscisic acid in the production of ethylene. Physiologia Plantarum 75:13–19.


## *Chapter 19 The Power of Movement in Plants*


Audus, L. J. 1979. Plant geosensors. Journal of Experimental Botany 30:1051–1073.

Balatti, Pedro A. and Jorge G. Willemöes. 1989. Role of ethylene in the geotropic response of Bermudagrass (*Cynodon dactylon* L. Pers.) stolons. Plant Physiology 91:1251–1254.

Ball, N. G. 1969. Nastic responses. Pages 277–300 *in* M. B. Wilkins (ed.), Physiology of Plant Growth and Development. McGraw-Hill, New York.

Bandurski, Robert S., A. Schulze, P. Dayanandan, and P. B. Kaufman. 1984. Response to gravity by *Zea mays* seedlings. I. Time course of the response. Plant Physiology 74:284–288.

Baskin, Tobias Isaac. 1986. Redistribution of growth during phototropism and nutation in the pea epicotyl. Planta 169:406–414.

Baskin, T. I., Winslow R. Briggs, and Moritoshi Iino. 1986. Can lateral redistribution of auxin account for phototropism in maize coleoptiles? Plant Physiol. 81:306–309.

Baskin, T. I. and Moritoshi Iino. 1987. An action spectrum in the blue and ultraviolet for phototropism in alfalfa. Photochemistry and Photobiology 46:127–136.

Baskin, T. I., M. Iino, P. B. Green, and W. R. Briggs. 1985. High-resolution measurement of growth during first positive phototropism in maize. Plant, Cell and Environment 8:595–603.

Bateson, A. and Francis Darwin. 1888. On a method of studying geotropism. Annals of Botany 2: 65–68.

Behrens, H. M., D. Gradmann, and A. Sievers. 1985. Membrane-potential responses following gravistimulation in roots of *Lepidium sativum* L. Planta 163:463–472.

Behrens, H. M., M. H. Weisenseel, and A. Sievers. 1982. Rapid changes in the pattern of electric current around the root tip of *Lepidium sativum* L. following gravistimulation. Plant Physiology 70:1079–1083.

Beyl, Caula A. and Cary A. Mitchell. 1983. Alteration of growth, exudation rate, and endogenous hormone profiles in mechanically dwarfed sunflower. J. Amer. Soc. Hort. Sci. 108:257–262.

Björkman, Olle and S. B. Powles. 1981. Leaf movement in the shade species *Oxalis oregana*. I. Response to light level and light quality. Annual Report of the Director, Dept. of Plant Biology, Stanford, Calif. Carnegie Institution of Washington Year Book, 80:59–62. (See also following article, Powles and Björkman, pages 63–66, and report in Science News (1981) 120:392.)

Björkman, T. and A. C. Leopold. 1987a. An electric current associated with gravity sensing in maize roots. Plant Physiol. 84:841–846.

Björkman, T. and A. C. Leopold. 1987b. Effect of inhibitors of auxin transport and of calmodulin on a gravisensing-dependent current in maize roots. Plant Physiol. 84:847–850.

Blaauw, A. H. 1909. Die Perzeption des Lichtes. (The perception of light.) Rec. Trav. Botica Neerlandica 5:209–272.

Blaauw, A. H. 1918. Licht und Wachstum III. (Light and growth III.) Mededelingen Landbouwhogeschool Wageningen 15:89–204.

Blaauw, O. H. and G. Blaauw-Jansen. 1970a. The phototropic responses of *Avena* coleoptiles. Acta Botica Neerlandica 19:755–763.

Blaauw, O. H. and G. Blaauw-Jansen. 1970b. Third positive (c-type) phototropism in the *Avena* coleoptile. Acta Botica Neerlandica 19:764–776.

Blake, T. J., R. P. Pharis, and D. M. Reid. 1980. Ethylene, gibberellins, auxin and the apical control of branch angle in a conifer, *Cupressus arizonica*. Planta 148:64–68.

Braam, Janet and Ronald W. Davis. 1990. Rain-, wind-, and touched-induced expression of calmodulin and calmodulin-related genes in *Arabidopsis*. Cell 60:357–364.

Brain, Robert D., J. A. Freeberg, C. V. Weiss, and Winslow R. Briggs. 1977. Blue light-induced absorbance changes in membrane fractions from corn and *Neurospora*. Plant Physiology 59:948–952.

Brauner, L. and F. Böck. 1963. Versuche zur Analyse der geotropischen Perzeption. II. Die Veränderung der osmotischen Saugkraft im Schwerefelt. (Experiments for the analysis of geotropic perception. II. Changes in osmotic suction force in the gravitational field.) Planta 56:416–437.

Brauner, L. and R. Diemer. 1971. Ueber den Einfluss der geotropischen Induktion auf den Wuchsstoffgehalt, die Wuchsstoffverteilung und die Wuchsstoffempfindlichkeit von *Helianthus*-Hypokotylen. (The influence of the geotropic induction on the content and the distribution of auxin in the hypocotyls of *Helianthus* and on their sensitivity to the growth substance.) Planta 97:337–353.

Brauner, L. and A. Hager. 1958. Versuche zur Analyse der geotropischen Perzeption. (Experiments to analyze geotropic perception.) Planta 51:115–147.

Briggs, Winslow R. 1963. Mediation of phototropic responses of corn coleoptiles by lateral transport of auxin. Plant Physiology 38:237–247.

Briggs, W. R. and T. I. Baskin. 1988. Phototropism in higher plants — controversies and caveats. Botanica Acta 101:133–139.

Briggs, W. R. and M. Iino. 1983. Blue-light-absorbing photoreceptors in plants. Philosophical Transactions of the Royal Society of London B303: 347–359.

Britz, Steven J. and Arthur W. Galston. 1982. Physiology of movements in stems of seedling *Pisum sativum* L. cv. Alaska. I. Experimental separation of nutation from gravitropism. Plant Physiol. 70:264–271.

Britz, Steven J. and Arthur W. Galston. 1983. Physiology of movements in the stems of seedling *Pisum sativum* L. cv Alaska. III. Phototropism in relation to gravitropism, nutation, and growth. Plant Physiol. 71:313–318.

Brock, Thomas G. and Peter B. Kaufman. 1988. Altered growth response to exogenous auxin and gibberellic acid by gravistimulation in pulvini of *Avena sativa*. Plant Physiol. 87:130–133.

Brown, Allan H. and David K. Chapman. 1984. Circumnutation observed without a significant gravitational force in spaceflight. Science 225: 230–232.

Brown, Allan H. and David K. Chapman. 1988. Kinetics of suppression of circumnutation by clinostatting favors modified internal oscillator model. Amer. J. Bot. 75(8):1247–1251.

Caspar, Timothy and Barbara G. Pickard. 1989. Gravitropism in a starchless mutant of *Arabidopsis*. Implications for the starch-statolith theory of gravity sensing. Planta 177:185–197.

Caspar, T., C. Somerville, and B. G. Pickard. 1985. Geotropic roots and shoots of a starch-free mutant of *Arabidopsis*. Supplement to Plant Physiol. 77(4):105.

Cholodny, N. 1926. Beiträge zur Analyse der geotropischen Reaktion. (Contributions to the analysis of the geotropic reaction.) Jahrb. Wiss. Bot. 65:447–459.

Ciesielski, Theophil. 1872. Untersuchungen über die Abwärtskrümmung der Wurzel. (Investigations on the downward bending of roots.) Beiträge zur Biologie der Pflanzen 1:1–30. (The Darwins cited Ciesielski's inaugural dissertation, Abwärtskrümmung der Wurzel, Breslau, 1871.)

Clifford, Paul E., D. S. Fensom, B. I. Munt, and W. D. McDowell, 1982. Lateral stress initiates bending responses in dandelion peduncles: A clue to geotropism? Canadian Journal of Botany 60:2671–2673.

Clifford, Paul E., D. M. Reid, and R. P. Pharis. 1983. Endogenous ethylene does not initiate but may modify geobending — a role for ethylene in autotropism. Plant, Cell and Environment 6: 433–436.

Curry, G. M. 1969. Phototropism. Pages 241–273 *in* M. B. Wilkins (ed.), The Physiology of Plant Growth and Development. McGraw-Hill, New York.

Darwin, Charles, assisted by Francis Darwin. 1880. The Power of Movement in Plants. Murray, London. (See also "Authorized Edition," 1896, Appleton, New York; reprinted by De Capo Press, New York, 1966.)

Davies, Eric and Anne Schuster. 1981. Intercellular communication in plants: Evidence for a rapidly generated, bidirectionally transmitted wound signal. Proceedings of the National Academy of Sciences 78:2422–2426.

Dayanandan, P., V. H. Frederick, V. D. Baldwin, and P. B. Kaufman. 1977. Structure of gravity-sensitive sheath and internodal pulvini in grass shoots. American Journal of Botany 64(10):1189–1199.

Dayanandan, P., F. V. Hebard, and P. B. Kaufman. 1976. Cell elongation in the grass pulvinus in response to geotropic stimulation and auxin application. Planta (Berlin) 131:245–252.

Dennison, David S. 1979. Phototropism. Pages 506–508 *in* W. Haupt and M. E. Feinleib (eds.), Physiology of Movements, Vol. 7 of A. Pirson and M. H. Zimmermann (eds.), Encyclopedia of Plant Physiology (New Series). Springer-Verlag, Berlin, Heidelberg, New York.

Dennison, David S. 1984. Phototropism. Pages 149–162 *in* M. B. Wilkins (ed.), Advanced Plant Physiology. Pitman, London and Marshfield, Mass.

Diehl, J. M., C. J. Gorter, G. Van Iterson, Jr., and A. Kleinhoonte. 1939. The influence of growth hormone on hypocotyls of *Helianthus* and the structure of their cell walls. Recueil travauxx botaniques Néerlandais 36:709–798.

Digby, John and Richard D. Firn. 1976. A critical assessment of the Cholodny-Went theory of shoot geotropism. Current Advances in Plant Science 8:953–960.

Dolk, H. E. 1930. Geotropie en groestof. (Geotropism and growth substances.) Dissertation, Utrecht, 1930. English translation by K. V. Thimann, Geotropism and the growth substance, in Rec. trav. bot. Néerl. 33:509–585, 1936.

du Buy, H. G. and E. Nuernbergk. 1934. Phototropismus und Wachstum der Pflanzen. (Phototropism and growth of plants.) II. Ergeb. Biol. 10:207–322.

Ehleringer, J. and I. Forseth. 1980. Solar tracking by plants. Science 210:1094–1098.

El-Antably, H. M. M. 1975. Redistribution of endogenous indole-acetic acid, abscisic acid and gibberellins in geotropically stimulated *Ribes nigrum* roots. Zeitschrift für Pflanzenphysiologie 76:400–410.

Esau, Katherine. 1977. Anatomy of Seed Plants. Wiley, New York.

Evans, Michael L., Randy Moore, and Karl-Heinz Hasenstein. 1986. How roots respond to gravity. Scientific American 255(6):112–119.

Evans, Michael L., Timothy J. Mulkey, and Mary Jo Vesper. 1980. Auxin action on proton influx in corn roots and its correlation with growth. Planta 148:510–512.

Feldman, L. J. 1981. Light-induced inhibitors from intact and cultured caps of *Zea* roots. Planta 153:471–475.

Firn, Richard D. and John Digby. 1980. The establishment of tropic curvatures in plants. Annual Review of Plant Physiology 31:131–148.

Franssen, J. M. and J. Bruinsma. 1981. Relationships between xanthoxin, phototropism, and elongation growth in the sunflower seedling *Helianthus annuus* L. Planta 151:365–370.

Franssen, J. M., Richard D. Firn, and John Digby. 1982. The role of the apex in the phototropic curvature of *Avena* coleoptiles: Positive curvature under conditions of continuous illumination. Planta 155:281–286.

Giridhar, G. and M. J. Jaffe. 1988. Thigmomorphogenesis: XXIII. Promotion of foliar senescence by mechanical perturbation of *Avena sativa* and four other species. Physiologia Plantarum 74:473–480.

Gould, F. W. 1968. Grass Systematics. McGraw-Hill, New York.

Haberlandt, G. 1902. Über die Statolithefunktion der Stärkekörner. (About the statolith function of starch grains.) Berichte der Deutschen Botanisches Gesellschaft 20:189–195.

Hammer, P. A., C. A. Mitchell, and T. C. Weiler. 1974. Height control in greenhouse chrysanthemum by mechanical stress. HortScience 9: 474–475.

Harrison, Marcia A. and Barbara G. Pickard. 1986. Evaluation of ethylene as a mediator of gravitropism by tomato hypocotyls. Plant Physiology 80:592–595.

Harrison, M. A. and B. G. Pickard. 1989. Auxin asymmetry during gravitropism by tomato hypocotyls. Plant Physiology 89:652–657.

Hart, J. W. 1990. Plant Tropisms and Other Growth Movements. Unwin Hyman, London.

Hasegawa, Koji, Masako Sakoda, and Johan Bruinsma. 1989. Revision of the theory of phototropism in plants: A new interpretation of a classical experiment. Planta 178:540–544.

Haupt, Wolfgang and M. E. Feinleib. 1979. Introduction. Pages 1–8 *in* W. Haupt and M. E. Feinleib (eds.), Physiology of Movements, Vol. 7 of A. Pirson and M. H. Zimmermann (eds.), Encyclopedia of Plant Physiology (New Series). Springer-Verlag, Berlin, Heidelberg, New York.

Heathcote, David G. and T. J. Aston. 1970. The physiology of plant nutation. Journal of Experimental Botany 21(69):997–1002.

Heslop-Harrison, Yokande. 1978. Carnivorous plants. Scientific American 238(2):104–115.

Hillman, S. K. and M. B. Wilkins. 1982. Gravity perception in decapped roots of *Zea mays*. Planta 155:267–271.

Hodick, Dieter and Andreas Sievers. 1989. On the mechanism of trap closure of Venus flytrap (*Dionaea muscipula* Ellis). Planta 179:32–42.

Houwink, A. L. 1935. The conduction of excitation in *Mimosa pudica*. Travaux botaniques néerlandais 32:51–91.
Iino, Moritoshi and Winslow R. Briggs. 1984. Growth distribution during first positive phototropic curvature of maize coleoptiles. Plant, Cell and Environment 7:97–104.
Iversen, T. H. 1969. Elimination of geotropic responsiveness in roots of cress (*Lepidium sativum*) by removal of statolith starch. Physiologia Plantarum 22:1251–1262.
Iversen, T. H. 1974. The roles of statoliths, auxin transport, and auxin metabolism in root geotropism. K. norske Vidensk, Selsk, Mus. Miscellanea 15:1–216.
Iversen, T. H. and P. Larsen. 1973. Movement of amyloplasts in the statocytes of geotropically stimulated roots. The pre-inversion effect. Physiologia Plantarum 28:172–181.
Jackson, M. B. and P. W. Barlow. 1981. Root geotropism and the role of growth regulators from the cap: A re-examination. Plant, Cell and Environment 4:107–123.
Jaffe, Mordecai J. 1973. Thigmomorphogenesis: The response of plant growth and development to mechanical stimulation. Planta 114:143–157.
Jaffe, M. J. 1976. Thigmomorphogenesis: A detailed characterization of the response of beans (*Phaseolus vulgaris* L.) to mechanical stimulation. Zeitschrift für Pflanzenphysiologie 77:437–453.
Jaffe, M. J. 1980. Morphogenetic responses of plants to mechanical stimuli or stress. BioScience 30(4):239–243.
Jaffe, M. J. and A. W. Galston. 1968. The physiology of tendrils. Annual Review of Plant Physiology 19:417–434.
Jaffe, M. J., H. Takahashi, and R. L. Biro. 1985. A pea mutant for the study of hydrotropism in roots. Science 230:445–447.
Jensen, William A. and Frank B. Salisbury. 1972. Botany: An Ecological Approach. Wadsworth, Belmont, Calif.
Jensen, W. A. and F. B. Salisbury. 1984. Botany, Second Edition. Wadsworth, Belmont, Calif.
Johnsson, Anders. 1971. Aspects on gravity-induced movements in plants. Quarterly Reviews of Biophysics 2(4):277–320.
Johnsson, A. 1979. Circumnutation. Pages 627–646 *in* W. Haupt and M. E. Feinleib (eds.), Physiology of Movements, Vol. 7 of A. Pirson and M. H. Zimmermann (eds.), Encyclopedia of Plant Physiology (New Series). Springer-Verlag, Berlin, Heidelberg, New York.
Juniper, Barrie E. 1976. Geotropism. Annual Review of Plant Physiology 27:385–406.
Juniper, Barrie E., S. Groves, B. Landua-Schachar, and L. J. Audus. 1966. Root cap and the perception of gravity. Nature (London) 209:93–94.
Kallas, Peter, Wolfram Meier-Augenstein, and Hermann Schildknecht. 1990. The structure-activity relationship of the turgorin PLMF 1 in the sensitive plant *Mimosa pudica* L.: *In vitro* binding of [$^{14}$C-carboxyl]-PLMF 1 to plasma membrane fractions from mimosa leaves and bioassays with PLMF 1-isomeric compounds. Journal of Plant Physiology 136:225–230.
Kaufman, P. B., R. P. Pharis, D. M. Reid, and F. D. Beall. 1985. Investigations into the possible regulation of negative gravitropic curvatures in intact *Avena sativa* plant and in isolated stem segments by ethylene and gibberellin. Physiologia Plantarum 65:237–244.
Kaufman, P. B., I. Song, and N. Ghosheh. 1986. Role of starch statoliths in the upward bending response in gravistimulated barley leaf-sheath pulvini. Plant Physiol. (Supp.) 80(4):8.
Kiss, John Z., Rainer Hertel, and Fred D. Sack. 1989. Amyloplasts are necessary for full gravitropic sensitivity in roots of *Arabidopsis thaliana*. Plants 177:198–206.
Lee, J. S., T. J. Mulkey, and M. L. Evans. 1983. Reversible loss of gravitropic sensitivity in maize roots after tip application of calcium chelators. Science 220:1375–1376.
MacDonald, Ian R. and James W. Hart. 1987. New light on the Cholodny-Went theory. Plant Physiology 84:568–570.
MacDonald, I. R., J. W. Hart, and Dennis C. Gordon. 1983. Analysis of growth during geotropic curvature in seedlings hypocotyls. Plant, Cell and Environment 6:401–406.
Mandoli, Dina F. and Winslow R. Briggs. 1982. Optical properties of etiolated plant tissues. Proceedings of the National Academy of Sciences USA 79:1902–1906.
Mandoli, D. F. and W. R. Briggs. 1983. Physiology and optics of plant tissues. What's New in Plant Physiology 14:13–16.
Mandoli, D. F. and W. R. Briggs. 1984. Fiber-optic plant tissues: Spectral dependence in dark-grown green tissues. Photochemistry and Photobiology 39(3):419–424.
Mauseth, James. 1988. Plant Anatomy. Benjamin/Cummings, Menlo Park, Calif.
Mertens, Rüdiger and Elmar W. Weiler. 1983. Kinetic studies on the redistribution of endogenous growth regulators in gravireacting plant organs. Planta 148:339–348.
Meyer, A. M. 1969. Versuche zur Trennung von 1. positiver and negativer Krümmung der *Avena* Koleoptile. (Experiments to separate first positive and negative curvature of the *Avena* coleoptile.) Zeitschrift für Pflanzenphysiologie 60:135–146.
Meyer, B. S. and D. B. Anderson. 1952. Plant Physiology, Second Edition. Van Nostrand, New York.
Mitchell, Cary A. 1977. Influence of mechanical stress on auxin-stimulated growth of excised pea stem sections. Physiologia Plantarum 41:129–134.
Mitchell, Cary A., Candace J. Severson, John A. Wott, and P. Allen Hammer. 1975. Seismomorphogenic regulation of plant growth. Journal of American Society of Horticultural Science 100(2):161–165.
Moran, Nava, Gerald Ehrenstein, Kunihiko Iwasa, Charles Mischke, Charles Bare, and Ruth L. Satter. 1988. Potassium channels in motor cells of *Samanea saman*, a patch-clamp study. Plant Physiology 88:643–648.
Mueller, Wesley J., Frank B. Salisbury, and P. Thomas Blotter. 1984. Gravitropism in higher plant shoots. II. Dimensional and pressure changes during stem bending. Plant Physiology 76:993–999.
Mulkey, Timothy J. and Michael L. Evans. 1981. Geotropism in corn roots: Evidence for its mediation by differential acid efflux. Science 212:70–71.
Mulkey, Timothy J. and Michael L. Evans. 1982. Suppression of asymmetric acid efflux and gravitropism in maize roots treated with auxin transport inhibitors or sodium orthovanadata. J. Plant Growth Regul. 1:259–265.
Mulkey, Timothy J., Konrad M. Kuzmanoff, and Michael L. Evans. 1981. The agar-dye method for visualizing acid efflux patterns during tropistic curvatures. What's New in Plant Physiology 12:9–12.
Muñoz, V. and W. L. Butler. 1975. Photoreceptor pigments for blue light in *Neurospora crassa*. Plant Physiology 55:421–426.
Neel, P. L. and R. W. Harris. 1971. Motion-induced inhibition of elongation and induction of dormancy in *Liquidambar*. Science 173:58–59.
Němec, B. 1901. Über die Wahrnehmung des Schwerkraftreizes bei den Pflanzen. (About the perception of gravity by plants.) Jahrb. Wiss. Bot. 36:80–178.
Pharis, R. P., R. L. Legge, M. Noma, P. B. Kaufman, N. S. Ghosheh, J. D. LaCroix, and K. Heller. 1981. Changes in endogenous gibberellins and the metabolism of $^{3}$H-GA$_4$ after geostimulation in shoots of the oat plant (*Avena sativa*). Plant Physiology 67:892–897.
Phillips, I. D. J. and W. Hartung. 1976. Longitudinal and lateral transport of [3,4-3H]gibberellin A1 and 3-indolyl(acetic acid-2-14C) in upright and geotropically responding green internode segments from *Helianthus annuus*. New Phytol. 76:1–9.
Pickard, B. G. 1973. Action potentials in higher plants. Botanical Review 39:172–201.
Pickard, B. G. 1985. Roles of hormones, protons and calcium in geotropism. Pages 193–281 *in* R. P. Pharis and D. M. Reid (eds.), Hormonal Regulation of Development III. Encyclopedia of Plant Physiology (New Series). Springer-Verlag, Berlin.
Pickard, B. G. and K. V. Thimann. 1964. Transport and distribution of auxin during tropistic response. II. The lateral migration of auxin in phototropism of coleoptiles. Plant Physiol. 39:341–350.
Ray, Thomas S., Jr. 1979. Slow-motion world of plant "behavior" visible in rain forest. Smithsonian 9(12):121–130.
Ricca, U. 1916a. Solutione di un probleme di fisiologia. La propagatione di stimulo vella *Mimosa*. Nuovo Giorn. bot. ital. N. S. 23:51–170.
Ricca, U. 1916b. Solution d'un probleme de physiologie. La propagation de stimules dans la Sensitive. Arch. ital. Biol. (Pisa) 65:219–232.
Rich, T. C. G., G. C. Whitelam, and H. Smith. 1987. Analysis of growth rates during phototropism: Modifications by separate light-growth responses. Plant, Cell and Environment 10: 303–311.
Roblin, G. 1982. Movements and bioelectrical events induced by photostimulation in the primary pulvinus of *Mimosa pudica*. Zeitschrift für Pflanzenphysiologie 106:299–303.
Rorabaugh, Patricia and F. B. Salisbury. 1989. Gravitropism in higher plant shoots. VI. Changing sensitivity to auxin in gravistimulated soybean hypocotyls. Plant Physiology 91:1329–1338.
Sack, F. and J. Kiss. 1988. Structural asymmetry in rootcap cells of wild type (WT) and starchless mutant (TC7) *Arabidopsis*. Plant Physiol. (Supp.) 86(4):29.
Salisbury, Frank B. 1963. The Flowering Process. Pergamon Press, Oxford, London, New York, Paris.
Salisbury, Frank B. and Ray M. Wheeler. 1981. Interpreting plant responses to clinostating. Plant Physiology 67:677–685.
Salisbury, Frank B., Linda Gillespie, and Patricia Rorabaugh. 1988. Gravitropism in higher plant shoots. V. Changing sensitivity to auxin. Plant Physiol. 88:1186–1194.
Samejima, Michikazu and Takao Sibaoka. 1980. Changes in the extracellular ion concentration in the main pulvinus of *Mimosa pudica* during rapid movement and recovery. Plant and Cell Physiology 21:467–479.
Satter, Ruth L. and Arthur W. Galston. 1981. Mechanisms of control of leaf movements. Annual Reviews of Plant Physiology 32:83–110.
Satter, R. L., M. J. Morse, Youngsook Lee, Richard C. Crain, Gary G. Coté, and Nava Moran. 1988. Light- and clock-controlled leaflet movements in *Samanea saman*: A physiological, biophysical and biochemical analysis. Botanica Acta 101:205–213.
Satter, R. L., D. D. Sabnis, and A. W. Galston. 1970. Phytochrome controlled nyctinasty in *Albizzi julibrissin*. I. Anatomy and fine structure of the pulvinule. American Journal of Botany 57: 374–381.
Schildknecht, Hermann. 1983. Turgorins, hormones of the endogenous daily rhythms of higher organized plants — detection, isolation, structure, synthesis, and activity. Angewandte Chemie Int. Edition English 22:695–710.
Schildknecht, Hermann. 1984. Turgorins — new chemical messengers for plant behaviour. Endeavour, New Series, 8(4):113–117.
Schopfer, P. 1984. Photomorphogenesis. Pages 380–407 *in* M. B. Wilkins (ed.), Advanced Plant Physiology. Pitman, London.
Schrempf, M., Ruth L. Satter, and Arthur W. Galston. 1976. Potassium-linked chloride fluxes during rhythmic leaf movement of *Albizzia julibrissin*. Plant Physiology 58:190–192.

Schwartz, Amnon and Dov Koller. 1978. Phototropic response to vectorial light in leaves of *Lavatera cretica* L. Plant Physiology 61:924–928.
Schwartz, A. and D. Koller. 1980. Role of the cotyledons in the phototropic response of *Lavatera cretica* seedlings. Plant Physiology 66:82–87.
Schwartz, A., Sarah Gilboa, and D. Koller. 1987. Photonastic control of leaf orientation in *Melilotus indicus* (Fabaceae). Plant Physiology 84:318–323.
Scurfield, G. 1973. Reaction wood: Its structure and function. Science 179:647–655.
Shackel, K. A. and A. E. Hall. 1979. Reversible leaflet movements in relation to drought adaptation of cowpeas, *Vigna unguiculata* L. Walp. Aust. J. Plant Physiol. 6:265–276.
Shaw, S. and M. B. Wilkins. 1973. The source and lateral transport of growth inhibitors in geotropically stimulated roots of *Zea mays* and *Pisum sativum*. Planta 109:11–26.
Shen-Miller, J. and R. R. Hinchman. 1974. Gravity sensing in plants: A critique of the statolith theory. BioScience 24:643–651.
Shen-Miller, J., P. Cooper, and S. A. Gordon. 1969. Phototropism and photoinhibition of basipolar transport of auxin in oat coleoptiles. Plant Physiol. 44: 491–496.
Sibaoka, Takao. 1969. Physiology of rapid movements in higher plants. Annual Review of Plant Physiology 20:165–184.
Sievers, Andreas, H. M. Behrens, T. J. Buckhout, and D. Gradmann. 1984. Can a $Ca^{2+}$ pump in the endoplasmic reticulum of the *Lepidium* root be the trigger for rapid changes in membrane potential after gravistimulation? Z. Pflanzenphysiol. 114:195–200.
Simons, P. J. 1981. The role of electricity in plant movements. New Phytologist 87:11–37.
Sinha, S. K. and S. Bose. 1988. Classical Research Papers in Plant Physiology from India. Society for Plant Physiology and Biochemistry, New Delhi. [Four of Prof. Bose's papers were published in this collection; original references are: J. Linnean Soc. 35:275–305 (1902); Phil. Trans. B204:63–97 (1914); Proc. Roy. Soc. 88B:483–507 (1915); Proc. Roy. Soc. 89B:213–231 (1916).]
Sliwinski, Julianne E. and Frank B. Salisbury. 1984. Gravitropism in higher plant shoots. III. Cell dimensions during gravitropic bending; perception of gravity. Plant Physiology 76:1000–1008.
Slocum, Robert D. and Stanley J. Roux. 1983. Cellular and subcellular localization of calcium in gravistimulated oat coleoptiles and its possible significance in the establishment of tropic curvature. Planta 157:481–492.
Song, I., C. R. Lee, T. G. Brock, and P. B. Kaufman. 1988. Do starch statoliths act as the gravisensors in cereal grass pulvini? Plant Physiology 86: 1155–1162.
Steyer, Brigitte. 1967. Die Dosis-Wirkungsrelationen bei geotroper und phototroper Reizung: Vergleich von Mono- mit Dicotyledonen. (The dose-response relations in geotropic and phototropic stimulation: Comparison of monocots with dicots.) Planta (Berl.) 77:277–286.
Strong, Donald R., Jr. and Thomas S. Ray, Jr. 1975. Host tree location behavior of a tropical vine (*Monstera gigantea*) by skototropism. Science 190:804–806.
Suzuki, T., N. Kondo, and T. Fujii. 1979. Distribution of growth regulators in relation to the light-induced geotropic responsiveness in *Zea* roots. Planta 145:323–329.
Tanada, Takuma and Christian Vinten-Johansen. 1980. Gravity induces fast electrical field change in soybean hypocotyls. Plant, Cell and Environment 3:127–130.
Thimann, K. V. and G. M. Curry. 1960. Phototropism and phototaxis. Pages 243–309 *in* M. Florkin and H. S. Mason (eds.), Comparative Biochemistry: A Comparative Treatise. Vol. I, Sources of Free Energy. Academic Press, New York.
Thimann, K. V. and C. L. Schneider. 1938. Differential growth in plant tissues. American Journal of Botany 25(8):627–641.
Tibbitts, T. W. and W. M. Hertzberg. 1978. Growth and epinasty of marigold plants maintained from emergence on horizontal clinostats. Plant Physiology 61:199–203.
Timell, T. E. 1986. Compression Wood in Gymnosperms. In Three Volumes. Springer-Verlag, Berlin, Heidelberg, New York, Tokyo.
Toriyama, H. 1955. Observational and experimental studies of sensitive plants. V. The development of the tannin vacuole of the motor cell of the pulvinus. Bot. Mag. 68:203–208.
Toriyama, H. 1962. Observational and experimental studies of sensitive plants. XV. The migration of potassium in the petiole of *Mimosa pudica*. Cytologia 27:431–442.
Umrath, Karl and G. Kastberger. 1983. Action potentials of the high-speed conduction in *Mimosa pudica* and *Neptunia plena*. Phyton 23:65–78.
Van Sambeek, Jerome W. and Barbara G. Pickard. 1976. Mediation of rapid electrical, metabolic, transpirational, and photosynthetic changes by factors released from wounds. III. Measurements of $CO_2$ and $H_2O$ flux. Canadian Journal of Botany 54:2662–2671.
Vierstra, R. D. and K. L. Poff. 1981. Role of carotenoids in the phototropic response of corn seedlings. Plant Physiology 68:798–801.
Volkmann, D. and A. Sievers. 1979. Graviperception in multicellular organs. Pages 573–600 *in* W. Haupt and M. E. Feinleib (eds.), Physiology of Movements, Vol. 7 of A. Pirson and M. H. Zimmermann (eds.), Encyclopedia of Plant Physiology (New Series). Springer-Verlag, Berlin, Heidelberg, New York.
von Sachs, Julius. 1873. Über das Wachstum der Haupt- und Nebenwurzel. (About the growth of primary and secondary roots.) Arb. Bot. Inst. Wurzburg 1:385–474.
von Sachs, J. 1882. Textbook of Botany (Second English Edition). Clarendon, Oxford, England. (See pages 796–808.)
Wainwright, C. M. 1977. Sun-tracking and related leaf movements in a desert lupine (*Lupinus arizonicus*). American Journal of Botany 64: 1032–1041.
Watanabe, S. and T. Sibaoka. 1983. Light- and auxin-induced leaflet opening in detached pinnae of *Mimosa pudica*. Plant and Cell Physiology 24: 641–647.
Weiler, E. W. 1984. Immunoassay of plant growth regulators. Ann. Rev. Plant Physiol. 35:85–95.
Went, F. W. 1926. On growth accelerating substances in the coleoptile of *Avena sativa*. Proc. K. Akad. Wet. Amsterdam 30:10–19.
Wheeler, Raymond M. and Frank B. Salisbury. 1979. Water spray as a convenient means of imparting mechanical stimulation of plants. HortScience 14(3):270–271.
Wheeler, Raymond M. and Frank B. Salisbury. 1981. Gravitropism in higher plant shoots. I. A role for ethylene. Plant Physiology 67:686–690.
Wheeler, R. M., R. G. White, and F. B. Salisbury. 1986. Gravitropism in higher plant shoots. IV. Further studies on participation of ethylene. Plant Physiol. 82:534–542.
White, Rosemary G. and Fred D. Sack. 1990. Actin microfilaments in presumptive statocytes of root caps and coleoptiles. American Journal of Botany 77:17–26.
Wilkins, Malcolm B. 1975. The role of the root cap in root geotropism. Current Advances in Plant Science 8:317–328.
Wilkins, M. B. 1979. Growth-control mechanisms in gravitropism. Pages 601–626 *in* W. Haupt and M. E. Feinleib (eds.), Physiology of Movements, Vol. 7 of A. Pirson and M. H. Zimmermann (eds.), Encyclopedia of Plant Physiology (New Series). Springer-Verlag, Berlin, Heidelberg, New York.
Wilkins, M. B. 1984. Gravitropism. Pages 163–185 *in* M. B. Wilkins (ed.), Advanced Plant Physiology. Pitman, London and Marshfield, Mass.
Wilkins, H. and R. L. Wain. 1974. The root cap and control of root elongation in *Zea mays* L. seedlings exposed to white light. Planta 121:1–8.
Williams, Stephen E. and Alan B. Bennett. 1982. Leaf closure in the Venus flytrap: An acid growth response. Science 218:1120–1122.
Wilson, Brayton F. and Robert R. Archer. 1977. Reaction wood: Induction and mechanical action. Annual Review of Plant Physiology 28:23–43.
Wright, Luann Z. and David L. Rayle. 1983. Evidence for a relationship between $H^+$ excretion and auxin in shoot gravitropism. Plant Physiology 72:99–104.
Yin, H. C. 1938. Diaphototropic movements of the leaves of *Malva neglecta*. American Journal of Botany 25:1–6.
Zawadski, Tadeusz and Kazimierz Trebacz. 1982. Action potentials in *Lupinus angustifolius* L. shoots. IV. Propagation of action potential in the stem after the application of mechanical block. Journal of Experimental Botany 33(132): 100–110.
Zimmermann, B. D. and W. R. Briggs. 1963. Phototropic dosage-response curves for oat coleoptiles. Plant Physiology 38:248–253.

### *Chapter 20 Photomorphogenesis*

Ballaré, C. L., R. A. Sánchez, A. L. Scopel, J. J. Casal, and C. M. Ghersa. 1987. Early detection of neighbour plants by phytochrome perception of spectral changes in reflected sunlight. Plant, Cell and Environment 10:551–557.
Ballaré, C. L., A. L. Scopel, and R. A. Sánchez. 1990. Far-red radiation reflected from adjacent leaves: An early signal of competition in plant canopies. Science 247:329–332.
Baskin, C. C. and J. M. Baskin. 1988. Germination ecophysiology of herbaceous plant species in a temperate region. American Journal of Botany 75:286–305.
Beale, S. I. 1990. Biosynthesis of the tetrapyrrole pigment precursor Δ-aminolevulinic acid, from glutamate. Plant Physiology 93:1273–1279.
Beggs, C. J., E. Wellmann, and H. Grisebach. 1986. Photocontrol of flavonoid biosynthesis. Pages 467–499 *in* R. E. Kendrick and G. H. M. Kronenberg (eds.), Photomorphogenesis in Plants. Martinus Nijhoff, Boston.
Bewley, J. D. and M. Black. 1982. Physiology and Biochemistry of Seeds, Vol. 2, Viability, Dormancy, and Environmental Control. Springer-Verlag, Berlin.
Bewley, J. D. and M. Black. 1985. Physiology of Seeds. Development and Germination. Plenum, London.
Björn, L. O. 1986. Introduction. Pages 3–16 *in* R. E. Kendrick and G. H. M. Kronenberg (eds.), Photomorphogenesis in Plants. Martinus Nijhoff, Boston.
Blaauw, O. H., G. Blaauw-Jansen, and W. J. van Leeuwen. 1968. An irreversible red-light-induced growth response in *Avena*. Planta 82:87–104.
Blowers, D. P. and A. J. Trewavas. 1989. Rapid cycling of autophosphorylation of a $Ca^{2+}$-calmodulin regulated plasma membrane located protein kinase from pea. Plant Physiology 90:1279–1285.
Borthwick, H. 1972. History of phytochrome. Pages 3–23 *in* K. Mitrakos and W. Shropshire, Jr. (eds.), Phytochrome. Academic Press, New York.
Borthwick, H. A., S. B. Hendricks, E. H. Toole, and V. K. Toole. 1954. Action of light on lettuce seed germination. Botanical Gazette 115:205–225.
Briggs, W. R. 1976. H. A. Borthwick and S. B. Hendricks — pioneers of photomorphogenesis. Pages 1–6 *in* H. Smith (ed.), Light and Plant Development. Butterworths, London.
Caldwell, M. M. 1981. Plant responses to solar ultraviolet radiation. Pages 170–197 *in* O. L. Lange, P. S. Nobel, C. B. Osmond, and H. Ziegler (eds.), Encyclopedia of Plant Physiology, New Series, Vol. 12A, Physiological Plant Ecology I. Springer-Verlag, Berlin.

Carpita, N. C., M. W. Nabors, C. W. Ross, and N. Petretic. 1979. Growth physics and water relations of red-light-induced germination in lettuce seeds. III. Changes in the osmotic and pressure potential in the embryonic axis of red- and far-red-treated seeds. Planta 144:217–224.

Carpita, N. C. and M. W. Nabors. 1981. Growth physics and water relations of red-light-induced germination in lettuce seeds. V. Promotion of elongation in the embryonic axes by gibberellin and phytochrome. Planta 152:131–136.

Casal, J. J. and H. Smith. 1989. The function, action and adaptive significance of phytochrome in light-grown plants. Plant, Cell and Environment 12:855–862.

Casal, J. J., V. A. Deregibus, and R. A. Sánchez. 1985. Variations in tiller dynamics and morphology in *Lolium multiflorum* Lam. vegetative and reproductive plants as affected by differences in red-far-red irradiation. Annals of Botany 56:553–559.

Colbert, J. T. 1988. Molecular biology of phytochrome. Plant, Cell and Environment 11:305–318.

Colbert, J. T. 1990. Regulation of type I phytochrome mRNA abundance. Physiologia Plantarum.

Cone, J. W. and R. E. Kendrick. 1986. Photocontrol of seed germination. Pages 443–466 *in* R. E. Kendrick and G. H. M. Kronenberg (eds.), Photomorphogenesis in Plants. Martinus Nijhoff, Boston.

Coohill, T. P. 1989. Ultraviolet action spectra (280 to 380 nm) and solar effectiveness spectra for higher plants. Photochemistry and Photobiology 50: 451–457.

Cordonnier, M-M. 1989. Yearly review: Monoclonal antibodies: Molecular probes for the study of phytochrome. Photochemistry and Photobiology 49:821–831.

Cosgrove, D. J. 1986. Selected responses. Pages 341–366 *in* R. E. Kendrick and G. H. M. Kronenberg (eds.), Photomorphogenesis in Plants. Martinus Nijhoff, Boston.

Cosgrove, D. J. 1988. Mechanism of rapid suppression of cell expansion in cucumber hypocotyls after blue-light irradiation. Planta 176:109–116.

Cresswell, E. G. and J. P. Grime. 1981. Induction of a light requirement during seed development and its ecological consequences. Nature 291:583–585.

Dale, J. E. 1988. The control of leaf expansion. Annual Review of Plant Physiology and Plant Molecular Biology 39:267–295.

De Greef, J. A. and H. Frédéricq. 1983. Photomorphogenesis and hormones. Pages 401–427 *in* W. Shropshire, Jr. and H. Mohr (eds.), Encyclopedia of Plant Physiology, New Series, Vol. 16A, Photomorphogenesis. Springer-Verlag, Berlin.

Deregibus, V. A., R. A. Sánchez, and J. Casal. 1983. Effects of light quality on tiller production in *Lolium* spp. Plant Physiology 72:900–902.

Downs, R. J. 1962. Photocontrol of growth and dormancy in woody plants. Pages 133–148 *in* T. T. Kozlowski (ed.), Tree Growth. Ronald Press.

Downs, R. J. and H. W. Siegelman. 1963. Photocontrol of anthocyanin synthesis in milo seedlings. Plant Physiology 38:25–30.

Dring, M. G. 1988. Photocontrol of development in algae. Annual Review of Plant Physiology and Molecular Biology 39:157–174.

Esashi, Y., R. Kuraishi, N. Tanaka, and S. Satoh. 1983. Transition from primary to secondary dormancy in cocklebur seeds. Plant, Cell and Environment 6:493–499.

Frankland, B. 1986. Perception of light quantity. Pages 219–235 *in* R. E. Kendrick and G. H. M. Kronenberg (eds.), Photomorphogenesis in Plants. Martinus Nijhoff, Boston.

Frankland, B. and R. Taylorson. 1983. Light control of seed germination. Pages 428–456 *in* W. Shropshire, Jr. and H. Mohr (eds.), Encyclopedia of Plant Physiology, New Series, Vol. 16A, Photomorphogenesis. Springer-Verlag, Berlin.

Furuya, M. 1987a. The history of phytochrome. I. Genesis (The Beltsville era: 1920–1963). Pages 3–8 *in* M. Furuya (ed.), Phytochrome and Photoregulation in Plants. Academic Press, New York.

Furuya, M. (Ed.). 1987b. Phytochrome and Photoregulation in Plants. Academic Press, New York.

Furuya, M. 1989. Molecular properties and biogenesis of phytochrome I and II. Advanced Biophysics 25:133–167.

Galland, P. and H. Senger. 1988. New trends in photobiology (invited review): The role of flavins as photoreceptors. Journal of Photochemistry and Photobiology, B: Biology 1:277–294.

Grime, J. P. 1979. Plant Strategies and Vegetation Processes. Wiley, London.

Grime, J. P. 1981. Plant strategies in shade. Pages 159–186 *in* H. Smith (ed.), Plants and the Daylight Spectrum. Academic Press, New York.

Groot, S. P. C. and C. M. Karssen. 1987. Gibberellins regulate seed germination in tomato by endosperm weakening: A study with gibberellin-deficient mutants. Planta 171:525–531.

Groot, S. P. C., B. Kieliszewska-Rokicka, E. Vermeer, and C. M. Karssen. 1988. Gibberellin-induced hydrolysis of endosperm cell walls in gibberellin-deficient tomato seeds prior to radicle protrusion. Planta 174:500–504.

Hahlbrock, K. and D. Scheel. 1989. Physiology and molecular biology of phenylpropanoid metabolism. Annual Review of Physiology and Plant Molecular Biology 40:347–369.

Hansjörg, A. W., H. A. W. Schneider-Poetsch, B. Braun, and W. Rüdiger. 1989. Phytochrome—all regions marked by a set of monoclonal antibodies reflect conformational changes. Planta 177: 511–514.

Hartmann, K. M. 1967. Ein Wirkungsspektrum der Photomorphogenese unter Hochenergiebedingungen und seine Interpretation auf der Basis des Phytochroms (Hypokotylwachstumshemmung bei *Lactuca sativa* L.). Zeitschrift für Naturforschung 22b:1172–1175.

Haupt, W. 1986. Photomovement. Pages 415–441 *in* R. E. Kendrick and G. H. M. Kronenberg (eds.), Photomorphogenesis in Plants. Martinus Nijhoff, Boston.

Haupt, W. 1987. Phytochrome control of intracellular movement. Pages 225–237 *in* M. Furuya (ed.), Phytochrome and Photoregulation in Plants. Academic Press, New York.

Hedden, P. and J. R. Lenton. 1988. Genetic and chemical approaches to the metabolic regulation and mode of action of gibberellins in plants. Pages 175–204 *in* Beltsville Symposia in Agricultural Resources 12, Biomechanisms Regulating Growth and Development. Kluwer Academic Publishers, Boston.

Holmes, M. G. and H. Smith. 1975. The function of phytochrome in plants growing in the natural environment. Nature 254:512–514.

Hoober, J. K. 1987. The molecular basis of chloroplast development. Pages 1–74 *in* M. D. Hatch and N. K. Boardman (eds.), The Biochemistry of Plants, Vol. 10, Photosynthesis. Academic Press, New York.

Hoshikawa, K. 1969. Underground organs of the seedlings and the systematics of gramineae. Botanical Gazette 130:192–203.

Ikuma, H. and K. V. Thimann. 1959. Photosensitive site in lettuce seeds. Science 13:568–569.

Inoue, Y. and K. Shibata. 1973. Light-induced chloroplast rearrangements and their action spectra as measured by absorption spectrophotometry. Planta 114:341–358.

Jabben, M., J. Shanklin, and R. D. Vierstra. 1989. Ubiquitin phytochrome conjugates. Journal of Biological Chemistry 264:4998–5005.

Jose, A. M. and E. Schäfer. 1978. Distorted phytochrome action spectra in green plants. Planta 138:25–28.

Kansara, M. S., J. Ramdas, and S. K. Srivastava. 1989. Phytochrome mediated photoregulation of NAD kinase in terminal buds of pea seedlings. Journal of Plant Physiology 134:603–607.

Karssen, C. M., S. Zagorski, J. Kepczynski, and S. P. C. Groot. 1989. Key role for endogenous gibberellins in the control of seed germination. Annals of Botany 63:71–80.

Kasemir, H. 1983. Light control of chlorophyll accumulation in higher plants. Pages 662–686 *in* W. Shropshire, Jr. and H. Mohr (eds.), Encyclopedia of Plant Physiology, New Series, Vol. 16B, Photomorphogenesis. Springer-Verlag, Berlin.

Kasperbauer, M. J. and D. L. Karlen. 1986. Light-mediated bioregulation of tillering and photosynthate partitioning in wheat. Physiologia Plantarum 66:159–163.

Kendrick, R. E. and M. E. Bossen. 1987. Photocontrol of ion fluxes and membrane properties in plants. Pages 215–224 *in* M. Furuya (ed.), Phytochrome and Photoregulation in Plants. Academic Press, New York.

Kendrick, R. E. and G. H. M. Kronenberg. (Eds.). 1986. Photomorphogenesis in Plants. Martinus Nijhoff, Boston.

Klein, W. H., L. Price, and K. Mitrakos. 1963. Light stimulated starch degradation in plastids and leaf morphogenesis. Photochemistry and Photobiology 2:233–240.

Koller, D. 1969. The physiology of dormancy and survival of plants in desert environments. Pages 449–469 *in* H. W. Woolhouse (ed.), Dormancy and Survival. Symposia for the Society of Experimental Biology, No. 23. Academic Press, New York.

Koornneef, M. and R. E. Kendrick. 1986. A genetic approach to photomorphogenesis. Pages 521–546 *in* R. E. Kendrick and G. H. M. Kronenberg (eds.), Photomorphogenesis in Plants. Martinus Nijhoff, Boston.

Kronenberg, G. H. M. and R. E. Kendrick. 1986. The physiology of action. Pages 99–114 *in* R. E. Kendrick and G. H. M. Kronenberg (eds.), Photomorphogenesis in Plants. Martinus Nijhoff, Boston.

Lang, G. A., J. D. Early, G. C. Martin, and R. L. Darnell. 1987. Endo-, para-, and ecodormancy: Physiological terminology and classification for dormancy research. HortScience 22:371–377.

Laskowski, M. J. and W. R. Briggs. 1989. Regulation of pea epicotyl elongation by blue light. Fluence-response relationship and growth distribution. Plant Physiology 89:293–298.

Lew, R. R., B. S. Serlin, C. L. Schauf, and M. E. Stockton. 1990. Red light regulates calcium-activated potassium channels in *Mougeotia* plasma membrane. Plant Physiology 92:822–830.

Link, G. 1988. Photocontrol of plastid gene expression. Plant, Cell and Environment 11: 329–338.

López-Figueroa, F., P. Lindemann, S. E. Braslavsky, K. Schaffner, H. A. W. Schneider-Poetsch, and W. Rüdiger. 1989. Detection of a phytochrome-like protein in macroalgae. Botanica Acta 102:178–180.

Mancinelli, A. L. 1980. Yearly review: The photoreceptors of the high irradiance responses of plant photomorphogenesis. Photochemistry and Photobiology 32:853–857.

Mancinelli, A. L. 1985. Light-dependent anthocyanin synthesis: A model system for the study of plant photomorphogenesis. The Botanical Review 51:107–157.

Mancinelli, A. L. 1989. Interaction between cryptochrome and phytochrome in higher plant photomorphogenesis. American Journal of Botany 76:143–154.

Mandoli, D. F. and W. R. Briggs. 1981. Phytochrome control of two low irradiance responses in etiolated oat seedlings. Plant Physiology 67: 733–739.

Marmé, D. 1989. The role of calcium and calmodulin in signal transduction. Pages 57–80 *in* W. F. Boss and D. J. Morré (eds.), Second Messengers in Plant Growth and Development. Alan R. Liss, New York.

Marrs, K. A. and L. S. Kaufman. 1989. Blue-light regulation of transcription for nuclear genes in pea. Proceedings of the National Academy of Sciences USA 86:4492–4495.

McCurdy, D. W. and L. H. Pratt. 1986. Immunogold electron microscopy of phytochrome in *Avena*: Identification of intracellular sites responsible for phytochrome sequestering and enhanced pelletability. Journal of Cell Biology 103: 2541–2550.

Mohr, H. 1957. Der Einfluss monochromatischer Strahlung auf das Langenwachstum des Hypocotyls und auf die Anthocyaninbildung bei Keimlingen von *Sinapis alba* (*Brassica alba* Boiss.). Planta 49:389–405.

Mohr, H. 1983. Pattern specification and realization in photomorphogenesis. Pages 338–357 *in* W. Shropshire, Jr. and H. Mohr (eds.), Encyclopedia of Plant Physiology, New Series, Vol. 16A, Photomorphogenesis. Springer-Verlag, Berlin.

Mohr, H. 1986. Coaction between pigment systems. Pages 547–564 *in* R. E. Kendrick and G. H. M. Kronenberg (eds.), Photomorphogenesis in Plants. Martinus Nijhoff, Boston.

Morgan, D. C. 1981. Shadelight quality effects on plant growth. Pages 205–222 *in* H. Smith (ed.), Plants and the Daylight Spectrum. Academic Press, New York.

Morgan, D. C. and H. Smith. 1976. Linear relationship between phytochrome photoequilibrium and growth in plants under simulated natural radiation. Nature 262:210–212.

Morse, M. J., R. C. Crain, G. G. Cote, and R. L. Satter. 1990. Light-signal transduction via accelerated inositol phospholipid turnover in *Samanea* pulvini. Pages 201–215 *in* D. J. Morré, W. F. Boss, and F. A. Loewus (eds.), Inositol Metabolism in Plants. Wiley-Liss, New York.

Moses, P. B. and N-H. Chua. 1988. Light switches for plant genes. Scientific American 258(4):88–93.

Moysset, L. and E. Simon. 1989. Role of calcium in phytochrome-controlled nyctinastic movements of *Albizzia lophantha* leaflets. Plant Physiology 90:1108–1114.

Nabors, M. W. and A. Lang. 1971. The growth physics and water relations of red-light induced germination in lettuce seeds. I. Embryos germinating in osmoticum. Planta 101:1–25.

Nagatani, A., R. E. Kendrick, M. Koornneef, and M. Furuya. 1989. Partial characterization of phytochrome I and II in etiolated and de-etiolated tissues of a photomorphogenetic mutant (*lh*) of cucumber (*Cucumis sativus* L.) and its isogenic wild type. Plant and Cell Physiology 30:685–690.

Nagatani, A., J. B. Reid, J. J. Ross, A. Dunnewijk, and M. Furuya. 1990. Internode length in *Pisum*. The response to light quality, and phytochrome Type I and II levels in *lv* plants. Journal of Plant Physiology 135:667–674.

Nagy, F., S. A. Kay, and N-H. Chua. 1988. Gene regulation by phytochrome. Trends in Genetics 4:37–42.

Okamuro, J. K. and R. B. Goldberg. 1989. Regulation of plant gene expression: General principles. Pages 1–81 *in* A. Marcus (ed.), The Biochemistry of Plants, Vol. 15, Molecular Biology. Academic Press, New York.

Park, M-H. and Q. Chae. 1989. Intracellular protein phosphorylation in oat (*Avena sativa* L.) protoplasts by phytochrome action. Biochemical and Biophysical Research Communications 162:9–14.

Parker, M. W., S. B. Hendricks, H. A. Borthwick, and F. W. Went. 1949. Spectral sensitivity for leaf and stem growth of etiolated pea seedlings and their similarity to action spectra for photoperiodism. American Journal of Botany 36:194–204.

Pollmann, L. and M. Wettern. 1989. The ubiquitin system in higher and lower plants — pathways in protein metabolism. Botanica Acta 102:21–30.

Pratt, L. H. 1986. Localization within the plant. Pages 61–81 *in* R. E. Kendrick and G. H. M. Kronenberg (eds.), Photomorphogenesis in Plants. Martinus Nijhoff, Boston.

Pratt, L. H., D. W. McCurdy, Y. Shimazaki, and M. M. Cordonnier. 1986. Immunodetection of phytochrome: Immunocytochemistry, immunoblotting and immunoquantitation. Modern Methods of Plant Analysis, New Series 4:51–74.

Psaras, G. 1984. On the structure of lettuce (*Lactuca sativa* L.) endosperm during germination. Annals of Botany 54:187–194.

Quail, P. H., C. Gatz, H. P. Hershey, A. M. Jones, J. L. Lissemore, B. M. Parks, R. A. Sharrock, R. F. Barker, K. Idler, M. G. Murray, M. Koornneef, and R. E. Kendrick. 1987. Molecular biology of phytochrome. Pages 23–37 *in* M. Furuya (ed.), Phytochrome and Photoregulation in Plants. Academic Press, New York.

Reid, J. B. 1987. The genetic control of growth via hormones. Pages 318–340 *in* P. J. Davies (ed.), Plant Hormones and Their Role in Plant Growth and Development. Martinus Nijhoff, Boston.

Reid, J. B. 1990. Phytohormone mutants in plant research. Journal of Plant Growth Regulation 9:97–111.

Rollin, P. 1972. Phytochrome control of seed germination. Pages 229–254 *in* K. Mitrakos and W. Shropshire, Jr. (eds.), Phytochrome. Academic Press, New York.

Roux, S. J. 1986. Phytochrome and membranes. Pages 115–136 *in* R. E. Kendrick and G. H. M. Kronenberg (eds.), Photomorphogenesis in Plants. Martinus Nijhoff, Boston.

Rüdiger, W. 1986. The chromophore. Pages 17–34 *in* R. E. Kendrick and G. H. M. Kronenberg (eds.), Photomorphogenesis in Plants. Martinus Nijhoff, Boston.

Rüdiger, W. 1987. Biochemistry of the phytochrome chromophore. Pages 127–137 *in* M. Furuya (ed.), Phytochrome and Photoregulation in Plants. Academic Press, New York.

Sánchez, R. A., L. De Miguel, and O. Mercuri. 1986. Phytochrome control of cellulase activity in *Datura ferox* L. seeds and its relationship with germination. Journal of Experimental Botany 37:1574–1580.

Sánchez, R. A., L. Sunell, J. M. Labavitch, and B. A. Bonner. 1990. Changes in the endosperm cell walls of two *Datura* species before radicle protrusion. Plant Physiology 93:89–97.

Schäfer, E. 1986. Primary action of phytochrome. Pages 279–288 *in* M. Furuya (ed.), Phytochrome and Photoregulation in Plants. Academic Press, New York.

Schäfer, E. and W. Haupt. 1983. Blue-light effects in phytochrome-mediated responses. Pages 722–744 *in* W. Shropshire, Jr. and H. Mohr (eds.), Encyclopedia of Plant Physiology, New Series, Vol. 16B, Photomorphogenesis. Springer-Verlag, Berlin.

Schäfer, E., T-U. Lassig, and P. Schopfer. 1982. Phytochrome-controlled extension growth of *Avena sativa* L. seedlings. II. Fluence rate response relationships and action spectra of mesocotyl and coleoptile responses. Planta 154:231–240.

Schopfer, P., K. H. Fidelak, and E. Schäfer. 1982. Phytochrome-controlled extension growth of *Avena sativa* L. seedlings. I. Kinetic characterization of mesocotyl, coleoptile, and leaf responses. Planta 154:224–230.

Scott, I. M. 1990. Plant hormone response mutants. Physiologia Plantarum 78:147–152.

Seitz, K. 1987. Light-dependent movement of chloroplasts in higher plant cells. Acta Physiologiae Plantarum 9:137–148.

Senger, H. (Ed.). 1987. Blue Light Responses, Vols. I and II. CRC Press, Boca Raton, Fla.

Senger, H. and E. D. Lipson. 1987. Problems and prospects of blue and ultraviolet light effects. Pages 315–331 *in* M. Furuya (ed.), Phytochrome and Photoregulation in Plants. Academic Press, New York.

Senger, H. and W. Schmidt. 1986. Cryptochrome and UV receptors. Pages 137–158 *in* R. E. Kendrick and G. H. M. Kronenberg (eds.), Photomorphogenesis in Plants. Martinus Nijhoff, Boston.

Sharrock, R. A. and P. H. Quail. 1989. Novel phytochrome sequences in *Arabidopsis thaliana*: Structure, evolution, and differential expression of a plant regulatory photoreceptor family. Genes and Development 3:1745–1757.

Short, T. W. and W. R. Briggs. 1990. Characterization of a rapid, blue light-mediated change in detectable phosphorylation of a plasma membrane protein from etiolated pea (*Pisum sativum* L.) seedlings. Plant Physiology 92:179–185.

Siegelman, H. W. and S. B. Hendricks. 1957. Photocontrol of anthocyanin formation in turnip and red cabbage seedlings. Plant Physiology 32:393–398.

Siegelman, H. W. and S. B. Hendricks. 1958. Photocontrol of anthocyanin synthesis in apple skin. Plant Physiology 33:185–190.

Simpson, J. and L. Herrera-Estrella. 1990. Light-regulated gene expression. CRC Critical Reviews in Plant Sciences 9:95–109.

Small, J. G. C., C. J. P. Spruit, G. Blaauw-Jansen, and O. H. Blaauw. 1979. Action spectra for light-induced germination in dormant lettuce seeds. II. Blue region. Planta 144:133–136.

Smith, H. 1972. The photocontrol of flavonoid synthesis. Pages 433–481 *in* K. Mitrakos and W. Shropshire, Jr. (eds.), Phytochrome. Academic Press, New York.

Smith, H. 1986. The light environment. Pages 187–217 *in* R. E. Kendrick and G. H. M. Kronenberg (eds.), Photomorphogenesis in Plants. Martinus Nijhoff, Boston.

Smith, H. and G. M. Jackson. 1987. Rapid phytochrome regulation of wheat seedling extension. Plant Physiology 84:1059–1062.

Smith, H. and G. Whitelam. 1987. Phytochrome action in the light-grown plant. Pages 289–303 *in* M. Furuya (ed.), Phytochrome and Photoregulation in Plants. Academic Press, New York.

Smith, H., J. J. Casal, and G. M. Jackson. 1990. Reflection signals and the perception by phytochrome of the proximity of neighbouring vegetation. Plant, Cell and Environment 13:73–78.

Smith, H. and G. C. Whitelam. 1990. Phytochrome, a family of photoreceptors with multiple physiological roles. Plant, Cell and Environment 13: 695–707.

Song, P-S., and I. Yamazaki. 1987. Structure-function relationship of the phytochrome chromophore. Pages 139–156 *in* M. Furuya (ed.), Phytochrome and Photoregulation in Plants. Academic Press, New York.

Spalding, E. P. and D. J. Cosgrove. 1988. Large plasma-membrane depolarization precedes rapid blue-light-induced growth inhibition in cucumber. Planta 178:407–410.

Tanada, T. 1968. A rapid photoreversible response of barley root tips in the presence of 3-indoleacetic acid. Proceedings of the National Academy of Sciences USA 59:376–380.

Tao, K. and A. W. Khan. 1979. Changes in the strength of lettuce endosperm during germination. Plant Physiology 63:126–128.

Thomas, B., S. E. Penn, and B. R. Jordan. 1989. Factors affecting phytochrome transcripts and apoprotein synthesis in germinating embryos of *Avena sativa* L. Journal of Experimental Botany 40:1299–1304.

Thomas, T. H. 1989. Gibberellin involvement in dormancy-break and germination of seeds of celery (*Apium graveolens* L.). Journal of Plant Growth Regulation 8:255–261.

Thompson, W. F. 1988. Photoregulation: Diverse gene responses in greening seedlings. Plant, Cell and Environment 11:319–328.

Thompson, W. F., L. S. Kaufman, B. A. Horwitz, A. D. Sagar, J. C. Watson, and W. R. Briggs. 1988. Patterns of phytochrome-induced gene expression in etiolated pea buds. Beltsville Symposia in Agricultural Research 12. Biomechanisms Regulating Growth and Development. Kluwer Academic Publishers, Boston.

Thomson, B. F. 1954. The effect of light on cell division and cell elongation in seedlings of oats and peas. American Journal of Botany 41:326–332.

Tokuhisha, J. G. and P. H. Quail. 1989. Phytochrome in green-tissue: Partial purification and characterization of the 118-kilodalton phytochrome species from light-grown *Avena sativa* L. Photochemistry and Photobiology 50:143–152.

Van Volkenburgh, E. 1987. Regulation of dicotyledonous leaf growth. Pages 193–201 *in* D. J. Cosgrove and D. P. Knievel (eds.), Physiology of Cell Expansion During Plant Growth. American Society of Plant Physiologists, Rockville, Md.

Vierstra, R. D. and P. H. Quail. 1983. Photochemistry of 124 kilodalton *Avena* phytochrome *in vitro*. Plant Physiology 72:264–267.

Vierstra, R. D. and P. H. Quail. 1986. The protein. Pages 35–60 *in* R. E. Kendrick and G. H. M. Kronenberg (eds.), Photomorphogenesis in Plants. Martinus Nijhoff, Boston.

Vince-Prue, D. 1975. Photoperiodism in Plants. McGraw-Hill, New York.

Vince-Prue, D. 1989. Review: The role of phytochrome in the control of flowering. Flowering Newsletter 8:3–14.

Virgin, H. I. 1989. An analysis of the light-induced unrolling of the grass leaf. Physiologia Plantarum 75:295–298.

Vogelmann, T. C. 1986. Light within the plant. Pages 307–337 *in* R. E. Kendrick and G. H. M. Kronenberg (eds.), Photomorphogenesis in Plants. Martinus Nijhoff, Boston.

Vogelmann, T. C. 1989. Yearly review: Penetration of light into plants. Photochemistry and Photobiology 50:895–902.

Warmbrodt, R. D., W. J. Van Der Woude, and W. O. Smith. 1989. Localization of phytochrome in *Secale cereale* L. by immunogold electron microscopy. Botanical Gazette 150:219–229.

Watson, J. C. 1989. Photoregulation of gene expression in plants. Pages 161–205 *in* S-D. Kung and C. J. Arntzen (eds.), Plant Biotechnology. Butterworths, Boston.

Wellmann, E. 1983. UV radiation in photomorphogenesis. Pages 745–756 *in* W. Shropshire, Jr. and H. Mohr (eds.), Encyclopedia of Plant Physiology, New Series, Vol. 16B, Photomorphogenesis. Springer-Verlag, Berlin.

Widell, K. O. and T. C. Vogelmann. 1988. Fiber optics studies of light gradients and spectral regime within *Lactuca sativa* achenes. Physiologia Plantarum 72:702–712.

Withrow, R. B., W. H. Klein, and V. Elstad. 1957. Action spectra of photomorphogenic induction and its photoinactivation. Plant Physiology 32:453–462.

Zhang, Y., C. W. Ross, and G. L. Orr. 1990. Effects of $P_{fr}$ on NAD kinase and nicotinamide coenzymes in lettuce seeds. Plant Physiology (supplement) 93:30.

## *Chapter 21 The Biological Clock: Rhythms of Life*

Aschoff, Jürgen and Rütger Wever. 1981. The circadian system of man. Pages 311–331 *in* J. Aschoff (ed.), Handbook of Behavioral Neurobiology. Vol. 4. Biological Rhythms. Plenum, New York.

Aschoff, J., K. Hoffman, H. Pohl, and R. Wever. 1975. Retrainment of circadian rhythms after phase-shifts of the Zeitgeber. Chronobiologia 2:23–78.

Behling, Ingeborg. 1929. Über das Zeitgedächtnis der Bienen. (On the time memory in honeybees.) Zeitschrift für Vergleichene Physiologie 9: 259–338.

Blakemore, Richard P. and Richard B. Frankel. 1981. Magnetic navigation in bacteria. Scientific American 245(6):58–65.

Brown, F. A., Jr. 1983. The biological clock phenomenon: Exogenous timing hypothesis. J. Interdiscipl. Cycle Res. 14(2):137–162.

Brown, Frank A., Jr. and Carol S. Chow. 1973a. Interorganismic and environmental influences through extremely weak electromagnetic fields. Biology Bulletin 144:437–461.

Brown, Frank A., Jr. and Carol S. Chow. 1973b. Lunar-correlated variations in water uptake by bean seeds. Biology Bulletin 145:265–278.

Brown, F. A., Jr. and H. M. Webb. 1948. Temperature relations of an endogenous daily rhythmicity in the fiddler crab, *Uca*. Physiologie Zoologie 21: 371–381.

Bünning, Erwin. 1960. Opening address: Biological clocks. Cold Spring Harbor Symposia on Quantitative Biology 15:1–9.

Bünning, Erwin. 1973. The Physiological Clock, Third Edition. Academic Press, London.

Bünning, Erwin. 1977. Fifty years of research in the wake of Wilhelm Pfeffer. Annual Review of Plant Physiology 28:1–22.

Bünning, E. and I. Moser. 1972. Influence of valinomycin on circadian leaf movements of *Phaseolus*. Proceedings of the National Academy of Sciences USA 69:2732–2733.

Bünning, E. and K. Stern. 1930. Über die tagesperiodischen Bewegungen der Primärblätter von *Phaseolus multiflorus* II. Die Bewegungen bei Termokonstanz. (The diurnal movements of the primary leaves of *Phaseolus multiflorus*. II. Movement at constant temperature.) Berichte der Deutschen Botanischen Gesellschaft 48:227–252.

Bünning, E., K. Stern, and R. Stoppel. 1930. Versuche über den Einfluss von Luftionen auf die Schlafbewegungen von *Phaseolus*. (Experiments on the influence of atmospheric ions on the sleep movements of *Phaseolus*.) Planta/Archiv für Wissenschaftliche Botanik 11(1):67–74.

Couderchet, Michel and Willard L. Koukkari. 1987. Daily variations in the sensitivity of soybean seedlings to low temperature. Chronobiology International 4(4):537–541.

Czeisler, Charles A., James S. Allan, Steven H. Strogatz, Joseph M. Ronda, Ramiro Sánchez, C. David Rios, Walter O. Freitag, Gary S. Richardson, and Richard E. Kronauer. 1986. Bright light resets the human circadian pacemaker independent of the timing of the sleep-wake cycle. Science 233:667–671.

Darwin, Charles, assisted by Francis Darwin. 1880. The Power of Movement in Plants. Murray, London; Appleton, New York. (See also "Authorized Edition," 1896, reprinted by De Capo Press, New York, 1966.)

de Mairan, J. 1729. Observation botanique. (Botanical observation.) Histoire de l'Academie Royale des Sciences, Paris.

Dowse, Harold B., Jeffrey C. Hall, and John M. Ringo. 1987. Circadian and ultradian rhythms in *period* mutants of *Drosophila melanogaster*. Behavior Genetics 17:19–35.

Ehret, Charles F. 1960. Action spectra and nucleic acid metabolism in circadian rhythms at the cellular level. Pages 149–158 *in* Cold Spring Harbor Symposia on Quantitative Biology, Vol. XXV, Biological Clocks. The Long Island Biological Association, Cold Spring Harbor, New York.

Englemann, Wolfgang and Martin Schrempf. 1980. Membrane models for circadian rhythms. Photochemical and Photobiological Reviews 5:49–86.

Enright, J. T. 1963. The tidal rhythm of activity of a sand-beach amphipod. Z. Vergl. Physiol. 46: 276–313.

Enright, J. T. 1982. Sleep movements of leaves: In defense of Darwin's interpretation. Oecologia 54:253–259.

Feldman, Jerry F. 1982. Genetic approaches to circadian clocks. Annual Review of Plant Physiology 33:583–608.

Feldman, Jerry F. 1983. Genetics of circadian clocks. BioScience 33:426–431.

Fryer, G. 1986. Lunar cycles in lake plankton. Nature (London) 322:306.

Galston, Arthur W. 1974. The unscientific method. Natural History, March, pp. 18–24.

Galston, A. W., A. A. Tuttle, and P. J. Penny. 1964. A kinetic study of growth movements and photomorphogenesis in etiolated pea seedlings. Am. J. Bot. 51:853–858.

Garner, W. W. and H. A. Allard. 1920. Effect of the relative length of day and night and other factors of the environment on growth and reproduction in plants. Journal of Agricultural Research 18: 553–606.

Garner, W. W. and H. A. Allard. 1931. Effect of abnormally long and short alterations of light and darkness on growth and development of plants. Journal of Agricultural Research 42:629–651.

Gorton, Holly L. and Ruth L. Satter. 1983. Circadian rhythmicity in leaf pulvini. BioScience 33: 451–456.

Goto, Ken, Danielle L. Laval-Martin, and Leland N. Edmunds, Jr. 1985. Biochemical modeling of an autonomously oscillatory circadian clock in *Euglena*. Science 228:1284–1288.

Grobbelaar, N., T. C. Huang, H. Y. Lin, and T. J. Chow. 1986. Dinitrogen-fixing endogenous rhythm in *Synechococcus* RF-1. FEMS Microbiology Letters 37:173–177.

Halberg, Franz, Franca Carandente, Germaine Cornelissen, and George S. Katinas. 1977. Glossary of chronobiology. Chronobiologia 4 (Supplement): 1–189.

Halberg, Franz, Erna Halberg, Cyrus P. Barnum, and John J. Bittner. 1959. Physiologic 24-hour periodicity in human beings and mice, the lighting regimen and daily routine. Pages 803–878 *in* R. B. Withrow (ed.), Photoperiodism and Related Phenomena in Plants and Animals. American Association for the Advancement of Science, Washington, D. C.

Harris, Philip J. C. and M. B. Wilkins. 1978a. Evidence of phytochrome involvement in the entrainment of the circadian rhythm of carbon dioxide metabolism in *Bryophyllum*. Planta 138:271–278.

Harris, Philip J. C. and M. B. Wilkins. 1978b. The circadian rhythm in *Bryophyllum* leaves: Phase control by radiant energy. Planta 143:323–328.

Hauenschild, C., A. Fischer, and D. K. Hoffman. 1968. Untersuchungen am pazifischen Palolowurm *Eunice viridis* (Polychaeta) im Samoa. Helg. Wiss. Meeresunters. 18:254–295. (Cited by Sweeney, 1987.)

Highkin, Harry R. and John B. Hanson. 1954. Possible interaction between light-dark cycles and endogenous daily rhythms on the growth of tomato plants. Plant Physiology 29:301–304.

Hillman, William S. 1976. Biological rhythms and physiological timing. Annual Review of Plant Physiology 27:159–179.

Jackson, F. Rob, Thaddeus A. Bargiello, Suk-Hyeon Yun, and Michael W. Young. 1986. Product of *per* locus of *Drosophila* shares homology with proteoglycans. Nature 320:185–188.

Janzen, Daniel H. 1976. Why bamboos wait so long to flower. Annual Review of Ecology and Systematics 7:347–391.

Jerebzoff, S. 1965. Manipulation of some oscillating systems in fungi by chemicals. Pages 183–189 *in* J. Aschoff (ed.), Circadian Clocks. North-Holland Pub., Amsterdam.

Johnson, C. H., James F. Roeber, and J. W. Hastings. 1984. Circadian changes in enzyme concentration account for rhythm of enzyme activity in *Gonyaulax*. Science 223:1428–1430.

King, Ann I., Michael S. Reid, and Brian D. Patterson. 1982. Diurnal changes in the chilling sensitivity of seedlings. Plant Physiology 70: 211–214.

Kleinhoonte, Anthonia. 1932. Untersuchungen über die autonomen Bewegungen der Primärblätter von *Canavalia ensiformis*. (Investigations on the autonomous leaf movements of the primary leaves of *Canavalia ensiformis*.) Jahrbuch für Wissenschaftliche Botanik 75:679–725.

Konopka, Ronald J. and Seymour Benzer. 1971. Clock mutants of *Drosophila melanogaster*. Proc. Natl. Acad. Sci. USA 68(9):2112–2116.

Koukkari, W. L. and S. B. Warde. 1985. Rhythms and their relations to hormones. Pages 37–77 *in* R. P. Pharis and D. M. Reid (eds.), Encyclopedia of Plant Physiology, New Series, Vol. 11. Hormonal Regulation of Development III. Role of Environmental Factors. Springer-Verlag, Berlin, Heidelberg.

Koukkari, W. L., C. Bingham, and S. H. Duke. 1987a. A special group of ultradian oscillations. Pages 29–33 *in* J. E. Pauly and L. E. Scheving (eds.), Advances in Chronobiology, Part A. Alan R. Liss, New York.

Koukkari, Willard L., Jeffrey L. Tate, and Susan B. Warde. 1987b. Chronobiology projects and laboratory exercises. Chronobiologia 14:405–442.

Kyriacou, C. P. and Jeffrey C. Hall. 1980. Circadian rhythm mutations in *Drosophila melanogaster* affect short-term fluctuations in the male's courtship song. Proc. Natl. Acad. Sci. USA 77(11)6729–6733.

Lloyd, David, Steven W. Edwards, and John C. Fry. 1982. Temperature-compensated oscillations in respiration and cellular protein content in synchronous cultures of *Acanthamoeba castellanii*. Proc. Natl. Acad. Sci. USA 79:3785–3788.

Lörcher, L. 1957. Die wirkung vershiedener Lichtqualitäten auf die endogene Tagesrhythmik von *Phaseolus*. (Effect of various light qualities on the endogenous diurnal rhythms of *Phaseolus*.) Zeitschrift für Botanik 46:209–241.

Luce, Gay Gaer. 1971. Biological Rhythms in Human and Animal Physiology. Dover, New York.

Lysek, G. 1978. Circadian rhythms. Pages 376–388 *in* John E. Smith and David R. Berry (eds.), The Filamentous Fungi, Vol. 3. Developmental Mycology. Wiley, New York.

Mackenzie, Jean. 1973. How biorhythms affect your life. Science Digest 74(2):18–22.

Mauseth, James D. 1988. Plant Anatomy. Benjamin/Cummings, Menlo Park, Calif.

Menaker, Michael. 1965. Circadian rhythms and photoperiodism in *Passer domesticus*. *In* Jürgen Aschoff (ed.), Circadian Clocks (Proceedings of the Feldafing Summer School, 7–18 September, 1964). North-Holland Publishing Co., Amsterdam.

Mitsui, A., S. Cao, A. Takahashi, and T. Arai. 1987. Growth synchrony and cellular parameters of the unicellular nitrogen-fixing marine cyanobacterium, *Synechococcus* sp. strain Miami BG 043511 under continuous illumination. Physiol. Plantarum 69:1–8.

Mitsui, A., S. Kumazawa, A. Takahashi, H. Ikemoto, S. Cao, and T. Arai. 1986. Strategy by which nitrogen-fixing unicellular cyanobacteria grow photoautotrophically. Nature 323:720–722.

Moore-Ede, Martin C., F. M. Sulzman, and C. A. Fuller. 1982. The Clocks That Time Us. Harvard University Press, Cambridge, Mass. and London.

Muñoz, Victor and Warren L. Butler. 1975. Photoreceptor pigment for blue light in *Neurospora crassa*. Plant Physiology 55:421–426.

Palmer, John D. 1975. Biological clocks of the tidal zone. Scientific American 232(2):70–79.

Palmer, J. D. 1990. The rhythmic lives of crabs. BioScience 40:352–358.

Palmer, J. D. and F. E. Round. 1967. Persistent vertical migration rhythms in the benthic microflora. VI. The tidal and diurnal nature of the rhythm in the diatom *Hantzschia virgata*. Biol. Bull. (Woods Hole, Mass.) 132:44–55.

Pengelley, E. T. and S. M. Asmundson. 1969. Free-running periods of endogenous circadian rhythms in the golden-mantled ground squirrel, *Citellus lateralis*. Comp. Biochem. Physiol. 30: 177–183.

Pfeffer, W. 1915. Beiträge zur Kenntnis der Entstehung der Schlafbewegungen. (Contributions to the knowledge of the origin of leaf movements.) Adhandl. Math. Phys. Kl. Kön. Sächs. Ges. d. Wiss. 34:1–154.

Piskorz-Binczycka, B., S. Jerebzoff, and S. Jerebzoff-Quintin. 1989. Asparagine and regulation of photoinduced rhythms in *Penicillium claviforme*. Physiologia Plantarum 76:315–318.

Pittendrigh, Colin S. 1954. On temperature independence in the clock system controlling emergence time in *Drosophila*. Proceedings of the National Academy of Sciences 40:1018–1029.

Pittendrigh, Colin S. 1967. On the mechanism of entrainment of a circadian rhythm by light cycles. Pages 277–297 *in* J. Aschoff (ed.), Circadian Clocks. North-Holland Publishing Co., Amsterdam.

Reinberg, A. 1971. La chronobiologie. Recherche 2:242–250.

Rodgers, C. W., R. L. Sprinkle, and F. H. Lindbert. 1974. Biorhythms: Three tests of the "critical days" hypothesis. International Journal of Chronobiology 2:215–310.

Salisbury, Frank B. 1963. Biological timing and hormone synthesis in flowering of *Xanthium*. Planta 59:518–534.

Salisbury, Frank B., George G. Spomer, Martha Sobral, and Richard T. Ward. 1968. Analysis of an alpine environment. Botanical Gazette 129(1): 16–32.

Satter, Ruth L. and Arthur W. Galston. 1981. Mechanisms of control of leaf movement. Annual Review of Plant Physiology 32:82–110.

Saunders, D. S. 1976. The biological clock of insects. Scientific American 234(2):114–121.

Schwemmle, Berthold. 1960. Thermoperiodic effects and circadian rhythms in flowering of plants. Pages 239–243 *in* Cold Spring Harbor Symposia on Quantitative Biology, Vol. XXV, Biological Clocks. Long Island Biological Association, Inc., Cold Spring Harbor, New York.

Schwintzer, C. R. 1971. Energy budgets and temperatures of nyctinastic leaves on freezing nights. Pl. Physiol., Lancaster 48:203–207.

Simon, Esther, Ruth L. Satter, and Arthur W. Galston. 1976. Circadian rhythmicity in excised *Samanea* pulvini. Plant Physiology 58:421–425.

Smith, Alan P. 1974. Bud temperature in relation to nyctinastic leaf movement in an Andean giant rosette plant. Biotropica 6(4):263–266.

Spruyt, E., L. Maes, J. P. Verbelen, E. Moereels, and J. A. De Greef. 1983. Circannual course of photomorphogenetic reactivity in etiolated bean seedlings. Photochemistry and Photobiology 37(4):471–473.

Sulzman, Frank M. 1983. Primate circadian rhythms. BioScience 33:445–450.

Sweeney, B. M. 1976. Pros and cons of the membrane model for circadian rhythms in the marine algae, *Gonyaulax* and *Acetabularia*. Pages 63–76 *in* P. J. De Coursey (ed.), Biological Rhythms in the Marine Environment. University of South Carolina Press, Columbia.

Sweeney, Beatrice M. 1983. Biological clocks — an introduction. BioScience 33:424–425. (See other articles in the same issue.)

Sweeney, Beatrice M. 1987. Rhythmic Phenomena in Plants, Second Edition. Academic Press, San Diego.

Thommen, G. 1973. Biorhythms: Is This Your Day? Avon Books, New York.

Thompson, Marcia J. and David W. Harsha. 1984. Our rhythms still follow the African sun. Psychology Today 18(1):50–54.

Turek, Fred W. and Susan Losee-Olson. 1986. A benzodiazepine used in the treatment of insomnia phase-shifts the mammalian circadian clock. Nature 321:167–168.

Walker, B. W. 1952. A guide to the grunion. Calif. Fish and Game 38:409–420.

Wever, Rütger A. 1979. The Circadian System of Man. Results of Experiments Under Temporal Isolation. Springer-Verlag, New York.

Wilkins, Malcolm B. and Philip J. C. Harris. 1975. Phytochrome and phase setting of endogenous rhythms. Pages 399–417 *in* H. Smith (ed.), Light and Plant Development. Butterworths, London and Boston.

Wilkinson, H. P. 1971. Leaf anatomy of various Anacardiaceae with special reference to the epidermis and some contributions to the taxonomy of the genus *Dracontomelon* Blume. Thesis. University of London.

Yoo, Keun Chang and Shunpei Uemoto. 1976. Studies on the physiology of bolting and flowering in *Raphanus sativus* L. II. Annual rhythm in readiness to flower in Japanese radish, cultivar "Wase-shijunichi." Plant & Cell Physiol. 17:863–865.

Young, Robert A. and Joseph R. Haun. 1961. Bamboo in the United States: Description, Culture, and Utilization. U. S. Department of Agriculture, Agriculture Handbook No. 193.

Zimmer, Rose. 1962. Phasenverschiebung und andere Störolichtwirkungen auf die endogen tagesperiodischen Blütenblattbewegungen. (Phase shift and other light-interruption effects on endogenous diurnal petal movements.) Planta 48:283–300.

Zimmerman, Natille H. and Michael Menaker. 1979. The pineal gland: The pacemaker within the circadian system of the house sparrow. Proc. Natl. Acad. Sci. USA 76:999–1003.

### *Chapter 22 Growth Responses to Temperature*

Altman, P. L. and Dorothy S. Dittmer. (Eds.). 1962. Growth, Including Reproduction and Development. Federation of the American Society for Experimental Biology, Washington, D. C.

Arias, I., P. M. Williams, and J. W. Bradbeer. 1976. Studies in seed dormancy. IX. The role of gibberellin biosynthesis and the release of bound gibberellin in the post-chilling accumulation of gibberellin in seeds of *Corylus avellana* L. Planta 131:135–139.

Baskin, Jerry M. and Carol C. Baskin. 1990. Temperature relations for bud growth in the root geophyte *pediomelum subacaule*, and ecological implications. Botanical Gazette. 151(4):506–509.

Bewley, J. Derek and Michael Black. 1982. Physiology and Biochemistry of Seeds in Relation to Germination. In Two Volumes. Springer-Verlag, Berlin, Heidelberg, New York.

Bewley, J. D. and M. Black. 1984. Seeds: Physiology of Development and Germination. Plenum, New York.

Caspari, E. W. and R. W. Marshak. 1965. The rise and fall of Lysenko. Science 149:275–278.

Chailakhyan, M. K. 1968. Internal factors of plant flowering. Annual Review of Plant Physiology 19:1–36.

Chapman, H. W. 1958. Tuberization in the potato plant. Physiol. Plant. 11:215–224.

Chouard, P. 1960. Vernalization and its relations to dormancy. Annual Review of Plant Physiology 11:191–238.

Crocker, W. and L. V. Barton. 1953. Physiology of Seeds. Chronica Botanica Co., Waltham, Mass.

Dell, B. 1980. Structure and function of the strophiolar plug in seeds of *Albizia lophantha*. American Journal of Botany 67(4):556–563.

Downs, R. J. and H. A. Borthwick. 1956. Effects of photoperiod on growth of trees. Botanical Gazette 117:310–326.

Earnshaw, M. J. 1981. Arrhenius plots of root respiration in some arctic plants. Arctic and Alpine Research 13:425–430.

Evenari, M. 1957. The physiological action and biological importance of germination inhibitors. Society of Experimental Biology Symposium 11:21–43.

Friend, D. J. C. and V. A. Helson. 1976. Thermoperiodic effects on the growth and photosynthesis of wheat and other crop plants. Botanical Gazette 137(1):75–84.

Gassner, G. 1918. Beiträge zur physiologischen Charakteristik Sommer und Winter annueller Gewächse insbesondere der Getreidepflanzen. (Contributions to the physiological characteristics of summer and winter annual growth, particularly with cereals.) Zeitschrift für Botanik 10: 417–430.

Hartsema, Annie M. 1961. Influence of temperatures on flower formation and flowering of bulbous and tuberous plants. Encyclopedia of Plant Physiol. 16:123–167.

Heide, Ole M. 1974. Growth and dormancy in Norway spruce ecotypes (*Picea abies*). I. Interaction of photoperiod and temperature. Physiologia Plantarum 30:1–12.

Henson, Cynthia A., Larry E. Schrader, and Stanley H. Duke. 1980. Effects of temperature on germination and mitochondrial dehydrogenases in two soybean (*Glycine max*) cultivars. Physiologia Plantarum 48:168–174.

Jann, R. C. and R. D. Amen. 1977. What is germination? Pages 7–28 *in* A. A. Khan (ed.), The Physiology and Biochemistry of Seed Dormancy and Germination. North-Holland Pub. Co., Amsterdam and New York.

Junttila, O. 1973. The mechanism of low temperature dormancy in mature seeds of *Syringa* species. Physiologia Plantarum 29:256–263.

Kadman-Zahavi, A., A. Horovitz, and Y. Ozer. 1984. Long-day induced dormancy in *Anemone coronaria* L. Annals of Botany 53:213–217.

Ketring, D. L. 1973. Germination inhibitors. Seed Science Technology 1(2):305–324.

Khan, A. A. 1977. Seed dormancy: Changing concepts and theories. Pages 29–50 *in* A. A. Khan (ed.), The Physiology and Biochemistry of Seed Dormancy and Germination. North-Holland Pub. Co., Amsterdam and New York.

Koda, Yasunori, El-Sayed A. Omer, Teruhiko Yoshihara, Haruki Shibata, Sadao Sakamura, and Yozo Okazawa. 1988. Isolation of a specific potato tuber-inducing substance from potato leaves. Plant Cell Physiology 29:1047–1051.

Koller, D. 1957. Germination-regulating mechanisms in some desert seeds. IV: *Atriplex dimorphostegia*. Ecology 38:1–13.

Kramer, Paul J. and T. T. Kozlowski. 1979. Physiology of Woody Plants. Academic Press, New York.

Lang, Anton. 1957. The effect of gibberellin upon flower formation. Proceedings of the National Academy of Sciences 43:709–711.

Lang, A. 1965a. Effects of some internal and external conditions on seed germination. Pages 849–893 *in* W. Ruhland (ed.), Encyclopedia of Plant Physiology, Vol. 15, Part 2. Springer-Verlag, Berlin.

Lang, A. 1965b. Physiology of flower initiation. Pages 1380–1536 *in* W. Ruhland (ed.), Encyclopedia of Plant Physiology, Vol. 15, Part 1. Springer-Verlag, Berlin.

Lang, G. A., J. D. Early, G. C. Martin, and R. L. Darnell. 1987. Endo-, para-, and ecodormancy: Physiological terminology and classification for dormancy research. HortScience 22:371–377.

Leopold, A. C. and Paul E. Kriedemann. 1975. Plant Growth and Development, Second Edition. McGraw-Hill, New York.

Lerman, J. C. and E. M. Cigliano. 1971. New carbon-14 evidence for six-hundred years old *Canna pacta* seed. Nature 232:568–570.

Mayer, A. M. 1974. Control of seed germination. Annual Review of Plant Physiology 25:167–193.

Mayer, A. M. and A. Poljakoff-Mayber. 1989. The Germination of Seeds, Fourth Edition. Pergamon Press, New York, London.

Melchers, G. 1937. Die Wirkung von Genen, tiefen Temperaturen und blühanden Pfropfpartnern auf die Blühreife von *Hyoscymas niger* L. (The effect of genes, low temperatures, and flowering graft partners on ripeness to flower of *Hyoscymas niger*.) Biologisches Zentralblatt 57:568–614.

Mingo-Castel, Angel M., Orrin E. Smith, and Junju Kumamoto. 1976. Studies on the carbon dioxide promotion and ethylene inhibition of tuberization in potato explants cultured *in vitro*. Plant Physiol. 57:480–485.

Nikolaeva, M. G. 1969. Physiology of Deep Dormancy in Seeds. Translated from Russian and published for the National Science Foundation by the Israel Program for Scientific Translations.

Osborne, Daphne. 1980. Senescence in seeds. Pages 13–37 *in* K. V. Thimann (ed.), Senescence in Plants. CRC Press, Boca Raton, Fla.

Perry, T. O. and H. Hellmers. 1973. Effects of abscisic acid on growth and dormancy of two races of red maple. Botanical Gazette 134:283–289.

Phillips, I. D. J., J. Miners, and J. R. Roddick. 1980. Effects of light and photoperiodic conditions on abscisic acid in leaves and roots of *Acer pseudoplatanus* L. Planta 149:118–122.

Porsild, A. E., C. R. Harington, and G. A. Mulligan. 1967. *Lupinus articus* Wats. grown from seeds of Pleistocene age. Science 148:113–114.

Priestly, David A. and Maarten A. Posthumus. 1982. Extreme longevity of lotus seeds from Plantien. Nature 299:148–149.

Purvis, O. N. 1961. The physiological analysis of vernalization. Encyclopedia of Plant Physiology 16:76–122.

Rees, A. R. 1972. The Growth of Bulbs. Academic Press, New York.

Roberts, E. H. and R. D. Smith. 1977. Dormancy and the pentose phosphate pathway. Pages 385–411 *in* A. A. Khan (ed.), The Physiology and Biochemistry of Seed Dormancy and Germination. North-Holland Pub. Co., Amsterdam and New York.

Salisbury, Frank B. 1963. The Flowering Process. Pergamon Press, Oxford, London, New York, Paris.

Salisbury, F. B. 1986. Dormancy terminology (letter). HortScience 21:185–186.

Samish, R. M. 1954. Dormancy in woody plants. Annual Review of Plant Physiology 5:183–204.

Thompson, H. C. 1953. Vernalization of growing plants. Pages 179–196 *in* W. E. Loomios (ed.), Growth and Differentiation in Plants. The Iowa State College Press, Ames.

Ulrich, Albert. 1955. Influence of night temperature and nitrogen nutrition on the growth, sucrose accumulation and leaf minerals of sugar beet plants. Plant Physiology 30:250–257.

Vegis, A. 1964. Dormancy in higher plants. Annual Review of Plant Physiology 15:185–224.

Vest, E. Dean. 1972. Shadscale and fungus: Desert partners. Pages 725–726 *in* W. A. Jensen and F. B. Salisbury, Botany, An Ecological Approach. Wadsworth, Belmont, Calif.

Villiers, T. A. 1972. Seed dormancy. Pages 219–281 *in* T. T. Kozlowski (ed.), Seed Biology, Vol. II. Academic Press, New York.

Vreugdenhil, Dick and Paul C. Struik. 1989. An integrated view of the hormonal regulation of tuber formation in potato (*Solanum tuberosum*). Physiol. Plant. 75:525–531.

Walton, D. C. 1977. Abscisic acid and seed germination. Pages 145–156 *in* A. A. Khan (ed.), The Physiology and Biochemistry of Seed Dormancy and Germination. North-Holland Pub. Co., Amsterdam and New York.

Walton, D. C. 1980. Biochemistry and physiology of abscisic acid. Annual Review of Plant Physiology 31:453–489.

Wareing, P. F. 1956. Photoperiodism in woody plants. Annual Review of Plant Physiology 7: 191–214.

Wellensiek, S. J. 1964. Dividing cells as the prerequisite for vernalization. Plant Physiology 39: 832–835.

Went, Frits W. 1957. The Experimental Control of Plant Growth. Chronica Botanica Co., Waltham, Mass.

Wood, W. M. L. 1953. Thermonasty in tulip and crocus flowers. Journal of Experimental Botany 4:65–77.

### *Chapter 23 Photoperiodism*

Atherton, J. G. 1987. Manipulation of Flowering. Butterworths, London.

Battle, R. W., J. K. Gaunt, and D. L. Laidman. 1976. The effect of photoperiod on endogenous γ-tocopherol and plastochromanol in leaves of *Xanthium strumarium* L. Biochemical Society of London Transactions 4:484–486.

Battle, R. W., D. L. Laidman, and J. K. Gaunt. 1977. The relationship between floral induction and γ-tocopherol concentrations in leaves of *Xanthium strumarium* L. Biochemical Society of London Transactions 5:322–324.

Bernier, Georges. 1988. The control of floral evocation and morphogenesis. Ann. Rev. Plant Physiol. Plant Mol. Biol. 39:175–219.

Bernier, G., J. Kinet, and R. M. Sachs. 1981. The physiology of flowering. Vol. I: The Initiation of Flowers. Vol. II: Transition to Reproductive Growth. CRC Press, Boca Raton, Fla.

Bismuth, Florence and Emile Miginiac. 1984. Influence of zeatin on flowering in root forming cuttings of *Anagallis arvensis* L. Plant & Cell Physiology 25:1073–1076.

Bjørnseth, Ian Petter. 1981. Effects of natural day-length and nutrition on the cessation of cambial activity in young plants of *Picea abies*. Mitteilungen der Forstlichen Bundesversuchsanstalt. Wien. 142. Heft: 167–176.

Black, M. and P. F. Wareing. 1955. Photoperiodism in the light inhibited seed of *Nemophila insignis*. J. Exp. Bot. 11:28–39.

Board, J. E. and J. R. Settimi. 1988. Photoperiod requirements for flowering and flower production in soybean. Agronomy Journal 80:518–525.

Bollig, I. 1977. Different circadian rhythms regulate photoperiodic flowering response and leaf movement in *Pharbitis nil* L. Choisy. Planta 135:137–142.

Bose, T. K. 1974. Effect of temperature and photoperiod on growth, flowering and seed formation in tossa jute. Indian Journal of Agricultural Science 44:32–35.

Bünning, Erwin. 1937. Die endonome Tagesrhythmik als Grundlage der photoperiodischen Reaktion. (The endogenous daily rhythm as the basis of the photoperiodic reaction.) Berichte der Deutschen Botanischen Gesellschaft 54: 590–607.

Cameron, J. Scott and Frank G. Dennis, Jr. 1986. The carbohydrate-nitrogen relationship and flowering/fruiting: Kraus and Draybill revisited. HortScience 21:1099–1102.

Carr, D. J. 1967. The relationship between florigen and the flower hormones. Pages 304–312 *in* J. F. Fredrick and E. M. Weyer (eds.), Plant Growth Regulators. Annals of the New York Academy of Sciences, Vol. 144.

Chailakhyan, Mikhail K. 1968. Internal factors of plant flowering. Annual Review of Plant Physiology 19:1–36.

Chailakhyan, M. K. and V. N. Khrianin. 1987. Sexuality in Plants and Its Hormonal Regulation. (Edited by K. V. Thimann, translated by Vanya Loroch.) Springer-Verlag, New York.

Chailakhyan, M. K., V. N. Lozhnikova. 1985. The florigen hypothesis and its substantiation by extraction of substances which induce flowering in plants. Fiziol. Rast. 32:1172–1181.

Claes, H. and A. Lang. 1947. Die Blütenbildung von *Hyoscyamus niger* in 48-stundigen Licht-Dunkel Zyklen mit aufgeteilten Lichtphasen. (Flower formation of *Hyoscyamus niger* in 48-hour light-dark cycles with partitioned light phases.) Zeitschrift für Naturforschung 2b:56–63.

Cleland, Charles F. 1978. The flowering enigma. BioScience 28:265–269.

Cleland, C. F. and A. Ajami. 1974. Identification of flower-inducing factor isolated from aphid honeydew as being salicylic acid. Plant Physiology 54:904–906. (See also 54:899–903.)

Cumming, Bruce G. 1959. Extreme sensitivity of germination and photoperiodic reaction in the genus *Chenopodium* (Tourn.) L. Nature 184: 1044–1045.

Cumming, Bruce G. 1969. *Chenopodium rubrum* L. and related species. Pages 156–185 *in* L. T. Evans (ed.), The Induction of Flowering, Some Case Histories. Macmillan of Australia, South Melbourne, Australia.

Deitzer, G. F., R. Hayes, and M. Jabben. 1979. Kinetics and time dependence of the effect of far-red light on the photoperiodic induction of flowering in winter barley. Plant Physiology 64:1015–1021.

De Proft, M., R. Van Dijck, L. Philippe, and J. A. De Greef. 1985. Hormonal regulation of flowering and apical dominance in bromeliad plants. 12th International Conference on Plant Growth Substances, Heidelberg. Page 93 (abstract).

Evans, L. T. (Ed.). 1969a. The Induction of Flowering, Some Case Histories. Macmillan of Australia, Victoria, Australia.

Evans, L. T. 1969b. The nature of flower induction. Pages 457–480 *in* L. T. Evans (ed.), The Induction of Flowering, Some Case Histories. Macmillan of Australia, Victoria, Australia.

Evans, L. T. 1975. Daylength and the Flowering of Plants. Benjamin/Cummings, Menlo Park, Calif.

Evans, L. T. 1987. Short day induction of inflorescence initiation in some winter wheat varieties. Australian Journal of Plant Physiology 14: 277–286.

Evans, L. T., A. Chu, Roderick W. King, Lewis N. Mander, and Richard P. Pharis. 1990. Gibberellin structure and florigenic activity in *Lolium temulentum*, a long-day plant. Planta 182:97–106.

Fischer, J. 1916. Zur Frage der Kohlensäureernährung der Pflanze. (On the question of the carbon dioxide nutrition of plants.) Gartenflora 65:232.

Friend, Douglas J., Monique Bodson, and Georges Bernier. 1984. Promotion of flowering in *Brassica campestris* L. cv Ceres by sucrose. Plant Physiol. 75:1085–1089.

Furuya, M. 1989. Molecular properties and biogenesis of phytochrome I and II. Advances in Biophysics 25:133–167.

Garner, W. W. and H. A. Allard. 1920. Effect of the relative length of day and night and other factors of the environment of growth and reproduction in plants. Journal of Agricultural Research 18:553–606.

Garner, W. W. and H. A. Allard. 1923. Further studies in photoperiodism, the response of plants to relative length of day and night. Journal of Agricultural Research 23:871–920.

Gibby, David D. and Frank B. Salisbury. 1971. Participation of long-day inhibition in flowering of *Xanthium strumarium* L. Plant Physiol. 47: 784–789.

Greulach, V. A. and Haesloop, J. G. 1958. Influence of gibberellin on *Xanthium* flowering as related to number of photoinductive cycles. Science 127:646.

Groenewald, E. G. and J. H. Visser. 1974. The effect of certain inhibitors of prostaglandin biosynthesis on flowering of *Pharbitis nil*. Zeitschrift für Pflanzenphysiologie 71:67–70.

Groenewald, E. G. and J. H. Visser. 1978. The effect of arachidonic acid, prostaglandins and inhibitors of prostaglandin synthetase, on the flowering of excised *Pharbitis nil* shoot apices under different photoperiods. Zeitschrift für Pflanzenphysiologie 88:423–429.

Groenewald, E. G., J. H. Visser, and N. Grobbelaar. 1983. The occurrence of prostaglandin (PG) $F_{2\alpha}$ in *Pharbitis nil* seedlings grown under short days or long days. South African Journal of Botany 2:82.

Halevy, Abraham H. (Ed.). 1985. Handbook of Flowering, Vols. I–V. CRC Press, Boca Raton, Fla.

Hamner, Karl C. 1963. Endogenous rhythms in controlled environments. Pages 215–232 *in* L. T. Evans (ed.), Environmental Control of Plant Growth. Academic Press, New York.

Hamner, K. C. 1969. *Glycine max* L. Merrill. Pages 62–89 *in* L. T. Evans (ed.), The Induction of Flowering, Some Case Histories. Macmillan of Australia, South Melbourne, Australia.

Hamner, K. C. and J. Bonner. 1938. Photoperiodism in relation to hormones as factors in floral initiation and development. Botanical Gazette 100:388.

Havelange, A. 1980. The quantitative ultrastructure of the meristematic cells of *Xanthium strumarium* during the transition to flowering. American Journal of Botany 67:1171–1178.

Hay, R. K. M. and O. M. Heide. 1983. Specific photoperiodic stimulation of dry matter production in a high-latitude cultivar of *Poa pratensis*. Physiologia Plantarum 47:135–142.

Heide, O. M. 1989. Environmental control of flowering and viviparous proliferation in seminiferous and viviparous arctic populations of two *Poa* species. Arctic and Alpine Research 21:305–315.

Hillman, W. S. 1979. Photoperiodism in plants and animals. J. J. Head (ed.), Carolina Biology Readers. Carolina Biology Supply Company, Burling, N. C.

Hodson, H. K. and K. C. Hamner. 1970. Floral inducing extract from *Xanthium*. Science 167: 384–385.

Hughes, J. E., D. C. Morgan, P. A. Lambton, C. R. Black, and H. Smith. 1984. Photoperiodic time signals during twilight. Plant Cell Environ. 7(4):269–278.

Ireland, C. R. and W. W. Schwabe. 1982a. Studies on the role of photosynthesis in the photoperiodic induction of flowering in the short-day plants *Kalanchoe blossfeldiana* Poellniz and *Xanthium pensylvanicum* Wallr. I. The requirement for $CO_2$ during photoperiodic induction. J. Exp. Bot. 33(135):738–747.

Ireland, C. R. and W. W. Schwabe. 1982b. Studies on the role of photosynthesis in the photoperiodic induction of flowering in the short-day plants *Kalanchoe blossfeldiana* Poellniz and *Xanthium pensylvanicum* Wallr. II. The effect of chemical inhibitors of photosynthesis. J. Exp. Bot. 33(135):748–760.

Jabben, Merten and Gerald F. Deitzer. 1979. Effects of the herbicide San 9789 on photomorphogenic responses. Plant Physiology 63:481–485.

Jacqmard, A., M. V. S. Raju, J-M. Kinet, and G. Bernier. 1976. The early action of the floral stimulus on mitotic activity and DNA synthesis in the apical meristem of *Xanthium strumarium*. American Journal of Botany 63:166–174.

Janistyn, Boris. 1982. Gas chromatographic-mass spectroscopic identification of prostaglandin $F_{2\alpha}$ in flowering *Kalanchoe blossfeldiana* v. Poelln. Planta 154:485–487.

Jordan, B. R., M. D. Partis, and B. Thomas. 1986. The biology and molecular biology of phytochrome. Pages 315–362 *in* B. J. Miflin (ed.), Oxford Survey of Plant Molecular and Cell Biology, Vol. 3. Oxford University Press, Oxford, England.

Junttila, Olavi. 1985. Experimental control of flowering and vivipary in timothy (*Phleum pratense*). Physiol. Plant. 63:35–42.

Junttila, Olavi and O. M. Heide. 1981. Shoot and needle growth in *Pinus sylvestris* as related to temperature in northern Fennoscandia. Forest Science 27(3):423–430.

Kadman-Zahavi, Avishag and Dovrat Peiper. 1987. Effects of moonlight on flower induction in *Pharbitis nil*, using a single dark period. Annals of Botany 6:621–623.

Kato, Y., N. Fukunharu, and R. Kobayashi. 1958. Stimulation of flower bud differentiation of conifers by gibberellin. Pages 67–68 *in* Abstracts of the Second Meeting of the Japan Gibberellin Research Association.

Kinet, Jean-Marie, Roy M. Sachs, and Georges Bernier. 1985. The Physiology of Flowering. III. The Development of Flowers. CRC Press, Boca Raton, Fla.

King, R. W. 1975. Multiple circadian rhythms regulate photoperiodic flowering responses in *Chenopodium rubrum*. Canadian Journal of Botany 53:2631–2638.

King, R. W. 1979. Photoperiodic time measurement and effects of temperature on flowering in *Chenopodium rubrum* L. Australian Journal of Plant Physiology 6:417–422.

King, R. W. and J. A. D. Zeevaart. 1973. Floral stimulus movement in *Perilla* and flower inhibition caused by noninduced leaves. Plant Physiology 51:727–738.

King, R., Lloyd Evans, Richard P. Pharis, and L. N. Mander. 1987. Gibberellins in relation to growth and flowering in *Pharbitis nil* Chois. Plant Physiology 84:1126–1131.

Klebs, G. 1904. Über Probleme der Entwicklung. (About problems of development.) Biologisches Centralblatt. 24(18):601–614. (Klebs published at least four other articles on the same topic and in the same journal during 1904; see Cameron and Dennis, 1986.)

Klebs, G. 1910. Alterations in the development and forms of plants as a result of environment. Proceedings of the Royal Society of London 82:547–558.

Klebs, G. 1918. Über die Blütenbildung bei *Sempervivum*. (On flower formation in *Sempervivum*.) Flora (Jena) 128:111–112.

Kraus, E. J. and H. R. Kraybill. 1918. Vegetation and reproduction with special reference to the tomato. Oregon Agricultural Experiment Station Bulletin 149:5.

Lang, Anton. 1965. Physiology of flower initiation. Pages 1380–1535 *in* W. Ruhland (ed.), Encyclopedia of Plant Physiology, Vol. 15. Springer-Verlag, Berlin.

Lang, A., M. Chailakhyan, and I. A. Frolova. 1977. Promotion and inhibition of flower formation in a day-neutral plant in grafts with a short-day plant and a long-day plant. Proceedings of the National Academy of Science USA 74:2412–2416.

Law, C. N. and Rachel Scarth. 1984. Genetics and its potential for understanding the action of light in flowering. Pages 193–209 *in* D. Vince-Prue, B. Thomas, and K. E. Cockshull (eds.), Light and the Flowering Process. Academic Press, London.

Lincoln, R. G., A. Cunningham, B. H. Carpenter, J. Alexander, and D. L. Mayfield. 1966. Florigenic acid from fungal cultures. Plant Physiology 41:1079–1080.

Lord, Elizabeth and Georges Bernier. 1989. Plant Reproduction: From Floral Induction to Pollination. American Society of Plant Physiologists, Rockville, Md.

Lumsden, Peter, Brian Thomas, and Daphne Vince-Prue. 1982. Photoperiodic control of flowering in dark-grown seedlings of *Pharbitis nil* Choisy. Plant Physiology 70:277–282.

McMillan, C. 1974. Photoperiodic adaptation of *Xanthium strumarium* in Europe, Asia Minor, and northern Africa. Canadian Journal of Botany 52:1779–1791.

Mooney, H. A. and W. D. Billings. 1961. Comparative physiological ecology of arctic and alpine populations of *Oxyria digyna*. Ecological Monographs 31:1–29.

Ogawa, Yukiyoshi. 1981. Stimulation of the flowering of *Pharbitis nil* Chois. by gibberellin $A_3$: Time dependent action at the apex. Plant & Cell Physiol. 22(4):675–681.

Olmsted, C. E. 1944. Growth and development in range grasses. IV. Photoperiodic responses in twelve geographic strains of side-oats grama. Botanical Gazette 106:46–74.

Papenfuss, H. D. and Frank B. Salisbury. 1967. Aspects of clock resetting in flowering of *Xanthium*. Plant Physiology 42:1562–1568.

Pharis, Richard P. and C. G. Kuo. 1977. Physiology of gibberellins in conifers. Canadian Journal of Forest Research 7(2):299–325.

Pharis, Richard P., Lloyd T. Evans, Roderick W. King, and Lewis N. Mander. 1989. Gibberellins and flowering in higher plants—differing structures yield highly specific effects. Pages 29–41 *in* Elizabeth Lord and Georges Bernier (eds.), Plant Reproduction: From Floral Induction to Pollination. American Society of Plant Physiologists Symposium Series Vol. 1, Amer. Soc. of Plant Physiol., 15501-A Monona Drive, Rockville, Md. 20855.

Pharis, R. P., R. W. King, L. T. Evans, and L. N. Mander. 1987a. Investigations on endogenous and applied gibberellins in relation to flower induction in the long-day plant, *Lolium temulentum*. Plant Physiology 84:1132–1138.

Pharis, R. P., Joe E. Webber, and Stephen D. Ross. 1987b. The promotion of flowering in forest trees by gibberellin $A_{4/7}$ and cultural treatments: A review of the possible mechanisms. Forest Ecology and Management 19:65–84.

Ramina, Angelo, Wesley P. Hackett, and Roy M. Sachs. 1979. Flowering in *Bougainvillea*. Plant Physiol. 64:810–813.

Rombach, J. 1986. Phytochrome in Norflurazon-treated seedlings of *Pharbitis nil*. Physiologia Plantarum 68:231–237.

Rood, Stewart B., Richard I. Buzzell, Lewis N. Mander, David Pearce, and Richard P. Pharis. 1988. Gibberellins: A phytohormonal basis for heterosis in maize. Science 241:1216–1218.

Ross, Stephen D., Mark P. Bollmann, Richard P. Pharis, and Geoffrey B. Sweet. 1984. Gibberellin $A_{4/7}$ and the promotion of flowering in *Pinus radiata*. Plant Physiol. 76:326–330.

Rowan, W. 1925. Relation of light to bird migration and developmental changes. Nature 115:494–495.

Sachs, R. M. 1978. Nutrient diversion: An hypothesis to explain the chemical control of flowering. HortScience 12:220–222.

Sachs, R. M. and W. P. Hackett. 1983. Source-sink relationships and flowering. Pages 263–272 *in* Werner J. Meudt (ed.), Beltsville Symposia in Agricultural Research, Vol. 6, Strategies of Plant Reproduction. Allanheld, Osmun, Totowa, N. J.

Salisbury, Frank B. 1955. The dual role of auxin in flowering. Plant Physiology 30:327–334.

Salisbury, Frank B. 1959. Influence of certain growth regulators on flowering of cocklebur. Page 381 *in* R. B. Withrow (ed.), Photoperiodism and Related Phenomena in Plants and Animals. American Association for the Advancement of Science, Washington, D. C.

Salisbury, Frank B. 1963a. The Flowering Process. Pergamon Press, Cambridge, New York.

Salisbury, Frank B. 1963b. Biological timing and hormone synthesis in flowering in *Xanthium*. Planta 49:518–534.

Salisbury, Frank B. 1965. Time measurement and the light period in flowering. Planta (Berl.) 66:1–26.

Salisbury, Frank B. 1981a. The twilight effect: Initiating dark measurement in photoperiodism of *Xanthium*. Plant Physiology 67:1230–1238.

Salisbury, Frank B. 1981b. Response to photoperiod. Pages 135–167 *in* O. L. Lange, P. S. Nobel, C. B. Osmond, and H. Ziegler (eds.), Physiological Plant Ecology I. Vol. 12A. *In* A. Pirson, M. H. Zimmermann (eds.), Encyclopedia of Plant Physiology, New Series. Springer-Verlag, Berlin, Heidelberg.

Salisbury, Frank B. 1982. Photoperiodism. Horticultural Reviews 4:66–105.

Salisbury, Frank B. 1989. The use of *Xanthium* in flowering research. Pages 153–214 *in* Roman Maksymowych (ed.), Analysis of Growth and Development of *Xanthium*. Cambridge University Press, Cambridge.

Salisbury, Frank B. and James Bonner. 1956. The reactions of the photoinductive dark period. Plant Physiology 31:141–147.

Salisbury, Frank B. and Alice Denney. 1974. Noncorrelation of leaf movements and photoperiodic clocks in *Xanthium strumarium* L. *Chronobiology*, Proceedings of the International Society of Chronobiology, Little Rock, Ark., Nov. 8–10, 1971. Igaku Shoin, Ltd.

Schwabe, W. W. 1971. Physiology of vegetative reproduction and flowering. Pages 233–411 *in* F. C. Steward (ed.), Plant Physiology — A Treatise, Vol. 7A. Academic Press, New York.

Takimoto, A. and K. Ikeda. 1961. Effect of twilight on photoperiodic induction in some short day plants. Plant Cell Physiology 2:213–229.

Tournois, J. 1914. Etudes sur la sexualité du houblon. (Studies on the sexuality of hops.) Annals des Sciences Naturelles (Botanique) 19:49–191.

Vince-Prue, Daphne. 1975. Photoperiodism in Plants. McGraw-Hill, London.

Vince-Prue, D. 1983. Photomorphogenesis and flowering. Pages 457–490 *in* W. Shropshire, Jr. and H. Mohr (eds.), Photomorphogenesis. Encyclopedia of Plant Physiology, New Series, Vol. 16. Springer-Verlag, Berlin, Heidelberg, New York.

Vince-Prue, D. 1989. The role of phytochrome in the control of flowering. Flowering Newsletter (Georges Bernier, editor), Issue No. 8, pp. 3–14.

Vince-Prue, Daphne and K. E. Cockshull. 1981. Photoperiodism and crop production. Pages 175–197 *in* C. B. Johnson (ed.), Physiological Process Limiting Plant Productivity. Butterworths, London.

Vince-Prue, D. and P. J. Lumsden. 1987. Inductive events in the leaves: Time measurement and photoperception in the short-day plant, *Pharbitis nil*. Pages 255–368 *in* J. G. Atherton (ed.), Manipulation of Flowering. Butterworths, London.

Vince-Prue, Daphne, Bryan Thomas, and K. E. Cockshull. (Eds.). 1984. Light and the Flowering Process. Academic Press, Orlando, Fla.

von Gaertner, Thekla and Ernst Braunroth. 1935. Über den Einfluß des Mondlichtes auf den Blühtermin der Lang- und Kurztagspflanzen. (About the influence of moonlight on flowering date of long- and short-day plants.) Botanisches Centralblatt Abt. A53:554–563.

Wellensiek, S. J. 1973. Genetics and flower formation of annual *Lunaria*. Netherland Journal of Agricultural Science 21:163–166.

Zeevaart, Jan A. D. 1976a. Physiology of flower formation. Annual Review of Plant Physiology 27:321–348.

Zeevaart, Jan A. D. 1976b. Phytohormones and flower formation. *In* D. S. Letham et al. (eds.), Plant Hormones and Related Compounds. ASP Biological and Medical, Amsterdam.

Zeevaart, Jan A. D. 1979. Perception, nature and complexity of transmitted signals. Pages 59–90 *in* La Physiologie de la Floraison. Éditions du Centre National de la Recherche Scientifique, Paris.

Zeevaart, Jan A. D. 1982. Transmission of the floral stimulus from a long-short-day plant, *Bryophyllum daigremontianum*, to the short-long-day plant *Echeveria harmsii*. Ann. Bot. 49:549–552.

### *Chapter 24 Molecular Genetics and the Plant Physiologist*

Bevan, M. 1984. Binary *Agrobacterium* vectors for plant transformation. Nucleic Acids Research 12:8711–8721.

Britten, R. J. and D. E. Kohne. 1968. Repeated sequences in DNA. Science 161:529–540.

Burnette, W. N. 1981. "Western blotting": Electrophoretic transfer of proteins from sodium dodecyl sulfate-polyacrylamide gels to unmodified nitrocellulose and detection with antibody and radioiodinated protein A. Analytical Biochemistry 112:195–203.

Capra, J. D. and A. B. Edmonson. 1977. The antibody combining site. Scientific American 236(1):50–59.

Cohen, S. N., A. C. Y. Chang, H. W. Boyer, and R. B. Helling. 1973. Construction of biologically functional bacterial plasmids in vitro. Proceedings of the National Academy of Science USA 70: 3240–3244.

Cox, K. H., D. V. DeLeon, L. M. Angerer, and R. C. Angerer. 1984. Detection of mRNAs in sea urchin embryos by in situ hybridization using asymmetric RNA probes. Developmental Biology 101:485–502.

Fedoroff, N. V. 1989. About maize transposable elements and development. Cell 56:181–191.

Fraley, R. T., S. G. Rogers, and R. B. Horsch. 1986. Genetic transformation in higher plants. CRC Critical Reviews in Plant Sciences 4:1–46.

Gilmartin, P. M., L. Sorokin, J. Memelink, and N. H. Chua. 1990. Molecular light switches for plant genes. The Plant Cell 2:369–378.

Goldberg, R. B., S. J. Barker, and L. Perez-Grau. 1989. Regulation of gene expression during plant embryogenesis. Cell 56:149–160.

Grunstein, M. and D. S. Hogness. 1975. Colony hybridization: A method for the isolation of cloned DNA that contains a specific gene. Proceedings of the National Academy of Sciences USA 72:3961–3965.

Hagen, G., A. Kleinschmidt, and T. J. Guilfoyle. 1984. Auxin-regulated gene expression in intact soybean hypocotyl and excised hypocotyl sections. Planta 162:147–153.

Higgins, T. J. V. 1984. Synthesis and regulation of major protein in seeds. Annual Review of Plant Physiology 35:191–221.

Jackson, D., R. Symons, and P. Berg. 1972. Biochemical method for inserting new genetic information into DNA of simian virus 40: Circular SV40 DNA molecules containing lambda phage genes and the galactose operon of *Escherichia coli*. Proceedings of the National Academy of Sciences USA 69:2904–2909.

Jefferson, R. A., T. A. Kavanagh, and M. W. Bevan. 1987. Gus fusions: Beta-glucuronidase as a sensitive and versatile gene fusion marker in higher plants. EMBO Journal 6:3901–3907.

Kelly, T. J. and H. O. Smith. 1970. A restriction enzyme from *Hemophilus influenzae*. II. Base sequence of the recognition site. Journal of Molecular Biology 51:393–409.

Key, J. L., J. Kimbel, E. Vierling, C-Y. Lin, R. T. Nagao, E. Czarneckaa, and F. Schöffl. 1985. Physiological and molecular analysis of the heat shock response in plants. Pages 327–348 *in* B. G. Atkinson and D. B. Walden (eds.), Changes in Eukaryotic Gene Expression in Response to Environmental Stress. Academic Press, Orlando, Fla.

Klee, H., R. Horsch, and S. Rogers. 1987. *Agrobacterium*-mediated plant transformation and its further applications to plant biology. Annual Review of Plant Physiology 38:467–486.

Klein, T. M., M. Fromm, A Weissinger, D. Tomes, S. Schaaf, M. Stetten, and J. C. Sanford. 1988. Transfer of foreign genes into intact maize cells with high-velocity microprojectiles. Proceedings of the National Academy of Sciences USA 85:4305–4309.

Kuhlemeier, C., P. J. Green, and N-H. Chua. 1987. Regulation of gene expression in higher plants. Annual Review of Plant Physiology 38:221–257.

Lin, C-Y., J. K. Roberts, and J. L. Key. 1984. Acquisition of thermotolerance in soybean seedlings: Synthesis and accumulation of heat shock proteins and their cellular localization. Plant Physiology 74:152–160.

Link, G. 1988. Photocontrol of plastid gene expression. Plant, Cell and Environment 11: 329–338.

Marmur, J. and P. Doty. 1962. Determination of the base composition of deoxyribonucleic acid from its thermal denaturation temperature. Journal of Molecular Biology 5:109–118.

Maxam, A. and W. Gilbert. 1977. A new method for sequencing DNA. Proceedings of the National Academy of Sciences USA 74:560–564.

McClintock, B. 1948. Mutable loci in maize. Carnegie Institution of Washington Year Book 47:155–169.

Medford, J. I., R. Horgan, Z. E. Sawi, and H. J. Klee. 1989. Alterations of endogenous cytokinins in transgenic plants using a chimeric isopentenyl transferase gene. 1:403–413.

Mitchell, P. J. and R. Tjian. 1989. Transcriptional regulation in mammalian cells by sequence-specific DNA binding proteins. Science 245: 371–378.

Moses, P. B. and N-H. Chua. 1988. Light switches for plant genes. Scientific American 258(4):88–93.

Murphy, T. M. and W. F. Thompson. 1988. Molecular Plant Development. Prentice-Hall, Englewood Cliffs, N. J.

Nagy, F., S. A. Kay, and N-H. Chua. 1988. Gene regulation by phytochrome. Trends in Genetics 4:37–42.

Okamuro, J. K. and R. B. Goldberg. 1989. Regulation of plant gene expression: General principles. Pages 1–81 *in* A. Marcus (ed.), The Biochemistry of Plants, Vol. 15, Molecular Biology. Academic Press, New York.

Powell, A. P., R. S. Nelson, B. H. N. De, S. G. Rogers, R. T. Fraley, and R. N. Beachy. 1986. Delay of disease development in transgenic plants that express the tobacco mosaic coat protein gene. Science 232:738–743.
Sanger, F., S. Nicklen, and A. R. Caulson. 1977. DNA sequencing with chain terminating inhibitors. Proceedings of the National Academy of Sciences USA 74:5463–5467.
Simpson, J. and L. Herrera-Estrella. 1990. Light-regulated gene expression. CRC Critical Reviews in Plant Sciences 9:95–109.
Skoog, F. and C. O. Miller. 1957. Chemical regulation of growth and organ formation in plant tissues cultured in vitro. Symposium of the Society for Experimental Biology 11:118–130.
Smith, C. J. S., C. F. Watson, J. Roy, C. R. Bird, P. C. Morris, W. Schuch, and D. Grierson. 1988. Antisense RNA inhibition of polygalacturonase gene expression in transgenic tomatoes. Nature 334:724–726.
Smith, H. O. and K. W. Wilcox. 1970. A restriction enzyme from *Hemophilus influenzae*. I. Purification and general properties. Journal of Molecular Biology 51:379–391.
Southern, E. M. 1975. Detection of specific sequences among DNA fragments separated by gel electrophoresis. Journal of Molecular Biology 98:503–517.
Tanksley, S. D., N. D. Young, A. H. Paterson, and M. W. Bonierbale. 1989. RFLP mapping in plant breeding: New tools for an old science. Bio-Technology 7:257–264.
Thomas, P. S. 1980. Hybridization of denatured RNA and small DNA fragments transferred to nitrocellulose. Proceedings of the National Academy of Sciences USA 77:5201–5205.
Thompson, W. F. 1988. Photoregulation: Diverse gene responses in greening seedlings. Plant, Cell and Environment 11:319–328.
Van der Krol, A. R., P. E. Lenting, J. Veenstra, I. M. Van der Meer, R. E. Kees, A.G. M. Gerats, J. N. M. Mol, and A. R. Stuitje. 1988. An antisense chalcone synthase gene in transgenic plants inhibits flower pigmentation. Nature 333: 866–869.
VarShavsky, A. 1987. Electrophoretic assay for DNA binding proteins. Methods Enzymol. 151:551–565.
Walker, J. C. and J. L. Key. 1982. Isolation of cloned cDNAs to auxin-responsive poly(A)$^+$ RNAs of elongating soybean hypocotyl. Proceedings of the National Academy of Sciences USA 79:185–189.
Watkins, P. C. 1988. Restriction fragment length polymorphism (RFLP): Applications in human chromosome mapping and genetic disease research. BioTechnology 6:310–320.
Watson, J. C. 1989. Photoregulation of gene expression in plants. Pages 161–205 *in* S-D. Kung and C. J. Arntzen (eds.), Plant Biotechnology. Butterworths, Boston.

### *Chapter 25 Topics in Environmental Physiology*

Amthor, J. S. 1989. Respiration and Crop Productivity. Springer-Verlag, New York.
Ballaré, Carlos, Ana L. Scopel, and Rodolfo A. Sánchez. 1990. Far-red radiation reflected from adjacent leaves: An early signal of competition in plant canopies. Science 247:329–331.
Barbour, M. G., J. H. Burk, and W. D. Pitts. 1987. Terrestrial Plant Ecology, Second Edition. Benjamin/Cummings, Menlo Park, Calif.
Barnes, P. W., P. W. Jordon, W. G. Gold, S. D. Flint, and M. M. Caldwell. 1988. Competition, morphology and canopy structure in wheat (*Triticum aestivum* L.) and wild oat (*Avena fatua* L.) exposed to enhanced ultraviolet-B radiation. Functional Ecology 2:319–330.
Bazzaz, Fakhri A., Nona R. Chiariello, Phyllis D. Coley, and Louis F. Pitelka. 1987. Allocating resources to reproduction and defense. BioScience 37:58–67.
Berry, Wade L. and Arthur Wallace. 1989. Zinc phytotoxicity: Physiological responses and diagnostic criteria for tissues and solutions. Soil Science 147:390–397.
Beyschlag, W., P. W. Barnes, S. D. Flint, and M. M. Caldwell. 1988. Enhanced UV-B irradiation has no effect on photosynthetic characteristics of wheat (*Triticum aestivum* L.) and wild oat (*Avena fatua* L.) under greenhouse and field conditions. Photosynthetica 22(4):516–525.
Bhargava, Suresh C. 1975. Photoperiodicity and seed germination in rice. Indian Journal of Agricultural Sciences 45:447–451.
Billings, W. D. 1970. Plants, Man and the Ecosystem, Second Edition. Wadsworth, Belmont, Calif.
Björkman, O. 1968. Further studies on differentiation of photosynthetic properties in sun and shade ecotypes of *Solidago virgaurea*. Physiologia Plantarum 21:84–99.
Björkman, O. 1981. Responses to different quantum flux densities. Pages 57–107 *in* O. L. Lange, P. S. Nobel, C. B. Osmond, and H. Ziegler (eds.), Encyclopedia of Plant Physiology, Vol. 12A, Physiological Plant Ecology I. Springer-Verlag, Berlin, Heidelberg, New York.
Blackman, F. F. 1905. Optima and limiting factors. Annals of Botany 14(74):281–295.
Boller, Thomas and Andres Wiemken. 1986. Dynamics of vacuolar compartmentation. Annual Review of Plant Physiology 37:137–164.
Bugbee, Bruce G. and F. B. Salisbury. 1988. Exploring the limits of crop productivity. I. Photosynthetic efficiency of wheat in high irradiance environments. Plant Physiology 88:869–878.
Bugbee, B. G. and F. B. Salisbury. 1989. Current and potential productivity of wheat for a controlled environment life support system. Advances in Space Research 9:5–15.
Caldwell, M. M. 1981. Plant response to solar ultraviolet radiation. Pages 170–194 *in* O. L. Lange, P. S. Nobel, C. B. Osmond, and H. Ziegler (eds.), Encyclopedia of Plant Physiology, Vol. 12A, Physiological Plant Ecology I. Springer-Verlag, Berlin, Heidelberg, New York.
Chapin III, F. Stuart, Arnold J. Bloom, Christopher B. Field, and Richard H. Waring. 1987. Plant responses to multiple environmental factors. BioScience 37:49–57.
Chazdon, R. L. and R. W. Pearcy. 1986. Photosynthetic responses to light variation in rain forest species. II. Carbon gain and light utilization during lightflecks. Oecologia 69:524–531.
Clausen, Jens, David D. Keck, and William M. Hiesey. 1940. Experimental Studies on the Nature of Species. I. Effect of Varied Environments on Western North American Plants. Carnegie Institution of Washington, Publication No. 520, Washington D. C.
Clements, Frederic E., Emmett V. Martin, and Frances L. Long. 1950. Adaptation and Origin in the Plant World. Chronica Botanica, Waltham, Mass.
Downs, R. J. and Borthwick, H. A. 1956. Effects of photoperiod on growth of trees. Botanical Gazette 117:310–326.
Ehleringer, James and Harold A. Mooney. 1984. Photosynthesis and productivity of desert and Mediterranean climate plants. Pages 205–231 *in* O. L. Lange, P. S. Nobel, C. B. Osmond, and H. Ziegler (eds.), Encyclopedia of Plant Physiology, New Series, Vol. 12D. Springer-Verlag, Berlin.
Gates, D. M. 1962. Energy Exchange in the Biosphere. Harper & Row, New York.
Highkin, H. R. and A. Lang. 1966. Residual effect of germination temperature on the growth of peas. Planta 68:94–98.
Keeley, J. E. and L. G. Busch. 1984. Carbon assimilation characteristics of the aquatic CAM plant *Isoetes howellii*. Plant Physiology 76:525–530.
Keeley, J. E., C. B. Osmond, and J. A. Raven. 1984. *Stylites*, a vascular land plant without stomata absorbs $CO_2$ via its roots. Nature 310:694–695. (The genus *Stylites* has been renamed *Isoetes*.)
Kigel, J., M. Ofir, and D. Koller. 1977. Control of the germination responses of *Amaranthus retroflexus* L. seeds by their parental photothermal environment. Journal of Experimental Botany 28: 1125–1136.
Koller, Dov. 1962. Preconditioning of germination in lettuce at time of fruit ripening. American Journal of Botany 49:841–844.
Lange, O. L., P. S. Nobel, C. B. Osmond, and H. Ziegler. 1981. Introduction: Perspectives in Ecological Plant Physiology. Pages 1–9 *in* O. L. Lange, P. S. Nobel, C. B. Osmond, and H. Ziegler (eds.), Encyclopedia of Plant Physiology, Vol. 12A, Physiological Plant Ecology I. Springer-Verlag, Berlin, Heidelberg, New York.
Liebig, Justus. 1841. Organic Chemistry in Its Applications to Agriculture and Physiology. John Owen, Cambridge. (First American Edition.)
Lockhart, James A. 1965. The analysis of interactions of physical and chemical factors on plant growth. Annual Review of Plant Physiology 16:37–52.
Marchand, Peter J. 1987. Life in the Cold. An Introduction to Winter Ecology. University Press of New England, Hanover and London.
Mayer, A. M. and A. Poljakoff-Mayber. 1989. The Germination of Seeds, Fourth Edition. Pergamon Press, New York, London.
Mohr, H. 1972. Lectures on Photomorphogenesis. Springer-Verlag, Berlin, Heidelberg, New York.
Mooney, Harold A. and S. L. Gulmon. 1979. Environmental and evolutionary constraints on the photosynthetic characteristics of higher plants. Pages 316–337 *in* O. T. Solbrig, S. Jain, G. B. Johnson, and P. H. Raven (eds.), Topics in Plant Population Biology. Columbia University Press, New York.
Mooney, H. A., Robert W. Pearcy, and James Ehleringer. 1987. Plant physiological ecology today. BioScience 37:18–20. (This is the introductory article of a special issue of BioScience devoted to physiological ecology.)
Morgan, D. C. and H. Smith. 1981. Non-photosynthetic responses to light quality. Pages 109–130 *in* O. L. Lange, P. S. Nobel, C. B. Osmond, and H. Ziegler (eds.), Encyclopedia of Plant Physiology, Vol. 12A, Physiological Plant Ecology I. Springer-Verlag, Berlin, Heidelberg, New York.
Muir, John. 1976. My First Summer in the Sierra. Houghton Mifflin, Boston. [This is a reprint of a book first published in 1886.]
Mulroy, T. W. and P. W. Rundel. 1977. Annual plants: Adaptations to desert environments. BioScience 27:109–114.
Osmond, C. B., M. P. Austin, J. A. Berry, W. D. Billings, J. S. Boyer, J. W. H. Cacey, P. S. Nobel, S. D. Smith, and W. E. Winner. 1987. Stress physiology and the distribution of plants. BioScience 37:38–48.
Pearcy, Robert W. 1988. Photosynthetic utilization of lightflecks by understory plants. Australian Journal of Plant Physiology 15:223–238.
Pearcy, R. W. 1990. Sunflecks and photosynthesis in plant canopies. Annual Review of Plant Physiology and Plant Molecular Biology 41: 421–453.
Pearcy, R. W., Olle Björkman, Martyn M. Caldwell, Jon E. Keeley, Russell K. Monson, and Boyd R. Strain. 1987. Carbon gain by plants in natural environments. BioScience 37:21–29.
Pourrat, Yvonne and Roger Jacques. 1975. The influence of photoperiodic conditions received by the mother plant on morphological and physiological characteristics of *Chenopodium polyspermum* L. seeds. Plant Science Letters 4: 273–279.
Raven, John A., Linda L. Handley, Jeffrey J. MacFarlane, Shona McInroy, Lewis McKenzie, Jennifer H. Richards, and Goran Samuelsson. 1988. The role of $CO_2$ uptake by roots and CAM in acquisition of inorganic C by plants of the isoetid life-form: A review, with new data on *Eriocaulon decangulare* L. New Phytologist 108:125–148.
Richardsen, S. and Frank B. Salisbury. 1977. Plant responses to the light penetrating snow. Ecology 58:1152–1158.

Salisbury, Frank B. 1975. Multiple factor effects on plants. Pages 501–520 *in* F. John Vernberg (ed.), Physiological Adaptation to the Environment. Intext Educational Publishers, New York.
Salisbury, Frank B. 1981a. Responses to photoperiod. Pages 135–167 *in* O. L. Lange, P. S. Nobel, C. B. Osmond, and H. Ziegler (eds.), Encyclopedia of Plant Physiology, Vol. 12A, Physiological Plant Ecology I. Springer-Verlag, Berlin, Heidelberg, New York.
Salisbury, Frank B. 1981b. The twilight effect: Initiating dark measurement in photoperiodism of *Xanthium*. Plant Physiology 67:1230–1238.
Salisbury, Frank B. 1982. Photoperiodism. Horticultural Reviews 4:66–105.
Salisbury, F. B. and Nicos G. Marinos. 1985. The ecological role of plant growth substances. Pages 707–766 *in* Richard P. Pharis and David M. Reid (eds.), Encyclopedia of Plant Physiology, Vol. 11, Hormonal Regulation of Development, III: Role of Environmental Factors. Springer-Verlag, Berlin, Heidelberg, New York.
Schopfer, Peter. 1969. Die Hemmung der Streckungswachstums durch Phytochrom—ein Stoffaufnahme erforderner Prozess? (Inhibition of elongation growth by phytochrome—a process requiring substrate uptake?) Planta (Berlin) 85:383–388.
Schulze, E-D., M. I. Fuchs, and M. Fuchs. 1977. Spatial distribution of photosynthetic capacity in a mountain spruce forest of Northern Germany. III. The significance of the evergreen habitat. Oecologia 30:239–248.
Schulze, E-D., R. H. Robichaux, J. Grace, P. W. Rundel, and J. R. Ehleringer. 1987. Plant water balance. BioScience 37:30–37.
Seemann, J. R., M. R. Badger, and J. A. Berry. 1984. Variations in specific activity of ribulose-1,5-bisphosphate carboxylase between species utilizing differing photosynthetic pathways. Plant Physiology 74:791–794.
Shelford, Victor E. 1913. Animal Communities in Temperate America. University of Chicago Press, Chicago.
Smith, Harry. 1983. Light quality, photoperception, and plant strategy. Annual Review of Plant Physiology 33:481–518.
Spomer, G. G. 1973. The concepts of "interaction" and "operational environment" in environmental analyses. Ecology 54(1):200–204.
Tieszen, L. L., P. C. Miller, M. C. Lewis, J. C. Mayo, W. C. Oechel, and F. S. Chapin. 1981. Processes of primary production in tundra. Pages 285–356 *in* L. C. Bliss, O. W. Heal, and J. J. Moore (eds.), Tundra Ecosystems: A Comparative Analysis. Cambridge University Press, Cambridge.
Turesson, G. 1922. The genotypic response of the plant species to the habitat. Hereditas 3:211–350.
Vince-Prue, D. 1975. Photoperiodism in Plants. McGraw-Hill, London.
Whittaker, R. H. 1975. Communities and Ecosystems, Second Edition. Macmillan, New York; Collier Macmillan, London.

## *Chapter 26 Stress Physiology*

Acevedo, Edmundo, Theodore C. Hsiao, and D. W. Henderson. 1971. Immediate and subsequent growth responses of maize leaves to changes in water status. Plant Physiology 48:631–636.
Allen, T. 1965. The Quest—A Report on Extraterrestrial Life. Chilton Books, Radnor, Pa.
Anderson, J. A., D. W. Buchanan, R. E. Stall, and C. B. Hall. 1982. Frost injury of tender plants increased by *Pseudomonas syringae* van Hall. Journal of the American Society of Horticultural Science 107:123–125.
Aragno, M. 1981. Responses of microorganisms to temperature. Pages 339–369 *in* O. L. Lange, P. S. Nobel, C. B. Osmond, and H. Ziegler (eds.), Encyclopedia of Plant Physiology, New Series, Vol. 12A, Physiological Plant Ecology. Springer-Verlag, Berlin, Heidelberg, New York.
Ashworth, E. N. 1984. Xylem development in *Prunus* flower buds and the relationship to deep supercooling. Plant Physiology 74:862–865.
Ball, Marilyn C. 1988. Salinity tolerance in the mangroves *Aegiceras corniculatum* and *Avicennia marina*. I. Water use in relation to growth, carbon partitioning, and salt balance. Australian Journal of Plant Physiology 15:447–464.
Barbour, Michael G. 1970. Is any angiosperm an obligate halophyte? American Midland Naturalist 84:106–119.
Barbour, M. G., Jack H. Burk, and Wanna D. Pitts. 1987. Terrestrial Plant Ecology, Second Edition. Benjamin/Cummings, Menlo Park, Calif.
Baross, J. A. and J. W. Deming. 1983. Growth of "black smoker" bacteria at temperatures of at least 250°C. Nature 303:423–426.
Bensen, Robert J., John S. Boyer, and John E. Mullet. 1988. Water deficit-induced changes in abscisic acid, growth, polysomes, and translatable RNA in soybean hypocotyls. Plant Physiol. 88:289–294.
Bewley, J. D. and J. E. Krochko. 1982. Desiccation tolerance. Pages 325–378 *in* O. L. Lange, P. S. Nobel, C. B. Osmond, and H. Ziegler (eds.), Encyclopedia of Plant Physiology, New Series, Vol. 12B, Physiological Plant Ecology II. Springer-Verlag, Berlin, Heidelberg, New York.
Bhagwat, Arvind A. and Shree Kumar Apte. 1989. Comparative analysis of proteins induced by heat shock, salinity, and osmotic stress in the nitrogen-fixing cyanobacterium *Anabaena* sp. strain L-31. Journal of Bacteriology 171(9):5187–5189.
Blumwald, Eduardo and Ronald J. Poole. 1987. Salt tolerance in suspension cultures of sugar beet. Induction of $Na^+/H^+$ antiport activity at the tonoplast by growth in salt. Plant Physiol. 83: 884–887.
Bodner, M. and E. Beck. 1987. Effect of supercooling and freezing on photosynthesis in freezing tolerant leaves of Afroalpine "giant rosette" plants. Oecologia (Berlin) 72:366–371.
Boyer, J. S. 1970. Leaf enlargement and metabolic rates in corn, soybean, and sunflower at various leaf water potentials. Plant Physiology 46: 233–235.
Bradford, K. J. and T. C. Hsiao. 1982. Physiological responses to moderate water stress. Pages 263–324 *in* O. L. Lange, P. S. Nobel, C. B. Osmond, and H. Ziegler (eds.), Physiological Plant Ecology II, Water Relations and Carbon Assimilation. *In* A. Pirson and M. H. Zimmermann (eds.), Encyclopedia of Plant Physiology, New Series, Vol. 12B. Springer-Verlag, Berlin, Heidelberg, New York.
Bray, Elizabeth A. 1988. Drought- and ABA-induced changes in polypeptide and mRNA accumulation in tomato leaves. Plant Physiol. 88:1210–1214.
Brock, T. D. 1978. Thermophilic Microorganisms and Life at High Temperatures. Springer-Verlag, Berlin, Heidelberg, New York.
Bunce, James A. 1988. Nonstomatal inhibition of photosynthesis by water stress. Reduction in photosynthesis at high transpiration rate without stomatal closure in field grown tomato. Photosynthesis Research 18:357–362.
Burke, J. J. and K. A. Orzech. 1988. The heat-shock response in higher plants: A biochemical model. Plant, Cell and Environment 11:441–444.
Burke, M. J., L. V. Gusta, H. A. Quammer, C. J. Weiser, and P. H. Li. 1976. Freezing and injury in plants. Annual Review of Plant Physiology 27:507–528.
Butcher, Russell D. and David E. Evans. 1987. Calcium transport by pea root membranes. I. Purification of membranes and characteristics of uptake. II. Effects of calmodulin and inhibitors. Planta 172:265–279.
Cheeseman, John M. 1988. Mechanisms of salinity tolerance in plants. Plant Physiol. 87:547–550.
Cramer, G. R., E. Epstein, and A. Läuchli. 1988. Kinetics of root elongation of maize in response to short-term exposure to NaCl and elevated calcium concentration. Jour. of Experimental Botany 39: 1513–1522.
Cramer, G. R., A. Läuchli, and Emanuel Epstein. 1986. Effects of NaCl and $CaCl_2$ on ion activities in complex nutrient solutions and root growth of cotton. Plant Physiol. 81:792–797.
Cramer, Grant R., André Läuchli, and Vito S. Polito. 1985. Displacement of $Ca^{2+}$ by $Na^+$ from the plasmalemma of root cells. Plant Physiol. 79: 207–211.
Crawford, R. M. M. 1989. Studies in Plant Survival. Ecological Case Histories of Plant Adaptation to Adversity. Blackwell, Oxford.
Daubenmire, Rex F. 1947. Plants and Environment. Wiley, New York.
DuPont, F. M. and J. B. Mudd. 1985. Acclimation to low temperature by microsomal membranes from tomato cell cultures. Plant Physiol. 77:74–78.
Edney, E. B. 1975. Absorption of water vapor from unsaturated air. Pages 77–97 *in* F. John Vernberg (ed.), Physiological Adaptation to the Environment. Intext Educational Publishers, New York.
Ehleringer, James R. and Tamsie A. Cooper. 1988. Correlations between carbon isotope ratio and microhabitat in desert plants. Oecologia 76: 562–566.
Ehleringer, J., Olle Björkman, and Harold A. Mooney. 1976. Leaf pubescence: Effects on absorptance and photosynthesis in a desert shrub. Science 192:376–377.
Evenari, M., E-D. Schulze, L. Kappen, U. Buschbom, and O. L. Lange. 1975. Adaptive mechanisms in desert plants. Pages 111–129 *in* F. John Vernberg (ed.), Physiological Adaptation to the Environment. Intext Educational Publishers, New York.
Fireman, M. and H. E. Hayward. 1952. Indicator significance of some shrubs in the Escalante Desert, Utah. Botanical Gazette 114:143–154.
Flores, H. E. and A. W. Galston. 1982. Polyamines and plant stress: Activation of putrescine biosynthesis by osmotic shock. Science 217: 1259–1260.
Flowers, T. J., P. F. Troke, and A. R. Yeo. 1977. The mechanism of salt tolerance in halophytes. Annual Review of Plant Physiology 28:89–121.
George, Milton F. and Michael J. Burke. 1984. Supercooling of tissue water to extreme low temperature in overwintering plants. TIBS 9: 211–214.
Giannini, John L. and Donald P. Briskin. 1989. The effect of assay composition, detergent solubilization and reconstitution on red beet (*Beta vulgaris* L.) plasma membrane $H^+$-ATPase kinetic properties. Plant Science 60:189–193.
Gilmour, Sarah J., Ravindra K. Hajela, and Michael F. Thomashow. 1988. Cold acclimation in *Arabidopsis thaliana*. Plant Physiol. 87:745–750.
Gimmler, Hartmut, Lothar Schneider, and Rosemarie Kaaden. 1989. The plasma membrane ATPase of *Dunaliella parva*. Zeitschrift für Naturforschung 44:128–138.
Graham, Douglas and Brian D. Patterson. 1982. Responses of plants to low, nonfreezing temperatures: Proteins, metabolism, and acclimation. Annual Review of Plant Physiology 33:347–372.
Greenway, H. and Rana Munns. 1980. Mechanisms of salt tolerance in nonhalophytes. Annual Review of Plant Physiology 30:149–190.
Gross, Dennis C., Edward L. Proebsting, Jr., and Heather Maccrindle-Zimmerman. 1988. Development, distribution, and characteristics of intrinsic, nonbacterial ice nuclei in *Prunus* wood. Plant Physiol. 88:915–922.
Guerrero, Felix D. and John E. Mullet. 1988. Reduction of turgor induces rapid changes in leaf translatable RNA. Plant Physiol. 88:401–408.
Guy, C. L. and D. Haskell. 1987. Induction of freezing tolerance in spinach is associated with the synthesis of cold acclimation induced proteins. Plant Physiol. 84:872–878.
Hanson, Andrew D. and William D. Hitz. 1982. Metabolic responses of mesophytes to plant water deficits. Annual Review of Plant Physiology 33:163–203.

Harvey, G. W., L. V. Gusta, D. C. Fork, and J. A. Berry. 1982. The relation between membrane lipid phase separation and frost tolerance of cereals and other cool climate plant species. Plant, Cell and Environment 5:241–244.
Hellebust, Johan A. 1976. Osmoregulation. Annual Review of Plant Physiology 27:485–505.
Hincha, Dirk K., Ulrich Heber, and Jürgen Schmitt. 1989. Freezing ruptures thylakoid membranes in leaves, and rupture can be prevented *in vitro* by cryoprotective proteins. Plant Physiology and Biochemistry 27(15):795–801.
Hsiao, T. C. 1973. Plant responses to water stress. Annual Review of Plant Physiology 24:519–570.
Huber, R., M. Kurr, Holger W. Jannasch, and K. O. Stetter. 1989. A novel group of abyssal methanogenic archaebacteria (*Methanopyrus*) growing at 110°C. Nature 342:833–834.
Hurkman, William J., Charlene K. Tanaka, and Frances M. DuPont. 1988. The effects of salt stress on polypeptides in membrane fractions from barley roots. Plant Physiol. 88:1263–1273.
Iraki, Naim M., Ray A. Bressan, P. M. Hasegawa, and Nicholas C. Carpita. 1989. Alteration of the physical and chemical structure of the primary cell wall of growth-limited plant cells adapted to osmotic stress. Plant Physiology 91:39–47.
Jackson, Michael B. 1985. Ethylene and responses of plants to soil waterlogging and submergence. Annual Review of Plant Physiology 36:145–174.
Jefferies, R. L. 1981. Osmotic adjustment and the response of halophytic plants to salinity. BioScience 31(1):42–46.
Jensen, W. A. and F. B. Salisbury. 1984. Botany, Second Edition. Wadsworth, Belmont, Calif.
Kaiser, Werner M. 1987. Effects of water deficit on photosynthetic capacity. Physiol. Plantarum 71:142–149.
Kappen, L. 1981. Ecological significance of resistance to high temperature. Pages 439–474 *in* O. L. Lange, P. S. Nobel, C. B. Osmond, and H. Ziegler (eds.), Encyclopedia of Plant Physiology, New Series, Vol. 12A, Physiological Plant Ecology. Springer-Verlag, Berlin, Heidelberg, New York.
Kappen, L. 1989. Field measurements of carbon dioxide exchange of the Antarctic lichen *Usnea sphacelata* in the frozen state. Antarctic Science 1(1):31–34.
Kasamo, Kunihiro. 1988. Response of tonoplast and plasma membrane ATPases in chilling-sensitive and -insensitive rice (*Oryza sativa* L.) culture cells to low temperature. Plant Cell Physiol. 29(7):1085–1094.
Key, Joe L., Janice Kimpel, Elizabeth Vierling, Chuyung Lin, Ronald T. Nagao, Eva Czarnecka, and Friedrich Schöffl. 1985. Physiological and molecular analysis of the heat shock response in plants. Pages 237–348 *in* B. G. Atkinson and D. B. Walden (eds.), Changes in Eukaryotic Gene Expression in Response to Environmental Stress. Academic Press, New York.
Kimpel, J. A. and J. L. Key. 1985. Heat shock in plants. Trends in Biochem. Science 117:353–357.
Körner, Christian and Walter Larcher. 1988. Plant life in cold climates. Pages 25–57 *in* S. P. Long and F. I. Woodward (eds.), Plants and Temperature, Symposia of the Society of Experimental Biology, No. 42. The Company of Biologists Limited, Cambridge.
Kramer, Paul J. 1980. Drought, stress, and the origin of adaptations. Pages 7–20 *in* N. C. Turner and P. J. Kramer (eds.), Adaptation of Plants to Water and High Temperature Stress. Wiley, New York.
Kramer, Paul J. 1983. Water Relations of Plants. Academic Press, New York and London.
Krause, G. Heinrich, S. Grafflage, S. Rumich-Bayer, and S. Somersalo. 1988. Effects of freezing on plant mesophyll cells. Pages 311–327 *in* S. P. Long and F. I. Woodward (eds.), Plants and Temperature. Symposia of the Society for Experimental Biology, No. 42. The Company of Biologists Limited, Cambridge.
Kriedemann, P. E. and B. R. Loveys. 1974. Hormonal mediation of plant responses to environmental stress. Pages 461–465 *in* R. L. Bieleski, A. R. Ferguson, and M. M. Creswell (eds.), Mechanisms of Regulation of Plant Growth, The Royal Society of New Zealand, Wellington.
Kruckeberg, A. R. 1954. The ecology of serpentine soils. III. Plant species in relation to serpentine soils. Ecology 35:267–274.
Kurth, Eva, Grant R. Cramer, André Läuchli, and Emanuel Epstein. 1986. Effects of NaCl and $CaCl_2$ on cell enlargement and cell production in cotton roots. Plant Physiol. 82:1102–1106.
LaHaye, P. A. and E. Epstein. 1969. Salt toleration by plants: Enhancement with calcium. Science 166:395–396.
Lalk, I. and K. Dörffling. 1985. Hardening, abscisic acid, proline and freezing resistance in two winter wheat varieties. Physiologia Pl. 63:287–292.
Larcher, Walter. 1983. Physiological Plant Ecology, Second Edition. Springer-Verlag, Berlin, Heidelberg, New York.
Larcher, Walter. 1987. Streβ bei Pflanzen. (Stress in plants.) Naturwissenschaften 74:158–167.
Larcher, W. and Helmut Bauer. 1981. Ecological significance of resistance to low temperature. Pages 403–437 *in* O. L. Lange, P. S. Nobel, C. B. Osmond, and H. Ziegler (eds.), Encyclopedia of Plant Physiology, New Series, Vol. 12A, Physiological Plant Ecology. Springer-Verlag, Berlin, Heidelberg, New York.
Larcher, W. and Monika Nagele. 1985. Induktionskinetic der Chlorophyllfluoreszenz unterkühlter und gefrorener Blätter von *Rhododendron ferrugineum* beim Übergang vom gefrierempfindlichen zum gefriertoleranten Zustand. (Induction kinetics of chlorophyll fluorescence of supercooled and frozen leaves of *Rhododendron ferrugineum* during the transition from the frost-sensitive to the frost-tolerant condition.) Sitzungsberichten der Österr. Akademie der Wissenschaften, Mathem.-naturw. Kl. I, 194:187–195.
Larcher, W. and Gilbert Neuner. 1989. Cold-induced sudden reversible lowering of *in vivo* chlorophyll fluorescence after saturating light pulses. Plant Physiol. 89:740–742.
Larcher, W., M. Holzner, and J. Pichler. 1989. Temperaturresistenz inneralpiner Trockenrasen. (Temperature resistance of graminoids from a dry valley of the Central Alps.) Flora 183:115–131.
Larcher, W., J. Wagner, and A. Thammathaworn. 1990. Effects of superimposed temperature stress on *in vivo* chlorophyll fluorescence of *Vigna unguiculata* under saline stress. Journal of Plant Physiology 136:92–102.
LeRudulier, D., A. R. Strom, A. M. Dandekar, L. T. Smith, and R. C. Valentine. 1984. Molecular biology of osmoregulation. Science 224: 1064–1068.
Levitt, J. 1962. A sulfhydryl-disulfide hypothesis of frost injury and resistance in plants. Journal of Theoretical Biology 3:355–391.
Levitt, J. 1972. Responses of Plants to Environmental Stresses. Academic Press, New York and London.
Levitt, Jacob. 1980. Response of Plants to Environmental Stresses, Second Edition, Vols. I and II. Academic Press, New York and London.
Li, Paul H. 1984. Subzero temperature stress physiology of herbaceous plants. Horticultural Reviews 6:373–416.
Lindow, Steven. 1983. The role of bacterial ice nucleation in frost injury to plants. Annual Review of Phytopathology 21:363–384.
Lindquist, S. and E. A. Craig. 1988. The heat-shock proteins. Annual Review of Genetics 22:631–677.
Ludwig, John A., Walter G. Whitford, and Joe M. Cornelius. 1989. Effects of water, nitrogen and sulfur amendments on cover, density and size of Chihuahuan Desert ephemerals. Journal of Arid Environments 16:35–42.
Lyons, J. M. 1973. Chilling injury in plants. Annual Review of Plant Physiology 24:445–466.
MacDougal, D. T. and E. S. Spaulding. 1910. The water-balance of succulent plants. Carnegie Institute Washington Publications 141:77.
Marchand, Peter J. 1987. Life in the Cold. University Press of New England, Hanover and London.
Michalowski, Christine B., Steven W. Olson, Mechtild Piepenbrock, Jürgen M. Schmitt, and Hans J. Bohnert. 1989. Time course of mRNA induction elicited by salt stress in the common ice plant (*Mesembryanthemum crystallinum*). Plant Physiol. 89:811–816.
Mooney, H. A., S. L. Gulmon, J. Ehleringer, and P. W. Rundel. 1980. Atmospheric water uptake by an Atacama desert shrub. Science 209:693–694.
Morgan, James M. 1984. Osmoregulation and water stress in higher plants. Annual Review of Plant Physiology 35:299–348.
Mussell, H. and R. Staples. 1979. Stress Physiology in Crop Plants. Wiley-Interscience, New York.
Neumann, Peter M., Elizabeth Van Volkenburgh, and Robert E. Cleland. 1988. Salinity stress inhibits bean leaf expansion by reducing turgor, not wall extensibility. Plant Physiol. 88:233–237.
Neuner, G. and W. Larcher. 1990. Determination of differences in chilling susceptibility of two soybean varieties by means of *in vivo* chlorophyll fluorescence measurements. J. of Agronomy and Crop Sci. 164(2):73–80.
Ougham, Helen J. and Catherine J. Howarth. 1988. Temperature shock proteins in plants. Pages 259–280 *in* S. P. Long and F. I. Woodward (eds.), Plants and Temperature. Symposia of the Society for Experimental Biology, No. 42. The Company of Biologists Limited, Cambridge.
Parker, J. 1963. Cold resistance in woody plants. Botanical Review 29:124–201.
Pearce, R. S. 1988. Extracellular ice and cell shape in frost-stressed cereal leaves: A low-temperature scanning-electron-microscopy study. Planta 175:313–324.
Pool, Robert. 1990. Pushing the envelope of life. Science 247:158–160.
Quader, H., A. Hofmann, and E. Schnepf. 1989. Reorganization of the endoplasmic reticulum in epidermal cells of onion bulb scales after cold stress: Involvement of cytoskeletal elements. Planta 177:273–280.
Quarrie, S. A. 1980. Genotypic differences in leaf water potential, abscisic acid and proline concentrations in spring wheat during drought stress. Annals of Botany 46:383–394.
Ramagopal, Subbanaidu. 1987a. Messenger RNA changes during drought stress in maize leaves. J. Plant. Physiol. 129:311–317.
Ramagopal, Subbanaidu. 1987b. Salinity stress induced tissue-specific proteins in barley seedlings. Plant Physiol. 84:324–331.
Ranieri, Annamaria, Rodolfo Bernardi, Paola Lanese, and Gian Franco Soldatini. 1989. Changes in free amino acid content and protein pattern of maize seedlings under water stress. Environmental and Experimental Botany 29(3): 351–357.
Raunkiaer, C. 1910. Statistik der Lebensformen als Grundlage für die biologische Pflanzengeographie. (Statistics of life forms as a basis for biological plant geography.) Beiheft Bot. Centralblatt 27II:171–206.
Richardson, S. G. and F. B. Salisbury. 1977. Plant responses to the light penetrating snow. Ecology 58:1152–1158.
Rouxel, Marie-France, Jai P. Singh, Nikos Beopoulos, Jean-Pierre Billard, and Robert Esnault. 1989. Effect of salinity stress on ribonucleolytic activities in glycophytic and halophytic plant species. J. Plant Physiol. 133:738–742.
Rundel, P. W. 1982. Water uptake by organs other than roots. Pages 111–128 *in* O. L. Lange, P. S. Nobel, C. B. Osmond, and H. Ziegler (eds.), Encyclopedia of Plant Physiology, New Series, Vol. 12B, Physiological Plant Ecology II. Springer-Verlag, Berlin, Heidelberg, New York.
Sachs, Martin M. and Tuan-Hua David Ho. 1986. Alteration of gene expression during environmental stress in plants. Annual Review of Plant Physiology 37:363–376.

Sakai, Akira and Walter Larcher. 1987. Frost Survival of Plants. Springer-Verlag, Berlin, Heidelberg.
Sakurai, Naoki and Susumu Kuraishi. 1988. Water potential and mechanical properties of the cell wall of hypocotyls of dark-grown squash (*Cucurbita maxima* Duch.) under water-stress conditions. Plant Cell Physiol. 29(8):1337–1343.
Salisbury, Frank B. 1964. Soil formation and vegetation on hydrothermally altered rock material in Utah. Ecology 45:1–9.
Salisbury, F. B. 1985. The Big Rock Candy Mountain. Utah Science 46:112–118.
Salisbury, F. B. and N. G. Marinos. 1985. The ecological role of plant growth substances. Pages 707–766 *in* R. P. Pharis and D. M. Reid (eds.), Encyclopedia of Plant Physiology, Vol. 11, Hormonal Regulation of Development, III: Role of Environmental Factors. Springer-Verlag, Berlin, Heidelberg, New York.
Salisbury, F. B. and George G. Spomer. 1964. Leaf temperatures of alpine plants in the field. Planta 60:497–505.
Santarius, Kurt A. and Engelbert Weis. 1988. Heat stress and membranes. Pages 97–112 *in* J. L. Harwood and T. J. Walton (eds.), Plant Membranes—Structure, Assembly and Function. The Biochemical Society, London.
Schachtman, D. P., A. J. Bloom, and J. Dvořák. 1989. Salt-tolerant *Triticum* x *Lophopyrum* derivatives limit the accumulation of sodium and chloride ions under saline-stress. Plant, Cell and Environment 12:47–55.
Schmitt, Andreas K., Helen S. J. Lee, and Ulrich Lüttge. 1988. The response of the $C_3$-CAM tree, *Clusia rosea*, to light and water stress. Journal of Experimental Botany 39:1581–1590.
Schulte, Paul J. and Park S. Nobel. 1989. Responses of a CAM plant to drought and rainfall: Capacitance and osmotic pressure influences on water movement. Journal of Experimental Botany 40:61–70.
Schulze, E-D. 1986. Carbon dioxide and water vapor exchange in response to drought in the atmosphere and in the soil. Annual Review of Plant Physiology 37:247–274.
Selye, Hans. 1936. A syndrome produced by Diverse Nocuous Agents. Nature 138(3479):32.
Selye, H. 1950. Stress and the general adaptation syndrome. British Medical Journal 1(June): 1383–1397.
Shantz, H. L. 1927. Drought resistance and soil moisture. Ecology 8:145–157.
Spomer, G. G. 1973. The concepts of "interaction" and "operational environment" in environmental analyses. Ecology 54(1):200–204.
Staples, R. C. and G. H. Toenniessen. (Eds.). 1984. Salinity Tolerance in Plants. Wiley, New York.
Steponkus, P. L. 1981. Responses to extreme temperatures. Cellular and subcellular bases. Pages 371–402 *in* O. L. Lange, P. S. Nobel, C. B. Osmond, and H. Ziegler (eds.), Encyclopedia of Plant Physiology, New Series, Vol. 12A, Physiological Plant Ecology. Springer-Verlag, Berlin, Heidelberg, New York.
Steponkus, Peter L. 1984. Role of the plasma membrane in freezing injury and cold acclimation. Annual Review of Plant Physiology 35:543–693.
Stewart, G. R. and J. A. Lee. 1974. The role of proline accumulation in halophytes. Planta 120:279–289.
Tietz, Dietmar and Arno Tietz. 1982. Streβ im Pflanzenreich. (Stress in the plant kingdom.) Biologie in unserer Zeit 12(4):113–119.
Tranquillini, W. 1982. Frost-drought and its ecological significance. Pages 379–400 *in* O. L. Lange, P. S. Nobel, C. B. Osmond, and H. Ziegler (eds.), Encyclopedia of Plant Physiology, New Series, Vol. 12B, Physiological Plant Ecology. Springer-Verlag, Berlin, Heidelberg, New York.
Troughton, J. and L. A. Donaldson. 1972. Probing Plant Structure. McGraw-Hill, New York.
Turner, Neil C. and Madelaine M. Jones. 1980. Turgor maintenance by osmotic adjustment: A review and evaluation. Pages 87–103 *in* N. C. Turner and P. J. Kramer (eds.), Adaptation of Plants to Water and High Temperature Stress. Wiley-Interscience, New York.
Turner, N. C. and Kramer, P. J. (Eds.). 1980. Adaptation of Plants to Water and High Temperature Stress. Wiley-Interscience, New York.
Ungar, I. A. 1977. The relationship between soil water potential and plant water potential in two inland halophytes under field conditions. Botanical Gazette 138:498–501.
Vu, Joseph C. V. and George Yelenosky. 1988. Water deficit and associated changes in some photosynthetic parameters in leaves of "Valencia" orange (*Citrus sinensis* [L.] Osbeck). Plant Physiol. 88:375–378.
Walton, D. C. 1980. Biochemistry and physiology of abscisic acid. Annual Review of Plant Physiology 31:453–489.
Weis, Engelbert and Joseph A. Berry. 1988. Plants and high temperature stress. Pages 329–346 *in* S. P. Long and F. I. Woodward (eds.), Plants and Temperature. Symposia of the Society for Experimental Biology, No. 42. The Company of Biologists Limited, Cambridge.
Weiser, C. J. 1970. Cold resistance and injury in woody plants. Science 169:1269–1278.
White, W. C. and C. J. Weiser. 1964. The relation of tissue desiccation, extreme cold, and rapid temperature fluctuations to winter injury of American Arborvitae. American Society for Horticultural Science 85:554–563.
Whittaker, R. H. 1954. The ecology of serpentine soils. IV. The vegetational response to serpentine soils. Ecology 35:278–288.
Yancey, P. H., M. E. Clark, S. C. Hand, R. D. Bowlus, and G. N. Somero. 1982. Living with water stress: Evolution of osmolyte systems. Science 217: 1214–1222.
Yoshida, S., T. Niki, and A. Sakai. 1979. Possible involvement of the tonoplast lesion in chilling injury of cultured plant cells. Pages 275–290 *in* J. M. Lyons, D. Graham, and J. K. Raison (eds.), Low Temperature Stress in Crop Plants. Academic Press, New York.

# Index of Species and Subjects

A lower-case "d" by the page number means that the term appears in **bold-face type** on that page and that the term is *defined* either formally or in the context of the discussion on that page. A lower-case "i" by a number means that the term appears in an *illustration* or *table* on that page (either in the illustration itself or in the caption).

# Author Index

A lower-case "i" by the page number means that an illustration on that page was provided by or based on work of the author shown. In a few cases, it means that a photograph of the author is shown on that page.